ANGELA SAINI presents science programmes for the BBC and her writing has appeared in *New Scientist*, the *Guardian*, *The Sunday Times*, *Nature*, *National Geographic* and *Wired*. She has a Masters in Engineering from Oxford University, where she was a chair of the student union's anti-racism committee, and she is a former fellow at the Massachusetts Institute of Technology. Angela's work has won a string of national and international awards. Her last book, *Inferior*, was *Physics World's* Book of the Year. Her first book, *Geek Nation*, was published by Hodder & Stoughton in 2011. In 2018 she was voted by her peers as one of the most respected journalists in the UK.

Praise for *Superior*:

'Roundly debunks racism's core lie – that inequality is to do with genetics, rather than political power' RENI EDDO-LODGE

'Deeply researched, masterfully written, and sorely needed, *Superior* is an exceptional work by one of the world's best science writers' ED YONG

'This is an essential book on an urgent topic by one of our most authoritative science writers' SATHNAM SANGHERA

'Saini takes readers on a walking tour through science, art, history, geography, nostalgia and personal revelation in order to unpack many of the most urgent debates about human origins, and about the origin myths of racial hierarchies. This beautifully written book will change the way you see the world' JONATHAN METZL

'She writes with a quiet and engaging dignity, drawing on her in-depth understanding of what science really has to say about race and genetics. An important, timely book' *New Statesman*

'Saini skilfully brings together interviews with historians, scientists and the objects of racial science themselves to paint a harrowing picture of the influence of race on science and vice versa' *Sunday Times*

'A rigorously researched and reported journey from the Enlightenment through 19th-century imperialism and 20th-century eugenics to the stealthy revival of race science in the 21st-century. Disturbing but written well enough to be entertaining at the same time' *Financial Times*

'A brilliant and devastating book' *Telegraph*

'It is Saini's attention to power that will make her book essential reading for anyone concerned with understanding the history of science's preoccupation with race'
Times Literary Supplement

'Exceptional and damning . . . Saini expertly chronicles the broader social forces that have reinvigorated race science . . . For such a weighty topic, *Superior* is a surprisingly easy-to-read blend of science reporting, cultural criticism and personal reflection' *Slate*

'This important, masterfully written book is a welcome reminder that science is political and we should, as Saini has done for us, follow the money to discover who benefits from prejudicial narratives. I urge you to read it' GAIA VINCE

'Angela Saini's incredible exploration of race science in *Superior* takes a hugely complex and dense subject and makes it accessible and engaging . . . shocking but essential reading, which unpicks the extent to which racial science studies have been used to justify extreme movements' *Vanity Fair*

'Saini's gaze is unrelenting, as she examines anthropologists, biologists, policy-makers and venerable institutions, who have extrapolated scientific principles and arguments to uphold views to justify racism embedded in our world . . . an important work at a time when the world is divided and dubious racist ideas are gaining wider currency quickly on the Internet, with consequences in the real world' SALIL TRIPATHI

'Angela Saini's *Superior* connects the dots, laying bare the history, continuity and connections of modern racist science, some more subtle than you might think. This is science journalism at its very best!' JONATHAN MARKS

'Angela Saini's investigative and narrative talents shine in *Superior*, her compelling look at racial biases in science past and present. The result is both a crystal-clear understanding of why race science is so flawed, and why science itself is so vulnerable to such deeply troubling fault lines in its approach to the world around us – and to ourselves' DEBORAH BLUM

'None have gone so deep under the skin of the subject as Angela Saini in *Superior*. In her deceptively relaxed writing style, Saini patiently leads readers through the intellectual minefields of "scientific" racism. She plainly exposes the conscious and unconscious biases that have led even some of our most illustrious scientists astray' MICHAEL BALTER

By the same author

Inferior:
The true power of women and the science that shows it

Geek Nation:
How Indian science is taking over the world

Superior

The Return of Race Science

Angela Saini

4th ESTATE • *London*

4th Estate
An imprint of HarperCollins*Publishers*
1 London Bridge Street
London SE1 9GF
www.4thEstate.co.uk

First published in Great Britain by 4th Estate in 2019
This 4th Estate paperback edition published in 2020

1

A catalogue record for this book is available from the British Library

ISBN 978-0-00-829386-4

Printed and bound by CPI Group (UK) Ltd, Croydon, CR0 4YY

MIX
Paper from
responsible sources
FSC™ C007454

This book is produced from independently certified FSC™ paper
to ensure responsible forest management

Find out more about HarperCollins and the environment at
www.harpercollins.co.uk/green

For my parents,
the only ancestors I need to know.

Contents

Prologue

'In the British Museum is where you can see 'em
The bones of African human beings'

– Fun-Da-Mental, 'English Breakfast'

I'M SURROUNDED BY DEAD PEOPLE, asking myself what I am.

Where I am is the British Museum. I've lived in London almost all my life and through the decades I've seen every gallery in the museum many times over. It was the place my husband took me on our first date, and years later, it was the first museum to which I brought my baby son. What draws me back here is the scale, the sheer quantity of artefacts, each seemingly older and more valuable than the last. I feel overwhelmed by it. But as I've learned, if you look carefully, there are secrets – secrets that undermine the grandeur, that offer a different narrative from the one the museum was built to tell.

When medical doctor, collector and slave owner Sir Hans Sloane bequeathed the British Museum's founding collection upon his death in 1753, an institution was established that would come to document the entire span of human culture, in time and space. The British Empire was growing, and in the museum you can still see how these Empire-builders envisioned their position in history. Britain framed itself as the heir to the great civilisations of Egypt, Greece, the Middle East and Rome.

The enormous colonnade at the entrance, completed in 1852, mimics the architecture of ancient Athens. The neo-classical style Londoners associate with this corner of the city owes itself to the fact that the British saw themselves as the cultural and intellectual successors of the Greeks and Romans.

Walk past the statues of Greek gods, their bodies considered the ideal of human physical perfection, and you're witness to this narrative. Walk past the white marble sculptures removed from the Parthenon in Athens even as they crumbled, and you begin to see the museum as a testament to the struggle for domination, for possession of the deep roots of civilisation itself. In 1798, when Napoleon conquered Egypt and a French army engineer uncovered the Rosetta Stone, allowing historians to translate Egyptian hieroglyphs for the first time, this priceless object was claimed for France. A few years after it was found, the British took it as a trophy and brought it here to the museum. They vandalised it with the words, 'Captured in Egypt by the British Army', which you can still see carved into one side. As historian Holger Hoock writes, 'the scale and quantity of the British Museum's collections owe much to the power and reach of the British military and imperial state.'

The museum served one story. Great Britain, this small island nation, had the might to take treasures, eight million exquisite objects from every corner of the globe, and transport them here. The inhabitants of Rapa Nui (Easter Island, as European explorers called it) built the enormous bust of Hoa Hakananai'a to capture the spirit of one of their ancestors, and the Aztecs carved the precious turquoise double-headed serpent as an emblem of authority, but in the nineteenth century both these jewels found their way here and here they've remained. To add insult to injury, they're just two of many, joining objects thousands of years older from Mesopotamia and the Indus

Valley. No single item in the museum is more important than the museum itself. All these jewels brought together like this have an obvious tale to tell, one constructed to remind us of Britain's place in the world. It's a testament to the audacity of power.

And this is why I'm at the museum once again. When I set out to write this book, I wanted to understand the biological facts around race. What does modern scientific evidence really tell us about human variation, and what do our differences mean? I read the genetic and medical literature, I investigated the history of the scientific ideas, I interviewed some of the leading researchers in their fields. What became clear was that biology can't answer this question, at least not fully. The key to understanding the meaning of race is understanding power. When you see how power has shaped the idea of race and continues to shape it, how it affects even the scientific facts, everything finally begins to make sense.

It was not long after the British Museum was founded that European scientists began to define what we now think of as race. In 1795, in the third edition of *On the Natural Varieties of Mankind*, German doctor Johann Friedrich Blumenbach described five human types: Caucasians, Mongolians, Ethiopians, Americans and Malays, elevating Caucasians – his own race – to the status of most beautiful of them all. Being precise, 'Caucasian' refers to people who live in the mountainous Caucasus region between the Black Sea to the west and the Caspian Sea to the east, but under Blumenbach's sweeping definition it encompassed everyone from Europe to India and North Africa. It was hardly scientific, even by the standards of his time, but his vague human taxonomy would nevertheless have lasting consequences. Caucasian is the polite word we still use today to describe white people of European descent.

The moment we were sifted into biological groups, placed in our respective galleries, was the beginning of the madness. Race feels so real and tangible now. We imagine that we know what we are, having forgotten that racial classification was always quite arbitrary. Take the case of Mostafa Hefny, an Egyptian immigrant to the United States who considers himself very firmly and very obviously black. According to the rules laid out by the US government in its 1997 Office of Management and Budget standards on race and ethnicity, people who originate in Europe, the Middle East and North Africa are officially classified as white, in the same way that Blumenbach would have categorised Hefny as Caucasian. So in 1997, aged forty-six, Hefny filed a lawsuit against the United States government to change his official racial classification from white to black. He points to his skin, which is darker than that of some self-identified black Americans. He points to his hair, which is black and curlier than that of some black Americans. To an everyday observer, he's a black man. Yet the authorities insist that he is white. His predicament still hasn't been resolved.

Hefny isn't alone. Much of the world's population falls through some crack or another when it comes to defining race. What we are, this hard measure of identity, so deep that it's woven into our skin and hair, a quality nobody can change, is harder to pin down than we think. My parents are from India, which means I'm variously described as Indian, Asian, or simply 'brown'. But when I grew up in south-east London in the 1990s, those of us who weren't white would often be categorised politically as black. The National Union of Journalists still considers me a 'Black member'. By Blumenbach's definition, being ancestrally north Indian makes me Caucasian. Like Mustafa Hefny then, I too am 'black', 'white' and other colours, depending on what you prefer.

We can draw lines across the world any way we choose, and in the history of race science, people have. What matters isn't where the lines are drawn, but what they mean. The meaning belongs to its time. And in Blumenbach's time, the power hierarchy had white people of European descent sitting at the top. They built their scientific story of the human species around this belief. They were the natural winners, they thought, the inevitable heirs of the great ancient civilisations nearby. They imagined that only Europe could have been the birthplace of modern science, that only the British could have built the railway network in India. Many still imagine that white Europeans have some innate edge, some superior set of genetic qualities that has propelled them to economic domination. They believe, as French President Nicolas Sarkozy said in 2007, that 'the tragedy of Africa is that the African has not fully entered into history . . . there is neither room for human endeavour nor the idea of progress.' The subtext is that history is over, the fittest have survived, and the victors have been decided.

But history is never over. There are objects in the British Museum that scream this truth silently, that betray the secret the museum tries to hide.

When you arrive for the first time it's almost impossible to notice them because they're so easily ignored by visitors in a rush to tick off every major treasure. You join the other fish in the shoal. But go upstairs to the Ancient Egypt galleries, to the plaster cast of a relief from the temple of Beit el-Wali in Lower Nubia, built by the pharaoh Ramesses II, who died in 1213 BCE. It's high near the ceiling, spanning almost the entire room. See the pharaoh depicted as an impressive figure on a chariot, wearing a tall blue headdress and brandishing a bow and arrow, his skin painted burnt ochre. He's ploughing into a legion of Nubians, dressed in leopard skins, some painted with black skin and some the same ochre as him. He sends their limbs into a tangle before they're finally conquered. As the relief shows, the Egyptians at that time believed themselves to be a superior people with the most advanced culture, imposing order on chaos. The racial hierarchy, if that's what you want to call it, looked this way in this time and place.

Then things changed. Downstairs on the ground floor is a granite sphinx from a century or two later, a reminder of the time when the Kushites, inhabitants of an ancient Nubian kingdom located in present-day Sudan, invaded Egypt. There was a new winner now, and the Ram Sphinx protecting King Taharqo – the black king of Egypt – illustrates how this conquering force took Egyptian culture and appropriated it. The Kushites built their own pyramids, the same way that the British would later replicate classical Greek architecture.

Through objects like this you can understand how power balances shift throughout history. They reveal a less simple version of the past, of who we are. And it's one that demands humility, warning us that power is fleeting. More importantly, they show that knowledge is not just an honest account of what we know, but has to be seen as something manipulated by those who happen to hold power when it is written.

The Ancient Egypt galleries of the British Museum are always the most crowded. As we walk past the ancient mummies in their glittering cases we don't always recognise that this is also a mausoleum. We're surrounded by the skeletons of real people who lived in a civilisation no less remarkable than the ones that followed or that went before. Every society that happens to be dominant comes to think of itself as being the best, deep down. The more powerful we become, the more our power begins to be framed as not only cultural but natural. We portray our enemies as ugly foreigners and our subordinates as inferior. We invent hierarchies, give meaning to our own categories. One day, a thousand years forward, in another museum, in another nation, these could be European bones encased in glass, what was once considered an advanced society replaced by a new one. A hundred years is nothing; everything can change within a millennium. No region or people has a claim on superiority.

Race is the counter-argument. Race is at its heart the belief that we are born different, deep inside our bodies, perhaps even in character and intellect, as well as in outward appearance. It's the notion that groups of people have certain innate qualities that are not only visible at the surface of their skins, but are intrinsic to their physical and mental capacities, that perhaps even help define the passage of progress, the success and failure of the nations our ancestors came from.

Notions of superiority and inferiority impact us in deep ways. I was told of an elderly man in Bangalore, south India, who ate his chapatis with a knife and fork because this was how the British ate. When my great-grandfather fought in the First World War for the British Empire and when my grandfather fought in the Second World War, their contributions were forgotten, like those of countless other Indian soldiers. They were considered not strictly equal to their white British counterparts. This is how it was. Generations of people in the twentieth century lived under colonial rule, apartheid and segregation, suffered violent racism and discrimination, because this is how it was. When boys from my school threw rocks at my sister and me when we were little, telling us to go home, this is how it was. I knew even as I bled that this is how it was. This is how it still is for many.

Race, shaped by power, has acquired a power of its own. We have so absorbed our classifications – the trend begun by scientists like Blumenbach – that we happily classify ourselves.

Many of those who visit the British Museum for the first time (I can tell you this from having spent hours watching them) come searching for their own place in these galleries. The Chinese tourists go straight to the Tang dynasty artefacts; the Greeks to the Parthenon marbles. The first time I came here, I made a beeline for the Indian galleries. My parents were born in India, as were their parents, and theirs before them, so this is where I imagined I would find the objects most relevant to my personal history. So many visitors have that same desire to know who their ancestors were, to know what *their* people achieved. We want to see ourselves in the past, forgetting that everything in the museum belongs to us all as human beings. We are each products of it all.

But, of course, that's not the lesson we take, because that's not what the museum was designed to tell us. Trapped inside glass cabinets, fixed to the floors, why are these objects in these rooms, and not where they were first made? Why do they live inside this museum in London, its neo-classical columns stretching into the wet, grey sky? Why are the bones of Africans here, and not where they were buried, in the magnificent tombs that were created for them, where they were supposed to live out eternity?

Because this is how power works. It takes, it claims and it keeps. It makes you believe that this is where they belong. It's designed to put you in your place.

The global power balance, as it played out in the eighteenth century, meant that treasures from all over the world could and would only end up in a museum like this, because Britain was one of the strongest nations at the time. It and other European powers were the latest colonisers, the most recent winners. So they gave themselves the right to take things. They gave themselves the right to document history their way, to define the scientific facts about humankind. European thinkers told us that their cultures were better, that they were the proprietors of thought and reason, and they married this with the notion that they belonged to a superior race. These became our realities.

The truth is something else.

1

Deep Time

Are we one human species, or aren't we?

Flanking a road dotted with the corpses of unlucky kangaroos, three hundred kilometres inland from the Western Australian city of Perth – and the other end of the world from where I call home – is what feels like a wilderness. Everything is alien to my eyes. Birds I've never seen before make sounds I've never heard. The dead branches of silvery trees, skeleton fingers, extend out of crumbly red soil. Gigantic rocks weathered over billions of years into soft pastel blobs resemble mossy spaceships. I imagine I've been transported to a galaxy beyond time, one in which humans have no place.

Except that inside a dark shelter beneath one undulating boulder are handprints.

Mulka's Cave is one of lots of ancient rock art sites dotted across Australia, but unique in this particular region for being so densely packed with images. I have to crouch to enter, navigating the darkness. One hand is all I see at first, stencilled within a spray of red ochre illuminated on the granite by a diffuse shaft of light. My eyes adjust, more hands appear. Infant hands and adult hands, hands on top of hands, hands all over the ceiling, hundreds of them in reds, yellows, oranges and whites. Becoming clearer in the half-light, it's as though they're pushing through the walls, willing for a high-five. There are parallel lines, too, maybe the vague outline of a dingo.

The images are hard to date. Some may be thousands of years old, others very recent. What is known is that the creation of rock art on this continent goes back to what in cultural terms feels like the dawn of time. Following excavation at the Madjedbebe rock shelter in Arnhem Land in northern Australia in 2017, it was conservatively estimated that modern humans had been present here for around 60,000 years – far longer than members of our species have lived in Europe, and long enough for people here to have witnessed an ice age, as well as the extinction of the giant mammals. And they may have been making art at the outset. At the Madjedbebe site, I'm told by one archaeologist who worked there, researchers found ochre 'crayons' worked down to a nub. At Lake Mungo in New South Wales, a site 42,000 years old, there is evidence of ceremonial burial, bodies sprinkled with ochre pigment that must have been transported there over hundreds of kilometres.

'Something like a handprint is likely to have many different meanings in different societies and even within a society,' says Benjamin Smith, a British-born rock art expert based at the University of Western Australia. It may be to signify place, possibly to assert that someone was here. But meaning is not always simple. The more experts like him have tried to decipher ancient art, wherever it is in the world, the more they've found themselves only scratching at the surface of systems of thought so deep that Western philosophical traditions can't contain them. In Australia, a rock isn't just a rock. The relationship that indigenous communities have with the land, even with inanimate natural objects, is practically boundless, everyone and everything intertwined.

What looks to me to be an alien wilderness isn't wild at all. It's a home that is more lived in than any I can imagine. Countless generations have absorbed and built upon knowledge of food sources and navigation. They have shaped the landscape sustainably over millennia, built a spiritual relationship with it, with its unique flora and fauna. As I learn slowly, in Aboriginal Australian thinking, the individual seems to melt away in the world around them. Time, space and object take on different dimensions. And none except those who have grown up immersed in this culture and place can quite understand it. I know that I could spend the rest of my life trying to fathom this and get no further than I am now, standing lonely in this cave.

We can't inhabit minds that aren't our own.

I was a teenager before I discovered that my mother might not actually know her own birth date. We were celebrating her birthday on the same day in October we always did when she told us in passing that her sisters thought she had actually been born in the summer. Pinning down dates hadn't been routine when she was growing up in India. It surprised me that she didn't care, and my surprise made her laugh. What mattered to her instead was her intricate web of family relationships, her place in society, her fate as mapped in the stars. And so I began to understand that the things we value are only what we know. I compare every city I visit with London, where I was born, for example. It's the centre of my universe.

For archaeologists interpreting the past, deciphering cultures that aren't their own is the challenge. 'Archaeologists have struggled for a long time to determine what it is, what is that unique trait, what makes us special,' says Smith, who as well as working in Australia has spent sixteen years at sites in South Africa. It's a job that has taken him to the cradles of humankind, rummaging through the remains of the beginning of our species. And this is a difficult business. It's surprisingly tough to date exactly when *Homo sapiens* emerged. Fossils of people who shared our facial features have been found from 300,000 to 100,000 years ago. Evidence of art, or at least the use of ochre, is reliably available in Africa far further back than 100,000 years, before some of our ancestors began venturing out of the continent and slowly populating other parts of the world, including Australia. 'It's one of the things that sets us apart as a species, the ability to make complex art,' he says.

But even if our ancestors were making art a hundred millennia ago, the world then was nothing like the world now. More than forty thousand years ago there weren't just modern humans, *Homo sapiens*, roaming the planet, but also archaic humans, including Neanderthals (sometimes called cavemen because their bones have been found in caves), who lived in Europe and parts of western and central Asia. And there were Denisovans, we now know, whose remains have been found in limestone caves in Siberia, their territory possibly spanning south-east Asia and Papua New Guinea. There were also at various times in the past many other kinds of human, most of which haven't yet been identified or named.

In the deep past we all shared the planet, even living alongside each other at certain times, in particular places. For some academics, this cosmopolitan moment in our ancient history lies at the heart of what it means to be modern. When we imagine these other humans, it's often as knuckle-dragging thugs. We must have had qualities that they didn't have, something that gave us an edge, the ability to survive and thrive as they went extinct. The word 'Neanderthal' has long been a term of abuse.

Dictionaries define it both as an extinct species of human that lived in ice-age Europe, and an uncivilised, uncouth man of low intelligence. Neanderthals and *Homo erectus* made stone tools like our own species, *Homo sapiens*, Smith explains, but as far as convincing evidence goes, he believes none had the same capacity to think symbolically, to talk in past and future tenses, to produce art quite like our own. These are the things that made us modern, that set us apart.

What separated 'us' from 'them' goes to the core of who we are. But it's not just a question for the past. Today, being human might seem so patently clear, so beyond need for clarification that it's hard to believe that not all that long ago it wasn't so. When archaeologists found fossils of other now-extinct human species in the late nineteenth and early twentieth centuries, they raised doubts about just how far all *Homo sapiens* living today really are the 'same'. Even as recently as the 1960s it wasn't controversial for a scientist to believe that modern humans may have evolved independently in different parts of the world from separate archaic forms. Indeed, some are still plagued by uncertainty over this question. Scientific debate around what makes a modern human a *modern* human is as contentious as it has ever been.

From our vantage point in the twenty-first century, this might sound absurd. The common, mainstream view is that we have shared origins, as described by the 'Out of Africa' hypothesis. Scientific data has confirmed in the last few decades that *Homo sapiens* evolved from a population of people in Africa before some of these people began migrating to the rest of the world around 100,000 years ago and adapting in small ways to their own particular environmental conditions. Within Africa, too, there was adaptation and change depending on where people lived. But overall, modern humans were then (and remain now) one species, *Homo sapiens*. We are special and we are united. It's nothing less than a scientific creed.

But this isn't a view shared universally within academia. It's not even the mainstream belief in certain countries. There are scientists who believe that, rather than modern humans migrating out of Africa relatively recently in evolutionary time, populations on each continent actually emerged into modernity separately from ancestors who lived there as far back as millions of years ago. In other words, different groups of people became human as we know it at different times in different places. A few go so far as to wonder whether, if different populations evolved separately into modern humans, maybe this could explain what we think of today as racial difference. And if that's the case, maybe the differences between 'races' run deeper than we realise.

*

In one early European account of indigenous Australians, the seventeenth-century English pirate and explorer William Dampier called them 'the miserablest people in the world'.

Dampier and the British colonists who followed him to the continent dismissed their new neighbours as savages who had been trapped in cultural stasis since migrating or emerging there, however long ago that was. Cultural researchers Kay Anderson, based at Western Sydney University, and Colin Perrin, an independent scholar, document the initial reaction of Europeans in Australia as one of sheer puzzlement. 'The non-cultivating Aborigine bewildered the early colonists,' they write. They didn't build houses, they didn't have agriculture, they didn't rear livestock. They couldn't figure out why these people, if they were equally human, hadn't 'improved' themselves by adopting these things. Why weren't they more like Europeans?

There was more to this than culture shock. Bewilderment – or rather, an unwillingness to try and understand the continent's original inhabitants – suited Europeans in the eighteenth century because it also served the belief that they were entering a territory they could justly claim for themselves. The landscape was thought to be no different from how it must have been in the beginning, because they couldn't recognise how it might have been changed. And if the land hadn't been cultivated, then by Western legal measures it was *terra nullius*, it didn't belong to anyone.

By the same token, if its inhabitants belonged to the past, to a time before modernity, their days were numbered. 'Indigenous Australians were considered to be primitive, a fossilised stage in human evolution,' I'm told by Billy Griffiths, a young Australian historian who has documented the story of archaeology in his country, challenging the narrative that once painted indigenous peoples as an evolutionary backwater. At least one early explorer even refused to believe they had created the rock art he saw. They were viewed as 'an earlier stage of western history, a living representative of an ancient form, a stepping stone'. From almost the first encounter, Aboriginal Australians were judged to have no history of their own, surviving in isolation as a flashback to how all humans might have lived before some became civilised. In 1958 the late distinguished Australian archaeologist John Mulvaney wrote that Victorians saw Australia as a 'museum of primeval humanity'. Even until the end of the twentieth century writers and scholars routinely called them 'Stone Age' people.

It's true that indigenous cultures have enduring connections to their ancestors, a continuation of traditions that go back millennia. 'The deep past is a living heritage,' Griffiths tells me. For Aboriginal Australians, 'it's something they feel in their bones . . . there are amazing stories of dramatic events that are preserved in oral histories, oral traditions, such as the rising of the seas at the end of the last ice age, and hills becoming islands, the eruption of volcanoes in western Victoria, even meteorites in different times.' But at the same time, this doesn't mean that ways of life have never changed. European colonisers failed to see this and it would take until the second half of the twentieth century for that view to be corrected.

'There was certainly little respect for the remarkable systems of understanding and land management that indigenous Australians had cultivated over millennia,' explains Griffiths. For thousands of years the land has been embedded with stories and songs, cultivated with digging sticks, fire and hand. 'While people have lived in Australia, there's been enormous environmental change as well as social change, political change, cultural change.' Their lives have never been static. In his 2014 book *Dark Emu, Black Seeds*, writer Bruce Pascoe argues, as other scholars have done, that this engagement with the land was so sophisticated and successful, including the harvesting of crops and fish, that it amounted to farming and agriculture.

But whatever they saw, the colonisers didn't value. For those raised in and around cities, industrialisation is still what represents civilisation. 'The idea of ranking, say, an industrial society higher than a hunter-gatherer society is absurd,' reminds Benjamin Smith. It's not easy to accept when you've grown up in a society that tells you concrete skyscrapers are the symbols of advanced culture, but when viewed from the perspective of deep time – across millennia rather than centuries, in the context of long historical trajectories – it becomes clearer. Empires and cities decline and fall. It is smaller, indigenous communities that have survived throughout, those whose societies date to many thousands rather than many hundreds of years. 'Archaeology shows us that all societies are incredibly sophisticated, they are just sophisticated in different ways,' Smith continues. 'These are the world's thinkers, and maybe they thought themselves into a better place. They have societies that have more leisure time than Western societies, lower suicide rates, higher standards of living in many ways, even though they don't have all of the technological sophistication.'

Respect for and pride in indigenous cultures has only started to build in the last few decades. And even now, there remains resistance among some non-indigenous Australians, especially as it has become clear from archaeological evidence that Aboriginal people have been occupying this territory not for just thousands of years, but for many tens of thousands. 'The mid-twentieth century revelation that people were here for that kind of depth of time . . . was received in many ways as a challenge to a settler nation with a very shallow history. There are cultural anxieties wrapped up in all of this,' says Griffiths. 'It challenges the legitimacy of white presence here.'

Among European colonists in the nineteenth century, there was a failure to engage with those they encountered, to accept them as the true inhabitants of the land, combined with a mercenary hastiness to write them off. Alongside the native people of Tierra del Fuego at the southernmost tip of South America, whose nakedness and apparent savagery had shocked biologist Charles Darwin when he saw them on his travels, indigenous Australians and Tasmanians were seen as occupying the lowest rungs in the human racial hierarchy. One observer described them as 'descending to the grave'. They were, Griffiths tells me, seen as doomed to go extinct. 'That was the dominant concept, that they would soon die out.

'There was a lot of talk of smoothing the pillow of a dying race.'

Smoothing the pillow was bloodthirsty work. Disease was the greatest killer, the forerunner of invasion. But starting in September 1794, six years after the First Fleet of British ships arrived in what would become Sydney, and continuing into the twentieth century, hundreds of massacres also helped to slowly and steadily shrink the indigenous population by around 80 per cent, according to some estimates. Many hundreds of thousands of people died, if not of smallpox and other illnesses shipped to Australia, then directly at the hands of individuals or gangs, and at other times of police. Equally harsh was the cultural genocide, adds Griffiths. There were bans on the practice of culture and use of language. 'Many people hid their identity, which also contributed to the decline in population.'

In 1869 the Australian government passed legislation allowing children to be forcibly taken away from their parents, particularly if they were of mixed heritage – described at the time as 'half-caste', 'quarter-caste' and smaller fractions. An official inquiry into the effects of this policy on the indelibly scarred 'Stolen Generations', finally published in 1997, is a catalogue of horrors. In Queensland and Western Australia, people were forced onto government settlements and missions, children removed from about the age of four and placed in dormitories, before being sent off to work at fourteen. 'Indigenous girls who became pregnant were sent back to the mission or dormitory to have their child. The removal process then repeated itself.'

By the 1930s, around half of Queensland's Aboriginal Australian population was living in institutions. Life was bleak, with high rates of illness and malnutrition; behaviour was strictly policed for fear that they would return to the 'immoral' ways of their home communities. Children were able to leave dormitories and missions only to provide cheap labour, the girls as domestic servants and the boys as farm labourers. They were considered mentally unsuited to any other kind of work. Historian Meg Parsons describes what happened as the 'remaking of Aboriginal bodies into suitable subjects and workers for White Queensland'.

Among those forced to live this way were the mother and grandmother of Gail Beck, an indigenous activist in Perth who was once a nurse but now works at the South West Aboriginal Land and Sea Council, fighting to reclaim land rights for her local community, the Noongar. When I visit her at her home in the picturesque port city of Fremantle, speaking to her as she cooks, awaiting a visit from the Aboriginal Australian side of her family, I find someone who has few ways to quantify the pain and loss.

Gail is sixty years old but her true family story is still fairly new to her. Until her thirties, she didn't even know she had any indigenous ancestry. She had been raised to believe she was Italian – a lie to explain her olive skin, her mother terrified that if she were told the truth, Gail might be taken away by the authorities as she herself had been. So she lived under a conspiracy of silence, shielded from the fact that her grandmother had been one of the Stolen Generations, a 'half-caste' taken from her family to live in a Catholic missionary home in 1911 at the age of two. There, she had been abused, physically, mentally, sexually. 'She was put out to service at thirteen. Didn't get paid, nothing like that. And she stayed there until she was an adult.' A similar fate befell Gail's mother, who was under the supervisory care of the nuns in the home from the day she was born, beaten and burned by them when she grew older. The Sisters of Mercy 'were very cruel people', Gail recounts.

Learning about her family's past, and having it confirmed by her grandmother's papers, was a bolt from the blue. 'I cried an ocean of tears.' At once, Gail gained a new identity, one that she was desperate to understand and build a connection to. It took her six years to find the part of her family that had been hidden from her, and she has devoted herself to absorbing their culture ever since. She shows me her blankets and pictures, adorned with the prints for which Aboriginal Australian artists have lately become famous. She has tried to learn an indigenous language, but it has been a struggle. She lives like most white Australians, in a nice house in a nice suburb, her knowledge of her great-grandmother's way of life, as it would have been, fragmentary.

'We are constantly in mourning, and people don't understand that,' she tells me. 'The young children that were lost, that doesn't just affect the nuclear family, that affects the community.' And this is perhaps the greatest tragedy of all, that the way of life she might have had, the knowledge and language she could have been raised with, the relationship to the local environment, all of this was trodden beneath the boot of what considered itself to be a superior race. After the arrival of the Europeans, even the creation of art sharply declined. It took until 1976 for Aboriginal people even to be able to gain legal rights over their land. Throughout, the victims had no choice. 'They weren't allowed to practise their culture, they weren't allowed to mix and they weren't allowed to speak their language.' Having been told they were inferior, that theirs was a life to be ashamed of, they adopted different ways of living – ways they were told were better.

'It was a real shameful thing.'

*

I don't cry easily. But in the car afterwards, I cry for Gail Beck. There is no scale of justice weighty enough to account for what happened. Not just for the abuse and the trauma, the children torn from their parents, the killings, but also for the lives that women and men like her didn't have the chance to live.

In recent decades, as scholars have tried to piece together the past and make sense of what happened, as they share with ordinary Australians in the long process of assessing the damage and its impact, we can see an overarching story about the definition of human difference. It shows us how people have drawn boundaries around other groups of people, and how far inside us and how far back in time the disparities are thought to stretch. These are the parameters of what we now call race.

That same day I meet with Martin Porr, a German-born archaeologist who works at the University of Western Australia, his work focusing on human origins. He feels, as do many archaeologists nowadays, that his is a profession weighed down by the baggage of colonialism. When the first European encounters with Australians happened, when the rules were drawn for how they should be treated, science and archaeology began to be woven in. And they have remained interwoven ever since. For Porr, this tale begins with the Enlightenment, at the birth of Western science. The Enlightenment reinforced the idea of human unity, of an essential biological quality that elevated humans above all other creatures. We live with that concept to this day, seeing it as positive and inclusive, a fact to be celebrated. There was a caveat, however. As Porr cautions, this modern universal way of framing human origins was constructed at a time when the world was a very different place, with far less understanding of other cultures. When European thinkers set the standard for what they considered a modern human, many built it around their own experiences and what they happened to value at that time.

A number of Enlightenment thinkers, including influential German philosophers Immanuel Kant and Georg Wilhelm

Friedrich Hegel, defined humanity without really having much of an idea how most of humanity lived or what it looked like. Those who lived in other lands, including the indigenous people of the New World and Australia, were often a mystery to them. 'A universal understanding of human origins was actually created at the time by white men in Europe who only had indirect access to information about other people in the world through the lens of colonialism,' explains Porr. So when they went out into the real world and encountered people who didn't look like them, who lived in ways they didn't choose to live, the first question they were forced to ask themselves was: Are they the same as *us*?

'If you define humanity in some universal sense, then it's very restrictive. And in the eighteenth century, that was totally Eurocentric. And of course, when you define it in that sense, then of course, so to speak, other people do not meet these standards,' Porr continues. Because of the narrow way Europeans had set their parameters of what constituted a human being, placing themselves as the paradigm, people of other cultures were almost guaranteed not to fit. They didn't necessarily share the same aesthetics, political systems or moral values, let alone food or habits. In universalising humanity, Enlightenment thinkers had inadvertently laid the foundations for dividing it.

And here lay the fatal error at the birth of modern science, one that would persist for centuries, and arguably persists to this day. It is a science of human origins, as British anthropologist Tim Ingold observes, that 'has written the essence of humanity in its own image, and that measures other people by how far they have come in living up to it'.

'When you look at these giants of the eighteenth century, Kant and Hegel, they were terribly racist. They were unbelievably racist!' says Porr. Kant stated in *Observations on the Feeling of the Beautiful and the Sublime* in 1764, 'The Negroes of Africa have by nature no feeling that rises above the trifling.' When he met a quick-witted carpenter, the man was quickly dismissed with the observation that 'this fellow was quite black from head to foot, a clear proof that what he said was stupid.' While a few Enlightenment thinkers did resist the idea of a racial hierarchy, many, including French philosopher Voltaire and Scottish philosopher David Hume, saw no contradiction between the values of liberty and fraternity and their belief that non-whites were innately inferior to whites.

By the nineteenth century, those who didn't live like Europeans were thought to have not yet fully realised their potential as human beings. Even now, notes Porr, when scientists discuss human origins, he still catches them describing *Homo sapiens* in what sound like nineteenth-century European economic terms, as being 'better' and 'faster' than other human species. There's an implicit assumption that higher productivity and more mastery over nature, the presence of settlements and cities, are the marks of human progress, even of evolution. The more we are superior to nature, the more we are superior as humans. It is a way of thinking that forces a ranking of people from closer to nature to more distant, from less developed to more, from worse to better. And history shows us that it's only a small leap from believing in cultural superiority to believing in biological superiority, that a group's achievements are due to their innate capacities.

What Europeans saw as shortcomings in other populations in the early nineteenth century quickly became conflated with how they looked. Cultural scholars Kay Anderson and Colin Perrin explain how, in that century, race came to be *everything*. One writer at the time noted that the natives of Australia differ 'from any other race of men in features, complexion, habits and language'. Their darker skin and different facial features became markers of their separateness, a sign of their permanent difference. Their perceived failure to cultivate the land, to domesticate animals and live in houses was taken as part and parcel of their appearance. And this had wider implications. Race, rather than history, could then be framed as the explanation for not only the Aboriginals' failure, but the failure of all non-white races to live up to the European ideal that Europeans had themselves defined. An Aboriginal Australian – just by having darker skin – could now be lumped together with a West African, for instance, despite being continents apart, with entirely different cultures and histories. Both were black, and this was all that mattered.

Whiteness became the visible measure of human modernity. It was an ideal that went so far as to become enshrined in Australian law. 'When Australia federated in 1901, when the states came together as a nation, one of the first pieces of legislation to pass through Parliament was the Immigration Restriction Act, which formed the basis of the White Australia policy. It sought to fuse the new nation together with whiteness by excluding non-European migration and attempting to assimilate and, ultimately, to eliminate Aboriginal and Torres Strait Islander identity,' explains Billy Griffiths. What happened to Gail Beck's family was one result of these attempts to remove the colour from Australia, in her case to drain it out of her mother's line over generations. 'There was this horrible language of "breeding out the colour" from full-bloods to half-castes to quarter-castes to octoroons,' Griffiths adds. The goal was to steadily replace one 'race' with another.

By the time this state-sanctioned ethnic cleansing was taking place, a crisis had already emerged within scientific circles. Since the Enlightenment, many European thinkers had united around the idea that humankind was one, that we all shared the same common capacities, the same spark of humanity that made it possible for even those of us condemned as 'miserable' to improve, with enough encouragement. Even if there was a racial hierarchy, even if there were lesser humans and greater ones, we were all still *human*. But in the nineteenth century, as Europeans encountered more people in other parts of the world, as they began to see the variety that exists across our species, and failed to 'improve' people the way they wanted to, some began to seriously doubt this cherished belief.

The passage of the nineteenth century saw some make an intellectual shift away from the original Enlightenment view of a single humanity with shared origins. Scientists ventured to wonder whether we all really did belong to the same species.

This wasn't just because of racism. Western scientists had been funnelled into a certain way of thinking about the world partly because of where they happened to be based. In the early days of archaeology, Europe was the reference point for subsequent research elsewhere. Before anyone was sure about humanity's African origins, human fossils in Europe provided the first data. According to John Shea, a professor of anthropology at Stony Brook University in New York, this created an indexing problem. 'If you have a series of observations, the first observations guide you more so than the latter ones. And our first observations about human evolution were based on an archaeological record in Europe.' The first movements out of Africa were eastwards, not westwards. This is why you see elephants in both Africa and Asia. Europe isn't where humans originated – indeed, being so inhospitable back then, it was one of the last places they migrated to, long after going to Australia. But since Europe was where the first archaeologists happened to live and work, this geographical outpost became the model for thinking about the past.

Some of the very oldest human sites in Europe bear evidence of fairly sophisticated cave art. So as a result of indexing, early archaeologists digging on their doorstep logically assumed that art and the ability to think using symbols and images must be a mark of human modernity, one of the features that make us special. But the first *Homo sapiens* arrived in Europe only around 45,000 years ago. When researchers then excavated far earlier sites in Africa, some as old as 200,000 years, they didn't always find the same evidence of symbolic thought and representational art. 'The archaeologists came up with a way to square this,' says Shea. 'They said, well, okay, you know these ancient Africans, Asians, they look morphologically modern but they aren't behaviourally modern. They're not quite right yet.' They decided that although such people *looked* like modern humans, for some reason they didn't *act* like them.

Rather than rethinking what it meant to be a modern human – perhaps taking out the requirement that *Homo sapiens* began making art immediately upon the emergence of our species – the rest of the world's history became a puzzle to be solved. It's a misstep that still has repercussions today. If art is what sets our species apart from Neanderthals and others, then at what point did we actually become our species? Was it 45,000 years ago when we see sophisticated cave art in Europe, or 100,000 years ago when, we now know, people used ochre for drawing? And if Neanderthals or other archaic humans turn out to show evidence of symbolic thought and to have made representational art, will we then have to call them modern too? 'Behavioural modernity is a diagnosis,' says Shea. All the archaeologists can think to do is 'rummage around looking for other evidence that will confirm this diagnosis of modernity'.

In the nineteenth century, such uncertainty around what constituted a modern human being was taken a leap further. If people weren't cultivating the land or living in brick houses, some asked, could they be considered modern? And if they weren't modern, were they even the same species?

Australia in all its alien strangeness posed a particular challenge to European thinkers. Anderson and Perrin argue that the discovery of the continent helped shatter the Enlightenment belief in human unity. After all, here was a remote place, with its own animals not seen elsewhere, kangaroos and koalas, and with its own plants, flowers and unusual landscape. 'Based on observations of the uniqueness of Australian flora and fauna' there were 'suspicions that the entire continent might have been the product of a separate creation,' they write. The humans of Australia were thought to be as strange as everything else there.

After the remains subsequently labelled Neanderthal were first identified in Germany's Neander Valley in 1856, Martin Porr and his colleague Jacqueline Matthews have noted, one of the first things anybody did was compare them to indigenous Australians. Five years later, English biologist Thomas Huxley, a champion of the work of Charles Darwin, described the skulls of Australians as being 'wonderfully near' those of the 'degraded type of the Neanderthal'. It was clear what they were insinuating. If any people on earth were going to have something in common with these now-extinct humans, European scientists assumed, it could only be the strange ones they called savages. Who else could it be but the people who were closest to nature, who had never fitted their definition of what a modern human was?

*

We are forever chasing our origins.

When we can't find what we want in the present, we go back, and back further still, until there at the dawn of time, we imagine we've found it. In the gloomy mists of the past, having squeezed ourselves back into the womb of humanity, we take a good look. Here it is, we say with satisfaction. Here is the root of our difference.

Once upon a time, scientists were convinced that Aboriginal Australians were further down the evolutionary ladder than other humans, perhaps closer to Neanderthals. In 2010 it turned out that Europeans are actually likely to have the largest metaphorical drop of Neanderthal blood. In January 2014 an international team of leading archaeologists, geneticists and anthropologists confirmed that humans outside Africa had bred with Neanderthals. Those of European and Asian ancestry have a very small but tangible presence of this now-extinct human in our lineage, up to 4 per cent of our DNA. People in Asia and Australia also bear traces of another archaic human, the Denisovans. There is likely to have been breeding with other kinds of humans as well. Neanderthals and Denisovans, too, mated with each other. In the deep past, it seems, they were pretty indiscriminate in their sexual partnerships.

'We're more complex than we initially thought,' explains John Shea. 'We initially thought there was either a lot of interbreeding or no interbreeding, and the truth is between those goalposts somewhere.'

The discovery had important consequences. It raked up a controversial, somewhat marginalised scientific theory that had been doing the rounds a few decades earlier. In April 1992 an article had been published in *Scientific American* magazine with the incendiary headline: 'The Multiregional Evolution of Humans'. The authors were Alan Thorne, a celebrated Australian anthropologist, who died in 2012, and Milford Wolpoff, a cheery American anthropologist based at the University of Michigan, where he still works today. Their hypothesis suggested that there was something deeper to human difference, that perhaps we hadn't come out of Africa as fully modern humans after all.

Although this notion had been mooted before, for Wolpoff, his ideas became cemented in the seventies. 'I travelled and I looked, I travelled and I looked, I travelled and I looked,' he tells me. 'And what I noticed was that in different regions, big regions – Europe, China, Australia, that is what I mean by regions, not small places – in different regions, it seemed to me there was a lot of similarity in fossils. They weren't the same and they all were evolving.'

Wolpoff's big realisation came in 1981 when he was working with a fossilised skull from Indonesia – one of Australia's closest neighbours, not far from its north coast – which was dated at roughly a million years and possibly older. A million years is an order of magnitude older than modern humans, hundreds of thousands of years before some of our ancestors first began to migrate out of Africa. It couldn't possibly be the ancestor of any living person. Yet Wolpoff says he was struck by the similarities he thought he could see between its facial structure and that of modern-day Australians. 'I had reconstructed a fossil that looked so much like a native Australian to me I almost dropped it,' he says. 'I propped it up on my lap with the face staring at me . . . when I turned it over on its side to get a good look at it, I was really surprised.'

Teaming up with Alan Thorne, who had done related research and shared his interpretation of the past, they came up with the theory that *Homo sapiens* evolved not only in Africa, but that some of the earlier ancestors of our species spread out of Africa and then independently evolved into modern humans, before mixing and interbreeding with other human groups to create the one single species we recognise today. In their article for *Scientific American*, which helped catapult their multiregional hypothesis into the mainstream, they wrote, 'some of the features that distinguish major human groups, such as Asians, Australian Aborigines and Europeans, evolved over a long period, roughly where these people are found today.'

They described these populations as 'types', judiciously steering clear of the word 'race'. 'A race in biology is a subspecies,' Wolpoff clarifies when I ask him about it. 'It's a part of a species that lives in its own geographic area, that has its own anatomy, its own morphology, and can integrate with other subspecies at the boundaries . . . There are no subspecies any more. There may have been subspecies in the past – that's something we argue about. But we do know there are no subspecies now.'

Many academics found Wolpoff and Thorne's idea unconvincing or offensive, or both. According to historian Billy Griffiths, the multiregional way of thinking about our origins, undercutting the fundamental belief that we are all human and nothing else, has echoes of an earlier intellectual tradition that viewed 'races' as separate species. 'Wherever we are in the world we look at the deep past and these immense spans of time through the lens of our present moment and our biases and what we want,' he tells me. 'Archaeology is a discipline that is saturated by colonialism, of course. It can't entirely escape its colonial roots.' Multiregionalism, while it was a response to the evidence available at the time, also carried echoes of the politics of colonialism and conquest. 'That's the ugly political legacy that dogs the multiregional hypothesis.'

Wolpoff has always been sensitive to the controversy. He faced down plenty of criticism when he and Alan Thorne published their work. 'We were the enemy,' he recalls. 'If we were right, there couldn't be a single recent origin for humans . . . They said, you're talking about the evolution of human races in separate places independently of each other.'

Their theory remains unproven. Academics in the West and in Africa today generally accept that humans became modern in Africa and then adapted to the environments where they happened to move to fairly recently in evolutionary time – and even these are only superficial adaptations such as skin colour, linked directly to survival. But not everyone everywhere agrees. In China, there's a belief among both the public and leading academics that Chinese ancestry goes back considerably further than the migration out of Africa. One of Wolpoff's collaborators, palaeontologist Wu Xinzhi at the Chinese Academy of Sciences, has argued that fossil evidence supports the notion that *Homo sapiens* evolved separately in China from earlier human species who were living there more than a million years ago, despite data showing that modern Chinese populations carry about as much of a genetic contribution from modern humans who left Africa as other non-African populations do.

'There are many people who are not happy with the idea of African origin,' says Eleanor Scerri, an archaeologist based at the University of Oxford who researches human origins. 'They have co-opted multiregionalism to make a claim that this is a simplistic idea, that races are real, and that people who have come from a particular area have always been there.' She tells me this appears to be prevalent not only in China, but also in Russia. 'There is no acceptance that they were ever African.'

While for some an unwillingness to accept African origins may be motivated by racism or nationalism, it isn't for all. There are those for whom it's simply a way of squaring old origin stories with modern science. In Australia, for instance, Billy

Griffiths tells me, many indigenous people favour the multiregional hypothesis because it sits closer to their own belief that they have been here from the very beginning. Indeed, this is an origin myth shared by cultures in many parts of the world. Until further evidence comes along (and maybe even after it does), the choice of theory may be driven as much by personal motivations as by data. The past can never be completely known, so the classic multiregional hypothesis persists despite its lack of support among experts. It has political power.

While classic multiregionalism seems unlikely to be the story of our past, the fact that we now know our ancestors bred with other kinds of archaic humans does have implications. It gives nourishment to those who would like to resurrect the multiregional hypothesis in full. It's a factual nugget that feeds fresh speculation about the roots of racial difference. Some dogged supporters of the multiregional hypothesis can rightly claim that at least one prediction made by Wolpoff and Thorne has turned out to be correct. The pair suggested that other now extinct humans such as Neanderthals either evolved into modern humans or interbred with them. And on interbreeding, we now know from genetic evidence, the pair got it right. Some of our ancestors did mate with Neanderthals, although their contribution to people's DNA today runs to just a few per cent, which means it couldn't have been particularly widespread. But it did happen.

When I ask Wolpoff if he feels vindicated by this, he laughs. 'You said vindicated. We said relief!'

Genetics has done the unthinkable, says rock art expert Benjamin Smith. 'The thing that has worried me is the way that genetics research has moved . . . We thought that we were basically all the same, whether you're a bushman in southern Africa, an Aboriginal Australian living in rural Western Australia, or someone like myself who is of European extraction. Everyone was telling us that we were all identical, all the modern science.' The latest discoveries appear to move the story back a little closer to the nineteenth-century account. 'This idea that some of us are more interbred with Neanderthals, some of us are more interbred with Denisovans . . . and Aboriginal Australians had quite a high proportion of Denisovan genetics, for example. That could lead us back to the nasty conclusion that we are all different,' he warns. 'I can see how it might be racialised.'

Indeed, when geneticists revealed the Neanderthal connection, personal ancestry testing companies were quick to sell services offering members of the public the opportunity to find out how much Neanderthal ancestry they might have, using data on genetic variants shared by both humans and Neanderthals – presumably in the expectation that this might mean something to everyday people. Maybe those having the test imagined they would have qualities in common with their extinct cousins.

The finding also had a peculiar effect on scientific research. Fairly soon after it was found to be modern-day Europeans who have the closer association to Neanderthals – not, as it turned out, Aboriginal Australians – the image of the Neanderthal underwent a dramatic makeover. When their remains were first discovered in 1856, the German naturalist Ernst Haeckel had suggested naming them 'Homo stupidus'. But in the twenty-first century, these same Neanderthals, the dictionary definition of simple-minded, loutish, uncivilised thugs, have become oddly rehabilitated.

Svante Pääbo, the director of the genetics department at the Max Planck Institute for Evolutionary Anthropology in Germany, who spearheaded some of the research that led to the discoveries of ancient interbreeding in the first place, was among those to marshal efforts to compare the genomes of

Neanderthals and *Homo sapiens*, in the search for what differs as well as what there is in common. This was accompanied by plenty of speculation from others. In 2018 a set of researchers in Switzerland and Germany suggested that Neanderthals actually had quite 'sophisticated cultural behaviour', prompting one British archaeologist to wonder out loud whether 'they were a lot more refined than previously thought'. An archaeologist in Spain claimed that modern humans and Neanderthals must have been 'cognitively indistinguishable'. A few even raised the possibility that Neanderthals could have been capable of symbolic thought, pointing to freshly discovered cave markings in Spain that appear to predate the arrival of modern humans (the finding has failed to convince rock art expert Benjamin Smith).

'Neanderthals are romanticised,' I'm told by John Shea. They're no longer around, and we don't have a great deal of evidence about what they were like or how they lived, which means they can be whatever we want them to be. 'We're free to project good qualities, things we admire, and the ideal on them.' In reality, whatever they were like, he says, 'the interbreeding thing is more like a symbolic thing for us than it is of evolutionary consequence.'

Yet researchers haven't been able to help themselves looking for evolutionary consequences. One team of scientists claimed that the tiny peppering of Neanderthal DNA may have given Europeans different immune systems from Africans. Another published paper linked Neanderthal DNA to a whole host of human differences, including 'skin tone and hair color, height, sleeping patterns, mood, and smoking status'. An American research group went so far as to try to link the amount of Neanderthal DNA people have with the shapes of their brains, implying that non-Africans may have some mental differences from Africans as a result of their interbreeding ancestors.

For more than a century the word 'Neanderthal' had been synonymous with low intelligence. In the space of a decade, once the genetic link to modern Europeans was suspected and then confirmed, that all changed. In the popular press, there was a flurry of excitement about our hitherto undervalued relatives. Headlines proclaimed that 'we haven't been giving Neanderthals enough credit' (*Popular Science*), that 'they were too smart for their own good' (*Telegraph*), that 'humans didn't outsmart the Neanderthals' (*Washington Post*). Meanwhile a piece in the *New Yorker* whimsically reflected on their apparent everyday similarity to humans, including the finding that they may have suffered from psoriasis. Poor things, they even itched like us. 'With each new discovery, the distance between them and us seems to narrow,' wrote the author. In the popular imagination, the family tree had gained a new member.

In January 2017, the *New York Times* asked: 'Neanderthals were people, too . . . Why did science get them so wrong?' This was indeed the big question. If the definition of 'people' had always included archaic humans, then why should Neanderthals so suddenly be accepted as 'people' now? And not just accepted, but elevated to the celebrity status of sadly deceased genius cousin? It wasn't so long ago that scientists had been reluctant to accept the full humanity even of Aboriginal Australians. Gail Beck's family had been denied their culture, treated in their own nation as unworthy of survival, their children ripped from them to be abused by strangers. In the nineteenth century, *they* had been lumped together with Neanderthals as evolutionary dead-ends, both destined for extinction. But now that kinship had been established between Europeans and Neanderthals, *now* we were all people? *Now* we had found our common ground?

If it had turned out that Aboriginal Australians were the

ones to possess that tiny bit of Neanderthal ancestry instead of Europeans, would our Neanderthal cousins have found themselves quite so remarkably reformed? Would they have been welcomed warmly with such tight hugs? It's hard not to see, in the public and scientific acceptance of Neanderthals as 'people like us', another manifestation of the Enlightenment habit of casting humanity in the European image. In this case, Neanderthals have been drawn into the circle of humankind by virtue of being just a little related to Europeans – forgetting that a century ago, it was their supposed resemblance to indigenous Australians that helped cast actual living human beings out of the circle.

*

Milford Wolpoff is clear with me that he doesn't think there is any biological basis to race, that there are no separate races, except as social categories. He comes across as honest and well meaning, and I believe him. But one obvious implication of his multiregional hypothesis is that if different populations became modern in their own way on their own territories, then maybe some became what we today recognise as human sooner than others. 'A modern human from China looks different than a modern human in Europe, not in the important ways, but in other ways,' he tells me. 'So did one become modern earlier than the other one?' Such a line of thinking opens a door for the politics of today to be projected onto the past, giving rise to racial speculation even if that's not what he intends.

There is still not enough evidence that any humans became modern outside Africa in the way that classic multiregional theory suggests. Even Wolpoff concedes that Africa must remain at the heart of the story. 'I will never say that all of modernity is African, but you've gotta think that most of it is' – even if only because in our deep past that's where most people lived. It is impossible to airbrush Africa out of the lineage of every living person. The genetic evidence we have to date confirms that some version of an 'Out of Africa' scenario must have happened.

But over time, the picture inside Africa has changed to incorporate the growing scientific realisation that our origins might have been a little fuzzier than we imagine. In the summer of 2018 Eleanor Scerri at the Institute of Archaeology in Oxford, together with a large international team of geneticists and anthropologists, published a scientific paper suggesting that rather than humans evolving from a single lineage that can be traced to a single small sub-Saharan African population, perhaps our ancestors were the product of many populations across a far wider area within Africa. These pan-African populations might have been isolated by distance or ecological barriers, and could therefore have been very different from one another. It is multiregionalism, if you like, but within one continent.

'Gradually we started to emerge from the occasional mixing of the populations that were spread around,' Scerri tells me. 'The characteristics that define us as a species don't appear in any single individual until much later. Before that, the characteristics of our species were distributed across the continent in different places at different times.' Modern humans, *Homo sapiens*, emerged from this 'mosaic'. 'We need to look at all of Africa to get a good picture of origins.' This version of our past still puts Africa at its centre, as the first home of our ancestors, but it also concedes that modern humans didn't appear suddenly in one place looking and sounding sophisticated, thinking symbolically and producing art. There was no sudden moment at which the first modern human emerged. The characteristics of us existed in various others before us.

'Humans evolved in Africa first,' agrees anthropologist John Shea. 'Not in just one garden of Eden, but among a broadly distributed population more or less like stops across a subway system. People were moving around along the rivers and coastlines.' In short, we are a product of longer periods of time and space, a mixture of qualities that incubated in Africa.

According to archaeologist Martin Porr in Australia, this version of the past is more plausible given the way that fossil evidence is scattered across the African continent. For him personally, it also resonates with indigenous Australian ways of defining what it means to be human. Up north in the Kimberley where he has done most of his work, he says, rock art is not thought of as just images upon rock. 'The rock is actually not a rock but it's a formation out of the dreamtime that is alive, that is in the living world, that people inhabit. And people themselves are part of that.' Human and object, object and environment, are not separated by hard divisions the way they are in Western philosophies.

'You can oscillate in and out of humanity just as objects and animals can oscillate between being human.' An inanimate object can take on human qualities, the way a doll does to a child. In that sense, too, Porr suggests that what made a being human in the past also oscillated.

'I think there's nothing essential about human beings at all.' This, he explains, is how he has come to think about our origins. Not that our evolutionary journey was one big leap, but that we are the gradual products of elements that already existed, in our African ancestors but also in Neanderthals, Denisovans and other archaic humans. Perhaps some of what we think of as purely human characteristics exist in other living creatures today, too.

It's a radically different way of thinking about what it means to be us, ditching the European Enlightenment view, and taking a cue instead from other cultures and older systems of thought. It's a challenge to researchers who have dedicated their careers to identifying the first modern humans and defining what they were like, chasing the tail of the Enlightenment philosophers who thought they already knew. Archaeologists are still trying to hunt down the earliest cave art, the earliest sign of symbolic, abstract thought that will signify the leap from a simpler primate to a sophisticated one, in the hope of pinpointing the magic moment at which *Homo sapiens* emerged, and where. Geneticists, too, hunt for magical ingredients in our genome, the ones that will indicate what makes us so remarkable. Increasingly the evidence suggests that it was never so simple.

'Very few people like looking at human origins from a post-colonial context, but there is a broader story,' says Porr. There are other ways of picturing humanity than as a uniquely special entity far removed from all other living things. Eleanor Scerri agrees that fresh scientific findings are forcing a rethink of what it means to be human. 'Popular science needs to get away from this idea that we originated, and that was *us*. There's never a time that we were not changing,' she says. 'The idea of these immutable forms, and that we originate in one place and that's who we are, that's where we're from.'

What does this mean for us today? If we can't agree on what makes a modern human, where does that leave the idea of universal humanity? If our origins aren't crystal clear then how do we know that we're all the same? What does it mean for race?

In a sense, it shouldn't be of any importance. How we choose to live and treat each other is a political and ethical matter, one that's already been decided by the fact that as a society we have chosen to call ourselves human and give every individual human rights. In reality, though, the tentacles of race reach into our minds and demand proof. If we are equally human, equally capable and equally modern, then there are those who need convincing before they grant full rights, freedoms and opportunities to those they have historically treated as inferior. They need to be convinced before they will commit to redressing the wrongs of the past, before they agree to affirmative action or decolonisation, before they fully dismantle the structures of race and racism. They're not about to give away their power for free.

And if we're honest, maybe we all need to be convinced. Many of us hold subtle prejudices, unconscious biases and stereotypes that reveal how we suspect we're not quite the same. We cling to race even when we know we shouldn't. A liberal, left-wing British friend of mine, of mixed Pakistani and white English ancestry, who has never been to Pakistan and has no deep ties to the country any more, told me recently that she believes there is something in her blood, something biological within her that makes her Pakistani. I feel this way occasionally about my Indian heritage. But where does culture end and ethnicity begin? Many of us who cherish our ethnic identities, whether on the political left, right or the centre, perhaps betray some commitment to the idea of racial difference.

This is the problem for science. When Enlightenment thinkers looked at the world around them, some took the politics of their day as the starting point. It was the lens through which they viewed all human difference. We do the same today. The facts only temper what we think we already know. Even when we study human origins, we don't actually start at the beginning. We begin at the end, with our assumptions as the basis for inquiry. We need to be persuaded before we cast aside our prior beliefs about who we are. The way new research is interpreted is always at the mercy of the old ideas.

'You can either use the present to explain the past. Or you can use the past to explain the present,' John Shea tells me. 'But you can't do both.' To make sense of the past – and of ourselves – is not a simple job of gathering together scientific data until we have the truth. It isn't just about how many fossils we have or how much genetic evidence. It's also about squaring the stories we have about who we are with the information we're given. Sometimes this information becomes slotted into the old stories, reinforcing them and giving them strength, even if it needs to be forced like a square peg into a round hole. Other times, we have to face the uncomfortable realisation that a story must be ditched and rewritten because however hard we try it no longer makes sense.

But the stories we're raised on, the tales, myths, legends, beliefs, even the old scientific orthodoxies, are how we frame everything we learn. The stories are our culture. They are the minds we inhabit. And that's where we have to start.

2

It's a Small World

How did scientists enter the story of race?

Once, a long time ago, I floated around the earth in the space of minutes.

I was on a ride at the Magic Kingdom in Walt Disney World, Florida, my little sisters and I perched alongside each other in a slow mechanical boat, buoyed by sugar. 'It's a Small World (After All)' chimed in tinny children's voices, while minuscule automata played out cultural stereotypes from different countries. From what I can recall, there were spinning Mexicans in sombreros and a ring of African dancers laughing alongside jungle animals. Indian dolls rocked their heads from side to side in front of the Taj Mahal. We sailed past, given just enough time to recognise each cultural stereotype, but not quite enough to take offence.

This long-forgotten vignette from my childhood is what comes back to me on the drizzly day I approach the eastern corner of the Bois de Vincennes woodland in Paris. I had heard that somewhere here I'd find the ruins of a set of enclosures in which humans were once kept – not as cruel punishment by the authorities, and not by some murderous psychopath. Apparently they were just ordinary, everyday people, kept here by everyday people, for the fascination of millions of other everyday people, for no other reason than where they happened to come from and what they happened to look like.

'Man is an animal suspended in webs of significance he himself has spun,' American anthropologist Clifford Geertz wrote in 1973. These webs are ours only until someone comes along to pull at the threads. The nineteenth century had marked an age of unprecedented movement and cultural contact, turning the world into a smaller place than it had ever been. It was less mysterious, perhaps, but no less fascinating. And people wanted to see it all. So in 1907 there was a grand Colonial Exposition on this overgrown site in Paris, within the Bois, in what was known as the Garden for Tropical Agriculture, recreating the different parts of the world in which France had its colonies.

Eight years earlier, the garden had been founded as a scientific project to see how crops in distant lands might be better cultivated, helping to bring in more income for colonisers back in Europe. This exposition went a step further. To exotic plants and flowers it added people, displaying them in houses vaguely typical of the ones they might have left behind, or at least how the French imagined them to be. There were five mini 'villages' in all, each designed to be as realistic as possible so visitors could experience what normal life was like for these foreigners. It was an Edwardian Disneyland, not with little dolls, but actual people. They transformed the tropical garden into nothing less than a human zoo.

'In Paris, there were many exhibitions with human zoos,' says French anthropologist Gilles Boëtsch, former president of the scientific council at the National Center for Scientific Research, who has studied their dark history. There was a circus element to it all, a cultural extravaganza. But there was also a genuine desire to showcase human diversity, to give a glimpse of life in the faraway colonies. According to some estimates, the 1907 Paris Exposition attracted two million visitors in the space of just six months – a hit with curious citizens who wanted to see the world in their backyard.

Wherever they were held, most evidence of human zoos has long disappeared, most likely deliberately forgotten. The Garden for Tropical Agriculture is one rare exception. That said, the French authorities don't appear to want to brag about it. It's tucked behind some quiet and well-to-do apartment blocks with barely any signposting. Greeting me as I enter is a Chinese arch that was once probably bright red, but has since faded to a dusty grey. As I walk under it down a gravel path, the place is peaceful but dilapidated. To my surprise, most of the buildings have survived the last century fairly intact, as though everything was abandoned immediately after the tourists left.

To one side is a weathered sculpture of a naked woman, reclining and covered in beads, her head gone, if it was ever there at all. A solitary jogger runs past.

For European scientists, zoos like this offered more than fleeting amusement value. They were a source of biological data, a laboratory stocked with captive human guinea pigs. 'They came to the human zoos to learn about the world,' explains Boëtsch. Escaping the bother of long sea voyages to the tropics, anatomists and anthropologists could conveniently pop down to their local colonial exhibition and sample from a selection of cultures in one place. Researchers measured head size, height, weight, colour of skin and eyes, and recorded the food these people ate, documenting their observations in dozens of scientific articles. With their notebooks, they set the parameters for modern race science.

Race itself was a fairly new idea. Some of the first known uses of the word date from as recently as the sixteenth century, but not in the way we use it now. Instead, at that time it referred to a group of people from common stock, like a family, a tribe, or perhaps – at a long stretch – a small nation. Even until the European Enlightenment in the eighteenth century, many still thought about physical difference as a permeable, shifting quantity. It was rooted in geography, perhaps explaining why people in hotter regions had darker skins. If those same people happened to move somewhere colder, it was assumed their skins would automatically lighten. A person could shift their identity by moving place or converting to another religion.

The notion that race was hard and fixed, a feature that people couldn't choose, an essence passed down to their children, came slowly, and in large part from Enlightenment science. Eighteenth-century Swedish botanist Carl Linnaeus, famous for classifying the natural world from the tiniest insects to the biggest beasts, turned his eye to humans. If flowers could be sorted by colour and shape, then perhaps we too could fall into groups. In the tenth edition of *Systema Naturae*, a catalogue published in 1758, he laid out the categories we still use today. He listed four main flavours of human, respectively corresponding to the Americas, Europe, Asia and Africa, and each easy to spot by their colours: red, white, yellow and black.

Categorising humans became a never-ending business. Every gentleman scholar (and they were almost exclusively men) drew up his own dividing lines, some going with as few as a couple of races, others with dozens or more. Many never saw the people they were describing, instead relying on second-hand accounts from travellers, or just hearsay. Linnaeus himself included two separate sub-categories within his *Systema Naturae* for monster-like and feral humans. However the lines were drawn, once defined, these 'races' rapidly became slotted into hierarchies based on the politics of the time, character conflated with appearance, political circumstance becoming biological fact. Linnaeus, for instance, described indigenous Americans (his 'red' race) as having straight black hair and wide nostrils, but also as 'subjugated', as though subjugation were in their nature.

And so it began. By the time human zoos were a popular attraction, when the ghostly enclosures of the Bois de Vincennes were not eerily empty as they are now, but full of performers – when I would have more likely been within a cage than outside it – the parameters of human difference had become hardened into what we recognise them as today.

Paris wasn't the only city to enjoy this breed of spectacle. Other European colonial powers hosted similar events. Indeed by the time of the 1907 Paris Exposition, human zoos had been around for more than a century. In 1853 a troupe of Zulus undertook a grand tour of Europe. And forty-three years before this an advertisement in London's *Morning Post* newspaper signalled the arrival of a woman who would go down in history as one of the most notorious of all racial freak shows, her story echoed by those to come. 'From the Banks of the River Gamtoos, on the Borders of Kaffraria, in the interior of South Africa, a most correct and perfect Specimen of that race of people,' it announced.

The 'Hottentot Venus', as she was described in the paper, was available for anyone to take a peek at, for a limited time only and at the cost of two shillings. Her real name was Saartjie Baartman and she was aged somewhere between twenty and thirty. What made her so fascinating were her enormous buttocks and elongated labia, considered by Europeans to be sexually grotesque. Calling her a 'Venus' was a joke at her expense. The *Morning Post* took pains to mention the expense shouldered by Boer farmer Hendric Cezar in transporting her all the way to Europe. He was banking on her body causing a scandal.

Baartman had been Cezar's servant in Africa, and by all accounts, she had come with him to Europe of her own free will. But it's unlikely that the life she endured as his travelling exhibit was what she expected. Her career was brief and humiliating. At each show, she was brought out of a cage to parade in front of visitors, who poked and pinched to check that she was real. Commentators in the press couldn't help but notice how unhappy she seemed, even remarking that if she felt ill or unwilling to perform, she was physically threatened. To add to the humiliation, she became, quite literally, the butt of jokes across the city, rendered in relentless caricature.

At the end of her run, Baartman ended up in Paris. She found herself at the mercy of celebrated French naturalist Georges Cuvier, a pioneer in the field of comparative anatomy, which aims to understand the physical differences between species. Like so many before him, he was spellbound by her – but his was an anatomist's fascination, one that drove him to undertake a detailed study of every bit of her body. When she died in 1815, just five years after being displayed in London, Cuvier dissected her, removing her brain and genitals and presenting them in jars to the French Academy of Sciences.

As far as Cuvier was concerned, this was just science and she was just another sample. The prodding, cutting, dehumanising fingers of researchers like him sought only to understand what made her and those like her different. What gave some of us dark skin and others light? Why did we have different hair, body shape, habits and language? If we were all one species, then why didn't we look and behave the same way? These were questions that had been asked before, but it was nineteenth-century scientists who turned the study of humans into the most gruesome art. People became objects, grouped together like museum exhibits. Any sense of common humanity was left at the door, replaced by the cold, hard tools of dissection and categorisation.

Following a lifetime of being relentlessly poked and prodded, Baartman remained on show for a hundred and fifty years after her death. Her abused body ended up at the Musée de l'Homme, the Museum of Man, looking out on the Eiffel Tower, a plaster cast of it still standing there until as recently as 1982. It was only in 2002, after a request from Nelson Mandela, that her remains were removed from Paris and finally returned to South Africa for burial.

*

'In the modern world we look to science as a rationalisation of political ideas,' I'm told by Jonathan Marks, a genial, generous professor of anthropology at the University of North Carolina at Charlotte. He is one of the most outspoken voices against scientific racism. Race science, he explains, emerged 'in the context of colonial political ideologies, of oppression and exploitation. It was a need to classify people, make them as homogeneous as possible.' By grouping people and dividing these groups, it was easier to control them.

It is no accident that modern ideas of race were formed during the heyday of European colonialism, when those in power had already decided on their superiority. By the nineteenth century, the possibility that races existed and some were inferior to others gave colonialism a moral kick in the drive for public support. The truth – that European nations were motivated by economic greed or power – was harder to swallow than the suggestion that the places they were colonising were too uncivilised and barbaric to matter, or that they were actually doing the savages a favour.

In the United States, the same tortured logic was used to justify slavery. The transatlantic trade in slaves officially ended in 1807 once the United Kingdom passed its Slave Trade Act, but the exploitation continued for far longer. The use of slave labour continued, people's bodies plundered both in life and death. Dead black slaves, for instance, were routinely stolen or sold for medical dissection. Daina Ramey Berry, professor of history at the University of Texas at Austin, has documented the economic value of slavery in the United States. She notes that there was a brisk trade in black corpses in the nineteenth century, some exhumed by their owners for a quick profit. It's ironic that much of our modern scientific understanding of human anatomy was built on the bodies of those who were considered at the time less than human.

'If you could say that the slavers were naturally distinct from the slaves, then you have essentially a moral argument in favour of slavery,' explains Jonathan Marks. Given this distinction, many feared that the abolition of slavery would set free the human zoo, unleashing chaos. In 1822 a group calling itself the American Colonisation Society bought land in West Africa to establish a colony named Liberia, now the Republic of Liberia, motivated largely by the desperate dread that freed black slaves would want to settle among them, with the same rights. Repatriation to the continent of their ancestors seemed like a convenient solution, ignoring the fact that after generations in slavery, most black Americans simply didn't have a tangible connection to it any more – let alone to a new country that their ancestors may never have seen.

Louis Agassiz, a Swiss naturalist who had been mentored by Georges Cuvier and moved to America in 1846, argued passionately against blacks being treated the same as whites. Shaken by such an intense physical disgust towards black domestic workers serving him food at a hotel that he almost couldn't eat there at all, he became convinced that separate races originated in different places, with different characters and intellectual abilities.

Enslavement was turned back on the slaves themselves. They were in this miserable, degrading position not because they had been forcibly enslaved, it was argued, but because it was their biological place in the universe. At a meeting of the British Association for the Advancement of Science in Plymouth in 1841, an American slave owner from Kentucky named Charles Caldwell had already claimed that Africans bore more of a resemblance to apes. In their 1854 book *Types of Mankind*, American physician Josiah Clark Nott and Egyptologist George Gliddon went so far as to sketch actual comparisons between the skulls of white and black people, alongside those of apes. While the typical European face was artfully modelled on classical sculpture, African faces were crude cartoons, exaggerating features that made it seem they had more in common with chimpanzees and gorillas.

Propelled by a belief that black people had their own unique diseases, Samuel Cartwright, a medical doctor practising in Louisiana and Mississippi, characterised in 1851 what he saw as a mental condition particular to black slaves, coining it 'drapetomania', or 'the disease causing Negroes to run away'. Harvard University historian Evelynn Hammonds, who teaches Cartwright's story to her students, laughs darkly when she recounts it. 'It makes sense to him, because if the natural state of the negro is to be a slave, then running away is going against their natural state. And therefore it's a disease.'

For Hammonds, another chilling aspect of Cartwright's work is the way in which he methodically described the sufferers of drapetomania. 'The colour of the skin is the main difference,' she reads for me from her notes, '. . . the membranes, the muscles, the tendons, all fluids and secretions, then the nerves, and the bile. There's a difference in the flesh. The bones are whiter and harder, the neck is shorter and more oblique.' Cartwright continues this way, couching racism in medical terminology. 'These kinds of observations turned into questions to be explored going forward. Since the 1850s, people have been trying to figure out if black bones are harder than white bones,' Hammonds explains. Cartwright's medical 'discoveries' were patently rooted in the desire to keep slaves enslaved, to maintain the status quo in the American South where he lived. In place of universal humanity came a self-serving version of the human story, in which racial difference became an excuse for treating people differently. Time and again, science provided the intellectual authority for racism, just as it had helped define race to begin with.

Race science became a pastime for non-scientists, too. French aristocrat and writer Count Arthur de Gobineau, in *An Essay on the Inequality of the Human Races*, published in 1853, proposed that there were three races, with what he saw as an obvious hierarchy between them: 'The negroid variety is the lowest, and stands at the foot of the ladder . . . His intellect will always move within a very narrow circle.' Pointing to the 'triangular' face shape of the 'yellow race', he explained that this was the opposite of the negroid variety. 'The yellow man has little physical energy, and is inclined to apathy . . . He tends to mediocrity in everything.' Neither could be a match for Gobineau's own race.

Reaching his predictable pinnacle, Gobineau added, 'We

come now to the white peoples. These are gifted with reflective energy, or rather with an energetic intelligence. They have a feeling for utility, but in a sense far wider and higher, more courageous and ideal, than the yellow races.' His work was a naked attempt to justify why those like him deserved the power and wealth they already had. This was the natural order of things, he argued. He didn't need hard evidence for his theories because there were plenty of people around him ready and willing to agree that they, too, belonged to a superior race.

It would be Gobineau's ideas that would later help reinforce the myth of racial purity and the creed of white supremacy. 'If the three great types had remained strictly separate, the supremacy would no doubt have always been in the hands of the finest of the white races, and the yellow and black varieties would have crawled forever at the feet of the lowest of the whites,' he wrote, promoting a notion of an imaginary 'Aryan' race. These glorious Aryans, he believed, had existed in India many centuries ago, speaking an ancestral Indo-European language, and had since spread across parts of the world, diluting their superior bloodline.

Myth and science coexisted, and both served politics. In the run-up to the passage in 1865 of the 13th Amendment, abolishing slavery in the United States, the race question wasn't resolved – it just became thornier. Although many Americans believed in emancipation on moral grounds, fewer were convinced that full equality would ever be possible, for the simple reason that groups weren't biologically the same. Even Presidents Thomas Jefferson and Abraham Lincoln believed that blacks were inherently inferior to whites. Jefferson, himself a slaveholder, agreed with those who thought that the best way to deal with freed slaves was to send them to a colony of their own. Freedom was framed as a gift bestowed on unfortunate black slaves by morally superior white leaders, rather than a reflection of a hope that everyone would one day live alongside each other as friends, colleagues and partners.

*

Not all scientists were quite so self-serving. For those who wanted to establish the facts about human difference, there were unanswered questions. The biggest puzzle was that there was no fleshed-out mechanism to account for how different races – if they were real – might have emerged. If each race was distinct, then where did they each come from, and why? Going by the Bible, as many Europeans did, one explanation for the existence of different races was that, after the big flood, Noah's children spread to different parts of the earth. How we truly originated, and how physical differences appeared between us, were anyone's guess.

In 1871 biologist Charles Darwin published *The Descent of Man*, sweeping away these religious creation myths and framing the human species as having had one common ancestor many millennia ago, evolving slowly like all other life on earth. Studying humans across the world, their emotions and expressions, he wrote, 'It seems improbable to me in the highest degree that so much similarity, or rather identity of structure, could have been acquired by independent means.' We are too alike in our basic responses, our smiles and tears, our blushes. On this alone, Darwin might have settled the race debate. He demonstrated that we could only have evolved from shared origins, that human races didn't emerge separately.

On a personal level, this was important to him. Darwin's family included influential abolitionists, his grandfathers Erasmus Darwin and Josiah Wedgwood. He himself had seen the brutality of slavery first-hand on his travels. When naturalist

Louis Agassiz in the United States spoke about human races having separate origins, Darwin wrote disparagingly in a letter that this must have come as comfort to slaveholding Southerners.

But this wasn't the last word on the subject. Darwin still struggled when it came to race. Like Abraham Lincoln, who was born on the same day, he opposed slavery but was also ambivalent on the question of whether black Africans and Australians were strictly equal to white Europeans on the evolutionary scale. He left open the possibility that, even though we could all be traced back to a common ancestor, that we were the same kind, populations may have diverged since then, producing levels of difference. As British anthropologist Tim Ingold notes, Darwin saw gradations between the 'highest men of the highest races and the lowest savages'. He suggested, for example, that the 'children of savages' have a stronger tendency to protrude their lips when they sulk than European children, because they are closer to the 'primordial condition', similar to chimps. Gregory Radick, historian and philosopher of science at the University of Leeds, observes that Darwin, even though he made such a bold and original contribution to the idea of racial unity, also seemed to be unembarrassed by his belief in an evolutionary hierarchy. Men were above women, and white races were above others.

In combination with the politics of the day, this was devastating. Uncertainty around the biological facts left more than enough room for ideology to be mixed with real science, fabricating fresh racial myths. Some argued that brown and yellow races were a bit higher up than black, while whites were the most evolved, and by implication, the most civilised and the most human. What was seen to be the success of the white races became couched in the language of the 'survival of the fittest', with the implication that the most 'primitive' peoples, as they were described, would inevitably lose the struggle for survival as the human race evolved. Rather than seeing evolution acting to make a species better adapted to its particular environment, Tim Ingold argues that Darwin himself began to frame evolution as an 'imperialist doctrine of progress'.

'In bringing the rise of science and civilisation within the compass of the same evolutionary process that had made humans out of apes, and apes out of creatures lower in the scale, Darwin was forced to attribute what he saw as the ascendancy of reason to hereditary endowment,' writes Ingold. 'For the theory to work, there had to be significant differences in such endowment between "tribes" or "nations".' For hunter-gatherers to live so differently from city-dwellers, the logic goes, it must be that their brains had not yet progressed to the same stage of evolution.

Adding fuel to this bonfire of flawed thinking (after all, we know that the brains of hunter-gatherers are no different from those of anyone else) were Darwin's supporters, some of whom happened to be fervent racists. The English biologist Thomas Henry Huxley, known as 'Darwin's Bulldog', argued that not all humans were equal. In an 1865 essay on the emancipation of black slaves, he wrote that the average white was 'bigger brained', adding, 'The highest places in the hierarchy of civilisation will assuredly not be within the reach of our dusky cousins.' For Huxley, freeing slaves was a morally good thing for white men to do, but the raw facts of biology made the idea of equal rights – for women as well as for black people – little more than an 'illogical delusion'. In Germany, meanwhile, Darwin's loudest cheerleader was Ernst Haeckel, who taught zoology at the University of Jena from 1862, and was a proud nationalist. He liked to draw connections between black Africans and primates, seeing them as a kind of living 'missing link' in the evolutionary chain that connected apes to white Europeans.

Darwinism did nothing to inhibit racism. Instead, ideas about the existence of different races and their relative superiority were merely repackaged in new theories. Science, or the lack of it, managed only to legitimise racism, rather than quash it. Whatever real and reasonable questions might have been asked about human difference were always tainted by power and money.

*

I pick my way through a tall thicket of bamboo and find an intricate wooden pagoda.

Further still inside the sunlit Garden for Tropical Agriculture is a Tunisian house, coated in thick green moss. If their histories were unknown to me, I might find the buildings in this quiet maze beautiful. They are grand and otherworldly, ethereal relics of foreign places as imagined by another age. But of course, I'm acutely aware that each was also once a kind-of home to real people like me, pulled from their lives thousands of miles away for the entertainment of paying visitors. As a reminder, through the smashed window of a Moroccan castle, complete with battlements and blue tiles, I'm caught off-guard by a glaring red face that must have been painted by vandals.

However beautiful they are, these aren't homes at all. They're gilded cages.

It's hard to imagine what life would have been like on the inside of the human zoos, looking out. The people kept here weren't slaves. They were paid, similar to actors under contract, but expected to dance, act, and carry out their everyday routines in public view. Their lives were live entertainment. They were objects first and people second. Little effort was made to help them feel comfortable in their temporary homes, much less to acclimatise them. After all, the whole point of the spectacle was to underscore just how different they were, to imagine that even in a cold climate they would choose to walk around in as few clothes as they wore in a hot one, that their behaviour couldn't change no matter where they lived. Visitors were made to believe that the cultural differences were woven into their bodies like stripes on a zebra. 'When there was a birth, it meant a new show,' Gilles Boëtsch from the National Center for Scientific Research tells me. People would flock to see the baby.

Science had created a distance between the viewers and the viewed, the colonisers and the colonised, the powerful and the powerless. For those confronted with people from foreign lands in this way, bizarrely out of context, referenced in a book or transplanted to some fake village in Paris, it only helped reinforce the notion that we were not all quite the same. For the spectators peering into their homes, the performers in human zoos must have been curiosities not just because they looked and behaved differently, but because control of their lives belonged to others who didn't look like them. The ones outside the cage were clothed, civilised and respectable while those inside were semi-naked, barbaric and subjugated.

'People are more readily perceived as inferior by nature when they are already seen as oppressed,' write American scholars Karen Fields and Barbara Fields in their 2012 book *Racecraft*. They explain how a sense of inevitability gets attached to a social routine until it becomes seen as natural. The idea of race didn't make people treat other people as subhuman. They were already treated as subhuman before race was invoked. But once it was invoked, the subjugation took on a new force.

There was something about treating human difference as a science that gave it a peculiar quality. The observation of humans turned humans into strange beasts. While the unimpeachable impression of scientific objectivity was maintained, somehow the gold standard of beauty and intelligence always turned out to be the scientist himself. His own race was safe in his hands. German naturalist Johann Blumenbach, for instance, idealised the Caucasian race to which he belonged, but described Ethiopians as being 'bandy-legged'. If legs were different, there was never any question that Caucasians might be the unusual ones. The creatures caged in the human zoos were those who had failed to reach the ideal of white European physical and mental perfection.

The scientific distance created by believing that racial hierarchies existed in nature, this uneven balance of power, allowed human zoos to treat their performers as less than equals, making life for them fatally precarious. According to Boëtsch, many died from pneumonia or tuberculosis. Concerns were expressed in the press. There were always protests, as there had been about Saartjie Baartman, but they made little difference.

In another example around the same time as the Paris Exposition, a Congolese 'pygmy' named Ota Benga, who had been brought to the United States to be displayed at the St Louis World's Fair, was put in the Monkey House at Bronx Zoo in New York, without shoes. Visitors loved him. 'Some of them poked him in the ribs, others tripped him up, all laughed at him,' the *New York Times* reported. He was eventually rescued by African American ministers, who found him a place in an orphanage. Ten years later, in despair because he couldn't return home to the Congo, he borrowed a revolver and shot himself through the heart.

As I stand among the weeds and crumbling former homes of Paris's human zoo, it's difficult to avoid concluding that the reason anyone pursued the scientific idea of race was not so much to understand the differences in our bodies, but to try to justify why we lead such different lives. Why else? Why would something as superficial as skin colour or body shape matter otherwise? What the scientists really wanted to know was why some people are enslaved and others free, why some prosper while others are poor, and why some civilisations have thrived while others haven't. Imagining themselves to be looking objectively at human variation, they sought answers in our bodies to questions that existed far outside them. Race science had sat, always, at the intersection of science and politics, of science and economics. Race wasn't just a tool for classifying physical difference, it was a way of measuring human progress, of placing judgement on the capacities and rights of others.

3

Scientific Priestcraft

Deciding that races could be improved, scientists looked for ways to improve their own

THE PAST IS BUILT of the things we choose to remember.

The Max Planck Society, with its headquarters in Munich, Germany, has an illustrious history. It has been the intellectual home of eighteen Nobel Prize winners, including the theoretical physicist Max Planck, after whom it's named. With an annual budget of 1.8 billion euros, its institutes employ more than 14,000 scientists, producing over 15,000 published scientific papers a year. By any standards, it's one of the most prestigious centres of science in the world. But in 1997 biologist Hubert Markl, then president of the Max Planck Society, made a decision that would threaten the reputation of his entire establishment. He wanted to scratch beneath its glorious history to reveal a secret that had been hidden for fifty years.

Before 1948, the Max Planck Society existed in a different incarnation, the Kaiser Wilhelm Society. Established in the German Empire in 1911, it was as important then as it is now, cementing Germany's place in modern scientific history. Even

Albert Einstein did some of his research at one of its institutes. But it was later, as the Nazis took power and began to act on their own scientific priorities, that things took a disturbing turn. We know that figures from within science and academia must have played a role in developing Adolf Hitler's ideology of racial hygiene, which argued that those of pure, 'Aryan' racial stock should be encouraged to breed, while others were gradually eliminated – an ideology that culminated in the Holocaust. It couldn't have been done without scientists, both to provide the theoretical framework for such an audacious experiment, and to carry out the job itself. On the practical side, there would have been those setting up concentration camps and gas chambers, as well as determining who should die. And then there were all the gruesome human experiments known to have been carried out on people who were eventually killed, plundering them for biological data.

There were rumours that staff from within the Kaiser Wilhelm Society had been involved, that they were maybe even party to murder and torture. In hindsight, they must have been. Under the regime, notes writer James Hawes, half the nation's doctors were Nazi party members. For a decade, German universities had taught racial theory.

But whatever went on was quietly forgotten after the Second World War. Although there was undoubtedly a story to be uncovered, it was thought wiser to leave it alone. By the Max Planck Society's own admission, it had a tradition of glossing over its ignominious past in favour of celebrating its greater scientific achievements. By the 1990s, however, there was too much pressure from the public to ignore that past any longer. And anyway, older members of staff who had been alive during the war – who might be affected by such revelations – had almost all died. The time had come. So Markl resolved to lift the lid, appointing an independent committee to investigate what German scientists at the Kaiser Wilhelm Society might have done during the war. It would be an investigation into the very darkest corners of race science. Younger researchers at the Max Planck Society justifiably worried whether the body of scientific work they had inherited might bear bloody stains.

They were right to worry. The past turned out to be dripping with blood. A few years after Markl launched the investigation, historians began publishing their findings, and they were devastating. Some had assumed that the Nazis were ignorant of or hostile towards science. Historical evidence proved this wasn't true. The Kaiser Wilhelm Society's scientists had willingly cooperated with the Nazi state, marrying academic interests and political expediency, helping to secure financial support and social standing for themselves. 'Such research not only literally built on the spoils of war, it also led scientists deep into the abysses of Nazi crimes,' wrote a reviewer. At least one prominent scientist helped draft and disseminate the legislation relating to racial ideology.

Those who weren't opportunistic were often complicit, displaying moral indifference when they could see inhumane or criminal acts happening right in front of them. When moves began in 1933 to expel Jewish scientists from the Kaiser Wilhelm Society (Einstein abandoned Germany that same year, leaving for a conference and wisely never returning), staff made little effort to stand in the way. At least two of its scientists and two other staff members ended up dying in concentration camps.

And then there were those who wholeheartedly supported the Nazis from the beginning. The work of Otmar von Verschuer, head of department at the Kaiser Wilhelm Institute of Anthropology, Human Heredity and Eugenics, makes for chilling reading. Until the war, von Verschuer was a widely respected academic, his research on twins as a means of understanding genetic inheritance funded for a few years by the Rockefeller Foundation in New York. He was once invited to speak at the Royal Society in London. But he was also, it transpired, an anti-Semite who openly praised Hitler and believed in a biological solution to what he saw as the Jewish threat to racial purity. According to American anthropologist Robert Wald Sussman, von Verschuer became one of the Nazis' race experts when it came to addressing the 'Jewish question', actively legitimising the regime's racial policies. One of his former students, the doctor Josef Mengele, went on to become infamous for his cruel experiments on twins and pregnant women at Auschwitz concentration camp. British writer Marek Kohn has documented in his 1995 book *The Race Gallery* that among the samples sent to von Verschuer from Auschwitz were 'pairs of eyes from twins . . . dissected after their murder . . . children's internal organs, corpses and the skeletons of murdered Jews'.

In 2001, the Max Planck Society at last accepted responsibility for historic crimes committed by its scientists. In its apology, the society admitted, 'Today it is safe to say that von Verschuer knew of the crimes being committed in Auschwitz and that he, together with some of his employees and colleagues, used them for his purposes.' Markl added in his speech, 'The Kaiser Wilhelm Society tolerated or even supported research among its ranks that cannot be justified on any ethical or moral grounds . . . I would like to apologise for the suffering of the victims of these crimes – the dead as well as the survivors – done in the name of science.'

This came too late for justice, of course. Those involved had died already. What was remarkable was that it had taken so long to root out the facts, to even find the will to do it. Scientists complicit with the regime had been skilled at covering their tracks, evidently. But maybe it was also easier for their colleagues to pretend that fellow scientists couldn't possibly have been active participants in murder and torture. Perhaps, they imagined, they were just bystanders, caught up in the mess while trying to get their work done.

The truth – that it is perfectly possible for prominent scientists to be racist, to murder, to abuse both people and knowledge – doesn't sit easily with the way we like to think about scientific research. We imagine that it's above politics, that it's a noble, rational and objective endeavour, untainted by feelings or prejudice. But if science is always so innocent, how is it that members of such a large and prestigious scientific organisation could have sold themselves to a murderous political regime as recently as the middle of the twentieth century?

The answer is simple: science is always shaped by the time and the place in which it is carried out. It ultimately sits at the mercy of the personal political beliefs of those carrying it out. In the case of some Nazi scientists, particular experiments may have been perfectly accurate and rigorous. They may even have produced good science, if goodness is measured in data and not human life. Other times, researchers didn't care about the truth or other people's lives, choosing instead to give the illusion of intellectual weight to a morally bankrupt ideology because it suited them.

Now, decades later, the horrors of the Second World War still have a warping effect on how we think about race science. Many of us choose to remember Nazi scientists like Otmar von Verschuer as some kind of uniquely evil exception, nothing like those who found themselves on the winning side of the war. The Holocaust and the twisted scientific rationale behind it are thought to belong to that time and place alone, purely the work of 'the bad guys'. But there was one question that went unanswered after the investigations into the bloodstained history of the Max Planck Society: Were scientists in the rest of the world so blameless?

To file away what happened during the war as aberrant, as something that could only have been done by the worst people under the worst circumstances, ignores the bigger truth. This was never a simple story of good versus evil. The well of scientific ideas from which Hitler and others in his regime drew their plans for 'racial hygiene', leading ultimately to genocide, didn't originate in Germany alone. They had been steadily supplied for more than a century by race scientists from all over the world, supported by well-respected intellectuals, aristocrats, political leaders and women and men of wealth.

Among the most influential of them all, as far as the Nazi regime was concerned, was a pair of statisticians working at 50 Gower Street, Bloomsbury – not in Germany, but in the famous old literary quarter of London.

*

'You have biologists who say there is no such thing as race, we need to get over it, forget it,' Subhadra Das tells me in an angry whisper. 'But then, if there is no such thing, why did you just say "race"? Where did that idea come from?'

Das is a curator of the University College London Medical and Science Collections, moonlighting occasionally as a stand-up comedian. Her dark wit betrays a fury fed by the things she's learned from her research. We're in the heart of Bloomsbury, recognisable by its peaceful garden squares and smart Georgian townhouses. Once a meeting point for artists and writers, including Virginia Woolf, it is still home to a large slice of London's universities and colleges. Outside, busy Gower Street is jam-packed with students heading for lectures, but where Das and I are it's library quiet. We're seated at a small table inside the Petrie Museum, named for Sir Flinders Petrie, an Egyptologist who, before he died in 1942, used to collect heads from around the world to shore up his ideas of racial superiority and inferiority.

'Scientists are socialised human beings who live within society, and their ideas are social constructions,' she continues. She wants me to hear this, setting the scene before she begins unfolding the packets of objects in front of us, which she has pulled from the archive. Among the first is a black-and-white photograph of a well-dressed older man, his bushy eyebrows resting in a canopy over his eyes, long white sideburns trailing down to his collar. Underneath is his autograph: it is the biologist Francis Galton, born in 1822, a younger cousin of Charles Darwin. Galton, she tells me, is the father of eugenics. He coined the term in 1883 from the Greek prefix '*eu*' for 'well' or 'good', to describe the idea of using social control to improve the health and intelligence of future generations.

Galton considered himself an expert on human difference, on the finer qualities that make a person better or worse. Not quite the genius that Darwin was, he certainly aspired to be. 'I find that talent is transmitted by inheritance in a very remarkable degree,' he had written in an essay titled 'Hereditary Character and Talent'. His idea drew on his cousin's theory of evolution by natural selection, that individuals in a population show a wide variety of characteristics, but those with the characteristics most suited to the environment will survive and breed, passing on those beneficial traits. Galton thought that a race of people could be more quickly improved if the most intelligent were encouraged to reproduce, while the stupidest weren't – the same way you might artificially breed a fatter cow or a redder apple. For him, this would speed up human evolution, driving the race closer to mental and physical perfection.

As an example, he drew on the fact that brilliant writers were often related to other brilliant writers. He noted that of 605 notable men who lived between 1453 and 1853, one in six were related. The ingredients for greatness must be heritable, he reasoned, choosing to overlook that being notable might also be a product of connections, privilege and wealth, which these men also had. 'If a twentieth part of the cost and pains were spent in measures for the improvement of the human race that is spent on the improvement of the breed of horses and cattle, what a galaxy of genius might we not create!' Galton dreamed of a 'utopia' of highly bred super people, and he made creating one his lifelong mission.

The first challenge would be to measure people's abilities, to build up a bank of data about who exactly were the most intelligent and who the least. In 1904 he convinced the University of London to set up the world's first Eugenics Record Office at 50 Gower Street, dedicated to measuring human differences, in the hope of understanding what kind of people Britain might want more of. University College London jumped at the chance, replying to his request within a week. After a short time the department became known as the Galton Laboratory for National Eugenics.

Eugenics is a word that's no longer used around here. Long after Galton's death, his laboratory was renamed the Department of Genetics, Evolution and Environment, housed in the Darwin Building. And this is where Subhadra Das steps in. Among the vast collection of objects she is responsible for at the university is Galton's archive, containing his personal photographs, equipment and papers, tracking the genesis and development of eugenics. She also looks after objects belonging to his close collaborator, mathematician Karl Pearson, who became the first professor of national eugenics in 1911 after Galton died. 'Pearson's greatest contribution, the thing that people remember him for, is founding the discipline of statistics. A lot of work on that was done with Galton. Galton, if you're going to bring his science down to anything in particular, is a statistician,' she tells me.

But before he settled down into science, Galton had been an explorer. He was lavishly funded by the estate of his father, who had made a fortune from supplying weapons that helped support the slave trade, and later from banking. An expedition in 1850 to Namibia, then known as Damaraland, earned Galton a medal from the Royal Geographical Society. Always proud of his appearance (there's a hand mirror and sewing kit among his possessions in the collection), he donned a white safari suit, becoming one of the first to cultivate what is now the classic image of the white European in Africa. 'If I say to you "African explorer", the picture that pops into your head? That's him,' Das tells me.

What was unusual about Galton was that travel failed to broaden his mind. His encounters with people in other countries didn't help him to see their common humanity. 'If anything, his racist assumptions were made stronger by his time in Africa.' As Galton told the Royal Society on his return, 'I saw enough of savage races to give me material to think about all the rest of my life.'

In London, racism combined in his scientific research with a passion for data. Galton was obsessed with measuring things, once using a sextant to size up an African woman's proportions from a distance. Another time, he came up with the mathematical formula for the perfect cup of tea. Through eugenics he saw a way of using what he thought he knew about human difference, shored up by Darwin's theories of natural selection, to systematically improve the quality of 'the British race'. 'Darwin said that humans are animals like any other animal. Galton said, well, if that's the case then we can breed them better,' Das explains. 'What he was concerned about was what he saw as the degeneration of the British race and how that could be prevented and improved.

'You have to call Galton a racist because the work that he did is fundamental in the story of scientific racism. So not only is he a racist, he is part of the way we invented racism, and the way that we think about it.'

*

Eugenics is a cold, calculated way of thinking about human life, reducing human beings to nothing but parts of the whole, either dragging down their race or pulling it up. It also assumes that almost all that we are is decided before we are born.

The origins of this idea – that everything is inherited, that it's in the genes – date back to the middle of the nineteenth century when Gregor Mendel, an Augustinian friar in Brno, Moravia, then part of the Austro-Hungarian Empire, became fascinated by plant hybrids. Working in the garden of his monastery, Mendel took seven strains of pea and bred them selectively until each one produced identical offspring every time. With these true-bred pea plants, he began to experiment, observing carefully to see what happened when different varieties were crossed. Nobody knew about genes at this point, and Mendel's paper on the topic published in 1866 went largely unnoticed within his lifetime. But his experimental finding that traits such as colour were being passed down the generations in certain patterns would form the linchpin of how geneticists in the following century thought about inheritance.

Once scientists understood that there were discrete packets of information in our cells that dictated how our bodies were built, and that we got these packets in roughly equal measure from each parent, the science of heredity finally took off. And it took almost no time for the political implications to be recognised. In 1905 the English biologist William Bateson, Mendel's principal populariser, predicted that it 'would soon provide power on a stupendous scale'.

Mendelism became a creed, an approach to thinking about human biology which suggested that it is largely set in motion as soon as an egg is fertilised, and that things then go on to work in fairly linear fashion. If you crossed one yellow-seeded pea plant with one green-seeded pea plant and you could predict which colours subsequent generations of pea plant would turn out to have, then it stood to reason that you might be able to predict how human children would look and behave based on the appearance and behaviour of their parents.

Through a narrow Mendelian lens, almost everything is planned by our genes. Environment counts for relatively little because we are at heart the products of chemical compounds mixing together. We are inevitable mixtures of our ancestors. Just as Bateson foresaw, this idea became the cornerstone of eugenics, the belief that better people could be bred by selecting better parents. 'Mendelism and determinism, the view that heredity is destiny, they go together,' says historian Gregory Radick, who has studied Mendel and his legacy.

But there was a problem with Mendel's pea plant research. At the beginning of the twentieth century Mendel's paper became the subject of ferocious debate, says Radick. 'Should the Mendelian view be the big generalisation around which you hang everything else? Or on the contrary, was it an interesting set of special cases?' When Mendel performed his experiments, he deliberately bred his peas to be reliable in every generation. Before he even began, he filtered out the aberrations, the random mutants, the messy spread of continuous variation you would normally see, so every generation bred as true as possible. Peas were either green or yellow. This allowed him to see a clear genetic signal through the noise, producing results that were far more perfect than nature would have provided.

Raphael Weldon, born in 1860, a professor at the University of Oxford with an interest in applying statistics to biology, spotted this dilemma and began campaigning for scientists to recognise the importance of environmental as well as genetic backgrounds when thinking about inheritance. 'What really bothered him about the emerging Mendelism was that it turned its back on what he regarded as the last twenty years of evidence from experimental embryology, whose message was that the effects a tissue has on a body depend radically on what it's interacting with, on what's around it,' explains Radick. Weldon's message was that variation matters, and that it is profoundly affected by context, be it neighbouring genes or the quality of air a person breathes. Everything can influence the direction of development, making nurture not some kind of afterthought tacked onto nature, but something embedded deep down in our bodies. 'Weldon was unusually sceptical.'

To prove his point, Weldon demonstrated how ordinary pea breeders couldn't come up with the same perfectly uniform peas as Mendel. Real peas are a multitude of colours between yellow and green. In the same way that our eyes aren't simply brown or blue or green, but a million different shades. Or that if a woman has a 'gene for breast cancer', it doesn't mean she will necessarily develop the disease. Or that a queen bee isn't born a queen; she is just another worker bee until she eats enough royal jelly. Between the gene and real life is not just the environment, but also random possibility. Comparing Mendel's peas with the real world, then, is like comparing a soap opera with real life. There is truth in there, but reality is a lot more complicated. Genes aren't Lego bricks or simple instruction manuals; they are interactive. They are enmeshed in a network of other genes, their immediate surroundings and the wider world, this ever-changing network producing a unique individual.

Sadly for Weldon, the ferocious debate for the soul of genetics ended prematurely in 1906 when he died of pneumonia, aged just forty-six. His manuscript went unfinished and unpublished. With less resistance than before, Mendel's ideas were gradually incorporated into biology textbooks, becoming the bedrock of modern genetics. Although Weldon's ideas have since slowly been reincorporated into scientific thinking, there still remains a strain of genetic determinism in both the scientific and the public imagination. Harvard biologist Richard Lewontin has called it the 'Central Dogma of Molecular Genetics'. It is a belief that all that we are is set in stone in the womb.

In the early twentieth century, before the advent of modern genetics but with Mendel's findings prominent in their minds, Francis Galton's theories seemed to make good sense to many. They had a logical appeal that stretched across the political spectrum. We associate eugenics today with the fascists who perpetrated the Holocaust, but before the 1930s, many on the left saw it as socially progressive. Galton himself was certainly not considered a crank. He was a fellow of the Royal Society, and an anthropometric laboratory he set up in 1884 to catalogue people's measurements enjoyed support from the British Medical Association. Eugenics belonged firmly to establishment science, and amongst intellectuals, it wasn't just mainstream, it was fashionable.

The fly in the ointment was how to carry it out. Galton

observed that the poor seemed to be outbreeding the rich, and he saw the poor as poor for the simple reason that they were congenitally unfit. Responsible action was necessary to address the problem and ensure genetic progress. On the one hand, the rich needed to step up their baby-making game. On the other, society's dregs, particularly those described as mentally feeble, physically weak, and criminal types, needed convincing to have fewer children. Managing reproduction was the linchpin of eugenics, even attracting a fan in women's rights activist and birth control pioneer Marie Stopes. To support her first clinic, Stopes founded the Society for Constructive Birth Control and Racial Progress. Philosopher Bertrand Russell, too, suggested that the state might improve the health of the population by fining the 'wrong' type of people for giving birth.

Eugenics was more than a theory, it was a plan in search of policymakers. Winston Churchill, then First Lord of the Admiralty, was welcomed as vice-president at the first International Eugenics Congress, held at the University of London in 1912. Other vice-presidents included the Lord Mayor of London and the Lord Chief Justice. Delegates came from all over Europe, Australia and the United States, including Harvard and Johns Hopkins University. The US state of Indiana had already passed the world's first involuntary sterilisation law in 1907, informed by eugenicists who argued that criminality, mental problems and poverty were hereditary. More than thirty other states soon followed, with enthusiastic public backing. By 1910 a Eugenics Record Office was established at Cold Spring Harbor on Long Island in New York, with support from oil industry magnate John D. Rockefeller and later funding by the Carnegie Institution of Washington.

A news item in the journal *Science* announced that one of the purposes of the new office in New York would be 'the study of miscegenation in the United States', the mixing and intermarriage of different racial groups. Its board of scientific directors included Alexander Graham Bell, inventor of the telephone, and the economist Irving Fisher. The hardware behind at least one of America's most ambitious eugenics projects came from none other than IBM, the same company that went on to supply the Nazi regime in Germany with the technology it needed to transport millions of victims to the concentration camps.

In the first decades of the twentieth century, all over the world, eugenics began to be conflated with nineteenth-century ideas about race. In Japan, Meiji-period thinker and politician Katō Hiroyuki used Darwinism to make the point that there was a struggle for survival between different nations. In China in 1905, the revolutionary Wang Jingwei argued that a state whose members were of a single race was stronger than one comprising multiple races. Other politicians advocated sterilisation as a means of human selection, and racial intermarriage to produce children with whiter skins. Historian Yuehtsen Juliette Chung has noted that during this time, 'China seemed to accept passively the notion of race as the West understood it.'

In India, too, European notions of racial superiority were easily absorbed by some, partly because they mirrored the country's existing caste system – itself a kind of racial hierarchy – but also because Germany's Aryan myth placed the noble race as having once lived in their region. The ideological quest for the true 'Aryans' remains alive in India, and Adolf Hitler's *Mein Kampf* is a bestseller in Indian bookshops. Each nation utilised the idea of race in its own ways, marrying it with science if it could be of use. Eugenics, then, became just another tool in what were longstanding power dynamics.

By 1914 the word 'eugenics' was being used with such abandon that it had almost became synonymous with being healthy, complained American eugenics professor Roswell H. Johnson in the *American Journal of Sociology*. 'A school for sex education is called a school of eugenics. Even a milk and ice station has been similarly designated,' he grumbled.

*

In its early days, particularly for its mainstream supporters, eugenics focused on improving racial stock by weeding out those seen to be at the margins of society, the feeble-minded, insane and disabled. But as time wore on, the umbrella inevitably expanded. Karl Pearson, who succeeded Galton as the main force behind eugenics when he died in 1911 and shared his views on race, believed that since other races than his own were inferior, intermixing was also dangerous to the health of the population. By this logic, the very existence of those other races represented something of a threat. 'Pearson's argument is that if you have uncontrolled immigration the welfare of British people is at stake,' Subhadra Das tells me.

At the time, despite the mainstream popularity of eugenics, some did notice the slippery slope. This is one reason why, despite all the support it attracted from politicians and intellectuals and how popular it became in other countries, eugenics never managed to gain a firm toehold in Britain and was not implemented by the government. British psychiatrist Henry Maudsley argued that privilege and upbringing could surely more accurately explain why some people were successful and others weren't. He noted that many remarkable people had unremarkable relatives. Another vocal critic was biologist Alfred Russel Wallace, who had come from humble beginnings to become an important and well-loved researcher, credited with formulating evolutionary theory at the same time as Darwin. 'The world does not want the eugenicist to set it straight,' he warned. 'Give the people good conditions, improve their environment, and all will tend towards the highest type. Eugenics is simply the meddlesome interference of an arrogant, scientific priestcraft.'

But it's important to remember that history might well have gone another way. Das pulls out another object from the archive. It's a narrow tin box, resembling a cigarette case but twice as long. It was brought to London by Karl Pearson, but had been designed by Eugen Fischer, a German scientist who had been director of the Kaiser Wilhelm Institute of Anthropology, Human Heredity and Eugenics. The box still bears Fischer's name. Inside is a neat row of thirty locks of artificial hair, ranging in colour from blonde (numbers 19 and 20) and light brunette in the centre, to bright red hair at one end and black Afro hair (number 30) at the other. At first glance it looks innocuous, like a colour chart you might find at the hairdresser's. But the disturbing story behind it is betrayed by the order in which the hair samples are placed. The most desirable colours and textures have been placed in the middle and the least acceptable at the margins. This simple little gauge tells a story of pure horror.

'Fischer used this device in Namibia in 1908 to establish the relative whiteness of mixed-race people,' reveals Das. In what is now remembered as the first genocide of the twentieth century, in the four years preceding 1908, Germany killed tens of thousands of Namibians as they rebelled against colonial rule. According to some estimates, up to 3,000 skulls belonging to those of the Herero ethnic group were sent back to Berlin to be studied by race scientists. 'Namibia was the first place that the Germans built a concentration camp. Depending on where your hair fell on the scale was the difference between life and death.' Similar methods would be used again, of course, a few decades later. Fischer's work would also go on to inform the Nuremberg Laws of 1935, outlawing intermarriage between Jews, blacks and other Germans. He became a member of the Nazi party in 1940.

Das takes out another box that belonged to Pearson, this time containing rows of glass eyes in different colours, framed in aluminium eyelids so eerily real that I fear one of them might blink. They are prosthetics of the kind that would have been fitted in patients who had eyes missing. In the context of eugenics, though, they served another purpose. 'This object, I have seen its twin brother on display in an exhibition about race hygiene in Germany at the Berlin Museum of Medical History at the Charité. This device was appropriated by Nazi scientists and, again, used to judge or measure race, particularly in Jewish people,' Das explains. 'You'll find photographs of Nazi scientists measuring people's heads, measuring people's noses, matching their eye colour.'

The eye and hair colour charts reveal just how slippery the dogged mantras of rationality and objectivity can be when it comes to studying human difference. 'Any scientist who claims that they are not politicised, or that they are asking questions out of pure curiosity, they are lying to themselves,' she continues. 'The structure in itself is fundamentally, structurally racist, because it has always been taken at its face. Never going back and taking apart those underpinnings.' What does it matter if one person has black hair and brown eyes, and another has blonde hair and blue eyes? Why not compare heights or weights or some other variable? These particular features matter only because they have political meaning attached to them.

In the United States, arguably the most racially charged place in the world at the time, evolutionary theory and eugenics came along at just the moment when intellectual racists could deploy them to full effect. Immigration into the US from countries considered to be undesirable had been curbed by the 1882 Chinese Exclusion Act, the country's first major law restricting immigrants. Twelve years later, three Harvard College graduates formed the Immigration Restriction League, arguing in favour of a literacy requirement for those who wanted to come to the US. The group's secretary, Prescott Farnsworth Hall, used Darwin's ideas of natural selection to caution against allowing into the country 'undesirable' immigrants who weren't 'kindred in habits, institutions and traditions to the original colonists'. In a lengthy racist tract in *The Annals of the American Academy of Political and Social Science* in 1904, he added, 'The doctrine is that the fittest survive; fittest for what? The fittest *to survive in the particular environment in which the organisms are placed*' (his emphasis).

By 1907 the Bellingham riots would see hundreds of white men, themselves recent arrivals from Europe, attack Indian immigrants living in the city of Bellingham in the state of Washington, blaming their 'filthy and immodest habits'. Reportedly, seven hundred Indians had to flee. The local *Bellingham Herald* complained, 'The Hindu is not a good citizen. It would require centuries to assimilate him, and this country need not take the trouble.'

It was against this backdrop that a new ideologue emerged. In 1916 a wealthy American law graduate named Madison Grant published a book that took eugenics to another level. Grant was known as a conservationist (as one of the co-founders of Bronx Zoo in New York, he had lobbied to put Congolese man Ota Benga on display among the apes there in 1906) but he wasn't a scientist. He recognised, however, the power of the language of science. In *The Passing of the Great Race:*

or The Racial Basis of European History, he revived the legacy of Count Arthur de Gobineau from the previous century, promoting the myth of Aryanism. Grant proposed that a blond, blue-eyed Nordic 'Master Race' represented the Aryans' true-life descendants.

Grant's racial hierarchy was geographically specific, consigning to inferiority everyone who wasn't northern European – including Italians and Greeks, who at that time were considered an undesirable immigrant group in the United States. He warned against racial intermixing in the belief that this would damage white racial purity even further. As casually as a biologist writing about plant hybrids, he wrote that a cross between any member of a European race and a Jew is a Jew.

In Grant, wealth and racism formed a toxic combination. Being descended from some of the first European colonists to settle in America, he counted himself – of course – among the descendants of Aryans, a noble race under threat. Openly in favour of both slavery and segregation, he made every possible effort to reduce immigration to the United States from anywhere but northern Europe. And he had powerful supporters, including soon-to-be President Theodore Roosevelt. In 1909 Grant became vice-president of the Immigration Restriction League, of which Roosevelt was a member. In 1921 Grant was the treasurer at the Second International Eugenics Conference in New York.

And yet it took only the slightest interrogation for both his historical and scientific evidence to be exposed as dodgy and self-serving. One reviewer raised an eyebrow at Grant's claim that Italian artists Dante, Raphael, Titian, Michelangelo and Leonardo da Vinci were of the Nordic type, and that – stretching the geographical parameters even further – so was Jesus. But the views of experts didn't matter to Grant's readers. His fake assertions were enough for those seeking some apparently intellectual support in their opposition to immigration.

Two parallel ideologies had by now become firmly intertwined in the minds of racists. First, the decades-old concept of the existence of a superior race. Second, informed by eugenics, the idea that unless checked, inferior races would outbreed superior ones. Human variation, before the eighteenth century a jelly-like set of loose generalisations, had now become a hard matter of progress and struggle. Grant's work was referenced by the Ku Klux Klan. It also became one of the inspirations behind the Immigration Act of 1924, which set quotas according to nationality aimed at decreasing immigration from southern and eastern Europe, including Italy, Greece and Poland, as well as effectively barring anyone from Asia.

His work even earned one lifelong fan in Germany. In a fawning letter to Grant about *The Passing of the Great Race*, Adolf Hitler wrote, 'The book is my Bible.'

*

It was all so long ago, we imagine that it's well and truly over now. We think of the horrors of the Holocaust and earlier genocides, of slavery and colonialism, of the many millions who were killed, of the twisted logic behind these actions, as belonging to another time. We imagine that the end of the Second World War spelled an abrupt end for race science. Eugenics is a dirty word. We're enlightened now. We're wiser.

But the story doesn't end quite so quickly. While they may have tempered their politics, race scientists didn't simply disappear after the war. Those who had built their work around eugenics and studying human difference, who staked their careers on it, simply found new avenues.

Take Otmar von Verschuer, who had plundered the tiny

bodies of Auschwitz victims for his studies of twins during the Holocaust: after being temporarily banned from teaching, in 1951 he became professor of human genetics at the University of Münster. Many scientists similarly changed tack, gently manoeuvring themselves out of eugenics into allied fields, such as genetics, that studied human difference in less controversial and more rigorous ways. Many stopped using the word 'race' altogether. Science learned at least one lesson, recognising that if human variation was to be studied, it had to try and at least appear to stay away from politics.

But the shift didn't happen abruptly. The Eugenics Record Office on Gower Street in London survived all the way through the war. There is still a Galton Professor of Genetics at University College London, funded by money Francis Galton left behind. What was the Eugenics Society became the Galton Institute in 1989. In 2016, the institute established the Artemis Trust, which according to its own promotional leaflet, handed to me at a conference, distributes grants of up to £15,000, partly with the aim of assisting in the provision of fertility control, and particularly to those from 'poorer communities'.

Subhadra Das tells me that a woman came to see her recently whose mother had worked in the Galton laboratory in the 1950s. Her job had been to study redheadedness in Wales. It was not until the 1960s that the word 'eugenics' could no longer be heard in these corridors. What helped kill it in the end wasn't just the war but also the fact that new research showed it probably wouldn't work. The way we inherit traits from our parents turns out to be more complicated than Galton imagined. There is no guarantee that two beautiful and brilliant parents will produce brilliant and beautiful kids. Genetics is more of a game of chance. The science of inheritance, once it was better understood, didn't support the idea that humans could breed themselves to perfection, whatever perfection meant. Complex psychological traits such as intelligence are not controlled by a mere few genes, and are also heavily influenced by environment and upbringing.

Yet it was decades before eugenics policies introduced to other parts of the world were abandoned. Only in 1974 did the American state of Indiana repeal legislation that had made it legal to sterilise those it considered undesirable. Investigations by reporter Corey Johnson in 2013 uncovered that doctors working for the California Department of Corrections and Rehabilitation had continued the practice, sterilising as many as 150 women inmates between 2006 and 2010, possibly by coercing them into having the procedure. In Japan, a Eugenic Protection Law introduced in 1948 to sterilise those with mental illness and physical disabilities and prevent the birth of 'inferior' offspring was repealed only in 1996. Victims of the legislation are still pushing for justice.

The process of self-examination, of experiencing regret and showing remorse – the kind attempted by the Max Planck Institute in 2001 – is slow. And it has been particularly slow in the places that found themselves on the winning side in the Second World War. In the decades after the war, scientists in Britain and the United States airbrushed away their pivotal role in race science and eugenics. They quietly moved into other fields, silently renamed their university departments, consigning to the past that dark chapter. History was rewritten by the victors.

According to Gavin Schaffer, a professor of British history at the University of Birmingham and author of *Racial Science and British Society, 1930–62*, 'It was much easier to point the finger at the horrible Nazis, and the same went for the scientists. This absence of introspection was rooted in the ability to point fingers at other people for being responsible for the perversion of science.'

The post-war narrative of good triumphing over evil glossed over the messier truth that, in fact, everyone should have had a finger pointing at themselves. Without ever really looking back to the past and asking how and where the idea of race had been constructed in the first place, why it had been relentlessly abused, without questioning the motives of scientists like Francis Galton, Karl Pearson and countless others, in this glaring 'absence of introspection', old ideas of race could never completely disappear. Even long after the war, scientific fascination with human variation remained tainted by a lingering belief that there might be something deeper about racial difference, that perhaps some races really are better than others.

Yes, there was good science that emerged out of the ashes. Biology did attempt to reform itself, to cast away the mistakes of the past and do a more precise and accurate job of understanding human variation. But at the same time, while the world around them changed, a few of the hardened old-school race scientists could still be found knocking about. 'Racist science continues, it just becomes more marginal,' Gavin Schaffer tells me. 'But there's no doubt that it does continue.'

4

Inside the Fold

After the war, intellectual racists forged new networks

THE SECOND WORLD WAR marked an unlucky turning point in the life of scientist Reginald Ruggles Gates.

Born in 1882, Gates was one of those well-to-do, gentlemanly race scientists who had been the norm in the nineteenth century. He was a colonial type, who believed that other races belonged to different human species, as well as a eugenicist and supporter of segregation in the United States. One particular obsession was what he saw as the danger of different races mixing with each other and having children. In his studies of mixed-race people, he kept skin colour charts that looked like paint swatches. Gates would almost certainly be considered a racist by modern standards, but at the time his views weren't uncommon. They certainly didn't get in the way of his career or his standing in society. He was successful and well respected.

To get a sense of who Gates was I've come to the Maughan Library at King's College London, to which his archive was bequeathed after he died in 1962. Here, in a vast nineteenth-century Gothic Revival building off Chancery Lane that was once Britain's Public Record Office, I leaf through his personal papers, slowly building a portrait. Sepia photographs show him to be smartly dressed, sporting a neatly clipped moustache. Gates had grown up in a wealthy family with thousands of acres of land across Nova Scotia in Canada before moving to Britain, where he was briefly married to Marie Stopes, a fellow member of the Eugenics Society. He enjoyed a career as a plant geneticist, becoming professor of botany at King's College in 1921, and later a fellow of the Royal Society.

He seems to have had a passion for travel, too, for understanding human difference across the world, his collection of scientific papers spanning almost every continent. In later life, he visited Cuba and Mexico to study 'mixed race' people, Japan and Australia to observe indigenous communities, and made a number of trips to India, a country that became a particular source of fascination. Browsing his personal collection of scientific papers in the library, I'm startled to discover there's even one on the blood groups of the Sainis in parts of Punjab – a study that may well have included relatives from my father's branch of the family.

Despite his racist beliefs, before the war Gates was riding high. Afterwards, though, it all changed. To his confusion and disappointment, he found himself left out in the cold by an establishment that had once welcomed him. His papers were rejected by scientific journals more often than they had ever been. And the reason was simple: deeply shaken by the genocidal use of eugenics by the Nazis in Germany, the world was turning its back on research that resembled their theory of racial hygiene. The enthusiasm for studying race that had once been almost fashionable was on its way out in scientific circles. Researchers who weren't wise enough to get with the new programme, who chose instead to cling to their unpalatable politics, as Gates did, found themselves flung from the warm centre of academic life to its chillier margins.

Yet he couldn't fathom it. 'What interests me about it is his incredulity,' I'm told by historian Gavin Schaffer. 'He seemed genuinely surprised.' In a sense, Gates was a man caught out by time. Francis Galton died in 1911 and Karl Pearson in 1936, before they could witness race science reach its most brutal peak. But others, such as Gates, lived long enough to see the political mood change – and then suffer the consequences when their own politics didn't.

At every opportunity, Gates refused to budge from his belief in racial superiority and inferiority. Wherever he found himself professionally hindered, he imagined himself to be the victim of some Jewish plot to derail his work. Schaffer recounts one especially bad experience in 1948 when Gates was working briefly at Howard University, the historically black college in Washington, DC. 'A petition was got up to remove him because of allegations that he was a racist – which he *was*. But he was stunned by that,' Schaffer says. 'He articulated his understanding of that as a manifestation of an international Jewish conspiracy, as opposed to just understanding that, in a historically black university, the kind of work that he did and the kind of things that he said were always going to be challenged.' Even when he agreed to leave Howard, Gates grumbled in private that only a few 'ignorant Negroes' were fit to be in a university at all.

Gates could never accept that the world was moving on, leaving those like him behind.

*

When it came to how the world thought about race, a wider political shift was underway. It was most clearly signposted in

1949 when more than a hundred scientists, anthropologists, diplomats and international policymakers met in Paris under the umbrella of the United Nations Educational, Scientific and Cultural Organisation, UNESCO, to redefine race. British-born American writer and anthropologist Ashley Montagu led the charge against scientific racism and its horrific legacy, taking his cue from a wave of social scientists who had already long argued that history, culture and environment were really behind what people thought of as racial difference.

'The word race is itself racist,' Montagu wrote in his influential 1942 book *Man's Most Dangerous Myth: The Fallacy of Race*. Both intellectually and culturally ahead of the curve, he explained in *American Anthropologist*, 'What a "race" is no one exactly seems to know, but everyone is most anxious to tell . . . The common definition . . . is based upon an arbitrary and superficial selection of external characters.' As anthropologists and geneticists were learning, individual variation within population groups, overlapping with other population groups, turned out to be so enormous that the boundaries of race made less and less sense. This was one reason why nobody had ever been able to agree on exactly how many races there were. Three, or four, or five, or several, there was never a consensus. The concept of race was as slippery as jelly, defying any effort to pin it down. In the end, academics had to concede that it probably wasn't an accurate or reliable way to think about human variation.

Montagu emphasised the likelihood that humans were genetically almost identical, and that in any case, our ancestral roots were certainly the same. Other anthropologists who had studied human diversity had already suggested that differences between humans were not only marginal, but also sat on continua, each so-called 'race' blurring into the next. What really made people and nations seem different was culture and language.

It was on the back of work like this that UNESCO, in July 1950, released its first statement on race, stressing unity between humans in a concerted effort to eradicate what it saw as the outcome of a 'fundamentally anti-rational system of thought'. It was meant to be the last word on the subject, to flush away racism once and for all. 'Scientists have reached general agreement in recognizing that mankind is one: that all men belong to the same species, *Homo sapiens*.'

The next few decades would be crucial to dismantling the idea that race was biologically real, and proving Montagu right. In 1972 a landmark paper exploring the true breadth of human biological diversity would appear in the annual edition of *Evolutionary Biology*. It was written by geneticist Richard Lewontin, who later became a professor at Harvard University. Dividing the planet up into seven human groups, based roughly on old-fashioned racial categories, Lewontin investigated just how much genetic diversity there was within each population, compared with the genetic diversity between them. What he found was that there was far more variation among people of the same 'race' than between the supposed races, concluding that around 85 per cent of all the genetic diversity we see sits within local populations, and 8 per cent more if you widen the net to continental populations. In total, around 90 per cent of the variation lies roughly within the old racial categories, not between them. There has been at least one critique of Lewontin's statistical method since then, but geneticists today overwhelmingly agree that although they may be able to use genomic data to roughly categorise people by the continent their ancestors came from (something we can often do equally well by sight), by far the biggest chunk of human genetic difference indeed lies within populations.

Lewontin's findings have been reinforced over time. An influential 2002 study published in *Science* by a team of American scientists, led by geneticist Noah Rosenberg, then at the University of Southern California, took genetic data from just over a thousand people around the world and showed that as much as 95 per cent of variation sits within the major population groups. Statistically this means that, although I look nothing like the white British woman who lives next door to me in my apartment building, it's perfectly possible for me to have more in common genetically with her than with my Indian-born neighbour who lives downstairs. Being of the same 'race' doesn't necessarily mean we are genetically more similar. In the long run, then, Ashley Montagu's position on race has been vindicated.

Mark Jobling, a respected professor of genetics at the University of Leicester, tells me that if there were a global catastrophe and all life were wiped out save just, say, Peruvians, 85 per cent of human genetic diversity would be safely retained. 'That just reflects the fact that we are a young species,' explains Jobling. Humans are relatively new, and being so new, we're still closely related to one another.

The greatest genetic diversity within *Homo sapiens* is found inside Africa, because this continent contains the oldest human communities. When some of our ancestors began to migrate into the rest of the world sometime between 50,000 and 100,000 years ago, the groups that moved were genetically less diverse than the ones left behind for the simple reason that they were made up of fewer people. The human variation we see across regions today is partly the result of this 'founder effect'.

Of course, groups of people have average physical differences, as a result of their biological and environmental histories. It has been estimated that 10,000 generations separate every single one of us from the original band of people in what is now Africa, but we vary in small ways because of the characteristics our ancestors happened to take with them as they migrated. As the small, sampled populations spread, bred and adapted to their local environments, they began to look more different from the relatives they left behind generations earlier and more like each other. And as small members of these groups themselves left for new territory, they would be slightly genetically different again because of a serial founder effect.

All this didn't happen in big clumps or clusters, but was rather more of a mesh, as people mated with those they encountered on the way, sometimes travelling further away and sometimes moving back. If everyone in the world had their genomes sequenced, adds Jobling, you wouldn't find hard borders between them, but gradients, with each small community blending into the next, the way hills blend into valleys. The racial categories we are used to seeing on census forms don't map onto the true picture of human variation.

The aim of the original 1950 UNESCO statement wasn't just to set out the science in a clear way, but to change the culture, to make people think differently about this idea they had lived with for so long, that had done incalculable damage to millions of lives. The statement emphasised that what we see as race is likely to be only a superficial variation on the same theme. Most of the visible variation is cultural. It tipped cold water over entrenched racial stereotypes, adding that there was no proof that groups of people differed in their innate mental characteristics, including intelligence and temperament.

This marked a crucial moment in history, a bold universal

attempt to reverse the deep-seated damage done by racism – and perpetuated by science – for at least two centuries. To some extent, it worked. Whether we realised it or not, all of us thought about race differently after that. Racism was no longer acceptable. Scientists and anthropologists by and large got behind UNESCO, and their work in the coming decades would in the main reflect that.

But this wasn't the end of it. And that's because not everyone was on board.

*

Despite the changing public mood about race, some researchers just couldn't bring themselves to ditch a body of work they had been cultivating for decades. Many didn't agree with UNESCO's claim that biology supported the idea of a universal brotherhood. A few couldn't accept that there were no mental differences between racial groups. And they weren't necessarily all racists. Some of them were respectable, eminent scientists at universities including Oxford and Cambridge, who simply wanted the statement to be revised with more scientific precision and qualification. But one of the most passionate voices of all belonged to Reginald Ruggles Gates.

'What Gates called for, time and again, was the objective continuing study of race . . . because he thought that his position was grounded in *true* science,' explains Gavin Schaffer. He and others felt that UNESCO was stepping outside the bounds of what biology could actually claim, that it was ignoring facts in favour of liberal, anti-racist politics. 'The biologists who countered it, what they wanted was the continuation of their own expertise, which they had asserted over twenty, thirty years. They agreed the Nazi state was completely wrong in the way it had used race, and that other political actors had been completely wrong, but they felt that the study of race would profit from further work. I think, to them in that period, they felt they wanted work on race to continue, and they felt that other people wanted it to stop.'

The pressure worked, at least to some extent. In 1951 UNESCO gathered a team of experts to publish a new statement, tempering its language to account for the lack of consensus around the biological facts. The changes were subtle but revealing. For instance, instead of saying that scientists had 'reached general agreement' that we were one human species, the revised statement was gently altered to say that scientists were 'generally agreed . . .' In short, it had to make clear that not every expert could accept even the most basic fact that we all belonged to the same species.

Despite concessions like these, Gates failed to keep race science alive in the same way as before. By now, it had been all but lifted out of the laboratory. The academic study of race no longer sat within the realms of biology. Whether all biologists liked it or not, by the second half of the twentieth century race belonged to the social sciences, to the study of culture and history. It was understood to be a social and political construction, not borne out by biology. Old-fashioned race researchers and eugenicists had to move on or be sidelined.

At that point, Schaffer explains, 'The biologists just go into themselves a bit. They go back to their work, they go back to their labs.' They moved into newer fields, such as genetics, evolutionary biology and psychology. For those who still wanted to study human variation, their research moved away from skin colour and hair texture and went to the molecular level. 'As long as you weren't hell-bent on the kind of politics that were going to call your position into threat, yes, why not.' The older, somewhat cruder and more controversial ways of studying human difference, using anatomy and twin studies, were treated with suspicion. By the 1950s the word 'race' was so unfashionable in scientific circles that it was barely used any more.

Historian Veronika Lipphardt at University College Freiburg, Germany, has noted that the 1950s saw new institutes dedicated to the study of human variation open across the world. There was one in Bombay, another at Columbia University in New York, and one at the Federal University of Paraná in Brazil. A politically correct scientific terminology emerged. Researchers began referring to 'populations', and occasionally 'ethnic groups'. But the departure from the old race science wasn't quite as complete as it might have been. Although the parameters of research had changed, racial categories were still alive in people's minds. They were still active in everyday life, playing out in the politics and racism of the real world. For scientists to suddenly stop thinking about humans in racial terms was impossible so long as everyone out there still thought about themselves and others that way. So they couldn't help but look for racial difference, to subconsciously force this way of thinking into their work.

One example is blood type. When genetics became the preferred way to talk about human variation, hard hereditary variables like blood type came under the spotlight. Categorising by blood sounded more mathematical, less wishy-washy, than talking about skin colour or hair texture. And in the process, blood became an obsession. It was already well known that the proportions of people with different blood types varied from population to population, because of a phenomenon known as genetic drift. In prehistory, as small founding communities of people migrated across the world, they took their own narrow subset of blood types with them. This is equivalent to your cousin, say, leaving home to set up a colony of her own. She may be closely related to you, but have a different blood type. As these communities grew bigger, their particular blood types became the common ones. For example my own, B+, is shared by around a third of people in India where my family are from, and less than a tenth of people in the United Kingdom, where I live. By studying which population groups have which blood types, researchers found they could open a window into how closely or distantly related these groups might be to each other.

In the post-war period, the distribution of blood types became a hot topic in anthropology journals. In the 1960s the World Health Organisation launched its own effort to document groups of people around the world, collecting data on skin pigmentation and hair form, but also on blood type, colour blindness and other genetic markers. When the blood of the Sainis of Punjab was collected and tested in 1961 by a pair of anthropologists at the University of Delhi, before the results were passed to Reginald Ruggles Gates in England, it was part of these bigger efforts. Similar tests were carried out in other Indian groups and castes. Ultimately, thousands of people from different communities would have been gauged in the same way. And the same happened all over the world. At least some of these scientists were searching for proof that race was genetically tangible, that evidence for deep racial differences could be found at the molecular level.

Gates was one of those. He just couldn't let go of his belief in the old, hard, biologically rooted racial gaps. And he would never change. His final work, entitled *The Emergence of Racial Genetics*, published posthumously in 1963, attempted to place the new genetics in the old framework of race. One reason he pressed on with his commitment to races as meaningful categories, argues Schaffer, is that he sincerely believed his own

research was objective, and those challenging him were the ones driven by ideology. He saw himself as the bearer of truth, held back by an anti-science political agenda that was mistakenly trying to impose racial equality on the world. He believed it even while he was receiving funding from segregationists in the United States.

Schaffer reminds me that it's important to understand the psychology behind this. When Gates complained that race research was being politicised, he was right. After all, following the Second World War and the bald brutality of the Holocaust, it would have been bizarre for any discussion of race not to be affected by politics. But what Gates failed to accept was that he was similarly affected by his own politics. 'People like him must also position themselves within that model. People who defend race historically have also done so for political reasons,' he explains. 'The science never becomes separated from political discourse.'

Gates wasn't a pseudoscientist, even if he was a bit of a crank. But his failure to gain professional recognition after the war wasn't just about his beliefs. Editors of some of the scientific journals rejecting his work warned him that his methods were becoming sloppy, relying on subjective interpretation rather than rigorous, intensive study. This slapdash approach may have been acceptable the previous century, but it didn't cut it in the world of modern science. By the end, both for his abhorrent views and his weak research, Gates had few supporters left in the scientific community. Reportedly, when his death was announced in 1962 at a meeting of American anthropologists, there were cheers.

That said, history doesn't move in a straight line. Ideas, even the worst ones, can go out of fashion in one century and come back in another. Those who imagined that the end of the war marked the abrupt death of race science were sadly mistaken.

*

In the final years before his death, when few scientific journals would touch his work, Gates decided to take matters into his own hands. If they wouldn't publish him, he would publish himself.

He and a handful of likeminded researchers, some on the very darkest margins of science – including former Nazi scientist Otmar von Verschuer (who died soon after Gates in a car accident, in 1969) and British eugenicist Roger Pearson (the last of the group still alive today, aged ninety, although he declined to give me an interview for reasons of ill health) – set up a journal of their own. Their aims were simple: to challenge what they saw as a politically correct, left-wing conspiracy around race and bring back some scientific objectivity. They named their brave new enterprise the *Mankind Quarterly*.

The founders of the journal regarded themselves as 'the defenders of the truth', says Schaffer; they even compared it to the gospels. But it would have been immediately clear to anyone who read it that the *Mankind Quarterly* was less impartial than it claimed to be. By 1960, when it launched, South Africa was in the midst of apartheid, the US civil rights movement was gaining momentum, and European colonies in Asia and Africa were winning independence. Race was high on the agenda everywhere, and the moral failures of the past were slowly being redressed. For racists who didn't welcome this shifting tide, now was the moment to assert their position. And the *Mankind Quarterly* was happy to oblige. It waded deep into the politics of the time, using science – even if only in a loose way – as its weapon of choice.

Recruiting truly respectable scientists to the cause was a challenge. But not an impossible one. The earliest editions included articles by Henry Garrett, a former president of the American Psychological Association and head of Columbia University's psychology department. Garrett was then one of the most powerful and eminent voices against desegregation in the United States. Most notably, in 1954, he had testified to stop the integration of black and white schools in the state of Virginia, when a trial known as Davis v. County School Board of Prince Edward County went up to the Supreme Court. Hundreds of students at an underfunded all-black school, with no gym or cafeteria, who were sometimes forced to study in an old school bus, fought against separate schools on the grounds that they were being disadvantaged because of their colour. The judge ruled against them. Under national pressure, the case was later combined as one of five brought before the Supreme Court in 1954 in the landmark action Brown v. Board of Education, which finally declared that separate schools for black and white students were unconstitutional.

Writing in the *Mankind Quarterly* in 1960, Garrett wasn't rolling back from his defeat, he was doubling down. Regardless of what the law said, for him, people of different races mixing with each other spelled certain disaster. 'The weak, disease-ridden population of modern Egypt offers dramatic evidence of the evil effects of a hybridization which has gone on for 5,000 years. In Brazil, coastal Bahia with its negroid mixtures is primitive and backward as compared with the relatively advanced civilization of white southern Brazil,' he wrote.

In another article for the journal in 1961, Garrett laid into academics, politicians and social reformers who didn't accept the 'common-sense' judgement that 'the Negro' was 'less intelligent and more indolent than the white'. Like so many scientific racists before him, Garrett argued that this was a matter of civilisational superiority. He claimed that Africans had never produced anything of great value. Could any African Negro, he charged, 'compare with the best of the European whites . . . for example, with Aristotle, Cicero, Thomas Acquinas [sic], Galileo, Voltaire, Goethe, Shakespeare or Newton?'

The twisting of facts to suit an ideological viewpoint would become a regular feature of the *Mankind Quarterly*. An especially cold-blooded article in 1966, this time by one of the editors, argued that Aboriginal Australians and Native Americans had been all but wiped out by European colonisers not out of greed or cruelty, but because it was a natural outcome of biology. 'If the conquered are markedly inferior to the conquerors . . . they will always remain an outcaste element at the bottom of the social structure.' Inter-racial conflict, the writer argued, was the product of natural selection, the fittest fighting to survive. Drawing parallels with the American civil rights movement, he added that it was virtually self-evident that racial integration would never work.

Articles like these didn't go unnoticed in the scientific community. Almost as soon as the *Mankind Quarterly* appeared, disgusted anthropologists sent in letters of complaint, accusing the journal of trying to make scientific racism respectable again. Slovene anthropologist Božo Škerlj, who had mistakenly joined the *Mankind Quarterly*'s advisory board only to be appalled by its 'ostensibly racialist editorial policy', entered into a public spat with the editors. Škerlj was particularly insulted by Gates's accusation that his mental outlook – and presumably his objectivity – was affected by the fact that he had been imprisoned in Dachau concentration camp during the war. Gates noted, revealingly, that he would never have considered Škerlj for the position in the first place had he known about his internment.

In *Science*, one of the world's leading journals, a reviewer called on scientists to take action. But it made no difference. The reason the *Mankind Quarterly* had been created at all was the lack of scientific approval for the kinds of ideas its founders wanted to publish. The editors didn't want or care about approval, they just needed a platform.

The other, deeper secret behind the *Mankind Quarterly* was that it had legs of its own. Support came indirectly from a reclusive, multimillionaire American textile heir with a vested political interest in what it was publishing. Wickliffe Draper was a diehard segregationist descended from a commanding officer in the Confederate Army on one side and the largest slaveholder in the state of Kentucky on the other. His family roots in America dated back to 1648, with enormous wealth and property steadily amassed over the centuries. In his 2002 book *The Funding of Scientific Racism*, William Tucker, emeritus professor of psychology at Rutgers University, details Draper's upbringing – one so privileged that a relatively weak academic record wasn't enough to stop him getting into Harvard University. There, as Tucker writes, he would have been exposed to those at the forefront of the American eugenics movement in the early twentieth century. An intellectual racist looking for ways to spend his inheritance, Draper would find in the *Mankind Quarterly* the perfect vehicle for his racist views.

In March 1937, Draper incorporated the Pioneer Fund, a private foundation aimed at disseminating information on human heredity and eugenics, providing race scientists who couldn't find backing anywhere else with the cash they needed to carry on. American anthropologist Robert Wald Sussman explains in his 2014 book *The Myth of Race* that 'Draper wanted to recruit scientific authorities with academic credentials and scholarly records who believed in the necessity of racial purity and that integration posed a threat to civilisation.' In short, he was trying to build a scholarly argument to defend segregation. During the war, his money also helped distribute a Nazi propaganda film about eugenics to US schools and churches. But it was after the war that the fund really came into its own. In 1959 he set up what he called the 'International Association for the Advancement of Ethnology and Eugenics', to produce and publish documents on race. The association's aim was to promote and distribute the *Mankind Quarterly*, helping turn it into one of the most important channels for race research in the world. William Tucker describes the association's original American directors as 'probably the most significant coterie of fascist intellectuals in the postwar United States and perhaps in the entire history of the country'.

The Pioneer's funding priority from the beginning was to back distinguished scientists, the more well known the better, along with racist ideologues. 'Grants to the former were intended to provide a façade of intellectual respectability for the latter, as well as results that could be used to justify their policies,' Tucker tells me. Cash gifts were routinely made to scientists who echoed Draper's political sentiments, while thousands of copies of the *Mankind Quarterly* containing their work were sent out to a list of American political conservatives. The science and the politics operated hand in glove.

Unsurprisingly then, according to Tucker, the journal made no concessions to political correctness. 'This was going to be a publication frankly written *by* racists *for* racists,' he writes. The target audience didn't appear to be the academic community at all, but racist movements searching for evidence that their prejudices might be rooted in scientific fact. 'Nothing seemed too bizarre or too repugnant to receive the *Mankind Quarterly*'s stamp of approval.' Among the lengthiest articles it published was one entitled 'The New Fanatics', slamming American intellectuals who used their authority to support equal rights for black people. Sussman has noted that the book review section was in essence a bulletin board for publications that had anything to do with eugenics, lavishing praise on new publications that were neo-Nazi, anti-Semitic or anti-black.

To those who read the journal, its editors' intentions would have been clear. Many of the mainstream scientists who did bother to read it saw straight through it. A scholarly review of the first three editions by the late British anthropologist Geoffrey Ainsworth Harrison, a president of the Royal Anthropological Institute, was scathing. He complained that one of the editors hadn't grasped the concepts of modern genetics, despite writing about them at length. He dismissed Henry Garrett's work, too, as full of inconsistencies. Although Harrison didn't rubbish the value of studying human variation altogether, he didn't see what the *Mankind Quarterly* did as academically useful. 'Few of the contributions have any merit whatsoever, and many are no more than incompetent attempts to rationalise irrational opinions . . . it is earnestly hoped that *The Mankind Quarterly* will succumb before it can further discredit anthropology and do more damage to mankind,' he concluded.

But that didn't happen. In fact, it kept going for many more decades, publishing scientists and not-quite-scientists at the margins of their fields, many of them bankrolled by Wickliffe Draper's Pioneer Fund. The fund stuck to its aims even after Draper died in 1972. In his will, he left $50,000 to Henry Garrett alone. Despite all the criticisms it faced when it was first published, despite the widespread expectation that it wouldn't last, the *Mankind Quarterly* never succumbed. Indeed, if you want to read it, it's still around today.

*

When I contact German biochemist Gerhard Meisenberg, the current editor-in-chief of the *Mankind Quarterly*, I don't expect to hear back from him. After all, this is a journal considered so inflammatory that the security filter of my home broadband provider won't even allow me to look at its website without changing my settings. So I'm surprised not only to hear from Meisenberg immediately, but that he seems perfectly happy to tell me whatever I want to know.

He advises me that he became editor of the *Mankind Quarterly* only within the last few years, and that his job is to 'start the seemingly hopeless task to salvage this run-down journal', betraying the possibility that he sees renewed interest for the ideas it publishes. At the same time, he warns me that he can only communicate with me through email – not because he considers the work he does to be inflammatory, but because it's tricky to reach him by any other means right now. Since 1984 he's worked at the Ross University School of Medicine, a for-profit private college based in Dominica, but he and his students recently found themselves kicked off the island by a hurricane. As a result, he's teaching from a rented cruise ship elsewhere in the Caribbean.

I can't call or see him, he says, but we can write to each other. What follows is a long and candid exchange.

'I can tell you about how modern races evolved,' he tells me in his first message. It's clear from the outset that he believes he has an understanding of the subject not shared by mainstream scientists. The reason he is so happy to talk is because I'm giving him the opportunity to enlighten me. 'One hang-up for academic definitions of race is that academics like precise definitions and precise boundaries between categories. They use only their left brain hemisphere. The right one is atrophied. For them, when categories into which they slice the world lack clear boundaries, they seem to assume that the categories are invalid.'

On school performance in the United States, he states, 'Jews tend to do very well, Chinese and Japanese pretty well, and Blacks and Hispanics not so well. The differences are small, but the most parsimonious explanation is that much and perhaps most of this is caused by genes.' There is no scientific evidence for this, it's just speculation. Nobody has ever found any genes linking ethnicity or race to school results. Like Henry Garrett half a century earlier, Meisenberg chooses to skip over the social, historical and economic aspects of racial inequality. Rather, he believes that scientific evidence that doesn't yet exist will explain the gaps eventually. He takes it as given that the answer must be biological.

'Of course we can use molecular genetics to figure out, for example, in what way intelligence-related genetic variants vary among different racial groups,' he suggests. 'This will answer the question about race differences in intelligence once and for all. Good riddance of a stupid debate!'

He continues in this vein, the occasional sensible observation punctuated by more bizarre ones. At one point he claims that 'Europeans became brighter since antiquity, but then became stupider again since the nineteenth century.' At another he asks, 'are different races genetically predisposed to think about the world in slightly different ways?'

Eventually I ask what attracted Meisenberg – neither a geneticist nor a psychologist, but a biochemist working at a medical school – to this scientific niche. What got him so interested in race?

'I am not particularly obsessed with race,' he replies. 'But I got interested in the subject in the context of the question of why some countries are rich and others are poor.' He notes that there is 'a 50-fold difference in per capita GDP between the poorest countries in the world and the advanced Western countries' and he believes that learning ability, which he sees as being tightly linked to intelligence, is what makes the difference to a country's economic success. For him, this learning ability is programmed into a person's – or a population's – DNA. 'In consequence, the question of whether there are genetic ability differences between people in different countries is perhaps the most fundamental question in development economics.'

He tells me what a shame it would be for what he calls 'low-IQ countries' such as Pakistan to lose their brightest citizens if they emigrate to the West. 'This cripples the poor countries and makes it impossible for them to catch up,' he laments. If some nations don't have the cognitive ability to catch up with Europeans and East Asians, they will 'get stuck somewhere on the lower rungs of the developmental path'. At a stroke, he condemns without evidence all the world outside Europe and parts of Asia to genetically inferior status.

Meisenberg certainly seems bothered by the social implications of what he sees as immutable racial difference. He expresses a fear that the racial stock of smarter, wealthier nations is under threat and that the problem urgently needs to be addressed, particularly by immigration control. 'Populations that get too bright and too rich invariably slip into sub-replacement fertility and slowly breed themselves out of existence,' he writes, 'while those that are stuck at a lower economic and cognitive level also get stuck . . . with continuing high fertility of the less educated sections of the population.' These could so easily be the words of an early twentieth-century eugenicist. He, too, believes that the inferior might outbreed the superior.

*

The mystery for me as a journalist is not that scientific racists exist. There have always been those with prejudice of every persuasion in academia, and possibly there will always be. The bigger puzzle is how someone like Gerhard Meisenberg – a professor at a private university in the Caribbean – manages to keep the *Mankind Quarterly* afloat, and finds researchers to write the articles that fill its pages. This requires networks, it requires coordination, and it requires funding. The fact is, across the world, old-style scientific race research of the kind his journal publishes is deliberately discouraged by science funding agencies and governments, not to mention deeply frowned upon within academia. It is as controversial as it was when the journal was founded. To do it independently, even if only online and for a small readership, one needs resources.

Following the money is where the trail begins. Attached to one of his early emails, Meisenberg sends me a paper that attempts to describe how racial categories work. It was published in one of the 2017 editions of the *Mankind Quarterly*, authored by someone named John Fuerst, from an organisation called the Ulster Institute for Social Research. I have heard of neither him nor it. When I do a check, the institute doesn't appear in the UK's list of officially recognised higher education bodies, describing itself instead as 'a think tank for the support of research on social issues and the publication of works by selected authors in this field'.

It turns out that since 2015, the Ulster Institute for Social Research has been publishing the *Mankind Quarterly*, along with a handful of books on race. On the *Mankind Quarterly* website, its address is given as a postal box in London. Meisenberg himself sits on its advisory council. How the institute is funded, he won't elaborate; he reveals only that it operates on a shoestring budget. I find one link between the institute and an offshore company based in the Bahamas. 'There certainly is no regular external funding from any outside source that I know of,' he tells me. 'I guess it's more a situation where someone may donate a larger sum, perhaps as part of a legacy . . . That's how most of these small foundations work.'

According to a report in the *Independent* newspaper in 1994, the institute received $50,000 the previous year from the Pioneer Fund. It was a period in which the fund, then based in Manhattan, New York, was particularly active. An investigation by the *Los Angeles Times* around the same time estimated that it was dispensing roughly a million dollars a year to academics, most of whom the newspaper claimed were looking for genetic differences between races. As well as giving grants to scientists, psychologist William Tucker has noted, between 1982 and 2000 the fund handed almost $1.5 million to lobbying groups in favour of immigration reform in the United States.

If the Ulster Institute for Social Research really is operating on a shoestring today, part of the reason may be that the Pioneer Fund seems to have since declined to a standstill. 'It is my strong sense that it is not nearly as influential as it once was, largely due to the deaths of the fund's key players,' William Tucker tells me. The Southern Poverty Law Center, a team of civil rights lawyers based in Montgomery, Alabama, inform me that the Pioneer Fund has been mostly silent of late. As far as they can tell, in the last decade it has been gradually emptied of all its assets. In summer 2018 the Associated Press investigated tax records to find that it had given out nearly $7.8 million between 1998 and 2016.

Wherever their funds come from now, it is clear that there is a small cadre of researchers, some with very few academic credentials, who are still publishing and citing each other's research through organs such as the Ulster Institute for Social

Research and the *Mankind Quarterly*. It is a small, self-contained network that operates on the margins of respectability. The same names crop up again and again. Richard Lynn, the assistant editor of the *Mankind Quarterly*, is also president of the institute (one of its books is a tribute to him in his eightieth year). Lynn told the *Independent on Sunday* newspaper in 1990 that he had received grants from the Pioneer Fund. In 2001, he even published a history of the fund, titled *The Science of Human Diversity*. In his own investigations, Robert Wald Sussman wrote that Lynn 'does very little science and the "science" he does is extremely poor'.

Edward Dutton from Oulu University in Finland, the author of at least two of the institute's books, including one on racial difference in sporting ability, is also a regular contributor to the *Mankind Quarterly*. Another contributor, Tatu Vanhanen, a recently deceased Finnish political scientist, co-authored a 2002 book with Richard Lynn titled *IQ and the Wealth of Nations*. Vanhanen was inspired to enter this area of research after reading up on evolutionary biology, interpreting it as a way to explain inter-ethnic conflict. He believed that political ideology could be used to serve people's genetic interests, by keeping them loyal to their own ethnic group. In one high-profile magazine interview in 2004 Vanhanen claimed that the average IQ of Finns was 97, while in Africa it was between 60 and 70. The comment went on to cause a national scandal because Vanhanen's son had just become Prime Minister of Finland.

Whatever the scientific merits of their work, the researchers inside this group were and still are undeniably tight-knit. Most of them are largely unknown outside their circle but highly prolific within it. They have managed to build a thin veneer of scientific credibility that comes from getting published and cited, almost entirely by publishing and citing one another. And they keep finding new outlets for their work. The latest addition to this alliance is *Open Differential Psychology*, an open-access online journal that claims to have been set up in 2014 by a Danish research fellow at the Ulster Institute of Social Research called Emil Kirkegaard. It includes Gerhard Meisenberg and John Fuerst among its reviewers, and its published papers so far include studies of IQ in Sudan, and of crime among Dutch immigrant groups.

Even with mutual support, those who write for the *Mankind Quarterly* rarely make much impact outside the shadowy recesses of the Internet. But there have been a handful of higher-profile figures among them. Until his death in 2012, one was Canadian psychologist John Philippe Rushton, a former head of the Pioneer Fund and professor at the University of Western Ontario. Rushton became notorious in academic circles for claiming that brain and genital size were inversely related, making black people better endowed but less intelligent than whites. Despite this, he was important enough to have his work read and reviewed by genuine scientists.

One review in particular shone a spotlight on the kind of work that still routinely appears in the *Mankind Quarterly*. When Rushton's book *Race, Evolution and Behaviour* was published in 1994, the University of Washington psychologist David Barash was stirred to write, 'Bad science and virulent racial prejudice drip like pus from nearly every page of this despicable book.' Rushton, he added, seemed to have collected scraps of unreliable evidence in 'the pious hope that by combining numerous little turds of variously tainted data, one can obtain a valuable result'. The reality, Barash concluded, is that 'the outcome is merely a larger than average pile of shit.'

5

Race Realists

Making racism respectable again

In 1985, American historian Barry Mehler had a dream.

Now in his seventies and a professor of humanities at Ferris State University in Michigan, studying genocide, Mehler's investigations had in the 1980s taken him into the murky territory of academia's extreme right-wing fringes. His focus was on the founders of the *Mankind Quarterly* and Wickliffe Draper's notorious Pioneer Fund, both of which were now known to have been helping keep fringe elements in science alive for decades. As Mehler worked, he found his waking life began to soak into his subconscious, colouring his sleep. In his dream – in truth, more of a nightmare – his son, then around two years old, was trapped in a runaway car hurtling down a hill towards oblivion.

'The traffic is going in both directions, and I am in the middle of the road desperately waving my hands trying to stop the flow of traffic in order to save the life of my son,' he recalls. 'It's a dream. It's a metaphor for how I felt.'

A few years before, prompted by historical research he had already carried out into early twentieth-century American eugenicists and their links to Nazi Germany, Mehler had begun looking into what had happened to these same scientists and others with similar worldviews once the Second World War was over. Many people assumed that the eugenicists had all but disappeared with the Nazi regime, and that race science was essentially finished at the same time. What Mehler learned instead was that the prejudice that had existed before the war – the fear of some kind of threat to the 'white race' – was still alive in a few small intellectual circles.

'I was really focused on the ideological continuity between the old and the new, and the fact that these ideologies were malicious and dangerous,' he explains. What worried him most of all as he did his investigations was that these people seemed now to be stepping outside their own limited cabal to penetrate not just mainstream academia, but also politics. Their target was nothing short of the highest echelons of the United States government.

One of the key figures in the network to have survived the old days was Roger Pearson. Pearson's career trajectory was very different from that of Reginald Ruggles Gates, fellow founder of the *Mankind Quarterly*. During the Second World War, Pearson had been an officer in the British Indian Army. In the 1950s he worked as managing director of a group of tea gardens in what was then known as East Pakistan, now Bangladesh. And it was around that time that he began publishing newsletters, printed in India, exploring issues of race, science and immigration. Very quickly, says Mehler, he connected with like-minded thinkers all over the world. 'He really was beginning to organise, institutionally organise, the remnants of the pre-war academic scholars who were doing work on eugenics and race. The war had disrupted all of their careers, and after the war they were trying to re-establish themselves. Establishing these institutional networks was essential for their rehabilitation,' he explains.

Pearson's newsletters and the *Mankind Quarterly* relied on being able to reach out to marginal figures from all over the world, people whose views were generally unacceptable in the societies in which they lived. It was a job, of course, that was being done before the benefit of the Internet and social media, before it was easy for likeminded people with extreme views to easily find each other. 'You have these people who come seemingly out of nowhere. It was just so amazing to me that they would be so well networked,' says Mehler.

One of Pearson's publications was the *Northlander*, which described itself as a monthly review of 'pan-Nordic affairs' – by which it meant, more broadly and euphemistically, matters supposedly of interest to white northern Europeans. Its very first edition in 1958 complained about the illegitimate children born thanks to the stationing of 'Negro' troops in Germany after the war, and about immigrants arriving in Britain from the West Indies. 'Britain resounds to the sound and sight of primitive peoples and of jungle rhythms,' Pearson warned. 'Why cannot we see the rot that is taking place in Britain herself?' On the following page, he printed a tribute to Charles Darwin. He had made it his goal to awaken people to what he saw as the existential threat of immigration and racial intermixing, referring to anti-racists as 'cosmopolitans'.

Within a couple of decades, Pearson ended up in Washington, DC, establishing publications there, too, including the *Journal of Indo-European Studies* in 1973 and the *Journal of Social, Political and Economic Studies* in 1975. 'And that's what piqued my curiosity,' Mehler continues. 'Really looking at people who are racist at a time when liberalism was the predominant ideology.' In April 1982 a letter even arrived for Pearson from the White House, bearing the signature of President Ronald Reagan, praising him for promoting scholars who supported 'a free enterprise economy, a firm and consistent foreign policy and a strong national defense'. Pearson used this endorsement to help raise funds and support for his publications. It was clear that he and those in his circle must have had access to the very peak of the US government. The *Independent on Sunday*'s 1990 investigation confirmed that Pearson received several grants from the Pioneer Fund around the same time.

Just as Mehler was carrying out his research, a soft-spoken civil servant in Washington, DC named Keith Hurt happened to be investigating the very same people in his spare time. When their paths crossed, Mehler and Hurt began combining their research. Hurt was then working for the Congressional Research Service, a branch of the Library of Congress that provides policy analysis to members of the House and Senate, which meant that he was keen to keep his identity private. Today, he tells me, he is free to talk on the record. 'I think I started out sort of naïvely,' he admits. 'I ran across some things that were disturbing, that I didn't expect. I didn't really understand that there were these structures and networks and associations of people that were attempting to keep alive a body of ideas that I had associated with at the very least the pre-civil rights movement in this country, and going back to the eugenics movement early in the last century. These ideas were still being developed and promulgated and promoted in discreet ways.'

Perhaps paradoxically, given the fierce nationalism of the figures involved, this also appeared to be a global network, spanning the United States and Europe at least, but also stretching to India and China. 'If you looked at the old *Mankind Quarterly*, it was a truly international journal, with contributors and editors from all over the world,' says Hurt. To this day, the *Mankind Quarterly* runs articles by many writers outside Europe, and its advisory board spans Russia, Japan, Saudi Arabia and Egypt.

'It was important to put together how the networks worked, where the funding came from, what the publications were, what the connections were. And what the connection is with right-wing political organisations,' adds Mehler. What he and Hurt were uncovering astounded them both. 'There was a network, an international network of these people who were not particularly well respected or regarded or even known outside of the network, but they had their own journals, their own publishing houses. They could review and comment upon each other's work,' explains Mehler. 'So it was almost like discovering this whole little world inside academia. And it was a rather nefarious world, of people whose origins went back to the Second World War.' What shocked them above all was the sheer professionalism of the operation, the slick ability of people with some of the most extreme views imaginable to connect to each other and communicate their views across thousands of miles. They were keeping scientific racism alive.

Mehler, who is Jewish, found it particularly disturbing. 'I have a lot of relatives who survived the Holocaust,' he tells me. 'When they flip the light switch and the light goes on, for them it's like "oh wow!". They are prepared for the world to collapse. They are prepared for things to cease to be normal very quickly because that was their experience.' I can hear the fear in his voice, an anxiety that political stability in even the strongest democracies sits ultimately on a precipice. 'I saw anti-Semitism. I was really alienated in American society. I was a person that felt that racism and anti-Semitism were predominant, and that the United States could easily become vicious, racist, and go back to its racist history when push came to shove, if people were threatened enough.' The past, he reminds me, is always capable of repeating itself.

It was around this time, as he and Hurt uncovered the network, that Mehler had his dream. 'I felt like I was desperately trying to prevent this from happening again . . . I thought that we were headed for more genocide.' The parallels between this far-right network of pseudoscientists and intellectuals and the rapid, devastating way in which eugenics research had been translated in Nazi Germany loomed large in Mehler's mind, terrifying him with the possibility that the brutal atrocities of the past could happen once more, that the ideological heart behind them was still beating.

*

Despite the urgency that both Barry Mehler and Keith Hurt felt, their investigations never made it into any high-profile publications. They appeared instead in a few small Jewish and left-wing newsletters, often with Hurt's name omitted or under an alias to protect his job at the Congressional Research Service. The lack of public interest reflected how many people assumed they no longer had anything to fear. Neo-Nazi political parties and white supremacists were thought to exist only on the irrelevant margins of real life.

'Race is such a difficult issue for Americans,' explains Hurt. 'People want to be optimistic. People want to believe that [racists] exist only in a sort of lunatic fringe, which is safely cabined off from the rest of society and that they have no consequences or implications for the future. That was the case then.' The world was thought to be moving in a liberal, more inclusive direction. Racists were thugs and skinheads, not men in power, not academics operating through covert networks.

Then in May 1988, Mehler and Hurt published an article in *The Nation*, a progressive weekly magazine, finally confirming that there might be reason to worry after all. It linked Ralph Scott, a professor of educational psychology at the University of Northern Iowa, to both the Pioneer Fund and the government. Their report claimed that Scott had used grants from the Pioneer Fund under a pseudonym in 1976 and 1977 to organise a national anti-busing campaign. Busing was a means of desegregating schools by transporting children from one area to another, and an important part of the civil rights movement. According to Mehler, some of Scott's money from the fund had also been used to sponsor a study in Mississippi looking at the physical and psychological traits of 'American Anglo-Saxon children'.

What turned the story into one of national significance was that in 1985 the Reagan administration appointed Scott the chair of the Iowa Advisory Commission on Civil Rights, a body whose express purpose was the enforcement of anti-discrimination legislation across the state. Just a few years earlier, Scott had brought, and later dropped, a lawsuit against three black civil rights activists who had described him as a racist. It was clear that even after taking up his influential post, Scott's views on racial difference hadn't changed. He continued to write pieces for the *Mankind Quarterly* and Pearson's *Journal of Social, Political and Economic Studies*. Indeed his most recent article for the *Mankind Quarterly* was published as recently as 2013. Scott, now an emeritus professor, refuses to give me any comments, or to confirm or deny Mehler's reports. But William Tucker has noted through his research that almost every one of his papers is a variation on the same theme: that integrated schools are holding back white students, and not improving achievement among black students, for the simple reason that the two groups are somehow genetically different.

So in 1985 here was Scott, a university professor known to be actively involved in blocking policies aimed at achieving desegregation, who had even been described as a racist, yet had somehow been made officially responsible for defending civil rights in his state. 'It was obviously alarming,' says Hurt. This also happened to come at a time when the Reagan administration was facing criticism for drastically cutting the Civil Rights Commission's budget, making his appointment look even more suspect. To outside observers, it was hard not to imagine that Scott had taken up the position as chair of the Commission to undermine it from within.

The month after Mehler and Hurt's article came out in *The Nation*, Ralph Scott resigned. At the time, although the story made some corners of the national press, it wasn't major headline news. 'It was largely dismissed, I would say, by people who in retrospect would probably admit [they] were mistaken to write it off,' says Hurt. Looking back on the case, in the context of today's politics, with the rise of far-right groups in Europe and the US, and of nationalism more globally, he believes that what they uncovered should have served as a warning. Scott was only an individual, but he operated within a larger network of intellectuals opposed to desegregation. 'What surprised me was how quickly and efficiently these groups worked,' adds Mehler. 'You would think it would be fringe people, and that they would remain on the fringe, and they would have difficulty raising funds and making contacts. That wasn't true at all. What surprised me was how quickly Roger Pearson went from Calcutta, India to Washington, DC to Ronald Reagan.'

Others within academia who have picked up the baton from Mehler and Hurt since then – including psychologist William Tucker at Rutgers University, who has carried out detailed investigations into the Pioneer Fund – have observed how well coordinated and resilient these networks have remained, even after key figures die off. When he set out to research the Pioneer Fund and its wealthy founder Wickliffe Draper in the late 1990s, Tucker tells me, 'I compiled a list of every academic or scientist I could think of who had been outspoken about racial differences and then searched the web or contacted their institution . . . Then I travelled to each of these places, fully expecting that some trips would be a waste of time and research money, because there would not be any Pioneer connection.' He was wrong. 'In fact, I never struck out. Every one of these persons had been contacted and usually supported either by Pioneer or by Draper.'

The Pioneer Fund may have since declined, but something important has entered to take its place. The global political landscape has changed, its focus moving away from the centre, making space once more for those at the extremes. The election of Republican President Donald Trump came at the same time as a growth in nationalist sentiment and far-right parties all over the world. For Hurt, the work he did three decades ago is prescient in today's political climate, not because Roger Pearson or Ralph Scott were ever particularly important figures in American politics during Reagan's time, but because they managed to gain close access to the government despite their views. Somehow, they both found a way to influence powerful people with their brand of intellectual racism.

'I've spent a lot of time in the last decade paying attention to the politics of immigration in this country, which are obviously related to all of that in intimate ways, and which dominate our politics in some ways today,' Hurt tells me. He believes what happened in the past can happen again. 'When Reagan came in, he didn't have established party networks of personnel that the establishment figures in the party had. So he cast a very wide net that included a very diverse range of people, including people like Ralph Scott. Scott wasn't, I think, representative of the central policy thrust of the administration, but there was a lot of carelessness at the beginning of the Reagan administration that allowed these people to step into positions of greater or lesser significance . . . He was symptomatic of a broader problem of entryism to the Republican Party by people like this.'

For those on the political extremes, it's a waiting game. As long as they can survive and maintain their networks, it's only a matter of time before society swings around and provides an entry point once more. The public assumed that scientific racism was dead, when in fact the racists were always active under the radar, suggests Hurt. 'I think there was a whole sequence of events between the late 1980s and the present in which these ideas, which have become pretty well established in the mainstream of American political culture, were step-by-step progressing, re-establishing themselves, eroding the norms of the post-civil rights environment.'

*

What difference does it make to science that a publication like the *Mankind Quarterly* exists today? In truth, barely any. Its work is so rarely read or cited by mainstream scientists that its impact factor (the measure used to judge the influence of a journal) hovers between 0 and a little more than 1. By contrast, the impact factor of a highly respected journal like *Nature* is more than 40. But then, of course, the *Mankind Quarterly* was never designed to be read by scientists or shape the future of research. It was always a platform for those seeking intellectual ballast for their political views. What is of concern, then, is what it represents. As a publication, it's a barometer for intellectual racism. Should it or its contributors become popular, then we will know that something is wrong. And in the last ten years, its impact factor has been on average higher than it was in the preceding decade.

At the same time, its editors have built a presence in other, more credible scientific journals. Assistant editor Richard Lynn, for example, today sits on the editorial advisory board of *Personality and Individual Differences*, produced by Elsevier, one of the world's largest scientific publishers, which counts the highly respected journals *The Lancet* and *Cell* among its titles. Among Lynn's papers was one in 2004 on 'The Intelligence of American Jews', arguing that 'Jews have a higher average level of verbal intelligence than non-Jewish whites.' Gerhard Meisenberg's work, looking at the links between intelligence, genetics and geography, has appeared in *Intelligence*, a psychology journal also published by Elsevier. He has authored at least eight articles in recent years, including one in 2010 on the average IQ of sub-Saharan Africans, and another in 2013 on the relationship between 'national intelligence' and economic success. When I look at who's on the editorial board of *Intelligence* in 2017, I find both Meisenberg and Lynn listed.

While journals are free to publish whatever they think is worthy, subject to peer review, the choice of who to appoint to an editorial board is important because those members help to shape its policy and scope. According to Elsevier's own online guidance for editors, they 'should be appointed from key research institutes'. Neither Lynn nor Meisenberg can claim that honour. A spokesperson for Elsevier, after repeated prompting, tells me that editorial board members 'are not involved in making decisions about which articles will be published. Their role is focused on reflecting the academic debate that takes place within the communities' domain that the journal serves.' Yet Elsevier's own website states that editorial board members 'review submitted manuscripts' and 'attract new authors and submissions'. The other implication of their brief statement is that the work of Lynn and Meisenberg, studying population-level differences – which some might equate with racial differences – in intelligence must now be a part of mainstream academic debate.

In 2017, when I call the current editor-in-chief of *Intelligence*, Richard Haier, an emeritus professor in the medical school at the University of California, Irvine, to find out how he feels about having editors from the *Mankind Quarterly* on the editorial board of his own journal, he sounds nervous. 'I struggled with this, frankly, when I became editor, and I consulted several people about this,' he admits. 'I decided that it's better to deal with these things with sunlight and by inclusion.' Throughout our conversation, he is uneasy, taking long pauses to choose his words. Keeping them inside the fold, he tells me finally, reflects his 'commitment to academic freedom'.

Haier reassures me that he has never met Meisenberg or Lynn. But he tells me he did personally know and defend the late Arthur Jensen, a professor of educational psychology at the University of California, Berkeley. In 1969 Jensen mooted in the *Harvard Educational Review* that gaps in intelligence test results between black and white students might be down to genetics. His article remains one of the most controversial psychology papers ever published. The *New York Times* reported in

1977 that the Pioneer Fund had been subsidising Jensen's work. An investigation published by the *Los Angeles Times* almost two decades later confirmed that by then grants to Jensen from the fund must have totalled more than a million dollars.

Haier continues, 'The area of the relationship between intelligence and group differences is probably the most incendiary area in the whole of psychology. And some of the people who work in that area have said incendiary things . . . I have read some quotes, indirect quotes, that disturb me, but throwing people off an editorial board for expressing an opinion really puts us in dicey area. I prefer to let the papers and the data speak for themselves.' He adds that he personally does believe there is something scientifically interesting about studying 'group differences' in intelligence. 'Scientific intelligence research has laboured under this cloud for fifty years, and it is my stated goal as editor to help bring intelligence research back into the mainstream, where it used to be.'

Even so, Haier seems to have been bothered by my inquiries. When I check the Elsevier website around a year later, at the end of 2018, both Gerhard Meisenberg and Richard Lynn have been removed from the editorial board of *Intelligence*.

If group-level or population-level differences in intelligence do need to come out from under the cloud of controversy, then what are the reasons that researchers might want to wade into this deeply divisive area of research? According to Richard Haier, one theme is shared by some of those who submit their work to *Intelligence*. 'I can tell you, and I'm not revealing anything secret here as editor, we receive a number of papers that try to speak to the relationship between intelligence and economic development, and group differences in general.' Sadly, it's not always of the highest quality, he admits. 'When I read many of those papers, they are substandard, and they never even get to peer review . . . We have had papers submitted that come up with some kind of result, and then the discussion section extrapolates from that result to immigration policy. Those papers never get as far as peer review in our journal because it's clear that there is an agenda.' As I recall, these were exactly the kind of extrapolations made by Gerhard Meisenberg in his emails to me.

Haier's admission betrays just how much this field remains plagued by dark politics. I can't help but think of the nineteenth-century race scientists who jumped to biological explanations for the inequality they saw in the world, who believed other races had been doomed to failure by nature because their brains were too small or their temperaments too weak. Confronted by slavery and colonialism, they skimmed over history and culture, preferring instead to look to biology for justification for this kind of exploitation. When researchers like Meisenberg today link economic development to intelligence, they imply that the vast inequality between the world's richest and poorest countries is rooted not just in the imbalance of power or historical circumstance, but in the innate weaknesses of the populations themselves. Racial injustice and inequality, in their minds, isn't injustice or inequality at all. It's there because the racial hierarchy is real.

*

'I think what we're experiencing now is a much more threatening environment,' says Keith Hurt. 'We're in a much worse situation than we were a couple of decades ago.' He believes that the kind of research once sponsored by the Pioneer Fund and still published by the *Mankind Quarterly* has now found fresh avenues of support. Scientific racism has come out of the shadows, at least partly because wider society has made room for it. 'Frankly, I think at this point the ideological stream that it was sustaining is now self-sustaining. There are other institutions, and a much, much broader culture that will sustain it.'

Over the last thirty years this broader culture that Hurt describes, which rails against 'political correctness' and calls for a greater diversity of political opinion and freedom of speech in academia as a disguise for the propagation of extreme right-wing views, has become stronger. In 2018 an investigation by the *London Student* newspaper revealed that *Mankind Quarterly* editors Richard Lynn and Gerhard Meisenberg had been organising and speaking at a series of small, invitation-only conferences held at University College London since 2014. The Associated Press reported later that a University of Arizona psychology professor with an interest in human behaviour and evolutionary psychology, who has also been on the editorial advisory board of the *Mankind Quarterly*, had used a grant from the Pioneer Fund to attend one such conference. Somehow the organisers had managed to secure a space within the university to discuss controversial issues around eugenics and intelligence, attracting one attendee who was later appointed (although he hastily withdrew) to head the Office for Students, the regulatory authority for the English higher education sector.

Race scientists working two decades ago are building a stronger presence once more. In 1994, in *The Bell Curve*, one of the most notorious bestsellers of the twentieth century, American political scientist Charles Murray and psychologist Richard Herrnstein suggested that black Americans were less intelligent than whites and Asians. A review at the time in the *New York Review of Books* observed that they cited five articles from the *Mankind Quarterly*, and no fewer than seventeen researchers who had contributed to the journal. Murray and Herrnstein went so far as to describe Richard Lynn as 'a leading scholar of racial and ethnic differences'. Although *The Bell Curve* was widely panned after its publication, an article in *American Behavioral Scientist* describing it as 'fascist ideology', in 2017 *Scientific American* noted that Charles Murray was enjoying 'an unfortunate resurgence'. Facing down protesters, he was being invited to give lectures on college campuses across the United States.

Another contributor to the *Mankind Quarterly* has become a key figure in the white supremacist movement. Yale-educated Jared Taylor, who belongs to a number of right-wing groups and think tanks, founded the magazine *American Renaissance* in 1990. William Tucker describes it to me as today being the true intellectual arm of the modern neo-Nazi movement. One phrase Taylor uses to defend racial segregation, for example, borrowed from the zoologist Raymond Hall writing in the first ever issue of the *Mankind Quarterly*, is that 'two subspecies of the same species do not occur in the same geographic area'. His brand of white supremacy draws from race science to lend itself the illusion of intellectual backbone. Taylor is in some ways the Wickliffe Draper of the twenty-first century.

Like Draper, Taylor has sought to make racism respectable again. His American Renaissance Foundation conferences – which anthropologist Robert Wald Sussman has described as 'a gathering place for white supremacists, white nationalists, white separatists, neo-Nazis, Ku Klux Klan members, Holocaust deniers, and eugenicists', and *The Journal of Blacks in Higher Education* has more succinctly dubbed 'a convocation of bigots' – have featured other intellectuals who have written for or edited the *Mankind Quarterly*. A visitor at the 1994 meeting reported that people didn't 'flinch from using terms such as "nigger" and "chink"'. Male attendees are expected to dress in smart business suits, to set themselves apart from the thuggish image most people associate with racists. They don't call themselves racists but 'race realists', a euphemism that reflects how they like to believe the scientific facts are on their side.

By the time of the 2016 US presidential election, Jared Taylor's place in American politics was more secure than ever. He even appeared in a television advert released by Hillary Clinton's campaign team to show the kind of support that rival Donald Trump's anti-immigration policies had among white nationalists. Explaining the increasing prominence of people such as Taylor during the election, one correspondent reflected in *Newsweek* magazine in 2017: 'These men have degrees from some of the nation's top universities . . . they are a well-read group who cloak their ideas about the intrinsic superiority of white men in selected passages of literature, history, philosophy and science.'

One regular speaker at American Renaissance Foundation conferences and another *Mankind Quarterly* contributor is Michael Levin, a professor of philosophy at the City University of New York (CUNY), which by its own mission as a large, relatively affordable urban university is committed to being accessible to under-represented minorities. He is the author of *Why Race Matters*, a book which went out of print after publication in 1997, but was republished with a new cover in 2016, together with a foreword by Jared Taylor. In 1986, Levin had written a letter to the *New York Times* arguing that it was legitimate for store owners to discriminate against all black people because they were more likely to be attacked by someone who was black. According to the Southern Poverty Law Center, Levin told the audience at a 1998 American Renaissance Foundation meeting, 'The two principal race differences that I see are race differences in intelligence and in motivation . . . It's no wonder there are very few black scientists . . . you have to have an IQ of 130 to be a successful research scientist.' According to an article in *The Journal of Blacks in Higher Education* in 1994, Levin had by then received more than $120,000 in grants from the Pioneer Fund.

'People like that are able to get funding,' Keith Hurt tells me. 'There will always be, unfortunately, men of wealth – and they are almost all *men* of wealth – who share these ideas and are willing to support them.' But another reason that scientific racists have more influence now is that the Internet and social media have given them simpler ways to access and grow their networks. 'Behind every racist joke is a scientific fact,' alt-right blogger Milo Yiannopoulos told a Bloomberg reporter in 2016. Among the cabal of 'race realists' operating online, as well as their followers, it's easy to spot an attitude of doggedness. They repeatedly insist that they are challenging the politically correct wider world by standing up for good science, that those who oppose them are irrational science-deniers.

Gerhard Meisenberg, for example, writes to me in our correspondence that 'some academics seem to believe that by simply claiming that race differences don't exist, we can prevent people from believing in them. Doesn't work like that . . . For example, if we tell people that black children are as smart as white children and it isn't true, there will be teachers and others who know first-hand that it isn't true. Also, if we tell people that it's because of some flaws in the school system and the flaws are repaired and it doesn't help one bit, not only do people get frustrated about all the wasted effort, but they also start distrusting the "scientists" who are telling them lies. Real people aren't postmodernists. They distinguish between truth and lies.' Meisenberg's tactic is simple: using people's gut prejudices and casual observations to undermine trust in mainstream science. If you *feel* it to be true, it must be.

This rhetoric around who has the genuine claim on the truth resonates today more than ever when the public on both sides of the political divide worry about fake news and media conspiracies. Yet it's a line that has been adopted by intellectual racists for decades. In 1998 an American corporate lawyer called Harry Weyher, president of the Pioneer Fund from 1958 until he died in 2002, was given space to write an eighteen-page editorial for the journal *Intelligence*. He used the opportunity to defend research supported by the fund. 'This is critical research by world-class scholars . . . Yet, if one were to believe some important segments of the media, this research was funded by an evil foundation and done by evil scientists, and is unfit for public dissemination.' Twenty years before Donald Trump was elected President, Weyher too laid into what he saw as an egalitarian orthodoxy and political correctness, going so far as to accuse the print and broadcast media of 'false reporting'.

*

'Why do we still have race science given everything that happened in the twentieth century?' I'm asked by American anthropologist Jonathan Marks, an academic many turn to for clarity when it comes to racism in science.

His answer is unequivocal: 'Because it is an important political issue. And there are powerful forces on the right that fund research into studying human differences with the goal of establishing those differences as a basis of inequalities.' Ultimately, politics is always a feature of the science, just as it was in the very beginning. Once there was the backdrop of slavery and colonialism, then it was immigration and segregation, and now it is the right-wing agenda of this age. Nativism remains an issue, but there is also a backlash against greater efforts to promote racial equality in multicultural societies. And just like before, the message of those with racist intentions is tailored gently, carefully sculpted to appeal to populist fears while at the same time sounding logical and reasonable. Communicating with me, for example, Gerhard Meisenberg uses the word 'culture' alongside the word 'race' as though they're fully interchangeable, understanding how much most people these days value and respect cultural boundaries, even if they don't recognise biological race.

'Without much selectivity in migration, all countries of the world become homogenised, not only in ability level but also in culture and everything else. Countries become more similar to each other,' writes Meisenberg. How tragic it would be to have the whole world look exactly the same. On a logical level he fails to explain how, if races are fixed and immutable in the way he thinks they are, we could all end up the same just by migrating. But this isn't the point. On the surface, heard quickly, his concerns sound almost sensible, in the same way that eugenics sounded so rational and attractive to progressive social reformers in the early twentieth century. Who back then could argue against the pursuit of a healthier, stronger population? Who today could argue against countries and groups maintaining their distinct cultures?

Race has always been an intrinsically political area of research, the idea itself born out of a certain world order. So it's small surprise that those doing this kind of research end up in the same place again and again, as they have for centuries. When they look for human variation, however objective they claim to be, they can't help but ask what the differences they think they see mean for society. This is why race science so often comes twinned with speculation as to the causes of economic and social inequality.

A common theme among today's race realists is their belief that, because racial differences exist, diversity and equal opportunity programmes – designed to make society fairer – are doomed to fail. 'As far as I understand, race-based policies of this kind were adopted by many American institutions since the 1960s or 70s, and were originally justified as something that is needed to make up for disadvantages that Blacks and some other minorities had suffered in the era before civil rights legislation,' Meisenberg tells me. 'Today, fifty years later, that old reasoning is no longer credible.' Rather than investing in these policies, then, he appears to argue that we should accept inequality as a biological fact. If an equal world isn't being forged fast enough, the race realists don't see it as a longer or tougher path than we imagined. Instead, it is a permanent natural roadblock created by the fact that, deep down, we're not the same.

'We have two nested fallacies here,' Jonathan Marks continues. The first is that the human species comes packaged up in a small number of discrete races, each with their own different traits. 'Second is the idea that there are innate explanations for political and economic inequality. And basically what you're doing there is saying that inequality exists, but it doesn't represent historical injustice. What these guys are trying to do is manipulate science to construct imaginary boundaries to social progress.'

For Marks, the prescription for this pathology is radical. In science, to ban any kind of research – if it can get funded and ethically approved – is a risk to academic freedom. He suggests instead that people who 'cannot handle the results shouldn't be studying it . . . We don't want racists working on human variation because that doesn't work. So it's not a question of "should this be studied?", it's a question of who should be studying it and how should they be credentialled, or how should they be vetted.' His argument is that if this field is always going to be affected by the politics of the scientists, then surely it makes sense to have people doing the research who aren't bent on division and destruction, whose aims don't lie at odds with what society as a whole has decided is morally acceptable.

To others, this sounds heavy-handed. Psychologist William Tucker, for example, whose painstaking work helped expose the Pioneer Fund and the scientists it has supported, supports the freedom of anyone to do any research – even the kind he abhors. 'People enjoy the right to take Pioneer's money; I would not like to see them deprived of that right. At the same time, I think that their decision to do so is awful and will do whatever is in my power to persuade them of the folly of such a course,' he tells me.

Fundamentally, though, the problem is not the science itself. If it were limited to academia, the kind of material that's published in the *Mankind Quarterly* and other likeminded publications would have next to no impact at all because mainstream scientists almost entirely ignore it. It just doesn't have enough scientific value. As for the ideology, most researchers today accept that what little we know about human variation can't be used to dictate how we treat people in the real world. It certainly can't be used to set policy. The problem is in how these ideas are used and abused in wider society, how much traction they can gain with the public and those in power. Nazi scientists carrying out their regime's programme of 'racial hygiene' had only a rickety scientific framework upon which to justify the destruction of millions. The same was true of those who used science to defend slavery, colonialism and segregation. And it applies today, too, to those on the political extremes. Nothing is more seductive than a nice string of data, a single bell curve, or a seemingly peer-reviewed scientific study. After all, it can't be racist if it is a 'fact'.

For those with a political ideology to sell, the science (such as it is) becomes a prop. The data itself doesn't matter so much as how it can be spun. Marks warns me, then, that those to really watch out for are the ones who claim to be uniquely free of bias, who tell you they have some kind of special, impartial claim on the truth. 'Whenever anybody tells you I am objective, I am apolitical, that is the time to watch your wallet. Because you're about to have your pocket picked.'

And he should know, because he almost had his pocket picked.

6

Human Biodiversity

How race was rebranded for the twenty-first century

It was 1998 or thereabouts that an invitation arrived in Jonathan Marks' email inbox. He doesn't recall the precise date because, until I asked him about it, he hadn't thought about it for years. What he does remember is that the sender was a then little-known American science journalist and former writer for the conservative *National Review* magazine, Steve Sailer. The invitation was for Marks to join a mailing list of people interested in the subject of human variation.

'I knew absolutely nothing about him,' Marks recalls. 'He just seemed to be someone who was organising something.'

At the time, Marks was teaching at the University of California, Berkeley, a few years after writing a popular textbook on race, genes and culture titled *Human Biodiversity*. Wedging these two words together, he had neatly coined a phrase to describe biological and social variation across the human race. Part of the reason he chose it, he admits, is that 'biodiversity' had become something of a buzzword. He never guessed it would cause any problems. And why would it? Diversity was being celebrated; both the glorious biodiversity of the natural world and the rainbow of cultural and physical diversity in human societies. It was the proud label of liberal anti-racists, of the good guys like himself.

'America's answer to the intolerant man is diversity,' Robert F. Kennedy had affirmed in 1964 at the dedication of an interfaith chapel in Georgia. 'United in diversity' was the motto of the European Union. It goes without saying that although different cultures have different things to offer, the Benetton breadth of human variation does nothing to undermine the general consensus that we're biologically pretty much all the same beneath the skin. This is a truth that's been universally acknowledged since the end of the Second World War. At least that's what Marks thought.

The invitation itself appeared perfectly innocent. The 1990s marked the early years of electronic mailing lists, and Steve Sailer apparently wanted to use one as a way of pulling together scientists, intellectuals and fellow journalists to start a private conversation about human difference. 'He said, hey, I'm interested in human variation, and I like your work. Let's get an email list together,' says Marks. 'It seemed pretty straightforward and harmless.' So he signed up. What intrigued him especially was that Sailer happened to be brandishing Marks' own neologism, calling his list the 'Human Biodiversity Discussion Group'.

Others joined in their dozens. By the summer of 1999, Sailer's roster of members was astounding. Along with prominent anthropologists like Marks, there was psychologist Steven Pinker, political scientist Francis Fukuyama and economist Paul Krugman. In hindsight, the large number of economists in the group might have been a warning. There in the mix, too, was the controversial author of *The Bell Curve*, political scientist Charles Murray. That should have been more of a red flag.

When Marks had talked about biological diversity, he meant the superficial variation we see right across the species, from individual to individual. 'There isn't really room for three, or four, or five biologically distinct kinds of people,' he tells me. He certainly didn't expect to see people reinforcing old-fashioned stereotypes of the kind that had long been debunked. That school of racism was long dead, he assumed. Yet here on this email list, something strange was happening. Observing the conversations that Sailer steered through the group, Marks noticed the term 'human biodiversity' being used differently from the way he had originally intended. Members were using it to refer to deep differences between human population groups.

Among the people added to the group that summer was Ron Unz, a Harvard graduate and founder of a financial services software company, who had recently run as the Republican candidate for governor of California. His introduction to other members was pasted alongside a 1994 article he had written in *The Wall Street Journal* titled 'How to Grab the Immigration Issue', observing the state of California's changing racial demographics. 'Conservatives and Anglos have become enormously angry and frustrated over the growth of crime, welfare, affirmative action and the general decay of their society,' he stated. Another addition to the group was Deepak Lal, a professor of international development at the University of California, Los Angeles, whose work explored the reasons the West happened to be economically more successful.

It dawned on Marks that Sailer's seemingly innocent email list was not so much about discussing science in an objective way, but more about tying together science and economics with existing racial stereotypes. One debate that sticks in Marks' mind today took place between himself and a journalist who claimed that black people were genetically endowed to be better at sports. Marks insisted that this was a scientifically shaky argument, not to mention one with dangerous political implications. The two experts clearly disagreed. But rather than help reach a consensus, 'Steve Sailer clearly took his side,' he tells me.

'At which point I realised, ah! This isn't an impartial scholarly discussion.'

Another time, Sailer defended a writer on the list who suggested that different ethnic groups had their own particular strengths, adding that Turks were born physically more powerful. This was, he suggested, one reason why affirmative action policies to hasten racial equality weren't a good idea. Different ethnicities should instead be encouraged to do what they do best. When Sailer talked about human biodiversity, he didn't appear to be using the phrase in a politically neutral way, but as a euphemism. He had spun the same language used by liberal anti-racists to celebrate human cultural diversity in order to build a new and ostensibly more acceptable language around biological race.

In email correspondence with me Sailer denies duping anyone, although he does admit that the group leant towards the heretical. But whatever his intentions, this wouldn't have been the first time someone had tried to subvert the idea of 'diversity'. In their conclusion to *The Bell Curve*, a book which just a few years earlier had claimed that black Americans were innately less intelligent than whites, Charles Murray and Richard Herrnstein similarly undermined the political push towards racial equality by arguing that biological differences between groups made it practically impossible. 'We are enthusiastic about diversity – the rich, unending diversity that free human beings generate as a matter of course, not the imposed diversity of group quotas,' they wrote. Every person in a diverse society had a valued place, they implied – just not the same place.

Realising his error, Marks left the group. Other members also denounced it as racist. 'If somebody said the same to me today I would probably be a little bit more suspicious,' Marks admits. 'I would look at who else is on the invite list. But even that can be misleading because he may be inviting people who are just as confused about this as I am.' At the time, it seemed harmless. Just a bunch of people with some marginal political ideas trying to convince others of the same.

Instead, Marks was about to find out just how prophetic the existence of the list was. 'That was my introduction to what became the alt-right.'

*

For those sucked into Sailer's electronic arena for the intellectual discussion of race, his email list was just a taste of the virulent racism that would later be seen far more often in shadowy corners of the Internet, then more openly on social media and right-wing websites, and finally in mainstream political discourse. Many more soon took hold of the phrase 'human biodiversity', giving it a life of its own online. Today it's nothing short of a mantra among self-styled race realists.

'Blogs have made the dissemination of wacko ideas much more efficient,' says Jonathan Marks. 'Actually in academia, we don't really know what to do with blogs because they have a really short shelf life. You forget about them a day after you read them, it's hard to cite them. But today, they are out there. They're these markers of people with wacko ideas.'

To be fair, few could have guessed that the email list was a precursor to something bigger. But as the group slowly became defunct, Steve Sailer's political convictions became increasingly obvious. He and other members of the list went on to become prominent conservative bloggers, writing frequently on race, genetics and intelligence. As a columnist for VDARE.com, an American website that describes itself as a news outlet for patriotic immigration reform (and, like the *Mankind Quarterly*, is automatically blocked by my Internet service provider), Sailer once argued passionately for the biological reality of race, stating that it was a fundamental aspect of the human condition. In 2009 the same right-wing website published Sailer's first book, about Barack Obama, titled *America's Half-Blood Prince*. In 2013 Ron Unz founded his own blogging platform, the *Unz Review*, as an alternative to the mainstream media, recruiting Sailer as one of his most prolific columnists.

Sailer truly rose to prominence, though, in the United States presidential election of 2016. Six years earlier, he had proposed focusing heavily on immigration to draw in white working-class voters as a single bloc. Having subverted the language of diversity through his email list, he did the same with identity politics. If ethnic minorities, such as black and Hispanic Americans, could assert their rights and defend their interests, he reasoned, then why not white voters who felt they were losing out to cheap immigrant labour and globalisation? It turned out to be an unexpectedly successful campaign strategy for Donald Trump.

'I think the big problem this country has is being politically correct,' Trump said in the first Republican Party presidential debate, in Ohio in 2015. Dog-whistle politics was reframed as a pushback to liberal elitism. A similar approach was adopted by some campaigners in the UK in favour of leaving Europe during the Brexit debate around the same time, emphasising that Turkey might join the European Union and flood Britain with millions of Turkish migrants.

An article in *New York* magazine in 2017 described Sailer's string of prophetic political insights as a new wave of populist thinking on the right, dubbing it 'Sailerism'. In the United States at least, he was credited with inventing a form of identity politics for disgruntled poorer whites, a group that had been neglected by politicians. But at the same time, political observers couldn't help but notice that his ideology often looked suspiciously like white nationalism.

For researcher Keith Hurt in Washington, DC, who has kept a close eye on the immigration debate ever since his investigations into right-wing intellectual racism in the 1980s, none of what happened during that time should have come as a surprise. 'The election of Trump made it impossible for many people to any longer overlook this stuff,' he tells me.

Hurt explains that the racist ideologies that existed at the start of the twentieth century, manifesting themselves in the eugenics movement and then in German nationalism, had survived by the end of it. The only difference was that those who held these views were later forced to keep them private. 'The post-World War Two ideological consensus rested on an implicit social contract that said, on the one hand, overt racist language will not be deployed in the public square in the way it was before. On the other hand, if this kind of talk is pushed out of the public square, society will refrain from making accusations of racism against all but the most extreme fringe.' And so the intellectual racists, the ideologues, communicated with each other through their own tight networks, disseminated their ideas through their own publications, some of which were so marginal and private that they were almost invisible to the outside world. It was easy for the public to assume they didn't exist, that the only genuine racists left were the ones they could see and hear, the skinheads and thugs.

When the time was right, however, political and intellectual racism slowly resurfaced into the mainstream. Sensing the rising tide, people like Steve Sailer waded in. 'They couldn't exist as they do now if there hadn't been the intellectual ideological continuity,' adds Hurt. They weren't starting from scratch, they were just repackaging the old ideas that they had been nursing the entire time, the same racial assumptions that had been around for decades.

But it all came as more of a surprise to academics like Jonathan Marks. 'I was working with the assumption that these guys were a lunatic fringe. If you had told me twenty years later that they would be part of a political mainstream wave, I would have said you are absolutely crazy. These guys are anti-science. These guys are positioning themselves against the empirical study of human variation and they are clearly ideologues for whom empirical evidence isn't important,' he laughs.

'But I think they were a lot cleverer than us professors.'

*

Offline, in mainstream academia and among respected scientists, a different debate was taking place. But it would similarly come to reshape the way people thought about race.

It was 1991, shortly after the launch of the multi-billion-dollar Human Genome Project, which was steadily working to build a map of the genetic data shared by our species for the first time by sequencing the entire length of human DNA. At the Stanford University School of Medicine in California, an influential geneticist by the name of Luigi Luca Cavalli-Sforza, then around seventy years old, spotted an opportunity. His own long and illustrious career had been dedicated to studying human variation across the world, work that had taken him from Italy to the United States and brought him to the forefront of a field known as population genetics.

Cavalli-Sforza and some of his colleagues – a small team of anthropologists and geneticists, based mostly in the United States – wondered whether the kind of data being collected by the Human Genome Project could also be used to pick through the fundamental differences between human groups, the genetic variations that make us who we are. Little did they guess that it would turn out to be one of the most controversial scientific initiatives of its time, leading to decades of debate about the biological reality of race.

Unlike Steve Sailer, Cavalli-Sforza and his team didn't come to their new project with race in mind. Indeed, quite the opposite. They were avowed anti-racists, fully signed up to the UNESCO statements on race in the 1950s, and firm believers that science would do nothing but prove racial stereotypes incorrect. They thought that their work might even help free the world from the scourge of racism. 'This was, at least on the American side, in general political terms, a quite left-wing group of people,' recalls Henry Greely, a professor at Stanford Law School who became involved in the initiative later on.

Cavalli-Sforza was used to receiving hate mail from people who disagreed with his outspoken belief that genetics didn't support old-fashioned notions of race. In 1973 he had publicly debated with William Shockley, a Stanford University physicist and joint Nobel Prize winner, who in later life became a notorious racist. Shockley believed that black Americans had intellectual shortcomings that were hereditary, and that black women should therefore be voluntarily sterilised. He was among the most prominent race theorists to receive support from the Pioneer Fund. When they met at Stanford, Cavalli-Sforza coolly demolished his claims, fact by fact.

Population genetics itself was born out of post-Second World War efforts to move away from traditional race science and eugenics. In the 1950s and 60s, when geneticists stopped talking about race, they turned their attention instead to 'populations', the 'human variation' between these populations, and the 'frequencies' of certain genes within these groups. It was a more rigorous, molecular, mathematical approach to studying human difference. And what population geneticists like Cavalli-Sforza quickly noticed was that there are no hard genetic boundaries around human groups, but rather continuous statistical variation, with a good deal of overlap. What differences there are exist along gradients, not borders. But that said, variation isn't random either. It can, depending on how you look at it, fall into clusters in which certain genes are statistically more common in some groups than in others.

Cavalli-Sforza and his peers became fascinated by these clusters. It was thought they would be more obvious in places where people had been geographically isolated for hundreds of years, living on islands or at the tops of mountains, mating almost exclusively within their own group through many generations. 'Primitive groups' were believed to be particularly distinct genetically because of the length of time they had spent away from others, shielding their genomes from the effects of intermixing. Since these people were so remote, their lineage conspicuous thanks to their long seclusion, maybe their genes could offer special insights into how humans evolved and adapted. Cavalli-Sforza was among those who believed that, by studying the genomes of these primitive groups, it might be possible to track historic patterns of migration. If they could see how gene frequencies varied in such groups and their neighbours, they could perhaps track where their ancestors lived in the distant past. In 1961 he became one of the first scientists to apply modern statistical methods to see how frequencies of major blood types varied among large human groups, creating a family tree to then show how these groups were related.

The search for what they thought would be genuine difference at the corners of the world took researchers to the remotest places imaginable. In 1964 the World Health Organisation picked out 'Eskimos' in the Arctic, Guayaki hunter-gatherers in the forests of eastern Paraguay, 'pygmies' in the Central African Republic, Aboriginal Australians, and the tribes of the Andaman Islands in India as potentially interesting 'primitive groups'. Scientists who couldn't make the trips would settle on immigrant groups in their own backyards, particularly Jewish and Roma communities, also known to be ancient and tight-knit. There was no point in studying very large human populations, say Africans or Europeans, because there was too much variation within them as a result of migration and mixing. Small, old indigenous communities were thought to be special, more distinct.

On the back of this research, Cavalli-Sforza formed a plan. He mooted the idea of using the same revolutionary gene sequencing technology as the Human Genome Project not just to map one genome, but by travelling the world to draw genetic data from lots of individuals of different ethnicities. This 'Human Genome Diversity Project', he and his collaborators announced in the journal *Genomics* in 1991, would 'supplement and strengthen findings from archaeology, linguistics, and history'. Through their work, genetics would sit alongside the study of culture, language and history to help paint the human story.

The Human Genome Diversity Project would initially zoom in on four to five hundred different small populations, especially geographically remote and apparently dwindling ones. Referring to these indigenous groups as 'isolates' – people including the Basques in Europe, the Kurds of eastern Turkey, and Native Americans – they expected their genetic data to provide clear signals about prehistoric migration and social structure. Calling for urgent funding from the world's governments, their argument was that this was a race against time to document the breadth of human variety before we all fell into the great big melting pot, when migration and assimilation would leave us all so genetically similar that such an effort might not be quite so worthwhile.

But what the scientists were proposing didn't sit easily with everyone. Some outside observers couldn't help but be a little uncomfortable. After all, it was hard not to wonder whether in the nineteenth century, this might have just been called race science. The Human Genome Diversity Project was technically more precise, of course, more scientific. It wasn't sampling skin and hair colour, or slotting people into racial hierarchies. It was using genetics. But in some ways it was hardly distinguishable from the study of human difference a hundred years earlier. The word 'race' had been prudently replaced by 'population', and 'racial difference' by 'human variation', but didn't it look suspiciously like the same old creature?

Then again, could it be called race science if the scientists involved were obviously anti-racists? As Henry Greely told me, they were left-wing liberals committed to stopping racism, whose public lives had been dedicated to fighting scientific racists and eugenicists. How could there be anything to worry about?

*

'They won't use the term "race",' I'm told by Joanna Radin in the department of history at Yale University, who has studied the progress of the Human Genome Diversity Project since 1991 to the present day. She notes that there was always a deliberate effort to avoid the word 'race', mostly because of its obvious political baggage, but also because the scientists didn't see genetic 'isolates' as being 'races' in the traditional sense. They were different enough to merit study, but not in the way racial difference had been described in the past. They were small groups, not large continental-scale ones. 'They don't necessarily map on to existing racial hierarchies,' she says.

Those behind the project insisted that their research was countering racial myths. Their stated intention was to replace ignorance and prejudice with hard scientific facts, and make it clear that we are one single human species, united in our common origins. Their plan was to look for difference not as racists had in the nineteenth century, to prove inferiority or superiority, but to use the tiny difference deep in our genomes to help build a picture of human migration. They would be like archaeologists, digging through our genes, looking for clues about our history. Their aim was simply to understand our past.

Given the unimpeachable political credentials of people like Luigi Luca Cavalli-Sforza, the scientists behind the Human Genome Diversity Project might have expected it to go forward without a hitch. All they needed was funding, and permission from their 'isolates', the indigenous communities whose blood and DNA they wanted to sample. But things weren't so easy. This wasn't the sixties any more, when foreign researchers could pick the communities they wanted to study and just rock up in the expectation that they would comply. Those communities had become more wary.

'Luca Cavalli-Sforza was an old school anthropologist,' geneticist Mark Jobling at the University of Leicester explains. 'I mean, I went to talks by him in the nineties where he would show old slides of him collecting DNA, blood samples in Africa from pygmy groups and offering glass beads and cigarettes in return, things like that.' This wasn't the way things were done now. This was the 1990s, the dawn of the Internet age, identity politics, and the fight for indigenous rights. Scientists who took an interest in indigenous communities now found that the same communities were taking an interest right back. They weren't as trusting as they had been in the past. And they were organised.

And they had good reason to be organised. Remote tribes and ethnic groups had been exploited throughout history, their land and cultural artefacts pillaged by Western colonisers, their bodies targets for unethical experimentation. Between 1946 and 1948, for example, the United States government ran secret experiments on thousands of people in Guatemala, deliberately exposing them to sexually transmitted diseases. Before and during the Second World War, British scientists had deliberately sent Indian soldiers fighting for Britain into gas chambers to study their response to mustard gas exposure. There had been a long and bloody tradition of scientists abusing other populations for their own ends, particularly populations deemed at the time to be racially inferior. But people were now prepared. Rights activists, alert to the risk of exploitation, were ready to defend the communities targeted by the Human Genome Diversity Project.

These activists warned of the possibility that DNA analysis might damage how the communities concerned chose to understand their past, reveal something profitable, or even be used as a weapon against them by racists. They weren't prepared to hand over their biological data, their blood and tissue, knowing that it might end up being misused. Even so, their resistance to join in the project baffled some of the scientists involved. As Radin explains, 'What had changed – and this is what caught the scientists by surprise – was that the indigenous groups they had imagined disappearing, that they didn't have to reckon with once they left, these purported isolates, were organising in indigenous movements. They had access to the web, they were in touch with activists.'

As the future of the project came increasingly under threat, Cavalli-Sforza and his colleagues called in Henry Greely from Stanford Law School to help navigate the ethical dilemmas and deal with critics. Greely, fascinated by the project and taking a personal liking to the people inside it, agreed. At the beginning, he had very little knowledge of the science itself. 'I knew how to spell DNA but that was about it.' But from the outset, he could see that there were likely to be problems. The scientists were of course aware that science did not have a great historical record when it came to race. 'They knew this was an issue . . . they knew that it had a bad past.' But they didn't see themselves as part of that past. As far as they saw it, 'they were the *break* from the past. They were the good guys. They were the ones who understood the rights-enhancing, equality-enhancing potential of genomics, and they were going to bring it to the world,' he says. Greely adds that this was the first place he came across the fact 'that all humanity is more similar to each other than a band of chimpanzees that lives in a particular region of Africa'.

But at the same time, they were oddly naïve about how the project might be seen from the outside. One scientist 'talked about the need to sample "isolates of historical interest", a term that indigenous populations did not care for,' Greely admits. 'It struck me that that was not likely to be well received because it's a very clinical, bloodless way of referring to people who are alive, and cultures that are living now. Historical interest is sort of something you find in a museum. It was tone deaf.'

'Naïveté is always easily diagnosed through the retrospectoscope,' he tells me.

Greely's job, to navigate the ethical quandaries posed by the Human Genome Diversity Project, turned out to be a poisoned chalice. Partly because of the political controversy surrounding the project, it didn't attract the funding they wanted. Yet the scientists struggled to understand why. 'Some were so comfortable with their own knowledge of their own moral bona fides that it was hard for them to imagine being attacked from the left. They would have imagined that any remaining *racists* would be attacking them, not that *they* would be attacked as being racists.' If anything, they were on the same side as their attackers. 'I was the most conservative person on the North American committee, and I'm a Carter-Clinton-Obama Democrat!'

The activists representing indigenous groups turned up the dial on their protests. At a meeting of the World Council of Indigenous Peoples in Guatemala in 1993, Greely found himself facing down the accusation that he was a CIA agent, intent on committing genocide, which was clearly untrue. 'We took some lumps. Some of them were deserved and some of them weren't.'

In an address at a special meeting of UNESCO in 1994, Cavalli-Sforza turned his focus to the charge of racism, insisting that his project would help combat prejudice, not perpetuate it. But in 1995 another political storm blew up when scientists funded by the United States National Institutes of Health tried to patent a virus-infected cell line from people belonging to the Hagahai tribe in the highlands of Papua New Guinea for the purpose of developing a new treatment for leukaemia. Activists accused them of stealing people's biological samples and attempting to profit from them.

It was in this charged political environment that Greely drew up a model ethical protocol, setting in stone that samples wouldn't be taken unless entire groups, not just individuals, had given their consent. But despite such concessions, there was a deeper problem. At the same time that the project claimed to be anti-racist, it was hard to escape the paradox that this was also all about finding out how people differed. If the genetic variation between us was already known to be trivial, then why embark on a multi-million-dollar international project to study it at all? In what way did this reinforce that we were all the same underneath?

For geneticist and critic Mark Jobling, the way the project was structured, deliberately going after isolated populations rather than scanning people all over the world wherever they happened to be, was what ultimately undermined it. 'How you define the population in the first place, these are culturally loaded things in themselves. So there was a lot of cultural discrimination in the original aims.' The isolated communities that scientists such as Cavalli-Sforza believed were unique were actually never all that isolated or unique, but they were treated as though they were.

When I raise this with him, Greely admits that there was an 'uneasy recognition that if your project is about looking for differences then it's sort of counterintuitive to say it's showing similarities'. Anti-racist as the scientists behind the project were, they had somehow fallen into the trap of treating groups of people as special and distinct, in the same way that racists do. They were still forcing humans into groups, even if they weren't calling those groups races. They were using similar intellectual frameworks to pre-war race scientists, but with fresh terminology.

That's not to suggest that Cavalli-Sforza was ever duplicitous or secretly racist. Perhaps he saw human differences as meaningful, but didn't want to focus on that uncomfortable aspect of his work in a world in which drawing attention to it could have political consequences. 'I don't think he believed himself to be engaged in a racist enterprise,' says Joanna Radin at Yale University. 'But I also don't think he really had any more sophisticated a sense of how this was going to fight racism than just being able to show we're all connected, we're all cousins or something.' Jobling adds that the anti-racist aims of the project weren't mirrored in the structure of the project itself. Anti-racism seemed to be more of a political ideal tagged on as an afterthought. 'They did have a slightly happy-clappy narrative to it, you know, joining up the whole human family kind of thing.'

In reality, it would have been perfectly possible to study human variation without grouping people. As Jobling explains, the divisions between us are so blurry that humans can theoretically be grouped any way you like. 'You could do a thought experiment where you just said we will take Kenyans, Swedes and Japanese, and will just proportion everybody into those three things.' If this were done, because we are all genetically connected to the average Kenyan, Swede or Japanese person, either directly or by historic migration, then everyone on earth could theoretically be fully grouped based on just these three nationalities. 'You could say that you were so many per cent Kenyan, so many per cent Swedish, and so many per cent Japanese.' This may seem meaningless, but actually it is no more so than dividing the world into black, brown, yellow, red and white. 'The definitions of those populations are cultural, and the choice of population is driven by expediency.'

Other geneticists have also warned against dividing up the world this way. It imposes a certain order on our species, ignoring the actual fuzziness. If the Human Genome Diversity Project had proposed sampling people more systematically, in a grid pattern across the globe perhaps, the true overlapping nature of human variation would be easier to see. Scientists would have been able to map gradual, continuous variation across regions, rather than tight knots centred on very small communities. It's hard not to imagine that this approach – which in fact was mooted at the time, but then discarded – might also have been a more effective way of fighting racism. But in the end, it wasn't the one the researchers chose.

Most governments, including that of the United States, proved unwilling to invest in the Human Genome Diversity Project. It never quite got off the ground in the way it was envisioned. To this day, it remains something of a cautionary tale. In hindsight, part of the problem was that the scientists, however well intentioned they were, failed to connect what they were doing with people's real-life experience of race, with the history and politics of this deadly idea. They thought they were above it all, when in fact they were always central to it.

*

Luigi Luca Cavalli-Sforza died in 2018 aged ninety-six. When I emailed him shortly before his death, he was retired and living in Milan. I found someone whose commitment to his science hadn't waned, and neither had his personal politics. 'There are simply no races in humankind,' he wrote to me. He remains a hero to biologists in his field, an inspiration, someone who had helped build his scientific discipline into one that today has enormous importance. It is impossible not to admire him. And yet, it is also difficult to read his work and come away convinced that his generation of scientists had fully abandoned race science after the Second World War. Although they had ditched race in name, it wasn't clear that they had necessarily shed it in practice.

In 2000, after almost a decade of controversy around the Human Genome Diversity Project, and with the project no closer to being realised, Cavalli-Sforza published a book titled *Genes, Peoples and Languages*. In it, he deftly sketched his grand plan for how genetics could be used to reconstruct human history. It's also a story, the book's back-page blurb added, that claimed to reveal 'the sheer unscientific absurdity of racism'.

In a section titled 'Why Classify Things?' Cavalli-Sforza wrote eloquently about taxonomy, and how humans have always felt compelled to categorise objects. Yet somehow, he managed to write this without any reference whatsoever to politics or social history. He never mentioned that humans were classified in large part because it was politically and economically useful to those who did it. He completely glossed over colonialism and slavery, and the ways in which they fundamentally shaped how European scientists thought about race in the nineteenth and early twentieth centuries (indeed, within his own lifetime). Instead, as he saw it, racism was merely a scientific idea that turned out to be incorrect. 'It seems wise to me,' his chapter concluded, 'to abandon any attempt at racial classification along the traditional lines.'

It's easy to miss the catch. And it lay in the final four words of his statement: *along the traditional lines*.

'A race is a group of individuals that we can recognize as biologically different from others,' he continued. Clearly, then, he hadn't abandoned the use of 'race' at all. Going by this statement, there may be no room for three, four or five old-fashioned racial types, with hard divisions between them – the way we usually think about race – but there could certainly be thousands of 'social groups' all over the globe, each characterised by its gene frequencies, and therefore having some biological distinctiveness to them. This is the obvious result of relatedness. People who are related are of course closer to each other genetically, and historically we have tended to live near our relatives, which is how clustering of genetic similarity happens. This means that even neighbouring towns may be genetically different from one another in some slight but statistically significant way. The small clusters produced by the fact that we don't mate completely at random could, by Cavalli-Sforza's definition, be considered 'races'. So according to his own definition, there *are* races, except the number of them is practically endless.

This is an old idea, which owes itself to early geneticists like Cavalli-Sforza himself who wanted to move the study of race away from vague generalisations and make it more precise. As early as the 1930s, when the field of population genetics was just emerging, evolutionary biologist Theodosius Dobzhansky – later an inspiration to Cavalli-Sforza – was the one to substitute the old-fashioned idea that races were fixed types with the more modern idea that they were populations sharing certain gene frequencies. Like Cavalli-Sforza, Dobzhansky was an outspoken anti-racist. But while being actively involved in anti-racist efforts within the scientific community, Dobzhansky also retained the concept of race. He just redefined it. The way it was redefined squared the circle of how it was possible for all humans to be practically the same while also being different. Under this designation, there's no contradiction in my having possibly more in common genetically with my white neighbour than with my Indian one. My population group as a whole (say, north Indians) will share genes in frequencies that her population group (say, white Britons) doesn't. In other words, if you want it to, race can exist, but you must remember that it's statistical. Not every individual will fit.

'They basically redistributed race,' argues Joanna Radin. According to her, the problem with this statistical 'population' approach to studying human difference is that even though it may look different in some ways, it hasn't fully discarded the baggage of the past. 'An interesting analogy would be colonialism,' she tells me. 'A colonial nation declares independence and they have to forge a new nation with the structures of the old colonial regime, and it's very, very hard to transcend that.' Even if the word 'race' isn't being used, the idea of race is still there, deep within the bedrock.

Canadian philosopher Lisa Gannett has similarly warned about the ethical limits of thinking about race in this new way. To some, it may not seem racist to think about average 'populations' rather than distinct 'types' of people. Certainly, early population geneticists such as Dobzhansky believed that racism was rooted in the assumption that within ethnic groups, people are all the same, whereas those like him believed that, within these groups, people are actually very different. But in the racist mind, as Gannett explains, it doesn't necessarily matter how differences are distributed, so long as they are there in some form or another. This conceptual loophole in population genetics – the fact that we're all different as individuals but that there is also some apparent order to this diversity – is what has since been seized upon by people with racist agendas. Gannett calls it 'statistical racism'.

The question all this raises is a slightly odd one: Is race still a problem if we redefine race? And even odder: Can science be racist if the people doing it are anti-racists? For Radin, intention does matter, but it doesn't fix the underlying problem. 'If you look at the UNESCO statements on race, people often think that they declared that race is a social construct and that race doesn't exist. But really what they did is try to constrict use of the term "race" to biologists who could be seen to use it responsibly, and not equate it with inequality.' In the hands of liberal, left-wing, anti-racist population geneticists, race was thought to be safe, because their politics were beyond reproach. 'I do think that their sense of virtue really emboldened them to feel like they were creating a transcendent mode of science, that might be able to leave race behind,' she explains.

In *Genes, Peoples and Languages*, Cavalli-Sforza added a humorous aside when talking about the fact that researchers who looked hard enough could spot average genetic distinctions between neighbouring populations, even at the village level: 'People in Pisa and Florence might be pleased that science had validated their ancient mutual distrust by demonstrating their genetic differences,' he quipped. But then, isn't this exactly what racism is? A dislike of others in the belief that they are biologically different? In the mind of the racist, it probably doesn't matter how big the groups happen to be, or if the differences are gradual or sharp. It presumably means equally little if it's all about gene frequencies or population averages, so long as the differences are real. If the people of Pisa and Florence could have their mutual distrust validated by population genetics, then why not the people of any other two places?

For Radin, the problem is obvious. It lies in the need to group in the first place, to separate even when that separation means having to zoom in on the very tiniest bits of the genome that might differ, and even then only on average. This need to separate, to treat people as different, is how race was invented. 'What happens is that you've got a large community of very well-meaning, self-described anti-racist scientists seeking to find a way to move beyond race into population genetics, which seems to be incredibly neutral. It's numbers, it's statistical, it's objective,' she says. 'What they have a more difficult time reckoning with is that even something like population genetics is a science done by people, working with the assumptions and the ideas that are available at the time.' Such scientists may believe themselves to be free of racism, but they can't help thinking about humans in racial terms.

In his correspondence with me, too, Cavalli-Sforza made a comment that betrayed this problem. He observed that interracial or mixed heritage relationships – relationships that in the early twentieth century, eugenicists feared might lead to offspring with strange physical and mental deformities – have turned out to be no bad thing. Miscegenation, as it was called in the past, is obviously no threat to human health. He wrote in his message that you only have to look at 'the beauty and vitality of hybrids, children of partners coming from genetically distant groups' for proof of this. It was his use of the phrases 'hybrid' and 'genetically distant' that disturbed me. This kind of language might have seemed at one time scientifically acceptable, but is it any more? It implies that human populations are like different breeds, even different species.

There is a grey area in which well-meaning people make what in other contexts might be considered incendiary statements, and we overlook them because we know these people are well-meaning. In reference to reproduction rates, Cavalli-Sforza's book stated, 'Europeans are largely at a standstill while populations in many developing countries are exploding; thus blonds and light-skinned people will decline in relative frequency.' If the superficial differences between us don't matter, then why should this? To the population geneticist, that people with blond hair are disappearing may be as much of a concern as the possibility that Native Americans might dwindle, or that Andaman Islanders living in the Bay of Bengal might be subsumed into the wider Indian population. To the anti-racist, objective scientist, there is no value judgement in this. It's a problem to science only because researchers are losing some interesting subjects of study, some statistical corners of human diversity, perhaps a few blue-eyed, blond gene combinations. But to someone with alternative politics, it might be seen instead as an argument in favour of racial purity, of preserving distinct population groups against the threat of miscegenation.

It's easy for academics to imagine that the language they use, and the frameworks they operate in, don't really matter. They are just words, not data. 'I think that in the real world what the scientists say has about as much influence as turning on a fan does on El Niño. It's what throwing a cobble into the English Channel has on Atlantic weather,' Henry Greely tells me near the end of our interview. 'We're just not that important.' But it does matter, because their frameworks and language contribute to our understanding of ourselves. If scientists call people of mixed ancestry 'hybrids', this implies that race is real because we are different enough to warrant using that word. If they talk about 'isolates', this sounds like there are groups who are more 'racially pure' than others. Dismantling the edifice of race is about more than just tweaking language, it is about fundamentally rewriting the way we think about human difference, to resist the urge to group people at all.

It takes some mental acrobatics to be an intellectual racist in the light of the scientific information we have today, but those who want to do it, will. Racists will find validation wherever they can, even if it means working a little harder than usual. And this is the reason that good scientists who do reliable research, ones who are also well-intentioned and anti-racist as Cavalli-Sforza was, can't afford to be cavalier or leave room for misinterpretation. There's an uncertain space between recognising that there is a gap in knowledge and actually filling that gap. It's a place where speculation thrives, where the racists reside. Racists adopted the same concepts as good scientists and the same language as anti-racists to claim that, if some average differences can be seen between certain groups then, by that logic, certain groups *might* be better on average than others at certain things. When Steve Sailer and his followers talk about 'human biodiversity', this is what they mean. This wolf in sheep's clothing is twenty-first century scientific racism.

We were told this might happen. A caustic report on the Human Genome Diversity Project released in 1995 by UNESCO's International Bioethics Committee sounded precisely that warning. It argued that the project could, whether it meant to or not, give racists some basis to believe certain groups were inferior or superior to others. In particular, the committee was concerned that by bringing genetics to the fore in telling the human story, people would ignore culture and history, and return to the kind of simplistic biological thinking that propelled the eugenics movement in the early twentieth century. It advised scientists to resist the temptation to use their work to shore up any kind of political ideology, whether racist or anti-racist. 'Racism,' it reminded them in case they had forgotten, is 'socially and politically constructed.'

Science is just a pawn in the bloody game.

*

Although the work itself never got off the ground, the concept behind the doomed Human Genome Diversity Project did survive. In the years that followed, other teams stepped in to achieve essentially the same outcome in other ways. The Center for the Study of Human Polymorphisms in Paris today keeps a bank of DNA samples from populations all over the world, ready for researchers who want to tap it. In 2002 the United States National Human Genome Research Institute introduced a $100 million initiative to study human variation. And in 2015 the United Kingdom launched its own project to make a genetic map of the people within its own borders, named People of the British Isles.

The project had one more unintended consequence. In 2005 the National Geographic Society in Washington, DC, the one behind the famous magazine and satellite TV channel, decided to dip a toe into the world of population genetics. Naming its effort the 'Genographic Project', it chose anthropologist, geneticist and television presenter Spencer Wells to lead it. Wells had spent a portion of the 1990s studying under Luigi Luca Cavalli-Sforza, seeing the difficulties around the Human Genome Diversity Project at close quarters. His solution to the controversy was simple. National Geographic would sell easy-to-use kits, helping people understand the history of migration that might be hidden in their DNA, and in the process build a data bank from their genetic information.

'We put together this consortium of scientists with the goal of sampling the world's DNA, and at the same time wanted to enable anybody, any member of the general public who was curious about their own genetic ancestry, to get themselves tested,' Wells tells me. The idea that people might want to spit into a cup and have their ancestry tested didn't seem at the time like a highly profitable venture; even the CEO of National Geographic warned Wells before the launch that nobody was going to spend a hundred bucks to test their DNA.

It turned out to be a money-spinner. 'The day we announced, we sold ten thousand. It had hit a hundred thousand by the end of the year,' he says. 'It launched the consumer genomics industry.' The trade was given a noticeable boost in 2006 when media legend Oprah Winfrey had her DNA tested for a television show, revealing ancestral links to people now living in Liberia, Cameroon and Zambia. She also turned out, unlike many black Americans, to share no recent ancestry at all with Europeans. 'I feel more connected to where I've come from,' she told the presenter, Henry Louis Gates Jr, Harvard professor of African American Studies, who wrote a book about Winfrey's experience.

Before long, companies such as 23andMe and AncestryDNA were selling their own kits, turning over billions of dollars. In 2018 it was announced that AncestryDNA alone had sold a total of around ten million kits around the world.

Spencer Wells has since left the Genographic Project, become the owner of a nightclub in Texas, and moved on to new genetic testing ventures. He tells me that the ancestry testing industry flourishes in the United States because of the rootlessness of so many of its citizens. 'In societies like the present-day US, where we have a lot of hyphenated Americans – Irish-Americans, Italian-Americans, Hispanic-Americans – people feel somewhat disconnected from the entity that comes before the hyphen. So they want to figure out who those people were. Who were the people who migrated to the US?' For African Americans in particular, many of whom were transported as slaves and ripped away from any connection to their families or homelands, the kits offer the only means they may have of tracing their genealogy. The psychological effect on the public of sequencing the human genome, of convincing them that our differences are identifiable in our DNA, is that this now appears to be a foolproof way to define who we are.

In reality, genetic testing is only an educated guess about where your relatives may have lived based on the data fed into the models in the first place. As geneticist Mark Jobling explained, it's possible to group people any way we like. Ancestry tests scan portions of people's genomes to find those who have genetically a little more in common, then pool them together. Theoretically, they could pool them using any measure, even as simple as those who live south of the equator and those who live north of it. But of course, companies most often use old-fashioned racial categories or nationalities. This means that if there are few or no DNA samples from people in the country your ancestors came from, you're stuck. One of the reasons Oprah is linked to Liberia, for example, may be that this is where former slaves were long ago 'repatriated' by white American leaders who couldn't bear the thought of such people living freely among them. Ancestry testing doesn't show you your past as much as it reveals the people you are distantly related to in the present, and even then only if they have had similar tests done. Oprah has some connection to people who now live in Liberia, but this is not necessarily her ancestral homeland, the place from which her relatives originally came.

Mark Thomas at University College London, a leading geneticist who has seen his own research recruited into these models, tells me he has always been sceptical of ancestry testing firms. 'It's not that they're cynical, it's not that they're nasty, it's not that they've got particularly racist agendas. They want to make money, and you make money by servicing people's prejudices.' Not just prejudices, but also people's natural desire to know *who they are*. In 2018 Thomas and his colleagues published a paper detailing how one particular firm based in Scotland, BritainsDNA, had threatened them with legal action after they challenged the firm's wide-ranging claims in the press – for instance that the actor Tom Conti is 'Saracen' in origin, and that Prince William has Indian ancestry. They didn't have the data to establish either claim.

What ancestry testing has done is take the work of well-meaning scientists, who only tried to do good in the world, and inadvertently helped reinforce the idea that race is real. Using their methods and data, an entire industry has achieved exactly the opposite of what scientists like Cavalli-Sforza once set out to do. The true way that variation works, the nuances, are rarely explained by those selling ancestry testing kits. Having seen how purely 'black' Oprah Winfrey really was, for example, white supremacists in the United States began using the very same tests to prove how 'white' they were, sometimes sending off vials of spit to various companies until one came back with the desired result, establishing beyond doubt that they were of nothing but European ancestry. By forcing people into categories, even if that means dividing our individual bodies into so much European, so much African, so much Asian and so on, the tests fortify the assumption that race is biologically meaningful. If it's possible to categorise, we assume, there must be something to the categories.

The irony is that as more research has been done into our origins since the launch of the Human Genome Project and consumer ancestry testing, it has only undermined these measures of identity. Within the last decade, as scientists have uncovered exponentially more genetic evidence about us and our ancestors, even they have been surprised by the results. Nothing has matched expectation. Our roots, it turns out, are very rarely where we think they are.

7

Roots

What race means now in the light of new scientific research

When she was growing up, my little sister was a diehard fan of Morrissey, frontman of The Smiths, genius songwriter, British cultural icon. For one of a handful of brown girls in a white working-class south-east London suburb, indie music spoke to that cold, lonely feeling of not quite being able to fit in. If the British National Party was marching outside our door, inside her headphones was a different British voice that she could relate to. He was a refuge from those who insisted that we all had to be the same.

But in an interview with a music magazine in 2007, Morrissey said something that couldn't help but trouble my sister, as well as other fans. 'Whatever England is now, it's not what it was and it's lamentable that we've lost so much,' he complained. He railed against high immigration, against what he saw as a change in the character of Britain. There was public outrage. She lost a hero. But as we in our family knew too well, out in the country as a whole there were many who felt this way. This was a debate that had been simmering for decades, occasionally stoked by national politics, making people anxious, wondering what it meant to be British.

A decade later, the pot bubbled over. A financial crisis and economic austerity, coupled with higher than usual rates of immigration from eastern Europe, helped fuel support for nationalists who wanted to cut the country free from the continent. In a referendum in 2016, the majority of voters agreed that leaving the European Union might be a good plan. They were promised a new dawn. The nation would stand alone, the way it had done during the days of Empire, riding the waves of unbridled trade and setting its own rules on who would be allowed into the country.

For visible immigrants, or children of immigrants like my sisters and I, watching this could sometimes feel like an out-of-body experience. The borough in which my parents lived, and where we grew up, was one of only five out of the thirty-two in London that voted to leave. As citizens, we had the right to vote to decide Britain's future, but we also knew that a sizeable slice of other voters wanted fewer of us there in the first place. A campaign poster showed legions of men with skin as brown as ours, queuing up against the slogan 'Breaking Point'. The far right was emboldened. Around the time of the referendum, reports of race-based crime rose, with a sharp spike in the kind of everyday racism that I last saw as a teenager.

Squaring your appearance with your nationality is one of the hardest things about being an ethnic minority. Not all, but some of those who voted to leave Europe wanted a return to their own particular vision of Britain. Skin colour mattered to them because it was a visible baseline, the reference. It was the way the British had always looked, from the beginning, before Empire, before Shakespeare, before kings and queens, before culture and values. Britain, as far as we were aware, has been forever white. So even though I was born in London, I speak the Queen's English, my dinners generally comprise meat and two veg (I probably eat a curry as often as most white Britons, and possibly less than some), and my wireless is set permanently to Radio 4, in their eyes it is my failure to be the right colour that truly undercuts my claim to Britishness.

What nobody could have predicted then was that, by an almost cosmic coincidence, at the very moment Britons were struggling to define their identity in the face of political turmoil, and particularly for those racists who saw Britain as a white nation first and foremost, some news was coming. They were about to be thrown a curve ball.

*

I saw it for myself at London's Natural History Museum in 2018, a package no fancier than a bunch of old bones.

The skeleton is laid out neatly in a small corner of the museum. Most of the visitors don't linger as I do. To be honest, it looks unremarkable. But this is the frame of one of the oldest dead bodies ever found in the country, dating from some 10,000 years ago. And it's full of secrets. Almost as soon as he was discovered in caves in Cheddar Gorge in Somerset in 1903, earning the name 'Cheddar Man', people began to wonder how he must have looked. They wanted to put a face to one of our early ancestors. Archaeologists could certainly guess that he was short by modern standards, that he probably had a good diet, and he may have been around twenty years old when he died. One speculative reconstruction showed him to be white-skinned, with rosy cheeks and a trailing brown moustache. But his actual appearance was a mystery.

The genomes of living people offer a limited and fuzzy picture of the past because of mixing and migration. We're just so similar. This is where the study of the bones of our distant ancestors and their ancient DNA has come in. It has achieved what the Human Genome Diversity Project couldn't. When it comes to tracking human migration patterns over thousands of years, even archaeology and linguistics can't provide the same historical data that ancient DNA can. By around 2010, genetic sequencing techniques had developed far enough to tease out highly reliable samples of DNA from ancient specimens (a bone just behind the ear turned out to be best) and use them to help reconstruct entire genomes of long-dead people. The use of this technique has mushroomed in the last decade. It has been credited with solving historical mysteries at a stroke. Thousands of skeletons from all over the world have been analysed already, and as the British public were about to learn in early 2018, Cheddar Man would be among them.

Scientists at the Natural History Museum and University College London revealed that he probably had blue eyes and curly hair – no great surprise here. But what came as a real shock to many was that his bones also carried genetic signatures of skin pigmentation more commonly found in sub-Saharan Africa. It was probable, then, that Cheddar Man would have had dark skin. So dark, in fact, that by today's standards he would be considered black. The revelation, along with a dramatic new reconstruction of his face markedly different from the original one, made front-page news and television bulletins.

'Hard cheese for the racist morons,' smirked a headline in the *Mirror*. 'Another racial panic for white supremacists,' announced the news website *Salon*.

Panic was indeed sparked. People experienced all the stages of grief. On far-right websites, a few immediately began doubting the scientific results – maybe, just maybe, the researchers had got it wrong. Some hopefully voiced the possibility that Cheddar Man hadn't been an actual Briton at all, but was just a passing visitor who happened to die here, like an unlucky tourist. Finally, there was acceptance. Some, especially those who for so long had believed that skin colour was the basic measure of Britishness, wondered if perhaps it was time to rethink national identity. If the original Britons were black, all bets were off.

*

Throughout the frenzy, there was one set of people for whom the news barely registered a flicker on their excitement dial. They weren't surprised at all.

'With the whole Cheddar Man thing, I was amazed initially at just how much press coverage it got,' I'm told by Mark Thomas, professor of evolutionary genetics at University College London, who worked on the finding. Leaning back in his chair, wearing stonewashed jeans and a grandad-collar shirt, Thomas is about as relatable as a professor comes. He is one of the world's leading experts on ancient DNA and, from this position of authority, he has a tendency to tell it how it is. For geneticists like him, the Cheddar Man discovery was unremarkable given what they already knew. They had more or less expected it. 'What I was even more amazed about was the *Daily Mail*-reading backlash,' he laughs.

Thomas had welcomed the outcome of the Cheddar Man tests as just another piece of evidence in a huge body of research. It took up a couple of sentences in his latest paper. Scientists had already known for a few years, from analysing the skeletons of other hunter-gatherer bones found in western Europe, that dark skin pigmentation could well have been common back then.

After all, light skin was likely an evolutionary adaptation, one that helped people living in northern climates absorb more vitamin D when there wasn't enough sunshine. The first human pioneers didn't arrive in Europe or Asia looking white, because they had originally migrated from Africa, where there was little or no survival advantage in having lighter skin pigmentation.

What researchers were a little less sure about was how quickly paler skin emerged, where and when. 'Over the last 10,000 years? Or over the last 40,000 years?' asks Thomas. One theory was that it developed very slowly and gradually, starting 40,000 years ago when modern humans first came to live in Europe. Another suggested it was a more recent phenomenon, contemporary perhaps with the advent of farming. Trading a hunter-gatherer lifestyle for settled agriculture would have limited people's diets and made it even more vital that they get the vitamin D they needed from the action of sunlight on their skin. Another is that light skin emerged elsewhere in the world, outside western Europe, and that the movement of people would have then introduced it to darker-skinned Europeans. Evidence as it stands indicates that, like Cheddar Man, many other pre-farming hunter-gatherers who lived in western Europe during his time and at least as recently as 7,000 years ago would have had light eyes, dark hair and dark skin. It was the first farmers to come into the region later from the east who brought with them their lighter skin and brown eyes.

So one thing was clear: Cheddar Man wasn't an exception in his time. People all over the world then didn't look anything like the way we look now. Not only this, they looked more different from each other than we do today.

'Differentiation between groups in different parts of the world would have been greater,' explains Thomas. The scientific explanation for this is genetic drift. Being in small groups as they were, every breakaway bunch of migrants as it moved began to look more and more different from the relatives they left behind as time passed. Since then, as groups have grown bigger and remixed with each other, populations across the world have become more homogenised. Today it's generally possible to know whether someone's ancestry may be rooted in, say, Asia or Europe or Africa by looking at them. But ten millennia ago, we would have struggled to do this. Appearance didn't map the way it does now, and physical features in some regions may have been dramatically different.

When Thomas and his team looked into the very earliest farmers in the Fertile Crescent, who lived in what is now Iran, and compared them with farmers in nearby Anatolia and the Aegean, they found to their surprise that the two were genetically very distinct from each other. 'They were as different as people from Ireland and Thailand today, more or less. I mean, of that order of magnitude.' Today, neighbouring populations tend to be much more similar. They've mingled and mated with each other, mostly dissolving away the gaps.

Our modern ideas of race are deeply connected with how we look. Our appearance is a shorthand for the stereotypes, a means of slotting people into groups and making judgements about them. The disbelief that met Cheddar Man's probable blackness arose because many among the British public couldn't help but assume that Britons had always looked a certain way, even in the distant past. They struggled to categorise him, forgetting that he existed thousands of years before our racial categories came about. He was proof that there couldn't be anything eternal or pure about race because once upon a time, not so very long ago in evolutionary terms, most of the people on earth didn't look like us. They were already human. They were, however distantly, *us*. But they looked different. The long lens of evolutionary history has a way of turning all you think you know on its head.

The picture becomes even more complex as we go further back in time. Geneticist Sarah Tishkoff at the University of Pennsylvania has carried out pioneering studies into skin colour variation across Africa, finding that the genetic variants – different forms of the same genes – associated with both dark and light skin have existed in Africa for a long time. The variants associated with light skin are common not only in Europe and sometimes in east Asia, but also in the San hunter-gatherers. 'These are the people in southern Africa who have the oldest genetic lineages in the world,' she says. This suggests that rather than evolving independently outside Africa, many of the gene variants associated with light skin may already have been there when people first migrated from the continent.

So not only did people with darker-pigmented skin occupy Europe, but even earlier, there were genetic variants for lighter-pigmented skin in Africa. Given the evidence so far, Tishkoff suggests that lighter-pigmented skin may even have been the ancestral state in the long-distant past. Underneath their bodies, chimpanzees – our closest genetic relatives – tend to be light skinned, their dark body hair providing protection against the sun. 'When our ancestors left the forest and went to the savannah, there would have been selection for better thermoregulation, so getting rid of body hair, increasing the number of sweat glands. And if you're decreasing the body hair, there would be selection for darker skin.' Darker skin could have been one of the adaptations to a new living environment within Africa.

Yet when scientific reconstructions are made of earlier human species, such as *Homo erectus*, they are almost always given dark skin. There's an assumption that our species began as black. 'I don't think that's necessarily the case, because both light and dark variants have been around for a really long time. And there could have been variation in Africa a million, two million years ago,' explains Tishkoff.

Even today, there is far more variation in Africa than the simplistic black–white model of race implies. 'I think many people don't recognise the large range in skin colour in Africa . . . The whole continent of sub-Saharan Africa is incredibly diverse genetically. It doesn't fit with a racial model, one homogeneous African race. There's a huge amount of variation amongst populations in Africa,' she adds. 'Skin colour is a terrible racial classifier. There really are no good biological classifiers for race.'

For the biologists who know this, skin colour begins to lose its meaning. 'I mean, it's skin pigmentation, you know! It's just so trivial,' reflects Mark Thomas. He found reaction to the new finding about Cheddar Man bizarre given the scientific facts. 'Obviously there are some idiot racists over there in the corner for whom it is important. But I think that if you base your identity on the pigmentation of some West Country bloke from 10,000 years ago then you really should rethink it. My own personal view is that today we over-privilege and fetishise the concept of identity.'

Thomas reminds me that the physical features we associate with race are poor proxies for overall genetic similarity. Even if one population tends to have darker skin and another lighter, that doesn't mean their genomes as a whole will have less in common than two populations with the same skin colour. Variations in physical appearance, whether it be skin pigmentation, ear shape, nose shape, whatever, says Thomas, makes the gaps between groups feel far larger than they really are genetically. In biological terms, the differences really do appear to be no more than skin-deep. It's an error to assume that the internal differences are as profound as the external ones appear. But it's an easy one to make. 'If we could see each other by looking at our genomes then, without a big computer, you would be hard pushed to work out whether somebody was from India or from Poland,' he explains. One might assume, for instance, that light-skinned Irish people would have very little in common genetically with darker-skinned south Indians. But that's not the case. 'There is relatively little genetic differentiation between southern India and Ireland. I mean, relatively similar ancestry components. But of course, the pigmentation differences are quite large, and so people assume that these people are massively different genetically.'

In that sense, how we look is misleading. 'Nature plays dirty tricks on us,' says Thomas.

It can play tricks on scientists, too. If data seems to suggest that populations are very different, for the most part it's because population geneticists are deliberately examining the small sections of our largely shared DNA that happen to differ. This is their job. 'We're zooming in. We're turning up the contrast on what are actually tiny little differences over extremely closely related populations,' he warns.

*

'The past is very surprising,' says David Reich, a geneticist in the ancient DNA laboratory at Harvard University. 'It's different from how most people picture the past in their heads.'

Reich is the most well-known person in this branch of science, at the forefront of using genetics to plot ancient migrations around the world. At the moment I happen to visit him, though, he has become embroiled in controversy for suggesting in the press that more work needs to be done to understand cognitive and psychological differences between population groups, a phrase that most people have interpreted as meaning 'racial differences'. His statement – a departure from the nearly seventy-year consensus that studying race isn't the business of biologists – has attracted angry emails from fellow academics. But he hasn't backed down. When I see him, I expect him to be defensive, maybe even brash.

I couldn't be more wrong. With his hands in his lap, so softly spoken that my voice recorder struggles to pick up every word, he surprises me with his gentleness. His half of his office is bare, save for a few drawings stuck to the plain white walls. He is unfailingly polite, pausing only to message his wife. The one clue to his global importance is the steady stream of students and researchers lining up to see him outside. One young man sits with his laptop at a bench in the corridor all day in the hope Reich may be able to spare him a minute or two later on.

Reich's lab is a powerhouse. It has scoured the world for skeletons that might provide genetic evidence of the past, and as Reich has noted, it churns out findings so quickly that the amount of data doubles faster than the time it takes for it to be published. Scientific journals simply can't keep up. But for him, this is more than a scientific gold rush. Genetics has a way of cutting through ancient historical questions in a way that nothing else can. His group, along with the lab of Mark Thomas and others across the world, helped confirm the hypothesis that farming emerged 10,000 years ago in the Near East – the region between Europe, Africa and Asia once spanned by the Ottoman Empire, and before that, by Mesopotamia – and that these farmers may have been a genetically varied group of humans who then helped spread agriculture to other parts. He is also fairly confident that natural selection has caused southern Europeans to be a little shorter on average than northern Europeans.

But it's the story of migration that is the most revealing. What we think of as 'indigenous' Europeans are, scientists like Reich now understand, the product of a number of migrations across the last 15,000 years, including from what is now the Middle East.

The British, in particular, have their own story to tell. 'Britons in the past didn't look like Britons today, and were genetically very unlike Britons today,' Reich explains. Whoever the first inhabitants of Britain were, their way of life was likely to have been replaced almost wholesale between around 5,000 and 4,000 years ago by a group of people coming through Europe from the steppe grassland that stretches between the Black Sea and the Caspian Sea. They are known by some anthropologists as 'Beaker folk' for their distinctive bell-shaped pottery. While artefacts of Beaker culture are found scattered all over Europe, the team here at Harvard has shown by studying the DNA of four hundred ancient Europeans that these people must have swept in and supplanted almost everyone who was living in Britain at the time as well.

How they did such a thing is unclear. They could have simply come in large numbers and bred with people who were already there. They may have been better equipped to survive in the environment, through resistance to certain diseases or by virtue of their technology. The pre-existing populations could have been collapsing already, as some data suggests. Whatever the explanation, their arrival changed not only the culture but also the way people looked. The steppe people with their Beaker culture had lighter skin. According to estimates drawn up recently by Reich and his colleagues, this Beaker invasion replaced around 90 per cent of Britain's gene pool. And all in the space of just a few centuries.

This means that light skin did not define Britons from the beginning. 'There's been a continuous process of skin lightening with big jumps that occurred at these migrations,' Reich suggests. 'So for example, when the first farmers came to Britain about 6,000 years ago there is a big change in the average hue of skin around then, predicted by the genetics. And then when this Beaker phenomenon spread into Britain, there was another big jump associated with that.'

While some of this confirms what archaeologists already suspected, it's more surprising just how much churn there has been in global patterns of migration throughout the ages. Reich himself was taught when he was younger that humans spread out from Africa, with little mixing once they started to split, like branches of a tree. Once they landed somewhere, people stayed put. That was the common assumption. But the evidence that's now emerging suggests something entirely different. 'It became very clear that the big large-scale mixture, migration, or gene flow, however you want to call it, is common and recurrent.'

The true human story, then, appears to be not of pure races rooted in one place for tens of thousands of years, but of constant mixing, with migration both one way and another. The cherished belief that people in certain places have looked the same way for millennia has to give way to the understanding that migration made the world a melting pot long before the last few centuries, long before the multicultural societies we have today. Our roots are not an orderly family tree but instead are tangled, according to Reich, more like a climbing plant on a trellis. Our ancestors branched out but then came back, and remixed, again and again throughout the past.

'I think this idea of indigeneity, and you being from a population that has been here for ages – I mean there may be populations that have better claims to that than others – but at some deep level the great majority of people in the world, if not everyone, is not derived directly from people who lived in the same place deep in the past.'

The British story is just one of thousands. For example, the Beaker folk were part of a far earlier, bigger and longer migration out of central Eurasia and into many different corners of the world of people associated with what archaeologists call the Yamnaya culture. They were pastoralists, raising and moving livestock, with wagons and horses that made them mobile in a way that may never have been seen before. Their diet was rich in meat and dairy. From roughly 7,000 years ago, across a span of two millennia, the Yamnaya (themselves a product of earlier migrations into the region they came from) trekked west and south-east, not only populating Europe but also going as far as north India. They brought the wheel and, it has been mooted, they also brought cannabis.

By 5,000 years ago, the Neolithic farming cultures of Europe had been largely replaced. Kumarasamy Thangaraj at the Centre for Cellular and Molecular Biology in Hyderabad tells me that around the same time, those of the Yamnaya culture came in from the north of India and mixed with the people who were already there. The Indian population was itself a mix of indigenous hunter-gatherers who had originally moved out of Africa many thousands of years earlier, and more recent farmers migrating from what is now Iran. All Indians, save the tiny community of Andaman Islanders who have been closed off from the Indian mainland for thousands of years, are a blend of these three ancestral populations.

Confirmed by genetics, these ancient connections can be spotted in the words we use. Linguists long ago saw similarities between European and Indian tongues, describing them together as Indo-European languages. Genetics has added more hard data to the history. Almost all Indians today are genetically connected to Europeans by their ancient ancestors who spread the Yamnaya culture, as well as the earlier farmers migrating from the Middle East.

*

'If you pay any attention to the discoveries coming out of science, they don't play into any sort of old systems of prejudice,' David Reich tells me.

Take Stonehenge, the mysterious prehistoric assembly of standing stones in south-west England, which attracts more than a million visitors every year. Within a few hundred years of its construction around 5,000 years ago, the Neolithic farmers who constructed it were largely gone. They were probably replaced by incoming folk who followed the Beaker culture, because within another thousand years, Reich's team could see little evidence of Neolithic ancestry in the genomes of ancient remains they were studying. Now, just pause to think about what this means: the symbol we associate with ancient Britain, the one thing that couldn't really be more authentically British, was built by people who are certainly not the main ancestors of those who consider themselves indigenous Brits today.

Cheddar Man and his relatives, too, who lived 10,000 years ago, couldn't have been from the same genetic population as Britons today, because like the builders of Stonehenge, they were replaced by farmers who spread across Europe from Anatolia. Cheddar Man and his people don't have any direct descendants – only bits of them exist now, explains Mark Thomas. What he means by 'bits' is that Cheddar Man and his relatives on the continent would have bred with whoever came into the region. So while his own particular population and their culture didn't survive intact, traces of them would have endured, either because they mixed to some extent with farmers coming into Britain, or because their continental relatives mixed with farmers spreading across Europe.

To know that this melting pot has been churning for thousands of years puts a fresh spin on the contemporary idea of race. 'I think that genetics and genomics have a wonderful opportunity to undermine these outdated and scientifically unsupported notions of race, ancestry, ethnicity and identity,' says Thomas. The feeling that there is a 'home' for us all, and that our bodies somehow reflect this, deeply and viscerally, begins to melt. The attachments we have to places and their relics, the ancient stories that tell us who 'our people' were, have to be rethought when we understand that 'our people' were actually migrants into a place occupied by others. The relics belong to them, not us. The place was theirs once, before others came along. We're all part of the churn.

What is even more mind-bending is that when you're looking this far back in time, ancestry expands to include almost everyone. 'Cheddar Man's people are to an extent the ancestors of just about everybody in Europe,' he explains. 'Indeed, it is possible that in his group are the ancestors of everybody in the world just about, maybe everybody in the world today.'

This may seem implausible, but it's just mathematics. The further back you go in time, the weaker your genetic link to your ancestors. Five generations ago, you would have as many as 32 possible ancestors contributing to your genetic makeup. Nine generations back, you could have 512, many of whom may have contributed next to nothing. Zoom back fifteen generations – still just the tiniest slice in recent human history – and you could have 32,768, assuming nobody was having babies with someone they were even distantly related to, which is unlikely.

Each of these could give you no more than the tiniest fraction of your DNA. Longer and longer ago, the theoretical number rises into the millions, and ultimately, to more people than were alive at the time. Of course, that is impossible, the only explanation being that we are all at least a little inbred.

Even if you could trace your lineage as far back as Cheddar Man, or more recently to Nebuchadnezzar or Cleopatra or any other figure from antiquity, you would probably be no more related to them than is a random person on the street. The more you zoom into the past, the more your ancestral history begins to overlap with that of everyone else on the planet. As Thomas notes, we only have to go back a few thousand years before we reach somebody who is the ancestor of everybody alive today. Go back a few thousand years more, and everybody who was alive is either the ancestor of everybody alive today (if they had descendants who survived), or nobody alive today (if they didn't). Hence Cheddar Man, if he had children and they had children, and so on until today, is both your relative and mine.

Race, nationality and ethnicity are not what we imagine them to be when seen from the deep past. They are ephemeral, real only in as much as we have made them real by living in the cultures we do, with the politics we have. David Reich tells me that he draws a sense of global kinship from his work on genetics. 'I have a personal way in which genetics is meaningful to me which doesn't involve my own ancestry,' he says, quietly. 'I think that one way of relating to the findings about genetics is that we're all related to each other, and we are all part of a broadly closely related group of people over the last couple of hundred thousand years, with a lot of complexity, and with a lot of mixtures and migrations and reticulations. And we're all part of that.'

But then his tone changes. Even after everything he's said, he doesn't dismiss the idea of race altogether.

*

David Reich is not a racist. But then neither does he adopt the staunch anti-racist position of the old-school population geneticists such as Luca Cavalli-Sforza, who bravely debated with the scientific racists of their time, wearing their politics on their sleeves. Reich respects Cavalli-Sforza, even writing about how much he has been an inspiration to him. But he confesses that he sees himself as apolitical.

The genetics of human variation are complicated and subtle, he tells me. And his own position on race is similarly subtle. Despite his research revealing the extent of interconnectedness between humans, the great uniting trellis of ancient migration, Reich still suspects there's something worth investigating about group difference. And he leaves open the possibility that this difference correlates with existing racial categories – categories that many academics would say were socially constructed, and not based in biology at all, except in very unreliable ways, such as along crude skin colour lines. 'There are real ancestry differences across populations that correlate to the social constructions we have,' he tells me firmly. 'We have to deal with that.'

He admits that some categories make no biological sense. For instance, 'Latino', as anyone from South America is referred to in the United States. 'Latinos is a crazy category that encompasses groups with different ancestry mixes ranging from Puerto Ricans, who have very little Native American ancestry, mostly African, a little European, to Mexicans, who have very little African ancestry and mostly Native American, European . . . It's a crazy category.' At the same time he thinks that some categories may have more biological meaning to them. African Americans are mostly West African in ancestry and white Americans tend to be European, both correlating to genuine population groups that were once separated at least partially by 70,000 years of human history. When it comes to West Africans and Europeans, he continues, 'there's a long time separating these two groups. Enough time for evolution to accumulate differences. We don't know very much about what those differences are because we're still at the beginning of collectively trying to identify biologically what differences do.'

Reich suggests that there may be more than superficial average differences between black and white Americans, possibly even cognitive and psychological ones, because before they arrived in the United States, each of these population groups had 70,000 years apart to adapt to its own different environment. He implies that natural selection may have acted on them differently within this timescale to produce changes that go further than skin deep. Reich adds, judiciously, that he doesn't think these differences will be large – only a fraction as big as the variation between individuals, just as biologist Richard Lewontin estimated in 1972. But he doesn't expect them to be non-existent either.

They are words I never expected to hear from a mainstream, respected geneticist. Reich is of course not a racist. Indeed, like Cavalli-Sforza, he believes that if race research is done, it will only further demolish racial prejudices. Scientists are concerned with fact, not fiction, his argument goes, and the facts we have so far are simply not in the racists' favour. The more good work that is done, the more it demolishes long-standing racial stereotypes, so by his logic there should be no barriers to doing yet more research, even if it feels risky. 'My feeling about this field has been that, broadly, it makes telling falsehoods more difficult. That's my feeling. It may be self-serving, but that's my feeling. And so I think these surprises, such as ancient Britons were very much more dark skinned than present people . . . I think this is broadly a force for combating prejudice, because it doesn't conform to anybody's pictures they had before.'

While he sees the racists as factually wrong, he also sees some anti-racists – those who insist that we are all exactly the same underneath – as not having the full facts either. 'It's a little bit painful to see very well-meaning people saying things that are contradicted by the science, because we want well-meaning people to say things that are correct,' he tells me. 'The way I see what's going on in this world right now, there are racist people that are just perpetrating falsehoods, and just representing the science in incorrect ways, tendentious ways in order to achieve certain goals. And then there's people whose perspective on the world I agree with who are actually saying things that are technically incorrect.'

Reich himself is technically correct that there *could* be more profound genetic differences between population groups than we are aware of at the moment. But to date, no scientific research has been able to show any average genetic differences between population groups that go further than the superficial and are linked to hard survival, such as skin colour or those that prevent a geographically linked disease. There is no gene or variant of any gene that has been found to exist in everyone of one 'race' and not in another. In London, Mark Thomas, who has collaborated with Reich, remains dismissive of the idea that race is useful to genetics. 'Most researchers, including geneticists, agree that "race" is a socially constructed category . . . There is no categorical imperative in biology, and no need or value in placing people in biological boxes. There are subtle genetic correlations with geographic origin, and physical traits, as well as medical ones, and understanding those correlations is important. But there are no hard borders, just gentle gradients,' he tells me. 'Unfortunately, that doesn't stop people "racialising" others, and perhaps that reflects our desire to categorise. Most categories are nonsense, although some may be useful. "Race" is useless, pernicious nonsense.'

The question of whether or not biological research into racial difference is useful still divides the scientific community. What seems to bother Reich above all is that the research simply isn't being done, at least not properly and not enough, so we just don't know how useful it might be. Part of the reason for this, of course, is the longstanding scientific taboo which has kept race off the table in mainstream genetics since the end of the Second World War – although certainly not in social science, which has built an enormous body of work on the topic. We have plenty of data on racial gaps in, for example, income, health and schooling in the UK and the United States. This is because race has been accepted by academics as a social reality, not a biological one. Race affects how we live, but not who we are genetically. Reich, however, appears to find this is unfair. 'We've been silenced by the great anxiety that we feel talking about these things, and by the history of abuse of genetics by people seventy years ago, or eighty years ago,' he says.

Reich is probably not the only scientist who would like to be free of 'the great anxiety' caused by eugenicists and scientific racists. But that freedom would have to come with responsibility. As the devastating mistakes of the nineteenth and twentieth centuries proved, race research never goes well when society is racist. And although Reich insists that biological data as it stands makes racism impossible, I'm not so sure.

Two days after I visit David Reich in his laboratory at Harvard, a party is held at the Cold Spring Harbor Laboratory, the world-class research institution on Long Island that was once the site of the United States Eugenics Record Office. The celebration marks the ninetieth birthday of James Watson, one of the legends of twentieth-century genetics, who helped discover the double helix structure of DNA alongside Rosalind Franklin, Francis Crick and Maurice Wilkins. In 1962 Watson, Crick and Wilkins were awarded the Nobel Prize. Watson went on to become the laboratory's director in 1968, and was crucial in helping to get it funding over the years and building its reputation. A Grammy award-winning pianist is invited to give a performance at the party, with no fewer than eight Nobel laureates among the four hundred guests.

Yet, for years, Watson has been known to hold racist and sexist views. He was famously derogatory about his former colleague Franklin, who did much of the experimental work that helped him make the discovery that led to his joint Nobel Prize. He told the *Sunday Times* newspaper in 2007 that he was 'inherently gloomy about the prospect of Africa', because 'all our social policies are based on the fact that their intelligence is the same as ours – whereas all the testing says not really.'

In 2010 David Reich witnessed Watson's racism first-hand at the Cold Spring Harbor Laboratory when they were both at a workshop on genetics and human history. Watson sidled up to him and asked him something along the lines of: 'When are you Jews going to figure out why you guys are so smart?' Reich was appalled. Watson openly compared Jewish people to Brahmins, high-caste Indians, who are known for being over-represented in universities and high-status jobs. Traditionally, they are India's educated, priestly class. Watson suggested that racial purity combined with millennia of selecting for scholarliness was the key to both Jewish and Brahmin success. He went on to make other racial slurs, about Indians being servile, a trait that suited British colonisers, and about the Chinese, whom he thought had been left genetically conformist by their society.

I wonder what Reich took away from this encounter. If understanding the scientific facts makes it so impossibly difficult to be racist, how does James Watson manage it?

Reich hesitates. 'Well, Watson is, you know, probably more sexist than he is racist,' he says awkwardly. 'I don't know. I don't know. He's like uncontrollable. It's impossible to control Jim Watson. He purposely wants to create, to annoy people, to scandalise people, so I don't know. You can't control everybody. I do think that. So, yeah, I don't know.'

There is a long pause, an uncomfortable half-shrug. 'I just don't know.'

8

Origin Stories

Why the scientific facts don't always matter

THE PAST IS PART OF THE PROBLEM. Not just the politics of the past, of the nineteenth-century theories of race that permeate our subconscious, the colonial hierarchy of races. But the deeper past. The problem is with how we build our ideas of who we are. When biologists try to understand ancient human migration, when they pick through our genomes and those of our distant ancestors, they are participating in age-old efforts to piece together our origin stories.

In China, it's believed that taming the flooding of the Yellow River many thousands of years ago, by a man named Gun and his son Yu, marked the dawn of Chinese agricultural civilisation. It's a legend that helped build national identity, serving a unifying purpose, lending a sense of superiority. Over the centuries, myths take on a life of their own, each generation recasting them to suit their needs until we can no longer tell the difference between myth and history. Before we know it, the glorious tales of our ancestors become our historical facts. Their ghosts become our icons. And of course we need to believe that our forebears were better than they really were, that they were nothing less than superhuman. The founding myth of Rome is of the abandoned baby twins Romulus and Remus, suckled by a she-wolf and rescued by a god. German nationalists told of a blond, rugged hero dubbed Hermann, who defeated the Romans and united Germany's disparate tribes. These figures have become woven into national identity, pulling people together in the belief in the cosmic power of their founders, cementing their particular claim on human civilisation.

So science is not enough. We need stories to assert our identity, even if they're held together with the tiniest grains of truth. There was indeed a German tribal chief named Arminius (from whom the legend of Hermann is constructed) who spearheaded a victory against the Romans two thousand years ago in what is known as the Battle of Teutoburg Forest. And in 2016 Chinese researchers confirmed that there really had been a giant flood around 1920 BCE. But legends must have been written around these facts until the people hearing them could no longer separate the two. The bloody realities became whitewashed over time, each iteration making the story cleaner, brighter and more dramatic than the original. And this was necessary, not just for the sake of gripping narrative, but also because it's tough to build national pride and a sense of superiority around a dirty history.

Anthropologist Jennifer Raff, based at the University of Kansas – in her words a 'middle-class white girl' – grew up on a powerful origin story of her own. 'Those of us in the United States have been taught this idea of American exceptionalism,' she says, 'that our country is the greatest country, and is founded on these wonderful beliefs, this freedom and equality and democracy.' It's a narrative that rests on the assumption that European pioneers in the seventeenth century filled a largely empty land with visions of a better society, deploying their unmatched skill and hard work to cultivate it. As in Australia, the indigenous inhabitants were framed as a dying race. If not gone, then definitely on their way out.

The subtext is that, without white Europeans, civilisation couldn't have flourished in North America. The United States was *theirs* to make.

It's not easy to square this popular founding myth with the more brutal historical facts. Of course, Native Americans weren't primitive, dwindling or scarce as the settlers liked to portray them. When the land turned out to be less vacant than they hoped, European colonisers made every effort to empty it. Thousands died on the 'Trail of Tears' – the forced relocation, following legislation passed in 1830, of several Native American tribes from their ancestral homelands in the south-eastern United States to designated Indian territories west of the Mississippi. Many more were killed by diseases brought by the migrants, to which Native Americans had no resistance. Deaths frequently went unrecorded, which means today we have only the vaguest estimates of how densely populated America really had been before. Genetic analysis published in 2011 suggests that the number of female Native Americans may have shrunk by half upon contact with Europeans five centuries ago.

We know this now. The founding myth becomes harder and harder to maintain. And yet it has strange ways of reasserting itself, even within academic circles, as Raff has found. Her work, trying to understand the distant past and the effects of race and migration, has shown her how easy it is for people, including respected scholars, to resist abandoning popular myths and racialised views of history even in the face of undeniable evidence. Indeed the myth of American exceptionalism is so pervasive that an entire scientific theory exists to explain it, weaving in archaeology and anthropology with the notion that Europeans are the ultimate bearers of human progress. It's known as the Solutrean hypothesis.

Crafted in earnest in the 1970s, the Solutrean hypothesis takes its name from archaeological evidence of certain tool-making techniques belonging to the Solutrean culture, which existed in parts of what is now France and Spain between 23,000 and 18,000 years ago. The Solutrean method of making blades by forming long, narrow flakes appears to be similar to that used in New Mexico by a culture known as the Clovis, which is thought to be some 13,000 years old. If the Clovis tools, which would have been used to kill such beasts as mammoths and bison, weren't developed independently, then the Solutreans might have brought them to the Americas first.

This version sits at odds with most other academic accounts of how the Americas came to be populated by our species. Geologists know that less than 15,000 years ago sea levels were low, allowing for a land bridge across the Bering Strait that would have joined modern-day Russia and Alaska. People could even have lived in the region between Asia and Alaska for an extended period of time before spreading further eastward. According to more recent research, there may have been waves of migration in both directions, with some people returning to Asia. So the most convincing account of what happened is this, that the first Americans came from the west, not the east, from Asia, not Europe.

The Solutrean hypothesis claims the opposite, that Europeans occupied the Americas long before the colonists of the seventeenth century, that in fact they were among the first people to live here. It suggests that those who became the native Americans must have reached there from Europe perhaps via Greenland and Canada, somehow crossing the treacherous

Atlantic during the last ice age, which ended around 11,700 years ago. Way back then, vast swaths of the planet would have been covered in sheets of ice, and sailing – or perhaps snowboarding – 6,000 kilometres across the Atlantic would have been a survival challenge of epic proportions. Yet those who defend this account believe that it was possible, especially if there was a continuous ice shelf across the ocean, which could have provided fresh water and food throughout the journey.

It's a theory at the very margins of science, but there remain a small number of American archaeologists who have staked their careers on it, publishing books on the hypothesis and clinging to the belief that more evidence will eventually prove them right. Among the most vocal is Bruce Bradley, usually based at the University of Exeter in England. It was in the 1970s that Bradley became aware of similarities between ancient stone tools dug up in northern Spain and those found in New Mexico. He couldn't believe that these similarities were mere coincidence.

'The basic underlying technology, the way stone tools are made, unless you understand how many detailed choices you have when you're making stone tools, things seem like they could happen accidentally . . . It's not just the blades, it's the way they made all the other tools. Virtually all of them have correspondences that are very, very striking between Solutrean and Clovis,' he tells me over the phone from Colorado.

The political implications of the Solutrean hypothesis are clear. If correct, it could be read as a suggestion that Europeans had a prior claim to the Americas, because their ancestors were already there many millenia before Columbus arrived in 1492. When they came later, then, they were only returning to a land that was already theirs. 'I see it as intimately tied up with the idea of Manifest Destiny,' explains Raff. This was a belief, particularly popular in the nineteenth century, that the European settlers who colonised what became known as the United States were somehow fated to expand their dominion across North America, that it was written into their history before they even arrived. It's a narrative they thought gave them a moral claim to the land, and later helped to square the inhuman treatment and murder of Native Americans with the squeaky-clean founding values of the United States.

That said, evidence for the Solutrean hypothesis is thin, and getting thinner all the time. One of the glaring snags is that the two cultures, the Clovis and the Solutreans, existed so many thousands of years apart in time. Nobody has discovered any ancient bones in Europe belonging to people who would have followed the Solutrean culture, only archaeological traces, such as the objects and art they left behind. So it's impossible as yet to connect modern-day Native Americans to Solutreans through their genomes.

Recent genetic evidence does show that almost all modern-day Native Americans have a shared lineage, and that this can be reliably traced to people who once lived in eastern Siberia. The 12,700-year-old remains of a Clovis boy found in Montana have shown him to be more closely related to all indigenous American populations than any other group. Raff explains that since ancient eastern Siberians were also related to the ancestors of modern-day east Asians, the obvious picture of migration is that the very earliest people to land in the Americas must have travelled through Asia and come by crossing the Pacific, not the Atlantic.Then again, archaeology is a field in which it's difficult to ever be completely certain of anything. New evidence can emerge at any time, overturning everything people thought they knew about the past. Science more broadly almost always leaves room for doubt as well. Proving something definitively wrong is tough because it requires you to look at every possibility in the universe, and then rule them all out. Sometimes this can be done – the earth is round and it rotates around the sun, we know that for sure. But when it comes to studies of the past, it's notoriously difficult. There's always the chance that a skull will turn up from under a ploughed field, or that a fresh scrap of archaeological evidence will bubble up from the Atlantic. Anything can happen. So, hypothetically, anything could be true.

This space for uncertainty, sometimes so small that you need a microscope to see it, is where the controversies live. And as far as Bruce Bradley is concerned, however controversial the Solutrean hypothesis, there's also a chance that he might be right, that history will dig up the evidence to vindicate him. 'Disproving is very, very difficult, and I don't even like the term prove and disprove. It's a matter of probabilities. Is this evidence more likely to indicate this than that? And that's the way we work all the time.'

Since he started working on the hypothesis a couple of decades ago, Bradley has come under sustained criticism from fellow archaeologists and geneticists. One team of researchers in the United States has even described his position as 'Solutreanism', implying that Bradley and those who share his ideas have crossed the line from science into ideology. Jennifer Raff insists that the lack of evidence linking Solutrean culture to Native Americans is about as clear as it could be. 'You would expect to see a bunch of other technologies,' she says. 'You would expect to see cave art of the same kind, you would expect to see settlements, and you don't see any of that.'

Geneticist Mark Thomas agrees that the theory has only the slightest likelihood, if any, of being correct. 'Let's be clear,' he says, when I raise the subject with him, sounding surprised that I'm even mentioning it. 'This is not a scientifically prevalent idea at all. If you are going to measure weight of argument in terms of word count, then maybe it seems prevalent. But this is very much like saying climate change is controversial because there are lots of words written saying there's no such thing as human-driven climate change. No.' For him, this is about a handful of researchers who have become so attached to an idea that they have embarked on a 'confirmation bias odyssey', as he calls it, scouring the world for evidence while neglecting whatever doesn't fit.

Despite my best efforts to put him at ease as we speak, Bradley does get combative, occasionally even raising his voice during our interview. 'I'm not trying to make anybody believe this hypothesis,' he tells me. 'I'm just putting the evidence out there and saying what we think it means.'

*

In a 2010 self-published novel, titled *White Apocalypse*, white-skinned Solutreans, having crossed the Atlantic and settled in North America, are slaughtered by savages who later cross the Bering Strait and become today's non-white Native Americans. The author, Kyle Bristow, a Detroit lawyer active on the political far right in the United States, makes fictional heroes of real-life archaeologists like Bruce Bradley, painting them as victims of a conspiracy by Native Americans and liberals, who don't want to face the apparent truth that the original Americans were white. Bristow's book has become popular in white supremacist circles. One review stated, 'This evidence could be the jolt whites need to awaken from our suicidal slumber.' When it was republished in 2013, he even included what were described as supplementary materials showing the validity of the Solutrean hypothesis.

For scholars such as Jennifer Raff, this comes as small surprise. The Solutrean hypothesis speaks to a nineteenth-century worldview that painted Europeans as the true inheritors of America, the only ones capable of civilising the continent. At the time, evidence of sophisticated technological cultures such as the Maya, Inca and Aztecs, which existed long before the Europeans arrived, were only further fuel for confusion. 'Since the beginning of the Americas, there has been this question of: Who are the Native Americans? Who are they? People actually wondered, are they *humans*? The first colonists did not really have a way to incorporate them into their biblical worldview,' Raff explains. 'After their humanity was more or less accepted, it then became this idea that, well, are they responsible for creating the culture, the very sophisticated technologies and art and monumental architecture that we see?'

In the shock of uncovering complex ancient civilisations in the New World, the first Europeans imagined elaborate ways in which they could have got there. 'It's so interesting to me when I look at ideas, alternative ideas, to explain the archaeology,' Raff continues. The Solutrean hypothesis, she says, is just the latest iteration. 'People are so desperate to find a non-mainstream answer to a lot of these issues. They won't just invoke Europeans, they will invoke aliens! They'll invoke people from Atlantis! Whatever they can find, as long as it's not Native Americans.' When *The Book of Mormon* was published in 1830, it claimed that Native Americans were descendants of the lost tribes of Israel who migrated to the Americas around 600 BCE and had been cursed with a dark skin for slaughtering their righteous relatives.

Of course, just because theories are exploited by the far right, this doesn't necessarily make them false. Raff is quick to add that while some supporters of the hypothesis may be motivated by racism, she doesn't believe that the researchers themselves are driven by this. 'They are good scientists and they are legitimate scientists. They are very well respected.' At the same time, though, she sees a doggedness in them that sets them apart. 'Everybody I've talked to who actually knows them personally tells me that you cannot change their minds. Nothing will change their minds. Nothing,' she adds. 'I wouldn't go so far as to call the Solutrean hypothesis pseudoscientific exactly. It did start out as a legitimate area of investigation, but I see it right now as being almost more ideological. I mean, people are not accepting any evidence against it. If you're pro-Solutrean, that's it.'

Of course, Bruce Bradley sees it differently. He tells me that Raff is 'deluded, to put it bluntly'. He believes he's been marginalised by the mainstream scientific consensus not because he is blindly clinging to a discredited idea, but because he is brave enough to challenge the academic orthodoxy. As far as he's concerned, his detractors are the ones motivated by bias. 'For me, it comes down to a lot of political stuff. When I first started promoting – not promoting, suggesting – this as a hypothesis, I was working in France and Spain, and different places over there. And I had very, very strong negative reactions from different people among colleagues in Spain. I think it's colonial guilt.'

Bradley insists that his work is just good archaeology. 'We've made it very clear all along that we're talking about our thing as a *hypothesis*,' he tells me, sounding more than a little worn down. 'People need to look up the definition of a hypothesis. A hypothesis cannot be right or wrong.'

The debate intersects with another recent controversy, that relating to one of the few ancient skeletons found in North America, known as Kennewick Man. Dated at around 8,500 years old, his middle-aged bones were discovered in 1996 by college students in Kennewick in the state of Washington.

Researchers at the time were quick to spot that his skull didn't look particularly like that of other modern-day Native Americans. In fact, one archaeologist described it as looking 'Caucasoid'. Of course, one way someone with Caucasoid features could have ended up here all that time ago was if some form of the Solutrean hypothesis was indeed correct, that the ancestors of Kennewick Man had travelled across the Atlantic from what is now Europe. A reconstruction of his face – cast in off-white, although nobody knew his real skin colour – even weirdly resembled the English actor, Patrick Stewart, best known for playing the captain of the starship *Enterprise* in *Star Trek: The Next Generation*.

Meanwhile, local Native American tribes rushed to claim him more plausibly as their own, insisting that he must have been a direct ancestor or related to their ancestors. In their historical legends, the land on which they lived had been their home since the beginning of time. They were products of it, not migrants to it. This narrative insisted that Kennewick Man simply couldn't belong to anyone else, and so, having been dug up and manhandled, he deserved a proper burial, conducted by the tribes. At the time, this call to return such remains to indigenous communities seemed irrational and emotional to some in the research community. A bitter court battle began, pitting scientists against Native Americans. It was about one origin story versus another.

The struggle over the remains of Kennewick Man wasn't just about identity or ritual. It was also wrapped up in a dark and brutal history that included scientific exploitation. In the nineteenth century Native American graves were often looted by anthropologists and hungry collectors, keen to claim their piece of this ancient culture before it disappeared, but with no respect for its traditions. Bones were rarely returned. The insults were not limited to artefacts and remains. As recently as 1990, blood samples from members of the Havasupai tribe who have lived in the Grand Canyon for centuries were taken by Arizona State University in the understanding that they would be used to study their risk of diabetes. In the end, without the permission of the Havasupai, the samples were also used to study other medical and mental disorders, including schizophrenia. The university agreed to pay $700,000 in compensation. When Native Americans defended the bones of their ancestors, then, they weren't only laying claim to their culture, they were also standing up for their rights over their own bodies.

Even so, in 2002 a judge finally ruled that the Kennewick bones weren't necessarily related to those of any modern tribes, in large part because researchers had declared that Kennewick Man didn't really look like the average Native American. As Kim TallBear in the Faculty of Native Studies at the University of Alberta has written, they privileged 'genome knowledge claims over indigenous knowledge claims'. With this, scientists were given a green light to study the skeleton.

Then slowly came the revelations.

Researchers in Denmark, led by Eske Willerslev, a pioneer in population genetics and ancient DNA, revealed in 2015 that Kennewick Man was after all more closely related to contemporary Native Americans than any other group. The tightest genetic link was found to be to the local Confederated Tribes of the Colville Reservation, which had originally claimed him as an ancestor. The indigenous groups had been right all along. He was one of their own, as much as it is possible to be when you are separated by millennia.

In February 2017, under legislation signed by President Barack Obama, Kennewick Man, now known by tribes as the 'Ancient One', was finally laid to rest in a traditional burial near the Columbia River, which runs through Washington state. The act of rewriting the story with something closer to the truth, and then returning his remains to the tribes, carried layers of significance. 'A wrong had finally been righted,' a spokesperson for one of the confederated tribes told the *Seattle Times* when the Ancient One was buried. A fresh forensic reconstruction showed a face starkly different from the first. Like Briton's Cheddar Man, whose facial reconstruction went from white to black in the space of a century, Kennewick Man, too, was completely different the second time around. Now he was given long hair and dark skin. The resemblance to Patrick Stewart was gone.

It was a lesson in how much culture and politics can shape how people read scientific evidence. It's an easy mistake to project contemporary racial parameters onto the past, explains Deborah Bolnick, an anthropological geneticist based at the University of Connecticut. She sees it happen in her own field all the time. 'If you see the genetic markers today that are found in western Europe, people will see those in the past and continue referring to them as western European, even if they're then also found in Siberia.' It's another example of an 'indexing problem', when the first available body of evidence influences subsequent thinking. Western researchers tend to have more access to European data because it's on their doorstep, so later discoveries elsewhere in the world are often interpreted relative to these.

Bolnick tells me of the example of a skeleton of a four-year-old boy discovered in south-central Siberia and thought to have been buried there some 24,000 years ago. In 2013 this became the oldest modern human genome yet sequenced, and scientists learned that he shared some genetic variants with people in western Europe. 'The way this got framed was: you have this individual in Siberia who has these western European genetic markers, and so maybe this means that there was a migration from western Europe to Siberia,' she tells me. In reality, the more parsimonious explanation, especially given the age of the skeleton, was that it was an east-to-west movement, not the other way. In other words, people in western Europe had *Siberian* genetic markers.

'Underlying assumptions and ideas definitely get embedded in ways that we don't even think about consciously, which can play out in the science,' Bolnick adds. 'We interpret data. We bring our perspectives, our framings to the data. You can use the same data to say many different things. I think modern genomic data provides the perfect example of that, because you can have different people who are all very smart and understand the data, who look at the same datasets and describe them in polar opposite directions.' It's impossible to escape our beliefs, our upbringing, our environment, even the pressure of wanting to be correct, when it comes to interpreting the facts. Our stories get in the way.

Indian historian Romila Thapar writes, 'In contemporary times we not only reconstruct the past but we also use it to give legitimacy to the way in which we order our own society.' Jennifer Raff believes this is quite clearly at play when it comes to the Solutrean hypothesis, just as it may have been when it came to understanding the true ancestry of Kennewick Man. There are powerful reasons why researchers may want to believe their own story is right, even when evidence declares otherwise.

The past is always at the mercy of the present.

*

'I remember during the Yugoslav Wars, I was in Paris.'

Kristian Kristiansen, a senior professor of archaeology at the University of Gothenburg in Sweden, works with ancient DNA expert Eske Willerslev, who carried out the investigation into Kennewick Man. Kristiansen is infectiously enthusiastic about this powerful new field of science. But as a longtime archaeologist, he also has a sober and measured perspective on the past. He agrees that leaps in genetics have the power to overturn everything we thought we knew about ourselves. They certainly challenge racial stereotypes by showing us just how much we have always mixed together throughout the past, and how much we have in common. But at the same time, he warns from his own personal experience that the political power of such insights has its limits.

It was in Paris in the nineties that he saw this for himself, during violent ethnic conflicts between Serbs and Croats in the struggle for independence in former Yugoslavia. 'I was a visiting researcher and I was living together with some expelled, you could say, archaeologists from Yugoslavia,' he says. The lives of ordinary people meant little in that place at that time. In the push for territory, a programme of ethnic cleansing, mainly of Bosnian Muslims, led to hundreds of thousands of civilians being forcibly displaced, women systematically raped, and murders so numerous and methodical that they rose to the level of genocide. Political leaders deliberately rewrote history to cast some ethnic groups as having a claim to certain tracts of land.

Reputable historians and archaeologists found themselves fighting an intellectual war against nationalist ideologues who wanted to justify their actions by promoting false versions of the past that suited their cause. 'And nobody wanted to listen,' says Kristiansen. 'That was the shocking thing. Nobody listened. And they published in newspapers, they did everything they could to get it across, but in the heat of the whole thing, they failed.' When push came to shove, truth became victim to politics. The facts mattered only if they suited the power-hungry agenda. 'Suddenly things can turn from left to right in a split second when politics changes,' he says, clicking his tongue.

This was nothing new. It had been seen before, most obviously earlier that same century when the Nazis pulled together a miscellany of racial theories to defend the genocide of millions of Jewish people and members of other groups during the Holocaust. Then, too, mainstream scientists and archaeologists found themselves marginalised and sacked while those whose ideas favoured the regime found themselves promoted and celebrated.

Bettina Arnold, a historian and professor of anthropology at the University of Wisconsin, Milwaukee, has researched just how gross these intellectual abuses were in the years leading up to the Second World War. After their country's humiliating defeat in the First World War, many Germans were looking for ways to rehabilitate their national pride, and the search for a more glorious prehistory was one means to that end. By promising to mend this collective feeling of bruised self-respect, the Nazis managed to gather public support. Slowly, they harnessed archaeological evidence that fitted their account of a great 'Germanic' past. At the same time, by proving that the German people had roots across Europe, they could lay moral claim to territory beyond their own borders. In their minds, they would expand to form an empire based on the ancient Germanic race, which they believed itself originally stemmed from noble, light-skinned Aryans, and was physically and mentally superior to all others.

Their intellectual framework came partly from linguist Gustaf Kossinna, who had been appointed a professor at the University of Berlin in 1902 and went on to become one of the country's most influential thinkers. By the time the Nazis came to power, Kossinna was dead. But the Third Reich had already nurtured his theories, seizing upon his argument that culture and ethnicity were wrapped up in each other. His ideas implied that when archaeologists uncovered evidence of shifting cultures, they were also seeing evidence of migration. So if they could find archaeological proof that the same cultures they could see in Germany had existed elsewhere, this would be proof that ancient Germans had lived there, that this was also part of *their* rightful territory. Archaeology, folklore and anthropology combined in service of this political idea.

The Nazis, says Arnold, were bent on 'proving that there was some kind of genetic – racial, essentially – commonality'. It was about expanding the boundaries of the traditional homeland using race as a rationale. This is not to say that the idea was welcomed, or even widely accepted. Kossinna was heavily criticised within his own lifetime for the quality of his work, most notably 'for the kind of cherry picking that he engaged in', she adds. 'You pick certain parts of material culture that support your arguments, you ignore those that don't. This is obviously a danger anyway in any archaeological interpretation. In his case, it was quite easy to pick holes in the arguments that he was making, and people did. Even his contemporaries did.'

Part of the reason that Kossinna was drawn to the Nazi party as they were beginning to claw their way to power in the early twentieth century is that he found support there he didn't necessarily have from his peers, reputable historians and archaeologists. 'He was a bit of a marginal figure early on in his career . . . He had been rejected by the mainstream cultural historians of the day. He had a hard time finding an academic job. There is a lot of personal bitterness,' she explains. But in the Nazi party, 'he found a niche and a place where he could matter, where his work was accepted and seen as important.'

Kossinna wasn't working for the party when he first developed his theories, but he was certainly motivated by an ethnically charged worldview that became useful later on. By the end, the party turned him into an icon, a founding father for the regime.

The politics suited him as much as he suited the politics. All the way up to his death in 1931, Gustaf Kossinna was firmly on board with Nazi ideology. Many of his publications make clear that he was aware of the political ramifications of the research he was conducting, says Arnold: 'He fully supported the idea that archaeology should be a handmaiden of the state.' Arnold has noted that in the first two years after Adolf Hitler came to power, eight new academic chairs in German prehistory were created. History was deliberately rewritten and appropriated by the party. The infamous swastika we associate with the Third Reich was employed after German archaeologists found the same prehistoric symbol on old German pottery. The 'SS' double lightning bolt that featured on Nazi uniforms was similarly adapted from an old Germanic rune.

Everything was recast through a political prism. Archaeologists writing for mainstream journals were replaced by those who toed the party line, and Germanic cultural influence on Western civilisation was intentionally exaggerated. In one bizarre instance of wishful thinking, the ancient Greeks were painted as ethnic Germans who had long ago somehow survived a natural catastrophe before developing a sophisticated culture of their own in southern Europe. Hermann the German, the tribal chief who had led his army to victory against the Romans almost two thousand years earlier, was dragged into service, too. Under the Nazis, his statue – erected in the Teutoburg Forest in the nineteenth century – became a focal point for nationalist pride, a reminder of a golden age of heroism.

Gustaf Kossinna remains a cautionary figure for archaeologists, as he does for academics more widely. The problem throughout, Arnold argues, is that archaeology – with its shortage of evidence and abundance of interpretation – has always lent itself to misinterpretation. The same may be said of other scientific fields, especially when data is thin on the ground and there are plenty of people desperate to speculate on the meaning and significance of what little there is. This has certainly been the problem with race science, and the study of human variation.

Kossinna is a reminder that shaping evidence around ideology, selecting specific results to suit a narrative, or even just a lack of care when it comes to interpreting or presenting data, can lead to disaster. What Kossinna did was no different from how scientific information was manipulated by anti-abolitionists in the American South in the nineteenth century, clinging tightly to their slaves, or by British imperialists who made the case for colonial rule by framing themselves as racially superior. But today it's a lesson taken seriously in Germany. When the two thousandth anniversary of the Battle of the Teutoburg Forest rolled around in 2009, the celebrations were sober, with a marked shortage of volunteers wanting to play the role of the Germans in a battlefield re-enactment. Most volunteers wanted to be Romans. Showing a distinct lack of nationalist fervour, even a spokesperson for the local museum told a reporter, 'I hope people in the future will take a closer look at history, question what they have learned and review the sources.'

'There is in Germany among my fellow archaeologists a really high sensitivity towards political misuse, towards simplification,' says Kristian Kristiansen, 'because they have seen the way that the Nazi regime constructed a false prehistory, by taking elements of the established prehistory and then twisting them.'

It was a sensitivity that could be seen in 2015 when Harvard geneticist David Reich was working on a paper for publication, examining evidence of the very same prehistoric culture that Kossinna had once described as Germanic. A German archaeologist who had supplied the team with skeletal samples was so concerned that the same conclusions might be reached about links between migration and cultural change as the Nazis had made that he and a number of other colleagues asked for their names be taken off the list of authors.

There is good reason to be cautious. In spring 2018 the prestigious science journal *Nature* issued an unusual editorial stating that, in a situation reminiscent of Gustaf Kossinna, 'Scholars are anxious because extremists are scrutinising the results of ancient-DNA studies and trying to use them for similar misleading ends.' It was the kind of warning that would have been unthinkable in a scientific journal a decade ago, but a new political climate combined with fresh discoveries emerging from human genetics was creating a crisis. 'They worry that DNA studies of groups described as Franks or Anglo-Saxons or Vikings will reify them . . .' People out there, the editorial suggested, are actively abusing science for racist purposes. In 2018 the *New York Times* reported that white nationalists had been seen 'chugging milk' at gatherings to demonstrate a genetic adaptation shared by many Europeans that allows adults to digest milk (a trait incidentally common to many non-white populations, too, who have historically also kept dairy cattle).

Kristiansen has witnessed this kind of racially motivated cherry-picking and distortion of scientific information for himself. 'Every time we publish, it goes into the global database. And what we can see is a lot of people are sitting out there that have all kinds of blogs where they go in and reanalyse data and see if they can falsify or get other results.' He suspects some of them may be fellow academics, but others seem to be enthusiastic amateurs. From what he can tell, they are deliberately scouring the genetic and archaeological data for evidence that fits in with their pet political or racial theories. In one memorable incident, he tells me that he was drawn into email correspondence with a respectable Canadian sociologist with a professorship at a public university who had cited his research. As they emailed each other, it slowly became clear to him that this man had views sympathetic to white supremacists.

'Everything can be twisted,' Kristiansen warns me. 'Everything.'

*

In the spring of 2018 a smattering of news reports began circulating in the Indian media which could have been lifted straight from Germany in the 1930s: the Indian government had set up a committee to rewrite history.

According to the reports, this was a decision that threatened to slam headlong against established scientific and historical facts, promoting a mythical version of history that painted India's dominant faith, Hinduism, as being central to its entire past. This particular origin story had been around for a century or so, enjoying varied levels of support. But it had become increasingly popular in recent decades, especially with the election of a conservative Hindu nationalist government in 2014. Now, it seemed, religious identity politics was being ratcheted into a higher gear.

Appointed by the Prime Minister, the twelve people on this new committee included a former senior official with the Archaeological Survey of India, along with the Minister of Culture, who was apparently keen to introduce a 'Hindu first' account of history into schools. Established facts about evolution, migration and genetics would be thrown out of the window in favour of a firmly religious narrative, one insisting that ancient Hindu texts are based in fact, not myth, and that those of other faiths have no claim to India.

In her book on contemporary identities in India, *The Past as Present*, historian Romila Thapar explains that the idea of a Hindu homeland has its roots in the struggle against British colonialism and efforts to construct a new national identity once independence was won. Just as in Germany following the First World War, politically motivated accounts of Hindu superiority have offered some Indians an opportunity to reclaim their self-respect, assert some collective pride and build a new sense of national identity. But in the process, India's ancient past, which is far from fully documented, has become a tool for projecting notions of technological and cultural superiority. Some of the members of the government committee to rewrite history, like other religious nationalists, believe that India belongs only to Hindus, even going so far as to suggest that Hindu Indians have no ancestry anywhere else, not even in Africa where our species originated. One member, a Sanskrit scholar, reportedly believes that Hindu culture is millions of years old, an order of magnitude older than the human species.

For religious nationalists, their ancestry and religion both tie them deeply to their land. Some have absorbed old European and American theories of an ancient, noble, pure-blooded Aryan race, and claim that these Aryans did indeed originate in India, living in the sophisticated cities of the Indus Valley Civilisation towards north-western India thousands of years ago. As the Nazis saw themselves, they see modern-day Hindus, particularly light-skinned, higher caste Hindus living mainly in northern India, as direct descendants of the Aryans. It's a connection thought to be not only timeless, but one that makes them superior to everyone else on the planet. As Thapar writes, the ideologues believe that 'the Aryans of India were not only indigenous but were the fountainhead of world civilization, and that all the achievements of human society had their origins in India and travelled out from India.'

It's a version of history that doesn't withstand much intellectual pressure. The oldest settlements to have been excavated in India, belonging to the Indus Valley Civilisation between 3,000 and 5,000 years ago, confirm that modern-day Indians must be products of different waves of migration from other parts of the world as we all are, some more closely related to Europeans in genetic terms, others less so, and everyone a mix. Hinduism and its cultures, too, have changed through time, and according to Thapar, bear little relation to the earliest civilisations.

But these facts don't always seem to matter, I'm told by Subir Sinha, a researcher at the School of Oriental and African Studies in London, who has been tracking the rise of religious extremism over the years. 'All I can say is that people who used to be scientific and rational at one time will now take a view that this [account of Hindu origins] is possible.'

While certain facts are deliberately ignored, at the same time there is a desperate desire to find others that do fit the ideology. The parallels with Gustaf Kossinna and Nazi Germany are striking. Indian archaeologists, for example, have been tasked with digging up evidence of places, people and events described in Hinduism's ancient texts. The legends include tales of demons, flying machines, monkey-headed and elephant gods. The nationalists say these weren't just beautiful allegories, but hard historical details. One example Sinha gives me is that of the mystical river known as the Saraswati, which sits at the centre of much of the action in one of Hinduism's scriptures. 'One of the first things the government did when it came to power this time was to set up a task force to identify the Saraswati River,' he tells me.

'There is a kind of will to truth. We will make this to be the truth if we try hard enough.'

Not just history and archaeology, but biology, too, are deployed to support the myth. While wilfully ignoring science on the one hand, they also 'care a lot about what the scientists are doing,' adds Sinha. When geneticists release new findings about human ancestry that don't sit well with the religious narrative, they are seen to pose an intellectual threat. It's a problem that has already landed at the door of Kumarasamy Thangaraj, India's leading population geneticist, based at the Centre for Cellular and Molecular Biology in Hyderabad. As one of the scientists who helped prove that modern Indians are the product of repeated migrations, he is well aware of the controversies surrounding the work of researchers like him. When he carried out research on Indian population genetics in collaboration with international colleagues, they decided to deliberately describe ancient migrant populations not as being African, Iranian or Middle Eastern in origin, as they might have for accuracy's sake. Instead, they called them 'ancestral north Indian' and 'ancestral south Indian', in an effort to be politically sensitive. With this wording, they avoided upsetting those who believe that Hindus spontaneously originated in India.

'It has not come to that level where I have to argue with them. People talk to themselves. They never fight back to me or oppose my findings, but that exists,' Thangaraj tells me, diplomatically.

Even so, Romila Thapar, Subir Sinha and other academics have expressed strong concerns about what they see happening in India. 'Most of the politics of connections with land and nature and "we are the true people" tends to be of a fascist, right-wing variety,' Sinha explains. 'They believe that civilisations should be based on a true homeland of righteous people, which have the same religion and language.' Religious minorities, particularly Muslims, have been picked out for persecution in this increasingly charged political environment. In the worst case, an eight-year-old Muslim girl living in the Indian-administered part of the state of Kashmir in north India was taken to a Hindu temple in early 2018 and gang-raped over a period of days before her dead body was dumped in a forest. Two government ministers attended a rally in support of the men accused of the crime.

The barbarity of how nationalism manifests itself may make it feel as though facts are peripheral, that they don't really matter, not when lives are at stake, not when young girls can be gang-raped for their faith. But for the religious nationalists, says Sinha, 'the past matters a lot for them to be confident in making the claims of greatness they want to make, claims to greatness in the world but also claims to land, power, claims to the right to show down people of other religions.' It gives them privilege over the truth, a version of their social structure that they can then sell to others. It throws weight behind the fists, it gives people the sense that what they're doing is morally justified, because this is the order of things as they see it. This is how the world was created, and they are only bringing some of that order back in place of chaos.

The nationalists must turn to the past for reassurance. The past is their problem.

But then, arguably, so is it for all of us, in smaller and bigger ways. When we study our genetic ancestry, aren't we also looking for clues about who we are, trying to reaffirm a story we have of ourselves? Why does it matter to some people that their ancestors were Vikings or Egyptian pharaohs? Does being related to Genghis Khan or Edward III make one person living today any different from the next? When we claim ethnic or racial pride, what are we doing but trying to piggyback on the achievements of those who went before us? It's not enough to be who we are now, to be good human beings in the present. The power of nationalism is that it calls to the part of us that doesn't want to accept being ordinary. It tells people that they are descended from greatness, that they have been genetically endowed with something special, something passed down to them over the generations. It attaches them to origin stories that have existed for hundreds of years, soaking into their subconscious, obscuring truth with the dazzling light of myth and legend. They are stories that shape even the convictions of world-famous geneticists like James Watson, who despite everything he has learned through science, clings to the belief that certain groups of people are simply born superior to the rest.

9

Caste

Are some races smarter than others?

On a smoggy January day I set out by taxi from the Indian capital, New Delhi, into the nearby state of Punjab to visit my extended family. It's a journey I've taken many times before, usually napping my way past the lush farmland and fruit sellers flanking the highway, stopping only at a *dhaba* for a butter-soaked lunch before returning to sleep. But this time I stay awake to study the faces I pass while I'm on the road, to watch the bodies jostling through the traffic.

I stare and I compare. The word 'brown' doesn't do any of us justice. Every possible skin colour is represented: ebony, paper white, yellowish, and countless other shades, along with almost every possible feature. India is unparalleled. The sheer span of the country and its environmental variety, from the sun-drenched beaches of the south to the snowy Himalayas of the north, seems to be mirrored in the physical diversity of its people.

An encounter the same morning had made me look with fresh eyes at this place I thought I already knew. I'd met geneticist Sridhar Sivasubbu in his scrupulously tidy office at the Institute of Genomics and Integrative Biology on Mathura Road in the heart of the city. Part of Sivasubbu's work is to investigate human genetic variation within India, with the aim of battling rare diseases. In a nation of more than a billion, he told me, no rare genetic illness is actually all that unusual. Rarity becomes a relative concept. Someone somewhere is comfortably beating the odds. But what fascinates him more than that is the variety. India is a microcosm, an entire hemisphere represented inside one country.

'We have something like fifty-five populations, major populations. Then there are minor populations within the country, and five linguistic groups are there,' he explained. Regions overlap genetically with other nations in south Asia and parts of the Middle East. The Andaman Islanders have close genetic affinities with Aboriginal Australians. This breadth of difference may explain why today India is the only remaining country in the world that has its own government-funded Anthropological Survey, designed to study the biological and cultural variation of its citizens.

But there's more to all this wondrous human diversity than meets the eye. One of the unsettling reasons that Indians exhibit such physical difference is that it's partly self-imposed – the culture demands it. Many centuries of marriage within fiercely tight-knit communities and a caste system that stretches back perhaps two millennia to keep privileged and non-privileged people apart, reinforced by the British under Empire, have deliberately separated populations from one another to this day.

Historian Romila Thapar has noted that there was always intermixing between population groups in India, that the strict divisions evident today were probably less strict in the past. But unlike neighbouring China, which though larger is not quite so ethnically diverse, freedom in India to marry and move between groups does seem to have been restrained. Millions still prefer to marry within their own religion, colour, caste and community, however shallow a pool of potential partners that might leave them with. And their preferences are policed not in law but by families, with regular cases of couples being attacked or killed for falling in love inappropriately. Inter-caste marriage was legalised in 1954, yet a survey in 2016 found that as many as 40 per cent of adults in Delhi who didn't belong to the lowest castes thought there should again be laws preventing it.

'You can find this in Hindus, you can find this in Christians, you can find this across India,' said Sivasubbu. 'So it's not about religion. It is about customs and marriage practices that have been passed down over generations. Indians tend to marry within the larger community that they live in, and in spite of all the few hundred years of knowledge that we have acquired, we still follow conservative and traditional marriage practices.'

My mother grew up with these values. Although she ended up married to a man of a different caste, religion and community, she retains a fatalistic view of life, steered by a society that has forever told her that everything is circumscribed, that has for hundreds of years kept people in their place. For those raised this way, social hierarchies feel knitted into their bodies. Their faith in the power of heredity is so strong that it overwhelms how they think even about themselves. They find nothing odd in a dynasty that remains in place over multiple generations, be it political, artistic or in business. Some extol the virtues of caste as giving everyone a valued place in society, ignoring that for people consigned to a life of cleaning toilets, it's little solace to be told that this is where they belong in the cosmic scheme of things. As a famous Indian economist belonging to a higher caste once told me, this is Indian culture, and it is unrealistic to expect change. It is regarded as given that people are fundamentally different, that they are born a certain way. Everyone becomes trapped in the net of their ancestral history.

Parallels have been drawn with the British class system, or race in the United States, but caste has features of both and of neither. It's ugly in its own way. At birth, you inherit your place in this social hierarchy and few transcend it. An 'untouchable' at the bottom will be given society's dirtiest work, existing in a permanent state of impurity, while those at the top of the ladder are a kind of aristocracy, favoured for jobs and education (people I meet as a journalist still like to drop into conversation that they are Brahmins, the highest, priestly caste, expecting this to carry some currency). The distant origins of all this are likely to have been partly strategic, to keep wealth and property within families. Some castes tend to encompass trades, creating generations of teachers, merchants or fishermen. Those at the peak of the system are apparently supposed to show benevolence to those below, but in reality there is considerable discrimination and violence.

Successive governments have brought in reservation policies, setting aside jobs and scholarships for disadvantaged castes. Even so, a report in 2014 by the international advocacy group Human Rights Watch found that teachers in some schools were still forcing lower-caste students to clean toilets and sit apart from everyone else. Top universities and colleges remain heavily populated by students of higher castes.

When scientist James Watson compared Brahmins to Jews, then, claiming that both had been bred for academic excellence, he was using both the language of caste and of science, seeing their differences as genetic qualities passed down over centuries. Their fortune is believed to lie beyond circumstance, to be part of what they are.

So Sivasubbu's comments overshadowed my thoughts as I rode to Punjab. In India, skin colour doesn't always faithfully betray someone's caste, but there's a lingering view that being fairer is better, that it denotes a higher status. The four main caste groups are sometimes even designated by colour – white, red, yellow and black – not unlike Europeans' classification of human races in the nineteenth century. 'Somebody very tall or very fair-coloured, obviously they would select an individual with similar features,' I had been told matter-of-factly by Indian population geneticist Kumarasamy Thangaraj. 'So there is a selection operating, not by natural but by man-made selection.' I recall when I was flat hunting in Delhi before starting my first job there as a reporter, I was asked to list my skin shade on the rental form. Having thought of myself only as 'brown' for most of my life, I had no idea what to write. The letting agent took a good look at me, and with a dirty smile scribbled, 'wheatish'. Colour takes on a new subtlety when every degree of pigmentation matters.

In both biological and social terms, India has long been a unique case study for scientists. This is why race scientist and eugenicist Reginald Ruggles Gates was so captivated by it. Here, systematic discrimination, the notion that groups of people are biologically pure and should be kept separate, that there are different breeds, isn't just an ideology. It's a living practice.

*

In the 1950s Indian geneticist L. D. Sanghvi wrote that people in his country 'are almost under an experimental environment . . . broken up into a large number of mutually exclusive groups, whose members are forbidden, by an inexorable social law, to marry outside their own group'.

When the idea of race research became unpalatable after the Second World War, Sanghvi was one of the first to turn to population genetics. In one nation, scientists like him believed they could explore what happens when human groups stay 'pure' – the kind of purity that nineteenth-century race scientists imagined might be possible, that Hitler wanted to see in Germany for the Nordic 'master race', and that white supremacists still want to see in Europe and the United States today. The grand social experiment that had already taken place in Indian society could reveal first-hand how the world would look if people mated only inside their narrow communities, selected for certain qualities over many generations.

It's remarkable just how widely Indians today believe that caste is deeply, biologically meaningful, that it has created exactly what it must have been intended to create: a social order reflected in biology, with the smartest and most gifted at the top, and others in various professions with their own skills below, as warriors and merchants, or as cleaners and servants. Even scientists think this way. 'The caste system, whatever has been practised for the last several generations, or several thousands of years, has a definite impact on everything,' Thangaraj told me. Character traits and abilities get passed down over generations, he implied. 'The offspring that's coming out of that founder are going to have such character. Usually, then, they become very unique features of that particular population.'

Sridhar Sivasubbu, too, suggested to me that people are biologically suited to the groups they are born into. By separating themselves for so long they have created genetic enclaves with particular talents, making caste not only a social reality but a biological one, too. 'Clearly certain communities have certain biological abilities which they are born with . . . we all look different and we each have our own strengths and abilities.' For him, the differences are so profound that castes are analogous to separate races, as population geneticist Luigi Luca Cavalli-Sforza might have defined them. 'You could call two groups, two peoples, completely different races and treat them as two separate entities. Or you could just celebrate both of them and say that they are different and each has the unique strengths and weaknesses.'

By way of an example, Sivasubbu pointed to Haryana, one of the states bordering New Delhi, which happens to be home to a disproportionate number of sportspeople, particularly wrestlers and runners. 'So, clearly they seem to have a better physique in terms of strength,' he suggested. Another example he offered was that of several tribal communities, which he claimed were naturally gifted at archery.

This casual speculation surprised me, coming as it did from a well-regarded geneticist. It demonstrated that more than half a century of research into human variation hasn't eliminated prejudice within science, wherever it's done. Old stereotypes are still being projected onto people, but perhaps in new ways. In Haryana there is certainly a long cultural tradition of wrestling in some families, for which people train their whole lives. But lifelong training could just as easily explain the prevalence of sportspeople as any innate ability. And if tribal communities happen to contain more skilled archers, this is most probably because they have traditionally been the ones to use bows and arrows, developing their skills through sheer hours of practice.

There's a slippery slope here, assuming that everyone in a particular community should be limited to certain paths in life. Indeed, social categories like these were harnessed and promoted by the British during colonial rule. My father's family, who were in the military and fought in both world wars for

Britain, were designated as one of the 'martial races'. They were seen by the British as physically and morally perfect soldiers. Yet it is a family tradition that has already disappeared. My father became an engineer, his brother a headmaster. Few of his siblings followed their ancestors into the military, and certainly none of their children.

Nevertheless, I can't help but ask myself whether Sivasubbu has a point. In 2018 scientists were amazed to discover that the nomadic Bajau people of south-east Asia, who live almost entirely at sea, surviving by free diving to hunt fish, have evolved an extraordinary ability to hold their breath underwater for long periods of time. The Bajau tend to have disproportionately larger spleens than neighbouring farmers, possibly helping them to keep up their blood oxygen levels when diving. There appears to be a measurable genetic difference between them and others, sharpened over many generations by living in an unusual environment.

This raises a question we don't like to ask out loud, but one that goes to the heart of the race debate. It is where race science began, with a belief that neglects history and jumps straight to the conclusion that the human zoo is like an animal zoo, each of us defined deep down by our stripes and spots. It follows from the offensive observation made by James Watson on the preponderance of Jewish intellectuals and Indian Brahmins in academia. Might it be possible, as Watson implied, for a group of people, isolated enough by time, space or culture, to become different? Could they, as eugenicists once suggested, evolve certain characteristics or abilities? Might they differ in their innate capacities?

Wandering down this road may be of scientific value, but it is risky, I know, and it's paved in blood. 'Could there be psychological differences between population groups?' I asked Thangaraj, tentatively. 'Differences in cognitive abilities?'

'That kind of thing is not known yet,' he replied. 'But I'm sure that everything has genetic basis.'

*

Back in London, I'm on the train on my way to a leafy corner in the south of the city, Denmark Hill.

The question of whether cognition, like skin colour or height, has a genetic basis is one of the most controversial in human biology. It's a grenade. And Robert Plomin is one of the few who have dared to handle it. A professor of behavioural genetics based at the southern campus of King's College London, he has dedicated his career to the search for the roots of intelligence, becoming one of the most divisive researchers in mainstream science. His work has far-reaching implications for how we think about human difference.

Tall, sporting a smart white beard and a crisp, pale blue shirt, Plomin is disarmingly charming in person, and rarely says anything to me that doesn't sound perfectly reasonable. I find him ruthlessly careful with his words. But there's a subtext. He moved to Britain from the United States in 1994, becoming known for brandishing the view that the cognitive differences between individuals can be accounted for largely in some way or another by genetics. The implication is that we are who we are, however we're raised. At the sharpest end, a few go so far as to take it to mean that the achievement gaps we see between large population groups (or races, as some might call them) may also just lie in their DNA.

It was in the 1970s, working at the University of Texas at Austin, that Plomin decided to dip his toe into the controversial field of behavioural genetics, eventually asking himself whether individual differences in intelligence might be heritable, and to what degree. It was perilous scientific territory with enormous social implications.

For him as a psychologist, 'it was still kind of forbidden to study genetics,' he tells me. 'It was dangerous professionally.' One obvious reason was the dark history of eugenics in the United States and Germany, as a result of which people were sterilised or killed in the belief that they would pass on their 'feeblemindedness' to their children. Another reason was that the handful who doggedly stuck with intelligence research after the war tended to say inflammatory things. In 1969 American educational psychologist Arthur Jensen, for example, claimed that black Americans had substantially lower IQs than white Americans, and that IQ was also significantly heritable. This implied that the black–white intelligence gaps seen in some tests weren't because blacks were socioeconomically worse off or discriminated against but were due to some innate genetic weakness. Plomin tells me he both personally knew and defended Jensen against his critics before he died. At the time, there was no genetic evidence to support Jensen, but he predicted that scientists would one day discover 'intelligence genes', which he believed would be 'found in populations in different proportions, somewhat like the distribution of blood types'. To compare 'intelligence genes' to blood types was telling because it was already known that blood types do vary in frequency between population groups. This would lead neatly to his next bit of speculation: that these 'intelligence genes' would also be found in lower frequencies in the black population than in the white.

Jensen's work went on to feature prominently in *The Bell Curve*, the controversial 1994 book by Richard Herrnstein and Charles Murray. Herrnstein, a psychologist like Jensen, had long insisted that intelligence was heavily heritable; he warned in 1971 that America was already slipping into a genetic caste system based on intelligence. Successful people, he claimed, were marrying each other and creating more successful children, while the unemployed languished because of their intellectual disadvantages, which 'may run in the genes of a family just as certainly as bad teeth do now'. Parallels with India's historic caste system were clear, except here it was being framed in purely biological terms. In Herrnstein's vision of the world, class, wealth and race overlapped because of biology, not because of history. The rich were rich because they were smarter, and their children were rich because they, too, were smarter. If certain racial or ethnic groups were poorer, then, it was their own fault.

In 2006 Arthur Jensen and Canadian psychologist John Philippe Rushton, who had been a pivotal figure in the Pioneer Fund and the *Mankind Quarterly*, published a brief commentary in the journal *Psychological Science*, repeating what they and Herrnstein had insisted their whole lives: that there was a racial gap of around fifteen IQ points and that it was unassailable.

But as they knew, even if this gap existed in IQ tests, proving that it was genetic and not the result of schooling, nutrition, upbringing or discrimination meant separating the effects of nature and nurture. This is the conundrum that has haunted human biology for more than a century. How do we know that differences between people are innate and not simply the product of social and cultural factors? Every human being is a unique product of biology and environment, which makes answering this question almost impossible.

Since there's no way to raise human genetic clones in a laboratory that we can run experiments on, the backbone of the psychological efforts to date have therefore been twin studies. By observing the similarities between identical twins, researchers have for decades thought they might be able to discern whether certain traits were potentially more heritable than others. But twin studies, too, are tainted by a toxic past. Josef Mengele, the notorious Nazi doctor who trawled concentration camps for involuntary subjects, had picked out young twins to deliberately amputate, mutilate and dissect. He carried out the kind of research that, for ethical reasons, is usually done on flies. And scientists learned nothing from it except just how far into hell a person would go to get his results.

For a while after the war, most researchers were wise to leave twins well enough alone. In 1979 Thomas Bouchard, a psychologist at the University of Minnesota, reignited the flame. Studying 100 pairs of twins who had been separated in infancy and then raised apart, he estimated that genetic factors accounted for approximately 70 per cent of their variance in IQ, implying that the bulk of the differences in intelligence we see between people who are healthy and well cared for are decided at birth. The rest lies in other factors in the environment, such as upbringing and schooling. Bouchard and his colleagues speculated that, at least in middle-class families living in industrialised societies, 'although parents may be able to affect their children's rate of cognitive skill acquisition, they may have relatively little influence on the ultimate level attained.' The implications of Bouchard's work, which had been financed initially by the Pioneer Fund, were obvious. If some people weren't doing so well at school – African American children, for example – it was nobody's fault but their own genes.

At the time, Bouchard was vilified, picketed, called a racist. Many decades have passed since then, but in some ways Robert Plomin has inherited this tarnished mantle. He has become part of the push to rehabilitate intelligence research, running his own studies using twin and sibling data to understand inheritance. The key is to see, as Bouchard did, whether twins raised in separate environments end up the same. And he believes that, more or less, they do.

*

'Everything is heritable,' Robert Plomin tells me straight out.

'In fact, I am not aware of anything reliably measured that's been shown not to be heritable in terms of psychology . . . Everything is moderately heritable.' Using studies of twins – especially those adopted into different families – his estimate for the portion of intelligence that is heritable sits at around 50 per cent. This may be far lower than the figure Bouchard came up with, but it is still pretty high. If half of our intelligence can be decided by our genes, then a large part of academic achievement may well be innate, immutable.

There are important caveats, however. First, measurement of intelligence is itself fraught, nobody fully satisfied that any IQ test can really do the job or, for that matter, that researchers have pinned down what intelligence really is and whether it can even be captured by a test. Theories about intelligence tend to be culturally loaded. Second, rates of heritability aren't the same for everyone. They depend critically on the environments of the people you're studying. Take a packet of seeds and shake half of them into a container filled with nutrient-rich soil, blessed with all the water and sunshine they need. Take the other half and put them in a container of poor soil with little water and light. In both pots, individual plants will grow to various heights, some taller, some smaller. The differences you see *within* each pot are largely hereditary because their conditions are the same. But in the first pot, each seed has been given the full opportunity to achieve its potential. In the second, they haven't been, so the plants will inevitably look smaller and scrappier. In this second pot, even the naturally strongest seed may not reach the same height as many of the plants in the more fortunate container. So the differences *between* the pots are not attributable to heredity.

Some traits, such as hair colour, are very strongly determined by our genes. Hair colour doesn't change depending on the environment, unless perhaps that environment is a hair salon. Even skin colour is to some extent affected by how we live. A group of paler-skinned children who play out in the sun all day will temporarily end up with darker pigmentation than paler children who don't, but this difference is purely due to environmental factors. So when scientists say that a trait such as height or intelligence is partly heritable, the only way they can know to what degree is by looking at people in the same normal, healthy environments, with few differences in how they are raised or treated by society. If people are deprived, it can obscure the genetic influence.

For instance, studies have shown that although their populations were until recently the same, North Koreans are today on average a little shorter than South Koreans. An alien landing on our planet with no knowledge of their histories might call it a racial difference, but it is purely down to their dramatically different economic circumstances. Height has very high heritability, but South Koreans have been prosperous and well fed and North Koreans haven't, so the differences in height between them are not genetic at all.

The problem with studies of adopted twins, as critics have noted, is that they usually involve children of fairly comfortable socioeconomic means. Even if the siblings are raised apart, they're unlikely to be at the poorest end of society, where the lack of good nutrition and a stable home life may be factors in their upbringing. They tend to go to fairly good schools. There's a risk, then, that these studies underestimate the role of environment across the true range of how people live.

But let's just assume for now that Robert Plomin and his twin studies are reliable, and intelligence is both measurable and highly heritable among children raised under normal conditions. Scientists like him must also be able to point to the genes responsible for the effects they claim to see. They need to be able to explain, step by step, how they get from this twin correlation to the genes and then to the brain. And to date, they haven't found anywhere near all the genes involved, let alone mechanisms for how they impact intelligence. In 2017 Plomin, along with a battalion of researchers in the Netherlands, Sweden and the United States, published the results of a study of nearly 80,000 people, which claimed to have found forty new genes linked to intelligence, bringing the total known to have such an effect to fifty-two. It was announced in the press as a breakthrough, but in reality these genes represented only a drop in the ocean. There are many, many more. 'We're not talking about a handful of genes, we're talking about thousands of genes of very small effect,' he admits. Half a century and millions of research dollars on, Arthur Jensen's prediction remains stubbornly unfulfilled.

Despite being unable to definitively isolate intelligence in human DNA, Plomin is proud of what he's achieved so far. He believes he is getting closer to an answer. 'Do you know, two years ago we could explain one per cent of the variance in intelligence with DNA? Now we can explain ten per cent! And it's only getting bigger. I would be amazed if at the end of the year we're not explaining fifteen per cent.' Supposing he does, though, he would still be left with the challenge of finding a single mechanism, one biological pathway, to explain how any of these genetic variations acts on the brain and leads to what we see as someone's general intelligence. We know, for instance, that X-linked mental retardation is a genetic condition, identifiable in a person's DNA, reliably leading to certain intellectual disabilities. There's a quantifiable link between the gene, inheritance and cognition. But for everyday intelligence, scientists don't have anything like this.

Psychologist Eric Turkheimer at the University of Virginia believes they will never find it. 'I've been around for a while,' he tells me. 'I've been in this field thirty years, and every single one of those thirty years, the biology people of one stripe or another have been saying, "I know we're not there yet but in five years, as soon as this next piece of technology is nailed, as soon as we have brain scans, as soon as the genome project is completed . . ." It's always right around the corner. And the reason I don't believe it is because I don't believe that's the way genetic causation works.'

Turkheimer compares intelligence to marriage. Psychologists know that if you have an identical twin who has been divorced, you are more likely to be divorced yourself. There is no suggestion that there's such a thing as a gene for divorce, because people understand this to be a complex outcome, influenced by countless factors, social as well as to do with personality and temperament. 'I think there are limits to how much we can understand something as complicated as divorce looking from the bottom up.' When it comes to intelligence, like most other complex traits, heritability depends crucially on context.

He and his colleagues have seen, for instance, that in studies of people with the lowest socioeconomic status, environment explains almost all the variation researchers see in IQ, with genes accounting for practically *nothing*. Children who are the most socially and economically disadvantaged have been shown to lose IQ points over their summer holidays, while the most advantaged ones gain knowledge and skills.

So for Turkheimer, it beggars belief that anyone should assume that the cognitive gaps psychologists now claim to see between groups in the United States must be biological. The effects of slavery and centuries of racism, in all its forms, are hard to quantify, but African Americans have undoubtedly suffered in ways that have left their mark on generations. 'Millions of people were kidnapped and thrown in the bottom of boats, and taken across the ocean, and a third of them died on the trip, and then thrown on plantations and enslaved for hundreds of years. And after that, treated with total discrimination. And now, *now* their IQs are a little lower? And we're saying it's in their *genes*?

'My feeling about that is give me a break.'

We know from the plant pot example that you can't contrast populations that live in different environments because the measurement of their heritabilities will be different. And there is no doubt that the social and economic circumstances of most black Americans remain significantly poorer than those of white Americans. The Institute for Women's Policy Research in Washington, DC, found that in 2017 the median weekly wage taken home by a full-time working white man was a third higher than that taken home by a black man. Compared with a black woman, it was almost 50 per cent higher. The wealth gap, which reflects capital accumulated over generations, is even starker. Research in 2017 showed that for lower and middle-income households, white families have four times as much wealth as black families. Across the board, from police brutality to quality of healthcare and schooling, black Americans are significantly worse off.

The logical consequence of insisting that IQ gaps between

races must be biologically determined is that nothing in human society can really be changed. In an age in which some like to believe that we have transcended the old rules of social inequality, when the playing field is supposed to be level, when women have the vote, when black Americans have civil rights and colonialism is over, they believe that biology is all that's left to explain the disparity that remains. Inequality, then, must be natural, the product of the survival of the fittest. Yet we still don't have the genetic evidence to prove any of this, says Turkheimer. All we have is the belief that the proof *will be* there somewhere in the genes. 'I don't see how we can get from where we are now to that kind of racial speculating that people like to do.'

*

Turkheimer lives in Charlottesville, Virginia. In August 2017 the city became the infamous backdrop to a Unite the Right rally, drawing together white nationalists, fascists and neo-Nazis from across the United States, brandishing swastikas and Confederate flags. Their march eventually escalated into violence, culminating in the death of peaceful anti-racism protester Heather Heyer, who was struck when a car drove into her and others who had braved the threat of violence to challenge the message of those on the far right. That day was described by many as a wake-up call to America. The dream of a post-racial society seemed more distant than ever.

'The synagogue we belong to is right next to the park where it all happened,' recalls Turkheimer. A letter from his rabbi told a disturbing story. 'It was Saturday morning, so there were services inside. They locked the doors while people marched up and down the street yelling, "Burn it down".'

In the days after this the scientific journal *Nature* felt the need to run an editorial reaffirming that science could not and should not be used to justify prejudice. It was a brief but remarkable statement, proving just how potent intellectual racism was seen to have become. 'This is not a new phenomenon,' it remarked about Charlottesville. 'But the recent worldwide rise of populist politics is again empowering disturbing opinions about gender and racial differences that seek to misuse science to reduce the status of both groups and individuals in a systematic way.'

Turkheimer explains that the problem is not in the data, which is so far either unclear or unsupportive of racism, but in the rampant speculation. If science could conclusively tell us that there was a biological difference in our DNA that made some groups smarter than others, then all bets would be off. There would be no more need for debate. 'But my point is I don't think we have that. And so what we're doing instead is speculating about our intuitions, speculating about people's dumb intuitions about Jews and blacks and whoever it is people like to speculate about.'

If bad intuition is the problem, it's a problem we all have. Intelligence is just as multifaceted as any other cognitive trait, but there's a widespread assumption that it is very heavily influenced by inherited natural ability. In reality, parents' IQ scores can only explain 15 per cent of the variance in their own children, admits Plomin. Exceptionally smart parents are likely to have children a little less smart than themselves because of a phenomenon known as regression to the mean, which works to bring everyone in a population back closer to the average. Very bright children are likelier to emerge from parents in the middle of the intelligence range, where most people live. This was precisely the statistical fact that made eugenics impossible.

As the *Nature* editorial noted, 'every individual is a potential exception.' Plomin himself is living evidence of just how possible it is for individuals to be unlike the rest of their family members. He was raised in a poor working-class family in Chicago, and neither of his parents was educated beyond school level. 'My sister is as different from me as you can be, in looks, in personality, and she was never interested in books. I would go to the library and get all these books, and she didn't go to university,' he tells me.

Today, there remains little doubt that there is at least some heritable component to what we perceive to be an individual's ability to reason and solve problems, to process complex ideas and generally figure things out. But it's the degree of flexibility in this hard-to-pin-down thing we call intelligence that still eats away at some in the scientific community, not to mention many on the outside who are interested in the politics of it all. The debate is reduced, as always, to a simple question of nature and nurture, to biology and the environment.

But it's not such a simple equation. Even identical twins can sometimes show very different abilities. In Thomas Bouchard's twin studies, he came across a pair of brothers who were raised in wildly different environments. One grew up to be an uneducated manual labourer and the other was highly educated. There was an IQ difference of twenty-four points between them. In the normal course of development, 'a ten or twelve-point difference between identical twins is not unusual,' Turkheimer tells me. 'So how much flexibility might there be in the system?'

*

In 1984 James Flynn, an intelligence researcher based at the University of Otago in New Zealand, raised a collective gasp in the scientific community when he announced that for nearly fifty years since 1932, American IQs had been rising at a rate of about three points a decade. Now known as the 'Flynn effect', this finding could have been interpreted as people getting significantly smarter with each passing generation. In fact, as Flynn recognised, people had simply become much more skilled at sitting IQ tests.

They were performing better, not because they had evolved mental capacities beyond those of their grandparents, but because the skills they had were being nurtured and sharpened now in ways they hadn't been before, by better and more education, more intellectually demanding jobs and hobbies. 'The period in question shows the radical malleability of IQ during a time of normal environmental change,' Flynn wrote at the end of his paper. Whatever link to intelligence that IQ tests measured saw a benefit from the cultural passage of time.

Comparing countries, he saw similar effects everywhere. Different versions of IQ tests are used across the world, with varying results. But between 1951 and 1975 Japan saw an IQ gain of more than twenty points. In Britain it was almost eight points between 1938 and 1979. Countries that were fully modern before the testing period began tended to show more modest gains than those that underwent significant social and economic change. Kenya and Caribbean nations made particularly big leaps. Scandinavian countries, on the other hand, showed a peak and then arguably even a little decline. Flynn proved that there must indeed be a great deal of flexibility in the system.

Mankind Quarterly editor Gerhard Meisenberg had told me that some countries are too cognitively challenged to prosper, that essentially they are poor because they are stupid. His only evidence was historical IQ test scores. Should anyone need it, the Flynn effect is some of the best proof yet that he is wrong. It shows that environment matters in IQ test results, even at a population level. In a paper published in *American Psychologist* in 2012, Flynn, Turkheimer and other experts suggested that, at this rate, the apparent 'IQ gap between developing and developed countries could close by the end of the 21st century'. Flynn has shown that the IQ performance of African Americans has risen faster than that of white Americans in the same time period. Between 1972 and 2002 they gained between seven and ten IQ points on 'non-Hispanic whites'.

Another easily overlooked fact in the American black–white IQ gap debate is that very few African Americans are quite as ancestrally African as we might think. In 1976, sociologist Robert Stuckert used United States census data to estimate that as many as a quarter of people listed as white might reasonably have some African ancestors, while as many as 80 per cent of black Americans were likely to have non-African ancestors. In 2015, geneticists at Harvard University and researchers from the ancestry testing company 23andMe investigated the heritage of more than 5,000 people who self-identified as African Americans, and found that on average almost a quarter of their ancestry appeared to be European.

If this is accurate, it should come as little surprise. The sexual exploitation of black women by their white owners was common during slavery. Founding Father Thomas Jefferson is thought to have fathered children by a slave in his household, Sally Hemings, who was herself of mixed ancestry. Of course, in the last century the United States has become more interwoven as the barriers between inter-racial relationships and marriage have been beaten down. But historically the 'one-drop rule' meant that anyone with even the smallest degree of African ancestry – one drop of blood – was classified as black. Today, anyone who looks at all 'black', however complex their heritage, is treated socially in the United States as *black*.

But if the biological portion of intelligence is rooted in a complex mixture of many thousands of genes, as biologists now agree it is, then it stands to reason that someone of mixed ancestry will have a mix of intelligence-linked genes from most if not all of their recent forebears. They wouldn't inherit genes from those with one skin shade and not another. And if that's the case, and there are indeed innate, genetic racial differences in intelligence, then logic dictates that they should show up in people of mixed ancestry. If, as Rushton and Jensen implied, black people are biologically less smart, then shouldn't African Americans with higher proportions of white European ancestry have slightly higher IQs?

As far back as 1936, a study of exactly this kind was published by American schoolteachers Paul Witty and Martin Jenkins. They picked sixty-three of the highest-performing black children in the Chicago public school system, and compared their IQs with the proportion of white ancestry they were thought to have according to their parents. Their results revealed no gap at all. Having more white ancestry didn't raise a child's IQ. Indeed the most remarkable student in the group, a girl with an IQ of 200, was reported to have no white ancestry whatsoever.

A similar piece of research was carried out in 1986, this time looking at black children who had been adopted into middle-class black and white families. Children who had one white parent and one black had around the same IQ as children with two black parents. What did make a difference to performance, though, was the family they were adopted into. The black and mixed-ancestry children adopted into white families had IQs thirteen points higher than those adopted into black families.

Psychologist Andrew Colman, originally from South Africa and now working at the University of Leicester in the UK, has interrogated the claims made by scientists who insist that innate intelligence gaps between 'races' are real. He believes that research like this strongly indicates that environmental factors could well account for the entire black–white IQ gap in the United States. Even for those who claim to show contradictory evidence, he notes, 'it is literally impossible to raise Black, White and mixed-race children in identical environments if racism itself is a significant environmental factor.' Being in the same school or even in the same family means little if society as a whole sees you as substandard.

Colman accuses researchers who cling to the idea, as one writer he cites once wrote, that 'negroes show some degree of genetic inferiority' of 'a form of self abuse'.

The United States is also a special case. What's interesting is how the debate over racial differences in IQ takes on a different flavour in other countries. In the United Kingdom, the group that achieves the lowest grades at GCSE level is white working-class boys, followed by white working-class girls. Scientists haven't jumped to claim that low intelligence is rooted in whiteness. There's no evidence that being white in the UK is a socially disadvantaging factor either, so by this logic it must be their socioeconomic status that's the problem. In the decade to 2016, some of the greatest progress in educational attainment was seen among Bangladeshi, black African and Chinese pupils. Girls have also historically tended to outperform boys, even though there is no average intelligence gap between the sexes. According to the founder of the Sutton Trust, which researches social mobility, it's clear that culture is at play here. There are social influences where class, ethnicity and gender intersect, and they all affect achievement.

This is a point that even Robert Plomin – whose work has been described in *Nature* as 'vintage genetic determinism' –

concedes. He acknowledges that studying group differences in intelligence is fraught because it is impossible to control adequately for the environmental effects, adding that he doesn't see any value in studying racial differences in intelligence. 'Based on what we know now, I don't see how you would do it. Certainly for black–white differences, people have tried for a very long time and I don't think we're any further along,' he tells me. 'We've had forty years of history of this, and it's just a lot of heat and no light.'

Heritability does matter, he insists before I leave his office. This position is, after all, his bread and butter. But he surprises me by adding that if you measure individual differences and find some people are a lot better than others, 'they all can achieve incredible levels of skill given culture.'

*

Sridhar Sivasubbu looked at me from behind his thick brush of a moustache.

I had asked him about his own community and what it meant to him. He told me that he came from a particular Tamil group in southern India known in ancient times for being warriors, and nowadays tending to join the military or police force. 'If you go back and read the history of my own community, largely people have been doing this for ages, so people just keep doing it. That's what they learned.' As a scientist, in his family he happened to be an outlier. I couldn't help but detect a hint of shame in his voice, or perhaps regret for the well-trodden path he had chosen not to take.

'People tend to follow what their ancestors did,' he mused. In a country where roots go so far back in time that truth and myth are often considered the same thing, connections to the past are not easily severed. They define how people live, forming the rigid framework of a sometimes precarious existence. 'Let's say we go back a few hundred years. You still had a set of beliefs, you still had a set of skills, trades that would be passed in the community. So you learned it, and you used that for livelihood. There will be communities that will end up growing a certain type of crop, in a certain type of region,' he explained. 'It helped them survive. You also had the bonding, friends, family, neighbours. So it helped you in times of crisis.'

But keeping close to a group has its costs. India's long history of community marriage has certainly left a mark in the genetic profiles of its inhabitants. These marks are most obvious when it comes to health. 'A genetic disease tends to remain in the community,' Sivasubbu told me. Studying particularly isolated communities has revealed stagnant pools. For the very smallest populations, their reluctance to marry outside the group can be deadly. The Parsis, descended from Persians who moved to India more than a thousand years ago, suffer such high rates of cancer that there have been fears they may disappear altogether.

He has helped identify other Indian families with rare genetic illnesses, including one in which siblings were covered in dried skin resembling scales on a snake, and another with a neurological disorder so severe that the parents wanted their children euthanised. These illnesses emerged for the simple reason that the communities concerned were too close-knit. People would end up unwittingly married to someone with a common recent ancestor, and if they happened to share the same recessive genes for a rare disease, abnormalities were more likely to pop up in their children. The same can be seen with Tay-Sachs, a genetic nerve disorder that is more common than average among Ashkenazi Jews, and in non-Jewish French Canadians living near the St Lawrence River. By isolating themselves, populations have formed tight bonds, but the smaller groups have also burdened themselves with higher genetic risks. Racial 'purity' comes at a high price.

I asked Sivasubbu whether he had married within his own community, or chosen to abandon that practice, knowing the risks. He told me that, despite fully understanding the genetic problems, his culture was so important to him that he found a wife from within his group. So long as he followed certain rules ensuring that he chose someone not too closely related to him, he believed it was better to stay tight to his community than to leave it. 'It was a very conscious decision, very conscious decision. It's going back to your roots, your beliefs,' he said, smiling.

For me, as a person of Indian ancestry raised in London, between two cultures and somewhat detached from both, this was something I had never fully understood. In that moment it struck me for the first time just how powerful culture can be. It can make us act against our better judgement, but it also anchors us in the world, in time and in place. Culture, and the safety and security it comes with, can have such a profound impact on behaviour over generations that to outsiders it may well appear to be genetic. It can seem to be woven into the fabric of a person, unshakeable, when in fact that same individual under other circumstances and raised in another place might behave completely differently.

When we see the effects of culture, we can't help but dream up biological mechanisms to explain it. The freshly unearthed travel diaries of Albert Einstein, written around 1922, have revealed that even he formed generalisations as he toured the world, despite being an anti-racist humanitarian. He described the Chinese as an 'industrious, filthy, obtuse people', adding, 'It would be a pity if these Chinese supplant all other races.'

Perhaps he imagined, like the builders of the old human zoos, that our differences run all the way from our habits into our bones, that to know one Chinese person is to know them all.

Biologists Marcus Feldman at Stanford University and Sohini Ramachandran at Brown University have suggested that the 'missing heritability' that scientists have for so long struggled to find in our DNA when it comes to intelligence and other complex traits may in fact be explained by the magic ingredient of culture. New scientific tools help us understand our genomes better, but they have only *reduced* the proportion of intelligence scientists now believe to be heritable. Feldman and Ramachandran ask the obvious: Why do scientists not look elsewhere for explanations?

In the same way that our parents pass on their genes to us, they also pass on their culture, their habits, their ways of thinking and doing things. And this can happen over generations. It is so sticky and persistent that it can seem biological to an observer. This is why measuring differences between groups, even over long periods of time, is laden with error. We are social beings, not just biological ones.

'The evolution of the Indian caste system is the perfect example of social determinism,' Indian biologist Rama Shankar Singh wrote in 2001. Studying the biological differences between castes, Singh started to understand it as a system defined not by evolved differences that made people better at different things, but as a set of barriers maintained by society for so long that it felt as if they were in the blood. Lives and choices were constrained by the invisible forces of culture, and so everything remained stratified. Everyone kept their position for fear of stepping out. We don't change easily.

But we can change. As societies do shift and inequalities finally flatten, then we start to see our assumptions overturned.

Stepping out of a rigid, unjust system can prove just how flexible we really are, just how far outside our genes our differences really may lie. In April 2018, a study was published into the performance of Indian bureaucrats hired as a result of affirmative action policies. The Indian Administrative Service, one of the largest and most powerful bureaucracies in the world, is also one of the toughest places to land a job. Of the 400,000 people who apply, 7,500 are invited to sit a gruelling exam, of which as few as a hundred or so will be offered a position. Controversially, half of these vacancies are then reserved for marginalised castes, whose slightly lower scores would usually disqualify them.

A common assumption has been that, even though they help redress social inequality, these quotas must have an impact on standards. If people are given a leg up to get in, then they surely can't be as good? Some believe that those born into lower castes are innately incapable of doing these high-status jobs well, regardless of their actual socioeconomic position. But when they investigated one particularly large sample project, American scholars Rikhil Bhavnani and Alexander Lee found no statistically significant difference in performance at all. 'Improvements in diversity can be obtained without efficiency losses,' they concluded. Caste had no impact. Indeed, the minority applicants who got through in the usual way, without the quota, tended to perform somewhat better than average.

10

Black Pills

Why racialised medicine doesn't work

'AN ENGLISHMAN TASTES THE Sweat of an African', reads the caption of the black and white engraving.

Dated 1725, it documents everyday life in the slave trading post of Calabar, West Africa. Part of a work that later became a kind of instructional guide for European seafarers, the picture shows a man being examined before he's sold into slavery and shipped to the New World. There is an obvious inhumanity to it, the flavour of a cattle market. But it's unnerving also for a strange act that's taking place in central focus. Here, the tall, well-built African slave dressed only in a loincloth crouches on his knees so that a skinny-legged, fully clothed Englishman with a sword hanging by his side can reach his face with an outstretched tongue to *lick his chin*.

In the following snapshot, a ship sets off with its cargo of slaves as families left behind mourn their loss, heads in their hands. The fatal transaction is complete. But there's still the memory of that odd lick. Academics have wondered just what the Englishman was doing. The accompanying description says he was confirming the slave's age and checking he wasn't sick, although it's not immediately clear how a lick would achieve this. We can only assume there was some method to it.

Centuries later, a young economist found this picture and came up with an alternative explanation. The licking, Roland Fryer at Harvard University has suggested, may have been to gauge the saltiness of the African's body, because being a little brinier might better equip him to handle the long sea voyage to the New World. It's an idea inspired by a scientific theory that claims black Americans, mostly the descendants of slaves, process salt differently from white Americans. The slave ships that transported people from places like Calabar to the New World would have seen immense loss of life along the way as a result of fluid depletion caused by dehydration, vomiting and diarrhoea, the theory goes. The proportion who naturally retained more salt would have fared better, producing a genetic bottleneck among the slaves who survived to the end.

This process of human evolution by natural selection on a rapid scale, acting only on black slaves, left them fundamentally different at the other end. Those who reached America were the saltiest.

Fast-forward to the present day. In the twenty-first century, this historical hypothesis has been commandeered to explain why black Americans today suffer persistently high blood pressure, commonly known as hypertension, at higher rates than other groups in the country. Hypertension is made worse by eating too much salt, but if black Americans naturally happen to retain more salt because of the legacy of the slave trade, then the theory goes that this might be the reason they suffer, rather than because of their diets. Their bodies, some scientists and doctors believe, just aren't the same. Others are more sceptical.

*

'Hypertension is probably the oldest chronic disease we know about,' says Richard Cooper, a 73-year-old public health researcher at Loyola University Medical School in Chicago. He has spent decades investigating blood pressure, stirred by his days as a medical student in eastern Arkansas when he saw patients dying of strokes while only in middle age. Part of the problem, he has found, is that it's a strangely nebulous disease. 'There are some people who argue that hypertension is not a disease, it's a condition, that it's a state, like anxiety,' he says. 'We don't really know initiating cause.' It isn't rooted in any one organ of the body, you can't run a scan or a biopsy for it. 'It just emerges out of the mists.' At the same time, it's a major killer. The World Health Organisation estimates that raised blood pressure accounts for nearly 13 per cent of all deaths worldwide. It is a simple thing to measure and it's a widespread problem all over the world. My mother happens to have it. In and of itself, it doesn't cause her any problems, but as the doctor reminds her, she needs to watch her sodium intake in order to bring down her blood pressure. Studies have shown that consuming just six fewer grams of salt a day could save two and a half million deaths from stroke and coronary heart disease worldwide a year. When I visit her for lunch, though, I can't help but notice how liberal she remains with the salt cellar. My mother doesn't need to remind me that it's a source of shame to serve Indian food under-seasoned. Habits don't change easily.

Despite how common it is in everyone, around the 1940s or 1950s American physicians began to notice more black patients than usual coming to them with hypertension. Today, in the United States, studies suggest that hypertension is almost twice as common in black Americans as it is in other groups. If you search the UK National Health Service website for the factors associated with high blood pressure, you'll see alongside salt, lack of exercise, too much alcohol, smoking and advanced age, one more: being of African or Caribbean descent. Hypertension is thought to be so powerfully correlated with blackness that UK clinical guidelines even recommend different drugs for black people and white people under the age of fifty-five.

Preventable heart disease and stroke resulting from high blood pressure are two to three times more likely to kill a black American than a white American. This mirrors death rates from other causes. The life expectancy of a black person born in the United States today is three and a half years lower than that of a white person. Almost every major cause of death and disability – even infant mortality – hits blacks harder if they live in the United States. So in a country like this, where black people die at disproportionate rates anyway, the mysterious quantity known as hypertension has long been a vehicle for racial speculation.

Cooper saw this speculation for himself growing up in Little Rock, Arkansas, and when he was at medical school in the 1960s. 'There were patients who were getting a transfusion and would say, "You're not going to give me black blood are you?" You were constantly confronted with that,' he recalls. At the time, wards weren't integrated and black Americans had obviously lower standards of healthcare. 'Growing up in Arkansas, race was very much apartheid.'

In some places, things haven't changed. Discrimination is a problem that extends all the way to the doctor's office even today. A hefty 432-page report published by the National Academy of Sciences in 2003 confirmed that evidence of racial and ethnic gaps in healthcare is 'remarkably consistent across a range of illnesses and healthcare services', despite surveys showing that most Americans believe blacks and whites receive the same quality of care. More recently, Roosa Tikkanen, a researcher at the Commonwealth Fund, a private healthcare foundation based in New York, found that black and minority patients in New York and Boston are disproportionately treated at poorer, less well-equipped public hospitals rather than in wealthier, high-quality teaching hospitals. She tells me that in some cases a public hospital may serve three times as many black patients as a private hospital located just a few blocks away.

Tikkanen suggests that institutional segregation and structural racism may be at play. Interviewing ambulance and emergency medical staff, she discovered that if someone was picked up from a low-income neighbourhood, in the Bronx or a certain part of Brooklyn, and that person happened to be from a minority background, ambulance personnel would by default tend to take them to a public hospital or a handful of so-called 'safety net' hospitals that have a history of welcoming poor and non-white patients. 'Even after you account for the fact that they have worse insurance, minority patients are disproportionately seen in the public system,' she adds.

Social disadvantage and inadequate medical care are obviously intimately linked to health and survival, yet in the case of black Americans, these factors remain curiously overlooked by researchers. In a paper published in the medical journal *The Lancet* in 2017, a raft of public health researchers including Mary Bassett, the New York City commissioner for health, warned that scientists were too often turning to biology to answer questions that could so clearly be better explained by social inequality. We know, for example, that 38 per cent of non-Hispanic black children in the United States live below the poverty line, compared with less than 15 per cent of children overall. Black people in poorer neighbourhoods live with worse levels of transportation, waste disposal and policing, and environmental hazards such as bus garages, sewage treatment plants and highways are more likely to be located near them. The areas where they live are also targets for cigarette and fast-food marketing.

'There is a rich social science literature conceptualising structural racism, but this research has not been adequately integrated into medical and scientific literature,' the authors of the study wrote. It's the elephant in the room. Of almost 48,000 articles they found on race and health, only 2,000 mentioned the word 'racism' even once.

It seems easier to believe that our bodies are different than to accept that our social circumstances are. The notion of black exceptionalism runs right through the history of American medicine, explains Cooper. 'There were very raw and basic racist ideas that were the norm in the medical school.' One common belief was that some diseases, such as tuberculosis and syphilis, manifested differently in blacks. Another was that even if black and white bodies were similar physiologically, black bodies had less value. In the Tuskegee experiment of 1932, the United States Public Health Service teamed up with researchers at the Tuskegee Institute in Alabama (at the time a college for black people) to track the effects of syphilis, deliberately denying patients antibiotics they knew could cure them. The men were observed until they died, their internal organs slowly ravaged. Only after ethical concerns were aired in the press did the study end, forty years later.

Hypertension is one of the conditions to have survived this era with its racial stereotypes intact. American doctors have continued to wonder whether the differences they see in rates of high blood pressure could be due to some intrinsic dissimilarity between races. They used to ask whether hypertension might even be a uniquely different illness in black people, connected somehow to skin pigmentation or testosterone levels, or to the heat and humidity of Africa.

As time passed and population studies were done, it turned out that people living in Africa, especially rural Africans, have the lowest levels of hypertension in the world. 'The people are skinny, they don't eat much salt, they're very active. There is no way that blood pressures cannot be low. It's just not possible. They don't have diabetes and they don't have hypertension,' says Cooper, who has carried out blood pressure studies on tens of thousands of people across Africa, Europe and the Americas. People in Nigeria and Ghana in West Africa, from where most black Americans can trace their ancestry, are known to have far lower blood pressure than those in other countries.

Topping the hypertension charts, exceeding levels elsewhere in the world, are in fact Finland, Germany and Russia. 'They have terrific hypertension,' adds Cooper. White North Americans and Canadians, meanwhile, tend to have lower levels than Europeans, including those in England, Spain and Italy. Hypertension, then, isn't a global problem for those with black skin, it's a local one. We know that black *Americans* have higher rates of hypertension on average than white *Americans*, and the same appears to be the case in Britain.

For doctors, the question has been: Why? The most obvious explanation would be the same reason that my mother has hypertension – diet, stress and lifestyle. 'Diet is the underlying cause of hypertension,' says Cooper. In Finland, for example, diets have traditionally been low in fruits and vegetables and high in fatty meat and salt. Food in the American South, traditionally associated with African Americans, is similarly rich in salt and fats. Cheap processed foods also have more added salt. Scientists seeking an easy explanation for differences in hypertension can find one in every kitchen.

But for some reason, the kitchen has been overlooked. In the 1980s, attempting to neatly square the circle of low hypertension in black Africans with high hypertension in black Americans, a doctor named Clarence Grim came up instead with what became known as the 'slavery hypertension hypothesis' – the theory, later championed by Harvard economist Roland Fryer, that black Americans are naturally predisposed to retain more salt because of a rapid process of natural selection on the slave ships that brought their ancestors to the New World.

Grim's was an evocative tale, giving the tragic brutality of slavery an extra poignancy. Sensitivity to salt, which had helped some through the brutal journey across the Atlantic, landed their unfortunate descendants in the twentieth century with the fatal scourge of hypertension. Western diets had damned them, and there was nothing they could do. The media loved it. Fans included Oprah Winfrey and the resident health expert on her talk show, Doctor Oz. With Fryer's apparent historical evidence – the picture of the slave being licked – supporting Grim's hypothesis, it all seemed to be tied up in a bow.

Not everyone was taken in by the story, though. Biologists raised eyebrows at the suggestion that evolutionary change could happen over such short timescales. Historian Philip Curtin, an expert on the African slave trade, argued that dehydration and salt depletion were not significant causes of death on slave ships. Neither, he noted, had there ever been a shortage of salt in West Africa to make people there particularly liable to retain salt. If anything, Curtin concluded, the historical evidence ran counter to the hypothesis rather than supporting it. Richard Cooper adds that salt sensitivity, if it is seen to be higher in black Americans, is likely to be a product of being primed over a lifetime for all the factors that give them hypertension in the first place, particularly diet. This is why other demographic groups who have higher hypertension, including men and the elderly, are also more sensitive to salt.

Even so, there was a widespread expectation among many medical researchers that harder proof would one day be found, most likely in our genes. If anything could settle the debate once and for all, it would be the glittering new science of genomics.

*

In 2009 researchers thought they had finally discovered the evidence they were looking for. A team led by scientists at the National Human Genome Research Institute in the United States took DNA samples from around a thousand people and discovered five genetic variants, or versions of the same gene, linked to blood pressure in African Americans. The effects were admittedly modest, but they seemed promising. In that moment it felt as though there really might be a host of tangible genetic differences between races that would help science get to the root of disproportionate ill health in black Americans.

The search for 'black genes' already had a precedent in the shape of sickle cell disease, a serious blood condition more prevalent in people who have ancestry in malaria-afflicted regions such as West Africa. The faulty gene indicative of sickle cell disease is known also to provide some resistance to malaria, which can otherwise be fatal. This is a watertight evolutionary explanation for why such a debilitating illness has persisted. But sickle cell disease also exists outside Africa, for example in Saudi Arabia and India, which means it can be found in people of different skin shades. Within the African continent, it isn't seen in high rates everywhere; it's not so prevalent, for example, in South Africa, where malaria is less of a problem. As the UK's National Institute for Health and Care Excellence states, the gene responsible for sickle cell is actually found in *all* ethnic groups.

In the United States these nuances were lost, mainly for demographic reasons. Many white Americans tend to be of European extraction, where sickle cell is rare, and black Americans tend to have West African roots, where it's more common, so it came to be seen as a 'black disease'. Once viewed this way, it reinforced existing assumptions about essential differences between blacks and whites. Two independent facts began to align in people's minds. First, that there may be different genes determining health according to race. Second, if black people suffer illness and death at higher rates than white people, could this then be genetic?

Hypertension, however, turned out not to be so straightforward. In 2012 another team of researchers, Clarence Grim among them, tried to replicate the 2009 study, this time with more than twice as many people. They failed. They just couldn't see the same correlations.

As scientists struggled to find the genetic evidence they thought must be out there, one team led by a researcher at the Harvard School of Public Health decided to look at factors other than race that might correlate with high blood pressure. They discovered that level of education, which often correlates with income and social class, was a far better predictor of hypertension in an individual than the percentage of African ancestry they had. Each year of education was associated with an extra half millimetre of mercury decrease in blood pressure readings. A year later, a study in Cuba showed that being black or white there made no difference to average blood pressure or hypertension. Others pointed out that living in an urban environment was strongly associated with blood pressure rises, as was being an immigrant or adopting a westernised lifestyle. Richard Cooper suggests, too, that the effects of chronic exposure to discrimination could well account for some of the black–white differences in blood pressure in the few countries where they're seen.

What is clear is that researchers are nowhere near understanding all the impacts of the social factors relating to health. One 2018 study even found a possible relationship between racism in a geographical area and the health of newborns in that area. Researchers saw, astonishingly, a direct correlation between the proportion of Google searches for the 'n' word in an area and the prevalence of black babies born prematurely or with a low birth weight. They noticed a similar heightened birth risk among women with Arab surnames in the six months after the 9/11 attacks.

For Cooper, the popularity of Grim's fairytale, the slavery hypertension hypothesis, betrays how some Americans would rather have a fanciful biological explanation for racial difference than a social one. 'It has been very hard to find genetic factors that affect salt sensitivity. What has been found is not more common in blacks than whites,' he concludes. 'There is no evidence of any significant selection across the African diaspora.' Despite the substantial resources poured into finding a gene, researchers have still found no association or mechanism that can fully account for a higher prevalence of hypertension in black Americans.

The desperate hunt for 'black genes' reveals just how deeply even well-meaning medical researchers believe that racial differences in health must be genetic, even when a goldmine of alternative explanations exists. 'It's a very useful window into people's thinking,' says Cooper. 'Below the surface, there is this very strong prior bias to believe in a mechanism like that. Racialised thinking is such a deep part, just like gendered thinking is such a deep part, of our psychology that we can't just by conscious effort free ourselves from it completely. It keeps popping up in ways when we're unprepared or not vigilant.'

People wanted so much for the hypothesis to be true, to be able to link the trauma of slavery to the trauma of black American deaths today, that they couldn't see past it to more mundane explanations. Hypertension, he says, is a case of science being retrofitted to accommodate race. The data, the theories, the facts themselves, are rotated and warped until they fit into a racial framework we can relate to. This is the power of race. It is the power to twist science to its own ends.

*

The list of drugs available to treat hypertension reads like a product catalogue from Willy Wonka's Chocolate Factory: there are beta-blockers to slow down your heart rate, alpha-blockers to relax blood vessels, diuretics to shift salt out of your body, ACE-inhibitors that work on the hormones, calcium-channel blockers, vasodilators, and more. And within each of these categories there are different brands, each competing fiercely for a piece of the lucrative blood pressure pie. They represent a global market worth in the region of £60 billion. It's small surprise that every pharmaceutical company wants to establish its unique selling point, to mark out its drug from the competition.

So it was with NitroMed, the marketers of BiDil, a pill that combined two different existing generic drugs: one that relaxed blood vessels and one that helped to combat heart failure, known to develop from hypertension. Around thirty years ago it became clear that mixing treatments in this way might allow patients to live longer, and BiDil was one of the earliest pills to do it. But there was a problem. The Food and Drug Administration, the gatekeeper for medicines that can be sold and marketed in the United States, refused to approve it because full clinical trials hadn't been done, so the company was unable to show exactly how well it worked. Its patent was running out, leaving the company in a jam. How could they get their pill approved as quickly and cheaply as possible?

The solution they came up with was unprecedented: they carried out a clinical trial in black patients only.

Studies already suggested that black patients tended to respond a little less effectively to ACE-inhibitors, for reasons that weren't yet fully understood. It's a fact so widely accepted that hypertension pills in the United States often specify different usage guidelines for different racial groups. In the United Kingdom, clinical advice goes so far as to state that white patients under the age of fifty-five should be given ACE-inhibitors to treat hypertension, while black patients shouldn't. BiDil wasn't an ACE-inhibitor, which meant it might make a promising new first drug of choice for black patients, says Jay Cohn, the pill's developer and a cardiologist based at the University of Minnesota.

Cohn's original tests on BiDil had shown that the small number of African American patients who were included in the trial (just forty-nine people) seemed to respond better to the pill than other groups did. By 2004 the results of a new trial, this time carried out on around a thousand patients, were published in the *New England Journal of Medicine*, confirming that BiDil, when taken in addition to existing medication, reduced mortality rates by 43 per cent. Given that the drug had already been shown to be effective in 1987, this outcome couldn't have been a complete surprise. But what was certainly different this time was that every single one of the patients tested was black.

Cohn tells me that this was a practical decision. 'We did not have adequate support to do a trial in the full population,' he admits. The cost of full-scale clinical trials can easily run into many millions of dollars. 'So we determined that maybe the best way to go would be to study the most responsive population, which was a self-designated black population.' This wasn't to say that BiDil didn't work in white patients, only that they didn't have the funds to do larger trials that included everyone. 'We would have needed a larger sample size to study a general population than we could get away with with a black population.' On the basis of this one group-specific trial, in June the following year the Food and Drug Administration approved BiDil as a medical treatment to be marketed solely to African Americans. It was the world's first black pill.

As soon as the decision was made, it divided people. Health campaigners and some high-profile groups such as the Association of Black Cardiologists welcomed it as a positive move, finally recognising the historically neglected medical needs of black Americans. Others, including many doctors, saw it as little more than a cynical marketing ploy to squeeze more profit out of a drug on which the patent was about to expire. NitroMed gained thirteen years of patent protection, with which it could sell the combination drug for as much as six times more than it could charge for the individual pills separately. The pharmaceutical goldmine had been excavated a little deeper.

That said, the going wasn't easy. NitroMed struggled to sell BiDil, in part because of scepticism among doctors, but also because of its eye-watering price. Since then, marketing rights have been sold to another firm, Arbor Pharmaceuticals in Atlanta, Georgia. Go to its website and you'll see black models on nearly every page, smiling and reassuringly healthy-looking. 'Although heart failure is on the rise all across America, it hits the African American community hardest,' it states. There's no doubt that BiDil is still being marketed as a black pill.

Yet, as I'm reminded by Jonathan Kahn at the Mitchell Hamline School of Law in Minnesota, who has tracked the case from the beginning, 'Race became relevant in the creation of this drug, not for medical reasons, but for legal and commercial reasons.' What's more, he explains, BiDil set a precedent. Seeing its success with the Food and Drug Administration, pharmaceutical firms began to file patent applications for other treatments that had been shown to work better in certain racial and ethnic groups. Looking at US patent applications filed between 2001 and 2005, the years before BiDil's approval, Kahn found that 65 mentioned race or ethnicity. Between 2006 and 2016, there were 384 that did.

Although no more explicitly race-specific drugs have been approved since BiDil, Kahn tells me that he has noticed an increased use of racial categories in drug labelling, the section that informs you about usage and dosage. 'You know, Asians respond differently from Caucasians, that kind of language,' he says. The website for beta-blocker Bystolic, for example, highlights how well it works in black and Hispanic patients. In the first half of 2008, Bystolic was the most heavily advertised drug in the United States. Drug labelling might seem of little importance, but he explains that it is a 'very powerful force for how medical professionals, and by extension anyone who's on these drugs, is being taught to think about the relationships of race to biology'. When you see a drug specified for use in a certain group, it implies that this group of people is biologically different from others.

In reality those working in medical research know that 'race' is hard to define, that it is a poor proxy for how human variation really works. But when there are few easy ways to distinguish people, it can feel as good as any. The ultimate aim for many in the medical profession is not to have racialised medicine, but personalised medicine, to be able to sequence an individual's genome and then tailor therapies to suit that individual. With personalised medicine, in principle, nobody will ever need to take a drug that doesn't work on them, or that gives them a bad reaction. But sequencing everyone's individual genome is expensive and ethically fraught, and we don't yet have all the data we need to analyse the results. Given these limitations, grouping people by race is seen as an imperfect but practical approximation. Most doctors and medical researchers will admit that it's a fudge, but they use it anyway. A proxy can save money and time, after all.

But really, how useful to medicine is the fudge of grouping by race? This is a question of statistics and demographics. Given all the medical data we have, how likely is it that an individual placed in a given racial group will benefit from a drug? And how likely is it that someone left out of that group and not given the drug could have benefited from it?

*

'Typically in my field, we don't just collect data and then show the data that we collected,' I'm told by Jay Kaufman, an epidemiologist and statistician based at McGill University in Canada. Data in its most raw form is rarely useful. Almost always, data has to be packaged, interpreted in some way to make sense.

He gives me a simple example. In 1970 the death rate in the city of Miami, Florida, was 8.92 per thousand people. In Alaska, at the north-west tip of the United States, that same year it was only 2.67. This is the unvarnished truth. If you happen to live in Miami, you are more likely to die than if you live in Alaska. But then, if you happen to live in Miami, you're also more likely to be retired. In 1970 the city, with a total population around

60 per cent of the whole of Alaska, had more than 92,000 people over the age of sixty-five, while Alaska had only around 2,000. If you're between fifteen and twenty-four in either of these places, the death rates are around the same.

Now take sickle cell disease. It was once suggested that only black infants in the United States be screened for the gene associated with sickle cell, because screening all infants would impose an unnecessary cost. Jay Kaufman and Richard Cooper worked together to dissect the statistics relating to sickle cell and found that, indeed, sickle cell trait prevalence in self-identified white Americans is only 250 per 100,000 members of the population, whereas in people who self-identify as black it is between 6,500 and 7,000. On this basis, it seems sensible to screen only black infants. On the other hand, there are many more white Americans than black Americans. The odds of a black newborn having the sickle trait may be 6.7 per cent, but the odds of *any* newborn having it are of the same order of magnitude, around 1.5 per cent. This is why US states today screen newborns universally, regardless of ethnicity or race.

When it comes to understanding the effectiveness of the 'black drug' BiDil, this kind of number-crunching hasn't been possible because the 2004 clinical trials included only self-identified black Americans and no white Americans. Statistical comparison just couldn't be done. What epidemiologists such as Kaufman have instead, though, are plenty of studies carried out over the decades on racial differences in response to common hypertension treatments. On the basis of these, for example, the National Institute for Health and Care Excellence in the United Kingdom recommends treating black people with calcium-channel blockers as the drug of first choice, rather than ACE-inhibitors or other alternatives.

So Kaufman and Cooper looked through all the published papers dealing with responses to blood pressure medication. Their aim was to figure out just how many individuals actually benefit from this racial distinction. They discovered that the perceived racial differences in drug response are in fact relatively small compared with differences within racial groups – exactly as anyone would expect given everything scientists know about the genetics of human variation. So while there might be statistically significant differences at a population level, this isn't always useful when it comes to treating any one individual patient. They found, for example, that regarding ACE-inhibitors, which are given to white patients under the age of fifty-five in the United Kingdom but not to black patients, data suggests that for 100 white people given the drug, 48 of them would fail to respond as hoped. Meanwhile if 100 black people were given this drug they are usually denied, 41 of them would benefit from it.

In this case, they conclude, assigning treatment by race is about as useful as flipping a coin.

Knowing that there are small average population-level differences in disease frequency, therefore, can be misleading when it comes to everyday treatment. Indeed, relying on those averages for guidance can even be life-threatening. American paediatrician Richard Garcia once described the case of a friend who as a child repeatedly failed to receive a correct diagnosis for cystic fibrosis because it was thought to be a white disease, and she was black. Only when a passing radiologist happened to spot her chest X-ray, without knowing to whom it belonged, was her condition instantly spotted. She had to wait until she was eight years old, and her colour had to be invisible, before she could be diagnosed.

For Kaufman, there are further problems when it comes to analysing the medical data on racial differences. Among the biggest pitfalls is statistical adjustment. 'The logic of these statistical adjustments is based on ideas from trials,' he explains. In experiments for new treatments, the gold standard is the double-blind randomised clinical trial, in which patients who are roughly the same in every other respect are randomly selected to receive either the treatment or a placebo but neither they nor the researchers know who has been given what. Scientists can then be sure that the effects they see are because of the drug and not something else. In real life, though, it's not always possible or ethical to carry out randomised trials, which is why adjustments are made after the fact, artificially removing the effects of variables such as age and weight.

When they do this, says Kaufman, researchers create what are in essence imaginary worlds where these things no longer matter. They are worlds made of manipulated data, illuminating a clear signal through the noise of reality. 'Once we start adjusting and describing some imaginary world, then the question is: What is that for? Why are we making an imaginary world? Which imaginary world are we going to make? What are we trying to learn from this imaginary world that we can't learn from the real world?' When it comes to racial data, the logic of adjustment is that if all the social and environmental aspects of racial difference can be removed and the data still shows a gap between groups, it must be biological.

In one study published in the *Journal of Allergy and Clinical Immunology* in July 2017, for example, Kaufman spotted an article by a large team of American medical doctors who claimed that the airways of black people become more inflamed than those of white people when they have asthma. In the United States, it's well known that black Americans carry the heavier burden of asthma. They're almost three times more likely to die from asthma-related causes than non-Hispanic white Americans. Black children are four times as likely as white children to be admitted to hospital for asthma. It's also well known that asthma is affected by the environment, including smoking, air pollution from busy roads and factories, and living among cockroaches, dust mites and mould. Yet this study claimed that there was also something intrinsic to black bodies that made black asthma patients suffer more severely, adding that they might even need their own therapies. Kaufman decided to pick through their figures.

The researchers' original data on airway inflammation showed no significant differences between white and black patients. So they chose to control for factors including lung function and degree of control of the disease, as well as body mass index, age and gender, until finally they came up with adjusted data that showed there *was* a small but significant difference between black and white patients, that black people's airways responded to asthma in a uniquely different way.

The lead author of the paper herself tells me that her study can't explain higher asthma rates among black people overall, only the severity of their condition. But as Kaufman explains, their adjustment can't even necessarily explain this. 'In a trial we would never adjust something that is affected by the exposure.' It's like testing a drug with a side-effect of weight gain. You can't judge the effects of that drug by adjusting for weight, because the drug itself causes patients to gain weight. In the asthma study 'they're saying let's imagine a world in which black people don't have a more severe disease, then what would their inflammatory response be? What use is that?' In Richard Cooper's words, the science is being retrofitted to accommodate race.

The logic of statistical adjustment holds true only if adjustment is actually possible. And racial difference is not a simple, measurable quantity like age or weight. The effects of racial discrimination, especially in a society as historically divided as the United States, run incalculably deep and wide.

'People get trained in schools to build models and make adjustments. This is the way we do things. And then they just apply it to race as though it's the same as a pill you would take. It's completely bizarre,' warns Kaufman. 'Most practitioners of medical research with medical degrees and basic science degrees don't really have much background in statistics. Many people with perfectly good intentions end up committing a lot of statistical errors because of lack of training and something we call wish bias, which is this idea that you want to find something interesting so you keep sifting through the data and fishing around until you find something interesting. That's a practice that generates many incorrect findings.'

Kaufman tells me he cannot understand why medical researchers persist in applying statistical methods to race when it's obvious that the methods cannot work the way they want them to, that they produce imaginary worlds of little or no use. 'It's epidemic in our literature that these adjustments are nonsensical. In economics, this approach that epidemiologists and biomedical people use doesn't fly at all.' He often sees scientists picking out a handful of variables to adjust for racial differences, without explaining why those variables were chosen and not others. At other times, he sees evidence of residual confounding, where the variables are measured poorly in the first place, making the final statistics even less reliable. This is especially true when it comes to adjusting for complex quantities, such as socioeconomic status.

The way data is collected and organised is also a problem. The habit of collecting data by racial or ethnic group has the unintended consequence of driving researchers to use it, hunting for gaps and trying any means possible to explain those gaps. Since 1993 the National Institutes of Health in the United States, the largest funder of medical research in the world, has had a general policy of requiring the clinical trials it supports to include women and minorities and also to collect data by 'race', across at least six categories. The purpose of this was never to look for differences between groups, but to ensure that medical studies include a broader spread of the population.

'All this is a record-keeping function that comports with other federal categories and guidelines, which are social categories,' explains Dorothy Roberts, professor of law and sociology at the University of Pennsylvania. 'It's the same categories as are used in the census to keep track of who is recruited into scientific studies. It's not a requirement for researchers to design their studies in any particular way.' But that's not the way the data always ends up being used.

In the attempt to understand racial disparities in health, social data sometimes gets treated as though it is biological data. Census categories are transformed into genetic groups. 'The US government provides millions and millions of dollars of grant money targeted to this question. If the government is giving you money, saying we want you to answer this question, and this is all people know how to do, then they're going to answer that question whichever way they know how. So that provides some incentive,' explains Jay Kaufman. It would be the same if funding agencies suddenly began collecting data by hair colour as well as by race and gender. You would be almost certain to see studies suggesting biological differences between people with brown hair and blonde hair. Just having the data invites comparisons.

As statisticians in this area know, pooling people into groups is always imperfect, and the larger the group, the more imperfect it becomes. In Britain, for example, it's common for health researchers to lump all south Asian communities into one convenient category. According to the National Health Service website, being south Asian is associated with a higher risk of cardiovascular disease, for instance. But this ignores the cultural and socioeconomic differences between Indians, Pakistanis and Bangladeshis, even within London alone. Rates of smoking tend to be high among Bangladeshis but very low among Indians, and smoking is a high risk factor for cardiovascular disease. Vegetarianism is common among many Indians, but rare among Pakistanis, and diet is also a crucial component of cardiovascular disease.

Hypertension is another condition considered to be slightly more prevalent among south Asians living in the UK. Yet although India, Pakistan and Bangladesh were one country until 1947, those of Indian origin tend to have higher blood pressure than those of Pakistani origin. Those of Bangladeshi heritage have lower levels than Britain's white population.

Although, with data as fuzzy and meaningless as this, adjusting for race or ethnicity can be a minefield, researchers routinely do it anyway. And sometimes with perplexing results, as Kaufman has found. One study he saw, for example, modelled hypertension and blood pressure for people in every country in the world in 2015 and in 1990. And 'so powerful was this model,' he wrote, 'that the authors even specified the mean blood pressure in 1990 for countries that did not even exist at that time.'

*

I ask Jay Cohn, the eminent cardiologist who invented BiDil, whether his drug, the world's first black pill, works well in patients who aren't black.

'Oh, of course it does! I use it all the time in white patients,' he replies. 'Everyone responds.'

Cohn has known this all along, and he has always been honest about it. His intentions were never to be racist, he tells me with a laugh. And I believe him. His goal was simply to get the drug approved any way he could. Labelling it as a 'black pill' was only ever driven by a commercial imperative. And in the end, this is business. Indeed, pharmaceuticals are big business.

But patients who are prescribed different drugs or who read the labels telling them that race matters may find it difficult to parse just how complex and subtle these facts are. When it comes to business, there are also those who may want to conceal them. In January 2017, independent news organisation ProPublica revealed that Tom Price, a Republican congressman nominated by Donald Trump to become head of the Department of Health and Human Services, had persistently lobbied on behalf of Arbor Pharmaceuticals, which owns the marketing rights to BiDil, to remove a certain study from a government website. Carried out in 2009 by heart researchers at the University of Colorado on more than 76,000 people, the study showed that across all the racial and ethnic groups investigated, the combination of drugs that made up BiDil was not associated with significant reductions in mortality or hospitalisation. The very first study on BiDil had looked at only forty-nine African Americans, and the second at just over a thousand. This was a far larger study and therefore likely to be more significant.

The US has a federal agency, the Agency for Healthcare Research and Quality, to help patients and doctors make informed choices about medical treatments. Yet according to the news report, one of Tom Price's aides had emailed the agency 'at least half a dozen times' to have the University of Colorado study removed. It turned out that Arbor Pharmaceuticals had previously donated to Price's campaign fund. The month after the ProPublica news report came out, Price was confirmed as Secretary of Health and Human Services, although he resigned before the year was out, having been criticised for his use of expensive chartered flights.

For now, BiDil is still on the market but its patent is approaching its expiration date. In the meantime, race has become a firmly and widely accepted variable in medicine, according to law professor Jonathan Kahn. He uses three clichés to describe what has happened in the last couple of decades. 'The road to hell is paved with good intentions,' is the first. 'I don't think the use of race in these contexts was nefariously plotted,' he explains. Almost everyone in the medical research community believed they were doing the right thing, curing as many people as possible in the most efficient way. The second is 'the law of unintended consequences, resulting from that. And finally, that it's creating an accident waiting to happen.' For Kahn, well-intentioned people have reintroduced race to medical science without fully understanding either their reasons or the consequences.

'It's not that the use of BiDil in black patients is immediately going to lead to the re-substantiation of slavery or scientific racism, but it's a step down the road of re-biologising race in a way that feeds deep strains of racism,' Kahn tells me. 'What we're seeing now in the US and again on the rise in Europe is a sort of ethno-nationalism. And any sort of indirect or direct approval or imprimatur of using race as a biological category becomes, no matter how well intended, dangerous in those contexts.'

And as the evidence shows, for all the studies that point to racial differences in health, the genetic evidence so far rarely tallies with them. Hypertension is just one case in point. Enormous experiments looking at the genomes of thousands of people have turned up little. Although hundreds of gene variants linked to blood pressure have been found, collectively they explain just 1 per cent or so of the variation we see, says Jay Kaufman. 'We've had a decade of genome-wide association studies now, we've spent billions and billions of dollars, and we still are at the position that it looks like ninety-seven per cent of the mortality disparity between blacks and whites in the United States has nothing to do with genes.'

This makes sense, he adds. It would be bizarre to imagine that black Americans are somehow so uniquely disadvantaged in biological terms that they would naturally die of this and almost everything at higher rates than everyone else.

Dorothy Roberts agrees that there's no logic in expecting black Americans to be so medically unusual. 'How could it possibly be that a group called black people, which first of all is defined differently around the world – it's been defined differently even within the United States, but the current definition in the United States is anyone with *any* discernible African ancestry – how could that hugely varied group, which could include someone with mostly European ancestry, someone with mostly Asian ancestry, someone with mostly Native American ancestry, how could it possibly be that that group for an innate biological reason could have a particular health outcome? That just doesn't make sense,' she says with a laugh. 'The most plausible, to me the only possible explanation could be because of inferior social conditions.'

'Race is a story we tell ourselves,' adds Richard Cooper. If you believe racial difference is biological, you will look for biological explanations. 'Everybody has a general belief in race and then they have stories about it, either something they've seen or something they've experienced. And the two reinforce each other. History, psychology, politics, we all have our belief systems and myths and things which a hundred years from now are not going to be true.'

In her book *The Social Life of DNA*, Columbia University sociologist Alondra Nelson notes that in eighteenth-century New York, the mortality rate of black infants was twice that of white. More than half the black population died in childhood, and those who lived saw their bodies driven to breaking point by the physical stresses of slavery. American medical experts claimed at the time that black people were naturally more robust, more resistant to diseases that killed others, including gallstones, tuberculosis, pneumonia and syphilis. It was a narrative that served slavery, allowing slaveholders to subject people to harsher labour and living conditions on the assumption that this couldn't harm them.

'So you have at one time a literature saying that black people are especially robust, and at another time you have a literature saying that black people are especially predisposed to illness,' says Kaufman. 'It's a contradiction, but each one serves its own purpose.'

He suggests that part of the reason Americans cling to the idea of black exceptionalism when it comes to health may be that, in some way, the idea lets society off the hook. It places the blame for inequality at the foot of biology. If poor health today is intrinsic to black bodies and nothing to do with racism, it's no one's fault. 'It says it's not our organisation of society that's somehow unfair or unjust or discriminatory. It's not that we treat people badly. It's not that we give people worse life chances,' he says. 'It's just that these people have some genetic defect and it's just the way they are.'

An interesting historical case in point is schizophrenia, a mental disorder for which people of black Caribbean ancestry living in the United Kingdom receive proportionately more diagnoses than white people, to the point where it has even been described as a 'black disease'. In recent years there has been a feverish search for the genes thought to be responsible, now that researchers widely accept that it may be heritable to varying degrees. In 2014 an enormous study involving more than 37,000 cases finally did find a number of genetic regions that may be associated with schizophrenia. But it turned out that even the most promising of these elevated risk by just 0.25 per cent. This gene variant turned up in 27 per cent of patients, but also in around 22 per cent of healthy subjects.

If schizophrenia is genetically more prevalent in one population group, then this clearly can't be a straightforward equation. Indeed environmental risk factors, such as living in an urban environment and being an immigrant, have already been shown to be at least as important to being diagnosed as any genetic links found so far. One study published in *Schizophrenia Bulletin* in 2012 found that patients with psychosis were almost three times as likely to have been exposed to adversity as children. That's not to say the disorder doesn't have a genetic component, but it does demonstrate that it can't be quantified by looking at genes in isolation. If there are racial differences in diagnoses, it may be life experiences, perhaps even the negative experiences resulting from racial discrimination, that tip some people over the edge while rescuing others. This is without even considering that schizophrenia diagnosis itself is known to be notoriously subjective.

If, even after all this, race is a factor, contrast the characterisation of schizophrenia as a 'black disease' today with an observation by Nazi scientist Otmar von Verschuer more than half a century ago. A year before the outbreak of the Second World War, he wrote, 'Schizophrenia is strikingly more frequent among Jews. According to statistics from Polish insane asylums, among insane Jews schizophrenia is twice as common as among insane Poles.' He went on: 'Since it is a matter of a hereditary disease . . . the more frequent occurrence of the disease in Jews must be viewed as a racial characteristic.'

At that moment in time in that particular place, then, it wasn't a black disease; it was a Jewish one.

11

The Illusionists

Down the rabbit hole of biological determinism

'ONCE UPON A TIME . . .'

At the turn of the millennium, excitement about the dizzying possibilities of genetics was still rife. People wondered whether gene therapy could someday cure cancer. Researchers imagined they would find genes for everything from being tall to being gay, whether we might even build designer babies by tinkering with our DNA. And two scientists working for the National Cancer Institute in the United States wrote a fairy tale.

Their protagonist is a well-meaning geneticist who one day begins to wonder why some people use chopsticks to eat their food and others don't. So of course, the hero does what all good experimentalists do: he rounds up several hundred students from his local university and asks them how often they each use chopsticks. Then he sensibly cross-references that data with their DNA and begins his hunt for a gene that shows some link between the two. Lo and behold, he finds it!

'One of the markers, located right in the middle of a region previously linked to several behavioural traits, showed a huge correlation to chopstick use,' the tale goes. He has discovered what he decides to call the 'successful-use-of-selected-hand-instruments' gene, neatly abbreviated to SUSHI. The magic spell is cast. The experiment is successfully replicated, the scientist's paper is published, and he lives happily ever after.

This might have been the end were it not for one fatal yet obvious flaw. It takes him as long as two years to hit upon the uncomfortable realisation that his research contains a mistake. The SUSHI gene he thought he had found just happened to occur in higher frequencies in Asian populations. So it wasn't the gene that made people better at using chopsticks; it was that people who used chopsticks for cultural reasons tended to share this one gene a little more often. He had fallen headlong into the trap of assuming that a link between the use of chopsticks and the gene was causal, when in fact it wasn't. The spell was lifted and the magic was gone.

Like all good fairy tales, there was a moral to this story. Although not everyone could see it.

*

In 2005 the hype around genetics had begun to fizzle out, to be slowly replaced with a healthier scepticism. Scientists began to wonder whether our bodies might not be quite as straightforward as they had thought. And then along came a young geneticist at the University of Chicago in the United States with an extraordinary claim.

Bruce Lahn's work was a shot in the arm for those who had always hoped that genes could explain everything, for the biological determinists who believed we were anything but blank slates, that much of what we are is decided on the day we're conceived. His claim was so bold it implied that maybe even the course of history could be decided by something as tiny as one gene.

Lahn had originally emigrated from China to study at Harvard University, and soon gained a reputation as a cocky maverick who didn't follow instructions, who did things his own way. A while after arriving in the United States, he changed his name to Bruce Lahn from Lan Tian in homage to the legendary actor and martial arts expert, Bruce Lee. Science journalist Michael Balter describes in a profile how once, when invited to go on a two-day hike with his colleagues, Lahn turned up with nothing but a jar of pickled eggs. 'He was kinda the whizz-kid, he was kinda the darling,' Balter recalls.

The whizz kid moved up the academic ladder at lightning speed. In 1999 he was named in *MIT Technology Review*'s list of innovators under thirty-five. Then in 2005 he published a pair of studies in the prestigious journal *Science* drawing a connection between a couple of genes and changes in human brain size. He and colleagues stated that as recently as 5,800 years ago (a mere heartbeat in evolutionary time), one genetic variant that was linked to the brain among other things had emerged and swept through populations as a result of evolution by natural selection. Their implication was that it bestowed some kind of survival advantage on our species, making our brains bigger and smarter. At the same time, he noted that this particular variant happened to be more common among people living in Europe, the Middle East, North Africa and parts of east Asia, but was curiously rare in the rest of Africa and in South America. Lahn speculated that perhaps 'the human brain is still undergoing rapid adaptive evolution' – although not for everyone in the same way.

His work caused a sensation. What set pulses racing above all was his observation that the timing of the spread of this gene variant seemed to coincide with the rise of the world's earliest civilisation in ancient Mesopotamia, which saw the emergence of the first highly sophisticated human cultures and written language. Lahn seemed to imply that the brains of different population groups might have evolved in different directions for the past five millennia, and that the groups with this special genetic difference may in consequence have become more sophisticated than others. In brief, Europeans, Middle Easterners and Asians had benefited from a cognitive boost, while Africans had languished – perhaps were still languishing – without it.

Racists ate it up and asked for second helpings. After all, here was hard scientific evidence that seemed to corroborate what all those nineteenth-century colonialists and twentieth-century contributors to the *Mankind Quarterly* had always claimed, that some nations were intellectually inferior to others. Their failure to prosper economically was rooted not in history, but in nature. 'There will be plenty more results where these came from,' predicted right-wing commentator John Derbyshire in the American conservative magazine *National Review*. Lahn also attracted support from the late Henry Harpending, a geneticist at the University of Utah and co-author of a controversial book arguing that biology could explain why Europeans conquered the Americas, and also that European Jews had evolved to be smarter on average than everyone else.

But there were problems with Lahn's findings. Even if his gene variants did show up with different frequencies in certain populations, it didn't necessarily mean that they provided those who had them with a *cognitive* advantage. The variants were known to be linked to organs other than the brain as well, so if natural selection was taking place, maybe this was nothing to do with intelligence. Maybe the genes conferred some advantage that wasn't related to the brain. The hypothesis needed more evidence.

Soon after the papers were published, controversial Canadian psychologist John Philippe Rushton ran IQ tests on hundreds of people to see if the gene variants really did make a difference to intelligence or brain size in those who possessed them. Try as he might (and we can reasonably assume that as head of the Pioneer Fund at the time, he tried his hardest), he couldn't find any evidence that they did. They neither increased head circumference nor general mental ability.

Before long, critics piled in from across the board, undermining every one of Lahn's scientific and historical assertions. For a start, the gene variant he described as emerging 5,800 years ago could actually have appeared within a time range as wide as 500 to 14,100 years ago, so it may not have coincided with any major historical events. Respected geneticist Sarah Tishkoff at the University of Pennsylvania, who had been a co-author of his papers, distanced herself from the suggestion that it might be linked to advances in human culture, as Lahn had suggested.

There were doubts, too, that Lahn's gene variants had seen any recent selection pressure at all. Tishkoff tells me that scientists today universally recognise intelligence as a highly complex trait, not only influenced by many genes but also likely to have evolved during the far longer portion of human history, ending around 10,000 years ago, when we were all mainly hunter-gatherers. 'There have been common selection pressures for intelligence,' she explains. 'People don't survive if they're not smart and able to communicate. There's no reason to think that there would be differential selection in different populations.

That doesn't mean somebody won't find something someday. Maybe it's possible, but I don't think there's any evidence right now that supports those claims.'

In the end, Lahn had no choice but to abandon this line of research. 'It was pretty damaging, because a lot of illustrious researchers either couldn't replicate his original findings or did not come to the same conclusions,' explains Michael Balter, who interviewed Lahn, his critics and his supporters at the time. *Science*, the journal that published his papers, came under attack for including the more speculative portions of his work in the first place.

To be fair to him, Lahn was partly a victim of how science works these days. The big discoveries have been made, so researchers often have little choice but to drill down into small, specific areas within biology. To make a name, they need themselves and the world to believe that this little thing they're studying is significant. According to Martin Yuille, a molecular biologist at the University of Manchester, 'If you're going to do an experiment you have to be reductionist. You have to look for one of the factors that is associated with a phenomenon, and you're tempted inevitably to try to think of that factor as being a cause, even though you know it is actually an association. So you're kind of driven to it.

'It is all too easy to exaggerate the role of the one variant of a gene that you might identify as associated with a trait . . . But you need to be modest.'

In this case, the world had seen the chopsticks fairy tale play out for real. In hindsight, it seems obvious that just because a genetic change in the brain may be more common in certain geographical populations than in others, that's no basis for claiming that it could be responsible for the fortunes of entire regions. Gerhard Meisenberg from the *Mankind Quarterly* made the same assumption when I interviewed him – that the innate abilities of a country's people are what define its success, even if we don't yet have scientific evidence for it. It's an idea that has underpinned racist thought for centuries. It assumes groups fall into ranks based on immutable biological features. It has the scent of the multi-regional hypothesis, implying that nature has taken different tracks, that some of us are more 'highly evolved' than others. By any measure, the intellectual leap that Lahn took was an irresponsible one. But then some people had thought he was cocky, that he did things his own way.

When I contact Lahn, now a professor of genetics at the University of Chicago, it has been more than a decade since his controversial papers were published. In 2009, undeterred by his failure, he wrote a piece in another top-flight journal, *Nature*, calling for the scientific community to be morally prepared for the possibility that they might find differences between populations, and to therefore embrace 'group diversity' in the way that societies already cherish cultural diversity. He argued that 'biological egalitarianism' won't be viable for much longer, implying that not all population groups are actually equal. He tells me that he's still 'open to the possibility that there may be genetic differences in intelligence between modern populations, just like there may be genetic differences in other biological traits between modern populations such as bodily measures, pigmentation, disease susceptibility and dietary adaptation.'

His hypothesis hasn't changed, even though he has no more evidence to support it than he did before. Yet Lahn sticks firmly to the line that he is guided by science, wherever this may take him. 'Before there is data, these are just possibilities,' he says.

'My nose follows the scientific method and data, not politics. I am willing to let the chips of data fall where they may, as any self-respecting scientist should.'

*

New York-based sociologist Barbara Katz Rothman has written: 'Genetics isn't just a science; it's a way of thinking . . . In this way of thinking, the seed contains all it could be. It is pure potential.'

For psychology professor Eric Turkheimer, the assumption that propels race research today in all its various forms is this deterministic pathology. 'There are people out there who think in a serious way that they're going to link up gene effects, the things you see in brain scans, the things you see on IQ tests,' he tells me. They are looking for that elusive mechanism, that magic formula which will allow them to take the genomes of people from Europe, or Africa, or China or India, or anywhere else, and prove beyond a shadow of a doubt that one population group really is smarter than another. It's all there in our bodies just waiting to be discovered.

'It's a racist hypothesis,' he adds.

In 2015 sociologists Carson Byrd at the University of Louisville and Victor Ray at the University of Tennessee investigated the belief of white Americans in genetic determinism. Having studied responses to the General Social Survey, which is carried out every two years to provide a snapshot of public attitudes, Byrd tells me they found that 'whites see racial difference in more biologically deterministic terms for blacks.' Yet they tend to view their own behaviour as more socially determined. If a black person, for instance, happens to be less smart, the interpretation is that they were born this way, whereas a white person's smartness or lack thereof is seen more as a product of outside factors such as schooling and hard work. 'So they give people a bit more leeway if they're white,' he explains.

It was also interesting to Byrd that even though the General Social Survey found that white conservatives were a little more biologically deterministic than white liberals, people with this view on both sides of the political spectrum shared the belief that policy measures such as affirmative action are needed to improve the lot of black Americans. There's a slippery slope here, he warns. 'The slipperiness is that they believe that because it's genetic, they can't help themselves, that it's innate, that they're going to be in a worse social position because of their race.' In other words, they want society to help black people, not because they believe we're all equal underneath, but because they believe we're not.

'Before it was something in the "blood" and now it's in our genes,' Byrd tells me. What has remained the same over the centuries is the racial stereotyping of black Americans. Rather than being seen as social or structural in origin, which it is, black disadvantage is conveniently rendered in the new scientific language of the day, which today is genetics. 'A lot of people have become enamoured with the science . . . the mystique of things that could be embedded within our genes.'

Stephan Palmié, an anthropologist at the University of Chicago, has argued that even now 'much genomic research proceeds from assumptions it culls from ostensibly "scientific" constructions of the past . . . and eventually restates them in the form of tabulations of [gene] frequencies.' Nineteenth-century ideas about race that have gone out of fashion take on an almost magical quality when they're freshly rewritten in the language of modern genetics. Today there is technical jargon, charts and numbers. Suddenly the old ideas seem shinier and more plausible than they did a moment ago. Suggest to anyone that the entire course of human history might have been decided by a single gene and they'll probably laugh. But that's exactly what Bruce Lahn did suggest in the pages of one of the most important journals in the world. For a moment, it felt possible because it was *new* science.

The belief that races have natural genetic propensities runs deep. One modern stereotype, for instance, is that of superior Asian cognitive ability. Race researchers, including Richard Lynn and John Philippe Rushton, have looked at academic test results in the United States and speculated that the smartest people in the world must be the Chinese, Japanese and other east Asians. When intelligence researcher James Flynn investigated the claim for work he published in 1991, he found that in fact they had the same average IQ as white Americans. Nevertheless, Asian Americans tended to score significantly higher on SAT college admission tests. They were also more likely to end up in professional, managerial and technical jobs. The edge they had was therefore a cultural one – more supportive parents or a stronger work ethic, maybe – endowed by their upbringing. They simply tended on average to work harder.

To anyone who has grown up as a member of an ethnic minority anywhere, especially those of us economic migrants who were told that we would have to work twice as hard to achieve the same as white people, this will come as small surprise. Among middle-class Indians living in the United Kingdom (the group my parents belong to) the weight of cultural pressure has generally been on children to become doctors, pharmacists, lawyers and accountants. These are professions that tend to be well respected and well paid, with no shortage of job opportunities and straightforward entry once you have the right qualifications. They are reliable routes into middle-class society. Medicine carries such immense prestige bias among immigrants and their children that, according to the most recent data gathered by the British Medical Association, around a quarter of all British doctors are Asian or British Asian. This is not because Indians make better medics, of course, but because culture acts as a silent funnel. In the same way, women get channelled into caring professions such as nursing because this is what society expects. Culture moulds people, even subconsciously, for certain lives and careers.

We forget that these stereotypes can change over time. Asian Americans are today considered a model minority. Yet more than a century ago, European race scientists saw Asians as biologically inferior, somewhere between themselves and what they referred to as the lowest races. In 1882 the United States passed the Chinese Exclusion Act to ban Chinese immigrant labourers because they were seen as undesirable citizens. Now that Japan has been highly prosperous for decades and India, China and South Korea are fast on the rise with their own wealthy elites, the stereotypes have shifted the other way. As people and nations prosper, the racial prejudices move target. Just as they always have.

*

'Think about what happened to all the old racial stereotypes,' Eric Turkheimer asks me.

'A hundred years ago, people were quite convinced that Greek people had low IQs. You know, people from southern Europe? Whatever happened to that? Did somebody do a big scientific study and check those Greek genes? No, nobody ever did that. It's just that time went on, Greek people overcame the disadvantages they faced a hundred years ago, and now they're fine and nobody thinks about it any more. And that's the way these things proceed. All we can do is wait for the world to change and what seemed like hardwired differences melt away and human flexibility just overwhelms it.'

But the waiting is hard. And as we wait, it remains all too easy for researchers to allow their assumptions about the world to muddy the lens through which they study it, and for the research they then produce to impact or reinforce racial stereotypes.

In 2011, Satoshi Kanazawa in the Department of Management at the London School of Economics, who writes widely on evolutionary psychology, speculated that black women are considered physically less attractive than women of other races. 'What accounts for the markedly lower average level of physical attractiveness among black women?' he blogged in *Psychology Today*, racking his brain. 'Black women are on average much heavier than nonblack women . . . However, this is not the reason black women are less physically attractive than nonblack women. Black women have lower average level of physical attractiveness net of BMI [body mass index]. Nor can the race difference in intelligence (and the positive association between intelligence and physical attractiveness) account for the race difference in physical attractiveness among women,' he continued, in the manner of a drunk uncle.

At a stroke, Kanazawa took it as a scientific given that black women are both less attractive, which is obviously a value judgement, and innately less intelligent, which is unproven. Presenting these two offensive statements unchallenged, he landed on the speculative conclusion that their unattractiveness, as he had now established it, might have something to do with their different 'levels of testosterone' from other women, again unproven. Kanazawa, whose published work has since looked at intelligence and homosexuality among other things, had his online post promptly pulled down under the weight of public and academic outrage. The London School of Economics banned him from publishing any more non-peer reviewed articles or blog posts for a year.

But how did it get published at all? When Kanazawa invoked race as a factor to explain why he perceived some women to be more attractive than others, he was performing a sleight of hand. He was diverting attention away from the underlying question of where his assumptions came from, or why he was asking this particular question. In so doing he shone the spotlight straight onto his racist conclusion. As soon as we, the audience, accepted his assumptions, they quickly transformed into a scientific question. For him, it was as legitimate as asking why apples fall down and not up, or why the sky is blue. Diverted, the publisher of his work failed to notice that his hypothesis was dripping with prejudice. It had no rigour to it at all.

American sociologist Karen Fields has compared use of the idea of race, like this, to witchcraft – it's the phenomenon that she calls 'racecraft'. Race is commonly described by scientists, politicians and race scholars as a social construct, as having no basis in biology. It's as biologically real as witches on broomsticks. And yet, writes Fields, she sees the same 'circular reasoning, prevalence of confirming rituals, barriers to disconfirming factual evidence, self-fulfilling prophecies . . .' as folk belief and superstition. It almost doesn't matter what anyone says because race feels as real to us as magic feels real to those who believe in it. It has been made real by overuse.

When Bruce Lahn, just four years after he was forced to retreat from his flawed research on intelligence genes, asked the scientific community to embrace 'group diversity', exactly what was he asking them to embrace? As he admitted to me himself, we don't yet have the data to tell us what the differences between populations are beyond the superficial, and even these superficial variations show enormous overlap. The chips haven't yet fallen. His plea is not for us to accept the science we have, but to accept in advance something we don't yet know. He is assuming that data will eventually confirm what he suspects, that there are cognitive differences between groups – and telling us to take his word for it. But how scientific is that? How close is it to being simply belief?

'I do science as if the truth mattered and your feelings about it didn't,' Satoshi Kanazawa states on his personal website, lacking ostensible remorse for his paper on black women. In 2018 he and a colleague at Westminster International University in Tashkent, Uzbekistan, published a paper in the *Journal of Biosocial Science*, produced by Cambridge University Press, asking why societies with 'higher average cognitive ability' have lower income inequality. Again, he started with the unproven assumption that populations have different cognitive abilities. Again, the editors failed to notice.

Among the very few researchers to have written on links between race, intelligence and the wealth of nations are Gerhard Meisenberg, Richard Lynn and Tatu Vanhanen, all intimately associated with the *Mankind Quarterly*. Together, they have claimed that Africans have an average IQ of about 70. When Dutch psychologist Jelte Wicherts investigated this figure, he found they could have arrived at it only by deliberately excluding the vast majority of data that actually shows African IQs to be higher. 'Lynn and Meisenberg's unsystematic methods are questionable and their results untrustworthy,' he concluded. Even so, Kanazawa cites heavily from their work in his own.

It's a problem that continues outside ivory towers and marginal journals. In 2013 a public policy researcher at a powerful conservative think tank, the Heritage Foundation in Washington, DC, was forced to resign after it was revealed that he had written a doctoral thesis while at Harvard University claiming that the average IQ of immigrants into the United States was lower than that of white Americans. Jason Richwine expressed the possibility that 'Hispanics' might never 'reach IQ parity with whites', ignoring that nobody considers 'Hispanics' a single genetic population group since they have such diverse ancestries. Most Argentinians, for instance, are of European ancestry, just like white Americans. Having created the illusion that Hispanics are a distinct biological race, Richwine made it real. He had performed a sleight of hand.

From this, he followed up with the suggestion that immigration policy should focus on attracting more intelligent people. Upon joining the think tank, he also happened to co-write a study which claimed that legalising the status of illegal immigrants, most of whom are Mexican and Central American, would result in an economic loss of trillions of dollars.

By the summer of 2018, as the result of a crackdown on illegal immigrants at the southern US border by the Trump administration, thousands of young children would be inhumanely separated from their parents by border patrol officers and held in metal cages in a warehouse in Texas. News reports described them sleeping under foil sheets, wailing in distress. When the children were finally returned, families feared the long-term psychological repercussions of their traumatic detainment. In January the same year, during a closed meeting on new immigration proposals held in the Oval Office, President Donald Trump had reportedly asked lawmakers, 'Why do we want all these people from shithole countries coming here?' He was referring to people from Haiti, El Salvador and Africa. He is believed to have added that the United States should be welcoming more immigrants from countries such as Norway.

✲

The notion that there are essential differences between population groups, that genetically 'shit' people come from 'shithole countries', may be an old one. But the science of inheritance helped propel these racially charged assumptions into modern intellectual thought. It is the concept of genetic determinism that has made some succumb to the illusion that every one of us has a racial destiny.

In reality, as science has advanced, it has only become clearer that human biology doesn't work this way. 'We can't sidestep the fundamental problem that biological systems are *systems*, they are collections of organisations of matter that interact with each other and each of their environments,' explains biologist Martin Yuille.

Evelynn Hammonds, historian of science at Harvard University, agrees that society has too long had a tendency to jump to the biological, to believe that our differences must be innate because how else do we explain them? 'When Jesse Owens won the medals in the 1936 Olympics, some people argued that he was not a full-blooded Negro, that he was actually mixed, and so they measured his whole body. That was 1936. They measured his whole body and made arguments like this: the thighbone is within the normal range of what the normal Negroid thighbone should be,' she says. 'That's always the hot button question at the end of the day. These people are different from each other, they're fundamentally different from each other, in disease, in athletic capacity and fundamentally in intelligence. That's the narrative. It acts as a kind of animating force under the surface.'

We can't help it. We keep looking back to race because of its familiarity. For so long, it has been the backdrop to our lives, the running narrative. We automatically translate the information our eyes and ears receive into the language of race, forgetting where that language came from. 'I think that scientists, they are trapped by the categories they use. They will either have to jettison it or find different ways of talking about this,' says Hammonds. 'They'll have to come to terms with that it has a social meaning.' This doesn't mean that racial categories shouldn't be used in medicine or in science more generally. But it does mean that those who use them should fully understand their significance, be able to define them, and know their history. They should at least know what race means.

For research published in 2007, anthropologist Duana Fullwiley, then at the Harvard School of Public Health, spent six months watching medical researchers in laboratories in California. Their job was to find genetic differences in how people responded to drugs. It was a fairly young, diverse, international team, not at all stuffy or old-fashioned. And she noticed all the scientists were routinely using racial categories not only to select their subjects, but to confidently pick out statistical differences between these racial groups. So, as Fullwiley observed, she asked each scientist she interviewed one simple question: 'How would you define race?'

Not one of them could answer her question confidently or clearly. The interviews were punctuated by long, awkward pauses and shy, embarrassed laughs. When pushed, some admitted that the concept of race made little sense, that the hard and fixed census categories actually didn't mean very much. One said, 'you can only judge race to a certain degree of confidence.' Another hesitated before admitting, 'I need to think more about it.'

Fullwiley concluded that most of the researchers 'were unsure of the meanings of the race categories that they used, yet they continue to assert that there is a biological basis to them, which they will soon corroborate'. Race was their bread and butter, the entire premise upon which they were doing their research, but they were unable to tell her what it was. Their work instead seemed to rest upon a hope that if they just persisted, they would eventually come to find scientific meaning in these categories. What they couldn't yet define would then be defined.

They appeared to believe that with enough data, with enough human guinea pigs, with enough science, they could take race – this imaginary, arbitrary set of categories invented by the powerful to control the weak – and somehow make it real.

Afterword

BARRY MEHLER'S WORDS ring in my ears.

'I have a lot of relatives who survived the Holocaust,' the historian told me. 'They are prepared for things to cease to be normal very quickly because that was their experience.'

I never imagined I might live through times that could make me feel this way, that could leave me dangling on a precipice afraid for my future. Politics is moving at such breakneck speed, taking such random turns, anything seems possible, and the worst of things feels likely. It's the suddenness of it all that makes it so strange. The cancerous surge in nationalism and racism around the world has taken many of us by surprise. I grew up in south-east London, not very far from where black teenager Stephen Lawrence was killed by racist thugs in 1993 while waiting for a bus. His murder left a mark on my generation. When we campaigned against racism, we knew there was a long way to go, but we were hopeful. And for a brief, sunlit moment the world really did seem to be changing. My son was born five years ago, when Barack Obama was President of the United States, and I dreamed then that he might grow up in a better society, perhaps even a post-racial one.

Things ceased to be normal very quickly. In the space of just a few years, far-right and anti-immigrant groups have become visible and powerful across Europe and the United States. In

Poland, nationalists march under the slogan 'Pure Poland, white Poland'. In Italy a right-wing leader rises to popularity on the promise to deport illegal immigrants and turn his back on refugees. White nationalists look to Russia under Vladimir Putin as a defender of 'traditional' values. In German elections in 2017 Alternative für Deutschland wins more than 12 per cent of the vote. Steve Bannon, the former chief strategist to Donald Trump, tells far-right nationalists in France in 2018, 'Let them call you racist, let them call you xenophobes, let them call you nativists. Wear it as a badge of honor.'

While it may be easy to blame white supremacists for this cancer, it's a brand of identity politics that has others in its grip, too. It's infecting people everywhere: Islamic fundamentalists in the Middle East and Pakistan, Hindu nationalists in India, Chinese scholars who turn their back on good science in favour of a worldview that paints the Chinese as having different evolutionary roots from everyone else. They may have different ideas and different histories, but their goal is the same: to assert difference for political gain. This is a twisted ideology that deliberately makes no appeal to a shared humanity, but instead rests on shadowy myths of belonging, on origin stories offering an umbrella to some but not others, sheltering them with false comfort. What nationalism stresses, as the late political scientist Ernst B. Haas wrote, is 'the individual's search for identity with strangers in an impersonal world'.

That desire to belong is powerful, I know. I was raised between cultures, and there's nothing quite so disruptive to your sense of belonging as not fully belonging anywhere, as being brown when everyone else is white in a place that notices these things. But don't be fooled. When they play on those feelings, when they tell you they can return you to a glorious past, offer a community of people just like you, who share your values and your dreams, a common history, they are selling you a myth. Enjoy your culture or religion, have pride in where you live or where your ancestors came from if you like, but don't imagine that these things give you any biological claim. Wear your identity lightly. Don't be sucked into believing that you are so different from others, that your rights have more value, that your blood is a different colour. There is no authenticity except that of personal experience.

The 'race realists', as they call themselves (perhaps because calling yourself a racist is still for now unpalatable even to most racists), work so hard to make the opposite case. They appeal to that dark corner of our souls that wants to believe human difference runs deep, making entire populations special, giving some nations an edge over others. And sadly, this is their moment. Whenever ugly politics become dominant, you can be sure that there are intellectuals and pseudo-intellectuals ready to jump on board. Those with dangerous ideas about 'human nature' and even more dangerous prescriptions for our problems are content to bide their time, knowing that the pendulum will swing their way eventually. Intellectual racism has always existed, and indeed for a chunk of history, it thrived. I believe it is still the toxic little seed at the heart of academia. However dead you might think it is, it needs only a little water, and now it's raining.

That said, what they're doing is also intellectually doomed. I've learned while writing this book that trying to force a biological understanding of race fails, often spectacularly, for the simple reason that it's history that has the answers. Science can't help you here. But then, perhaps the 'race realists' know this. Maybe they know that if we truly want an end to racism, we need to understand the past, to have more equitable education and healthcare, to end discrimination in work and institutions, to be a little more open with our hearts and maybe also with our borders. Maybe they know that the answers are not in our blood, but in us. They are in our actions, in the choices we make, and in the ways in which we treat each other. Maybe their insistent banging of the drum, their increasing violence and anger, is simply to mask the fact that they don't want to make these concessions.

There are plenty of ignorant racists, but the problem is not ignorance alone. The problem is that, even when people know the facts, not all of them actually want an end to racial inequality or even to the idea of race. Some would rather things stayed the way they are, or went backwards. And this means that those committed to the biological reality of race won't back down if the data proves them wrong. There's no incentive for them to admit intellectual defeat. They will simply keep reaching for fresher, more elaborate theories when the old ones fail. If skin colour can't explain racial inequality, then maybe the structure of our brains and bodies will. If not anatomy, then maybe our genes. When then this, too, throws up nothing of value, they will reach for the next thing. All this intellectual jumping through hoops to maintain the status quo. All this to prove what they have always really wanted to know: that they are superior.

Well, keep reaching, keep reaching, keep reaching. One day there will be nothing left to reach for.

Acknowledgements

THIS IS THE BOOK I have wanted to write since I was ten years old, and I have poured my soul into it. I'm grateful that my editors at Fourth Estate and Beacon Press, Louise Haines and Amy Caldwell, didn't hesitate in commissioning it, and that my publicists Michelle Kane and Caitlin Meyer have been my champions throughout. My agents Peter Tallack, Tisse Takagi and Louisa Pritchard were equally supportive. I am very fortunate to have such a loyal, kind-hearted team around me.

I would also like to deeply thank Jon Marks, Eric Turkheimer, Bill Tucker, Jay Kaufman, Subhadra Das, Jennifer Raff, Greg Radick and Billy Griffiths for their generous assistance. My friend, the archaeologist Tim Power, guided me through the British Museum and helped me see the past from a different perspective. I would also like to thank my sister Rima, herself a scholar of race and politics, for her critical feedback, my parents-in-law Neena and Pammi for their kind help at home, and my husband Mukul for his love and patience. When we started dating, Mukul was a fan of the band Fun-Da-Mental and he would often recite the two lines of lyrics I included at the beginning of this book.

In my business, there are real friends – the ones who care – and there's everyone else. My deepest gratitude is reserved for Peter Wrobel, a true friend who, out of the goodness of his heart, scoured the entire manuscript for errors.

This is the second book I have written since my son, Aneurin, was born, and he has made writing both a pleasure. He lights up my days. I don't know what the future holds for him, but I hope he never has to face the struggles that his parents or grandparents did. I hope he understands that how we look, our ancestry, where we come from or where we live are not the most important markers of identity. What makes us who we are is the content of our characters, as Martin Luther King reminded us. And this is entirely in our hands.

Don't forget that, my baby.

References

Prologue

QUOTE AT TOP: Song lyrics by Dennis Webb taken from 'English Breakfast', *Seize the Time* album by Fun-Da-Mental. With deepest thanks to Aki Nawaz and Dennis Webb for their kind permission to reproduce the words.

Factual details about the objects in the British Museum are taken from their labels, with guidance from archaeologist Tim Power.

Bartlett, John. '"Stolen friend": Rapa Nui Seek Return of Moai Statue'. BBC News, 18 November 2018, https://www.bbc.co.uk/news/world-latin-america-46222276 (accessed 19 November 2018)

Blumenbach, Johann Friedrich. *De generis humani varietate native*. Gottingae: Vandenhoek et Ruprecht, 1795

Freedman, Bernard J. 'Caucasian'. *British Medical Journal* (Clinical Research Edition) 288, no. 6418 (3 March 1984), 696–8

Hoock, Holger. 'The British State and the Anglo-French Wars over Antiquities, 1798–1858'. *The Historical Journal* 50, no. 1 (March 2007), 49–72

'Petition: Cultural Genocide: U.S. Government Forces Egyptian Nubians to be Classified as White and Not Black', MoveOn.org, https://petitions.moveon.org/sign/justice-for-an-indigenous (accessed 19 November 2018)

'Sir Hans Sloane', British Museum, https://www.britishmuseum.org/about_us/the_museums_story/general_history/sir_hans_sloane.aspx (accessed 19 November 2018)
Tharoor, Shashi. *Inglorious Empire: What the British Did to India*. London: Hurst & Company, 2017

1 Deep Time

Adcock, Gregory J. et al. 'Mitochondrial DNA Sequences in Ancient Australians: Implications for Modern Human Origins'. *Proceedings of the National Academy of Sciences* 98, no. 2 (January 2001), 537–42
Allen, Harry. 'The Past in the Present? Archaeological Narratives and Aboriginal History', in *Long History, Deep Time: Deepening Histories of Place*, ed. Ann McGrath and Mary Anne Jebb. Australia: ANU Press, 2015, 176–202
Anderson, Kay and Perrin, Colin. '"The Miserablest People in the World": Race, Humanism and the Australian Aborigine'. *Australian Journal of Anthropology* 18, no. 1 (2007), 18–39
Anderson, Kay and Perrin, Colin. 'How Race Became Everything: Australia and Polygenism'. *Ethnic and Racial Studies* 31, no. 5 (2008), 962–90
Athreya, Sheela. 'Picking a Bone with Evolutionary Essentialism', *Anthropology News*, 18 September 2018, http://www.anthropology-news.org/index.php/2018/09/18/picking-a-bone-with-evolutionary-essentialism/ (accessed 15 October 2018)
'Bringing Them Home: Report of the National Inquiry into the Separation of Aboriginal and Torres Strait Islander Children from Their Families', Commonwealth of Australia, 1997
ChangZhu, Jin et al. 'The Homo Sapiens Cave Hominin site of Mulan Mountain, Jiangzhou District, Chongzuo, Guangxi with Emphasis on its Age'. *Chinese Science Bulletin* 54, no. 21 (November 2009), 3848
Clarkson, Chris et al. 'Human Occupation of Northern Australia by 65,000 Years Ago'. *Nature* 547 (20 July 2017), 306–10

'Colonial Frontier Massacres in Central and Eastern Australia 1788–1930', University of Newcastle, Australia, https://c21ch.newcastle.edu.au/colonialmassacres/timeline.php (accessed 7 September 2018)

Dannemann, Michael and Kelso, Janet. 'The Contribution of Neanderthals to Phenotypic Variation in Modern Humans'. *American Journal of Human Genetics* 101, no. 4 (5 October 2017), 578–89

Gibbons, Ann. 'Who Were the Denisovans?' *Science* 333, no. 6046 (26 August 2011), 1084–7

Gosden, Chris. 'Race and Racism in Archaeology: Introduction'. *World Archaeology* 38, no. 1 (2006), 1–7

Gregory, Michael D. et al. 'Neanderthal-Derived Genetic Variation Shapes Modern Human Cranium and Brain'. *Scientific Reports* 7, no. 6308 (December 2017)

Griffiths, Billy. *Deep Time Dreaming*. Carleton, Victoria: Black Inc. Books, 2018

Gunn, R. G. 'Mulka's Cave Aboriginal rock art site: its context and content'. *Records of the Western Australian Museum* 23, 2006, 19–41

Hublin, Jean-Jacques et al. 'New fossils from Jebel Irhoud, Morocco and the Pan-African origin of *Homo sapiens*'. *Nature* 546 (8 June 2017), 289–92

Ingold, Tim. 'Beyond Biology and Culture: The Meaning of Evolution in a Relational World'. *Social Anthropology* 12, no. 2 (June 2004), 209–21

Kaplan, Sarah. 'Humans Didn't Outsmart the Neanderthals. We Just Outlasted Them', *Washington Post*, 1 November 2017. www.washingtonpost.com/news/speaking-of-science/wp/2017/11/01/humans-didnt-outsmart-the-neanderthals-we-just-outlasted-them/?utm_term=.4a78010999de (accessed 16 October 2018)

Karakostis, Fotios Alexandros et al. 'Evidence for Precision Grasping in Neandertal Daily Activities'. *Science Advances* 4, no. 9 (26 September 2018)

Kolbert, Elizabeth, 'Our Neanderthals, Ourselves', *New Yorker*, 12 February 2015, https://www.newyorker.com/news/daily-comment/neanderthals (accessed 16 October 2018)

Krause, Johannes et al. 'The Complete Mitochondrial DNA Genome of an Unknown Hominin from Southern Siberia'. *Nature* 464 (8 April 2010), 894–7

Marris, Emma. 'News: Neanderthal Artists Made Oldest-known Cave Paintings'. *Nature* (22 February 2018)

Milks, Annemieke. 'We Haven't Been Giving Neanderthals Enough Credit.' *Popular Science*, 26 June 2018, www.popsci.com/neanderthal-hunting-spears (accessed 16 October 2018)

Mills, Charles W. *The Racial Contract*. Ithaca, NY: Cornell University Press, 1997

Mooallem, Jon, 'Neanderthals Were People, Too'. *New York Times*, 11 January 2017, https://www.nytimes.com/2017/01/11/magazine/neanderthals-were-people-too.html (accessed 16 October 2018)

'Neanderthals Thought Like We Do', Max-Planck-Gesellschaft, 22 February 2018, https://www.mpg.de/11948095/neandertals-cave-art (accessed 11 October 2018)

'Neanderthals Were Too Smart for Their Own Good'. *The Telegraph*, 18 November 2011, www.telegraph.co.uk/news/science/science-news/8898321/Neanderthals-were-too-smart-for-their-own-good.html (accessed 16 October 2018)

Noonuccal, Oodgeroo. 'Stone Age'. In *The Dawn Is at Hand*, London: Marion Boyars, 1992

Parsons, Meg. 'Creating a Hygienic Dorm: The Refashioning of Aboriginal Women and Children and the Politics of Racial Classification in Queensland 1920s–40s'. *Health and History* 14, no. 2 (2012), 112–39

Pascoe, Bruce. *Dark Emu, Black Seeds: Agriculture or Accident*. Broome, Western Australia: Magabala Books, 2014

Porr, Martin. 'Essential Questions: Modern Humans and the Capacity for Modernity', in *Southern Asia, Australia and the Search for Human Origins*, ed. Robin Dennell and Martin Porr. Cambridge: Cambridge University Press, 2014, 257–64

Porr, Martin and Matthews, Jacqueline M. 'Post-colonialism, Human Origins and the Paradox of Modernity. *Antiquity* 91, no. 358 (August 2017), 1058–68

Prüfer, Kay et al. 'The Complete Genome Sequence of a Neanderthal from the Altai Mountains'. *Nature* 505 (2 January 2014), 43–9

Qiu, Jane. 'The Forgotten Continent'. *Nature* 535 (14 July 2016), 218–20

Quach, Hélène et al. 'Genetic Adaptation and Neandertal Admixture Shaped the Immune System of Human Populations'. *Cell* 167, no. 3 (20 October 2016), 643–56

Scerri, Eleanor M.L. et al. 'Did Our Species Evolve in Subdivided Populations across Africa, and Why Does It Matter?' *Trends in Ecology & Evolution* 33, no. 8 (August 2018), 582–94

Thorne, Alan G. and Wolpoff, Milford H. 'Regional Continuity in Australasian Pleistocene Hominid Evolution'. *American Journal of Physical Anthropology* 55, no. 3 (July 1981), 337–49

Thorne, Alan G. and Wolpoff, Milford H. 'The Multiregional Evolution of Humans'. *Scientific American* 266, no. 4 (April 1992), 76–83

van der Kaars, Sander et al. 'Humans Rather than Climate the Primary Cause of Pleistocene Megafaunal Extinction in Australia'. *Nature Communications* 8, no. 14142 (20 January 2017)

Webb, Steve. *Made in Africa: Hominin Explorations and the Australian Skeletal Evidence*. Cambridge, MA: Academic Press, 2018

Wenban-Smith, Frances, 'Neanderthals Were No Brutes – Research Reveals they May Have Been Precision Workers', *The Conversation*, 26 September 2018, https://theconversation.com/neanderthals-were-no-brutes-research-reveals-they-may-have-been-precision-workers-103858?utm_medium=Social&utm_source=Twitter#Echobox=1538043704 (accessed 11 October 2018)

2 It's a Small World

'The African-American Mosaic', Library of Congress, http://www.loc.gov/exhibits/african/afam002.html (accessed 12 October 2017)

Berry, Daina Ramey. *The Price for Their Pound of Flesh: The Value of the Enslaved, from Womb to Grave, in the Building of a Nation*. Boston, MA: Beacon Press, 2017

Cartwright, Samuel A. 'Diseases and Peculiarities of the Negro Race'. *De Bow's Review*, Southern and Western States 11, New Orleans, 1851

Desmond, Adrian and Moore, James. *Darwin's Sacred Cause: How a Hatred of Slavery Shaped Darwin's Views on Human Evolution*. New York: Houghton Mifflin Harcourt, 2009

Douglas, Bronwen. 'Climate to Crania: Science and the Racialization of Human Difference', in *Foreign Bodies: Oceania and the Science of Race 1750–1940*, ed. Bronwen Douglas and Chris Ballard. Canberra: ANU Press, 2008

Fields, Karen E. and Fields, Barbara J. *Racecraft: The Soul of Inequality in American Life*, Reprint edn. London and New York: Verso Books, 2014

Geertz, Clifford. *The Interpretation of Cultures*. New York: Basic Books, 1973

Gobineau, Arthur, comte de. *The Inequality of Human Races*. 1853; English translation by Adrian Collins. London: William Heinemann, 1915

Gould, Stephen Jay. *The Mismeasure of Man*. New York: W. W. Norton, 1981

'The Hottentot Venus Is Going Home'. *Journal of Blacks in Higher Education*, no. 35 (Spring 2002), 63

Huxley, Thomas Henry. *Collected Essays: Volume 3, Science and Education*. London: Macmillan, 1893

Keller, Mitch. 'The Scandal at the Zoo'. *New York Times*, 6 August 2006

Lewis, Bernard. 'The Historical Roots of Racism'. *The American Scholar* 67, no. 1 (Winter 1998), 17–25
Malik, Kenan. *Strange Fruit: Why Both Sides are Wrong in the Race Debate*. London: Oneworld, 2008
Marks, Jonathan. *Is Science Racist?* Cambridge: Polity Press, 2017
O'Brien, Conor Cruise. 'Thomas Jefferson: Radical and Racist'. *The Atlantic*, October 1996
Radick, Gregory. 'Darwin and Humans', in *The Cambridge Encyclopedia of Darwin and Evolutionary Thought*, ed. Michael Ruse. Cambridge: Cambridge University Press, 2013, 173–81
Radick, Gregory. 'How and Why Darwin Got Emotional about Race', in *Historicizing Humans: Deep Time, Evolution and Race in Nineteenth Century British Sciences*, ed. Efram Sera-Shriar. Pittsburgh: University of Pittsburgh Press, 2018
Solly, S., Moojen. Geo and Lindfors, Bernth. 'Courting the Hottentot Venus'. *Africa: Rivista trimestrale di studi e documentazione dell'Istituto italiano per l'Africa e l'Oriente*. Vol. 40, no. 1 (March 1985), 133–48
Swanton, John R. 'Review: The Inequality of Human Races by Arthur De Gobineau and Adrian Collins'. *American Anthropologist*, New Series 18, no. 3 (July–September 1916), 429–31
Zeitoun, Charline. 'In the Days of Human Zoos'. CNRS, 22 November 2016, https://news.cnrs.fr/articles/in-the-days-of-human-zoos (accessed 9 October 2017)

3 Scientific Priestcraft

A. B. S. 'Reviewed Work: The Passing of the Great Race: Or the Racial Basis of European History by Madison Grant'. *American Historical Review* 22, no. 4 (July 1917), 842–4
Black, Edwin. *IBM and the Holocaust: The Strategic Alliance between Nazi Germany and America's Most Powerful Corporation*. New York: Crown, 2001

Boulter, Michael. 'The Rise of Eugenics, 1901–14', in *Bloomsbury Scientists: Science and Art in the Wake of Darwin*. London: UCL Press, 2017, 102–14
Burke, Jason and Oltermann, Philip. 'Germany moves to atone for "forgotten genocide" in Namibia'. *The Guardian*, 25 December 2016
Chung, Yuehtsen Juliette. 'Better Science and Better Race?: Social Darwinism and Chinese Eugenics'. *Isis* 105, no. 4 (December 2014), 793–802
Durant, John R. 'Scientific Naturalism and Social Reform in the Thought of Alfred Russel Wallace'. *British Journal for the History of Science* 12, no. 1 (March 1979), 31–58
'The Eugenics Record Office', *Science* 37, no. 954 (11 April 1913), 553–4
'First International Eugenics Congress, London, July 24th to July 30th, 1912, University of London, South Kensington: Programme and time table', https://archive.org/stream/b22439833/b22439833_djvu.txt (accessed 12 January 2018)
Gabriel, Elliott. 'National Hygiene and "Inferior Offspring": Japan's Eugenics Victims Demand Justice'. *Mint Press News*, 19 March 2018, https://www.mintpressnews.com/inferior-offspring-japans-eugenics-victims-demand-justice/239078/ (accessed 26 March 2018)
Galton, Francis. 'Hereditary Character and Talent'. *Macmillan's Magazine* 12 (1865), 157–66
Grant, Madison. *The Passing of the Great Race: Or the Racial Basis of European History*. New York: Charles Scribner's Sons, 1916
Hall, Prescott F. 'Selection of Immigration'. *Annals of the American Academy of Political and Social Science* 24 (July 1904), 169–84
Hawes, James. *The Shortest History of Germany*. London: Old Street, 2017
Heim, Susanne, Sachse, Carola and Walker, Mark. *The Kaiser Wilhelm Society under National Socialism*. Cambridge: Cambridge University Press, 2009

'History of the Kaiser Wilhelm Society under National Socialism', Max Planck Society, https://www.mpg.de/9811513/kws-under-national-socialism (accessed 22 October 2017)

'International Eugenics Congress', *The Scientific Monthly* 12, no. 4 (April 1921), 383–4

Johnson, Corey. 'California Was Sterilising its Female Prisoners as Late as 2010'. *The Guardian*, 8 November 2013

Johnson, Roswell H. 'Eugenics and So-Called Eugenics'. *American Journal of Sociology* 20, no. 1 (July 1914), 98–103

Kohn, Marek. *The Race Gallery: The Return of Racial Science*. London: Jonathan Cape, 1995

Lee, Jonathan H. X. *History of Asian Americans: Exploring Diverse Roots*. Santa Barbara, CA: Greenwood, 2015

Lewontin, Richard C. 'Biological Determinism'. *Tanner Lectures on Human Values* 4 (1983), 147–83

Mukherjee, Siddhartha. *The Gene: An Intimate History*. New York: Scribner, 2016

Müller-Hill, Benno. 'The Blood from Auschwitz and the Silence of the Scholars'. *History and Philosophy of the Life Sciences* 21, no. 3 (1999), 331–65

Priemel, Kim Christian. 'Review: The Kaiser Wilhelm Society under National Socialism by Susanne Heim, Carola Sachse and Mark Walker'. *Journal of Modern History* 83, no. 1 (March 2011), 216–18

Schaffer, Gavin. '"Like a Baby with a Box of Matches": British Scientists and the Concept of "Race" in the Inter-War Period'. *British Journal for the History of Science* 38, no. 3 (September 2005), 307–24

Sussman, Robert Wald. *The Myth of Race: The Troubling Persistence of an Unscientific Idea*. Cambridge, MA and London: Harvard University Press, 2014

Tuffs, Annette. 'German Research Society Apologises to Victims of Nazis'. *British Medical Journal* (16 June 2001), 322

Weigmann, Katrin. 'In the Name of Science'. *EMBO Reports* 2, no. 10 (2001), 871–5

Whitman, James Q. *Hitler's American Model: The United States and the Making of Nazi Race Law*. Princeton, NJ and Oxford: Princeton University Press, 2017

4 Inside the Fold

With thanks to King's College London for access to their special collections.

Barash, David. 'Review: Race, Evolution and Behavior'. *Animal Behaviour* 49, no. 4 (April 1995), 1131–3

Campbell, Michael C. and Tishkoff, Sarah A. 'African Genetic Diversity: Implications for Human Demographic History, Modern Human Origins, and Complex Disease Mapping'. *Annual Review of Genomics and Human Genetics* 9 (2008), 403–33

Dutton, Edward. 'Obituary: Tatu Vanhanen 1929–2015'. *Mankind Quarterly* 56, no. 2 (2015), 225–32

Edwards, A. W. F. 'Human Genetic Diversity: Lewontin's Fallacy'. *Bioessays* 25, no. 8 (August 2003), 798–801

Garrett, Henry E. 'Klineberg's Chapter on Race and Psychology'. *Mankind Quarterly* (July 1960), 15–22

Garrett, Henry E. 'The Equalitarian Dogma'. *Mankind Quarterly* (April 1961), 253–7

Gayre, R. 'The Dilemma of Inter-Racial Relations'. *Mankind Quarterly* 6, no. 4 (April–June 1966)

Genoves, Santiago. 'Racism and "The Mankind Quarterly"'. *Science* 134, no. 3493 (8 December 1961), 1928–32

Harrison, Geoffrey Ainsworth. 'Reviewed Work: The Mankind Quarterly by R. Gayre'. *Man* 61 (September 1961), 163–4

International Consortium of Investigative Journalists Offshore Leaks Database, https://offshoreleaks.icij.org/nodes/21000166 (accessed 19 December 2017)

Kelsey, Tim. 'Ulster University Took Grant from Fund Backing Whites'. *The Independent*, 9 January 1994

Kunzelman, Michael. 'University Accepted $458K from Eugenics

Fund'. Associated Press, 25 August 2018, https://apnews.com/a9791e6174374437b3bbe17af8b76215 (accessed 27 August 2018)

Lewontin, Richard Charles. 'The Apportionment of Human Diversity'. *Evolutionary Biology* 6 (1972) 381–98

Lipphardt, Veronika. 'From "Races" to "Isolates" and "Endogamous Communities": Human Genetics and the Notion of Human Diversity in the 1950s', in *Human Heredity in the Twentieth Century*, ed. Bernd Gausemeier, Staffan Müller-Wille and Edmund Ramsden. London: Pickering & Chatto, 2013, 55–68

Miller, Adam. 'The Pioneer Fund: Bankrolling the Professors of Hate'. *Journal of Blacks in Higher Education*, no. 6 (Winter 1994–5), 58–61 (a version of this article originally appeared in the *Los Angeles Times* in 1994)

Montagu, Ashley. *Man's Most Dangerous Myth: The Fallacy of Race*. New York: Columbia University Press, 1942

Montagu, Ashley. 'The Genetical Theory of Race, and Anthropological Method'. *American Anthropologist* 44, no. 3 (July–September 1942)

Radick, Gregory. 'Beyond the "Mendel-Fisher Controversy"'. *Science* 350, no. 6257 (9 October 2015), 159–60

Radick, Gregory. 'Presidential Address: Experimenting with the Scientific Past'. *British Journal for the History of Science* 49, no. 2 (June 2016), 153–72

'Reginald Ruggles Gates Collection', King's College London, https://www.kcl.ac.uk/library/archivespec/special-collections/Individualcollections/rugglesgates.aspx (accessed 2 November 2017)

Rosenberg, Noah A. et al. 'Genetic Structure of Human Populations'. *Science* 298, no. 5602 (20 December 2002), 2381–5

Schaffer, Gavin. '"Scientific" Racism Again?: Reginald Gates, the "Mankind Quarterly" and the Question of "Race" in Science after the Second World War'. *Journal of American Studies* 41, no. 2 (August 2007), 253–78

Selcer, Perrin. 'Beyond the Cephalic Index: Negotiating Politics to Produce UNESCO's Scientific Statements on Race'. *Current Anthropology* 53, Supplement 5 (April 2012)

Silverman, Rachel. 'The Blood Group "Fad" in Post-War Racial Anthropology'. *Kroeber Anthropological Society Papers* 84 (2000), 11–27

Singh, Indera P. and Singh, Darshan. 'The Study of ABO Blood Groups of Sainis of Punjab'. *American Journal of Physical Anthropology* 19, no. 3 (September 1961), 223–6

Škerlj, Božo. 'The Mankind Quarterly'. Man 60 (November 1960), 172–3

Sussman, Robert Wald. *The Myth of Race: The Troubling Persistence of an Unscientific Idea*. Cambridge, MA and London: Harvard University Press, 2014

Tucker, William H. *The Funding of Scientific Racism: Wickliffe Draper and the Pioneer Fund*. Urbana: University of Illinois Press, 2002

UNESCO, 'The Race Concept: Results of an Inquiry', in *The Race Question in Modern Science*. Paris: UNESCO, 1952

Witherspoon, D. J. et al. 'Genetic Similarities Within and Between Human Populations'. *Genetics* 176, no. 1 (May 2007), 351–9

5 Race Realists

Burleigh, Nina. 'Steve Bannon, Jared Taylor and the Radical Right's Ivy League Pedigree'. *Newsweek*, 23 March 2017, http://www.newsweek.com/bannon-spencer-trump-alt-right-breitbart-infowars-yale-gottfried-oathkeepers-572585 (accessed 16 November 2017)

'A Convocation of Bigots: The 1998 American Renaissance Conference'. *Journal of Blacks in Higher Education*, no. 21 (Autumn 1998), 120–4

'Intelligence Editorial Board', Elsevier, https://www.journals.elsevier.com/intelligence/editorial-board (accessed 10 November 2017 and 20 November 2018)

Jensen, Arthur. 'How Much Can We Boost IQ and Scholastic Achievement?' *Harvard Educational Review* 39, no. 1 (April 1969), 1–123

Kelsey, Tim, and Rowe, Trevor. 'Academics "were funded by racist American trust"'. *Independent on Sunday*, 4 March 1990

Lane, Charles. 'The Tainted Sources of "The Bell Curve"'. *New York Review of Books*, 1 December 1994

Levin, Michael. Letter to the Editor: 'Howard Beach Turns a Beam on Racial Tensions'. *New York Times*, 11 January 1987

Lynn, Richard. 'The Intelligence of American Jews'. *Personality and Individual Differences* 36, no. 1 (January 2004), 201–6

Lynn, Richard and Meisenberg, Gerhard. Review article: 'The Average IQ of Sub-Saharan Africans: Comments on Wicherts, Dolan, and van der Maas'. *Intelligence* 38, no. 1 (January–February 2010), 21–9

Mehler, Barry. 'The New Eugenics: Academic Racism in the U.S. Today'. *Science for the People* 15, no. 3 (May–June 1983), 18–23

Mehler, Barry. 'Rightist on the Rights Panel'. *The Nation* (7 May 1988), 640–1

'Michael Levin', Southern Poverty Law Center, https://www.splcenter.org/fighting-hate/extremist-files/individual/michael-levin (accessed 13 November 2017)

Murray, Charles A. and Herrnstein, Richard. *The Bell Curve: Intelligence and Class Structure in American Life*. 1994; New York: Free Press Paperbacks, 1996

Nelson, Louis. 'Clinton Ad Ties Trump to KKK, White Supremacists'. POLITICO, 25 August 2016, https://www.politico.com/story/2016/08/clinton-ad-kkk-trump-227404 (accessed 28 November 2017)

Pearson, Roger. 'Immigration into Britain'. *Northlander* 1, no. 1 (April 1958), 2

Rosenthal, Steven J. 'The Pioneer Fund: Financier of Fascist Research'. *American Behavioral Scientist* 39, no. 1 (September–October 1995), 44–61

Santiago, Frank. 'Rights Official Has Racial "Purity" Links'. *Des Moines Register* (28 February 1988)

Schulson, Michael. 'Race, Science, and Razib Khan'. Undark, 28 February 2017, https://undark.org/article/race-science-razib-khan-racism/ (accessed 20 November 2018)

Scott, Ralph. 'Arthur Jensen: A Latter-Day "Enemy of the People"?' *Mankind Quarterly* 53, no. 3–4 (2013)

Siegel, Eric. 'The Real Problem with Charles Murray and "The Bell Curve"'. *Scientific American*, https://blogs.scientificamerican.com/voices/the-real-problem-with-charles-murray-and-the-bell-curve/, 12 April 2017 (accessed 6 November 2017)

Stein, Joel. 'Milo Yiannopoulos is the Pretty, Monstrous Face of the Alt-Right'. Bloomberg, 15 September 2016, https://www.bloomberg.com/features/2016-america-divided/milo-yiannopoulos/ (accessed 20 November 2018)

Stolarski, Maciej, Zajenkowski, Marcin and Meisenberg, Gerhard. 'National Intelligence and Personality: Their Relationships and Impact on National Economic Success'. *Intelligence* 41, no. 2 (March–April 2013), 94–101

Sussman, Robert Wald. 'America's Virulent Racists: The Sick Ideas and Perverted "science" of the American Renaissance Foundation'. *Salon*, 11 October 2014, https://www.salon.com/2014/10/11/americas_virulent_racists_the_sick_ideas_and_perverted_science_of_the_american_renaissance_foundation/ (accessed 13 November 2017)

van der Merwe, Ben. 'Exposed: London's Eugenics Conference and its Neo-Nazi Links. *London Student*, 10 January 2018, http://londonstudent.coop/news/2018/01/10/exposed-london-eugenics-conferences-neo-nazi-links/amp/?__twitter_impression=true (accessed 15 January 2018)

Weyher, Harry F. 'The Pioneer Fund, the Behavioral Sciences, and the Media's False Stories'. *Intelligence* 26, no. 4 (1998), 319–36

6 Human Biodiversity

'Ancestry Names Margo Georgiadis Chief Executive Officer', Ancestry.com, 19 April 2018, https://www.ancestry.com/corporate/newsroom/press-releases/ancestry-names-margo-georgiadis-chief-executive-officer (accessed 20 November 2018)

Boodman, Eric. 'White Nationalists Are Flocking to Genetic Ancestry Tests. Some Don't Like what they Find'. *STAT News*, 16 August 2017, https://www.statnews.com/2017/08/16/white-nationalists-genetic-ancestry-test/ (accessed 7 March 2018)

Cavalli-Sforza, Luigi Luca. *Genes, Peoples, and Languages*. London: Penguin, 2001

Cavalli-Sforza, Luca. 'The Human Genome Diversity Project: An Address Delivered to a Special Meeting of UNESCO', Paris (12 September 1994)

Evans, Rob. 'Military Scientists Tested Mustard Gas on Indians'. *The Guardian*, 1 September 2007, https://www.theguardian.com/uk/2007/sep/01/india.military (accessed 3 August 2018)

Gannett, Lisa. 'Racism and Human Genome Diversity Research: The Ethical Limits of "Population Thinking"'. *Philosophy of Science* 68, no. 3 (2001), Supplement S479–S492

Gates, Henry Louis. *Finding Oprah's Roots: Finding Your Own*. New York: Crown, 2007

Kennett, Debbie A. et al. 'The Rise and Fall of BritainsDNA: A Tale of Misleading Claims, Media Manipulation and Threats to Academic Freedom'. *Genealogy* 2, no. 4 (2 November 2018)

Lipphardt, Veronika. '"Geographical Distribution Patterns of Various Genes": Genetic Studies of Human Variation after 1945'. *Studies in History and Philosophy of Biological and Biomedical Sciences* 47 (September 2014), 50–61

MacDougald, Park and Willick, Jason. 'The Man Who Invented Identity Politics for the New Right'. *New York*, 30 April 2017, http://nymag.com/daily/intelligencer/2017/04/steve-sailer-invented-identity-politics-for-the-alt-right.html (accessed 4 January 2018)

Malik, Kenan. *Strange Fruit: Why Both Sides Are Wrong in the Race Debate*. Oxford: Oneworld, 2008

Marks, Jonathan. *Human Biodiversity: Genes, Race, and History*. New York: Aldine de Gruyter, 1995

Murray, Charles A. and Herrnstein, Richard. *The Bell Curve: Intelligence and Class Structure in American Life*. 1994; New York: Free Press, 1996

Olson, Steve. 'The Genetic Archaeology of Race'. *The Atlantic*, April 2001, https://www.theatlantic.com/magazine/archive/2001/04/the-genetic-archaeology-of-race/302180/ (accessed 14 February 2018)

'Oprah Winfrey's Surprising DNA Test', Ancestry.com, https://blogs.ancestry.com/cm/the-surprising-facts-oprah-winfrey-learned-about-her-dna/ (accessed 7 March 2018)

Radin, Joanna. 'Human Genome Diversity Project: History', in *International Encyclopedia of the Social & Behavioral Sciences*, 2nd edn, vol. 11, ed. James D. Wright. Oxford: Elsevier, 2015, 306–10

Reardon, Jenny. *Race to the Finish: Identity and Governance in an Age of Genomics*. Princeton, NJ: Princeton University Press, 2004

Reardon, Jenny, 'Finding Oprah's Roots, Losing the World: Beyond the Liberal AntiRacist Genome'. Unpublished paper presented at Berkeley Workshop on Environmental Politics, 23 October 2009

Taubes, Gary. 'Scientists Attacked for "Patenting" Pacific Tribe'. *Science* 270, no. 5239 (17 November 1995), 1112

UNESCO IBC Working Group on Population Genetics. *Draft Report: Bioethics and Human Population Genetics Research*. Third Session of the IBC, 27–29 September 1995

Walker, Matthew. 'First, Do Harm'. *Nature* 482 (9 February 2012), 148–52

WHO Scientific Group on Research in Population Genetics of Primitive Groups, *Research in population genetics of primitive groups: Report of a WHO Scientific Group [meeting held in Geneva from 27 November to 3 December 1962]*. Geneva: World Health Organization, 1964

7 Roots

Allentoft, Morten E. et al. 'Population Genomics of Bronze Age Eurasia'. *Nature* 522, no. 7555 (June 2015), 167–72

Brace, Selina et al. 'Population Replacement in Early Neolithic Britain'. *bioRxiv*, 18 February 2018, https://www.biorxiv.org/content/early/2018/02/18/267443

Crawford, Nicholas G. et al. 'Loci associated with skin pigmentation identified in African populations'. *Science*, published online 12 October 2017

Devega, Chauncey. 'Cheddar Man is "black"! Another racial panic for white supremacists'. *Salon*, 12 February 2018, https://www.salon.com/amp/cheddar-man-is-black-another-racial-panic-for-white-supremacists (accessed 20 November 2018)

Hunt-Grubbe, Charlotte. 'The Elementary DNA of Dr Watson'. *Sunday Times*, 14 October 2007

Kristiansen, Kristian et al. 'Re-theorising Mobility and the Formation of Culture and Language among the Corded Ware Culture in Europe'. *Antiquity* 91, no. 356 (April 2017), 334–47

Long, Tegwen et al. 'Cannabis in Eurasia: Origin of Human Use and Bronze Age Trans-continental Connections'. *Vegetation History and Archaeobotany* 26, no. 2 (March 2017), 245–58

Mathieson, Iain et al. 'Genome-wide Patterns of Selection in 230 Ancient Eurasians'. *Nature* 528, no. 7583 (December 2015), 499–503

Mathieson, Iain et al. 'The Genomic History of Southeastern Europe. *Nature*, 21 February 2018, https://www.nature.com/articles/nature25778

Moorjani, Priya et al. 'Genetic Evidence for Recent Population Mixture in India'. *American Journal of Human Genetics* 93, no. 3 (September 2013), 422–38

'Morrissey: Big mouth strikes again', *New Musical Express*, December 2007

Narasimhan, Vagheesh M. et al. 'The Genomic Formation of South and Central Asia'. *bioRxiv*, 31 March 2018, https://www.biorxiv.org/content/early/2018/03/31/292581

'Nobelist Jim Watson is honored on his 90th birthday', Cold Spring Harbour Laboratory website, 11 April 2018, https://www.cshl.edu/nobelist-jim-watson-honored-90th-birthday/ (accessed 18 April 2018)

Olalde, Iñigo et al. 'Derived Immune and Ancestral Pigmentation Alleles in a 7,000-year-old Mesolithic European'. *Nature* 507, no. 7491 (13 March 2014), 225–8

Olalde, Iñigo et al. 'The Beaker Phenomenon and the Genomic Transformation of Northwest Europe'. *Nature* 555 (8 March 2018), 190–6

Pasha-Robinson, Lucy. 'One in Three Black, Asian or Minority Ethnic People Racially Abused Since Brexit, Study Reveals'. *The Independent*, 17 March 2017, http://www.independent.co.uk/news/uk/home-news/one-three-black-asian-minority-ethnic-bame-racism-abuse-assault-brexit-hate-crime-tuc-study-a7634231.html (accessed 6 March 2018)

Reade, Brian. 'Dark-skinned Cheddar Man is Hard Cheese for the Racist Morons of the far right'. *Mirror*, 10 February 2018, https://www.mirror.co.uk/news/uk-news/dark-skinned-cheddar-man-hard-11999683 (accessed 20 November 2018)

Reich, David. *Who We Are and How We Got Here*. Oxford: Oxford University Press, 2018

Rutherford, Adam. *A Brief History of Everyone Who Ever Lived: The Stories in Our Genes*. London: Weidenfeld and Nicolson, 2016

University College London press release, 'Face of first Brit revealed', 7 February 2018, https://www.ucl.ac.uk/news/newsarticles/0218/070218-Face-of-cheddar-man-revealed (accessed 10 March 2018)

Watson, James. *The Double Helix: A Personal Account of the Discovery of the Structure of* DNA. New York: Atheneum, 1968

8 Origin Stories

Arnold, Bettina. 'The Past as Propaganda: Totalitarian Archaeology in Nazi Germany'. *Antiquity* 64, No 244 (September 1990), 464–78

Arnold, Bettina. '"Arierdämmerung": Race and Archaeology in Nazi Germany'. *World Archaeology* 38, no. 1 (March 2006), 8–31

'Asifa Bano: The child rape and murder that has Kashmir on edge', BBC News website, 12 April 2018, http://www.bbc.co.uk/news/world-asia-india-43722714 (accessed 26 April 2018)

Bardill, Jessica et al. 'Advancing the Ethics of Paleogenomics'. *Science* 360, no. 6387 (27 April 2018), 384–5

Colavito, Jason, 'White Nationalists and the Solutrean Hypothesis', blog, 31 January 2014, http://www.jasoncolavito.com/blog/white-nationalists-and-the-solutrean-hypothesis (accessed 22 March 2018)

Crossland, David. 'Germany Recalls Myth That Created the Nation'. *Spiegel Online* (28 August 2009), http://www.spiegel.de/international/germany/battle-of-the-teutoburg-forest-germany-recalls-myth-that-created-the-nation-a-644913.html (accessed 28 April 2018)

'Editorial: Use and abuse of ancient DNA'. *Nature* 555 (March 2018), 559

Friese, Kai. '4500-year-old DNA from Rakhigarhi Reveals Evidence that Will Unsettle Hindutva Nationalists'. *India Today*, 31 August 2018, https://www.indiatoday.in/amp/magazine/cover-story/story/20180910-rakhigarhi-dna-study-findings-indus-valley-civilisation-1327247-2018-08-31 (accessed 5 September 2018)

Ghose, Tia. 'Ancient Kennewick Man Finally Laid to Rest'. *Live Science*, 22 February 2017, https://www.livescience.com/57977-kennewick-man-reburied.html (accessed 20 March 2018)

Gibbons, Ann. 'There's No Such Thing as a "Pure" European – or Anyone Else'. *Science* (15 May 2017)

Green, Sara Jean. '"A Wrong had Finally Been Righted": Tribes Bury Remains of Ancient Ancestor Known as Kennewick Man'. *Seattle Times*, 19 February 2017

Harmon, Amy. 'Indian Tribe Wins Fight to Limit Research of its DNA'. *New York Times*, 21 April 2010

Harmon, Amy. 'Geneticists See Work Distorted for Racist Ends'. *New York Times*, 18 October 2018

Jain, Rupam and Lasseter, Tom. 'Special Report – By Rewriting History, Hindu Nationalists Aim to Assert their Dominance over India'. Reuters, 6 March 2018, https://www.reuters.com/investigates/special-report/india-modi-culture/ (accessed 23 April 2018)

Jha, D. N. 'Against Communalising History'. *Social Scientist* 26, no. 9/10 (September–October 1998), 52–62

Lindo, John et al. 'Ancient Individuals from the North American Northwest Coast Reveal 10,000 Years of Regional Genetic Continuity'. *Proceedings of the National Academy of Sciences of the United States of America* 114, no. 16 (April 2017), 4093–8

Mees, Bernard. 'Hitler and Germanentum'. *Journal of Contemporary History* 39, no. 2 (April 2004), 255–70

'Nephi's Neighbors: Book of Mormon Peoples and Pre-Columbian Populations', Fair Mormon website, https://www.fairmormon.org/conference/august-2003/nephis-neighbors-book-of-mormon-peoples-and-pre-columbian-populations (accessed 23 March 2018)

O'Brien, Michael J. et al. 'On Thin Ice: Problems with Stanford and Bradley's Proposed Solutrean Colonisation of North America'. *Antiquity* 88, no. 340 (May 2014), 606–13

O'Brien, Michael J. et al. 'Solutreanism', *Antiquity* 88, no. 340 (January 2015), 622–4

O'Fallon, Brendan D. and Fehren-Schmitz, Lars. 'Native Americans Experienced a Strong Population Bottleneck Coincident with European Contact.' *Proceedings of the National Academy of Sciences*, Vol. 108, no. 51 (December 2011), 20444–8

Oppenheimer, Stephen Bradley, Bruce and Stanford, Dennis. 'Solutrean Hypothesis: Genetics, the Mammoth in the Room'. *World Archaeology* 46, no. 5 (October 2014), 752–74

Pillalamarri, Akhilesh. 'When History Gets Political: India's Grand "Aryan" Debate and the Indus Valley Civilization'. *The Diplomat*, 18 August 2016, https://thediplomat.com/ 2016/08/ when-history-gets-political-indias-grand-aryan-debate-and-the-indus-valley-civilization/ (accessed 29 April 2018)

Preston, Douglas. 'The Kennewick Man Finally Freed to Share His Secrets'. *Smithsonian Magazine*, September 2014

Raff, Jennifer. 'Rejecting the Solutrean Hypothesis'. *The Guardian*, 21 February 2018, https://www.theguardian.com/science/2018/feb/21/rejecting-the-solutrean-hypothesis-the-first-peoples-in-the-americas-were-not-from-europe (accessed 19 March 2018)

Raff, Jennifer A. and Bolnick, Deborah A. 'Does Mitochondrial Haplogroup X Indicate Ancient Trans-Atlantic Migration to the Americas? A Critical Re-Evaluation'. *PaleoAmerica* 1, no. 4 (November 2015), 297–304

Raghavan, Maanasa et al, 'Upper Palaeolithic Siberian Genome Reveals Dual Ancestry of Native Americans'. *Nature* 505 (January 2014), 87–91

Rasmussen, Morten et al. The Genome of a Late Pleistocene Human from a Clovis Burial Site in Western Montana. *Nature* 506 (February 2014), 225–9

Sharma, Manimugdha S. 'Faking History Starts Online'. *The Times of India*, 19 November 2017, https://timesofindia.indiatimes.com/home/sunday-times/faking-history-starts-online/articleshow/61705453.cms (accessed 29 April 2018)

Srivastava, Sushil. 'The Abuse of History: A Study of the White Papers on Ayodhya'. *Social Scientist* 22, no. 5/6 (May–June 1994), 39–51

Stanford, Dennis and Bradley, Bruce. 'Reply to O'Brien et al'. *Antiquity* 88, no. 340, (January 2015), 614–21

TallBear, Kim. *Native American DNA: Tribal Belonging and the False Promise of Genetic Science*. Minneapolis: University of Minnesota Press, 2013

TallBear, Kim. 'Genomic Articulations of Indigeneity'. *Social Studies of Science* 43, no. 4 (August 2013), 509–33

Thapar, Romila. *The Past as Present*. New Delhi: Aleph, 2014

Thapar, Romila. 'Can Genetics Help Us Understand Indian Social History?' *Cold Spring Harbor Perspectives in Biology* 6, no. 11 (November 2014)

Wu, Qinglong et al. 'Outburst Flood at 1920 BCE Supports Historicity of China's Great Flood and the Xia dynasty'. *Science* 353, no. 6299 (August 2016), 579–82

Zimmer, Carl. 'New Study Links Kennewick Man to Native Americans'. *New York Times*, 19 June 2015

9 Caste

Bhavnani, Rikhil R. and Lee, Alexander. 'Does Affirmative Action Worsen Bureaucratic Performance? Evidence from the Indian Administrative Service'. April 2018, https://faculty.polisci.wisc.edu/bhavnani/wp-content/uploads/2018/04/aa.pdf (accessed 10 May 2018)

Bouchard, Thomas J. et al. 'Sources of Human Psychological Differences: The Minnesota Study of Twins Reared Apart'. *Science* 250, no. 4978 (12 October 1990), 223–8

Bryc, Katarzyna et al. 'The Genetic Ancestry of African Americans, Latinos, and European Americans across the United States'. *American Journal of Human Genetics* 96, no. 1 (January 2015), 37–53

Chen, Jieming et al. 'Genetic Structure of the Han Chinese

Population Revealed by Genome-wide SNP Variation.' *American Journal of Human Genetics* 85, no. 6 (December 2009), 775–85

Coffey, Diane et al. 'Explicit Prejudice'. *Economic and Political Weekly* 53, no. 1 (6 January 2018)

Colman, Andrew. 'Race Differences in IQ: Hans Eysenck's Contribution to the Debate in the Light of Subsequent Research'. *Personality and Individual Differences* 103, (September 2016), 182–9

Comfort, Nathaniel. 'Books & Arts: Genetic Determinism Redux'. *Nature* 561 (27 September 2018), 461–3

Cullinane, Carl and Kirby, Philip. Research Brief: 'Class Differences: Ethnicity and Disadvantage'. The Sutton Trust, November 2016, https://www.suttontrust.com/wp-content/uploads/2016/11/Class-differences-report_References-available-online.pdf

Dickens, William T. and Flynn, James R. 'Black Americans Reduce the Racial IQ Gap'. *Psychological Science* 17, no. 10 (1 October 2006), 913–20

'Editorial: Against Bigotry'. *Nature* 548 (17 August 2017), 259

Edson, Lee. 'Jensenism: The Theory that I.Q. is Largely Determined by the Genes'. *New York Times Magazine* (31 August 1969), 10

Einstein, Albert. *The Travel Diaries of Albert Einstein: The Far East, Palestine, and Spain, 1922–1923*, ed. Ze'ev Rosenkranz. Princeton, NJ: Princeton University Press, 2018

'Father Kills Daughter for Marrying Outside their Caste in Maharashtra', *Hindustan Times*, 6 April 2017, http://www.hindustantimes.com/india-news/father-allegedly-kills-daughter-for-marrying-outside-their-caste-in-maharashtra/story-WKLKo54zk5IXznDJtThxnI.html (accessed 9 January 2018)

Feldman, Marcus W. and Ramachandran, Sohini. 'Missing Compared to What? Revisiting Heritability, Genes and Culture'. *Philosophical Transactions of the Royal Society B* 373, no. 1743 (5 April 2018)

Flynn, James R. 'The Mean IQ of Americans: Massive Gains 1932 to 1978'. *Psychological Bulletin* 95, no. 1 (1984), 29–51

Flynn, James R. 'Massive IQ Gains in 14 Nations: What IQ Tests Really Measure'. *Psychological Bulletin* 101, no. 2 (1987), 171–91

Gates-Coon, Rebecca. 'The Children of Sally Hemings'. Library of Congress, May 2002, https://www.loc.gov/loc/lcib/0205/hemings.html (accessed 11 May 2018)

Hart, Betty and Risley, Todd R. *Meaningful Differences in the Everyday Experience of Young American Children*. Baltimore: Paul H. Brookes, 1995

Hegewisch, Ariane and Williams-Baron, Emma. 'The Gender Wage Gap: 2017 Earnings Differences by Race and Ethnicity'. Institute for Women's Policy Research, 7 March 2018, https://iwpr.org/publications/gender-wage-gap-2017-race-ethnicity/ (accessed 10 May 2018)

Holden, Constance. 'Identical Twins Reared Apart'. *Science* 207, no. 4437 (21 March 1980), 1323–5 and 1327–8

Human Rights Watch, '"They Say We're Dirty" – Denying an Education to India's Marginalized', Report, 22 April 2014, https://www.hrw.org/report/2014/04/22/they-say-were-dirty/denying-education-indias-marginalized (accessed 9 January 2018)

Ilardo, Melissa A. et al. 'Physiological and Genetic Adaptations to Diving in Sea Nomads', *Cell* 173, no. 3 (April 2018), 569–80

Kochhar, Rakesh and Cilluffo, Anthony. 'How Wealth Inequality Has Changed in the U.S. Since the Great Recession, by Race, Ethnicity and Income'. Pew Research Center, 1 November 2017, http://www.pewresearch.org/fact-tank/2017/11/01/how- wealth-inequality-has-changed-in-the-u-s-since-the-great-recession-by-race-ethnicity-and-income/ (accessed 15 November 2018)

Moore, Elsie G. 'Family Socialization and the IQ Test Performance of Traditionally and Transracially Adopted Black

Children'. *Developmental Psychology* 22, no. 3 (May 1986), 317–26

Nisbett, Richard E. et al. 'Intelligence: New Findings and Theoretical Developments'. *American Psychologist* 67, no. 2 (February–March 2012), 130–59

Plomin, Robert, *Blueprint: How DNA Makes Us Who We Are*, London: Allen Lane, 2018

Plomin, Robert and Deary, Ian J. 'Genetics and Intelligence Differences: Five Special Findings'. *Molecular Psychiatry* 20 (2015), 98–108

Plomin, Robert and von Stumm, Sophie. 'The New Genetics of Intelligence'. *Nature Reviews Genetics* 19 (March 2018), 148–59

Ropers, H.-Hilger and Hamel, Ben C. J. 'X-linked Mental Retardation'. *Nature Reviews Genetics* 6 (January 2005), 46–57

Rushton, J. Philippe and Jensen, Arthur R. 'The Totality of Available Evidence Shows the Race IQ Gap Still Remains'. *Psychological Science* 17, no. 10 (October 2006), 921–2

Schwekendiek, Daniel Jong. 'Height and Weight Differences between North and South Korea'. *Journal of Biosocial Science* 41, no. 1 (January 2009), 51–5

Singh, Rama Shankar. 'The Indian Caste System, Human Diversity and Genetic Determinism', in *Thinking about Evolution: Historical, Philosophical, and Political Perspectives, Volume* 2, ed. Rama Shankar Singh et al. Cambridge: Cambridge University Press, 2000

Sniekers, Suzanne et al. 'Genome-Wide Association Meta-analysis Of 78,308 Individuals Identifies New Loci and Genes Influencing Human Intelligence'. *Nature Genetics*. 49, no. 7 (July 2017), 1107–12

Stevens, William K. 'Doctor Foresees an I.Q. Caste System'. *New York Times*, 29 August 1971

Witty, P. A. and Jenkins, M. A. 'Intra-Race Testing and Negro Intelligence'. *Journal of Psychology: Interdisciplinary and Applied* 1 (1936), 179–92

10 Black Pills

Adeyemo, Adebowale et al. 'A Genome-Wide Association Study of Hypertension and Blood Pressure in African Americans'. *PLoS Genetics* 5, no. 7, (17 July 2009)

Arias, Elizabeth et al. 'United States Life Tables'. National Vital Statistics Reports of the United States Centers for Disease Control and Prevention 66, no. 4 (August 2017)

'Asthma and African Americans', US Department of Health and Human Services Office of Minority Health website, https://minorityhealth.hhs.gov/omh/browse.aspx?lvl=4&lvlid=15 (accessed 10 July 2018)

Bailey, Zinzi D. et al. 'Structural Racism and Health Inequities in the USA: Evidence and Interventions. *The Lancet* 389, no. 10077 (8 April 2017), 1453–63

'Cardiovascular disease', NHS, https://www.nhs.uk/conditions/cardiovascular-disease/ (accessed 9 July 2018)

Chae, David H. et al. 'Area Racism and Birth Outcomes among Blacks in the United States. *Social Science & Medicine* 199 (February 2018), 49–55

Cooper, Richard S. 'Race in Biological and Biomedical Research'. *Cold Spring Harbor Perspectives in Medicine* 3, no. 11 (November 2013)

Cooper, Richard S. and Rotimi, Charles. 'Hypertension in Blacks'. *American Journal of Hypertension* 10, no. 7 (1 July 1997), 804–12

Cooper, Richard S. et al. 'An International Comparative Study of Blood Pressure in Populations of European vs. African Descent. *BioMed Central Medicine* 3, no. 2 (5 January 2005)

Cooper, Richard S. et al. 'Elevated Hypertension Risk for African-Origin Populations in Biracial Societies: Modeling the Epidemiologic Transition Study'. *Journal of Hypertension* 33, no. 3 (March 2015), 473–81

Curtin, Philip D. 'The Slavery Hypothesis for Hypertension

among African Americans: the Historical Evidence'. *American Journal of Public Health* 82, no. 12 (December 1992), 1681–6
Dubner, Stephen J. 'Toward a Unified Theory of Black America'. *New York Times Magazine* (20 March 2005)
Faturechi, Robert, 'When a Study Cast Doubt on a Heart Pill, the Drug Company Turned to Tom Price', *ProPublica*, 19 January 2017, https://www.propublica.org/article/when-a-study-cast-doubt-on-heart-pill-the-drug-company-turned-to-tom-price (accessed 18 July 2018)
Ferdinand, Keith C. et al. 'Disparities in hypertension and Cardiovascular Disease in Blacks: The Critical Role of Medication Adherence'. *Journal of Clinical Hypertension* 19, no. 10 (October 2017), 1015–24
Garcia, Richard S. 'The Misuse of Race in Medical Diagnosis'. *Pediatrics* 111, no. 5 (May 2004), 1394–5
'Global Health Observatory Data: Raised Blood Pressure', World Health Organization website, http://www.who.int/gho/ncd/risk_factors/blood_pressure_prevalence_text/en/ (accessed 10 July 2018)
Grim, Clarence E. and Robinson, Miguel. 'Commentary: Salt, Slavery and Survival: Hypertension in the African Diaspora'. *Epidemiology* 14, no. 1 (January 2003), 120–2
Hammermeister, Karl E. et al. 'Effectiveness of Hydralazine/Isosorbide Dinitrate in Racial/Ethnic Subgroups with Heart Failure'. *Clinical Therapeutics* 31, no. 3 (March 2009), 632–43
He, Feng J. et al. 'WASH – World Action on Salt and Health'. *Kidney International* 78, no. 8 (2 October 2010), 745–53
'Health A-Z: High blood pressure (hypertension)', National Health Service, https://www.nhs.uk/conditions/high-blood-pressure-hypertension/ (accessed 28 June 2018)
'Hypertension in adults: diagnosis and management: Clinical guideline', National Institute for Health and Care Excellence, published August 2011 and last updated November 2016, https://www.nice.org.uk/guidance/cg127/chapter/1-Guidance

#initiating-and-monitoring-antihypertensive-drug-treatment-including-blood-pressure-targets-2 (accessed 10 July 2018)

Kahn, Jonathan D. 'Race in a Bottle'. *Scientific American* 297, no. 2 (August 2007), 40–5

Kahn, Jonathan D. 'Beyond BiDil: the Expanding Embrace of Race in Biomedical Research and Product Development'. *St. Louis University Journal of Health Law & Policy* 3 (2009), 61–92

Kahn, Jonathan D. 'Revisiting Racial Patents in an Era of Precision Medicine'. *Case Western Reserve Law Review* 67, no. 4 (2017), 1153–69

Kalinowski, Leszek et al. 'Race-specific Differences in Endothelial Function: Predisposition of African Americans to Vascular Diseases'. *Circulation* 109, no. 21, 1 June 2004, 2511–17

Kaufman, Jay S. 'Statistics, Adjusted Statistics, and Maladjusted Statistics'. *American Journal of Law & Medicine* 43 (May 2017), 193–208

Kaufman, Jay S. and Cooper, Richard S. 'Use of Racial and Ethnic Identity in Medical Evaluations and Treatments', in *What's the Use of Race? Modern Governance and the Biology of Difference*, ed. Ian Whitmarsh and David S, Jones. Cambridge, MA: MIT Press, 2010

Kaufman, Jay S. and Hall, Susan A. 'The Slavery Hypertension Hypothesis: Dissemination and Appeal of a Modern Race Theory'. *Epidemiology* 14, no. 1 (January 2003), 111–18

Kaufman, Jay S. et al. 'Socioeconomic Status and Health in Blacks and Whites: The Problem of Residual Confounding and the Resiliency of Race'. *Epidemiology* 8, no. 6 (November 1997), 621–8

Khan, Jawad M. and Beevers, Gareth D. 'Management of Hypertension in Ethnic Minorities'. *Heart* 91, no. 8 (2005), 1105–9

Kidambi, Srividya et al. 'Non-replication Study of a Genome-

Wide Association Study for Hypertension and Blood Pressure in African Americans'. *BioMed Central Medical Genetics* 13, no. 27 (April 2012)

Nelson, Alondra. *The Social Life of DNA: Race, Reparations and Reconciliation after the Genome*. Boston, MA: Beacon Press, 2016

'NIH Policy and Guidelines on The Inclusion of Women and Minorities as Subjects in Clinical Research', United States National Institutes of Health, https://grants.nih.gov/grants/funding/women_min/guidelines.htm (accessed 20 August 2018)

Non, Amy L. et al. 'Education, Genetic Ancestry, and Blood Pressure in African Americans and Whites'. *American Journal of Public Health* 102, no. 8 (August 2012), 1559–65

Nyenhuis, Sharmilee M. et al. 'Race is Associated with Differences in Airway Inflammation in Patients with Asthma'. *Journal of Allergy and Clinical Immunology* 140, no. 1 (January 2017), 257–65

Obasogie, Osagie K. 'Oprah's Unhealthy Mistake'. *Los Angeles Times*, 17 May 2007

Ordúñez, Pedro et al. 'Blacks and Whites in Cuba Have Equal Prevalence of Hypertension: Confirmation from a New Population Survey'. *BioMed Central Public Health* 13, no. 169 (2013)

Roberts, Dorothy. *Fatal Invention: How Science, Politics, and Big Business Re-create Race in the Twenty-first Century*. New York: The New Press, 2012

Sankar, Pamela and Kahn, Jonathan. 'BiDil: Race Medicine or Race Marketing?' *Health Affairs* (July–December 2005), W5-455–63

Smedley, Brian D. et al. *Unequal Treatment: Confronting Racial and Ethnic Disparities in Health Care*. Institute of Medicine (US) Committee on Understanding and Eliminating Racial and Ethnic Disparities in Health Care. Washington, DC: National Academies Press, 2003

Taylor, Anne L. et al. 'Combination of Isosorbide Dinitrate and Hydralazine in Blacks with Heart Failure'. *New England Journal of Medicine* 351, no. 20 (11 November 2004), 2049–57

Tomson, Joseph and Lip, Gregory Y. H. 'Blood Pressure Demographics: Nature or Nurture . . . Genes or Environment?' *BioMed Central Medicine* 3, no. 3 (7 January 2005)

Transparency Market Research, 'Global Cardiovascular Drugs Market: Incessantly Rising Cases of Hypertension and Hyperlipidemia to Fuel Market Growth', 7 March 2018, https://www.prnewswire.com/news-releases/global-cardiovascular-drugs-market-incessantly-rising-cases-of-hypertension-and-hyperlipidemia-to-fuel-market-growth-says-tmr-676097783.html (accessed 17 July 2018)

'U.S. Public Health Service Syphilis Study at Tuskegee', Centers for Disease Control and Prevention, https://www.cdc.gov/tuskegee/index.html (accessed 3 July 2018)

Wolf-Maier, Katharina et al. 'Hypertension Prevalence and Blood Pressure Levels in 6 European Countries, Canada, and the United States'. *Journal of the American Medical Association* 289, no. 18 (14 May 2003), 2363–9

11 The Illusionists

Balter, Michael. 'Brain Man Makes Waves with Claims of Recent Human Evolution'. *Science*, New Series 314, no. 5807 (22 December 2006), 1871, 1873

Balter, Michael. 'Links between Brain Genes, Evolution, and Cognition Challenged'. *Science*, New Series 314, no. 5807 (22 December 2006), 1872

Balter, Michael. 'Schizophrenia's Unyielding Mysteries'. *Scientific American* (May 2017), 55–61

British Medical Association, 'Trend of Growing Numbers of BME Doctors in the Profession Continues', 14 May 2018, https://www.bma.org.uk/about-us/equality-diversity-and-inclusion/equality-lens/trend-2 (accessed 7 June 2018)

Byrd, W. Carson and Ray, Victor E. 'Ultimate Attribution in the Genetic Era: White Support for Genetic Explanations of Racial Difference and Policies'. *Annals of the American Academy of Political and Social Sciences* 661, no. 1 (1 September 2015), 212–35

Cochran, Gregory and Harpending, Henry. *The 10,000 Year Explosion: How Civilization Accelerated Human Evolution*. New York: Basic Books, 2009

Derbyshire. John. 'Evolution of the Brain'. *National Review*, 9 September 2005, https://www.nationalreview.com/corner/evolution-brain-john-derbyshire/ (accessed 6 June 2018)

Evans, Patrick D. et al. 'Microcephalin, a Gene Regulating Brain Size, Continues to Evolve Adaptively in Humans'. *Science*, New Series 309, no. 5741 (9 September 2005), 1717–20

Flynn, James R. *Asian Americans: Achievement beyond IQ*. Hillsdale, NJ: Lawrence Erlbaum, 1991

Fullwiley, Duana. 'Race and Genetics: Attempts to Define the Relationship'. *BioSocieties* 2, no. 2 (2007), 221–37

Hamer, Dean H. and Sirota, Lev. 'Beware the Chopsticks Gene'. *Molecular Psychiatry* 5, no. 1 (February 2000), 11–13

Kanazawa, Satoshi. 'Why Are Black Women Less Physically Attractive than Other Women?' *Psychology Today*, 15 May 2011, available online at http://tishushu.tumblr.com/post/ 5548905092/here-is-the-psychology-today-article-by (accessed 6 June 2018)

Kanazawa, Satoshi and Salahodjaev, Raufhon. 'Why Do Societies with Higher Average Cognitive Ability Have Lower Income Inequality? The Role of Redistributive Policies'. *Journal of Biosocial Science* 50, no. 3 (May 2018), 347–64

Lahn, Bruce and Ebenstein, Lanny. 'Let's Celebrate Human Genetic Diversity'. *Nature* 461, no. 461 (8 October 2009), 726–8

Matthews, Dylan. 'Heritage Study Co-author Opposed Letting in Immigrants with Low IQs'. *Washington Post*, 8 May 2013, https://www.washingtonpost.com/news/wonk/wp/2013/05/08/heritage-study-co-author-opposed-letting-in-immigrants-with-low-iqs/?utm_term=.cf882f04806f (accessed 6 June 2018)

Mekel-Bobrov, Nitzan et al. 'Ongoing Adaptive Evolution of ASPM, a Brain Size Determinant in Homo sapiens'. *Science*, New Series 309, no. 5741 (9 September 2005), 1720–2

Palmié, Stephan. 'Genomics, Divination, "Racecraft"'. *American Ethnologist* 34, no. 2 (May 2007), 205–22

Pinto, Rebecca et al. 'Schizophrenia in Black Caribbeans Living in the UK: an Exploration of Underlying Causes of the High Incidence Rate. *British Journal of General Practice* 58, no. 551 (June 2008), 429–34

Rothman, Barbara Katz. *The Book of Life: A Personal and Ethical Guide to Race, Normality and the Human Gene Study: A Personal and Ethical Guide to Race, Normality, and the Implications of the Human Genome Project*. Boston: Beacon Press, 2001

Rushton, J. Philippe et al. 'No evidence that polymorphisms of brain regulator genes Microcephalin and ASPM are associated with general mental ability, head circumference or altruism'. *Biology Letters* 3, no. 2 (2007), 157–60

'US migrant children cry for separated parents on audio', BBC News, 19 June 2018, https://www.bbc.co.uk/news/world-us-canada-44531187 (accessed 19 June 2018)

Varese, Filippo et al. 'Childhood Adversities Increase the Risk of Psychosis: A Meta-analysis of Patient-Control, Prospective- and Cross-sectional Cohort Studies'. *Schizophrenia Bulletin* 38, no. 4 (18 June 2012), 661–71

von Verschuer, Otmar. *Racial Biology of the Jews*, trans. from German by Charles E. Weber, Ph.D. Reedy, WV: Liberty Bell, 1983

Watkins, Eli and Phillip, Abby. 'Trump decries immigrants from "shithole countries" coming to US'. CNN Politics, 12 January 2018, https://edition.cnn.com/2018/01/11/politics/immigrants-shithole-countries-trump/index.html (accessed 14 June 2018)

Wicherts, Jelte M. et al. 'The Dangers of Unsystematic Selection Methods and the Representativeness of 46 Samples of African Test-takers. *Intelligence* 38 (2010), 30–7

Afterword

Bremmer, Ian. 'These 5 Countries Show How the European Far-Right Is Growing in Power'. *Time*, 13 September 2018, http://time.com/5395444/europe-far-right-italy-salvini-sweden-france-germany/ (accessed 20 November 2018)

Haas, Ernst B. 'Review: What is Nationalism and Why Should We Study it?' *International Organization* 40, no. 3 (Summer 1986), 707–44

'Poland Independence: Huge Crowds March Amid Far-right Row', BBC News, 11 November 2018, https://www.bbc.co.uk/news/world-europe-46172662 (accessed 20 November 2018)

Porter, Tom. 'Charlottesville's Alt-right Leaders Have a Passion for Vladimir Putin'. *Newsweek*, 16 August 2017, https://www.newsweek.com/leaders-charlottesvilles-alt-right-protest-all-have-ties-russian-fascist-651384 (accessed 21 November 2018)

Sommer, Allison Kaplan. 'The Global Anti-globalist: Steve Bannon Comes Out as Proud "Racist" on His European Comeback Tour'. *Haaretz*, 11 March 2018, https://www.haaretz.com/us-news/.premium-on-european-tour-steve-bannon-comes-out-as-a-proud-racist-1.5890885 (accessed 20 November 2018)

Stille, Alexander. 'How Matteo Salvini Pulled Italy to the Far Right'. *The Guardian*, 9 August 2018, https://www.theguardian.com/news/2018/aug/09/how-matteo-salvini-pulled-italy-to-the-far-right (accessed 20 November 2018)

Index

The Oxford Spanish Language Programme

The inauguration of the Oxford Spanish Language Programme marked the start of a new age of Spanish dictionaries. The Programme has produced, with unrivalled clarity and authority, the only dictionaries to present the full wealth of Spanish from both sides of the Atlantic and across 24 different Spanish-speaking countries and regions.

• The Bank of Spanish

Drawing on The Bank of Spanish, a vast electronic databank of up-to-date, authentic language in use, these dictionaries provide a more accurate and complete picture of *real* language than has ever been possible before. The Bank shapes every dictionary entry and translation to meet the needs of today's users, highlighting important constructions, illustrating difficult meanings, and focusing attention on common usage.

• The richest choice of words

In-depth coverage of over 24 different regional varieties of Spanish with special emphasis on modern idioms and colloquial usage are distinctive elements of the Oxford Spanish Language Programme. Words and phrases restricted to particular areas of the Spanish-speaking world are precisely labelled for country or wider region, from Spain to Chile to Mexico, from Central America to the River Plate. In addition, variant pronunciations and the register of words, from formal right through to taboo, are signalled throughout.

• The Spanish Literary Heritage

A wide range of vocabulary and usage found in the literary heritage of the Spanish-speaking world has been analysed and described by the editors of the Programme to assist readers and students of Spanish literature.

• The British National Corpus

Each English entry is shaped by direct evidence from the British National Corpus, an unrivalled balanced collection of 100 million words of text representing every kind of writing and speech in English.

Total Language Accessibility

Oxford's unparalleled reputation in the field of dictionary publishing is founded on more than 150 years of experience. Each dictionary in the range bears the Oxford hallmarks of integrity and authority. The Oxford Spanish dictionaries are an integral part of this tradition and offer an unequalled range of carefully-designed benefits to ensure maximum language accessibility.

• Rapid access design

Oxford's new quick-access page designs and typography have been specially created to ensure exceptional clarity and accessibility. Entries are written in clear, jargon-free language without confusing abbreviations.

• Unrivalled practical help

Extended treatment of the core vocabulary offers the user step-by-step guidance on how to translate a given word correctly. Unrivalled practical grammatical help has been built into every dictionary within the range. Thousands of examples, drawn from the evidence of the Bank of Spanish, are carefully chosen to illustrate the many different nuances of meaning and context.

• Supplementary Information

All the dictionaries in the Oxford range offer valuable additional help and information, including verb tables, thematic vocabulary boxes, political and cultural information, guides to effective communication (how to write letters, CVs, book holidays, or take minutes), pronunciation guidance, and colour texts for easy access.

• The best range in the world

Oxford provides Spanish dictionaries for all levels of user, from advanced to beginner. In addition, Oxford also publishes a wealth of Spanish reference titles, including guides to Spanish grammar, usage, verbs, correspondence and core vocabulary. Whatever type of dictionary—whether for children, native speakers, university students or learners; on paper or CD-ROM; in English, French, Spanish, German, Italian, Russian, Japanese, Latin, Greek, Arabic, Turkish, Portuguese, Hungarian, Hindi, Gujarati or Chinese—Oxford offers the most trusted range available anywhere in the world today.

The Oxford
Quick Reference Spanish Dictionary

SPANISH–ENGLISH
ENGLISH–SPANISH

ESPAÑOL–INGLÉS
INGLÉS–ESPAÑOL

Christine Lea

Spanish in Context
prepared by Michael Britton
and Carol Styles

OXFORD UNIVERSITY PRESS
1998

Oxford University Press, Great Clarendon Street, Oxford OX2 6DP

Oxford New York
Athens Auckland Bangkok Bogota Bombay
Buenos Aires Calcutta Cape Town Dar es Salaam
Delhi Florence Hong Kong Istanbul Karachi
Kuala Lumpur Madras Madrid Melbourne
Mexico City Nairobi Paris Singapore
Taipei Tokyo Toronto Warsaw

and associated companies in
Berlin Ibadan

This edition first published 1998
Originally published as The Oxford Paperback Spanish Dictionary, 1993

British Library Cataloguing in Publication Data
Data available

Library of Congress Cataloging-in-Publication Data
Lea, Christine. [Oxford paperback Spanish dictionary]
The Oxford quick reference Spanish dictionary : Spanish–English, English–Spanish = español–inglés, inglés–español / Christine Lea.
Originally published as: The Oxford paperback Spanish dictionary.
Oxford [England] ; New York : Oxford University Press, 1994.
1. Spanish language—Dictionaries—English. 2. English language—Dictionaries—Spanish. I. Title.
PC4640.L44 1998 463'.21—dc21 97–40950

ISBN 0–19–860185–9

10 9 8 7 6 5 4 3 2 1

Printed in Great Britain by
Mackays of Chatham plc
Chatham, Kent

Contents • Índice

Preface to *The Oxford Quick Reference Spanish Dictionary*

The Oxford Quick Reference Spanish Dictionary is the latest addition to the Oxford Spanish Dictionary range. It is specifically designed for beginners of Spanish as an affordable, accessible dictionary with valuable additional help provided by the unique *Spanish in Context* supplement in the middle of the book. *Spanish in Context* is designed to help you build your knowledge of grammar and vocabulary, and provides valuable practice in dealing with everyday situations and conversations.

Foreword

This dictionary has been written with speakers of both English and Spanish in mind and contains the most useful words and expressions of the English and Spanish languages of today. Wide coverage of culinary and motoring terms has been included to help the tourist.

Common abbreviations, names of countries, and other useful geographical names are included.

English pronunciation is given by means of the International Phonetic Alphabet. It is shown for all headwords and for those derived words whose pronunciation is not easily deduced from that of a headword. The rules for pronunciation of Spanish are given on page x.

I should like to thank particularly Mary-Carmen Beaven, whose comments have been invaluable. I would also like to acknowledge the help given to me unwittingly by Dr M. Janes and Mrs J. Andrews, whose French and Italian Minidictionaries have served as models for the present work.

C. A. L

Prólogo

Este diccionario de Oxford se escribió tanto para los hispanohablantes como para los angloparlantes y contiene las palabras y frases más corrientes de ambas lenguas de hoy. Se incluyen muchos términos culinarios y de movilismo que pueden servir al turista.

Las abreviaturas más corrientes, los nombres de países, y otros términos geográficos figuran en este diccionario.

La pronunciación inglesa sigue el Alfabeto Fonético Internacional. Se incluye para cada palabra clave y todas las derivadas cuya pronunciación no es fácil de deducir a partir de la palabra clave. Las reglas de la pronunciación española se encuentran en la página x.

Quisiera reconocer la ayuda de Mary-Carmen Beaven cuyas observaciones me han sido muy valiosas. También quiero agradecerles al Dr. M. Janes y a la Sra. J. Andrews cuyos minidiccionarios del francés y del italiano me han servido de modelo para el presente.

C. A. L

Introduction

The swung dash (~) is used to replace a headword or that part of a headword preceding the vertical bar (|). In both English and Spanish only irregular plurals are given. Normally Spanish nouns and adjectives ending in an unstressed vowel form the plural by adding s (e.g. *libro, libros*). Nouns and adjectives ending in a stressed vowel or a consonant add *es* (e.g. *rubí, rubíes*; *pared, paredes*). An accent on the final syllable is not required when *es* is added (e.g. *nación, naciones*). Final *z* becomes *ces* (e.g. *vez, veces*). Spanish nouns and adjectives ending in *o* form the feminine by changing the final *o* to *a* (e.g. *hermano, hermana*). Most Spanish nouns and adjectives ending in anything other than final *o* do not have a separate feminine form with the exception of those denoting nationality etc.; these add *a* to the masculine singular form (e.g. *español, española*). An accent on the final syllable is then not required (e.g. *inglés, inglesa*). Adjectives ending in *án*, *ón*, or *or* behave like those denoting nationality with the following exceptions: *inferior, mayor, mejor, menor, peor, superior*, where the feminine has the same form as the masculine. Spanish verb tables will be found in the appendix.

The Spanish alphabet

In Spanish *ch, ll* and *ñ* are considered separate letters and in the Spanish–English section, therefore, they will be found after *cu*, *luî* and *ny* respectively.

Introducción

La tilde (~) se emplea para sustituir a la palabra cabeza de artículo o aquella parte de tal palabra que precede a la barra vertical (|). Tanto en inglés como en español se dan los plurales solamente si son irregulares. Para formar el plural regular en inglés se añade la letra *s* al sustantivo singular, pero se añade *es* cuando se trata de una palabra que termina en *ch, sh, s, ss, us, x, o, z* (p.ej. *sash, sashes*). En el caso de una palabra que termine en *y* precedida por una consonante, la *y* se cambia en *ies* (p.ej. *baby, babies*). Para formar el tiempo pasado y el participio pasado se añade *ed* al infinitivo de los verbos regulares ingleses (p.ej. *last, lasted*). En el caso de los verbos ingleses que terminan en *e* muda se añade sólo la *d* (p.ej. *move, moved*) . En el caso de los verbos ingleses que terminan en *y* hay que cambiar la *y* por *ied* (p.ej. *carry, carried*). Los verbos irregulares se encuentran en el diccionario por orden alfabético remitidos al infinitivo, y también en la lista en el apéndice.

Pronunciation of Spanish

Vowels

a between pronunciation of *a* in English *cat* and *arm*

e like *e* in English *bed*

i like *ee* in English *see* but a little shorter

o like *o* in English *hot* but a little longer

u like *oo* in English *too*

y when a vowel, like Spanish **i**

Consonants

b (1) in initial position or after nasal consonant, like English *b*
(2) in other positions, between English *b* and English *v*

c (1) before **e** or **i**, like *th* in English *thin*
(2) in other positions, like *c* in English *cat*

ch like *ch* in English *chip*

d (1) in initial position, after nasal consonants and after **l**, like English **d**
(2) in other positions, like *th* in English *this*

f like English *f*

g (1) before **e** or **i**, like *ch* in Scottish *loch*
(2) in initial position, like g in English *get*
(3) in other positions, like (2) but a little softer

h silent in Spanish but see also **ch**

j like *ch* in Scottish *loch*

k like English *k*

l like English *l* but see also **ll**

ll like *lli* in English *million*

m like English *m*

n like English *n*

ñ like *ni* in English *opinion*

p like English *p*

q like English *k*

r rolled or trilled

s like *s* in English *sit*

t like English *t*

v (1) in initial position or after nasal consonant, like English *b*
(2) in other positions, between English *b* and English *v*

w like Spanish **b** or **v**

x like English *x*

y like English *y*

z like *th* in English *thin*

Pronunciación Inglesa

Símbolos fonéticos

Vocales y diptongos

iː	s*ee*	ɔː	s*aw*	əɪ	p*a*ge	ɔɪ	j*oi*n
ɪ	s*i*t	ʊ	p*u*t	əʊ	h*o*me	ɪə	n*ea*r
e	t*e*n	uː	t*oo*	aɪ	f*i*ve	eə	h*air*
æ	h*a*t	ʌ	c*u*p	aɪə	f*ire*	ʊə	p*oor*
ɑ	*ar*m	ɜː	f*ur*	aʊ	n*ow*		
ɒ	g*o*t	ə	*a*go	aʊə	fl*our*		

Consonantes

p	*p*en	tʃ	*ch*in	s	*s*o	n	*n*o
b	*b*ad	dʒ	*J*une	z	*z*oo	ŋ	si*ng*
t	*t*ea	f	*f*all	ʃ	*sh*e	l	*l*eg
d	*d*ip	v	*v*oice	ʒ	mea*s*ure	r	*r*ed
k	*c*at	θ	*th*in	h	*h*ow	j	*y*es
g	*g*ot	ð	*th*en	m	*m*an	w	*w*et

Abbreviations / Abreviaturas

adjective	*a*	adjetivo
abbreviation	*abbr /abrev*	abreviatura
administration	*admin*	administración
adverb	*adv*	adverbio
American	*Amer*	americano
anatomy	*anat*	anatomía
architecture	*archit /arquit*	arquitectura
definite article	*art def*	artículo definido
indefinite article	*art indef*	artículo indefinido
astrology	*astr*	astrología
motoring	*auto*	automóvil
auxiliary	*aux*	auxiliar
aviation	*aviat /aviac*	aviación
biology	*biol*	biología
botany	*bot*	botánica
commerce	*com*	comercio
conjunction	*conj*	conjunción
cooking	*culin*	cocina
electricity	*elec*	electricidad
school	*escol*	enseñanza
Spain	*Esp*	España
feminine	*f*	femenino
familiar	*fam*	familiar
figurative	*fig*	figurado
philosophy	*fil*	filosofía
photography	*foto*	fotografía
geography	*geog*	geografía
geology	*geol*	geología
grammar	*gram*	gramática
humorous	*hum*	humorístico
interjection	*int*	interjección
interrogative	*inter*	interrogativo
invariable	*invar*	invariable
legal, law	*jurid*	jurídico
Latin American	*LAm*	latinoamericano
language	*lang*	lengua(je)
masculine	*m*	masculino
mathematics	*mat(h)*	matemáticas
mechanics	*mec*	mecánica
medicine	*med*	medicina

military	*mil*	militar
music	*mus*	música
mythology	*myth*	mitología
noun	*n*	nombre
nautical	*naut*	náutica
oneself	*o. s.*	uno mismo, se
proprietary term	*P*	marca registrada
pejorative	*pej*	peyorativo
philosophy	*phil*	filosofía
photography	*photo*	fotografía
plural	*pl*	plural
politics	*pol*	política
possessive	*poss*	posesivo
past participle	*pp*	participio pasado
prefix	*pref*	prefijo
preposition	*prep*	preposición
present participle	*pres p*	participio de presente
pronoun	*pron*	pronombre
psychology	*psych*	psicología
past tense	*pt*	tiempo pasado
railroad	*rail*	ferrocarril
relative	*rel*	relativo
religion	*relig*	religión
school	*schol*	enseñanza
singular	*sing*	singular
slang	*sl*	argot
someone	*s. o.*	alguien
something	*sth*	algo
technical	*tec*	técnico
television	*TV*	televisión
university	*univ*	universidad
auxiliary verb	*v aux*	verbo auxiliar
verb	*vb*	verbo
intransitive verb	*vi*	verbo intransitivo
pronominal verb	*vpr*	verbo pronominal
transitive verb	*vt*	verbo transitivo
transitive & intransitive verb	*vti*	verbo transitivo e intransitivo

Proprietary terms

This dictionary includes some words which have, or are asserted to have, proprietary status as trademarks. Their inclusion does not imply that they have acquired for legal purposes a non-proprietary or general significance, nor any other judgement concerning their legal status. In cases where the editorial staff have some evidence that a word has proprietary status this is indicated in the entry for that word by the letter (P), but no judgement concerning the legal status of such words is made or implied thereby.

Marcas registradas

Este diccionario incluye algunas palabras que son o pretendenser marcas registradas. No debe atri buirse ningún valor jurídico ni a la presencia ni a la ausencia de tal designación.

ESPAÑOL-INGLÉS
SPANISH-ENGLISH

A

a *prep* in, at; (*dirección*) to; (*tiempo*) at; (*hasta*) to, until; (*fecha*) on; (*más tarde*) later; (*medio*) by; (*precio*) for, at. **~ 5 km** 5 km away. **¿~ cuántos estamos?** what's the date? **~l día siguiente** the next day. **~ la francesa** in the French fashion. **~ las 2** at 2 o'clock. **~ los 25 años** (*edad*) at the age of 25; (*después de*) after 25 years. **~ no ser por** but for. **~ que** I bet. **~ 28 de febrero** on the 28th of February

ábaco *m* abacus

abad *m* abbot

abadejo *m* (*pez*) cod

abad|esa *f* abbess. **~ía** *f* abbey

abajo *adv* (down) below; (*dirección*) down(wards); (*en casa*) downstairs. ● *int* down with. **calle ~** down the street. **el ~ firmante** the undersigned. **escaleras ~** downstairs. **la parte de ~** the bottom part. **los de ~** those at the bottom. **más ~** below.

abalanzarse [10] *vpr* rush towards

abalorio *m* glass bead

abanderado *m* standard-bearer

abandon|ado *adj* abandoned; (*descuidado*) neglected; ‹*personas*› untidy. **~ar** *vt* leave ‹*un lugar*›; abandon ‹*personas, cosas*›. ● *vi* give up. **~arse** *vpr* give in; (*descuidarse*) let o.s. go. **~o** *m* abandonment; (*estado*) abandon

abani|car [7] *vt* fan. **~co** *m* fan. **~queo** *m* fanning

abarata|miento *m* reduction in price. **~r** *vt* reduce. **~rse** *vpr* ‹*precios*› come down

abarca *f* sandal

abarcar [7] *vt* put one's arms around, embrace; (*comprender*) embrace; (*LAm, acaparar*) monopolize

abarquillar *vt* warp. **~se** *vpr* warp

abarrotar *vt* overfill, pack full

abarrotes *mpl* (*LAm*) groceries

abast|ecer [11] *vt* supply. **~ecimiento** *m* supply; (*acción*) supplying. **~o** *m* supply. **dar ~o a** supply

abati|do *a* depressed. **~miento** *m* depression. **~r** *vt* knock down, demolish; (*fig, humillar*) humiliate. **~rse** *vpr* swoop (**sobre** on); (*ponerse abatido*) get depressed

abdica|ción *f* abdication. **~r** [7] *vt* give up. ● *vi* abdicate

abdom|en *m* abdomen. **~inal** *a* abdominal

abec|é *m* (*fam*) alphabet, ABC. **~edario** *m* alphabet

abedul *m* birch (tree)

abej|a *f* bee. **~arrón** *m* bumble-bee. **~ón** *m* drone. **~orro** *m* bumble-bee; (*insecto coleóptero*) cockchafer

aberración *f* aberration

abertura *f* opening

abet|al *m* fir wood. **~o** *m* fir (tree)

abierto *pp véase* **abrir.** ● *a* open

abigarra|do *a* multi-coloured; (*fig, mezclado*) mixed. **~miento** *m* variegation

abigeato *m* (*Mex*) rustling

abism|al *a* abysmal; (*profundo*) deep. **~ar** *vt* throw into an abyss; (*fig, abatir*) humble. **~arse** *vpr* be absorbed (**en** in), be lost (**en** in). **~o** *m* abyss; (*fig, diferencia*) world of difference

abizcochado *a* spongy

abjura|ción *f* abjuration. **~r** *vt* forswear. ● *vi*. **~r de** forswear

abland a|miento *m* softening. **~r** *vt* soften. **~rse** *vpr* soften

ablución *f* ablution

abnega|ción *f* self-sacrifice. **~do** *a* self-sacrificing

aboba|do *a* silly. **~miento** *m* silliness

aboca|do *a* ‹*vino*› medium. **~r** [7] *vt* pour out

abocetar *vt* sketch

abocinado *a* trumpet-shaped

abochornar *vt* suffocate; (*fig, avergonzar*) embarrass. **~se** *vpr* feel embarassed; ‹*plantas*› wilt

abofetear *vt* slap

aboga|cía *f* legal profession. **~do** *m* lawyer; (*notario*) solicitor; (*en el tribunal*) barrister, attorney (*Amer*). **~r** [12] *vi* plead

abolengo *m* ancestry

aboli|ción *f* abolition. **~cionismo** *m* abolitionism. **~cionista** *m & f* abolitionist. **~r** [24] *vt* abolish

abolsado *a* baggy

abolla|dura *f* dent. **~r** *vt* dent

abomba|do *a* convex; (*Arg, borracho*) drunk. **~r** *vt* make convex. **~rse** *vpr* (*LAm, corromperse*) start to rot, go bad

abomina|ble *a* abominable. **~ción** *f* abomination. **~r** *vt* detest. ● *vi*. **~r de** detest

abona|ble *a* payable. **~do** *a* paid. ● *m* subscriber

abonanzar *vi* ‹*tormenta*› abate; ‹*tiempo*› improve

abon|ar *vt* pay; (*en agricultura*) fertilize. **~aré** *m* promissory note. **~arse** *vpr* subscribe. **~o** *m* payment; (*estiércol*) fertilizer; (*a un periódico*) subscription

aborda|ble *a* reasonable; ‹*persona*› approachable. **~je** *m* boarding. **~r** *vt* tackle ‹*un asunto*›; approach ‹*una persona*›; (*naut*) come alongside

aborigen *a & m* native

aborrascarse [7] *vpr* get stormy

aborrec|er [11] *vt* hate; (*exasperar*) annoy. **~ible** *a* loathsome. **~ido** *a* hated. **~imiento** *m* hatred

aborregado *a* ‹*cielo*› mackerel

abort|ar *vi* have a miscarriage. **~ivo** *a* abortive. **~o** *m* miscarriage; (*voluntario*) abortion; (*fig, monstruo*) abortion. **hacerse ~ar** have an abortion

abotaga|miento *m* swelling. **~rse** [12] *vpr* swell up

abotonar *vt* button (up)

aboveda|do *a* vaulted. **~r** *vt* vault

abra *f* cove

abracadabra *m* abracadabra

abrasa|dor *a* burning. **~r** *vt* burn; (*fig, consumir*) consume. **~rse** *vpr* burn

abrasi|ón *f* abrasion; (*geología*) erosion. **~vo** *a* abrasive

abraz|adera *f* bracket. **~ar** *vt* [10] embrace; (*encerrar*) enclose. **~arse** *vpr* embrace. **~o** *m* hug. **un fuerte ~o de** (*en una carta*) with best wishes from

abrecartas *m* paper-knife

ábrego *m* south wind

abrelatas *m invar* tin opener (*Brit*), can opener

abreva|dero *m* watering place. **~r** *vt* water ‹*animales*›. **~rse** *vpr* ‹*animales*› drink

abrevia|ción *f* abbreviation; (*texto abreviado*) abridged text. **~do** *a* brief; ‹*texto*› abridged. **~r** *vt* abbreviate; abridge ‹*texto*›; cut short ‹*viaje etc*›. ● *vi* be brief. **~tura** *f* abbreviation

abrig|ada *f* shelter. **~adero** *m* shelter. **~ado** *a* ‹*lugar*› sheltered; ‹*personas*› well wrapped up. **~ar** [12] *vt* shelter; cherish ‹*esperanza*›; harbour ‹*duda, sospecha*›. **~arse** *vpr* (take) shelter; (*con ropa*) wrap up. **~o** *m* (over)coat; (*lugar*) shelter

abril *m* April. **~eño** *a* April

abrillantar *vt* polish

abrir [*pp* **abierto**] *vt/i* open. **~se** *vpr* open; (*extenderse*) open out; ‹*el tiempo*› clear

abrocha|dor *m* buttonhook. **~r** *vt* do up; (*con botones*) button up

abrojo *m* thistle

abroncar [7] *vt* (*fam*) tell off; (*abuchear*) boo; (*avergonzar*) shame. **~se** *vpr* be ashamed; (*enfadarse*) get annoyed

abroquelarse *vpr* shield o.s.

abruma|dor *a* overwhelming. **~r** *vt* overwhelm

abrupto *a* steep; (*áspero*) harsh

abrutado *a* brutish

absceso *m* abscess

absentismo *m* absenteeism

ábside *m* apse

absintio *m* absinthe

absolución *f* (*relig*) absolution; (*jurid*) acquittal

absolut|amente *adv* absolutely, completely. **~ismo** *m* absolutism. **~ista** *a & m & f* absolutist. **~o** *a* absolute. **~orio** *a* of acquittal. **en ~o** (*de manera absoluta*) absolutely; (*con sentido negativo*) (not) at all

absolver [2, *pp* **absuelto**] *vt* (*relig*) absolve; (*jurid*) acquit

absor|bente *a* absorbent; (*fig, interesante*) absorbing. **~ber** *vt* absorb. **~ción** *f* absorption. **~to** *a* absorbed

abstemio *a* teetotal. ● *m* teetotaller

absten|ción *f* abstention. **~erse** [40] *vpr* abstain, refrain (**de** from)
abstinen|cia *f* abstinence. **~te** *a* abstinent
abstra|cción *f* abstraction. **~cto** *a* abstract. **~er** [41] *vt* abstract. **~erse** *vpr* be lost in thought. **~ído** *a* absent-minded
abstruso *a* abstruse
absuelto *a* (*relig*) absolved; (*jurid*) acquitted
absurdo *a* absurd. ● *m* absurd thing
abuche|ar *vt* boo. **~o** *m* booing
abuel|a *f* grandmother. **~o** *m* grandfather. **~os** *mpl* grandparents
ab|ulia *f* lack of willpower. **~úlico** *a* weak-willed
abulta|do *a* bulky. **~miento** *m* bulkiness. **~r** *vt* enlarge; (*hinchar*) swell; (*fig, exagerar*) exaggerate. ● *vi* be bulky
abunda|ncia *f* abundance. **~nte** *a* abundant, plentiful. **~r** *vi* be plentiful. **nadar en la ~ncia** be rolling in money
aburguesa|miento *m* conversion to a middle-class way of life. **~rse** *vpr* become middle-class
aburri|do *a* (*con estar*) bored; (*con ser*) boring. **~miento** *m* boredom; (*cosa pesada*) bore. **~r** *vt* bore. **~rse** *vpr* be bored, get bored
abus|ar *vi* take advantage. **~ar de la bebida** drink too much. **~ivo** *a* excessive. **~o** *m* abuse. **~ón** *a* (*fam*) selfish
abyec|ción *f* wretchedness. **~to** *a* abject
acá *adv* here; (*hasta ahora*) until now. **~ y allá** here and there. **de ~ para allá** to and fro. **de ayer ~** since yesterday
acaba|do *a* finished; (*perfecto*) perfect; (*agotado*) worn out. ● *m* finish. **~miento** *m* finishing; (*fin*) end. **~r** *vt/i* finish. **~rse** *vpr* finish; (*agotarse*) run out; (*morirse*) die. **~r con** put an end to. **~r de** (+ *infinitivo*) have just (+ *pp*). **~ de llegar** he has just arrived. **~r por** (+ *infinitivo*) end up (+ *gerundio*). **¡se acabó!** that's it!
acabóse *m*. **ser el ~** be the end, be the limit
acacia *f* acacia
acad|emia *f* academy. **~émico** *a* academic
acaec|er [11] *vi* happen. **~imiento** *m* occurrence
acalora|damente *adv* heatedly. **~do** *a* heated. **~miento** *m* heat. **~r** *vt* warm up; (*fig, excitar*) excite. **~rse** *vpr* get hot; (*fig, excitarse*) get excited
acallar *vt* silence
acampanado *a* bell-shaped
acampar *vi* camp
acanala|do *a* grooved. **~dura** *f* groove. **~r** *vt* groove
acantilado *a* steep. ● *m* cliff
acanto *m* acanthus
acapara|r *vt* hoard; (*monopolizar*) monopolize. **~miento** *m* hoarding; (*monopolio*) monopolizing
acaracolado *a* spiral
acaricia|dor *a* caressing. **~r** *vt* caress; (*rozar*) brush; ‹*proyectos etc*› have in mind
ácaro *m* mite
acarre|ar *vt* transport; ‹*desgracias etc*› cause. **~o** *m* transport
acartona|do *a* ‹*persona*› wizened. **~rse** *vpr* (*ponerse rígido*) go stiff; ‹*persona*› become wizened
acaso *adv* maybe, perhaps. ● *m* chance. **~ llueva mañana** perhaps it will rain tomorrow. **al ~** at random. **por si ~** in case
acata|miento *m* respect (**a** for). **~r** *vt* respect
acatarrarse *vpr* catch a cold, get a cold
acaudalado *a* well off
acaudillar *vt* lead
acceder *vi* agree; (*tener acceso*) have access
acces|ibilidad *f* accessibility. **~ible** *a* accessible; ‹*persona*› approachable. **~o** *m* access, entry; (*med, ataque*) attack; (*llegada*) approach
accesorio *a & m* accessory
accidentado *a* ‹*terreno*› uneven; (*agitado*) troubled; ‹*persona*› injured
accident|al *a* accidental. **~arse** *vpr* have an accident. **~e** *m* accident
acci|ón *f* (*incl jurid*) action; (*hecho*) deed. **~onar** *vt* work. ● *vi* gesticulate. **~onista** *m & f* shareholder
acebo *m* holly (tree)
acebuche *m* wild olive tree
acecinar *vt* cure ‹*carne*›. **~se** *vpr* become wizened
acech|ar *vt* spy on; (*aguardar*) lie in wait for. **~o** *m* spying. **al ~o** on the look-out

acedera *f* sorrel
acedía *f* (*pez*) plaice; (*acidez*) heartburn
aceit|ar *vt* oil; (*culin*) add oil to. **~e** *m* oil; (*de oliva*) olive oil. **~era** *f* oil bottle; (*para engrasar*) oilcan. **~ero** *a* oil. **~oso** *a* oily
aceitun|a *f* olive. **~ado** *a* olive. **~o** *m* olive tree
acelera|ción *f* acceleration. **~damente** *adv* quickly. **~dor** *m* accelerator. **~r** *vt* accelerate; (*fig*) speed up, quicken
acelga *f* chard
ac|émila *f* mule; (*como insulto*) ass (*fam*). **~emilero** *m* muleteer
acendra|do *a* pure. **~r** *vt* purify; refine ‹*metales*›
acensuar *vt* tax
acent|o *m* accent; (*énfasis*) stress. **~uación** *f* accentuation. **~uar** [21] *vt* stress; (*fig*) emphasize. **~uarse** *vpr* become noticeable
aceña *f* water-mill
acepción *f* meaning, sense
acepta|ble *a* acceptable. **~ción** *f* acceptance; (*aprobación*) approval. **~r** *vt* accept
acequia *f* irrigation channel
acera *f* pavement (*Brit*), sidewalk (*Amer*)
acerado *a* steel; (*fig*, *mordaz*) sharp
acerca de *prep* about
acerca|miento *m* approach; (*fig*) reconciliation. **~r** [7] *vt* bring near. **~rse** *vpr* approach
acería *f* steelworks
acerico *m* pincushion
acero *m* steel. **~ inoxidable** stainless steel
acérrimo *a* (*fig*) staunch
acert|ado *a* right, correct; (*apropiado*) appropriate. **~ar** [1] *vt* hit ‹*el blanco*›; (*adivinar*) get right, guess. ● *vi* get right. **~ar a** happen to. **~ar con** hit on. **~ijo** *m* riddle
acervo *m* pile; (*bienes*) common property
acetato *m* acetate
acético *a* acetic
acetileno *m* acetylene
acetona *m* acetone
aciago *a* unlucky
aciano *m* cornflower
ac|íbar *m* aloes; (*planta*) aloe; (*fig*, *amargura*) bitterness. **~ibarar** *vt* add aloes to; (*fig*, *amargar*) embitter
acicala|do *a* dressed up, overdressed. **~r** *vt* dress up. **~rse** *vpr* get dressed up
acicate *m* spur
acid|ez *f* acidity. **~ificar** [7] *vt* acidify. **~ificarse** *vpr* acidify
ácido *a* sour. ● *m* acid
acierto *m* success; (*idea*) good idea; (*habilidad*) skill
aclama|ción *f* acclaim; (*aplausos*) applause. **~r** *vt* acclaim; (*aplaudir*) applaud
aclara|ción *f* explanation. **~r** *vt* lighten ‹*colores*›; (*explicar*) clarify; (*enjuagar*) rinse. ● *vi* ‹*el tiempo*› brighten up. **~rse** *vpr* become clear. **~torio** *a* explanatory
aclimata|ción *f* acclimatization, acclimation (*Amer*). **~r** *vt* acclimatize, acclimate (*Amer*). **~rse** *vpr* become acclimatized, become acclimated (*Amer*)
acné *m* acne
acobardar *vt* intimidate. **~se** *vpr* get frightened
acocil *m* (*Mex*) freshwater shrimp
acod|ado *a* bent. **~ar** *vt* (*doblar*) bend; (*agricultura*) layer. **~arse** *vpr* lean on (**en** on). **~o** *m* layer
acog|edor *a* welcoming; ‹*ambiente*› friendly. **~er** [14] *vt* welcome; (*proteger*) shelter; (*recibir*) receive. **~erse** *vpr* take refuge. **~ida** *f* welcome; (*refugio*) refuge
acogollar *vi* bud. **~se** *vpr* bud
acolcha|do *a* quilted. **~r** *vt* quilt, pad
acólito *m* acolyte; (*monaguillo*) altar boy
acomet|edor *a* aggressive; (*emprendedor*) enterprising. **~er** *vt* attack; (*emprender*) undertake; (*llenar*) fill. **~ida** *f* attack. **~ividad** *f* aggression; (*iniciativa*) enterprise
acomod|able *a* adaptable. **~adizo** *a* accommodating. **~ado** *a* well off. **~ador** *m* usher. **~adora** *f* usherette. **~amiento** *m* suitability. **~ar** *vt* arrange; (*adaptar*) adjust. ● *vi* be suitable. **~arse** *vpr* settle down; (*adaptarse*) conform. **~aticio** *a* accommodating. **~o** *m* position
acompaña|do *a* accompanied; (*concurrido*) busy. **~miento** *m* accompaniment. **~nta** *f* companion. **~nte** *m* companion; (*mus*) accompanist. **~r** *vt* accompany; (*adjuntar*) enclose. **~rse** *vpr* (*mus*) accompany o.s.

acompasa|do *a* rhythmic. **~r** *vt* keep in time; (*fig, ajustar*) adjust
acondiciona|do *a* equipped. **~miento** *m* conditioning. **~r** *vt* fit out; (*preparar*) prepare
acongojar *vt* distress. **~se** *vpr* get upset
acónito *m* aconite
aconseja|ble *a* advisable. **~do** *a* advised. **~r** *vt* advise. **~rse** *vpr* take advice. **~rse con** consult
aconsonantar *vt/i* rhyme
acontec|er [11] *vi* happen. **~imiento** *m* event
acopi|ar *vt* collect. **~o** *m* store
acopla|do *a* coordinated. **~miento** *m* coupling; (*elec*) connection. **~r** *vt* fit; (*elec*) connect; (*rail*) couple
acoquina|miento *m* intimidation. **~r** *vt* intimidate. **~rse** *vpr* be intimidated
acoraza|do *a* armour-plated. ● *m* battleship. **~r** [10] *vt* armour
acorazonado *a* heart-shaped
acorcha|do *a* spongy. **~rse** *vpr* go spongy; ‹*parte del cuerpo*› go to sleep
acord|ado *a* agreed. **~ar** [2] *vt* agree (upon); (*decidir*) decide; (*recordar*) remind. **~e** *a* in agreement; (*mus*) harmonious. ● *m* chord
acorde|ón *m* accordion. **~onista** *m & f* accordionist
acordona|do *a* ‹*lugar*› cordoned off. **~miento** *m* cordoning off. **~r** *vt* tie, lace; (*rodear*) surround, cordon off
acorrala|miento *m* (*de animales*) rounding up; (*de personas*) cornering. **~r** *vt* round up ‹*animales*›; corner ‹*personas*›
acorta|miento *m* shortening. **~r** *vt* shorten; (*fig*) cut down
acos|ar *vt* hound; (*fig*) pester. **~o** *m* pursuit; (*fig*) pestering
acostar [2] *vt* put to bed; (*naut*) bring alongside. ● *vi* (*naut*) reach land. **~se** *vpr* go to bed; (*echarse*) lie down; (*Mex, parir*) give birth
acostumbra|do *a* (*habitual*) usual. **~do a** used to, accustomed to. **~r** *vt* get used. **me ha acostumbrado a levantarme por la noche** he's got me used to getting up at night. ● *vi*. **~r (a)** be accustomed to. **acostumbro comer a la una** I usually have lunch at one o'clock. **~rse** *vpr* become accustomed, get used
acota|ción *f* (*nota*) marginal note; (*en el teatro*) stage direction; (*cota*) elevation mark. **~do** *a* enclosed. **~r** *vt* mark out ‹*terreno*›; (*anotar*) annotate
ácrata *a* anarchistic. ● *m & f* anarchist
acre *m* acre. ● *a* ‹*olor*› pungent; ‹*sabor*› sharp, bitter
acrecenta|miento *m* increase. **~r** [1] *vt* increase. **~rse** *vpr* increase
acrec|er [11] *vt* increase. **~imiento** *m* increase
acredita|do *a* reputable; (*pol*) accredited. **~r** *vt* prove; accredit ‹*representante diplomático*›; (*garantizar*) guarantee; (*autorizar*) authorize. **~rse** *vpr* make one's name
acreedor *a* worthy (**a** of). ● *m* creditor
acribillar *vt* (*a balazos*) riddle (**a** with); (*a picotazos*) cover (**a** with); (*fig, a preguntas etc*) pester (**a** with)
acrimonia *f* (*de sabor*) sharpness; (*de olor*) pungency; (*fig*) bitterness
acrisola|do *a* pure; (*fig*) proven. **~r** *vt* purify; (*confirmar*) prove
acritud *f* (*de sabor*) sharpness; (*de olor*) pungency; (*fig*) bitterness
acr|obacia *f* acrobatics. **~obacias aéreas** aerobatics. **~óbata** *m & f* acrobat. **~obático** *a* acrobatic. **~obatismo** *m* acrobatics
acrónimo *m* acronym
acróstico *a & m* acrostic
acta *f* minutes; (*certificado*) certificate
actinia *f* sea anemone
actitud *f* posture, position; (*fig*) attitude, position
activ|ación *f* speed-up. **~amente** *adv* actively. **~ar** *vt* activate; (*acelerar*) speed up. **~idad** *f* activity. **~o** *a* active. ● *m* assets
acto *m* act; (*ceremonia*) ceremony. **en el ~** immediately
act|or *m* actor. **~riz** *f* actress
actuación *f* action; (*conducta*) behaviour; (*theat*) performance
actual *a* present; ‹*asunto*› topical. **~idad** *f* present. **~idades** *fpl* current affairs. **~ización** *f* modernization. **~izar** [10] *vt* modernize. **~mente** *adv* now, at the present time. **en la ~idad** nowadays
actuar [21] *vt* work. ● *vi* act. **~ como, ~ de** act as

actuario *m* clerk of the court. **~ (de seguros)** actuary
acuarel|a *f* watercolour. **~ista** *m & f* watercolourist
acuario *m* aquarium. **A~** Aquarius
acuartela|do *a* quartered. **~miento** *m* quartering. **~r** *vt* quarter, billet; (*mantener en cuartel*) confine to barracks
acuático *a* aquatic
acuci|ador pressing. **~ar** *vt* urge on; (*dar prisa a*) hasten. **~oso** *a* keen
acuclillarse *vpr* crouch down, squat down
acuchilla|do *a* slashed; ‹*persona*› stabbed. **~r** *vt* slash; stab ‹*persona*›; (*alisar*) smooth
acudir *vi*. **~ a** go to, attend; keep ‹*una cita*›; (*en auxilio*) go to help
acueducto *m* aqueduct
acuerdo *m* agreement. ● *vb véase* **acordar**. **¡de ~!** OK! **de ~ con** in accordance with. **estar de ~** agree. **ponerse de ~** agree
acuesto *vb véase* **acostar**
acuidad *f* acuity, sharpness
acumula|ción *f* accumulation. **~dor** *a* accumulative. ● *m* accumulator. **~r** *vt* accumulate. **~rse** *vpr* accumulate
acunar *vt* rock
acuña|ción *f* minting, coining. **~r** *vt* mint, coin
acuos|idad *f* wateriness. **~o** *a* watery
acupuntura *f* acupuncture
acurrucarse [7] *vpr* curl up
acusa|ción *f* accusation. **~do** *a* accused; (*destacado*) marked. ● *m* accused. **~dor** *a* accusing. ● *m* accuser. **~r** *vt* accuse; (*mostrar*) show; (*denunciar*) denounce. **~rse** *vpr* confess; (*notarse*) become marked. **~torio** *a* accusatory
acuse *m*. **~ de recibo** acknowledgement of receipt
acus|ica *m & f* (*fam*) telltale. **~ón** *a & m* telltale
acústic|a *f* acoustics. **~o** *a* acoustic
achacar [7] *vt* attribute
achacoso *a* sickly
achaflanar *vt* bevel
achantar *vt* (*fam*) intimidate. **~se** *vpr* hide; (*fig*) back down
achaparrado *a* stocky
achaque *m* ailment
achares *mpl* (*fam*). **dar ~** make jealous
achata|miento *m* flattening. **~r** *vt* flatten
achica|do *a* childish. **~r** [7] *vt* make smaller; (*fig, empequeñecer, fam*) belittle; (*naut*) bale out. **~rse** *vpr* become smaller; (*humillarse*) be humiliated
achicopalado *a* (*Mex*) depressed
achicoria *f* chicory
achicharra|dero *m* inferno. **~nte** *a* sweltering. **~r** *vt* burn; (*fig*) pester. **~rse** *vpr* burn
achispa|do *a* tipsy. **~rse** *vpr* get tipsy
achocolatado *a* (chocolate-)brown
achuch|ado *a* (*fam*) hard. **~ar** *vt* jostle, push. **~ón** *m* shove, push
achulado *a* cocky
adagio *m* adage, proverb; (*mus*) adagio
adalid *m* leader
adamascado *a* damask
adapta|ble *a* adaptable. **~ción** *f* adaptation. **~dor** *m* adapter. **~r** *vt* adapt; (*ajustar*) fit. **~rse** *vpr* adapt o.s.
adecentar *vt* clean up. **~se** *vpr* tidy o.s. up
adecua|ción *f* suitability. **~damente** *adv* suitably. **~do** *a* suitable. **~r** *vt* adapt, make suitable
adelant|ado *a* advanced; ‹*niño*› precocious; ‹*reloj*› fast. **~amiento** *m* advance(ment); (*auto*) overtaking. **~ar** *vt* advance, move forward; (*acelerar*) speed up; put forward ‹*reloj*›; (*auto*) overtake. ● *vi* advance, go forward; ‹*reloj*› gain, be fast. **~arse** *vpr* advance, move forward; ‹*reloj*› gain; (*auto*) overtake. **~e** *adv* forward. ● *int* come in!; (*¡siga!*) carry on! **~o** *m* advance; (*progreso*) progress. **más ~e** (*lugar*) further on; (*tiempo*) later on. **pagar por ~ado** pay in advance.
adelfa *f* oleander
adelgaza|dor *a* slimming. **~miento** *m* slimming. **~r** [10] *vt* make thin. ● *vi* lose weight; (*adrede*) slim. **~rse** *vpr* lose weight; (*adrede*) slim
ademán *m* gesture. **ademanes** *mpl* (*modales*) manners. **en ~ de** as if to
además *adv* besides; (*también*) also. **~ de** besides
adentr|arse *vpr*. **~ en** penetrate into; study thoroughly ‹*tema etc*›. **~o** *adv* in(side). **mar ~o** out at sea. **tierra ~o** inland
adepto *m* supporter

aderez|ar [10] *vt* flavour ‹*bebidas*›; (*condimentar*) season; dress ‹*ensalada*›. **~o** *m* flavouring; (*con condimentos*) seasoning; (*para ensalada*) dressing
adeud|ar *vt* owe. **~o** *m* debit
adhe|rencia *f* adhesion; (*fig*) adherence. **~rente** *a* adherent. **~rir** [4] *vt* stick on. ● *vi* stick. **~rirse** *vpr* stick; (*fig*) follow. **~sión** *f* adhesion; (*fig*) support. **~sivo** *a* & *m* adhesive
adici|ón *f* addition. **~onal** *a* additional. **~onar** *vt* add
adicto *a* devoted. ● *m* follower
adiestra|do *a* trained. **~miento** *m* training. **~r** *vt* train. **~rse** *vpr* practise
adinerado *a* wealthy
adiós *int* goodbye!; (*al cruzarse con alguien*) hello!
adit|amento *m* addition; (*accesorio*) accessory. **~ivo** *m* additive
adivin|ación *f* divination; (*por conjeturas*) guessing. **~ador** *m* fortune-teller. **~anza** *f* riddle. **~ar** *vt* foretell; (*acertar*) guess. **~o** *m* fortune-teller
adjetivo *a* adjectival. ● *m* adjective
adjudica|ción *f* award. **~r** [7] *vt* award. **~rse** *vpr* appropriate. **~tario** *m* winner of an award
adjunt|ar *vt* enclose. **~o** *a* enclosed; (*auxiliar*) assistant. ● *m* assistant
adminículo *m* thing, gadget
administra|ción *f* administration; (*gestión*) management. **~dor** *m* administrator; (*gerente*) manager. **~dora** *f* administrator; manageress. **~r** *vt* administer. **~tivo** *a* administrative
admira|ble *a* admirable. **~ción** *f* admiration. **~dor** *m* admirer. **~r** *vt* admire; (*asombrar*) astonish. **~rse** *vpr* be astonished. **~tivo** *a* admiring
admi|sibilidad *f* admissibility. **~sible** *a* acceptable. **~sión** *f* admission; (*aceptación*) acceptance. **~tir** *vt* admit; (*aceptar*) accept
adobar *vt* (*culin*) pickle; (*fig*) twist
adobe *m* sun-dried brick. **~ra** *f* mould for making (sun-dried) bricks
adobo *m* pickle
adocena|do *a* common. **~rse** *vpr* become common
adoctrinamiento *m* indoctrination
adolecer [11] *vi* be ill. **~ de** suffer with
adolescen|cia *f* adolescent. **~te** *a* & *m* & *f* adolescent
adonde *conj* where
adónde *adv* where?
adop|ción *f* adoption. **~tar** *vt* adopt. **~tivo** *a* adoptive; ‹*patria*› of adoption
adoqu|ín *m* paving stone; (*imbécil*) idiot. **~inado** *m* paving. **~inar** *vt* pave
adora|ble *a* adorable. **~ción** *f* adoration. **~dor** *a* adoring. ● *n* worshipper. **~r** *vt* adore
adormec|edor *a* soporific; ‹*droga*› sedative. **~er** [11] *vt* send to sleep; (*fig, calmar*) calm, soothe. **~erse** *vpr* fall asleep; ‹*un miembro*› go to sleep. **~ido** *a* sleepy; ‹*un miembro*› numb. **~imiento** *m* sleepiness; (*de un miembro*) numbness
adormidera *f* opium poppy
adormilarse *vpr* doze
adorn|ar *vt* adorn (**con, de** with). **~o** *m* decoration
adosar *vt* lean (**a** against)
adqui|rido *a* acquired. **~rir** [4] *vt* acquire; (*comprar*) buy. **~sición** *f* acquisition; (*compra*) purchase. **~sitivo** *a* acquisitive. **poder** *m* **~sitivo** purchasing power
adrede *adv* on purpose
adrenalina *f* adrenalin
adscribir [*pp* **adscrito**] *vt* appoint
aduan|a *f* customs. **~ero** *a* customs. ● *m* customs officer
aducir [47] *vt* allege
adueñarse *vpr* take possession
adul|ación *f* flattery. **~ador** *a* flattering. ● *m* flatterer. **~ar** *vt* flatter
ad|ulteración *f* adulteration. **~ulterar** *vt* adulterate. ● *vi* commit adultery. **~ulterino** *a* adulterous. **~ulterio** *m* adultery. **~últera** *f* adulteress. **~últero** *a* adulterous. ● *m* adulterer
adulto *a* & *m* adult, grown-up
adusto *a* severe, harsh
advenedizo *a* & *m* upstart
advenimiento *m* advent, arrival; (*subida al trono*) accession
adventicio *a* accidental
adverbi|al *a* adverbial. **~o** *m* adverb
advers|ario *m* adversary. **~idad** *f* adversity. **~o** *a* adverse, unfavourable
advert|encia *f* warning; (*prólogo*) foreword. **~ido** *a* informed. **~ir** [4] *vt* warn; (*notar*) notice

adviento *m* Advent
advocación *f* dedication
adyacente *a* adjacent
aéreo *a* air; (*photo*) aerial; ‹*ferrocarril*› overhead; (*fig*) flimsy
aeróbica *f* aerobics
aerodeslizador *m* hovercraft
aerodinámic|a *f* aerodynamics. **~o** *a* aerodynamic
aeródromo *m* aerodrome, airdrome (*Amer*)
aero|espacial *a* aerospace. **~faro** *m* beacon. **~lito** *m* meteorite. **~nauta** *m* & *f* aeronaut. **~náutica** *f* aeronautics. **~náutico** *a* aeronautical. **~nave** *f* airship. **~puerto** *m* airport. **~sol** *m* aerosol
afab|ilidad *f* affability. **~le** *a* affable
afamado *a* famous
af|án *m* hard work; (*deseo*) desire. **~anar** *vt* (*fam*) pinch. **~anarse** *vpr* strive (**en, por** to). **~anoso** *a* laborious
afea|miento *m* disfigurement. **~r** *vt* disfigure, make ugly; (*censurar*) censure
afección *f* disease
afecta|ción *f* affectation. **~do** *a* affected. **~r** *vt* affect
afect|ísimo *a* affectionate. **~ísimo amigo** (*en cartas*) my dear friend. **~ividad** *f* emotional nature. **~ivo** *a* sensitive. **~o** *m* (*cariño*) affection. ● *a*. **~o a** attached to. **~uosidad** *f* affection. **~uoso** *a* affectionate. **con un ~uoso saludo** (*en cartas*) with kind regards. **suyo ~ísimo** (*en cartas*) yours sincerely
afeita|do *m* shave. **~dora** *f* electric razor. **~r** *vt* shave. **~rse** *vpr* (have a) shave
afelpado *a* velvety
afemina|do *a* effeminate. ● *m* effeminate person. **~miento** *m* effeminacy. **~rse** *vpr* become effeminate
aferrar [1] *vt* grasp
afgano *a* & *m* Afghan
afianza|miento *m* (*reforzar*) strengthening; (*garantía*) guarantee. **~rse** [10] *vpr* become established
afici|ón *f* liking; (*conjunto de aficionados*) fans. **~onado** *a* keen (**a** on), fond (**a** of). ● *m* fan. **~onar** *vt* make fond. **~onarse** *vpr* take a liking to. **por ~ón** as a hobby
afila|do *a* sharp. **~dor** *m* knife-grinder. **~dura** *f* sharpening. **~r** *vt* sharpen. **~rse** *vpr* get sharp; (*ponerse flaco*) grow thin
afilia|ción *f* affiliation. **~do** *a* affiliated. **~rse** *vpr* become a member (**a** of)
afiligranado *a* filigreed; (*fig*) delicate
afín *a* similar; (*próximo*) adjacent; ‹*personas*› related
afina|ción *f* refining; (*auto, mus*) tuning. **~do** *a* finished; (*mus*) in tune. **~r** *vt* refine; (*afilar*) sharpen; (*acabar*) finish; (*auto, mus*) tune. ● *vi* be in tune. **~rse** *vpr* become more refined
afincarse [7] *vpr* settle
afinidad *f* affinity; (*parentesco*) relationship
afirma|ción *f* affirmation. **~r** *vt* make firm; (*asentir*) affirm. **~rse** *vpr* steady o.s.; (*confirmar*) confirm. **~tivo** *a* affirmative
aflic|ción *f* affliction. **~tivo** *a* distressing
afligi|do *a* distressed. ● *m* afflicted. **~r** [14] *vt* distress. **~rse** *vpr* grieve
afloja|miento *m* loosening. **~r** *vt* loosen; (*relajar*) ease. ● *vi* let up
aflora|miento *m* outcrop. **~r** *vi* appear on the surface
aflu|encia *f* flow. **~ente** *a* flowing. ● *m* tributary. **~ir** [17] *vi* flow (**a** into)
af|onía *f* hoarseness. **~ónico** *a* hoarse
aforismo *m* aphorism
aforo *m* capacity
afortunado *a* fortunate, lucky
afrancesado *a* francophile
afrent|a *f* insult; (*vergüenza*) disgrace. **~ar** *vt* insult. **~oso** *a* insulting
África *f* Africa. **~ del Sur** South Africa
africano *a* & *m* African
afrodisíaco *a* & *m*, **afrodisiaco** *a* & *m* aphrodisiac
afrontar *vt* bring face to face; (*enfrentar*) face, confront
afuera *adv* out(side). **¡~!** out of the way! **~s** *fpl* outskirts.
agachar *vt* lower. **~se** *vpr* bend over
agalla *f* (*de los peces*) gill. **~s** *fpl* (*fig*) guts
agarrada *f* row
agarrader|a *f* (*LAm*) handle. **~o** *m* handle. **tener ~as** (*LAm*), **tener ~os** have influence

agarr|ado *a* (*fig, fam*) mean. **~ador** *a* (*Arg*) ‹*bebida*› strong. **~ar** *vt* grasp; (*esp LAm*) take, catch. ● *vi* ‹*plantas*› take root. **~arse** *vpr* hold on; (*reñirse, fam*) fight. **~ón** *m* tug; (*LAm, riña*) row

agarrota|miento *m* tightening; (*auto*) seizing up. **~r** *vt* tie tightly; ‹*el frío*› stiffen; garotte ‹*un reo*›. **~rse** *vpr* go stiff; (*auto*) seize up

agasaj|ado *m* guest of honour. **~ar** *vt* look after well. **~o** *m* good treatment

ágata *f* agate

agavilla|dora *f* (*máquina*) binder. **~r** *vt* bind

agazaparse *vpr* hide

agencia *f* agency. **~ de viajes** travel agency. **~ inmobiliaria** estate agency (*Brit*), real estate agency (*Amer*). **~r** *vt* find. **~rse** *vpr* find (out) for o.s.

agenda *f* notebook

agente *m* agent; (*de policía*) policeman. **~ de aduanas** customs officer. **~ de bolsa** stockbroker

ágil *a* agile

agilidad *f* agility

agita|ción *f* waving; (*de un líquido*) stirring; (*intranquilidad*) agitation. **~do** *a* ‹*el mar*› rough; (*fig*) agitated. **~dor** *m* (*pol*) agitator

agitanado *a* gypsy-like

agitar *vt* wave; shake ‹*botellas etc*›; stir ‹*líquidos*›; (*fig*) stir up. **~se** *vpr* wave; ‹*el mar*› get rough; (*fig*) get excited

aglomera|ción *f* agglomeration; (*de tráfico*) traffic jam. **~r** *vt* amass. **~rse** *vpr* form a crowd

agn|osticismo *m* agnosticism. **~óstico** *a & m* agnostic

agobi|ador *a* ‹*trabajo*› exhausting; ‹*calor*› oppressive. **~ante** *a* ‹*trabajo*› exhausting; ‹*calor*› oppressive. **~ar** *vt* weigh down; (*fig, abrumar*) overwhelm. **~o** *m* weight; (*cansancio*) exhaustion; (*opresión*) oppression

agolpa|miento *m* (*de gente*) crowd; (*de cosas*) pile. **~rse** *vpr* crowd together

agon|ía *f* death throes; (*fig*) agony. **~izante** *a* dying; ‹*luz*› failing. **~izar** [10] *vi* be dying

agor|ar [16] *vt* prophesy. **~ero** *a* of ill omen. ● *m* soothsayer

agostar *vt* wither

agosto *m* August. **hacer su ~** feather one's nest

agota|do *a* exhausted; ‹*libro*› out of print. **~dor** *a* exhausting. **~miento** *m* exhaustion. **~r** *vt* exhaust. **~rse** *vpr* be exhausted; ‹*libro*› go out of print

agracia|do *a* attractive; (*que tiene suerte*) lucky. **~r** make attractive

agrada|ble *a* pleasant, nice. **~r** *vi* please. **esto me ~** I like this

agradec|er [11] *vt* thank ‹*persona*›; be grateful for ‹*cosa*›. **~ido** *a* grateful. **~imiento** *m* gratitude. **¡muy ~ido!** thanks a lot!

agrado *m* pleasure; (*amabilidad*) friendliness

agrandar *vt* enlarge; (*fig*) exaggerate. **~se** *vpr* get bigger

agrario *a* agrarian, land; ‹*política*› agricultural

agrava|miento *m* worsening. **~nte** *a* aggravating. ● *f* additional problem. **~r** *vt* aggravate; (*aumentar el peso*) make heavier. **~rse** *vpr* get worse

agravi|ar *vt* offend; (*perjudicar*) wrong. **~arse** *vpr* be offended. **~o** *m* offence

agraz *m*. **en ~** prematurely

agredir [24] *vt* attack. **~ de palabra** insult

agrega|do *m* aggregate; (*funcionario diplomático*) attaché. **~r** [12] *vt* add; (*unir*) join; appoint ‹*persona*›

agremiar *vt* form into a union. **~se** *vpr* form a union

agres|ión *f* aggression; (*ataque*) attack. **~ividad** *f* aggressiveness. **~ivo** aggressive. **~or** *m* aggressor

agreste *a* country

agria|do *a* (*fig*) embittered. **~r** [*regular, o raramente* 20] *vt* sour. **~rse** *vpr* turn sour; (*fig*) become embittered

agr|ícola *a* agricultural. **~icultor** *a* agricutural. ● *m* farmer. **~icultura** *f* agriculture, farming

agridulce *a* bitter-sweet; (*culin*) sweet-and-sour

agriera *f* (*LAm*) heartburn

agrietar *vt* crack. **~se** *vpr* crack; ‹*piel*› chap

agrimens|or *m* surveyor. **~ura** *f* surveying

agrio *a* sour; (*fig*) sharp. **~s** *mpl* citrus fruits

agronomía *f* agronomy

agropecuario *a* farming

agrupa|ción *f* group; (*acción*) grouping. **~r** *vt* group. **~rse** *vpr* form a group

agua *f* water; (*lluvia*) rain; (*marea*) tide; (*vertiente del tejado*) slope. **~ abajo** downstream. **~ arriba** upstream. **~ bendita** holy water. **~ caliente** hot water. **estar entre dos ~s** sit on the fence. **hacer ~** (*naut*) leak. **nadar entre dos ~s** sit on the fence

aguacate *m* avocado pear; (*árbol*) avocado pear tree

aguacero *m* downpour, heavy shower

agua *f* **corriente** running water

aguachinarse *vpr* (*Mex*) ‹*cultivos*› be flooded

aguada *f* watering place; (*naut*) drinking water; (*acuarela*) water-colour

agua *f* **de colonia** eau-de-Cologne

aguad|o *a* watery. **~ucho** *m* refreshment kiosk

agua: ~ dulce fresh water. **~fiestas** *m & f invar* spoil-sport, wet blanket. **~ fría** cold water. **~fuerte** *m* etching

aguaje *m* spring tide

agua: ~mala *f*, **~mar** *m* jellyfish

aguamarina *f* aquamarine

agua: ~miel *f* mead. **~ mineral con gas** fizzy mineral water. **~ mineral sin gas** still mineral water. **~nieve** *f* sleet

aguanoso *a* watery; ‹*tierra*› waterlogged

aguant|able *a* bearable. **~aderas** *fpl* patience. **~ar** *vt* put up with, bear; (*sostener*) support. ● *vi* hold out. **~arse** *vpr* restrain o.s. **~e** *m* patience; (*resistencia*) endurance

agua: ~pié *m* watery wine. **~ potable** drinking water. **~r** [15] *vt* water down. **~ salada** salt water.

aguardar *vt* wait for. ● *vi* wait

agua: ~rdiente *m* (cheap) brandy. **~rrás** *m* turpentine, turps (*fam*). **~turma** *f* Jerusalem artichoke. **~zal** *m* puddle

agud|eza *f* sharpness; (*fig, perspicacia*) insight; (*fig, ingenio*) wit. **~izar** [10] *vt* sharpen. **~izarse** *vpr* ‹*enfermedad*› get worse. **~o** *a* sharp; ‹*ángulo, enfermedad*› acute; ‹*voz*› high-pitched

agüero *m* omen. **ser de buen ~** augur well

aguij|ada *f* goad. **~ar** *vt* (*incl fig*) goad. **~ón** *m* point of a goad. **~onazo** *m* prick. **~onear** *vt* goad

águila *f* eagle; (*persona perspicaz*) astute person

aguileña *f* columbine

aguil|eño *a* aquiline. **~ucho** *m* eaglet

aguinaldo *m* Christmas box

aguja *f* needle; (*del reloj*) hand; (*arquit*) steeple. **~s** *fpl* (*rail*) points

agujer|ear *vt* make holes in. **~o** *m* hole

agujetas *fpl* stiffness. **tener ~** be stiff

agujón *m* hairpin

agusanado *a* full of maggots

agutí *m* (*LAm*) guinea pig

aguza|do *a* sharp. **~miento** *m* sharpening. **~r** [10] *vt* sharpen

ah *int* ah!, oh!

aherrojar *vt* (*fig*) oppress

ahí *adv* there. **de ~ que** so that. **por ~** over there; (*aproximadamente*) thereabouts

ahija|da *f* god-daughter, godchild. **~do** *m* godson, godchild. **~r** *vt* adopt

ahínco *m* enthusiasm; (*empeño*) insistence

ahíto *a* full up

ahog|ado *a* (*en el agua*) drowned; (*asfixiado*) suffocated. **~ar** [12] *vt* (*en el agua*) drown; (*asfixiar*) suffocate; put out ‹*fuego*›. **~arse** *vpr* (*en el agua*) drown; (*asfixiarse*) suffocate. **~o** *m* breathlessness; (*fig, angustia*) distress; (*apuro*) financial trouble

ahondar *vt* deepen; ● *vi* go deep. **~ en** (*fig*) examine in depth. **~se** *vpr* get deeper

ahora *adv* now; (*hace muy poco*) just now; (*dentro de poco*) very soon. **~ bien** but. **~ mismo** right now. **de ~ en adelante** from now on, in future. **por ~** for the time being

ahorca|dura *f* hanging. **~r** [7] *vt* hang. **~rse** *vpr* hang o.s.

ahorita *adv* (*fam*) now. **~ mismo** right now

ahorquillar *vt* shape like a fork

ahorr|ador *a* thrifty. **~ar** *vt* save. **~arse** *vpr* save o.s. **~o** *m* saving; (*cantidad ahorrada*) savings. **~os** *mpl* savings

ahuecar [7] *vt* hollow; fluff up ‹*colchón*›; deepen ‹*la voz*›; (*marcharse, fam*) clear off (*fam*)

ahuizote *m* (*Mex*) bore
ahulado *m* (*LAm*) oilskin
ahuma|do *a* (*culin*) smoked; (*de colores*) smoky. **~r** *vt* (*culin*) smoke; (*llenar de humo*) fill with smoke. ● *vi* smoke. **~rse** *vpr* become smoky; ‹*comida*› acquire a smoky taste; (*emborracharse, fam*) get drunk
ahusa|do *a* tapering. **~rse** *vpr* taper
ahuyentar *vt* drive away; banish ‹*pensamientos etc*›
airado *a* annoyed
aire *m* air; (*viento*) breeze; (*corriente*) draught; (*aspecto*) appearance; (*mus*) tune, air. **~ación** *f* ventilation. **~ acondicionado** air-conditioned. **~ar** *vt* air; (*ventilar*) ventilate; (*fig, publicar*) make public. **~arse** *vpr*. **salir para ~arse** go out for some fresh air. **al ~ libre** in the open air. **darse ~s** give o.s. airs
airón *m* heron
airos|amente *adv* gracefully. **~o** *a* draughty; (*fig*) elegant
aisla|do *a* isolated; (*elec*) insulated. **~dor** *a* (*elec*) insulating. ● *m* (*elec*) insulator. **~miento** *m* isolation; (*elec*) insulation. **~nte** *a* insulating. **~r** [23] *vt* isolate; (*elec*) insulate
ajajá *int* good! splendid!
ajar *vt* crumple; (*estropear*) spoil
ajedre|cista *m & f* chess-player. **~z** *m* chess. **~zado** *a* chequered, checked
ajenjo *m* absinthe
ajeno *a* (*de otro*) someone else's; (*de otros*) other people's; (*extraño*) alien
ajetre|arse *vpr* be busy. **~o** *m* bustle
ají *m* (*LAm*) chilli; (*salsa*) chilli sauce
aj|iaceite *m* garlic sauce. **~ilimójili** *m* piquant garlic sauce. **~illo** *m* garlic. **al ~illo** cooked with garlic. **~o** *m* garlic. **~o-a-rriero** *m* cod in garlic sauce
ajorca *f* bracelet
ajuar *m* furnishings; (*de novia*) trousseau
ajuma|do *a* (*fam*) drunk. **~rse** *vpr* (*fam*) get drunk
ajust|ado *a* right; ‹*vestido*› tight. **~ador** *m* fitter. **~amiento** *m* fitting; (*adaptación*) adjustment; (*acuerdo*) agreement; (*de una cuenta*) settlement. **~ar** *vt* fit; (*adaptar*) adapt; (*acordar*) agree; settle ‹*una cuenta*›; (*apretar*) tighten. ● *vi* fit. **~arse** *vpr* fit; (*adaptarse*) adapt o.s.; (*acordarse*) come to an agreement. **~e** m fitting; (*adaptación*) *f* adjustment; (*acuerdo*) agreement; (*de una cuenta*) settlement
ajusticiar *vt* execute
al = **a**; **el**
ala *f* wing; (*de sombrero*) brim; (*deportes*) winger
alaba|ncioso *a* boastful. **~nza** *f* praise. **~r** *vt* praise. **~rse** *vpr* boast
alabastro *m* alabaster
álabe *m* (*paleta*) paddle; (*diente*) cog
alabe|ar *vt* warp. **~arse** *vpr* warp. **~o** *m* warping
alacena *f* cupboard (*Brit*), closet (*Amer*)
alacrán *m* scorpion
alacridad *f* alacrity
alado *a* winged
alambi|cado *a* distilled; (*fig*) subtle. **~camiento** *m* distillation; (*fig*) subtlety. **~car** [7] *vt* distil. **~que** *m* still
alambr|ada *f* wire fence; (*de alambre de espinas*) barbed wire fence. **~ar** *vt* fence. **~e** *m* wire. **~e de espinas** barbed wire. **~era** *f* fireguard
alameda *f* avenue; (*plantío de álamos*) poplar grove
álamo *m* poplar. **~ temblón** aspen
alano *m* mastiff
alarde *m* show. **~ar** *vi* boast
alarga|dera *f* extension. **~do** *a* long. **~dor** *m* extension. **~miento** *m* lengthening. **~r** [12] *vt* lengthen; stretch out ‹*mano etc*›; (*dar*) give, pass. **~rse** *vpr* lengthen, get longer
alarido *m* shriek
alarm|a *f* alarm. **~ante** *a* alarming. **~ar** *vt* alarm, frighten. **~arse** *vpr* be alarmed. **~ista** *m & f* alarmist
alba *f* dawn
albacea *m* executor. ● *f* executrix
albacora (*culin*) tuna(-fish)
albahaca *f* basil
albanés *a & m* Albanian
Albania *f* Albania
albañal *m* sewer, drain
albañil *m* bricklayer. **~ería** *f* (*arte*) bricklaying
albarán *m* delivery note
albarda *f* packsaddle; (*Mex*) saddle. **~r** *vt* saddle
albaricoque *m* apricot. **~ro** *m* apricot tree
albatros *m* albatross
albedrío *m* will. **libre ~** free will

albéitar *m* veterinary surgeon (*Brit*), veterinarian (*Amer*), vet (*fam*)
alberca *f* tank, reservoir
alberg|ar [12] *vt* (*alojar*) put up; ‹*viviendas*› house; (*dar asilo*) shelter. **~arse** *vpr* stay; (*refugiarse*) shelter. **~ue** *m* accommodation; (*refugio*) shelter. **~ue de juventud** youth hostel
albóndiga *f* meatball, rissole
albor *m* dawn. **~ada** *f* dawn; (*mus*) dawn song. **~ear** *vi* dawn
albornoz *m* (*de los moros*) burnous; (*para el baño*) bathrobe
alborot|adizo *a* excitable. **~ado** *a* excited; (*aturdido*) hasty. **~ador** *a* rowdy. ● *m* trouble-maker. **~ar** *vt* disturb, upset. ● *vi* make a racket. **~arse** *vpr* get excited; ‹*el mar*› get rough. **~o** *m* row, uproar
alboroz|ado *a* overjoyed. **~ar** [10] *vt* make laugh; (*regocijar*) make happy. **~arse** *vpr* be overjoyed. **~o** *m* joy
albufera *f* lagoon
álbum *m* (*pl* **~es** *o* **~s**) album
alcachofa *f* artichoke
alcald|e *m* mayor. **~esa** *f* mayoress. **~ía** *f* mayoralty; (*oficina*) mayor's office
álcali *m* alkali
alcalino *a* alkaline
alcance *m* reach; (*de arma, telescopio etc*) range; (*déficit*) deficit
alcancía *f* money-box
alcantarilla *f* sewer; (*boca*) drain
alcanzar [10] *vt* (*llegar a*) catch up; (*coger*) reach; catch ‹*un autobús*›; ‹*bala etc*› strike, hit. ● *vi* reach; (*ser suficiente*) be enough. **~ a** manage
alcaparra *f* caper
alcaucil *m* artichoke
alcayata *f* hook
alcazaba *f* fortress
alcázar *m* fortress
alcoba *f* bedroom
alcoh|ol *m* alcohol. **~ol desnaturalizado** methylated spirits, meths (*fam*). **~ólico** *a* & *m* alcoholic. **~olímetro** *m* breathalyser (*Brit*). **~olismo** *m* alcoholism. **~olizarse** [10] *vpr* become an alcoholic
Alcorán *m* Koran
alcornoque *m* cork-oak; (*persona torpe*) idiot
alcuza *f* (olive) oil bottle
aldaba *f* door-knocker. **~da** *f* knock at the door
alde|a *f* village. **~ano** *a* village; (*campesino*) rustic, country. **~huela** *f* hamlet
alea|ción *f* alloy. **~r** *vt* alloy
aleatorio *a* uncertain
alecciona|dor *a* instructive. **~miento** *m* instruction. **~r** *vt* instruct
aledaños *mpl* outskirts
alega|ción *f* allegation; (*Arg, Mex, disputa*) argument. **~r** [12] *vt* claim; (*jurid*) allege. ● *vi* (*LAm*) argue. **~to** *m* plea
aleg|oría *f* allegory. **~órico** *a* allegorical
alegr|ar *vt* make happy; (*avivar*) brighten up. **~arse** *vpr* be happy; (*emborracharse*) get merry. **~e** *a* happy; (*achispado*) merry, tight. **~emente** *adv* happily. **~ía** *f* happiness. **~ón** *m* sudden joy, great happiness
aleja|do *a* distant. **~miento** *m* removal; (*entre personas*) estrangement; (*distancia*) distance. **~r** *vt* remove; (*ahuyentar*) get rid of; (*fig, apartar*) separate. **~rse** *vpr* move away
alela|do *a* stupid. **~r** *vt* stupefy. **~rse** *vpr* be stupefied
aleluya *m* & *f* alleluia
alemán *a* & *m* German
Alemania *f* Germany. **~ Occidental** (*historia*) West Germany. **~ Oriental** (*historia*) East Germany
alenta|dor *a* encouraging. **~r** [1] *vt* encourage. ● *vi* breathe
alerce *m* larch
al|ergia *f* allergy. **~érgico** *a* allergic
alero *m* (*del tejado*) eaves
alerón *m* aileron
alerta *adv* alert, on the alert. **¡~!** look out! **~r** *vt* alert
aleta *f* wing; (*de pez*) fin
aletarga|do *a* lethargic. **~miento** *m* lethargy. **~r** [12] *vt* make lethargic. **~rse** *vpr* become lethargic
alet|azo *m* (*de un ave*) flap of the wings; (*de un pez*) flick of the fin. **~ear** *vi* flap its wings, flutter. **~eo** *m* flapping (of the wings)
aleve *a* treacherous
alevín *m* young fish
alevos|ía *f* treachery. **~o** *a* treacherous
alfab|ético *a* alphabetical. **~etizar** [10] *vt* alphabetize; teach to read

and write ‹*a uno*›. **~eto** *m* alphabet. **~eto Morse** Morse code
alfalfa *f* lucerne (*Brit*), alfalfa (*Amer*)
alfar *m* pottery. **~ería** *f* pottery. **~ero** *m* potter
alféizar *m* window-sill
alferecía *f* epilepsy
alférez *m* second lieutenant
alfil *m* (*en ajedrez*) bishop
alfile|r *m* pin. **~razo** *m* pinprick. **~tero** *m* pin-case
alfombr|a *f* (*grande*) carpet; (*pequeña*) rug, mat. **~ar** *vt* carpet. **~illa** *f* rug, mat; (*med*) German measles
alforja *f* saddle-bag
algas *fpl* seaweed
algarabía *f* (*fig, fam*) gibberish, nonsense
algarada *f* uproar
algarrob|a *f* carob bean. **~o** *m* carob tree
algazara *f* uproar
álgebra *f* algebra
algebraico *a* algebraic
álgido *a* (*fig*) decisive
algo *pron* something; (*en frases interrogativas*) anything. ● *adv* rather. **¿~ más?** is there anything else? **¿quieres tomar algo?** (*de beber*) would you like a drink?; (*de comer*) would you like something to eat?
algod|ón *m* cotton. **~ón de azúcar** candy floss (*Brit*), cotton candy (*Amer*). **~onero** *a* cotton. ● *m* cotton plant. **~ón hidrófilo** cotton wool
alguacil *m* bailiff
alguien *pron* someone, somebody; (*en frases interrogativas*) anyone, anybody
alguno *a* (*delante de nombres masculinos en singular* **algún**) some; (*en frases interrogativas*) any; (*pospuesto al nombre en frases negativas*) at all. **no tiene idea alguna** he hasn't any idea at all. ● *pron* one; (*en plural*) some; (*alguien*) someone. **alguna que otra vez** from time to time. **algunas veces, alguna vez** sometimes
alhaja *f* piece of jewellery; (*fig*) treasure. **~r** *vt* deck with jewels; (*amueblar*) furnish
alharaca *f* fuss
alhelí *m* wallflower
alheña *f* privet
alhucema *f* lavender
alia|do *a* allied. ● *m* ally. **~nza** *f* alliance; (*anillo*) wedding ring. **~r** [20] *vt* combine. **~rse** *vpr* be combined; (*formar una alianza*) form an alliance
alias *adv* & *m* alias
alicaído *a* (*fig, débil*) weak; (*fig, abatido*) depressed
alicates *mpl* pliers
aliciente *m* incentive; (*de un lugar*) attraction
alien|ado *a* mentally ill. **~ista** *m* & *f* psychiatrist
aliento *m* breath; (*ánimo*) courage
aligera|miento *m* lightening; (*alivio*) alleviation. **~r** *vt* make lighter; (*aliviar*) alleviate, ease; (*apresurar*) quicken
alij|ar *vt* (*descargar*) unload; smuggle ‹*contrabando*›. **~o** *m* unloading; (*contrabando*) contraband
alimaña *f* vicious animal
aliment|ación *f* food; (*acción*) feeding. **~ar** *vt* feed; (*nutrir*) nourish. ● *vi* be nourishing. **~arse** *vpr* feed (**con, de** on). **~icio** *a* nourishing. **~o** *m* food. **~os** *mpl* (*jurid*) alimony. **productos** *mpl* **~icios** foodstuffs
alimón. al ~ *adv* jointly
alinea|ción *f* alignment; (*en deportes*) line-up. **~r** *vt* align, line up
aliñ|ar *vt* (*culin*) season. **~o** *m* seasoning
alioli *m* garlic sauce
alisar *vt* smooth
alisios *apl.* **vientos** *mpl* **~** trade winds
aliso *m* alder (tree)
alista|miento *m* enrolment. **~r** *vt* put on a list; (*mil*) enlist. **~rse** *vpr* enrol; (*mil*) enlist
aliteración *f* alliteration
alivi|ador *a* comforting. **~ar** *vt* lighten; relieve ‹*dolor, etc*›; (*hurtar, fam*) steal, pinch (*fam*). **~arse** *vpr* ‹*dolor*› diminish; ‹*persona*› get better. **~o** *m* relief
aljibe *m* tank
alma *f* soul; (*habitante*) inhabitant
almac|én *m* warehouse; (*LAm, tienda*) grocer's shop; (*de un arma*) magazine. **~enes** *mpl* department store. **~enaje** *m* storage; (*derechos*) storage charges. **~enamiento** *m* storage; (*mercancías almacenadas*) stock. **~enar** *vt* store; stock up with

‹*provisiones*›. **~enero** *m* (*Arg*) shopkeeper. **~enista** *m & f* shopkeeper
almádena *f* sledge-hammer
almanaque *m* almanac
almeja *f* clam
almendr|a *f* almond. **~ado** *a* almond-shaped. **~o** *m* almond tree
almiar *m* haystack
alm|íbar *m* syrup. **~ibarado** *a* syrupy. **~ibarar** *vt* cover in syrup
almid|ón *m* starch. **~onado** *a* starched; (*fig*, *estirado*) starchy
alminar *m* minaret
almirant|azgo *m* admiralty. **~e** *m* admiral
almirez *m* mortar
almizcle *m* musk
almohad|a *f* cushion; (*de la cama*) pillow; (*funda*) pillowcase. **~illa** *f* small cushion; (*acerico*) pincushion. **~ón** *m* large pillow, bolster. **consultar con la ~a** sleep on it
almorranas *fpl* haemorrhoids, piles
alm|orzar [2 & 10] *vt* (*a mediodía*) have for lunch; (*desayunar*) have for breakfast. ● *vi* (*a mediodía*) have lunch; (*desayunar*) have breakfast. **~uerzo** *m* (*a mediodía*) lunch; (*desayuno*) breakfast
alocado *a* scatter-brained
alocución *f* address, speech
aloja|do *m* (Mex) lodger, guest. **~miento** *m* accommodation. **~r** *vt* put up. **~rse** *vpr* stay
alondra *f* lark
alpaca *f* alpaca
alpargat|a *f* canvas shoe, espadrille. **~ería** *f* shoe shop
Alpes *mpl* Alps
alpin|ismo *m* mountaineering, climbing. **~ista** *m & f* mountaineer, climber. **~o** *a* Alpine
alpiste *m* birdseed
alquil|ar *vt* (*tomar en alquiler*) hire ‹*vehículo*›, rent ‹*piso, casa*›; (*dar en alquiler*) hire (out) ‹*vehículo*›, rent (out) ‹*piso, casa*›. **~arse** *vpr* ‹*casa*› be let; ‹*vehículo*› be on hire. **se alquila** to let (*Brit*), for rent (*Amer*). **~er** *m* (*acción de alquilar un piso etc*) renting; (*acción de alquilar un vehículo*) hiring; (*precio por el que se alquila un piso etc*) rent; (*precio por el que se alquila un vehículo*) hire charge. **de ~er** for hire
alquimi|a *f* alchemy. **~sta** *m* alchemist
alquitara *f* still. **~r** *vt* distil
alquitr|án *m* tar. **~anar** *vt* tar
alrededor *adv* around. **~ de** around; (*con números*) about. **~es** *mpl* surroundings; (*de una ciudad*) outskirts
alta *f* discharge
altamente *adv* highly
altaner|ía *f* (*orgullo*) pride. **~o** *a* proud, haughty
altar *m* altar
altavoz *m* loudspeaker
altera|bilidad *f* changeability. **~ble** *a* changeable. **~ción** *f* change, alteration. **~do** *a* changed, altered; (*perturbado*) disturbed. **~r** *vt* change, alter; (*perturbar*) disturb; (*enfadar*) anger, irritate. **~rse** *vpr* change, alter; (*agitarse*) get upset; (*enfadarse*) get angry; ‹*comida*› go off
alterca|do *m* argument. **~r** [7] *vi* argue
altern|ado *a* alternate. **~ador** *m* alternator. **~ante** *a* alternating. **~ar** *vt/i* alternate. **~arse** *vpr* take turns. **~ativa** *f* alternative. **~ativo** *a* alternating. **~o** *a* alternate
alteza *f* height. **A~** (*título*) Highness
altibajos *mpl* (*de terreno*) unevenness; (*fig*) ups and downs
altiplanicie *f* high plateau
altísimo *a* very high. ● *m*. **el A~** the Almighty
altisonante *a*, **altísono** *a* pompous
altitud *f* height; (*aviat, geog*) altitude
altiv|ez *f* arrogance. **~o** *a* arrogant
alto *a* high; ‹*persona*› tall; ‹*voz*› loud; (*fig*, *elevado*) lofty; (*mus*) ‹*nota*› high(-pitched); (*mus*) ‹*voz, instrumento*› alto; ‹*horas*› early. **tiene 3 metros de ~** it is 3 metres high. ● *adv* high; (*de sonidos*) loud(ly). ● *m* height; (*de un edificio*) high floor; (*viola*) viola; (*voz*) alto; (*parada*) stop. ● *int* halt!, stop! **en lo ~ de** on the top of
altoparlante *m* (*esp LAm*) loudspeaker
altruis|mo *m* altruism. **~ta** *a* altruistic. ● *m & f* altruist
altura *f* height; (*altitud*) altitude; (*de agua*) depth; (*fig*, *cielo*) sky. **a estas ~s** at this stage. **tiene 3 metros de ~** it is 3 metres high
alubia *f* French bean
alucinación *f* hallucination
alud *m* avalanche

aludi|do *a* in question. **darse por ~do** take it personally. **no darse por ~do** turn a deaf ear. **~r** *vi* mention

alumbra|do *a* lit; (*achispado, fam*) tipsy. ● *m* lighting. **~miento** *m* lighting; (*parto*) childbirth. **~r** *vt* light. ● *vi* give birth. **~rse** *vpr* (*emborracharse*) get tipsy

aluminio *m* aluminium (*Brit*), aluminum (*Amer*)

alumno *m* pupil; (*univ*) student

aluniza|je *m* landing on the moon. **~r** [10] *vi* land on the moon

alusi|ón *f* allusion. **~vo** *a* allusive

alverja *f* vetch; (*LAm, guisante*) pea

alza *f* rise. **~cuello** *m* clerical collar, dog-collar (*fam*). **~da** *f* (*de caballo*) height; (*jurid*) appeal. **~do** *a* raised; ‹*persona*› fraudulently bankrupt; (*Mex, soberbio*) vain; ‹*precio*› fixed. **~miento** *m* raising; (*aumento*) rise, increase; (*pol*) revolt. **~r** [10] *vt* raise, lift (up); raise ‹*precios*›. **~rse** *vpr* rise; (*ponerse en pie*) stand up; (*pol*) revolt; (*quebrar*) go fraudulently bankrupt; (*apelar*) appeal

allá *adv* there. **¡~ él!** that's his business. **~ fuera** out there. **~ por el 1970** around about 1970. **el más ~** the beyond. **más ~** further on. **más ~ de** beyond. **por ~** over there

allana|miento *m* levelling; (*de obstáculos*) removal. **~miento de morada** burglary. **~r** *vt* level; remove ‹*obstáculos*›; (*fig*) iron out ‹*dificultades etc*›; burgle ‹*una casa*›. **~rse** *vpr* level off; (*hundirse*) fall down; (*ceder*) submit (**a** to)

allega|do *a* close. ● *m* relation. **~r** [12] vt collect

allí *adv* there; (*tiempo*) then. **~ donde** wherever. **~ fuera** out there. **por ~** over there

ama *f* lady of the house. **~ de casa** housewife. **~ de cría** wet-nurse. **~ de llaves** housekeeper

amab|ilidad *f* kindness. **~le** *a* kind; (*simpático*) nice

amado *a* dear. **~r** *m* lover

amaestra|do *a* trained; (*en circo*) performing. **~miento** *m* training. **~r** *vt* train

amag|ar [12] *vt* (*amenazar*) threaten; (*mostrar intención de*) show signs of. ● *vi* threaten; ‹*algo bueno*› be in the offing. **~o** *m* threat; (*señal*) sign; (*med*) sympton

amalgama *f* amalgam. **~r** *vt* amalgamate

amamantar *vt* breast-feed

amancebarse *vpr* live together

amanecer *m* dawn. ● *vi* dawn; ‹*persona*› wake up. **al ~** at dawn, at daybreak

amanera|do *a* affected. **~miento** *m* affectation. **~rse** *vpr* become affected

amanezca *f* (*Mex*) dawn

amansa|dor *m* tamer. **~miento** *m* taming. **~r** *vt* tame; break in ‹*un caballo*›; soothe ‹*dolor etc*›. **~rse** *vpr* calm down

amante *a* fond. ● *m & f* lover

amañ|ar *vt* arrange. **~o** *m* scheme

amapola *f* poppy

amar *vt* love

amara|je *m* landing on the sea; (*de astronave*) splash-down. **~r** *vt* land on the sea; ‹*astronave*› splash down

amarg|ado *a* embittered. **~ar** [12] *vt* make bitter; embitter ‹*persona*›. **~arse** *vpr* get bitter. **~o** *a* bitter. ● *m* bitterness. **~ura** *f* bitterness

amariconado *a* effeminate

amarill|ear *vi* go yellow. **~ento** *a* yellowish; ‹*tez*› sallow. **~ez** *f* yellow; (*de una persona*) paleness. **~o** *a & m* yellow

amarra *f* mooring rope. **~s** *fpl* (*fig, fam*) influence. **~do** *a* (*LAm*) mean. **~r** *vt* moor; (*atar*) tie. ● *vi* (*empollar, fam*) study hard, swot (*fam*)

amartillar *vt* cock (*arma de fuego*)

amas|ar *vt* knead; (*fig, tramar, fam*) concoct, cook up (*fam*). **~ijo** *m* dough; (*acción*) kneading; (*fig, mezcla, fam*) hotchpotch

amate *m* (*Mex*) fig tree

amateur *a & m & f* amateur

amatista *f* amethyst

amazona *f* Amazon; (*mujer varonil*) mannish woman; (*que monta a caballo*) horsewoman

Amazonas *m*. **el río ~** the Amazon

ambages *mpl* circumlocutions. **sin ~** in plain language

ámbar *m* amber

ambarino *a* amber

ambici|ón *f* ambition. **~onar** *vt* strive after. **~onar ser** have an ambition to be. **~oso** *a* ambitious. ● *m* ambitious person

ambidextro *a* ambidextrous. ● *m* ambidextrous person

ambient|ar *vt* give an atmosphere to. **~arse** *vpr* adapt o.s. **~e** *m* atmosphere; (*medio*) environment

ambig|uamente *adv* ambiguously. **~üedad** *f* ambiguity. **~uo** *a* ambiguous; (*fig, afeminado, fam*) effeminate

ámbito *m* ambit

ambos *a & pron* both. **~ a dos** both (of them)

ambulancia *f* ambulance; (*hospital móvil*) field hospital

ambulante *a* travelling

ambulatorio *m* out-patients' department

amedrentar *vt* frighten, scare. **~se** *vpr* be frightened

amén *m* amen. ● *int* amen! **en un decir ~** in an instant

amenaza *f* threat. **~dor** *a*, **~nte** *a* threatening. **~r** [10] *vt* threaten

amen|idad *f* pleasantness. **~izar** [10] *vt* brighten up. **~o** *a* pleasant

América *f* America. **~ Central** Central America. **~ del Norte** North America. **~ del Sur** South America. **~ Latina** Latin America

american|a *f* jacket. **~ismo** *m* Americanism. **~ista** *m & f* Americanist. **~o** *a* American

amerindio *a & m & f* Amerindian, American Indian

ameriza|je *m* landing on the sea; (*de astronave*) splash-down. **~r** [10] *vt* land on the sea; ‹*astronave*› splash down

ametralla|dora *f* machine-gun. **~r** *vt* machine-gun

amianto *m* asbestos

amig|a *f* friend; (*novia*) girl-friend; (*amante*) lover. **~able** *a* friendly. **~ablemente** *adv* amicably. **~rse** [12] *vpr* live together

am|ígdala *f* tonsil. **~igdalitis** *f* tonsillitis

amigo *a* friendly. ● *m* friend; (*novio*) boy-friend; (*amante*) lover. **ser ~ de** be fond of. **ser muy ~s** be good friends

amilanar *vt* frighten, scare. **~se** *vpr* be frightened

aminorar *vt* lessen; slow down ‹*velocidad*›

amist|ad *f* friendship. **~ades** *mpl* friends. **~osamente** *adv* amicably. **~oso** *a* friendly

amnesia *f* amnesia

amnist|ía *f* amnesty. **~iar** [20] *vt* grant an amnesty to

amo *m* master; (*dueño*) owner; (*jefe*) boss; (*cabeza de familia*) head of the family

amodorra|miento *m* sleepiness. **~rse** *vpr* get sleepy

amojonar *vt* mark out

amola|dor *m* knife-grinder. **~r** [2] *vt* sharpen; (*molestar, fam*) annoy

amoldar *vt* mould; (*acomodar*) fit

amonedar *vt* coin, mint

amonesta|ción *f* rebuke, reprimand; (*de una boda*) banns. **~r** *vt* rebuke, reprimand; (*anunciar la boda*) publish the banns

amoniaco *m*, **amoníaco** *m* ammonia

amontillado *m* Amontillado, pale dry sherry

amontona|damente *adv* in a heap. **~miento** *m* piling up. **~r** *vt* pile up; (*fig, acumular*) accumulate. **~rse** *vpr* pile up; ‹*gente*› crowd together; (*amancebarse, fam*) live together

amor *m* love. **~es** *mpl* (*relaciones amorosas*) love affairs. **con mil ~es**, **de mil ~es** with (the greatest of) pleasure. **hacer el ~** make love. **por (el) ~ de Dios** for God's sake

amorata|do *a* purple; (*de frío*) blue. **~rse** *vpr* go black and blue

amorcillo *m* Cupid

amordazar [10] *vt* gag; (*fig*) silence

amorfo *a* amorphous, shapeless

amor: ~ío *m* affair. **~oso** *a* loving; ‹*cartas*› love

amortajar *vt* shroud

amortigua|dor *a* deadening. ● *m* (*auto*) shock absorber. **~miento** *m* deadening; (*de la luz*) dimming. **~r** [15] *vt* deaden ‹*ruido*›; dim ‹*luz*›; cushion ‹*golpe*›; tone down ‹*color*›

amortiza|ble *a* redeemable. **~ción** *f* (*de una deuda*) repayment; (*recuperación*) redemption. **~r** [10] *vt* repay ‹*una deuda*›

amoscarse [7] *vpr* (*fam*) get cross, get irritated

amostazarse [10] *vpr* get cross

amotina|do *a & m* insurgent, rebellious. **~miento** *m* riot; (*mil*) mutiny. **~r** *vt* incite to riot. **~rse** *vpr* rebel; (*mil*) mutiny

ampar|ar *vt* help; (*proteger*) protect. **~arse** *vpr* seek protection; (*de la lluvia*) shelter. **~o** *m* protection; (*de la lluvia*) shelter. **al ~o de** under the protection of

amperio *m* ampere, amp (fam)

amplia|ción *f* extension; (*photo*) enlargement. **~r** [20] *vt* enlarge, extend; (*photo*) enlarge

amplifica|ción *f* amplification. **~dor** *m* amplifier. **~r** [7] amplify

ampli|o *a* wide; (*espacioso*) spacious; ‹*ropa*› loose-fitting. **~tud** *f* extent; (*espaciosidad*) spaciousness; (*espacio*) space

ampolla *f* (*med*) blister; (*frasco*) flask; (*de medicamento*) ampoule, phial

ampuloso *a* pompous

amputa|ción *f* amputation; (*fig*) deletion. **~r** *vt* amputate; (*fig*) delete

amueblar *vt* furnish

amuinar *vt* (*Mex*) annoy

amuralla|do *a* walled. **~r** *vt* build a wall around

anacardo *m* (*fruto*) cashew nut

anaconda *f* anaconda

anacr|ónico *a* anachronistic. **~onismo** *m* anachronism

ánade *m & f* duck

anagrama *m* anagram

anales *mpl* annals

analfabet|ismo *m* illiteracy. **~o** *a & m* illiterate

analgésico *a & m* analgesic, pain-killer

an|álisis *m invar* analysis. **~álisis de sangre** blood test. **~alista** *m & f* analyst. **~alítico** *a* analytical. **~alizar** [10] *vt* analyze

an|alogía *f* analogy. **~álogo** *a* analogous

ananás *m* pineapple

anaquel *m* shelf

anaranjado *a* orange

an|arquía *f* anarchy. **~árquico** *a* anarchic. **~arquismo** *m* anarchism. **~arquista** *a* anarchistic. ● *m & f* anarchist

anatema *m* anathema

anat|omía *f* anatomy. **~ómico** *a* anatomical

anca *f* haunch; (*parte superior*) rump; (*nalgas, fam*) bottom. **~s** *fpl* **de rana** frogs' legs

ancestral *a* ancestral

anciano *a* elderly, old. ● *m* elderly man, old man; (*relig*) elder. **los ~s** old people

ancla *f* anchor. **~dero** *m* anchorage. **~r** *vi* anchor, drop anchor. **echar ~s** anchor. **levar ~s** weigh anchor

áncora *f* anchor; (*fig*) refuge

ancho *a* wide; ‹*ropa*› loose-fitting; (*fig*) relieved; (*demasiado grande*) too big; (*ufano*) smug. ● *m* width; (*rail*) gauge. **a mis anchas, a sus anchas etc** comfortable, relaxed. **quedarse tan ancho** behave as if nothing has happened. **tiene 3 metros de ~** it is 3 metres wide

anchoa *f* anchovy

anchura *f* width; (*medida*) measurement

andaderas *fpl* baby-walker

andad|or *a* good at walking. ● *m* baby-walker. **~ura** *f* walking; (*manera de andar*) walk

Andalucía *f* Andalusia

andaluz *a & m* Andalusian

andamio *m* platform. **~s** *mpl* scaffolding

andar [25] *vt* (*recorrer*) cover, go. ● *vi* walk; ‹*máquina*› go, work; (*estar*) be; (*moverse*) move. ● *m* walk. **¡anda!** go on! come on! **~iego** *a* fond of walking; (*itinerante*) wandering. **~ por** be about. **~se** *vpr* (*marcharse*) go away

andén *m* platform; (*de un muelle*) quayside; (*LAm, acera*) pavement (*Brit*), sidewalk (*Amer*)

Andes *mpl* Andes

andino *a* Andean

Andorra *f* Andorra

andrajo *m* rag. **~so** *a* ragged

andurriales *mpl* (*fam*) out-of-the-way place

anduve *vb véase* **andar**

anécdota *f* anecdote

anega|dizo *a* subject to flooding. **~r** [12] *vt* flood. **~rse** *vpr* be flooded, flood

anejo *a* attached. ● *m* annexe; (*de libro etc*) appendix

an|emia *f* anaemia. **~émico** *a* anaemic

anest|esia *f* anaesthesia. **~ésico** *a & m* anaesthetic. **~esista** *m & f* anaesthetist

anex|ión *f* annexation. **~ionar** *vt* annex. **~o** *a* attached. ● *m* annexe

anfibio *a* amphibious. ● *m* amphibian

anfiteatro *m* amphitheatre; (*en un teatro*) upper circle

anfitri|ón *m* host. **~ona** *f* hostess

ángel *m* angel; (*encanto*) charm

angelical *a*, **angélico** *a* angelic

angina *f*. **~ de pecho** angina (pectoris). **tener ~s** have tonsillitis

anglicano *a & m* Anglican

anglicismo *m* Anglicism

anglófilo *a & m* Anglophile

anglo|hispánico *a* Anglo-Spanish. **~sajón** *a & m* Anglo-Saxon

angosto *a* narrow
anguila *f* eel
angula *f* elver, baby eel
angular *a* angular
ángulo *m* angle; (*rincón, esquina*) corner; (*curva*) bend
anguloso *a* angular
angusti|a *f* anguish. **~ar** *vt* distress; (*inquietar*) worry. **~arse** *vpr* get distressed; (*inquietarse*) get worried. **~oso** *a* anguished; (*que causa angustia*) distressing
anhel|ante *a* panting; (*deseoso*) longing. **~ar** *vt* (+ *nombre*) long for; (+ *verbo*) long to. ● *vi* pant. **~o** *m* (*fig*) yearning. **~oso** *a* panting; (*fig*) eager
anidar *vi* nest
anill|a *f* ring. **~o** *m* ring. **~o de boda** wedding ring
ánima *f* soul
anima|ción *f* (*de personas*) life; (*de cosas*) liveliness; (*bullicio*) bustle; (*en el cine*) animation. **~do** *a* lively; ‹*sitio etc*› busy. **~dor** *m* compère, host
animadversión *f* ill will
animal *a* animal; (*fig, torpe, fam*) stupid. ● *m* animal; (*fig, idiota, fam*) idiot; (*fig, bruto, fam*) brute
animar *vt* give life to; (*dar ánimo*) encourage; (*dar vivacidad*) liven up. **~se** *vpr* (*decidirse*) decide; (*ponerse alegre*) cheer up. **¿te animas a venir al cine?** do you fancy coming to the cinema?
ánimo *m* soul; (*mente*) mind; (*valor*) courage; (*intención*) intention. **¡~!** come on!, cheer up! **dar ~s** encourage
animosidad *f* animosity
animoso *a* brave; (*resuelto*) determined
aniquila|ción *f* annihilation. **~miento** *m* annihilation. **~r** *vt* annihilate; (*acabar con*) ruin. **~rse** *vpr* deteriorate
anís *m* aniseed; (*licor*) anisette
aniversario *m* anniversary
ano *m* anus
anoche *adv* last night, yesterday evening
anochecer [11] *vi* get dark; ‹*persona*› be at dusk. **anochecí en Madrid** I was in Madrid at dusk. ● *m* nightfall, dusk. **al ~** at nightfall
anodino *a* indifferent
an|omalía *f* anomaly. **~ómalo** *a* anomalous
an|onimato *m* anonymity. **~ónimo** *a* anonymous; ‹*sociedad*› limited. ● *m* anonymity; (*carta*) anonymous letter
anormal *a* abnormal; (*fam*) stupid, silly. **~idad** *f* abnormality
anota|ción *f* noting; (*acción de poner notas*) annotation; (*nota*) note. **~r** *vt* (*poner nota*) annotate; (*apuntar*) make a note of
anquilosa|miento *m* paralysis. **~r** *vt* paralyze. **~rse** *vpr* become paralyzed
ansi|a *f* anxiety, worry; (*anhelo*) yearning. **~ar** [20 *o regular*] *vt* long for. **~edad** *f* anxiety. **~oso** *a* anxious; (*deseoso*) eager
antag|ónico *a* antagonistic. **~onismo** *m* antagonism. **~onista** *m & f* antagonist
antaño *adv* in days gone by
antártico *a & m* Antarctic
ante *prep* in front of, before; (*en comparación con*) compared with; (*frente a peligro, enemigo*) in the face of; (*en vista de*) in view of. ● *m* (*piel*) suede. **~anoche** *adv* the night before last. **~ayer** *adv* the day before yesterday. **~brazo** *m* forearm
ante... *pref* ante...
antece|dente *a* previous. ● *m* antecedent. **~dentes** *mpl* history, background. **~dentes penales** criminal record. **~der** *vt* precede. **~sor** *m* predecessor; (*antepasado*) ancestor
antedicho *a* aforesaid
antelación *f* advance. **con ~** in advance
antemano *adv*. **de ~** beforehand
antena *f* antenna; (*radio, TV*) aerial
anteojeras *fpl* blinkers
anteojo *m* telescope. **~s** *mpl* (*gemelos*) opera glasses; (*prismáticos*) binoculars; (*LAm, gafas*) glasses, spectacles
ante: ~pasados *mpl* forebears, ancestors. **~pecho** *m* rail; (*de ventana*) sill. **~poner** [34] *vt* put in front (**a** of); (*fig*) put before, prefer. **~proyecto** *m* preliminary sketch; (*fig*) blueprint. **~puesto** *a* put before
anterior *a* previous; (*delantero*) front, fore. **~idad** *f*. **con ~idad** previously. **~mente** *adv* previously
antes *adv* before; (*antiguamente*) in days gone by; (*mejor*) rather; (*primero*) first. **~ de** before. **~ de ayer**

the day before yesterday. **~ de que** + *subj* before. **~ de que llegue** before he arrives. **cuanto ~, lo ~ posible** as soon as possible

antesala *f* anteroom; (*sala de espera*) waiting-room. **hacer ~** wait (to be received)

anti... *pref* anti...

anti: ~aéreo *a* anti-aircraft. **~biótico** *a & m* antibiotic. **~ciclón** *m* anticyclone

anticip|ación *f* anticipation. **con ~ación** in advance. **con media hora de ~ación** half an hour early. **~adamente** *adv* in advance. **~ado** *a*. **por ~ado** in advance. **~ar** *vt* bring forward; advance ‹*dinero*›. **~arse** *vpr* be early. **~o** *m* (*dinero*) advance; (*fig*) foretaste

anti: ~concepcional *a & m* contraceptive. **~conceptivo** *a & m* contraceptive. **~congelante** *m* antifreeze

anticua|do *a* old-fashioned. **~rio** *m* antique dealer. **~rse** *vpr* go out of date

anticuerpo *m* antibody

antídoto *m* antidote

anti: ~estético *a* ugly. **~faz** *m* mask. **~gás** *a invar*. **careta ~gás** gas mask

antig|ualla *f* old relic. **~uamente** *adv* formerly; (*hace mucho tiempo*) long ago. **~üedad** *f* antiquity; (*objeto*) antique; (*en un empleo*) length of service. **~uo** *a* old, ancient. **chapado a la ~ua** old-fashioned

antílope *m* antelope

Antillas *fpl* West Indies

antinatural *a* unnatural

antip|atía *f* dislike; (*cualidad de antipático*) unpleasantness. **~ático** *a* unpleasant, unfriendly

anti: ~semita *m & f* anti-Semite. **~semítico** *a* anti-Semitic. **~semitismo** *m* anti-Semitism. **~séptico** *a & m* antiseptic. **~social** *a* antisocial

antítesis *f invar* antithesis

antoj|adizo *a* capricious. **~arse** *vpr* fancy. **se le ~a un caramelo** he fancies a sweet. **~o** *m* whim; (*de embarazada*) craving

antología *f* anthology

antorcha *f* torch

antro *m* cavern; (*fig*) dump, hole. **~ de perversión** den of iniquity

antropófago *m* cannibal

antrop|ología *f* anthropology. **~ólogo** *m & f* anthropologist

anua|l *a* annual. **~lidad** *f* annuity. **~lmente** *adv* yearly. **~rio** *m* yearbook

anudar *vt* tie, knot; (*fig, iniciar*) begin; (*fig, continuar*) resume. **~se** *vpr* get into knots. **~se la voz** get a lump in one's throat

anula|ción *f* annulment, cancellation. **~r** *vt* annul, cancel. • *a* ‹*dedo*› ring. • *m* ring finger

Anunciación *f* Annunciation

anunci|ante *m & f* advertiser. **~ar** *vt* announce; advertise ‹*producto comercial*›; (*presagiar*) be a sign of. **~arse** *vpr* promise to be. **~o** *m* announcement; (*para vender algo*) advertisement, advert (*fam*); (*cartel*) poster

anzuelo *m* (fish)hook; (*fig*) bait. **tragar el ~** be taken in, fall for it

añadi|do *a* added. **~dura** *f* addition. **~r** *vt* add. **por ~dura** besides

añejo *a* ‹*vino*› mature; ‹*jamón etc*› cured

añicos *mpl* bits. **hacer ~** (*romper*) smash (to pieces); (*dejar cansado*) wear out

añil *m* indigo

año *m* year. **~ bisiesto** leap year. **~ nuevo** new year. **al ~** per year, a year. **¿cuántos ~s tiene? tiene 5 ~s** how old is he? he's 5 (years old). **el ~ pasado** last year. **el ~ que viene** next year. **entrado en ~s** elderly. **los ~s 60** the sixties

añora|nza *f* nostalgia. **~r** *vt* miss. • *vi* pine

apabullar *vt* crush; (*fig*) intimidate

apacentar [1] *vt* graze. **~se** *vpr* graze

apacib|ilidad *f* gentleness; (*calma*) peacefulness. **~le** *a* gentle; ‹*tiempo*› mild

apacigua|dor *a* pacifying. **~miento** *m* appeasement. **~r** [15] *vt* pacify; (*calmar*) calm; relieve ‹*dolor etc*›. **~rse** *vpr* calm down

apadrina|miento *m* sponsorship. **~r** *vt* sponsor; be godfather to ‹*a un niño*›; (*en una boda*) be best man for

apaga|dizo *a* slow to burn. **~do** *a* extinguished; ‹*color*› dull; ‹*aparato eléctrico*› off; ‹*persona*› lifeless; ‹*sonido*› muffled. **~r** [12] *vt* put out ‹*fuego, incendio*›; turn off, switch off ‹*aparato eléctrico*›; quench ‹*sed*›; muffle ‹*sonido*›. **~rse** *vpr* ‹*fuego*› go

out; ‹*luz*› go out; ‹*sonido*› die away; (*fig*) pass away
apagón *m* blackout
apalabrar *vt* make a verbal agreement; (*contratar*) engage. **~se** *vpr* come to a verbal agreement
apalanca|miento *m* leverage. **~r** [7] *vt* (*levantar*) lever up; (*abrir*) lever open
apalea|miento *m* (*de grano*) winnowing; (*de alfombras, frutos, personas*) beating. **~r** *vt* winnow ‹*grano*›; beat ‹*alfombras, frutos, personas*›; (*fig*) be rolling in ‹*dinero*›
apantallado *a* (*Mex*) stupid
apañ|ado *a* handy. **~ar** *vt* (*arreglar*) fix; (*remendar*) mend; (*agarrar*) grasp, take hold of. **~arse** *vpr* get along, manage. **¡estoy ~ado!** that's all I need!
aparador *m* sideboard
aparato *m* apparatus; (*máquina*) machine; (*teléfono*) telephone; (*rad, TV*) set; (*ostentación*) show, pomp. **~samente** *adv* ostentatiously; (*impresionante*) spectacularly. **~sidad** *f* ostentation. **~so** *a* showy, ostentatious; ‹*caída*› spectacular
aparca|miento *m* car park (*Brit*), parking lot (*Amer*). **~r** [7] *vt/i* park
aparea|miento *m* pairing off. **~r** *vt* pair off; mate ‹*animales*›. **~rse** *vpr* match; ‹*animales*› mate
aparecer [11] *vi* appear. **~se** *vpr* appear
aparej|ado *a* ready; (*adecuado*) fitting. **llevar ~ado, traer ~ado** mean, entail. **~o** *m* preparation; (*avíos*) equipment
aparent|ar *vt* (*afectar*) feign; (*parecer*) look. ● *vi* show off. **~a 20 años** she looks like she's 20. **~e** *a* apparent; (*adecuado, fam*) suitable
apari|ción *f* appearance; (*visión*) apparition. **~encia** *f* appearance; (*fig*) show. **cubrir las ~encias** keep up appearances
apartad|ero *m* lay-by; (*rail*) siding. **~o** *a* separated; (*aislado*) isolated. ● *m* (*de un texto*) section. **~o (de correos)** post-office box, PO box
apartamento *m* flat (*Brit*), apartment
apart|amiento *m* separation; (*LAm, piso*) flat (*Brit*), apartment; (*aislamiento*) seclusion. **~ar** *vt* separate; (*quitar*) remove. **~arse** *vpr* leave; abandon ‹*creencia*›; (*quitarse de en medio*) get out of the way; (*aislarse*) cut o.s. off. **~e** *adv* apart; (*por separado*) separately; (*además*) besides. ● *m* aside; (*párrafo*) new paragraph. **~e de** apart from. **dejar ~e** leave aside. **eso ~e** apart from that
apasiona|do *a* passionate; (*entusiasta*) enthusiastic; (*falto de objetividad*) biassed. ● *m* lover (**de** of). **~miento** *m* passion. **~r** *vt* excite. **~rse** *vpr* get excited (**de, por** about), be mad (**de, por** about); (*ser parcial*) become biassed
ap|atía *f* apathy. **~ático** *a* apathetic
apea|dero *m* (*rail*) halt. **~r** *vt* fell ‹*árbol*›; (*disuadir*) dissuade; overcome ‹*dificultad*›; sort out ‹*problema*›. **~rse** *vpr* (*de un vehículo*) get off
apechugar [12] *vi* push (with one's chest). **~ con** put up with
apedrear *vt* stone
apeg|ado *a* attached. **~o** *m* (*fam*) affection. **tener ~o a** be fond of
apela|ción *f* appeal. **~r** appeal; (*recurrir*) resort (**a** to)
apelmazar [10] *vt* compress
apellid|ar *vt* call. **~arse** *vpr* be called. **¿cómo te apellidas?** what's your surname? **~o** *m* surname
apenar *vt* pain. **~se** *vpr* grieve
apenas *adv* hardly, scarcely; (*enseguida que*) as soon as. **~ si** (*fam*) hardly
ap|éndice *m* (*med*) appendix; (*fig*) appendage; (*de un libro*) appendix. **~endicitis** *f* appendicitis
apercibi|miento *m* warning. **~r** *vt* warn (**de** of, about); (*amenazar*) threaten. **~rse** *vpr* prepare; (*percatarse*) provide o.s. (**de** with)
apergaminado *a* ‹*piel*› wrinkled
aperitivo *m* (*bebida*) aperitif; (*comida*) appetizer
aperos *mpl* agricultural equipment
apertura *f* opening
apesadumbrar *vt* upset. **~se** *vpr* be upset
apestar *vt* stink out; (*fastidiar*) pester. ● *vi* stink (**a** of)
apet|ecer [11] *vt* long for; (*interesar*) appeal to. **¿te ~ece una copa?** do you fancy a drink? do you feel like a drink?. ● *vi* be welcome. **~ecible** *a* attractive. **~ito** *m* appetite; (*fig*) desire. **~itoso** *a* tempting
apiadarse *vpr* feel sorry (**de** for)

ápice *m* (*nada, en frases negativas*) anything. **no ceder un ~** not give an inch
apicult|or *m* bee-keeper. **~ura** *f* bee-keeping
apilar *vt* pile up
apiñar *vt* pack in. **~se** *vpr* ‹*personas*› crowd together; ‹*cosas*› be packed tight
apio *m* celery
apisonadora *f* steamroller
aplacar [7] *vt* placate; relieve ‹*dolor*›
aplanar *vt* smooth. **~se** *vpr* become smooth; ‹*persona*› lose heart
aplasta|nte *a* overwhelming. **~r** *vt* crush. **~rse** *vpr* flatten o.s.
aplatanarse *vpr* become lethargic
aplau|dir *vt* clap, applaud; (*fig*) applaud. **~so** *m* applause; (*fig*) praise
aplaza|miento *m* postponement. **~r** [10] *vt* postpone; defer ‹*pago*›
aplebeyarse *vpr* lower o.s.
aplica|ble *a* applicable. **~ción** *f* application. **~do** *a* ‹*persona*› diligent. **~r** [7] *vt* apply; (*fijar*) attach. **~rse** *vpr* apply o.s.
aplom|ado *a* self-confident; (*vertical*) vertical. **~o** *m* (self-) confidence, aplomb; (*verticalidad*) verticality
apocado *a* timid
Apocalipsis *f* Apocalypse
apocalíptico *a* apocalyptic
apoca|miento *m* diffidence. **~r** [7] *vt* belittle ‹*persona*›. **~rse** *vpr* feel small
apodar *vt* nickname
apodera|do *m* representative. **~r** *vt* authorize. **~rse** *vpr* seize
apodo *m* nickname
apogeo *m* (*fig*) height
apolilla|do *a* moth-eaten. **~rse** *vpr* get moth-eaten
apolítico *a* non-political
apología *f* defence
apoltronarse *vpr* get lazy
apoplejía *f* stroke
apoquinar *vt/i* (*fam*) fork out
aporrear *vt* hit, thump; beat up ‹*persona*›
aporta|ción *f* contribution. **~r** *vt* contribute
aposent|ar *vt* put up, lodge. **~o** *m* room, lodgings
apósito *m* dressing
aposta *adv* on purpose
apostar[1] [2] *vt/i* bet
apostar[2] *vt* station. **~se** *vpr* station o.s.
apostilla *f* note. **~r** *vt* add notes to
apóstol *m* apostle
apóstrofo *m* apostrophe
apoy|ar *vt* lean (**en** against); (*descansar*) rest; (*asentar*) base; (*reforzar*) support. **~arse** *vpr* lean, rest. **~o** *m* support
apreci|able *a* appreciable; (*digno de estima*) worthy. **~ación** *f* appreciation; (*valoración*) appraisal. **~ar** *vt* value; (*estimar*) appreciate. **~ativo** *a* appreciative. **~o** *m* appraisal; (*fig*) esteem
aprehensión *f* capture
apremi|ante *a* urgent, pressing. **~ar** *vt* urge; (*obligar*) compel; (*dar prisa a*) hurry up. ● *vi* be urgent. **~o** *m* urgency; (*obligación*) obligation
aprender *vt/i* learn. **~se** *vpr* learn (by heart)
aprendiz *m* apprentice. **~aje** *m* apprenticeship
aprensi|ón *f* apprehension; (*miedo*) fear. **~vo** *a* apprehensive, fearful
apresa|dor *m* captor. **~miento** *m* capture. **~r** *vt* seize; (*prender*) capture
aprestar *vt* prepare. **~se** *vpr* prepare
apresura|damente *adv* hurriedly, in a hurry. **~do** *a* in a hurry; (*hecho con prisa*) hurried. **~miento** *m* hurry. **~r** *vt* hurry. **~rse** *vpr* hurry
apret|ado *a* tight; (*difícil*) difficult; (*tacaño*) stingy, mean. **~ar** [1] *vt* tighten; press ‹*botón*›; squeeze ‹*persona*›; (*comprimir*) press down. ● *vi* be too tight. **~arse** *vpr* crowd together. **~ón** *m* squeeze. **~ón de manos** handshake
aprieto *m* difficulty. **verse en un ~** be in a tight spot
aprisa *adv* quickly
aprisionar *vt* imprison
aproba|ción *f* approval. **~r** [2] *vt* approve (of); pass ‹*examen*›. ● *vi* pass
apropia|do *a* appropriate. **~rse** *vpr*. **~rse de** appropriate, take
aprovecha|ble *a* usable. **~do** *a* (*aplicado*) diligent; (*ingenioso*) resourceful; (*egoísta*) selfish; (*económico*) thrifty. **~miento** *m* advantage; (*uso*) use. **~r** *vt* take advantage of; (*utilizar*) make use of. ● *vi* be useful. **~rse** *vpr* make the

most of it. **~rse de** take advantage of. **¡que aproveche!** enjoy your meal!
aprovisionar *vt* supply (**con, de** with)
aproxima|ción *f* approximation; (*proximidad*) closeness; (*en la lotería*) consolation prize. **~damente** *adv* roughly, approximately. **~do** *a* approximate, rough. **~r** *vt* bring near; (*fig*) bring together ‹*personas*›. **~rse** *vpr* come closer, approach
apt|itud *f* suitability; (*capacidad*) ability. **~o** *a* (*capaz*) capable; (*adecuado*) suitable
apuesta *f* bet
apuesto *m* smart. ● *vb véase* **apostar**
apunta|ción *f* note. **~do** *a* sharp. **~dor** *m* prompter
apuntalar *vt* shore up
apunt|amiento *m* aiming; (*nota*) note. **~ar** *vt* aim ‹*arma*›; (*señalar*) point at; (*anotar*) make a note of, note down; (*sacar punta*) sharpen; (*en el teatro*) prompt. **~arse** *vpr* put one's name down; score ‹*triunfo, tanto etc*›. **~e** *m* note; (*bosquejo*) sketch. **tomar ~s** take notes
apuñalar *vt* stab
apur|adamente *adv* with difficulty. **~ado** *a* difficult; (*sin dinero*) hard up; (*agotado*) exhausted; (*exacto*) precise, carefully done. **~ar** *vt* exhaust; (*acabar*) finish; drain ‹*vaso etc*›; (*fastidiar*) annoy; (*causar vergüenza*) embarrass. **~arse** *vpr* worry; (*esp LAm, apresurarse*) hurry up. **~o** *m* tight spot, difficult situation; (*vergüenza*) embarrassment; (*estrechez*) hardship, want; (*esp LAm, prisa*) hurry
aquejar *vt* trouble
aquel *a* (*f* **aquella,** *mpl* **aquellos,** *fpl* **aquellas**) that; (*en plural*) those; (*primero de dos*) former
aquél *pron* (*f* **aquélla,** *mpl* **aquéllos,** *fpl* **aquéllas**) that one; (*en plural*) those; (*primero de dos*) the former
aquello *pron* that; (*asunto*) that business
aquí *adv* here. **de ~** from here. **de ~ a 15 días** in a fortnight's time. **de ~ para allí** to and fro. **de ~ que** so that. **hasta ~** until now. **por ~** around here
aquiescencia *f* acquiescence
aquietar *vt* calm (down)
aquí: ~ fuera out here. **~ mismo** right here
árabe *a* & *m* & *f* Arab; (*lengua*) Arabic
Arabia *f* Arabia. **~ saudita, ~ saudí** Saudi Arabia
arábigo *a* Arabic
arado *m* plough. **~r** *m* ploughman
Aragón *m* Aragon
aragonés *a* & *m* Aragonese
arancel *m* tariff. **~ario** *a* tariff
arandela *f* washer
araña *f* spider; (*lámpara*) chandelier
arañar *vt* scratch
arar *vt* plough
arbitra|je *m* arbitration; (*en deportes*) refereeing. **~r** *vt/i* arbitrate; (*en fútbol etc*) referee; (*en tenis etc*) umpire
arbitr|ariedad *f* arbitrariness. **~ario** *a* arbitrary. **~io** *m* (free) will; (*jurid*) decision, judgement
árbitro *m* arbitrator; (*en fútbol etc*) referee; (*en tenis etc*) umpire
árbol *m* tree; (*eje*) axle; (*palo*) mast
arbol|ado *m* trees. **~adura** *f* rigging. **~eda** *f* wood
árbol: ~ genealógico family tree. **~ de navidad** Christmas tree
arbusto *m* bush
arca *f* (*caja*) chest. **~ de Noé** Noah's ark
arcada *f* arcade; (*de un puente*) arches; (*náuseas*) retching
arca|ico *a* archaic. **~ísmo** *m* archaism
arcángel *m* archangel
arcano *m* mystery. ● *a* mysterious, secret
arce *m* maple (tree)
arcén *m* (*de autopista*) hard shoulder; (*de carretera*) verge
arcilla *f* clay
arco *m* arch; (*de curva*) arc; (*arma, mus*) bow. **~ iris** *m* rainbow
archipiélago *m* archipelago
archiv|ador *m* filing cabinet. **~ar** *vt* file (away). **~o** *m* file; (*de documentos históricos*) archives
arder *vt/i* burn; (*fig, de ira*) seethe. **~se** *vpr* burn (up). **estar que arde** be very tense. **y va que arde** and that's enough
ardid *m* trick, scheme
ardiente *a* burning. **~mente** *adv* passionately
ardilla *f* squirrel

ardor *m* heat; (*fig*) ardour. **~ del estómago** *m* heartburn. **~oso** *a* burning
arduo *a* arduous
área *f* area
arena *f* sand; (*en deportes*) arena; (*en los toros*) (bull)ring. **~l** *m* sandy area
arenga *f* harangue. **~r** [12] *vt* harangue
aren|isca *f* sandstone. **~isco** *a*, **~oso** *a* sandy
arenque *m* herring. **~ ahumado** kipper
argamasa *f* mortar
Argel *m* Algiers. **~ia** *f* Algeria
argelino *a* & *m* Algerian
argentado *a* silver-plated
Argentina *f*. **la ~** Argentina
argentin|ismo *m* Argentinism. **~o** *a* silvery; (*de la Argentina*) Argentinian, Argentine. ● *m* Argentinian
argolla *f* ring
argot *m* slang
argucia *f* sophism
argüir [19] *vt* (*deducir*) deduce; (*probar*) prove, show; (*argumentar*) argue; (*echar en cara*) reproach. ● *vi* argue
argument|ación *f* argument. **~ador** *a* argumentative. **~ar** *vt/i* argue. **~o** *m* argument; (*de libro, película etc*) story, plot; (*resumen*) synopsis
aria *f* aria
aridez *f* aridity, dryness
árido *a* arid, dry. ● *m*. **~s** *mpl* dry goods
Aries *m* Aries
arisco *a* ‹*persona*› unsociable; ‹*animal*› vicious
arist|ocracia *f* aristocracy. **~ócrata** *m* & *f* aristocrat. **~ocrático** *a* aristocratic
aritmética *f* arithmetic
arma *f* arm, weapon; (*sección*) section. **~da** *f* navy; (*flota*) fleet. **~ de fuego** firearm. **~do** *a* armed (**de** with). **~dura** *f* armour; (*de gafas etc*) frame; (*tec*) framework. **~mento** *m* arms, armaments; (*acción de armar*) armament. **~r** *vt* arm (**de** with); (*montar*) put together. **~r un lío** kick up a fuss. **La A~da Invencible** the Armada
armario *m* cupboard; (*para ropa*) wardrobe. **~ ropero** wardrobe
armatoste *m* monstrosity, hulk (*fam*)
armazón *m* & *f* frame(work)
armer|ía *f* gunsmith's shop; (*museo*) war museum. **~o** *m* gunsmith
armiño *m* ermine
armisticio *m* armistice
armonía *f* harmony
armónica *f* harmonica, mouth organ
armoni|oso harmonious. **~zación** *f* harmonizing. **~zar** [10] *vt* harmonize. ● *vi* harmonize; ‹*personas*› get on well (**con** with); ‹*colores*› go well (**con** with)
arnés *m* armour. **arneses** *mpl* harness
aro *m* ring, hoop; (*Arg, pendiente*) ear-ring
arom|a *m* aroma; (*de vino*) bouquet. **~ático** *a* aromatic. **~atizar** [10] *vt* perfume; (*culin*) flavour
arpa *f* harp
arpado *a* serrated
arpía *f* harpy; (*fig*) hag
arpillera *f* sackcloth, sacking
arpista *m* & *f* harpist
arp|ón *m* harpoon. **~onar** *vt*, **~onear** *vt* harpoon
arque|ar *vt* arch, bend. **~arse** *vpr* arch, bend. **~o** *m* arching, bending
arque|ología *f* archaeology. **~ológico** *a* archaeological. **~ólogo** *m* archaeologist
arquería *f* arcade
arquero *m* archer; (*com*) cashier
arqueta *f* chest
arquetipo *m* archetype; (*prototipo*) prototype
arquitect|o *m* architect. **~ónico** *a* architectural. **~ura** *f* architecture
arrabal *m* suburb; (*LAm, tugurio*) slum. **~es** *mpl* outskirts. **~ero** *a* suburban; (*de modales groseros*) common
arracima|do *a* in a bunch; (*apiñado*) bunched together. **~rse** *vpr* bunch together
arraiga|damente *adv* firmly. **~r** [12] *vi* take root. **~rse** *vpr* take root; (*fig*) settle
arran|cada *f* sudden start. **~car** [7] *vt* pull up ‹*planta*›; extract ‹*diente*›; (*arrebatar*) snatch; (*auto*) start. ● *vi* start. **~carse** *vpr* start. **~que** *m* sudden start; (*auto*) start; (*de emoción*) outburst
arras *fpl* security
arrasa|dor *a* overwhelming, devastating. **~r** *vt* level, smooth; raze to the ground ‹*edificio etc*›; (*llenar*) fill to the brim. ● *vi* ‹*el cielo*› clear.

~rse *vpr* ‹*el cielo*› clear; ‹*los ojos*› fill with tears; (*triunfar*) triumph

arrastr|ado *a* (*penoso*) wretched. **~ar** *vt* pull; (*rozar contra el suelo*) drag (along); give rise to ‹*consecuencias*›. ● *vi* trail on the ground. **~arse** *vpr* crawl; (*humillarse*) grovel. **~e** *m* dragging; (*transporte*) haulage. **estar para el ~e** (*fam*) have had it, be worn out. **ir ~ado** be hard up

arrayán *m* myrtle

arre *int* gee up! **~ar** *vt* urge on; give ‹*golpe*›

arrebañar *vt* scrape together; scrape clean ‹*plato etc*›

arrebat|ado *a* enraged; (*irreflexivo*) impetuous; ‹*cara*› flushed. **~ar** *vt* snatch (away); ‹*el viento*› blow away; (*fig*) win (over); captivate ‹*corazón etc*›. **~arse** *vpr* get carried away. **~o** *m* (*de cólera etc*) fit; (*éxtasis*) extasy

arrebol *m* red glow

arreciar *vi* get worse, increase

arrecife *m* reef

arregl|ado *a* neat; (*bien vestido*) well-dressed; (*moderado*) moderate. **~ar** *vt* arrange; (*poner en orden*) tidy up; sort out ‹*asunto, problema etc*›; (*reparar*) mend. **~arse** *vpr* (*ponerse bien*) improve; (*prepararse*) get ready; (*apañarse*) manage, make do; (*ponerse de acuerdo*) come to an agreement. **~árselas** manage, get by. **~o** *m* (*incl mus*) arrangement; (*acción de reparar*) repair; (*acuerdo*) agreement; (*orden*) order. **con ~o a** according to

arrellanarse *vpr* lounge, sit back

arremangar [12] *vt* roll up ‹*mangas*›; tuck up ‹*falda*›. **~se** *vpr* roll up one's sleeves

arremet|er *vt/i* attack. **~ida** *f* attack

arremolinarse *vpr* mill about

arrenda|dor *m* (*que da en alquiler*) landlord; (*que toma en alquiler*) tenant. **~miento** *m* renting; (*contrato*) lease; (*precio*) rent. **~r** [1] *vt* (*dar casa en alquiler*) let; (*dar cosa en alquiler*) hire out; (*tomar en alquiler*) rent. **~tario** *m* tenant

arreos *mpl* harness

arrepenti|miento *m* repentance, regret. **~rse** [4] *vpr*. **~rse de** be sorry, regret; repent ‹*pecados*›

arrest|ar *vt* arrest, detain; (*encarcelar*) imprison. **~o** *m* arrest; (*encarcelamiento*) imprisonment

arriar [20] *vt* lower ‹*bandera, vela*›; (*aflojar*) loosen; (*inundar*) flood. **~se** *vpr* be flooded

arriba *adv* (up) above; (*dirección*) up(wards); (*en casa*) upstairs. ● *int* up with; (*¡levántate!*) up you get!; (*¡ánimo!*) come on! **¡~ España!** long live Spain! **~ mencionado** aforementioned. **calle ~** up the street. **de ~ abajo** from top to bottom. **de 100 pesetas para ~** more than 100 pesetas. **escaleras ~** upstairs. **la parte de ~** the top part. **los de ~** those at the top. **más ~** above

arribar *vi* ‹*barco*› reach port; (*esp LAm, llegar*) arrive

arribista *m & f* self-seeking person, arriviste

arribo *m* (*esp LAm*) arrival

arriero *m* muleteer

arriesga|do *a* risky. **~r** [12] *vt* risk; (*aventurar*) venture. **~rse** *vpr* take a risk

arrim|ar *vt* bring close(r); (*apartar*) move out of the way ‹*cosa*›; (*apartar*) push aside ‹*persona*›. **~arse** *vpr* come closer, approach; (*apoyarse*) lean (**a** on). **~o** *m* support. **al ~o de** with the support of

arrincona|do *a* forgotten. **~rse** *vt* put in a corner; (*perseguir*) corner; (*arrumbar*) put aside; (*apartar a uno*) leave out, ignore. **~rse** *vpr* become a recluse

arriscado *a* ‹*terreno*› uneven

arrobar *vt* entrance. **~se** *vpr* be enraptured

arrocero *a* rice

arrodillarse *vpr* kneel (down)

arrogan|cia *f* arrogance; (*orgullo*) pride. **~te** *a* arrogant; (*orgulloso*) proud

arrogarse [12] *vpr* assume

arroj|ado *a* brave. **~ar** *vt* throw; (*dejar caer*) drop; (*emitir*) give off, throw out; (*producir*) produce. ● *vi* (*esp LAm, vomitar*) be sick. **~arse** *vpr* throw o.s. **~o** *m* courage

arrolla|dor *a* overwhelming. **~r** *vt* roll (up); (*atropellar*) run over; ‹*ejército*› crush; ‹*agua*› sweep away; (*tratar sin respeto*) have no respect for

arropar *vt* wrap up; (*en la cama*) tuck up; (*fig, amparar*) protect. **~se** *vpr* wrap (o.s.) up

arroy|o *m* stream; (*de una calle*) gutter; (*fig, de lágrimas*) flood; (*fig, de sangre*) pool. **poner en el ~o** throw into the street. **~uelo** *m* small stream
arroz *m* rice. **~al** *m* rice field. **~ con leche** rice pudding
arruga *f* (*en la piel*) wrinkle, line; (*en tela*) crease. **~r** [12] *vt* wrinkle; crumple ‹*papel*›; crease ‹*tela*›. **~rse** *vpr* ‹*la piel*› wrinkle, get wrinkled; ‹*tela*› crease, get creased
arruinar *vt* ruin; (*destruir*) destroy. **~se** *vpr* ‹*persona*› be ruined; ‹*edificio*› fall into ruins
arrullar *vt* lull to sleep. ● *vi* ‹*palomas*› coo. **~se** *vpr* bill and coo
arrumaco *m* caress; (*zalamería*) flattery
arrumbar *vt* put aside
arsenal *m* (*astillero*) shipyard; (*de armas*) arsenal; (*fig*) store
arsénico *m* arsenic
arte *m en singular, f en plural* art; (*habilidad*) skill; (*astucia*) cunning. **bellas ~s** fine arts. **con ~** skilfully. **malas ~s** trickery. **por amor al ~** for nothing, for love
artefacto *m* device
arter|amente *adv* artfully. **~ía** *f* cunning
arteria *f* artery; (*fig, calle*) main road
artero *a* cunning
artesan|al *a* craft. **~ía** *f* handicrafts. **~o** *m* artisan, craftsman. **objeto** *m* **de ~ía** hand-made article
ártico *a & m* Arctic
articula|ción *f* joint; (*pronunciación*) articulation. **~damente** *adv* articulately. **~do** *a* articulated; ‹*lenguaje*› articulate. **~r** *vt* articulate
articulista *m & f* columnist
artículo *m* article. **~s** *mpl* (*géneros*) goods. **~ de exportación** export commodity. **~ de fondo** editorial, leader
artificial *a* artificial
artificiero *m* bomb-disposal expert
artificio *m* (*habilidad*) skill; (*dispositivo*) device; (*engaño*) trick. **~so** *a* clever; (*astuto*) artful
artilugio *m* gadget
artiller|ía *f* artillery. **~o** *m* artilleryman, gunner
artimaña *f* trap
art|ista *m & f* artist; (*en espectáculos*) artiste. **~ísticamente** *adv* artistically. **~ístico** *a* artistic
artr|ítico *a* arthritic. **~itis** *f* arthritis
arveja *f* vetch; (*LAm, guisante*) pea
arzobispo *m* archbishop
as *m* ace
asa *f* handle
asad|o *a* roast(ed). ● *m* roast (meat), joint. **~o a la parrilla** grilled. **~o al horno** (*sin grasa*) baked; (*con grasa*) roast. **~or** *m* spit. **~ura** *f* offal
asalariado *a* salaried. ● *m* employee
asalt|ante *m* attacker; (*de un banco*) robber. **~ar** *vt* storm ‹*fortaleza*›; attack ‹*persona*›; raid ‹*banco etc*›; (*fig*) ‹*duda*› assail; (*fig*) ‹*idea etc*› cross one's mind. **~o** *m* attack; (*en boxeo*) round
asamble|a *f* assembly; (*reunión*) meeting; (*congreso*) conference. **~ísta** *m & f* member of an assembly
asapán *m* (*Mex*) flying squirrel
asar *vt* roast; (*fig, acosar*) pester (**a** with). **~se** *vpr* be very hot. **~ a la parrilla** grill. **~ al horno** (*sin grasa*) bake; (*con grasa*) roast
asbesto *m* asbestos
ascendencia *f* descent
ascend|ente *a* ascending. **~er** [1] *vt* promote. ● *vi* go up, ascend; ‹*cuenta etc*› come to, amount to; (*ser ascendido*) be promoted. **~iente** *m & f* ancestor; (*influencia*) influence
ascens|ión *f* ascent; (*de grado*) promotion. **~ional** *a* upward. **~o** *m* ascent; (*de grado*) promotion. **día** *m* **de la A~ión** Ascension Day
ascensor *m* lift (*Brit*), elevator (*Amer*). **~ista** *m & f* lift attendant (*Brit*), elevator operator (*Amer*)
asc|eta *m & f* ascetic. **~ético** *a* ascetic
asco *m* disgust. **dar ~** be disgusting; (*fig, causar enfado*) be infuriating. **estar hecho un ~** be disgusting. **hacer ~s de algo** turn up one's nose at sth. **me da ~ el ajo** I can't stand garlic. **¡qué ~!** how disgusting! **ser un ~** be a disgrace
ascua *f* ember. **estar en ~s** be on tenterhooks
asea|damente *adv* cleanly. **~do** *a* clean; (*arreglado*) neat. **~r** *vt* (*lavar*) wash; (*limpiar*) clean; (*arreglar*) tidy up
asedi|ar *vt* besiege; (*fig*) pester. **~o** *m* siege
asegura|do *a & m* insured. **~dor** *m* insurer. **~r** *vt* secure, make safe; (*decir*) assure; (*concertar un seguro*)

insure; (*preservar*) safeguard. **~rse** *vpr* make sure
asemejarse *vpr* be alike
asenta|da *f*. **de una ~da** at a sitting. **~do** *a* situated; (*arraigado*) established. **~r** [1] *vt* place; (*asegurar*) settle; (*anotar*) note down. • *vi* be suitable. **~rse** *vpr* settle; (*estar situado*) be situated
asenti|miento *m* consent. **~r** [4] *vi* agree (**a** to). **~r con la cabeza** nod
aseo *m* cleanliness. **~s** *mpl* toilets
asequible *a* obtainable; ‹*precio*› reasonable; ‹*persona*› approachable
asesin|ar *vt* murder; (*pol*) assassinate. **~ato** *m* murder; (*pol*) assassination. **~o** *m* murderer; (*pol*) assassin
asesor *m* adviser, consultant. **~amiento** *m* advice. **~ar** *vt* advise. **~arse** *vpr*. **~arse con/de** consult. **~ía** *f* consultancy; (*oficina*) consultant's office
asestar *vt* aim ‹*arma*›; strike ‹*golpe etc*›; (*disparar*) fire
asevera|ción *f* assertion. **~r** *vt* assert
asfalt|ado *a* asphalt. **~ar** *vt* asphalt. **~o** *m* asphalt
asfixia *f* suffocation. **~nte** *a* suffocating. **~r** *vt* suffocate. **~rse** *vpr* suffocate
así *adv* so; (*de esta manera*) like this, like that. • *a* such. **~ ~, ~ asá, ~ asado** so-so. **~ como** just as. **~... como** both... and. **~ pues** so. **~ que** so; (*enseguida*) as soon as. **~ sea** so be it. **~ y todo** even so. **aun ~** even so. **¿no es ~?** isn't that right? **y ~ (sucesivamente)** and so on
Asia *f* Asia
asiático *a* & *m* Asian
asidero *m* handle; (*fig*, *pretexto*) excuse
asidu|amente *adv* regularly. **~idad** *f* regularity. **~o** *a* & *m* regular
asiento *m* seat; (*situación*) site. **~ delantero** front seat. **~ trasero** back seat. **tome Vd ~** please take a seat
asigna|ción *f* assignment; (*sueldo*) salary. **~r** *vt* assign; allot ‹*porción, tiempo etc*›
asignatura *f* subject. **~ pendiente** (*escol*) failed subject; (*fig*) matter still to be resolved
asil|ado *m* inmate. **~ado político** refugee. **~o** *m* asylum; (*fig*) shelter; (*de ancianos etc*) home. **~o de huérfanos** orphanage. **pedir ~o político** ask for political asylum
asimétrico *a* asymmetrical
asimila|ción *f* assimilation. **~r** *vt* assimilate. **~rse** *vpr* be assimilated. **~rse a** resemble
asimismo *adv* in the same way, likewise
asir [45] *vt* grasp. **~se** *vpr* grab hold (**a, de** of)
asist|encia *f* attendance; (*gente*) people (present); (*en un teatro etc*) audience; (*ayuda*) assistance. **~encia médica** medical care. **~enta** *f* assistant; (*mujer de la limpieza*) charwoman. **~ente** *m* assistant. **~ente social** social worker. **~ido** *a* assisted. **~ir** *vt* assist, help; ‹*un médico*› treat. • *vi*. **~ir a** attend, be present at
asm|a *f* asthma. **~ático** *a* & *m* asthmatic
asn|ada *f* (*fig*) silly thing. **~o** *m* donkey; (*fig*) ass
asocia|ción *f* association; (*com*) partnership. **~do** *a* associated; ‹*miembro etc*› associate. • *m* associate. **~r** *vt* associate; (*com*) take into partnership. **~rse** *vpr* associate; (*com*) become a partner
asolador *a* destructive
asolar[1] [1] *vt* destroy. **~se** *vpr* be destroyed
asolar[2] *vt* dry up ‹*plantas*›
asoma|da *f* brief appearance. **~r** *vt* show. • *vi* appear, show. **~rse** *vpr* ‹*persona*› lean out (**a, por** of); ‹*cosa*› appear
asombr|adizo *a* easily frightened. **~ar** *vt* (*pasmar*) amaze; (*sorprender*) surprise. **~arse** *vpr* be amazed; (*sorprenderse*) be surprised. **~o** *m* amazement, surprise. **~osamente** *adv* amazingly. **~oso** *a* amazing, astonishing
asomo *m* sign. **ni por ~** by no means
asonada *f* mob; (*motín*) riot
aspa *f* cross, X-shape; (*de molino*) (windmill) sail. **~do** *a* X-shaped
aspaviento *m* show, fuss. **~s** *mpl* gestures. **hacer ~s** make a big fuss
aspecto *m* look, appearance; (*fig*) aspect
aspereza *f* roughness; (*de sabor etc*) sourness
áspero *a* rough; ‹*sabor etc*› bitter
aspersión *f* sprinkling

aspiración *f* breath; (*deseo*) ambition

aspirador *a* suction. **~a** *f* vacuum cleaner

aspira|nte *m* candidate. **~r** *vt* breathe in; ‹*máquina*› suck up. ● *vi* breathe in; ‹*máquina*› suck. **~r a** aspire to

aspirina *f* aspirin

asquear *vt* sicken. ● *vi* be sickening. **~se** *vpr* be disgusted

asqueros|amente *adv* disgustingly. **~idad** *f* filthiness. **~o** *a* disgusting

asta *f* spear; (*de la bandera*) flagpole; (*mango*) handle; (*cuerno*) horn. **a media ~** at half-mast. **~do** *a* horned

asterisco *m* asterisk

astilla *f* splinter. **~s** *fpl* firewood. **~r** *vt* splinter. **hacer ~s** smash. **hacerse ~s** shatter

astillero *m* shipyard

astringente *a* & *m* astringent

astro *m* star

astr|ología *f* astrology. **~ólogo** *m* astrologer

astrona|uta *m* & *f* astronaut. **~ve** *f* spaceship

astr|onomía *f* astronomy. **~onómico** *a* astronomical. **~ónomo** *m* astronomer

astu|cia *f* cleverness; (*ardid*) cunning. **~to** *a* astute; (*taimado*) cunning

asturiano *a* & *m* Asturian

Asturias *fpl* Asturias

asueto *m* time off, holiday

asumir *vt* assume

asunción *f* assumption. **A~** Assumption

asunto *m* subject; (*cuestión*) matter; (*de una novela*) plot; (*negocio*) business. **~s** *mpl* **exteriores** foreign affairs. **el ~ es que** the fact is that

asusta|dizo *a* easily frightened. **~r** *vt* frighten. **~rse** *vpr* be frightened

ataca|nte *m* & *f* attacker. **~r** [7] *vt* attack

atad|ero *m* rope; (*cierre*) fastening; (*gancho*) hook. **~ijo** *m* bundle. **~o** *a* tied; (*fig*) timid. ● *m* bundle. **~ura** *f* tying; (*cuerda*) string

ataj|ar *vi* take a short cut. **~o** *m* short cut; (*grupo*) bunch. **echar por el ~o** take the easy way out

atalaya *f* watch-tower; (*fig*) vantage point

atañer [22] *vt* concern

ataque *m* attack; (*med*) fit, attack. **~ al corazón** heart attack. **~ de nervios** hysterics

atar *vt* tie (up). **~se** *vpr* get tied up

atardecer [11] *vi* get dark. ● *m* dusk. **al ~** at dusk

atarea|do *a* busy. **~rse** *vpr* work hard

atasc|adero *m* (*fig*) stumbling block. **~ar** [7] *vt* block; (*fig*) hinder. **~arse** *vpr* get stuck; ‹*tubo etc*› block. **~o** *m* obstruction; (*auto*) traffic jam

ataúd *m* coffin

atav|iar [20] *vt* dress up. **~iarse** *vpr* dress up, get dressed up. **~ío** *m* dress, attire

atemorizar [10] *vt* frighten. **~se** *vpr* be frightened

Atenas *fpl* Athens

atenazar [10] *vt* (*fig*) torture; ‹*duda, miedo*› grip

atención *f* attention; (*cortesía*) courtesy, kindness; (*interés*) interest. **¡~!** look out! **~ a** beware of. **llamar la ~** attract attention, catch the eye. **prestar ~** pay attention

atender [1] *vt* attend to; heed ‹*consejo etc*›; (*cuidar*) look after. ● *vi* pay attention

atenerse [40] *vpr* abide (**a** by)

atentado *m* offence; (*ataque*) attack. **~ contra la vida de uno** attempt on s.o.'s life

atentamente *adv* attentively; (*con cortesía*) politely; (*con amabilidad*) kindly. **le saluda ~** (*en cartas*) yours faithfully

atentar *vi* commit an offence. **~ contra la vida de uno** make an attempt on s.o.'s life

atento *a* attentive; (*cortés*) polite; (*amable*) kind

atenua|nte *a* extenuating. ● *f* extenuating circumstance. **~r** [21] *vt* attenuate; (*hacer menor*) diminish, lessen. **~rse** *vpr* weaken

ateo *a* atheistic. ● *m* atheist

aterciopelado *a* velvety

aterido *a* frozen (stiff), numb (with cold)

aterra|dor *a* terrifying. **~r** *vt* terrify. **~rse** *vpr* be terrified

aterriza|je *m* landing. **~je forzoso** emergency landing. **~r** [10] *vt* land

aterrorizar [10] *vt* terrify

atesorar *vt* hoard

atesta|do *a* packed, full up. ● *m* sworn statement. **~r** *vt* fill up, pack; (*jurid*) testify

atestiguar [15] *vt* testify to; (*fig*) prove
atiborrar *vt* fill, stuff. **~se** *vpr* stuff o.s.
ático *m* attic
atilda|do *a* elegant, neat. **~r** *vt* put a tilde over; (*arreglar*) tidy up. **~rse** *vpr* smarten o.s. up
atina|damente *adv* rightly. **~do** *a* right; (*juicioso*) wise, sensible. **~r** *vt/i* hit upon; (*acertar*) guess right
atípico *a* exceptional
atiplado *a* high-pitched
atirantar *vt* tighten
atisb|ar *vt* spy on; (*vislumbrar*) make out. **~o** *m* spying; (*indicio*) hint, sign
atizar [10] *vt* poke; give ‹*golpe*›; (*fig*) stir up; arouse, excite ‹*pasión etc*›
atlántico *a* Atlantic. **el (océano) A~** the Atlantic (Ocean)
atlas *m* atlas
atl|eta *m & f* athlete. **~ético** *a* athletic. **~etismo** *m* athletics
atm|ósfera *f* atmosphere. **~osférico** *a* atmospheric
atolondra|do *a* scatter-brained; (*aturdido*) bewildered. **~miento** *m* bewilderment; (*irreflexión*) thoughtlessness. **~r** *vt* bewilder; (*pasmar*) stun. **~rse** *vpr* be bewildered
atolladero *m* bog; (*fig*) tight corner
at|ómico *a* atomic. **~omizador** *m* atomizer. **~omizar** [10] *vt* atomize
átomo *m* atom
atónito *m* amazed
atonta|do *a* bewildered; (*tonto*) stupid. **~r** *vt* stun. **~rse** *vpr* get confused
atormenta|dor *a* tormenting. ● *m* tormentor. **~r** *vt* torture. **~rse** *vpr* worry, torment o.s.
atornillar *vt* screw on
atosigar [12] *vt* pester
atracadero *m* quay
atracador *m* bandit
atracar [7] *vt* (*amarrar*) tie up; (*arrimar*) bring alongside; rob ‹*banco, persona*›. ● *vi* ‹*barco*› tie up; ‹*astronave*› dock. **~se** *vpr* stuff o.s. **(de** with)
atracci|ón *f* attraction. **~ones** *fpl* entertainment, amusements
atrac|o *m* hold-up, robbery. **~ón** *m*. **darse un ~ón** stuff o.s.
atractivo *a* attractive. ● *m* attraction; (*encanto*) charm
atraer [41] *vt* attract
atragantarse *vpr* choke (**con** on). **la historia se me atraganta** I can't stand history
atranc|ar [7] *vt* bolt ‹*puerta*›; block up ‹*tubo etc*›. **~arse** *vpr* get stuck; ‹*tubo*› get blocked. **~o** *m* difficulty
atrapar *vt* trap; (*fig*) land ‹*empleo etc*›; catch ‹*resfriado*›
atrás *adv* behind; (*dirección*) backwards); (*tiempo*) previously, before. ● *int* back! **dar un paso ~** step backwards. **hacia ~, para ~** backwards
atras|ado *a* behind; ‹*reloj*› slow; (*con deudas*) in arrears; ‹*país*› backward. **llegar ~ado** arrive late. **~ar** *vt* slow down; (*retrasar*) put back; (*demorar*) delay, postpone. ● *vi* ‹*reloj*› be slow. **~arse** *vpr* be late; ‹*reloj*› be slow; (*quedarse atrás*) be behind. **~o** *m* delay; (*de un reloj*) slowness; (*de un país*) backwardness. **~os** *mpl* arrears
atravesa|do *a* lying across; (*bizco*) cross-eyed; (*fig, malo*) wicked. **~r** [1] *vt* cross; (*traspasar*) go through; (*poner transversalmente*) lay across. **~rse** *vpr* lie across; (*en la garganta*) get stuck, stick; (*entrometerse*) interfere
atrayente *a* attractive
atrev|erse *vpr* dare. **~erse con** tackle. **~ido** *a* daring, bold; (*insolente*) insolent. **~imiento** *m* daring, boldness; (*descaro*) insolence
atribución *f* attribution. **atribuciones** *fpl* authority
atribuir [17] *vt* attribute; confer ‹*función*›. **~se** *vpr* take the credit for
atribular *vt* afflict. **~se** *vpr* be distressed
atribut|ivo *a* attributive. **~o** *m* attribute; (*símbolo*) symbol
atril *m* lectern; (*mus*) music stand
atrincherar *vt* fortify with trenches. **~se** *vpr* entrench (o.s.)
atrocidad *f* atrocity. **decir ~es** make silly remarks. **¡qué ~!** how terrible!
atrochar *vi* take a short cut
atrojarse *vpr* (*Mex*) be cornered
atrona|dor *a* deafening. **~r** [2] *vt* deafen
atropell|adamente *adv* hurriedly. **~ado** *a* hasty. **~ar** *vt* knock down, run over; (*empujar*) push aside; (*maltratar*) bully; (*fig*) outrage, insult. **~arse** *vpr* rush. **~o** *m* (*auto*) accident; (*fig*) outrage

atroz *a* atrocious; (*fam*) huge. **~mente** *adv* atrociously, awfully

atuendo *m* dress, attire

atufar *vt* choke; (*fig*) irritate. **~se** *vpr* be overcome; (*enfadarse*) get cross

atún *m* tuna (fish)

aturdi|do *a* bewildered; (*irreflexivo*) thoughtless. **~r** *vt* bewilder, stun; ‹*ruido*› deafen. **~rse** *vpr* be stunned; (*intentar olvidar*) try to forget

atur(r)ullar *vt* bewilder

atusar *vt* smooth; trim ‹*pelo*›

auda|cia *f* boldness, audacity. **~z** *a* bold

audib|ilidad *f* audibility. **~le** *a* audible

audición *f* hearing; (*concierto*) concert

audiencia *f* audience; (*tribunal*) court

auditor *m* judge-advocate; (*de cuentas*) auditor

auditorio *m* audience; (*sala*) auditorium

auge *m* peak; (*com*) boom

augur|ar *vt* predict; ‹*cosas*› augur. **~io** *m* omen. **~ios** *mpl*. **con nuestros ~ios para** with our best wishes for

augusto *a* august

aula *f* class-room; (*univ*) lecture room

aulaga *f* gorse

aull|ar [23] *vi* howl. **~ido** *m* howl

aument|ar *vt* increase; put up ‹*precios*›; magnify ‹*imagen*›; step up ‹*producción, voltaje*›. ● *vi* increase. **~arse** *vpr* increase. **~ativo** *a* & *m* augmentative. **~o** *m* increase; (*de sueldo*) rise

aun *adv* even. **~ así** even so. **~ cuando** although. **más ~** even more. **ni ~** not even

aún *adv* still, yet. **~ no ha llegado** it still hasn't arrived, it hasn't arrived yet

aunar [23] *vt* join. **~se** *vpr* join together

aunque *conj* although, (even) though

aúpa *int* up! **de ~** wonderful

aureola *f* halo

auricular *m* (*de teléfono*) receiver. **~es** *mpl* headphones

aurora *f* dawn

ausen|cia *f* absence. **~tarse** *vpr* leave. **~te** *a* absent. ● *m* & *f* absentee; (*jurid*) missing person. **en ~ de** in the absence of

auspicio *m* omen. **bajo los ~s de** sponsored by

auster|idad *f* austerity. **~o** *a* austere

austral *a* southern. ● *m* (*unidad monetaria argentina*) austral

Australia *m* Australia

australiano *a* & *m* Australian

Austria *f* Austria

austriaco, austríaco *a* & *m* Austrian

aut|enticar [7] authenticate. **~enticidad** *f* authenticity. **~éntico** *a* authentic

auto *m* sentence; (*auto, fam*) car. **~s** *mpl* proceedings

auto... *pref* auto...

auto|ayuda *f* self-help. **~biografía** *f* autobiography. **~biográfico** *a* autobiographical. **~bombo** *m* self-glorification

autobús *m* bus. **en ~** by bus

autocar *m* coach (*Brit*), (long-distance) bus (*Amer*)

aut|ocracia *f* autocracy. **~ócrata** *m* & *f* autocrat. **~ocrático** *a* autocratic

autóctono *a* autochthonous

auto: ~determinación *f* self-determination. **~defensa** *f* self-defence. **~didacto** *a* self-taught. ● *m* autodidact. **~escuela** *f* driving school. **~giro** *m* autogiro

autógrafo *m* autograph

automación *f* automation

autómata *m* robot

autom|ático *a* automatic. ● *m* press-stud. **~atización** *f* automation. **~atizar** [10] *vt* automate

automotor *a* (*f* **automotriz**) self-propelled. ● *m* diesel train

autom|óvil *a* self-propelled. ● *m* car. **~ovilismo** *m* motoring. **~ovilista** *m* & *f* driver, motorist

aut|onomía *f* autonomy. **~onómico** *a*, **~ónomo** *a* autonomous

autopista *f* motorway (*Brit*), freeway (*Amer*)

autopsia *f* autopsy

autor *m* author. **~a** *f* author(ess)

autori|dad *f* authority. **~tario** *a* authoritarian. **~tarismo** *m* authoritarianism

autoriza|ción *f* authorization. **~damente** *adv* officially. **~do** *a* authorized, offical; ‹*opinión etc*› authoritative. **~r** [10] *vt* authorize

auto: ~rretrato *m* self-portrait. **~servicio** *m* self-service restaurant. **~stop** *m* hitch-hiking. **hacer ~stop** hitch-hike
autosuficien|cia *f* self-sufficiency. **~te** *a* self-sufficient
autovía *f* dual carriageway
auxili|ar *a* assistant; ‹*servicios*› auxiliary. ● *m* assistant. ● *vt* help. **~o** *m* help. **¡~o!** help! **~os espirituales** last rites. **en ~o de** in aid of. **pedir ~o** shout for help. **primeros ~os** first aid
Av. *abrev* (*Avenida*) Ave, Avenue
aval *m* guarantee
avalancha *f* avalanche
avalar *vt* guarantee
avalorar *vt* enhance; (*fig*) encourage
avance *m* advance; (*en el cine*) trailer; (*balance*) balance; (*de noticias*) early news bulletin. **~ informativo** publicity hand-out
avante *adv* (*esp LAm*) forward
avanza|do *a* advanced. **~r** [10] *vt* move forward. ● *vi* advance
avar|icia *f* avarice. **~icioso** *a*, **~iento** *a* greedy; (*tacaño*) miserly. **~o** *a* miserly. ● *m* miser
avasalla|dor *a* overwhelming. **~r** *vt* dominate
Avda. *abrev* (*Avenida*) Ave, Avenue
ave *f* bird. **~ de paso** (*incl fig*) bird of passage. **~ de presa**, **~ de rapiña** bird of prey
avecinarse *vpr* approach
avecindarse *vpr* settle
avejentarse *vpr* age
avellan|a *f* hazel-nut. **~o** *m* hazel (tree)
avemaría *f* Hail Mary. **al ~** at dusk
avena *f* oats
avenar *vt* drain
avenida *f* (*calle*) avenue; (*de río*) flood
avenir [53] *vt* reconcile. **~se** *vpr* come to an agreement
aventaja|do *a* outstanding. **~r** *vt* surpass
aventar [1] *vt* fan; winnow ‹*grano etc*›; ‹*viento*› blow away
aventur|a *f* adventure; (*riesgo*) risk. **~a amorosa** love affair. **~ado** *a* risky. **~ar** *vt* risk. **~arse** *vpr* dare. **~a sentimental** love affair. **~ero** *a* adventurous. ● *m* adventurer
avergonza|do *a* ashamed; (*embarazado*) embarrassed. **~r** [10 & 16] *vt* shame; (*embarazar*) embarrass. **~rse** *vpr* be ashamed; (*embarazarse*) be embarrassed
aver|ía *f* (*auto*) breakdown; (*daño*) damage. **~iado** *a* broken down; ‹*fruta*› damaged, spoilt. **~iar** [20] *vt* damage. **~iarse** *vpr* get damaged; ‹*coche*› break down
averigua|ble *a* verifiable. **~ción** *f* verification; (*investigación*) investigation; (*Mex, disputa*) argument. **~dor** *m* investigator. **~r** [15] *vt* verify; (*enterarse de*) find out; (*investigar*) investigate. ● *vi* (*Mex*) quarrel
aversión *f* aversion **(a, hacia, por** for)
avestruz *m* ostrich
aviación *f* aviation; (*mil*) air force
aviado *a* (*Arg*) well off. **estar ~** be in a mess
aviador *m* (*aviat*) member of the crew; (*piloto*) pilot; (*Arg, prestamista*) money-lender; (*Arg, de minas*) mining speculator
aviar [20] *vt* get ready, prepare; (*arreglar*) tidy; (*reparar*) repair; (*LAm, prestar dinero*) lend money; (*dar prisa*) hurry up. **~se** *vpr* get ready. **¡avíate!** hurry up!
av|ícula *a* poultry. **~icultor** *m* poultry farmer. **~icultura** *f* poultry farming
avidez *f* eagerness, greed
ávido *a* eager, greedy
avieso *a* (*maligno*) wicked
avinagra|do *a* sour. **~r** *vt* sour; (*fig*) embitter. **~rse** *vpr* go sour; (*fig*) become embittered
avío *m* preparation. **~s** *mpl* provisions; (*utensilios*) equipment
avi|ón *m* aeroplane (*Brit*), airplane (*Amer*). **~oneta** *f* light aircraft
avis|ado *a* wise. **~ar** *vt* warn; (*informar*) notify, inform; call ‹*medico etc*›. **~o** *m* warning; (*anuncio* notice. **estar sobre ~o** be on the alert. **mal ~ado** ill-advised. **sin previo ~o** without notice
avisp|a *f* wasp. **~ado** *a* sharp. **~ero** *m* wasps' nest; (*fig*) mess. **~ón** *m* hornet
avistar *vt* catch sight of
avitualla|miento *m* supplying. **~r** *vt* provision
avivar *vt* stoke up ‹*fuego*›; brighten up ‹*color*›; arouse ‹*interés, pasión*›; intensify ‹*dolor*›. **~se** *vpr* revive; (*animarse*) cheer up
axila *f* axilla, armpit

axiom|a *m* axiom. **~ático** *a* axiomatic

ay *int* (*de dolor*) ouch!; (*de susto*) oh!; (*de pena*) oh dear! **~ de** poor. **¡~ de tí!** poor you!

aya *f* governess, child's nurse

ayer *adv* yesterday. ● *m* past. **antes de ~** the day before yesterday. **~ por la mañana** yesterday morning. **~ (por la) noche** last night

ayo *m* tutor

ayote *m* (*Mex*) pumpkin

ayuda *f* help, aid. **~ de cámara** valet. **~nta** *f*, **~nte** *m* assistant; (*mil*) adjutant. **~nte técnico sanitario (ATS)** nurse. **~r** *vt* help

ayun|ar *vi* fast. **~as** *fpl*. **estar en ~as** have had no breakfast; (*fig, fam*) be in the dark. **~o** *m* fasting

ayuntamiento *m* town council, city council; (*edificio*) town hall

azabache *m* jet

azad|a *f* hoe. **~ón** *m* (large) hoe

azafata *f* air hostess

azafrán *m* saffron

azahar *m* orange blossom

azar *m* chance; (*desgracia*) misfortune. **al ~** at random. **por ~** by chance

azararse *vpr* go wrong; (*fig*) get flustered

azaros|amente *adv* hazardously. **~o** *a* hazardous, risky; ‹*persona*› unlucky

azoga|do *a* restless. **~rse** [12] *vpr* be restless

azolve *m* (*Mex*) obstruction

azora|do *a* flustered, excited, alarmed. **~miento** *m* confusion, embarrassment. **~r** *vt* embarrass; (*aturdir*) alarm. **~rse** *vpr* get flustered, be alarmed

Azores *fpl* Azores

azot|aina *f* beating. **~ar** *vt* whip, beat. **~e** *m* whip; (*golpe*) smack; (*fig, calamidad*) calamity

azotea *f* flat roof. **estar mal de la ~** be mad

azteca *a* & *m* & *f* Aztec

az|úcar *m* & *f* sugar. **~ucarado** *a* sweet. **~ucarar** *vt* sweeten. **~ucarero** *m* sugar bowl

azucena *f* (white) lily

azufre *m* sulphur

azul *a* & *m* blue. **~ado** *a* bluish. **~ de lavar** (washing) blue. **~ marino** navy blue

azulejo *m* tile

azuzar *vt* urge on, incite

B

bab|a *f* spittle. **~ear** *vi* drool, slobber; ‹*niño*› dribble. **caerse la ~a** be delighted

babel *f* bedlam

babe|o *m* drooling; (*de un niño*) dribbling. **~ro** *m* bib

Babia *f*. **estar en ~** have one's head in the clouds

babieca *a* stupid. ● *m* & *f* simpleton

babor *m* port. **a ~** to port, on the port side

babosa *f* slug

babosada *f* (*Mex*) silly remark

babos|ear *vt* slobber over; ‹*niño*› dribble over. **~eo** *m* drooling; (*de niño*) dribbling. **~o** *a* slimy; (*LAm, tonto*) silly

babucha *f* slipper

babuino *m* baboon

baca *f* luggage rack

bacaladilla *f* small cod

bacalao *m* cod

bacon *m* bacon

bacteria *f* bacterium

bache *m* hole; (*fig*) bad patch

bachillerato *m* school-leaving examination

badaj|azo *m* stroke (of a bell). **~o** *m* clapper; (*persona*) chatterbox

bagaje *m* baggage; (*animal*) beast of burden; (*fig*) knowledge

bagatela *f* trifle

Bahamas *fpl* Bahamas

bahía *f* bay

bail|able *a* dance. **~ador** *a* dancing. ● *m* dancer. **~aor** *m* Flamenco dancer. **~ar** *vt/i* dance. **~arín** dancer. **~arina** *f* dancer; (*de baile clásico*) ballerina. **~e** *m* dance. **~e de etiqueta** ball. **ir a ~ar** go dancing

baja *f* drop, fall; (*mil*) casualty. **~ por maternidad** maternity leave. **~da** *f* slope; (*acto de bajar*) descent. **~mar** *m* low tide. **~r** *vt* lower; (*llevar abajo*) get down; bow ‹*la cabeza*›. **~r la escalera** go downstairs. ● *vi* go down; ‹*temperatura, precio*› fall. **~rse** *vpr* bend down. **~r(se) de** get out of ‹*coche*›; get off ‹*autobús, caballo, tren, bicicleta*›. **dar(se) de ~** take sick leave

bajeza *f* vile deed

bajío *m* sandbank

bajo *a* low; (*de estatura*) short, small; ‹*cabeza, ojos*› lowered; (*humilde*) humble, low; (*vil*) vile, low; ‹*color*› pale; ‹*voz*› low; (*mus*) deep. ● *m* lowland; (*bajío*) sandbank; (*mus*) bass. ● *adv* quietly; ‹*volar*› low. ● *prep* under; (*temperatura*) below. **~ la lluvia** in the rain. **los ~s fondos** the low district. **por lo ~** under one's breath; (*fig*) in secret

bajón *m* drop; (*de salud*) decline; (*com*) slump

bala *f* bullet; (*de algodón etc*) bale. **~ perdida** stray bullet. **como una ~** like a shot

balada *f* ballad

baladí *a* trivial

baladrón *a* boastful

baladron|ada *f* boast. **~ear** *vi* boast

balan|ce *m* swinging; (*de una cuenta*) balance; (*documento*) balance sheet. **~cear** *vt* balance. ● *vi* hesitate. **~cearse** *vpr* swing; (*vacilar*) hesitate. **~ceo** *m* swinging. **~za** *f* scales; (*com*) balance

balar *vi* bleat

balaustrada *f* balustrade, railing(s); (*de escalera*) banisters

balay *m* (*LAm*) wicker basket

balazo *m* (*disparo*) shot; (*herida*) bullet wound

balboa *f* (*unidad monetaria panameña*) balboa

balbuc|ear *vt/i* stammer; ‹*niño*› babble. **~eo** *m* stammering; (*de niño*) babbling. **~iente** *a* stammering; ‹*niño*› babbling. **~ir** [24] *vt/i* stammer; ‹*niño*› babble

balc|ón *m* balcony. **~onada** *f* row of balconies. **~onaje** *m* row of balconies

balda *f* shelf

baldado *a* disabled, crippled; (*rendido*) shattered. ● *m* disabled person, cripple

baldaquín *m*, **baldaquino** *m* canopy

baldar *vt* cripple

balde *m* bucket. **de ~** free (of charge). **en ~** in vain. **~ar** *vt* wash down

baldío *a* ‹*terreno*› waste; (*fig*) useless

baldosa *f* (floor) tile; (*losa*) flagstone

balduque *m* (*incl fig*) red tape

balear *a* Balearic. ● *m* native of the Balearic Islands. **las Islas** *fpl* **B~es** the Balearics, the Balearic Islands

baleo *m* (*LAm, tiroteo*) shooting; (*Mex, abanico*) fan

balido *m* bleat; (*varios sonidos*) bleating

bal|ín *m* small bullet. **~ines** *mpl* shot

balística *f* ballistics

baliza *f* (*naut*) buoy; (*aviat*) beacon

balneario *m* spa; (*con playa*) seaside resort. ● *a*. **estación** *f* **balnearia** spa; (*con playa*) seaside resort

balompié *m* football (*Brit*), soccer

bal|ón *m* ball, football. **~oncesto** *m* basketball. **~onmano** *m* handball. **~volea** *m* volleyball

balotaje *m* (*LAm*) voting

balsa *f* (*de agua*) pool; (*plataforma flotante*) raft

bálsamo *m* balsam; (*fig*) balm

balsón *m* (*Mex*) stagnant water

baluarte *m* (*incl fig*) bastion

balumba *f* mass, mountain

ballena *f* whale

ballesta *f* crossbow

ballet /ba'le/ (*pl* **ballets** uba'le/) *m* ballet

bambole|ar *vi* sway; ‹*mesa etc*› wobble. **~arse** *vpr* sway; ‹*mesa etc*› wobble. **~o** *m* swaying; (*de mesa etc*) wobbling

bambú *m* (*pl* **bambúes**) bamboo

banal *a* banal. **~idad** *f* banality

banan|a *f* (*esp LAm*) banana. **~o** *m* (*LAm*) banana tree

banast|a *f* large basket. **~o** *m* large round basket

banc|a *f* banking; (*en juegos*) bank; (*LAm, asiento*) bench. **~ario** *a* bank, banking. **~arrota** *f* bankruptcy. **~o** *m* (*asiento*) bench; (*com*) bank; (*bajío*) sandbank. **hacer ~arrota, ir a la ~arrota** go bankrupt

banda *f* (*incl mus, radio*) band; (*grupo*) gang, group; (*lado*) side. **~da** *f* (*de aves*) flock; (*de peces*) shoal. **~ de sonido, ~ sonora** sound-track

bandeja *f* tray; (*LAm, plato*) serving dish. **servir algo en ~ a uno** hand sth to s.o. on a plate

bandera *f* flag; (*estandarte*) banner, standard

banderill|a *f* banderilla. **~ear** *vt* stick the banderillas in. **~ero** *m* banderillero

banderín *m* pennant, small flag, banner

bandido *m* bandit

bando *m* edict, proclamation; (*partido*) faction. **~s** *mpl* banns. **pasarse al otro ~** go over to the other side
bandolero *m* bandit
bandolina *f* mandolin
bandoneón *m* large accordion
banjo *m* banjo
banquero *m* banker
banqueta *f* stool; (*LAm, acera*) pavement (*Brit*), sidewalk (*Amer*)
banquete *m* banquet; (*de boda*) wedding reception. **~ar** *vt/i* banquet
banquillo *m* bench; (*jurid*) dock; (*taburete*) footstool
bañ|ado *m* (*LAm*) swamp. **~ador** *m* (*de mujer*) swimming costume; (*de hombre*) swimming trunks. **~ar** *vt* bathe, immerse; bath ‹*niño*›; (*culin, recubrir*) coat. **~arse** *vpr* go swimming, have a swim; (*en casa*) have a bath. **~era** *f* bath, bath-tub. **~ero** *m* life-guard. **~ista** *m & f* bather. **~o** *m* bath; (*en piscina, mar etc*) swim; (*bañera*) bath, bath-tub; (*capa*) coat(ing)
baptisterio *m* baptistery; (*pila*) font
baquet|a *f* (*de fusil*) ramrod; (*de tambor*) drumstick. **~ear** *vt* bother. **~eo** *m* nuisance, bore
bar *m* bar
barahúnda *f* uproar
baraja *f* pack of cards. **~r** *vt* shuffle; juggle, massage ‹*cifras etc*›. ● *vi* argue (**con** with); (*enemistarse*) fall out (**con** with). **~s** *fpl* argument. **jugar a la ~** play cards. **jugar a dos ~s, jugar con dos ~s** be deceitful, indulge in double-dealing
baranda *f*, **barandal** *m*, **barandilla** *f* handrail; (*de escalera*) banisters
barat|a *f* (*Mex*) sale. **~ija** *f* trinket. **~illo** *m* junk shop; (*géneros*) cheap goods. **~o** *a* cheap. ● *m* sale. ● *adv* cheap(ly). **~ura** *f* cheapness
baraúnda *f* uproar
barba *f* chin; (*pelo*) beard. **~do** *a* bearded
barbacoa *f* barbecue; (*Mex, carne*) barbecued meat
bárbaramente *adv* savagely; (*fig*) tremendously
barbari|dad *f* barbarity; (*fig*) outrage; (*mucho, fam*) awful lot (*fam*). **¡qué ~dad!** how awful! **~e** *f* barbarity; (*fig*) ignorance. **~smo** *m* barbarism
bárbaro *a* barbaric, cruel; (*bruto*) uncouth; (*estupendo, fam*) terrific (*fam*). ● *m* barbarian. **¡qué ~!** how marvellous!
barbear *vt* (*afeitar*) shave; (*Mex, lisonjear*) fawn on
barbecho *m* fallow
barber|ía *f* barber's (shop). **~o** *m* barber; (*Mex, adulador*) flatterer
barbi|lampiño *a* beardless; (*fig*) inexperienced, green. **~lindo** *m* dandy
barbilla *f* chin
barbitúrico *m* barbiturate
barbo *m* barbel. **~ de mar** red mullet
barbot|ar *vt/i* mumble. **~ear** *vt/i* mumble. **~eo** *m* mumbling
barbudo *a* bearded
barbullar *vi* jabber
barca *f* (small) boat. **~ de pasaje** ferry. **~je** *m* fare. **~za** *f* barge
Barcelona *f* Barcelona
barcelonés *a* of Barcelona, from Barcelona. ● *m* native of Barcelona
barco *m* boat; (*navío*) ship. **~ cisterna** tanker. **~ de vapor** steamer. **~ de vela** sailing boat. **ir en ~** go by boat
bario *m* barium
barítono *m* baritone
barman *m* (*pl* **barmans**) barman
barniz *m* varnish; (*para loza etc*) glaze; (*fig*) veneer. **~ar** [10] *vt* varnish; glaze ‹*loza etc*›
bar|ométrico *a* barometric. **~ómetro** *m* barometer
bar|ón *m* baron. **~onesa** *f* baroness
barquero *m* boatman
barra *f* bar; (*pan*) French bread; (*de oro o plata*) ingot; (*palanca*) lever. **~ de labios** lipstick. **no pararse en ~s** stop at nothing
barrabasada *f* mischief, prank
barraca *f* hut; (*vivienda pobre*) shack, shanty
barranco *m* ravine, gully; (*despeñadero*) cliff, precipice
barre|dera *f* road-sweeper. **~dura** *f* rubbish. **~minas** *m invar* mine-sweeper
barren|a *f* drill, bit. **~ar** *vt* drill. **~o** *m* large (mechanical) drill. **entrar en ~a** ‹*avión*› go into a spin
barrer *vt* sweep; (*quitar*) sweep aside
barrera *f* barrier. **~ del sonido** sound barrier
barriada *f* district

barrica *f* barrel
barricada *f* barricade
barrido *m* sweeping
barrig|a *f* (pot-)belly. **~ón** *a*, **~udo** *a* pot-bellied
barril *m* barrel. **~ete** *m* keg, small barrel
barrio *m* district, area. **~bajero** *a* vulgar, common. **~s bajos** poor quarter, poor area. **el otro ~** (*fig, fam*) the other world
barro *m* mud; (*arcilla*) clay; (*arcilla cocida*) earthenware
barroco *a* Baroque. ● *m* Baroque style
barrote *m* heavy bar
barrunt|ar *vt* sense, have a feeling. **~e** *m*, **~o** *m* sign; (*presentimiento*) feeling
bartola *f*. **tenderse a la ~, tumbarse a la ~** take it easy
bártulos *mpl* things. **liar los ~** pack one's bags
barullo *m* uproar; (*confusión*) confusion. **a ~** galore
basa *f*, **basamento** *m* base; (*fig*) basis
basar *vt* base. **~se** *vpr*. **~se en** be based on
basc|a *f* crowd. **~as** *fpl* nausea. **~osidad** *f* filth. **la ~a** the gang
báscula *f* scales
bascular *vi* tilt
base *f* base; (*fig*) basis, foundation. **a ~ de** thanks to; (*mediante*) by means of; (*en una receta*) as the basic ingredient(s). **a ~ de bien** very well. **partiendo de la ~ de, tomando como ~** on the basis of
básico *a* basic
basílica *f* basilica
basilisco *m* basilisk. **hecho un ~** furious
basta *f* tack, tacking stitch
bastante *a* enough; (*varios*) quite a few, quite a lot of. ● *adv* rather, fairly; (*mucho tiempo*) long enough; (*suficiente*) enough; (*Mex, muy*) very
bastar *vi* be enough. **¡basta!** that's enough! **basta decir que** suffice it to say that. **basta y sobra** that's more than enough
bastardilla *f* italics. **poner en ~** italicize
bastardo *m* bastard; (*fig, vil*) mean, base
bastidor *m* frame; (*auto*) chassis. **~es** *mpl* (*en el teatro*) wings. **entre ~es** behind the scenes
bastión *f* (*incl fig*) bastion
basto *a* coarse. **~s** *mpl* (*naipes*) clubs
bast|ón *m* walking stick. **empuñar el ~ón** take command. **~onazo** *m* blow with a stick
basur|a *f* rubbish, garbage (*Amer*); (*en la calle*) litter. **~ero** *m* dustman (*Brit*), garbage collector (*Amer*); (*sitio*) rubbish dump; (*recipiente*) dustbin (*Brit*), garbage can (*Amer*). **cubo** *m* **de la ~a** dustbin (*Brit*), garbage can (*Amer*)
bata *f* dressing-gown; (*de médico etc*) white coat. **~ de cola** Flamenco dress
batall|a *f* battle. **~a campal** pitched battle. **~ador** *a* fighting. ● *m* fighter. **~ar** *vi* battle, fight. **~ón** *m* battalion. ● *a*. **cuestión** *f* **batallona** vexed question. **de ~a** everyday
batata *f* sweet potato
bate *m* bat. **~ador** *m* batter; (*cricket*) batsman
batería *f* battery; (*mus*) percussion. **~ de cocina** kitchen utensils, pots and pans
batido *a* beaten; ‹*nata*› whipped. ● *m* batter; (*bebida*) milk shake. **~ra** *f* beater. **~ra eléctrica** mixer
batín *m* dressing-gown
batir *vt* beat; (*martillar*) hammer; mint ‹*monedas*›; whip ‹*nata*›; (*derribar*) knock down. **~ el récord** break the record. **~ palmas** clap. **~se** *vpr* fight
batuta *f* baton. **llevar la ~** be in command, be the boss
baúl *m* trunk; (*LAm, auto*) boot (*Brit*), trunk (*Amer*)
bauti|smal *a* baptismal. **~smo** *m* baptism, christening. **~sta** *a* & *m* & *f* Baptist. **~zar** [10] *vt* baptize, christen
baya *f* berry
bayeta *f* (floor-)cloth
bayoneta *f* bayonet. **~zo** *m* (*golpe*) bayonet thrust; (*herida*) bayonet wound
baza *f* (*naipes*) trick; (*fig*) advantage. **meter ~** interfere
bazar *m* bazaar
bazofia *f* leftovers; (*basura*) rubbish
beat|itud *f* (*fig*) bliss. **~o** *a* blessed; (*de religiosidad afectada*) sanctimonious
bebé *m* baby
beb|edero *m* drinking trough; (*sitio*) watering place. **~edizo** *a*

drinkable. • *m* potion; (*veneno*) poison. **~edor** *a* drinking. • *m* heavy drinker. **~er** *vt/i* drink. **dar de ~er a uno** give s.o. a drink. **~ida** *f* drink. **~ido** *a* tipsy, drunk
beca *f* grant, scholarship. **~rio** *m* scholarship holder, scholar
becerro *m* calf
befa *f* jeer, taunt. **~r** *vt* scoff at. **~rse** *vpr*. **~rse de** scoff at. **hacer ~ de** scoff at
beige /beis, bes/ *a & m* beige
béisbol *m* baseball
beldad *f* beauty
belén *m* crib, nativity scene; (*barullo*) confusion
belga *a & m & f* Belgian
Bélgica *f* Belgium
bélico *a*, **belicoso** *a* warlike
beligerante *a* belligerent
bella|co *a* wicked. • *m* rogue. **~quear** *vi* cheat. **~quería** *f* dirty trick
bell|eza *f* beauty. **~o** *a* beautiful. **~as artes** *fpl* fine arts
bellota *f* acorn
bemol *m* flat. **tener (muchos) ~es** be difficult
bencina *f* (*Arg*, *gasolina*) petrol (*Brit*), gasoline (*Amer*)
bend|ecir [46 *pero imperativo* **bendice**, *futuro, condicional y pp regulares*] *vt* bless. **~ición** *f* blessing. **~ito** *a* blessed, holy; (*que tiene suerte*) lucky; (*feliz*) happy
benefactor *m* benefactor. **~a** *f* benefactress
benefic|encia *f* (*organización pública*) charity. **~iar** *vt* benefit. **~iarse** *vpr* benefit. **~iario** *m* beneficiary; (*de un cheque etc*) payee. **~io** *m* benefit; (*ventaja*) advantage; (*ganancia*) profit, gain. **~ioso** *a* beneficial, advantageous
benéfico *a* beneficial; (*de beneficencia*) charitable
benemérito *a* worthy
beneplácito *m* approval
ben|evolencia *f* benevolence. **~évolo** *a* benevolent
bengala *f* flare. **luz** *f* **de B~** flare
benign|idad *f* kindness; (*falta de gravedad*) mildness. **~o** *a* kind; (*moderado*) gentle, mild; ‹*tumor*› benign
beodo *a* drunk
berberecho *m* cockle
berenjena *f* aubergine (*Brit*), egg-plant. **~l** *m* (*fig*) mess
bermejo *a* red
berr|ear *vi* ‹*animales*› low, bellow; ‹*niño*› howl; (*cantar mal*) screech. **~ido** *m* bellow; (*de niño*) howl; (*de cantante*) screech
berrinche *m* temper; (*de un niño*) tantrum
berro *m* watercress
berza *f* cabbage
besamel(a) *f* white sauce
bes|ar *vt* kiss; (*rozar*) brush against. **~arse** *vpr* kiss (each other); (*tocarse*) touch each other. **~o** *m* kiss
bestia *f* beast; (*bruto*) brute; (*idiota*) idiot. **~ de carga** beast of burden. **~l** *a* bestial, animal; (*fig, fam*) terrific. **~lidad** *f* bestiality; (*acción brutal*) horrid thing
besugo *m* sea-bream. **ser un ~** be stupid
besuquear *vt* cover with kisses
betún *m* bitumen; (*para el calzado*) shoe polish
biberón *m* feeding-bottle
Biblia *f* Bible
bíblico *a* biblical
bibliografía *f* bibliography
biblioteca *f* library; (*librería*) book-case. **~ de consulta** reference library. **~ de préstamo** lending library. **~rio** *m* librarian
bicarbonato *m* bicarbonate. **~ sódico** bicarbonate of soda
bici *f* (*fam*) bicycle, bike (*fam*). **~cleta** *f* bicycle. **ir en ~cleta** go by bicycle, cycle. **montar en ~cleta** ride a bicycle
bicolor *a* two-colour
bicultural *a* bicultural
bicho *m* (*animal*) small animal, creature; (*insecto*) insect. **~ raro** odd sort. **cualquier ~ viviente, todo ~ viviente** everyone
bidé *m*, **bidet** *m* bidet
bidón *m* drum, can
bien *adv* (**mejor**) well; (*muy*) very, quite; (*correctamente*) right; (*de buena gana*) willingly. • *m* good; (*efectos*) property; (*provecho*) advantage, benefit. **¡~!** fine!, OK!, good! **~... (o) ~** either... or. **~ que** although. **¡está ~!** fine! alright! **más ~** rather. **¡muy ~!** good! **no ~** as soon as. **¡qué ~!** marvellous!, great! (*fam*). **si ~** although
bienal *a* biennial
bien: ~aventurado *a* fortunate. **~estar** *m* well-being. **~hablado** *a* well-spoken. **~hechor** *m* benefactor.

~hechora *f* benefactress. **~intencionado** *a* well-meaning
bienio *m* two years, two year-period
bien: ~quistar *vt* reconcile. **~quistarse** *vpr* become reconciled. **~quisto** *a* well-liked
bienvenid|a *f* welcome. **~o** *a* welcome. **¡~o!** welcome! **dar la ~a a uno** welcome s.o.
bife *m* (*Arg*), **biftek** *m* steak
bifurca|ción *f* fork, junction. **~rse** [7] *vpr* fork
b|igamia *f* bigamy. **~ígamo** *a* bigamous. • *m & f* bigamist
bigot|e *m* moustache. **~udo** *a* with a big moustache
bikini *m* bikini; (*culin*) toasted cheese and ham sandwich
bilingüe *a* bilingual
billar *m* billiards
billete *m* ticket; (*de banco*) note (*Brit*), bill (*Amer*). **~ de banco** banknote. **~ de ida y vuelta** return ticket (*Brit*), round-trip ticket (*Amer*). **~ sencillo** single ticket (Brit), one-way ticket (*Amer*). **~ro** *m*, **~ra** *f* wallet, billfold (*Amer*)
billón *m* billion (*Brit*), trillion (*Amer*)
bimbalete *m* (*Mex*) swing
bi|mensual *a* fortnightly, twice-monthly. **~mestral** *a* two-monthly. **~motor** *a* twin-engined. • *m* twin-engined plane
binocular *a* binocular. **~es** *mpl* binoculars
biodegradable *a* biodegradable
bi|ografía *f* biography. **~ográfico** *a* biographical. **~ógrafo** *m* biographer
bi|ología *f* biology. **~ológico** *a* biological. **~ólogo** *m* biologist
biombo *m* folding screen
biopsia *f* biopsy
bioquímic|a *f* biochemistry; (*persona*) biochemist. **~o** *m* biochemist
bípedo *m* biped
biplano *m* biplane
biquini *m* bikini
birlar *vt* (*fam*) steal, pinch (*fam*)
birlibirloque *m*. **por arte de ~** (as if) by magic
Birmania *f* Burma
birmano *a & m* Burmese
biromen *m* (*Arg*) ball-point pen
bis *m* encore. • *adv* twice. **¡~!** encore! **vivo en el 3 ~** I live at 3A
bisabuel|a *f* great-grandmother. **~o** *m* great-grandfather. **~os** *mpl* great-grandparents
bisagra *f* hinge
bisar *vt* encore
bisbise|ar *vt* whisper. **~o** *m* whisper(ing)
bisemanal *a* twice-weekly
bisiesto *a* leap. **año** *m* **~** leap year
bisniet|a *f* great-granddaughter. **~o** *m* great-grandson. **~os** *mpl* great-grandchildren
bisonte *m* bison
bisté *m*, **bistec** *m* steak
bisturí *m* scalpel
bisutería *f* imitation jewellery, costume jewellery
bizco *a* cross-eyed. **quedarse ~** be dumbfounded
bizcocho *m* sponge (cake); (*Mex, galleta*) biscuit
bizquear *vi* squint
blanc|a *f* white woman; (*mus*) minim. **~o** *a* white; ‹*tez*› fair. • *m* white; (*persona*) white man; (*intervalo*) interval; (*espacio*) blank; (*objetivo*) target. **~o de huevo** white of egg, egg-white. **dar en el ~o** hit the mark. **dejar en ~o** leave blank. **pasar la noche en ~o** have a sleepless night. **~o y negro** black and white. **~ura** *f* whiteness. **~uzco** *a* whitish
blandir [24] *vt* brandish
bland|o *a* soft; ‹*carácter*› weak; (*cobarde*) cowardly; ‹*palabras*› gentle, tender. **~ura** *f* softness. **~uzco** *a* softish
blanque|ar *vt* whiten; white-wash ‹*paredes*›; bleach ‹*tela*›. • *vi* turn white; (*presentarse blanco*) look white. **~cino** *a* whitish. **~o** *m* whitening
blasfem|ador *a* blasphemous. • *m* blasphemer. **~ar** *vi* blaspheme. **~ia** *f* blasphemy. **~o** *a* blasphemous. • *m* blasphemer
blas|ón *m* coat of arms; (*fig*,) honour, glory. **~onar** *vt* emblazon. • *vi* boast (**de** of, about)
bledo *m* nothing. **me importa un ~, no se me da un ~** I couldn't care less
blinda|je *m* armour. **~r** *vt* armour
bloc *m* (*pl* **blocs**) pad
bloque *m* block; (*pol*) bloc. **~ar** *vt* block; (*mil*) blockade; (*com*) freeze. **~o** *m* blockade; (*com*) freezing. **en ~** en bloc
blusa *f* blouse
boato *m* show, ostentation

bob|ada *f* silly thing. **~alicón** *a* stupid. **~ería** *f* silly thing. **decir ~adas** talk nonsense

bobina *f* bobbin, reel; (*foto*) spool; (*elec*) coil

bobo *a* silly, stupid. ● *m* idiot, fool

boca *f* mouth; (*fig, entrada*) entrance; (*de cañón*) muzzle; (*agujero*) hole. **~ abajo** face down. **~ arriba** face up. **a ~ de jarro** point-blank. **con la ~ abierta** dumbfounded

bocacalle *f* junction. **la primera ~ a la derecha** the first turning on the right

bocad|illo *m* sandwich; (*comida ligera, fam*) snack. **~o** *m* mouthful; (*mordisco*) bite; (*de caballo*) bit

boca: ~jarro. a ~jarro point-blank. **~manga** *f* cuff

bocanada *f* puff; (*de vino etc*) mouthful

bocaza *f invar*, **bocazas** *f invar* big-mouth

boceto *m* outline, sketch

bocina *f* horn. **~zo** *m* toot, blast. **tocar la ~** sound one's horn

bock *m* beer mug

bocha *f* bowl. **~s** *fpl* bowls

bochinche *m* uproar

bochorno *m* sultry weather; (*fig, vergüenza*) embarrassment. **~so** *a* oppressive; (*fig*) embarrassing. **¡qué ~!** how embarrassing!

boda *f* marriage; (*ceremonia*) wedding

bodeg|a *f* cellar; (*de vino*) wine cellar; (*almacén*) warehouse; (*de un barco*) hold. **~ón** *m* cheap restaurant; (*pintura*) still life

bodoque *m* pellet; (*tonto, fam*) thickhead

bofes *mpl* lights. **echar los ~** slog away

bofet|ada *f* slap; (*fig*) blow. **dar una ~ada a uno** slap s.o. in the face. **darse de ~adas** clash. **~ón** *m* punch

boga *m & f* rower; (*hombre*) oarsman; (*mujer*) oarswoman; (*moda*) fashion. **estar en ~** be in fashion, be in vogue. **~da** *f* stroke (of the oar). **~dor** rower, oarsman. **~r** [12] *vt* row. **~vante** *m* (*crustáceo*) lobster

Bogotá *f* Bogotá

bogotano *a* from Bogotá. ● *m* native of Bogotá

bohemio *a & m* Bohemian

bohío *m* (*LAm*) hut

boicot *m* (*pl* **boicots**) boycott. **~ear** *vt* boycott. **~eo** *m* boycott. **hacer el ~** boycott

boina *f* beret

boîte /bwat/ *m* night-club

bola *f* ball; (*canica*) marble; (*naipes*) slam; (*betún*) shoe polish; (*mentira*) fib; (*Mex, reunión desordenada*) rowdy party. **~ del mundo** (*fam*) globe. **contar ~s** tell fibs. **dejar que ruede la ~** let things take their course. **meter ~s** tell fibs

bolas *fpl* (*LAm*) bolas

boleada *f* (*Mex*) polishing of shoes

boleadoras (*LAm*) *fpl* bolas

bolera *f* bowling alley

bolero *m* (*baile, chaquetilla*) bolero; (*fig, mentiroso, fam*) liar; (*Mex, limpiabotas*) bootblack

boletín *m* bulletin; (*publicación periódica*) journal; (*escolar*) report. **~ de noticias** news bulletin. **~ de precios** price list. **~ informativo** news bulletin. **~ meteorológico** weather forecast

boleto *m* (*esp LAm*) ticket

boli *m* (*fam*) Biro (P), ball-point pen

boliche *m* (*juego*) bowls; (*bolera*) bowling alley

bolígrafo *m* Biro (P), ball-point pen

bolillo *m* bobbin; (*Mex, panecillo*) (bread) roll

bolívar *m* (*unidad monetaria venezolana*) bolívar

Bolivia *f* Bolivia

boliviano *a* Bolivian. ● *m* Bolivian; (*unidad monetaria de Bolivia*) boliviano

bolo *m* skittle

bolsa *f* bag; (*monedero*) purse; (*LAm, bolsillo*) pocket; (*com*) stock exchange; (*cavidad*) cavity. **~ de agua caliente** hot-water bottle

bolsillo *m* pocket; (*monedero*) purse. **de ~** pocket

bolsista *m & f* stockbroker

bolso *m* (*de mujer*) handbag

boll|ería *f* baker's shop. **~ero** *m* baker. **~o** *m* roll; (*con azúcar*) bun; (*abolladura*) dent; (*chichón*) lump; (*fig, jaleo, fam*) fuss

bomba *f* bomb; (*máquina*) pump; (*noticia*) bombshell. **~ de aceite** (*auto*) oil pump. **~ de agua** (*auto*) water pump. **~ de incendios** fire-engine. **pasarlo ~** have a marvellous time

bombach|as *fpl* (*LAm*) knickers, pants. **~o** *m* (*esp Mex*) baggy trousers, baggy pants (*Amer*)

bombarde|ar *vt* bombard; (*mil*) bomb. **~o** *m* bombardment; (*mil*) bombing. **~ro** *m* (*avión*) bomber
bombazo *m* explosion
bombear *vt* pump; (*mil*) bomb
bombero *m* fireman. **cuerpo** *m* **de ~s** fire brigade (*Brit*), fire department (*Amer*)
bombilla *f* (light) bulb; (*LAm, para maté*) pipe for drinking maté; (*Mex, cucharón*) ladle
bombín *m* pump; (*sombrero, fam*) bowler (hat) (*Brit*), derby (*Amer*)
bombo *m* (*tambor*) bass drum. **a ~ y platillos** with a lot of fuss
bomb|ón *m* chocolate. **ser un ~ón** be a peach. **~ona** *f* container. **~onera** *f* chocolate box
bonachón *a* easygoing; (*bueno*) good-natured
bonaerense *a* from Buenos Aires. • *m* native of Buenos Aires
bonanza *f* (*naut*) fair weather; (*prosperidad*) prosperity. **ir en ~** (*naut*) have fair weather; (*fig*) go well
bondad *f* goodness; (*amabilidad*) kindness. **tenga la ~ de** would you be kind enough to. **~osamente** *adv* kindly. **~oso** *a* kind
bongo *m* (*LAm*) canoe
boniato *m* sweet potato
bonito *a* nice; (*mono*) pretty. **¡muy ~!, ¡qué ~!** that's nice!, very nice!. • *m* bonito
bono *m* voucher; (*título*) bond. **~ del Tesoro** government bond
boñiga *f* dung
boqueada *f* gasp. **dar las ~s** be dying
boquerón *m* anchovy
boquete *m* hole; (*brecha*) breach
boquiabierto *a* open-mouthed; (*fig*) amazed, dumbfounded. **quedarse ~** be amazed
boquilla *f* mouthpiece; (*para cigarillos*) cigarette-holder; (*filtro de cigarillo*) tip
borboll|ar *vi* bubble. **~ón** *m* bubble. **hablar a ~ones** gabble. **salir a ~ones** gush out
borbot|ar *vt* bubble. **~ón** *m* bubble. **hablar a ~ones** gabble. **salir a ~ones** gush out
bordado *a* embroidered. • *m* embroidery. **quedar ~, salir ~** come out very well
bordante *m* (*Mex*) lodger
bordar *vt* embroider; (*fig, fam*) do very well
bord|e *m* edge; (*de carretera*) side; (*de plato etc*) rim; (*de un vestido*) hem. **~ear** *vt* go round the edge of; (*fig*) border on. **~illo** *m* kerb. **al ~e de** on the edge of; (*fig*) on the brink of
bordo *m* board. **a ~** on board
borinqueño *a & m* Puerto Rican
borla *f* tassel
borra *f* flock; (*pelusa*) fluff; (*sedimento*) sediment
borrach|era *f* drunkenness. **~ín** *m* drunkard. **~o** *a* drunk. • *m* drunkard; (*temporalmente*) drunk. **estar ~o** be drunk. **ni ~o** never in a million years. **ser ~o** be a drunkard
borrador *m* rough copy; (*libro*) rough notebook
borradura *f* crossing-out
borrajear *vt/i* scribble
borrar *vt* rub out; (*tachar*) cross out
borrasc|a *f* storm. **~oso** *a* stormy
borreg|o *m* year-old lamb; (*fig*) simpleton; (*Mex, noticia falsa*) hoax. **~uil** *a* meek
borric|ada *f* silly thing. **~o** *m* donkey; (*fig, fam*) ass
borrón *m* smudge; (*fig, imperfección*) blemish; (*de una pintura*) sketch. **~ y cuenta nueva** let's forget about it!
borroso *a* blurred; (*fig*) vague
bos|caje *m* thicket. **~coso** *a* wooded. **~que** *m* wood, forest. **~quecillo** *m* copse
bosquej|ar *vt* sketch. **~o** *m* sketch
bosta *f* dung
bostez|ar [10] *vi* yawn. **~o** *m* yawn
bota *f* boot; (*recipiente*) leather wine bottle
botadero *m* (*Mex*) ford
botánic|a *f* botany. **~o** *a* botanical. • *m* botanist
botar *vt* launch. • *vi* bounce. **estar que bota** be hopping mad
botarat|ada *f* silly thing. **~e** *m* idiot
bote *m* bounce; (*golpe*) blow; (*salto*) jump; (*sacudida*) jolt; (*lata*) tin, can; (*vasija*) jar; (*en un bar*) jar for tips; (*barca*) boat. **~ salvavidas** lifeboat. **de ~ en ~** packed
botell|a *f* bottle. **~ita** *f* small bottle
botica *f* chemist's (shop) (*Brit*), drugstore (*Amer*). **~rio** *m* chemist (*Brit*), druggist (*Amer*)
botija *f*, **botijo** *m* earthenware jug
botín *m* half boot; (*despojos*) booty; (*LAm, calcetín*) sock

botiquín *m* medicine chest; (*de primeros auxilios*) first aid kit
bot|ón *m* button; (*yema*) bud. **~onadura** *f* buttons. **~ón de oro** buttercup. **~ones** *m invar* bellboy (*Brit*), bellhop (*Amer*)
botulismo *m* botulism
boutique /bu'tik/ *m* boutique
bóveda *f* vault
boxe|ador *m* boxer. **~ar** *vi* box. **~o** *m* boxing
boya *f* buoy; (*corcho*) float. **~nte** *a* buoyant
bozal *m* (*de perro etc*) muzzle; (*de caballo*) halter
bracear *vi* wave one's arms; (*nadar*) swim, crawl
bracero *m* labourer. **de ~** (*fam*) arm in arm
braga *f* underpants, knickers; (*cuerda*) rope. **~dura** *f* crotch. **~s** *fpl* knickers, pants. **~zas** *m invar* (*fam*) henpecked man
bragueta *f* flies
braille /breil/ *m* Braille
bram|ar *vi* roar; ‹*vaca*› moo; ‹*viento*› howl. **~ido** *m* roar
branquia *f* gill
bras|a *f* hot coal. **a la ~a** grilled. **~ero** *m* brazier; (*LAm, hogar*) hearth
Brasil *m*. **el ~** Brazil
brasile|ño *a* & *m* Brazilian. **~ro** *a* & *m* (*LAm*) Brazilian
bravata *f* boast
bravío *a* wild; ‹*persona*› coarse, uncouth
brav|o *a* brave; ‹*animales*› wild; ‹*mar*› rough. **¡~!** *int* well done! bravo! **~ura** *f* ferocity; (*valor*) courage
braz|a *f* fathom. **nadar a ~a** do the breast-stroke. **~ada** *f* waving of the arms; (*en natación*) stroke; (*cantidad*) armful. **~ado** *m* armful. **~al** *m* arm-band. **~alete** *m* bracelet; (*brazal*) arm-band. **~o** *m* arm; (*de animales*) foreleg; (*rama*) branch. **~o derecho** right-hand man. **a ~o** by hand. **del ~o** arm in arm
brea *f* tar, pitch
brear *vt* ill-treat
brécol *m* broccoli
brecha *f* gap; (*mil*) breach; (*med*) gash. **estar en la ~** be in the thick of it
brega *f* struggle. **~r** [12] *vi* struggle; (*trabajar mucho*) work hard, slog away. **andar a la ~** work hard
breña *f*, **breñal** *m* scrub
Bretaña *f* Brittany. **Gran ~** Great Britain
breve *a* short. **~dad** *f* shortness. **en ~** soon, shortly. **en ~s momentos** soon
brez|al *m* moor. **~o** *m* heather
brib|ón *m* rogue, rascal. **~onada** *f*, **~onería** *f* dirty trick
brida *f* bridle. **a toda ~** at full speed
bridge /britʃ/ *m* bridge
brigada *f* squad; (*mil*) brigade. **general de ~** brigadier (*Brit*), brigadier-general (*Amer*)
brill|ante *a* brilliant. ● *m* diamond. **~antez** *f* brilliance. **~ar** *vi* shine; (*centellear*) sparkle. **~o** *m* shine; (*brillantez*) brilliance; (*centelleo*) sparkle. **dar ~o, sacar ~o** polish
brinc|ar [7] *vi* jump up and down. **~o** *m* jump. **dar un ~o** jump. **estar que brinca** be hopping mad. **pegar un ~o** jump
brind|ar *vt* offer. ● *vi*. **~ar por** toast, drink a toast to. **~is** *m* toast
br|ío *m* energy; (*decisión*) determination. **~ioso** *a* spirited; (*garboso*) elegant
brisa *f* breeze
británico *a* British. ● *m* Briton, British person
brocado *m* brocade
bróculi *m* broccoli
brocha *f* paintbrush; (*para afeitarse*) shaving-brush
broche *m* clasp, fastener; (*joya*) brooch; (*Arg, sujetapapeles*) paper-clip
brocheta *f* skewer
brom|a *f* joke. **~a pesada** practical joke. **~ear** *vi* joke. **~ista** *a* fun-loving. ● *m* & *f* joker. **de ~a, en ~a** in fun. **ni de ~a** never in a million years
bronca *f* row; (*reprensión*) telling-off
bronce *m* bronze. **~ado** *a* bronze; (*por el sol*) tanned, sunburnt. **~ar** *vt* tan ‹*piel*›. **~arse** *vpr* get a suntan
bronco *a* rough
bronquitis *f* bronchitis
broqueta *f* skewer
brot|ar *vi* ‹*plantas*› bud, sprout; (*med*) break out; ‹*líquido*› gush forth; ‹*lágrimas*› well up. **~e** *m* bud, shoot; (*med*) outbreak; (*de líquido*) gushing; (*de lágrimas*) welling-up
bruces *mpl*. **de ~** face down(wards). **caer de ~** fall flat on one's face

bruj|a *f* witch. ● *a* (*Mex*) penniless. **~ear** *vi* practise witchcraft. **~ería** *f* witchcraft. **~o** *m* wizard, magician; (*LAm*) medicine man

brújula *f* compass

brum|a *f* mist; (*fig*) confusion. **~oso** *a* misty, foggy

bruñi|do *m* polish. **~r** [22] *vt* polish

brusco *a* (*repentino*) sudden; ‹*persona*› brusque

Bruselas *fpl* Brussels

brusquedad *f* abruptness

brut|al *a* brutal. **~alidad** *f* brutality; (*estupidez*) stupidity. **~o** *a* (*estúpido*) stupid; (*tosco*) rough, uncouth; ‹*peso, sueldo*› gross

bucal *a* oral

buce|ar *vi* dive; (*fig*) explore. **~o** *m* diving

bucle *m* curl

budín *m* pudding

budis|mo *m* Buddhism. **~ta** *m & f* Buddhist

buen *véase* **bueno**

buenamente *adv* easily; (*voluntariamente*) willingly

buenaventura *f* good luck; (*adivinación*) fortune. **decir la ~ a uno, echar la ~ a uno** tell s.o.'s fortune

bueno *a* (*delante de nombre masculino en singular* **buen**) good; (*apropiado*) fit; (*amable*) kind; ‹*tiempo*› fine. ● *int* well!; (*de acuerdo*) OK!, very well! **¡buena la has hecho!** you've gone and done it now! **¡buenas noches!** good night! **¡buenas tardes!** (*antes del atardecer*) good afternoon!; (*después del atardecer*) good evening! **¡~s días!** good morning! **estar de buenas** be in a good mood. **por las buenas** willingly

Buenos Aires *m* Buenos Aires

buey *m* ox

búfalo *m* buffalo

bufanda *f* scarf

bufar *vi* snort. **estar que bufa** be hopping mad

bufete *m* (*mesa*) writing-desk; (*despacho*) lawyer's office

bufido *m* snort; (*de ira*) outburst

buf|o *a* comic. **~ón** *a* comical. ● *m* buffoon. **~onada** *f* joke

bugle *m* bugle

buharda *f*, **buhardilla** *f* attic; (*ventana*) dormer window

búho *m* owl

buhoner|ía *f* pedlar's wares. **~o** *m* pedlar

buitre *m* vulture

bujía *f* candle; (*auto*) spark(ing)-plug

bula *f* bull

bulbo *m* bulb

bulevar *m* avenue, boulevard

Bulgaria *f* Bulgaria

búlgaro *a & m* Bulgarian

bulo *m* hoax

bulto *m* (*volumen*) volume; (*tamaño*) size; (*forma*) shape; (*paquete*) package; (*protuberancia*) lump. **a ~** roughly

bulla *f* uproar; (*muchedumbre*) crowd

bullicio *m* hubbub; (*movimiento*) bustle. **~so** *a* bustling; (*ruidoso*) noisy

bullir [22] *vt* stir, move. ● *vi* boil; (*burbujear*) bubble; (*fig*) bustle

buñuelo *m* doughnut; (*fig*) mess

BUP *abrev* (*Bachillerato Unificado Polivalente*) secondary school education

buque *m* ship, boat

burbuj|a *f* bubble. **~ear** *vi* bubble; ‹*vino*› sparkle. **~eo** *m* bubbling

burdel *m* brothel

burdo *a* rough, coarse; ‹*excusa*› clumsy

burgu|és *a* middle-class, bourgeois. ● *m* middle-class person. **~esía** *f* middle class, bourgeoisie

burla *f* taunt; (*broma*) joke; (*engaño*) trick. **~dor** *a* mocking. ● *m* seducer. **~r** *vt* trick, deceive; (*seducir*) seduce. **~rse** *vpr*. **~rse de** mock, make fun of

burlesco *a* funny

burlón *a* mocking

bur|ocracia *f* civil service. **~ócrata** *m & f* civil servant. **~ocrático** *a* bureaucratic

burro *m* donkey; (*fig*) ass

bursátil *a* stock-exchange

bus *m* (*fam*) bus

busca *f* search. **a la ~ de** in search of. **en ~ de** in search of

busca: ~pié *m* feeler. **~pleitos** *m invar* (*LAm*) trouble-maker

buscar [7] *vt* look for. ● *vi* look. **buscársela** ask for it. **ir a ~ a uno** fetch s.o.

buscarruidos *m invar* trouble-maker

buscona *f* prostitute

busilis *m* snag

búsqueda *f* search

busto *m* bust

butaca *f* armchair; (*en el teatro etc*) seat
butano *m* butane
buzo *m* diver
buzón *m* postbox (*Brit*), mailbox (*Amer*)

C

Cu *abrev* (*Calle*) St, Street, Rd, Road
cabal *a* exact; (*completo*) complete. **no estar en sus ~es** not be in one's right mind
cabalga|dura *f* mount, horse. **~r** [12] *vt* ride. • *vi* ride, go riding. **~ta** *f* ride; (*desfile*) procession
cabalmente *adv* completely; (*exactamente*) exactly
caballa *f* mackerel
caballada *f* (*LAm*) stupid thing
caballeresco *a* gentlemanly. **literatura** *f* **caballeresca** books of chivalry
caballer|ía *f* mount, horse. **~iza** *f* stable. **~izo** *m* groom
caballero *m* gentleman; (*de orden de caballería*) knight; (*tratamiento*) sir. **~samente** *adv* like a gentleman. **~so** *a* gentlemanly
caballete *m* (*del tejado*) ridge; (*de la nariz*) bridge; (*de pintor*) easel
caballito *m* pony. **~ del diablo** dragonfly. **~ de mar** sea-horse. **los ~s** (*tiovivo*) merry-go-round
caballo *m* horse; (*del ajedrez*) knight; (*de la baraja española*) queen. **~ de vapor** horsepower. **a ~** on horseback
cabaña *f* hut
cabaret /kaba're/ *m* (*pl* **cabarets** /kaba're/) night-club
cabece|ar *vi* nod; (*para negar*) shake one's head. **~o** *m* nodding, nod; (*acción de negar*) shake of the head
cabecera *f* (*de la cama, de la mesa*) head; (*en un impreso*) heading
cabecilla *m* leader
cabell|o *m* hair. **~os** *mpl* hair. **~udo** *a* hairy
caber [28] *vi* fit (**en** into). **los libros no caben en la caja** the books won't fit into the box. **no cabe duda** there's no doubt
cabestr|illo *m* sling. **~o** *m* halter
cabeza *f* head; (*fig, inteligencia*) intelligence. **~da** *f* butt; (*golpe recibido*) blow; (*saludo, al dormirse*) nod. **~zo** *m* butt; (*en fútbol*) header. **andar de ~** have a lot to do. **dar una ~da** nod off
cabida *f* capacity; (*extensión*) area. **dar ~ a** leave room for, leave space for
cabina *f* (*de avión*) cabin, cockpit; (*electoral*) booth; (*de camión*) cab. **~ telefónica** telephone box (*Brit*), telephone booth (*Amer*)
cabizbajo *a* crestfallen
cable *m* cable
cabo *m* end; (*trozo*) bit; (*mil*) corporal; (*mango*) handle; (*geog*) cape; (*naut*) rope. **al ~** eventually. **al ~ de una hora** after an hour. **de ~ a rabo** from beginning to end. **llevar(se) a ~** carry out
cabr|a *f* goat. **~a montesa** *f* mountain goat. **~iola** *f* jump, skip. **~itilla** *f* kid. **~ito** *m* kid
cabrón *m* cuckold
cabuya *f* (*LAm*) pita, agave
cacahuate *m* (*Mex*), **cacahuete** *m* peanut
cacao *m* (*planta y semillas*) cacao; (*polvo*) cocoa; (*fig*) confusion
cacare|ar *vt* boast about. • *vi* ‹*gallo*› crow; ‹*gallina*› cluck. **~o** *m* (*incl fig*) crowing; (*de gallina*) clucking
cacería *f* hunt
cacerola *f* casserole, saucepan
caciqu|e *m* cacique, Indian chief; (*pol*) cacique, local political boss. **~il** *a* despotic. **~ismo** *m* caciquism, despotism
caco *m* pickpocket, thief
cacof|onía *f* cacophony. **~ónico** *a* cacophonous
cacto *m* cactus
cacumen *m* acumen
cacharro *m* earthenware pot; (*para flores*) vase; (*coche estropeado*) wreck; (*cosa inútil*) piece of junk; (*chisme*) thing. **~s** *mpl* pots and pans
cachear *vt* frisk
cachemir *m*, **cachemira** *f* cashmere
cacheo *m* frisking
cachetada *f* (*LAm*), **cachete** *m* slap
cachimba *f* pipe
cachiporra *f* club, truncheon. **~zo** *m* blow with a club
cachivache *m* thing, piece of junk
cacho *m* bit, piece; (*LAm, cuerno*) horn; (*miga*) crumb
cachondeo *m* (*fam*) joking, joke
cachorro *m* (*perrito*) puppy; (*de otros animales*) young

cada *a invar* each, every. **~ uno** each one, everyone. **uno de ~ cinco** one in five
cadalso *m* scaffold
cadáver *m* corpse. **ingresar ~** be dead on arrival
cadena *f* chain; (*TV*) channel. **~ de fabricación** production line. **~ de montañas** mountain range. **~ perpetua** life imprisonment
cadencia *f* cadence, rhythm
cadera *f* hip
cadete *m* cadet
caduc|ar [7] *vi* expire. **~idad** *f*. **fecha** *f* **de ~idad** sell-by date. **~o** *a* decrepit
cae|dizo *a* unsteady. **~r** [29] *vi* fall. **~rse** *vpr* fall (over). **dejar ~r** drop. **estar al ~r** be about to happen. **este vestido no me ~ bien** this dress doesn't suit me. **hacer ~r** knock over. **Juan me ~ bien** I get on well with Juan. **su cumpleaños cayó en Martes** his birthday fell on a Tuesday
café *m* coffee; (*cafetería*) café. ● *a*. **color ~** coffee-coloured. **~ con leche** white coffee. **~ cortado** coffee with a little milk. **~ (solo)** black coffee
cafe|ína *f* caffeine. **~tal** *m* coffee plantation. **~tera** *f* coffee-pot. **~tería** *f* café. **~tero** *a* coffee
caíd|a *f* fall; (*disminución*) drop; (*pendiente*) slope. **~o** *a* fallen; (*abatido*) dejected. ● *m* fallen
caigo *vb véase* **caer**
caimán *m* cayman, alligator
caj|a *f* box; (*grande*) case; (*de caudales*) safe; (*donde se efectúan los pagos*) cash desk; (*en supermercado*) check-out. **~a de ahorros** savings bank. **~a de caudales**, **~a fuerte** safe. **~a postal de ahorros** post office savings bank. **~a registradora** till. **~ero** *m* cashier. **~etilla** *f* packet. **~ita** *f* small box. **~ón** *m* large box; (*de mueble*) drawer; (*puesto de mercado*) stall. **ser de ~ón** be a matter of course
cal *m* lime
cala *f* cove
calaba|cín *m* marrow; (*fig, idiota, fam*) idiot. **~za** *f* pumpkin; (*fig, idiota, fam*) idiot
calabozo *m* prison; (*celda*) cell
calado *a* soaked. ● *m* (*naut*) draught. **estar ~ hasta los huesos** be soaked to the skin
calamar *m* squid
calambre *m* cramp
calami|dad *f* calamity, disaster. **~toso** *a* calamitous, disastrous
calar *vt* soak; (*penetrar*) pierce; (*fig, penetrar*) see through; sample ‹*fruta*›. **~se** *vpr* get soaked; ‹*zapatos*› leak; (*auto*) stall
calavera *f* skull
calcar [7] *vt* trace; (*fig*) copy
calceta *f*. **hacer ~** knit
calcetín *m* sock
calcinar *vt* burn
calcio *m* calcium
calco *m* tracing. **~manía** *f* transfer. **papel** *m* **de ~** tracing-paper
calcula|dor *a* calculating. **~dora** *f* calculator. **~dora de bolsillo** pocket calculator. **~r** *vt* calculate; (*suponer*) reckon, think
cálculo *m* calculation; (*fig*) reckoning
caldea|miento *m* heating. **~r** *vt* heat, warm. **~rse** *vpr* get hot
calder|a *f* boiler; (*Arg, para café*) coffee-pot; (*Arg, para té*) teapot. **~eta** *f* small boiler
calderilla *f* small change, coppers
calder|o *m* small boiler. **~ón** *m* large boiler
caldo *m* stock; (*sopa*) soup, broth. **poner a ~ a uno** give s.o. a dressing-down
calefacción *f* heating. **~ central** central heating
caleidoscopio *m* kaleidoscope
calendario *m* calendar
caléndula *f* marigold
calenta|dor *m* heater. **~miento** *m* heating; (*en deportes*) warm-up. **~r** [1] *vt* heat, warm. **~rse** *vpr* get hot, warm up
calentur|a *f* fever, (high) temperature. **~iento** *a* feverish
calibr|ar *vt* calibrate; (*fig*) measure. **~e** *m* calibre; (*diámetro*) diameter; (*fig*) importance
calidad *f* quality; (*función*) capacity. **en ~ de** as
cálido *a* warm
calidoscopio *m* kaleidoscope
caliente *a* hot, warm; (*fig, enfadado*) angry
califica|ción *f* qualification; (*evaluación*) assessment; (*nota*) mark. **~r** [7] *vt* qualify; (*evaluar*) assess; mark ‹*examen etc*›. **~r de** describe as, label. **~tivo** *a* qualifying. ● *m* epithet

caliz|a *f* limestone. **~o** *a* lime
calm|a *f* calm. **¡~a!** calm down! **~ante** *a & m* sedative. **~ar** *vt* calm, soothe. ● *vi* ‹*viento*› abate. **~arse** *vpr* calm down; ‹*viento*› abate. **~oso** *a* calm; (*flemático, fam*) phlegmatic. **en ~a** calm. **perder la ~a** lose one's composure
calor *m* heat, warmth. **hace ~** it's hot. **tener ~** be hot
caloría *f* calorie
calorífero *m* heater
calumni|a *f* calumny; (*oral*) slander; (*escrita*) libel. **~ar** *vt* slander; (*por escrito*) libel. **~oso** *a* slanderous; ‹*cosa escrita*› libellous
caluros|amente *adv* warmly. **~o** *a* warm
calv|a *f* bald patch. **~ero** *m* clearing. **~icie** *f* baldness. **~o** *a* bald; ‹*terreno*› barren
calza *f* (*fam*) stocking; (*cuña*) wedge
calzada *f* road
calza|do *a* wearing shoes. ● *m* footwear, shoe. **~dor** *m* shoehorn. **~r** [10] *vt* put shoes on; (*llevar*) wear. ● *vi* wear shoes. ● *vpr* put on. **¿qué número calza Vd?** what size shoe do you take?
calz|ón *m* shorts; (*ropa interior*) knickers, pants. **~ones** *mpl* shorts. **~oncillos** *mpl* underpants
calla|do *a* quiet. **~r** *vt* silence; keep ‹*secreto*›; hush up ‹*asunto*›. ● *vi* be quiet, keep quiet, shut up (*fam*). **~rse** *vpr* be quiet, keep quiet, shut up (*fam*). **¡cállate!** be quiet! shut up! (*fam*)
calle *f* street, road; (*en deportes, en autopista*) lane. **~ de dirección única** one-way street. **~ mayor** high street, main street. **abrir ~** make way
callej|a *f* narrow street. **~ear** *vi* wander about the streets. **~ero** *a* street. ● *m* street plan. **~ón** *m* alley. **~uela** *f* back street, side street. **~ón sin salida** cul-de-sac
call|ista *m & f* chiropodist. **~o** *m* corn, callus. **~os** *mpl* tripe. **~oso** *a* hard, rough
cama *f* bed. **~ de matrimonio** double bed. **~ individual** single bed. **caer en la ~** fall ill. **guardar ~** be confined to bed
camada *f* litter; (*fig, de ladrones*) gang
camafeo *m* cameo
camaleón *m* chameleon
cámara *f* room; (*de reyes*) royal chamber; (*fotográfica*) camera; (*de armas, pol*) chamber. **~ fotográfica** camera. **a ~ lenta** in slow motion
camarada *f* colleague; (*amigo*) companion
camarer|a *f* chambermaid; (*de restaurante etc*) waitress; (*en casa*) maid. **~o** *m* waiter
camarín *m* dressing-room; (*naut*) cabin
camarón *m* shrimp
camarote *m* cabin
cambi|able *a* changeable; (*com etc*) exchangeable. **~ante** *a* variable. **~ar** *vt* change; (*trocar*) exchange. ● *vi* change. **~ar de idea** change one's mind. **~arse** *vpr* change. **~o** *m* change; (*com*) exchange rate; (*moneda menuda*) (small) change. **~sta** *m & f* money-changer. **en ~o** on the other hand
camelia *f* camellia
camello *m* camel
camilla *f* stretcher; (*sofá*) couch
camina|nte *m* traveller. **~r** *vt* cover. ● *vi* travel; (*andar*) walk; ‹*río, astros etc*› move. **~ta** *f* long walk
camino *m* road; (*sendero*) path, track; (*dirección, medio*) way. **~ de** towards, on the way to. **abrir ~** make way. **a medio ~, a la mitad del ~** half-way. **de ~** on the way. **ponerse en ~** set out
cami|ón *m* lorry; (*Mex, autobús*) bus. **~onero** *m* lorry-driver. **~oneta** *f* van
camis|a *f* shirt; (*de un fruto*) skin. **~a de dormir** nightdress. **~a de fuerza** strait-jacket. **~ería** *f* shirt shop. **~eta** *f* T-shirt; (*ropa interior*) vest. **~ón** *m* nightdress
camorra *f* (*fam*) row. **buscar ~** look for trouble, pick a quarrel
camote *m* (*LAm*) sweet potato
campamento *m* camp
campan|a *f* bell. **~ada** *f* stroke of a bell; (*de reloj*) striking. **~ario** *m* bell tower, belfry. **~eo** *m* peal of bells. **~illa** *f* bell. **~udo** *a* bell-shaped; ‹*estilo*› bombastic
campaña *f* countryside; (*mil, pol*) campaign. **de ~** (*mil*) field
campe|ón *a & m* champion. **~onato** *m* championship
campes|ino *a* country. ● *m* peasant. **~tre** *a* country
camping /'kampin/ *m* (*pl* **campings** /'kampin/) camping; (*lugar*) campsite. **hacer ~** go camping

campiña *f* countryside
campo *m* country; (*agricultura, fig*) field; (*de tenis*) court; (*de fútbol*) pitch; (*de golf*) course. **~santo** *m* cemetery
camufla|do *a* camouflaged. **~je** *m* camouflage. **~r** *vt* camouflage
cana *f* grey hair, white hair. **echar una ~ al aire** have a fling. **peinar ~s** be getting old
Canadá *m*. **el ~** Canada
canadiense *a & m* Canadian
canal *m* (*incl TV*) channel; (*artificial*) canal; (*del tejado*) gutter. **~ de la Mancha** English Channel. **~ de Panamá** Panama Canal. **~ón** *m* (*horizontal*) gutter; (*vertical*) drain-pipe
canalla *f* rabble. ● *m* (*fig, fam*) swine. **~da** *f* dirty trick
canapé *m* sofa, couch; (*culin*) canapé
Canarias *fpl*. **(las islas) ~** the Canary Islands, the Canaries
canario *a* of the Canary Islands. ● *m* native of the Canary Islands; (*pájaro*) canary
canast|a *f* (large) basket. **~illa** *f* small basket; (*para un bebé*) layette. **~illo** *m* small basket. **~o** *m* (large) basket
cancela *f* gate
cancela|ción *f* cancellation . **~r** *vt* cancel; write off ‹*deuda*›; (*fig*) forget
cáncer *m* cancer. **C~** Cancer
canciller *m* chancellor; (*LAm, ministro de asuntos exteriores*) Minister of Foreign Affairs
canci|ón *f* song. **~ón de cuna** lullaby. **~onero** *m* song-book. **¡siempre la misma ~ón!** always the same old story!
cancha *f* (*de fútbol*) pitch, ground; (*de tenis*) court
candado *m* padlock
candel|a *f* candle. **~ero** *m* candlestick. **~illa** *f* candle
candente *a* (*rojo*) red-hot; (*blanco*) white-hot; (*fig*) burning
candidato *m* candidate
candidez *f* innocence; (*ingenuidad*) naïvety
cándido *a* naïve
candil *m* oil-lamp; (*Mex, araña*) chandelier. **~ejas** *fpl* footlights
candinga *m* (*Mex*) devil
candor *m* innocence; (*ingenuidad*) naïvety. **~oso** *a* innocent; (*ingenuo*) naïve
canela *f* cinnamon. **ser ~** be beautiful
cangrejo *m* crab. **~ de río** crayfish
canguro *m* kangaroo; (*persona*) baby-sitter
can|íbal *a & m* cannibal. **~ibalismo** *m* cannibalism
canica *f* marble
canijo *m* weak
canino *a* canine. ● *m* canine (tooth)
canje *m* exchange. **~ar** *vt* exchange
cano *a* grey-haired
canoa *f* canoe; (*con motor*) motor boat
canon *m* canon
can|ónigo *m* canon. **~onizar** [10] *vt* canonize
canoso *a* grey-haired
cansa|do *a* tired. **~ncio** *m* tiredness. **~r** *vt* tire; (*aburrir*) bore. ● *vi* be tiring; (*aburrir*) get boring. **~rse** *vpr* get tired
cantábrico *a* Cantabrian. **el mar ~** the Bay of Biscay
canta|nte *a* singing. ● *m* singer; (*en óperas*) opera singer. **~or** *m* Flamenco singer. **~r** *vt/i* sing. ● *m* singing; (*canción*) song; (*poema*) poem. **~rlas claras** speak frankly
cántar|a *f* pitcher. **~o** *m* pitcher. **llover a ~os** pour down
cante *m* folk song. **~ flamenco, ~ jondo** Flamenco singing
cantera *f* quarry
cantidad *f* quantity; (*número*) number; (*de dinero*) sum. **una ~ de** lots of
cantilena *f*, **cantinela** *f* song
cantimplora *f* water-bottle
cantina *f* canteen; (*rail*) buffet
canto *m* singing; (*canción*) song; (*borde*) edge; (*de un cuchillo*) blunt edge; (*esquina*) corner; (*piedra*) pebble. **~ rodado** boulder. **de ~** on edge
cantonés *a* Cantonese
cantor *a* singing. ● *m* singer
canturre|ar *vt/i* hum. **~o** *m* humming
canuto *m* tube
caña *f* stalk, stem; (*planta*) reed; (*vaso*) glass; (*de la pierna*) shin. **~ de azúcar** sugar-cane. **~ de pescar** fishing-rod
cañada *f* ravine; (*camino*) track
cáñamo *m* hemp. **~ índio** cannabis
cañ|ería *f* pipe; (*tubería*) piping. **~o** *m* pipe, tube; (*de fuente*) jet. **~ón** *m* pipe, tube; (*de órgano*) pipe; (*de*

chimenea) flue; (*arma de fuego*) cannon; (*desfiladero*) canyon. **~onazo** *m* gunshot. **~onera** *f* gunboat
caoba *f* mahogany
ca|os *m* chaos. **~ótico** *a* chaotic
capa *f* cloak; (*de pintura*) coat; (*culin*) coating; (*geol*) stratum, layer
capacidad *f* capacity; (*fig*) ability
capacitar *vt* qualify, enable; (*instruir*) train
caparazón *m* shell
capataz *m* foreman
capaz *a* capable, able; (*espacioso*) roomy. **~ para** which holds, with a capacity of
capazo *m* large basket
capcioso *a* sly, insidious
capellán *m* chaplain
caperuza *f* hood; (*de pluma*) cap
capilla *f* chapel; (*mus*) choir
capita *f* small cloak, cape
capital *a* capital, very important. ● *m* (*dinero*) capital. ● *f* (*ciudad*) capital; (*LAm, letra*) capital (letter). **~ de provincia** county town
capitali|smo *m* capitalism. **~sta** *a* & *m* & *f* capitalist. **~zar** [10] *vt* capitalize
capit|án *m* captain. **~anear** *vt* lead, command; (*un equipo*) captain
capitel *m* (*arquit*) capital
capitulaci|ón *f* surrender; (*acuerdo*) agreement. **~ones** *fpl* marriage contract
capítulo *m* chapter. **~s matrimoniales** marriage contract
capó *m* bonnet (*Brit*), hood (*Amer*)
capón *m* (*pollo*) capon
caporal *m* chief, leader
capota *f* (*de mujer*) bonnet; (*auto*) folding top, sliding roof
capote *m* cape
Capricornio *m* Capricorn
capricho *m* whim. **~so** *a* capricious, whimsical. **a ~** capriciously
cápsula *f* capsule
captar *vt* harness ‹*agua*›; grasp ‹*sentido*›; hold ‹*atención*›; win ‹*confianza*›; (*radio*) pick up
captura *f* capture. **~r** *vt* capture
capucha *f* hood
capullo *m* bud; (*de insecto*) cocoon
caqui *m* khaki
cara *f* face; (*de una moneda*) obverse; (*de un objeto*) side; (*aspecto*) look, appearance; (*descaro*) cheek. **~ a** towards; (*frente a*) facing. **~ a ~** face to face. **~ o cruz** heads or tails. **dar la ~** face up to. **hacer ~ a** face. **no volver la ~ atrás** not look back. **tener ~ de** look, seem to be. **tener ~ para** have the face to. **tener mala ~** look ill. **volver la ~** look the other way
carabela *f* caravel, small light ship
carabina *f* rifle; (*fig, señora, fam*) chaperone
Caracas *m* Caracas
caracol *m* snail; (*de pelo*) curl. **¡~es!** Good Heavens! **escalera** *f* **de ~** spiral staircase
carácter *m* (*pl* **caracteres**) character. **con ~ de, por su ~ de** as
característic|a *f* characteristic; (*LAm, teléfonos*) dialling code. **~o** *a* characteristic, typical
caracteriza|do *a* characterized; (*prestigioso*) distinguished. **~r** [10] *vt* characterize
cara: ~ dura cheek, nerve. **~dura** *m* & *f* cheeky person, rotter (*fam*)
caramba *int* good heavens!, goodness me!
carámbano *m* icicle
caramelo *m* sweet (*Brit*), candy (*Amer*); (*azúcar fundido*) caramel
carancho *m* (*Arg*) vulture
carapacho *m* shell
caraqueño *a* from Caracas. ● *m* native of Caracas
carátula *f* mask; (*fig, teatro*) theatre; (*Mex, esfera del reloj*) face
caravana *f* caravan; (*fig, grupo*) group; (*auto*) long line, traffic jam
caray *int* (*fam*) good heavens!, goodness me!
carb|ón *m* coal; (*papel*) carbon (paper); (*para dibujar*) charcoal. **~oncillo** *m* charcoal. **~onero** *a* coal. ● *m* coal-merchant. **~onizar** [10] *vt* (*fig*) burn (to a cinder). **~ono** *m* carbon
carburador *m* carburettor
carcajada *f* burst of laughter. **reírse a ~s** roar with laughter. **soltar una ~** burst out laughing
cárcel *m* prison, jail; (*en carpintería*) clamp
carcel|ario *a* prison. **~ero** *a* prison. ● *m* prison officer
carcom|a *f* woodworm. **~er** *vt* eat away; (*fig*) undermine. **~erse** *vpr* be eaten away; (*fig*) waste away
cardenal *m* cardinal; (*contusión*) bruise
cárdeno *a* purple

cardiaco, **cardíaco** *a* cardiac, heart. ● *m* heart patient
cardinal *a* cardinal
cardiólogo *m* cardiologist, heart specialist
cardo *m* thistle
carear *vt* bring face to face ‹*personas*›; compare ‹*cosas*›
carecer [11] *vi*. ~ **de** lack. ~ **de sentido** not to make sense
caren|cia *f* lack. ~**te** *a* lacking
carero *a* expensive
carestía *f* (*precio elevado*) high price; (*escasez*) shortage
careta *f* mask
carey *m* tortoiseshell
carga *f* load; (*fig*) burden; (*acción*) loading; (*de barco*) cargo; (*obligación*) obligation. ~**do** *a* loaded; (*fig*) burdened; ‹*tiempo*› heavy; ‹*hilo*› live; ‹*pila*› charged. ~**mento** *m* load; (*acción*) loading; (*de un barco*) cargo. ~**nte** *a* demanding. ~**r** [12] *vt* load; (*fig*) burden; (*mil, elec*) charge; fill ‹*pluma etc*›; (*fig, molestar, fam*) annoy. ● *vi* load. ~**r con** pick up. ~**rse** *vpr* (*llenarse*) fill; ‹*cielo*› become overcast; (*enfadarse, fam*) get cross. **llevar la** ~ **de algo** be responsible for sth
cargo *m* load; (*fig*) burden; (*puesto*) post; (*acusación*) accusation, charge; (*responsabilidad*) charge. **a** ~ **de** in the charge of. **hacerse** ~ **de** take responsibility for. **tener a su** ~ be in charge of
carguero *m* (*Arg*) beast of burden; (*naut*) cargo ship
cari *m* (*LAm*) grey
cariacontecido *a* crestfallen
caria|do *a* decayed. ~**rse** *vpr* decay
caribe *a* Caribbean. **el mar** *m* **C**~ the Caribbean (Sea)
caricatura *f* caricature
caricia *f* caress
caridad *f* charity. **¡por** ~**!** for goodness sake!
caries *f invar* (dental) decay
carilampiño *a* clean-shaven
cariño *m* affection; (*caricia*) caress. ~ **mío** my darling. ~**samente** *adv* tenderly, lovingly; (*en carta*) with love from. ~**so** *a* affectionate. **con mucho** ~ (*en carta*) with love from. **tener** ~ **a** be fond of. **tomar** ~ **a** take a liking to. **un** ~ (*en carta*) with love from
carism|a *m* charisma. ~**ático** *a* charismatic
caritativo *a* charitable
cariz *m* look
carlinga *f* cockpit
carmesí *a* & *m* crimson
carmín *m* (*de labios*) lipstick; (*color*) red
carnal *a* carnal; ‹*pariente*› blood, full. **primo** ~ first cousin
carnaval *m* carnival. ~**esco** *a* carnival. **martes** *m* **de** ~ Shrove Tuesday
carne *f* (*incl de frutos*) flesh; (*para comer*) meat. ~ **de cerdo** pork. ~ **de cordero** lamb. ~ **de gallina** gooseflesh. ~ **picada** mince. ~ **de ternera** veal. ~ **de vaca** beef. **me pone la** ~ **de gallina** it gives me the creeps. **ser de** ~ **y hueso** be only human
carné *m* card; (*cuaderno*) notebook. ~ **de conducir** driving licence (*Brit*), driver's license (*Amer*). ~ **de identidad** identity card.
carnero *m* sheep; (*culin*) lamb
carnet /kar'ne/ *m* card; (*cuaderno*) notebook. ~ **de conducir** driving licence (*Brit*), driver's license (*Amer*). ~ **de identidad** identity card
carnicer|ía *f* butcher's (shop); (*fig*) massacre. ~**o** *a* carnivorous; (*fig, cruel*) cruel, savage. ● *m* butcher; (*animal*) carnivore
carnívoro *a* carnivorous. ● *m* carnivore
carnoso *a* fleshy
caro *a* dear. ● *adv* dear, dearly. **costar** ~ **a uno** cost s.o. dear
carpa *f* carp; (*tienda*) tent
carpeta *f* file, folder. ~**zo** *m*. **dar** ~**zo a** shelve, put on one side
carpinter|ía *f* carpentry. ~**o** *m* carpinter, joiner
carraspe|ar *vi* clear one's throat. ~**ra** *f*. **tener** ~**ra** have a frog in one's throat
carrera *f* run; (*prisa*) rush; (*concurso*) race; (*recorrido, estudios*) course; (*profesión*) profession, career
carreta *f* cart. ~**da** *f* cart-load
carrete *m* reel; (*película*) 35mm film
carretera *f* road. ~ **de circunvalación** bypass, ring road. ~ **nacional** A road (*Brit*), highway (*Amer*). ~ **secundaria** B road (*Brit*), secondary road (*Amer*)
carret|illa *f* trolley; (*de una rueda*) wheelbarrow; (*de bebé*) baby-walker. ~**ón** *m* small cart

carril *m* rut; (*rail*) rail; (*de autopista etc*) lane
carrillo *m* cheek; (*polea*) pulley
carrizo *m* reed
carro *m* cart; (*LAm, coche*) car. **~ de asalto, ~ de combate** tank
carrocería *f* (*auto*) bodywork; (*taller*) car repairer's
carroña *f* carrion
carroza *f* coach, carriage; (*en desfile de fiesta*) float
carruaje *m* carriage
carrusel *m* merry-go-round
carta *f* letter; (*documento*) document; (*lista de platos*) menu; (*lista de vinos*) list; (*geog*) map; (*naipe*) card. **~ blanca** free hand. **~ de crédito** credit card
cartearse *vpr* correspond
cartel *m* poster; (*de escuela etc*) wall-chart. **~era** *f* hoarding; (*en periódico*) entertainments. **~ito** *m* notice. **de ~** celebrated. **tener ~** be a hit, be successful
cartera *f* wallet; (*de colegial*) satchel; (*para documentos*) briefcase
cartería *f* sorting office
carterista *m & f* pickpocket
cartero *m* postman, mailman (*Amer*)
cartílago *m* cartilage
cartilla *f* first reading book. **~ de ahorros** savings book. **leerle la ~ a uno** tell s.o. off
cartón *m* cardboard
cartucho *m* cartridge
cartulina *f* thin cardboard
casa *f* house; (*hogar*) home; (*empresa*) firm; (*edificio*) building. **~ de correos** post office. **~ de huéspedes** boarding-house. **~ de socorro** first aid post. **amigo** *m* **de la ~** family friend. **ir a ~** go home. **salir de ~** go out
casad|a *f* married woman. **~o** *a* married. • *m* married man. **los recién ~os** the newly-weds
casamentero *m* matchmaker
casa|miento *m* marriage; (*ceremonia*) wedding. **~r** *vt* marry. • *vi* get married. **~rse** *vpr* get married
cascabel *m* small bell. **~eo** *m* jingling
cascada *f* waterfall
cascado *a* broken; ‹*voz*› harsh
cascanueces *m invar* nutcrackers
cascar [7] *vt* break; crack ‹*frutos secos*›; (*pegar*) beat. • *vi* (*fig, fam*) chatter, natter (*fam*). **~se** *vpr* crack
cáscara *f* (*de huevo, frutos secos*) shell; (*de naranja*) peel; (*de plátano*) skin
casco *m* helmet; (*de cerámica etc*) piece, fragment; (*cabeza*) head; (*de barco*) hull; (*envase*) empty bottle; (*de caballo*) hoof; (*de una ciudad*) part, area
cascote *m* rubble
caserío *m* country house; (*conjunto de casas*) hamlet
casero *a* home-made; (*doméstico*) domestic, household; (*amante del hogar*) home-loving; ‹*reunión*› family. • *m* owner; (*vigilante*) caretaker
caseta *f* small house, cottage. **~ de baño** bathing hut
caset(t)e *m & f* cassette
casi *adv* almost, nearly; (*en frases negativas*) hardly. **~ ~** very nearly. **~ nada** hardly any. **¡~ nada!** is that all! **~ nunca** hardly ever
casilla *f* small house; (*cabaña*) hut; (*de mercado*) stall; (*en ajedrez etc*) square; (*departamento de casillero*) pigeon-hole
casillero *m* pigeon-holes
casimir *m* cashmere
casino *m* casino; (*sociedad*) club
caso *m* case; (*atención*) notice. **~ perdido** hopeless case. **~ urgente** emergency. **darse el ~ (de) que** happen. **el ~ es que** the fact is that. **en ~ de** in the event of. **en cualquier ~** in any case, whatever happens. **en ese ~** in that case. **en todo ~** in any case. **en último ~** as a last resort. **hacer ~ de** take notice of. **poner por ~** suppose
caspa *f* dandruff
cáspita *int* good heavens!, goodness me!
casquivano *a* scatter-brained
cassette *m & f* cassette
casta *f* (*de animal*) breed; (*de persona*) descent
castaña *f* chestnut
castañet|a *f* click of the fingers. **~ear** *vi* ‹*dientes*› chatter
castaño *a* chestnut, brown. • *m* chestnut (tree)
castañuela *f* castanet
castellano *a* Castilian. • *m* (*persona*) Castilian; (*lengua*) Castilian, Spanish. **~parlante** *a* Castilian-speaking, Spanish-speaking. **¿habla Vd ~?** do you speak Spanish?

castidad *f* chastity
castig|ar [12] *vt* punish; (*en deportes*) penalize. **~o** *m* punishment; (*en deportes*) penalty
Castilla *f* Castille. **~ la Nueva** New Castille. **~ la Vieja** Old Castille
castillo *m* castle
cast|izo *a* true; ‹*lengua*› pure. **~o** *a* pure
castor *m* beaver
castra|ción *f* castration. **~r** *vt* castrate
castrense *m* military
casual *a* chance, accidental. **~idad** *f* chance, coincidence. **~mente** *adv* by chance. **dar la ~idad** happen. **de ~idad, por ~idad** by chance. **¡qué ~idad!** what a coincidence!
cataclismo *m* cataclysm
catador *m* taster; (*fig*) connoisseur
catalán *a* & *m* Catalan
catalejo *m* telescope
catalizador *m* catalyst
cat|alogar [12] *vt* catalogue; (*fig*) classify. **~álogo** *m* catalogue
Cataluña *f* Catalonia
catamarán *m* catamaran
cataplúm *int* crash! bang!
catapulta *f* catapult
catar *vt* taste, try
catarata *f* waterfall, falls; (*med*) cataract
catarro *m* cold
cat|ástrofe *m* catastrophe. **~astrófico** *a* catastrophic
catecismo *m* catechism
catedral *f* cathedral
catedrático *m* professor; (*de instituto*) teacher, head of department
categ|oría *f* category; (*clase*) class. **~órico** *a* categorical. **de ~oría** important. **de primera ~oría** first-class
catinga *f* (*LAm*) bad smell
catita *f* (*Arg*) parrot
catoche *m* (*Mex*) bad mood
cat|olicismo *m* catholicism. **~ólico** *a* (Roman) Catholic. ● *m* (Roman) Catholic
catorce *a* & *m* fourteen
cauce *m* river bed; (*fig, artificial*) channel
caución *f* caution; (*jurid*) guarantee
caucho *m* rubber
caudal *m* (*de río*) flow; (*riqueza*) wealth. **~oso** *a* ‹*río*› large
caudillo *m* leader, caudillo
causa *f* cause; (*motivo*) reason; (*jurid*) lawsuit. **~r** *vt* cause. **a ~ de, por ~ de** because of
cáustico *a* caustic
cautel|a *f* caution. **~arse** *vpr* guard against. **~osamente** *adv* warily, cautiously. **~oso** *a* cautious, wary
cauterizar [10] *vt* cauterize; (*fig*) apply drastic measures to
cautiv|ar *vt* capture; (*fig, fascinar*) captivate. **~erio** *m*, **~idad** *f* captivity. **~o** *a* & *m* captive
cauto *a* cautious
cavar *vt/i* dig
caverna *f* cave, cavern
caviar *m* caviare
cavidad *f* cavity
cavil|ar *vi* ponder, consider. **~oso** *a* worried
cayado *m* (*de pastor*) crook; (*de obispo*) crozier
caza *f* hunting; (*una expedición*) hunt; (*animales*) game. ● *m* fighter. **~dor** *m* hunter. **~dora** *f* jacket. **~ mayor** big game hunting. **~ menor** small game hunting. **~r** [10] *vt* hunt; (*fig*) track down; (*obtener*) catch, get. **andar a (la) ~ de** be in search of. **dar ~** chase, go after
cazo *m* saucepan; (*cucharón*) ladle. **~leta** *f* (small) saucepan
cazuela *f* casserole
cebada *f* barley
ceb|ar *vt* fatten (up); (*con trampa*) bait; prime ‹*arma de fuego*›. **~o** *m* bait; (*de arma de fuego*) charge
ceboll|a *f* onion. **~ana** *f* chive. **~eta** *f* spring onion. **~ino** *m* chive
cebra *f* zebra
cece|ar *vi* lisp. **~o** *m* lisp
cedazo *m* sieve
ceder *vt* give up. ● *vi* give in; (*disminuir*) ease off; (*fallar*) give way, collapse. **ceda el paso** give way
cedilla *f* cedilla
cedro *m* cedar
cédula *f* document; (*ficha*) index card
CE(E) *abrev* (*Comunidad* (*Económica*) *Europea*) E(E)C, European (Economic) Community
cefalea *f* severe headache
ceg|ador *a* blinding. **~ar** [1 & 12] *vt* blind; (*tapar*) block up. **~arse** *vpr* be blinded (**de** by). **~ato** *a* short-sighted. **~uera** *f* blindness
ceja *f* eyebrow
cejar *vi* move back; (*fig*) give way
celada *f* ambush; (*fig*) trap
cela|dor *m* (*de niños*) monitor; (*de cárcel*) prison warder; (*de museo etc*) attendant. **~r** *vt* watch

celda *f* cell
celebra|ción *f* celebration. **~r** *vt* celebrate; (*alabar*) praise. **~rse** *vpr* take place
célebre *a* famous; (*fig*, *gracioso*) funny
celebridad *f* fame; (*persona*) celebrity
celeridad *f* speed
celest|e *a* heavenly. **~ial** *a* heavenly. **azul ~e** sky-blue
celibato *m* celibacy
célibe *a* celibate
celo *m* zeal. **~s** *mpl* jealousy. **dar ~s** make jealous. **papel** *m* **~** adhesive tape, Sellotape (P). **tener ~s** be jealous
celofán *m* cellophane
celoso *a* enthusiastic; (*que tiene celos*) jealous
celta *a* Celtic. ● *m & f* Celt
céltico *a* Celtic
célula *f* cell
celular *a* cellular
celuloide *m* celluloid
celulosa *f* cellulose
cellisca *f* sleetstorm
cementerio *m* cemetery
cemento *m* cement; (*hormigón*) concrete; (*LAm*, *cola*) glue
cena *f* dinner; (*comida ligera*) supper. **~duría** *f* (*Mex*) restaurant
cenag|al *m* marsh, bog; (*fig*) tight spot. **~oso** *a* muddy
cenar *vt* have for dinner; (*en cena ligera*) have for supper. ● *vi* have dinner; (*tomar cena ligera*) have supper
cenicero *m* ashtray
cenit *m* zenith
ceniz|a *f* ash. **~o** *a* ashen. ● *m* jinx
censo *m* census. **~ electoral** electoral roll
censura *f* censure; (*de prensa etc*) censorship. **~r** *vt* censure; censor ‹*prensa etc*›
centavo *a & m* hundredth; (*moneda*) centavo
centell|a *f* flash; (*chispa*) spark. **~ar** *vi*, **~ear** *vi* sparkle. **~eo** *m* sparkle, sparkling
centena *f* hundred. **~r** *m* hundred. **a ~res** by the hundred
centenario *a* centenary; ‹*persona*› centenarian. ● *m* centenary; (*persona*) centenarian
centeno *m* rye
centésim|a *f* hundredth. **~o** *a* hundredth; (*moneda*) centésimo
cent|ígrado *a* centigrade, Celsius. **~igramo** *m* centigram. **~ilitro** *m* centilitre. **~ímetro** *m* centimetre
céntimo *a* hundredth. ● *m* cent
centinela *f* sentry
centolla *f*, **centollo** *m* spider crab
central *a* central. ● *f* head office. **~ de correos** general post office. **~ eléctrica** power station. **~ nuclear** nuclear power station. **~ telefónica** telephone exchange. **~ismo** *m* centralism. **~ita** *f* switchboard
centraliza|ción *f* centralization. **~r** [10] *vt* centralize
centrar *vt* centre
céntrico *a* central
centrífugo *a* centrifugal
centro *m* centre. **~ comercial** shopping centre
Centroamérica *f* Central America
centroamericano *a & m* Central American
centuplicar [7] *vt* increase a hundredfold
ceñi|do *a* tight. **~r** [5 & 22] *vt* surround, encircle; ‹*vestido*› be a tight fit. **~rse** *vpr* limit o.s. (**a** to)
ceñ|o *m* frown. **~udo** *a* frowning. **fruncir el ~o** frown
cepill|ar *vt* brush; (*en carpintería*) plane. **~o** *m* brush; (*en carpintería*) plane. **~o de dientes** toothbrush
cera *f* wax
cerámic|a *f* ceramics; (*materia*) pottery; (*objeto*) piece of pottery. **~o** *a* ceramic
cerca *f* fence. ● *adv* near, close. **~s** *mpl* foreground. **~ de** *prep* near; (*con números, con tiempo*) nearly. **de ~** from close up, closely
cercado *m* enclosure
cercan|ía *f* nearness, proximity. **~ías** *fpl* outskirts. **tren** *m* **de ~ías** local train. **~o** *a* near, close. **C~o Oriente** *m* Near East
cercar [7] *vt* fence in, enclose; ‹*gente*› surround, crowd round; (*asediar*) besiege
cerciorar *vt* convince. **~se** *vpr* make sure, find out
cerco *m* (*grupo*) circle; (*cercado*) enclosure; (*asedio*) siege
Cerdeña *f* Sardinia
cerdo *m* pig; (*carne*) pork
cereal *m* cereal
cerebr|al *a* cerebral. **~o** *m* brain; (*fig*, *inteligencia*) intelligence, brains

ceremoni|a *f* ceremony. **~al** *a* ceremonial. **~oso** *a* ceremonious, stiff
céreo *a* wax
cerez|a *f* cherry. **~o** cherry tree
cerill|a *f* match. **~o** *m* (*Mex*) match
cern|er [1] *vt* sieve. **~erse** *vpr* hover; (*fig*, *amenazar*) hang over. **~idor** *m* sieve
cero *m* nought, zero; (*fútbol*) nil (*Brit*), zero (*Amer*); (*tenis*) love; (*persona*) nonentity. **partir de ~** start from scratch
cerquillo *m* (*LAm*, *flequillo*) fringe
cerquita *adv* very near
cerra|do *a* shut, closed; (*espacio*) shut in, enclosed; ‹*cielo*› overcast; ‹*curva*› sharp. **~dura** *f* lock; (*acción de cerrar*) shutting, closing. **~jero** *m* locksmith. **~r** [1] *vt* shut, close; (*con llave*) lock; (*con cerrojo*) bolt; (*cercar*) enclose; turn off ‹*grifo*›; block up ‹*agujero etc*›. ● *vi* shut, close. **~rse** *vpr* shut, close; ‹*herida*› heal. **~r con llave** lock
cerro *m* hill. **irse por los ~s de Úbeda** ramble on
cerrojo *m* bolt. **echar el ~** bolt
certamen *m* competition, contest
certero *a* accurate
certeza *f*, **certidumbre** *f* certainty
certifica|do *a* ‹*carta etc*› registered. ● *m* certificate; (*carta*) registered letter. **~r** [7] *vt* certify; register ‹*carta etc*›
certitud *f* certainty
cervato *m* fawn
cerve|cería *f* beerhouse, bar; (*fábrica*) brewery. **~za** *f* beer. **~za de barril** draught beer. **~za de botella** bottled beer
cesa|ción *f* cessation, suspension. **~nte** *a* out of work. **~r** *vt* stop. ● *vi* stop, cease; (*dejar un empleo*) give up. **sin ~r** incessantly
cesáreo *a* Caesarian. **operación** *f* **cesárea** Caesarian section
cese *m* cessation; (*de un empleo*) dismissal
césped *m* grass, lawn
cest|a *f* basket. **~ada** *f* basketful. **~o** *m* basket. **~o de los papeles** wastepaper basket
cetro *m* sceptre; (*fig*) power
cianuro *m* cyanide
ciática *f* sciatica
cibernética *f* cibernetics
cicatriz *f* scar. **~ación** *f* healing. **~ar** [10] *vt*/*i* heal. **~arse** *vpr* heal
ciclamino *m* cyclamen
cíclico *a* cyclic(al)
ciclis|mo *m* cycling. **~ta** *m* & *f* cyclist
ciclo *m* cycle; (*LAm*, *curso*) course
ciclomotor *m* moped
ciclón *m* cyclone
ciclostilo *m* cyclostyle, duplicating machine
ciego *a* blind. ● *m* blind man, blind person. **a ciegas** in the dark
cielo *m* sky; (*relig*) heaven; (*persona*) darling. **¡~s!** good heavens!, goodness me!
ciempiés *m invar* centipede
cien *a* a hundred. **~ por ~** (*fam*) completely, one hundred per cent. **me pone a ~** it drives me mad
ciénaga *f* bog, swamp
ciencia *f* science; (*fig*) knowledge. **~s** *fpl* (*univ etc*) science. **~s empresariales** business studies. **saber a ~ cierta** know for a fact, know for certain
cieno *m* mud
científico *a* scientific. ● *m* scientist
ciento *a* & *m* (*delante de nombres, y numerales a los que multiplica* **cien**) a hundred, one hundred. **por ~** per cent
cierne *m* blossoming. **en ~** in blossom; (*fig*) in its infancy
cierre *m* fastener; (*acción de cerrar*) shutting, closing. **~ de cremallera** zip, zipper (*Amer*)
cierro *vb véase* **cerrar**
cierto *a* certain; (*verdad*) true. **estar en lo ~** be right. **lo ~ es que** the fact is that. **no es ~** that's not true. **¿no es ~?** right? **por ~** certainly, by the way. **si bien es ~ que** although
ciervo *m* deer
cifra *f* figure, number; (*cantidad*) sum. **~do** *a* coded. **~r** *vt* code; (*resumir*) summarize. **en ~** code, in code
cigala *f* (Norway) lobster
cigarra *f* cicada
cigarr|illo *m* cigarette. **~o** *m* (*cigarillo*) cigarette; (*puro*) cigar
cigüeña *f* stork
cil|índrico *a* cylindrical. **~indro** *m* cylinder; (*Mex*, *organillo*) barrel organ
cima *f* top; (*fig*) summit
címbalo *m* cymbal
cimbrear *vt* shake. **~se** *vpr* sway
cimentar [1] *vt* lay the foundations of; (*fig*, *reforzar*) strengthen
cimer|a *f* crest. **~o** *a* highest

cimiento *m* foundations; (*fig*) source. **desde los ~s** from the very beginning
cinc *m* zinc
cincel *m* chisel. **~ar** *vt* chisel
cinco *a & m* five
cincuent|a *a & m* fifty; (*quincuagésimo*) fiftieth. **~ón** *a* about fifty
cine *m* cinema. **~matografiar** [20] *vt* film
cinético *a* kinetic
cínico *a* cynical; (*desvergonzado*) shameless. ● *m* cynic
cinismo *m* cynicism; (*desvergüenza*) shamelessness
cinta *f* band; (*adorno de pelo etc*) ribbon; (*película*) film; (*magnética*) tape; (*de máquina de escribir etc*) ribbon. **~ aisladora, ~ aislante** insulating tape. **~ magnetofónica** magnetic tape. **~ métrica** tape measure
cintur|a *f* waist. **~ón** *m* belt. **~ón de seguridad** safety belt. **~ón salvavidas** lifebelt
ciprés *m* cypress (tree)
circo *m* circus
circuito *m* circuit; (*viaje*) tour. **~ cerrado** closed circuit. **corto ~** short circuit
circula|ción *f* circulation; (*vehículos*) traffic. **~r** *a* circular. ● *vt* circulate. ● *vi* circulate; ‹*líquidos*› flow; (*conducir*) drive; ‹*autobús etc*› run
círculo *m* circle. **~ vicioso** vicious circle. **en ~** in a circle
circunci|dar *vt* circumcise. **~sión** *f* circumcision
circunda|nte *a* surrounding. **~r** *vt* surround
circunferencia *f* circumference
circunflejo *m* circumflex
circunscri|bir [*pp* **circunscrito**] *vt* confine. **~pción** *f* (*distrito*) district. **~pción electoral** constituency
circunspecto *a* wary, circumspect
circunstan|cia *f* circumstance. **~te** *a* surrounding. ● *m* bystander. **los ~tes** those present
circunvalación *f*. **carretera** *f* **de ~** bypass, ring road
cirio *m* candle
ciruela *f* plum. **~ claudia** greengage. **~ damascena** damson
ciru|gía *f* surgery. **~jano** *m* surgeon
cisne *m* swan
cisterna *f* tank, cistern
cita *f* appointment; (*entre chico y chica*) date; (*referencia*) quotation. **~ción** *f* quotation; (*jurid*) summons. **~do** *a* aforementioned. **~r** *vt* make an appointment with; (*mencionar*) quote; (*jurid*) summons. **~rse** *vpr* arrange to meet
cítara *f* zither
ciudad *f* town; (*grande*) city. **~anía** *f* citizenship; (*habitantes*) citizens. **~ano** *a* civic ● *m* citizen, inhabitant; (*habitante de ciudad*) city dweller
cívico *a* civic
civil *a* civil. ● *m* civil guard. **~idad** *f* politeness
civiliza|ción *f* civilization. **~r** [10] *vt* civilize. **~rse** *vpr* become civilized
civismo *m* community spirit
cizaña *f* (*fig*) discord
clam|ar *vi* cry out, clamour. **~or** *m* cry; (*griterío*) noise, clamour; (*protesta*) outcry. **~oroso** *a* noisy
clandestin|idad *f* secrecy. **~o** *a* clandestine, secret
clara *f* (*de huevo*) egg white
claraboya *f* skylight
clarear *vi* dawn; (*aclarar*) brighten up. **~se** *vpr* be transparent
clarete *m* rosé
claridad *f* clarity; (*luz*) light
clarifica|ción *f* clarification. **~r** [7] *vt* clarify
clarín *m* bugle
clarinet|e *m* clarinet; (*músico*) clarinettist. **~ista** *m & f* clarinettist
clarividen|cia *f* clairvoyance; (*fig*) far-sightedness. **~te** *a* clairvoyant; (*fig*) far-sighted
claro *a* (*con mucha luz*) bright; (*transparente, evidente*) clear; ‹*colores*› light; ‹*líquido*› thin. ● *m* (*en bosque etc*) clearing; (*espacio*) gap. ● *adv* clearly. ● *int* of course! **~ de luna** moonlight. **¡~ que sí!** yes of course! **¡~ que no!** of course not!
clase *f* class; (*aula*) classroom. **~ media** middle class. **~ obrera** working class. **~ social** social class. **dar ~s** teach. **toda ~ de** all sorts of
clásico *a* classical; (*fig*) classic. ● *m* classic
clasifica|ción *f* classification; (*deportes*) league. **~r** [7] *vt* classify; (*seleccionar*) sort
claudia *f* greengage
claudicar [7] (*ceder*) give in; (*cojear*) limp

claustro *m* cloister; (*univ*) staff
claustrof|obia *f* claustrophobia. **~óbico** *a* claustrophobic
cláusula *f* clause
clausura *f* closure; (*ceremonia*) closing ceremony. **~r** *vt* close
clava|do *a* fixed; (*con clavo*) nailed. **~r** *vt* knock in ‹*clavo*›; (*introducir a mano*) stick; (*fijar*) fix; (*juntar*) nail together. **es ~do a su padre** he's the spitting image of his father
clave *f* key; (*mus*) clef; (*clavicémbalo*) harpsichord
clavel *m* carnation
clavicémbalo *m* harpsichord
clavícula *f* collar bone, clavicle
clavija *f* peg; (*elec*) plug
clavo *m* nail; (*culin*) clove
claxon *m* (*pl* **claxons** /ˈklakson/) horn
clemen|cia *f* clemency, mercy. **~te** *a* clement, merciful
clementina *f* tangerine
cleptómano *m* kleptomaniac
cler|ecía *f* priesthood. **~ical** *a* clerical
clérigo *m* priest
clero *m* clergy
cliché *m* cliché; (*foto*) negative
cliente *m* & *f* client, customer; (*de médico*) patient. **~la** *f* clientele, customers; (*de médico*) patients, practice
clim|a *m* climate. **~ático** *a* climatic. **~atizado** *a* air-conditioned. **~atológico** *a* climatological
clínic|a *f* clinic. **~o** *a* clinical. • *m* clinician
clip *m* (*pl* **clips**) clip
clo *m* cluck. **hacer ~ ~** cluck
cloaca *f* drain, sewer
cloque|ar *vi* cluck. **~o** *m* clucking
cloro *m* chlorine
club *m* (*pl* **clubs** *o* **clubes**) club
coacci|ón *f* coercion, compulsion. **~onar** *vt* coerce, compel
coagular *vt* coagulate; clot ‹*sangre*›; curdle ‹*leche*›. **~se** *vpr* coagulate; ‹*sangre*› clot; ‹*leche*› curdle
coalición *f* coalition
coartada *f* alibi
coartar *vt* hinder; restrict ‹*libertad etc*›
cobard|e *a* cowardly. • *m* coward. **~ía** *f* cowardice
cobaya *f*, **cobayo** *m* guinea pig
cobert|era *f* (*tapadera*) lid. **~izo** *m* lean-to, shelter. **~or** *m* bedspread; (*manta*) blanket. **~ura** *f* covering
cobij|a *f* (*LAm, ropa de cama*) bedclothes; (*Mex, manta*) blanket. **~ar** *vt* shelter. **~arse** *vpr* shelter, take shelter. **~o** *m* shelter
cobra *f* cobra
cobra|dor *m* conductor. **~dora** *f* conductress. **~r** *vt* collect; (*ganar*) earn; charge ‹*precio*›; cash ‹*cheque*›; (*recuperar*) recover. • *vi* be paid. **~rse** *vpr* recover
cobre *m* copper; (*mus*) brass (instruments)
cobro *m* collection; (*de cheque*) cashing; (*pago*) payment. **ponerse en ~** go into hiding. **presentar al ~** cash
cocada *f* (*LAm*) sweet coconut
cocaína *f* cocaine
cocción *f* cooking; (*tec*) baking, firing
cocear *vt/i* kick
coc|er [2 & 9] *vt/i* cook; (*hervir*) boil; (*en horno*) bake. **~ido** *a* cooked. • *m* stew
cociente *m* quotient. **~ intelectual** intelligence quotient, IQ
cocin|a *f* kitchen; (*arte de cocinar*) cookery, cuisine; (*aparato*) cooker. **~a de gas** gas cooker. **~a eléctrica** electric cooker. **~ar** *vt/i* cook. **~ero** *m* cook
coco *m* coconut; (*árbol*) coconut palm; (*cabeza*) head; (*duende*) bogeyman. **comerse el ~** think hard
cocodrilo *m* crocodile
cocotero *m* coconut palm
cóctel *m* (*pl* **cóctels** *o* **cócteles**) cocktail; (*reunión*) cocktail party
coche *m* car (*Brit*), motor car (*Brit*), automobile (*Amer*); (*de tren*) coach, carriage. **~-cama** sleeper. **~ fúnebre** hearse. **~ra** *f* garage; (*de autobuses*) depot. **~ restaurante** dining-car. **~s de choque** dodgems
cochin|ada *f* dirty thing. **~o** *a* dirty, filthy. • *m* pig
cod|azo *m* nudge (with one's elbow); (*Mex, aviso secreto*) tip-off. **~ear** *vt/i* elbow, nudge
codici|a *f* greed. **~ado** *a* coveted, sought after. **~ar** *vt* covet. **~oso** *a* greedy (**de** for)
código *m* code. **~ de la circulación** Highway Code
codo *m* elbow; (*dobladura*) bend. **hablar por los ~s** talk too much. **hasta los ~s** up to one's neck
codorniz *m* quail
coeducación *f* coeducation

coerción *f* coercion
coetáneo *a & m* contemporary
coexist|encia *f* coexistence. **~ir** *vi* coexist
cofradía *f* brotherhood
cofre *m* chest
coger [14] *vt* (*España*) take; catch ‹*tren, autobús, pelota, catarro*›; (*agarrar*) take hold of; (*del suelo*) pick up; pick ‹*frutos etc*›. ● *vi* (*caber*) fit. **~se** *vpr* trap, catch
cogollo *m* (*de lechuga etc*) heart; (*fig, lo mejor*) cream; (*fig, núcleo*) centre
cogote *m* back of the neck
cohech|ar *vt* bribe. **~o** *m* bribery
coherente *a* coherent
cohesión *f* cohesion
cohete *m* rocket; (*Mex, pistola*) pistol
cohibi|ción *f* inhibition. **~r** *vt* restrict; inhibit ‹*persona*›. **~rse** *vpr* feel inhibited; (*contenerse*) restrain o.s.
coincid|encia *f* coincidence. **~ente** *a* coincidental. **~ir** *vt* coincide. **dar la ~encia** happen
coje|ar *vt* limp; ‹*mueble*› wobble. **~ra** *f* lameness
coj|ín *m* cushion. **~inete** *m* small cushion. **~inete de bolas** ball bearing
cojo *a* lame; ‹*mueble*› wobbly. ● *m* lame person
col *f* cabbage. **~es de Bruselas** Brussel sprouts
cola *f* tail; (*fila*) queue; (*para pegar*) glue. **a la ~** at the end. **hacer ~** queue (up). **tener ~, traer ~** have serious consequences
colabora|ción *f* collaboration. **~dor** *m* collaborator. **~r** *vi* collaborate
colada *f* washing. **hacer la ~** do the washing
colador *m* strainer
colapso *m* collapse; (*fig*) stoppage
colar [2] *vt* strain ‹*líquidos*›; (*lavar*) wash; pass ‹*moneda falsa etc*›. ● *vi* ‹*líquido*› seep through; (*fig*) be believed, wash (*fam*). **~se** *vpr* slip; (*no hacer caso de la cola*) jump the queue; (*en fiesta*) gatecrash; (*meter la pata*) put one's foot in it
colch|a *f* bedspread. **~ón** *m* mattress. **~oneta** *f* mattress
colear *vi* wag its tail; ‹*asunto*› not be resolved. **vivito y coleando** alive and kicking
colecci|ón *f* collection; (*fig, gran número de*) a lot of. **~onar** *vt* collect. **~onista** *m & f* collector
colecta *f* collection
colectiv|idad *f* community. **~o** *a* collective. ● *m* (*Arg*) minibus
colector *m* (*en las alcantarillas*) main sewer
colega *m & f* colleague
colegi|al *m* schoolboy. **~ala** *f* schoolgirl. **~o** *m* private school; (*de ciertas profesiones*) college. **~o mayor** hall of residence
colegir [5 & 14] *vt* gather
cólera *f* cholera; (*ira*) anger, fury. **descargar su ~** vent one's anger. **montar en ~** fly into a rage
colérico *a* furious, irate
colesterol *m* cholesterol
coleta *f* pigtail
colga|nte *a* hanging. ● *m* pendant. **~r** [2 & 12] *vt* hang; hang out ‹*colada*›; hang up ‹*abrigo etc*›. ● *vi* hang; (*teléfono*) hang up, ring off. **~rse** *vpr* hang o.s. **dejar a uno ~do** let s.o. down
cólico *m* colic
coliflor *m* cauliflower
colilla *f* cigarette end
colina *f* hill
colinda|nte *a* adjacent. **~r** *vt* border (**con** on)
colisión *f* collision, crash; (*fig*) clash
colmar *vt* fill to overflowing; (*fig*) fulfill. **~ a uno de amabilidad** overwhelm s.o. with kindness
colmena *f* beehive, hive
colmillo *m* eye tooth, canine (tooth); (*de elefante*) tusk; (*de otros animales*) fang
colmo *m* height. **ser el ~** be the limit, be the last straw
coloca|ción *f* positioning; (*empleo*) job, position. **~r** [7] *vt* put, place; (*buscar empleo*) find work for. **~rse** *vpr* find a job
Colombia *f* Colombia
colombiano *a & m* Colombian
colon *m* colon
colón *m* (*unidad monetaria de Costa Rica y El Salvador*) colón
Colonia *f* Cologne
coloni|a *f* colony; (*agua de colonia*) eau-de-Cologne; (*LAm, barrio*) suburb. **~a de verano** holiday camp. **~al** *a* colonial. **~ales** *mpl* imported foodstuffs; (*comestibles en general*) groceries. **~alista** *m & f* colonialist. **~zación** *f* colonization. **~zar** [10] colonize
coloqui|al *a* colloquial. **~o** *m* conversation; (*congreso*) conference

color *m* colour. **~ado** *a* (*rojo*) red. **~ante** *m* colouring. **~ar** *vt* colour. **~ear** *vt/i* colour. **~ete** *m* rouge. **~ido** *m* colour. **de ~** colour. **en ~** (*fotos, película*) colour
colosal *a* colossal; (*fig, magnífico, fam*) terrific
columna *f* column; (*fig, apoyo*) support
columpi|ar *vt* swing. **~arse** *vpr* swing. **~o** *m* swing
collar *m* necklace; (*de perro etc*) collar
coma *f* comma. ● *m* (*med*) coma
comadre *f* midwife; (*madrina*) godmother; (*vecina*) neighbour. **~ar** *vi* gossip
comadreja *f* weasel
comadrona *f* midwife
comand|ancia *f* command. **~ante** *m* commander. **~o** *m* command; (*soldado*) commando
comarca *f* area, region
comba *f* bend; (*juguete*) skipping-rope. **~r** *vt* bend. **~rse** *vpr* bend. **saltar a la ~** skip
combat|e *m* fight; (*fig*) struggle. **~iente** *m* fighter. **~ir** *vt/i* fight
combina|ción *f* combination; (*bebida*) cocktail; (*arreglo*) plan, scheme; (*prenda*) slip. **~r** *vt* combine; (*arreglar*) arrange; (*armonizar*) match, go well with. **~rse** *vpr* combine; (*ponerse de acuerdo*) agree (**para** to)
combustible *m* fuel
comedia *f* comedy; (*cualquier obra de teatro*) play. **hacer la ~** pretend
comedi|do *a* reserved. **~rse** [5] *vpr* be restrained
comedor *m* dining-room; (*restaurante*) restaurant; (*persona*) glutton. **ser buen ~** have a good appetite
comensal *m* companion at table, fellow diner
comentar *vt* comment on; (*anotar*) annotate. **~io** *m* commentary; (*observación*) comment; (*fam*) gossip. **~ista** *m & f* commentator
comenzar [1 & 10] *vt/i* begin, start
comer *vt* eat; (*a mediodía*) have for lunch; (*corroer*) eat away; (*en ajedrez*) take. ● *vi* eat; (*a mediodía*) have lunch. **~se** *vpr* eat (up). **dar de ~ a** feed
comerci|al *a* commercial. **~ante** *m* trader; (*de tienda*) shopkeeper. **~ar** *vt* trade (**con, en** in); (*con otra persona*) do business. **~o** *m* commerce; (*actividad*) trade; (*tienda*) shop; (*negocio*) business
comestible *a* edible. **~s** *mpl* food. **tienda de ~s** grocer's (shop) (*Brit*), grocery (*Amer*)
cometa *m* comet. ● *f* kite
comet|er *vt* commit; make ‹*falta*›. **~ido** *m* task
comezón *m* itch
comicastro *m* poor actor, ham (*fam*)
comicios *mpl* elections
cómico *a* comic(al). ● *m* comic actor; (*cualquier actor*) actor
comida *f* food; (*a mediodía*) lunch. **hacer la ~** prepare the meals
comidilla *f* topic of conversation. **ser la ~ del pueblo** be the talk of the town
comienzo *m* beginning, start. **a ~s de** at the beginning of
comil|ón *a* greedy. **~ona** *f* feast
comillas *fpl* inverted commas
comino *m* cumin. **(no) me importa un ~** I couldn't care less
comisar|ía *f* police station. **~io** *m* commissioner; (*deportes*) steward. **~io de policía** police superintendent
comisión *f* assignment; (*comité*) commission, committee; (*com*) commission
comisura *f* corner. **~ de los labios** corner of the mouth
comité *m* committee
como *adv* like, as. ● *conj* as; (*en cuanto*) as soon as. **~ quieras** as you like. **~ sabes** as you know. **~ si** as if
cómo *a* how? **¿~?** I beg your pardon? **¿~ está Vd?** how are you? **¡~ no!** (*esp LAm*) of course! **¿~ son?** what are they like? **¿~ te llamas?** what's your name? **¡y ~!** and how!
cómoda *f* chest of drawers
comodidad *f* comfort. **a su ~** at your convenience
cómodo *a* comfortable; (*útil*) handy
comoquiera *conj*. **~ que** since. **~ que sea** however it may be
compacto *a* compact; (*denso*) dense; ‹*líneas etc*› close
compadecer [11] *vt* feel sorry for. **~se** *vpr*. **~se de** feel sorry for
compadre *m* godfather; (*amigo*) friend
compañ|ero *m* companion; (*de trabajo*) colleague; (*amigo*) friend. **~ía** *f* company. **en ~ía de** with

compara|ble *a* comparable. **~ción** *f* comparison. **~r** *vt* compare. **~tivo** *a & m* comparative. **en ~ción con** in comparison with, compared with
comparecer [11] *vi* appear
comparsa *f* group; (*en el teatro*) extra
compartimiento *m* compartment
compartir *vt* share
compás *m* (*instrumento*) (pair of) compasses; (*ritmo*) rhythm; (*división*) bar (*Brit*), measure (*Amer*); (*naut*) compass. **a ~** in time
compasi|ón *f* compassion, pity. **tener ~ón de** feel sorry for. **~vo** *a* compassionate
compatib|ilidad *f* compatibility. **~le** *a* compatible
compatriota *m & f* compatriot
compeler *vt* compel, force
compendi|ar *vt* summarize. **~o** *m* summary
compenetración *f* mutual understanding
compensa|ción *f* compensation. **~ción por despido** redundancy payment. **~r** *vt* compensate
competen|cia *f* competition; (*capacidad*) competence; (*terreno*) field, scope. **~te** *a* competent; (*apropiado*) appropriate, suitable
competi|ción *f* competition. **~dor** *m* competitor. **~r** [5] *vi* compete
compilar *vt* compile
compinche *m* accomplice; (*amigo, fam*) friend, mate (*fam*)
complac|encia *f* pleasure; (*indulgencia*) indulgence. **~er** [32] *vt* please; (*prestar servicio*) help. **~erse** *vpr* have pleasure, be pleased. **~iente** *a* helpful; ‹*marido*› complaisant
complej|idad *f* complexity. **~o** *a & m* complex
complement|ario *a* complementary. **~o** *m* complement; (*gram*) object, complement
complet|ar *vt* complete. **~o** *a* complete; (*lleno*) full; (*perfecto*) perfect
complexión *f* disposition; (*constitución*) constitution
complica|ción *f* complication. **~r** [7] *vt* complicate; involve ‹*persona*›. **~rse** *vpr* become complicated
cómplice *m* accomplice
complot *m* (*pl* **complots**) plot
compon|ente *a* component. • *m* component; (*culin*) ingredient; (*miembro*) member. **~er** [34] *vt* make up; (*mus, literatura etc*) write, compose; (*reparar*) mend; (*culin*) prepare; (*arreglar*) restore; settle ‹*estómago*›; reconcile ‹*diferencias*›. **~erse** *vpr* be made up; (*arreglarse*) get ready. **~érselas** manage
comporta|miento *m* behaviour. **~r** *vt* involve. **~rse** *vpr* behave. **~rse como es debido** behave properly. **~rse mal** misbehave
composi|ción *f* composition. **~tor** *m* composer
compostelano *a* from Santiago de Compostela. • *m* native of Santiago de Compostela
compostura *f* composition; (*arreglo*) repair; (*culin*) condiment; (*comedimiento*) composure
compota *f* stewed fruit
compra *f* purchase. **~ a plazos** hire purchase. **~dor** *m* buyer; (*en una tienda*) customer. **~r** *vt* buy. **~venta** *f* dealing. **hacer la ~, ir a la ~, ir de ~s** do the shopping, go shopping. **negocio** *m* **de ~venta** second-hand shop
compren|der *vt* understand; (*incluir*) include. **~sible** *a* understandable. **~sión** *f* understanding. **~sivo** *a* understanding; (*que incluye*) comprehensive
compresa *f* compress; (*de mujer*) sanitary towel
compr|esión *f* compression. **~imido** *a* compressed. • *m* pill, tablet. **~imir** *vt* compress; keep back ‹*lágrimas*›; (*fig*) restrain
comproba|nte *m* (*recibo*) receipt. **~r** *vt* check; (*confirmar*) confirm
compromet|er *vt* compromise; (*arriesgar*) endager. **~erse** *vpr* compromise o.s.; (*obligarse*) agree to. **~ido** *a* ‹*situación*› awkward, embarrassing
compromiso *m* obligation; (*apuro*) predicament; (*cita*) appointment; (*acuerdo*) agreement. **sin ~** without obligation
compuesto *a* compound; ‹*persona*› smart. • *m* compound
compungido *a* sad, sorry
computador *m*, **computadora** *f* computer
computar *vt* calculate
cómputo *m* calculation
comulgar [12] *vi* take Communion
común *a* common. • *m* community. **en ~** in common. **por lo ~** generally

comunal *a* municipal, communal

comunica|ción *f* communication. **~do** *m* communiqué. **~do a la prensa** press release. **~r** [7] *vt/i* communicate; pass on ‹*enfermedad, información*›. **~rse** *vpr* communicate; ‹*enfermedad*› spread. **~tivo** *a* communicative. **está ~ndo** (*al teléfono*) it's engaged, the line's engaged

comunidad *f* community. **~ de vecinos** residents' association. **C~ (Económica) Europea** European (Economic) Community. **en ~** together

comunión *f* communion; (*relig*) (Holy) Communion

comunis|mo *m* communism. **~ta** *a & m & f* communist

comúnmente *adv* generally, usually

con *prep* with; (*a pesar de*) in spite of; (+ *infinitivo*) by. **~ decir la verdad** by telling the truth. **~ que** so. **~ tal que** as long as

conato *m* attempt

concatenación *f* chain, linking

cóncavo *a* concave

concebir [5] *vt/i* conceive

conceder *vt* concede, grant; award ‹*premio*›; (*admitir*) admit

concej|al *m* councillor. **~o** *m* town council

concentra|ción *f* concentration. **~do** *m* concentrated. **~r** *vt* concentrate. **~rse** *vpr* concentrate

concep|ción *f* conception. **~to** *m* concept; (*opinión*) opinion. **bajo ningún ~to** in no way. **en mi ~to** in my view. **por ningún ~to** in no way

concerniente *a* concerning. **en lo ~ a** with regard to

concertar [1] *vt* (*mus*) harmonize; (*coordinar*) coordinate; (*poner de acuerdo*) agree. ● *vi* be in tune; (*fig*) agree. **~se** *vpr* agree

concertina *f* concertina

concesión *f* concession

conciencia *f* conscience; (*conocimiento*) consciousness. **~ción** *f* awareness. **~ limpia** clear conscience. **~ sucia** guilty conscience. **a ~ de que** fully aware that. **en ~** honestly. **tener ~ de** be aware of. **tomar ~ de** become aware of

concienzudo *a* conscientious

concierto *m* concert; (*acuerdo*) agreement; (*mus, composición*) concerto

concilia|ble *a* reconcilable. **~ción** *f* reconciliation. **~r** *vt* reconcile. **~r el sueño** get to sleep. **~rse** *vpr* gain

concilio *m* council

conciso *m* concise

conciudadano *m* fellow citizen

conclu|ir [17] *vt* finish; (*deducir*) conclude. ● *vi* finish, end. **~irse** *vpr* finish, end. **~sión** *f* conclusion. **~yente** *a* conclusive

concord|ancia *f* agreement. **~ar** [2] *vt* reconcile. ● *vi* agree. **~e** *a* in agreement. **~ia** *f* harmony

concret|amente *adv* specifically, to be exact. **~ar** *vt* make specific. **~arse** *vpr* become definite; (*limitarse*) confine o.s. **~o** *a* concrete; (*determinado*) specific, particular. ● *m* (*LAm, hormigón*) concrete. **en ~o** definite; (*concretamente*) to be exact; (*en resumen*) in short

concurr|encia *f* coincidence; (*reunión*) crowd, audience. **~ido** *a* crowded, busy. **~ir** *vi* meet; (*asistir*) attend; (*coincidir*) coincide; (*contribuir*) contribute; (*en concurso*) compete

concurs|ante *m & f* competitor, contestant. **~ar** *vi* compete, take part. **~o** *m* competition; (*concurrencia*) crowd; (*ayuda*) help

concha *f* shell; (*carey*) tortoiseshell

condado *m* county

conde *m* earl, count

condena *f* sentence. **~ción** *f* condemnation. **~do** *m* convict. **~r** *vt* condemn; (*jurid*) convict

condensa|ción *f* condensation. **~r** *vt* condense. **~rse** *vpr* condense

condesa *f* countess

condescende|ncia *f* condescension; (*tolerancia*) indulgence. **~r** [1] *vi* agree; (*dignarse*) condescend

condici|ón *f* condition; (*naturaleza*) nature. **~onado** *a*, **~onal** *a* conditional. **~onar** *vt* condition. **a ~ón de (que)** on the condition that

condiment|ar *vt* season. **~o** *m* condiment

condolencia *f* condolence

condominio *m* joint ownership

condón *m* condom

condonar *vt* (*perdonar*) reprieve; cancel ‹*deuda*›

conducir [47] *vt* drive ‹*vehículo*›; carry ‹*electricidad, gas, agua etc*›. ● *vi* drive; (*fig, llevar*) lead. **~se** *vpr* behave. **¿a qué conduce?** what's the point?

conducta *f* behaviour
conducto *m* pipe, tube; (*anat*) duct. **por ~ de** through
conductor *m* driver; (*jefe*) leader; (*elec*) conductor
conduzco *vb véase* **conducir**
conectar *vt/i* connect; (*enchufar*) plug in
conejo *m* rabbit
conexión *f* connection
confabularse *vpr* plot
confecci|ón *f* making; (*prenda*) ready-made garment. **~ones** *fpl* clothing, clothes. **~onado** *a* ready-made. **~onar** *vt* make
confederación *f* confederation
conferencia *f* conference; (*al teléfono*) long-distance call; (*univ etc*) lecture. **~ cumbre**, **~ en la cima**, **~ en la cumbre** summit conference. **~nte** *m & f* lecturer
conferir [4] *vt* confer; award ‹*premio*›
confes|ar [1] *vt/i* confess. **~arse** *vpr* confess. **~ión** *f* confession. **~ional** *a* confessional. **~ionario** *m* confessional. **~or** *m* confessor
confeti *m* confetti
confia|do *a* trusting; (*seguro de sí mismo*) confident. **~nza** *f* trust; (*en sí mismo*) confidence; (*intimidad*) familiarity. **~r** [20] *vt* entrust. • *vi* trust. **~rse** *vpr* put one's trust in
confiden|cia *f* confidence, secret. **~cial** *a* confidential. **~te** *m & f* close friend; (*de policía*) informer
configuración *f* configuration, shape
conf|ín *m* border. **~inar** *vt* confine; (*desterrar*) banish. • *vi* border (**con** on). **~ines** *mpl* outermost parts
confirma|ción *f* confirmation. **~r** *vt* confirm
confiscar [7] *vt* confiscate
confit|ería *f* sweet-shop (*Brit*), candy store (*Amer*). **~ura** *f* jam
conflagración *f* conflagration
conflicto *m* conflict
confluencia *f* confluence
conforma|ción *f* conformation, shape. **~r** *vt* (*acomodar*) adjust. • *vi* agree. **~rse** *vpr* conform
conform|e *a* in agreement; (*contento*) happy, satisfied; (*según*) according (**con** to). • *conj* as. • *int* OK! **~e a** in accordance with, according to. **~idad** *f* agreement; (*tolerancia*) resignation. **~ista** *m & f* conformist
conforta|ble *a* comfortable. **~nte** *a* comforting. **~r** *vt* comfort
confronta|ción *f* confrontation; (*comparación*) comparison. **~r** *vt* confront; (*comparar*) compare
confu|ndir *vt* blur; (*equivocar*) mistake, confuse; (*perder*) lose; (*mezclar*) mix up, confuse. **~ndirse** *vpr* become confused; (*equivocarse*) make a mistake. **~sión** *f* confusion; (*vergüenza*) embarrassment. **~so** *a* confused; (*avergonzado*) embarrassed
congela|do *a* frozen. **~dor** *m* freezer. **~r** *vt* freeze
congeniar *vi* get on
congesti|ón *f* congestion. **~onado** *a* congested. **~onar** *vt* congest. **~onarse** *vpr* become congested
congoja *f* distress
congraciar *vt* win over. **~se** *vpr* ingratiate o.s.
congratular *vt* congratulate
congrega|ción *f* gathering; (*relig*) congregation. **~rse** [12] *vpr* gather, assemble
congres|ista *m & f* delegate, member of a congress. **~o** *m* congress, conference. **C~o de los Diputados** House of Commons
cónico *a* conical
conífer|a *f* conifer. **~o** *a* coniferous
conjetura *f* conjecture, guess. **~r** *vt* conjecture, guess
conjuga|ción *f* conjugation. **~r** [12] *vt* conjugate
conjunción *f* conjunction
conjunto *a* joint. • *m* collection; (*mus*) band; (*ropa*) suit, outfit. **en ~** altogether
conjura *f*, **conjuración** *f* conspiracy
conjurar *vt* plot, conspire
conmemora|ción *f* commemoration. **~r** *vt* commemorate. **~tivo** *a* commemorative
conmigo *pron* with me
conminar *vt* threaten; (*avisar*) warn
conmiseración *f* commiseration
conmo|ción *f* shock; (*tumulto*) upheaval; (*terremoto*) earthquake. **~cionar** *vt* shock. **~ cerebral** concussion. **~ver** [2] *vt* shake; (*emocionar*) move
conmuta|dor *m* switch. **~r** *vt* exchange
connivencia *f* connivance
connota|ción *f* connotation. **~r** *vt* connote
cono *m* cone

conoc|edor *a & m* expert. **~er** [11] *vt* know; (*por primera vez*) meet; (*reconocer*) recognize, know. **~erse** *vpr* know o.s.; ‹*dos personas*› know each other; (*notarse*) be obvious. **dar a ~er** make known. **darse a ~er** make o.s. known. **~ido** *a* well-known. ● *m* acquaintance. **~imiento** *m* knowledge; (*sentido*) consciousness; (*conocido*) acquaintance. **perder el ~imiento** faint. **se ~e que** apparently. **tener ~imiento de** know about

conozco *vb véase* **conocer**

conque *conj* so

conquense *a* from Cuenca. ● *m* native of Cuenca

conquista *f* conquest. **~dor** *a* conquering. ● *m* conqueror; (*de América*) conquistador; (*fig*) lady-killer. **~r** *vt* conquer, win

consabido *a* well-known

consagra|ción *f* consecration. **~r** *vt* consecrate; (*fig*) devote. **~rse** *vpr* devote o.s.

consanguíneo *m* blood relation

consciente *a* conscious

consecución *f* acquisition; (*de un deseo*) realization

consecuen|cia *f* consequence; (*firmeza*) consistency. **~te** *a* consistent. **a ~cia de** as a result of. **en ~cia, por ~cia** consequently

consecutivo *a* consecutive

conseguir [5 & 13] *vt* get, obtain; (*lograr*) manage; achieve ‹*objetivo*›

conseja *f* story, fable

consej|ero *m* adviser; (*miembro de consejo*) member. **~o** *m* advice; (*pol*) council. **~o de ministros** cabinet

consenso *m* assent, consent

consenti|do *a* ‹*niño*› spoilt. **~miento** *m* consent. **~r** [4] *vt* allow. ● *vi* consent. **~rse** *vpr* break

conserje *m* porter, caretaker. **~ría** *f* porter's office

conserva *f* preserves; (*mermelada*) jam, preserve; (*en lata*) tinned food. **~ción** *f* conservation; (*de alimentos*) preservation; (*de edificio*) maintenance. **en ~** preserved

conservador *a & m* (*pol*) conservative

conservar *vt* keep; preserve ‹*alimentos*›. **~se** *vpr* keep; ‹*costumbre etc*› survive

conservatorio *m* conservatory

considera|ble *a* considerable. **~ción** *f* consideration; (*respeto*) respect. **~do** *a* considered; (*amable*) considerate; (*respetado*) respected. **~r** *vt* consider; (*respetar*) respect. **de ~ción** considerable. **de su ~ción** (*en cartas*) yours faithfully. **tomar en ~ción** take into consideration

consigna *f* order; (*rail*) left luggage office (*Brit*), baggage room (*Amer*); (*eslogan*) slogan

consigo *pron* (*él*) with him; (*ella*) with her; (*Ud, Uds*) with you; (*uno mismo*) with o.s.

consiguiente *a* consequent. **por ~** consequently

consist|encia *f* consistency. **~ente** *a* consisting (**en** of); (*firme*) solid. **~ir** *vi* consist (**en** of); (*deberse*) be due (**en** to)

consola|ción *f* consolation. **~r** [2] *vt* console, comfort

consolidar *vt* consolidate. **~se** *vpr* consolidate

consomé *m* clear soup, consommé

consonan|cia *f* consonance. **~te** *a* consonant. ● *f* consonant

consorcio *m* consortium

consorte *m & f* consort

conspicuo *a* eminent; (*visible*) visible

conspira|ción *f* conspiracy. **~dor** *m* conspirator. **~r** *vi* conspire

constan|cia *f* constancy. **~te** *a* constant

constar *vi* be clear; (*figurar*) appear, figure; (*componerse*) consist. **hacer ~** point out. **me consta que** I'm sure that. **que conste que** believe me

constatar *vt* check; (*confirmar*) confirm

constelación *f* constellation

consternación *f* consternation

constipa|do *m* cold. ● *a*. **estar ~do** have a cold. **~rse** *vpr* catch a cold

constitu|ción *f* constitution; (*establecimiento*) setting up. **~cional** *a* constitutional. **~ir** [17] *vt* constitute; (*formar*) form; (*crear*) set up, establish. **~irse** *vpr* set o.s. up (**en** as); (*presentarse*) appear. **~tivo** *a*, **~yente** *a* constituent

constreñir [5 & 22] *vt* force, oblige; (*restringir*) restrain

constricción *f* constriction

constru|cción *f* construction. **~ctor** *m* builder. **~ir** [17] *vt* construct; build ‹*edificio*›

consuelo *m* consolation, comfort
consuetudinario *a* customary
cónsul *m* consul
consula|do *m* consulate. **~r** *a* consular
consult|a *f* consultation. **~ar** *vt* consult. **~orio** *m* surgery. **~orio sentimental** problem page. **horas** *fpl* **de ~a** surgery hours. **obra** *f* **de ~a** reference book
consumar *vt* complete; commit ‹*crimen*›; consummate ‹*matrimonio*›
consum|ición *f* consumption; (*bebida*) drink; (*comida*) food. **~ido** *a* ‹*persona*› skinny, wasted; ‹*frutas*› shrivelled. **~idor** *m* consumer. **~ir** *vt* consume. **~irse** *vpr* ‹*persona*› waste away; ‹*cosa*› wear out; (*quedarse seco*) dry up. **~ismo** *m* consumerism. **~o** *m* consumption
contab|ilidad *f* book-keeping; (*profesión*) accountancy. **~le** *m* & *f* accountant
contacto *m* contact. **ponerse en ~ con** get in touch with
contado *a* counted. **~s** *apl* few. **~r** *m* meter; (*LAm*, *contable*) accountant. **al ~** cash
contagi|ar *vt* infect ‹*persona*›; pass on ‹*enfermedd*›; (*fig*) contaminate. **~o** *m* infection. **~oso** *a* infectious
contamina|ción *f* contamination, pollution. **~r** *vt* contaminate, pollute
contante *a*. **dinero** *m* **~** cash
contar [2] *vt* count; tell ‹*relato*›. ● *vi* count. **~ con** rely on, count on. **~se** *vpr* be included (**entre** among); (*decirse*) be said
contempla|ción *f* contemplation. **~r** *vt* look at; (*fig*) contemplate. **sin ~ciones** unceremoniously
contemporáneo *a* & *m* contemporary
contend|er [1] *vi* compete. **~iente** *m* & *f* competitor
conten|er [40] *vt* contain; (*restringir*) restrain. **~erse** *vpr* restrain o.s. **~ido** *a* contained. ● *m* contents
content|ar *vt* please. **~arse** *vpr*. **~arse de** be satisfied with, be pleased with. **~o** *a* (*alegre*) happy; (*satisfecho*) pleased
contesta|ción *f* answer. **~dor** *m*. **~ automático** answering machine. **~r** *vt/i* answer; (*replicar*) answer back
contexto *m* context
contienda *f* struggle
contigo *pron* with you
contiguo *a* adjacent
continen|cia *f* continence. **~tal** *a* continental. **~te** *m* continent
contingen|cia *f* contingency. **~te** *a* contingent. ● *m* contingent; (*cuota*) quota
continu|ación *f* continuation. **~ar** [21] *vt* continue, resume. ● *vi* continue. **~ará** (*en revista*, *TV etc*) to be continued. **~idad** *f* continuity. **~o** *a* continuous; (*muy frecuente*) continual. **a ~ación** immediately after. **corriente** *f* **~a** direct current
contorno *m* outline; (*geog*) contour. **~s** *mpl* surrounding area
contorsión *f* contortion
contra *adv* & *prep* against. ● *m* cons. **en ~** against
contraalmirante *m* rear-admiral
contraata|car [7] *vt/i* counter-attack. **~que** *m* counter-attack
contrabajo *m* double-bass; (*persona*) double-bass player
contrabalancear *vt* counterbalance
contraband|ista *m* & *f* smuggler. **~o** *m* contraband
contracción *f* contraction
contrachapado *m* plywood
contrad|ecir [46] *vt* contradict. **~icción** *f* contradiction. **~ictorio** *a* contradictory
contraer [41] *vt* contract. **~ matrimonio** marry. **~se** *vpr* contract; (*limitarse*) limit o.s.
contrafuerte *m* buttress
contragolpe *m* backlash
contrahecho *a* fake; ‹*moneda*› counterfeit; ‹*persona*› hunchbacked
contraindicación *f* contraindication
contralto *m* alto. ● *f* contralto
contramano. **a ~** in the wrong direction
contrapartida *f* compensation
contrapelo. **a ~** the wrong way
contrapes|ar *vt* counterbalance. **~o** *m* counterbalance
contraponer [34] oppose; (*comparar*) compare
contraproducente *a* counterproductive
contrari|ar [20] *vt* oppose; (*molestar*) annoy. **~edad** *f* obstacle; (*disgusto*) annoyance. **~o** *a* contrary; ‹*dirección*› opposite; ‹*persona*› opposed. **al ~o** on the contrary. **al**

~o de contrary to. **de lo ~o** otherwise. **en ~o** against. **llevar la ~a** contradict. **por el ~o** on the contrary

contrarrestar *vt* counteract

contrasentido *m* contradiction

contraseña *f* secret mark; (*palabra*) password

contrast|ar *vt* check, verify. ● *vi* contrast. **~e** *m* contrast; (*en oro, plata etc*) hallmark

contratar *vt* sign a contract for; engage ‹*empleados*›

contratiempo *m* setback; (*accidente*) accident

contrat|ista *m & f* contractor. **~o** *m* contract

contraven|ción *f* contravention. **~ir** [53] *vi*. **~ir a** contravene

contraventana *f* shutter

contribu|ción *f* contribution; (*tributo*) tax. **~ir** [17] *vt/i* contribute. **~yente** *m & f* contributor; (*que paga impuestos*) taxpayer

contrincante *m* rival, opponent

contrito *a* contrite

control *m* control; (*inspección*) check. **~ar** *vt* control; (*examinar*) check

controversia *f* controversy

contundente *a* ‹*arma*› blunt; ‹*argumento etc*› convincing

conturbar *vt* perturb

contusión *f* bruise

convalec|encia *f* convalescence. **~er** [11] *vi* convalesce. **~iente** *a & m & f* convalescent

convalidar *vt* confirm; recognize ‹*título*›

convenc|er [9] *vt* convince. **~imiento** *m* conviction

convenci|ón *f* convention. **~onal** *a* conventional

conveni|encia *f* convenience; (*aptitud*) suitability. **~encias (sociales)** conventions. **~ente** *a* suitable; (*aconsejable*) advisable; (*provechoso*) useful, advantageous. **~o** *m* agreement. **~r** [53] *vt* agree. ● *vi* agree; (*ser conveniente*) be convenient for, suit; (*ser aconsejable*) be advisable

convento *m* (*de monjes*) monastery; (*de monjas*) convent

convergente *a* converging

converger [14] *vi*, **convergir** [14] *vi* converge

conversa|ción *f* conversation. **~r** *vi* converse, talk

conver|sión *f* conversion. **~so** *a* converted. ● *m* convert. **~tible** *a* convertible. **~tir** [4] *vt* convert. **~tirse** *vpr* be converted

convexo *a* convex

convic|ción *f* conviction. **~to** *a* convicted

convida|do *m* guest. **~r** *vt* invite. **te convido a un helado** I'll treat you to an ice-cream

convincente *a* convincing

convite *m* invitation; (*banquete*) banquet

conviv|encia *f* coexistence. **~ir** *vi* live together

convocar [7] *vt* convene ‹*reunión*›; summon ‹*personas*›

convoy *m* convoy; (*rail*) train; (*vinagrera*) cruet

convulsión *f* convulsion; (*fig*) upheaval

conyugal *a* conjugal; (*vida*› married

cónyuge *m* spouse. **~s** *mpl* (married) couple

coñac *m* (*pl* **coñacs**) brandy

coopera|ción *f* co-operation. **~r** *vi* co-operate. **~tiva** *f* co-operative. **~tivo** *a* co-operative

coord|enada *f* coordinate. **~inación** *f* co-ordination. **~inar** *vt* co-ordinate

copa *f* glass; (*deportes, fig*) cup. **~s** *fpl* (*naipes*) hearts. **tomar una ~** have a drink

copia *f* copy. **~ en limpio** fair copy. **~r** *vt* copy. **sacar una ~** make a copy

copioso *a* copious; ‹*lluvia, nevada etc*› heavy

copla *f* verse; (*canción*) song

copo *m* flake. **~ de nieve** snowflake. **~s de maíz** cornflakes

coquet|a *f* flirt; (*mueble*) dressing-table. **~ear** *vi* flirt. **~eo** *m* flirtation. **~o** *a* flirtatious

coraje *m* courage; (*rabia*) anger. **dar ~** make mad, make furious

coral *a* choral. ● *m* (*materia, animal*) coral

Corán *m* Koran

coraza *f* (*naut*) armour-plating; (*de tortuga*) shell

coraz|ón *m* heart; (*persona*) darling. **~onada** *f* hunch; (*impulso*) impulse. **sin ~ón** heartless. **tener buen ~ón** be good-hearted

corbata *f* tie, necktie (*esp Amer*). **~ de lazo** bow tie

corcova *f* hump. **~do** *a* hunchbacked
corchea *f* quaver
corchete *m* fastener, hook and eye; (*gancho*) hook; (*paréntesis*) square bracket
corcho *m* cork
cordel *m* cord, thin rope
cordero *m* lamb
cordial *a* cordial, friendly. ● *m* tonic. **~idad** *f* cordiality, warmth
cordillera *f* mountain range
córdoba *m* (*unidad monetaria de Nicaragua*) córdoba
Córdoba *f* Cordova
cordón *m* string; (*de zapatos*) lace; (*cable*) flex; (*fig*) cordon. **~ umbilical** umbilical cord
corear *vt* chant
coreografía *f* choreography
corista *m* & *f* member of the chorus. ● *f* (*bailarina*) chorus girl
cornet|a *f* bugle. **~ín** *m* cornet
Cornualles *m* Cornwall
cornucopia *f* cornucopia
cornudo *a* horned. ● *m* cuckold
coro *m* chorus; (*relig*) choir
corona *f* crown; (*de flores*) wreath, garland. **~ción** *f* coronation. **~r** *vt* crown
coronel *m* colonel
coronilla *f* crown. **estar hasta la ~** be fed up
corporación *f* corporation
corporal *a* corporal
corpulento *a* stout
corpúsculo *m* corpuscle
corral *m* pen. **aves** *fpl* **de ~** poultry
correa *f* strap; (*de perro*) lead; (*cinturón*) belt
correc|ción *f* correction; (*represión*) rebuke; (*cortesía*) good manners. **~to** *a* correct; (*cortés*) polite
corre|dizo *a* running. **nudo ~dizo** slip knot. **puerta** *f* **~diza** sliding door. **~dor** *m* runner; (*pasillo*) corridor; (*agente*) agent, broker. **~dor automovilista** racing driver
corregir [5 & 14] *vt* correct; (*reprender*) rebuke
correlaci|ón *f* correlation. **~onar** *vt* correlate
correo *m* courier; (*correos*) post, mail; (*tren*) mail train. **~s** *mpl* post office. **echar al ~** post
correr *vt* run; (*viajar*) travel; draw ‹*cortinas*›. ● *vi* run; ‹*agua*, *electricidad etc*› flow; ‹*tiempo*› pass. **~se** *vpr* (*apartarse*) move along; (*pasarse*) go too far; ‹*colores*› run. **~se una juerga** have a ball
correspond|encia *f* correspondence. **~er** *vi* correspond; (*ser adecuado*) be fitting; (*contestar*) reply; (*pertenecer*) belong; (*incumbir*) fall to. **~erse** *vpr* (*amarse*) love one another. **~iente** *a* corresponding
corresponsal *m* correspondent
corrid|a *f* run. **~a de toros** bullfight. **~o** *a* ‹*peso*› good; (*continuo*) continuous; (*avergonzado*) embarrassed. **de ~a** from memory
corriente *a* ‹*agua*› running; ‹*monedas*, *publicación*, *cuenta*, *año etc*› current; (*ordinario*) ordinary. ● *f* current; (*de aire*) draught; (*fig*) tendency. ● *m* current month. **al ~** (*al día*) up-to-date; (*enterado*) aware
corr|illo *m* small group, circle. **~o** *m* circle
corroborar *vt* corroborate
corroer [24 & 37] *vt* corrode; (*geol*) erode; (*fig*) eat away. **~se** *vpr* corrode
corromper *vt* rot ‹*madera*›; turn bad ‹*alimentos*›; (*fig*) corrupt. ● *vi* (*fam*) stink. **~se** *vpr* ‹*madera*› rot; ‹*alimentos*› go bad; (*fig*) be corrupted
corrosi|ón *f* corrosion. **~vo** *a* corrosive
corrupción *f* (*de madera etc*) rot; (*soborno*) bribery; (*fig*) corruption
corsé *m* corset
cortacésped *m invar* lawn-mower
cortad|o *a* cut; ‹*leche*› sour; (*avergonzado*) embarrassed; (*confuso*) confused. ● *m* coffee with a little milk. **~ura** *f* cut
corta|nte *a* sharp; ‹*viento*› biting; ‹*frío*› bitter. **~r** *vt* cut; (*recortar*) cut out; (*aislar*, *detener*) cut off; (*interrumpir*) cut in. ● *vi* cut. **~rse** *vpr* cut o.s.; ‹*leche etc*› curdle; (*al teléfono*) be cut off; (*fig*) be embarrassed, become tongue-tied. **~rse el pelo** have one's hair cut. **~rse las uñas** cut one's nails
cortaúñas *m invar* nail-clippers
corte *m* cutting; (*de instrumento cortante*) cutting edge; (*de corriente*) cut; (*de prendas de vestir*) cut; (*de tela*) length. ● *f* court. **~ de luz** power cut. **~ y confección** dressmaking. **hacer la ~** court. **las C~s** the Spanish parliament

cortej|ar *vt* court. **~o** *m* (*de rey etc*) entourage. **~o fúnebre** cortège, funeral procession. **~o nupcial** wedding procession

cortés *a* polite

cortesan|a *f* courtesan. **~o** *m* courtier

cortesía *f* courtesy

corteza *f* bark; (*de naranja etc*) peel, rind; (*de pan*) crust

cortijo *m* farm; (*casa*) farmhouse

cortina *f* curtain

corto *a* short; (*escaso*) scanty; (*apocado*) shy. **~circuito** *m* short circuit. **~ de alcances** dim, thick. **~ de oído** hard of hearing. **~ de vista** short-sighted. **a la corta o a la larga** sooner or later. **quedarse ~** fall short; (*miscalcular*) under-estimate

Coruña *f*. **La ~** Corunna

corvo *a* bent

cosa *f* thing; (*asunto*) business; (*idea*) idea. **~ de** about. **como si tal ~** just like that; (*como si no hubiera pasado nada*) as if nothing had happened. **decirle a uno cuatro ~s** tell s.o. a thing or two. **lo que son las ~s** much to my surprise

cosaco *a & m* Cossack

cosech|a *f* harvest; (*de vino*) vintage. **~ar** *vt* harvest. **~ero** *m* harvester

coser *vt/i* sew. **~se** *vpr* stick to s.o. **eso es ~ y cantar** it's as easy as pie

cosmético *a & m* cosmetic

cósmico *a* cosmic

cosmonauta *m & f* cosmonaut

cosmopolita *a & m & f* cosmopolitan

cosmos *m* cosmos

cosquillas *fpl* ticklishness. **buscar a uno las ~** provoke s.o. **hacer ~** tickle. **tener ~** be ticklish

costa *f* coast. **a ~ de** at the expense of. **a toda ~** at any cost

costado *m* side

costal *m* sack

costar [2] *vt/i* cost. **~ caro** be expensive. **cueste lo que cueste** at any cost

Costa Rica *f* Costa Rica

costarricense *a & m*, **costarriqueño** *a & m* Costa Rican

coste *m* cost. **~ar** *vt* pay for; (*naut*) sail along the coast

costero *a* coastal

costilla *f* rib; (*chuleta*) chop

costo *m* cost. **~so** *a* expensive

costumbre *f* custom, habit. **de ~** *a* usual. ● *adv* usually

costur|a *f* sewing; (*línea*) seam; (*confección*) dressmaking. **~era** *f* dressmaker. **~ero** *m* sewing box

cotejar *vt* compare

cotidiano *a* daily

cotille|ar *vt* gossip. **~o** *m* gossip

cotiza|ción *f* quotation, price. **~r** [10] *vt* (*en la bolsa*) quote. ● *vi* pay one's subscription. **~rse** *vpr* fetch; (*en la bolsa*) stand at; (*fig*) be valued

coto *m* enclosure; (*de caza*) preserve. **~ de caza** game preserve

cotorr|a *f* parrot; (*urraca*) magpie; (*fig*) chatterbox. **~ear** *vi* chatter

coyuntura *f* joint; (*oportunidad*) opportunity; (*situación*) situation; (*circunstancia*) occasion, juncture

coz *f* kick

cráneo *m* skull

cráter *m* crater

crea|ción *f* creation. **~dor** *a* creative. ● *m* creator. **~r** *vt* create

crec|er [11] *vi* grow; (*aumentar*) increase. **~ida** *f* (*de río*) flood. **~ido** *a* ‹*persona*› grown-up; ‹*número*› large, considerable; ‹*plantas*› fully-grown. **~iente** *a* growing; ‹*luna*› crescent. **~imiento** *m* growth

credencial *a* credential. **~es** *fpl* credentials

credibilidad *f* credibility

crédito *m* credit. **digno de ~** reliable, trustworthy

credo *m* creed. **en un ~** in a flash

crédulo *a* credulous

cre|encia *f* belief. **~er** [18] believe; (*pensar*) think. **~o que no** I don't think so, I think not. **~o que sí** I think so. ● *vi* believe. **~erse** *vpr* consider o.s. **no me lo ~o** I don't believe it. **~íble** *a* credible. **¡ya lo ~o!** I should think so!

crema *f* cream; (*culin*) custard. **~ bronceadora** sun-tan cream

cremación *f* cremation; (*de basura*) incineration

cremallera *f* zip, zipper (*Amer*)

crematorio *m* crematorium; (*de basura*) incinerator

crepitar *vi* crackle

crepúsculo *m* twilight

crescendo *m* crescendo

cresp|o *a* frizzy. **~ón** *m* crêpe

cresta *f* crest; (*tupé*) toupee; (*geog*) ridge

Creta *f* Crete

cretino *m* cretin

creyente *m* believer

cría *f* breeding; (*animal*) baby animal
cria|da *f* maid, servant. **~dero** *m* nursery. **~do** *a* brought up. ● *m* servant. **~dor** *m* breeder. **~nza** *f* breeding. **~r** [20] *vt* suckle; grow ‹*plantas*›; breed ‹*animales*›; (*educar*) bring up. **~rse** *vpr* grow up
criatura *f* creature; (*niño*) baby
crim|en *m* crime. **~inal** *a* & *m* & *f* criminal
crin *m* mane; (*relleno*) horsehair
crinolina *f* crinoline
crío *m* child
criollo *a* & *m* Creole
cripta *f* crypt
crisantemo *m* chrysanthemum
crisis *f* crisis
crisol *m* melting-pot
crispar *vt* twitch; (*irritar*, *fam*) annoy. **~ los nervios a uno** get on s.o.'s nerves
cristal *m* crystal; (*vidrio*) glass; (*de una ventana*) pane of glass. **~ de aumento** magnifying glass. **~ino** *a* crystalline; (*fig*) crystal-clear. **~izar** [10] crystallize. **limpiar los ~es** clean the windows
cristian|amente *adv* in a Christian way. **~dad** *f* Christianity. **~ismo** *m* Christianity. **~o** *a* & *m* Christian
Cristo *m* Christ
cristo *m* crucifix
criterio *m* criterion; (*opinión*) opinion
cr|ítica *f* criticism; (*reseña*) review. **~iticar** [7] *vt* criticize. **~ítico** *a* critical. ● *m* critic
croar *vi* croak
crom|ado *a* chromium-plated. **~o** *m* chromium, chrome
cromosoma *m* chromosome
crónic|a *f* chronicle; (*de periódico*) news. **~o** *a* chronic
cronista *m* & *f* reporter
cronol|ogía *f* chronology. **~ógico** *a* chronological
cron|ometraje *m* timing. **~ometrar** *vt* time. **~ómetro** *m* chronometer; (*en deportes*) stop-watch
croquet /'kroket/ *m* croquet
croqueta *f* croquette
cruce *m* crossing; (*de calles, de carreteras*) crossroads; (*de peatones*) (pedestrian) crossing
crucial *a* cross-shaped; (*fig*) crucial
crucifi|car [7] *vt* crucify. **~jo** *m* crucifix. **~xión** *f* crucifiction
crucigrama *m* crossword (puzzle)
crudo *a* raw; (*fig*) crude. **petróleo** *m* **~** crude oil
cruel *a* cruel. **~dad** *f* cruelty
cruji|do *m* (*de seda, de hojas secas etc*) rustle; (*de muebles etc*) creak. **~r** *vi* ‹*seda, hojas secas etc*› rustle; ‹*muebles etc*› creak
cruz *f* cross; (*de moneda*) tails. **~ gamada** swastika. **la C~ Roja** the Red Cross
cruzada *f* crusade
cruzar [10] *vt* cross; (*poner de un lado a otro*) lay across. **~se** *vpr* cross; (*pasar en la calle*) pass
cuaderno *m* exercise book; (*para apuntes*) notebook
cuadra *f* (*caballeriza*) stable; (*LAm, manzana*) block
cuadrado *a* & *m* square
cuadragésimo *a* fortieth
cuadr|ar *vt* square. ● *vi* suit; (*estar de acuerdo*) agree. **~arse** *vpr* (*mil*) stand to attention; (*fig*) dig one's heels in. **~ilátero** *a* quadrilateral. ● *m* quadrilateral; (*boxeo*) ring
cuadrilla *f* group; (*pandilla*) gang
cuadro *m* square; (*pintura*) painting; (*de obra de teatro, escena*) scene; (*de jardín*) bed; (*de números*) table; (*de mando etc*) panel; (*conjunto del personal*) staff. **~ de distribución** switchboard. **a ~s, de ~s** check. **en ~** in a square. **¡qué ~!, ¡vaya un ~!** what a sight!
cuadrúpedo *m* quadruped
cuádruple *a* & *m* quadruple
cuajar *vt* thicken; clot ‹*sangre*›; curdle ‹*leche*›; (*llenar*) fill up. ● *vi* ‹*nieve*› settle; (*fig*, *fam*) work out. **cuajado de** full of. **~se** *vpr* coagulate; ‹*sangre*› clot; ‹*leche*› curdle. **~ón** *m* clot
cual *pron*. **el ~, la ~ etc** (*animales y cosas*) that, which; (*personas, sujeto*) who, that; (*personas, objeto*) whom. ● *adv* as, like. ● *a* such as. **~ si** as if. **~... tal** like... like. **cada ~** everyone. **por lo ~** because of which
cuál *pron* which
cualidad *f* quality; (*propiedad*) property
cualquiera *a* (*delante de nombres* **cualquier**, *pl* **cualesquiera**) any. ● *pron* (*pl* **cualesquiera**) anyone, anybody; (*cosas*) whatever, whichever. **un ~** a nobody
cuando *adv* when. ● *conj* when; (*aunque*) even if. **~ más** at the most.

~ menos at the least. **~ no** if not. **aun ~** even if. **de ~ en ~** from time to time

cuándo *adv & conj* when. **¿de ~ acá?**, **¿desde ~?** since when?

cuant|ía *f* quantity; (*extensión*) extent. **~ioso** *a* abundant

cuanto *a* as much... as, as many... as. ● *pron* as much as, as many as. ● *adv* as much as. **~ más, mejor** the more the merrier. **en ~** as soon as. **en ~ a** as for. **por ~** since. **unos ~s** a few, some

cuánto *a* (*interrogativo*) how much?; (*interrogativo en plural*) how many?; (*exclamativo*) what a lot of! ● *pron* how much?; (*en plural*) how many? ● *adv* how much. **¿~ tiempo?** how long? **¡~ tiempo sin verte!** it's been a long time! **¿a ~?** how much? **¿a ~s estamos?** what's the date today? **un Sr. no sé ~s** Mr So-and-So

cuáquero *m* Quaker

cuarent|a *a & m* forty; (*cuadragésimo*) fortieth. **~ena** *f* (about) forty; (*med*) quarantine. **~ón** *a* about forty

cuaresma *f* Lent

cuarta *f* (*palmo*) span

cuartear *vt* quarter, divide into four; (*zigzaguear*) zigzag. **~se** *vpr* crack

cuartel *m* (*mil*) barracks. **~ general** headquarters. **no dar ~** show no mercy

cuarteto *m* quartet

cuarto *a* fourth. ● *m* quarter; (*habitación*) room. **~ de baño** bathroom. **~ de estar** living room. **~ de hora** quarter of an hour. **estar sin un ~** be broke. **menos ~** (a) quarter to. **y ~** (a) quarter past

cuarzo *m* quartz

cuatro *a & m* four. **~cientos** *a & m* four hundred

Cuba *f* Cuba

cuba: ~libre *m* rum and Coke (P). **~no** *a & m* Cuban

cúbico *a* cubic

cubículo *m* cubicle

cubiert|a *f* cover, covering; (*de la cama*) bedspread; (*techo*) roof; (*neumático*) tyre; (*naut*) deck. **~o** *a* covered; ‹*cielo*› overcast. ● *m* place setting, cutlery; (*comida*) meal. **a ~o** under cover. **a ~o de** safe from

cubis|mo *m* cubism. **~ta** *a & m & f* cubist

cubil *m* den, lair. **~ete** *m* bowl; (*molde*) mould; (*para echar los dados*) cup

cubo *m* bucket; (*en geometría y matemáticas*) cube

cubrecama *m* bedspread

cubrir *vt* [*pp* **cubierto**] cover; ‹*sonido*› drown; fill ‹*vacante*›. **~se** *vpr* cover o.s.; (*ponerse el sombrero*) put on one's hat; ‹*el cielo*› cloud over, become overcast

cucaracha *f* cockcroach

cuclillas. en ~ *adv* squatting

cuclillo *m* cuckoo

cuco *a* shrewd; (*mono*) pretty, nice. ● *m* cuckoo; (*insecto*) grub

cucurucho *m* cornet

cuchar|a *f* spoon. **~ada** *f* spoonful. **~adita** *f* teaspoonful. **~illa** *f*, **~ita** *f* teaspoon. **~ón** *m* ladle

cuchiche|ar *vi* whisper. **~o** *m* whispering

cuchill|a *f* large knife; (*de carnicero*) cleaver; (*hoja de afeitar*) razor blade. **~ada** *f* slash; (*herida*) knife wound. **~o** *m* knife

cuchitril *m* pigsty; (*fig*) hovel

cuello *m* neck; (*de camisa*) collar. **cortar el ~ a uno** cut s.o.'s throat

cuenc|a *f* hollow; (*del ojo*) (eye) socket; (*geog*) basin. **~o** *m* hollow; (*vasija*) bowl

cuenta *f* count; (*acción de contar*) counting; (*factura*) bill; (*en banco, relato*) account; (*asunto*) affair; (*de collar etc*) bead. **~ corriente** current account, checking account (*Amer*). **ajustar las ~s** settle accounts. **caer en la ~ de que** realize that. **darse ~ de** realize. **en resumidas ~s** in short. **por mi ~** for myself. **tener en ~**, **tomar en ~** bear in mind

cuentakilómetros *m invar* milometer

cuent|ista *m & f* story-writer; (*de mentiras*) fibber. **~o** *m* story; (*mentira*) fib, tall story. ● *vb véase* **contar**

cuerda *f* rope; (*más fina*) string; (*mus*) string. **~ floja** tightrope. **dar ~ a** wind up ‹*un reloj*›

cuerdo *a* ‹*persona*› sane; ‹*acción*› sensible

cuern|a *f* horns. **~o** *m* horn

cuero *m* leather; (*piel*) skin; (*del grifo*) washer. **~ cabelludo** scalp. **en ~s (vivos)** stark naked

cuerpo *m* body

cuervo *m* crow

cuesta *f* slope, hill. **~ abajo** downhill. **~ arriba** uphill. **a ~s** on one's back
cuesti|ón *f* matter; (*altercado*) quarrel; (*dificultad*) trouble. **~onario** *m* questionnaire
cueva *f* cave; (*sótano*) cellar
cuida|do *m* care; (*preocupación*) worry; (*asunto*) affair. **¡~do!** (be) careful! **~doso** *a* careful. **~dosamente** *adv* carefully. **~r** *vt* look after. ● *vi*. **~r de** look after. **~rse** *vpr* look after o.s. **~rse de** be careful to. **tener ~do** be careful
culata *f* (*de arma de fuego*) butt; (*auto*) cylinder head. **~zo** *m* recoil
culebra *f* snake
culebrón *m* (*LAm*) soap opera
culinario *a* culinary
culmina|ción *f* culmination. **~r** *vi* culminate
culo *m* (*fam*) bottom. **ir de ~** go downhill
culpa *f* fault; (*jurid*) guilt. **~bilidad** *f* guilt. **~ble** *a* guilty. ● *m* culprit. **~r** *vt* blame (**de** for). **echar la ~** blame. **por ~ de** because of. **tener la ~ de** be to blame for
cultiv|ar *vt* farm; grow ‹*plantas*›; (*fig*) cultivate. **~o** *m* farming; (*de plantas*) growing
cult|o *a* ‹*tierra etc*› cultivated; ‹*persona*› educated. ● *m* cult; (*homenaje*) worship. **~ura** *f* culture. **~ural** *a* cultural
culturismo *m* body-building
cumbre *f* summit; (*fig*) height
cumpleaños *m invar* birthday
cumplido *a* perfect; (*grande*) large; (*cortés*) polite. ● *m* compliment. **~r** *a* reliable. **de ~** courtesy. **por ~** out of politeness
cumplim|entar *vt* carry out; (*saludar*) pay a courtesy call to; (*felicitar*) congratulate. **~iento** *m* carrying out, execution
cumplir *vt* carry out; observe ‹*ley*›; serve ‹*condena*›; reach ‹*años*›; keep ‹*promesa*›. ● *vi* do one's duty. **~se** *vpr* expire; (*realizarse*) be fulfilled. **hoy cumple 3 años** he's 3 (years old) today. **por ~** as a mere formality
cumulativo *a* cumulative
cúmulo *m* pile, heap
cuna *f* cradle; (*fig, nacimiento*) birthplace
cundir *vi* spread; (*rendir*) go a long way
cuneta *f* gutter
cuña *f* wedge
cuñad|a *f* sister-in-law. **~o** *m* brother-in-law
cuño *m* stamp. **de nuevo ~** new
cuota *f* quota; (*de sociedad etc*) subscription, fees
cupe *vb véase* **caber**
cupé *m* coupé
Cupido *m* Cupid
cupo *m* cuota
cupón *m* coupon
cúpula *f* dome
cura *f* cure; (*tratamiento*) treatment. ● *m* priest. **~ble** *a* curable. **~ción** *f* healing. **~ndero** *m* faith-healer. **~r** *vt* (*incl culin*) cure; dress ‹*herida*›; (*tratar*) treat; (*fig*) remedy; tan ‹*pieles*›. ● *vi* ‹*persona*› get better; ‹*herida*› heal; (*fig*) be cured. **~rse** *vpr* get better
curios|ear *vi* pry; (*mirar*) browse. **~idad** *f* curiosity; (*limpieza*) cleanliness. **~o** *a* curious; (*raro*) odd, unusual; (*limpio*) clean
curriculum vitae *m* curriculum vitae
cursar *vt* send; (*estudiar*) study
cursi *a* pretentious, showy. ● *m* affected person
cursillo *m* short course
cursiva *f* italics
curso *m* course; (*univ etc*) year. **en ~** under way; ‹*año etc*› current
curtir *vt* tan; (*fig*) harden. **~se** *vpr* become tanned; (*fig*) become hardened
curv|a *f* curve; (*de carretera*) bend. **~o** *a* curved
cúspide *f* peak
custodi|a *f* care, safe-keeping. **~ar** *vt* take care of. **~o** *a & m* guardian
cutáneo *a* skin. **enfermedad** *f* **cutánea** skin disease
cutícula *f* cuticle
cutis *m* skin, complexion
cuyo *pron* (*de persona*) whose, of whom; (*de cosa*) whose, of which. **en ~ caso** in which case

CH

chabacano *a* common; ‹*chiste etc*› vulgar. ● *m* (*Mex, albaricoque*) apricot
chabola *f* shack. **~s** *fpl* shanty town
chacal *m* jackal
chacota *f* fun. **echar a ~** make fun of

chacra *f* (*LAm*) farm
cháchara *f* chatter
chacharear *vt* (*Mex*) sell. ● *vi* chatter
chafar *vt* crush. **quedar chafado** be nonplussed
chal *m* shawl
chalado *a* (*fam*) crazy
chalé *m* house (with a garden), villa
chaleco *m* waistcoat, vest (*Amer*). ~ **salvavidas** life-jacket
chalequear *vt* (*Arg, Mex*) trick
chalet *m* (*pl* **chalets**) house (with a garden), villa
chalón *m* (*LAm*) shawl
chalote *m* shallot
chalupa *f* boat
chamac|a *f* (*esp Mex*) girl. **~o** *m* (*esp Mex*) boy
chamagoso *a* (*Mex*) filthy
chamarr|a *f* sheepskin jacket. **~o** *m* (*LAm*) coarse blanket
chamba *f* (*fam*) fluke; (*Mex, empleo*) job. **por ~** by fluke
champán *m*, **champaña** *m* champagne
champiñón *m* mushroom
champú *m* (*pl* **champúes** *o* **champús**) shampoo
chamuscar [7] *vt* scorch; (*Mex, vender*) sell cheaply
chance *m* (*esp LAm*) chance
chanclo *m* clog; (*de caucho*) rubber overshoe
chancho *m* (*LAm*) pig
chanchullo *m* swindle, fiddle (*fam*)
chandal *m* tracksuit
chanquete *m* whitebait
chantaj|e *m* blackmail. **~ista** *m & f* blackmailer
chanza *f* joke
chapa *f* plate, sheet; (*de madera*) plywood; (*de botella*) metal top. **~do** *a* plated. **~do a la antigua** old-fashioned. **~do de oro** gold-plated
chaparrón *m* downpour. **llover a chaparrones** pour (down), rain cats and dogs
chapotear *vi* splash
chapuce|ar *vt* botch; (*Mex, engañar*) deceive. **~ro** *a* ‹*persona*› careless; ‹*cosas*› shoddy. ● *m* careless worker
chapurrar *vt*, **chapurrear** *vt* speak badly, speak a little; mix ‹*licores*›
chapuza *f* botched job, mess; (*de poca importancia*) odd job
chaqueta *f* jacket. **cambiar la ~** change sides
chaquetero *m* turncoat
charada *f* charade
charc|a *f* pond, pool. **~o** *m* puddle, pool. **cruzar el ~o** cross the water; (*ir a América*) cross the Atlantic
charla *f* chat; (*conferencia*) talk. **~dor** *a* talkative. **~r** *vi* (*fam*) chat
charlatán *a* talkative. ● *m* chatterbox; (*curandero*) charlatan
charol *m* varnish; (*cuero*) patent leather
chárter *a* charter
chascar [7] *vt* crack ‹*látigo*›; click ‹*lengua*›; snap ‹*dedos*›. ● *vi* ‹*látigo*› crack; (*con la lengua*) click one's tongue; ‹*los dedos*› snap
chascarrillo *m* joke, funny story
chasco *m* disappointment; (*broma*) joke; (*engaño*) trick
chasis *m* (*auto*) chassis
chasqu|ear *vt* crack ‹*látigo*›; click ‹*lengua*›; snap ‹*dedos*›. ● *vi* ‹*látigo*› crack; (*con la lengua*) click one's tongue; ‹*los dedos*› snap. **~ido** *m* crack; (*de la lengua*) click; (*de los dedos*) snap
chatarra *f* scrap iron; (*fig*) scrap
chato *a* ‹*nariz*› snub; ‹*persona*› snub-nosed; ‹*objetos*› flat. ● *m* wine glass; (*niño, mujer, fam*) dear, darling; (*hombre, fam*) mate (*fam*)
chaval *m* (*fam*) boy, lad. **~a** *f* girl, lass
che *int* (*Arg*) listen!, hey!
checo *a & m* Czech. **la república** *f* **Checa** the Czech Republic
checoslovaco *a & m* (*history*) Czechoslovak
Checoslovaquia *f* (*history*) Czechoslovakia
chelín *m* shilling
chelo *a* (*Mex, rubio*) fair
cheque *m* cheque. **~ de viaje** traveller's cheque. **~ra** *f* cheque-book
chica *f* girl; (*criada*) maid, servant
chicano *a & m* Chicano, Mexican-American
chicle *m* chewing-gum
chico *a* (*fam*) small. ● *m* boy. **~s** *mpl* children
chicoleo *m* compliment
chicoria *f* chicory
chicharra *f* cicada; (*fig*) chatterbox
chicharrón *m* (*de cerdo*) crackling; (*fig*) sunburnt person
chichón *m* bump, lump
chifla|do *a* (*fam*) crazy, daft. **~r** *vt* (*fam*) drive crazy. **~rse** *vpr* be mad (**por** about). **le chifla el chocolate**

he's mad about chocolate. **le tiene chiflado esa chica** he's crazy about that girl
Chile *m* Chile
chile *m* chilli
chileno *a & m* Chilean
chill|ar *vi* scream, shriek; ‹*gato*› howl; ‹*ratón*› squeak; ‹*cerdo*› squeal. **~ido** *m* scream, screech; (*de gato etc*) howl. **~ón** *a* noisy; ‹*colores*› loud; ‹*sonido*› shrill
chimenea *f* chimney; (*hogar*) fireplace
chimpancé *m* chimpanzee
China *f* China
chinch|ar *vt* (*fam*) annoy, pester. **~e** *m* drawing-pin (*Brit*), thumbtack (*Amer*); (*insecto*) bedbug; (*fig*) nuisance. **~eta** *f* drawing-pin (*Brit*), thumbtack (*Amer*)
chinela *f* slipper
chino *a & m* Chinese
Chipre *m* Cyprus
chipriota *a & m & f* Cypriot
chiquillo *a* childish. ● *m* child, kid (*fam*)
chiquito *a* small, tiny. ● *m* child, kid (*fam*)
chiribita *f* spark. **estar que echa ~s** be furious
chirimoya *f* custard apple
chiripa *f* fluke. **por ~** by fluke
chirivía *f* parsnip
chirri|ar *vi* creak; ‹*pájaro*› chirp. **~do** *m* creaking; (*al freir*) sizzling; (*de pájaros*) chirping
chis *int* sh!, hush!; (*para llamar a uno, fam*) hey!, psst!
chism|e *m* gadget, thingumajig (*fam*); (*chismorreo*) piece of gossip. **~es** *mpl* things, bits and pieces. **~orreo** *m* gossip. **~oso** *a* gossipy. ● *m* gossip
chispa *f* spark; (*gota*) drop; (*gracia*) wit; (*fig*) sparkle. **estar que echa ~(s)** be furious
chispea|nte *a* sparkling. **~r** *vi* spark; (*lloviznar*) drizzle; (*fig*) sparkle
chisporrotear *vt* throw out sparks; ‹*fuego*› crackle; ‹*aceite*› sizzle
chistar *vi* speak. **sin ~** without saying a word
chiste *m* joke, funny story. **hacer ~ de** make fun of. **tener ~** be funny
chistera *f* (*fam*) top hat, topper (*fam*)
chistoso *a* funny
chiva|r *vi* inform ‹*policía*›; ‹*niño*› tell. **~tazo** *m* tip-off. **~to** *m* informer; (*niño*) telltale
chivo *m* kid, young goat
choca|nte *a* surprising; ‹*persona*› odd. **~r** [7] *vt* clink ‹*vasos*›; shake ‹*la mano*›. ● *vi* collide, hit. **~r con, ~r contra** crash into. **lo ~nte es que** the surprising thing is that
chocolate *m* chocolate. **tableta** *f* **de ~** bar of chocolate
choch|ear *vi* be senile. **~o** *a* senile; (*fig*) soft
chófer *m* chauffeur; (*conductor*) driver
cholo *a & m* (*LAm*) half-breed
chopo *m* poplar
choque *m* collision; (*fig*) clash; (*eléctrico*) shock; (*auto, rail etc*) crash, accident; (*sacudida*) jolt
chorizo *m* salami
chorr|ear *vi* gush forth; (*fig*) be dripping. **~o** *m* jet, stream; (*caudal pequeño*) trickle; (*fig*) stream. **a ~os** (*fig*) in abundance. **hablar a ~os** jabber
chovinis|mo *m* chauvinism. **~ta** *a* chauvinistic. ● *m & f* chauvinist
choza *f* hut
chubas|co *m* squall, heavy shower; (*fig*) bad patch. **~quero** *m* raincoat, anorak
chuchería *f* trinket; (*culin*) sweet
chufa *f* tiger nut
chuleta *f* chop
chulo *a* insolent; (*vistoso*) showy. ● *m* ruffian; (*rufián*) pimp
chumbo *m* prickly pear; (*fam*) bump. **higo** *m* **~** prickly pear
chup|ada *f* suck; (*al cigarro etc*) puff. **~ado** *a* skinny; (*fácil, fam*) very easy. **~ar** *vt* suck, lick; puff at ‹*cigarro etc*›; (*absorber*) absorb. **~arse** *vpr* lose weight. **~ete** *m* dummy (*Brit*), pacifier (*Amer*)
churro *m* fritter; (*fam*) mess. **me salió un ~** I made a mess of it
chusco *a* funny
chusma *f* riff-raff
chutar *vi* shoot. **¡va que chuta!** it's going well!

D

dactilógrafo *m* typist
dado *m* dice. ● *a* given; ‹*hora*› gone. **~ que** since, given that
dalia *f* dahlia
daltoniano *a* colour-blind

dama *f* lady; (*en la corte*) lady-in-waiting. **~s** *fpl* draughts (*Brit*), checkers (*Amer*)
damasco *m* damask
danés *a* Danish. ● *m* Dane; (*idioma*) Danish
danza *f* dance; (*acción*) dancing; (*enredo*) affair. **~r** [10] *vt/i* dance
dañ|ado *a* damaged. **~ar** *vt* damage; harm ‹*persona*›. **~ino** *a* harmful. **~o** *m* damage; (*a una persona*) harm. **~oso** *a* harmful. **~os y perjuicios** damages. **hacer ~o a** harm; hurt ‹*persona*›. **hacerse ~o** hurt o.s.
dar [26] *vt* give; (*producir*) yield; strike ‹*la hora*›. ● *vi* give. **da igual** it doesn't matter. **¡dale!** go on! **da lo mismo** it doesn't matter. **~ a** ‹*ventana*› look on to; ‹*edificio*› face. **~ a luz** give birth. **~ con** meet ‹*persona*›; find ‹*cosa*›; **~ de cabeza** fall flat on one's face. **~ por** assume; (+ *infinitivo*) decide. **~se** *vpr* give o.s. up; (*suceder*) happen. **dárselas de** make o.s. out to be. **~se por** consider o.s. **¿qué más da?** it doesn't matter!
dardo *m* dart
dársena *f* dock
datar *vt* date. ● *vi*. **~ de** date from
dátil *m* date
dato *m* fact. **~s** *mpl* data, information
de *prep* of; (*procedencia*) from; (*suposición*) if. **~ día** by day. **~ dos en dos** two by two. **~ haberlo sabido** if I (you, he etc) had known. **~ niño** as a child. **el libro ~ mi amigo** my friend's book. **las 2 ~ la madrugada** 2 (o'clock) in the morning. **un puente ~ hierro** an iron bridge. **soy ~ Loughborough** I'm from Loughborough
deambular *vi* stroll
debajo *adv* underneath. **~ de** underneath, under. **el de ~** the one underneath. **por ~** underneath. **por ~ de** below
debat|e *m* debate. **~ir** *vt* debate
deber *vt* owe. ● *vi* have to, must. ● *m* duty. **~es** *mpl* homework. **~se** *vpr*. **~se a** be due to. **debo marcharme** I must go, I have to go
debido *a* due; (*correcto*) proper. **~ a** due to. **como es ~** as is proper. **con el respeto ~** with due respect
débil *a* weak; ‹*ruido*› faint; ‹*luz*› dim
debili|dad *f* weakness. **~tar** *vt* weaken. **~tarse** *vpr* weaken, get weak
débito *m* debit; (*deuda*) debt
debutar *vi* make one's debut
década *f* decade
deca|dencia *f* decline. **~dente** *a* decadent. **~er** [29] *vi* decline; (*debilitarse*) weaken. **~ído** *a* depressed. **~imiento** *m* decline, weakening
decano *m* dean; (*miembro más antiguo*) senior member
decantar *vt* decant ‹*vino etc*›
decapitar *vt* behead
decena *f* ten; (*aproximadamente*) about ten
decencia *f* decency, honesty
decenio *m* decade
decente *a* ‹*persona*› respectable, honest; ‹*cosas*› modest; (*limpio*) clean, tidy
decepci|ón *f* disappointment. **~onar** *vt* disappoint
decibelio *m* decibel
decidi|do *a* decided; ‹*persona*› determined, resolute. **~r** *vt* decide; settle ‹*cuestión etc*›. ● *vi* decide. **~rse** *vpr* make up one's mind
decimal *a & m* decimal
décimo *a & m* tenth. ● *m* (*de lotería*) tenth part of a lottery ticket
decimo: ~ctavo *a & m* eighteenth. **~cuarto** *a & m* fourteenth. **~nono** *a & m*, **~noveno** *a & m* nineteenth. **~quinto** *a & m* fifteenth. **~séptimo** *a & m* seventeenth. **~sexto** *a & m* sixteenth. **~tercero** *a & m*, **~tercio** *a & m* thirteenth
decir [46] *vt* say; (*contar*) tell. ● *m* saying. **~se** *vpr* be said. **~ que no** say no. **~ que sí** say yes. **dicho de otro modo** in other words. **dicho y hecho** no sooner said than done. **¿dígame?** can I help you? **¡dígame!** (*al teléfono*) hello! **digamos** let's say. **es ~** that is to say. **mejor dicho** rather. **¡no me digas!** you don't say!, really! **por así ~, por ~lo así** so to speak, as it were. **querer ~** mean. **se dice que** it is said that, they say that
decisi|ón *f* decision. **~vo** *a* decisive
declamar *vt* declaim
declara|ción *f* statement. **~ción de renta** income tax return. **~r** *vt/i* declare. **~rse** *vpr* declare o.s.; (*epidemia etc*) break out
declina|ción *f* (*gram*) declension. **~r** *vt/i* decline; ‹*salud*› deteriorate

declive *m* slope; (*fig*) decline. **en ~** sloping
decolorar *vt* discolour, fade. **~se** *vpr* become discoloured, fade
decora|ción *f* decoration. **~do** *m* (*en el teatro*) set. **~dor** *m* decorator. **~r** *vt* decorate. **~tivo** *a* decorative
decoro *m* decorum; (*respeto*) respect. **~so** *a* proper; (*modesto*) modest; ‹*profesión*› honourable
decrecer [11] *vi* decrease, diminish; ‹*aguas*› subside
decrépito *a* decrepit
decret|ar *vt* decree. **~o** *m* decree
dedal *m* thimble
dedica|ción *f* dedication. **~r** [7] *vt* dedicate; devote ‹*tiempo*›. **~toria** *f* dedication, inscription
ded|il *m* finger-stall. **~illo** *m*. **al ~illo** at one's fingertips. **~o** *m* finger; (*del pie*) toe. **~o anular** ring finger. **~ corazón** middle finger. **~o gordo** thumb. **~o índice** index finger. **~o meñique** little finger. **~o pulgar** thumb
deduc|ción *f* deduction. **~ir** [47] *vt* deduce; (*descontar*) deduct
defect|o *m* fault, defect. **~uoso** *a* defective
defen|der [1] *vt* defend. **~sa** *f* defence. **~sivo** *a* defensive. **~sor** *m* defender. **abogado** *m* **~sor** defence counsel
deferen|cia *f* deference. **~te** *a* deferential
deficien|cia *f* deficiency. **~cia mental** mental handicap. **~te** *a* deficient; (*imperfecto*) defective. **~te mental** mentally handicapped
déficit *m invar* deficit
defini|ción *f* definition. **~do** *a* defined. **~r** *vt* define; (*aclarar*) clarify. **~tivo** *a* definitive. **en ~tiva** (*en resumen*) in short
deflación *f* deflation
deform|ación *f* deformation; (*TV etc*) distortion. **~ar** *vt* deform; (*TV etc*) distort. **~arse** *vpr* go out of shape. **~e** *a* deformed; (*feo*) ugly
defraudar *vt* cheat; (*decepcionar*) disappoint; evade ‹*impuestos etc*›
defunción *f* death
degenera|ción *f* degeneration; (*moral*) degeneracy. **~do** *a* degenerate. **~r** *vi* degenerate
deglutir *vt/i* swallow
degollar [16] *vt* cut s.o.'s throat; (*fig, arruinar*) ruin
degradar *vt* degrade. **~se** *vpr* lower o.s.
degusta|ción *f* tasting. **~r** *vt* taste
dehesa *f* pasture
dei|dad *f* deity. **~ficar** [7] *vt* deify
deja|ción *f* surrender. **~dez** *f* abandon; (*pereza*) laziness. **~do** *a* negligent. **~r** *vt* leave; (*abandonar*) abandon; (*prestar*) lend; (*permitir*) let. **~r aparte**, **~r a un lado** leave aside. **~r de** stop. **no ~r de** not fail to
dejo *m* aftertaste; (*tonillo*) accent
del = **de**; **el**
delantal *m* apron
delante *adv* in front; (*enfrente*) opposite. **~ de** in front of. **de ~** front
delanter|a *f* front; (*de teatro etc*) front row; (*ventaja*) advantage. **coger la ~a** get ahead. **~o** *a* front. ● *m* forward. **llevar la ~a** be ahead
delat|ar *vt* denounce. **~or** *m* informer
delega|ción *f* delegation; (*sucursal*) branch. **~do** *m* delegate; (*com*) agent, representative. **~r** [12] *vt* delegate
deleit|ar *vt* delight. **~e** *m* delight
deletéreo *a* deleterious
deletre|ar *vt* spell (out). **~o** *m* spelling
deleznable *a* brittle, crumbly; ‹*argumento etc*› weak
delfín *m* dolphin
delgad|ez *f* thinness. **~o** *a* thin; (*esbelto*) slim. **~ucho** *a* skinny
delibera|ción *f* deliberation. **~r** *vt* discuss, decide. ● *vi* deliberate
delicad|eza *f* delicacy; (*fragilidad*) frailty; (*tacto*) tact. **~o** *a* delicate; (*sensible* sensitive; (*discreto*) tactful, discreet. **falta de ~eza** tactlessness
delici|a *f* delight. **~oso** *a* delightful; ‹*sabor etc*› delicious; (*gracioso, fam*) funny
delimitar *vt* delimit
delincuen|cia *f* delinquency. **~te** *a & m* delinquent
delinea|nte *m* draughtsman. **~r** *vt* outline; (*dibujar*) draw
delinquir [8] *vi* commit an offence
delir|ante *a* delirious. **~ar** *vi* be delirious; (*fig*) talk nonsense. **~io** *m* delirium; (*fig*) frenzy
delito *m* crime, offence
delta *f* delta
demacrado *a* emaciated

demagogo *m* demagogue
demanda *f*. **en ~ de** asking for; (*en busca de*) in search of. **~nte** *m & f* (*jurid*) plaintiff. **~r** *vt* (*jurid*) bring an action against
demarca|ción *f* demarcation. **~r** [7] *vt* demarcate
demás *a* rest of the, other. ● *pron* rest, others. **lo ~** the rest. **por ~** useless; (*muy*) very. **por lo ~** otherwise
demasía *f* excess; (*abuso*) outrage; (*atrevimiento*) insolence. **en ~** too much
demasiado *a* too much; (*en plural*) too many. ● *adv* too much; (*con adjetivo*) too
demen|cia *f* madness. **~te** *a* demented, mad
dem|ocracia *f* democracy. **~ócrata** *m & f* democrat. **~ocrático** *a* democratic
demol|er [2] *vt* demolish. **~ición** *f* demolition
demonio *m* devil, demon. **¡~s!** hell! **¿cómo ~s?** how the hell? **¡qué ~s!** what the hell!
demora *f* delay. **~r** *vt* delay. ● *vi* stay on. **~rse** *vpr* be a long time
demostra|ción *f* demonstration, show. **~r** [2] *vt* demonstrate; (*mostrar*) show; (*probar*) prove. **~tivo** *a* demonstrative
denegar [1 & 12] *vt* refuse
deng|oso *a* affected, finicky. **~ue** *m* affectation
denigrar *vt* denigrate
denomina|ción *f* denomination. **~do** *a* called. **~dor** *m* denominator. **~r** *vt* name
denotar *vt* denote
dens|idad *f* density. **~o** *a* dense, thick
denta|dura *f* teeth. **~dura postiza** denture, false teeth. **~l** *a* dental
dentera *f*. **dar ~ a uno** set s.o.'s teeth on edge; (*dar envidia*) make s.o. green with envy
dentífrico *m* toothpaste
dentista *m & f* dentist
dentro *adv* inside; (*de un edificio*) indoors. **~ de** in. **~ de poco** soon. **por ~** inside
denuncia *f* report; (*acusación*) accusation. **~r** *vt* report (a crime); ‹*periódico etc*› denounce; (*indicar*) indicate
departamento *m* department; (*Arg, piso*) flat (*Brit*), apartment (*Amer*)
dependencia *f* dependence; (*sección*) section; (*sucursal*) branch
depender *vi* depend (**de** on)
dependient|a *f* shop assistant. **~e** *a* dependent (**de** on). ● *m* employee; (*de oficina*) clerk; (*de tienda*) shop assistant
depila|ción *f* depilation. **~r** *vt* depilate. **~torio** *a* depilatory
deplora|ble *a* deplorable. **~r** *vt* deplore, regret
deponer [34] *vt* remove from office. ● *vi* give evidence
deporta|ción *f* deportation. **~r** *vt* deport
deport|e *m* sport. **~ista** *m* sportsman. ● *f* sportswoman. **~ivo** *a* sports. ● *m* sports car. **hacer ~e** take part in sports
deposición *f* deposition; (*de un empleo*) removal from office
dep|ositador *m* depositor. **~ositante** *m & f* depositor. **~ositar** *vt* deposit; (*poner*) put, place. **~ósito** *m* deposit; (*conjunto de cosas*) store; (*almacén*) warehouse; (*mil*) depot; (*de líquidos*) tank
deprava|ción *f* depravity. **~do** *a* depraved. **~r** *vt* deprave. **~rse** *vpr* become depraved
deprecia|ción *f* depreciation. **~r** *vt* depreciate. **~rse** *vpr* depreciate
depresión *f* depression
deprim|ente *a* depressing. **~ido** *a* depressed. **~ir** *vt* depress. **~irse** *vpr* get depressed
depura|ción *f* purification; (*pol*) purging. **~r** *vt* purify; (*pol*) purge
derech|a *f* (*mano*) right hand; (*lado*) right. **~ista** *a* right-wing. ● *m & f* right-winger. **~o** *a* right; (*vertical*) upright; (*recto*) straight. ● *adv* straight. ● *m* right; (*ley*) law; (*lado*) right side. **~os** *mpl* dues. **~os de autor** royalties. **a la ~a** on the right; (*hacia el lado derecho*) to the right. **todo ~o** straight on
deriva *f* drift. **a la ~** drifting, adrift
deriva|ción *f* derivation; (*cambio*) diversion. **~do** *a* derived. ● *m* derivative, by-product. **~r** *vt* derive; (*cambiar la dirección de*) divert. ● *vi*. **~r de** derive from, be derived from. **~rse** *vpr* be derived
derram|amiento *m* spilling. **~amiento de sangre** bloodshed. **~ar** *vt* spill; (*verter*) pour; shed ‹*lágrimas*›. **~arse** *vpr* spill. **~e** *m* spilling; (*pérdida*) leakage; (*cantidad perdida*)

spillage; (*med*) discharge; (*med, de sangre*) haemorrhage
derretir [5] vt melt. **~se** *vpr* melt; (*enamorarse*) fall in love (**por** with)
derriba|do *a* fallen down. **~r** *vt* knock down; bring down, overthrow ‹*gobierno etc*›. **~rse** *vpr* fall down
derrocar [7] *vt* bring down, overthrow ‹*gobierno etc*›
derroch|ar *vt* squander. **~e** *m* waste
derrot|a *f* defeat; (*rumbo*) course. **~ar** *vt* defeat. **~ado** *a* defeated; ‹*vestido*› shabby. **~ero** *m* course
derrumba|miento *m* collapse. **~r** *vt* (*derribar*) knock down. **~rse** *vpr* collapse
desaborido *a* tasteless; ‹*persona*› dull
desabotonar *vt* unbutton, undo. ● *vi* bloom. **~se** *vpr* come undone
desabrido *a* tasteless; ‹*tiempo*› unpleasant; ‹*persona*› surly
desabrochar *vt* undo. **~se** *vpr* come undone
desacat|ar *vt* have no respect for. **~o** *m* disrespect
desac|ertado *a* ill-advised; (*erróneo*) wrong. **~ertar** [1] *vt* be wrong. **~ierto** *m* mistake
desaconseja|ble *a* inadvisable. **~do** *a* unwise, ill-advised. **~r** *vt* advise against, dissuade
desacorde *a* discordant
desacostumbra|do *a* unusual. **~r** *vt* give up
desacreditar *vt* discredit
desactivar *vt* defuse
desacuerdo *m* disagreement
desafiar [20] *vt* challenge; (*afrontar*) defy
desafilado *a* blunt
desafina|do *a* out of tune. **~r** *vi* be out of tune. **~rse** *vpr* go out of tune
desafío *m* challenge; (*combate*) duel
desaforado *a* ‹*comportamiento*› outrageous; (*desmedido*) excessive; ‹*sonido*› loud; (*enorme*) huge
desafortunad|amente *adv* unfortunately. **~o** *a* unfortunate
desagrada|ble *a* unpleasant. **~r** *vt* displease. ● *vi* be unpleasant. **me ~ el sabor** I don't like the taste
desagradecido *a* ungrateful
desagrado *m* displeasure. **con ~** unwillingly
desagravi|ar *vt* make amends to. **~o** *m* amends; (*expiación*) atonement
desagregar [12] *vt* break up. **~se** *vpr* disintegrate
desagüe *m* drain; (*acción*) drainage. **tubo** *m* **de ~** drain-pipe
desaguisado *a* illegal. ● *m* offence; (*fam*) disaster
desahog|ado *a* roomy; (*adinerado*) well-off; (*fig, descarado, fam*) impudent. **~ar** [12] *vt* relieve; vent ‹*ira*›. **~arse** *vpr* (*desfogarse*) let off steam. **~o** *m* comfort; (*alivio*) relief
desahuci|ar *vt* deprive of hope; give up hope for ‹*enfermo*›; evict ‹*inquilino*›. **~o** *m* eviction
desair|ado *a* humiliating; ‹*persona*› humiliated, spurned. **~ar** *vt* snub ‹*persona*›; disregard ‹*cosa*›. **~e** *m* rebuff
desajuste *m* maladjustment; (*avería*) breakdown
desal|entador *a* disheartening. **~entar** [1] *vt* (*fig*) discourage. **~iento** *m* discouragement
desaliño *m* untidiness, scruffiness
desalmado *a* wicked
desalojar *vt* eject ‹*persona*›; evacuate ‹*sitio*›. ● *vi* move (house)
desampar|ado *a* helpless; (*abandonado*) abandoned. **~ar** *vt* abandon. **~o** *m* helplessness; (*abandono*) abandonment
desangelado *a* insipid, dull
desangrar *vt* bleed. **~se** *vpr* bleed
desanima|do *a* down-hearted. **~r** *vt* discourage. **~rse** *vpr* lose heart
desánimo *m* discouragement
desanudar *vt* untie
desapacible *a* unpleasant; ‹*sonido*› harsh
desapar|ecer [11] *vi* disappear; ‹*efecto*› wear off. **~ecido** *a* disappeared. ● *m* missing person. **~ecidos** *mpl* missing. **~ición** *f* disappearance
desapasionado *a* dispassionate
desapego *m* indifference
desapercebido *a* unnoticed
desaplicado *a* lazy
desaprensi|ón *f* unscrupulousness. **~vo** *a* unscrupulous
desaproba|ción *f* disapproval. **~r** [2] *vt* disapprove of; (*rechazar*) reject.
desaprovecha|do *a* wasted; ‹*alumno*› lazy. **~r** *vt* waste
desarm|ar *vt* disarm; (*desmontar*) take to pieces. **~e** *m* disarmament
desarraig|ado *a* rootless. **~ar** [12] *vt* uproot; (*fig, erradicar*) wipe out. **~o** *m* uprooting; (*fig*) eradication

desarregl|ado *a* untidy; (*desordenado*) disorderly. **~ar** *vt* mess up; (*deshacer el orden*) make untidy. **~o** *m* disorder; (*de persona*) untidiness

desarroll|ado *a* (well-) developed. **~ar** *vt* develop; (*desenrollar*) unroll, unfold. **~arse** *vpr* (*incl foto*) develop; (*desenrollarse*) unroll; ‹*suceso*› take place. **~o** *m* development

desarrugar [12] *vt* smooth out

desarticular *vt* dislocate ‹*hueso*›; (*fig*) break up

desaseado *a* dirty; (*desordenado*) untidy

desasirse [45] *vpr* let go (**de** of)

desasos|egar [1 & 12] *vt* disturb. **~egarse** *vpr* get uneasy. **~iego** *m* anxiety; (*intranquilidad*) restlessness

desastr|ado *a* scruffy. **~e** *m* disaster. **~oso** *a* disastrous

desata|do *a* untied; (*fig*) wild. **~r** *vt* untie; (*fig, soltar*) unleash. **~rse** *vpr* come undone

desatascar [7] *vt* pull out of the mud; unblock ‹*tubo etc*›

desaten|ción *f* inattention; (*descortesía*) discourtesy. **~der** [1] *vt* not pay attention to; neglect ‹*deber etc*›. **~to** *a* inattentive; (*descortés*) discourteous

desatin|ado *a* silly. **~o** *m* silliness; (*error*) mistake

desatornillar *vt* unscrew

desatracar [7] *vt/i* cast off

desautorizar [10] *vt* declare unauthorized; (*desmentir*) deny

desavenencia *f* disagreement

desayun|ar *vt* have for breakfast. ● *vi* have breakfast. **~o** *m* breakfast

desazón *m* (*fig*) anxiety

desbandarse *vpr* (*mil*) disband; (*dispersarse*) disperse

desbarajust|ar *vt* throw into confusion. **~e** *m* confusion

desbaratar *vt* spoil

desbloquear *vt* unfreeze

desbocado *a* ‹*vasija etc*› chipped; ‹*caballo*› runaway; ‹*persona*› foul-mouthed

desborda|nte *a* overflowing. **~r** *vt* go beyond; (*exceder*) exceed. ● *vi* overflow. **~rse** *vpr* overflow

descabalgar [12] *vi* dismount

descabellado *a* crazy

descabezar [10] *vt* behead

descafeinado *a* decaffeinated. ● *m* decaffeinated coffee

descalabr|ar *vt* injure in the head; (*fig*) damage. **~o** *m* disaster

descalificar [7] *vt* disqualify; (*desacreditar*) discredit

descalz|ar [10] *vt* take off ‹*zapato*›. **~o** *a* barefoot

descaminar *vt* misdirect; (*fig*) lead astray

descamisado *a* shirtless; (*fig*) shabby

descampado *a* open. ● *m* open ground

descans|ado *a* rested; ‹*trabajo*› easy. **~apiés** *m* footrest. **~ar** *vt/i* rest. **~illo** *m* landing. **~o** *m* rest; (*descansillo*) landing; (*en deportes*) half-time; (*en el teatro etc*) interval

descapotable *a* convertible

descarado *a* insolent, cheeky; (*sin vergüenza*) shameless

descarg|a *f* unloading; (*mil, elec*) discharge. **~ar** [12] *vt* unload; (*mil, elec*) discharge, shock; deal ‹*golpe etc*›. ● *vi* flow into. **~o** *m* unloading; (*recibo*) receipt; (*jurid*) evidence

descarnado *a* scrawny, lean; (*fig*) bare

descaro *m* insolence, cheek; (*cinismo*) nerve, effrontery

descarriar [20] *vt* misdirect; (*fig*) lead astray. **~se** *vpr* go the wrong way; ‹*res*› stray; (*fig*) go astray

descarrila|miento *m* derailment. **~r** *vi* be derailed. **~se** *vpr* be derailed

descartar *vt* discard; (*rechazar*) reject. **~se** *vpr* discard

descascarar *vt* shell

descen|dencia *f* descent; (*personas*) descendants. **~dente** *a* descending. **~der** [1] *vt* lower, get down; go down ‹*escalera etc*›. ● *vi* go down; (*provenir*) be descended (**de** from). **~diente** *m & f* descendent. **~so** *m* descent; (*de temperatura, fiebre etc*) fall, drop

descentralizar [10] *vt* decentralize

descifrar *vt* decipher; decode ‹*clave*›

descolgar [2 & 12] *vt* take down; pick up ‹*el teléfono*›. **~se** *vpr* let o.s. down; (*fig, fam*) turn up

descolorar *vt* discolour, fade

descolori|do *a* discoloured, faded; ‹*persona*› pale. **~r** *vt* discolour, fade

descomedido *a* rude; (*excesivo*) excessive, extreme

descomp|ás *m* disproportion. **~asado** *a* disproportionate
descomp|oner [34] *vt* break down; decompose ‹*substancia*›; distort ‹*rasgos*›; (*estropear*) break; (*desarreglar*) disturb, spoil. **~onerse** *vpr* decompose; ‹*persona*› lose one's temper. **~osición** *f* decomposition; (*med*) diarrhoea. **~ostura** *f* breaking; (*de un motor*) breakdown; (*desorden*) disorder. **~uesto** *a* broken; (*podrido*) decomposed; (*encolerizado*) angry. **estar ~uesto** have diarrhoea
descomunal *a* (*fam*) enormous
desconc|ertante *a* disconcerting. **~ertar** [1] *vt* disconcert; (*dejar perplejo*) puzzle. **~ertarse** *vpr* be put out, be disconcerted; ‹*mecanismo*› break down. **~ierto** *m* confusion
desconectar *vt* disconnect
desconfia|do *a* distrustful. **~nza** *f* distrust, suspicion. **~r** [20] *vi*. **~r de** not trust; (*no creer*) doubt
descongelar *vt* defrost; (*com*) unfreeze
desconoc|er [11] *vt* not know, not recognize. **~ido** *a* unknown; (*cambiado*) unrecognizable. ● *m* stranger. **~imiento** *m* ignorance
desconsidera|ción *f* lack of consideration. **~do** *a* inconsiderate
descons|olado *a* distressed. **~olar** [2] *vt* distress. **~olarse** *vpr* despair. **~uelo** *m* distress; (*tristeza*) sadness
desconta|do *a*. **dar por ~do** take for granted. **por ~do** of course. **~r** [2] *vt* discount
descontent|adizo *a* hard to please. **~ar** *vt* displease. **~o** *a* unhappy (**de** about), discontented (**de** with). ● *m* discontent
descontrolado *a* uncontrolled
descorazonar *vt* discourage. **~se** *vpr* lose heart
descorchar *vt* uncork
descorrer *vt* draw ‹*cortina*›. **~ el cerrojo** unbolt the door
descort|és *a* rude, discourteous. **~esía** *f* rudeness
descos|er *vt* unpick. **~erse** *vpr* come undone. **~ido** *a* unstitched; (*fig*) disjointed. **como un ~ido** a lot
descoyuntar *vt* dislocate
descrédito *m* disrepute. **ir en ~ de** damage the reputation of
descreído *a* unbelieving
descremar *vt* skim
descri|bir [*pp* **descrito**] *vt* describe. **~pción** *f* description. **~ptivo** *a* descriptive
descuartizar [10] *vt* cut up
descubierto *a* discovered; (*no cubierto*) uncovered; (*expuesto*) exposed; ‹*cielo*› clear; (*sin sombrero*) bareheaded. ● *m* overdraft; (*déficit*) deficit. **poner al ~** expose
descubri|miento *m* discovery. **~r** [*pp* **descubierto**] *vt* discover; (*quitar lo que cubre*) uncover; (*revelar*) reveal; unveil ‹*estatua*›. **~rse** *vpr* be discovered; ‹*cielo*› clear; (*quitarse el sombrero*) take off one's hat
descuento *m* discount
descuid|ado *a* careless; ‹*aspecto etc*› untidy; (*desprevenido*) unprepared. **~ar** *vt* neglect. ● *vi* not worry. **~arse** *vpr* be careless; (*no preocuparse*) not worry. **¡~a!** don't worry! **~o** *m* carelessness; (*negligencia*) negligence. **al ~o** nonchalantly. **estar ~ado** not worry, rest assured
desde *prep* (*lugar etc*) from; (*tiempo*) since, from. **~ hace poco** for a short time. **~ hace un mes** for a month. **~ luego** of course. **~ Madrid hasta Barcelona** from Madrid to Barcelona. **~ niño** since childhood
desdecir [46, *pero imperativo* **desdice,** *futuro y condicional regulares*] *vi*. **~ de** be unworthy of; (*no armonizar*) not match. **~se** *vpr*. **~ de** take back ‹*palabras etc*›; go back on ‹*promesa*›
desd|én *m* scorn. **~eñable** *a* contemptible. **~eñar** *vt* scorn. **~eñoso** *a* scornful
desdicha *f* misfortune. **~do** *a* unfortunate. **por ~** unfortunately
desdoblar *vt* straighten; (*desplegar*) unfold
desea|ble *a* desirable. **~r** *vt* want; wish ‹*algo a uno*›. **de ~r** desirable. **le deseo un buen viaje** I hope you have a good journey. **¿qué desea Vd?** can I help you?
desecar [7] *vt* dry up
desech|ar *vt* throw out. **~o** *m* rubbish
desembalar *vt* unpack
desembarazar [10] *vt* clear. **~se** *vpr* free o.s.
desembarca|dero *m* landing stage. **~r** [7] *vt* unload. ● *vi* disembark
desembocа|dura *f* (*de río*) mouth; (*de calle*) opening. **~r** [7] *vi*. **~r en**

‹*río*› flow into; ‹*calle*› join; (*fig*) lead to, end in
desembols|ar *vt* pay. **~o** *m* payment
desembragar [12] *vi* declutch
desembrollar *vt* unravel
desembuchar *vi* tell, reveal a secret
desemejan|te *a* unlike, dissimilar. **~za** *f* dissimilarity
desempapelar *vt* unwrap
desempaquetar *vt* unpack, unwrap
desempat|ar *vi* break a tie. **~e** *m* tie-breaker
desempeñ|ar *vt* redeem; play ‹*papel*›; hold ‹*cargo*›; perform, carry out ‹*deber etc*›. **~arse** *vpr* get out of debt. **~o** *m* redemption; (*de un papel, de un cargo*) performance
desemple|ado *a* unemployed. ● *m* unemployed person. **~o** *m* unemployment. **los ~ados** *mpl* the unemployed
desempolvar *vt* dust; (*fig*) unearth
desencadenar *vt* unchain; (*fig*) unleash. **~se** *vpr* break loose; ‹*guerra etc*› break out
desencajar *vt* dislocate; (*desconectar*) disconnect. **~se** *vpr* become distorted
desencant|ar *vt* disillusion. **~o** *m* disillusionment
desenchufar *vt* unplug
desenfad|ado *a* uninhibited. **~ar** *vt* calm down. **~arse** *vpr* calm down. **~o** *m* openness; (*desenvoltura*) assurance
desenfocado *a* out of focus
desenfren|ado *a* unrestrained. **~arse** *vpr* rage. **~o** *m* licentiousness
desenganchar *vt* unhook
desengañ|ar *vt* disillusion. **~arse** *vpr* be disillusioned; (*darse cuenta*) realize. **~o** *m* disillusionment, disappointment
desengrasar *vt* remove the grease from. ● *vi* lose weight
desenla|ce *m* outcome. **~zar** [10] *vt* undo; solve ‹*problema*›
desenmarañar *vt* unravel
desenmascarar *vt* unmask
desenojar *vt* calm down. **~se** *vpr* calm down
desenred|ar *vt* unravel. **~arse** *vpr* extricate o.s. **~o** *m* denoument
desenrollar *vt* unroll, unwind
desenroscar [7] *vt* unscrew
desentenderse [1] *vpr* want nothing to do with; (*afectar ignorancia*) pretend not to know. **hacerse el desentendido** (*fingir no oír*) pretend not to hear
desenterrar [1] *vt* exhume; (*fig*) unearth
desenton|ar *vi* be out of tune; ‹*colores*› clash. **~o** *m* rudeness
desentrañar *vt* work out
desenvoltura *f* ease; (*falta de timidez*) confidence; (*descaro*) insolence
desenvolver [2, *pp* **desenvuelto**] *vt* unwrap; expound ‹*idea etc*›. **~se** *vpr* act with confidence
deseo *m* wish, desire. **~so** *a* desirous. **arder en ~s de** long for. **buen ~** good intentions. **estar ~so de** be eager to
desequilibr|ado *a* unbalanced. **~io** *m* imbalance
des|erción *f* desertion; (*pol*) defection. **~ertar** *vt* desert. **~értico** *a* desert-like. **~ertor** *m* deserter
desespera|ción *f* despair. **~do** *a* desperate. **~nte** *a* infuriating. **~r** *vt* drive to despair. ● *vi* despair (**de** of). **~rse** *vpr* despair
desestimar *vt* (*rechazar*) reject
desfachat|ado *a* brazen, impudent. **~ez** *f* impudence
desfalc|ar [7] *vt* embezzle. **~o** *m* embezzlement
desfallec|er [11] *vt* weaken. ● *vi* get weak; (*desmayarse*) faint. **~imiento** *m* weakness
desfas|ado *a* ‹*persona*› out of place, out of step; ‹*máquina etc*› out of phase. **~e** *m* jet-lag. **estar ~ado** have jet-lag
desfavor|able *a* unfavourable. **~ecer** [11] *vt* ‹*ropa*› not suit
desfigurar *vt* disfigure; (*desdibujar*) blur; (*fig*) distort
desfiladero *m* pass
desfil|ar *vi* march (past). **~e** *m* procession, parade. **~e de modelos** fashion show
desfogar [12] *vt* vent (**en**, **con** on). **~se** *vpr* let off steam
desgajar *vt* tear off; (*fig*) uproot ‹*persona*›. **~se** *vpr* come off
desgana *f* (*falta de apetito*) lack of appetite; (*med*) weakness, faintness; (*fig*) unwillingness
desgarr|ador *a* heart-rending. **~ar** *vt* tear; (*fig*) break ‹*corazón*›. **~o** *m* tear, rip; (*descaro*) insolence. **~ón** *m* tear

desgast|ar *vt* wear away; wear out ‹*ropa*›. **~arse** *vpr* wear away; ‹*ropa*› be worn out; ‹*persona*› wear o.s. out. **~e** *m* wear

desgracia *f* misfortune; (*accidente*) accident; (*mala suerte*) bad luck. **~damente** *adv* unfortunately. **~do** *a* unlucky; (*pobre*) poor; (*desagradable*) unpleasant. ● *m* unfortunate person, poor devil (*fam*). **~r** *vt* spoil. **caer en ~** fall from favour. **estar en ~** be unfortunate. **por ~** unfortunately. **¡qué ~!** what a shame!

desgranar *vt* shell ‹*guisantes etc*›

desgreñado *a* ruffled, dishevelled

desgua|ce *m* scrapyard. **~zar** [10] *vt* scrap

deshabitado *a* uninhabited

deshabituarse [21] *vpr* get out of the habit

deshacer [31] *vt* undo; strip ‹*cama*›; unpack ‹*maleta*›; (*desmontar*) take to pieces; break ‹*trato*›; (*derretir*) melt; (*en agua*) dissolve; (*destruir*) destroy; (*estropear*) spoil; (*derrotar*) defeat. **~se** *vpr* come undone; (*descomponerse*) fall to pieces; (*derretirse*) melt. **~se de algo** get rid of sth. **~se en lágrimas** burst into tears. **~se por hacer algo** go out of one's way to do sth

deshelar [1] *vt* thaw. **~se** *vpr* thaw

desheredar *vt* disinherit

deshidratar *vt* dehydrate. **~se** *vpr* become dehydrated

deshielo *m* thaw

deshilachado *a* frayed

deshincha|do *a* ‹*neumático*› flat. **~r** *vt* deflate. **~rse** *vpr* go down

deshollina|dor *m* (chimney-)sweep. **~r** *vt* sweep ‹*chimenea*›

deshon|esto *a* dishonest; (*obsceno*) indecent. **~or** *m*, **~ra** *f* disgrace. **~rar** *vt* dishonour

deshora *f*. **a ~** (*a hora desacostumbrada*) at an unusual time; (*a hora inoportuna*) at an inconvenient time; (*a hora avanzada*) very late

deshuesar *vt* bone ‹*carne*›; stone ‹*fruta*›

desidia *f* laziness

desierto *a* deserted. ● *m* desert

designa|ción *f* designation. **~r** *vt* designate; (*fijar*) fix

desigual *a* unequal; ‹*terreno*› uneven; (*distinto*) different. **~dad** *f* inequality

desilusi|ón *f* disappointment; (*pérdida de ilusiones*) disillusionment. **~onar** *vt* disappoint; (*quitar las ilusiones*) disillusion. **~onarse** *vpr* become disillusioned

desinfecta|nte *m* disinfectant. **~r** *vt* disinfect

desinfestar *vt* decontaminate

desinflar *vt* deflate. **~se** *vpr* go down

desinhibido *a* uninhibited

desintegra|ción *f* disintegration. **~r** *vt* disintegrate. **~rse** *vpr* disintegrate

desinter|és *m* impartiality; (*generosidad*) generosity. **~esado** *a* impartial; (*liberal*) generous

desistir *vi*. **~ de** give up

desleal *a* disloyal. **~tad** *f* disloyalty

desleír [51] *vt* thin down, dilute

deslenguado *a* foul-mouthed

desligar [12] *vt* untie; (*separar*) separate; (*fig, librar*) free. **~se** *vpr* break away; (*de un compromiso*) free o.s.

deslizar [10] *vt* slide, slip. **~se** *vpr* slide, slip; ‹*tiempo*› slide by, pass; (*fluir*) flow

deslucido *a* tarnished; (*gastado*) worn out; (*fig*) undistinguished

deslumbrar *vt* dazzle

deslustrar *vt* tarnish

desmadr|ado *a* unruly. **~arse** *vpr* get out of control. **~e** *m* excess

desmán *m* outrage

desmandarse *vpr* get out of control

desmantelar *vt* dismantle; (*despojar*) strip

desmañado *a* clumsy

desmaquillador *m* make-up remover

desmay|ado *a* unconscious. **~ar** *vi* lose heart. **~arse** *vpr* faint. **~o** *m* faint; (*estado*) unconsciousness; (*fig*) depression

desmedido *a* excessive

desmedrarse *vpr* waste away

desmejorarse *vpr* deteriorate

desmelenado *a* dishevelled

desmembrar *vt* (*fig*) divide up

desmemoriado *a* forgetful

desmentir [4] *vt* deny. **~se** *vpr* contradict o.s.; (*desdecirse*) go back on one's word

desmenuzar [10] *vt* crumble; chop ‹*carne etc*›

desmerecer [11] *vt* be unworthy of. ● *vi* deteriorate

desmesurado *a* excessive; (*enorme*) enormous

desmigajar *vt*, **desmigar** [12] *vt* crumble

desmonta|ble *a* collapsible. **~r** *vt* (*quitar*) remove; (*desarmar*) take to pieces; (*derribar*) knock down; (*allanar*) level. ● *vi* dismount

desmoralizar [10] *vt* demoralize

desmoronar *vt* wear away; (*fig*) make inroads into. **~se** *vpr* crumble

desmovilizar [10] *vt/i* demobilize

desnatar *vt* skim

desnivel *m* unevenness; (*fig*) difference, inequality

desnud|ar *vt* strip; undress, strip ‹*persona*›. **~arse** *vpr* get undressed. **~ez** *f* nudity. **~o** *a* naked; (*fig*) bare. ● *m* nude

desnutri|ción *f* malnutrition. **~do** *a* undernourished

desobed|ecer [11] *vt* disobey. **~iencia** *f* disobedience. **~iente** *a* disobedient

desocupa|do *a* ‹*asiento etc*› vacant, free; (*sin trabajo*) unemployed; (*ocioso*) idle. **~r** *vt* vacate

desodorante *m* deodorant

desoír [50] *vt* take no notice of

desola|ción *f* desolation; (*fig*) distress. **~do** *a* desolate; ‹*persona*› sorry, sad. **~r** *vt* ruin; (*desconsolar*) distress

desollar *vt* skin; (*fig, criticar*) criticize; (*fig, hacer pagar demasiado, fam*) fleece

desorbitante *a* excessive

desorden *m* disorder, untidiness; (*confusión*) confusion. **~ado** *a* untidy. **~ar** *vt* disarrange, make a mess of

desorganizar [10] *vt* disorganize; (*trastornar*) disturb

desorienta|do *a* confused. **~r** *vt* disorientate. **~rse** *vpr* lose one's bearings

desovar *vi* ‹*pez*› spawn; ‹*insecto*› lay eggs

despabila|do *a* wide awake; (*listo*) quick. **~r** *vt* (*despertar*) wake up; (*avivar*) brighten up. **~rse** *vpr* wake up; (*avivarse*) brighten up. **¡despabílate!** get a move on!

despaci|o *adv* slowly. ● *int* easy does it! **~to** *adv* slowly

despach|ar *vt* finish; (*tratar con*) deal with; (*vender*) sell; (*enviar*) send; (*despedir*) send away; issue ‹*billete*›. ● *vi* hurry up. **~arse** *vpr* get rid; (*terminar*) finish. **~o** *m* dispatch; (*oficina*) office; (*venta*) sale; (*del teatro*) box office

despampanante *a* stunning

desparejado *a* odd

desparpajo *m* confidence; (*descaro*) impudence

desparramar *vt* scatter; spill ‹*líquidos*›; squander ‹*fortuna*›

despavorido *a* terrified

despectivo *a* disparaging; ‹*sentido etc*› pejorative

despecho *m* spite. **a ~ de** in spite of. **por ~** out of spite

despedazar [10] *vt* tear to pieces

despedi|da *f* goodbye, farewell. **~da de soltero** stag-party. **~r** [5] *vt* say goodbye, see off; dismiss ‹*empleado*›; evict ‹*inquilino*›; (*arrojar*) throw; give off ‹*olor etc*›. **~rse** *vpr*. **~rse de** say goodbye to

despeg|ado *a* cold, indifferent. **~ar** [12] *vt* unstick. ● *vi* ‹*avión*› take off. **~o** *m* indifference. **~ue** *m* take-off

despeinar *vt* ruffle the hair of

despeja|do *a* clear; ‹*persona*› wide awake. **~r** *vt* clear; (*aclarar*) clarify. ● *vi* clear. **~rse** *vpr* (*aclararse*) become clear; ‹*cielo*› clear; ‹*tiempo*› clear up; ‹*persona*› liven up

despellejar *vt* skin

despensa *f* pantry, larder

despeñadero *m* cliff

desperdici|ar *vt* waste. **~o** *m* waste. **~os** *mpl* rubbish. **no tener ~o** be good all the way through

desperezarse [10] *vpr* stretch

desperfecto *m* flaw

desperta|dor *m* alarm clock. **~r** [1] *vt* wake up; (*fig*) awaken. **~rse** *vpr* wake up

despiadado *a* merciless

despido *m* dismissal

despierto *a* awake; (*listo*) bright

despilfarr|ar *vt* waste. **~o** *m* squandering; (*gasto innecesario*) extravagance

despista|do *a* (*con estar*) confused; (*con ser*) absent-minded. **~r** *vt* throw off the scent; (*fig*) mislead. **~rse** *vpr* go wrong; (*fig*) get confused

despiste *m* swerve; (*error*) mistake; (*confusión*) muddle

desplaza|do *a* out of place. **~miento** *m* displacement; (*de opinión etc*) swing, shift. **~r** [10] *vt* displace. **~rse** *vpr* travel

despl|egar [1 & 12] *vt* open out; spread ‹*alas*›; (*fig*) show. **~iegue** *m* opening; (*fig*) show
desplomarse *vpr* lean; (*caerse*) collapse
desplumar *vt* pluck; (*fig, fam*) fleece
despobla|do *m* deserted area. **~r** [2] *vt* depopulate
despoj|ar *vt* deprive ‹*persona*›; strip ‹*cosa*›. **~o** *m* plundering; (*botín*) booty. **~os** *mpl* left-overs; (*de res*) offal; (*de ave*) giblets
desposado *a & m* newly-wed
déspota *m & f* despot
despreci|able *a* despicable; ‹*cantidad*› negligible. **~ar** *vt* despise; (*rechazar*) scorn. **~o** *m* contempt
desprend|er *vt* remove; give off ‹*olor*›. **~erse** *vpr* fall off; (*fig*) part with; (*deducirse*) follow. **~imiento** *m* loosening; (*generosidad*) generosity
despreocupa|ción *f* carelessness. **~do** *a* unconcerned; (*descuidado*) careless. **~rse** *vpr* not worry
desprestigiar *vt* discredit
desprevenido *a* unprepared. **coger a uno ~** catch s.o. unawares
desproporci|ón *f* disproportion. **~onado** *a* disproportionate
despropósito *m* irrelevant remark
desprovisto *a*. **~ de** lacking, without
después *adv* after, afterwards; (*más tarde*) later; (*a continuación*) then. **~ de** after. **~ de comer** after eating. **~ de todo** after all. **~ que** after. **poco ~** soon after. **una semana ~** a week later
desquiciar *vt* (*fig*) disturb
desquit|ar *vt* compensate. **~arse** *vpr* make up for; (*vengarse*) take revenge. **~e** *m* compensation; (*venganza*) revenge
destaca|do *a* outstanding. **~r** [7] *vt* emphasize. • *vi* stand out. **~rse** *vpr* stand out
destajo *m* piece-work. **hablar a ~** talk nineteen to the dozen
destap|ar *vt* uncover; open ‹*botella*›. **~arse** *vpr* reveal one's true self. **~e** *m* (*fig*) permissiveness
destartalado *a* ‹*habitación*› untidy; ‹*casa*› rambling
destell|ar *vi* sparkle. **~o** *m* sparkle; (*de estrella*) twinkle; (*fig*) glimmer
destemplado *a* out of tune; (*agrio*) harsh; ‹*tiempo*› unsettled; ‹*persona*› out of sorts
desteñir [5 & 22] *vt* fade; (*manchar*) discolour. • *vi* fade. **~se** *vpr* fade; ‹*color*› run
desterra|do *m* exile. **~r** [1] *vt* banish
destetar *vt* wean
destiempo *m*. **a ~** at the wrong moment
destierro *m* exile
destil|ación *f* distillation. **~ar** *vt* distil. **~ería** *f* distillery
destin|ar *vt* destine; (*nombrar*) appoint. **~atario** *m* addressee. **~o** *m* (*uso*) use, function; (*lugar*) destination; (*empleo*) position; (*suerte*) destiny. **con ~o a** going to, bound for. **dar ~o a** find a use for
destitu|ción *f* dismissal. **~ir** [17] *vt* dismiss
destornilla|dor *m* screwdriver. **~r** *vt* unscrew
destreza *f* skill
destripar *vt* rip open
destroz|ar [10] *vt* ruin; (*fig*) shatter. **~o** *m* destruction. **causar ~os, hacer ~os** ruin
destru|cción *f* destruction. **~ctivo** *a* destructive. **~ir** [17] *vt* destroy; demolish ‹*edificio*›
desunir *vt* separate
desus|ado *a* old-fashioned; (*insólito*) unusual. **~o** *m* disuse. **caer en ~o** become obsolete
desvaído *a* pale; (*borroso*) blurred; ‹*persona*› dull
desvalido *a* needy, destitute
desvalijar *vt* rob; burgle ‹*casa*›
desvalorizar [10] *vt* devalue
desván *m* loft
desvanec|er [11] *vt* make disappear; tone down ‹*colores*›; (*borrar*) blur; (*fig*) dispel. **~erse** *vpr* disappear; (*desmayarse*) faint. **~imiento** *m* (*med*) fainting fit
desvariar [20] *vi* be delirious; (*fig*) talk nonsense
desvel|ar *vt* keep awake. **~arse** *vpr* stay awake, have a sleepless night. **~o** *m* insomnia, sleeplessness
desvencijar *vt* break; (*agotar*) exhaust
desventaja *f* disadvantage
desventura *f* misfortune. **~do** *a* unfortunate
desverg|onzado *a* impudent, cheeky. **~üenza** *f* impudence, cheek
desvestirse [5] *vpr* undress
desv|iación *f* deviation; (*auto*) diversion. **~iar** [20] *vt* deflect, turn aside.

~iarse *vpr* be deflected; (*del camino*) make a detour; (*del tema*) stray. **~ío** *m* diversion; (*frialdad*) *f* indifference

desvivirse *vpr* long (**por** for); (*afanarse*) strive, do one's utmost

detall|ar *vt* relate in detail. **~e** *m* detail; (*fig*) gesture. **~ista** *m & f* retailer. **al ~e** in detail; (*al por menor*) retail. **con todo ~e** in great detail. **en ~es** in detail. **¡qué ~e!** how thoughtful!

detect|ar *vt* detect. **~ive** *m* detective

deten|ción *f* stopping; (*jurid*) arrest; (*en la cárcel*) detention. **~er** [40] *vt* stop; (*jurid*) arrest; (*encarcelar*) detain; (*retrasar*) delay. **~erse** *vpr* stop; (*entretenerse*) spend a lot of time. **~idamente** *adv* carefully. **~ido** *a* (*jurid*) under arrest; (*minucioso*) detailed. ● *m* prisoner

detergente *a & m* detergent

deterior|ar *vt* damage, spoil. **~arse** *vpr* deteriorate. **~o** *m* damage

determina|ción *f* determination; (*decisión*) decison. **~nte** *a* decisive. **~r** *vt* determine; (*decidir*) decide; (*fijar*) fix. **tomar una ~ción** make a decision

detestar *vt* detest

detonar *vi* explode

detrás *adv* behind; (*en la parte posterior*) on the back. **~ de** behind. **por ~** on the back; (*detrás de*) behind

detrimento *m* detriment. **en ~ de** to the detriment of

detrito *m* debris

deud|a *f* debt. **~or** *m* debtor

devalua|ción *f* devaluation. **~r** [21] *vt* devalue

devanar *vt* wind

devasta|dor *a* devastating. **~r** *vt* devastate

devoción *f* devotion

devol|ución *f* return; (*com*) repayment, refund. **~ver** [5] (*pp* **devuelto**) *vt* return; (*com*) repay, refund; restore ‹*edificio etc*›. ● *vi* be sick

devorar *vt* devour

devoto *a* devout; ‹*amigo etc*› devoted. ● *m* enthusiast

di *vb véase* **dar**

día *m* day. **~ de fiesta** (public) holiday. **~ del santo** saint's day. **~ festivo** (public) holiday. **~ hábil**, **~ laborable** working day. **al ~** up to date. **al ~ siguiente** (on) the following day. **¡buenos ~s!** good morning! **dar los buenos ~s** say good morning. **de ~** by day. **el ~ de hoy** today. **el ~ de mañana** tomorrow. **en pleno ~** in broad daylight. **en su ~** in due course. **todo el santo ~** all day long. **un ~ de estos** one of these days. **un ~ sí y otro no** every other day. **vivir al ~** live from hand to mouth

diab|etes *f* diabetes. **~ético** *a* diabetic

diab|lo *m* devil. **~lura** *f* mischief. **~ólico** *a* diabolical

diácono *m* deacon

diadema *f* diadem

diáfano *a* diaphanous

diafragma *m* diaphragm

diagn|osis *f* diagnosis. **~osticar** [7] *vt* diagnose. **~óstico** *a* diagnostic

diagonal *a & f* diagonal

diagrama *m* diagram

dialecto *m* dialect

diálisis *f* dialysis

di|alogar [12] *vi* talk. **~álogo** *m* dialogue

diamante *m* diamond

diámetro *m* diameter

diana *f* reveille; (*blanco*) bull's-eye

diapasón *m* (*para afinar*) tuning fork

diapositiva *f* slide, transparency

diari|amente *adv* every day. **~o** *a* daily. ● *m* newspaper; (*libro*) diary. **a ~o** daily. **~o hablado** (*en la radio*) news bulletin. **de ~o** everyday, ordinary

diarrea *f* diarrhoea

diatriba *f* diatribe

dibuj|ar *vt* draw. **~o** *m* drawing. **~os animados** cartoon (film)

diccionario *m* dictionary

diciembre *m* December

dictado *m* dictation

dictad|or *m* dictator. **~ura** *f* dictatorship

dictamen *m* opinion; (*informe*) report

dictar *vt* dictate; pronounce ‹*sentencia etc*›

dich|a *f* happiness. **~o** *a* said; (*susodicho*) aforementioned. ● *m* saying. **~oso** *a* happy; (*afortunado*) fortunate. **~o y hecho** no sooner said than done. **mejor ~o** rather. **por ~a** fortunately

didáctico *a* didactic

dieci|nueve *a & m* nineteen. **~ocho** *a & m* eighteen. **~séis** *a & m* sixteen. **~siete** *a & m* seventeen
diente *m* tooth; (*de tenedor*) prong; (*de ajo*) clove. **~ de león** dandelion. **hablar entre ~s** mumble
diesel /'disel/ *a* diesel
diestr|a *f* right hand. **~o** *a* (*derecho*) right; (*hábil*) skillful
dieta *f* diet
diez *a & m* ten
diezmar *vt* decimate
difama|ción *f* (*con palabras*) slander; (*por escrito*) libel. **~r** *vt* (*hablando*) slander; (*por escrito*) libel
diferen|cia *f* difference; (*desacuerdo*) disagreement. **~ciar** *vt* differentiate between. ● *vi* differ. **~ciarse** *vpr* differ. **~te** *a* different
difer|ido *a* (*TV etc*) recorded. **~ir** [4] *vt* postpone, defer. ● *vi* differ
dif|ícil *a* difficult. **~icultad** *f* difficulty; (*problema*) problem. **~icultar** *vt* make difficult
difteria *f* diphtheria
difundir *vt* spread; (*TV etc*) broadcast. **~se** *vpr* spread
difunto *a* late, deceased. ● *m* deceased
difusión *f* spreading
dige|rir [4] *vt* digest. **~stión** *f* digestion. **~stivo** *a* digestive
digital *a* digital; (*de los dedos*) finger
dignarse *vpr* deign. **dígnese Vd** be so kind as
dign|atario *m* dignitary. **~idad** *f* dignity; (*empleo*) office. **~o** *a* worthy; (*apropiado*) appropriate
digo *vb véase* **decir**
digresión *f* digression
dije *vb véase* **decir**
dila|ción *f* delay. **~tación** *f* dilation, expansion. **~tado** *a* extensive; ‹*tiempo*› long. **~tar** *vt* expand; (*med*) dilate; (*prolongar*) prolong. **~tarse** *vpr* expand; (*med*) dilate; (*extenderse*) extend. **sin ~ción** immediately
dilema *m* dilemma
diligen|cia *f* diligence; (*gestión*) job; (*historia*) stagecoach. **~te** *a* diligent
dilucidar *vt* explain; solve ‹*misterio*›
diluir [17] *vt* dilute
diluvio *m* flood
dimensión *f* dimension; (*tamaño*) size
diminut|ivo *a & m* diminutive. **~o** *a* minute
dimi|sión *f* resignation. **~tir** *vt/i* resign
Dinamarca *f* Denmark
dinamarqués *a* Danish. ● *m* Dane
din|ámica *f* dynamics. **~ámico** *a* dynamic. **~amismo** *m* dynamism
dinamita *f* dynamite
dínamo *m*, **dinamo** *m* dynamo
dinastía *f* dynasty
dineral *m* fortune
dinero *m* money. **~ efectivo** cash. **~ suelto** change
dinosaurio *m* dinosaur
diócesis *f* diocese
dios *m* god. **~a** *f* goddess. **¡D~ mío!** good heavens! **¡gracias a D~!** thank God! **¡válgame D~!** bless my soul!
diploma *m* diploma
diplomacia *f* diplomacy
diplomado *a* qualified
diplomático *a* diplomatic. ● *m* diplomat
diptongo *m* diphthong
diputa|ción *f* delegation. **~ción provincial** county council. **~do** *m* delegate; (*pol, en España*) member of the Cortes; (*pol, en Inglaterra*) Member of Parliament; (*pol, en Estados Unidos*) congressman
dique *m* dike
direc|ción *f* direction; (*señas*) address; (*los que dirigen*) management; (*pol*) leadership. **~ción prohibida** no entry. **~ción única** one-way. **~ta** *f* (*auto*) top gear. **~tiva** *f* directive, guideline. **~tivo** *m* executive. **~to** *a* direct; ‹*línea*› straight; ‹*tren*› through. **~tor** *m* director; (*mus*) conductor; (*de escuela etc*) headmaster; (*de periódico*) editor; (*gerente*) manager. **~tora** *f* (*de escuela etc*) headmistress. **en ~to** (*TV etc*) live. **llevar la ~ción de** direct
dirig|ente *a* ruling. ● *m & f* leader; (*de empresa*) manager. **~ible** *a & m* dirigible. **~ir** [14] *vt* direct; (*mus*) conduct; run ‹*empresa etc*›; address ‹*carta etc*›. **~irse** *vpr* make one's way; (*hablar*) address
discernir [1] *vt* distinguish
disciplina *f* discipline. **~r** *vt* discipline. **~rio** *a* disciplinary
discípulo *m* disciple; (*alumno*) pupil
disco *m* disc; (*mus*) record; (*deportes*) discus; (*de teléfono*) dial; (*auto*) lights; (*rail*) signal
disconforme *a* not in agreement
discontinuo *a* discontinuous

discord|ante *a* discordant. **~e** *a* discordant. **~ia** *f* discord
discoteca *f* discothèque, disco (*fam*); (*colección de discos*) record library
discreción *f* discretion
discrepa|ncia *f* discrepancy; (*desacuerdo*) disagreement. **~r** *vi* differ
discreto *a* discreet; (*moderado*) moderate; ‹*color*› subdued
discrimina|ción *f* discrimination. **~r** *vt* (*distinguir*) discriminate between; (*tratar injustamente*) discriminate against
disculpa *f* apology; (*excusa*) excuse. **~r** *vt* excuse, forgive. **~rse** *vpr* apologize. **dar ~s** make excuses. **pedir ~s** apologize
discurrir *vt* think up. ● *vi* think (**en** about); ‹*tiempo*› pass
discurs|ante *m* speaker. **~ar** *vi* speak (**sobre** about). **~o** *m* speech
discusión *f* discussion; (*riña*) argument. **eso no admite ~** there can be no argument about that
discuti|ble *a* debatable. **~r** *vt* discuss; (*argumentar*) argue about; (*contradecir*) contradict. ● *vi* discuss; (*argumentar*) argue
disec|ar [7] *vt* dissect; stuff ‹*animal muerto*›. **~ción** *f* dissection
disemina|ción *f* dissemination. **~r** *vt* disseminate, spread
disentería *f* dysentery
disenti|miento *m* dissent, disagreement. **~r** [4] *vi* disagree (**de** with) (**en** on)
diseñ|ador *m* designer. **~ar** *vt* design. **~o** *m* design; (*fig*) sketch
disertación *f* dissertation
disfraz *m* disguise; (*vestido*) fancy dress. **~ar** [10] *vt* disguise. **~arse** *vpr*. **~arse de** disguise o.s. as
disfrutar *vt* enjoy. ● *vi* enjoy o.s. **~ de** enjoy
disgregar [12] *vt* disintegrate
disgust|ar *vt* displease; (*molestar*) annoy. **~arse** *vpr* get annoyed, get upset; ‹*dos personas*› fall out. **~o** *m* annoyance; (*problema*) trouble; (*repugnancia*) disgust; (*riña*) quarrel; (*dolor*) sorrow, grief
disiden|cia *f* disagreement, dissent. **~te** *a* & *m* & *f* dissident
disímil *a* (*LAm*) dissimilar
disimular *vt* conceal. ● *vi* pretend
disipa|ción *f* dissipation; (*de dinero*) squandering. **~r** *vt* dissipate; (*derrochar*) squander
diskette *m* floppy disk
dislocarse [7] *vpr* dislocate
disminu|ción *f* decrease. **~ir** [17] *vi* diminish
disociar *vt* dissociate
disolver [2, *pp* **disuelto**] *vt* dissolve. **~se** *vpr* dissolve
disonante *a* dissonant
dispar *a* different
disparar *vt* fire. ● *vi* shoot (**contra** at)
disparat|ado *a* absurd. **~ar** *vi* talk nonsense. **~e** *m* silly thing; (*error*) mistake. **decir ~es** talk nonsense. **¡qué ~e!** how ridiculous! **un ~e** (*mucho*, *fam*) a lot, an awful lot (*fam*)
disparidad *f* disparity
disparo *m* (*acción*) firing; (*tiro*) shot
dispensar *vt* distribute; (*disculpar*) excuse. **¡Vd dispense!** forgive me
dispers|ar *vt* scatter, disperse. **~arse** *vpr* scatter, disperse. **~ión** *f* dispersion. **~o** *a* scattered
dispon|er [34] *vt* arrange; (*preparar*) prepare. ● *vi*. **~er de** have; (*vender etc*) dispose of. **~erse** *vpr* get ready. **~ibilidad** *f* availability. **~ible** *a* available
disposición *f* arrangement; (*aptitud*) talent; (*disponibilidad*) disposal; (*jurid*) order, decree. **~ de ánimo** frame of mind. **a la ~ de** at the disposal of. **a su ~** at your service
dispositivo *m* device
dispuesto *a* ready; (*hábil*) clever; (*inclinado*) disposed; (*servicial*) helpful
disputa *f* dispute. **~r** *vt* dispute. ● *vi*. **~r por** argue about; (*competir para*) compete for. **sin ~** undoubtedly
distan|cia *f* distance. **~ciar** *vt* space out; (*en deportes*) outdistance. **~ciarse** *vpr* ‹*dos personas*› fall out. **~te** *a* distant. **a ~cia** from a distance. **guardar las ~cias** keep one's distance
distar *vi* be away; (*fig*) be far. **dista 5 kilómetros** it's 5 kilometres away
distin|ción *f* distinction. **~guido** *a* distinguished; (*en cartas*) Honoured. **~guir** [13] *vt/i* distinguish. **~guirse** *vpr* distinguish o.s.; (*diferenciarse*) differ; (*verse*) be visible. **~tivo** *a* distinctive. ● *m* badge. **~to** *a* different; (*claro*) distinct
distorsión *f* distortion; (*med*) sprain
distra|cción *f* amusement; (*descuido*) absent-mindedness, inattention. **~er** [41] *vt* distract; (*divertir*)

amuse; embezzle ‹*fondos*›. ● *vi* be entertaining. **~erse** *vpr* amuse o.s.; (*descuidarse*) not pay attention. **~ido** *a* amusing; (*desatento*) absent-minded

distribu|ción *f* distribution. **~idor** *m* distributor, agent. **~idor automático** vending machine. **~ir** [17] *vt* distribute

distrito *m* district

disturbio *m* disturbance

disuadir *vt* dissuade

diurético *a & m* diuretic

diurno *a* daytime

divagar [12] *vi* (*al hablar*) digress

diván *m* settee, sofa

diverg|encia *f* divergence. **~ente** *a* divergent. **~ir** [14] *vi* diverge

diversidad *f* diversity

diversificar [7] *vt* diversify

diversión *f* amusement, entertainment; (*pasatiempo*) pastime

diverso *a* different

diverti|do *a* amusing; (*que tiene gracia*) funny; (*agradable*) enjoyable. **~r** [4] *vt* amuse, entertain. **~rse** *vpr* enjoy o.s.

dividir *vt* divide; (*repartir*) share out

divin|idad *f* divinity. **~o** *a* divine

divisa *f* emblem. **~s** *fpl* foreign exchange

divisar *vt* make out

divis|ión *f* division. **~or** *m* divisor. **~orio** *a* dividing

divorci|ado *a* divorced. ● *m* divorcee. **~ar** *vt* divorce. **~arse** *vpr* get divorced. **~o** *m* divorce

divulgar [12] *vt* divulge; (*propagar*) spread. **~se** *vpr* become known

do *m* C; (*solfa*) doh

dobl|adillo *m* hem; (*de pantalón*) turn-up (*Brit*), cuff (*Amer*). **~ado** *a* double; (*plegado*) folded; ‹*película*› dubbed. **~ar** *vt* double; (*plegar*) fold; (*torcer*) bend; turn ‹*esquina*›; dub ‹*película*›. ● *vi* turn; ‹*campana*› toll. **~arse** *vpr* double; (*encorvarse*) bend; (*ceder*) give in. **~e** *a* double. ● *m* double; (*pliegue*) fold. **~egar** [12] *vt* (*fig*) force to give in. **~egarse** *vpr* give in. **el ~e** twice as much

doce *a & m* twelve. **~na** *f* dozen. **~no** *a* twelfth

docente *a* teaching. ● *m & f* teacher

dócil *a* obedient

doct|o *a* learned. **~or** *m* doctor. **~orado** *m* doctorate. **~rina** *f* doctrine

document|ación *f* documentation, papers. **~al** *a & m* documentary. **~ar** *vt* document. **~arse** *vpr* gather information. **~o** *m* document. **D~o Nacional de Identidad** national identity card

dogm|a *m* dogma. **~ático** *a* dogmatic

dólar *m* dollar

dol|er [2] *vi* hurt, ache; (*fig*) grieve. **me duele la cabeza** my head hurts. **le duele el estómago** he has a pain in his stomach. **~erse** *vpr* regret; (*quejarse*) complain. **~or** *m* pain; (*sordo*) ache; (*fig*) sorrow. **~oroso** *a* painful. **~or de cabeza** headache. **~or de muelas** toothache

domar *vt* tame; break in ‹*caballo*›

dom|esticar [7] *vt* domesticate. **~éstico** *a* domestic. ● *m* servant

domicilio *m* home. **a ~** at home. **servicio a ~** home delivery service

domina|ción *f* domination. **~nte** *a* dominant; ‹*persona*› domineering. **~r** *vt* dominate; (*contener*) control; (*conocer*) have a good knowledge of. ● *vi* dominate; (*destacarse*) stand out. **~rse** *vpr* control o.s.

domin|go *m* Sunday. **~guero** *a* Sunday. **~ical** *a* Sunday

dominio *m* authority; (*territorio*) domain; (*fig*) good knowledge

dominó *m* (*juego*) dominoes

don *m* talent, gift; (*en un sobre*) Mr. **~ Pedro** Pedro. **tener ~ de lenguas** have a gift for languages. **tener ~ de gentes** have a way with people

donación *f* donation

donaire *m* grace, charm

dona|nte *m* (*de sangre*) donor. **~r** *vt* donate

doncella *f* (*criada*) maid

donde *adv* where

dónde *adv* where? **¿hasta ~?** how far? **¿por ~?** whereabouts?; (*¿por qué camino?*) which way? **¿a ~ vas?** where are you going? **¿de ~ eres?** where are you from?

dondequiera *adv* anywhere; (*en todas partes*) everywhere. **~ que** wherever. **por ~** everywhere

doña *f* (*en un sobre*) Mrs. **~ María** María

dora|do *a* golden; (*cubierto de oro*) gilt. **~dura** *f* gilding. **~r** *vt* gilt; (*culin*) brown

dormi|lón *m* sleepyhead. ● *a* lazy. **~r** [6] *vt* send to sleep. ● *vi* sleep. **~rse** *vpr* go to sleep. **~tar** *vi* doze. **~torio** *m* bedroom. **~r la siesta**

have an afternoon nap, have a siesta. **echarse a dormir** go to bed
dors|al *a* back. ● *m* (*en deportes*) number. **~o** *m* back
dos *a* & *m* two. **~cientos** *a* & *m* two hundred. **cada ~ por tres** every five minutes. **de ~ en ~** in twos, in pairs. **en un ~ por tres** in no time. **los dos, las dos** both (of them)
dosi|ficar [7] *vt* dose; (*fig*) measure out. **~s** *f* dose
dot|ado *a* gifted. **~ar** *vt* give a dowry; (*proveer*) endow (**de** with). **~e** *m* dowry
doy *vb véase* **dar**
dragar [12] *vt* dredge
drago *m* dragon tree
dragón *m* dragon
dram|a *m* drama; (*obra de teatro*) play. **~ático** *a* dramatic. **~atizar** [10] *vt* dramatize. **~aturgo** *m* playwright
drástico *a* drastic
droga *f* drug. **~dicto** *m* drug addict. **~do** *a* drugged. ● *m* drug addict. **~r** [12] *vt* drug. **~rse** *vpr* take drugs. **~ta** *m* & *f* (*fam*) drug addict
droguería *f* hardware shop (*Brit*), hardware store (*Amer*)
dromedario *m* dromedary
ducha *f* shower. **~rse** *vpr* have a shower
dud|a *f* doubt. **~ar** *vt/i* doubt. **~oso** *a* doubtful; (*sospechoso*) dubious. **poner en ~a** question. **sin ~a (alguna)** without a doubt
duelo *m* duel; (*luto*) mourning
duende *m* imp
dueñ|a *f* owner, proprietress; (*de una pensión*) landlady. **~o** *m* owner, proprietor; (*de una pensión*) landlord
duermo *vb véase* **dormir**
dul|ce *a* sweet; ‹*agua*› fresh; (*suave*) soft, gentle. ● *m* sweet. **~zura** *f* sweetness; (*fig*) gentleness
duna *f* dune
dúo *m* duet, duo
duodécimo *a* & *m* twelfth
duplica|do *a* in duplicate. ● *m* duplicate. **~r** [7] *vt* duplicate. **~rse** *vpr* double
duque *m* duke. **~sa** *f* duchess
dura|ción *f* duration, length. **~dero** *a* lasting
durante *prep* during, in; (*medida de tiempo*) for. **~ todo el año** all year round
durar *vi* last
durazno *m* (*LAm, fruta*) peach
dureza *f* hardness, toughness; (*med*) hard patch
durmiente *a* sleeping
duro *a* hard; (*culin*) tough; (*fig*) harsh. ● *adv* hard. ● *m* five-peseta coin. **ser ~ de oído** be hard of hearing

E

e *conj* and
ebanista *m* & *f* cabinet-maker
ébano *m* ebony
ebri|edad *f* drunkenness. **~o** *a* drunk
ebullición *f* boiling
eccema *m* eczema
eclesiástico *a* ecclesiastical. ● *m* clergyman
eclipse *m* eclipse
eco *m* echo. **hacer(se) ~** echo
ecolog|ía *f* ecology. **~ista** *m* & *f* ecologist
economato *m* cooperative store
econ|omía *f* economy; (*ciencia*) economics. **~ómicamente** *adv* economically. **~ómico** *a* economic(al); (*no caro*) inexpensive. **~omista** *m* & *f* economist. **~omizar** [10] *vt/i* economize
ecuación *f* equation
ecuador *m* equator. **el E~** Ecuador
ecuánime *a* level-headed; (*imparcial*) impartial
ecuanimidad *f* equanimity
ecuatoriano *a* & *m* Ecuadorian
ecuestre *a* equestrian
echar *vt* throw; post ‹*carta*›; give off ‹*olor*›; pour ‹*líquido*›; sprout ‹*hojas etc*›; (*despedir*) throw out; dismiss ‹*empleado*›; (*poner*) put on; put out ‹*raices*›; show ‹*película*›. **~se** *vpr* throw o.s.; (*tumbarse*) lie down. **~ a** start. **~ a perder** spoil. **~ de menos** miss. **~se atrás** (*fig*) back down. **echárselas de** feign
edad *f* age. **~ avanzada** old age. **E~ de Piedra** Stone Age. **E~ Media** Middle Ages. **¿qué ~ tiene?** how old is he?
edición *f* edition; (*publicación*) publication
edicto *m* edict
edific|ación *f* building. **~ante** *a* edifying. **~ar** [7] *vt* build; (*fig*) edify. **~io** *m* building; (*fig*) structure

Edimburgo *m* Edinburgh
edit|ar *vt* publish. **~or** *a* publishing. ● *m* publisher. **~orial** *a* editorial. ● *m* leading article. ● *f* publishing house
edredón *m* eiderdown
educa|ción *f* upbringing; (*modales*) (good) manners; (*enseñanza*) education. **~do** *a* polite. **~dor** *m* teacher. **~r** [7] *vt* bring up; (*enseñar*) educate. **~tivo** *a* educational. **bien ~do** polite. **falta de ~ción** rudeness, bad manners. **mal ~do** rude
edulcorante *m* sweetener
EE.UU. *abrev* (*Estados Unidos*) USA, United States (of America)
efect|ivamente *adv* really; (*por supuesto*) indeed. **~ivo** *a* effective; (*auténtico*) real; ‹*empleo*› permanent. ● *m* cash. **~o** *m* effect; (*impresión*) impression. **~os** *mpl* belongings; (*com*) goods. **~uar** [21] *vt* carry out, effect; make ‹*viaje, compras etc*›. **en ~o** in fact; (*por supuesto*) indeed
efervescente *a* effervescent; ‹*bebidas*› fizzy
efica|cia *f* effectiveness; (*de persona*) efficiency. **~z** *a* effective; ‹*persona*› efficient
eficien|cia *f* efficiency. **~te** *a* efficient
efigie *f* effigy
efímero *a* ephemeral
efluvio *m* outflow
efusi|ón *n* effusion. **~vo** *a* effusive; ‹*gracias*› warm
Egeo *m*. **mar ~** Aegean Sea
égida *f* aegis
egipcio *a* & *m* Egyptian
Egipto *m* Egypt
ego|céntrico *a* egocentric. ● *m* egocentric person. **~ísmo** *m* selfishness. **~ísta** *a* selfish. ● *m* selfish person
egregio *a* eminent
egresar *vi* (*LAm*) leave; (*univ*) graduate
eje *m* axis; (*tec*) axle
ejecu|ción *f* execution; (*mus etc*) performance. **~tante** *m* & *f* executor; (*mus etc*) performer. **~tar** *vt* carry out; (*mus etc*) perform; (*matar*) execute
ejecutivo *m* director, manager
ejempl|ar *a* exemplary. ● *m* (*ejemplo*) example, specimen; (*libro*) copy; (*revista*) issue, number. **~ificar** [7] *vt* exemplify. **~o** *m* example. **dar ~o** set an example. **por ~o** for example. **sin ~** unprecedented
ejerc|er [9] *vt* exercise; practise ‹*profesión*›; exert ‹*influencia*›. ● *vi* practise. **~icio** *m* exercise; (*de una profesión*) practice. **~itar** *vt* exercise. **~itarse** *vpr* exercise. **hacer ~icios** take exercise
ejército *m* army
el *art def m* (*pl* **los**) the. ● *pron* (*pl* **los**) the one. **~ de Antonio** Antonio's. **~ que** whoever, the one
él *pron* (*persona*) he; (*persona con prep*) him; (*cosa*) it. **el libro de ~** his book
elabora|ción *f* processing; (*fabricación*) manufacture. **~r** *vt* process; manufacture ‹*producto*›; (*producir*) produce
el|asticidad *f* elasticity. **~ástico** *a* & *m* elastic
elec|ción *f* choice; (*de político etc*) election. **~ciones** *fpl* (*pol*) election. **~tor** *m* voter. **~torado** *m* electorate. **~toral** *a* electoral
electrici|dad *f* electricity. **~sta** *m* & *f* electrician
eléctrico *a* electric; (*de la electricidad*) electrical
electrificar [7] *vt*, **electrizar** [10] *vt* electrify
electrocutar *vt* electrocute
electrodo *m* electrode
electrodoméstico *a* electrical household. **~s** *mpl* electrical household appliances
electrólisis *f* electrolysis
electrón *m* electron
electrónic|a *f* electronics. **~o** *a* electronic
elefante *m* elephant
elegan|cia *f* elegance. **~te** *a* elegant
elegía *f* elegy
elegi|ble *a* eligible. **~do** *a* chosen. **~r** [5 & 14] *vt* choose; (*por votación*) elect
element|al *a* elementary. **~o** *m* element; (*persona*) person, bloke (*fam*). **~os** *mpl* (*nociones*) basic principles
elenco *m* (*en el teatro*) cast
eleva|ción *f* elevation; (*de precios*) rise, increase; (*acción*) raising. **~dor** *m* (*LAm*) lift. **~r** *vt* raise; (*promover*) promote
elimina|ción *f* elimination. **~r** *vt* eliminate. **~toria** *f* preliminary heat
el|ipse *f* ellipse. **~íptico** *a* elliptical

élite /e'lit, e'lite/ *f* elite
elixir *m* elixir
elocución *f* elocution
elocuen|cia *f* eloquence. **~te** *a* eloquent
elogi|ar *vt* praise. **~o** *m* praise
elote *m* (*Mex*) corn on the cob
eludir *vt* avoid, elude
ella *pron* (*persona*) she; (*persona con prep*) her; (*cosa*) it. **~s** *pron pl* they; (*con prep*) them. **el libro de ~** her book. **el libro de ~s** their book
ello *pron* it
ellos *pron pl* they; (*con prep*) them. **el libro de ~** their book
emaciado *a* emaciated
emana|ción *f* emanation. **~r** *vi* emanate (**de** from); (*originarse*) originate (**de** from, in)
emancipa|ción *f* emancipation. **~do** *a* emancipated. **~r** *vt* emancipate. **~rse** *vpr* become emancipated
embadurnar *vt* smear
embajad|a *f* embassy. **~or** *m* ambassador
embalar *vt* pack
embaldosar *vt* tile
embalsamar *vt* embalm
embalse *m* dam; (*pantano*) reservoir
embaraz|ada *a* pregnant. ● *f* pregnant woman. **~ar** [10] *vt* hinder. **~o** *m* hindrance; (*de mujer*) pregnancy. **~oso** *a* awkward, embarrassing
embar|cación *f* boat. **~cadero** *m* jetty, pier. **~car** [7] *vt* embark ‹*personas*›; ship ‹*mercancías*›. **~carse** *vpr* embark. **~carse en** (*fig*) embark upon
embargo *m* embargo; (*jurid*) seizure. **sin ~** however
embarque *m* loading
embarullar *vt* muddle
embaucar [7] *vt* deceive
embeber *vt* absorb; (*empapar*) soak. ● *vi* shrink. **~se** *vpr* be absorbed
embelesar *vt* delight. **~se** *vpr* be delighted
embellecer [11] *vt* embellish
embesti|da *f* attack. **~r** [5] *vt/i* attack
emblema *m* emblem
embobar *vt* amaze
embobecer [11] *vt* make silly. **~se** *vpr* get silly
embocadura *f* (*de un río*) mouth
emboquillado *a* tipped
embolsar *vt* pocket
emborrachar *vt* get drunk. **~se** *vpr* get drunk
emborrascarse [7] *vpr* get stormy
emborronar *vt* blot
embosca|da *f* ambush. **~rse** [7] *vpr* lie in wait
embotar *vt* blunt; (*fig*) dull
embotella|miento *m* (*de vehículos*) traffic jam. **~r** *vt* bottle
embrague *m* clutch
embriag|ar [12] *vt* get drunk; (*fig*) intoxicate; (*fig, enajenar*) enrapture. **~arse** *vpr* get drunk. **~uez** *f* drunkenness; (*fig*) intoxication
embrión *m* embryo
embroll|ar *vt* mix up; involve ‹*personas*›. **~arse** *vpr* get into a muddle; *en un asunto*) get involved. **~o** *m* tangle; (*fig*) muddle. **~ón** *m* troublemaker
embromar *vt* make fun of; (*engañar*) fool
embruja|do *a* bewitched; ‹*casa etc*› haunted. **~r** *vt* bewitch
embrutecer [11] *vt* brutalize
embuchar *vt* wolf ‹*comida*›
embudo *m* funnel
embuste *m* lie. **~ro** *a* deceitful. ● *m* liar
embuti|do *m* (*culin*) sausage. **~r** *vt* stuff
emergencia *f* emergency; (*acción de emerger*) emergence. **en caso de ~** in case of emergency
emerger [14] *vi* appear, emerge; ‹*submarino*› surface
emigra|ción *f* emigration. **~nte** *m & f* emigrant. **~r** *vi* emigrate
eminen|cia *f* eminence. **~te** *a* eminent
emisario *m* emissary
emis|ión *f* emission; (*de dinero*) issue; (*TV etc*) broadcast. **~or** *a* issuing; (*TV etc*) broadcasting. **~ora** *f* radio station
emitir *vt* emit; let out ‹*grito*›; (*TV etc*) broadcast; (*expresar*) express; (*poner en circulación*) issue
emoci|ón *f* emotion; (*excitación*) excitement. **~onado** *a* moved. **~onante** *a* exciting; (*conmovedor*) moving. **~onar** *vt* excite; (*conmover*) move. **~onarse** *vpr* get excited; (*conmoverse*) be moved. **¡qué ~ón!** how exciting!
emotivo *a* emotional; (*conmovedor*) moving
empacar [7] *vt* (*LAm*) pack

empacho *m* indigestion; (*vergüenza*) embarrassment

empadronar *vt* register. **~se** *vpr* register

empalagoso *a* sickly; (*demasiado amable*) ingratiating; (*demasiado sentimental*) mawkish

empalizada *f* fence

empalm|ar *vt* connect, join. ● *vi* meet. **~e** *m* junction; (*de trenes*) connection

empanad|a *f* (savoury) pie. **~illa** *f* (small) pie. **~o** *a* fried in breadcrumbs

empanizado *a* (*Mex*) fried in breadcrumbs

empantanar *vt* flood. **~se** *vpr* become flooded; (*fig*) get bogged down

empañar *vt* mist; dull ‹*metales etc*›; (*fig*) tarnish. **~se** *vpr* ‹*cristales*› steam up

empapar *vt* soak; (*absorber*) soak up. **~se** *vpr* be soaked

empapela|do *m* wallpaper. **~r** *vt* paper; (*envolver*) wrap (in paper)

empaquetar *vt* package; pack together ‹*personas*›

emparedado *m* sandwich

emparejar *vt* match; (*nivelar*) make level. **~se** *vpr* pair off

empast|ar *vt* fill ‹*muela*›. **~e** *m* filling

empat|ar *vi* draw. **~e** *m* draw

empedernido *a* inveterate; (*insensible*) hard

empedrar [1] *vt* pave

empeine *m* instep

empeñ|ado *a* in debt; (*decidido*) determined; (*acalorado*) heated. **~ar** *vt* pawn; pledge ‹*palabras*›; (*principiar*) start. **~arse** *vpr* (*endeudarse*) get into debt; (*meterse*) get involved; (*estar decidido a*) insist (**en** on). **~o** *m* pledge; (*resolución*) determination. **casa de ~s** pawnshop

empeorar *vt* make worse. ● *vi* get worse. **~se** *vpr* get worse

empequeñecer [11] *vt* dwarf; (*fig*) belittle

empera|dor *m* emperor. **~triz** *f* empress

empezar [1 & 10] *vt/i* start, begin. **para ~** to begin with

empina|do *a* upright; ‹*cuesta*› steep. **~r** *vt* raise. **~rse** *vpr* ‹*persona*› stand on tiptoe; ‹*animal*› rear

empírico *a* empirical

emplasto *m* plaster

emplaza|miento *m* (*jurid*) summons; (*lugar*) site. **~r** [10] *vt* summon; (*situar*) site

emple|ado *m* employee. **~ar** *vt* use; employ ‹*persona*›; spend ‹*tiempo*›. **~arse** *vpr* be used; ‹*persona*› be employed. **~o** *m* use; (*trabajo*) employment; (*puesto*) job

empobrecer [11] *vt* impoverish. **~se** *vpr* become poor

empolvar *vt* powder

empoll|ar *vt* incubate ‹*huevos*›; (*estudiar, fam*) swot up (*Brit*), grind away at (*Amer*). ● *vi* ‹*ave*› sit; ‹*estudiante*› swot (*Brit*), grind away (*Amer*). **~ón** *m* swot

emponzoñar *vt* poison

emporio *m* emporium; (*LAm, almacén*) department store

empotra|do *a* built-in, fitted. **~r** *vt* fit

emprendedor *a* enterprising

emprender *vt* undertake; set out on ‹*viaje etc*›. **~la con uno** pick a fight with s.o.

empresa *f* undertaking; (*com*) company, firm. **~rio** *m* impresario; (*com*) contractor

empréstito *m* loan

empuj|ar *vt* push; press ‹*botón*›. **~e** *m* push, shove; (*fig*) drive. **~ón** *m* push, shove

empuñar *vt* grasp; take up ‹*pluma, espada*›

emular *vt* emulate

emulsión *f* emulsion

en *prep* in; (*sobre*) on; (*dentro*) inside, in; (*con dirección*) into; (*medio de transporte*) by. **~ casa** at home. **~ coche** by car. **~ 10 días** in 10 days. **de pueblo ~ pueblo** from town to town

enagua *f* petticoat

enajena|ción *f* alienation; (*éxtasis*) rapture. **~r** *vt* alienate; (*volver loco*) drive mad; (*fig, extasiar*) enrapture. **~ción mental** insanity

enamora|do *a* in love. ● *m* lover. **~r** *vt* win the love of. **~rse** *vpr* fall in love (**de** with)

enan|ito *m* dwarf. **~o** *a & m* dwarf

enardecer [11] *vt* inflame. **~se** *vpr* get excited (**por** about)

encabeza|miento *m* heading; (*de periódico*) headline. **~r** [10] *vt* introduce ‹*escrito*›; (*poner título a*) entitle; head ‹*una lista*›; lead ‹*revolución etc*›; (*empadronar*) register

encadenar *vt* chain; (*fig*) tie down
encaj|ar *vt* fit; fit together ‹*varias piezas*›. ● *vi* fit; (*estar de acuerdo*) tally. **~arse** *vpr* squeeze into. **~e** *m* lace; (*acción de encajar*) fitting
encajonar *vt* box; (*en sitio estrecho*) squeeze in
encalar *vt* whitewash
encallar *vt* run aground; (*fig*) get bogged down
encaminar *vt* direct. **~se** *vpr* make one's way
encandilar *vt* (*pasmar*) bewilder; (*estimular*) stimulate
encanecer [11] *vi* go grey
encant|ado *a* enchanted; (*hechizado*) bewitched; ‹*casa etc*› haunted. **~ador** *a* charming. ● *m* magician. **~amiento** *m* magic. **~ar** *vt* bewitch; (*fig*) charm, delight. **~o** *m* magic; (*fig*) delight. **¡~ado!** pleased to meet you! **me ~a la leche** I love milk
encapotado *a* ‹*cielo*› overcast
encapricharse *vpr*. **~ con** take a fancy to
encarar *vt* face. **~se** *vpr*. **~se con** face
encarcelar *vt* imprison
encarecer [11] *vt* put up the price of; (*alabar*) praise. ● *vi* go up
encarg|ado *a* in charge. ● *m* manager, attendant, person in charge. **~ar** [12] *vt* entrust; (*pedir*) order. **~arse** *vpr* take charge (**de** of). **~o** *m* job; (*com*) order; (*recado*) errand. **hecho de ~o** made to measure
encariñarse *vpr*. **~ con** take to, become fond of
encarna|ción *f* incarnation. **~do** *a* incarnate; (*rojo*) red. ● *m* red
encarnizado *a* bitter
encarpetar *vt* file; (*LAm, dar carpetazo*) shelve
encarrilar *vt* put back on the rails; (*fig*) direct, put on the right road
encasillar *vt* pigeonhole
encastillarse *vpr*. **~ en** (*fig*) stick to
encauzar [10] *vt* channel
encend|edor *m* lighter. **~er** [1] *vt* light; (*pegar fuego a*) set fire to; switch on, turn on ‹*aparato eléctrico*›; (*fig*) arouse. **~erse** *vpr* light; (*prender fuego*) catch fire; (*excitarse*) get excited; (*ruborizarse*) blush. **~ido** *a* lit; ‹*aparato eléctrico*› on; (*rojo*) bright red. ● *m* (*auto*) ignition

encera|do *a* waxed. ● *m* (*pizarra*) blackboard. **~r** *vt* wax
encerr|ar [1] *vt* shut in; (*con llave*) lock up; (*fig, contener*) contain. **~ona** *f* trap
encía *f* gum
encíclica *f* encyclical
enciclop|edia *f* encyclopaedia. **~édico** *a* enclyclopaedic
encierro *m* confinement; (*cárcel*) prison
encima *adv* on top; (*arriba*) above. **~ de** on, on top of; (*sobre*) over; (*además de*) besides, as well as. **por ~** on top; (*a la ligera*) superficially. **por ~ de todo** above all
encina *f* holm oak
encinta *a* pregnant
enclave *m* enclave
enclenque *a* weak; (*enfermizo*) sickly
encog|er [14] *vt* shrink; (*contraer*) contract. **~erse** *vpr* shrink. **~erse de hombros** shrug one's shoulders. **~ido** *a* shrunk; (*fig, tímido*) timid
encolar *vt* glue; (*pegar*) stick
encolerizar [10] *vt* make angry. **~se** *vpr* get angry, lose one's temper
encomendar [1] *vt* entrust
encomi|ar *vt* praise. **~o** *m* praise
encono *m* bitterness, ill will
encontra|do *a* contrary, conflicting. **~r** [2] *vt* find; (*tropezar con*) meet. **~rse** *vpr* meet; (*hallarse*) be. **no ~rse** feel uncomfortable
encorvar *vt* bend, curve. **~se** *vpr* stoop
encrespado *a* ‹*pelo*› curly; ‹*mar*› rough
encrucijada *f* crossroads
encuaderna|ción *f* binding. **~dor** *m* bookbinder. **~r** *vt* bind
encuadrar *vt* frame
encub|ierto *a* hidden. **~rir** [*pp* **encubierto**] *vt* hide, conceal; shelter ‹*delincuente*›
encuentro *m* meeting; (*colisión*) crash; (*en deportes*) match; (*mil*) skirmish
encuesta *f* survey; (*investigación*) inquiry
encumbra|do *a* eminent. **~r** *vt* (*fig, elevar*) exalt. **~rse** *vpr* rise
encurtidos *mpl* pickles
encharcar [7] *vt* flood. **~se** *vpr* be flooded
enchuf|ado *a* switched on. **~ar** *vt* plug in; fit together ‹*tubos etc*›. **~e** *m* socket; (*clavija*) plug; (*de tubos*

etc) joint; (*fig*, *empleo*, *fam*) cushy job; (*influencia*, *fam*) influence. **tener ~e** have friends in the right places
endeble *a* weak
endemoniado *a* possessed; (*malo*) wicked
enderezar [10] *vt* straighten out; (*poner vertical*) put upright (again); (*fig*, *arreglar*) put right, sort out; (*dirigir*) direct. **~se** *vpr* straighten out
endeudarse *vpr* get into debt
endiablado *a* possessed; (*malo*) wicked
endomingarse [12] *vpr* dress up
endosar *vt* endorse ‹*cheque etc*›; (*fig*, *fam*) lumber
endrogarse [12] *vpr* (*Mex*) get into debt
endulzar [10] *vt* sweeten; (*fig*) soften
endurecer [11] *vt* harden. **~se** *vpr* harden; (*fig*) become hardened
enema *m* enema
enemi|go *a* hostile. ● *m* enemy. **~stad** *f* enmity. **~star** *vt* make an enemy of. **~starse** *vpr* fall out (**con** with)
en|ergía *f* energy. **~érgico** *a* ‹*persona*› lively; ‹*decisión*› forceful
energúmeno *m* madman
enero *m* January
enervar *vt* enervate
enésimo *a* nth, umpteenth (*fam*)
enfad|adizo *a* irritable. **~ado** *a* cross, angry. **~ar** *vt* make cross, anger; (*molestar*) annoy. **~arse** *vpr* get cross. **~o** *m* anger; (*molestia*) annoyance
énfasis *m invar* emphasis, stress. **poner ~** stress, emphasize
enfático *a* emphatic
enferm|ar *vi* fall ill. **~edad** *f* illness. **~era** *f* nurse. **~ería** *f* sick bay. **~ero** *m* (male) nurse. **~izo** *a* sickly. **~o** *a* ill. ● *m* patient
enflaquecer [11] *vt* make thin. ● *vi* lose weight
enfo|car [7] *vt* shine on; focus ‹*lente etc*›; (*fig*) consider. **~que** *m* focus; (*fig*) point of view
enfrascarse [7] *vpr* (*fig*) be absorbed
enfrentar *vt* face, confront; (*poner frente a frente*) bring face to face. **~se** *vpr*. **~se con** confront; (*en deportes*) meet
enfrente *adv* opposite. **~ de** opposite. **de ~** opposite
enfria|miento *m* cooling; (*catarro*) cold. **~r** [20] *vt* cool (down); (*fig*) cool down. **~rse** *vpr* go cold; (*fig*) cool off
enfurecer [11] *vt* infuriate. **~se** *vpr* lose one's temper; ‹*mar*› get rough
enfurruñarse *vpr* sulk
engalanar *vt* adorn. **~se** *vpr* dress up
enganchar *vt* hook; hang up ‹*ropa*›. **~se** *vpr* get caught; (*mil*) enlist
engañ|ar *vt* deceive, trick; (*ser infiel*) be unfaithful. **~arse** *vpr* be wrong, be mistaken; (*no admitir la verdad*) deceive o.s. **~o** *m* deceit, trickery; (*error*) mistake. **~oso** *a* deceptive; ‹*persona*› deceitful
engarzar [10] *vt* string ‹*cuentas*›; set ‹*joyas*›; (*fig*) link
engatusar *vt* (*fam*) coax
engendr|ar *vt* breed; (*fig*) produce. **~o** *m* (*monstruo*) monster; (*fig*) brainchild
englobar *vt* include
engomar *vt* glue
engordar *vt* fatten. ● *vi* get fatter, put on weight
engorro *m* nuisance
engranaje *m* (*auto*) gear
engrandecer [11] *vt* (*enaltecer*) exalt, raise
engrasar *vt* grease; (*con aceite*) oil; (*ensuciar*) make greasy
engreído *a* arrogant
engrosar [2] *vt* swell. ● *vi* ‹*persona*› get fatter; ‹*río*› swell
engullir [22] *vt* gulp down
enharinar *vt* sprinkle with flour
enhebrar *vt* thread
enhorabuena *f* congratulations. **dar la ~** congratulate
enigm|a *m* enigma. **~ático** *a* enigmatic
enjabonar *vt* soap; (*fig*, *fam*) butter up
enjalbegar [12] *vt* whitewash
enjambre *m* swarm
enjaular *vt* put in a cage
enjuag|ar [12] *vt* rinse (out). **~atorio** *m* mouthwash. **~ue** *m* rinsing; (*para la boca*) mouthwash
enjugar [12] *vt* dry; (*limpiar*) wipe; cancel ‹*deuda*›
enjuiciar *vt* pass judgement on
enjuto *a* ‹*persona*› skinny
enlace *m* connection; (*matrimonial*) wedding
enlatar *vt* tin, can

enlazar [10] *vt* tie together; (*fig*) relate, connect
enlodar *vt*, **enlodazar** [10] *vt* cover in mud
enloquecer [11] *vt* drive mad. ● *vi* go mad. **~se** *vpr* go mad
enlosar *vt* (*con losas*) pave; (*con baldosas*) tile
enlucir [11] *vt* plaster
enluta|do *a* in mourning. **~r** *vt* dress in mourning; (*fig*) sadden
enmarañar *vt* tangle (up), entangle; (*confundir*) confuse. **~se** *vpr* get into a tangle; (*confundirse*) get confused
enmarcar [7] *vt* frame
enmascarar *vt* mask. **~se de** masquerade as
enm|endar *vt* correct. **~endarse** *vpr* mend one's way. **~ienda** *f* correction; (*de ley etc*) amendment
enmohecerse [11] *vpr* (*con óxido*) go rusty; (*con hongos*) go mouldy
enmudecer [11] *vi* be dumbstruck; (*callar*) say nothing
ennegrecer [11] *vt* blacken
ennoblecer [11] *vt* ennoble; (*fig*) add style to
enoj|adizo *a* irritable. **~ado** *a* angry, cross. **~ar** *vt* make cross, anger; (*molestar*) annoy. **~arse** *vpr* get cross. **~o** *m* anger; (*molestia*) annoyance. **~oso** *a* annoying
enorgullecerse [11] *vpr* be proud
enorm|e *a* enormous; (*malo*) wicked. **~emente** *adv* enormously. **~idad** *f* immensity; (*atrocidad*) enormity. **me gusta una ~idad** I like it enormously
enrabiar *vt* infuriate
enraizar [10 & 20] *vi* take root
enrarecido *a* rarefied
enrasar *vt* make level
enred|adera *f* creeper. **~adero** *a* climbing. **~ar** *vt* tangle (up), entangle; (*confundir*) confuse; (*comprometer a uno*) involve, implicate; (*sembrar la discordia*) cause trouble between. ● *vi* get up to mischief. **~ar con** fiddle with, play with. **~arse** *vpr* get into a tangle; (*confundirse*) get confused; ‹*persona*› get involved. **~o** *m* tangle; (*fig*) muddle, mess
enrejado *m* bars
enrevesado *a* complicated
enriquecer [11] *vt* make rich; (*fig*) enrich. **~se** *vpr* get rich
enrojecer [11] *vt* turn red, redden. **~se** *vpr* ‹*persona*› go red, blush
enrolar *vt* enlist
enrollar *vt* roll (up); wind ‹*hilo etc*›
enroscar [7] *vt* coil; (*atornillar*) screw in
ensalad|a *f* salad. **~era** *f* salad bowl. **~illa** *f* Russian salad. **armar una ~a** make a mess
ensalzar [10] *vt* praise; (*enaltecer*) exalt
ensambladura *f*, **ensamblaje** *m* (*acción*) assembling; (*efecto*) joint
ensamblar *vt* join
ensanch|ar *vt* widen; (*agrandar*) enlarge. **~arse** *vpr* get wider. **~e** *m* widening; (*de ciudad*) new district
ensangrentar [1] *vt* stain with blood
ensañarse *vpr*. **~ con** treat cruelly
ensartar *vt* string ‹*cuentas etc*›
ensay|ar *vt* test; rehearse ‹*obra de teatro etc*›. **~arse** *vpr* rehearse. **~o** *m* test, trial; (*composición literaria*) essay
ensenada *f* inlet, cove
enseña|nza *f* education; (*acción de enseñar*) teaching. **~nza media** secondary education. **~r** *vt* teach; (*mostrar*) show
enseñorearse *vpr* take over
enseres *mpl* equipment
ensillar *vt* saddle
ensimismarse *vpr* be lost in thought
ensoberbecerse [11] *vpr* become conceited
ensombrecer [11] *vt* darken
ensordecer [11] *vt* deafen. ● *vi* go deaf
ensortijar *vt* curl ‹*pelo etc*›
ensuciar *vt* dirty. **~se** *vpr* get dirty
ensueño *m* dream
entablar *vt* (*empezar*) start
entablillar *vt* put in a splint
entalegar [12] *vt* put into a bag; (*fig*) hoard
entallar *vt* fit ‹*un vestido*›. ● *vi* fit
entarimado *m* parquet
ente *m* entity, being; (*persona rara, fam*) odd person; (*com*) firm, company
entend|er [1] *vt* understand; (*opinar*) believe, think; (*querer decir*) mean. ● *vi* understand. **~erse** *vpr* make o.s. understood; (*comprenderse*) be understood. **~er de** know all about. **~erse con** get on with. **~ido** *a* understood; (*enterado*) well-informed. ● *interj* agreed!, OK! (*fam*). **~imiento** *m* understanding.

a mi ~er in my opinion. **dar a ~er** hint. **no darse por ~ido** pretend not to understand, turn a deaf ear

entenebrecer [11] *vt* darken. **~se** *vpr* get dark

enterado *a* well-informed; (*que sabe*) aware. **no darse por ~** pretend not to understand, turn a deaf ear

enteramente *adv* entirely, completely

enterar *vt* inform. **~se** *vpr*. **~se de** find out about, hear of. **¡entérate!** listen! **¿te enteras?** do you understand?

entereza *f* (*carácter*) strength of character

enternecer [11] *vt* (*fig*) move, touch. **~se** *vpr* be moved, be touched

entero *a* entire, whole; (*firme*) firm. **por ~** entirely, completely

enterra|dor *m* gravedigger. **~r** [1] *vt* bury

entibiar *vt* cool. **~se** *vpr* cool down; (*fig*) cool off

entidad *f* entity; (*organización*) organization; (*com*) company

entierro *m* burial; (*ceremonia*) funeral

entona|ción *f* intonation; (*fig*) arrogance. **~r** *vt* intone. ● *vi* (*mus*) be in tune; ‹*colores*› match. **~rse** *vpr* (*fortalecerse*) tone o.s. up; (*engreírse*) be arrogant

entonces *adv* then. **en aquel ~, por aquel ~** at that time, then

entontecer [11] *vt* make silly. **~se** *vpr* get silly

entornar *vt* half close; leave ajar ‹*puerta*›

entorpecer [11] *vt* ‹*frío etc*› numb; (*dificultar*) hinder

entra|da *f* entrance; (*acceso*) admission, entry; (*billete*) ticket; (*de datos, tec*) input. **~do** *a*. **~do en años** elderly. **ya ~da la noche** late at night. **~nte** *a* next, coming. **dar ~da a** (*admitir*) admit. **de ~da** right away.

entraña *f* (*fig*) heart. **~s** *fpl* entrails; (*fig*) heart. **~ble** *a* ‹*cariño etc*› deep; ‹*amigo*› close. **~r** *vt* involve

entrar *vt* put; (*traer*) bring. ● *vi* go in, enter; (*venir*) come in, enter; (*empezar*) start, begin. **no ~ ni salir en** have nothing to do with

entre *prep* (*de dos personas o cosas*) between; (*más de dos*) among(st)

entreab|ierto *a* half-open. **~rir** [*pp* **entreabierto**] *vt* half open

entreacto *m* interval

entrecano *a* ‹*pelo*› greying; ‹*persona*› who is going grey

entrecejo *m* forehead. **arrugar el ~, fruncir el ~** frown

entrecerrar [1] *vt* (*Amer*) half close

entrecortado *a* ‹*voz*› faltering; ‹*respiración*› laboured

entrecruzar [10] *vt* intertwine

entrega *f* handing over; (*de mercancías etc*) delivery; (*de novela etc*) instalment; (*dedicación*) commitment. **~r** [12] *vt* hand over, deliver, give. **~rse** *vpr* surrender, give o.s. up; (*dedicarse*) devote o.s. (**a** to)

entrelazar [10] *vt* intertwine

entremés *m* hors-d'oeuvre; (*en el teatro*) short comedy

entremet|er *vt* insert. **~erse** *vpr* interfere. **~ido** *a* interfering

entremezclar *vt* mix

entrena|dor *m* trainer. **~miento** *m* training. **~r** *vt* train. **~rse** *vpr* train

entrepierna *f* crotch

entresacar [7] *vt* pick out

entresuelo *m* mezzanine

entretanto *adv* meanwhile

entretejer *vt* interweave

entreten|er [40] *vt* entertain, amuse; (*detener*) delay, keep; (*mantener*) keep alive, keep going. **~erse** *vpr* amuse o.s.; (*tardar*) delay, linger. **~ido** *a* entertaining. **~imiento** *m* entertainment; (*mantenimiento*) upkeep

entrever [43] *vt* make out, glimpse

entrevista *f* interview; (*reunión*) meeting. **~rse** *vpr* have an interview

entristecer [11] *vt* sadden, make sad. **~se** *vpr* be sad

entromet|erse *vpr* interfere. **~ido** *a* interfering

entroncar [7] *vi* be related

entruchada *f*, **entruchado** *m* (*fam*) plot

entumec|erse [11] *vpr* go numb. **~ido** *a* numb

enturbiar *vt* cloud

entusi|asmar *vt* fill with enthusiasm; (*gustar mucho*) delight. **~asmarse** *vpr*. **~asmarse con** get enthusiastic about; (*ser aficionado a*) be mad about, love. **~asmo** *m* enthusiasm. **~asta** *a* enthusiastic.

● *m & f* enthusiast. **~ástico** *a* enthusiastic
enumera|ción *f* count, reckoning. **~r** *vt* enumerate
enuncia|ción *f* enunciation. **~r** *vt* enunciate
envainar *vt* sheathe
envalentonar *vt* encourage. **~se** *vpr* be brave, pluck up courage
envanecer [11] *vt* make conceited. **~se** *vpr* be conceited
envas|ado *a* tinned. ● *m* packaging. **~ar** *vt* package; (*en latas*) tin, can; (*en botellas*) bottle. **~e** *m* packing; (*lata*) tin, can; (*botella*) bottle
envejec|er [11] *vt* make old. ● *vi* get old, grow old. **~erse** *vpr* get old, grow old. **~ido** *a* aged, old
envenenar *vt* poison
envergadura *f* (*alcance*) scope
envés *m* wrong side
envia|do *a* sent. ● *m* representative; (*de la prensa*) correspondent. **~r** *vt* send
enviciar *vt* corrupt
envidi|a *f* envy; (*celos*) jealousy. **~able** *a* enviable. **~ar** *vt* envy, be envious of. **~oso** *a* envious. **tener ~a a** envy
envilecer [11] *vt* degrade
envío *m* sending, dispatch; (*de mercancías*) consignment; (*de dinero*) remittance. **~ contra reembolso** cash on delivery. **gastos** *mpl* **de envío** postage and packing (costs)
enviudar *vi* ‹*mujer*› become a widow, be widowed; ‹*hombre*› become a widower, be widowed
env|oltura *f* wrapping. **~olver** [2, *pp* **envuelto**] *vt* wrap; (*cubrir*) cover; (*fig, acorralar*) corner; (*fig, enredar*) involve; (*mil*) surround. **~olvimiento** *m* involvement. **~uelto** *a* wrapped (up)
enyesar *vt* plaster; (*med*) put in plaster
enzima *f* enzyme
épica *f* epic
epicentro *m* epicentre
épico *a* epic
epid|emia *f* epidemic. **~émico** *a* epidemic
epil|epsia *f* epilepsy. **~éptico** *a* epileptic
epílogo *m* epilogue
episodio *m* episode
epístola *f* epistle
epitafio *m* epitaph
epíteto *m* epithet
epítome *m* epitome
época *f* age; (*período*) period. **hacer ~** make history, be epoch-making
equidad *f* equity
equilátero *a* equilateral
equilibr|ar *vt* balance. **~io** *m* balance; (*de balanza*) equilibrium. **~ista** *m & f* tightrope walker
equino *a* horse, equine
equinoccio *m* equinox
equipaje *m* luggage (*esp Brit*), baggage (*esp Amer*); (*de barco*) crew
equipar *vt* equip; (*de ropa*) fit out
equiparar *vt* make equal; (*comparar*) compare
equipo *m* equipment; (*en deportes*) team
equitación *f* riding
equivale|ncia *f* equivalence. **~nte** *a* equivalent. **~r** [42] *vi* be equivalent; (*significar*) mean
equivoca|ción *f* mistake, error. **~do** *a* wrong. **~r** [7] *vt* mistake. **~rse** *vpr* be mistaken, be wrong, make a mistake. **~rse de** be wrong about. **~rse de número** dial the wrong number. **si no me equivoco** if I'm not mistaken
equívoco *a* equivocal; (*sospechoso*) suspicious. ● *m* ambiguity; (*juego de palabras*) pun; (*doble sentido*) double meaning
era *f* era. ● *vb véase* **ser**
erario *m* treasury
erección *f* erection; (*fig*) establishment
eremita *m* hermit
eres *vb véase* **ser**
erguir [48] *vt* raise. **~ la cabeza** hold one's head high. **~se** *vpr* straighten up
erigir [14] *vt* erect. **~se** *vpr* set o.s. up (**en** as)
eriza|do *a* prickly. **~rse** [10] *vpr* stand on end
erizo *m* hedgehog; (*de mar*) sea urchin. **~ de mar, ~ marino** sea urchin
ermita *f* hermitage. **~ño** *m* hermit
erosi|ón *f* erosion. **~onar** *vt* erode
er|ótico *a* erotic. **~otismo** *m* eroticism
errar [1, *la* **i** *inicial se escribe* **y**) *vt* miss. ● *vi* wander; (*equivocarse*) make a mistake, be wrong
errata *f* misprint
erróneo *a* erroneous, wrong
error *m* error, mistake. **estar en un ~** be wrong, be mistaken

eructar *vi* belch
erudi|ción *f* learning, erudition. **~to** *a* learned
erupción *f* eruption; (*med*) rash
es *vb véase* **ser**
esa *a véase* **ese**
ésa *pron véase* **ése**
esbelto *a* slender, slim
esboz|ar [10] *vt* sketch, outline. **~o** *m* sketch, outline
escabeche *m* pickle. **en ~** pickled
escabroso *a* ‹*terreno*› rough; ‹*asunto*› difficult; (*atrevido*) crude
escabullirse [22] *vpr* slip away
escafandra *f*, **escafandro** *m* diving-suit
escala *f* scale; (*escalera de mano*) ladder; (*de avión*) stopover. **~da** *f* climbing; (*pol*) escalation. **~r** *vt* scale; break into ‹*una casa*›. ● *vi* (*pol*) escalate. **hacer ~ en** stop at. **vuelo sin ~s** non-stop flight
escaldar *vt* scald
escalera *f* staircase, stairs; (*de mano*) ladder. **~ de caracol** spiral staircase. **~ de incendios** fire escape. **~ mecánica** escalator. **~ plegable** step-ladder
escalfa|do *a* poached. **~r** *vt* poach
escalinata *f* flight of steps
escalofrío *m* shiver
escal|ón *m* step; (*de escalera interior*) stair; (*de escala*) rung. **~onar** *vt* spread out
escalope *m* escalope
escam|a *f* scale; (*de jabón*) flake; (*fig*) suspicion. **~oso** *a* scaly
escamotear *vt* make disappear; (*robar*) steal, pinch (*fam*); disregard ‹*dificultad*›
escampar *vi* stop raining
esc|andalizar [10] *vt* scandalize, shock. **~andalizarse** *vpr* be shocked. **~ándalo** *m* scandal; (*alboroto*) uproar. **~andaloso** *a* scandalous; (*alborotador*) noisy
Escandinavia *f* Scandinavia
escandinavo *a* & *m* Scandinavian
escaño *m* bench; (*pol*) seat
escapa|da *f* escape; (*visita*) flying visit. **~do** *a* in a hurry. **~r** *vi* escape. **~rse** *vpr* escape; ‹*líquido, gas*› leak. **dejar ~r** let out
escaparate *m* (shop) window. **ir de ~s** go window-shopping
escapatoria *f* (*fig, fam*) way out
escape *m* (*de gas, de líquido*) leak; (*fuga*) escape; (*auto*) exhaust
escarabajo *m* beetle
escaramuza *f* skirmish
escarbar *vt* scratch; pick ‹*dientes, herida etc*›; (*fig, escudriñar*) delve (**en** into)
escarcha *f* frost. **~do** *a* ‹*fruta*› crystallized
escarlat|a *a invar* scarlet. **~ina** *f* scarlet fever
escarm|entar [1] *vt* punish severely. ● *vi* learn one's lesson. **~iento** *m* punishment; (*lección*) lesson
escarn|ecer [11] *vt* mock. **~io** *m* ridicule
escarola *f* endive
escarpa *f* slope. **~do** *a* steep
escas|ear *vi* be scarce. **~ez** *f* scarcity, shortage; (*pobreza*) poverty. **~o** *a* scarce; (*poco*) little; (*insuficiente*) short; (*muy justo*) barely
escatimar *vt* be sparing with
escayola *f* plaster. **~r** *vt* put in plaster
escena *f* scene; (*escenario*) stage. **~rio** *m* stage; (*en el cine*) scenario; (*fig*) scene
escénico *a* scenic
escenografía *f* scenery
esc|epticismo *m* scepticism. **~éptico** *a* sceptical. ● *m* sceptic
esclarecer [11] *vt* (*fig*) throw light on, clarify
esclavina *f* cape
esclav|itud *f* slavery. **~izar** [10] *vt* enslave. **~o** *m* slave
esclerosis *f* sclerosis
esclusa *f* lock
escoba *f* broom
escocer [2 & 9] *vt* hurt. ● *vi* sting
escocés *a* Scottish. ● *m* Scotsman
Escocia *f* Scotland
escog|er [14] *vt* choose, select. **~ido** *a* chosen; (*de buena calidad*) choice
escolar *a* school. ● *m* schoolboy. ● *f* schoolgirl. **~idad** *f* schooling
escolta *f* escort
escombros *mpl* rubble
escond|er *vt* hide. **~erse** *vpr* hide. **~idas. a ~idas** secretly. **~ite** *m* hiding place; (*juego*) hide-and-seek. **~rijo** *m* hiding place
escopeta *f* shotgun. **~zo** *m* shot
escoplo *m* chisel
escoria *f* slag; (*fig*) dregs
Escorpión *m* Scorpio
escorpión *m* scorpion
escot|ado *a* low-cut. **~adura** *f* low neckline. **~ar** *vt* cut out. ● *vi* pay

one's share. **~e** *m* low neckline. **ir a ~e, pagar a ~e** share the expenses

escozor *m* pain

escri|bano *m* clerk. **~biente** *m* clerk. **~bir** [*pp* **escrito**] *vt/i* write. **~bir a máquina** type. **~birse** *vpr* write to each other; (*deletrearse*) be spelt. **~to** *a* written. ● *m* writing; (*documento*) document. **~tor** *m* writer. **~torio** *m* desk; (*oficina*) office. **~tura** *f* (hand)writing; (*documento*) document; (*jurid*) deed. **¿cómo se escribe...?** how do you spell...? **poner por ~to** put into writing

escr|úpulo *m* scruple; (*escrupulosidad*) care, scrupulousness. **~uloso** *a* scrupulous

escrut|ar *vt* scrutinize; count ‹*votos*›. **~inio** *m* count. **hacer el ~inio** count the votes

escuadr|a *f* (*instrumento*) square; (*mil*) squad; (*naut*) fleet. **~ón** *m* squadron

escuálido *a* skinny; (*sucio*) squalid

escuchar *vt* listen to. ● *vi* listen

escudilla *f* bowl

escudo *m* shield. **~ de armas** coat of arms

escudriñar *vt* examine

escuela *f* school. **~ normal** teachers' training college

escueto *a* simple

escuincle *m* (*Mex*, *perro*) stray dog; (*Mex*, *muchacho*, *fam*) child, kid (*fam*)

escul|pir *vt* sculpture. **~tor** *m* sculptor. **~tora** *f* sculptress. **~tura** *f* sculpture; (*en madera*) carving

escupir *vt/i* spit

escurr|eplatos *m invar* plate-rack. **~idizo** *a* slippery. **~ir** *vt* drain; wring out ‹*ropa*›. ● *vi* drip; (*ser resbaladizo*) be slippery. **~irse** *vpr* slip

ese *a* (*f* **esa**, *mpl* **esos**, *fpl* **esas**) that; (*en plural*) those

ése *pron* (*f* **ésa**, *mpl* **ésos**, *fpl* **ésas**) that one; (*en plural*) those; (*primero de dos*) the former. **ni por ésas** on no account

esencia *f* essence. **~l** *a* essential. **lo ~l** the main thing

esf|era *f* sphere; (*de reloj*) face. **~érico** *a* spherical

esfinge *f* sphinx

esf|orzarse [2 & 10] *vpr* make an effort. **~uerzo** *m* effort

esfumarse *vpr* fade away; ‹*persona*› vanish

esgrim|a *f* fencing. **~ir** *vt* brandish; (*fig*) use

esguince *m* swerve; (*med*) sprain

eslab|ón *m* link. **~onar** *vt* link (together)

eslavo *a* Slav, Slavonic

eslogan *m* slogan

esmalt|ar *vt* enamel; varnish ‹*uñas*›; (*fig*) adorn. **~e** *m* enamel. **~ de uñas, ~e para las uñas** nail varnish (*Brit*), nail polish (*Amer*)

esmerado *a* careful

esmeralda *f* emerald

esmerarse *vpr* take care (**en** over)

esmeril *m* emery

esmero *m* care

esmoquin *m* dinner jacket, tuxedo (*Amer*)

esnob *a invar* snobbish. ● *m & f* (*pl* **esnobs**) snob. **~ismo** *m* snobbery

esnórkel *m* snorkel

eso *pron* that. **¡~ es!** that's it! **~ mismo** exactly. **¡~ no!** certainly not! **¡~ sí!** of course. **a ~ de** about. **en ~** at that moment. **¿no es ~?** isn't that right? **por ~** therefore. **y ~ que** although

esos *a pl véase* **ese**

ésos *pron pl véase* **ése**

espabila|do *a* bright. **~r** *vt* snuff ‹*vela*›; (*avivar*) brighten up; (*despertar*) wake up. **~rse** *vpr* wake up; (*apresurarse*) hurry up

espaci|al *a* space. **~ar** *vt* space out. **~o** *m* space. **~oso** *a* spacious

espada *f* sword. **~s** *fpl* (*en naipes*) spades

espagueti *m* spaghetti

espald|a *f* back. **~illa** *f* shoulder-blade. **a ~as de uno** behind s.o.'s back. **a las ~as** on one's back. **tener las ~as anchas** be broad-shouldered. **volver la ~a a uno, volver las ~as a uno** give s.o. the cold shoulder

espant|ada *f* stampede. **~adizo** *a* timid, timorous. **~ajo** *m*, **~apájaros** *m inv* scarecrow. **~ar** *vt* frighten; (*ahuyentar*) frighten away. **~arse** *vpr* be frightened; (*ahuyentarse*) be frightened away. **~o** *m* terror; (*horror*) horror. **~oso** *a* frightening; (*muy grande*) terrible. **¡qué ~ajo!** what a sight!

España *f* Spain

español *a* Spanish. ● *m* (*persona*) Spaniard; (*lengua*) Spanish. **los**

~es the Spanish. **~izado** *a* Hispanicized

esparadrapo *m* sticking-plaster, plaster (*Brit*)

esparci|do *a* scattered; (*fig*) widespread. **~r** [9] *vt* scatter; (*difundir*) spread. **~rse** *vpr* be scattered; (*difundirse*) spread; (*divertirse*) enjoy o.s.

espárrago *m* asparagus

esparto *m* esparto (grass)

espasm|o *m* spasm. **~ódico** *a* spasmodic

espátula *f* spatula; (*en pintura*) palette knife

especia *f* spice

especial *a* special. **~idad** *f* speciality (*Brit*), specialty (*Amer*). **~ista** *a* & *m* & *f* specialist. **~ización** *f* specialization. **~izar** [10] *vt* specialize. **~izarse** *vpr* specialize. **~mente** *adv* especially. **en ~** especially

especie *f* kind, sort; (*en biología*) species; (*noticia*) piece of news. **en ~** in kind

especifica|ción *f* specification. **~r** [7] *vt* specify

específico *a* specific

espect|áculo *m* sight; (*diversión*) entertainment, show. **~ador** *m* & *f* spectator. **~acular** *a* spectacular

espectro *m* spectre; (*en física*) spectrum

especula|ción *f* speculation. **~dor** *m* speculator. **~r** *vi* speculate. **~tivo** *a* speculative

espej|ismo *m* mirage. **~o** *m* mirror. **~o retrovisor** (*auto*) rear-view mirror

espeleólogo *m* potholer

espeluznante *a* horrifying

espera *f* wait. **sala** *f* **de ~** waiting room

espera|nza *f* hope. **~r** *vt* hope; (*aguardar*) wait for; (*creer*) expect. ● *vi* hope; (*aguardar*) wait. **~r en uno** trust in s.o. **en ~ de** awaiting. **espero que no** I hope not. **espero que sí** I hope so

esperma *f* sperm

esperpento *m* fright; (*disparate*) nonsense

espes|ar *vt* thicken. **~arse** *vpr* thicken. **~o** *a* thick; ‹*pasta etc*› stiff. **~or** *m*, **~ura** *f* thickness; (*bot*) thicket

espetón *m* spit

esp|ía *f* spy. **~iar** [20] *vt* spy on. ● *vi* spy

espiga *f* (*de trigo etc*) ear

espina *f* thorn; (*de pez*) bone; (*dorsal*) spine; (*astilla*) splinter; (*fig, dificultad*) difficulty. **~ dorsal** spine

espinaca *f* spinach

espinazo *m* spine

espinilla *f* shin; (*med*) blackhead

espino *m* hawthorn. **~ artificial** barbed wire. **~so** *a* thorny; ‹*pez*› bony; (*fig*) difficult

espionaje *m* espionage

espiral *a* & *f* spiral

espirar *vt/i* breathe out

esp|iritismo *m* spiritualism. **~iritoso** *a* spirited. **~iritista** *m* & *f* spiritualist. **~íritu** *m* spirit; (*mente*) mind; (*inteligencia*) intelligence. **~iritual** *a* spiritual. **~iritualismo** *m* spiritualism

espita *f* tap, faucet (*Amer*)

espl|éndido *a* splendid; ‹*persona*› generous. **~endor** *m* splendour

espliego *m* lavender

espolear *vt* (*fig*) spur on

espoleta *f* fuse

espolvorear *vt* sprinkle

esponj|a *f* sponge; (*tejido*) towelling. **~oso** *a* spongy. **pasar la ~a** forget about it

espont|aneidad *f* spontaneity. **~áneo** *a* spontaneous

esporádico *a* sporadic

espos|a *f* wife. **~as** *fpl* handcuffs. **~ar** *vt* handcuff. **~o** *m* husband. **los ~os** the couple

espuela *f* spur; (*fig*) incentive. **dar de ~s** spur on

espum|a *f* foam; (*en bebidas*) froth; (*de jabón*) lather. **~ar** *vt* skim. ● *vi* foam; ‹*bebidas*› froth; ‹*jabón*› lather. **~oso** *a* ‹*vino*› sparkling. **echar ~a** foam, froth

esqueleto *m* skeleton

esquem|a *m* outline. **~ático** *a* sketchy

esqu|í *m* (*pl* **esquís**) ski; (*el deporte*) skiing. **~iador** *m* skier. **~iar** [20] *vi* ski

esquilar *vt* shear

esquimal *a* & *m* Eskimo

esquina *f* corner

esquirol *m* blackleg

esquiv|ar *vt* avoid. **~o** *a* aloof

esquizofrénico *a* & *m* schizophrenic

esta *a véase* **este**

ésta *pron véase* **éste**

estab|ilidad *f* stability. **~ilizador** *m* stabilizer. **~ilizar** [10] *vt* stabilize. **~le** *a* stable

establec|er [11] *vt* establish. **~erse** *vpr* settle; (*com*) start a business. **~imiento** *m* establishment
establo *m* cowshed
estaca *f* stake; (*para apalear*) stick. **~da** *f* (*cerca*) fence
estación *f* station; (*del año*) season; (*de vacaciones*) resort. **~ de servicio** service station
estaciona|miento *m* parking. **~r** *vt* station; (*auto*) park. **~rio** *a* stationary
estadio *m* stadium; (*fase*) stage
estadista *m* statesman. ● *f* stateswoman
estadístic|a *f* statistics. **~o** *a* statistical
estado *m* state. **~ civil** marital status. **~ de ánimo** frame of mind. **~ de cuenta** bank statement. **~ mayor** (*mil*) staff. **en buen ~** in good condition. **en ~ (interesante)** pregnant
Estados Unidos *mpl* United States
estadounidense *a* American, United States. ● *m & f* American
estafa *f* swindle. **~r** *vt* swindle
estafeta *f* (*oficina de correos*) (sub-) post office
estala|ctita *f* stalactite. **~gmita** *f* stalagmite
estall|ar *vi* explode; ‹*olas*› break; ‹*guerra, epidemia etc*› break out; (*fig*) burst. **~ar en llanto** burst into tears. **~ar de risa** burst out laughing. **~ido** *m* explosion; (*de guerra, epidemia etc*) outbreak; (*de risa etc*) outburst
estamp|a *f* print; (*aspecto*) appearance. **~ado** *a* printed. ● *m* printing; (*tela*) cotton print. **~ar** *vt* stamp; (*imprimir*) print. **dar a la ~a** (*imprimir*) print; (*publicar*) publish. **la viva ~a** the image
estampía. de ~ía suddenly
estampido *m* explosion
estampilla *f* stamp; (*Mex*) (postage) stamp
estanca|do *a* stagnant. **~miento** *m* stagnation. **~r** [7] *vt* stem; (*com*) turn into a monopoly
estanci|a *f* stay; (*Arg, finca*) ranch, farm; (*cuarto*) room. **~ero** *m* (*Arg*) farmer
estanco *a* watertight. ● *m* tobacconist's (shop)
estandarte *m* standard, banner
estanque *m* lake; (*depósito de agua*) reservoir
estanquero *m* tobacconist
estante *m* shelf. **~ría** *f* shelves; (*para libros*) bookcase
estañ|o *m* tin. **~adura** *f* tin-plating
estar [27] *vi* be; (*quedarse*) stay; (*estar en casa*) be in. **¿estamos?** alright? **estamos a 29 de noviembre** it's the 29th of November. **~ para** be about to. **~ por** remain to be; (*con ganas de*) be tempted to; (*ser partidario de*) be in favour of. **~se** *vpr* stay. **¿cómo está Vd?, ¿cómo estás?** how are you?
estarcir [9] *vt* stencil
estatal *a* state
estático *a* static; (*pasmado*) dumbfounded
estatua *f* statue
estatura *f* height
estatut|ario *a* statutory. **~o** *m* statute
este *m* east; (*viento*) east wind. ● *a* (*f* **esta,** *mpl* **estos,** *fpl* **estas**) this; (*en plural*) these. ● *int* (*LAm*) well, er
éste *pron* (*f* **ésta,** *mpl* **éstos,** *fpl* **éstas**) this one, (*en plural*) these; (*segundo de dos*) the latter
estela *f* wake; (*arquit*) carved stone
estera *f* mat; (*tejido*) matting
est|éreo *a* stereo. **~ereofónico** *a* stereo, stereophonic
esterilla *f* mat
estereotip|ado *a* stereotyped. **~o** *m* stereotype
est|éril *a* sterile; ‹*mujer*› infertile; ‹*terreno*› barren. **~erilidad** *f* sterility; (*de mujer*) infertility; (*de terreno*) barrenness
esterlina *a* sterling. **libra** *f* **~** pound sterling
estético *a* aesthetic
estevado *a* bow-legged
estiércol *m* dung; (*abono*) manure
estigma *m* stigma. **~s** *mpl* (*relig*) stigmata
estilarse *vpr* be used
estil|ista *m & f* stylist. **~izar** [10] *vt* stylize. **~o** *m* style. **por el ~o** of that sort
estilográfica *f* fountain pen
estima *f* esteem. **~do** *a* esteemed. **~do señor** (*en cartas*) Dear Sir. **~r** *vt* esteem; have great respect for ‹*persona*›; (*valorar*) value; (*juzgar*) think
est|imulante *a* stimulating. ● *m* stimulant. **~imular** *vt* stimulate; (*incitar*) incite. **~ímulo** *m* stimulus
estipular *vt* stipulate

estir|ado *a* stretched; ‹*persona*› haughty. **~ar** *vt* stretch; (*fig*) stretch out. **~ón** *m* pull, tug; (*crecimiento*) sudden growth
estirpe *m* stock
estival *a* summer
esto *pron neutro* this; (*este asunto*) this business. **en ~** at this point. **en ~ de** in this business of. **por ~** therefore
estofa *f* class. **de baja ~** ‹*gente*› low-class
estofa|do *a* stewed. ● *m* stew. **~r** *vt* stew
estoic|ismo *m* stoicism. **~o** *a* stoical. ● *m* stoic
estómago *m* stomach. **dolor** *m* **de ~** stomach-ache
estorb|ar *vt* hinder, obstruct; (*molestar*) bother, annoy. ● *vi* be in the way. **~o** *m* hindrance; (*molestia*) nuisance
estornino *m* starling
estornud|ar *vi* sneeze. **~o** *m* sneeze
estos *a mpl véase* **este**
éstos *pron mpl véase* **éste**
estoy *vb véase* **estar**
estrabismo *m* squint
estrado *m* stage; (*mus*) bandstand
estrafalario *a* outlandish
estrag|ar [12] *vt* devastate. **~o** *m* devastation. **hacer ~os** devastate
estragón *m* tarragon
estrambótico *a* outlandish
estrangula|ción *f* strangulation. **~dor** *m* strangler; (*auto*) choke. **~miento** *m* blockage; (*auto*) bottleneck. **~r** *vt* strangle
estraperlo *m* black market. **comprar algo de ~** buy sth on the black market
estratagema *f* stratagem
estrateg|a *m & f* strategist. **~ia** *f* strategy
estratégic|amente *adv* strategically. **~o** *a* strategic
estrato *m* stratum
estratosfera *f* stratosphere
estrech|ar *vt* make narrower; take in ‹*vestido*›; (*apretar*) squeeze; hug ‹*persona*›. **~ar la mano a uno** shake hands with s.o. **~arse** *vpr* become narrower; (*apretarse*) squeeze up. **~ez** *f* narrowness; (*apuro*) tight spot; (*falta de dinero*) want. **~o** *a* narrow; ‹*vestido etc*› tight; (*fig, íntimo*) close. ● *m* straits. **~o de miras, de miras ~as** narrow-minded
estregar [1 & 12] *vt* rub
estrella *f* star. **~ de mar**, **~mar** *m* starfish
estrellar *vt* smash; fry ‹*huevos*›. **~se** *vpr* smash; (*fracasar*) fail. **~se contra** crash into
estremec|er [11] *vt* shake. **~erse** *vpr* tremble (**de** with). **~imiento** *m* shaking
estren|ar *vt* use for the first time; wear for the first time ‹*vestido etc*›; show for the first time ‹*película*›. **~arse** *vpr* make one's début; ‹*película*› have its première; ‹*obra de teatro*› open. **~o** *m* first use; (*de película*) première; (*de obra de teatro*) first night
estreñi|do *a* constipated. **~miento** *m* constipation
estr|épito *m* din. **~epitoso** *a* noisy; (*fig*) resounding
estreptomicina *f* streptomycin
estrés *m* stress
estría *f* groove
estribar *vt* rest (**en** on); (*consistir*) lie (**en** in)
estribillo *m* refrain; (*muletilla*) catchphrase
estribo *m* stirrup; (*de vehículo*) step; (*contrafuerte*) buttress. **perder los ~s** lose one's temper
estribor *m* starboard
estricto *a* strict
estridente *a* strident, raucous
estrofa *f* strophe
estropajo *m* scourer. **~so** *a* ‹*carne etc*› tough; ‹*persona*› slovenly
estropear *vt* spoil; (*romper*) break. **~se** *vpr* be damaged; ‹*fruta etc*› go bad; (*fracasar*) fail
estructura *f* structure. **~l** *a* structural
estruendo *m* din; (*de mucha gente*) uproar. **~so** *a* deafening
estrujar *vt* squeeze; (*fig*) drain
estuario *m* estuary
estuco *m* stucco
estuche *m* case
estudi|ante *m & f* student. **~antil** *a* student. **~ar** *vt* study. **~o** *m* study; (*de artista*) studio. **~oso** *a* studious
estufa *f* heater; (*LAm*) cooker
estupefac|ción *f* astonishment. **~iente** *a* astonishing. ● *m* narcotic. **~to** *a* astonished
estupendo *a* marvellous; (*hermoso*) beautiful
est|upidez *f* stupidity; (*acto*) stupid thing. **~úpido** *a* stupid

estupor *m* amazement
esturión *m* sturgeon
estuve *vb véase* **estar**
etapa *f* stage. **hacer ~ en** break the journey at. **por ~s** in stages
etc *abrev* (*etcétera*) etc
etcétera *adv* et cetera
éter *m* ether
etéreo *a* ethereal
etern|amente *adv* eternally. **~idad** *f* eternity. **~izar** [10] *vt* drag out. **~izarse** *vpr* be interminable. **~o** *a* eternal
étic|a *f* ethics. **~o** *a* ethical
etimología *f* etymology
etiqueta *f* ticket, tag; (*ceremonial*) etiquette. **de ~** formal
étnico *a* ethnic
eucalipto *m* eucalyptus
eufemismo *m* euphemism
euforia *f* euphoria
Europa *f* Europe
europe|o *a & m* European. **~izar** [10] *vt* Europeanize
eutanasia *f* euthanasia
evacua|ción *f* evacuation. **~r** [21 *o regular*] *vt* evacuate
evadir *vt* avoid. **~se** *vpr* escape
evaluar [21] *vt* evaluate
evang|élico *a* evangelical. **~elio** *m* gospel. **~elista** *m & f* evangelist
evapora|ción *f* evaporation. **~r** *vi* evaporate. **~rse** *vpr* evaporate; (*fig*) disappear
evasi|ón *f* evasion; (*fuga*) escape. **~vo** *a* evasive
evento *m* event. **a todo ~** at all events
eventual *a* possible. **~idad** *f* eventuality
eviden|cia *f* evidence. **~ciar** *vt* show. **~ciarse** *vpr* be obvious. **~te** *a* obvious. **~temente** *adv* obviously. **poner en ~cia** show; (*fig*) make a fool of
evitar *vt* avoid; (*ahorrar*) spare
evocar [7] *vt* evoke
evoluci|ón *f* evolution. **~onado** *a* fully-developed. **~onar** *vi* evolve; (*mil*) manoevre
ex *pref* ex-, former
exacerbar *vt* exacerbate
exact|amente *adv* exactly. **~itud** *f* exactness. **~o** *a* exact; (*preciso*) accurate; (*puntual*) punctual. **¡~!** exactly!. **con ~itud** exactly
exagera|ción *f* exaggeration. **~do** *a* exaggerated. **~r** *vt/i* exaggerate
exalta|do *a* exalted; (*fanático*) fanatical. **~r** *vt* exalt. **~rse** *vpr* get excited
exam|en *m* examination; (*escol, univ*) exam(ination). **~inador** *m* examiner. **~inar** *vt* examine. **~inarse** *vpr* take an exam
exánime *a* lifeless
exaspera|ción *f* exasperation. **~r** *vt* exasperate. **~rse** *vpr* get exasperated
excava|ción *f* excavation. **~dora** *f* digger. **~r** *vt* excavate
excede|ncia *f* leave of absence. **~nte** *a & m* surplus. **~r** *vi* exceed. **~rse** *vpr* go too far. **~rse a sí mismo** excel o.s.
excelen|cia *f* excellence; (*tratamiento*) Excellency. **~te** *a* excellent
exc|entricidad *f* eccentricity. **~éntrico** *a & m* eccentric
excepci|ón *f* exception. **~onal** *a* exceptional. **a ~ón de, con ~ón de** except (for)
except|o *prep* except (for). **~uar** [21] *vt* except
exces|ivo *a* excessive. **~o** *m* excess. **~o de equipaje** excess luggage (*esp Brit*), excess baggage (*esp Amer*)
excita|ble *a* excitable. **~ción** *f* excitement. **~nte** *a* exciting. ● *m* stimulant. **~r** *vt* excite; (*incitar*) incite. **~rse** *vpr* get excited
exclama|ción *f* exclamation. **~r** *vi* exclaim
exclu|ir [17] *vt* exclude. **~sión** *f* exclusion. **~siva** *f* sole right; (*en la prensa* exclusive (story). **~sive** *adv* exclusive; (*exclusivamente*) exclusively. **~sivo** *a* exclusive
excomu|lgar [12] *vt* excommunicate. **~nión** *f* excommunication
excremento *m* excrement
exculpar *vt* exonerate; (*jurid*) acquit
excursi|ón *f* excursion, trip. **~onista** *m & f* day-tripper. **ir de ~ón** go on an excursion
excusa *f* excuse; (*disculpa*) apology. **~r** *vt* excuse. **presentar sus ~s** apologize
execra|ble *a* loathsome. **~r** *vt* loathe
exento *a* exempt; (*libre*) free
exequias *fpl* funeral rites
exhala|ción *f* shooting star. **~r** *vt* exhale, breath out; give off ‹*olor etc*›. **~rse** *vpr* hurry. **como una ~ción** at top speed

exhaust|ivo *a* exhaustive. **~o** *a* exhausted
exhibi|ción *f* exhibition. **~cionista** *m & f* exhibitionist. **~r** *vt* exhibit
exhortar *vt* exhort (**a** to)
exhumar *vt* exhume; (*fig*) dig up
exig|encia *f* demand. **~ente** *a* demanding. **~ir** [14] *vt* demand. **tener muchas ~encias** be very demanding
exiguo *a* meagre
exil|(i)ado *a* exiled. • *m* exile. **~(i)arse** *vpr* go into exile. **~io** *m* exile
eximio *a* distinguished
eximir *vt* exempt; (*liberar*) free
existencia *f* existence. **~s** *fpl* stock
existencial *a* existential. **~ismo** *m* existentialism
exist|ente *a* existing. **~ir** *vi* exist
éxito *m* success. **no tener ~** fail. **tener ~** be successful
exitoso *a* successful
éxodo *m* exodus
exonerar *vt* (*de un empleo*) dismiss; (*de un honor etc*) strip
exorbitante *a* exorbitant
exorci|smo *m* exorcism. **~zar** [10] *vt* exorcise
exótico *a* exotic
expan|dir *vt* expand; (*fig*) spread. **~dirse** *vpr* expand. **~sión** *f* expansion. **~sivo** *a* expansive
expatria|do *a & m* expatriate. **~r** *vt* banish. **~rse** *vpr* emigrate; (*exiliarse*) go into exile
expectativa *f*. **estar a la ~** be on the lookout
expedición *f* dispatch; (*cosa expedida*) shipment; (*mil, científico etc*) expedition
expediente *m* expedient; (*jurid*) proceedings; (*documentos*) record, file
expedi|r [5] *vt* dispatch, send; issue ‹*documento*›. **~to** *a* clear
expeler *vt* expel
expende|dor *m* dealer. **~dor automático** vending machine. **~duría** *f* shop; (*de billetes*) ticket office. **~r** *vt* sell
expensas *fpl*. **a ~ de** at the expense of. **a mis ~** at my expense
experiencia *f* experience
experiment|al *a* experimental. **~ar** *vt* test, experiment with; (*sentir*) experience. **~o** *m* experiment
experto *a & m* expert
expiar [20] *vt* atone for
expirar *vi* expire; (*morir*) die
explana|da *f* levelled area; (*paseo*) esplanade. **~r** *vt* level
explayar *vt* extend. **~se** *vpr* spread out, extend; (*hablar*) be long-winded; (*confiarse*) confide (**a** in)
expletivo *m* expletive
explica|ción *f* explanation. **~r** [7] *vt* explain. **~rse** *vpr* understand; (*hacerse comprender*) explain o.s. **no me lo explico** I can't understand it
explícito *a* explicit
explora|ción *f* exploration. **~dor** *m* explorer; (*muchacho*) boy scout. **~r** *vt* explore. **~torio** *a* exploratory
explosi|ón *f* explosion; (*fig*) outburst. **~onar** *vt* blow up. **~vo** *a & m* explosive
explota|ción *f* working; (*abuso*) exploitation. **~r** *vt* work ‹*mina*›; farm ‹*tierra*›; (*abusar*) exploit. • *vi* explode
expone|nte *m* exponent. **~r** [34] *vt* expose; display ‹*mercancías*›; (*explicar*) expound; exhibit ‹*cuadros etc*›; (*arriesgar*) risk. • *vi* hold an exhibition. **~rse** *vpr* run the risk (**a** of)
exporta|ción *f* export. **~dor** *m* exporter. **~r** *vt* export
exposición *f* exposure; (*de cuadros etc*) exhibition; (*en escaparate etc*) display; (*explicación*) exposition, explanation
expresamente *adv* specifically
expres|ar *vt* express. **~arse** *vpr* express o.s. **~ión** *f* expression. **~ivo** *a* expressive; (*cariñoso*) affectionate
expreso *a* express. • *m* express messenger; (*tren*) express
exprimi|dor *m* squeezer. **~r** *vt* squeeze; (*explotar*) exploit
expropiar *vt* expropriate
expuesto *a* on display; ‹*lugar etc*› exposed; (*peligroso*) dangerous. **estar ~ a** be liable to
expuls|ar *vt* expel; throw out ‹*persona*›; send off ‹*jugador*›. **~ión** *f* expulsion
expurgar [12] *vt* expurgate
exquisit|o *a* exquisite. **~amente** *adv* exquisitely
extasiar [20] *vt* enrapture
éxtasis *m invar* ecstasy
extático *a* ecstatic
extend|er [1] *vt* spread (out); draw up ‹*documento*›. **~erse** *vpr* spread;

‹*paisaje etc*› extend, stretch; (*tenderse*) stretch out. **~ido** *a* spread out; (*generalizado*) widespread; ‹*brazos*› outstretched

extens|amente *adv* widely; (*detalladamente*) in full. **~ión** *f* extension; (*amplitud*) expanse; (*mus*) range. **~o** *a* extensive

extenuar [21] *vt* exhaust

exterior *a* external, exterior; (*del extranjero*) foreign; ‹*aspecto etc*› outward. ● *m* exterior; (*países extranjeros*) abroad. **~izar** [10] *vt* show

extermin|ación *f* extermination. **~ar** *vt* exterminate. **~io** *m* extermination

externo *a* external; ‹*signo etc*› outward. ● *m* day pupil

extin|ción *f* extinction. **~guir** [13] *vt* extinguish. **~guirse** *vpr* die out; ‹*fuego*› go out. **~to** *a* extinguished; ‹*raza etc*› extinct. **~tor** *m* fire extinguisher

extirpa|r *vt* uproot; extract ‹*muela etc*›; remove ‹*tumor*›. **~ción** *f* (*fig*) eradication

extorsi|ón *f* (*fig*) inconvenience. **~onar** *vt* inconvenience

extra *a invar* extra; (*de buena calidad*) good-quality; ‹*huevos*› large. **paga** *f* **~** bonus

extrac|ción *f* extraction; (*de lotería*) draw. **~to** *m* extract

extradición *f* extradition

extraer [41] *vt* extract

extranjero *a* foreign. ● *m* foreigner; (*países*) foreign countries. **del ~** from abroad. **en el ~**, **por el ~** abroad

extrañ|ar *vt* surprise; (*encontrar extraño*) find strange; (*LAm, echar de menos*) miss; (*desterrar*) banish. **~arse** *vpr* be surprised (**de** at); ‹*2 personas*› grow apart. **~eza** *f* strangeness; (*asombro*) surprise. **~o** *a* strange. ● *m* stranger

extraoficial *a* unofficial

extraordinario *a* extraordinary. ● *m* (*correo*) special delivery; (*plato*) extra dish; (*de periódico etc*) special edition. **horas** *fpl* **extraordinarias** overtime

extrarradio *m* suburbs

extrasensible *a* extra-sensory

extraterrestre *a* extraterrestrial. ● *m* alien

extravagan|cia *f* oddness, eccentricity. **~te** *a* odd, eccentric

extravertido *a* & *m* extrovert

extrav|iado *a* lost; ‹*lugar*› isolated. **~iar** [20] *vt* lose. **~iarse** *vpr* get lost; ‹*objetos*› be missing. **~ío** *m* loss

extremar *vt* overdo. **~se** *vpr* make every effort

extremeño *a* from Extremadura. ● *m* person from Extremadura

extrem|idad *f* extremity. **~idades** *fpl* extremities. **~ista** *a* & *m* & *f* extremist. **~o** *a* extreme. ● *m* end; (*colmo*) extreme. **en ~o** extremely. **en último ~o** as a last resort

extrovertido *a* & *m* extrovert

exuberan|cia *f* exuberance. **~te** *a* exuberant

exulta|ción *f* exultation. **~r** *vi* exult

eyacular *vt/i* ejaculate

F

fa *m* F; (*solfa*) fah

fabada *f* Asturian stew

fábrica *f* factory. **marca** *f* **de ~** trade mark

fabrica|ción *f* manufacture. **~ción en serie** mass production. **~nte** *m* & *f* manufacturer. **~r** [7] *vt* manufacture; (*inventar*) fabricate

fábula *f* fable; (*mentira*) story, lie; (*chisme*) gossip

fabuloso *a* fabulous

facci|ón *f* faction. **~ones** *fpl* (*de la cara*) features

faceta *f* facet

fácil *a* easy; (*probable*) likely; ‹*persona*› easygoing

facili|dad *f* ease; (*disposición*) aptitude. **~dades** *fpl* facilities. **~tar** *vt* facilitate; (*proporcionar*) provide

fácilmente *adv* easily

facistol *m* lectern

facón *m* (*Arg*) gaucho knife

facsímil(e) *m* facsimile

factible *a* feasible

factor *m* factor

factoría *f* agency; (*esp LAm, fábrica*) factory

factura *f* bill, invoice; (*hechura*) manufacture. **~r** *vt* (*hacer la factura*) invoice; (*cobrar*) charge; (*en ferrocarril*) register (*Brit*), check (*Amer*)

faculta|d *f* faculty; (*capacidad*) ability; (*poder*) power. **~tivo** *a* optional

facha *f* (*aspecto, fam*) look

fachada *f* façade; (*fig*, *apariencia*) show
faena *f* job. **~s domésticas** housework
fagot *m* bassoon; (*músico*) bassoonist
faisán *m* pheasant
faja *f* (*de tierra*) strip; (*corsé*) corset; (*mil etc*) sash
fajo *m* bundle; (*de billetes*) wad
falang|e *f* (*política española*) Falange. **~ista** *m & f* Falangist
falda *f* skirt; (*de montaña*) side
fálico *a* phallic
fals|ear *vt* falsify, distort. **~edad** *f* falseness; (*mentira*) lie, falsehood. **~ificación** *f* forgery. **~ificador** *m* forger. **~ificar** [7] *vt* forge. **~o** *a* false; (*equivocado*) wrong; (*falsificado*) fake
falt|a *f* lack; (*ausencia*) absence; (*escasez*) shortage; (*defecto*) fault, defect; (*culpa*) fault; (*error*) mistake; (*en fútbol etc*) foul; (*en tenis*) fault. **~ar** *vi* be lacking; (*estar ausente*) be absent. **~o** *a* lacking (**de** in). **a ~a de** for lack of. **echar en ~a** miss. **hacer ~a** be necessary. **me hace ~a** I need. **¡no ~aba más!** don't mention it! (*naturalmente*) of course! **sacar ~as** find fault
falla *f* (*incl geol*) fault. **~r** *vi* fail; (*romperse*) break, give way; ‹*motor*, *tiro etc*› miss. **sin ~r** without fail
fallec|er [11] *vi* die. **~ido** *a* late. ● *m* deceased
fallido *a* vain; (*fracasado*) unsuccessful
fallo *m* failure; (*defecto*) fault; (*jurid*) sentence
fama *f* fame; (*reputación*) reputation. **de mala ~** of ill repute. **tener ~ de** have the reputation of
famélico *a* starving
familia *f* family. **~ numerosa** large family. **~r** *a* familiar; (*de la familia*) family; (*sin ceremonia*) informal. **~ridad** *f* familiarity. **~rizarse** [10] *vpr* become familiar (**con** with)
famoso *a* famous
fanático *a* fanatical. ● *m* fanatic
fanfarr|ón *a* boastful. ● *m* braggart. **~onada** *f* boasting; (*dicho*) boast. **~onear** *vi* show off
fango *m* mud. **~so** *a* muddy
fantas|ear *vi* daydream; (*imaginar*) fantasize. **~ía** *f* fantasy. **de ~** fancy
fantasma *m* ghost
fantástico *a* fantastic
fantoche *m* puppet
faringe *f* pharynx
fardo *m* bundle
farfullar *vi* jabber, gabble
farmac|eútico *a* pharmaceutical. ● *m* chemist (*Brit*), pharmacist, druggist (*Amer*). **~ia** *f* (*ciencia*) pharmacy; (*tienda*) chemist's (shop) (*Brit*), pharmacy, drugstore (*Amer*)
faro *m* lighthouse; (*aviac*) beacon; (*auto*) headlight
farol *m* lantern; (*de la calle*) street lamp. **~a** *f* street lamp. **~ita** *f* small street lamp
farsa *f* farce
fas *adv*. **por ~ o por nefas** rightly or wrongly
fascículo *m* instalment
fascina|ción *f* fascination. **~r** *vt* fascinate
fascis|mo *m* fascism. **~ta** *a & m & f* fascist
fase *f* phase
fastidi|ar *vt* annoy; (*estropear*) spoil. **~arse** *vpr* (*aguantarse*) put up with it; (*hacerse daño*) hurt o.s. **~o** *m* nuisance; (*aburrimiento*) boredom. **~oso** *a* annoying. **¡para que te ~es!** so there! **¡qué ~o!** what a nuisance!
fatal *a* fateful; (*mortal*) fatal; (*pésimo*, *fam*) terrible. **~idad** *f* fate; (*desgracia*) misfortune. **~ista** *m & f* fatalist
fatig|a *f* fatigue. **~as** *fpl* troubles. **~ar** [12] *vt* tire. **~arse** *vpr* get tired. **~oso** *a* tiring
fatuo *a* fatuous
fauna *f* fauna
fausto *a* lucky
favor *m* favour. **~able** *a* favourable. **a ~ de**, **en ~ de** in favour of. **haga el ~ de** would you be so kind as to, please. **por ~** please
favorec|edor *a* flattering. **~er** [11] *vt* favour; ‹*vestido*, *peinado etc*› suit. **~ido** *a* favoured
favorit|ismo *m* favouritism. **~o** *a & m* favourite
faz *f* face
fe *f* faith. **dar ~ de** certify. **de buena ~** in good faith
fealdad *f* ugliness
febrero *m* February
febril *a* feverish
fecund|ación *f* fertilization. **~ación artificial** artificial insemination. **~ar** *vt* fertilize. **~o** *a* fertile; (*fig*) prolific

fecha *f* date. **~r** *vt* date. **a estas ~s** now; (*todavía*) still. **hasta la ~** so far. **poner la ~** date
fechoría *f* misdeed
federa|ción *f* federation. **~l** *a* federal
feísimo *a* hideous
felici|dad *f* happiness. **~dades** *fpl* best wishes; (*congratulaciones*) congratulations. **~tación** *f* congratulation. **~tar** *vt* congratulate. **~tarse** *vpr* be glad
feligr|és *m* parishioner. **~esía** *f* parish
felino *a* & *m* feline
feliz *a* happy; (*afortunado*) lucky. **¡Felices Pascuas!** Happy Christmas! **¡F~ Año Nuevo!** Happy New Year!
felpudo *a* plush. ● *m* doormat
femeni|l *a* feminine. **~no** *a* feminine; (*biol, bot*) female. ● *m* feminine. **~nidad** *f* femeninity. **~sta** *a* & *m* & *f* feminist
fen|omenal *a* phenomenal. **~ómeno** *m* phenomenon; (*monstruo*) freak
feo *a* ugly; (*desagradable*) nasty; (*malo*) bad
féretro *m* coffin
feria *f* fair; (*verbena*) carnival; (*descanso*) holiday; (*Mex, cambio*) change. **~do** *a*. **día ~do** holiday
ferment|ación *f* fermentation. **~ar** *vt/i* ferment. **~o** *m* ferment
fero|cidad *f* ferocity. **~z** *a* fierce; ‹*persona*› savage
férreo *a* iron. **vía férrea** railway (*Brit*), railroad (*Amer*)
ferreter|ía *f* ironmonger's (shop) (*Brit*), hardware store (*Amer*). **~o** *m* ironmonger (*Brit*), hardware dealer (*Amer*)
ferro|bús *m* local train. **~carril** *m* railway (*Brit*), railroad (*Amer*). **~viario** *a* rail. ● *m* railwayman (*Brit*), railroad worker (*Amer*)
fértil *a* fertile
fertili|dad *f* fertility. **~zante** *m* fertilizer. **~zar** [10] *vt* fertilize
férvido *a* fervent
ferv|iente *a* fervent. **~or** *m* fervour
festej|ar *vt* celebrate; entertain ‹*persona*›; court ‹*novia etc*›; (*Mex, golpear*) beat. **~o** *m* entertainment; (*celebración*) celebration
festiv|al *m* festival. **~idad** *f* festivity. **~o** *a* festive; (*humorístico*) humorous. **día ~o** feast day, holiday
festonear *vt* festoon
fétido *a* stinking
feto *m* foetus
feudal *a* feudal
fiado *m*. **al ~** on credit. **~r** *m* fastener; (*jurid*) guarantor
fiambre *m* cold meat
fianza *f* (*dinero*) deposit; (*objeto*) surety. **bajo ~** on bail. **dar ~** pay a deposit
fiar [20] *vt* guarantee; (*vender*) sell on credit; (*confiar*) confide. ● *vi* trust. **~se** *vpr*. **~se de** trust
fiasco *m* fiasco
fibra *f* fibre; (*fig*) energy. **~ de vidrio** fibreglass
fic|ción *f* fiction. **~ticio** *a* fictitious; (*falso*) false
fich|a *f* token; (*tarjeta*) index card; (*en los juegos*) counter. **~ar** *vt* file. **~ero** *m* card index. **estar ~ado** have a (police) record
fidedigno *a* reliable
fidelidad *f* faithfulness. **alta ~** hi-fi (*fam*), high fidelity
fideos *mpl* noodles
fiebre *f* fever. **~ del heno** hay fever. **tener ~** have a temperature
fiel *a* faithful; ‹*memoria, relato etc*› reliable. ● *m* believer; (*de balanza*) needle. **los ~es** the faithful
fieltro *m* felt
fier|a *f* wild animal; (*persona*) brute. **~o** *a* fierce; (*cruel*) cruel. **estar hecho una ~a** be furious
fierro *m* (*LAm*) iron
fiesta *f* party; (*día festivo*) holiday. **~s** *fpl* celebrations. **~ nacional** bank holiday (*Brit*), national holiday
figura *f* figure; (*forma*) shape; (*en obra de teatro*) character; (*en naipes*) court-card. **~r** *vt* feign; (*representar*) represent. ● *vi* figure; (*ser importante*) be important. **~rse** *vpr* imagine. **¡figúrate!** just imagine! **~tivo** *a* figurative
fij|ación *f* fixing. **~ar** *vt* fix; stick ‹*sello*›; post ‹*cartel*›. **~arse** *vpr* settle; (*fig, poner atención*) notice. **¡fíjate!** just imagine! **~o** *a* fixed; (*firme*) stable; ‹*persona*› settled. **de ~o** certainly
fila *f* line; (*de soldados etc*) file; (*en el teatro, cine etc*) row; (*cola*) queue. **ponerse en ~** line up
filamento *m* filament
fil|antropía *f* philanthropy. **~antrópico** *a* philanthropic. **~ántropo** *m* philanthropist

filarmónico *a* philharmonic
filat|elia *f* stamp collecting, philately. **~élico** *a* philatelic. ● *m* stamp collector, philatelist
filete *m* fillet
filfa *f* (*fam*) hoax
filial *a* filial. ● *f* subsidiary
filigrana *f* filigree (work); (*en papel*) watermark
Filipinas *fpl*. **las (islas)** ~ the Philippines
filipino *a* Philippine, Filipino
filmar *vt* film
filo *m* edge; (*de hoja*) cutting edge; (*Mex*, *hambre*) hunger. **al ~ de las doce** at exactly twelve o'clock. **dar ~ a**, **sacar ~ a** sharpen
filología *f* philology
filón *m* vein; (*fig*) gold-mine
fil|osofía *f* philosophy. **~osófico** *a* philosophical. **~ósofo** *m* philosopher
filtr|ar *vt* filter. **~arse** *vpr* filter; ‹*dinero*› disappear. **~o** *m* filter; (*bebida*) philtre
fin *m* end; (*objetivo*) aim. **~ de semana** weekend. **a ~ de** in order to. **a ~ de cuentas** all things considered. **a ~ de que** in order that. **a ~es de** at the end of. **al ~** finally. **al ~ y al cabo** after all. **dar ~ a** end. **en ~** in short. **poner ~ a** end. **por ~** finally. **sin ~** endless
final *a* final, last. ● *m* end. ● *f* final. **~idad** *f* aim. **~ista** *m & f* finalist. **~izar** [10] *vt/i* end. **~mente** *adv* finally
financi|ar *vt* finance. **~ero** *a* financial. ● *m* financier
finca *f* property; (*tierras*) estate; (*LAm*, *granja*) farm
finés *a* Finnish. ● *m* Finn; (*lengua*) Finnish
fingi|do *a* false. **~r** [14] *vt* feign; (*simular*) simulate. ● *vi* pretend. **~rse** *vpr* pretend to be
finito *a* finite
finlandés *a* Finnish. ● *m* (*persona*) Finn; (*lengua*) Finnish
Finlandia *f* Finland
fin|o *a* fine; (*delgado*) slender; (*astuto*) shrewd; ‹*sentido*› keen; (*cortés*) polite; ‹*jerez*› dry. **~ura** *f* fineness; (*astucia*) shrewdness; (*de sentido*) keenness; (*cortesía*) politeness
fiordo *m* fiord
firma *f* signature; (*empresa*) firm
firmamento *m* firmament
firmar *vt* sign
firme *a* firm; (*estable*) stable, steady; ‹*persona*› steadfast. ● *m* (*pavimento*) (road) surface. ● *adv* hard. **~za** *f* firmness. **de ~** hard. **en ~** firm, definite
fisc|al *a* fiscal. ● *m & f* public prosecutor. **~o** *m* treasury
fisg|ar [12] *vt* pry into ‹*asunto*›; spy on ‹*persona*›. ● *vi* pry. **~ón** *a* prying. ● *m* busybody
físic|a *f* physics. **~o** *a* physical. ● *m* physique; (*persona*) physicist
fisi|ología *f* physiology. **~ológico** *a* physiological. **~ólogo** *m* physiologist
fisioterap|euta *m & f* physiotherapist. **~ia** *f* physiotherapy. **~ista** *m & f* (*fam*) physiotherapist
fisonom|ía *f* physiognomy, face. **~ista** *m & f*. **ser buen ~ista** be good at remembering faces
fisura *f* (*Med*) fracture
fláccido *a* flabby
flaco *a* thin, skinny; (*débil*) weak
flagelo *m* scourge
flagrante *a* flagrant. **en ~** red-handed
flamante *a* splendid; (*nuevo*) brand-new
flamenco *a* flamenco; (*de Flandes*) Flemish. ● *m* (*música etc*) flamenco
flan *m* crème caramel
flaqueza *f* thinness; (*debilidad*) weakness
flash *m* flash
flato *m*, **flatulencia** *f* flatulence
flaut|a *f* flute. ● *m & f* (*músico*) flautist, flutist (*Amer*). **~ín** *m* piccolo. **~ista** *m & f* flautist, flutist (*Amer*)
fleco *m* fringe
flecha *f* arrow
flem|a *f* phlegm. **~ático** *a* phlegmatic
flequillo *m* fringe
fletar *vt* charter
flexib|ilidad *f* flexibility. **~le** *a* flexible. ● *m* flex, cable
flirte|ar *vi* flirt. **~o** *m* flirting
floj|ear *vi* ease up. **~o** *a* loose; (*poco fuerte*) weak; ‹*viento*› light; (*perezoso*) lazy
flor *f* flower; (*fig*) cream. **~a** *f* flora. **~al** *a* floral. **~ecer** [11] *vi* flower, bloom; (*fig*) flourish. **~eciente** *a* (*fig*) flourishing. **~ero** *m* flower vase. **~ido** *a* flowery; (*selecto*) select; ‹*lenguaje*› florid. **~ista** *m & f* florist

flota *f* fleet
flot|ador *m* float. **~ar** *vi* float. **~e** *m*. **a ~e** afloat
flotilla *f* flotilla
fluctua|ción *f* fluctuation. **~r** [21] *vi* fluctuate
flu|idez *f* fluidity; (*fig*) fluency. **~ido** *a* fluid; (*fig*) fluent. ● *m* fluid. **~ir** [17] *vi* flow. **~jo** *m* flow. **~o y reflujo** ebb and flow
fluorescente *a* fluorescent
fluoruro *m* fluoride
fluvial *a* river
fobia *f* phobia
foca *f* seal
foc|al *a* focal. **~o** *m* focus; (*lámpara*) floodlight; (*LAm, bombilla*) light bulb
fogón *m* (*cocina*) cooker
fogoso *a* spirited
folio *m* leaf
folkl|ore *m* folklore. **~órico** *a* folk
follaje *m* foliage
follet|ín *m* newspaper serial. **~o** *m* pamphlet
follón *m* (*lío*) mess; (*alboroto*) row
fomentar *vt* foment, stir up
fonda *f* (*pensión*) boarding-house
fondo *m* bottom; (*parte más lejana*) bottom, end; (*de escenario, pintura etc*) background; (*profundidad*) depth. **~s** *mpl* funds, money. **a ~** thoroughly. **en el ~** deep down
fonétic|a *f* phonetics. **~o** *a* phonetic
fono *m* (*LAm, del teléfono*) earpiece
fontaner|ía plumbing. **~o** *m* plumber
footing /'futin/ *m* jogging
forastero *a* alien. ● *m* stranger
forceje|ar *vi* struggle. **~o** *m* struggle
fórceps *m invar* forceps
forense *a* forensic
forjar *vt* forge
forma *f* form, shape; (*horma*) mould; (*modo*) way; (*de zapatero*) last. **~s** *fpl* conventions. **~ción** *f* formation; (*educación*) training. **dar ~ a** shape; (*expresar*) formulate. **de ~ que** so (that). **de todas ~s** anyway. **estar en ~** be in good form. **guardar ~s** keep up appearances
formal *a* formal; (*de fiar*) reliable; (*serio*) serious. **~idad** *f* formality; (*fiabilidad*) reliability; (*seriedad*) seriousness
formar *vt* form; (*hacer*) make; (*enseñar*) train. **~se** *vpr* form; (*desarrollarse*) develop
formato *m* format
formidable *a* formidable; (*muy grande*) enormous; (*muy bueno, fam*) marvellous
fórmula *f* formula; (*receta*) recipe
formular *vt* formulate; make ‹*queja etc*›; (*expresar*) express
fornido *a* well-built
forraje *m* fodder. **~ar** *vt/i* forage
forr|ar *vt* (*en el interior*) line; (*en el exterior*) cover. **~o** *m* lining; (*cubierta*) cover. **~o del freno** brake lining
fortale|cer [11] *vt* strengthen. **~za** *f* strength; (*mil*) fortress; (*fuerza moral*) fortitude
fortificar [7] *vt* fortify
fortuito *a* fortuitous. **encuentro** *m* **~** chance meeting
fortuna *f* fortune; (*suerte*) luck. **por ~** fortunately
forz|ado *a* hard. **~ar** [2 & 10] *vt* force. **~osamente** *adv* necessarily. **~oso** *a* inevitable; (*necesario*) necessary
fosa *f* grave
fosfato *m* phosphate
fósforo *m* phosphorus; (*cerilla*) match
fósil *a & m* fossil
fosilizarse [10] *vpr* fossilize
foso *m* ditch
foto *f* photo, photograph. **sacar ~s** take photographs
fotocopia *f* photocopy. **~dora** *f* photocopier. **~r** *vt* photocopy
fotogénico *a* photogenic
fot|ografía *f* photography; (*foto*) photograph. **~ografiar** [20] *vt* photograph. **~ográfico** *a* photographic. **~ógrafo** *m* photographer. **sacar ~ografías** take photographs
foyer *m* foyer
frac *m* (*pl* **fraques** *o* **fracs**) tails
fracas|ar *vi* fail. **~o** *m* failure
fracción *f* fraction; (*pol*) faction
fractura *f* fracture. **~r** *vt* fracture, break. **~rse** *vpr* fracture, break
fragan|cia *f* fragrance. **~te** *a* fragrant
fragata *f* frigate
fr|ágil *a* fragile; (*débil*) weak. **~agilidad** *f* fragility; (*debilidad*) weakness
fragment|ario *a* fragmentary. **~o** *m* fragment
fragor *m* din
fragoso *a* rough

fragua *f* forge. **~r** [15] *vt* forge; (*fig*) concoct. ● *vi* harden

fraile *m* friar; (*monje*) monk

frambuesa *f* raspberry

francés *a* French. ● *m* (*persona*) Frenchman; (*lengua*) French

Francia *f* France

franco *a* frank; (*com*) free. ● *m* (*moneda*) franc

francotirador *m* sniper

franela *f* flannel

franja *f* border; (*fleco*) fringe

franque|ar *vt* clear; stamp ‹*carta*›; overcome ‹*obstáculo*›. **~o** *m* stamping; (*cantidad*) postage

franqueza *f* frankness; (*familiaridad*) familiarity

franquis|mo *m* General Franco's regime; (*política*) Franco's policy. **~ta** *a* pro-Franco

frasco *m* small bottle

frase *f* phrase; (*oración*) sentence. **~ hecha** set phrase

fratern|al *a* fraternal. **~idad** *f* fraternity

fraud|e *m* fraud. **~ulento** *a* fraudulent

fray *m* brother, friar

frecuen|cia *f* frequency. **~tar** *vt* frequent. **~te** *a* frequent. **con ~cia** frequently

frega|dero *m* sink. **~r** [1 & 12] *vt* scrub; wash up ‹*los platos*›; mop ‹*el suelo*›; (*LAm*, *fig*, *molestar*, *fam*) annoy

freír [51, *pp* **frito**] *vt* fry; (*fig*, *molestar*, *fam*) annoy. **~se** *vpr* fry; ‹*persona*› be very hot, be boiling (*fam*)

frenar *vt* brake; (*fig*) check

fren|esí *m* frenzy. **~ético** *a* frenzied

freno *m* (*de caballería*) bit; (*auto*) brake; (*fig*) check

frente *m* front. ● *f* forehead. **~ a** opposite; (*en contra de*) opposed to. **~ por ~** opposite; (*en un choque*) head-on. **al ~** at the head; (*hacia delante*) forward. **arrugar la ~** frown. **de ~** forward. **hacer ~ a** face ‹*cosa*›; stand up to ‹*persona*›

fresa *f* strawberry

fresc|a *f* fresh air. **~o** *a* (*frío*) cool; (*nuevo*) fresh; (*descarado*) cheeky. ● *m* fresh air; (*frescor*) coolness; (*mural*) fresco; (*persona*) impudent person. **~or** *m* coolness. **~ura** *f* freshness; (*frío*) coolness; (*descaro*) cheek. **al ~o** in the open air. **hacer ~o** be cool. **tomar el ~o** get some fresh air

fresno *m* ash (tree)

friable *a* friable

frialdad *f* coldness; (*fig*) indifference

fricci|ón *f* rubbing; (*fig*, *tec*) friction; (*masaje*) massage. **~onar** *vt* rub

frigidez *f* coldness; (*fig*) frigidity

frígido *a* frigid

frigorífico *m* refrigerator, fridge (*fam*)

fríjol *m* bean. **~es refritos** (*Mex*) purée of black beans

frío *a* & *m* cold. **coger ~** catch cold. **hacer ~** be cold

frisar *vi*. **~ en** be getting on for, be about

frito *a* fried; (*exasperado*) exasperated. **me tiene ~** I'm sick of him

fr|ivolidad *f* frivolity. **~ívolo** *a* frivolous

fronda *f* foliage

fronter|a *f* frontier; (*fig*) limit. **~izo** *a* frontier. **~o** *a* opposite

frontón *m* pelota court

frotar *vt* rub; strike ‹*cerilla*›

fructífero *a* fruitful

frugal *a* frugal

fruncir [9] *vt* gather ‹*tela*›; wrinkle ‹*piel*›

fruslería *f* trifle

frustra|ción *f* frustration. **~r** *vt* frustrate. **~rse** *vpr* (*fracasar*) fail. **quedar ~do** be disappointed

frut|a *f* fruit. **~ería** *f* fruit shop. **~ero** *a* fruit. ● *m* fruiterer; (*recipiente*) fruit bowl. **~icultura** *f* fruit-growing. **~illa** *f* (*LAm*) strawberry. **~o** *m* fruit

fucsia *f* fuchsia

fuego *m* fire. **~s artificiales** fireworks. **a ~ lento** on a low heat. **tener ~** have a light

fuente *f* fountain; (*manantial*) spring; (*plato*) serving dish; (*fig*) source

fuera *adv* out; (*al exterior*) outside; (*en otra parte*) away; (*en el extranjero*) abroad. ● *vb véase* **ir** *y* **ser**. **~ de** outside; (*excepto*) except for, besides. **por ~** on the outside

fuerte *a* strong; ‹*color*› bright; ‹*sonido*› loud; ‹*dolor*› severe; (*duro*) hard; (*grande*) large; ‹*lluvia*, *nevada*› heavy. ● *m* fort; (*fig*) strong point. ● *adv* hard; (*con hablar etc*) loudly; (*mucho*) a lot

fuerza *f* strength; (*poder*) power; (*en física*) force; (*mil*) forces. **~ de**

voluntad will-power. **a ~ de** by dint of, by means of. **a la ~** by necessity. **por ~** by force; (*por necesidad*) by necessity. **tener ~s para** have the strength to
fuese *vb véase* **ir** *y* **ser**
fug|a *f* flight, escape; (*de gas etc*) leak; (*mus*) fugue. **~arse** [12] *vpr* flee, escape. **~az** *a* fleeting. **~itivo** *a & m* fugitive. **ponerse en ~a** take to flight
fui *vb véase* **ir** *y* **ser**
fulano *m* so-and-so. **~, mengano y zutano** Tom, Dick and Harry
fulgor *m* brilliance; (*fig*) splendour
fulminar *vt* strike by lightning; (*fig, mirar*) look daggers at
fuma|dor *a* smoking. ● *m* smoker. **~r** *vt/i* smoke. **~rse** *vpr* smoke; (*fig, gastar*) squander. **~rada** *f* puff of smoke. **~r en pipa** smoke a pipe. **prohibido ~r** no smoking
funámbulo *m* tightrope walker
funci|ón *f* function; (*de un cargo etc*) duties; (*de teatro*) show, performance. **~onal** *a* functional. **~onar** *vi* work, function. **~onario** *m* civil servant. **no ~ona** out of order
funda *f* cover. **~ de almohada** pillowcase
funda|ción *f* foundation. **~mental** *a* fundamental. **~mentar** *vt* lay the foundations of; (*fig*) base. **~mento** *m* foundation. **~r** *vt* found; (*fig*) base. **~rse** *vpr* be based
fundi|ción *f* melting; (*de metales*) smelting; (*taller*) foundry. **~r** *vt* melt; smelt ‹*metales*›; cast ‹*objeto*›; blend ‹*colores*›; (*fusionar*) merge. **~rse** *vpr* melt; (*unirse*) merge
fúnebre *a* funeral; (*sombrío*) gloomy
funeral *a* funeral. ● *m* funeral. **~es** *mpl* funeral
funicular *a & m* funicular
furg|ón *m* van. **~oneta** *f* van
fur|ia *f* fury; (*violencia*) violence. **~ibundo** *a* furious. **~ioso** *a* furious. **~or** *m* fury
furtivo *a* furtive
furúnculo *m* boil
fuselaje *m* fuselage
fusible *m* fuse
fusil *m* gun. **~ar** *vt* shoot
fusión *f* melting; (*unión*) fusion; (*com*) merger
fútbol *m* football
futbolista *m* footballer
fútil *a* futile
futur|ista *a* futuristic. ● *m & f* futurist. **~o** *a & m* future

G

gabán *m* overcoat
garbardina *f* raincoat; (*tela*) gabardine
gabinete *m* (*pol*) cabinet; (*en museo etc*) room; (*de dentista, médico etc*) consulting room
gacela *f* gazelle
gaceta *f* gazette
gachas *fpl* porridge
gacho *a* drooping
gaélico *a* Gaelic
gafa *f* hook. **~s** *fpl* glasses, spectacles. **~s de sol** sun-glasses
gaf|ar *vt* hook; (*fam*) bring bad luck to. **~e** *m* jinx
gaita *f* bagpipes
gajo *m* (*de naranja, nuez etc*) segment
gala|s *fpl* finery, best clothes. **estar de ~** be dressed up. **hacer ~ de** show off
galán *m* (*en el teatro*) male lead; (*enamorado*) lover
galante *a* gallant. **~ar** *vt* court. **~ría** *f* gallantry
galápago *m* turtle
galardón *m* reward
galaxia *f* galaxy
galeón *m* galleon
galera *f* galley
galería *f* gallery
Gales *m* Wales. **país de ~** Wales
gal|és *a* Welsh. ● *m* Welshman; (*lengua*) Welsh. **~esa** *f* Welshwoman
galgo *m* greyhound
Galicia *f* Galicia
galimatías *m invar* (*fam*) gibberish
galón *m* gallon; (*cinta*) braid; (*mil*) stripe
galop|ar *vi* gallop. **~e** *m* gallop
galvanizar [10] *vt* galvanize
gallard|ía *f* elegance. **~o** *a* elegant
gallego *a & m* Galician
galleta *f* biscuit (*Brit*), cookie (*Amer*)
gall|ina *f* hen, chicken; (*fig, fam*) coward. **~o** *m* cock
gama *f* scale; (*fig*) range
gamba *f* prawn (*Brit*), shrimp (*Amer*)
gamberro *m* hooligan
gamuza *f* (*piel*) chamois leather

gana *f* wish, desire; (*apetito*) appetite. **de buena ~** willingly. **de mala ~** reluctantly. **no me da la ~** I don't feel like it. **tener ~s de** (+ *infinitivo*) feel like (+ *gerundio*)
ganad|ería *f* cattle raising; (*ganado*) livestock. **~o** *m* livestock. **~o de cerda** pigs. **~o lanar** sheep. **~o vacuno** cattle
ganar *vt* earn; (*en concurso, juego etc*) win; (*alcanzar*) reach; (*aventajar*) beat. ● *vi* (*vencer*) win; (*mejorar*) improve. **~se la vida** earn a living. **salir ganando** come out better off
ganch|illo *m* crochet. **~o** *m* hook. **~oso** *a*, **~udo** *a* hooked. **echar el ~o a** hook. **hacer ~illo** crochet. **tener ~o** be very attractive
gandul *a* & *m* & *f* good-for-nothing
ganga *f* bargain; (*buena situación*) easy job, cushy job (*fam*)
gangrena *f* gangrene
gans|ada *f* silly thing. **~o** *m* goose
gañi|do *m* yelping. **~r** [22] *vi* yelp
garabat|ear *vt/i* (*garrapatear*) scribble. **~o** *m* (*garrapato*) scribble
garaj|e *m* garage. **~ista** *m* & *f* garage attendant
garant|e *m* & *f* guarantor. **~ía** *f* guarantee. **~ir** [24] *vt* (*esp LAm*), **~izar** [10] *vt* guarantee
garapiñado *a*. **almendras** *fpl* **garapiñadas** sugared almonds
garbanzo *m* chick-pea
garbo *m* poise; (*de escrito*) style. **~so** *a* elegant
garfio *m* hook
garganta *f* throat; (*desfiladero*) gorge; (*de botella*) neck
gárgaras *fpl*. **hacer ~** gargle
gargarismo *m* gargle
gárgola *f* gargoyle
garita *f* hut; (*de centinela*) sentry box
garito *m* gambling den
garra *f* (*de animal*) claw; (*de ave*) talon
garrafa *f* carafe
garrapata *f* tick
garrapat|ear *vi* scribble. **~o** *m* scribble
garrote *m* club, cudgel; (*tormento*) garrotte
gárrulo *a* garrulous
garúa *f* (*LAm*) drizzle
garza *f* heron
gas *m* gas. **con ~** fizzy. **sin ~** still
gasa *f* gauze
gaseosa *f* lemonade
gasfitero *m* (*Arg*) plumber
gas|óleo *m* diesel. **~olina** *f* petrol (*Brit*), gasoline (*Amer*), gas (*Amer*). **~olinera** *f* petrol station (*Brit*), gas station (*Amer*); (*lancha*) motor boat. **~ómetro** *m* gasometer
gast|ado *a* spent; ‹*vestido etc*› worn out. **~ador** *m* spendthrift. **~ar** *vt* spend; (*consumir*) use; (*malgastar*) waste; wear ‹*vestido etc*›; crack ‹*broma*›. ● *vi* spend. **~arse** *vpr* wear out. **~o** *m* expense; (*acción de gastar*) spending
gástrico *a* gastric
gastronomía *f* gastronomy
gat|a *f* cat. **a ~as** on all fours. **~ear** *vi* crawl
gatillo *m* trigger; (*de dentista*) (dental) forceps
gat|ito *m* kitten. **~o** *m* cat. **dar ~o por liebre** take s.o. in
gaucho *a* & *m* Gaucho
gaveta *f* drawer
gavilla *f* sheaf; (*de personas*) band, gang
gaviota *f* seagull
gazpacho *m* gazpacho, cold soup
géiser *m* geyser
gelatina *f* gelatine; (*jalea*) jelly
gelignita *f* gelignite
gema *f* gem
gemelo *m* twin. **~s** *mpl* (*anteojos*) binoculars; (*de camisa*) cuff-links. **G~s** Gemini
gemido *m* groan
Géminis *mpl* Gemini
gemir [5] *vi* groan; ‹*animal*› whine, howl
gen *m*, **gene** *m* gene
geneal|ogía *f* genealogy. **~ógico** *a* genealogical. **árbol** *m* **~ógico** family tree
generación *f* generation
general *a* general; (*corriente*) common. ● *m* general. **~ísimo** *m* generalissimo, supreme commander. **~ización** *f* generalization. **~izar** [10] *vt/i* generalize. **~mente** *adv* generally. **en ~** in general. **por lo ~** generally
generar *vt* generate
género *m* type, sort; (*biol*) genus; (*gram*) gender; (*producto*) product. **~s de punto** knitwear. **~ humano** mankind
generos|idad *f* generosity. **~o** *a* generous; ‹*vino*› full-bodied
génesis *m* genesis
genétic|a *f* genetics. **~o** *a* genetic

genial *a* brilliant; (*agradable*) pleasant
genio *m* temper; (*carácter*) nature; (*talento, persona*) genius
genital *a* genital. **~es** *mpl* genitals
gente *f* people; (*nación*) nation; (*familia, fam*) family; (*Mex, persona*) person
gentil *a* charming; (*pagano*) pagan. **~eza** *f* elegance; (*encanto*) charm; (*amabilidad*) kindness
gentío *m* crowd
genuflexión *f* genuflection
genuino *a* genuine
ge|ografía *f* geography. **~ográfico** *a* geographical. **~ógrafo** *m* geographer
ge|ología *f* geology. **~ólogo** *m* geologist
geom|etría *f* geometry. **~étrico** *a* geometrical
geranio *m* geranium
geren|cia *f* management. **~te** *m* manager
geriatría *f* geriatrics
germánico *a & m* Germanic
germen *m* germ
germicida *f* germicide
germinar *vi* germinate
gestación *f* gestation
gesticula|ción *f* gesticulation. **~r** *vi* gesticulate; (*hacer muecas*) grimace
gesti|ón *f* step; (*administración*) management. **~onar** *vt* take steps to arrange; (*dirigir*) manage
gesto *m* expression; (*ademán*) gesture; (*mueca*) grimace
Gibraltar *m* Gibraltar
gibraltareño *a & m* Gibraltarian
gigante *a* gigantic. ● *m* giant. **~sco** *a* gigantic
gimn|asia *f* gymnastics. **~asio** *m* gymnasium, gym (*fam*). **~asta** *m & f* gymnast. **~ástica** *f* gymnastics
gimotear *vi* whine
ginebra *f* gin
Ginebra *f* Geneva
ginec|ología *f* gynaecology. **~ólogo** *m* gynaecologist
gira *f* excursion; (*a varios sitios*) tour
girar *vt* spin; (*por giro postal*) transfer. ● *vi* rotate, go round; ‹*camino etc*› turn
girasol *m* sunflower
gir|atorio *a* revolving. **~o** *m* turn; (*com*) draft; (*locución*) expression. **~o postal** postal order
giroscopio *m* gyroscope
gis *m* chalk
gitano *a & m* gypsy
glacia|l *a* icy. **~r** *m* glacier
gladiador *m* gladiator
glándula *f* gland
glasear *vt* glaze; (*culin*) ice
glicerina *f* glycerine
glicina *f* wisteria
glob|al *a* global; (*fig*) overall. **~o** *m* globe; (*aeróstato, juguete*) balloon
glóbulo *m* globule; (*med*) corpuscle
gloria *f* glory. **~rse** *vpr* boast (**de** about)
glorieta *f* bower; (*auto*) roundabout (*Brit*), (traffic) circle (*Amer*)
glorificar [7] *vt* glorify
glorioso *a* glorious
glosario *m* glossary
glot|ón *a* gluttonous. ● *m* glutton. **~onería** *f* gluttony
glucosa *f* glucose
gnomo /'nomo/ *m* gnome
gob|ernación *f* government. **~ernador** *a* governing. ● *m* governor. **~ernante** *a* governing. **~ernar** [1] *vt* govern; (*dirigir*) manage, direct. **~ierno** *m* government; (*dirección*) management, direction. **~ierno de la casa** housekeeping. **Ministerio** *m* **de la G~ernación** Home Office (*Brit*), Department of the Interior (*Amer*)
goce *m* enjoyment
gol *m* goal
golf *m* golf
golfo *m* gulf; (*niño*) urchin; (*holgazán*) layabout
golondrina *f* swallow
golos|ina *f* titbit; (*dulce*) sweet. **~o** *a* fond of sweets
golpe *m* blow; (*puñetazo*) punch; (*choque*) bump; (*de emoción*) shock; (*acceso*) fit; (*en fútbol*) shot; (*en golf, en tenis, de remo*) stroke. **~ar** *vt* hit; (*dar varios golpes*) beat; (*con mucho ruido*) bang; (*con el puño*) punch. ● *vi* knock. **~ de estado** coup d'etat. **~ de fortuna** stroke of luck. **~ de mano** raid. **~ de vista** glance. **~ militar** military coup. **de ~** suddenly. **de un ~** at one go
gom|a *f* rubber; (*para pegar*) glue; (*anillo*) rubber band; (*elástico*) elastic. **~a de borrar** rubber. **~a de pegar** glue. **~a espuma** foam rubber. **~ita** *f* rubber band
gongo *m* gong
gord|a *f* (*Mex*) thick tortilla. **~iflón** *m* (*fam*), **~inflón** *m* (*fam*) fatty. **~o** *a* ‹*persona*› fat; ‹*carne*› fatty;

(*grande*) large, big. ● *m* first prize. **~ura** *f* fatness; (*grasa*) fat
gorila *f* gorilla
gorje|ar *vi* chirp. **~o** *m* chirping
gorra *f* cap
gorrión *m* sparrow
gorro *m* cap; (*de niño*) bonnet
got|a *f* drop; (*med*) gout. **~ear** *vi* drip. **~eo** *m* dripping. **~era** *f* leak. **ni ~a** nothing
gótico *a* Gothic
gozar [10] *vt* enjoy. ● *vi*. **~ de** enjoy. **~se** *vpr* enjoy
gozne *m* hinge
gozo *m* pleasure; (*alegría*) joy. **~so** *a* delighted
graba|ción *f* recording. **~do** *m* engraving, print; (*en libro*) illustration. **~r** *vt* engrave; record ‹*discos etc*›
gracejo *m* wit
graci|a *f* grace; (*favor*) favour; (*humor*) wit. **~as** *fpl* thanks. **¡~as!** thank you!, thanks! **~oso** *a* funny. ● *m* fool, comic character. **dar las ~as** thank. **hacer ~a** amuse; (*gustar*) please. **¡muchas ~as!** thank you very much! **tener ~a** be funny
grad|a *f* step; (*línea*) row; (*de anfiteatro*) tier. **~ación** *f* gradation. **~o** *m* degree; (*escol*) year (*Brit*), grade (*Amer*); (*voluntad*) willingness
gradua|ción *f* graduation; (*de alcohol*) proof. **~do** *m* graduate. **~l** *a* gradual. **~r** [21] *vt* graduate; (*medir*) measure; (*univ*) confer a degree on. **~rse** *vpr* graduate
gráfic|a *f* graph. **~o** *a* graphic. ● *m* graph
grajo *m* rook
gram|ática *f* grammar. **~atical** *a* grammatical
gramo *m* gram, gramme (*Brit*)
gramófono *m* record-player, gramophone (*Brit*), phonograph (*Amer*)
gran *a véase* **grande**
grana *f* (*color*) scarlet
granada *f* pomegranate; (*mil*) grenade
granate *m* garnet
Gran Bretaña *f* Great Britain
grande *a* (*delante de nombre en singular* **gran**) big, large; (*alto*) tall; (*fig*) great. ● *m* grandee. **~za** *f* greatness
grandioso *a* magnificent
granel *m*. **a ~** in bulk; (*suelto*) loose; (*fig*) in abundance
granero *m* barn
granito *m* granite; (*grano*) small grain
graniz|ado *m* iced drink. **~ar** [10] *vi* hail. **~o** *m* hail
granj|a *f* farm. **~ero** *m* farmer
grano *m* grain; (*semilla*) seed; (*de café*) bean; (*med*) spot. **~s** *mpl* cereals
granuja *m & f* rogue
gránulo *m* granule
grapa *f* staple
gras|a *f* grease; (*culin*) fat. **~iento** *a* greasy
gratifica|ción *f* (*propina*) tip; (*de sueldo*) bonus. **~r** [7] *vt* (*dar propina*) tip
gratis *adv* free
gratitud *f* gratitude
grato *a* pleasant; (*bienvenido*) welcome
gratuito *a* free; (*fig*) uncalled for
grava *f* gravel
grava|men *m* obligation. **~r** *vt* tax; (*cargar*) burden
grave *a* serious; (*pesado*) heavy; ‹*sonido*› low; ‹*acento*› grave. **~dad** *f* gravity
gravilla *f* gravel
gravita|ción *f* gravitation. **~r** *vi* gravitate; (*apoyarse*) rest (**sobre** on); (*fig, pesar*) weigh (**sobre** on)
gravoso *a* onerous; (*costoso*) expensive
graznar *vi* ‹*cuervo*› caw; ‹*pato*› quack
Grecia *f* Greece
gregario *a* gregarious
greguería *f* uproar
gremio *m* union
greñ|a *f* mop of hair. **~udo** *a* unkempt
gresca *f* uproar; (*riña*) quarrel
griego *a & m* Greek
grieta *f* crack
grifo *m* tap, faucet (*Amer*); (*animal fantástico*) griffin
grilletes *mpl* shackles
grillo *m* cricket; (*bot*) shoot. **~s** *mpl* shackles
grima *f*. **dar ~** annoy
gringo *m* (*LAm*) Yankee (*fam*), American
gripe *f* flu (*fam*), influenza
gris *a* grey. ● *m* grey; (*policía, fam*) policeman
grit|ar *vt* shout (for); (*como protesta*) boo. ● *vi* shout. **~ería** *f*, **~erío** *m*

uproar. **~o** *m* shout; (*de dolor, sorpresa*) cry; (*chillido*) scream. **dar ~s** shout
grosella *f* redcurrant. **~ negra** blackcurrant
groser|ía *f* coarseness; (*palabras etc*) coarse remark. **~o** *a* coarse; (*descortés*) rude
grosor *m* thickness
grotesco *a* grotesque
grúa *f* crane
grues|a *f* gross. **~o** *a* thick; ‹*persona*› fat, stout. ● *m* thickness; (*fig*) main body
grulla *f* crane
grumo *m* clot; (*de leche*) curd
gruñi|do *m* grunt; (*fig*) grumble. **~r** [22] *vi* grunt; ‹*perro*› growl; (*refunfuñar*) grumble
grupa *f* hindquarters
grupo *m* group
gruta *f* grotto
guacamole *m* (*Mex*) avocado purée
guadaña *f* scythe
guagua *f* trifle; (*esp LAm, autobús, fam*) bus
guante *m* glove
guapo *a* good-looking; ‹*chica*› pretty; (*elegante*) smart
guarapo *m* (*LAm*) sugar cane liquor
guarda *m & f* guard; (*de parque etc*) keeper. ● *f* protection. **~barros** *m invar* mudguard. **~bosque** *m* gamekeeper. **~costas** *m invar* coastguard vessel. **~dor** *a* careful. ● *m* keeper. **~espaldas** *m invar* bodyguard. **~meta** *m invar* goalkeeper. **~r** *vt* keep; (*vigilar*) guard; (*proteger*) protect; (*reservar*) save, keep. **~rse** *vpr* be on one's guard. **~rse de** (+ *infinitivo*) avoid (+ *gerundio*). **~rropa** *m* wardrobe; (*en local público*) cloakroom. **~vallas** *m invar* (*LAm*) goalkeeper
guardería *f* nursery
guardia *f* guard; (*custodia*) care. ● *f* guard. **G~ Civil** Civil Guard. **~ municipal** policeman. **~ de tráfico** traffic policeman. **estar de ~** be on duty. **estar en ~** be on one's guard. **montar la ~** mount guard
guardián *m* guardian; (*de parque etc*) keeper; (*de edificio*) caretaker
guardilla *f* attic
guar|ecer [11] (*albergar*) give shelter to. **~ecerse** *vpr* take shelter. **~ida** *f* den, lair; (*de personas*) hideout
guarn|ecer [11] *vt* provide; (*adornar*) decorate; (*culin*) garnish. **~ición** *m* decoration; (*de caballo*) harness; (*culin*) garnish; (*mil*) garrison; (*de piedra preciosa*) setting
guarro *m* pig
guasa *f* joke; (*ironía*) irony
guaso *a* (*Arg*) coarse
guasón *a* humorous. ● *m* joker
Guatemala *f* Guatemala
guatemalteco *a* from Guatemala. ● *m* person from Guatemala
guateque *m* party
guayaba *f* guava; (*dulce*) guava jelly
guayabera *f* (*Mex*) shirt
gubernamental *a*, **gubernativo** *a* governmental
güero *a* (*Mex*) fair
guerr|a *f* war; (*método*) warfare. **~a civil** civil war. **~ear** *vi* wage war. **~ero** *a* war; (*belicoso*) fighting. ● *m* warrior. **~illa** *f* band of guerillas. **~illero** *m* guerilla. **dar ~a** annoy
guía *m & f* guide. ● *f* guidebook; (*de teléfonos*) directory; (*de ferrocarriles*) timetable
guiar [20] *vt* guide; (*llevar*) lead; (*auto*) drive. **~se** *vpr* be guided (**por** by)
guij|arro *m* pebble. **~o** *m* gravel
guillotina *f* guillotine
guind|a *f* morello cherry. **~illa** *f* chilli
guiñapo *m* rag; (*fig, persona*) reprobate
guiñ|ar *vt/i* wink. **~o** *m* wink. **hacer ~os** wink
gui|ón *m* hyphen, dash; (*de película etc*) script. **~onista** *m & f* scriptwriter
guirnalda *f* garland
güiro *m* (*LAm*) gourd
guisa *f* manner, way. **a ~ de** as. **de tal ~** in such a way
guisado *m* stew
guisante *m* pea. **~ de olor** sweet pea
guis|ar *vt/i* cook. **~o** *m* dish
güisqui *m* whisky
guitarr|a *f* guitar. **~ista** *m & f* guitarist
gula *f* gluttony
gusano *m* worm; (*larva de mosca*) maggot
gustar *vt* taste. ● *vi* please. **¿te gusta?** do you like it? **me gusta el vino** I like wine
gusto *m* taste; (*placer*) pleasure. **~so** *a* tasty; (*agradable*) pleasant. **a ~** comfortable. **a mi ~** to my liking.

buen ~ (good) taste. **con mucho ~** with pleasure. **dar ~** please. **mucho ~** pleased to meet you

gutural *a* guttural

H

ha *vb véase* **haber**

haba *f* broad bean; (*de café etc*) bean

Habana *f*. **la ~** Havana

haban|era *f* habanera, Cuban dance. **~ero** *a* from Havana. ● *m* person from Havana. **~o** *m* (*puro*) Havana

haber *v aux* [30] have. ● *v. impersonal* (*presente s & pl* **hay**, *imperfecto s & pl* **había**, *pretérito s & pl* **hubo**) be. **hay 5 bancos en la plaza** there are 5 banks in the square. **hay que hacerlo** it must be done, you have to do it. **he aquí** here is, here are. **no hay de qué** don't mention it, not at all. **¿qué hay?** (*¿qué pasa?*) what's the matter?; (*¿qué tal?*) how are you?

habichuela *f* bean

hábil *a* skilful; (*listo*) clever; (*adecuado*) suitable

habilidad *f* skill; (*astucia*) cleverness

habilita|ción *f* qualification. **~r** *vt* qualify

habita|ble *a* habitable. **~ción** *f* room; (*casa etc*) dwelling; (*cuarto de dormir*) bedroom; (*en biología*) habitat. **~ción de matrimonio**, **~ción doble** double room. **~ción individual** , **~ción sencilla** single room. **~do** *a* inhabited. **~nte** *m* inhabitant. **~r** *vt* live in. ● *vi* live

hábito *m* habit

habitual *a* usual, habitual; ‹*cliente*› regular. **~mente** *adv* usually

habituar [21] *vt* accustom. **~se** *vpr*. **~se a** get used to

habla *f* speech; (*idioma*) language; (*dialecto*) dialect. **al ~** (*al teléfono*) speaking. **ponerse al ~ con** get in touch with. **~dor** *a* talkative. ● *m* chatterbox. **~duría** *f* rumour. **~durías** *fpl* gossip. **~nte** *a* speaking. ● *m & f* speaker. **~r** *vt* speak. ● *vi* speak, talk (**con** to). **~rse** *vpr* speak. **¡ni ~r!** out of the question! **se ~ español** Spanish spoken

hacedor *m* creator, maker

hacendado *m* landowner; (*LAm*) farmer

hacendoso *a* hard-working

hacer [31] *vt* do; (*fabricar, producir etc*) make; (*en matemáticas*) make, be. ● *v impersonal* (*con expresiones meteorológicas*) be; (*con determinado periodo de tiempo*) ago. **~se** *vpr* become; (*acostumbrarse*) get used (**a** to); (*estar hecho*) be made. **~ de** act as. **~se a la mar** put to sea. **~se el sordo** pretend to be deaf. **hace buen tiempo** it's fine weather. **hace calor** it's hot. **hace frío** it's cold. **hace poco** recently. **hace 7 años** 7 years ago. **hace sol** it's sunny. **hace viento** it's windy. **¿qué le vamos a ~?** what are we going to do?

hacia *prep* towards; (*cerca de*) near; (*con tiempo*) at about. **~ abajo** down(wards). **~ arriba** up(wards). **~ las dos** at about two o'clock

hacienda *f* country estate; (*en LAm*) ranch; (*LAm, ganado*) livestock; (*pública*) treasury. **Ministerio** *m* **de H~** Ministry of Finance; (*en Gran Bretaña*) Exchequer; (*en Estados Unidos*) Treasury. **ministro** *m* **de H~** Minister of Finance; (*en Gran Bretaña*) Chancellor of the Exchequer; (*en Estados Unidos*) Secretary of the Treasury

hacinar *vt* stack

hacha *f* axe; (*antorcha*) torch

hachís *m* hashish

hada *f* fairy. **cuento** *m* **de ~s** fairy tale

hado *m* fate

hago *vb véase* **hacer**

Haití *m* Haiti

halag|ar [12] *vt* flatter. **~üeño** *a* flattering

halcón *m* falcon

hálito *m* breath

halo *m* halo

hall /xol/ *m* hall

halla|r *vt* find; (*descubrir*) discover. **~rse** *vpr* be. **~zgo** *m* discovery

hamaca *f* hammock; (*asiento*) deck-chair

hambr|e *f* hunger; (*de muchos*) famine. **~iento** *a* starving. **tener ~e** be hungry

Hamburgo *m* Hamburg

hamburguesa *f* hamburger

hamp|a *f* underworld. **~ón** *m* thug

handicap /'xandikap/ *m* handicap

hangar *m* hangar

haragán *a* lazy, idle. ● *m* layabout
harap|iento *a* in rags. **~o** *m* rag
harina *f* flour
harpa *f* harp
hart|ar *vt* satisfy; (*fastidiar*) annoy. **~arse** *vpr* (*comer*) eat one's fill; (*cansarse*) get fed up (**de** with). **~azgo** *m* surfeit. **~o** *a* full; (*cansado*) tired; (*fastidiado*) fed up (**de** with). ● *adv* enough; (*muy*) very. **~ura** *f* surfeit; (*abundancia*) plenty; (*de deseo*) satisfaction
hasta *prep* as far as; (*con tiempo*) until, till; (*Mex*) not until. ● *adv* even. **¡~ la vista!** goodbye!, see you! (*fam*). **¡~ luego!** see you later! **¡~ mañana!** see you tomorrow! **¡~ pronto!** see you soon!
hast|iar [20] *vt* annoy; (*cansar*) weary, tire; (*aburrir*) bore. **~iarse** *vpr* get fed up (**de** with). **~ío** *m* weariness; (*aburrimiento*) boredom; (*asco*) disgust
hat|illo *m* bundle (of belongings); (*ganado*) small flock. **~o** *m* belongings; (*ganado*) flock, herd
haya *f* beech (tree). ● *vb véase* **haber**
Haya *f*. **la ~** the Hague
haz *m* bundle; (*de trigo*) sheaf; (*de rayos*) beam
hazaña *f* exploit
hazmerreír *m* laughing-stock
he *vb véase* **haber**
hebdomadario *a* weekly
hebilla *f* buckle
hebra *f* thread; (*fibra*) fibre
hebreo *a* Hebrew; (*actualmente*) Jewish. ● *m* Hebrew; (*actualmente*) Jew; (*lengua*) Hebrew
hecatombe *m* (*fig*) disaster
hechi|cera *f* witch. **~cería** *f* witchcraft. **~cero** *a* magic. ● *m* wizard. **~zar** [10] *vt* cast a spell on; (*fig*) fascinate. **~zo** *m* witchcraft; (*un acto de brujería*) spell; (*fig*) fascination
hech|o *pp de* **hacer**. ● *a* mature; (*terminado*) finished; ‹*vestidos etc*› ready-made; (*culin*) done. ● *m* fact; (*acto*) deed; (*cuestión*) matter; (*suceso*) event. **~ura** *f* making; (*forma*) form; (*del cuerpo*) build; (*calidad de fabricación*) workmanship. **de ~o** in fact
hed|er [1] *vi* stink. **~iondez** *f* stench. **~iondo** *a* stinking, smelly. **~or** *m* stench
hela|da *f* freeze; (*escarcha*) frost. **~dera** *f* (*LAm*) refrigerator, fridge (*Brit, fam*). **~dería** *f* ice-cream shop. **~do** *a* frozen; (*muy frío*) very cold. ● *m* ice-cream. **~dora** *f* freezer. **~r** [1] *vt* freeze. **~rse** *vpr* freeze
helecho *m* fern
hélice *f* spiral; (*propulsor*) propeller
heli|cóptero *m* helicopter. **~puerto** *m* heliport
hembra *f* female; (*mujer*) woman
hemisferio *m* hemisphere
hemorragia *f* haemorrhage
hemorroides *fpl* haemorrhoids, piles
henchir [5] *vt* fill. **~se** *vpr* stuff o.s.
hend|er [1] *vt* split. **~idura** *f* crack, split; (*geol*) fissure
heno *m* hay
heráldica *f* heraldry
herb|áceo *a* herbaceous. **~olario** *m* herbalist. **~oso** *a* grassy
hered|ad *f* country estate. **~ar** *vt/i* inherit. **~era** *f* heiress. **~ero** *m* heir. **~itario** *a* hereditary
herej|e *m* heretic. **~ía** *f* heresy
herencia *f* inheritance; (*fig*) heritage
heri|da *f* injury. **~do** *a* injured, wounded. ● *m* injured person. **~r** [4] *vt* injure, wound; (*fig*) hurt. **~rse** *vpr* hurt o.s. **los ~dos** the injured; (*cantidad*) the number of injured
herman|a *f* sister. **~a política** sister-in-law. **~astra** *f* stepsister. **~astro** *m* stepbrother. **~dad** *f* brotherhood. **~o** *m* brother. **~o político** brother-in-law. **~os gemelos** twins
hermético *a* hermetic; (*fig*) watertight
hermos|o *a* beautiful; (*espléndido*) splendid; ‹*hombre*› handsome. **~ura** *f* beauty
hernia *f* hernia
héroe *m* hero
hero|ico *a* heroic. **~ína** *f* heroine; (*droga*) heroin. **~ismo** *m* heroism
herr|adura *f* horseshoe. **~amienta** *f* tool. **~ería** *f* smithy. **~ero** *m* blacksmith. **~umbre** *f* rust
herv|idero *m* (*manantial*) spring; (*fig*) hotbed; (*multitud*) throng. **~ir** [4] *vt/i* boil. **~or** *m* boiling; (*fig*) ardour
heterogéneo *a* heterogeneous
heterosexual *a & m & f* heterosexual
hex|agonal *a* hexagonal. **~ágono** *m* hexagon
hiato *m* hiatus

hiberna|ción *f* hibernation. **~r** *vi* hibernate
hibisco *m* hibiscus
híbrido *a* & *m* hybrid
hice *vb véase* **hacer**
hidalgo *m* nobleman
hidrata|nte *a* moisturizing. **~r** *vt* hydrate; ‹*crema etc*› moisturize. **crema** *f* **~nte** moisturizing cream
hidráulico *a* hydraulic
hidroavión *m* seaplane
hidroeléctrico *a* hydroelectric
hidrófilo *a* absorbent
hidr|ofobia *f* rabies. **~ófobo** *a* rabid
hidrógeno *m* hydrogen
hidroplano *m* seaplane
hiedra *f* ivy
hiel *f* (*fig*) bitterness
hielo *m* ice; (*escarcha*) frost; (*fig*) coldness
hiena *f* hyena; (*fig*) brute
hierba *f* grass; (*culin*, *med*) herb. **~buena** *f* mint. **mala ~** weed; (*gente*) bad people, evil people
hierro *m* iron
hígado *m* liver
higi|ene *f* hygiene. **~énico** *a* hygienic
hig|o *m* fig. **~uera** *f* fig tree
hij|a *f* daughter. **~a política** daughter-in-law. **~astra** *f* step-daughter. **~astro** *m* stepson. **~o** *m* son. **~o político** son-in-law. **~s** *mpl* sons; (*chicos y chicas*) children
hilar *vt* spin. **~ delgado** split hairs
hilaridad *f* laughter, hilarity
hilera *f* row; (*mil*) file
hilo *m* thread; (*elec*) wire; (*de líquido*) trickle; (*lino*) linen
hilv|án *m* tacking. **~anar** *vt* tack; (*fig*, *bosquejar*) outline
himno *m* hymn. **~ nacional** anthem
hincapié *m*. **hacer ~ en** stress, insist on
hincar [7] *vt* drive in. **~se** *vpr* sink into. **~se de rodillas** kneel down
hincha *f* (*fam*) grudge; (*aficionado*, *fam*) fan
hincha|do *a* inflated; (*med*) swollen; ‹*persona*› arrogant. **~r** *vt* inflate, blow up. **~rse** *vpr* swell up; (*fig*, *comer mucho*, *fam*) gorge o.s. **~zón** *f* swelling; (*fig*) arrogance
hindi *m* Hindi
hindú *a* Hindu
hiniesta *f* (*bot*) broom
hinojo *m* fennel
hiper... *pref* hyper...
hiper|mercado *m* hypermarket. **~sensible** *a* hypersensitive. **~tensión** *f* high blood pressure
hípico *a* horse
hipn|osis *f* hypnosis. **~ótico** *a* hypnotic. **~otismo** *m* hypnotism. **~otizador** *m* hypnotist. **~otizar** [10] *vt* hypnotize
hipo *m* hiccup. **tener ~** have hiccups
hipocondríaco *a* & *m* hypochondriac
hip|ocresía *f* hypocrisy. **~ócrita** *a* hypocritical. ● *m* & *f* hypocrite
hipodérmico *a* hypodermic
hipódromo *m* racecourse
hipopótamo *m* hippopotamus
hipoteca *f* mortgage. **~r** [7] *vt* mortgage
hip|ótesis *f invar* hypothesis. **~otético** *a* hypothetical
hiriente *a* offensive, wounding
hirsuto *a* shaggy
hirviente *a* boiling
hispánico *a* Hispanic
hispano... *pref* Spanish
Hispanoamérica *f* Spanish America
hispano|americano *a* Spanish American. **~hablante** *a*, **~parlante** *a* Spanish-speaking
hist|eria *f* hysteria. **~érico** *a* hysterical. **~erismo** *m* hysteria
hist|oria *f* history; (*cuento*) story. **~oriador** *m* historian. **~órico** *a* historical. **~orieta** *f* tale; (*con dibujos*) strip cartoon. **pasar a la ~oria** go down in history
hito *m* milestone
hizo *vb véase* **hacer**
hocico *m* snout; (*fig*, *de enfado*) grimace
hockey *m* hockey. **~ sobre hielo** ice hockey
hogar *m* hearth; (*fig*) home. **~eño** *a* home; ‹*persona*› home-loving
hogaza *f* large loaf
hoguera *f* bonfire
hoja *f* leaf; (*de papel*, *metal etc*) sheet; (*de cuchillo*, *espada etc*) blade. **~ de afeitar** razor blade. **~lata** *f* tin. **~latería** *f* tinware. **~latero** *m* tinsmith
hojaldre *m* puff pastry, flaky pastry
hojear *vt* leaf through; (*leer superficialmente*) glance through
hola *int* hello!
Holanda *f* Holland
holand|és *a* Dutch. ● *m* Dutchman; (*lengua*) Dutch. **~esa** *f* Dutchwoman

holg|ado *a* loose; (*fig*) comfortable. **~ar** [2 & 12] *vt* (*no trabajar*) not work, have a day off; (*sobrar*) be unnecessary. **~azán** *a* lazy. ● *m* idler. **~ura** *f* looseness; (*fig*) comfort; (*en mecánica*) play. **huelga decir que** needless to say

holocausto *m* holocaust

hollín *m* soot

hombre *m* man; (*especie humana*) man(kind). ● *int* Good Heavens!; (*de duda*) well. **~ de estado** statesman. **~ de negocios** businessman. **~ rana** frogman. **el ~ de la calle** the man in the street

hombr|era *f* epaulette; (*almohadilla*) shoulder pad. **~o** *m* shoulder

hombruno *a* masculine

homenaje *m* homage; (*fig*) tribute. **rendir ~ a** pay tribute to

home|ópata *m* homoeopath. **~opatía** *f* homoeopathy. **~opático** *a* homoeopathic

homicid|a *a* murderous. ● *m & f* murderer. **~io** *m* murder

homogéneo *a* homogeneous

homosexual *a & m & f* homosexual. **~idad** *f* homosexuality

hond|o *a* deep. **~onada** *f* hollow. **~ura** *f* depth

Honduras *fpl* Honduras

hondureño *a & m* Honduran

honest|idad *f* decency. **~o** *a* proper

hongo *m* fungus; (*culin*) mushroom; (*venenoso*) toadstool

hon|or *m* honour. **~orable** *a* honourable. **~orario** *a* honorary. **~orarios** *mpl* fees. **~ra** *f* honour; (*buena fama*) good name. **~radez** *f* honesty. **~rado** *a* honest. **~rar** *vt* honour. **~rarse** *vpr* be honoured

hora *f* hour; (*momento determinado, momento oportuno*) time. **~ avanzada** late hour. **~ punta** rush hour. **~s** *fpl* **de trabajo** working hours. **~s** *fpl* **extraordinarias** overtime. **a estas ~s** now. **¿a qué ~?** at what time? when? **de ~ en ~** hourly. **de última ~** last-minute. **en buena ~** at the right time. **media ~** half an hour. **¿qué ~ es?** what time is it? **¿tiene Vd ~?** can you tell me the time?

horario *a* time; (*cada hora*) hourly. ● *m* timetable. **a ~** (*LAm*) on time

horca *f* gallows

horcajadas. a ~ astride

horchata *f* tiger-nut milk

horda *f* horde

horizont|al *a & f* horizontal. **~e** *m* horizon

horma *f* mould; (*para fabricar calzado*) last; (*para conservar forma del calzado*) shoe-tree

hormiga *f* ant

hormigón *m* concrete

hormigue|ar *vt* tingle; (*bullir*) swarm. **me ~a la mano** I've got pins and needles in my hand. **~o** *m* tingling; (*fig*) anxiety

hormiguero *m* anthill; (*de gente*) swarm

hormona *f* hormone

horn|ada *f* batch. **~ero** *m* baker. **~illo** *m* cooker. **~o** *m* oven; (*para ladrillos, cerámica etc*) kiln; (*tec*) furnace

horóscopo *m* horoscope

horquilla *f* pitchfork; (*para el pelo*) hairpin

horr|endo *a* awful. **~ible** *a* horrible. **~ipilante** *a* terrifying. **~or** *m* horror; (*atrocidad*) atrocity. **~orizar** [10] *vt* horrify. **~orizarse** *vpr* be horrified. **~oroso** *a* horrifying. **¡qué ~or!** how awful!

hort|aliza *f* vegetable. **~elano** *m* market gardener. **~icultura** *f* horticulture

hosco *a* surly; ‹*lugar*› gloomy

hospeda|je *m* lodging. **~r** *vt* put up. **~rse** *vpr* lodge

hospital *m* hospital

hospital|ario *m* hospitable. **~idad** *f* hospitality

hostal *m* boarding-house

hostería *f* inn

hostia *f* (*relig*) host; (*golpe, fam*) punch

hostigar [12] *vt* whip; (*fig, excitar*) urge; (*fig, molestar*) pester

hostil *a* hostile. **~idad** *f* hostility

hotel *m* hotel. **~ero** *a* hotel. ● *m* hotelier

hoy *adv* today. **~ (en) día** nowadays. **~ mismo** this very day. **~ por ~** for the time being. **de ~ en adelante** from now on

hoy|a *f* hole; (*sepultura*) grave. **~o** *m* hole; (*sepultura*) grave. **~uelo** *m* dimple

hoz *f* sickle; (*desfiladero*) pass

hube *vb véase* **haber**

hucha *f* money box

hueco *a* hollow; (*vacío*) empty; (*esponjoso*) spongy; (*resonante*) resonant. ● *m* hollow

huelg|a *f* strike. **~a de brazos caídos** sit-down strike. **~a de celo** work-to-rule. **~a de hambre** hunger strike. **~uista** *m & f* striker. **declarar la ~a, declararse en ~a** come out on strike
huelo *vb véase* **oler**
huella *f* footprint; (*de animal, vehículo etc*) track. **~ dactilar, ~ digital** fingerprint
huérfano *a* orphaned. ● *m* orphan. **~ de** without
huero *a* empty
huert|a *f* market garden (*Brit*), truck farm (*Amer*); (*terreno de regadío*) irrigated plain. **~o** *m* vegetable garden; (*de árboles frutales*) orchard
huesa *f* grave
hueso *m* bone; (*de fruta*) stone. **~so** *a* bony
huésped *m* guest; (*que paga*) lodger; (*animal*) host
huesudo *a* bony
huev|a *f* roe. **~era** *f* eggcup. **~o** *m* egg. **~o duro** hard-boiled egg. **~o escalfado** poached egg. **~o estrellado, ~o frito** fried egg. **~o pasado por agua** boiled egg. **~os revueltos** scrambled eggs
hui|da *f* flight, escape. **~dizo** *a* (*tímido*) shy; (*fugaz*) fleeting. **~r** [17] *vt/i* flee, run away; (*evitar*) avoid
huipil *m* (*Mex*) embroidered smock
huitlacoche *m* (*Mex*) edible black fungus
hule *m* oilcloth, oilskin
human|idad *f* mankind; (*fig*) humanity. **~idades** *fpl* humanities. **~ismo** *m* humanism. **~ista** *m & f* humanist. **~itario** *a* humanitarian. **~o** *a* human; (*benévolo*) humane. ● *m* human (being)
hum|areda *f* cloud of smoke. **~ear** *vi* smoke; (*echar vapor*) steam
humed|ad *f* dampness (*en meteorología*) humidity. **~ecer** [11] *vt* moisten. **~ecerse** *vpr* become moist
húmedo *a* damp; ‹*clima*› humid; (*mojado*) wet
humi|ldad *f* humility. **~lde** *a* humble. **~llación** *f* humiliation. **~llar** *vt* humiliate. **~llarse** *vpr* humble o.s.
humo *m* smoke; (*vapor*) steam; (*gas nocivo*) fumes. **~s** *mpl* conceit
humor *m* mood, temper; (*gracia*) humour. **~ismo** *m* humour. **~ista** *m & f* humorist. **~ístico** *a* humorous. **estar de mal ~** be in a bad mood
hundi|do *a* sunken. **~miento** *m* sinking. **~r** *vt* sink; destroy ‹*edificio*›. **~rse** *vpr* sink; ‹*edificio*› collapse
húngaro *a & m* Hungarian
Hungría *f* Hungary
huracán *m* hurricane
huraño *a* unsociable
hurg|ar [12] *vt* poke; (*fig*) stir up. **~ón** *m* poker
hurón *m* ferret. ● *a* unsociable
hurra *int* hurray!
hurraca *f* magpie
hurtadillas. a ~ stealthily
hurt|ar *vt* steal. **~o** *m* theft; (*cosa robada*) stolen object
husmear *vt* sniff out; (*fig*) pry into
huyo *vb véase* **huir**

I

Iberia *f* Iberia
ibérico *a* Iberian
ibero *a & m* Iberian
íbice *m* ibex, mountain goat
Ibiza *f* Ibiza
iceberg /iθ'ber/ *m* iceberg
icono *m* icon
ictericia *f* jaundice
ida *f* outward journey; (*salida*) departure. **de ~ y vuelta** return (*Brit*), round-trip (*Amer*)
idea *f* idea; (*opinión*) opinion. **cambiar de ~** change one's mind. **no tener la más remota ~, no tener la menor ~** not have the slightest idea, not have a clue (*fam*)
ideal *a* ideal; (*imaginario*) imaginary. ● *m* ideal. **~ista** *m & f* idealist. **~izar** [10] *vt* idealize
idear *vt* think up, conceive; (*inventar*) invent
ídem *pron & adv* the same
idéntico *a* identical
identi|dad *f* identity. **~ficación** *f* identification. **~ficar** [7] *vt* identify. **~ficarse** *vpr*. **~ficarse con** identify with
ideol|ogía *f* ideology. **~ógico** *a* ideological
idílico *a* idyllic
idilio *m* idyll
idiom|a *m* language. **~ático** *a* idiomatic

idiosincrasia *f* idiosyncrasy
idiot|a *a* idiotic. ● *m & f* idiot. **~ez** *f* idiocy
idiotismo *m* idiom
idolatrar *vt* worship; (*fig*) idolize
ídolo *m* idol
idóneo *a* suitable (**para** for)
iglesia *f* church
iglú *m* igloo
ignición *f* ignition
ignomini|a *f* ignominy, disgrace. **~oso** *a* ignominious
ignora|ncia *f* ignorance. **~nte** *a* ignorant. ● *m* ignoramus. **~r** *vt* not know, be unaware of
igual *a* equal; (*mismo*) the same; (*similar*) like; (*llano*) even; (*liso*) smooth. ● *adv* easily. ● *m* equal. **~ que** (the same) as. **al ~ que** the same as. **da ~, es ~** it doesn't matter
igual|ar *vt* make equal; (*ser igual*) equal; (*allanar*) level. **~arse** *vpr* be equal. **~dad** *f* equality. **~mente** *adv* equally; (*también*) also, likewise; (*respuesta de cortesía*) the same to you
ijada *f* flank
ilegal *a* illegal
ilegible *a* illegible
ilegítimo *a* illegitimate
ileso *a* unhurt
ilícito *a* illicit
ilimitado *a* unlimited
ilógico *a* illogical
ilumina|ción *f* illumination; (*alumbrado*) lighting; (*fig*) enlightenment. **~r** *vt* light (up); (*fig*) enlighten. **~rse** *vpr* light up
ilusi|ón *f* illusion; (*sueño*) dream; (*alegría*) joy. **~onado** *a* excited. **~onar** *vt* give false hope. **~onarse** *vpr* have false hopes. **hacerse ~ones** build up one's hopes. **me hace ~ón** I'm thrilled; I'm looking forward to ‹*algo en el futuro*›
ilusionis|mo *m* conjuring. **~ta** *m & f* conjurer
iluso *a* easily deceived. ● *m* dreamer. **~rio** *a* illusory
ilustra|ción *f* learning; (*dibujo*) illustration. **~do** *a* learned; (*con dibujos*) illustrated. **~r** *vt* explain; (*instruir*) instruct; (*añadir dibujos etc*) illustrate. **~rse** *vpr* acquire knowledge. **~tivo** *a* illustrative
ilustre *a* illustrious
imagen *f* image; (*TV etc*) picture
imagina|ble *a* imaginable. **~ción** *f* imagination. **~r** *vt* imagine. **~rse** *vpr* imagine. **~rio** *m* imaginary. **~tivo** *a* imaginative
imán *m* magnet
imantar *vt* magnetize
imbécil *a* stupid. ● *m & f* imbecile, idiot
imborrable *a* indelible; ‹*recuerdo etc*› unforgettable
imbuir [17] *vt* imbue (**de** with)
imita|ción *f* imitation. **~r** *vt* imitate
impacien|cia *f* impatience. **~tarse** *vpr* lose one's patience. **~te** *a* impatient; (*intranquilo*) anxious
impacto *m* impact
impar *a* odd
imparcial *a* impartial. **~idad** *f* impartiality
impartir *vt* impart
impasible *a* impassive
impávido *a* fearless; (*impasible*) impassive
impecable *a* impeccable
impedi|do *a* disabled. **~menta** *f* (*esp mil*) baggage. **~mento** *m* hindrance. **~r** [5] *vt* prevent; (*obstruir*) hinder
impeler *vt* drive
impenetrable *a* impenetrable
impenitente *a* unrepentant
impensa|ble *a* unthinkable. **~do** *a* unexpected
imperar *vi* reign
imperativo *a* imperative; ‹*persona*› imperious
imperceptible *a* imperceptible
imperdible *m* safety pin
imperdonable *a* unforgivable
imperfec|ción *f* imperfection. **~to** *a* imperfect
imperial *a* imperial. ● *f* upper deck. **~ismo** *m* imperialism
imperio *m* empire; (*poder*) rule; (*fig*) pride. **~so** *a* imperious
impermeable *a* waterproof. ● *m* raincoat
impersonal *a* impersonal
impertérrito *a* undaunted
impertinen|cia *f* impertinence. **~te** *a* impertinent
imperturbable *a* imperturbable
ímpetu *m* impetus; (*impulso*) impulse; (*impetuosidad*) impetuosity
impetuos|idad *f* impetuosity; (*violencia*) violence. **~o** *a* impetuous; (*violento*) violent

impío *a* ungodly; ‹*acción*› irreverent
implacable *a* implacable
implantar *vt* introduce
implica|ción *f* implication. **~r** [7] *vt* implicate; (*significar*) imply
implícito *a* implicit
implora|ción *f* entreaty. **~r** *vt* implore
imponderable *a* imponderable; (*inapreciable*) invaluable
impon|ente *a* imposing; (*fam*) terrific. **~er** [34] *vt* impose; (*requerir*) demand; deposit ‹*dinero*›. **~erse** *vpr* be imposed; (*hacerse obedecer*) assert o.s.; (*hacerse respetar*) command respect. **~ible** *a* taxable
impopular *a* unpopular. **~idad** *f* unpopularity
importa|ción *f* import; (*artículo*) import. **~dor** *a* importing. ● *m* importer
importa|ncia *f* importance; (*tamaño*) size. **~nte** *a* important; (*en cantidad*) considerable. **~r** *vt* import; (*valer*) cost. ● *vi* be important, matter. **¡le importa...?** would you mind...? **no ~** it doesn't matter
importe *m* price; (*total*) amount
importun|ar *vt* bother. **~o** *a* troublesome; (*inoportuno*) inopportune
imposib|ilidad *f* impossibility. **~le** *a* impossible. **hacer lo ~le** do all one can
imposición *f* imposition; (*impuesto*) tax
impostor *m* & *f* impostor
impotable *a* undrinkable
impoten|cia *f* impotence. **~te** *a* powerless, impotent
impracticable *a* impracticable; (*intransitable*) unpassable
impreca|ción *f* curse. **~r** [7] *vt* curse
imprecis|ión *f* vagueness. **~o** *a* imprecise
impregnar *vt* impregnate; (*empapar*) soak; (*fig*) cover
imprenta *f* printing; (*taller*) printing house, printer's
imprescindible *a* indispensable, essential
impresi|ón *f* impression; (*acción de imprimir*) printing; (*tirada*) edition; (*huella*) imprint. **~onable** *a* impressionable. **~onante** *a* impressive; (*espantoso*) frightening. **~onar** *vt* impress; (*conmover*) move; (*foto*) expose. **~onarse** *vpr* be impressed; (*conmover*) be moved
impresionis|mo *m* impressionism. **~ta** *a* & *m* & *f* impressionist
impreso *a* printed. ● *m* printed paper, printed matter. **~ra** *f* printer
imprevis|ible *a* unforseeable. **~to** *a* unforeseen
imprimir [*pp* **impreso**] *vt* impress; print ‹*libro etc*›
improbab|ilidad *f* improbability. **~le** *a* unlikely, improbable
improcedente *a* unsuitable
improductivo *a* unproductive
improperio *m* insult. **~s** *mpl* abuse
impropio *a* improper
improvis|ación *f* improvisation. **~adamente** *adv* suddenly. **~ado** *a* improvised. **~ar** *vt* improvise. **~o** *a*. **de ~o** suddenly
impruden|cia *f* imprudence. **~te** *a* imprudent
impuden|cia *f* impudence. **~te** *a* impudent
imp|údico *a* immodest; (*desvergonzado*) shameless. **~udor** *m* immodesty; (*desvergüenza*) shamelessness
impuesto *a* imposed. ● *m* tax. **~ sobre el valor añadido** VAT, value added tax
impugnar *vt* contest; (*refutar*) refute
impulsar *vt* impel
impuls|ividad *f* impulsiveness. **~ivo** *a* impulsive. **~o** *m* impulse
impun|e *a* unpunished. **~idad** *f* impunity
impur|eza *f* impurity. **~o** *a* impure
imputa|ción *f* charge. **~r** *vt* attribute; (*acusar*) charge
inacabable *a* interminable
inaccesible *a* inaccessible
inaceptable *a* unacceptable
inacostumbrado *a* unaccustomed
inactiv|idad *f* inactivity. **~o** *a* inactive
inadaptado *a* maladjusted
inadecuado *a* inadequate; (*inapropiado*) unsuitable
inadmisible *a* inadmissible; (*intolerable*) intolerable
inadvert|ido *a* unnoticed. **~encia** *f* inadvertence
inagotable *a* inexhaustible
inaguantable *a* unbearable; ‹*persona*› insufferable
inaltera|ble unchangeable; ‹*color*› fast; ‹*carácter*› calm. **~do** *a* unchanged

inanimado *a* inanimate
inaplicable *a* inapplicable
inapreciable *a* imperceptible
inapropiado *a* inappropriate
inarticulado *a* inarticulate
inasequible *a* out of reach
inaudito *a* unheard-of
inaugura|ción *f* inauguration. **~l** *a* inaugural. **~r** *vt* inaugurate
inca *a* Incan. ● *m* & *f* Inca. **~ico** *a* Incan
incalculable *a* incalculable
incandescen|cia *f* incandescence. **~te** *a* incandescent
incansable *a* tireless
incapa|cidad *f* incapacity. **~citar** *vt* incapacitate. **~z** *a* incapable
incauto *a* unwary; (*fácil de engañar*) gullible
incendi|ar *vt* set fire to. **~arse** *vpr* catch fire. **~ario** *a* incendiary. ● *m* arsonist. **~o** *m* fire
incentivo *m* incentive
incertidumbre *f* uncertainty
incesante *a* incessant
incest|o *m* incest. **~uoso** *a* incestuous
inciden|cia *f* incidence; (*incidente*) incident. **~tal** *a* incidental. **~te** *m* incident
incidir *vi* fall; (*influir*) influence
incienso *m* incense
incierto *a* uncertain
incinera|ción *f* incineration; (*de cadáveres*) cremation. **~dor** *m* incinerator. **~r** *vt* incinerate; cremate ‹*cadáver*›
incipiente *a* incipient
incisión *f* incision
incisivo *a* incisive. ● *m* incisor
incitar *vt* incite
incivil *a* rude
inclemen|cia *f* harshness. **~te** *a* harsh
inclina|ción *f* slope; (*de la cabeza*) nod; (*fig*) inclination. **~r** *vt* incline. **~rse** *vpr* lean; (*encorvarse*) stoop; (*en saludo*) bow; (*fig*) be inclined. **~rse a** (*parecerse*) resemble
inclu|ido *a* included; ‹*precio*› inclusive; (*en cartas*) enclosed. **~ir** [17] *vt* include; (*en cartas*) enclose. **~sión** *f* inclusion. **~sive** *adv* inclusive. **hasta el lunes ~sive** up to and including Monday. **~so** *a* included; (*en cartas*) enclosed. ● *adv* including; (*hasta*) even
incógnito *a* unknown. **de ~** incognito
incoheren|cia *f* incoherence. **~te** *a* incoherent
incoloro *a* colourless
incólume *a* unharmed
incomestible *a*, **incomible** *a* uneatable, inedible
incomodar *vt* inconvenience; (*molestar*) bother. **~se** *vpr* trouble o.s.; (*enfadarse*) get angry
incómodo *a* uncomfortable; (*inoportuno*) inconvenient
incomparable *a* imcomparable
incompatib|ilidad *f* incompatibility. **~le** *a* incompatible
incompeten|cia *f* incompetence. **~te** *a* incompetent
incompleto *a* incomplete
incompren|dido *a* misunderstood. **~sible** *a* incomprehensible. **~sión** *f* incomprehension
incomunicado *a* isolated; ‹*preso*› in solitary confinement
inconcebible *a* inconceivable
inconciliable *a* irreconcilable
inconcluso *a* unfinished
incondicional *a* unconditional
inconfundible *a* unmistakable
incongruente *a* incongruous
inconmensurable *a* (*fam*) enormous
inconscien|cia *f* unconsciousness; (*irreflexión*) recklessness. **~te** *a* unconscious; (*irreflexivo*) reckless
inconsecuente *a* inconsistent
inconsiderado *a* inconsiderate
inconsistente *a* insubstantial
inconsolable *a* unconsolable
inconstan|cia *f* inconstancy. **~te** *a* changeable; ‹*persona*› fickle
incontable *a* countless
incontaminado *a* uncontaminated
incontenible *a* irrepressible
incontestable *a* indisputable
incontinen|cia *f* incontinence. **~te** *a* incontinent
inconvenien|cia *f* disadvantage. **~te** *a* inconvenient; (*inapropiado*) inappropriate; (*incorrecto*) improper. ● *m* difficulty; (*desventaja*) drawback
incorpora|ción *f* incorporation. **~r** *vt* incorporate; (*culin*) mix. **~rse** *vpr* sit up; join ‹*sociedad, regimiento etc*›
incorrecto *a* incorrect; ‹*acción*› improper; (*descortés*) discourteous
incorregible *a* incorrigible
incorruptible *a* incorruptible
incrédulo *a* incredulous

increíble *a* incredible
increment|ar *vt* increase. **~o** *m* increase
incriminar *vt* incriminate
incrustar *vt* encrust
incuba|ción *f* incubation. **~dora** *f* incubator. **~r** *vt* incubate; (*fig*) hatch
incuestionable *a* unquestionable
inculcar [7] *vt* inculcate
inculpar *vt* accuse; (*culpar*) blame
inculto *a* uncultivated; ‹*persona*› uneducated
incumplimiento *m* non-fulfilment; (*de un contrato*) breach
incurable *a* incurable
incurrir *vi*. **~ en** incur; fall into ‹*error*›; commit ‹*crimen*›
incursión *f* raid
indaga|ción *f* investigation. **~r** [12] *vt* investigate
indebido *a* undue
indecen|cia *f* indecency. **~te** *a* indecent
indecible *a* inexpressible
indecis|ión *f* indecision. **~o** *a* undecided
indefenso *a* defenceless
indefini|ble *a* indefinable. **~do** *a* indefinite
indeleble *a* indelible
indelicad|eza *f* indelicacy. **~o** *a* indelicate; (*falto de escrúpulo*) unscrupulous
indemn|e *a* undamaged; ‹*persona*› unhurt. **~idad** *f* indemnity. **~izar** [10] *vt* indemnify, compensate
independ|encia *f* independence. **~iente** *a* independent
independizarse [10] *vpr* become independent
indescifrable *a* indecipherable, incomprehensible
indescriptible *a* indescribable
indeseable *a* undesirable
indestructible *a* indestructible
indetermina|ble *a* indeterminable. **~do** *a* indeterminate
India *f*. **la ~** India. **las ~s** *fpl* the Indies
indica|ción *f* indication; (*sugerencia*) suggestion. **~ciones** *fpl* directions. **~dor** *m* indicator; (*tec*) gauge. **~r** [7] *vt* show, indicate; (*apuntar*) point at; (*hacer saber*) point out; (*aconsejar*) advise. **~tivo** *a* indicative. ● *m* indicative; (*al teléfono*) dialling code
índice *m* indication; (*dedo*) index finger; (*de libro*) index; (*catálogo*) catalogue; (*aguja*) pointer
indicio *m* indication, sign; (*vestigio*) trace
indiferen|cia *f* indifference. **~te** *a* indifferent. **me es ~te** it's all the same to me
indígena *a* indigenous. ● *m & f* native
indigen|cia *f* poverty. **~te** *a* needy
indigest|ión *f* indigestion. **~o** *a* undigested; (*difícil de digerir*) indigestible
indign|ación *f* indignation. **~ado** *a* indignant. **~ar** *vt* make indignant. **~arse** *vpr* be indignant. **~o** *a* unworthy; (*despreciable*) contemptible
indio *a & m* Indian
indirect|a *f* hint. **~o** *a* indirect
indisciplina *f* lack of discipline. **~do** *a* undisciplined
indiscre|ción *f* indiscretion. **~to** *a* indiscreet
indiscutible *a* unquestionable
indisoluble *a* indissoluble
indispensable *a* indispensable
indisp|oner [34] *vt* (*enemistar*) set against. **~onerse** *vpr* fall out; (*ponerse enfermo*) fall ill. **~osición** *f* indisposition. **~uesto** *a* indisposed
indistinto *a* indistinct
individu|al *a* individual; ‹*cama*› single. **~alidad** *f* individuality. **~alista** *m & f* individualist. **~alizar** [10] *vt* individualize. **~o** *a & m* individual
índole *f* nature; (*clase*) type
indolen|cia *f* indolence. **~te** *a* indolent
indoloro *a* painless
indomable *a* untameable
indómito *a* indomitable
Indonesia *f* Indonesia
inducir [47] *vt* induce; (*deducir*) infer
indudable *a* undoubted. **~mente** *adv* undoubtedly
indulgen|cia *f* indulgence. **~te** *a* indulgent
indult|ar *vt* pardon; exempt (*de un pago etc*). **~o** *m* pardon
industria *f* industry. **~l** *a* industrial. ● *m* industrialist. **~lización** *f* industrialization. **~lizar** [10] *vt* industrialize
industriarse *vpr* do one's best
industrioso *a* industrious

inédito *a* unpublished; (*fig*) unknown
ineducado *a* impolite
inefable *a* inexpressible
ineficaz *a* ineffective
ineficiente *a* inefficient
inelegible *a* ineligible
ineludible *a* inescapable, unavoidable
inept|itud *f* ineptitude. **~o** *a* inept
inequívoco *a* unequivocal
iner|cia *f* inertia
inerme *a* unarmed; (*fig*) defenceless
inerte *a* inert
inesperado *a* unexpected
inestable *a* unstable
inestimable *a* inestimable
inevitable *a* inevitable
inexacto *a* inaccurate; (*incorrecto*) incorrect; (*falso*) untrue
inexistente *a* non-existent
inexorable *a* inexorable
inexper|iencia *f* inexperience. **~to** *a* inexperienced
inexplicable *a* inexplicable
infalible *a* infallible
infam|ar *vt* defame. **~atorio** *a* defamatory. **~e** *a* infamous; (*fig, muy malo, fam*) awful. **~ia** *f* infamy
infancia *f* infancy
infant|a *f* infanta, princess. **~e** *m* infante, prince; (*mil*) infantryman. **~ería** *f* infantry. **~il** *a* (*de niño*) child's; (*como un niño*) infantile
infarto *m* coronary (thrombosis)
infatigable *a* untiring
infatua|ción *f* conceit. **~rse** *vpr* get conceited
infausto *a* unlucky
infec|ción *f* infection. **~cioso** *a* infectious. **~tar** *vt* infect. **~tarse** *vpr* become infected. **~to** *a* infected; (*fam*) disgusting
infecundo *a* infertile
infeli|cidad *f* unhappiness. **~z** *a* unhappy
inferior *a* inferior. ● *m & f* inferior. **~idad** *f* lower; (*calidad*) inferiority
inferir [4] *vt* infer; (*causar*) cause
infernal *a* infernal, hellish
infestar *vt* infest; (*fig*) inundate
infi|delidad *f* unfaithfulness. **~el** *a* unfaithful
infierno *m* hell
infiltra|ción *f* infiltration. **~rse** *vpr* infiltrate
ínfimo *a* lowest
infini|dad *f* infinity. **~tivo** *m* infinitive. **~to** *a* infinite. ● *m* infinite; (*en matemáticas*) infinity. **una ~dad de** countless
inflación *f* inflation; (*fig*) conceit
inflama|ble *a* (in)flammable. **~ción** *f* inflammation. **~r** *vt* set on fire; (*fig, med*) inflame. **~rse** *vpr* catch fire; (*med*) become inflamed
inflar *vt* inflate; (*fig, exagerar*) exaggerate
inflexi|ble *a* inflexible. **~ón** *f* inflexion
infligir [14] *vt* inflict
influ|encia *f* influence. **~enza** *f* flu (*fam*), influenza. **~ir** [17] *vt/i* influence. **~jo** *m* influence. **~yente** *a* influential
informa|ción *f* information. **~ciones** *fpl* (*noticias*) news; (*de teléfonos*) directory enquiries. **~dor** *m* informant
informal *a* informal; (*incorrecto*) incorrect
inform|ante *m & f* informant. **~ar** *vt/i* inform. **~arse** *vpr* find out. **~ática** *f* information technology. **~ativo** *a* informative
informe *a* shapeless. ● *m* report; (*información*) information
infortun|ado *a* unfortunate. **~io** *m* misfortune
infracción *f* infringement
infraestructura *f* infrastructure
infranqueable *a* impassable; (*fig*) insuperable
infrarrojo *a* infrared
infrecuente *a* infrequent
infringir [14] *vt* infringe
infructuoso *a* fruitless
infundado *a* unfounded
infu|ndir *vt* instil. **~sión** *f* infusion
ingeniar *vt* invent
ingenier|ía *f* engineering. **~o** *m* engineer
ingenio *m* ingenuity; (*agudeza*) wit; (*LAm, de azúcar*) refinery. **~so** *a* ingenious
ingenu|idad *f* ingenuousness. **~o** *a* ingenuous
ingerir [4] *vt* swallow
Inglaterra *f* England
ingle *f* groin
ingl|és *a* English. ● *m* Englishman; (*lengua*) English. **~esa** *f* Englishwoman
ingrat|itud *f* ingratitude. **~o** *a* ungrateful; (*desagradable*) thankless
ingrediente *m* ingredient

ingres|ar *vt* deposit. • *vi*. **~ar en** come in, enter; join ‹*sociedad*›. **~o** *m* entry; (*en sociedad, hospital etc*) admission. **~os** *mpl* income

inh|ábil *a* unskillful; (*no apto*) unfit. **~abilidad** *f* unskillfulness

inhabitable *a* uninhabitable

inhala|ción *f* inhalation. **~dor** *m* inhaler. **~r** *vt* inhale

inherente *a* inherent

inhibi|ción *f* inhibition. **~r** *vt* inhibit

inhospitalario *a*, **inhóspito** *a* inhospitable

inhumano *a* inhuman

inicia|ción *f* beginning. **~l** *a* & *f* initial. **~r** *vt* initiate; (*comenzar*) begin, start. **~tiva** *f* initiative

inicio *m* beginning

inicuo *a* iniquitous

inigualado *a* unequalled

ininterrumpido *a* continuous

injer|encia *f* interference. **~ir** [4] *vt* insert. **~irse** *vpr* interfere

injert|ar *vt* graft. **~to** *m* graft

injuri|a *f* insult; (*ofensa*) offence. **~ar** *vt* insult. **~oso** *a* offensive

injust|icia *f* injustice. **~o** *a* unjust

inmaculado *a* immaculate

inmaduro *a* unripe; ‹*persona*› immature

inmediaciones *fpl* neighbourhood

inmediat|amente *adv* immediately. **~o** *a* immediate; (*contiguo*) next

inmejorable *a* excellent

inmemorable *a* immemorial

inmens|idad *f* immensity. **~o** *a* immense

inmerecido *a* undeserved

inmersión *f* immersion

inmigra|ción *f* immigration. **~nte** *a* & *m* immigrant. **~r** *vt* immigrate

inminen|cia *f* imminence. **~te** *a* imminent

inmiscuirse [17] *vpr* interfere

inmobiliario *a* property

inmoderado *a* immoderate

inmodesto *a* immodest

inmolar *vt* sacrifice

inmoral *a* immoral. **~idad** *f* immorality

inmortal *a* immortal. **~izar** [10] *vt* immortalize

inmóvil *a* immobile

inmueble *a*. **bienes ~s** property

inmund|icia *f* filth. **~icias** *fpl* rubbish. **~o** *a* filthy

inmun|e *a* immune. **~idad** *f* immunity. **~ización** *f* immunization. **~izar** [10] *vt* immunize

inmuta|ble *a* unchangeable. **~rse** *vpr* turn pale

innato *a* innate

innecesario *a* unnecessary

innegable *a* undeniable

innoble *a* ignoble

innova|ción *f* innovation. **~r** *vt/i* innovate

innumerable *a* innumerable

inocen|cia *f* innocence. **~tada** *f* practical joke. **~te** *a* innocent. **~tón** *a* naïve

inocuo *a* innocuous

inodoro *a* odourless. • *m* toilet

inofensivo *a* inoffensive

inolvidable *a* unforgettable

inoperable *a* inoperable

inopinado *a* unexpected

inoportuno *a* untimely; (*incómodo*) inconvenient

inorgánico *a* inorganic

inoxidable *a* stainless

inquebrantable *a* unbreakable

inquiet|ar *vt* worry. **~arse** *vpr* get worried. **~o** *a* worried; (*agitado*) restless. **~ud** *f* anxiety

inquilino *m* tenant

inquirir [4] *vt* enquire into, investigate

insaciable *a* insatiable

insalubre *a* unhealthy

insanable *a* incurable

insatisfecho *a* unsatisfied; (*descontento*) dissatisfied

inscri|bir [*pp* **inscrito**] *vt* inscribe; (*en registro etc*) enrol, register. **~birse** *vpr* register. **~pción** *f* inscription; (*registro*) registration

insect|icida *m* insecticide. **~o** *m* insect

insegur|idad *f* insecurity. **~o** *a* insecure; (*dudoso*) uncertain

insemina|ción *f* insemination. **~r** *vt* inseminate

insensato *a* senseless

insensible *a* insensitive; (*med*) insensible; (*imperceptible*) imperceptible

inseparable *a* inseparable

insertar *vt* insert

insidi|a *f* trap. **~oso** *a* insidious

insigne *a* famous

insignia *f* badge; (*bandera*) flag

insignificante *a* insignificant

insincero *a* insincere

insinua|ción *f* insinuation. **~nte** *a* insinuating. **~r** [21] *vt* insinuate. **~rse** *vpr* ingratiate o.s. **~rse en** creep into

insípido *a* insipid
insist|encia *f* insistence. **~ente** *a* insistent. **~ir** *vi* insist; (*hacer hincapié*) stress
insolación *f* sunstroke
insolen|cia *f* rudeness, insolence. **~te** *a* rude, insolent
insólito *a* unusual
insoluble *a* insoluble
insolven|cia *f* insolvency. **~te** *a* & *m* & *f* insolvent
insomn|e *a* sleepless. **~io** *m* insomnia
insondable *a* unfathomable
insoportable *a* unbearable
insospechado *a* unexpected
insostenible *a* untenable
inspec|ción *f* inspection. **~cionar** *vt* inspect. **~tor** *m* inspector
inspira|ción *f* inspiration. **~r** *vt* inspire. **~rse** *vpr* be inspired
instala|ción *f* installation. **~r** *vt* install. **~rse** *vpr* settle
instancia *f* request
instant|ánea *f* snapshot. **~áneo** *a* instantaneous; ‹*café etc*› instant. **~e** *m* instant. **a cada ~e** constantly. **al ~e** immediately
instar *vt* urge
instaura|ción *f* establishment. **~r** *vt* establish
instiga|ción *f* instigation. **~dor** *m* instigator. **~r** [12] *vt* instigate; (*incitar*) incite
instint|ivo *a* instinctive. **~o** *m* instinct
institu|ción *f* institution. **~cional** *a* institutional. **~ir** [17] *vt* establish. **~to** *m* institute; (*escol*) (secondary) school. **~triz** *f* governess
instru|cción *f* instruction. **~ctivo** *a* instructive. **~ctor** *m* instructor. **~ir** [17] *vt* instruct; (*enseñar*) teach
instrument|ación *f* instrumentation. **~al** *a* instrumental. **~o** *m* instrument; (*herramienta*) tool
insubordina|ción *f* insubordination. **~r** *vt* stir up. **~rse** *vpr* rebel
insuficien|cia *f* insufficiency; (*inadecuación*) inadequacy. **~te** *a* insufficient
insufrible *a* insufferable
insular *a* insular
insulina *f* insulin
insulso *a* tasteless; (*fig*) insipid
insult|ar *vt* insult. **~o** *m* insult
insuperable *a* insuperable; (*excelente*) excellent
insurgente *a* insurgent
insurrec|ción *f* insurrection. **~to** *a* insurgent
intacto *a* intact
intachable *a* irreproachable
intangible *a* intangible
integra|ción *f* integration. **~l** *a* integral; (*completo*) complete; ‹*pan*› wholemeal (*Brit*), wholewheat (*Amer*). **~r** *vt* make up
integridad *f* integrity; (*entereza*) wholeness
íntegro *a* complete; (*fig*) upright
intelect|o *m* intellect. **~ual** *a* & *m* & *f* intellectual
inteligen|cia *f* intelligence. **~te** *a* intelligent
inteligible *a* intelligible
intemperancia *f* intemperance
intemperie *f* bad weather. **a la ~** in the open
intempestivo *a* untimely
intenci|ón *f* intention. **~onado** *a* deliberate. **~onal** *a* intentional. **bien ~onado** well-meaning. **mal ~onado** malicious. **segunda ~ón** duplicity
intens|idad *f* intensity. **~ificar** [7] *vt* intensify. **~ivo** *a* intensive. **~o** *a* intense
intent|ar *vt* try. **~o** *m* intent; (*tentativa*) attempt. **de ~o** intentionally
intercalar *vt* insert
intercambio *m* exchange
interceder *vt* intercede
interceptar *vt* intercept
intercesión *f* intercession
interdicto *m* ban
inter|és *m* interest; (*egoísmo*) self-interest. **~esado** *a* interested; (*parcial*) biassed; (*egoísta*) selfish. **~esante** *a* interesting. **~esar** *vt* interest; (*afectar*) concern. ● *vi* be of interest. **~esarse** *vpr* take an interest (**por** in)
interfer|encia *f* interference. **~ir** [4] *vi* interfere
interino *a* temporary; ‹*persona*› acting. ● *m* stand-in; (*médico*) locum
interior *a* interior. ● *m* inside. **Ministerio** *m* **del I~** Home Office (*Brit*), Department of the Interior (*Amer*)
interjección *f* interjection
interlocutor *m* speaker
interludio *m* interlude
intermediario *a* & *m* intermediary
intermedio *a* intermediate. ● *m* interval

interminable *a* interminable
intermitente *a* intermittent. ● *m* indicator
internacional *a* international
intern|ado *m* (*escol*) boarding-school. **~ar** *vt* intern; (*en manicomio*) commit. **~arse** *vpr* penetrate. **~o** *a* internal; (*escol*) boarding. ● *m* (*escol*) boarder
interpelar *vt* appeal
interponer [34] *vt* interpose. **~se** *vpr* intervene
int|erpretación *f* interpretation. **~erpretar** *vt* interpret. **~érprete** *m* interpreter; (*mus*) performer
interroga|ción *f* question; (*acción*) interrogation; (*signo*) question mark. **~r** [12] *vt* question. **~tivo** *a* interrogative
interru|mpir *vt* interrupt; (*suspender*) stop. **~pción** *f* interruption. **~ptor** *m* switch
intersección *f* intersection
interurbano *a* inter-city; ‹*conferencia*› long-distance
intervalo *m* interval; (*espacio*) space. **a ~s** at intervals
interven|ir [53] *vt* control; (*med*) operate on. ● *vi* intervene; (*participar*) take part. **~tor** *m* inspector; (*com*) auditor
intestino *m* intestine
intim|ar *vi* become friendly. **~idad** *f* intimacy
intimidar *vt* intimidate
íntimo *a* intimate. ● *m* close friend
intitular *vt* entitle
intolera|ble *a* intolerable. **~nte** *a* intolerant
intoxicar [7] *vt* poison
intranquil|izar [10] *vt* worry. **~o** *a* worried
intransigente *a* intransigent
intransitable *a* impassable
intransitivo *a* intransitive
intratable *a* intractable
intrépido *a* intrepid
intriga *f* intrigue. **~nte** *a* intriguing. **~r** [12] *vt/i* intrigue
intrincado *a* intricate
intrínseco *a* intrinsic
introduc|ción *f* introduction. **~ir** [47] *vt* introduce; (*meter*) insert. **~irse** *vpr* get into; (*entrometerse*) interfere
intromisión *f* interference
introvertido *a* & *m* introvert
intrus|ión *f* intrusion. **~o** *a* intrusive. ● *m* intruder
intui|ción *f* intuition. **~r** [17] *vt* sense. **~tivo** *a* intuitive
inunda|ción *f* flooding. **~r** *vt* flood
inusitado *a* unusual
in|útil *a* useless; (*vano*) futile. **~utilidad** *f* uselessness
invadir *vt* invade
inv|alidez *f* invalidity; (*med*) disability. **~álido** *a* & *m* invalid
invaria|ble *a* invariable. **~do** *a* unchanged
invas|ión *f* invasion. **~or** *a* invading. ● *m* invader
invectiva *f* invective
invencible *a* invincible
inven|ción *f* invention. **~tar** *vt* invent
inventario *m* inventory
invent|iva *f* inventiveness. **~ivo** *a* inventive. **~or** *m* inventor
invernadero *m* greenhouse
invernal *a* winter
inverosímil *a* improbable
inversión *f* inversion; (*com*) investment
inverso *a* inverse; (*contrario*) opposite. **a la inversa** the other way round
invertebrado *a* & *m* invertebrate
inverti|do *a* inverted; (*homosexual*) homosexual. ● *m* homosexual. **~r** [4] *vt* reverse; (*volcar*) turn upside down; (*com*) invest; spend ‹*tiempo*›
investidura *f* investiture
investiga|ción *f* investigation; (*univ*) research. **~dor** *m* investigator. **~r** [12] *vt* investigate
investir [5] *vt* invest
inveterado *a* inveterate
invicto *a* unbeaten
invierno *m* winter
inviolable *a* inviolate
invisib|ilidad *f* invisibility. **~le** *a* invisible
invita|ción *f* invitation. **~do** *m* guest. **~r** *vt* invite. **te invito a una copa** I'll buy you a drink
invoca|ción *f* invocation. **~r** [7] *vt* invoke
involuntario *a* involuntary
invulnerable *a* invulnerable
inyec|ción *f* injection. **~tar** *vt* inject
ion *m* ion
ir [49] *vi* go; ‹*ropa*› (*convenir*) suit. ● *m* going. **~se** *vpr* go away. **~ a hacer** be going to do. **~ a pie** walk. **~ de paseo** go for a walk. **~ en coche** go by car. **no me va ni me viene** it's all the same to me. **no**

vaya a ser que in case. **¡qué va!** nonsense! **va mejorando** it's gradually getting better. **¡vamos!**, **¡vámonos!** come on! let's go! **¡vaya!** fancy that! **¡vete a saber!** who knows? **¡ya voy!** I'm coming!
ira *f* anger. **~cundo** *a* irascible
Irak *m* Iraq
Irán *m* Iran
iraní *a* & *m* & *f* Iranian
iraquí *a* & *m* & *f* Iraqi
iris *m* (*anat*) iris; (*arco iris*) rainbow
Irlanda *f* Ireland
irland|és *a* Irish. ● *m* Irishman; (*lengua*) Irish. **~esa** *f* Irishwoman
ir|onía *f* irony. **~ónico** *a* ironic
irracional *a* irrational
irradiar *vt/i* radiate
irrazonable *a* unreasonable
irreal *a* unreal. **~idad** *f* unreality
irrealizable *a* unattainable
irreconciliable *a* irreconcilable
irreconocible *a* unrecognizable
irrecuperable *a* irretrievable
irreducible *a* irreducible
irreflexión *f* impetuosity
irrefutable *a* irrefutable
irregular *a* irregular. **~idad** *f* irregularity
irreparable *a* irreparable
irreprimible *a* irrepressible
irreprochable *a* irreproachable
irresistible *a* irresistible
irresoluto *a* irresolute
irrespetuoso *a* disrespectful
irresponsable *a* irresponsible
irrevocable *a* irrevocable
irriga|ción *f* irrigation. **~r** [12] *vt* irrigate
irrisorio *a* derisive; (*insignificante*) ridiculous
irrita|ble *a* irritable. **~ción** *f* irritation. **~r** *vt* irritate. **~rse** *vpr* get annoyed
irrumpir *vi* burst (**en** in)
irrupción *f* irruption
isla *f* island. **las I~s Británicas** the British Isles
Islam *m* Islam
islámico *a* Islamic
islandés *a* Icelandic. ● *m* Icelander; (*lengua*) Icelandic
Islandia *f* Iceland
isleño *a* island. ● *m* islander
Israel *m* Israel
israelí *a* & *m* Israeli
istmo /'ismo/ *m* isthmus
Italia *f* Italy
italiano *a* & *m* Italian
itinerario *a* itinerary
IVA *abrev* (*impuesto sobre el valor añadido*) VAT, value added tax
izar [10] *vt* hoist
izquierd|a *f* left(-hand); (*pol*) left (-wing). **~ista** *m* & *f* leftist. **~o** *a* left. **a la ~a** on the left; (*con movimiento*) to the left

J

ja *int* ha!
jabalí *m* wild boar
jabalina *f* javelin
jab|ón *m* soap. **~onar** *vt* soap. **~onoso** *a* soapy
jaca *f* pony
jacinto *m* hyacinth
jacta|ncia *f* boastfulness; (*acción*) boasting. **~rse** *vpr* boast
jadea|nte *a* panting. **~r** *vi* pant
jaez *m* harness
jaguar *m* jaguar
jalea *f* jelly
jaleo *m* row, uproar. **armar un ~** kick up a fuss
jalón *m* (*LAm, tirón*) pull; (*Mex, trago*) drink
Jamaica *f* Jamaica
jamás *adv* never; (*en frases afirmativas*) ever
jamelgo *m* nag
jamón *m* ham. **~ de York** boiled ham. **~ serrano** cured ham
Japón *m*. **el ~** Japan
japonés *a* & *m* Japanese
jaque *m* check. **~ mate** checkmate
jaqueca *f* migraine. **dar ~** bother
jarabe *m* syrup
jardín *m* garden. **~ de la infancia** kindergarten, nursery school
jardiner|ía *f* gardening. **~o** *m* gardener
jarocho *a* (*Mex*) from Veracruz
jarr|a *f* jug. **~o** *m* jug. **echar un ~o de agua fría a** throw cold water on. **en ~as** with hands on hips
jaula *f* cage
jauría *f* pack of hounds
jazmín *m* jasmine
jef|a *f* boss. **~atura** *f* leadership; (*sede*) headquarters. **~e** *m* boss; (*pol etc*) leader. **~e de camareros** head waiter. **~e de estación** stationmaster. **~e de ventas** sales manager
jengibre *m* ginger

jeque *m* sheikh
jer|arquía *f* hierarchy. **~árquico** *a* hierarchical
jerez *m* sherry. **al ~** with sherry
jerga *f* coarse cloth; (*argot*) jargon
jerigonza *f* jargon; (*galimatías*) gibberish
jeringa *f* syringe; (*LAm, molestia*) nuisance. **~r** [12] *vt* (*fig, molestar, fam*) annoy
jeroglífico *m* hieroglyph(ic)
jersey *m* (*pl* **jerseys**) jersey
Jerusalén *m* Jerusalem
Jesucristo *m* Jesus Christ. **antes de ~** BC, before Christ
jesuita *a* & *m* & *f* Jesuit
Jesús *m* Jesus. ● *int* good heavens!; (*al estornudar*) bless you!
jícara *f* small cup
jilguero *m* goldfinch
jinete *m* rider, horseman
jipijapa *f* straw hat
jirafa *f* giraffe
jirón *m* shred, tatter
jitomate *m* (*Mex*) tomato
jocoso *a* funny, humorous
jorna|da *f* working day; (*viaje*) journey; (*etapa*) stage. **~l** *m* day's wage; (*trabajo*) day's work. **~lero** *m* day labourer
joroba *f* hump. **~do** *a* hunchbacked. ● *m* hunchback. **~r** *vt* annoy
jota *f* letter J; (*danza*) jota, popular dance; (*fig*) iota. **ni ~** nothing
joven (*pl* **jóvenes**) *a* young. ● *m* young man, youth. ● *f* young woman, girl
jovial *a* jovial
joy|a *f* jewel. **~as** *fpl* jewellery. **~ería** *f* jeweller's (shop). **~ero** *m* jeweller; (*estuche*) jewellery box
juanete *m* bunion
jubil|ación *f* retirement. **~ado** *a* retired. **~ar** *vt* pension off. **~arse** *vpr* retire. **~eo** *m* jubilee
júbilo *m* joy
jubiloso *a* jubilant
judaísmo *m* Judaism
judía *f* Jewish woman; (*alubia*) bean. **~ blanca** haricot bean. **~ escarlata** runner bean. **~ verde** French bean
judicial *a* judicial
judío *a* Jewish. ● *m* Jewish man
judo *m* judo
juego *m* game; (*de niños, tec*) play; (*de azar*) gambling; (*conjunto*) set. ● *vb véase* **jugar**. **estar en ~** be at stake. **estar fuera de ~** be offside. **hacer ~** match
juerga *f* spree
jueves *m* Thursday
juez *m* judge. **~ de instrucción** examining magistrate. **~ de línea** linesman
juga|dor *m* player; (*en juegos de azar*) gambler. **~r** [3] *vt* play. ● *vi* play; (*a juegos de azar*) gamble; (*apostar*) bet. **~rse** *vpr* risk. **~r al fútbol** play football
juglar *m* minstrel
jugo *m* juice; (*de carne*) gravy; (*fig*) substance. **~so** *a* juicy; (*fig*) substantial
juguet|e *m* toy. **~ear** *vi* play. **~ón** *a* playful
juicio *m* judgement; (*opinión*) opinion; (*razón*) reason. **~so** *a* wise. **a mi ~** in my opinion
juliana *f* vegetable soup
julio *m* July
junco *m* rush, reed
jungla *f* jungle
junio *m* June
junt|a *f* meeting; (*consejo*) board, committee; (*pol*) junta; (*tec*) joint. **~ar** *vt* join; (*reunir*) collect. **~arse** *vpr* join; ‹*gente*› meet. **~o** *a* joined; (*en plural*) together. **~o a** next to. **~ura** *f* joint. **por ~o** all together
jura|do *a* sworn. ● *m* jury; (*miembro de jurado*) juror. **~mento** *m* oath. **~r** *vt/i* swear. **~r en falso** commit perjury. **jurárselas a uno** have it in for s.o. **prestar ~mento** take the oath
jurel *m* (type of) mackerel
jurídico *a* legal
juris|dicción *f* jurisdiction. **~prudencia** *f* jurisprudence
justamente *a* exactly; (*con justicia*) fairly
justicia *f* justice
justifica|ción *f* justification. **~r** [7] *vt* justify
justo *a* fair, just; (*exacto*) exact; ‹*ropa*› tight. ● *adv* just. **~ a tiempo** just in time
juven|il *a* youthful. **~tud** *f* youth; (*gente joven*) young people
juzga|do *m* (*tribunal*) court. **~r** [12] *vt* judge. **a ~r por** judging by

K

kilo *m*, **kilogramo** *m* kilo, kilogram
kil|ometraje *m* distance in kilometres, mileage. **~ométrico** *a* (*fam*) endless. **~ómetro** *m* kilometre. **~ómetro cuadrado** square kilometre
kilovatio *m* kilowatt
kiosco *m* kiosk

L

la *m* A; (*solfa*) lah. ● *art def f* the. ● *pron* (*ella*) her; (*Vd*) you; (*ello*) it. **~ de** the one. **~ de Vd** your one, yours. **~ que** whoever, the one
laberinto *m* labyrinth, maze
labia *f* glibness
labio *m* lip
labor *f* work; (*tarea*) job. **~able** *a* working. **~ar** *vi* work. **~es** *fpl* **de aguja** needlework. **~es** *fpl* **de ganchillo** crochet. **~es** *fpl* **de punto** knitting. **~es** *fpl* **domésticas** housework
laboratorio *m* laboratory
laborioso *a* laborious
laborista *a* Labour. ● *m & f* member of the Labour Party
labra|do *a* worked; ‹*madera*› carved; ‹*metal*› wrought; ‹*tierra*› ploughed. **~dor** *m* farmer; (*obrero*) labourer. **~nza** *f* farming. **~r** *vt* work; carve ‹*madera*›; cut ‹*piedra*›; till ‹*la tierra*›; (*fig, causar*) cause
labriego *m* peasant
laca *f* lacquer
lacayo *m* lackey
lacerar *vt* lacerate
lacero *m* lassoer; (*cazador*) poacher
lacio *a* straight; (*flojo*) limp
lacón *m* shoulder of pork
lacónico *a* laconic
lacra *f* scar
lacr|ar *vt* seal. **~e** *m* sealing wax
lactante *a* breast-fed
lácteo *a* milky. **productos** *mpl* **~s** dairy products
ladear *vt/i* tilt. **~se** *vpr* lean
ladera *f* slope
ladino *a* astute
lado *m* side. **al ~** near. **al ~ de** at the side of, beside. **los de al ~** the next door neighbours. **por otro ~** on the other hand. **por todos ~s** on all sides. **por un ~** on the one hand
ladr|ar *vi* bark. **~ido** *m* bark
ladrillo *m* brick; (*de chocolate*) block
ladrón *a* thieving. ● *m* thief
lagart|ija *f* (small) lizard. **~o** *m* lizard
lago *m* lake
lágrima *f* tear
lagrimoso *a* tearful
laguna *f* small lake; (*fig, omisión*) gap
laico *a* lay
lamé *m* lamé
lamedura *f* lick
lament|able *a* lamentable, pitiful. **~ar** *vt* be sorry about. **~arse** *vpr* lament; (*quejarse*) complain. **~o** *m* moan
lamer *vt* lick; ‹*olas etc*› lap
lámina *f* sheet; (*foto*) plate; (*dibujo*) picture
lamina|do *a* laminated. **~r** *vt* laminate
lámpara *f* lamp; (*bombilla*) bulb; (*lamparón*) grease stain. **~ de pie** standard lamp
lamparón *m* grease stain
lampiño *a* clean-shaven, beardless
lana *f* wool. **~r** *a*. **ganado** *m* **~r** sheep. **de ~** wool(len)
lanceta *f* lancet
lancha *f* boat. **~ motora** *f* motor boat. **~ salvavidas** lifeboat
lanero *a* wool(len)
langost|a *f* (*crustáceo marino*) lobster; (*insecto*) locust. **~ino** *m* prawn
languide|cer [11] *vi* languish. **~z** *f* languor
lánguido *a* languid; (*decaído*) listless
lanilla *f* nap; (*tela fina*) flannel
lanudo *a* woolly
lanza *f* lance, spear
lanza|llamas *m invar* flame-thrower. **~miento** *m* throw; (*acción de lanzar*) throwing; (*de proyectil, de producto*) launch. **~r** [10] *vt* throw; (*de un avión*) drop; launch ‹*proyectil, producto*›. **~rse** *vpr* fling o.s.
lapicero *m* (propelling) pencil
lápida *f* memorial tablet. **~ sepulcral** tombstone
lapidar *vt* stone
lápiz *m* pencil; (*grafito*) lead. **~ de labios** lipstick
Laponia *f* Lapland

lapso *m* lapse
larg|a *f.* **a la ~a** in the long run. **dar ~as** put off. **~ar** [12] *vt* slacken; (*dar, fam*) give; (*fam*) deal ‹*bofetada etc*›. **~arse** *vpr* (*fam*) go away, clear off (*fam*). **~o** *a* long; (*demasiado*) too long. ● *m* length. **¡~o!** go away! **~ueza** *f* generosity. **a lo ~o** lengthwise. **a lo ~o de** along. **tener 100 metros de ~o** be 100 metres long
laring|e *f* larynx. **~itis** *f* laryngitis
larva *f* larva
las *art def fpl* the. ● *pron* them. **~ de** those, the ones. **~ de Vd** your ones, yours. **~ que** whoever, the ones
lascivo *a* lascivious
láser *m* laser
lástima *f* pity; (*queja*) complaint. **dar ~** be pitiful. **ella me da ~** I feel sorry for her. **¡qué ~!** what a pity!
lastim|ado *a* hurt. **~ar** *vt* hurt. **~arse** *vpr* hurt o.s. **~ero** *a* doleful. **~oso** *a* pitiful
lastre *m* ballast
lata *f* tinplate; (*envase*) tin (*esp Brit*), can; (*molestia, fam*) nuisance. **dar la ~** be a nuisance. **¡qué ~!** what a nuisance!
latente *a* latent
lateral *a* side, lateral
latido *m* beating; (*cada golpe*) beat
latifundio *m* large estate
latigazo *m* (*golpe*) lash; (*chasquido*) crack
látigo *m* whip
latín *m* Latin. **saber ~** (*fam*) not be stupid
latino *a* Latin. **L~américa** *f* Latin America. **~americano** *a & m* Latin American
latir *vi* beat; ‹*herida*› throb
latitud *f* latitude
latón *m* brass
latoso *a* annoying; (*pesado*) boring
laucha *f* (*Arg*) mouse
laúd *m* lute
laudable *a* laudable
laureado *a* honoured; (*premiado*) prize-winning
laurel *m* laurel; (*culin*) bay
lava *f* lava
lava|ble *a* washable. **~bo** *m* wash-basin; (*retrete*) toilet. **~dero** *m* sink, wash-basin. **~do** *m* washing. **~do de cerebro** brainwashing. **~do en seco** dry-cleaning. **~dora** *f* washing machine. **~ndería** *f* laundry. **~ndería automática** launderette, laundromat (*esp Amer*). **~parabrisas** *m invar* windscreen washer (*Brit*), windshield washer (*Amer*). **~platos** *m & f invar* dishwasher; (*Mex, fregadero*) sink. **~r** *vt* wash. **~r en seco** dry-clean. **~rse** *vpr* have a wash. **~rse las manos** (*incl fig*) wash one's hands. **~tiva** *f* enema. **~vajillas** *m & f inv* dishwasher
lax|ante *a & m* laxative. **~o** *a* loose
laz|ada *f* bow. **~o** *m* knot; (*lazada*) bow; (*fig, vínculo*) tie; (*cuerda con nudo corredizo*) lasso; (*trampa*) trap
le *pron* (*acusativo, él*) him; (*acusativo, Vd*) you; (*dativo, él*) (to) him; (*dativo, ella*) (to) her; (*dativo, ello*) (to) it; (*dativo, Vd*) (to) you
leal *a* loyal; (*fiel*) faithful. **~tad** *f* loyalty; (*fidelidad*) faithfulness
lebrel *m* greyhound
lección *f* lesson; (*univ*) lecture
lect|or *m* reader; (*univ*) language assistant. **~ura** *f* reading
leche *f* milk; (*golpe*) bash. **~ condensada** condensed milk. **~ desnatada** skimmed milk. **~ en polvo** powdered milk. **~ra** *f* (*vasija*) milk jug. **~ría** *f* dairy. **~ro** *a* milk, dairy. ● *m* milkman. **~ sin desnatar** whole milk. **tener mala ~** be spiteful
lecho *m* bed
lechoso *a* milky
lechuga *f* lettuce
lechuza *f* owl
leer [18] *vt/i* read
legación *f* legation
legado *m* legacy; (*enviado*) legate
legajo *m* bundle, file
legal *a* legal. **~idad** *f* legality. **~izar** [10] *vt* legalize; (*certificar*) authenticate. **~mente** *adv* legally
legar [12] *vt* bequeath
legendario *a* legendary
legible *a* legible
legi|ón *f* legion. **~onario** *m* legionary
legisla|ción *f* legislation. **~dor** *m* legislator. **~r** *vi* legislate. **~tura** *f* legislature
leg|itimidad *f* legitimacy. **~ítimo** *a* legitimate; (*verdadero*) real
lego *a* lay; (*ignorante*) ignorant. ● *m* layman
legua *f* league
legumbre *f* vegetable
lejan|ía *f* distance. **~o** *a* distant
lejía *f* bleach
lejos *adv* far. **~ de** far from. **a lo ~** in the distance. **desde ~** from a distance, from afar

lelo *a* stupid
lema *m* motto
lencería *f* linen; (*de mujer*) lingerie
lengua *f* tongue; (*idioma*) language. **irse de la ~** talk too much. **morderse la ~** hold one's tongue. **tener mala ~** have a vicious tongue
lenguado *m* sole
lenguaje *m* language
lengüeta *f* (*de zapato*) tongue
lengüetada *f*, **lengüetazo** *m* lick
lente *f* lens. **~s** *mpl* glasses. **~s de contacto** contact lenses
lentej|a *f* lentil. **~uela** *f* sequin
lentilla *f* contact lens
lent|itud *f* slowness. **~o** *a* slow
leñ|a *f* firewood. **~ador** *m* woodcutter. **~o** *m* log
Leo *m* Leo
le|ón *m* lion. **León** Leo. **~ona** *f* lioness
leopardo *m* leopard
leotardo *m* thick tights
lepr|a *f* leprosy. **~oso** *m* leper
lerdo *a* dim; (*torpe*) clumsy
les *pron* (*acusativo*) them; (*acusativo, Vds*) you; (*dativo*) (to) them; (*dativo, Vds*) (to) you
lesbia(na) *f* lesbian
lesbiano *a*, **lesbio** *a* lesbian
lesi|ón *f* wound. **~onado** *a* injured. **~onar** *vt* injure; (*dañar*) damage
letal *a* lethal
letanía *f* litany
let|árgico *a* lethargic. **~argo** *m* lethargy
letr|a *f* letter; (*escritura*) handwriting; (*de una canción*) words, lyrics. **~a de cambio** bill of exchange. **~a de imprenta** print. **~ado** *a* learned. **~ero** *m* notice; (*cartel*) poster
letrina *f* latrine
leucemia *f* leukaemia
levadizo *a*. **puente** *m* **~** drawbridge
levadura *f* yeast. **~ en polvo** baking powder
levanta|miento *m* lifting; (*sublevación*) uprising. **~r** *vt* raise, lift; (*construir*) build; (*recoger*) pick up; (*separar*) take off. **~rse** *vpr* get up; (*ponerse de pie*) stand up; (*erguirse, sublevarse*) rise up
levante *m* east; (*viento*) east wind. **L~** Levant
levar *vt* weigh ‹*ancla*›. • *vi* set sail
leve *a* light; ‹*enfermedad etc*› slight; (*de poca importancia*) trivial. **~dad** *f* lightness; (*fig*) slightness
léxico *m* vocabulary
lexicografía *f* lexicography
ley *f* law; (*parlamentaria*) act. **plata** *f* **de ~** sterling silver
leyenda *f* legend
liar [20] *vt* tie; (*envolver*) wrap up; roll ‹*cigarillo*›; (*fig, confundir*) confuse; (*fig, enredar*) involve. **~se** *vpr* get involved
libanés *a* & *m* Lebanese
Líbano *m*. **el ~** Lebanon
libel|ista *m* & *f* satirist. **~o** *m* satire
libélula *f* dragonfly
libera|ción *f* liberation. **~dor** *a* liberating. • *m* liberator
liberal *a* & *m* & *f* liberal. **~idad** *f* liberality. **~mente** *adv* liberally
liber|ar *vt* free. **~tad** *f* freedom. **~tad de cultos** freedom of worship. **~tad de imprenta** freedom of the press. **~tad provisional** bail. **~tar** *vt* free. **en ~tad** free
libertino *m* libertine
Libia *f* Libya
libido *m* libido
libio *a* & *m* Libyan
libra *f* pound. **~ esterlina** pound sterling
Libra *f* Libra
libra|dor *m* (*com*) drawer. **~r** *vt* free; (*de un peligro*) rescue. **~rse** *vpr* free o.s. **~rse de** get rid of
libre *a* free; ‹*aire*› open; (*en natación*) freestyle. **~ de impuestos** tax-free. • *m* (*Mex*) taxi
librea *f* livery
libr|ería *f* bookshop (*Brit*), bookstore (*Amer*); (*mueble*) bookcase. **~ero** *m* bookseller. **~eta** *f* notebook. **~o** *m* book. **~o de a bordo** logbook. **~o de bolsillo** paperback. **~o de ejercicios** exercise book. **~o de reclamaciones** complaints book
licencia *f* permission; (*documento*) licence. **~do** *m* graduate. **~ para manejar** (*LAm*) driving licence. **~r** *vt* (*mil*) discharge; (*echar*) dismiss. **~tura** *f* degree
licencioso *a* licentious
liceo *m* (*esp LAm*) (secondary) school
licita|dor *m* bidder. **~r** *vt* bid for
lícito *a* legal; (*permisible*) permissible
licor *m* liquid; (*alcohólico*) liqueur
licua|dora *f* liquidizer. **~r** [21] liquefy
lid *f* fight. **en buena ~** by fair means

líder *m* leader
liderato *m*, **liderazgo** *m* leadership
lidia *f* bullfighting; (*lucha*) fight; (*LAm, molestia*) nuisance. **~r** *vt/i* fight
liebre *f* hare
lienzo *m* linen; (*del pintor*) canvas; (*muro, pared*) wall
liga *f* garter; (*alianza*) league; (*mezcla*) mixture. **~dura** *f* bond; (*mus*) slur; (*med*) ligature. **~mento** *m* ligament. **~r** [12] *vt* tie; (*fig*) join; (*mus*) slur. • *vi* mix. **~r con** (*fig*) pick up. **~rse** *vpr* (*fig*) commit o.s.
liger|eza *f* lightness; (*agilidad*) agility; (*rapidez*) swiftness; (*de carácter*) fickleness. **~o** *a* light; (*rápido*) quick; (*ágil*) agile; (*superficial*) superficial; (*de poca importancia*) slight. • *adv* quickly. **a la ~a** lightly, superficially
liguero *m* suspender belt
lija *f* dogfish; (*papel de lija*) sandpaper. **~r** *vt* sand
lila *f* lilac
Lima *f* Lima
lima *f* file; (*fruta*) lime. **~duras** *fpl* filings. **~r** *vt* file (down)
limbo *m* limbo
limita|ción *f* limitation. **~do** *a* limited. **~r** *vt* limit. **~r con** border on. **~tivo** *a* limiting
límite *m* limit. **~ de velocidad** speed limit
limítrofe *a* bordering
limo *m* mud
lim|ón *m* lemon. **~onada** *f* lemonade
limosn|a *f* alms. **~ear** *vi* beg. **pedir ~a** beg
limpia *f* cleaning. **~botas** *m invar* bootblack. **~parabrisas** *m inv* windscreen wiper (*Brit*), windshield wiper (*Amer*). **~pipas** *m invar* pipe-cleaner. **~r** *vt* clean; (*enjugar*) wipe
limpi|eza *f* cleanliness; (*acción de limpiar*) cleaning. **~eza en seco** dry-cleaning. **~o** *a* clean; ‹*cielo*› clear; (*fig, honrado*) honest. • *adv* fairly. **en ~o** (*com*) net. **jugar ~o** play fair
linaje *m* lineage; (*fig, clase*) kind
lince *m* lynx
linchar *vt* lynch
lind|ante *a* bordering (**con** on). **~ar** *vi* border (**con** on). **~e** *f* boundary. **~ero** *m* border
lindo *a* pretty, lovely. **de lo ~** (*fam*) a lot
línea *f* line. **en ~s generales** in broad outline. **guardar la ~** watch one's figure
lingote *m* ingot
lingü|ista *m & f* linguist. **~ística** *f* linguistics. **~ístico** *a* linguistic
lino *m* flax; (*tela*) linen
linóleo *m*, **linóeum** *m* lino, linoleum
linterna *f* lantern; (*de bolsillo*) torch, flashlight (*Amer*)
lío *m* bundle; (*jaleo*) fuss; (*embrollo*) muddle; (*amorío*) affair
liquen *m* lichen
liquida|ción *f* liquidation; (*venta especial*) (clearance) sale. **~r** *vt* liquify; (*com*) liquidate; settle ‹*cuenta*›
líquido *a* liquid; (*com*) net. • *m* liquid
lira *f* lyre; (*moneda italiana*) lira
líric|a *f* lyric poetry. **~o** *a* lyric(al)
lirio *m* iris. **~ de los valles** lily of the valley
lirón *m* dormouse; (*fig*) sleepyhead. **dormir como un ~** sleep like a log
Lisboa *f* Lisbon
lisia|do *a* disabled. **~r** *vt* disable; (*herir*) injure
liso *a* smooth; ‹*pelo*› straight; ‹*tierra*› flat; (*sencillo*) plain
lisonj|a *f* flattery. **~eador** *a* flattering. • *m* flatterer. **~ear** *vt* flatter. **~ero** *a* flattering
lista *f* stripe; (*enumeración*) list; (*de platos*) menu. **~ de correos** poste restante. **~do** *a* striped. **a ~s** striped
listo *a* clever; (*preparado*) ready
listón *m* ribbon; (*de madera*) strip
lisura *f* smoothness
litera *f* (*en barco*) berth; (*en tren*) sleeper; (*en habitación*) bunk bed
literal *a* literal
litera|rio *a* literary. **~tura** *f* literature
litig|ar [12] *vi* dispute; (*jurid*) litigate. **~io** *m* dispute; (*jurid*) litigation
litografía *f* (*arte*) lithography; (*cuadro*) lithograph
litoral *a* coastal. • *m* coast
litro *m* litre
lituano *a & m* Lithuanian
liturgia *f* liturgy
liviano *a* fickle, inconstant
lívido *a* livid
lizo *m* warp thread

lo *art def neutro*. **~ importante** what is important, the important thing. ● *pron* (*él*) him; (*ello*) it. **~ que** what(ever), that which

loa *f* praise. **~ble** *a* praiseworthy. **~r** *vt* praise

lobo *m* wolf

lóbrego *a* gloomy

lóbulo *m* lobe

local *a* local. ● *m* premises; (*lugar*) place. **~idad** *f* locality; (*de un espectáculo*) seat; (*entrada*) ticket. **~izar** [10] *vt* localize; (*encontrar*) find, locate

loción *f* lotion

loco *a* mad; (*fig*) foolish. ● *m* lunatic. **~ de alegría** mad with joy. **estar ~ por** be crazy about. **volverse ~** go mad

locomo|ción *f* locomotion. **~tora** *f* locomotive

locuaz *a* talkative

locución *f* expression

locura *f* madness; (*acto*) crazy thing. **con ~** madly

locutor *m* announcer

locutorio *m* (*de teléfono*) telephone booth

lod|azal *m* quagmire. **~o** *m* mud

logaritmo *m* logarithm, log

lógic|a *f* logic. **~o** *a* logical

logística *f* logistics

logr|ar *vt* get; win ‹*premio*›. **~ hacer** manage to do. **~o** *m* achievement; (*de premio*) winning; (*éxito*) success

loma *f* small hill

lombriz *f* worm

lomo *m* back; (*de libro*) spine; (*doblez*) fold. **~ de cerdo** loin of pork

lona *f* canvas

loncha *f* slice; (*de tocino*) rasher

londinense *a* from London. ● *m* Londoner

Londres *m* London

loneta *f* thin canvas

longánimo *a* magnanimous

longaniza *f* sausage

longev|idad *f* longevity. **~o** *a* long-lived

longitud *f* length; (*geog*) longitude

lonja *f* slice; (*de tocino*) rasher; (*com*) market

lord *m* (*pl* **lores**) lord

loro *m* parrot

los *art def mpl* the. ● *pron* them. **~ de Antonio** Antonio's. **~ que** whoever, the ones

losa *f* slab; (*baldosa*) flagstone. **~ sepulcral** tombstone

lote *m* share

lotería *f* lottery

loto *m* lotus

loza *f* crockery

lozano *a* fresh; ‹*vegetación*› lush; ‹*persona*› lively

lubri(fi)ca|nte *a* lubricating. ● *m* lubricant. **~r** [7] *vt* lubricate

lucero *m* (*estrella*) bright star; (*planeta*) Venus

lucid|ez *f* lucidity. **~o** *a* splendid

lúcido *a* lucid

luciérnaga *f* glow-worm

lucimiento *m* brilliance

lucir [11] *vt* (*fig*) show off. ● *vi* shine; ‹*lámpara*› give off light; ‹*joya*› sparkle. **~se** *vpr* (*fig*) shine, excel

lucr|ativo *a* lucrative. **~o** *m* gain

lucha *f* fight. **~dor** *m* fighter. **~r** *vi* fight

luego *adv* then; (*más tarde*) later. ● *conj* therefore. **~ que** as soon as. **desde ~** of course

lugar *m* place. **~ común** cliché. **~eño** *a* village. **dar ~ a** give rise to. **en ~ de** instead of. **en primer ~** in the first place. **hacer ~** make room. **tener ~** take place

lugarteniente *m* deputy

lúgubre *a* gloomy

lujo *m* luxury. **~so** *a* luxurious. **de ~** de luxe

lujuria *f* lust

lumbago *m* lumbago

lumbre *f* fire; (*luz*) light. **¿tienes ~?** have you got a light?

luminoso *a* luminous; (*fig*) brilliant

luna *f* moon; (*de escaparate*) window; (*espejo*) mirror. **~ de miel** honeymoon. **~r** *a* lunar. ● *m* mole. **claro de ~** moonlight. **estar en la ~** be miles away

lunes *m* Monday. **cada ~ y cada martes** day in, day out

lupa *f* magnifying glass

lúpulo *m* hop

lustr|abotas *m inv* (*LAm*) bootblack. **~ar** *vt* shine, polish. **~e** *m* shine; (*fig, esplendor*) splendour. **~oso** *a* shining. **dar ~e a, sacar ~e a** polish

luto *m* mourning. **estar de ~** be in mourning

luxación *f* dislocation

Luxemburgo *m* Luxemburg

luz *f* light; (*electricidad*) electricity. **luces** *fpl* intelligence. **~ antiniebla**

(*auto*) fog light. **a la ~ de** in the light of. **a todas luces** obviously. **dar a ~** give birth. **hacer la ~ sobre** shed light on. **sacar a la ~** bring to light

LL

llaga *f* wound; (*úlcera*) ulcer
llama *f* flame; (*animal*) llama
llamada *f* call; (*golpe*) knock; (*señal*) sign
llama|do *a* known as. **~miento** *m* call. **~r** *vt* call; (*por teléfono*) ring (up). ● *vi* call; (*golpear en la puerta*) knock; (*tocar el timbre*) ring. **~rse** *vpr* be called. **~r por teléfono** ring (up), telephone. **¿cómo te ~s?** what's your name?
llamarada *f* blaze; (*fig*) blush; (*fig, de pasión etc*) outburst
llamativo *a* loud, gaudy
llamear *vi* blaze
llan|eza *f* simplicity. **~o** *a* flat, level; ‹*persona*› natural; (*sencillo*) plain. ● *m* plain
llanta *f* (*auto*) (wheel) rim; (*LAm, neumático*) tyre
llanto *m* weeping
llanura *f* plain
llave *f* key; (*para tuercas*) spanner; (*grifo*) tap (*Brit*), faucet (*Amer*); (*elec*) switch. **~ inglesa** monkey wrench. **~ro** *m* key-ring. **cerrar con ~** lock. **echar la ~** lock up
llega|da *f* arrival. **~r** [12] *vi* arrive, come; (*alcanzar*) reach; (*bastar*) be enough. **~rse** *vpr* come near; (*ir*) go (round). **~r a** (*conseguir*) manage to. **~r a saber** find out. **~r a ser** become
llen|ar *vt* fill (up); (*rellenar*) fill in. **~o** *a* full. ● *m* (*en el teatro etc*) full house. **de ~** completely
lleva|dero *a* tolerable. **~r** *vt* carry; (*inducir, conducir*) lead; (*acompañar*) take; wear ‹*ropa*›; (*traer*) bring. **~rse** *vpr* run off with ‹*cosa*›. **~rse bien** get on well together. **¿cuánto tiempo ~s aquí?** how long have you been here? **llevo 3 años estudiando inglés** I've been studying English for 3 years
llor|ar *vi* cry; ‹*ojos*› water. **~iquear** *vi* whine. **~iqueo** *m* whining. **~o** *m* crying. **~ón** *a* whining. ● *m* cry-baby. **~oso** *a* tearful
llov|er [2] *vi* rain. **~izna** *f* drizzle. **~iznar** *vi* drizzle
llueve *vb véase* **llover**
lluvi|a *f* rain; (*fig*) shower. **~oso** *a* rainy; ‹*clima*› wet

M

maca *f* defect; (*en fruta*) bruise
macabro *a* macabre
macaco *a* (*LAm*) ugly. ● *m* macaque (monkey)
macadam *m*, **macadán** *m* Tarmac (*P*)
macanudo *a* (*fam*) great
macarrón *m* macaroon. **~es** *mpl* macaroni
macerar *vt* macerate
maceta *f* mallet; (*tiesto*) flowerpot
macilento *a* wan
macizo *a* solid. ● *m* mass; (*de plantas*) bed
macrobiótico *a* macrobiotic
mácula *f* stain
macuto *m* knapsack
mach /mak/ *m*. **(número de) ~** Mach (number)
machac|ar [7] *vt* crush. ● *vi* go on (**en** about). **~ón** *a* boring. ● *m* bore
machamartillo. a ~ *adv* firmly
machaqueo *m* crushing
machet|azo *m* blow with a machete; (*herida*) wound from a machete. **~e** *m* machete
mach|ista *m* male chauvinist. **~o** *a* male; (*varonil*) macho
machón *m* buttress
machucar [7] *vt* crush; (*estropear*) damage
madeja *f* skein
madera *m* (*vino*) Madeira. ● *f* wood; (*naturaleza*) nature. **~ble** *a* yielding timber. **~je** *m*, **~men** *m* woodwork
madero *m* log; (*de construcción*) timber
madona *f* Madonna
madr|astra *f* stepmother. **~e** *f* mother. **~eperla** *f* mother-of-pearl. **~eselva** *f* honeysuckle
madrigal *m* madrigal
madriguera *f* den; (*de liebre*) burrow
madrileño *a* of Madrid. ● *m* person from Madrid
madrina *f* godmother; (*en una boda*) chief bridesmaid
madroño *m* strawberry-tree

madrug|ada *f* dawn. **~ador** *a* who gets up early. ● *m* early riser. **~ar** [12] *vi* get up early. **~ón** *m*. **darse un ~ón** get up very early

madur|ación *f* maturing; (*de fruta*) ripening. **~ar** *vt/i* mature; ‹*fruta*› ripen. **~ez** *f* maturity; (*de fruta*) ripeness. **~o** *a* mature; ‹*fruta*› ripe

maestr|a *f* teacher. **~ía** *f* skill. **~o** *m* master. **~a, ~o (de escuela)** schoolteacher

mafia *f* Mafia

magdalena *f* madeleine, small sponge cake

magia *f* magic

mágico *a* magic; (*maravilloso*) magical

magín *m* (*fam*) imagination

magisterio *m* teaching (profession); (*conjunto de maestros*) teachers

magistrado *m* magistrate; (*juez*) judge

magistral *a* teaching; (*bien hecho*) masterly; ‹*lenguaje*› pedantic

magistratura *f* magistracy

magn|animidad *f* magnanimity. **~ánimo** *a* magnanimous

magnate *m* magnate

magnesia *f* magnesia. **~ efervescente** milk of magnesia

magnético *a* magnetic

magneti|smo *m* magnetism. **~zar** [10] *vt* magnetize

magnetofón *m*, **magnetófono** *m* tape recorder

magnificencia *f* magnificence

magnífico *a* magnificent

magnitud *f* magnitude

magnolia *f* magnolia

mago *m* magician. **los (tres) reyes ~s** the Magi

magr|a *f* slice of ham. **~o** *a* lean; ‹*tierra*› poor; ‹*persona*› thin

magulla|dura *f* bruise. **~r** *vt* bruise

mahometano *a & m* Muhammadan

maíz *m* maize, corn (*Amer*)

majada *f* sheepfold; (*estiércol*) manure; (*LAm*) flock of sheep

majader|ía *f* silly thing. **~o** *m* idiot; (*mano del mortero*) pestle. ● *a* stupid

majador *m* crusher

majagranzas *m* idiot

majar *vt* crush; (*molestar*) bother

majest|ad *f* majesty. **~uoso** *a* majestic

majo *a* nice

mal *adv* badly; (*poco*) poorly; (*difícilmente*) hardly; (*equivocadamente*) wrongly. ● *a* see **malo**. ● *m* evil; (*daño*) harm; (*enfermedad*) illness. **~ que bien** somehow (or other). **de ~ en peor** worse and worse. **hacer ~ en** be wrong to. **¡menos ~!** thank goodness!

malabar *a*. **juegos ~es** juggling. **~ismo** *m* juggling. **~ista** *m & f* juggler

malaconsejado *a* ill-advised

malacostumbrado *a* with bad habits

malagueño *a* of Málaga. ● *m* person from Málaga

malamente *adv* badly; (*fam*) hardly enough

malandanza *f* misfortune

malapata *m & f* nuisance

malaria *f* malaria

Malasia *f* Malaysia

malasombra *m & f* clumsy person

malavenido *a* incompatible

malaventura *f* misfortune. **~do** *a* unfortunate

malayo *a* Malay(an)

malbaratar *vt* sell off cheap; (*malgastar*) squander

malcarado *a* ugly

malcasado *a* unhappily married; (*infiel*) unfaithful

malcomer *vi* eat poorly

malcriad|eza *f* (*LAm*) bad manners. **~o** *a* ‹*niño*› spoilt

maldad *f* evil; (*acción*) wicked thing

maldecir [46 *pero imperativo* **maldice**, *futuro y condicional regulares*, *pp* **maldecido** *o* **maldito**] *vt* curse. ● *vi* speak ill (**de** of); (*quejarse*) complain (**de** about)

maldici|ente *a* backbiting; (*que blasfema*) foul-mouthed. **~ón** *f* curse

maldit|a *f* tongue. **¡~a sea!** damn it! **~o** *a* damned. ● *m* (*en el teatro*) extra

maleab|ilidad *f* malleability. **~le** *a* malleable

malea|nte *a* wicked. ● *m* vagrant. **~r** *vt* damage; (*pervertir*) corrupt. **~rse** *vpr* be spoilt; (*pervertirse*) be corrupted

malecón *m* breakwater; (*rail*) embankment; (*para atracar*) jetty

maledicencia *f* slander

maleficio *m* curse

maléfico *a* evil

malestar *m* indisposition; (*fig*) uneasiness

malet|a *f* (suit)case; (*auto*) boot, trunk (*Amer*); (*LAm, lío de ropa*)

bundle; (*LAm*, *de bicicleta*) saddlebag. **hacer la ~a** pack one's bags. ● *m & f* (*fam*) bungler. **~ero** *m* porter; (*auto*) boot, trunk (*Amer*). **~ín** *m* small case
malevolencia *f* malevolence
malévolo *a* malevolent
maleza *f* weeds; (*matorral*) undergrowth
malgasta|dor *a* wasteful. ● *m* spendthrift. **~r** *vt* waste
malgeniado *a* (*LAm*) bad-tempered
malhablado *a* foul-mouthed
malhadado *a* unfortunate
malhechor *m* criminal
malhumorado *a* bad-tempered
malici|a *f* malice. **~arse** *vpr* suspect. **~as** *fpl* (*fam*) suspicions. **~oso** *a* malicious
malign|idad *f* malice; (*med*) malignancy. **~o** *a* malignant; ‹*persona*› malicious
malintencionado *a* malicious
malmandado *a* disobedient
malmirado *a* (*con estar*) disliked; (*con ser*) inconsiderate
malo *a* (*delante de nombre masculino en singular* **mal**) bad; (*enfermo*) ill. **~ de** difficult. **estar de malas** be out of luck; (*malhumorado*) be in a bad mood. **lo ~ es que** the trouble is that. **ponerse a malas con uno** fall out with s.o. **por las malas** by force
malogr|ar *vt* waste; (*estropear*) spoil. **~arse** *vpr* fall through. **~o** *m* failure
maloliente *a* smelly
malparto *m* miscarriage
malpensado *a* nasty, malicious
malquerencia *f* dislike
malquist|ar *vt* set against. **~arse** *vpr* fall out. **~o** *a* disliked
malsano *a* unhealthy; (*enfermizo*) sickly
malsonante *a* ill-sounding; (*grosero*) offensive
malta *f* malt; (*cerveza*) beer
maltés *a & m* Maltese
maltratar *vt* ill-treat
maltrecho *a* battered
malucho *a* (*fam*) poorly
malva *f* mallow. **(color de) ~** *a invar* mauve
malvado *a* wicked
malvavisco *m* marshmallow
malvender *vt* sell off cheap
malversa|ción *f* embezzlement. **~dor** *a* embezzling. ● *m* embezzler. **~r** *vt* embezzle
Malvinas *fpl*. **las islas ~** the Falkland Islands
malla *f* mesh. **cota de ~** coat of mail
mallo *m* mallet
Mallor|ca *f* Majorca. **~quín** *a & m* Majorcan
mama *f* teat; (*de mujer*) breast
mamá *f* mum(my)
mama|da *f* sucking. **~r** *vt* suck; (*fig*) grow up with; (*engullir*) gobble
mamario *a* mammary
mamarrach|adas *fpl* nonsense. **~o** *m* clown; (*cosa ridícula*) (ridiculous) sight
mameluco *a* Brazilian half-breed; (*necio*) idiot
mamífero *a* mammalian. ● *m* mammal
mamola *f*. **hacer la ~** chuck (under the chin); (*fig*) make fun of
mamotreto *m* notebook; (*libro voluminoso*) big book
mampara *f* screen
mamporro *m* blow
mampostería *f* masonry
mamut *m* mammoth
maná *f* manna
manada *f* herd; (*de lobos*) pack. **en ~** in crowds
manager /'manaʒer/ *m* manager
mana|ntial *m* spring; (*fig*) source. **~r** *vi* flow; (*fig*) abound. ● *vt* run with
manaza *f* big hand; (*sucia*) dirty hand. **ser un ~s** be clumsy
manceb|a *f* concubine. **~ía** *f* brothel. **~o** *m* youth; (*soltero*) bachelor
mancera *f* plough handle
mancilla *f* stain. **~r** *vt* stain
manco *a* (*de una mano*) one-handed; (*de las dos manos*) handless; (*de un brazo*) one-armed; (*de los dos brazos*) armless
mancomún *adv*. **de ~** jointly
mancomun|adamente *adv* jointly. **~ar** *vt* unite; (*jurid*) make jointly liable. **~arse** *vpr* unite. **~idad** *f* union
mancha *f* stain
Mancha *f*. **la ~** la Mancha (region of Spain). **el canal de la ~** the English Channel
mancha|do *a* dirty; ‹*animal*› spotted. **~r** *vt* stain. **~rse** *vpr* get dirty
manchego *a* of la Mancha. ● *m* person from la Mancha
manchón *m* large stain

manda *f* legacy

manda|dero *m* messenger. **~miento** *m* order; (*relig*) commandment. **~r** *vt* order; (*enviar*) send; (*gobernar*) rule. ● *vi* be in command. **¿mande?** (*esp LAm*) pardon?

mandarín *m* mandarin

mandarin|a *f* (*naranja*) mandarin; (*lengua*) Mandarin. **~o** *m* mandarin tree

mandat|ario *m* attorney. **~o** *m* order; (*jurid*) power of attorney

mandíbula *f* jaw

mandil *m* apron

mandioca *f* cassava

mando *m* command; (*pol*) term of office. **~ a distancia** remote control. **los ~s** the leaders

mandolina *f* mandolin

mandón *a* bossy

manducar [7] *vt* (*fam*) stuff oneself with

manecilla *f* needle; (*de reloj*) hand

manej|able *a* manageable. **~ar** *vt* handle; (*fig*) manage; (*LAm, conducir*) drive. **~arse** *vpr* behave. **~o** *m* handling; (*intriga*) intrigue

manera *f* way. **~s** *fpl* manners. **de ~ que** so (that). **de ninguna ~** not at all. **de otra ~** otherwise. **de todas ~s** anyway

manga *f* sleeve; (*tubo de goma*) hose(pipe); (*red*) net; (*para colar*) filter

mangante *m* beggar; (*fam*) scrounger

mangle *m* mangrove

mango *m* handle; (*fruta*) mango

mangonear *vt* boss about. ● *vi* (*entrometerse*) interfere

manguera *f* hose(pipe)

manguito *m* muff

manía *f* mania; (*antipatía*) dislike

maniaco *a*, **maníaco** *a* maniac(al). ● *m* maniac

maniatar *vt* tie s.o.'s hands

maniático *a* maniac(al); (*fig*) crazy

manicomio *m* lunatic asylum

manicura *f* manicure; (*mujer*) manicurist

manido *a* stale; ‹*carne*› high

manifesta|ción *f* manifestation; (*pol*) demonstration. **~nte** *m* demonstrator. **~r** [1] *vi* manifest; (*pol*) state. **~rse** *vpr* show; (*pol*) demonstrate

manifiesto *a* clear; ‹*error*› obvious; ‹*verdad*› manifest. ● *m* manifesto

manilargo *a* light-fingered

manilla *f* bracelet; (*de hierro*) handcuffs

manillar *m* handlebar(s)

maniobra *f* manoeuvring; (*rail*) shunting; (*fig*) manoeuvre. **~r** *vt* operate; (*rail*) shunt. ● *vi* manoeuvre. **~s** *fpl* (*mil*) manoeuvres

manipula|ción *f* manipulation. **~r** *vt* manipulate

maniquí *m* dummy. ● *f* model

manirroto *a* extravagant. ● *m* spendthrift

manita *f* little hand

manivela *f* crank

manjar *m* (*special*) dish

mano *f* hand; (*de animales*) front foot; (*de perros, gatos*) front paw. **~ de obra** work force. **¡~s arriba!** hands up! **a ~** by hand; (*próximo*) handy. **de segunda ~** second hand. **echar una ~** lend a hand. **tener buena ~ para** be good at

manojo *m* bunch

manose|ar *vt* handle; (*fig*) overwork. **~o** *m* handling

manotada *f*, **manotazo** *m* slap

manote|ar *vi* gesticulate. **~o** *m* gesticulation

mansalva. a ~ *adv* without risk

mansarda *f* attic

mansedumbre *f* gentleness; (*de animal*) tameness

mansión *f* stately home

manso *a* gentle; ‹*animal*› tame

manta *f* blanket. **~ eléctrica** electric blanket. **a ~ (de Dios)** a lot

mantec|a *f* fat; (*LAm*) butter. **~ado** *m* bun; (*helado*) ice-cream. **~oso** *a* greasy

mantel *m* tablecloth; (*del altar*) altar cloth. **~ería** *f* table linen

manten|er [40] *vt* support; (*conservar*) keep; (*sostener*) maintain. **~erse** *vpr* remain. **~ de/con** live off. **~imiento** *m* maintenance

mantequ|era *f* butter churn. **~ería** *f* dairy. **~illa** *f* butter

mantilla *f* mantilla

manto *m* cloak

mantón *m* shawl

manual *a & m* manual

manubrio *m* crank

manufactura *f* manufacture; (*fábrica*) factory

manuscrito *a* handwritten. ● *m* manuscript

manutención *f* maintenance

f apple. ~r m (apple) ...omile tea; (vino) ...herry
day after ... the morning
mañoso *a* clever; (*astuto*) crafty
mapa *m* map. **~mundi** *m* map of the world
mapache *m* racoon
mapurite *m* skunk
maqueta *f* scale model
maquiavélico *a* machiavellian
maquilla|je *m* make-up. **~r** *vt* make up. **~rse** *vpr* make up
máquina *f* machine; (*rail*) engine. **~ de escribir** typewriter. **~ fotográfica** camera
maquin|ación *f* machination. **~al** *a* mechanical. **~aria** *f* machinery. **~ista** *m & f* operator; (*rail*) engine driver
mar *m & f* sea. **alta ~** high seas. **la ~ de** (*fam*) lots of
maraña *f* thicket; (*enredo*) tangle; (*embrollo*) muddle
maravedí *m* (*pl* **maravedís, maravedises**) maravedi, old Spanish coin
maravill|a *f* wonder. **~ar** *vt* astonish. **~arse** *vpr* be astonished (**con** at). **~oso** *a* marvellous, wonderful. **a ~a, a las mil ~as** marvellously. **contar/decir ~as de** speak wonderfully of. **hacer ~as** work wonders
marbete *m* label
marca *f* mark; (*de fábrica*) trademark; (*deportes*) record. **~do** *a* marked. **~dor** *m* marker; (*deportes*) scoreboard. **~r** [7] *vt* mark; (*señalar*) show; (*anotar*) note down; score ‹*un gol*›; dial ‹*número de teléfono*›. ● *vi* score. **de ~** brand name; (*fig*) excellent. **de ~ mayor** (*fam*) first-class
marcial *a* martial
marciano *a & m* Martian
marco *m* frame; (*moneda alemana*) mark; (*deportes*) goal-posts
marcha *f* (*incl mus*) march; (*auto*) gear; (*curso*) course. **a toda ~** at full speed. **dar/hacer ~ atrás** put into reverse. **poner en ~** start; (*fig*) set in motion
marchante *m* (*f* **marchanta**) dealer; (*LAm*, *parroquiano*) client
marchar *vi* go; (*funcionar*) work, go. **~se** *vpr* go away, leave
marchit|ar *vt* wither. **~arse** *vpr* wither. **~o** *a* withered
...rea *f* tide. **~do** *a* sick; (*en el mar*) ...; (*aturdido*) dizzy; (*bo-*...) ... **~r** *vt* sail, navigate; ... annoy. **~rse** *vpr* feel ... be seasick; (*estar* ...) (...*irse la cabeza*) get slightly drunk ... worse)
marejada *f* swell; (*fig*) wave
maremagno *m* (*de cosas*) sea; (*de gente*) (noisy) crowd
mareo *m* sickness; (*en el mar*) seasickness; (*aturdimiento*) dizziness; (*fig*, *molestia*) nuisance
marfil *m* ivory. **~eño** *a* ivory. **torre** *f* **de ~** ivory tower
margarina *f* margarine
margarita *f* pearl; (*bot*) daisy
marg|en *m* margin; (*borde*) edge, border; (*de un río*) bank; (*de un camino*) side; (*nota marginal*) marginal note. **~inado** *a* on the edge. ● *m* outcast. **~inal** *a* marginal. **~inar** *vt* (*excluir*) exclude; (*dejar márgenes*) leave margins; (*poner notas*) write notes in the margin. **al ~en** (*fig*) outside
mariachi (*Mex*) *m* (*música popular de Jalisco*) Mariachi; (*conjunto popular*) Mariachi band
mariano *a* Marian
marica *f* (*hombre afeminado*) sissy; (*urraca*) magpie
maricón *m* homosexual, queer (*sl*)
marid|aje *m* married life; (*fig*) harmony. **~o** *m* husband
mariguana *f*, **marihuana** *f* marijuana
marimacho *m* mannish woman
marimandona *f* bossy woman
marimba *f* (type of) drum; (*LAm*, *especie de xilofón*) marimba
marimorena *f* (*fam*) row
marin|a *f* coast; (*cuadro*) seascape; (*conjunto de barcos*) navy; (*arte de navegar*) seamanship. **~era** *f* seamanship; (*conjunto de marineros*) crew. **~ero** *a* marine; ‹*barco*› seaworthy. ● *m* sailor. **~o** *a* marine. **~a de guerra** navy. **~a mercante** merchant navy. **a la ~era** in tomato

and garlic sauce. **azul ~o** navy blue
marioneta *f* puppet. **~s** *fpl* puppet show
maripos|a *f* butterfly. **~ear** *vi* be fickle; (*galantear*) flirt. **~n** *m* fl
~a nocturna moth
mariquita *f* ladybird, (*Amer*)
marisabidilla *f* know
mariscador *m* shel
mariscal *m*
maris|co *m*
~quero

...sh. ...pesca mar- ...marerman; (*persona* ...iscos) seafood seller
...*a* marital
marítimo *a* maritime; ‹*ciudad etc*› coastal, seaside
maritornes *f* uncouth servant
marmit|a *f* pot. **~ón** *m* kitchen boy
mármol *m* marble
marmol|era *f* marblework, marbles. **~ista** *m & f* marble worker
marmóreo *a* marble
marmota *f* marmot
maroma *f* rope; (*LAm, función de volatines*) tightrope walking
marqu|és *m* marquess. **~esa** *f* marchioness. **~esina** *f* glass canopy
marquetería *f* marquetry
marrajo *a* ‹*toro*› vicious; ‹*persona*› cunning. ● *m* shark
marran|a *f* sow. **~ada** *f* filthy thing; (*cochinada*) dirty trick. **~o** *a* filthy. ● *m* hog
marrar *vt* (*errar*) miss; (*fallar*) fail
marrón *a & m* brown
marroquí *a & m & f* Moroccan. ● *m* (*tafilete*) morocco
marrubio *m* (*bot*) horehound
Marruecos *m* Morocco
marruller|ía *f* cajolery. **~o** *a* cajoling. ● *m* cajoler
marsopa *f* porpoise
marsupial *a & m* marsupial
marta *f* marten
martajar *vt* (*Mex*) grind ‹*maíz*›
Marte *m* Mars
martes *m* Tuesday
martill|ada *f* blow with a hammer. **~ar** *vt* hammer. **~azo** *m* blow with a hammer. **~ear** *vt* hammer. **~eo** *m* hammering. **~o** *m* hammer
martín *m* **pescador** kingfisher
martinete *m* (*macillo del piano*) hammer; (*mazo*) drop hammer
martingala *f* (*ardid*) trick
mártir *m & f* martyr

mas...*ativo*) more; ...st. **~ caro** dearer. ...more curious. **el ~ caro** ...dearest; (*de dos*) the dearer. **el ~ curioso** the most curious; (*de dos*) the more curious. ● *conj* and, plus. ● *m* plus (sign). **~ bien** rather. **~ de** (*cantidad indeterminada*) more than. **~ o menos** more or less. **~ que** more than. **~ y ~** more and more. **a lo ~** at (the) most. **de ~** too many. **es ~** moreover. **no ~** no more
masa *f* dough; (*cantidad*) mass; (*física*) mass. **en ~** en masse
masacre *f* massacre
masaj|e *m* massage. **~ista** *m* masseur. ● *f* masseuse
masca|da *f* (*LAm*) plug of tobacco. **~dura** *f* chewing. **~r** [7] *vt* chew
máscara *f* mask; (*persona*) masked figure/person
mascar|ada *f* masquerade. **~illa** *f* mask. **~ón** *m* (large) mask
mascota *f* mascot
masculin|idad *f* masculinity. **~o** *a* masculine; ‹*sexo*› male. ● *m* masculine
mascullar [3] *vt* mumble
masilla *f* putty
masivo *a* massive, large-scale
mas|ón *m* (free)mason. **~onería** *f* (free)masonry. **~ónico** *a* masonic
masoquis|mo *m* masochism. **~ta** *a* masochistic. ● *m & f* masochist
mastate *m* (*Mex*) loincloth
mastelero *m* topmast
mastica|ción *f* chewing. **~r** [7] *vt* chew; (*fig*) chew over
mástil *m* mast; (*palo*) pole; (*en instrumentos de cuerda*) neck
mastín *m* mastiff
mastitis *f* mastitis
mastodonte *m* mastodon
mastoides *a & f* mastoid
mastuerzo *m* cress
masturba|ción *f* masturbation. **~rse** *vpr* masturbate
mata *f* grove; (*arbusto*) bush
matad|ero *m* slaughterhouse. **~or** *a* killing. ● *m* killer; (*torero*) matador
matadura *f* sore

matamoscas *m invar* fly swatter

mata|nza *f* killing. **~r** *vt* kill ‹*personas*›; slaughter ‹*reses*›. **~rife** *m* butcher. **~rse** *vpr* commit suicide; (*en un acidente*) be killed. **estar a ~r con uno** be deadly enemies with s.o.

matarratas *m invar* cheap liquor

matasanos *m invar* quack

matasellos *m invar* postmark

match *m* match

mate *a* matt, dull; ‹*sonido*› dull. ● *m* (*ajedrez*) (check)mate; (*LAm, bebida*) maté

matemátic|as *fpl* mathematics, maths (*fam*), math (*Amer, fam*). **~o** *a* mathematical. ● *m* mathematician

materia *f* matter; (*material*) material. **~ prima** raw material. **en ~ de** on the question of

material *a* & *m* material. **~idad** *f* material nature. **~ismo** *m* materialism. **~ista** *a* materialistic. ● *m* & *f* materialist. **~izar** [10] *vt* materialize. **~izarse** *vpr* materialize. **~mente** *adv* materially; (*absolutamente*) absolutely

matern|al *a* maternal; (*como de madre*) motherly. **~idad** *f* motherhood; (*casa de maternidad*) maternity home. **~o** *a* motherly; ‹*lengua*› mother

matin|al *a* morning. **~ée** *m* matinée

matiz *m* shade. **~ación** *f* combination of colours. **~ar** [10] *vt* blend ‹*colores*›; (*introducir variedad*) vary; (*teñir*) tinge (**de** with)

matojo *m* bush

mat|ón *m* bully. **~onismo** *m* bullying

matorral *m* scrub; (*conjunto de matas*) thicket

matra|ca *f* rattle. **~quear** *vt* rattle; (*dar matraca*) pester. **dar ~ca** pester. **ser un(a) ~ca** be a nuisance

matraz *m* flask

matriarca|do *m* matriarchy. **~l** *a* matriarchal

matr|ícula *f* (*lista*) register, list; (*acto de matricularse*) registration; (*auto*) registration number. **~icular** *vt* register. **~icularse** *vpr* enrol, register

matrimoni|al *a* matrimonial. **~o** *m* marriage; (*pareja*) married couple

matritense *a* from Madrid

matriz *f* matrix; (*anat*) womb, uterus

matrona *f* matron; (*partera*) midwife

Matusalén *m* Methuselah. **más viejo que ~** as old as Methuselah

matute *m* smuggling. **~ro** *m* smuggler

matutino *a* morning

maula *f* piece of junk

maull|ar *vi* miaow. **~ido** *m* miaow

mauritano *a* & m Mauritanian

mausoleo *m* mausoleum

maxilar *a* maxillary. **hueso ~** jaw(bone)

máxima *f* maxim

máxime *adv* especially

máximo *a* maximum; (*más alto*) highest. ● *m* maximum

maya *f* daisy; (*persona*) Maya Indian

mayestático *a* majestic

mayo *m* May; (*palo*) maypole

mayólica *f* majolica

mayonesa *f* mayonnaise

mayor *a* (*más grande, comparativo*) bigger; (*más grande, superlativo*) biggest; (*de edad, comparativo*) older; (*de edad, superlativo*) oldest; (*adulto*) grown-up; (*principal*) main, major; (*mus*) major. ● *m* & *f* boss; (*adulto*) adult. **~al** *m* foreman; (*pastor*) head shepherd. **~azgo** *m* entailed estate. **al por ~** wholesale

mayordomo *m* butler

mayor|ía *f* majority. **~ista** *m* & *f* wholesaler. **~mente** *adv* especially

mayúscul|a *f* capital (letter). **~o** *a* capital; (*fig, grande*) big

maza *f* mace

mazacote *m* hard mass

mazapán *m* marzipan

mazmorra *f* dungeon

mazo *m* mallet; (*manojo*) bunch

mazorca *f*. **~ de maíz** corn on the cob

me *pron* (*acusativo*) me; (*dativo*) (to) me; (*reflexivo*) (to) myself

meandro *m* meander

mecánic|a *f* mechanics. **~o** *a* mechanical. ● *m* mechanic

mecani|smo *m* mechanism. **~zación** *f* mechanization. **~zar** [10] *vt* mechanize

mecanograf|ía *f* typing. **~iado** *a* typed, typewritten. **~iar** [20] *vt* type

mecanógrafo *m* typist

mecate *m* (*LAm*) (*pita*) rope

mecedora *f* rocking chair

mecenazgo *m* patronage

mecer [9] *vt* rock; swing ‹*columpio*›. **~se** *vpr* rock; (*en un columpio*) swing

mecha *f* (*de vela*) wick; (*de mina*) fuse

mechar *vt* stuff, lard

mechero *m* (cigarette) lighter

mechón *m* (*de pelo*) lock

medall|a *f* medal. **~ón** *m* medallion; (*relicario*) locket

media *f* stocking; (*promedio*) average

mediación *f* mediation

mediado *a* half full; ‹*trabajo etc*› halfway through. **a ~s de marzo** in the middle of March

mediador *m* mediator

medialuna *f* croissant

median|amente *adv* fairly. **~era** *f* party wall. **~ero** *a* ‹*muro*› party. **~a** *f* average circumstances. **~o** *a* average, medium; (*mediocre*) mediocre

medianoche *f* midnight; (*culin*) small sandwich

mediante *prep* through, by means of

mediar *vi* mediate; (*llegar a la mitad*) be halfway (**en** through)

mediatizar [10] *vt* annex

medic|ación *f* medication. **~amento** *m* medicine. **~ina** *f* medicine. **~inal** *a* medicinal. **~inar** *vt* administer medicine

medición *f* measurement

médico *a* medical. ● *m* doctor. **~ de cabecera** GP, general practitioner

medid|a *f* measurement; (*unidad*) measure; (*disposición*) measure, step; (*prudencia*) moderation. **~or** *m* (*LAm*) meter. **a la ~a** made to measure. **a ~a que** as. **en cierta ~a** to a certain point

mediero *m* share-cropper

medieval *a* medieval. **~ista** *m & f* medievalist

medio *a* half (a); (*mediano*) average. **~ litro** half a litre. ● *m* middle; (*manera*) means; (*en deportes*) half(-back). **en ~** in the middle (**de** of). **por ~ de** through

mediocr|e *a* (*mediano*) average; (*de escaso mérito*) mediocre. **~idad** *f* mediocrity

mediodía *m* midday, noon; (*sur*) south

medioevo *m* Middle Ages

Medio Oriente *m* Middle East

medir [5] *vt* medir; weigh up ‹*palabras etc*›. ● *vi* measure, be. **~se** *vpr* (*moderarse*) be moderate

medita|bundo *a* thoughtful. **~ción** *f* meditation. **~r** *vt* think about. ● *vi* meditate

Mediterráneo *m* Mediterranean

mediterráneo *a* Mediterranean

médium *m & f* medium

medrar *vi* thrive

medroso *a* (*con estar*) frightened; (*con ser*) fearful

médula *f* marrow

medusa *f* jellyfish

mefítico *a* noxious

mega... *pref* mega...

megáfono *m* megaphone

megal|ítico *a* megalithic. **~ito** *m* megalith

megal|omanía *f* megalomania. **~ómano** *m* megalomaniac

mejicano *a & m* Mexican

Méjico *m* Mexico

mejido *a* ‹*huevo*› beaten

mejilla *f* cheek

mejillón *m* mussel

mejor *a & adv* (*comparativo*) better; (*superlativo*) best. **~a** *f* improvement. **~able** *a* improvable. **~amiento** *m* improvement. **~ dicho** rather. **a lo ~** perhaps. **tanto ~** so much the better

mejorana *f* marjoram

mejorar *vt* improve, better. ● *vi* get better

mejunje *m* mixture

melanc|olía *f* melancholy. **~ólico** *a* melancholic

melaza *f* molasses, treacle (*Amer*)

melen|a *f* long hair; (*de león*) mane. **~udo** *a* long-haired

melifluo *a* mellifluous

melillense *a* of/from Melilla. ● *m* person from Melilla

melindr|e *m* (*mazapán*) sugared marzipan cake; (*masa frita con miel*) honey fritter. **~oso** *a* affected

melocot|ón *m* peach. **~onero** *m* peach tree

mel|odía *f* melody. **~ódico** *a* melodic. **~odioso** *a* melodious

melodram|a *m* melodrama. **~áticamente** *adv* melodramatically. **~ático** *a* melodramatic

melómano *m* music lover

mel|ón *m* melon; (*bobo*) fool. **~onada** *f* something stupid

meloncillo *m* (*animal*) mongoose

melos|idad *f* sweetness. **~o** *a* sweet

mella *f* notch. **~do** *a* jagged. **~r** *vt* notch
mellizo *a & m* twin
membran|a *f* membrane. **~oso** *a* membranous
membrete *m* letterhead
membrill|ero *m* quince tree. **~o** *m* quince
membrudo *a* burly
memez *f* something silly
memo *a* stupid. ● *m* idiot
memorable *a* memorable
memorando *m*, **memorándum** *m* notebook; (*nota*) memorandum
memoria *f* memory; (*informe*) report; (*tesis*) thesis. **~s** *fpl* (*recuerdos personales*) memoirs. **de ~** from memory
memorial *m* memorial. **~ista** *m* amanuensis
memor|ión *m* good memory. **~ista** *a* having a good memory. **~ístico** *a* memory
mena *f* ore
menaje *m* furnishings
menci|ón *f* mention. **~onado** *a* aforementioned. **~onar** *vt* mention
menda|cidad *f* mendacity. **~z** *a* lying
mendi|cante *a & m* mendicant. **~cidad** *f* begging. **~gar** [12] *vt* beg (for). ● *vi* beg. **~go** *m* beggar
mendrugo *m* (*pan*) hard crust; (*zoquete*) blockhead
mene|ar *vt* move, shake. **~arse** *vpr* move, shake. **~o** *m* movement, shake
menester *m* need. **~oso** *a* needy. **ser ~** be necessary
menestra *f* stew
menestral *m* artesan
mengano *m* so-and-so
mengua *f* decrease; (*falta*) lack; (*descrédito*) discredit. **~do** *a* miserable; (*falto de carácter*) spineless. **~nte** *a* decreasing; ‹*luna*› waning; ‹*marea*› ebb. ● *f* (*del mar*) ebb tide; (*de un río*) low water. **~r** [15] *vt/i* decrease, diminish
meningitis *f* meningitis
menisco *m* meniscus
menjurje *m* mixture
menopausia *f* menopause
menor *a* (*más pequeño, comparativo*) smaller; (*más pequeño, superlativo*) smallest; (*más joven, comparativo*) younger; (*más joven*) youngest; (*mus*) minor. ● *m & f* (*menor de edad*) minor. **al por ~** retail
Menorca *f* Minorca
menorquín *a & m* Minorcan
menos *a* (*comparativo*) less; (*comparativo, con plural*) fewer; (*superlativo*) least; (*superlativo, con plural*) fewest. ● *adv* (*comparativo*) less; (*superlativo*) least. ● *prep* except. **~cabar** *vt* lessen; (*fig, estropear*) damage. **~cabo** *m* lessening. **~preciable** *a* contemptible. **~preciar** *vt* despise. **~precio** *m* contempt. **a ~ que** unless. **al ~** at least. **ni mucho ~** far from it. **por lo ~** at least
mensaje *m* message. **~ro** *m* messenger
menso *a* (*Mex*) stupid
menstru|ación *f* menstruation. **~al** *a* menstrual. **~ar** [21] *vi* menstruate. **~o** *m* menstruation
mensual *a* monthly. **~idad** *f* monthly pay
ménsula *f* bracket
mensurable *a* measurable
menta *f* mint
mental *a* mental. **~idad** *f* mentality. **~mente** *adv* mentally
mentar [1] *vt* mention, name
mente *f* mind
mentecato *a* stupid. ● *m* idiot
mentir [4] *vi* lie. **~a** *f* lie. **~oso** *a* lying. ● *m* liar. **de ~ijillas** for a joke
mentís *m invar* denial
mentol *m* menthol
mentor *m* mentor
menú *m* menu
menudear *vi* happen frequently
menudencia *f* trifle
menudeo *m* retail trade
menudillos *mpl* giblets
menudo *a* tiny; ‹*lluvia*› fine; (*insignificante*) insignificant. **~s** *mpl* giblets. **a ~** often
meñique *a* ‹*dedo*› little. ● *m* little finger
meollo *m* brain; (*médula*) marrow; (*parte blanda*) soft part; (*fig, inteligencia*) brains
meramente *adv* merely
mercachifle *m* hawker; (*fig*) profiteer
mercader *m* (*LAm*) merchant
mercado *m* market. **M~ Común** Common Market. **~ negro** black market
mercan|cía *f* article. **~cías** *fpl* goods, merchandise. **~te** *a & m*

merchant. **~til** *a* mercantile, commercial. **~tilismo** *m* mercantilism
mercar [7] *vt* buy
merced *f* favour. **su/vuestra ~** your honour
mercenario *a* & *m* mercenary
mercer|ía *f* haberdashery, notions (*Amer*). **~o** *m* haberdasher
mercurial *a* mercurial
Mercurio *m* Mercury
mercurio *m* mercury
merec|edor *a* deserving. **~er** [11] *vt* deserve. ● *vi* be deserving. **~idamente** *adv* deservedly. **~ido** *a* well deserved. **~imiento** *m* (*mérito*) merit
merend|ar [1] vt have as an afternoon snack. ● *vi* have an afternoon snack. **~ero** *m* snack bar; (*lugar*) picnic area
merengue *m* meringue
meretriz *f* prostitute
mergo *m* cormorant
meridian|a *f* (*diván*) couch. **~o** *a* midday; (*fig*) dazzling. ● *m* meridian
meridional *a* southern. ● *m* southerner
merienda *f* afternoon snack
merino *a* merino
mérito *m* merit; (*valor*) worth
meritorio *a* meritorious. ● *m* unpaid trainee
merlo *m* black wrasse
merluza *f* hake
merma *f* decrease. **~r** *vt/i* decrease, reduce
mermelada *f* jam
mero *a* mere; (*Mex, verdadero*) real. ● *adv* (*Mex, precisamente*) exactly; (*Mex, verdaderamente*) really. ● *m* grouper
merode|ador *a* marauding. ● *m* marauder. **~ar** *vi* maraud. **~o** *m* marauding
merovingio *a* & *m* Merovingian
mes *m* month; (*mensualidad*) monthly pay
mesa *f* table; (*para escribir o estudiar*) desk. **poner la ~** lay the table
mesana *f* (*palo*) mizen-mast
mesarse *vpr* tear at one's hair
mesenterio *m* mesentery
meseta *f* plateau; (*descansillo*) landing
mesiánico *a* Messianic
Mesías *m* Messiah
mesilla *f* small table. **~ de noche** bedside table
mesón *m* inn
mesoner|a *f* landlady. **~o** *m* landlord
mestiz|aje *m* crossbreeding. **~o** *a* ‹*persona*› half-caste; ‹*animal*› cross-bred. ● *m* (*persona*) half-caste; (*animal*) cross-breed
mesura *f* moderation. **~do** *a* moderate
meta *f* goal; (*de una carrera*) finish
metabolismo *m* metabolism
metacarpiano *m* metacarpal
metafísic|a *f* metaphysics. **~o** *a* metaphysical
met|áfora *f* metaphor. **~afórico** *a* metaphorical
met|al *m* metal; (*instrumentos de latón*) brass; (*de la voz*) timbre. **~álico** *a* ‹*objeto*› metal; ‹*sonido*› metallic. **~alizarse** [10] *vpr* (*fig*) become mercenary
metal|urgia *f* metallurgy. **~úrgico** *a* metallurgical
metam|órfico *a* metamorphic. **~orfosear** *vt* transform. **~orfosis** *f* metamorphosis
metano *m* methane
metatarsiano *m* metatarsal
metátesis *f invar* metathesis
metedura *f*. **~ de pata** blunder
mete|órico *a* meteoric. **~orito** *m* meteorite. **~oro** *m* meteor. **~orología** *f* meteorology. **~orológico** meteorological. **~orólogo** *m* meteorologist
meter *vt* put, place; (*ingresar*) deposit; score ‹*un gol*›; (*enredar*) involve; (*causar*) make. **~se** *vpr* get; (*entrometerse*) meddle. **~se con uno** pick a quarrel with s.o.
meticulos|idad *f* meticulousness. **~o** *a* meticulous
metido *m* reprimand. ● *a*. **~ en años** getting on. **estar muy ~ con uno** be well in with s.o.
metilo *m* methyl
metódico *a* methodical
metodis|mo *m* Methodism. **~ta** *a* & *m* & *f* Methodist
método *m* method
metodología *f* methodology
metomentodo *m* busybody
metraje *m* length. **de largo ~** ‹*película*› feature
metrall|a *f* shrapnel. **~eta** *f* submachine gun
métric|a *f* metrics. **~o** *a* metric; ‹*verso*› metrical

metro *m* metre; (*tren*) underground, subway (*Amer*). **~ cuadrado** cubic metre
metrónomo *m* metronome
metr|ópoli *f* metropolis. **~opolitano** *a* metropolitan. ● *m* metropolitan; (*tren*) underground, subway (*Amer*)
mexicano *a* & *m* (*LAm*) Mexican
México *m* (*LAm*) Mexico. **~ D. F.** Mexico City
mezcal *m* (*Mex*) (type of) brandy
mezc|la *f* (*acción*) mixing; (*substancia*) mixture; (*argamasa*) mortar. **~lador** *m* mixer. **~lar** *vt* mix; shuffle ‹*los naipes*›. **~larse** *vpr* mix; (*intervenir*) interfere. **~olanza** *f* mixture
mezquin|dad *f* meanness. **~o** *a* mean; (*escaso*) meagre. ● *m* mean person
mezquita *f* mosque
mi *a* my. ● *m* (*mus*) E; (*solfa*) mi
mí *pron* me
miaja *f* crumb
miasma *m* miasma
miau *m* miaow
mica *f* (*silicato*) mica; (*Mex*, *embriaguez*) drunkenness
mico *m* (long-tailed) monkey
micro... *pref* micro...
microbio *m* microbe
micro: ~biología *f* microbiology. **~cosmo** *m* microcosm. **~film(e)** *m* microfilm
micrófono *m* microphone
micrómetro *m* micrometer
microonda *f* microwave. **horno** *m* **de ~s** microwave oven
microordenador *m* microcomputer
microsc|ópico *a* microscopic. **~opio** *m* microscope
micro: ~surco *m* long-playing record. **~taxi** *m* minicab
miedo *m* fear. **~so** *a* fearful. **dar ~** frighten. **morirse de ~** be scared to death. **tener ~** be frightened
miel *f* honey
mielga *f* lucerne, alfalfa (*Amer*)
miembro *m* limb; (*persona*) member
mientras *conj* while. ● *adv* meanwhile. **~ que** whereas. **~ tanto** in the meantime
miércoles *m* Wednesday. **~ de ceniza** Ash Wednesday
mierda *f* (*vulgar*) shit
mies *f* corn, grain (*Amer*)
miga *f* crumb; (*fig*, *meollo*) essence. **~jas** *fpl* crumbs. **~r** [12] *vt* crumble
migra|ción *f* migration. **~torio** *a* migratory
mijo *m* millet
mil *a* & *m* a/one thousand. **~es de** thousands of. **~ novecientos noventa y dos** nineteen ninety-two. **~ pesetas** a thousand pesetas
milagro *m* miracle. **~so** *a* miraculous
milano *m* kite
mildeu *m*, **mildiu** *m* mildew
milen|ario *a* millenial. **~io** *m* millennium
milenrama *f* milfoil
milésimo *a* & m thousandth
mili *f* (*fam*) military service
milicia *f* soldiering; (*gente armada*) militia
mili|gramo *m* milligram. **~litro** *m* millilitre
milímetro *m* millimetre
militante *a* militant
militar *a* military. ● *m* soldier. **~ismo** *m* militarism. **~ista** *a* militaristic. ● *m* & *f* militarist. **~izar** [10] *vt* militarize
milonga *f* (*Arg*, *canción*) popular song; (*Arg*, *baile*) popular dance
milord *m*. **vivir como un ~** live like a lord
milpies *m invar* woodlouse
milla *f* mile
millar *m* thousand. **a ~es** by the thousand
mill|ón *m* million. **~onada** *f* fortune. **~onario** *m* millionaire. **~onésimo** *a* & *m* millionth. **un ~n de libros** a million books
mimar *vt* spoil
mimbre *m* & *f* wicker. **~arse** *vpr* sway. **~ra** *f* osier. **~ral** *m* osier-bed
mimetismo *m* mimicry
mímic|a *f* mime. **~o** *a* mimic
mimo *m* mime; (*a un niño*) spoiling; (*caricia*) caress
mimosa *f* mimosa
mina *f* mine. **~r** *vt* mine; (*fig*) undermine
minarete *m* minaret
mineral *m* mineral; (*mena*) ore. **~ogía** *f* mineralogy. **~ogista** *m* & *f* mineralogist
miner|ía *f* mining. **~o** *a* mining. ● *m* miner
mini... *pref* mini...
miniar *vt* paint in miniature
miniatura *f* miniature
minifundio *m* smallholding
minimizar [10] *vt* minimize

mínim|o *a* & *m* minimum. **~um** *m* minimum
minino *m* (*fam*) cat, puss (*fam*)
minio *m* red lead
minist|erial *a* ministerial. **~erio** *m* ministry. **~ro** *m* minister
minor|ación *f* diminution. **~a** *f* minority. **~idad** *f* minority. **~ista** *m* & *f* retailer
minuci|a *f* trifle. **~osidad** *f* thoroughness. **~oso** *a* thorough; (*con muchos detalles*) detailed
minué *m* minuet
minúscul|a *f* small letter, lower case letter. **~o** *a* tiny
minuta *f* draft; (*menú*) menu
minut|ero *m* minute hand. **~o** *m* minute
mío *a* & *pron* mine. **un amigo ~** a friend of mine
miop|e *a* short-sighted. ● *m* & *f* short-sighted person. **~ía** *f* short-sightedness
mira *f* sight; (*fig, intención*) aim. **~da** *f* look. **~do** *a* thought of; (*comedido*) considerate; (*cirunspecto*) circumspect. **~dor** *m* windowed balcony; (*lugar*) viewpoint. **~miento** *m* consideration. **~r** *vt* look at; (*observar*) watch; (*considerar*) consider. **~r fijamente a** stare at. ● *vi* look; ‹*edificio etc*› face. **~rse** *vpr* ‹*personas*› look at each other. **a la ~** on the lookout. **con ~s a** with a view to. **echar una ~da a** glance at
mirilla *f* peephole
miriñaque *m* crinoline
mirlo *m* blackbird
mirón *a* nosey. ● *m* nosey-parker; (*espectador*) onlooker
mirra *f* myrrh
mirto *m* myrtle
misa *f* mass
misal *m* missal
mis|antropía *f* misanthropy. **~antrópico** *a* misanthropic. **~ántropo** *m* misanthropist
miscelánea *f* miscellany; (*Mex, tienda*) corner shop
miser|able *a* very poor; (*lastimoso*) miserable; (*tacaño*) mean. **~ia** *f* extreme poverty; (*suciedad*) squalor
misericordi|a *f* pity; (*piedad*) mercy. **~oso** *a* merciful
mísero *a* very poor; (*lastimoso*) miserable; (*tacaño*) mean
misil *m* missile
misi|ón *f* mission. **~onal** *a* missionary. **~onero** *m* missionary
misiva *f* missive
mism|amente *adv* just. **~ísimo** *a* very same. **~o** *a* same; (*después de pronombre personal*) myself, yourself, himself, herself, itself, ourselves, yourselves, themselves; (enfático) very. ● *adv* right. **ahora ~** right now. **aquí ~** right here
mis|oginia *f* misogyny. **~ógino** *m* misogynist
misterio *m* mystery. **~so** *a* mysterious
místic|a *f* mysticism. **~o** *a* mystical
mistifica|ción *f* falsification; (*engaño*) trick. **~r** [7] vt falsify; (*engañar*) deceive
mitad *f* half; (*centro*) middle
mítico *a* mythical
mitiga|ción *f* mitigation. **~r** [12] *vt* mitigate; quench ‹*sed*›; relieve ‹*dolor etc*›
mitin *m* meeting
mito *m* myth. **~logía** *f* mythology. **~lógico** *a* mythological
mitón *m* mitten
mitote *m* (*LAm*) Indian dance
mitra *f* mitre. **~do** *m* prelate
mixteca *f* (*Mex*) southern Mexico
mixt|o *a* mixed. ● *m* passenger and goods train; (*cerilla*) match. **~ura** *f* mixture
mnemotécnic|a *f* mnemonics. **~o** *a* mnemonic
moaré *m* moiré
mobiliario *m* furniture
moblaje *m* furniture
moca *m* mocha
moce|dad *f* youth. **~ro** *m* young people. **~tón** *m* strapping lad
moción *f* motion
moco *m* mucus
mochales *a invar*. **estar ~** be round the bend
mochila *f* rucksack
mocho *a* blunt. ● *m* butt end
mochuelo *m* little owl
moda *f* fashion. **~l** *a* modal. **~les** *mpl* manners. **~lidad** *f* kind. **de ~** in fashion
model|ado *m* modelling. **~ador** *m* modeller. **~ar** *vt* model; (*fig, configurar*) form. **~o** *m* model
modera|ción *f* moderation. **~do** *a* moderate. **~r** *vt* moderate; reduce ‹*velocidad*›. **~rse** *vpr* control oneself

modern|amente *adv* recently. **~idad** *f* modernity. **~ismo** *m* modernism. **~ista** *m & f* modernist. **~izar** [10] *vt* modernize. **~o** *a* modern

modest|ia *f* modesty. **~o** *a* modest

modicidad *f* reasonableness

módico *a* moderate

modifica|ción *f* modification. **~r** [7] *vt* modify

modismo *m* idiom

modist|a *f* dressmaker. **~o** *m & f* designer

modo *m* manner, way; (*gram*) mood; (*mus*) mode. **~ de ser** character. **de ~ que** so that. **de ningún ~** certainly not. **de todos ~s** anyhow

modorr|a *f* drowsiness. **~o** *a* drowsy

modoso *a* well-behaved

modula|ción *f* modulation. **~dor** *m* modulator. **~r** *vt* modulate

módulo *m* module

mofa *f* mockery. **~rse** *vpr*. **~rse de** make fun of

mofeta *f* skunk

moflet|e *m* chubby cheek. **~udo** *a* with chubby cheeks

mogol *m* Mongol. **el Gran M~** the Great Mogul

moh|ín *m* grimace. **~ino** *a* sulky. **hacer un ~ín** pull a face

moho *m* mould; (*óxido*) rust. **~so** *a* mouldy; ‹*metales*› rusty

moisés *m* Moses basket

mojado *a* damp, wet

mojama *f* salted tuna

mojar *vt* wet; (*empapar*) soak; (*humedecer*) moisten, dampen. ● *vi*. **~ en** get involved in

mojicón *m* blow in the face; (*bizcocho*) sponge cake

mojiganga *f* masked ball; (*en el teatro*) farce

mojigat|ería *f* hypocrisy. **~o** *m* hypocrite

mojón *m* boundary post; (*señal*) signpost

molar *m* molar

mold|e *m* mould; (*aguja*) knitting needle. **~ear** *vt* mould, shape; (*fig*) form. **~ura** *f* moulding

mole *f* mass, bulk. ● *m* (*Mex, guisado*) (Mexican) stew with chili sauce

mol|écula *f* molecule. **~ecular** *a* molecular

mole|dor *a* grinding. ● *m* grinder; (*persona*) bore. **~r** [2] grind; (*hacer polvo*) pulverize

molest|ar *vt* annoy; (*incomodar*) bother. **¿le ~a que fume?** do you mind if I smoke? **no ~ar** do not disturb. ● *vi* be a nuisance. **~arse** *vpr* bother; (*ofenderse*) take offence. **~ia** *f* bother, nuisance; (*inconveniente*) inconvenience; (*incomodidad*) discomfort. **~o** *a* annoying; (*inconveniente*) inconvenient; (*ofendido*) offended

molicie *f* softness; (*excesiva comodidad*) easy life

molido *a* ground; (*fig, muy cansado*) worn out

molienda *f* grinding

molin|ero *m* miller. **~ete** *m* toy windmill. **~illo** *m* mill; (*juguete*) toy windmill. **~o** *m* (water) mill. **~o de viento** windmill

molusco *m* mollusc

mollar *a* soft

molleja *f* gizzard

mollera *f* (*de la cabeza*) crown; (*fig, sesera*) brains

moment|áneamente *adv* momentarily; (*por el momento*) right now. **~áneo** *a* momentary. **~o** *m* moment; (*mecánica*) momentum

momi|a *f* mummy. **~ficación** *f* mummification. **~ficar** [7] *vt* mummify. **~ficarse** *vpr* become mummified

momio *a* lean. ● *m* bargain; (*trabajo*) cushy job

monaca|l *a* monastic. **~to** *m* monasticism

monada *f* beautiful thing; (*de un niño*) charming way; (*acción tonta*) silliness

monaguillo *m* altar boy

mon|arca *m & f* monarch. **~arquía** *f* monarchy. **~árquico** *a* monarchic(al). **~arquismo** *m* monarchism

mon|asterio *m* monastery. **~ástico** *a* monastic

monda *f* pruning; (*peladura*) peel

mond|adientes *m invar* toothpick. **~adura** *f* pruning; (*peladura*) peel. **~ar** *vt* peel ‹*fruta etc*›; dredge ‹*un río*›. **~o** *a* (*sin pelo*) bald; (*sin dinero*) broke; (*sencillo*) plain

mondongo *m* innards

moned|a *f* coin; (*de un país*) currency. **~ero** *m* minter; (*portamonedas*) purse

monetario *a* monetary
mongol *a & m* Mongolian
mongolismo *m* Down's syndrome
monigote *m* weak character; (*muñeca*) rag doll; (*dibujo*) doodle
monises *mpl* money, dough (*fam*)
monitor *m* monitor
monj|a *f* nun. **~e** *m* monk. **~il** *a* nun's; (*como de monja*) like a nun
mono *m* monkey; (*sobretodo*) overalls. ● *a* pretty
mono... *pref* mono...
monocromo *a & m* monochrome
monóculo *m* monocle
mon|ogamia *f* monogamy. **~ógamo** *a* monogamous
monografía *f* monograph
monograma *m* monogram
monol|ítico *a* monolithic. **~ito** *m* monolith
mon|ologar [12] *vi* soliloquize. **~ólogo** *m* monologue
monoman|ía *f* monomania. **~iaco** *m* monomaniac
monoplano *m* monoplane
monopoli|o *m* monopoly. **~zar** [10] *vt* monopolize
monos|ilábico *a* monosyllabic. **~ílabo** *m* monosyllable
monoteís|mo *m* monotheism. **~ta** *a* monotheistic. ● *m & f* monotheist
mon|otonía *f* monotony. **~ótono** *a* monotonous
monseñor *m* monsignor
monserga *f* boring talk
monstruo *m* monster. **~sidad** *f* monstrosity. **~so** *a* monstrous
monta *f* mounting; (*valor*) value
montacargas *m invar* service lift
monta|do *a* mounted. **~dor** *m* fitter. **~je** *m* assembly; (*cine*) montage; (*teatro*) staging, production
montañ|a *f* mountain. **~ero** *a* mountaineer. **~és** *a* mountain. ● *m* highlander. **~ismo** *m* mountaineering. **~oso** *a* mountainous. **~a rusa** big dipper
montaplatos *m invar* service lift
montar *vt* ride; (*subirse*) get on; (*ensamblar*) assemble; cock ‹*arma*›; set up ‹*una casa, un negocio*›. ● *vi* ride; (*subirse a*) mount. **~ a caballo** ride a horse
montaraz *a* ‹*animales*› wild; ‹*personas*› mountain
monte *m* (*montaña*) mountain; (*terreno inculto*) scrub; (*bosque*) forest. **~ de piedad** pawn-shop. **ingeniero** *m* **de ~s** forestry expert
montepío *m* charitable fund for dependents
monter|a *f* cloth cap. **~o** *m* hunter
montés *a* wild
Montevideo *m* Montevideo
montevideano *a* & m Montevidean
montículo *m* hillock
montón *m* heap, pile. **a montones** in abundance, lots of
montuoso *a* hilly
montura *f* mount; (*silla*) saddle
monument|al *a* monumental; (*fig, muy grande*) enormous. **~o** *m* monument
monzón *m & f* monsoon
moñ|a *f* hair ribbon. **~o** *m* bun
moque|o *m* runny nose. **~ro** *m* handkerchief
moqueta *f* fitted carpet
moquillo *m* distemper
mora *f* mulberry; (*zarzamora*) blackberry
morada *f* dwelling
morado *a* purple
morador *m* inhabitant
moral *m* mulberry tree. ● *f* morals. ● *a* moral. **~eja** *f* moral. **~idad** *f* morality. **~ista** *m & f* moralist. **~izador** *a* moralizing. ● *m* moralist. **~izar** [10] *vt* moralize
morapio *m* (*fam*) cheap red wine
morar *vi* live
moratoria *f* moratorium
morbidez *f* softness
mórbido *a* soft; (*malsano*) morbid
morbo *m* illness. **~sidad** *f* morbidity. **~so** *a* unhealthy
morcilla *f* black pudding
morda|cidad *f* bite. **~z** *a* biting
mordaza *f* gag
mordazmente *adv* bitingly
morde|dura *f* bite. **~r** [2] *vt* bite; (*fig, quitar porciones a*) eat into; (*denigrar*) gossip about. ● *vi* bite
mordis|car [7] *vt* nibble (at). ● *vi* nibble. **~co** *m* bite. **~quear** *vt* nibble (at)
morelense *a* (*Mex*) from Morelos. ● *m & f* person from Morelos
morena *f* (*geol*) moraine
moreno *a* dark; (*de pelo obscuro*) dark-haired; (*de raza negra*) negro
morera *f* mulberry tree
morería *f* Moorish lands; (*barrio*) Moorish quarter
moretón *m* bruise
morfema *m* morpheme
morfin|a *f* morphine. **~ómano** *a* morphine. ● *m* morphine addict

morfol|ogía *f* morphology. **~ógico** *a* morphological
moribundo *a* moribund
morillo *m* andiron
morir [6] (*pp* **muerto**) *vi* die; (*fig, extinguirse*) die away; (*fig, terminar*) end. **~se** *vpr* die. **~se de hambre** starve to death; (*fig*) be starving. **se muere por una flauta** she's dying to have a flute
moris|co *a* Moorish. ● *m* Moor. **~ma** *f* Moors
morm|ón *m & f* Mormon. **~ónico** *a* Mormon. **~onismo** *m* Mormonism
moro *a* Moorish. ● *m* Moor
moros|idad *f* dilatoriness. **~o** *a* dilatory
morrada *f* butt; (*puñetazo*) punch
morral *m* (*mochila*) rucksack; (*del cazador*) gamebag; (*para caballos*) nosebag
morralla *f* rubbish
morrillo *m* nape of the neck
morriña *f* homesickness
morro *m* snout
morrocotudo *a* (*esp Mex*) (*fam*) terrific (*fam*)
morsa *f* walrus
mortaja *f* shroud
mortal *a & m & f* mortal. **~idad** *f* mortality. **~mente** *adv* mortally
mortandad *f* death toll
mortecino *a* failing; ‹*color*› faded
mortero *m* mortar
mortífero *a* deadly
mortifica|ción *f* mortification. **~r** [7] *vt* (*med*) damage; (*atormentar*) plague; (*humillar*) humiliate. **~rse** *vpr* (*Mex*) feel embarassed
mortuorio *a* death
morueco *m* ram
moruno *a* Moorish
mosaico *a* of Moses, Mosaic. ● *m* mosaic
mosca *f* fly. **~rda** *f* blowfly. **~rdón** *m* botfly; (*mosca de cuerpo azul*) bluebottle
moscatel *a* muscatel
moscón *m* botfly; (*mosca de cuerpo azul*) bluebottle
moscovita *a & m & f* Muscovite
Moscú *m* Moscow
mosque|arse *vpr* get cross. **~o** *m* resentment
mosquete *m* musket. **~ro** *m* musketeer
mosquit|ero *m* mosquito net. **~o** *m* mosquito; (*mosca pequeña*) fly, gnat
mostacho *m* moustache
mostachón *m* macaroon
mostaza *f* mustard
mosto *m* must
mostrador *m* counter
mostrar [2] *vt* show. **~se** *vpr* (show oneself to) be. **se mostró muy amable** he was very kind
mostrenco *a* ownerless; ‹*animal*› stray; (*torpe*) thick; (*gordo*) fat
mota *f* spot, speck
mote *m* nickname; (*lema*) motto
motea|do *a* speckled. **~r** *vt* speckle
motejar *vt* call
motel *m* motel
motete *m* motet
motín *m* riot; (*rebelión*) uprising; (*de tropas*) mutiny
motiv|ación *f* motivation. **~ar** *vt* motivate; (*explicar*) explain. **~o** *m* reason. **con ~o de** because of
motocicl|eta *f* motor cycle, motor bike (*fam*). **~ista** *m & f* motorcyclist
motón *m* pulley
motonave *f* motor boat
motor *a* motor. ● *m* motor, engine. **~a** *f* motor boat. **~ de arranque** starter motor
motoris|mo *m* motorcycling. **~ta** *m & f* motorist; (*de una moto*) motorcyclist
motorizar [10] *vt* motorize
motriz *af* motive, driving
move|dizo *a* movable; (*poco firme*) unstable; ‹*persona*› fickle. **~r** [2] *vt* move; shake ‹*la cabeza*›; (*provocar*) cause. **~rse** *vpr* move; (*darse prisa*) hurry up. **arenas** *fpl* **~dizas** quicksand
movi|ble *a* movable. **~do** *a* moved; (*foto*) blurred; (*inquieto*) fidgety
móvil *a* movable. ● *m* motive
movili|dad *f* mobility. **~zación** *f* mobilization. **~zar** [10] *vt* mobilize
movimiento *m* movement, motion; (*agitación*) bustle
moza *f* girl; (*sirvienta*) servant, maid. **~lbete** *m* young lad
mozárabe *a* Mozarabic. ● *m & f* Mozarab
moz|o *m* boy, lad. **~uela** *f* young girl. **~uelo** *m* young boy/lad
muaré *m* moiré
mucam|a *f* (*Arg*) servant. **~o** *m* (*Arg*) servant
mucos|idad *f* mucus. **~o** *a* mucous
muchach|a *f* girl; (*sirvienta*) servant, maid. **~o** *m* boy, lad; (*criado*) servant

muchedumbre *f* crowd
muchísimo *a* very much. ● *adv* a lot
mucho *a* much (*pl* **many),** a lot of. ● *pron* a lot; (*personas*) many (people). ● *adv* a lot, very much; (*de tiempo*) long, a long time. **ni ~ menos** by no means. **por ~ que** however much
muda *f* change of clothing; (*de animales*) moult. **~ble** *a* changeable; ‹*personas*› fickle. **~nza** *f* change; (*de casa*) removal. **~r** *vt/i* change. **~rse** (*de ropa*) change one's clothes; (*de casa*) move (house)
mudéjar *a* & *m* & *f* Mudéjar
mud|ez *f* dumbness. **~o** *a* dumb; (*callado*) silent
mueble *a* movable. ● *m* piece of furniture
mueca *f* grimace, face. **hacer una ~** pull a face
muela *f* (*diente*) tooth; (*diente molar*) molar; (*piedra de afilar*) grindstone; (*piedra de molino*) millstone
muelle *a* soft. ● *m* spring; (*naut*) wharf; (*malecón*) jetty
muérdago *m* mistletoe
muero *vb véase* **morir**
muert|e *f* death; (*homicidio*) murder. **~o** *a* dead; (*matado, fam*) killed; ‹*colores*› pale. ● *m* dead person; (*cadáver*) body, corpse
muesca *f* nick; (*ranura*) slot
muestra *f* sample; (*prueba*) proof; (*modelo*) model; (*seal*) sign. **~rio** *m* collection of samples
muestro *vb véase* **mostrar**
muevo *vb véase* **mover**
mugi|do *m* moo. **~r** [14] *vi* moo; (*fig*) roar
mugr|e *m* dirt. **~iento** *a* dirty, filthy
mugrón *m* sucker
muguete *m* lily of the valley
mujer *f* woman; (*esposa*) wife. ● *int* my dear! **~iego** *a* ‹*hombre*› fond of the women. **~il** *a* womanly. **~ío** *m* (crowd of) women. **~zuela** *f* prostitute
mújol *m* mullet
mula *f* mule; (*Mex*) unsaleable goods. **~da** *f* drove of mules
mulato *a* & *m* mulatto
mulero *m* muleteer
mulet|a *f* crutch; (*fig*) support; (*toreo*) stick with a red flag
mulo *m* mule
multa *f* fine. **~r** *vt* fine
multi... *pref* multi...
multicolor *a* multicolour(ed)
multicopista *m* copying machine
multiforme *a* multiform
multilateral *a* multilateral
multilingüe *a* multilingual
multimillonario *m* multimillionaire
múltiple *a* multiple
multiplic|ación *f* multiplication. **~ar** [7] *vt* multiply. **~arse** *vpr* multiply; (*fig*) go out of one's way. **~idad** *f* multiplicity
múltiplo *a* & *m* multiple
multitud *f* multitude, crowd. **~inario** *a* multitudinous
mulli|do *a* soft. ● *m* stuffing. **~r** [22] *vt* soften
mund|ano *a* wordly; (*de la sociedad elegante*) society. ● *m* socialite. **~ial** *a* world-wide. **la segunda guerra ~ial** the Second World War. **~illo** *m* world, circles. **~o** *m* world. **~ología** *f* worldly wisdom. **todo el ~o** everybody
munición *f* ammunition; (*provisiones*) supplies
municip|al *a* municipal. **~alidad** *f* municipality. **~io** *m* municipality; (*ayuntamiento*) town council
mun|ificencia *f* munificence. **~ífico** *a* munificent
muñe|ca *f* (*anat*) wrist; (*juguete*) doll; (*maniquí*) dummy. **~co** *m* boy doll. **~quera** *f* wristband
muñón *m* stump
mura|l *a* mural, wall. ● *m* mural. **~lla** *f* (city) wall. **~r** *vt* wall
murciélago *m* bat
murga *f* street band; (*lata*) bore, nuisance. **dar la ~** bother, be a pain (*fam*)
murmullo *m* (*de personas*) whisper(ing), murmur(ing); (*del agua*) rippling; (*del viento*) sighing, rustle
murmura|ción *f* gossip. **~dor** *a* gossiping. ● *m* gossip. **~r** *vi* murmur; (*hablar en voz baja*) whisper; (*quejarse en voz baja*) mutter; (*criticar*) gossip
muro *m* wall
murri|a *f* depression. **~o** *a* depressed
mus *m* card game
musa *f* muse
musaraña *f* shrew
muscula|r *a* muscular. **~tura** *f* muscles
músculo *m* muscle
musculoso *a* muscular
muselina *f* muslin

museo *m* museum. **~ de arte** art gallery
musgaño *m* shrew
musgo *m* moss. **~so** *a* mossy
música *f* music
musical *a & m* musical
músico *a* musical. ● *m* musician
music|ología *f* musicology. **~ólogo** *m* musicologist
musitar *vt/i* mumble
muslímico *a* Muslim
muslo *m* thigh
mustela *a* weasel
musti|arse *vpr* wither, wilt. **~o** *a* ‹*plantas*› withered; ‹*cosas*› faded; ‹*personas*› gloomy; (*Mex, hipócrita*) hypocritical
musulmán *a & m* Muslim
muta|bilidad *f* mutability. **~ción** *f* change; (*en biología*) mutation
mutila|ción *f* mutilation. **~do** *a* crippled. ● *m* cripple. **~r** *vt* mutilate; cripple, maim ‹*persona*›
mutis *m* (*en el teatro*) exit. **~mo** *m* silence
mutu|alidad *f* mutuality; (*asociación*) friendly society. **~amente** *adv* mutually. **~o** *a* mutual
muy *adv* very; (*demasiado*) too

N

nab|a *f* swede. **~o** *m* turnip
nácar *m* mother-of-pearl
nac|er [11] *vi* be born; ‹*huevo*› hatch; ‹*planta*› sprout. **~ido** *a* born. **~iente** *a* ‹*sol*› rising. **~imiento** *m* birth; (*de río*) source; (*belén*) crib. **dar ~imiento a** give rise to. **lugar** *m* **de ~imiento** place of birth. **recien ~ido** newborn. **volver a ~er** have a narrow escape
naci|ón *f* nation. **~onal** *a* national. **~onalidad** *f* nationality. **~onalismo** *m* nationalism. **~onalista** *m & f* nationalist. **~onalizar** [10] *vt* nationalize. **~onalizarse** *vpr* become naturalized
nada *pron* nothing, not anything. ● *adv* not at all. **¡~ de eso!** nothing of the sort! **antes de ~** first of all. **¡de ~!** (*después de 'gracias'*) don't mention it! **para ~** (not) at all. **por ~ del mundo** not for anything in the world
nada|dor *m* swimmer. **~r** *vi* swim
nadería *f* trifle
nadie *pron* no one, nobody
nado *adv*. **a ~** swimming
nafta *f* (*LAm, gasolina*) petrol, (*Brit*), gas (*Amer*)
nailon *m* nylon
naipe *m* (playing) card. **juegos** *mpl* **de ~s** card games
nalga *f* buttock. **~s** *fpl* bottom
nana *f* lullaby
Nápoles *m* Naples
naranj|a *f* orange. **~ada** *f* orangeade. **~al** *m* orange grove. **~o** *m* orange tree
narcótico *a & m* narcotic
nariz *f* nose; (*orificio de la nariz*) nostril. **¡narices!** rubbish!
narra|ción *f* narration. **~dor** *m* narrator. **~r** *vt* tell. **~tivo** *a* narrative
nasal *a* nasal
nata *f* cream
natación *f* swimming
natal *a* birth; ‹*pueblo etc*› home. **~idad** *f* birth rate
natillas *fpl* custard
natividad *f* nativity
nativo *a & m* native
nato *a* born
natural *a* natural. ● *m* native. **~eza** *f* nature; (*nacionalidad*) nationality; (*ciudadanía*) naturalization. **~eza muerta** still life. **~idad** *f* naturalness. **~ista** *m & f* naturalist. **~izar** [10] *vt* naturalize. **~izarse** *vpr* become naturalized. **~mente** *adv* naturally. ● *int* of course!
naufrag|ar [12] *vi* ‹*barco*› sink; ‹*persona*› be shipwrecked; (*fig*) fail. **~io** *m* shipwreck
náufrago *a* shipwrecked. ● *m* shipwrecked person
náusea *f* nausea. **dar ~s a uno** make s.o. feel sick. **sentir ~s** feel sick
nauseabundo *a* sickening
náutico *a* nautical
navaja *f* penknife; (*de afeitar*) razor. **~zo** *m* slash
naval *a* naval
Navarra *f* Navarre
nave *f* ship; (*de iglesia*) nave. **~ espacial** spaceship. **quemar las ~s** burn one's boats
navega|ble *a* navigable; ‹*barco*› seaworthy. **~ción** *f* navigation. **~nte** *m & f* navigator. **~r** [12] *vi* sail; ‹*avión*› fly
Navid|ad *f* Christmas. **~eño** *a* Christmas. **en ~ades** at Christmas. **¡feliz ~ad!** Happy Christmas! **por ~ad** at Christmas

navío *m* ship
nazi *a* & *m* & *f* Nazi
neblina *f* mist
nebuloso *a* misty; (*fig*) vague
necedad *f* foolishness. **decir ~es** talk nonsense. **hacer una ~** do sth stupid
necesari|amente *adv* necessarily. **~o** *a* necessary
necesi|dad *f* necessity; (*pobreza*) poverty. **~dades** *fpl* hardships. **por ~dad** (out) of necessity. **~tado** *a* in need (**de** of); (*pobre*) needy. **~tar** *vt* need. • *vi*. **~tar de** need
necio *a* silly. • *m* idiot
necrología *f* obituary column
néctar *m* nectar
nectarina *f* nectarine
nefasto *a* unfortunate, ominous
nega|ción *f* negation; (*desmentimiento*) denial; (*gram*) negative. **~do** *a* incompetent. **~r** [1 & 12] *vt* deny; (*rehusar*) refuse. **~rse** *vpr*. **~rse a** refuse. **~tiva** *f* negative; (*acción*) denial; (*acción de rehusar*) refusal. **~tivo** *a* & *m* negative
negligen|cia *f* negligence. **~te** *a* negligent
negoci|able *a* negotiable. **~ación** *f* negotiation. **~ante** *m* & *f* dealer. **~ar** *vt/i* negotiate. **~ar en** trade in. **~o** *m* business; (*com, trato*) deal. **~os** *mpl* business. **hombre** *m* **de ~os** businessman
negr|a *f* Negress; (*mus*) crotchet. **~o** *a* black; ‹*persona*› Negro. • *m* (*color*) black; (*persona*) Negro. **~ura** *f* blackness. **~uzco** *a* blackish
nene *m* & *f* baby, child
nenúfar *m* water lily
neo... *pref* neo...
neocelandés *a* from New Zealand. • *m* New Zealander
neolítico *a* Neolithic
neón *m* neon
nepotismo *m* nepotism
nervio *m* nerve; (*tendón*) sinew; (*bot*) vein. **~sidad** *f*, **~sismo** *m* nervousness; (*impaciencia*) impatience. **~so** *a* nervous; (*de temperamento*) highly-strung. **crispar los ~s a uno** (*fam*) get on s.o.'s nerves. **ponerse ~so** get excited
neto *a* clear; ‹*verdad*› simple; (*com*) net
neumático *a* pneumatic. • *m* tyre
neumonía *f* pneumonia
neuralgia *f* neuralgia
neur|ología *f* neurolgy. **~ólogo** *m* neurologist
neur|osis *f* neurosis. **~ótico** *a* neurotic
neutr|al *a* neutral. **~alidad** *f* neutrality. **~alizar** [10] *vt* neutralize. **~o** *a* neutral; (*gram*) neuter
neutrón *m* neutron
neva|da *f* snowfall. **~r** [1] *vi* snow. **~sca** *f* blizzard
nevera *f* fridge (*Brit, fam*), refrigerator
nevisca *f* light snowfall. **~r** [7] *vi* snow lightly
nexo *m* link
ni *conj* nor, neither; (*ni siquiera*) not even. **~... ~** neither... nor. **~ que** as if. **~ siquiera** not even
Nicaragua *f* Nicaragua
nicaragüense *a* & *m* & *f* Nicaraguan
nicotina *f* nicotine
nicho *m* niche
nido *m* nest; (*de ladrones*) den; (*escondrijo*) hiding-place
niebla *f* fog; (*neblina*) mist. **hay ~** it's foggy
niet|a *f* granddaughter. **~o** *m* grandson. **~os** *mpl* grandchildren
nieve *f* snow; (*LAm, helado*) ice-cream
Nigeria *f* Nigeria. **~no** *a* Nigerian
niki *m* T-shirt
nilón *m* nylon
nimbo *m* halo
nimi|edad *f* triviality. **~o** *a* insignificant
ninfa *f* nymph
ninfea *f* water lily
ningún *véase* **ninguno**
ninguno *a* (*delante de nombre masculino en singular* **ningún**) no, not any. • *pron* none; (*persona*) no-one, nobody; (*de dos*) neither. **de ninguna manera, de ningún modo** by no means. **en ninguna parte** nowhere
niñ|a *f* (little) girl. **~ada** *f* childish thing. **~era** *f* nanny. **~ería** *f* childish thing. **~ez** *f* childhood. **~o** *a* childish. • *m* (little) boy. **de ~o** as a child. **desde ~o** from childhood
níquel *m* nickel
níspero *m* medlar
nitidez *f* clearness
nítido *a* clear; (*foto*) sharp
nitrato *m* nitrate
nítrico *a* nitric
nitrógeno *m* nitrogen

nivel *m* level; (*fig*) standard. **~ar** *vt* level. **~arse** *vpr* become level. **~ de vida** standard of living

no *adv* not; (*como respuesta*) no. **¿~?** isn't it? **~ más** only. **¡a que ~!** I bet you don't! **¡cómo ~!** of course! **Felipe ~ tiene hijos** Felipe has no children. **¡que ~!** certainly not!

nob|iliario *a* noble. **~le** *a* & *m* & *f* noble. **~leza** *f* nobility

noción *f* notion. **nociones** *fpl* rudiments

nocivo *a* harmful

nocturno *a* nocturnal; ‹*clase*› evening; ‹*tren etc*› night. ● *m* nocturne

noche *f* night. **~ vieja** New Year's Eve. **de ~** at night. **hacer ~** spend the night. **media ~** midnight. **por la ~** at night

Nochebuena *f* Christmas Eve

nodo *m* (*Esp, película*) newsreel

nodriza *f* nanny

nódulo *m* nodule

nogal *m* walnut(-tree)

nómada *a* nomadic. ● *m* & *f* nomad

nombr|adía *f* fame. **~ado** *a* famous; (*susodicho*) aforementioned. **~amiento** *m* appointment. **~ar** *vt* appoint; (*citar*) mention. **~e** *m* name; (*gram*) noun; (*fama*) renown. **~e de pila** Christian name. **en ~e de** in the name of. **no tener ~e** be unspeakable. **poner de ~e** call

nomeolvides *m invar* forget-me-not

nómina *f* payroll

nomina|l *a* nominal. **~tivo** *a* & *m* nominative. **~tivo a** ‹*cheque etc*› made out to

non *a* odd. ● *m* odd number

nonada *f* trifle

nono *a* ninth

nordeste *a* ‹*región*› north-eastern; ‹*viento*› north-easterly. ● *m* north-east

nórdico *a* northern. ● *m* northerner

noria *f* water-wheel; (*en una feria*) ferris wheel

norma *f* rule

normal *a* normal. ● *f* teachers' training college. **~idad** normality (*Brit*), normalcy (*Amer*). **~izar** [10] *vt* normalize. **~mente** *adv* normally, usually

Normandía *f* Normandy

noroeste *a* ‹*región*› north-western; ‹*viento*› north-westerly. ● *m* north-west

norte *m* north; (*viento*) north wind; (*fig, meta*) aim

Norteamérica *f* (North) America

norteamericano *a* & *m* (North) American

norteño *a* northern. ● *m* northerner

Noruega *f* Norway

noruego *a* & *m* Norwegian

nos *pron* (*acusativo*) us; (*dativo*) (to) us; (*reflexivo*) (to) ourselves; (*recíproco*) (to) each other

nosotros *pron* we; (*con prep*) us

nost|algia *f* nostalgia; (*de casa, de patria*) homesickness. **~álgico** *a* nostalgic

nota *f* note; (*de examen etc*) mark. **~ble** *a* notable. **~ción** *f* notation. **~r** *vt* notice; (*apuntar*) note down. **de mala ~** notorious. **de ~** famous. **digno de ~** notable. **es de ~r** it should be noted. **hacerse ~r** stand out

notario *m* notary

notici|a *f* (piece of) news. **~as** *fpl* news. **~ario** *m* news. **~ero** *a* news. **atrasado de ~as** behind the times. **tener ~as de** hear from

notifica|ción *f* notification. **~r** [7] *vt* notify

notori|edad *f* notoriety. **~o** *a* well-known; (*evidente*) obvious

novato *m* novice

novecientos *a* & *m* nine hundred

noved|ad *f* newness; (*noticia*) news; (*cambio*) change; (*moda*) latest fashion. **~oso** *a* (*LAm*) novel. **sin ~ad** no news

novel|a *f* novel. **~ista** *m* & *f* novelist

noveno *a* ninth

novent|a *a* & *m* ninety; (*nonagésimo*) ninetieth. **~ón** *a* & *m* ninety-year-old

novia *f* girlfriend; (*prometida*) fiancée; (*en boda*) bride. **~zgo** *m* engagement

novicio *m* novice

noviembre *m* November

novilunio *m* new moon

novill|a *f* heifer. **~o** *m* bullock. **hacer ~os** play truant

novio *m* boyfriend; (*prometido*) fiancé; (*en boda*) bridegroom. **los ~s** the bride and groom

novísimo *a* very new

nub|arrón *m* large dark cloud. **~e** *f* cloud; (*de insectos etc*) swarm. **~lado** *a* cloudy, overcast. ● *m*

cloud. **~lar** *vt* cloud. **~larse** *vpr* become cloudy. **~loso** *a* cloudy
nuca *f* back of the neck
nuclear *a* nuclear
núcleo *m* nucleus
nudillo *m* knuckle
nudis|mo *m* nudism. **~ta** *m & f* nudist
nudo *m* knot; (*de asunto etc*) crux. **~so** *a* knotty. **tener un ~ en la garganta** have a lump in one's throat
nuera *f* daughter-in-law
nuestro *a* our; (*pospuesto al sustantivo*) of ours. • *pron* ours. **~ coche** our car. **un coche ~** a car of ours
nueva *f* (piece of) news. **~s** *fpl* news. **~mente** *adv* newly; (*de nuevo*) again
Nueva York *f* New York
Nueva Zelanda *f*, **Nueva Zelandia** *f* (*LAm*) New Zealand
nueve *a & m* nine
nuevo *a* new. **de ~** again
nuez *f* nut; (*del nogal*) walnut; (*anat*) Adam's apple. **~ de Adán** Adam's apple. **~ moscada** nutmeg
nul|idad *f* incompetence; (*persona, fam*) nonentity. **~o** *a* useless; (*jurid*) null and void
num|eración *f* numbering. **~eral** *a & m* numeral. **~erar** *vt* number. **~érico** *a* numerical
número *m* number; (*arábigo, romano*) numeral; (*de zapatos etc*) size. **sin ~** countless
numeroso *a* numerous
nunca *adv* never, not ever. **~ (ja)más** never again. **casi ~** hardly ever. **más que ~** more than ever
nupcia|l *a* nuptial. **~s** *fpl* wedding. **banquete ~l** wedding breakfast
nutria *f* otter
nutri|ción *f* nutrition. **~do** *a* nourished, fed; (*fig*) large; ‹*aplausos*› loud; ‹*fuego*› heavy. **~r** *vt* nourish, feed; (*fig*) feed. **~tivo** *a* nutritious. **valor** *m* **~tivo** nutritional value
nylon *m* nylon

Ñ

ña *f* (*LAm, fam*) Mrs
ñacanina *f* (*Arg*) poisonous snake
ñame *m* yam
ñapindá *m* (*Arg*) mimosa
ñato (*LAm*) snub-nosed
ño *m* (*LAm, fam*) Mr
ñoñ|ería *f*, **~ez** *f* insipidity. **~o** *a* insipid; (*tímido*) bashful; (*quisquilloso*) prudish
ñu *m* gnu

O

o *conj* or. **~ bien** rather. **~... ~** either... or. **~ sea** in other words
oasis *m invar* oasis
obcecar [7] *vt* blind
obed|ecer [11] *vt/i* obey. **~iencia** *f* obedience. **~iente** *a* obedient
obelisco *m* obelisk
obertura *f* overture
obes|idad *f* obesity. **~o** *a* obese
obispo *m* bishop
obje|ción *f* objection. **~tar** *vt/i* object
objetiv|idad *f* objectivity. **~o** *a* objective. • *m* objective; (*foto etc*) lens
objeto *m* object
objetor *m* objector. **~ de conciencia** conscientious objector
oblicuo *a* oblique; ‹*mirada*› sidelong
obliga|ción *f* obligation; (*com*) bond. **~do** *a* obliged; (*forzoso*) obligatory; **~r** [12] *vt* force, oblige. **~rse** *vpr*. **~rse a** undertake to. **~torio** *a* obligatory
oboe *m* oboe; (*músico*) oboist
obra *f* work; (*de teatro*) play; (*construcción*) building. **~ maestra** masterpiece. **en ~s** under construction. **por ~ de** thanks to. **~r** *vt* do; (*construir*) build
obrero *a* labour; ‹*clase*› working. • *m* workman; (*en fábrica*) worker
obscen|idad *f* obscenity. **~o** *a* obscene
obscu... *véase* **oscu...**
obsequi|ar *vt* lavish attention on. **~ar con** give, present with. **~o** *m* gift, present; (*agasajo*) attention. **~oso** *a* obliging. **en ~o de** in honour of
observa|ción *f* observation; (*objeción*) objection. **~dor** *m* observer. **~ncia** *f* observance. **~nte** *a* observant. **~r** *vt* observe; (*notar*) notice. **~rse** *vpr* be noted. **~torio** *m* observatory. **hacer una ~ción** make a remark

obses|ión *f* obsession. **~ionar** *vt* obsess. **~ivo** *a* obsessive. **~o** *a* obsessed

obst|aculizar [10] *vt* hinder. **~áculo** *m* obstacle

obstante. no ~ *adv* however, nevertheless. ● *prep* in spite of

obstar *vi*. **~ para** prevent

obstétrico *a* obstetric

obstina|ción *f* obstinacy. **~do** *a* obstinate. **~rse** *vpr* be obstinate. **~rse en** (+ *infintivo*) persist in (+ *gerundio*)

obstru|cción *f* obstruction. **~ir** [17] *vt* obstruct

obtener [40] *vt* get, obtain

obtura|dor *m* (*foto*) shutter. **~r** *vt* plug; fill ‹*muela etc*›

obtuso *a* obtuse

obviar *vt* remove

obvio *a* obvious

oca *f* goose

ocasi|ón *f* occasion; (*oportunidad*) opportunity; (*motivo*) cause. **~onal** *a* chance. **~onar** *vt* cause. **aprovechar la ~ón** take the opportunity. **con ~ón de** on the occasion of. **de ~ón** bargain; (*usado*) second-hand. **en ~ones** sometimes. **perder una ~ón** miss a chance

ocaso *m* sunset; (*fig*) decline

occident|al *a* western. ● *m & f* westerner. **~e** *m* west

océano *m* ocean

ocio *m* idleness; (*tiempo libre*) leisure time. **~sidad** *f* idleness. **~so** *a* idle; (*inútil*) pointless

oclusión *f* occlusion

octano *m* octane. **índice** *m* **de ~** octane number, octane rating

octav|a *f* octave. **~o** *a & m* eighth

octogenario *a & m* octogenarian, eighty-year-old

oct|ogonal *a* octagonal. **~ógono** *m* octagon

octubre *m* October

oculista *m & f* oculist, optician

ocular *a* eye

ocult|ar *vt* hide. **~arse** *vpr* hide. **~o** *a* hidden; (*secreto*) secret

ocupa|ción *f* occupation. **~do** *a* occupied; ‹*persona*› busy. **~nte** *m* occupant. **~r** *vt* occupy. **~rse** *vpr* look after

ocurr|encia *f* occurrence, event; (*idea*) idea; (*que tiene gracia*) witty remark. **~ir** *vi* happen. **~irse** *vpr* occur. **¿qué ~e?** what's the matter? **se me ~e que** it occurs to me that

ochent|a *a & m* eighty. **~ón** *a & m* eighty-year-old

ocho *a & m* eight. **~cientos** *a & m* eight hundred

oda *f* ode

odi|ar *vt* hate. **~o** *m* hatred. **~oso** *a* hateful

odisea *f* odyssey

oeste *m* west; (*viento*) west wind

ofen|der *vt* offend; (*insultar*) insult. **~derse** *vpr* take offence. **~sa** *f* offence. **~siva** *f* offensive. **~sivo** *a* offensive

oferta *f* offer; (*en subasta*) bid; (*regalo*) gift. **~s de empleo** situations vacant. **en ~** on (special) offer

oficial *a* official. ● *m* skilled worker; (*funcionario*) civil servant; (*mil*) officer. **~a** *f* skilled (woman) worker

oficin|a *f* office. **~a de colocación** employment office. **~a de Estado** government office. **~a de turismo** tourist office. **~ista** *m & f* office worker. **horas** *fpl* **de ~a** business hours

oficio *m* job; (*profesión*) profession; (*puesto*) post. **~so** *a* (*no oficial*) unofficial

ofrec|er [11] *vt* offer; give ‹*fiesta, banquete etc*›; (*prometer*) promise. **~erse** *vpr* ‹*persona*› volunteer; ‹*cosa*› occur. **~imiento** *m* offer

ofrenda *f* offering. **~r** *vt* offer

ofusca|ción *f* blindness; (*confusión*) confusion. **~r** [7] *vt* blind; (*confundir*) confuse. **~rse** *vpr* be dazzled

ogro *m* ogre

oí|ble *a* audible. **~da** *f* hearing. **~do** *m* hearing; (*anat*) ear. **al ~do** in one's ear. **de ~das** by hearsay. **de ~do** by ear. **duro de ~do** hard of hearing

oigo *vb véase* **oír**

oír [50] *vt* hear. **~ misa** go to mass. **¡oiga!** listen!; (*al teléfono*) hello!

ojal *m* buttonhole

ojalá *int* I hope so! ● *conj* if only

ojea|da *f* glance. **~r** *vt* eye; (*para inspeccionar*) see; (*ahuyentar*) scare away. **dar una ~da a**, **echar una ~da a** glance at

ojeras *fpl* (*del ojo*) bags

ojeriza *f* ill will. **tener ~ a** have a grudge against

ojete *m* eyelet

ojo *m* eye; (*de cerradura*) keyhole; (*de un puente*) span. **¡~!** careful!
ola *f* wave
olé *int* bravo!
olea|da *f* wave. **~je** *m* swell
óleo *m* oil; (*cuadro*) oil painting
oleoducto *m* oil pipeline
oler [2, *las formas que empezarían por* **ue** *se escriben* **hue**] *vt* smell; (*curiosear*) pry into; (*descubrir*) discover. ● *vi* smell (**a** of)
olfat|ear *vt* smell, sniff; (*fig*) sniff out. **~o** *m* (sense of) smell; (*fig*) intuition
olimpiada *f*, **olimpíada** *f* Olympic games, Olympics
olímpico *a* ‹*juegos*› Olympic
oliv|a *f* olive; (*olivo*) olive tree. **~ar** *m* olive grove. **~o** *m* olive tree
olmo *m* elm (tree)
olor *m* smell. **~oso** *a* sweet-smelling
olvid|adizo *a* forgetful. **~ar** *vt* forget. **~arse** *vpr* forget; (*estar olvidado*) be forgotten. **~o** *m* oblivion; (*acción de olvidar*) forgetfulness. **se me ~ó** I forgot
olla *f* pot, casserole; (*guisado*) stew. **~ a/de presión**, **~ exprés** pressure cooker. **~ podrida** Spanish stew
ombligo *m* navel
ominoso *a* awful, abominable
omi|sión *f* omission; (*olvido*) forgetfulness. **~tir** *vt* omit
ómnibus *a* omnibus
omnipotente *a* omnipotent
omóplato *m*, **omoplato** *m* shoulder blade
once *a* & *m* eleven
ond|a *f* wave. **~a corta** short wave. **~a larga** long wave. **~ear** *vi* wave; ‹*agua*› ripple. **~ulación** *f* undulation; (*del pelo*) wave. **~ular** *vi* wave. **longitud** *f* **de ~a** wavelength
oneroso *a* onerous
ónice *m* onyx
onomástico *a*. **día ~**, **fiesta onomástica** name-day
ONU *abrev* (*Organización de las Naciones Unidas*) UN, United Nations
onza *f* ounce
opa *a* (*LAm*) stupid
opaco *a* opaque; (*fig*) dull
ópalo *m* opal
opción *f* option
ópera *f* opera
opera|ción *f* operation; (*com*) transaction. **~dor** *m* operator; (*cirujano*) surgeon; (*TV*) cameraman. **~r** *vt* operate on; work ‹*milagro etc*›. ● *vi* operate; (*com*) deal. **~rse** *vpr* occur; (*med*) have an operation. **~torio** *a* operative
opereta *f* operetta
opin|ar *vi* think. **~ión** *f* opinion. **la ~ión pública** public opinion
opio *m* opium
opone|nte *a* opposing. ● *m* & *f* opponent. **~r** *vt* oppose; offer ‹*resistencia*›; raise ‹*objeción*›. **~rse** *vpr* be opposed; ‹*dos personas*› oppose each other
oporto *m* port (wine)
oportun|idad *f* opportunity; (*cualidad de oportuno*) timeliness. **~ista** *m* & *f* opportunist. **~o** *a* opportune; (*apropiado*) suitable
oposi|ción *f* opposition. **~ciones** *fpl* competition, public examination. **~tor** *m* candidate
opres|ión *f* oppression; (*ahogo*) difficulty in breathing. **~ivo** *a* oppressive. **~o** *a* oppressed. **~or** *m* oppressor
oprimir *vt* squeeze; press ‹*botón etc*›; ‹*ropa*› be too tight for; (*fig*) oppress
oprobio *m* disgrace
optar *vi* choose. **~ por** opt for
óptic|a *f* optics; (*tienda*) optician's (shop). **~o** *a* optic(al). ● *m* optician
optimis|mo *m* optimism. **~ta** *a* optimisitic. ● *m* & *f* optimist
opuesto *a* opposite; (*enemigo*) opposed
opulen|cia *f* opulence. **~to** *a* opulent
oración *f* prayer; (*discurso*) speech; (*gram*) sentence
oráculo *m* oracle
orador *m* speaker
oral *a* oral
orar *vi* pray
oratori|a *f* oratory. **~o** *a* oratorical. ● *m* (*mus*) oratorio
orbe *m* orb
órbita *f* orbit
orden *m* & *f* order; (*Mex*, *porción*) portion. **~ado** *a* tidy. **~ del día** agenda. **órdenes** *fpl* **sagradas** Holy Orders. **a sus órdenes** (*esp Mex*) can I help you? **en ~** in order. **por ~** in turn
ordenador *m* computer
ordena|nza *f* order. ● *m* (*mil*) orderly. **~r** *vt* put in order; (*mandar*) order; (*relig*) ordain
ordeñar *vt* milk
ordinal *a* & *m* ordinal

ordinario *a* ordinary; (*grosero*) common
orear *vt* air
orégano *m* oregano
oreja *f* ear
orfanato *m* orphanage
orfebre *m* goldsmith, silversmith
orfeón *m* choral society
orgánico *a* organic
organigrama *m* flow chart
organillo *m* barrel-organ
organismo *m* organism
organista *m & f* organist
organiza|ción *f* organization. **~dor** *m* organizer. **~r** [10] *vt* organize. **~rse** *vpr* get organized
órgano *m* organ
orgasmo *m* orgasm
orgía *f* orgy
orgullo *m* pride. **~so** *a* proud
orientación *f* direction
oriental *a & m & f* oriental
orientar *vt* position. **~se** *vpr* point; ‹*persona*› find one's bearings
oriente *m* east. **O~ Medio** Middle East
orificio *m* hole
orig|en *m* origin. **~inal** *a* original; (*excéntrico*) odd. **~inalidad** *f* originality. **~inar** *vt* give rise to. **~inario** *a* original; (*nativo*) native. **dar ~en a** give rise to. **ser ~inario de** come from
orilla *f* (*del mar*) shore; (*de río*) bank; (*borde*) edge
orín *m* rust
orina *f* urine. **~l** *m* chamber-pot. **~r** *vi* urinate
oriundo *a*. **~ de** ‹*persona*› (originating) from; ‹*animal etc*› native to
orla *f* border
ornamental *a* ornamental
ornitología *f* ornithology
oro *m* gold. **~s** *mpl* Spanish card suit. **~ de ley** 9 carat gold. **hacerse de ~** make a fortune. **prometer el ~ y el moro** promise the moon
oropel *m* tinsel
orquesta *f* orchestra. **~l** *a* orchestral. **~r** *vt* orchestrate
orquídea *f* orchid
ortiga *f* nettle
ortodox|ia *f* orthodoxy. **~o** *a* orthodox
ortografía *f* spelling
ortop|edia *f* orthopaedics. **~édico** *a* orthopaedic
oruga *f* caterpillar
orzuelo *m* sty
os *pron* (*acusativo*) you; (*dativo*) (to) you; (*reflexivo*) (to) yourselves; (*recíproco*) (to) each other
osad|ía *f* boldness. **~o** *a* bold
oscila|ción *f* swinging; (*de precios*) fluctuation; (*tec*) oscillation. **~r** *vi* swing; ‹*precio*› fluctuate; (*tec*) oscillate; (*fig, vacilar*) hesitate
oscur|ecer [11] *vi* darken; (*fig*) obscure. **~ecerse** *vpr* grow dark; (*nublarse*) cloud over. **~idad** *f* darkness; (*fig*) obscurity. **~o** *a* dark; (*fig*) obscure. **a ~as** in the dark
óseo *a* bony
oso *m* bear. **~ de felpa, ~ de peluche** teddy bear
ostensible *a* obvious
ostent|ación *f* ostentation. **~ar** *vt* show off; (*mostrar*) show. **~oso** *a* ostentatious
osteoartritis *f* osteoarthritis
oste|ópata *m & f* osteopath. **~opatía** *f* osteopathy
ostión *m* (*esp Mex*) oyster
ostra *f* oyster
ostracismo *m* ostracism
Otan *abrev* (*Organización del Tratado del Atlántico Norte*) NATO, North Atlantic Treaty Organization
otear *vt* observe; (*escudriñar*) scan, survey
otitis *f* inflammation of the ear
otoño *m* autumn (*Brit*), fall (*Amer*)
otorga|miento *m* granting; (*documento*) authorization. **~r** [12] *vt* give; (*jurid*) draw up
otorrinolaringólogo *m* ear, nose and throat specialist
otro *a* other; (*uno más*) another. • *pron* another (one); (*en plural*) others; (*otra persona*) someone else. **el ~** the other. **el uno al ~** one another, each other
ovación *f* ovation
oval *a* oval
óvalo *m* oval
ovario *m* ovary
oveja *f* sheep; (*hembra*) ewe
overol *m* (*LAm*) overalls
ovino *a* sheep
ovillo *m* ball. **hacerse un ~** curl up
OVNI *abrev* (*objeto volante no identificado*) UFO, unidentified flying object
ovulación *f* ovulation

oxida|ción *f* rusting. **~r** *vi* rust. **~rse** *vpr* go rusty
óxido *m* oxide
oxígeno *m* oxygen
oye *vb véase* **oír**
oyente *a* listening. ● *m & f* listener
ozono *m* ozone

P

pabellón *m* bell tent; (*edificio*) building; (*de instrumento*) bell; (*bandera*) flag
pabilo *m* wick
paceño *a* from La Paz. ● *m* person from La Paz
pacer [11] *vi* graze
pacien|cia *f* patience. **~te** *a & m & f* patient
pacificar [7] *vt* pacify; reconcile ‹*dos personas*›. **~se** *vpr* calm down
pacífico *a* peaceful. **el (Océano** *m* **) P~** the Pacific (Ocean)
pacifis|mo *m* pacifism. **~ta** *a & m & f* pacifist
pact|ar *vi* agree, make a pact. **~o** *m* pact, agreement
pachucho *a* ‹*fruta*› overripe; ‹*persona*› poorly
padec|er [11] *vt/i* suffer (**de** from); (*soportar*) bear. **~imiento** *m* suffering; (*enfermedad*) ailment
padrastro *m* stepfather
padre *a* (*fam*) great. ● *m* father. **~s** *mpl* parents
padrino *m* godfather; (*en boda*) best man
padrón *m* census
paella *f* paella
paga *f* pay, wages. **~ble** *a*, **~dero** *a* payable
pagano *a & m* pagan
pagar [12] *vt* pay; pay for ‹*compras*›. ● *vi* pay. **~é** *m* IOU
página *f* page
pago *m* payment
pagoda *f* pagoda
país *m* country; (*región*) region. **~ natal** native land. **el P~ Vasco** the Basque Country. **los P~es Bajos** the Low Countries
paisa|je *m* countryside. **~no** *a* of the same country. ● *m* compatriot
paja *f* straw; (*fig*) nonsense
pajarera *f* aviary
pájaro *m* bird. **~ carpintero** woodpecker
paje *m* page
Pakistán *m*. **el ~** Pakistan
pala *f* shovel; (*laya*) spade; (*en deportes*) bat; (*de tenis*) racquet
palabr|a *f* word; (*habla*) speech. **~ota** *f* swear-word. **decir ~otas** swear. **pedir la ~a** ask to speak. **soltar ~otas** swear. **tomar la ~a** (begin to) speak
palacio *m* palace; (*casa grande*) mansion
paladar *m* palate
paladino *a* clear; (*público*) public
palanca *f* lever; (*fig*) influence. **~ de cambio (de velocidades)** gear lever (*Brit*), gear shift (*Amer*)
palangana *f* wash-basin
palco *m* (*en el teatro*) box
Palestina *f* Palestine
palestino *a & m* Palestinian
palestra *f* (*fig*) arena
paleta *f* (*de pintor*) palette; (*de albañil*) trowel
paleto *m* yokel
paliativo *a & m* palliative
palide|cer [11] *vi* turn pale. **~z** *f* paleness
pálido *a* pale
palillo *m* small stick; (*de dientes*) toothpick
palique *m*. **estar de ~** be chatting
paliza *f* beating
palizada *f* fence; (*recinto*) enclosure
palma *f* (*de la mano*) palm; (*árbol*) palm (tree); (*de dátiles*) date palm. **~s** *fpl* applause. **~da** *f* slap. **~das** *fpl* applause. **dar ~(da)s** clap. **tocar las ~s** clap
palmera *f* date palm
palmo *m* span; (*fig, pequeña cantidad*) small amount. **~ a ~** inch by inch
palmote|ar *vi* clap, applaud. **~o** *m* clapping, applause
palo *m* stick; (*del teléfono etc*) pole; (*mango*) handle; (*de golf*) club; (*golpe*) blow; (*de naipes*) suit; (*mástil*) mast
paloma *f* pigeon, dove
palomitas *fpl* popcorn
palpa|ble *a* palpable. **~r** *vt* feel
palpita|ción *f* palpitation. **~nte** *a* throbbing. **~r** *vi* throb; (*latir*) beat
palta *f* (*LAm*) avocado pear
pal|údico *a* marshy; (*de paludismo*) malarial. **~udismo** *m* malaria
pamp|a *f* pampas. **~ear** *vi* (*LAm*) travel across the pampas. **~ero** *a* of the pampas

pan *m* bread; (*barra*) loaf. **~ integral** wholemeal bread (*Brit*), wholewheat bread (*Amer*). **~ tostado** toast. **~ rallado** breadcrumbs. **ganarse el ~** earn one's living
pana *f* corduroy
panacea *f* panacea
panader|ía *f* bakery; (*tienda*) baker's (shop). **~o** *m* baker
panal *m* honeycomb
Panamá *f* Panama
panameño *a* & *m* Panamanian
pancarta *f* placard
panda *m* panda; (*pandilla*) gang
pander|eta *f* (small) tambourine. **~o** *m* tambourine
pandilla *f* gang
panecillo *m* (bread) roll
panel *m* panel
panfleto *m* pamphlet
pánico *m* panic
panor|ama *m* panorama. **~ámico** *a* panoramic
panqué *m* (*LAm*) pancake
pantaletas *fpl* (*LAm*) underpants, knickers
pantal|ón *m* trousers. **~ones** *mpl* trousers. **~ón corto** shorts. **~ón tejano**, **~ón vaquero** jeans
pantalla *f* screen; (*de lámpara*) (lamp)shade
pantano *m* marsh; (*embalse*) reservoir. **~so** *a* boggy
pantera *f* panther
pantomima *f* pantomime
pantorrilla *f* calf
pantufla *f* slipper
panucho *m* (*Mex*) stuffed tortilla
panz|a *f* belly. **~ada** *f* (*hartazgo*, *fam*) bellyful; (*golpe*, *fam*) blow in the belly. **~udo** *a* fat, pot-bellied
pañal *m* nappy (*Brit*), diaper (*Amer*)
pañ|ería *f* draper's (shop). **~o** *m* material; (*de lana*) woollen cloth; (*trapo*) cloth. **~o de cocina** dishcloth; (*para secar*) tea towel. **~o higiénico** sanitary towel. **en ~os menores** in one's underclothes
pañuelo *m* handkerchief; (*de cabeza*) scarf
papa *m* pope. ● *f* (*esp LAm*) potato. **~s francesas** (*LAm*) chips
papá *m* dad(dy). **~s** *mpl* parents. **P~ Noel** Father Christmas
papada *f* (*de persona*) double chin
papado *m* papacy
papagayo *m* parrot
papal *a* papal
papanatas *m inv* simpleton
paparrucha *f* (*tontería*) silly thing
papaya *f* pawpaw
papel *m* paper; (*en el teatro etc*) role. **~ carbón** carbon paper. **~ celofán** celophane paper. **~ de calcar** carbon paper. **~ de embalar**, **~ de envolver** wrapping paper. **~ de plata** silver paper. **~ de seda** tissue paper. **~era** *f* waste-paper basket. **~ería** *f* stationer's (shop). **~eta** *f* ticket; (*para votar*) paper. **~ higiénico** toilet paper. **~ pintado** wallpaper. **~ secante** blotting paper. **blanco como el ~** as white as a sheet. **desempeñar un ~**, **hacer un ~** play a role
paperas *fpl* mumps
paquebote *m* packet (boat)
paquete *m* packet; (*paquebote*) packet (boat); (*Mex*, *asunto difícil*) difficult job. **~ postal** parcel
paquistaní *a* & *m* Pakistani
par *a* equal; ‹*número*› even. ● *m* couple; (*dos cosas iguales*) pair; (*igual*) equal; (*título*) peer. **a la ~** at the same time; ‹*monedas*› at par. **al ~ que** at the same time. **a ~es** two by two. **de ~ en ~** wide open. **sin ~** without equal
para *prep* for; (*hacia*) towards; (*antes del infinitivo*) (in order) to. **~ con** to(wards). **¿~ qué?** why? **~ que** so that
parabienes *mpl* congratulations
parábola *f* (*narración*) parable
parabrisas *m inv* windscreen (*Brit*), windshield (*Amer*)
paraca *f* (*LAm*) strong wind (from the Pacific)
paraca|ídas *m inv* parachute. **~idista** *m* & *f* parachutist; (*mil*) paratrooper
parachoques *m inv* bumper (*Brit*), fender (*Amer*); (*rail*) buffer
parad|a *f* (*acción*) stopping; (*sitio*) stop; (*de taxis*) rank; (*mil*) parade. **~ero** *m* whereabouts; (*alojamiento*) lodging. **~o** *a* stationary; ‹*obrero*› unemployed; (*lento*) slow. **dejar ~o** confuse. **tener mal ~ero** come to a sticky end
paradoja *f* paradox
parador *m* state-owned hotel
parafina *f* paraffin
par|afrasear *vt* paraphrase. **~áfrasis** *f inv* paraphrase
paraguas *m inv* umbrella
Paraguay *m* Paraguay
paraguayo *a* & *m* Paraguayan

paraíso *m* paradise; (*en el teatro*) gallery

paralel|a *f* parallel (line). **~as** *fpl* parallel bars. **~o** *a* & *m* parallel

par|álisis *f inv* paralysis. **~alítico** *a* paralytic. **~alizar** [10] *vt* paralyse

paramilitar *a* paramilitary

páramo *m* barren plain

parang|ón *m* comparison. **~onar** *vt* compare

paraninfo *m* hall

paranoi|a *f* paranoia. **~co** *a* paranoiac

parapeto *m* parapet; (*fig*) barricade

parapléjico *a* & *m* paraplegic

parar *vt/i* stop. **~se** *vpr* stop. **sin ~** continuously

pararrayos *m inv* lightning conductor

parásito *a* parasitic. • *m* parasite

parasol *m* parasol

parcela *f* plot. **~r** *vt* divide into plots

parcial *a* partial. **~idad** *f* prejudice; (*pol*) faction. **a tiempo ~** part-time

parco *a* sparing, frugal

parche *m* patch

pardo *a* brown

parear *vt* pair off

parec|er *m* opinion; (*aspecto*) appearance. • *vi* [11] seem; (*asemejarse*) look like; (*aparecer*) appear. **~erse** *vpr* resemble, look like. **~ido** *a* similar. • *m* similarity. **al ~er** apparently. **a mi ~er** in my opinion. **bien ~ido** good-looking. **me ~e** I think. **¿qué te parece?** what do you think? **según ~e** apparently

pared *f* wall. **~ón** *m* thick wall; (*de ruinas*) standing wall. **~ por medio** next door. **llevar al ~ón** shoot

parej|a *f* pair; (*hombre y mujer*) couple; (*la otra persona*) partner. **~o** *a* alike, the same; (*liso*) smooth

parente|la *f* relations. **~sco** *m* relationship

paréntesis *m inv* parenthesis; (*signo ortográfico*) bracket. **entre ~** (*fig*) by the way

paria *m* & *f* outcast

paridad *f* equality

pariente *m* & *f* relation, relative

parihuela *f*, **parihuelas** *fpl* stretcher

parir *vt* give birth to. • *vi* have a baby, give birth

París *m* Paris

parisiense *a* & *m* & *f*, **parisino** *a* & *m* Parisian

parking /'parkin/ *m* car park (*Brit*), parking lot (*Amer*)

parlament|ar *vi* discuss. **~ario** *a* parliamentary. • *m* member of parliament (*Brit*), congressman (*Amer*). **~o** *m* parliament

parlanchín *a* talkative. • *m* chatterbox

parmesano *a* Parmesan

paro *m* stoppage; (*desempleo*) unemployment; (*pájaro*) tit

parodia *f* parody. **~r** *vt* parody

parpadear *vi* blink; ‹*luz*› flicker; ‹*estrella*› twinkle

párpado *m* eyelid

parque *m* park. **~ de atracciones** funfair. **~ infantil** children's playground. **~ zoológico** zoo, zoological gardens

parqué *m* parquet

parquedad *f* frugality; (*moderación*) moderation

parra *f* grapevine

párrafo *m* paragraph

parrilla *f* grill; (*LAm*, *auto*) radiator grill. **~da** *f* grill. **a la ~** grilled

párroco *m* parish priest

parroquia *f* parish; (*iglesia*) parish church. **~no** *m* parishioner; (*cliente*) customer

parsimoni|a *f* thrift. **~oso** *a* thrifty

parte *m* message; (*informe*) report. • *f* part; (*porción*) share; (*lado*) side; (*jurid*) party. **dar ~** report. **de mi ~** for me. **de ~ de** from. **¿de ~ de quién?** (*al teléfono*) who's speaking? **en cualquier ~** anywhere. **en gran ~** largely. **en ~** partly. **en todas ~s** everywhere. **la mayor ~** the majority. **ninguna ~** nowhere. **por otra ~** on the other hand. **por todas ~s** everywhere

partera *f* midwife

partición *f* sharing out

participa|ción *f* participation; (*noticia*) notice; (*de lotería*) lottery ticket. **~nte** *a* participating. • *m* & *f* participant. **~r** *vt* notify. • *vi* take part

participio *m* participle

partícula *f* particle

particular *a* particular; ‹*clase*› private. • *m* matter. **~idad** *f* peculiarity. **~izar** [10] *vt* distinguish; (*detallar*) give details about. **en ~** in particular. **nada de ~** nothing special

partida *f* departure; (*en registro*) entry; (*documento*) certificate; (*juego*) game; (*de gente*) group. **mala ~** dirty trick

partidario *a & m* partisan. **~ de** keen on

parti|do *a* divided. ● *m* (*pol*) party; (*encuentro*) match, game; (*equipo*) team. **~r** *vt* divide; (*romper*) break; (*repartir*) share; crack ‹*nueces*›. ● *vi* leave; (*empezar*) start. **~rse** *vpr* (*romperse*) break; (*dividirse*) split. **a ~r de** (starting) from

partitura *f* (*mus*) score

parto *m* birth; (*fig*) creation. **estar de ~** be in labour

párvulo *m*. **colegio de ~s** nursery school

pasa *f* raisin. **~ de Corinto** currant. **~ de Esmirna** sultana

pasa|ble *a* passable. **~da** *f* passing; (*de puntos*) row. **~dero** *a* passable. **~dizo** *m* passage. **~do** *a* past; ‹*día, mes etc*› last; (*anticuado*) old-fashioned; ‹*comida*› bad, off. **~do mañana** the day after tomorrow. **~dor** *m* bolt; (*de pelo*) hair-slide; (*culin*) strainer. **de ~da** in passing. **el lunes ~do** last Monday

pasaje *m* passage; (*naut*) crossing; (*viajeros*) passengers. **~ro** *a* passing. ● *m* passenger

pasamano(s) *m* handrail; (*barandilla de escalera*) banister(s)

pasamontañas *m inv* Balaclava (helmet)

pasaporte *m* passport

pasar *vt* pass; (*poner*) put; (*filtrar*) strain; spend ‹*tiempo*›; (*tragar*) swallow; show ‹*película*›; (*tolerar*) tolerate, overlook; give ‹*mensaje, enfermedad*›. ● *vi* pass; (*suceder*) happen; (*ir*) go; (*venir*) come; ‹*tiempo*› go by. **~ de** have no interest in. **~se** *vpr* pass; (*terminarse*) be over; ‹*flores*› wither; ‹*comida*› go bad; spend ‹*tiempo*›; (*excederse*) go too far. **~lo bien** have a good time. **~ por alto** leave out. **como si no hubiese pasado nada** as if nothing had happened. **lo que pasa es que** the fact is that. **pase lo que pase** whatever happens. **¡pase Vd!** come in!, go in! **¡que lo pases bien!** have a good time! **¿qué pasa?** what's the matter?, what's happening?

pasarela *f* footbridge; (*naut*) gangway

pasatiempo *m* hobby, pastime

pascua *f* (*fiesta de los hebreos*) Passover; (*de Resurrección*) Easter; (*Navidad*) Christmas. **~s** *fpl* Christmas. **hacer la ~ a uno** mess things up for s.o. **¡y santas ~s!** and that's that!

pase *m* pass

pase|ante *m & f* passer-by. **~ar** *vt* take for a walk; (*exhibir*) show off. ● *vi* go for a walk; (*en coche etc*) go for a ride. **~arse** *vpr* go for a walk; (*en coche etc*) go for a ride. **~o** *m* walk; (*en coche etc*) ride; (*calle*) avenue. **~o marítimo** promenade. **dar un ~o** go for a walk. **¡vete a ~o!** (*fam*) go away!, get lost! (*fam*)

pasillo *m* passage

pasión *f* passion

pasiv|idad *f* passiveness. **~o** *a* passive

pasm|ar *vt* astonish. **~arse** *vpr* be astonished. **~o** *m* astonishment. **~oso** *a* astonishing

paso *a* ‹*fruta*› dried ● *m* step; (*acción de pasar*) passing; (*huella*) footprint; (*manera de andar*) walk; (*camino*) way through; (*entre montañas*) pass; (*estrecho*) strait(s). **~ a nivel** level crossing (*Brit*), grade crossing (*Amer*). **~ de cebra** Zebra crossing. **~ de peatones** pedestrian crossing. **~ elevado** flyover. **a cada ~** at every turn. **a dos ~s** very near. **al ~ que** at the same time as. **a ~ lento** slowly. **ceda el ~** give way. **de ~** in passing. **de ~ por** on the way through. **prohibido el ~** no entry

pasodoble *m* (*baile*) pasodoble

pasota *m & f* drop-out

pasta *f* paste; (*masa*) dough; (*dinero, fam*) money. **~s** *fpl* pasta; (*pasteles*) pastries. **~ de dientes**, **~ dentífrica** toothpaste

pastar *vt/i* graze

pastel *m* cake; (*empanada*) pie; (*lápiz*) pastel. **~ería** *f* cakes; (*tienda*) cake shop, confectioner's

paste(u)rizar [10] *vt* pasteurize

pastiche *m* pastiche

pastilla *f* pastille; (*de jabón*) bar; (*de chocolate*) piece

pastinaca *f* parsnip

pasto *m* pasture; (*hierba*) grass; (*Mex, césped*) lawn. **~r** *m* shepherd; (*relig*) minister. **~ral** *a* pastoral

pata *f* leg; (*pie*) paw, foot. **~s arriba** upside down. **a cuatro ~s** on all fours. **meter la ~** put one's foot in it. **tener mala ~** have bad luck

pataca *f* Jerusalem artichoke

pata|da *f* kick. **~lear** *vt* stamp; ‹*niño pequeño*› kick

pataplum *int* crash!
patata *f* potato. **~s fritas** chips (*Brit*), French fries (*Amer*). **~s fritas (a la inglesa)** (potato) crisps (*Brit*), potato chips (*Amer*)
patent|ar *vt* patent. **~e** *a* obvious. ● *f* licence. **~e de invención** patent
patern|al *a* paternal; ‹*cariño etc*› fatherly. **~idad** *f* paternity. **~o** *a* paternal; ‹*cariño etc*› fatherly
patético *a* moving
patillas *fpl* sideburns
patín *m* skate; (*juguete*) scooter
pátina *f* patina
patina|dero *m* skating rink. **~dor** *m* skater. **~je** *m* skating. **~r** *vi* skate; (*deslizarse*) slide. **~zo** *m* skid; (*fig, fam*) blunder
patio *m* patio. **~ de butacas** stalls (*Brit*), orchestra (*Amer*)
pato *m* duck
patol|ogía *f* pathology. **~ógico** *a* pathological
patoso *a* clumsy
patraña *f* hoax
patria *f* native land
patriarca *m* patriarch
patrimonio *m* inheritance; (*fig*) heritage
patri|ota *a* patriotic. ● *m & f* patriot. **~ótico** *a* patriotic. **~otismo** *m* patriotism
patrocin|ar *vt* sponsor. **~io** *m* sponsorship
patr|ón *m* patron; (*jefe*) boss; (*de pensión etc*) landlord; (*modelo*) pattern. **~onato** *m* patronage; (*fundación*) trust, foundation
patrulla *f* patrol; (*fig, cuadrilla*) group. **~r** *vt/i* patrol
paulatinamente *adv* slowly
pausa *f* pause. **~do** *a* slow
pauta *f* guideline
paviment|ar *vt* pave. **~o** *m* pavement
pavo *m* turkey. **~ real** peacock
pavor *m* terror. **~oso** *a* terrifying
payas|ada *f* buffoonery. **~o** *m* clown
paz *f* peace. **La P~** La Paz
peaje *m* toll
peatón *m* pedestrian
pebet|a *f* (*LAm*) little girl. **~e** *m* little boy
peca *f* freckle
peca|do *m* sin; (*defecto*) fault. **~dor** *m* sinner. **~minoso** *a* sinful. **~r** [7] *vi* sin
pecoso *a* freckled
pectoral *a* pectoral; (*para la tos*) cough
peculiar *a* peculiar, particular. **~idad** *f* peculiarity
pech|era *f* front. **~ero** *m* bib. **~o** *m* chest; (*de mujer*) breast; (*fig, corazón*) heart. **~uga** *f* breast. **dar el ~o** breast-feed ‹*a un niño*›; (*afrontar*) confront. **tomar a ~o** take to heart
pedagogo *m* teacher
pedal *m* pedal. **~ear** *vi* pedal
pedante *a* pedantic
pedazo *m* piece, bit. **a ~s** in pieces. **hacer ~s** break to pieces. **hacerse ~s** fall to pieces
pedernal *m* flint
pedestal *m* pedestal
pedestre *a* pedestrian
pediatra *m & f* paediatrician
pedicuro *m* chiropodist
pedi|do *m* order. **~r** [5] *vt* ask (for); (*com, en restaurante*) order. ● *vi* ask. **~r prestado** borrow
pegadizo *a* sticky; (*mus*) catchy
pegajoso *a* sticky
pega|r [12] *vt* stick (on); (*coser*) sew on; give ‹*enfermedad etc*›; (*juntar*) join; (*golpear*) hit; (*dar*) give. ● *vi* stick. **~rse** *vpr* stick; (*pelearse*) hit each other. **~r fuego a** set fire to. **~tina** *f* sticker
pein|ado *m* hairstyle. **~ar** *vt* comb. **~arse** *vpr* comb one's hair. **~e** *m* comb. **~eta** *f* ornamental comb
p.ej. *abrev* (*por ejemplo*) e.g., for example
pela|do *a* ‹*fruta*› peeled; ‹*cabeza*› bald; ‹*número*› exactly; ‹*terreno*› barren. ● *m* bare patch. **~dura** *f* (*acción*) peeling; (*mondadura*) peelings
pela|je *m* (*de animal*) fur; (*fig, aspecto*) appearance. **~mbre** *m* (*de animal*) fur; (*de persona*) thick hair
pelar *vt* cut the hair; (*mondar*) peel; (*quitar el pellejo*) skin
peldaño *m* step; (*de escalera de mano*) rung
pelea *f* fight; (*discusión*) quarrel. **~r** *vi* fight. **~rse** *vpr* fight
peletería *f* fur shop
peliagudo *a* difficult, tricky
pelícano *m*, **pelicano** *m* pelican
película *f* film (*esp Brit*), movie (*Amer*). **~ de dibujos (animados)** cartoon (film). **~ en colores** colour film
peligro *m* danger; (*riesgo*) risk. **~so** *a* dangerous. **poner en ~** endanger

pelirrojo *a* red-haired

pelma *m & f*, **pelmazo** *m* bore, nuisance

pel|o *m* hair; (*de barba o bigote*) whisker. **~ón** *a* bald; (*rapado*) with very short hair. **no tener ~os en la lengua** be outspoken. **tomar el ~o a uno** pull s.o.'s leg

pelota *f* ball; (*juego vasco*) pelota. **~ vasca** pelota. **en ~(s)** naked

pelotera *f* squabble

pelotilla *f*. **hacer la ~ a** ingratiate o.s. with

peluca *f* wig

peludo *a* hairy

peluquer|ía *f* (*de mujer*) hairdresser's; (*de hombre*) barber's. **~o** *m* (*de mujer*) hairdresser; (*de hombre*) barber

pelusa *f* down; (*celos, fam*) jealousy

pelvis *f* pelvis

pella *f* lump

pelleja *f*, **pellejo** *m* skin

pellizc|ar [7] *vt* pinch. **~o** *m* pinch

pena *f* sadness; (*dificultad*) difficulty. **~ de muerte** death penalty. **a duras ~s** with difficulty. **da ~ que** it's a pity that. **me da ~ que** I'm sorry that. **merecer la ~** be worthwhile. **¡qué ~!** what a pity! **valer la ~** be worthwhile

penacho *m* tuft; (*fig*) plume

penal *a* penal; (*criminal*) criminal. ● *m* prison. **~idad** *f* suffering; (*jurid*) penalty. **~izar** [10] *vt* penalize

penalty *m* penalty

penar *vt* punish. ● *vi* suffer. **~ por** long for

pend|er *vi* hang. **~iente** *a* hanging; ‹*terreno*› sloping; ‹*cuenta*› outstanding; (*fig*) ‹*asunto etc*› pending. ● *m* earring. ● *f* slope

pendón *m* banner

péndulo *a* hanging. ● *m* pendulum

pene *m* penis

penetra|nte *a* penetrating; ‹*sonido*› piercing; ‹*herida*› deep. **~r** *vt* penetrate; (*fig*) pierce; (*entender*) understand. ● *vi* penetrate; (*entrar*) go into

penicilina *f* penicillin

pen|ínsula *f* peninsula. **península Ibérica** Iberian Peninsula. **~insular** *a* peninsular

penique *m* penny

peniten|cia *f* penitence; (*castigo*) penance. **~te** *a & m & f* penitent

penoso *a* painful; (*difícil*) difficult

pensa|do *a* thought. **~dor** *m* thinker. **~miento** *m* thought. **~r** [1] *vt* think; (*considerar*) consider. ● *vi* think. **~r en** think about. **~tivo** *a* thoughtful. **bien ~do** all things considered. **cuando menos se piensa** when least expected. **menos ~do** least expected. **¡ni ~rlo!** certainly not! **pienso que sí** I think so

pensi|ón *f* pension; (*casa de huéspedes*) guest-house. **~ón completa** full board. **~onista** *m & f* pensioner; (*huésped*) lodger; (*escol*) boarder

pentágono *m* pentagon

pentagrama *m* stave

Pentecostés *m* Whitsun; (*fiesta judía*) Pentecost

penúltimo *a & m* penultimate, last but one

penumbra *f* half-light

penuria *f* shortage

peñ|a *f* rock; (*de amigos*) group; (*club*) club. **~ón** *m* rock. **el peñón de Gibraltar** The Rock (of Gibraltar)

peón *m* labourer; (*en ajedrez*) pawn; (*en damas*) piece; (*juguete*) (spinning) top

peonía *f* peony

peonza *f* (spinning) top

peor *a* (*comparativo*) worse; (*superlativo*) worst. ● *adv* worse. **~ que ~** worse and worse. **lo ~** the worst thing. **tanto ~** so much the worse

pepin|illo *m* gherkin. **~o** *m* cucumber. **(no) me importa un ~o** I couldn't care less

pepita *f* pip

pepitoria *f* fricassee

pequeñ|ez *f* smallness; (*minucia*) trifle. **~ito** *a* very small, tiny. **~o** *a* small, little. **de ~o** as a child. **en ~o** in miniature

pequinés *m* (*perro*) Pekingese

pera *f* (*fruta*) pear. **~l** *m* pear (tree)

percance *m* setback

percatarse *vpr*. **~ de** notice

perc|epción *f* perception. **~eptible** *a* perceptible. **~eptivo** *a* perceptive. **~ibir** *vt* perceive; earn ‹*dinero*›

percusión *f* percussion

percutir *vt* tap

percha *f* hanger; (*de aves*) perch. **de ~** off the peg

perde|dor *a* losing. ● *m* loser. **~r** [1] *vt* lose; (*malgastar*) waste; miss ‹*tren etc*›. ● *vi* lose; ‹*tela*› fade. **~rse** *vpr* get lost; (*desparecer*) disappear;

(*desperdiciarse*) be wasted; (*estropearse*) be spoilt. **echar(se) a ~r** spoil
pérdida *f* loss; (*de líquido*) leak; (*de tiempo*) waste
perdido *a* lost
perdiz *f* partridge
perd|ón *m* pardon, forgiveness. ● *int* sorry! **~onar** *vt* excuse, forgive; (*jurid*) pardon. **¡~one (Vd)!** sorry! **pedir ~ón** apologize
perdura|ble *a* lasting. **~r** *vi* last
perece|dero *a* perishable. **~r** [11] *vi* perish
peregrin|ación *f* pilgrimage. **~ar** *vi* go on a pilgrimage; (*fig*, fam) travel. **~o** *a* strange. ● *m* pilgrim
perejil *m* parsley
perengano *m* so-and-so
perenne *a* everlasting; (*bot*) perennial
perentorio *a* peremptory
perez|a *f* laziness. **~oso** *a* lazy
perfec|ción *f* perfection. **~cionamiento** *m* perfection; (*mejora*) improvement. **~cionar** *vt* perfect; (*mejorar*) improve. **~cionista** *m & f* perfectionist. **~tamente** *adv* perfectly. ● *int* of course! **~to** *a* perfect; (*completo*) complete. **a la ~ción** perfectly, to perfection
perfidia *f* treachery
pérfido *a* treacherous
perfil *m* profile; (*contorno*) outline; **~es** *mpl* (*fig*, *rasgos*) features. **~ado** *a* (*bien terminado*) well-finished. **~ar** *vt* draw in profile; (*fig*) put the finishing touches to
perfora|ción *f* perforation. **~do** *m* perforation. **~dora** *f* punch. **~r** *vt* pierce, perforate; punch ‹*papel*, *tarjeta etc*›
perfum|ar *vt* perfume. **~arse** *vpr* put perfume on. **~e** *m* perfume, scent. **~ería** *f* perfumery
pergamino *m* parchment
pericia *f* expertise
pericón *m* popular Argentinian dance
perif|eria *f* (*de población*) outskirts. **~érico** *a* peripheral
perilla *f* (*barba*) goatee
perímetro *m* perimeter
periódico *a* periodic(al). ● *m* newspaper
periodis|mo *m* journalism. **~ta** *m & f* journalist
período *m*, **periodo** *m* period
periquito *m* budgerigar
periscopio *m* periscope
perito *a & m* expert
perju|dicar [7] *vt* harm; (*desfavorecer*) not suit. **~dicial** *a* harmful. **~icio** *m* harm. **en ~icio de** to the detriment of
perjur|ar *vi* perjure o.s. **~io** *m* perjury
perla *f* pearl. **de ~s** *adv* very well. ● *a* excellent
permane|cer [11] *vi* remain. **~ncia** *f* permanence; (*estancia*) stay. **~nte** *a* permanent. ● *f* perm
permeable *a* permeable
permi|sible *a* permissible. **~sivo** *a* permissive. **~so** *m* permission; (*documento*) licence; (*mil etc*) leave. **~so de conducción**, **~so de conducir** driving licence (*Brit*), driver's license (*Amer*). **~tir** *vt* allow, permit. **~tirse** *vpr* be allowed. **con ~so** excuse me. **¿me ~te?** may I?
permutación *f* exchange; (*math*) permutation
pernicioso *a* pernicious; ‹*persona*› wicked
pernio *m* hinge
perno *m* bolt
pero *conj* but. ● *m* fault; (*objeción*) objection
perogrullada *f* platitude
perol *m* pan
peronista *m & f* follower of Juan Perón
perorar *vi* make a speech
perpendicular *a & f* perpendicular
perpetrar *vt* perpetrate
perpetu|ar [21] *vt* perpetuate. **~o** *a* perpetual
perplej|idad *f* perplexity. **~o** *a* perplexed
perr|a *f* (*animal*) bitch; (*moneda*) coin, penny (*Brit*), cent (*Amer*); (*rabieta*) tantrum. **~era** *f* kennel. **~ería** *f* (*mala jugada*) dirty trick; (*palabra*) harsh word. **~o** *a* awful ● *m* dog. **~o corredor** hound. **~o de aguas** spaniel. **~o del hortelano** dog in the manger. **~o galgo** greyhound. **de ~os** awful. **estar sin una ~a** be broke
persa *a & m & f* Persian
perse|cución *f* pursuit; (*tormento*) persecution. **~guir** [5 & 13] *vt* pursue; (*atormentar*) persecute
persevera|ncia *f* perseverance. **~nte** *a* persevering. **~r** *vi* persevere
persiana *f* (Venetian) blind

persist|encia *f* persistence. **~ente** *a* persistent. **~ir** *vi* persist

person|a *f* person. **~as** *fpl* people. **~aje** *m* (*persona importante*) important person; (*de obra literaria*) character. **~al** *a* personal; (*para una persona*) single. ● *m* staff. **~alidad** *f* personality. **~arse** *vpr* appear in person. **~ificar** [7] *vt* personify. **~ificación** *f* personification

perspectiva *f* perspective

perspica|cia *f* shrewdness; (*de vista*) keen eye-sight. **~z** *a* shrewd; ‹*vista*› keen

persua|dir *vt* persuade. **~sión** *f* persuasion. **~sivo** *a* persuasive

pertenecer [11] *vi* belong

pertinaz *a* persistent

pertinente *a* relevant

perturba|ción *f* disturbance. **~r** *vt* perturb

Perú *m*. **el ~** Peru

peruano *a* & *m* Peruvian

perver|sión *f* perversion. **~so** *a* perverse. ● *m* pervert. **~tir** [4] *vt* pervert

pervivir *vi* live on

pesa *f* weight. **~dez** *f* weight; (*de cabeza etc*) heaviness; (*lentitud*) sluggishness; (*cualidad de fastidioso*) tediousness; (*cosa fastidiosa*) bore, nuisance

pesadilla *f* nightmare

pesad|o *a* heavy; (*lento*) slow; (*duro*) hard; (*aburrido*) boring, tedious. **~umbre** *f* (*pena*) sorrow

pésame *m* sympathy, condolences

pesar *vt/i* weigh. ● *m* sorrow; (*remordimiento*) regret. **a ~ de (que)** in spite of. **me pesa que** I'm sorry that. **pese a (que)** in spite of

pesario *m* pessary

pesca *f* fishing; (*peces*) fish; (*pescado*) catch. **~da** *f* hake. **~dería** *f* fish shop. **~dilla** *f* whiting. **~do** *m* fish. **~dor** *a* fishing. ● *m* fisherman. **~r** [7] *vt* catch. ● *vi* fish. **ir de ~** go fishing

pescuezo *m* neck

pesebre *m* manger

pesero *m* (*Mex*) minibus taxi

peseta *f* peseta; (*Mex*) twenty-five centavos

pesimis|mo *m* pessimism. **~ta** *a* pessimistic. ● *m* & *f* pessimist

pésimo *a* very bad, awful

peso *m* weight; (*moneda*) peso. **~ bruto** gross weight. **~ neto** net weight. **a ~** by weight. **de ~** influential

pesquero *a* fishing

pesquisa *f* inquiry

pestañ|a *f* eyelash. **~ear** *vi* blink. **sin ~ear** without batting an eyelid

pest|e *f* plague; (*hedor*) stench. **~icida** *m* pesticide. **~ilencia** *f* pestilence; (*hedor*) stench

pestillo *m* bolt

pestiño *m* pancake with honey

petaca *f* tobacco case; (*LAm, maleta*) suitcase

pétalo *m* petal

petardo *m* firework

petición *f* request; (*escrito*) petition. **a ~ de** at the request of

petirrojo *m* robin

petrificar [7] *vt* petrify

petr|óleo *m* oil. **~olero** *a* oil. ● *m* oil tanker. **~olífero** *a* oil-bearing

petulante *a* arrogant

peyorativo *a* pejorative

pez *f* fish; (*substancia negruzca*) pitch. **~ espada** swordfish

pezón *m* nipple; (*bot*) stalk

pezuña *f* hoof

piada *f* chirp

piadoso *a* compassionate; (*devoto*) devout

pian|ista *m* & *f* pianist. **~o** *m* piano. **~o de cola** grand piano

piar [20] *vi* chirp

pib|a *f* (*LAm*) little girl. **~e** *m* (*LAm*) little boy

picad|illo *m* mince; (*guiso*) stew. **~o** *a* perforated; ‹*carne*› minced; (*ofendido*) offended; ‹*mar*› choppy; ‹*diente*› bad. **~ura** *f* bite, sting; (*de polilla*) moth hole

picante *a* hot; ‹*palabras etc*› cutting

picaporte *m* door-handle; (*aldaba*) knocker

picar [7] *vt* prick, pierce; ‹*ave*› peck; ‹*insecto, pez*› bite; ‹*avispa*› sting; (*comer poco*) pick at; mince ‹*carne*›. ● *vi* prick; ‹*ave*› peck; ‹*insecto, pez*› bite; ‹*sol*› scorch; ‹*sabor fuerte*› be hot. **~ alto** aim high

picard|ear *vt* corrupt. **~ía** *f* wickedness; (*travesura*) naughty thing

picaresco *a* roguish; ‹*literatura*› picaresque

pícaro *a* villainous; ‹*niño*› mischievous. ● *m* rogue

picatoste *m* toast; (*frito*) fried bread

picazón *f* itch

pico *m* beak; (*punta*) corner; (*herramienta*) pickaxe; (*cima*) peak.

~**tear** *vt* peck; (*comer*, *fam*) pick at. **y** ~ (*con tiempo*) a little after; (*con cantidad*) a little more than
picudo *a* pointed
pich|ona *f* (*fig*) darling; ~**ón** *m* pigeon
pido *vb véase* **pedir**
pie *m* foot; (*bot*, *de vaso*) stem. ~ **cuadrado** square foot. **a cuatro** ~**s** on all fours. **al** ~ **de la letra** literally. **a** ~ on foot. **a** ~**(s) juntillas** (*fig*) firmly. **buscarle tres** ~**s al gato** split hairs. **de** ~ standing (up). **de** ~**s a cabeza** from head to foot. **en** ~ standing (up). **ponerse de/en** ~ stand up
piedad *f* pity; (*relig*) piety
piedra *f* stone; (*de mechero*) flint; (*granizo*) hailstone
piel *f* skin; (*cuero*) leather. **artículos de** ~ leather goods
pienso *vb véase* **pensar**
pierdo *vb véase* **perder**
pierna *f* leg. **estirar las** ~**s** stretch one's legs
pieza *f* piece; (*parte*) part; (*obra teatral*) play; (*moneda*) coin; (*habitación*) room. ~ **de recambio** spare part
pífano *m* fife
pigment|ación *f* pigmentation. ~**o** *m* pigment
pigmeo *a* & *m* pygmy
pijama *m* pyjamas
pila *f* (*montón*) pile; (*recipiente*) basin; (*eléctrica*) battery. ~ **bautismal** font
píldora *f* pill
pilot|ar *vt* pilot. ~**o** *m* pilot
pilla|je *m* pillage. ~**r** *vt* pillage; (*alcanzar*, *agarrar*) catch; (*atropellar*) run over
pillo *a* wicked. • *m* rogue
pim|entero *m* (*vasija*) pepper-pot. ~**entón** *m* paprika, cayenne pepper. ~**ienta** *f* pepper. ~**iento** *m* pepper. **grano** *m* **de** ~**ienta** peppercorn
pináculo *m* pinnacle
pinar *m* pine forest
pincel *m* paintbrush. ~**ada** *f* brushstroke. **la última** ~**ada** (*fig*) the finishing touch
pinch|ar *vt* pierce, prick; puncture ‹*neumático*›; (*fig*, *incitar*) push; (*med*, *fam*) give an injection to. ~**azo** *m* prick; (*en neumático*) puncture. ~**itos** *mpl* kebab(s); (*tapas*) savoury snacks. ~**o** *m* point
ping|ajo *m* rag. ~**o** *m* rag
ping-pong *m* table tennis, ping-pong
pingüino *m* penguin
pino *m* pine (tree)
pint|a *f* spot; (*fig*, *aspecto*) appearance. ~**ada** *f* graffiti. ~**ar** *vt* paint. ~**arse** *vpr* put on make-up. ~**or** *m* painter. ~**or de brocha gorda** painter and decorator. ~**oresco** *a* picturesque. ~**ura** *f* painting. **no** ~**a nada** (*fig*) it doesn't count. **tener** ~**a de** look like
pinza *f* (clothes-)peg (*Brit*), (clothes-) pin (*Amer*); (*de cangrejo etc*) claw. ~**s** *fpl* tweezers
pinzón *m* chaffinch
piñ|a *f* pine cone; (*ananás*) pineapple; (*fig*, *grupo*) group. ~**ón** *m* (*semilla*) pine nut
pío *a* pious; ‹*caballo*› piebald. • *m* chirp. **no decir (ni)** ~ not say a word
piocha *f* pickaxe
piojo *m* louse
pionero *m* pioneer
pipa *f* pipe; (*semilla*) seed; (*de girasol*) sunflower seed
pipián *m* (*LAm*) stew
pique *m* resentment; (*rivalidad*) rivalry. **irse a** ~ sink
piqueta *f* pickaxe
piquete *m* picket
piragua *f* canoe
pirámide *f* pyramid
pirata *m* & *f* pirate
Pirineos *mpl* Pyrenees
piropo *m* (*fam*) compliment
piruet|a *f* pirouette. ~**ear** *vi* pirouette
pirulí *m* lollipop
pisa|da *f* footstep; (*huella*) footprint. ~**papeles** *m invar* paperweight. ~**r** *vt* tread on; (*apretar*) press; (*fig*) walk over. • *vi* tread. **no** ~**r el césped** keep off the grass
piscina *f* swimming pool; (*para peces*) fish-pond
Piscis *m* Pisces
piso *m* floor; (*vivienda*) flat (*Brit*), apartment (*Amer*); (*de zapato*) sole
pisotear *vt* trample (on)
pista *f* track; (*fig*, *indicio*) clue. ~ **de aterrizaje** runway. ~ **de baile** dance floor. ~ **de hielo** skating-rink. ~ **de tenis** tennis court
pistacho *m* pistachio (nut)
pisto *m* fried vegetables
pistol|a *f* pistol. ~**era** *f* holster. ~**ero** *m* gunman

pistón *m* piston
pit|ar *vt* whistle at. ● *vi* blow a whistle; (*auto*) sound one's horn. **~ido** *m* whistle
pitill|era *f* cigarette case. **~o** *m* cigarette
pito *m* whistle; (*auto*) horn
pitón *m* python
pitorre|arse *vpr*. **~arse de** make fun of. **~o** *m* teasing
pitorro *m* spout
pivote *m* pivot
pizarr|a *f* slate; (*encerrado*) blackboard. **~ón** *m* (*LAm*) blackboard
pizca *f* (*fam*) tiny piece; (*de sal*) pinch. **ni ~** not at all
pizz|a *f* pizza. **~ería** *f* pizzeria
placa *f* plate; (*conmemorativa*) plaque; (*distintivo*) badge
pláceme *m* congratulations
place|ntero *a* pleasant. **~r** [32] *vt* please. **me ~** I like. ● *m* pleasure
plácido *a* placid
plaga *f* plague; (*fig, calamidad*) disaster; (*fig, abundancia*) glut. **~r** [12] *vt* fill
plagi|ar *vt* plagiarize. **~o** *m* plagiarism
plan *m* plan; (*med*) course of treatment. **a todo ~** on a grand scale. **en ~ de** as
plana *f* (*llanura*) plain; (*página*) page. **en primera ~** on the front page
plancha *f* iron; (*lámina*) sheet. **~do** *m* ironing. **~r** *vt/i* iron. **a la ~** grilled. **tirarse una ~** put one's foot in it
planeador *m* glider
planear *vt* plan. ● *vi* glide
planeta *m* planet. **~rio** *a* planetary. ● *m* planetarium
planicie *f* plain
planifica|ción *f* planning. **~r** [7] *vt* plan
planilla *f* (*LAm*) list
plano *a* flat. ● *m* plane; (*de ciudad*) plan. **primer ~** foreground; (*foto*) close-up
planta *f* (*anat*) sole; (*bot, fábrica*) plant; (*plano*) ground plan; (*piso*) floor. **~ baja** ground floor (*Brit*), first floor (*Amer*)
planta|ción *f* plantation. **~do** *a* planted. **~r** *vt* plant; deal ‹*golpe*›. **~r en la calle** throw out. **~rse** *vpr* stand; (*fig*) stand firm. **bien ~do** good-looking
plantear *vt* (*exponer*) expound; (*causar*) create; raise ‹*cuestión*›
plantilla *f* insole; (*modelo*) pattern; (*personal*) personnel
plaqué *m* plate
plasma *m* plasma
plástico *a* & *m* plastic
plata *f* silver; (*fig, dinero, fam*) money. **~ de ley** sterling silver. **~ alemana** nickel silver
plataforma *f* platform
plátano *m* plane (tree); (*fruta*) banana; (*platanero*) banana tree
platea *f* stalls (*Brit*), orchestra (*Amer*)
plateado *a* silver-plated; (*color de plata*) silver
pl|ática *f* chat, talk. **~aticar** [7] *vi* chat, talk
platija *f* plaice
platillo *m* saucer; (*mus*) cymbal. **~ volante** flying saucer
platino *m* platinum. **~s** *mpl* (*auto*) points
plato *m* plate; (*comida*) dish; (*parte de una comida*) course
platónico *a* platonic
plausible *a* plausible; (*loable*) praiseworthy
playa *f* beach; (*fig*) seaside
plaza *f* square; (*mercado*) market; (*sitio*) place; (*empleo*) job. **~ de toros** bullring
plazco *vb véase* **placer**
plazo *m* period; (*pago*) instalment; (*fecha*) date. **comprar a ~s** buy on hire purchase (*Brit*), buy on the installment plan (*Amer*)
plazuela *f* little square
pleamar *f* high tide
plebe *f* common people. **~yo** *a* & *m* plebeian
plebiscito *m* plebiscite
plectro *m* plectrum
plega|ble *a* pliable; ‹*silla etc*› folding. **~r** [1 & 12] *vt* fold. **~rse** *vpr* bend; (*fig*) give way
pleito *m* (court) case; (*fig*) dispute
plenilunio *m* full moon
plen|itud *f* fullness; (*fig*) height. **~o** *a* full. **en ~o día** in broad daylight. **en ~o verano** at the height of the summer
pleuresía *f* pleuresy
plieg|o *m* sheet. **~ue** *m* fold; (*en ropa*) pleat
plinto *m* plinth
plisar *vt* pleat

plom|ero *m* (*esp LAm*) plumber. **~o** *m* lead; (*elec*) fuse. **de ~o** lead

pluma *f* feather; (*para escribir*) pen. **~ estilográfica** fountain pen. **~je** *m* plumage

plúmbeo *a* leaden

plum|ero *m* feather duster; (*para plumas, lapices etc*) pencil-case. **~ón** *m* down

plural *a* & *m* plural. **~idad** *f* plurality; (*mayoría*) majority. **en ~** in the plural

pluriempleo *m* having more than one job

plus *m* bonus

pluscuamperfecto *m* pluperfect

plusvalía *f* appreciation

plut|ocracia *f* plutocracy. **~ócrata** *m* & *f* plutocrat. **~ocrático** *a* plutocratic

plutonio *m* plutonium

pluvial *a* rain

pobla|ción *f* population; (*ciudad*) city, town; (*pueblo*) village. **~do** *a* populated. ● *m* village. **~r** [2] *vt* populate; (*habitar*) inhabit. **~rse** *vpr* get crowded

pobre *a* poor. ● *m* & *f* poor person; (*fig*) poor thing. **¡~cito!** poor (little) thing! **¡~ de mí!** poor (old) me! **~za** *f* poverty

pocilga *f* pigsty

poción *f* potion

poco *a* not much, little; (*en plural*) few; (*unos*) a few. ● *m* (a) little. ● *adv* little, not much; (*con adjetivo*) not very; (*poco tiempo*) not long. **~ a ~** little by little, gradually. **a ~ de** soon after. **dentro de ~** soon. **hace ~** not long ago. **poca cosa** nothing much. **por ~** (*fam*) nearly

podar *vt* prune

poder [33] *vi* be able. **no pudo venir** he couldn't come. **¿puedo hacer algo?** can I do anything? **¿puedo pasar?** may I come in? ● *m* power. **~es** *mpl* **públicos** authorities. **~oso** *a* powerful. **en el ~** in power. **no ~ con** not be able to cope with; (*no aguantar*) not be able to stand. **no ~ más** be exhausted; (*estar harto de algo*) not be able to manage any more. **no ~ menos que** not be able to help. **puede que** it is possible that. **puede ser** it is possible. **¿se puede ...?** may I ...?

podrido *a* rotten

po|ema *m* poem. **~esía** *f* poetry; (*poema*) poem. **~eta** *m* poet. **~ético** *a* poetic

polaco *a* Polish. ● *m* Pole; (*lengua*) Polish

polar *a* polar. **estrella ~** polestar

polarizar [10] *vt* polarize

polca *f* polka

polea *f* pulley

pol|émica *f* controversy. **~émico** *a* polemic(al). **~emizar** [10] *vi* argue

polen *m* pollen

policía *f* police (force); (*persona*) policewoman. ● *m* policeman. **~co** *a* police; ‹*novela etc*› detective

policlínica *f* clinic, hospital

policromo, polícromo *a* polychrome

polideportivo *m* sports centre

poliéster *m* polyester

poliestireno *m* polystyrene

polietileno *m* polythene

pol|igamia *f* polygamy. **~ígamo** *a* polygamous

polígloto *m* & *f* polyglot

polígono *m* polygon

polilla *f* moth

polio(mielitis) *f* polio(myelitis)

pólipo *m* polyp

politécnic|a *f* polytechnic. **~o** *a* polytechnic

polític|a *f* politics. **~o** *a* political; ‹*pariente*› -in-law. ● *m* politician. **padre** *m* **~o** father-in-law

póliza *f* document; (*de seguros*) policy

polo *m* pole; (*helado*) ice lolly (*Brit*); (*juego*) polo. **~ helado** ice lolly (*Brit*). **~ norte** North Pole

Polonia *f* Poland

poltrona *f* armchair

polución *f* (*contaminación*) pollution

polv|areda *f* cloud of dust; (*fig, escándalo*) scandal. **~era** *f* compact. **~o** *m* powder; (*suciedad*) dust. **~os** *mpl* powder. **en ~o** powdered. **estar hecho ~o** be exhausted. **quitar el ~o** dust

pólvora *f* gunpowder; (*fuegos artificiales*) fireworks

polvor|iento *a* dusty. **~ón** *m* Spanish Christmas shortcake

poll|ada *f* brood. **~era** *f* (*para niños*) baby-walker; (*LAm, falda*) skirt. **~ería** *f* poultry shop. **~o** *m* chicken; (*gallo joven*) chick

pomada *f* ointment

pomelo *m* grapefruit

pómez *a*. **piedra** *f* **~** pumice stone

pomp|a *f* bubble; (*esplendor*) pomp. **~as fúnebres** funeral. **~oso** *a* pompous; (*espléndido*) splendid

pómulo *m* cheek; (*hueso*) cheekbone

poncha|do *a* (*Mex*) punctured, flat. **~r** *vt* (*Mex*) puncture

ponche *m* punch

poncho *m* poncho

ponderar *vt* (*alabar*) speak highly of

poner [34] *vt* put; put on ‹*ropa, obra de teatro, TV etc*›; (*suponer*) suppose; lay ‹*la mesa, un huevo*›; (*hacer*) make; (*contribuir*) contribute; give ‹*nombre*›; show ‹*película, interés*›; open ‹*una tienda*›; equip ‹*una casa*›. • *vi* lay. **~se** *vpr* put o.s.; (*volverse*) get; put on ‹*ropa*›; ‹*sol*› set. **~ con** (*al teléfono*) put through to. **~ en claro** clarify. **~ por escrito** put into writing. **~ una multa** fine. **~se a** start to. **~se a mal con uno** fall out with s.o. **pongamos** let's suppose

pongo *vb véase* **poner**

poniente *m* west; (*viento*) west wind

pont|ificado *m* pontificate. **~ifical** *a* pontifical. **~ificar** [7] *vi* pontificate. **~ífice** *m* pontiff

pontón *m* pontoon

popa *f* stern

popelín *m* poplin

popul|acho *m* masses. **~ar** *a* popular; ‹*lenguaje*› colloquial. **~aridad** *f* popularity. **~arizar** [10] *vt* popularize. **~oso** *a* populous

póquer *m* poker

poquito *m* a little bit. • *adv* a little

por *prep* for; (*para*) (in order) to; (*a través de*) through; (*a causa de*) because of; (*como agente*) by; (*en matemática*) times; (*como función*) as; (*en lugar de*) instead of. **~ la calle** along the street. **~ mí** as for me, for my part. **~ si** in case. **~ todo el país** throughout the country. **50 kilómetros ~ hora** 50 kilometres per hour

porcelana *f* china

porcentaje *m* percentage

porcino *a* pig. • *m* small pig

porción *f* portion; (*de chocolate*) piece

pordiosero *m* beggar

porf|ía *f* persistence; (*disputa*) dispute. **~iado** *a* persistent. **~iar** [20] *vi* insist. **a ~ía** in competition

pormenor *m* detail

pornogr|afía *f* pornography. **~áfico** *a* pornographic

poro *m* pore. **~so** *a* porous

poroto *m* (*LAm, judía*) bean

porque *conj* because; (*para que*) so that

porqué *m* reason

porquería *f* filth; (*basura*) rubbish; (*grosería*) dirty trick

porra *f* club; (*culin*) fritter

porrón *m* wine jug (with a long spout)

portaaviones *m invar* aircraft-carrier

portada *f* façade; (*de libro*) title page

portador *m* bearer

porta|equipaje(s) *m invar* boot (*Brit*), trunk (*Amer*); (*encima del coche*) roof-rack. **~estandarte** *m* standard-bearer

portal *m* hall; (*puerta principal*) main entrance; (*soportal*) porch

porta|lámparas *m invar* socket. **~ligas** *m invar* suspender belt. **~monedas** *m invar* purse

portarse *vpr* behave

portátil *a* portable

portavoz *m* megaphone; (*fig, persona*) spokesman

portazgo *m* toll

portazo *m* bang. **dar un ~** slam the door

porte *m* transport; (*precio*) carriage. **~ador** *m* carrier

portento *m* marvel

porteño *a* (*de Buenos Aires*) from Buenos Aires. • *m* person from Buenos Aires

porter|ía *f* caretaker's lodge, porter's lodge; (*en deportes*) goal. **~o** *m* caretaker, porter; (*en deportes*) goalkeeper. **~o automático** intercom (*fam*)

portezuela *f* small door; (*auto*) door

pórtico *m* portico

portill|a *f* gate; (*en barco*) porthole. **~o** *m* opening

portorriqueño *a* Puerto Rican

Portugal *m* Portugal

portugués *a & m* Portuguese

porvenir *m* future

posada *f* guest house; (*mesón*) inn

posaderas *fpl* (*fam*) bottom

posar *vt* put. • *vi* ‹*pájaro*› perch; ‹*modelo*› sit. **~se** *vpr* settle

posdata *f* postscript

pose|edor *m* owner. **~er** [18] *vt* have, own; (*saber*) know well. **~ído** *a* possessed. **~sión** *f* possession. **~sionar** *vt*. **~sionar de** hand over. **~sionarse** *vpr*. **~sionarse de** take possession of. **~sivo** *a* possessive

posfechar *vt* postdate

posguerra *f* post-war years
posib|ilidad *f* possibility. **~le** *a* possible. **de ser ~le** if possible. **en lo ~le** as far as possible. **hacer todo lo ~le para** do everything possible to. **si es ~le** if possible
posición *f* position
positivo *a* positive
poso *m* sediment
posponer [34] *vt* put after; (*diferir*) postpone
posta *f*. **a ~** on purpose
postal *a* postal. ● *f* postcard
poste *m* pole
postergar [12] *vt* pass over; (*diferir*) postpone
posteri|dad *f* posterity. **~or** *a* back; (*ulterior*) later. **~ormente** *adv* later
postigo *m* door; (*contraventana*) shutter
postizo *a* false, artificial. ● *m* hairpiece
postra|do *a* prostrate. **~r** *vt* prostrate. **~rse** *vpr* prostrate o.s.
postre *m* dessert, sweet (*Brit*). **de ~** for dessert
postular *vt* postulate; collect ‹*dinero*›
póstumo *a* posthumous
postura *f* position, stance
potable *a* drinkable; ‹*agua*› drinking
potaje *m* vegetable stew
potasio *m* potassium
pote *m* jar
poten|cia *f* power. **~cial** *a* & *m* potential. **~te** *a* powerful. **en ~cia** potential
potingue *m* (*fam*) concoction
potr|a *f* filly. **~o** *m* colt; (*en gimnasia*) horse. **tener ~a** be lucky
pozo *m* well; (*hoyo seco*) pit; (*de mina*) shaft
pozole *m* (*Mex*) stew
práctica *f* practice; (*destreza*) skill. **en la ~** in practice. **poner en ~** put into practice
practica|ble *a* practicable. **~nte** *m* & *f* nurse. **~r** [7] *vt* practise; play ‹*deportes*›; (*ejecutar*) carry out
práctico *a* practical; (*diestro*) skilled. ● *m* practitioner
prad|era *f* meadow; (*terreno grande*) prairie. **~o** *m* meadow
pragmático *a* pragmatic
preámbulo *m* preamble
precario *a* precarious
precaución *f* precaution; (*cautela*) caution. **con ~** cautiously
precaver *vt* guard against
precede|ncia *f* precedence; (*prioridad*) priority. **~nte** *a* preceding. ● *m* precedent. **~r** *vt/i* precede
precepto *m* precept. **~r** *m* tutor
precia|do *a* valuable; (*estimado*) esteemed. **~rse** *vpr* boast
precinto *m* seal
precio *m* price. **~ de venta al público** retail price. **al ~ de** at the cost of. **no tener ~** be priceless. **¿qué ~ tiene?** how much is it?
precios|idad *f* value; (*cosa preciosa*) beautiful thing. **~o** *a* precious; (*bonito*) beautiful. **¡es una ~idad!** it's beautiful!
precipicio *m* precipice
precipita|ción *f* precipitation. **~damente** *adv* hastily. **~do** *a* hasty. **~r** *vt* hurl; (*acelerar*) accelerate; (*apresurar*) hasten. **~rse** *vpr* throw o.s.; (*correr*) rush; (*actuar sin reflexionar*) act rashly
precis|amente *a* exactly. **~ar** *vt* require; (*determinar*) determine. **~ión** *f* precision; (*necesidad*) need. **~o** *a* precise; (*necesario*) necessary
preconcebido *a* preconceived
precoz *a* early; ‹*niño*› precocious
precursor *m* forerunner
predecesor *m* predecessor
predecir [46]; *o* [46, *pero imperativo* **predice**, *futuro y condicional regulares*] *vt* foretell
predestina|ción *f* predestination. **~r** *vt* predestine
prédica *f* sermon
predicamento *m* influence
predicar [7] *vt/i* preach
predicción *f* prediction; (*del tiempo*) forecast
predilec|ción *f* predilection. **~to** *a* favourite
predisponer [34] *vt* predispose
predomin|ante *a* predominant. **~ar** *vt* dominate. ● *vi* predominate. **~io** *m* predominance
preeminente *a* pre-eminent
prefabricado *a* prefabricated
prefacio *m* preface
prefect|o *m* prefect. **~ura** *f* prefecture
prefer|encia *f* preference. **~ente** *a* preferential. **~ible** *a* preferable. **~ido** *a* favourite. **~ir** [4] *vt* prefer. **de ~encia** preferably
prefigurar *vt* foreshadow
prefij|ar *vt* fix beforehand; (*gram*) prefix. **~o** *m* prefix; (*telefónico*) dialling code

preg|ón *m* announcement. **~onar** *vt* announce

pregunta *f* question. **~r** *vt/i* ask. **~rse** *vpr* wonder. **hacer ~s** ask questions

prehistórico *a* prehistoric

preju|icio *m* prejudice. **~zgar** [12] *vt* prejudge

prelado *m* prelate

preliminar *a & m* preliminary

preludio *m* prelude

premarital *a*, **prematrimonial** *a* premarital

prematuro *a* premature

premedita|ción *f* premeditation. **~r** *vt* premeditate

premi|ar *vt* give a prize to; (*recompensar*) reward. **~o** *m* prize; (*recompensa*) reward; (*com*) premium. **~o gordo** first prize

premonición *f* premonition

premura *f* urgency; (*falta*) lack

prenatal *a* antenatal

prenda *f* pledge; (*de vestir*) article of clothing, garment; (*de cama etc*) linen. **~s** *fpl* (*cualidades*) talents; (*juego*) forfeits. **~r** *vt* captivate. **~rse** *vpr* be captivated (**de** by); (*enamorarse*) fall in love (**de** with)

prender *vt* capture; (*sujetar*) fasten. ● *vi* catch; (*arraigar*) take root. **~se** *vpr* (*encenderse*) catch fire

prensa *f* press. **~r** *vt* press

preñado *a* pregnant; (*fig*) full

preocupa|ción *f* worry. **~do** *a* worried. **~r** *vt* worry. **~rse** *vpr* worry. **~rse de** look after. **¡no te preocupes!** don't worry!

prepara|ción *f* preparation. **~do** *a* prepared. ● *m* preparation. **~r** *vt* prepare. **~rse** *vpr* get ready. **~tivo** *a* preparatory. ● *m* preparation. **~torio** *a* preparatory

preponderancia *f* preponderance

preposición *f* preposition

prepotente *a* powerful; (*fig*) presumptuous

prerrogativa *f* prerogative

presa *f* (*acción*) capture; (*cosa*) catch; (*embalse*) dam

presagi|ar *vt* presage. **~o** *m* omen; (*premonición*) premonition

présbita *a* long-sighted

presb|iteriano *a & m* Presbyterian. **~iterio** *m* presbytery. **~ítero** *m* priest

prescindir *vi*. **~ de** do without; (*deshacerse de*) dispense with

prescri|bir (*pp* **prescrito**) *vt* prescribe. **~pción** *f* prescription

presencia *f* presence; (*aspecto*) appearance. **~r** *vt* be present at; (*ver*) witness. **en ~ de** in the presence of

presenta|ble *a* presentable. **~ción** *f* presentation; (*aspecto*) appearance; (*de una persona a otra*) introduction. **~dor** *m* presenter. **~r** *vt* present; (*ofrecer*) offer; (*hacer conocer*) introduce; show ‹*película*›. **~rse** *vpr* present o.s.; (*hacerse conocer*) introduce o.s.; (*aparecer*) turn up

presente *a* present; (*este*) this. ● *m* present. **los ~s** those present. **tener ~** remember

presenti|miento *m* presentiment; (*de algo malo*) foreboding. **~r** [4] *vt* have a presentiment of

preserva|ción *f* preservation. **~r** *vt* preserve. **~tivo** *m* condom

presiden|cia *f* presidency; (*de asamblea*) chairmanship. **~cial** *a* presidential. **~ta** *f* (woman) president. **~te** *m* president; (*de asamblea*) chairman. **~te del gobierno** leader of the government, prime minister

presidi|ario *m* convict. **~o** *m* prison

presidir *vt* preside over

presilla *f* fastener

presi|ón *f* pressure. **~onar** *vt* press; (*fig*) put pressure on. **a ~ón** under pressure. **hacer ~ón** press

preso *a* under arrest; (*fig*) stricken. ● *m* prisoner

presta|do *a* (*a uno*) lent; (*de uno*) borrowed. **~mista** *m & f* moneylender. **pedir ~do** borrow

préstamo *m* loan; (*acción de pedir prestado*) borrowing

prestar *vt* lend; give ‹*ayuda etc*›; pay ‹*atención*›. ● *vi* lend

prestidigita|ción *f* conjuring. **~dor** *m* magician

prestigio *m* prestige. **~so** *a* prestigious

presu|mido *a* presumptuous. **~mir** *vt* presume. ● *vi* be conceited. **~nción** *f* presumption. **~nto** *a* presumed. **~ntuoso** *a* presumptuous

presup|oner [34] *vt* presuppose. **~uesto** *m* budget

presuroso *a* quick

preten|cioso *a* pretentious. **~der** *vt* try to; (*afirmar*) claim; (*solicitar*) apply for; (*cortejar*) court. **~dido** *a* so-called. **~diente** *m* pretender; (*a* *una mujer*) suitor. **~sión** *f* pretension; (*aspiración*) aspiration

pretérito *m* preterite, past

pretexto *m* pretext. **a ~ de** on the pretext of

prevalec|er [11] *vi* prevail. **~iente** *a* prevalent

prevalerse [42] *vpr* take advantage

preven|ción *f* prevention; (*prejuicio*) prejudice. **~ido** *a* ready; (*precavido*) cautious. **~ir** [53] *vt* prepare; (*proveer*) provide; (*precaver*) prevent; (*advertir*) warn. **~tivo** *a* preventive

prever [43] *vt* foresee; (*prepararse*) plan

previo *a* previous

previs|ible *a* predictable. **~ión** *f* forecast; (*prudencia*) prudence. **~ión de tiempo** weather forecast. **~to** *a* foreseen

prima *f* (*pariente*) cousin; (*cantidad*) bonus

primario *a* primary

primate *m* primate; (*fig, persona*) important person

primavera *f* spring. **~l** *a* spring

primer *a véase* **primero**

primer|a *f* (*auto*) first (gear); (*en tren etc*) first class. **~o** *a* (*delante de nombre masculino en singular* **primer**) first; (*principal*) main; (*anterior*) former; (*mejor*) best. ● *n* (the) first. ● *adv* first. **~a enseñanza** primary education. **a ~os de** at the beginning of. **de ~a** first-class

primitivo *a* primitive

primo *m* cousin; (*fam*) fool. **hacer el ~** be taken for a ride

primogénito *a & m* first-born, eldest

primor *m* delicacy; (*cosa*) beautiful thing

primordial *a* basic

princesa *f* princess

principado *m* principality

principal *a* principal. ● *m* (*jefe*) head, boss (*fam*)

príncipe *m* prince

principi|ante *m & f* beginner. **~ar** *vt/i* begin, start. **~o** *m* beginning; (*moral, idea*) principle; (*origen*) origin. **al ~o** at first. **a ~o(s) de** at the beginning of. **dar ~o a** start. **desde el ~o** from the outset. **en ~o** in principle. **~os** *mpl* (*nociones*) rudiments

pring|oso *a* greasy. **~ue** *m* dripping; (*mancha*) grease mark

prior *m* prior. **~ato** *m* priory

prioridad *f* priority

prisa *f* hurry, haste. **a ~** quickly. **a toda ~** (*fam*) as quickly as possible. **correr ~** be urgent. **darse ~** hurry (up). **de ~** quickly. **tener ~** be in a hurry

prisi|ón *f* prison; (*encarcelamiento*) imprisonment. **~onero** *m* prisoner

prism|a *m* prism. **~áticos** *mpl* binoculars

priva|ción *f* deprivation. **~do** *a* (*particular*) private. **~r** *vt* deprive (**de** of); (*prohibir*) prevent (**de** from). ● *vi* be popular. **~tivo** *a* exclusive (**de** to)

privilegi|ado *a* privileged; (*muy bueno*) exceptional. **~o** *m* privilege

pro *prep* for. ● *m* advantage. ● *pref* pro-. **el ~ y el contra** the pros and cons. **en ~ de** on behalf of. **los ~s y los contras** the pros and cons

proa *f* bows

probab|ilidad *f* probability. **~le** *a* probable, likely. **~lemente** *adv* probably

proba|dor *m* fitting-room. **~r** [2] *vt* try; try on ‹*ropa*›; (*demostrar*) prove. ● *vi* try. **~rse** *vpr* try on

probeta *f* test-tube

problem|a *m* problem. **~ático** *a* problematic

procaz *a* insolent

proced|encia *f* origin. **~ente** *a* (*razonable*) reasonable. **~ente de** (coming) from. **~er** *m* conduct. ● *vi* proceed. **~er contra** start legal proceedings against. **~er de** come from. **~imiento** *m* procedure; (*sistema*) process; (*jurid*) proceedings

procesador *m*. **~ de textos** word processor

procesal *a*. **costas ~es** legal costs

procesamiento *m* processing. **~ de textos** word-processing

procesar *vt* prosecute

procesión *f* procession

proceso *m* process; (*jurid*) trial; (*transcurso*) course

proclama *f* proclamation. **~ción** *f* proclamation. **~r** *vt* proclaim

procrea|ción *f* procreation. **~r** *vt* procreate

procura|dor *m* attorney, solicitor. **~r** *vt* try; (*obtener*) get; (*dar*) give

prodigar [12] *vt* lavish. **~se** *vpr* do one's best

prodigio *m* prodigy; (*milagro*) miracle. **~ioso** *a* prodigious
pródigo *a* prodigal
produc|ción *f* production. **~ir** [47] *vt* produce; (*causar*) cause. **~irse** *vpr* (*aparecer*) appear; (*suceder*) happen. **~tivo** *a* productive. **~to** *m* product. **~tor** *m* producer. **~to derivado** by-product. **~tos agrícolas** farm produce. **~tos de belleza** cosmetics. **~tos de consumo** consumer goods
proeza *f* exploit
profan|ación *f* desecration. **~ar** *vt* desecrate. **~o** *a* profane
profecía *f* prophecy
proferir [4] *vt* utter; hurl ‹*insultos etc*›
profes|ar *vt* profess; practise ‹*profesión*›. **~ión** *f* profession. **~ional** *a* professional. **~or** *m* teacher; (*en universidad etc*) lecturer. **~orado** *m* teaching profession; (*conjunto de profesores*) staff
prof|eta *m* prophet. **~ético** *a* prophetic. **~etizar** [10] *vt/i* prophesize
prófugo *a* & *m* fugitive
profund|idad *f* depth. **~o** *a* deep; (*fig*) profound
profus|ión *f* profusion. **~o** *a* profuse. **con ~ión** profusely
progenie *f* progeny
programa *m* programme; (*de ordenador*) program; (*de estudios*) curriculum. **~ción** *f* programming; (*TV etc*) programmes; (*en periódico*) TV guide. **~r** *vt* programme; program ‹*ordenador*›. **~dor** *m* computer programmer
progres|ar *vi* (make) progress. **~ión** *f* progression. **~ista** *a* progressive. **~ivo** *a* progressive. **~o** *m* progress. **hacer ~os** make progress
prohibi|ción *f* prohibition. **~do** *a* forbidden. **~r** *vt* forbid. **~tivo** *a* prohibitive
prójimo *m* fellow man
prole *f* offspring
proletari|ado *m* proletariat. **~o** *a* & *m* proletarian
prol|iferación *f* proliferation. **~iferar** *vi* proliferate. **~ífico** *a* prolific
prolijo *a* long-winded, extensive
prólogo *m* prologue
prolongar [12] *vt* prolong; (*alargar*) lengthen. **~se** *vpr* go on
promedio *m* average
prome|sa *f* promise. **~ter** *vt/i* promise. **~terse** *vpr* ‹*novios*› get engaged. **~térselas muy felices** have high hopes. **~tida** *f* fiancée. **~tido** *a* promised; ‹*novios*› engaged. ● *m* fiancé
prominen|cia *f* prominence. **~te** *a* prominent
promiscu|idad *f* promiscuity. **~o** *a* promiscuous
promoción *f* promotion
promontorio *m* promontory
promo|tor *m* promoter. **~ver** [2] *vt* promote; (*causar*) cause
promulgar [12] *vt* promulgate
pronombre *m* pronoun
pron|osticar [7] *vt* predict. **~óstico** *m* prediction; (*del tiempo*) forecast; (*med*) prognosis
pront|itud *f* quickness. **~o** *a* quick; (*preparado*) ready. ● *adv* quickly; (*dentro de poco*) soon; (*temprano*) early. ● *m* urge. **al ~o** at first. **de ~o** suddenly. **por lo ~o** for the time being; (*al menos*) anyway. **tan ~o como** as soon as
pronuncia|ción *f* pronunciation. **~miento** *m* revolt. **~r** *vt* pronounce; deliver ‹*discurso*›. **~rse** *vpr* be pronounced; (*declarase*) declare o.s.; (*sublevarse*) rise up
propagación *f* propagation
propaganda *f* propaganda; (*anuncios*) advertising
propagar [12] *vt/i* propagate. **~se** *vpr* spread
propano *m* propane
propasarse *vpr* go too far
propens|ión *f* inclination. **~o** *a* inclined
propiamente *adv* exactly
propici|ar *vt* (*provocar*) cause, bring about. **~o** *a* favourable
propie|dad *f* property; (*posesión*) possession. **~tario** *m* owner
propina *f* tip
propio *a* own; (*característico*) typical; (*natural*) natural; (*apropiado*) proper. **de ~** on purpose. **el médico ~** the doctor himself
proponer [34] *vt* propose. **~se** *vpr* propose
proporci|ón *f* proportion. **~onado** *a* proportioned. **~onal** *a* proportional. **~onar** *vt* proportion; (*facilitar*) provide
proposición *f* proposition
propósito *m* intention. **a ~** (*adrede*) on purpose; (*de paso*) incidentally.

a ~ de with regard to. **de ~** on purpose

propuesta *f* proposal

propuls|ar *vt* propel; (*fig*) promote. **~ión** *f* propulsion. **~ión a chorro** jet propulsion

prórroga *f* extension

prorrogar [12] *vt* extend

prorrumpir *vi* burst out

prosa *f* prose. **~ico** *a* prosaic

proscri|bir (*pp* **proscrito**) *vt* banish; (*prohibido*) ban. **~to** *a* banned. ● *m* exile; (*persona*) outlaw

prosecución *f* continuation

proseguir [5 & 13] *vt/i* continue

prospección *f* prospecting

prospecto *m* prospectus

prosper|ar *vi* prosper. **~idad** *f* prosperity; (*éxito*) success

próspero *a* prosperous. **¡P~ Año Nuevo!** Happy New Year!

prostit|ución *f* prostitution. **~uta** *f* prostitute

protagonista *m & f* protagonist

prote|cción *f* protection. **~ctor** *a* protective. ● *m* protector; (*patrocinador*) patron. **~ger** [14] *vt* protect. **~gida** *f* protegée. **~gido** *a* protected. ● *m* protegé

proteína *f* protein

protesta *f* protest; (*declaración*) protestation

protestante *a & m & f* (*relig*) Protestant

protestar *vt/i* protest

protocolo *m* protocol

protuberan|cia *f* protuberance. **~te** *a* protuberant

provecho *m* benefit. **¡buen ~!** enjoy your meal! **de ~** useful. **en ~ de** to the benefit of. **sacar ~ de** benefit from

proveer [18] (*pp* **proveído** *y* **provisto**) *vt* supply, provide

provenir [53] *vi* come (**de** from)

proverbi|al *a* proverbial. **~o** *m* proverb

providencia *f* providence. **~l** *a* providential

provincia *f* province. **~l** *a*, **~no** *a* provincial

provisi|ón *f* provision; (*medida*) measure. **~onal** *a* provisional

provisto *a* provided (**de** with)

provoca|ción *f* provocation. **~r** [7] *vt* provoke; (*causar*) cause. **~tivo** *a* provocative

próximamente *adv* soon

proximidad *f* proximity

próximo *a* next; (*cerca*) near

proyec|ción *f* projection. **~tar** *vt* hurl; cast ‹*luz*›; show ‹*película*›. **~til** *m* missile. **~to** *m* plan. **~to de ley** bill. **~tor** *m* projector. **en ~to** planned

pruden|cia *f* prudence. **~nte** *a* prudent, sensible

prueba *f* proof; (*examen*) test; (*de ropa*) fitting. **a ~** on trial. **a ~ de** proof against. **a ~ de agua** waterproof. **en ~ de** in proof of. **poner a ~** test

pruebo *vb véase* **probar**

psicoan|álisis *f* psychoanalysis. **~alista** *m & f* psychoanalyst. **~alizar** [10] *vt* psychoanalyse

psicodélico *a* psychedelic

psic|ología *f* psychology. **~ológico** *a* psychological. **~ólogo** *m* psychologist

psicópata *m & f* psychopath

psicosis *f* psychosis

psique *f* psyche

psiqui|atra *m & f* psychiatrist. **~atría** *f* psychiatry. **~átrico** *a* psychiatric

psíquico *a* psychic

ptas, pts *abrev* (*pesetas*) pesetas

púa *f* sharp point; (*bot*) thorn; (*de erizo*) quill; (*de peine*) tooth; (*mus*) plectrum

pubertad *f* puberty

publica|ción *f* publication. **~r** [7] *vt* publish; (*anunciar*) announce

publici|dad *f* publicity; (*com*) advertising. **~tario** *a* advertising

público *a* public. ● *m* public; (*de espectáculo etc*) audience. **dar al ~** publish

puchero *m* cooking pot; (*guisado*) stew. **hacer ~s** (*fig, fam*) pout

pude *vb véase* **poder**

púdico *a* modest

pudiente *a* rich

pudín *m* pudding

pudor *m* modesty. **~oso** *a* modest

pudrir (*pp* **podrido**) *vt* rot; (*fig, molestar*) annoy. **~se** *vpr* rot

puebl|ecito *m* small village. **~o** *m* town; (*aldea*) village; (*nación*) nation, people

puedo *vb véase* **poder**

puente *m* bridge; (*fig, fam*) long weekend. **~ colgante** suspension bridge. **~ levadizo** drawbridge. **hacer ~** (*fam*) have a long weekend

puerco *a* filthy; (*grosero*) coarse. ● *m* pig. **~ espín** porcupine

pueril *a* childish
puerro *m* leek
puerta *f* door; (*en deportes*) goal; (*de ciudad*) gate. **~ principal** main entrance. **a ~ cerrada** behind closed doors
puerto *m* port; (*fig*, *refugio*) refuge; (*entre montañas*) pass. **~ franco** free port
Puerto Rico *m* Puerto Rico
puertorriqueño *a & m* Puerto Rican
pues *adv* (*entonces*) then; (*bueno*) well. • *conj* since
puest|a *f* setting; (*en juegos*) bet. **~a de sol** sunset. **~a en escena** staging. **~a en marcha** starting. **~o** *a* put; (*vestido*) dressed. • *m* place; (*empleo*) position, job; (*en mercado etc*) stall. • *conj*. **~o que** since. **~o de socorro** first aid post
pugna *f* fight. **~r** *vt* fight
puja *f* effort; (*en subasta*) bid. **~r** *vt* struggle; (*en subasta*) bid
pulcro *a* neat
pulga *f* flea; (*de juego*) tiddly-wink. **tener malas ~s** be bad-tempered
pulga|da *f* inch. **~r** *m* thumb; (*del pie*) big toe
puli|do *a* neat. **~mentar** *vt* polish. **~mento** *m* polishing; (*substancia*) polish. **~r** *vt* polish; (*suavizar*) smooth
pulm|ón *m* lung. **~onar** *a* pulmonary. **~onía** *f* pneumonia
pulpa *f* pulp
pulpería *f* (*LAm*) grocer's shop (*Brit*), grocery store (*Amer*)
púlpito *m* pulpit
pulpo *m* octopus
pulque *m* (*Mex*) pulque, alcoholic Mexican drink
pulsa|ción *f* pulsation. **~dor** *a* pulsating. • *m* button. **~r** *vt* (*mus*) play
pulsera *f* bracelet; (*de reloj*) strap
pulso *m* pulse; (*muñeca*) wrist; (*firmeza*) steady hand; (*fuerza*) strength; (*fig*, *tacto*) tact. **tomar el ~ a uno** take s.o.'s pulse
pulular *vi* teem with
pulveriza|dor *m* (*de perfume*) atomizer. **~r** [10] *vt* pulverize; atomize ‹*líquido*›
pulla *f* cutting remark
pum *int* bang!
puma *m* puma
puna *f* puna, high plateau
punitivo *a* punitive
punta *f* point; (*extremo*) tip; (*clavo*) (small) nail. **estar de ~** be in a bad mood. **estar de ~ con uno** be at odds with s.o. **ponerse de ~ con uno** fall out with s.o.. **sacar ~ a** sharpen; (*fig*) find fault with
puntada *f* stitch
puntal *m* prop, support
puntapié *m* kick
puntear *vt* mark; (*mus*) pluck
puntera *f* toe
puntería *f* aim; (*destreza*) markmanship
puntiagudo *a* sharp, pointed
puntilla *f* (*encaje*) lace. **de ~s** on tiptoe
punto *m* point; (*señal*) dot; (*de examen*) mark; (*lugar*) spot, place; (*de taxis*) stand; (*momento*) moment; (*punto final*) full stop (*Brit*), period (*Amer*); (*puntada*) stitch; (*de tela*) mesh. **~ de admiración** exclamation mark. **~ de arranque** starting point. **~ de exclamación** exclamation mark. **~ de interrogación** question mark. **~ de vista** point of view. **~ final** full stop. **~ muerto** (*auto*) neutral (gear). **~ y aparte** full stop, new paragraph (*Brit*), period, new paragraph (*Amer*). **~ y coma** semicolon. **a ~** on time; (*listo*) ready. **a ~ de** on the point of. **de ~** knitted. **dos ~s** colon. **en ~** exactly. **hacer ~** knit. **hasta cierto ~** to a certain extent
puntuación *f* punctuation; (*en deportes, acción*) scoring; (*en deportes, número de puntos*) score
puntual *a* punctual; (*exacto*) accurate. **~idad** *f* punctuality; (*exactitud*) accuracy
puntuar [21] *vt* punctuate. • *vi* score
punza|da *f* prick; (*dolor*) pain; (*fig*) pang. **~nte** *a* sharp. **~r** [10] *vt* prick
puñado *m* handful. **a ~s** by the handful
puñal *m* dagger. **~ada** *f* stab
puñ|etazo *m* punch. **~o** *m* fist; (*de ropa*) cuff; (*mango*) handle. **de su ~o (y letra)** in his own handwriting
pupa *f* spot; (*en los labios*) cold sore. **hacer ~** hurt. **hacerse ~** hurt o.s.
pupila *f* pupil
pupitre *m* desk
puquío *m* (*Arg*) spring
puré *m* purée; (*sopa*) thick soup. **~ de patatas** mashed potato
pureza *f* purity
purga *f* purge. **~r** [12] *vt* purge. **~torio** *m* purgatory

purifica|ción *f* purification. **~r** [7] *vt* purify
purista *m & f* purist
puritano *a* puritanical. ● *m* puritan
puro *a* pure; ‹*cielo*› clear; (*fig*) simple. ● *m* cigar. **de ~** so. **de pura casualidad** by sheer chance
púrpura *f* purple
purpúreo *a* purple
pus *m* pus
puse *vb véase* **poner**
pusilánime *a* cowardly
pústula *f* spot
puta *f* whore
putrefacción *f* putrefaction
pútrido *a* rotten, putrid

Q

que *pron rel* (*personas, sujeto*) who; (*personas, complemento*) whom; (*cosas*) which, that. ● *conj* that. **¡~ tengan Vds buen viaje!** have a good journey! **¡que venga!** let him come! **~ venga o no venga** whether he comes or not. **a que** I bet. **creo que tiene razón** I think (that) he is right. **de ~** from which. **yo ~ tú** if I were you
qué *a* (*con sustantivo*) what; (*con a o adv*) how. ● *pron* what. **¡~ bonito!** how nice. **¿en ~ piensas?** what are you thinking about?
quebra|da *f* gorge; (*paso*) pass. **~dizo** *a* fragile. **~do** *a* broken; (*com*) bankrupt. ● *m* (*math*) fraction. **~dura** *f* fracture; (*hondonada*) gorge. **~ntar** *vt* break; (*debilitar*) weaken. **~nto** *m* (*pérdida*) loss; (*daño*) damage. **~r** [1] *vt* break. ● *vi* break; (*com*) go bankrupt. **~rse** *vpr* break
quechua *a & m & f* Quechuan
queda *f* curfew
quedar *vi* stay, remain; (*estar*) be; (*faltar, sobrar*) be left. **~ bien** come off well. **~se** *vpr* stay. **~ con** arrange to meet. **~ en** agree to. **~ en nada** come to nothing. **~ por** (+ *infinitivo*) remain to be (+ *pp*)
quehacer *m* job. **~es domésticos** household chores
quej|a *f* complaint; (*de dolor*) moan. **~arse** *vpr* complain (**de** about); (*gemir*) moan. **~ido** *m* moan. **~oso** *a* complaining
quema|do *a* burnt; (*fig, fam*) bitter. **~dor** *m* burner. **~dura** *f* burn. **~r** *vt* burn; (*prender fuego a*) set fire to. ● *vi* burn. **~rse** *vpr* burn o.s.; (*consumirse*) burn up; (*con el sol*) get sunburnt. **~rropa** *adv.* **a ~rropa** point-blank
quena *f* Indian flute
quepo *vb véase* **caber**
queque *m* (*Mex*) cake
querella *f* (*riña*) quarrel, dispute; (*jurid*) charge
quer|er [35] *vt* want; (*amar*) love; (*necesitar*) need. **~er decir** mean. **~ido** *a* dear; (*amado*) loved. ● *m* darling; (*amante*) lover. **como quiera que** since; (*de cualquier modo*) however. **cuando quiera que** whenever. **donde quiera** wherever. **¿quieres darme ese libro?** would you pass me that book? **quiere llover** it's trying to rain. **¿quieres un helado?** would you like an ice-cream? **quisiera ir a la playa** I'd like to go to the beach. **sin ~er** without meaning to
queroseno *m* kerosene
querubín *m* cherub
ques|adilla *f* cheesecake; (*Mex, empanadilla*) pie. **~o** *m* cheese. **~o de bola** Edam cheese
quiá *int* never!, surely not!
quicio *m* frame. **sacar de ~ a uno** infuriate s.o.
quiebra *f* break; (*fig*) collapse; (*com*) bankruptcy
quiebro *m* dodge
quien *pron rel* (*sujeto*) who; (*complemento*) whom
quién *pron interrogativo* (*sujeto*) who; (*tras preposición*) whom. **¿de ~?** whose. **¿de ~ son estos libros?** whose are these books?
quienquiera *pron* whoever
quiero *vb véase* **querer**
quiet|o *a* still; (*inmóvil*) motionless; ‹*carácter etc*› calm. **~ud** *f* stillness
quijada *f* jaw
quilate *m* carat
quilla *f* keel
quimera *f* (*fig*) illusion
químic|a *f* chemistry. **~o** *a* chemical. ● *m* chemist
quincalla *f* hardware; (*de adorno*) trinket
quince *a & m* fifteen. **~ días** a fortnight. **~na** *f* fortnight. **~nal** *a* fortnightly
quincuagésimo *a* fiftieth

quiniela *f* pools coupon. **~s** *fpl* (football) pools
quinientos *a & m* five hundred
quinino *m* quinine
quinqué *m* oil-lamp; (*fig, fam*) shrewdness
quinquenio *m* (period of) five years
quinta *f* (*casa*) villa
quintaesencia *f* quintessence
quintal *m* a hundred kilograms
quinteto *m* quintet
quinto *a & m* fifth
quiosco *m* kiosk; (*en jardín*) summerhouse; (*en parque etc*) bandstand
quirúrgico *a* surgical
quise *vb véase* **querer**
quisque *pron.* **cada ~** (*fam*) (absolutely) everybody
quisquill|a *f* trifle; (*camarón*) shrimp. **~oso** *a* irritable; (*chinchorrero*) fussy
quita|manchas *m invar* stain remover. **~nieves** *m invar* snow plough. **~r** *vt* remove, take away; take off ‹*ropa*›; (*robar*) steal. **~ndo** (*a excepción de, fam*) apart from. **~rse** *vpr* be removed; take off ‹*ropa*›. **~rse de** (*no hacerlo más*) stop. **~rse de en medio** get out of the way. **~sol** *m invar* sunshade
Quito *m* Quito
quizá(s) *adv* perhaps
quórum *m* quorum

R

rábano *m* radish. **~ picante** horseradish. **me importa un ~** I couldn't care less
rabi|a *f* rabies; (*fig*) rage. **~ar** *vi* (*de dolor*) be in great pain; (*estar enfadado*) be furious; (*fig, tener ganas, fam*) long. **~ar por algo** long for sth. **~ar por hacer algo** long to do sth. **~eta** *f* tantrum. **dar ~a** infuriate
rabino *m* Rabbi
rabioso *a* rabid; (*furioso*) furious; ‹*dolor etc*› violent
rabo *m* tail
racial *a* racial
racimo *m* bunch
raciocinio *m* reason; (*razonamiento*) reasoning
ración *f* share, ration; (*de comida*) portion
racional *a* rational. **~izar** [10] *vt* rationalize
racionar *vt* (*limitar*) ration; (*repartir*) ration out
racis|mo *m* racism. **~ta** *a* racist
racha *f* gust of wind; (*fig*) spate
radar *m* radar
radiación *f* radiation
radiactiv|idad *f* radioactivity. **~o** *a* radioactive
radiador *m* radiator
radial *a* radial
radiante *a* radiant
radical *a & m & f* radical
radicar [7] *vi* (*estar*) be. **~ en** (*fig*) lie in
radio *m* radius; (*de rueda*) spoke; (*elemento metálico*) radium. • *f* radio
radioactiv|idad *f* radioactivity. **~o** *a* radioactive
radio|difusión *f* broadcasting. **~emisora** *f* radio station. **~escucha** *m & f* listener
radiografía *f* radiography
radi|ología *f* radiology. **~ólogo** *m* radiologist
radioterapia *f* radiotherapy
radioyente *m & f* listener
raer [36] *vt* scrape off
ráfaga *f* (*de viento*) gust; (*de luz*) flash; (*de ametralladora*) burst
rafia *f* raffia
raído *a* threadbare
raigambre *f* roots; (*fig*) tradition
raíz *f* root. **a ~ de** immediately after. **echar raíces** (*fig*) settle
raja *f* split; (*culin*) slice. **~r** *vt* split. **~rse** *vpr* split; (*fig*) back out
rajatabla. a ~ vigorously
ralea *f* sort
ralo *a* sparse
ralla|dor *m* grater. **~r** *vt* grate
rama *f* branch. **~je** *m* branches. **~l** *m* branch. **en ~** raw
rambla *f* gully; (*avenida*) avenue
ramera *f* prostitute
ramifica|ción *f* ramification. **~rse** [7] *vpr* branch out
ramilla *f* twig
ramillete *m* bunch
ramo *m* branch; (*de flores*) bouquet
rampa *f* ramp, slope
ramplón *a* vulgar
rana *f* frog. **ancas** *fpl* **de ~** frogs' legs. **no ser ~** not be stupid
rancio *a* rancid; ‹*vino*› old; (*fig*) ancient

ranch|ero *m* cook; (*LAm, jefe de rancho*) farmer. **~o** *m* (*LAm*) ranch, farm
rango *m* rank
ranúnculo *m* buttercup
ranura *f* groove; (*para moneda*) slot
rapar *vt* shave; crop ‹*pelo*›
rapaz *a* rapacious; ‹*ave*› of prey. ● *m* bird of prey
rapidez *f* speed
rápido *a* fast, quick. ● *adv* quickly. ● *m* (*tren*) express. **~s** *mpl* rapids
rapiña *f* robbery. **ave** *f* **de ~** bird of prey
rapsodia *f* rhapsody
rapt|ar *vt* kidnap. **~o** *m* kidnapping; (*de ira etc*) fit; (*éxtasis*) ecstasy
raqueta *f* racquet
raramente *adv* seldom, rarely
rarefacción *f* rarefaction
rar|eza *f* rarity; (*cosa rara*) oddity. **~o** *a* rare; (*extraño*) odd. **es ~o que** it is strange that. **¡qué ~o!** how strange!
ras *m*. **a ~ de** level with
rasar *vt* level; (*rozar*) graze
rasca|cielos *m invar* skyscraper. **~dura** *f* scratch. **~r** [7] *vt* scratch; (*raspar*) scrape
rasgar [12] *vt* tear
rasgo *m* stroke. **~s** *mpl* (*facciones*) features
rasguear *vt* strum; (*fig, escribir*) write
rasguñ|ar *vt* scratch. **~o** *m* scratch
raso *a* (*llano*) flat; (*liso*) smooth; ‹*cielo*› clear; ‹*cucharada etc*› level; ‹*vuelo etc*› low. ● *m* satin. **al ~** in the open air. **soldado** *m* **~** private
raspa *f* (*de pescado*) backbone
raspa|dura *f* scratch; (*acción*) scratching. **~r** *vt* scratch; (*rozar*) scrape
rastr|a *f* rake. **a ~as** dragging. **~ear** *vt* track. **~eo** *m* dragging. **~ero** *a* creeping; ‹*vuelo*› low. **~illar** *vt* rake. **~illo** *m* rake. **~o** *m* rake; (*huella*) track; (*señal*) sign. **el R~o** the flea market in Madrid. **ni ~o** not a trace
rata *f* rat
rate|ar *vt* steal. **~ría** *f* pilfering. **~ro** *m* petty thief
ratifica|ción *f* ratification. **~r** [7] *vt* ratify
rato *m* moment, short time. **~s libres** spare time. **a ~s** at times. **hace un ~** a moment ago. **¡hasta otro ~!** (*fam*) see you soon! **pasar mal ~** have a rough time
rat|ón *m* mouse. **~onera** *f* mousetrap; (*madriguera*) mouse hole
raud|al *m* torrent; (*fig*) floods. **~o** *a* swift
raya *f* line; (*lista*) stripe; (*de pelo*) parting. **~r** *vt* rule. ● *vi* border (**con** on). **a ~s** striped. **pasar de la ~** go too far
rayo *m* ray; (*descarga eléctrica*) lightning. **~s X** X-rays
raza *f* race; (*de animal*) breed. **de ~** ‹*caballo*› thoroughbred; ‹*perro*› pedigree
raz|ón *f* reason. **a ~ón de** at the rate of. **perder la ~ón** go out of one's mind. **tener ~ón** be right. **~onable** *a* reasonable. **~onamiento** *m* reasoning. **~onar** *vt* reason out. ● *vi* reason
re *m* D; (*solfa*) re
reac|ción *f* reaction. **~cionario** *a* & *m* reactionary. **~ción en cadena** chain reaction. **~tor** *m* reactor; (*avión*) jet
real *a* real; (*de rey etc*) royal. ● *m* real, old Spanish coin
realce *m* relief; (*fig*) splendour
realidad *f* reality; (*verdad*) truth. **en ~** in fact
realis|mo *m* realism. **~ta** *a* realistic. ● *m* & *f* realist; (*monárquico*) royalist
realiza|ción *f* fulfilment. **~r** [10] *vt* carry out; make ‹*viaje*›; achieve ‹*meta*›; (*vender*) sell. **~rse** *vpr* ‹*plan etc*› be carried out; ‹*sueño, predicción etc*› come true; ‹*persona*› fulfil o.s.
realzar [10] *vt* (*fig*) enhance
reanima|ción *f* revival. **~r** *vt* revive. **~rse** *vpr* revive
reanudar *vt* resume; renew ‹*amistad*›
reaparecer [11] *vi* reappear
rearm|ar *vt* rearm. **~e** *m* rearmament
reavivar *vt* revive
rebaja *f* reduction. **~do** *a* ‹*precio*› reduced. **~r** *vt* lower. **en ~s** in the sale
rebanada *f* slice
rebaño *m* herd; (*de ovejas*) flock
rebasar *vt* exceed; (*dejar atrás*) leave behind
rebatir *vt* refute
rebel|arse *vpr* rebel. **~de** *a* rebellious. ● *m* rebel. **~día** *f* rebelliousness. **~ión** *f* rebellion
reblandecer [11] *vt* soften

rebosa|nte *a* overflowing. **~r** *vi* overflow; (*abundar*) abound
rebot|ar *vt* bounce; (*rechazar*) repel. ● *vi* bounce; ‹*bala*› ricochet. **~e** *m* bounce, rebound. **de ~e** on the rebound
rebozar [10] *vt* wrap up; (*culin*) coat in batter
rebullir [22] *vi* stir
rebusca|do *a* affected. **~r** [7] *vt* search thoroughly
rebuznar *vi* bray
recabar *vt* claim
recado *m* errand; (*mensaje*) message. **dejar ~** leave a message
reca|er [29] *vi* fall back; (*med*) relapse; (*fig*) fall. **~ída** *f* relapse
recalcar [7] *vt* squeeze; (*fig*) stress
recalcitrante *a* recalcitrant
recalentar [1] *vt* (*de nuevo*) reheat; (*demasiado*) overheat
recamar *vt* embroider
recámara *f* small room; (*de arma de fuego*) chamber; (*LAm, dormitorio*) bedroom
recambio *m* change; (*de pluma etc*) refill. **~s** *mpl* spare parts. **de ~** spare
recapitula|ción *f* summing up. **~r** *vt* sum up
recarg|ar [12] *vt* overload; (*aumentar*) increase; recharge ‹*batería*›. **~o** *m* increase
recat|ado *a* modest. **~ar** *vt* hide. **~arse** *vpr* hide o.s. away; (*actuar discretamente*) act discreetly. **~o** *m* prudence; (*modestia*) modesty. **sin ~arse, sin ~o** openly
recauda|ción *f* (*cantidad*) takings. **~dor** *m* tax collector. **~r** *vt* collect
recel|ar *vt/i* suspect. **~o** *m* distrust; (*temor*) fear. **~oso** *a* suspicious
recepci|ón *f* reception. **~onista** *m & f* receptionist
receptáculo *m* receptacle
recept|ivo *a* receptive. **~or** *m* receiver
recesión *f* recession
receta *f* recipe; (*med*) prescription
recib|imiento *m* (*acogida*) welcome. **~ir** *vt* receive; (*acoger*) welcome. ● *vi* entertain. **~irse** *vpr* graduate. **~o** *m* receipt. **acusar ~o** acknowledge receipt
reci|én *adv* recently; ‹*casado, nacido etc*› newly. **~ente** *a* recent; (*culin*) fresh
recinto *m* enclosure
recio *a* strong; ‹*voz*› loud. ● *adv* hard; (*en voz alta*) loudly
recipiente *m* (*persona*) recipient; (*cosa*) receptacle
recíproco *a* reciprocal. **a la recíproca** vice versa
recita|l *m* recital; (*de poesías*) reading. **~r** *vt* recite
reclama|ción *f* claim; (*queja*) complaint. **~r** *vt* claim. ● *vi* appeal
reclinar *vi* lean. **~se** *vpr* lean
reclu|ir [17] *vt* shut away. **~sión** *f* seclusion; (*cárcel*) prison. **~so** *m* prisoner
recluta *m* recruit. ● *f* recruitment. **~miento** *m* recruitment; (*conjunto de reclutas*) recruits. **~r** *vt* recruit
recobrar *vt* recover. **~se** *vpr* recover
recodo *m* bend
recog|er [14] *vt* collect; pick up ‹*cosa caída*›; (*cosechar*) harvest; (*dar asilo*) shelter. **~erse** *vpr* withdraw; (*ir a casa*) go home; (*acostarse*) go to bed. **~ida** *f* collection; (*cosecha*) harvest. **~ido** *a* withdrawn; (*pequeño*) small
recolección *f* harvest
recomenda|ción *f* recommendation. **~r** [1] *vt* recommend; (*encomendar*) entrust
recomenzar [1 & 10] *vt/i* start again
recompensa *f* reward. **~r** *vt* reward
recomponer [34] *vt* mend
reconcilia|ción *f* reconciliation. **~r** *vt* reconcile. **~rse** *vpr* be reconciled
recóndito *a* hidden
reconoc|er [11] *vt* recognize; (*admitir*) acknowledge; (*examinar*) examine. **~imiento** *m* recognition; (*admisión*) acknowledgement; (*agradecimiento*) gratitude; (*examen*) examination
reconozco *vb véase* **reconocer**
reconquista *f* reconquest. **~r** *vt* reconquer; (*fig*) win back
reconsiderar *vt* reconsider
reconstitu|ir [17] *vt* reconstitute. **~yente** *m* tonic
reconstru|cción *f* reconstruction. **~ir** [17] *vt* reconstruct
récord /'rekor/ *m* record. **batir un ~** break a record
recordar [2] *vt* remember; (*hacer acordar*) remind; (*Lam, despertar*) wake up. ● *vi* remember. **que yo recuerde** as far as I remember. **si mal no recuerdo** if I remember rightly

recorr|er *vt* tour ‹*país*›; (*pasar por*) travel through; cover ‹*distancia*›; (*registrar*) look over. **~ido** *m* journey; (*itinerario*) route

recort|ado *a* jagged. **~ar** *vt* cut (out). **~e** *m* cutting (out); (*de periódico etc*) cutting

recoser *vt* mend

recostar [2] *vt* lean. **~se** *vpr* lie back

recoveco *m* bend; (*rincón*) nook

recre|ación *f* recreation. **~ar** *vt* recreate; (*divertir*) entertain. **~arse** *vpr* amuse o.s. **~ativo** *a* recreational. **~o** *m* recreation; (*escol*) break

recrimina|ción *f* recrimination. **~r** *vt* reproach

recrudecer [11] *vi* increase, worsen, get worse

recta *f* straight line

rect|angular *a* rectangular; ‹*triángulo*› right-angled. **~ángulo** *a* rectangular; ‹*triángulo*› right-angled. ● *m* rectangle

rectifica|ción *f* rectification. **~r** [7] *vt* rectify

rect|itud *f* straightness; (*fig*) honesty. **~o** *a* straight; (*fig, justo*) fair; (*fig, honrado*) honest. ● *m* rectum. **todo ~o** straight on

rector *a* governing. ● *m* rector

recuadro *m* (*en periódico*) box

recubrir [*pp* **recubierto**] *vt* cover

recuerdo *m* memory; (*regalo*) souvenir. ● *vb véase* **recordar**. **~s** *mpl* (*saludos*) regards

recupera|ción *f* recovery. **~r** *vt* recover. **~rse** *vpr* recover. **~r el tiempo perdido** make up for lost time

recur|rir *vi*. **~rir a** resort to ‹*cosa*›; turn to ‹*persona*›. **~so** *m* resort; (*medio*) resource; (*jurid*) appeal. **~sos** *mpl* resources

recusar *vt* refuse

rechaz|ar [10] *vt* repel; reflect ‹*luz*›; (*no aceptar*) refuse; (*negar*) deny. **~o** *m*. **de ~o** on the rebound; (*fig*) consequently

rechifla *f* booing; (*burla*) derision

rechinar *vi* squeak; ‹*madera etc*› creak; ‹*dientes*› grind

rechistar *vt* murmur. **sin ~** without saying a word

rechoncho *a* stout

red *f* network; (*malla*) net; (*para equipaje*) luggage rack; (*fig, engaño*) trap

redac|ción *f* editing; (*conjunto de redactores*) editorial staff; (*oficina*) editorial office; (*escol, univ*) essay. **~tar** *vt* write. **~tor** *m* writer; (*de periódico*) editor

redada *f* casting; (*de policía*) raid

redecilla *f* small net; (*para el pelo*) hairnet

rededor *m*. **al ~, en ~** around

reden|ción *f* redemption. **~tor** *a* redeeming

redil *f* sheepfold

redimir *vt* redeem

rédito *m* interest

redoblar *vt* redouble; (*doblar*) bend back

redoma *f* flask

redomado *a* sly

redond|a *f* (*de imprenta*) roman (type); (*mus*) semibreve (*Brit*), whole note (*Amer*). **~amente** *adv* (*categóricamente*) flatly. **~ear** *vt* round off. **~el** *m* circle; (*de plaza de toros*) arena. **~o** *a* round; (*completo*) complete. ● *m* circle. **a la ~a** around. **en ~o** round; (*categóricamente*) flatly

reduc|ción *f* reduction. **~ido** *a* reduced; (*limitado*) limited; (*pequeño*) small; ‹*precio*› low. **~ir** [47] *vt* reduce. **~irse** *vpr* be reduced; (*fig*) amount

reduje *vb véase* **reducir**

redundan|cia *f* redundancy. **~te** *a* redundant

reduplicar [7] *vt* (*aumentar*) redouble

reduzco *vb véase* **reducir**

reedificar [7] *vt* reconstruct

reembols|ar *vt* reimburse. **~o** *m* repayment. **contra ~o** cash on delivery

reemplaz|ar [10] *vt* replace. **~o** *m* replacement

reemprender *vt* start again

reenviar [20] *vt*, **reexpedir** [5] *vt* forward

referencia *f* reference; (*información*) report. **con ~ a** with reference to. **hacer ~ a** refer to

referéndum *m* (*pl* **referéndums**) referendum

referir [4] *vt* tell; (*remitir*) refer. **~se** *vpr* refer. **por lo que se refiere a** as regards

refiero *vb véase* **referir**

refilón. **de ~** obliquely

refin|amiento *m* refinement. **~ar** *vt* refine. **~ería** *f* refinery

reflector *m* reflector; (*proyector*) searchlight
reflej|ar *vt* reflect. **~o** *a* reflected; (*med*) reflex. ● *m* reflection; (*med*) reflex; (*en el pelo*) highlights
reflexi|ón *f* reflection. **~onar** *vi* reflect. **~vo** *a* ‹*persona*› thoughtful; (*gram*) reflexive. **con ~ón** on reflection. **sin ~ón** without thinking
reflujo *m* ebb
reforma *f* reform. **~s** *fpl* (*reparaciones*) repairs. **~r** *vt* reform. **~rse** *vpr* reform
reforzar [2 & 10] *vt* reinforce
refrac|ción *f* refraction. **~tar** *vt* refract. **~tario** *a* heat-resistant
refrán *m* saying
refregar [1 & 12] *vt* rub
refrenar *vt* rein in ‹*caballo*›; (*fig*) restrain
refrendar *vt* endorse
refresc|ar [7] *vt* refresh; (*enfriar*) cool. ● *vi* get cooler. **~arse** *vpr* refresh o.s.; (*salir*) go out for a walk. **~o** *m* cold drink. **~os** *mpl* refreshments
refrigera|ción *f* refrigeration; (*aire acondicionado*) air-conditioning. **~r** *vt* refrigerate. **~dor** *m*, **~dora** *f* refrigerator
refuerzo *m* reinforcement
refugi|ado *m* refugee. **~arse** *vpr* take refuge. **~o** *m* refuge, shelter
refulgir [14] *vi* shine
refundir *vt* (*fig*) revise, rehash
refunfuñar *vi* grumble
refutar *vt* refute
regadera *f* watering-can; (*Mex, ducha*) shower
regala|damente *adv* very well. **~do** *a* as a present, free; (*cómodo*) comfortable. **~r** *vt* give; (*agasajar*) treat very well. **~rse** *vpr* indulge o.s.
regaliz *m* liquorice
regalo *m* present, gift; (*placer*) joy; (*comodidad*) comfort
regañ|adientes. **a ~adientes** reluctantly. **~ar** *vt* scold. ● *vi* moan; (*dos personas*) quarrel. **~o** *m* (*reprensión*) scolding
regar [1 & 12] *vt* water
regata *f* regatta
regate *m* dodge; (*en deportes*) dribbling. **~ar** *vt* haggle over; (*economizar*) economize on. ● *vi* haggle; (*en deportes*) dribble. **~o** *m* haggling; (*en deportes*) dribbling
regazo *m* lap
regencia *f* regency
regenerar *vt* regenerate
regente *m & f* regent; (*director*) manager
régimen *m* (*pl* **regímenes**) rule; (*pol*) regime; (*med*) diet. **~ alimenticio** diet
regimiento *m* regiment
regio *a* royal
regi|ón *f* region. **~onal** *a* regional
regir [5 & 14] *vt* rule; govern ‹*país*›; run ‹*colegio, empresa*›. ● *vi* apply, be in force
registr|ado *a* registered. **~ador** *m* recorder; (*persona*) registrar. **~ar** *vt* register; (*grabar*) record; (*examinar*) search. **~arse** *vpr* register; (*darse*) be reported. **~o** *m* (*acción de registrar*) registration; (*libro*) register; (*cosa anotada*) entry; (*inspección*) search. **~o civil** (*oficina*) register office
regla *f* ruler; (*norma*) rule; (*menstruación*) period, menstruation. **~mentación** *f* regulation. **~mentar** *vt* regulate. **~mentario** *a* obligatory. **~mento** *m* regulations. **en ~** in order. **por ~ general** as a rule
regocij|ar *vt* delight. **~arse** *vpr* be delighted. **~o** *m* delight. **~os** *mpl* festivities
regode|arse *vpr* be delighted. **~o** *m* delight
regordete *a* chubby
regres|ar *vi* return. **~ión** *f* regression. **~ivo** *a* backward. **~o** *m* return
reguer|a *f* irrigation ditch. **~o** *m* irrigation ditch; (*señal*) trail
regula|dor *m* control. **~r** *a* regular; (*mediano*) average; (*no bueno*) so-so. ● *vt* regulate; (*controlar*) control. **~ridad** *f* regularity. **con ~ridad** regularly. **por lo ~r** as a rule
rehabilita|ción *f* rehabilitation; (*en un empleo etc*) reinstatement. **~r** *vt* rehabilitate; (*al empleo etc*) reinstate
rehacer [31] *vt* redo; (*repetir*) repeat; (*reparar*) repair. **~se** *vpr* recover
rehén *m* hostage
rehogar [12] *vt* sauté
rehuir [17] *vt* avoid
rehusar *vt/i* refuse
reimpr|esión *f* reprinting. **~imir** (*pp* **reimpreso**) *vt* reprint
reina *f* queen. **~do** *m* reign. **~nte** *a* ruling; (*fig*) prevailing. **~r** *vi* reign; (*fig*) prevail

reincidir *vi* relapse, repeat an offence

reino *m* kingdom. **R~ Unido** United Kingdom

reinstaurar *vt* restore

reintegr|ar *vt* reinstate ‹*persona*›; refund ‹*cantidad*›. **~arse** *vpr* return. **~o** *m* refund

reír [51] *vi* laugh. **~se** *vpr* laugh. **~se de** laugh at. **echarse a ~** burst out laughing

reivindica|ción *f* claim. **~r** [7] *vt* claim; (*restaurar*) restore

rej|a *f* grille, grating. **~illa** *f* grille, grating; (*red*) luggage rack; (*de mimbre*) wickerwork. **entre ~as** behind bars

rejuvenecer [11] *vt/i* rejuvenate. **~se** *vpr* be rejuvenated

relaci|ón *f* relation(ship); (*relato*) tale; (*lista*) list. **~onado** *a* concerning. **~onar** *vt* relate (**con** to). **~onarse** *vpr* be connected. **bien ~onado** well-connected. **con ~ón a**, **en ~ón a** in relation to. **hacer ~ón a** refer to

relaja|ción *f* relaxation; (*aflojamiento*) slackening. **~do** *a* loose. **~r** *vt* relax; (*aflojar*) slacken. **~rse** *vpr* relax

relamerse *vpr* lick one's lips

relamido *a* overdressed

rel|ámpago *m* (flash of) lightning. **~ampaguear** *vi* thunder; (*fig*) sparkle

relatar *vt* tell, relate

relativ|idad *f* relativity. **~o** *a* relative. **en lo ~o a** in relation to

relato *m* tale; (*informe*) report

relegar [12] *vt* relegate. **~ al olvido** forget about

relev|ante *a* outstanding. **~ar** *vt* relieve; (*substituir*) replace. **~o** *m* relief. **carrera** *f* **de ~os** relay race

relieve *m* relief; (*fig*) importance. **de ~** important. **poner de ~** emphasize

religi|ón *f* religion. **~osa** *f* nun. **~oso** *a* religious. ● *m* monk

relinch|ar *vi* neigh. **~o** *m* neigh

reliquia *f* relic

reloj *m* clock; (*de bolsillo o pulsera*) watch. **~ de caja** grandfather clock. **~ de pulsera** wrist-watch. **~ de sol** sundial. **~ despertador** alarm clock. **~ería** *f* watchmaker's (shop). **~ero** *m* watchmaker

reluci|ente *a* shining. **~r** [11] *vi* shine; (*destellar*) sparkle

relumbrar *vi* shine

rellano *m* landing

rellen|ar *vt* refill; (*culin*) stuff; fill in ‹*formulario*›. **~o** *a* full up; (*culin*) stuffed. ● *m* filling; (*culin*) stuffing

remach|ar *vt* rivet; (*fig*) drive home. **~e** *m* rivet

remangar [12] *vt* roll up

remanso *m* pool; (*fig*) haven

remar *vi* row

remat|ado *a* (*total*) complete; ‹*niño*› very naughty. **~ar** *vt* finish off; (*agotar*) use up; (*com*) sell off cheap. **~e** *m* end; (*fig*) finishing touch. **de ~e** completely

remedar *vt* imitate

remedi|ar *vt* remedy; (*ayudar*) help; (*poner fin a*) put a stop to; (*fig, resolver*) solve. **~o** *m* remedy; (*fig*) solution. **como último ~o** as a last resort. **no hay más ~o** there's no other way. **no tener más ~o** have no choice

remedo *m* imitation

rem|endar [1] *vt* repair. **~iendo** *m* patch; (*fig, mejora*) improvement

remilg|ado *a* fussy; (*afectado*) affected. **~o** *m* fussiness; (*afectación*) affectation

reminiscencia *f* reminiscence

remirar *vt* look again at

remisión *f* sending; (*referencia*) reference; (*perdón*) forgiveness

remiso *a* remiss

remit|e *m* sender's name and address. **~ente** *m* sender. **~ir** *vt* send; (*referir*) refer. ● *vi* diminish

remo *m* oar

remoj|ar *vt* soak; (*fig, fam*) celebrate. **~o** *m* soaking. **poner a ~o** soak

remolacha *f* beetroot. **~ azucarera** sugar beet

remolcar [7] *vt* tow

remolino *m* swirl; (*de aire etc*) whirl; (*de gente*) throng

remolque *m* towing; (*cabo*) towrope; (*vehículo*) trailer. **a ~** on tow. **dar ~ a** tow

remontar *vt* mend. **~se** *vpr* soar; (*con tiempo*) go back to

rémora *f* (*fig*) hindrance

remord|er [2] (*fig*) worry. **~imiento** *m* remorse. **tener ~imientos** feel remorse

remoto *a* remote

remover [2] *vt* move; stir ‹*líquido*›; turn over ‹*tierra*›; (*quitar*) remove; (*fig, activar*) revive

remozar [10] *vt* rejuvenate ‹*persona*›; renovate ‹*edificio etc*›

remunera|ción *f* remuneration. **~r** *vt* remunerate

renac|er [11] *vi* be reborn; (*fig*) revive. **~imiento** *m* rebirth. **R~** Renaissance

renacuajo *m* tadpole; (*fig*) tiddler

rencilla *f* quarrel

rencor *m* bitterness. **~oso** *a* (*estar*) resentful; (*ser*) spiteful. **guardar ~ a** have a grudge against

rendi|ción *f* surrender. **~do** *a* submissive; (*agotado*) exhausted

rendija *f* crack

rendi|miento *m* efficiency; (*com*) yield. **~r** [5] *vt* yield; (*vencer*) defeat; (*agotar*) exhaust; pay ‹*homenaje*›. ● *vi* pay; (*producir*) produce. **~rse** *vpr* surrender

renega|do *a & m* renegade. **~r** [1 & 12] *vt* deny. ● *vi* grumble. **~r de** renounce ‹*fe etc*›; disown ‹*personas*›

RENFE *abrev* (*Red Nacional de los Ferrocarriles Españoles*) Spanish National Railways

renglón *m* line; (*com*) item. **a ~ seguido** straight away

reno *m* reindeer

renombr|ado *a* renowned. **~e** *m* renown

renova|ción *f* renewal; (*de edificio*) renovation; (*de cuarto*) decorating. **~r** *vt* renew; renovate ‹*edificio*›; decorate ‹*cuarto*›

rent|a *f* income; (*alquiler*) rent; (*deuda*) national debt. **~able** *a* profitable. **~ar** *vt* produce, yield; (*LAm, alquilar*) rent, hire. **~a vitalicia** (life) annuity. **~ista** *m & f* person of independent means

renuncia *f* renunciation. **~r** *vi*. **~r a** renounce, give up

reñi|do *a* hard-fought. **~r** [5 & 22] *vt* tell off. ● *vi* quarrel. **estar ~do con** be incompatible with ‹*cosas*›; be on bad terms with ‹*personas*›

reo *m & f* culprit; (*jurid*) accused. **~ de Estado** person accused of treason. **~ de muerte** prisoner sentenced to death

reojo. **mirar de ~** look out of the corner of one's eye at; (*fig*) look askance at

reorganizar [10] *vt* reorganize

repanchigarse [12] *vpr*, **repantigarse** [12] *vpr* sprawl out

repar|ación *f* repair; (*acción*) repairing; (*fig, compensación*) reparation. **~ar** *vt* repair; (*fig*) make amends for; (*notar*) notice. ● *vi*. **~ar en** notice; (*hacer caso de*) pay attention to. **~o** *m* fault; (*objeción*) objection. **poner ~os** raise objections

repart|ición *f* division. **~idor** *m* delivery man. **~imiento** *m* distribution. **~ir** *vt* distribute, share out; deliver ‹*cartas, leche etc*›; hand out ‹*folleto, premio*›. **~o** *m* distribution; (*de cartas, leche etc*› delivery; (*actores*) cast

repas|ar *vt* go over; check ‹*cuenta*›; revise ‹*texto*›; (*leer a la ligera*) glance through; (*coser*) mend. ● *vi* go back. **~o** *m* revision; (*de ropa*) mending. **dar un ~o** look through

repatria|ción *f* repatriation. **~r** *vt* repatriate

repecho *m* steep slope

repele|nte *a* repulsive. **~r** *vt* repel

repensar [1] *vt* reconsider

repent|e. **de ~** suddenly. **~ino** *a* sudden

repercu|sión *f* repercussion. **~tir** *vi* reverberate; (*fig*) have repercussions (**en** on)

repertorio *m* repertoire; (*lista*) index

repeti|ción *f* repetition; (*mus*) repeat. **~damente** *adv* repeatedly. **~r** [5] *vt* repeat; (*imitar*) copy; ● *vi*. **~r de** have a second helping of. **¡que se repita!** encore!

repi|car [7] *vt* ring ‹*campanas*›. **~que** *m* peal

repisa *f* shelf. **~ de chimenea** mantlepiece

repito *vb véase* **repetir**

replegarse [1 & 12] *vpr* withdraw

repleto *a* full up

réplica *a* answer; (*copia*) replica

replicar [7] *vi* answer

repliegue *m* crease; (*mil*) withdrawal

repollo *m* cabbage

reponer [34] *vt* replace; revive ‹*obra de teatro*›; (*contestar*) reply. **~se** *vpr* recover

report|aje *m* report. **~ero** *m* reporter

repos|ado *a* quiet; (*sin prisa*) unhurried. **~ar** *vi* rest. **~arse** *vpr* settle. **~o** *m* rest

repost|ar *vt* replenish; refuel ‹*avión*›; fill up ‹*coche etc*›. **~ería** *f* cake shop

repren|der *vt* reprimand. **~sible** *a* reprehensible

represalia *f* reprisal. **tomar ~s** retaliate

representa|ción *f* representation; (*en el teatro*) performance. **en ~ción de** representing. **~nte** *m* representative; (*actor*) actor. ● *f* representative; (*actriz*) actress. **~r** *vt* represent; perform ‹*obra de teatro*›; play ‹*papel*›; (*aparentar*) look. **~rse** *vpr* imagine. **~tivo** *a* representative

represi|ón *f* repression. **~vo** *a* repressive

reprimenda *f* reprimand

reprimir *vt* supress. **~se** *vpr* stop o.s.

reprobar [2] *vt* condemn; reproach ‹*persona*›

réprobo *a* & *m* reprobate

reproch|ar *vt* reproach. **~e** *m* reproach

reproduc|ción *f* reproduction. **~ir** [47] *vt* reproduce. **~tor** *a* reproductive

reptil *m* reptile

rep|ública *f* republic. **~ublicano** *a* & *m* republican

repudiar *vt* repudiate

repuesto *m* store; (*auto*) spare (part). **de ~** in reserve

repugna|ncia *f* disgust. **~nte** *a* repugnant. **~r** *vt* disgust

repujar *vt* emboss

repuls|a *f* rebuff. **~ión** *f* repulsion. **~ivo** *a* repulsive

reputa|ción *f* reputation. **~do** *a* reputable. **~r** *vt* consider

requebrar [1] *vt* flatter

requemar *vt* scorch; (*culin*) burn; tan ‹*piel*›

requeri|miento *m* request; (*jurid*) summons. **~r** [4] *vt* need; (*pedir*) ask

requesón *m* cottage cheese

requete... *pref* extremely

requiebro *m* compliment

réquiem *m* (*pl* **réquiems**) *m* requiem

requis|a *f* inspection; (*mil*) requisition. **~ar** *vt* requisition. **~ito** *m* requirement

res *f* animal. **~ lanar** sheep. **~ vacuna** (*vaca*) cow; (*toro*) bull; (*buey*) ox. **carne de ~** (*Mex*) beef

resabido *a* well-known; ‹*persona*› pedantic

resabio *m* (unpleasant) after-taste; (*vicio*) bad habit

resaca *f* undercurrent; (*después de beber alcohol*) hangover

resaltar *vi* stand out. **hacer ~** emphasize

resarcir [9] *vt* repay; (*compensar*) compensate. **~se** *vpr* make up for

resbal|adizo *a* slippery. **~ar** *vi* slip; (*auto*) skid; ‹*líquido*› trickle. **~arse** *vpr* slip; (*auto*) skid; ‹*líquido*› trickle. **~ón** *m* slip; (*de vehículo*) skid

rescat|ar *vt* ransom; (*recuperar*) recapture; (*fig*) recover. **~e** *m* ransom; (*recuperación*) recapture; (*salvamento*) rescue

rescindir *vt* cancel

rescoldo *m* embers

resecar [7] *vt* dry up; (*med*) remove. **~se** *vpr* dry up

resenti|do *a* resentful. **~miento** *m* resentment. **~rse** *vpr* feel the effects; (*debilitarse*) be weakened; (*ofenderse*) take offence (**de** at)

reseña *f* account; (*en periódico*) report, review. **~r** *vt* describe; (*en periódico*) report on, review

resero *m* (*Arg*) herdsman

reserva *f* reservation; (*provisión*) reserve(s). **~ción** *f* reservation. **~do** *a* reserved. **~r** *vt* reserve; (*guardar*) keep, save. **~rse** *vpr* save o.s. **a ~ de** except for. **a ~ de que** unless. **de ~** in reserve

resfria|do *m* cold; (*enfriamiento*) chill. **~r** *vt*. **~r a uno** give s.o. a cold. **~rse** *vpr* catch a cold; (*fig*) cool off

resguard|ar *vt* protect. **~arse** *vpr* protect o.s.; (*fig*) take care. **~o** *m* protection; (*garantía*) guarantee; (*recibo*) receipt

resid|encia *f* residence; (*univ*) hall of residence, dormitory (*Amer*); (*de ancianos etc*) home. **~encial** *a* residential. **~ente** *a* & *m* & *f* resident. **~ir** *vi* reside; (*fig*) lie

residu|al *a* residual. **~o** *m* remainder. **~os** *mpl* waste

resigna|ción *f* resignation. **~damente** *adv* with resignation. **~r** *vt* resign. **~rse** *vpr* resign o.s. (**a**, **con** to)

resina *f* resin

resist|encia *f* resistence. **~ente** *a* resistent. **~ir** *vt* resist; (*soportar*) bear. ● *vi* resist. **oponer ~encia a** resist

resma *f* ream

resobado *a* trite

resol|ución *f* resolution; (*solución*) solution; (*decisión*) decision. **~ver**

[2] (*pp* **resuelto**) resolve; solve ‹*problema etc*›. **~verse** *vpr* be solved; (*resultar bien*) work out; (*decidirse*) make up one's mind
resollar [2] *vi* breathe heavily. **sin ~** without saying a word
resona|ncia *f* resonance. **~nte** *a* resonant; (*fig*) resounding. **~r** [2] *vi* resound. **tener ~ncia** cause a stir
resopl|ar *vi* puff; (*por enfado*) snort; (*por cansancio*) pant. **~ido** *m* heavy breathing; (*de enfado*) snort; (*de cansancio*) panting
resorte *m* spring. **tocar (todos los) ~s** (*fig*) pull strings
respald|ar *vt* back; (*escribir*) endorse. **~arse** *vpr* lean back. **~o** *m* back
respect|ar *vi* concern. **~ivo** *a* respective. **~o** *m* respect. **al ~o** on the matter. **(con) ~o a** as regards. **en/por lo que ~ a** as regards
respet|able *a* respectable. ● *m* audience. **~ar** *vt* respect. **~o** *m* respect. **~uoso** *a* respectful. **de ~o** best. **faltar al ~o a** be disrespectful to. **hacerse ~ar** command respect
respingo *m* start
respir|ación *f* breathing; (*med*) respiration; (*ventilación*) ventilation. **~ador** *a* respiratory. **~ar** *vi* breathe; (*fig*) breathe a sigh of relief. **no ~ar** (*no hablar*) not say a word. **~o** *m* breathing; (*fig*) rest
respland|ecer [11] *vi* shine. **~eciente** *a* shining. **~or** *m* brilliance; (*de llamas*) glow
responder *vi* answer; (*replicar*) answer back; (*fig*) reply, respond. **~ de** answer for
responsab|ilidad *f* responsibility. **~le** *a* responsible. **hacerse ~le de** assume responsibilty for
respuesta *f* reply, answer
resquebra|dura *f* crack. **~jar** *vt* crack. **~jarse** *vpr* crack
resquemor *m* (*fig*) uneasiness
resquicio *m* crack; (*fig*) possibility
resta *f* subtraction
restablecer [11] *vt* restore. **~se** *vpr* recover
restallar *vi* crack
restante *a* remaining. **lo ~** the rest
restar *vt* take away; (*substraer*) subtract. ● *vi* be left
restaura|ción *f* restoration. **~nte** *m* restaurant. **~r** *vt* restore
restitu|ción *f* restitution. **~ir** [17] *vt* return; (*restaurar*) restore
resto *m* rest, remainder; (*en matemática*) remainder. **~s** *mpl* remains; (*de comida*) leftovers
restorán *m* restaurant
restregar [1 & 12] *vt* rub
restri|cción *f* restriction. **~ngir** [14] *vt* restrict, limit
resucitar *vt* resuscitate; (*fig*) revive. ● *vi* return to life
resuelto *a* resolute
resuello *m* breath; (*respiración*) breathing
resulta|do *m* result. **~r** *vi* result; (*salir*) turn out; (*ser*) be; (*ocurrir*) happen; (*costar*) come to
resum|en *m* summary. **~ir** *vt* summarize; (*recapitular*) sum up; (*abreviar*) abridge. **en ~en** in short
resur|gir [14] *vi* reappear; (*fig*) revive. **~gimiento** *m* resurgence. **~rección** *f* resurrection
retaguardia *f* (*mil*) rearguard
retahíla *f* string
retal *m* remnant
retama *f*, **retamo** *m* (*LAm*) broom
retar *vt* challenge
retardar *vt* slow down; (*demorar*) delay
retazo *m* remnant; (*fig*) piece, bit
retemblar [1] *vi* shake
rete... *pref* extremely
reten|ción *f* retention. **~er** [40] *vt* keep; (*en la memoria*) retain; (*no dar*) withhold
reticencia *f* insinuation; (*reserva*) reticence, reluctance
retina *f* retina
retintín *m* ringing. **con ~** (*fig*) sarcastically
retir|ada *f* withdrawal. **~ado** *a* secluded; (*jubilado*) retired. **~ar** *vt* move away; (*quitar*) remove; withdraw ‹*dinero*›; (*jubilar*) pension off. **~arse** *vpr* draw back; (*mil*) withdraw; (*jubilarse*) retire; (*acostarse*) go to bed. **~o** *m* retirement; (*pensión*) pension; (*lugar apartado*) retreat
reto *m* challenge
retocar [7] *vt* retouch
retoño *m* shoot
retoque *m* (*acción*) retouching; (*efecto*) finishing touch
retorc|er [2 & 9] *vt* twist; wring ‹*ropa*›. **~erse** *vpr* get twisted up; (*de dolor*) writhe. **~imiento** *m* twisting; (*de ropa*) wringing
retóric|a *f* rhetoric; (*grandilocuencia*) grandiloquence. **~o** *m* rhetorical

retorn|ar *vt/i* return. **~o** *m* return

retortijón *m* twist; (*de tripas*) stomach cramp

retoz|ar [10] *vi* romp, frolic. **~ón** *a* playful

retractar *vt* retract. **~se** *vpr* retract

retra|er [41] *vt* retract. **~erse** *vpr* withdraw. **~ído** *a* retiring

retransmitir *vt* relay

retras|ado *a* behind; ‹*reloj*› slow; (*poco desarrollado*) backward; (*anticuado*) old-fashioned; (*med*) mentally retarded. **~ar** *vt* delay; put back ‹*reloj*›; (*retardar*) slow down. ● *vi* fall behind; ‹*reloj*› be slow. **~arse** *vpr* be behind; ‹*reloj*› be slow. **~o** *m* delay; (*poco desarrollo*) backwardness; (*de reloj*) slowness. **~os** *mpl* arrears. **con 5 minutos de ~o** 5 minutes late. **traer ~o** be late

retrat|ar *vt* paint a portrait of; (*foto*) photograph; (*fig*) protray. **~ista** *m & f* portrait painter. **~o** *m* portrait; (*fig, descripción*) description. **ser el vivo ~o de** be the living image of

retreparse *vpr* lean back

retreta *f* retreat

retrete *m* toilet

retribu|ción *f* payment. **~ir** [17] *vt* pay

retroce|der *vi* move back; (*fig*) back down. **~so** *m* backward movement; (*de arma de fuego*) recoil; (*med*) relapse

retrógrado *a & m* (*pol*) reactionary

retropropulsión *f* jet propulsion

retrospectivo *a* retrospective

retrovisor *m* rear-view mirror

retumbar *vt* echo; ‹*trueno etc*› boom

reuma *m*, **reúma** *m* rheumatism

reum|ático *a* rheumatic. **~atismo** *m* rheumatism

reuni|ón *f* meeting; (*entre amigos*) reunion. **~r** [23] *vt* join together; (*recoger*) gather (together). **~rse** *vpr* join together; ‹*personas*› meet

rev|álida *f* final exam. **~alidar** *vt* confirm; (*escol*) take an exam in

revancha *f* revenge. **tomar la ~** get one's own back

revela|ción *f* revelation. **~do** *m* developing. **~dor** *a* revealing. **~r** *vt* reveal; (*foto*) develop

revent|ar [1] *vi* burst; (*tener ganas*) be dying to. **~arse** *vpr* burst. **~ón** *m* burst; (*auto*) puncture

reverbera|ción *f* (*de luz*) reflection; (*de sonido*) reverberation. **~r** *vi* ‹*luz*› be reflected; ‹*sonido*› reverberate

reveren|cia *f* reverence; (*muestra de respeto*) bow; (*muestra de respeto de mujer*) curtsy. **~ciar** *vt* revere. **~do** *a* respected; (*relig*) reverend. **~te** *a* reverent

revers|ible *a* reversible. **~o** *m* reverse

revertir [4] *vi* revert

revés *m* wrong side; (*desgracia*) misfortune; (*en deportes*) backhand. **al ~** the other way round; (*con lo de arriba abajo*) upside down; (*con lo de dentro fuera*) inside out

revesti|miento *m* coating. **~r** [5] *vt* cover; put on ‹*ropa*›; (*fig*) take on

revis|ar *vt* check; overhaul ‹*mecanismo*›; service ‹*coche etc*›. **~ión** *f* check(ing); (*inspección*) inspection; (*de coche etc*) service. **~or** *m* inspector

revist|a *f* magazine; (*inspección*) inspection; (*artículo*) review; (*espectáculo*) revue. **~ero** *m* critic; (*mueble*) magazine rack. **pasar ~a a** inspect

revivir *vi* come to life again

revocar [7] *vt* revoke; whitewash ‹*pared*›

revolcar [2 & 7] *vt* knock over. **~se** *vpr* roll

revolotear *vi* flutter

revoltijo *m*, **revoltillo** *m* mess. **~ de huevos** scrambled eggs

revoltoso *a* rebellious; ‹*niño*› naughty

revoluci|ón *f* revolution. **~onar** *vt* revolutionize. **~onario** *a & m* revolutionary

revolver [2, *pp* **revuelto**] *vt* mix; stir ‹*líquido*›; (*desordenar*) mess up; (*pol*) stir up. **~se** *vpr* turn round. **~se contra** turn on

revólver *m* revolver

revoque *m* (*con cal*) whitewashing

revuelo *m* fluttering; (*fig*) stir

revuelt|a *f* turn; (*de calle etc*) bend; (*motín*) revolt; (*conmoción*) disturbance. **~o** *a* mixed up; ‹*líquido*› cloudy; ‹*mar*› rough; ‹*tiempo*› unsettled; ‹*huevos*› scrambled

rey *m* king. **~es** *mpl* king and queen

reyerta *f* quarrel

rezagarse [12] *vpr* fall behind

rez|ar [10] *vt* say. ● *vi* pray; (*decir*) say. **~o** *m* praying; (*oración*) prayer

rezongar [12] *vi* grumble

rezumar *vt/i* ooze
ría *f* estuary
riachuelo *m* stream
riada *f* flood
ribera *f* bank
ribete *m* border; (*fig*) embellishment
ricino *m*. **aceite de ~** castor oil
rico *a* rich; (*culin, fam*) delicious. • *m* rich person
rid|ículo *a* ridiculous. **~iculizar** [10] *vt* ridicule
riego *m* watering; (*irrigación*) irrigation
riel *m* rail
rienda *f* rein
riesgo *m* risk. **a ~ de** at the risk of. **correr (el) ~ de** run the risk of
rifa *f* raffle. **~r** *vt* raffle. **~rse** *vpr* (*fam*) quarrel over
rifle *m* rifle
rigidez *f* rigidity; (*fig*) inflexibility
rígido *a* rigid; (*fig*) inflexible
rig|or *m* strictness; (*exactitud*) exactness; (*de clima*) severity. **~uroso** *a* rigorous. **de ~or** compulsory. **en ~or** strictly speaking
rima *f* rhyme. **~r** *vt/i* rhyme
rimbombante *a* resounding; ‹*lenguaje*› pompous; (*fig, ostentoso*) showy
rimel *m* mascara
rincón *m* corner
rinoceronte *m* rhinoceros
riña *f* quarrel; (*pelea*) fight
riñ|ón *m* kidney. **~onada** *f* loin; (*guiso*) kidney stew
río *m* river; (*fig*) stream. • *vb véase* **reír**. **~ abajo** downstream. **~ arriba** upstream
rioja *m* Rioja wine
riqueza *f* wealth; (*fig*) richness. **~s** *fpl* riches
riquísimo *a* delicious
risa *f* laugh. **desternillarse de ~** split one's sides laughing. **la ~** laughter
risco *m* cliff
ris|ible *a* laughable. **~otada** *f* guffaw
ristra *f* string
risueño *a* smiling; (*fig*) happy
rítmico *a* rhythmic(al)
ritmo *m* rhythm; (*fig*) rate
rit|o *m* rite; (*fig*) ritual. **~ual** *a* & *m* ritual. **de ~ual** customary
rival *a* & *m* & *f* rival. **~idad** *f* rivalry. **~izar** [10] *vi* rival
riz|ado *a* curly. **~ar** [10] *vt* curl; ripple ‹*agua*›. **~o** *m* curl; (*en agua*) ripple. **~oso** *a* curly
róbalo *m* bass
robar *vt* steal ‹*cosa*›; rob ‹*persona*›; (*raptar*) kidnap
roble *m* oak (tree)
roblón *m* rivet
robo *m* theft; (*fig, estafa*) robbery
robot (*pl* **robots**) *m* robot
robust|ez *f* strength. **~o** *a* strong
roca *f* rock
roce *m* rubbing; (*toque ligero*) touch; (*señal*) mark; (*fig, entre personas*) contact
rociar [20] *vt* spray
rocín *m* nag
rocío *m* dew
rodaballo *m* turbot
rodado *m* (*Arg, vehículo*) vehicle
rodaja *f* disc; (*culin*) slice
roda|je *m* (*de película*) shooting; (*de coche*) running in. **~r** [2] *vt* shoot ‹*película*›; run in ‹*coche*›; (*recorrer*) travel. • *vi* roll; ‹*coche*› run; (*hacer una película*) shoot
rode|ar *vt* surround. **~arse** *vpr* surround o.s. (**de** with). **~o** *m* long way round; (*de ganado*) round-up. **andar con ~os** beat about the bush. **sin ~os** plainly
rodill|a *f* knee. **~era** *f* knee-pad. **de ~as** kneeling
rodillo *m* roller; (*culin*) rolling-pin
rododendro *m* rhododendron
rodrigón *m* stake
roe|dor *m* rodent. **~r** [37] *vt* gnaw
rogar [2 & 12] *vt/i* ask; (*relig*) pray. **se ruega a los Sres pasajeros...** passengers are requested.... **se ruega no fumar** please do not smoke
roj|ete *m* rouge. **~ez** *f* redness. **~izo** *a* reddish. **~o** *a* & *m* red. **ponerse ~o** blush
roll|izo *a* round; ‹*persona*› plump. **~o** *m* roll; (*de cuerda*) coil; (*culin, rodillo*) rolling-pin; (*fig, pesadez, fam*) bore
romance *a* Romance. • *m* Romance language; (*poema*) romance. **hablar en ~** speak plainly
rom|ánico *a* Romanesque; ‹*lengua*› Romance. **~ano** *a* & *m* Roman. **a la ~ana** (*culin*) (deep-)fried in batter
rom|anticismo *m* romanticism. **~ántico** *a* romantic
romería *f* pilgrimage
romero *m* rosemary
romo *a* blunt; ‹*nariz*› snub; (*fig, torpe*) dull
rompe|cabezas *m invar* puzzle; (*con tacos de madera*) jigsaw (puzzle).

~**nueces** *m invar* nutcrackers. ~**olas** *m invar* breakwater

romp|er (*pp* **roto**) *vt* break; break off ‹*relaciones etc*›. ● *vi* break; ‹*sol*› break through. ~**erse** *vpr* break. ~**er a** burst out. ~**imiento** *m* (*de relaciones etc*) breaking off

ron *m* rum

ronc|ar [7] *vi* snore. ~**o** *a* hoarse

roncha *f* lump; (*culin*) slice

ronda *f* round; (*patrulla*) patrol; (*carretera*) ring road. ~**lla** *f* group of serenaders; (*invención*) story. ~**r** *vt/i* patrol

rondón. de ~ unannounced

ronquedad *f*, **ronquera** *f* hoarseness

ronquido *m* snore

ronronear *vi* purr

ronzal *m* halter

roñ|a *f* (*suciedad*) grime. ~**oso** *a* dirty; (*oxidado*) rusty; (*tacaño*) mean

rop|a *f* clothes, clothing. ~**a blanca** linen; (*ropa interior*) underwear. ~**a de cama** bedclothes. ~**a hecha** ready-made clothes. ~**a interior** underwear. ~**aje** *m* robes; (*excesivo*) heavy clothing. ~**ero** *m* wardrobe

ros|a *a invar* pink. ● *f* rose; (*color*) pink. ~**áceo** *a* pink. ~**ado** *a* rosy. ● *m* (*vino*) rosé. ~**al** *m* rose-bush

rosario *m* rosary; (*fig*) series

rosbif *m* roast beef

rosc|a *f* coil; (*de tornillo*) thread; (*de pan*) roll. ~**o** *m* roll

rosetón *m* rosette

rosquilla *f* doughnut; (*oruga*) grub

rostro *m* face

rota|ción *f* rotation. ~**tivo** *a* rotary

roto *a* broken

rótula *f* kneecap

rotulador *m* felt-tip pen

rótulo *m* sign; (*etiqueta*) label

rotundo *a* emphatic

rotura *f* break

roturar *vt* plough

roza *f* groove. ~**dura** *f* scratch

rozagante *a* showy

rozar [10] *vt* rub against; (*ligeramente*) brush against; (*ensuciar*) dirty; (*fig*) touch on. ~**se** *vpr* rub; (*con otras personas*) mix

Rte. *abrev* (*Remite*(*nte*)) sender

rúa *f* (small) street

rubéola *f* German measles

rubí *m* ruby

rubicundo *a* ruddy

rubio *a* ‹*pelo*› fair; ‹*persona*› fair-haired; ‹*tabaco*› Virginian

rubor *m* blush; (*fig*) shame. ~**izado** *a* blushing; (*fig*) ashamed. ~**izar** [10] *vt* make blush. ~**izarse** *vpr* blush

rúbrica *f* red mark; (*de firma*) flourish; (*título*) heading

rudeza *f* roughness

rudiment|al *a* rudimentary. ~**os** *mpl* rudiments

rudo *a* rough; (*sencillo*) simple

rueda *f* wheel; (*de mueble*) castor; (*de personas*) ring; (*culin*) slice. ~ **de prensa** press conference

ruedo *m* edge; (*redondel*) arena

ruego *m* request; (*súplica*) entreaty. ● *vb véase* **rogar**

rufi|án *m* pimp; (*granuja*) villain. ~**anesco** *a* roguish

rugby *m* Rugby

rugi|do *m* roar. ~**r** [14] *vi* roar

ruibarbo *m* rhubarb

ruido *m* noise; (*alboroto*) din; (*escándalo*) commotion. ~**so** *a* noisy; (*fig*) sensational

ruin *a* despicable; (*tacaño*) mean

ruina *f* ruin; (*colapso*) collapse

ruindad *f* meanness

ruinoso *a* ruinous

ruiseñor *m* nightingale

ruleta *f* roulette

rulo *m* (*culin*) rolling-pin; (*del pelo*) curler

Rumania *f* Romania

rumano *a & m* Romanian

rumba *f* rumba

rumbo *m* direction; (*fig*) course; (*fig, generosidad*) lavishness. ~**so** *a* lavish. **con** ~ **a** in the direction of. **hacer** ~ **a** head for

rumia|nte *a & m* ruminant. ~**r** *vt* chew; (*fig*) chew over. ● *vi* ruminate

rumor *m* rumour; (*ruido*) murmur. ~**earse** *vpr* be rumoured. ~**oso** *a* murmuring

runr|ún *m* rumour; (*ruido*) murmur. ~**unearse** *vpr* be rumoured

ruptura *f* break; (*de relaciones etc*) breaking off

rural *a* rural

Rusia *f* Russia

ruso *a & m* Russian

rústico *a* rural; (*de carácter*) coarse. **en rústica** paperback

ruta *f* route; (*camino*) road; (*fig*) course

rutilante *a* shining

rutina *f* routine. ~**rio** *a* routine

S

S.A. *abrev* (*Sociedad Anónima*) Ltd, Limited, plc, Public Limited Company
sábado *m* Saturday
sabana *f* (*esp LAm*) savannah
sábana *f* sheet
sabandija *f* bug
sabañón *m* chilblain
sabático *a* sabbatical
sab|elotodo *m* & *f invar* know-all (*fam*). **~er** [38] *vt* know; (*ser capaz de*) be able to, know how to; (*enterarse de*) learn. ● *vi*. **~er a** taste of. **~er** *m* knowledge. **~ido** *a* well-known. **~iduría** *f* wisdom; (*conocimientos*) knowledge. **a ~er si** I wonder if. **¡haberlo ~ido!** if only I'd known! **hacer ~er** let know. **no sé cuántos** what's-his-name. **para que lo sepas** let me tell you. **¡qué sé yo!** how should I know? **que yo sepa** as far as I know. **¿~es nadar?** can you swim? **un no sé qué** a certain sth. **¡yo qué sé!** how should I know?
sabiendas. **a ~** knowingly; (*a propósito*) on purpose
sabio *a* learned; (*prudente*) wise
sabor *m* taste, flavour; (*fig*) flavour. **~ear** *vt* taste; (*fig*) savour
sabot|aje *m* sabotage. **~eador** *m* saboteur. **~ear** *vt* sabotage
sabroso *a* tasty; (*fig, substancioso*) meaty
sabueso *m* (*perro*) bloodhound; (*fig, detective*) detective
saca|corchos *m invar* corkscrew. **~puntas** *m invar* pencil-sharpener
sacar [7] *vt* take out; put out ‹*parte del cuerpo*›; (*quitar*) remove; take ‹*foto*›; win ‹*premio*›; get ‹*billete, entrada etc*›; withdraw ‹*dinero*›; reach ‹*solución*›; draw ‹*conclusión*›; make ‹*copia*›. **~ adelante** bring up ‹*niño*›; carry on ‹*negocio*›
sacarina *f* saccharin
sacerdo|cio *m* priesthood. **~tal** *a* priestly. **~te** *m* priest
saciar *vt* satisfy
saco *m* bag; (*anat*) sac; (*LAm, chaqueta*) jacket; (*de mentiras*) pack. **~ de dormir** sleeping-bag
sacramento *m* sacrament
sacrific|ar [7] *vt* sacrifice. **~arse** *vpr* sacrifice o.s. **~io** *m* sacrifice
sacr|ilegio *m* sacrilege. **~ílego** *a* sacrilegious
sacro *a* sacred, holy. **~santo** *a* sacrosanct
sacudi|da *f* shake; (*movimiento brusco*) jolt, jerk; (*fig*) shock. **~da eléctrica** electric shock. **~r** *vt* shake; (*golpear*) beat; (*ahuyentar*) chase away. **~rse** *vpr* shake off; (*fig*) get rid of
sádico *a* sadistic. ● *m* sadist
sadismo *m* sadism
saeta *f* arrow; (*de reloj*) hand
safari *m* safari
sagaz *a* shrewd
Sagitario *m* Sagittarius
sagrado *a* sacred, holy. ● *m* sanctuary
Sahara *m*, **Sáhara** /'saxara/ *m* Sahara
sainete *m* short comedy
sal *f* salt
sala *f* room; (*en teatro*) house. **~ de espectáculos** concert hall, auditorium. **~ de espera** waiting-room. **~ de estar** living-room. **~ de fiestas** nightclub
sala|do *a* salty; ‹*agua del mar*› salt; (*vivo*) lively; (*encantador*) cute; (*fig*) witty. **~r** *vt* salt
salario *m* wages
salazón *f* (*carne*) salted meat; (*pescado*) salted fish
salchich|a *f* (pork) sausage. **~ón** *m* salami
sald|ar *vt* pay ‹*cuenta*›; (*vender*) sell off; (*fig*) settle. **~o** *m* balance; (*venta*) sale; (*lo que queda*) remnant
salero *m* salt-cellar
salgo *vb véase* **salir**
sali|da *f* departure; (*puerta*) exit, way out; (*de gas, de líquido*) leak; (*de astro*) rising; (*com, posibilidad de venta*) opening; (*chiste*) witty remark; (*fig*) way out. **~da de emergencia** emergency exit. **~ente** *a* projecting; (*fig*) outstanding. **~r** [52] *vi* leave; (*de casa etc*) go out; ‹*revista etc*› be published; (*resultar*) turn out; ‹*astro*› rise; (*aparecer*) appear. **~rse** *vpr* leave; ‹*recipiente, líquido etc*› leak. **~r adelante** get by. **~rse con la suya** get one's own way
saliva *f* saliva
salmo *m* psalm
salm|ón *m* salmon. **~onete** *m* red mullet
salmuera *f* brine
salón *m* lounge, sitting-room. **~ de actos** assembly hall. **~ de fiestas** dancehall

salpica|dero *m* (*auto*) dashboard. **~dura** *f* splash; (*acción*) splashing. **~r** [7] *vt* splash; (*fig*) sprinkle

sals|a *f* sauce; (*para carne asada*) gravy; (*fig*) spice. **~a verde** parsley sauce. **~era** *f* sauce-boat

salt|amontes *m invar* grasshopper. **~ar** *vt* jump (over); (*fig*) miss out. • *vi* jump; (*romperse*) break; ‹*líquido*› spurt out; (*desprenderse*) come off; ‹*pelota*› bounce; (*estallar*) explode. **~eador** *m* highwayman. **~ear** *vt* rob; (*culin*) sauté. • *vi* skip through

saltimbanqui *m* acrobat

salt|o *m* jump; (*al agua*) dive. **~o de agua** waterfall. **~ón** *a* ‹*ojos*› bulging. • *m* grasshopper. **a ~os** by jumping; (*fig*) by leaps and bounds. **de un ~o** with one jump

salud *f* health; (*fig*) welfare. • *int* cheers! **~able** *a* healthy

salud|ar *vt* greet, say hello to; (*mil*) salute. **~o** *m* greeting; (*mil*) salute. **~os** *mpl* best wishes. **le ~a atentamente** (*en cartas*) yours faithfully

salva *f* salvo; (*de aplausos*) thunders

salvación *f* salvation

salvado *m* bran

Salvador *m*. **El ~** El Salvador

salvaguardia *f* safeguard

salvaje *a* ‹*planta, animal*› wild; (*primitivo*) savage. • *m & f* savage

salvamanteles *m invar* table-mat

salva|mento *m* rescue. **~r** *vt* save, rescue; (*atraversar*) cross; (*recorrer*) travel; (*fig*) overcome. **~rse** *vpr* save o.s. **~vidas** *m invar* lifebelt. **chaleco** *m* **~vidas** life-jacket

salvia *f* sage

salvo *a* safe. • *adv & prep* except (for). **~ que** unless. **~conducto** *m* safe-conduct. **a ~** out of danger. **poner a ~** put in a safe place

samba *f* samba

San *a* Saint, St. **~ Miguel** St Michael

sana|r *vt* cure. • *vi* recover. **~torio** *m* sanatorium

sanci|ón *f* sanction. **~onar** *vt* sanction

sancocho *m* (*LAm*) stew

sandalia *f* sandal

sándalo *m* sandalwood

sandía *f* water melon

sandwich /'sambitʃ/ *m* (*pl* **sandwichs, sandwiches**) sandwich

sanear *vt* drain

sangr|ante *a* bleeding; (*fig*) flagrant. **~ar** *vt/i* bleed. **~e** *f* blood. **a ~e fría** in cold blood

sangría *f* (*bebida*) sangria

sangriento *a* bloody

sangu|ijuela *f* leech. **~íneo** *a* blood

san|idad *f* health. **~itario** *a* sanitary. **~o** *a* healthy; (*seguro*) sound. **~o y salvo** safe and sound. **cortar por lo ~o** settle things once and for all

santiamén *m*. **en un ~** in an instant

sant|idad *f* sanctity. **~ificar** [7] *vt* sanctify. **~iguar** [15] *vt* make the sign of the cross over. **~iguarse** *vpr* cross o.s. **~o** *a* holy; (*delante de nombre*) Saint, St. • *m* saint; (*día*) saint's day, name day. **~uario** *m* sanctuary. **~urrón** *a* sanctimonious, hypocritical

sañ|a *f* fury; (*crueldad*) cruelty. **~oso** *a*, **~udo** *a* furious

sapo *m* toad; (*bicho, fam*) small animal, creature

saque *m* (*en tenis*) service; (*en fútbol*) throw-in; (*inicial en fútbol*) kick-off

saque|ar *vt* loot. **~o** *m* looting

sarampión *m* measles

sarape *m* (*Mex*) blanket

sarc|asmo *m* sarcasm. **~ástico** *a* sarcastic

sardana *f* Catalonian dance

sardina *f* sardine

sardo *a & m* Sardinian

sardónico *a* sardonic

sargento *m* sergeant

sarmiento *m* vine shoot

sarpullido *m* rash

sarta *f* string

sartén *f* frying-pan (*Brit*), fry-pan (*Amer*)

sastre *m* tailor. **~ría** *f* tailoring; (*tienda*) tailor's (shop)

Satanás *m* Satan

satánico *a* satanic

satélite *m* satellite

satinado *a* shiny

sátira *f* satire

satírico *a* satirical. • *m* satirist

satisf|acción *f* satisfaction. **~acer** [31] *vt* satisfy; (*pagar*) pay; (*gustar*) please; meet ‹*gastos, requisitos*›. **~acerse** *vpr* satisfy o.s.; (*vengarse*) take revenge. **~actorio** *a* satisfactory. **~echo** *a* satisfied. **~echo de sí mismo** smug

satura|ción *f* saturation. **~r** *vt* saturate

Saturno *m* Saturn

sauce *m* willow. **~ llorón** weeping willow

saúco *m* elder
savia *f* sap
sauna *f* sauna
saxofón *m*, **saxófono** *m* saxophone
saz|ón *f* ripeness; (*culin*) seasoning. **~onado** *a* ripe; (*culin*) seasoned. **~onar** *vt* ripen; (*culin*) season. **en ~ón** in season
se *pron* (*él*) him; (*ella*) her; (*Vd*) you; (*reflexivo, él*) himself; (*reflexivo, ella*) herself; (*reflexivo, ello*) itself; (*reflexivo, uno*) oneself; (*reflexivo, Vd*) yourself; (*reflexivo, ellos, ellas*) themselves; (*reflexivo, Vds*) yourselves; (*recíproco*) (to) each other. **~ dice** people say, they say, it is said (**que** that). **~ habla español** Spanish spoken
sé *vb véase* **saber** *y* **ser**
sea *vb véase* **ser**
sebo *m* tallow; (*culin*) suet
seca|dor *m* drier; (*de pelo*) hairdrier. **~nte** *a* drying. ● *m* blotting-paper. **~r** [7] *vt* dry. **~rse** *vpr* dry; ‹*río etc*› dry up; ‹*persona*› dry o.s.
sección *f* section
seco *a* dry; ‹*frutos, flores*› dried; (*flaco*) thin; ‹*respuesta*› curt; (*escueto*) plain. **a secas** just. **en ~** (*bruscamente*) suddenly. **lavar en ~** dry-clean
secre|ción *f* secretion. **~tar** *vt* secrete
secretar|ía *f* secretariat. **~io** *m* secretary
secreto *a* & *m* secret
secta *f* sect. **~rio** *a* sectarian
sector *m* sector
secuela *f* consequence
secuencia *f* sequence
secuestr|ar *vt* confiscate; kidnap ‹*persona*›; hijack ‹*avión*›. **~o** *m* seizure; (*de persona*) kidnapping; (*de avión*) hijack(ing)
secular *a* secular
secundar *vt* second, help. **~io** *a* secondary
sed *f* thirst. ● *vb véase* **ser**. **tener ~** be thirsty. **tener ~ de** (*fig*) be hungry for
seda *f* silk
sedante *a* & *m*, **sedativo** *a* & *m* sedative
sede *f* seat; (*relig*) see
sedentario *a* sedentary
sedici|ón *f* sedition. **~oso** *a* seditious
sediento *a* thirsty
sediment|ar *vi* deposit. **~arse** *vpr* settle. **~o** *m* sediment
seduc|ción *f* seduction. **~ir** [47] *vt* seduce; (*atraer*) attract. **~tor** *a* seductive. ● *m* seducer
sega|dor *m* harvester. **~dora** *f* harvester, mower. **~r** [1 & 12] *vt* reap
seglar *a* secular. ● *m* layman
segmento *m* segment
segoviano *m* person from Segovia
segrega|ción *f* segregation. **~r** [12] *vt* segregate
segui|da *f*. **en ~da** immediately. **~do** *a* continuous; (*en plural*) consecutive. ● *adv* straight; (*después*) after. **todo ~do** straight ahead. **~dor** *a* following. ● *m* follower. **~r** [5 & 13] *vt* follow. (*continuar*) continue
según *prep* according to. ● *adv* it depends; (*a medida que*) as
segundo *a* second. ● *m* second; (*culin*) second course
segur|amente *adv* certainly; (*muy probablemente*) surely. **~idad** *f* safety; (*certeza*) certainty; (*aplomo*) confidence. **~idad en sí mismo** self-confidence. **~idad social** social security. **~o** *a* safe; (*cierto*) certain, sure; (*firme*) secure; (*de fiar*) reliable. ● *adv* for certain. ● *m* insurance; (*dispositivo de seguridad*) safety device. **~o de sí mismo** self-confident. **~o de terceros** third-party insurance
seis *a* & *m* six. **~cientos** *a* & *m* six hundred
seísmo *m* earthquake
selec|ción *f* selection. **~cionar** *vt* select, choose. **~tivo** *a* selective. **~to** *a* selected; (*fig*) choice
selva *f* forest; (*jungla*) jungle
sell|ar *vt* stamp; (*cerrar*) seal. **~o** *m* stamp; (*en documento oficial*) seal; (*fig, distintivo*) hallmark
semáforo *m* semaphore; (*auto*) traffic lights; (*rail*) signal
semana *f* week. **~l** *a* weekly. **~rio** *a* & *m* weekly. **S~ Santa** Holy Week
semántic|a *f* semantics. **~o** *a* semantic
semblante *m* face; (*fig*) look
sembrar [1] *vt* sow; (*fig*) scatter
semeja|nte *a* similar; (*tal*) such. ● *m* fellow man; (*cosa*) equal. **~nza** *f* similarity. **~r** *vi* seem. **~rse** *vpr* look alike. **a ~nza de** like. **tener ~nza con** resemble

semen *m* semen. **~tal** *a* stud. ● *m* stud animal

semestr|al *a* half-yearly. **~e** *m* six months

semibreve *m* semibreve (*Brit*), whole note (*Amer*)

semic|ircular *a* semicircular. **~írculo** *m* semicircle

semicorchea *f* semiquaver (*Brit*), sixteenth note (*Amer*)

semifinal *f* semifinal

semill|a *f* seed. **~ero** *m* nursery; (*fig*) hotbed

seminario *m* (*univ*) seminar; (*relig*) seminary

sem|ita *a* Semitic. ● *m* Semite. **~ítico** *a* Semitic

sémola *f* semolina

senado *m* senate; (*fig*) assembly. **~r** *m* senator

sencill|ez *f* simplicity. **~o** *a* simple; (*uno solo*) single

senda *f*, **sendero** *m* path

sendos *apl* each

seno *m* bosom. **~ materno** womb

sensaci|ón *f* sensation. **~onal** *a* sensational

sensat|ez *f* good sense. **~o** *a* sensible

sensi|bilidad *f* sensibility. **~ble** *a* sensitive; (*notable*) notable; (*lamentable*) lamentable. **~tivo** *a* ‹*órgano*› sense

sensual *a* sensual. **~idad** *f* sensuality

senta|do *a* sitting (down). **dar algo por ~do** take something for granted. **~r** [1] *vt* place; (*establecer*) establish. ● *vi* suit; (*de medidas*) fit; ‹*comida*› agree with. **~rse** *vpr* sit (down); ‹*sedimento*› settle

sentencia *f* saying; (*jurid*) sentence. **~r** *vt* sentence

sentido *a* deeply felt; (*sincero*) sincere; (*sensible*) sensitive. ● *m* sense; (*dirección*) direction. **~ común** common sense. **~ del humor** sense of humour. **~ único** one-way. **doble ~** double meaning. **no tener ~** not make sense. **perder el ~** faint. **sin ~** unconscious; ‹*cosa*› senseless

sentim|ental *a* sentimental. **~iento** *m* feeling; (*sentido*) sense; (*pesar*) regret

sentir [4] *vt* feel; (*oír*) hear; (*lamentar*) be sorry for. ● *vi* feel; (*lamentarse*) be sorry. ● *m* (*opinión*) opinion. **~se** *vpr* feel. **lo siento** I'm sorry

seña *f* sign. **~s** *fpl* (*dirección*) address; (*descripción*) description

señal *f* sign; (*rail etc*) signal; (*telefónico*) tone; (*com*) deposit. **~ado** *a* notable. **~ar** *vt* signal; (*poner señales en*) mark; (*apuntar*) point out; ‹*manecilla, aguja*› point to; (*determinar*) fix. **~arse** *vpr* stand out. **dar ~es de** show signs of. **en ~ de** as a token of

señero *a* alone; (*sin par*) unique

señor *m* man; (*caballero*) gentleman; (*delante de nombre propio*) Mr; (*tratamiento directo*) sir. **~a** *f* lady, woman; (*delante de nombre propio*) Mrs; (*esposa*) wife; (*tratamiento directo*) madam. **~ial** *a* ‹*casa*› stately. **~ita** *f* young lady; (*delante de nombre propio*) Miss; (*tratamiento directo*) miss. **~ito** *m* young gentleman. **el ~ alcalde** the mayor. **el ~** Mr. **muy ~ mío** Dear Sir. **¡no ~!** certainly not! **ser ~ de** be master of, control

señuelo *m* lure

sepa *vb véase* **saber**

separa|ción *f* separation. **~do** *a* separate. **~r** *vt* separate; (*apartar*) move away; (*de empleo*) dismiss. **~rse** *vpr* separate; ‹*amigos*› part. **~tista** *a* & *m* & *f* separatist. **por ~do** separately

septentrional *a* north(ern)

séptico *a* septic

septiembre *m* September

séptimo *a* seventh

sepulcro *m* sepulchre

sepult|ar *vt* bury. **~ura** *f* burial; (*tumba*) grave. **~urero** *m* gravedigger

sequ|edad *f* dryness. **~ía** *f* drought

séquito *m* entourage; (*fig*) aftermath

ser [39] *vi* be. ● *m* being. **~ de** be made of; (*provenir de*) come from; (*pertenecer a*) belong to. **~ humano** human being. **a no ~ que** unless. **¡así sea!** so be it! **es más** what is more. **lo que sea** anything. **no sea que, no vaya a ~ que** in case. **o sea** in other words. **sea lo que fuere** be that as it may. **sea... sea** either... or. **siendo así que** since. **soy yo** it's me

seren|ar *vt* calm down. **~arse** *vpr* calm down; ‹*tiempo*› clear up. **~ata** *f* serenade. **~idad** *f* serenity. **~o** *a* ‹*cielo*› clear; ‹*tiempo*› fine; (*fig*) calm. ● *m* night watchman. **al ~o** in the open

seri|al *m* serial. **~e** *f* series. **fuera de ~e** (*fig, extraordinario*) special. **producción** *f* **en ~** mass production
seri|edad *f* seriousness. **~o** *a* serious; (*confiable*) reliable. **en ~o** seriously. **poco ~o** frivolous
sermón *m* sermon
serp|enteante *a* winding. **~entear** *vi* wind. **~iente** *f* snake. **~iente de cascabel** rattlesnake
serrano *a* mountain; ‹*jamón*› cured
serr|ar [1] *vt* saw. **~ín** *m* sawdust. **~ucho** *m* (hand)saw
servi|cial *a* helpful. **~cio** *m* service; (*conjunto*) set; (*aseo*) toilet. **~cio a domicilio** delivery service. **~dor** *m* servant. **~dumbre** *f* servitude; (*criados*) servants, staff. **~l** *a* servile. **su (seguro) ~dor** (*en cartas*) yours faithfully
servilleta *f* serviette, (table) napkin
servir [5] *vt* serve; (*ayudar*) help; (*en restaurante*) wait on. ● *vi* serve; (*ser útil*) be of use. **~se** *vpr* help o.s. **~se de** use. **no ~ de nada** be useless. **para ~le** at your service. **sírvase sentarse** please sit down
sesear *vi* pronounce the Spanish *c* as an *s*
sesent|a *a* & *m* sixty. **~ón** *a* & *m* sixty-year-old
seseo *m* pronunciation of the Spanish *c* as an *s*
sesg|ado *a* slanting. **~o** *m* slant; (*fig, rumbo*) turn
sesión *f* session; (*en el cine*) showing; (*en el teatro*) performance
ses|o *m* brain; (*fig*) brains. **~udo** *a* inteligent; (*sensato*) sensible
seta *f* mushroom
sete|cientos *a* & *m* seven hundred. **~nta** *a* & *m* seventy. **~ntón** *a* & *m* seventy-year-old
setiembre *m* September
seto *m* fence; (*de plantas*) hedge. **~ vivo** hedge
seudo... *pref* pseudo...
seudónimo *m* pseudonym
sever|idad *f* severity. **~o** *a* severe; ‹*disciplina, profesor etc*› strict
Sevilla *f* Seville
sevillan|as *fpl* popular dance from Seville. **~o** *m* person from Seville
sexo *m* sex
sext|eto *m* sextet. **~o** *a* sixth
sexual *a* sexual. **~idad** *f* sexuality
si *m* (*mus*) B; (*solfa*) te. ● *conj* if; (*dubitativo*) whether. **~ no** or else. **por ~ (acaso)** in case
sí *pron reflexivo* (*él*) himself; (*ella*) herself; (*ello*) itself; (*uno*) oneself; (*Vd*) yourself; (*ellos, ellas*) themselves; (*Vds*) yourselves; (*recíproco*) each other
sí *adv* yes. ● *m* consent
Siamés *a* & *m* Siamese
Sicilia *f* Sicily
sida *m* Aids
siderurgia *f* iron and steel industry
sidra *f* cider
siega *f* harvesting; (*época*) harvest time
siembra *f* sowing; (*época*) sowing time
siempre *adv* always. **~ que** if. **como ~** as usual. **de ~** (*acostumbrado*) usual. **lo de ~** the same old story. **para ~** for ever
sien *f* temple
siento *vb véase* **sentar** *y* **sentir**
sierra *f* saw; (*cordillera*) mountain range
siervo *m* slave
siesta *f* siesta
siete *a* & *m* seven
sífilis *f* syphilis
sifón *m* U-bend; (*de soda*) syphon
sigilo *m* secrecy
sigla *f* initials, abbreviation
siglo *m* century; (*época*) time, age; (*fig, mucho tiempo, fam*) ages; (*fig, mundo*) world
significa|ción *f* meaning; (*importancia*) significance. **~do** *a* (*conocido*) well-known. ● *m* meaning. **~r** [7] *vt* mean; (*expresar*) express. **~rse** *vpr* stand out. **~tivo** *a* significant
signo *m* sign. **~ de admiración** exclamation mark. **~ de interrogación** question mark
sigo *vb véase* **seguir**
siguiente *a* following, next. **lo ~** the following
sílaba *f* syllable
silb|ar *vt/i* whistle. **~ato** *m*, **~ido** *m* whistle
silenci|ador *m* silencer. **~ar** *vt* hush up. **~o** *m* silence. **~oso** *a* silent
sílfide *f* sylph
silicio *m* silicon
silo *m* silo
silueta *f* silhouette; (*dibujo*) outline
silvestre *a* wild
sill|a *f* chair; (*de montar*) saddle; (*relig*) see. **~a de ruedas** wheelchair. **~ín** *m* saddle. **~ón** *m* armchair

simb|ólico *a* symbolic(al). **~olismo** *m* symbolism. **~olizar** [10] *vt* symbolize
símbolo *m* symbol
sim|etría *f* symmetry. **~étrico** *a* symmetric(al)
simiente *f* seed
similar *a* similar
simp|atía *f* liking; (*cariño*) affection; (*fig*, *amigo*) friend. **~ático** *a* nice, likeable; (*amable*) kind. **~atizante** *m* & *f* sympathizer. **~atizar** [10] *vi* get on (well together). **me es ~ático** I like
simpl|e *a* simple; (*mero*) mere. **~eza** *f* simplicity; (*tontería*) stupid thing; (*insignificancia*) trifle. **~icidad** *f* simplicity. **~ificar** [7] *vt* simplify. **~ón** *m* simpleton
simposio *m* symposium
simula|ción *f* simulation. **~r** *vt* feign
simultáneo *a* simultaneous
sin *prep* without. **~ que** without
sinagoga *f* synagogue
sincer|idad *f* sincerity. **~o** *a* sincere
síncopa *f* (*mus*) syncopation
sincopar *vt* syncopate
sincronizar [10] *vt* synchronize
sindica|l *a* (trade-)union. **~lista** *m* & *f* trade-unionist. **~to** *m* trade union
síndrome *m* syndrome
sinfín *m* endless number
sinf|onía *f* symphony. **~ónico** *a* symphonic
singular *a* singular; (*excepcional*) exceptional. **~izar** [10] *vt* single out. **~izarse** *vpr* stand out
siniestro *a* sinister; (*desgraciado*) unlucky. ● *m* disaster
sinnúmero *m* endless number
sino *m* fate. ● *conj* but; (*salvo*) except
sínodo *m* synod
sinónimo *a* synonymous. ● *m* synonym
sinrazón *f* wrong
sintaxis *f* syntax
síntesis *f invar* synthesis
sint|ético *a* synthetic. **~etizar** [10] *vt* synthesize; (*resumir*) summarize
síntoma *f* sympton
sintomático *a* symptomatic
sinton|ía *f* (*en la radio*) signature tune. **~izar** [10] *vt* (*con la radio*) tune (in)
sinuoso *a* winding
sinvergüenza *m* & *f* scoundrel
sionis|mo *m* Zionism. **~ta** *m* & *f* Zionist
siquiera *conj* even if. ● *adv* at least. **ni ~** not even
sirena *f* siren
Siria *f* Syria
sirio *a* & *m* Syrian
siroco *m* sirocco
sirvienta *f*, **sirviente** *m* servant
sirvo *vb véase* **servir**
sise|ar *vt*/*i* hiss. **~o** *m* hissing
sísmico *a* seismic
sismo *m* earthquake
sistem|a *m* system. **~ático** *a* systematic. **por ~a** as a rule
sitiar *vt* besiege; (*fig*) surround
sitio *m* place; (*espacio*) space; (*mil*) siege. **en cualquier ~** anywhere
situa|ción *f* position. **~r** [21] *vt* situate; (*poner*) put; (*depositar*) deposit. **~rse** *vpr* be successful, establish o.s.
slip /es'lip/ *m* (*pl* **slips** /es'lip/) underpants, briefs
slogan /es'logan/ *m* (*pl* **slogans** /es'logan/) slogan
smoking /es'mokin/ *m* (*pl* **smokings** /es'mokin/) dinner jacket (*Brit*), tuxedo (*Amer*)
sobaco *m* armpit
sobar *vt* handle; knead ‹*masa*›
soberan|ía *f* sovereignty. **~o** *a* sovereign; (*fig*) supreme. ● *m* sovereign
soberbi|a *f* pride; (*altanería*) arrogance. **~o** *a* proud; (*altivo*) arrogant
soborn|ar *vt* bribe. **~o** *m* bribe
sobra *f* surplus. **~s** *fpl* leftovers. **~do** *a* more than enough. **~nte** *a* surplus. **~r** *vi* be left over; (*estorbar*) be in the way. **de ~** more than enough
sobrasada *f* Majorcan sausage
sobre *prep* on; (*encima de*) on top of; (*más o menos*) about; (*por encima de*) above; (*sin tocar*) over; (*además de*) on top of. ● *m* envelope. **~cargar** [12] *vt* overload. **~coger** [14] *vt* startle. **~cogerse** *vpr* be startled. **~cubierta** *f* dust cover. **~dicho** *a* aforementioned. **~entender** [1] *vt* understand, infer. **~entendido** *a* implicit. **~humano** *a* superhuman. **~llevar** *vt* bear. **~mesa** *f*. **de ~mesa** after-dinner. **~natural** *a* supernatural. **~nombre** *m* nickname. **~pasar** *vt* exceed. **~poner** [34] *vt* superimpose; (*fig*, *anteponer*) put before. **~ponerse** *vpr* overcome. **~pujar** *vt* surpass. **~saliente** *a* (*fig*) outstanding. ● *m* excellent mark. **~salir** [52] *vi* stick out;

(*fig*) stand out. **~saltar** *vt* startle. **~salto** *m* fright. **~sueldo** *m* bonus. **~todo** *m* overall; (*abrigo*) overcoat. **~ todo** above all, especially. **~venir** [53] *vi* happen. **~viviente** *a* surviving. ● *m & f* survivor. **~vivir** *vi* survive. **~volar** *vt* fly over

sobriedad *f* restraint

sobrin|a *f* niece. **~o** *m* nephew

sobrio *a* moderate, sober

socarr|ón *a* sarcastic; (*taimado*) sly. **~onería** *f* sarcasm

socavar *vt* undermine

soci|able *a* sociable. **~al** *a* social. **~aldemocracia** *f* social democracy. **~aldemócrata** *m & f* social democrat. **~alismo** *m* socialsim. **~alista** *a & m & f* socialist. **~alizar** [10] *vt* nationalize. **~edad** *f* society; (*com*) company. **~edad anónima** limited company. **~o** *m* member; (*com*) partner. **~ología** *f* sociology. **~ólogo** *m* sociologist

socorr|er *vt* help. **~o** *m* help

soda *f* (*bebida*) soda (water)

sodio *m* sodium

sofá *m* sofa, settee

sofistica|ción *f* sophistication. **~do** *a* sophisticated. **~r** [7] *vt* adulterate

sofoca|ción *f* suffocation. **~nte** *a* (*fig*) stifling. **~r** [7] *vt* suffocate; (*fig*) stifle. **~rse** *vpr* suffocate; (*ruborizarse*) blush

soga *f* rope

soja *f* soya (bean)

sojuzgar [12] *vt* subdue

sol *m* sun; (*luz solar*) sunlight; (*mus*) G; (*solfa*) soh. **al ~** in the sun. **día *m* de ~** sunny day. **hace ~, hay ~** it is sunny. **tomar el ~** sunbathe

solamente *adv* only

solapa *f* lapel; (*de bolsillo etc*) flap. **~do** *a* sly. **~r** *vt/i* overlap

solar *a* solar. ● *m* plot

solariego *a* ‹*casa*› ancestral

solaz *m* relaxation

soldado *m* soldier. **~ raso** private

solda|dor *m* welder; (*utensilio*) soldering iron. **~r** [2] *vt* weld, solder

solea|do *a* sunny. **~r** *vt* put in the sun

soledad *f* solitude; (*aislamiento*) loneliness

solemn|e *a* solemn. **~idad** *f* solemnity; (*ceremonia*) ceremony

soler [2] *vi* be in the habit of. **suele despertarse a las 6** he usually wakes up at 6 o'clock

sol|icitar *vt* request; apply for ‹*empleo*›; attract ‹*atención*›. **~ícito** *a* solicitous. **~icitud** *f* (*atención*) concern; (*petición*) request; (*para un puesto*) application

solidaridad *f* solidarity

solid|ez *f* solidity; (*de color*) fastness. **~ificar** [7] *vt* solidify. **~ificarse** *vpr* solidify

sólido *a* solid; ‹*color*› fast; (*robusto*) strong. ● *m* solid

soliloquio *m* soliloquy

solista *m & f* soloist

solitario *a* solitary; (*aislado*) lonely. ● *m* recluse; (*juego, diamante*) solitaire

solo *a* (*sin compañía*) alone; (*aislado*) lonely; (*único*) only; (*mus*) solo; ‹*café*› black. ● *m* solo; (*juego*) solitaire. **a solas** alone

sólo *adv* only. **~ que** only. **aunque ~ sea** even if it is only. **con ~ que** if; (*con tal que*) as long as. **no ~... sino también** not only... but also... **tan ~** only

solomillo *m* sirloin

solsticio *m* solstice

soltar [2] *vt* let go of; (*dejar caer*) drop; (*dejar salir, decir*) let out; give ‹*golpe etc*›. **~se** *vpr* come undone; (*librarse*) break loose

solter|a *f* single woman. **~o** *a* single. ● *m* bachelor. **apellido *m* de ~a** maiden name

soltura *f* looseness; (*agilidad*) agility; (*en hablar*) ease, fluency

solu|ble *a* soluble. **~ción** *f* solution. **~cionar** *vt* solve; settle ‹*huelga, asunto*›

solvent|ar *vt* resolve; settle ‹*deuda*›. **~e** *a & m* solvent

sollo *m* sturgeon

solloz|ar [10] *vi* sob. **~o** *m* sob

sombr|a *f* shade; (*imagen oscura*) shadow. **~eado** *a* shady. **a la ~a** in the shade

sombrero *m* hat. **~ hongo** bowler hat

sombrío *a* sombre

somero *a* shallow

someter *vt* subdue; subject ‹*persona*›; (*presentar*) submit. **~se** *vpr* give in

somn|oliento *a* sleepy. **~ífero** *m* sleeping-pill

somos *vb véase* **ser**

son *m* sound. ● *vb véase* **ser**

sonámbulo *m* sleepwalker

sonar [2] *vt* blow; ring ‹*timbre*›. ● *vi* sound; ‹*timbre, teléfono etc*› ring; ‹*reloj*› strike; (*pronunciarse*) be pronounced; (*mus*) play; (*fig, ser conocido*) be familiar. **~se** *vpr* blow one's nose. **~ a** sound like
sonata *f* sonata
sonde|ar *vt* sound; (*fig*) sound out. **~o** *m* sounding; (*fig*) poll
soneto *m* sonnet
sónico *a* sonic
sonido *m* sound
sonoro *a* sonorous; (*ruidoso*) loud
sonr|eír [51] *vi* smile. **~eírse** *vpr* smile. **~iente** *a* smiling. **~isa** *f* smile
sonroj|ar *vt* make blush. **~arse** *vpr* blush. **~o** *m* blush
sonrosado *a* rosy, pink
sonsacar [7] *vt* wheedle out
soñ|ado *a* dream. **~ador** *m* dreamer. **~ar** [2] *vi* dream (**con** of). **¡ni ~arlo!** not likely! **(que) ni ~ado** marvellous
sopa *f* soup
sopesar *vt* (*fig*) weigh up
sopl|ar *vt* blow; blow out ‹*vela*›; blow off ‹*polvo*›; (*inflar*) blow up. ● *vi* blow. **~ete** *m* blowlamp. **~o** *m* puff; (*fig, momento*) moment
soporífero *a* soporific. ● *m* sleeping-pill
soport|al *m* porch. **~ales** *mpl* arcade. **~ar** *vt* support; (*fig*) bear. **~e** *m* support
soprano *f* soprano
sor *f* sister
sorb|er *vt* suck; sip ‹*bebida*›; (*absorber*) absorb. **~ete** *m* sorbet, water-ice. **~o** *m* swallow; (*pequeña cantidad*) sip
sord|amente *adv* silently, dully. **~era** *f* deafness
sórdido *a* squalid; (*tacaño*) mean
sordo *a* deaf; (*silencioso*) quiet. ● *m* deaf person. **~mudo** *a* deaf and dumb. **a la sorda, a sordas** on the quiet. **hacerse el ~** turn a deaf ear
sorna *f* sarcasm. **con ~** sarcastically
soroche *m* (*LAm*) mountain sickness
sorpre|ndente *a* surprising. **~nder** *vt* surprise; (*coger desprevenido*) catch. **~sa** *f* surprise
sorte|ar *vt* draw lots for; (*rifar*) raffle; (*fig*) avoid. ● *vi* draw lots; (*con moneda*) toss up. **~o** *m* draw; (*rifa*) raffle; (*fig*) avoidance
sortija *f* ring; (*de pelo*) ringlet
sortilegio *m* witchcraft; (*fig*) spell
sos|egado *a* calm. **~egar** [1 & 12] *vt* calm. ● *vi* rest. **~iego** *m* calmness. **con ~iego** calmly
soslayo. al ~, de ~ sideways
soso *a* tasteless; (*fig*) dull
sospech|a *f* suspicion. **~ar** *vt/i* suspect. **~oso** *a* suspicious. ● *m* suspect
sost|én *m* support; (*prenda femenina*) bra (*fam*), brassière. **~ener** [40] *vt* support; (*sujetar*) hold; (*mantener*) maintain; (*alimentar*) sustain. **~enerse** *vpr* support o.s.; (*continuar*) remain. **~enido** *a* sustained; (*mus*) sharp. ● *m* (*mus*) sharp
sota *f* (*de naipes*) jack
sótano *m* basement
sotavento *m* lee
soto *m* grove; (*matorral*) thicket
soviético *a* (*historia*) Soviet
soy *vb véase* **ser**
Sr *abrev* (*Señor*) Mr. **~a** *abrev* (*Señora*) Mrs. **~ta** *abrev* (*Señorita*) Miss
su *a* (*de él*) his; (*de ella*) her; (*de ello*) its; (*de uno*) one's; (*de Vd*) your; (*de ellos, de ellas*) their; (*de Vds*) your
suav|e *a* smooth; (*fig*) gentle; ‹*color, sonido*› soft. **~idad** *f* smoothness, softness. **~izar** [10] *vt* smooth, soften
subalimentado *a* underfed
subalterno *a* secondary; ‹*persona*› auxiliary
subarrendar [1] *vt* sublet
subasta *f* auction; (*oferta*) tender. **~r** *vt* auction
sub|campeón *m* runner-up. **~consciencia** *f* subconscious. **~consciente** *a* & *m* subconscious. **~continente** *m* subcontinent. **~desarrollado** *a* under-developed. **~director** *m* assistant manager
súbdito *m* subject
sub|dividir *vt* subdivide. **~estimar** *vt* underestimate. **~gerente** *m* & *f* assistant manager
subi|da *f* ascent; (*aumento*) rise; (*pendiente*) slope. **~do** *a* ‹*precio*› high; ‹*color*› bright; ‹*olor*› strong. **~r** *vt* go up; (*poner*) put; (*llevar*) take up; (*aumentar*) increase. ● *vi* go up. **~r a** get into ‹*coche*›; get on ‹*autobús, avión, barco, tren*›; (*aumentar*) increase. **~rse** *vpr* climb up. **~rse a** get on ‹*tren etc*›
súbito *a* sudden. ● *adv* suddenly. **de ~** suddenly

subjetivo *a* subjective
subjuntivo *a & m* subjunctive
subleva|ción *f* uprising. **~r** *vt* incite to rebellion. **~rse** *vpr* rebel
sublim|ar *vt* sublimate. **~e** *a* sublime
submarino *a* underwater. ● *m* submarine
subordinado *a & m* subordinate
subrayar *vt* underline
subrepticio *a* surreptitious
subsanar *vt* remedy; overcome ‹*dificultad*›
subscri|bir *vt* (*pp* **subscrito**) sign. **~birse** *vpr* subscribe. **~pción** *f* subscription
subsidi|ario *a* subsidiary. **~o** *m* subsidy. **~o de paro** unemployment benefit
subsiguiente *a* subsequent
subsist|encia *f* subsistence. **~ir** *vi* subsist; (*perdurar*) survive
substanci|a *f* substance. **~al** *a* important. **~oso** *a* substantial
substantivo *m* noun
substitu|ción *f* substitution. **~ir** [17] *vt/i* substitute. **~to** *a & m* substitute
substraer [41] *vt* take away
subterfugio *m* subterfuge
subterráneo *a* underground. ● *m* (*bodega*) cellar; (*conducto*) underground passage
subtítulo *m* subtitle
suburb|ano *a* suburban. ● *m* suburban train. **~io** *m* suburb; (*en barrio pobre*) slum
subvenci|ón *f* grant. **~onar** *vt* subsidize
subver|sión *f* subversion. **~sivo** *a* subversive. **~tir** [4] *vt* subvert
subyugar [12] *vt* subjugate; (*fig*) subdue
succión *f* suction
suce|der *vi* happen; (*seguir*) follow; (*substituir*) succeed. **~dido** *m* event. **lo ~dido** what happened. **~sión** *f* succession. **~sivo** *a* successive; (*consecutivo*) consecutive. **~so** *m* event; (*incidente*) incident. **~sor** *m* successor. **en lo ~sivo** in future. **lo que ~de es que** the trouble is that. **¿qué ~de?** what's the matter?
suciedad *f* dirt; (*estado*) dirtiness
sucinto *a* concise; ‹*prenda*› scanty
sucio *a* dirty; (*vil*) mean; ‹*conciencia*› guilty. **en ~** in rough
sucre *m* (*unidad monetaria del Ecuador*) sucre
suculento *a* succulent
sucumbir *vi* succumb
sucursal *f* branch (office)
Sudáfrica *m & f* South Africa
sudafricano *a & m* South African
Sudamérica *f* South America
sudamericano *a & m* South American
sudar *vt* work hard for. ● *vi* sweat
sud|este *m* south-east; (*viento*) south-east wind. **~oeste** *m* south-west; (*viento*) south-west wind
sudor *m* sweat
Suecia *f* Sweden
sueco *a* Swedish. ● *m* (*persona*) Swede; (*lengua*) Swedish. **hacerse el ~** pretend not to hear
suegr|a *f* mother-in-law. **~o** *m* father-in-law. **mis ~os** my in-laws
suela *f* sole
sueldo *m* salary
suelo *m* ground; (*dentro de edificio*) floor; (*tierra*) land. ● *vb véase* **soler**
suelto *a* loose; (*libre*) free; (*sin pareja*) odd; ‹*lenguaje*› fluent. ● *m* (*en periódico*) item; (*dinero*) change
sueño *m* sleep; (*ilusión*) dream. **tener ~** be sleepy
suero *m* serum; (*de leche*) whey
suerte *f* luck; (*destino*) fate; (*azar*) chance. **de otra ~** otherwise. **de ~ que** so. **echar ~s** draw lots. **por ~** fortunately. **tener ~** be lucky
suéter *m* jersey
suficien|cia *f* sufficiency; (*presunción*) smugness; (*aptitud*) suitability. **~te** *a* sufficient; (*presumido*) smug. **~temente** *adv* enough
sufijo *m* suffix
sufragio *m* (*voto*) vote
sufri|do *a* ‹*persona*› long-suffering; ‹*tela*› hard-wearing. **~miento** *m* suffering. **~r** *vt* suffer; (*experimentar*) undergo; (*soportar*) bear. ● *vi* suffer
suge|rencia *f* suggestion. **~rir** [4] *vt* suggest. **~stión** *f* suggestion. **~stionable** *a* impressionable. **~stionar** *vt* influence. **~stivo** *a* (*estimulante*) stimulating; (*atractivo*) attractive
suicid|a *a* suicidal. ● *m & f* suicide; (*fig*) maniac. **~arse** *vpr* commit suicide. **~io** *m* suicide
Suiza *f* Switzerland
suizo *a* Swiss. ● *m* Swiss; (*bollo*) bun

suje|ción *f* subjection. **~tador** *m* fastener; (*de pelo, papeles etc*) clip; (*prenda femenina*) bra (*fam*), brassière. **~tapapeles** *m invar* paper-clip. **~tar** *vt* fasten; (*agarrar*) hold; (*fig*) restrain. **~tarse** *vr* subject o.s.; (*ajustarse*) conform. **~to** *a* fastened; (*susceptible*) subject. • *m* individual

sulfamida *f* sulpha (drug)

sulfúrico *a* sulphuric

sult|án *m* sultan. **~ana** *f* sultana

suma *f* sum; (*total*) total. **en ~** in short. **~mente** *adv* extremely. **~r** *vt* add (up); (*fig*) gather. • *vi* add up. **~rse** *vpr*. **~rse a** join in

sumario *a* brief. • *m* summary; (*jurid*) indictment

sumergi|ble *m* submarine. • *a* submersible. **~r** [14] *vt* submerge

sumidero *m* drain

suministr|ar *vt* supply. **~o** *m* supply; (*acción*) supplying

sumir *vt* sink; (*fig*) plunge

sumis|ión *f* submission. **~o** *a* submissive

sumo *a* greatest; (*supremo*) supreme. **a lo ~** at the most

suntuoso *a* sumptuous

supe *vb véase* **saber**

superar *vt* surpass; (*vencer*) overcome; (*dejar atrás*) get past. **~se** *vpr* excel o.s.

supercheria *f* swindle

superestructura *f* superstructure

superfici|al *a* superficial. **~e** *f* surface; (*extensión*) area. **de ~e** surface

superfluo *a* superfluous

superhombre *m* superman

superintendente *m* superintendent

superior *a* superior; (*más alto*) higher; (*mejor*) better; ‹*piso*› upper. • *m* superior. **~idad** *f* superiority

superlativo *a & m* superlative

supermercado *m* supermarket

supersónico *a* supersonic

supersticí|ón *f* superstition. **~oso** *a* superstitious

supervis|ión *f* supervision. **~or** *m* supervisor

superviviente *a* surviving. • *m & f* survivor

suplantar *vt* supplant

suplement|ario *a* supplementary. **~o** *m* supplement

suplente *a & m & f* substitute

súplica *f* entreaty; (*petición*) request

suplicar [7] *vt* beg

suplicio *m* torture

suplir *vt* make up for; (*reemplazar*) replace

supo|ner [34] *vt* suppose; (*significar*) mean; (*costar*) cost. **~sición** *f* supposition

supositorio *m* suppository

suprem|acía *f* supremacy. **~o** *a* supreme; ‹*momento etc*› critical

supr|esión *f* suppression. **~imir** *vt* suppress; (*omitir*) omit

supuesto *a* supposed. • *m* assumption. **~ que** if. **¡por ~!** of course!

sur *m* south; (*viento*) south wind

surc|ar [7] *vt* plough. **~o** *m* furrow; (*de rueda*) rut; (*en la piel*) wrinkle

surgir [14] *vi* spring up; (*elevarse*) loom up; (*aparecer*) appear; ‹*dificultad, oportunidad*› arise, crop up

surrealis|mo *m* surrealism. **~ta** *a & m & f* surrealist

surti|do *a* well-stocked; (*variado*) assorted. • *m* assortment, selection. **~dor** *m* (*de gasolina*) petrol pump (*Brit*), gas pump (*Amer*). **~r** *vt* supply; have ‹*efecto*›. **~rse** *vpr* provide o.s. (**de** with)

susceptib|ilidad *f* susceptibility; (*sensibilidad*) sensitivity. **~le** *a* susceptible; (*sensible*) sensitive

suscitar *vt* provoke; arouse ‹*curiosidad, interés, sospechas*›

suscr... *véase* **subscr...**

susodicho *a* aforementioned

suspen|der *vt* hang (up); (*interrumpir*) suspend; (*univ etc*) fail. **~derse** *vpr* stop. **~sión** *f* suspension. **~so** *a* hanging; (*pasmado*) amazed; (*univ etc*) failed. • *m* fail. **en ~so** pending

suspicaz *a* suspicious

suspir|ar *vi* sigh. **~o** *m* sigh

sust... *véase* **subst...**

sustent|ación *f* support. **~ar** *vt* support; (*alimentar*) sustain; (*mantener*) maintain. **~o** *m* support; (*alimento*) sustenance

susto *m* fright. **caerse del ~** be frightened to death

susurr|ar *vi* ‹*persona*› whisper; ‹*agua*› murmur; ‹*hojas*› rustle. **~o** *m* (*de persona*) whisper; (*de agua*) murmur; (*de hojas*) rustle

sutil *a* fine; (*fig*) subtle. **~eza** *f* fineness; (*fig*) subtlety

suyo *a & pron* (*de él*) his; (*de ella*) hers; (*de ello*) its; (*de uno*) one's; (*de Vd*) yours; (*de ellos, de ellas*) theirs;

(*de Vds*) yours. **un amigo ~** a friend of his, a friend of theirs, etc

T

taba *f* (*anat*) ankle-bone; (*juego*) jacks
tabac|alera *f* (state) tobacconist. **~alero** *a* tobacco. **~o** *m* tobacco; (*cigarillos*) cigarettes; (*rapé*) snuff
tabalear *vi* drum (with one's fingers)
Tabasco *m* Tabasco (**P**)
tabern|a *f* bar. **~ero** *m* barman; (*dueño*) landlord
tabernáculo *m* tabernacle
tabique *m* (thin) wall
tabl|a *f* plank; (*de piedra etc*) slab; (*estante*) shelf; (*de vestido*) pleat; (*lista*) list; (*índice*) index; (*en matemática etc*) table. **~ado** *m* platform; (*en el teatro*) stage. **~ao** *m* place where flamenco shows are held. **~as reales** backgammon. **~ero** *m* board. **~ero de mandos** dashboard. **hacer ~a rasa de** disregard
tableta *f* tablet; (*de chocolate*) bar
tabl|illa *f* small board. **~ón** *m* plank. **~ón de anuncios** notice board (*esp Brit*), bulletin board (*Amer*)
tabú *m* taboo
tabular *vt* tabulate
taburete *m* stool
tacaño *a* mean
tacita *f* small cup
tácito *a* tacit
taciturno *a* taciturn; (*triste*) miserable
taco *m* plug; (*LAm, tacón*) heel; (*de billar*) cue; (*de billetes*) book; (*fig, lío, fam*) mess; (*Mex, culin*) filled tortilla
tacógrafo *m* tachograph
tacón *m* heel
táctic|a *f* tactics. **~o** *a* tactical
táctil *a* tactile
tacto *m* touch; (*fig*) tact
tacuara *f* (*Arg*) bamboo
tacurú *m* (small) ant
tacha *f* fault; (*clavo*) tack. **poner ~s a** find fault with. **sin ~** flawless
tachar *vt* (*borrar*) rub out; (*con raya*) cross out. **~ de** accuse of
tafia *f* (*LAm*) rum
tafilete *m* morocco
tahúr *m* card-sharp
Tailandia *f* Thailand
tailandés *a* & *m* Thai
taimado *a* sly
taj|ada *f* slice. **~ante** *a* sharp. **~o** *m* slash; (*fig, trabajo, fam*) job; (*culin*) chopping block. **sacar ~ada** profit
Tajo *m* Tagus
tal *a* such; (*ante sustantivo en singular*) such a. ● *pron* (*persona*) someone; (*cosa*) such a thing. ● *adv* so; (*de tal manera*) in such a way. **~ como** the way. **~ cual** (*tal como*) the way; (*regular*) fair. **~ para cual** (*fam*) two of a kind. **con ~ que** as long as. **¿qué ~?** how are you? **un ~** a certain
taladr|ar *vt* drill. **~o** *m* drill; (*agujero*) drill hole
talante *m* mood. **de buen ~** willingly
talar *vt* fell; (*fig*) destroy
talco *m* talcum powder
talcualillo *a* (*fam*) so so
talega *f*, **talego** *m* sack
talento *m* talent
TALGO *m* high-speed train
talismán *m* talisman
tal|ón *m* heel; (*recibo*) counterfoil; (*cheque*) cheque. **~onario** *m* receipt book; (*de cheques*) cheque book
talla *f* carving; (*grabado*) engraving; (*de piedra preciosa*) cutting; (*estatura*) height; (*medida*) size; (*palo*) measuring stick; (*Arg, charla*) gossip. **~do** *a* carved. ● *m* carving. **~dor** *m* engraver
tallarín *m* noodle
talle *m* waist; (*figura*) figure; (*medida*) size
taller *m* workshop; (*de pintor etc*) studio
tallo *m* stem, stalk
tamal *m* (*LAm*) tamale
tamaño *a* (*tan grande*) so big a; (*tan pequeño*) so small a. ● *m* size. **de ~ natural** life-size
tambalearse *vpr* ‹*persona*› stagger; ‹*cosa*› wobble
también *adv* also, too
tambor *m* drum. **~ del freno** brake drum. **~ilear** *vi* drum
Támesis *m* Thames
tamiz *m* sieve. **~ar** [10] *vt* sieve
tampoco *adv* nor, neither, not either
tampón *m* tampon; (*para entintar*) ink-pad
tan *adv* so. **tan... ~** as... as

tanda *f* group; (*capa*) layer; (*de obreros*) shift
tangente *a & f* tangent
Tánger *m* Tangier
tangible *a* tangible
tango *m* tango
tanque *m* tank; (*camión, barco*) tanker
tante|ar *vt* estimate; (*ensayar*) test; (*fig*) weigh up. ● *vi* score. **~o** *m* estimate; (*prueba*) test; (*en deportes*) score
tanto *a* (*en singular*) so much; (*en plural*) so many; (*comparación en singular*) as much; (*comparación en plural*) as many. ● *pron* so much; (*en plural*) so many. ● *adv* so much; (*tiempo*) so long. ● *m* certain amount; (*punto*) point; (*gol*) goal. **~ como** as well as; (*cantidad*) as much as. **~ más... cuanto que** all the more... because. **~ si... como si** whether... or. **a ~s de** sometime in. **en ~, entre ~** meanwhile. **en ~ que** while. **entre ~** meanwhile. **estar al ~ de** be up to date with. **hasta ~ que** until. **no es para ~** it's not as bad as all that. **otro ~** the same; (*el doble*) as much again. **por (lo) ~** so. **un ~** *adv* somewhat
tañer [22] *vt* play
tapa *f* lid; (*de botella*) top; (*de libro*) cover. **~s** *fpl* savoury snacks
tapacubos *m invar* hub-cap
tapa|dera *f* cover, lid; (*fig*) cover. **~r** *vt* cover; (*abrigar*) wrap up; (*obturar*) plug; put the top on ‹*botella*›
taparrabo(s) *m invar* loincloth; (*bañador*) swimming-trunks
tapete *m* (*de mesa*) table cover; (*alfombra*) rug
tapia *f* wall. **~r** *vt* enclose
tapicería *f* tapestry; (*de muebles*) upholstery
tapioca *f* tapioca
tapiz *m* tapestry. **~ar** [10] *vt* hang with tapestries; upholster ‹*muebles*›
tap|ón *m* stopper; (*corcho*) cork; (*med*) tampon; (*tec*) plug. **~onazo** *m* pop
taqu|igrafía *f* shorthand. **~ígrafo** *m* shorthand writer
taquill|a *f* ticket office; (*archivador*) filing cabinet; (*fig, dinero*) takings. **~ero** *m* clerk, ticket seller. ● *a* box-office
tara *f* (*peso*) tare; (*defecto*) defect
taracea *f* marquetry
tarántula *f* tarantula
tararear *vt/i* hum
tarda|nza *f* delay. **~r** *vi* take; (*mucho tiempo*) take a long time. **a más ~r** at the latest. **sin ~r** without delay
tard|e *adv* late. ● *f* (*antes del atardecer*) afternoon; (*después del atardecer*) evening. **~e o temprano** sooner or later. **~ío** *a* late. **de ~e en ~e** from time to time. **por la ~e** in the afternoon
tardo *a* (*torpe*) slow
tarea *f* task, job
tarifa *f* rate, tariff
tarima *f* platform
tarjeta *f* card. **~ de crédito** credit card. **~ postal** postcard
tarro *m* jar
tarta *f* cake; (*torta*) tart. **~ helada** ice-cream gateau
tartamud|ear *vi* stammer. **~o** *a* stammering. ● *m* stammerer. **es ~o** he stammers
tártaro *m* tartar
tarugo *m* chunk
tasa *f* valuation; (*precio*) fixed price; (*índice*) rate. **~r** *vt* fix a price for; (*limitar*) ration; (*evaluar*) value
tasca *f* bar
tatarabuel|a *f* great-great-grandmother. **~o** *m* great-great-grandfather
tatua|je *m* (*acción*) tattooing; (*dibujo*) tattoo. **~r** [21] *vt* tattoo
taurino *a* bullfighting
Tauro *m* Taurus
tauromaquia *f* bullfighting
tax|i *m* taxi. **~ímetro** *m* taxi meter. **~ista** *m & f* taxi-driver
tayuyá *m* (*Arg*) water melon
taz|a *f* cup. **~ón** *m* bowl
te *pron* (*acusativo*) you; (*dativo*) (to) you; (*reflexivo*) (to) yourself
té *m* tea. **dar el ~** bore
tea *f* torch
teatr|al *a* theatre; (*exagerado*) theatrical. **~alizar** [10] *vt* dramatize. **~o** *m* theatre; (*literatura*) drama. **obra** *f* **~al** play
tebeo *m* comic
teca *f* teak
tecla *f* key. **~do** *m* keyboard. **tocar la ~, tocar una ~** pull strings
técnica *f* technique
tecn|icismo *m* technicality
técnico *a* technical. ● *m* technician
tecnol|ogía *f* technology. **~ógico** *a* technological

tecolote *m* (*Mex*) owl
tecomate *m* (*Mex*) earthenware cup
tech|ado *m* roof. **~ar** *vt* roof. **~o** *m* (*interior*) ceiling; (*exterior*) roof. **~umbre** *f* roofing. **bajo ~ado** indoors
teja *f* tile. **~do** *m* roof. **a toca ~** cash
teje|dor *m* weaver. **~r** *vt* weave; (*hacer punto*) knit
tejemaneje *m* (*fam*) fuss; (*intriga*) scheming
tejido *m* material; (*anat, fig*) tissue. **~s** *mpl* textiles
tejón *m* badger
tela *f* material; (*de araña*) web; (*en líquido*) skin
telar *m* loom. **~es** *mpl* textile mill
telaraña *f* spider's web, cobweb
tele *f* (*fam*) television
tele|comunicación *f* telecommunication. **~diario** *m* television news. **~dirigido** *a* remote-controlled. **~férico** *m* cable-car; (*tren*) cable-railway
tel|efonear *vt/i* telephone. **~efónico** *a* telephone. **~efonista** *m & f* telephonist. **~éfono** *m* telephone. **al ~éfono** on the phone
tel|egrafía *f* telegraphy. **~egrafiar** [20] *vt* telegraph. **~egráfico** *a* telegraphic. **~égrafo** *m* telegraph
telegrama *m* telegram
telenovela *f* television soap opera
teleobjetivo *m* telephoto lens
telep|atía *f* telepathy. **~ático** *a* telepathic
telesc|ópico *a* telescopic. **~opio** *m* telescope
telesilla *m* ski-lift, chair-lift
telespectador *m* viewer
telesquí *m* ski-lift
televi|dente *m & f* viewer. **~sar** *vt* televise. **~sión** *f* television. **~sor** *m* television (set)
télex *m* telex
telón *m* curtain. **~ de acero** (*historia*) Iron Curtain
tema *m* subject; (*mus*) theme
tembl|ar [1] *vi* shake; (*de miedo*) tremble; (*de frío*) shiver; (*fig*) shudder. **~or** *m* shaking; (*de miedo*) trembling; (*de frío*) shivering. **~or de tierra** earthquake. **~oroso** *a* trembling
temer *vt* be afraid (of). ● *vi* be afraid. **~se** *vpr* be afraid
temerario *a* reckless
tem|eroso *a* frightened. **~ible** *a* fearsome. **~or** *m* fear
témpano *m* floe
temperamento *m* temperament
temperatura *f* temperature
temperie *f* weather
tempest|ad *f* storm. **~uoso** *a* stormy. **levantar ~ades** (*fig*) cause a storm
templ|ado *a* moderate; (*tibio*) warm; ‹*clima, tiempo*› mild; (*valiente*) courageous; (*listo*) bright. **~anza** *f* moderation; (*de clima o tiempo*) mildness. **~ar** *vt* temper; (*calentar*) warm; (*mus*) tune. **~e** *m* tempering; (*temperatura*) temperature; (*humor*) mood
templ|ete *m* niche; (*pabellón*) pavilion. **~o** *m* temple
tempora|da *f* time; (*época*) season. **~l** *a* temporary. ● *m* (*tempestad*) storm; (*período de lluvia*) rainy spell
tempran|ero *a* ‹*frutos*› early. **~o** *a & adv* early. **ser ~ero** be an early riser
tena|cidad *f* tenacity
tenacillas *fpl* tongs
tenaz *a* tenacious
tenaza *f*, **tenazas** *fpl* pliers; (*para arrancar clavos*) pincers; (*para el fuego, culin*) tongs
tende|ncia *f* tendency. **~nte** *a*. **~nte a** aimed at. **~r** [1] *vt* spread (out); hang out ‹*ropa a secar*›; (*colocar*) lay. ● *vi* have a tendency (**a** to). **~rse** *vpr* stretch out
tender|ete *m* stall. **~o** *m* shopkeeper
tendido *a* spread out; ‹*ropa*› hung out; ‹*persona*› stretched out. ● *m* (*en plaza de toros*) front rows. **~s** *mpl* (*ropa lavada*) washing
tendón *m* tendon
tenebroso *a* gloomy; (*turbio*) shady
tenedor *m* fork; (*poseedor*) holder
tener [40] *vt* have (got); (*agarrar*) hold; be ‹*años, calor, celos, cuidado, frío, ganas, hambre, miedo, razón, sed etc*›. **¡ten cuidado!** be careful! **tengo calor** I'm hot. **tiene 3 años** he's 3 (years old). **~se** *vpr* stand up; (*considerarse*) consider o.s., think o.s. **~ al corriente**, **~ al día** keep up to date. **~ 2 cm de largo** be 2 cms long. **~ a uno por** consider s.o. **~ que** have (got) to. **tenemos que comprar pan** we've got to buy some bread. **¡ahí tienes!** there you are! **no ~ nada que ver con** have nothing to do with. **¿qué tienes?** what's the

matter (with you)? **¡tenga!** here you are!
tengo *vb véase* **tener**
teniente *m* lieutenant. **~ de alcalde** deputy mayor
tenis *m* tennis. **~ta** *m & f* tennis player
tenor *m* sense; (*mus*) tenor. **a este ~** in this fashion
tens|ión *f* tension; (*presión*) pressure; (*arterial*) blood pressure; (*elec*) voltage; (*de persona*) tenseness. **~o** *a* tense
tentación *f* temptation
tentáculo *m* tentacle
tenta|dor *a* tempting. **~r** [1] *vt* feel; (*seducir*) tempt
tentativa *f* attempt
tenue *a* thin; ‹*luz, voz*› faint
teñi|do *m* dye. **~r** [5 & 22] *vt* dye; (*fig*) tinge (**de** with). **~rse** *vpr* dye one's hair
te|ología *f* theology. **~ológico** *a* theological. **~ólogo** *m* theologian
teorema *m* theorem
te|oría *f* theory. **~órico** *a* theoretical
tepache *m* (*Mex*) (alcoholic) drink
tequila *f* tequila
TER *m* high-speed train
terap|éutico *a* therapeutic. **~ia** *f* therapy
tercer *a véase* **tercero**. **~a** *f* (*auto*) third (gear). **~o** *a* (*delante de nombre masculino en singular* **tercer**) third. ● *m* third party
terceto *m* trio
terciar *vi* mediate. **~ en** join in. **~se** *vpr* occur
tercio *m* third
terciopelo *m* velvet
terco *a* obstinate
tergiversar *vt* distort
terma|l *a* thermal. **~s** *fpl* thermal baths
termes *m invar* termite
térmico *a* thermal
termina|ción *f* ending; (*conclusión*) conclusion. **~l** *a & m* terminal. **~nte** *a* categorical. **~r** *vt* finish, end. **~rse** *vpr* come to an end. **~r por** end up
término *m* end; (*palabra*) term; (*plazo*) period. **~ medio** average. **~ municipal** municipal district. **dar ~ a** finish off. **en último ~** as a last resort. **estar en buenos ~s con** be on good terms with. **llevar a ~** carry out. **poner ~ a** put an end to. **primer ~** foreground
terminología *f* terminology
termita *f* termite
termo *m* Thermos flask (P), flask
termómetro *m* thermometer
termo|nuclear *a* thermonuclear. **~sifón** *m* boiler. **~stato** *m* thermostat
terner|a *f* (*carne*) veal. **~o** *m* calf
ternura *f* tenderness
terquedad *f* stubbornness
terracota *f* terracotta
terrado *m* flat roof
terraplén *m* embankment
terrateniente *m & f* landowner
terraza *f* terrace; (*terrado*) flat roof
terremoto *m* earthquake
terre|no *a* earthly. ● *m* land; (*solar*) plot; (*fig*) field. **~stre** *a* earthly; (*mil*) ground
terr|ible *a* terrible. **~iblemente** *adv* awfully. **~ífico** *a* terrifying
territori|al *a* territorial. **~o** *m* territory
terrón *m* (*de tierra*) clod; (*culin*) lump
terror *m* terror. **~ífico** *a* terrifying. **~ismo** *m* terrorism. **~ista** *m & f* terrorist
terr|oso *a* earthy; (*color*) brown. **~uño** *m* land; (*patria*) native land
terso *a* polished; ‹*piel*› smooth
tertulia *f* social gathering, get-together (*fam*). **~r** *vi* (*LAm*) get together. **estar de ~** chat. **hacer ~** get together
tesi|na *f* dissertation. **~s** *f inv* thesis; (*opinión*) theory
tesón *m* perseverance
tesor|ería *f* treasury. **~ero** *m* treasurer. **~o** *m* treasure; (*tesorería*) treasury; (*libro*) thesaurus
testa *f* (*fam*) head. **~ferro** *m* figurehead
testa|mento *m* will. **T~mento** (*relig*) Testament. **~r** *vi* make a will
testarudo *a* stubborn
testículo *m* testicle
testi|ficar [7] *vt/i* testify. **~go** *m* witness. **~go de vista**, **~go ocular**, **~go presencial** eyewitness. **~monio** *m* testimony
teta *f* nipple; (*de biberón*) teat
tétanos *m* tetanus
tetera *f* (*para el té*) teapot; (*Mex, biberón*) feeding-bottle
tetilla *f* nipple; (*de biberón*) teat
tétrico *a* gloomy

textil *a & m* textile
text|o *m* text. **~ual** *a* textual
textura *f* texture
teyú *m* (*Arg*) iguana
tez *f* complexion
ti *pron* you
tía *f* aunt; (*fam*) woman
tiara *f* tiara
tibio *a* lukewarm. **ponerle ~ a uno** insult s.o.
tiburón *m* shark
tic *m* tic
tiempo *m* time; (*atmosférico*) weather; (*mus*) tempo; (*gram*) tense; (*en deportes*) half. **a su ~** in due course. **a ~** in time. **¿cuánto ~?** how long? **hace buen ~** the weather is fine. **hace ~** some time ago. **mucho ~** a long time. **perder el ~** waste time. **¿qué ~ hace?** what is the weather like?
tienda *f* shop; (*de campaña*) tent. **~ de comestibles**, **~ de ultramarinos** grocer's (shop) (*Brit*), grocery store (*Amer*)
tiene *vb véase* **tener**
tienta. **a ~s** gropingly. **andar a ~s** grope one's way
tiento *m* touch; (*de ciego*) blind person's stick; (*fig*) tact
tierno *a* tender; (*joven*) young
tierra *f* land; (*planeta, elec*) earth; (*suelo*) ground; (*geol*) soil, earth. **caer por ~** (*fig*) crumble. **por ~** overland, by land
tieso *a* stiff; (*firme*) firm; (*engreído*) conceited; (*orgulloso*) proud
tiesto *m* flowerpot
tifoideo *a* typhoid
tifón *m* typhoon
tifus *m* typhus; (*fiebre tifoidea*) typhoid (fever); (*en el teatro*) people with complimentary tickets
tigre *m* tiger
tijera *f*, **tijeras** *fpl* scissors; (*de jardín*) shears
tijeret|a *f* (*insecto*) earwig; (*bot*) tendril. **~ear** *vt* snip
tila *f* lime(-tree); (*infusión*) lime tea
tild|ar *vt*. **~ar de** (*fig*) call. **~e** *m* tilde
tilín *m* tinkle. **hacer ~** appeal
tilingo *a* (*Arg, Mex*) silly
tilma *f* (*Mex*) poncho
tilo *m* lime(-tree)
timar *vt* swindle
timbal *m* drum; (*culin*) timbale, meat pie
timbiriche *m* (*Mex*) (alcoholic) drink
timbr|ar *vt* stamp. **~e** *m* (*sello*) stamp; (*elec*) bell; (*sonido*) timbre. **tocar el ~e** ring the bell
timidez *f* shyness
tímido *a* shy
timo *m* swindle
timón *m* rudder; (*fig*) helm
tímpano *m* kettledrum; (*anat*) eardrum. **~s** *mpl* (*mus*) timpani
tina *f* tub. **~ja** *f* large earthenware jar
tinglado *m* (*fig*) intrigue
tinieblas *fpl* darkness; (*fig*) confusion
tino *f* (*habilidad*) skill; (*moderación*) moderation; (*tacto*) tact
tint|a *f* ink. **~e** *m* dyeing; (*color*) dye; (*fig*) tinge. **~ero** *m* ink-well. **de buena ~a** on good authority
tint|ín *m* tinkle; (*de vasos*) chink, clink. **~inear** *vi* tinkle; ‹*vasos*› chink, clink
tinto *a* ‹*vino*› red
tintorería *f* dyeing; (*tienda*) dry cleaner's
tintura *f* dyeing; (*color*) dye; (*noción superficial*) smattering
tío *m* uncle; (*fam*) man. **~s** *mpl* uncle and aunt
tiovivo *m* merry-go-round
típico *a* typical
tipo *m* type; (*persona, fam*) person; (*figura de mujer*) figure; (*figura de hombre*) build; (*com*) rate
tip|ografía *f* typography. **~ográfico** *a* typographic(al). **~ógrafo** *m* printer
típula *f* crane-fly, daddy-long-legs
tique *m*, **tíquet** *m* ticket
tiquete *m* (*LAm*) ticket
tira *f* strip. **la ~ de** lots of
tirabuzón *m* corkscrew; (*de pelo*) ringlet
tirad|a *f* distance; (*serie*) series; (*de libros etc*) edition. **~o** *a* (*barato*) very cheap; (*fácil, fam*) very easy. **~or** *m* (*asa*) handle; (*juguete*) catapult (*Brit*), slingshot (*Amer*). **de una ~a** at one go
tiran|ía *f* tyranny. **~izar** [10] *vt* tyrannize. **~o** *a* tyrannical. • *m* tyrant
tirante *a* tight; (*fig*) tense; ‹*relaciones*› strained. • *m* shoulder strap. **~s** *mpl* braces (*esp Brit*), suspenders (*Amer*)
tirar *vt* throw; (*desechar*) throw away; (*derribar*) knock over; give ‹*golpe, coz etc*›; (*imprimir*) print. • *vi* (*disparar*) shoot. **~se** *vpr*

throw o.s.; (*tumbarse*) lie down. **~ a** tend to (be); (*parecerse a*) resemble. **~ de** pull; (*atraer*) attract. **a todo ~** at the most. **ir tirando** get by

tirita *f* sticking-plaster, plaster (*Brit*)

tirit|ar *vi* shiver. **~ón** *m* shiver

tiro *m* throw; (*disparo*) shot; (*alcance*) range. **~ a gol** shot at goal. **a ~** within range. **errar el ~** miss. **pegarse un ~** shoot o.s.

tiroides *m* thyroid (gland)

tirón *m* tug. **de un ~** in one go

tirote|ar *vt* shoot at. **~o** *m* shooting

tisana *f* herb tea

tisis *f* tuberculosis

tisú *m* (*pl* **tisus**) tissue

títere *m* puppet. **~ de guante** glove puppet. **~s** *mpl* puppet show

titilar *vi* quiver; ‹*estrella*› twinkle

titiritero *m* puppeteer; (*acróbata*) acrobat; (*malabarista*) juggler

titube|ante *a* shaky; (*fig*) hesistant. **~ar** *vi* stagger; ‹*cosa*› be unstable; (*fig*) hesitate. **~o** *m* hesitation

titula|do *a* ‹*libro*› entitled; ‹*persona*› qualified. **~r** *m* headline; (*persona*) holder. ● *vt* call. **~rse** *vpr* be called

título *m* title; (*persona*) titled person; (*académico*) qualification; (*univ*) degree; (*de periódico etc*) headline; (*derecho*) right. **a ~ de** as, by way of

tiza *f* chalk

tiz|nar *vt* dirty. **~ne** *m* soot. **~ón** *m* half-burnt stick; (*fig*) stain

toall|a *f* towel. **~ero** *m* towel-rail

tobillo *m* ankle

tobogán *m* slide; (*para la nieve*) toboggan

tocadiscos *m invar* record-player

toca|do *a* (*con sombrero*) wearing. ● *m* hat. **~dor** *m* dressing-table. **~dor de señoras** ladies' room. **~nte** *a* touching. **~r** [7] *vt* touch; (*mus*) play; ring ‹*timbre*›; (*mencionar*) touch on; ‹*barco*› stop at. ● *vi* knock; (*corresponder a uno*) be one's turn. **~rse** *vpr* touch each other; (*cubrir la cabeza*) cover one's head. **en lo que ~ a, en lo ~nte a** as for. **estar ~do (de la cabeza)** be mad. **te ~ a ti** it's your turn

tocateja. **a ~** cash

tocayo *m* namesake

tocino *m* bacon

tocólogo *m* obstetrician

todavía *adv* still, yet. **~ no** not yet

todo *a* all; (*entero*) the whole; (*cada*) every. ● *adv* completely, all. ● *m* whole. ● *pron* everything, all; (*en plural*) everyone. **~ el día** all day. **~ el mundo** everyone. **~ el que** anyone who. **~ incluido** all in. **~ lo contrario** quite the opposite. **~ lo que** anything which. **~s los días** every day. **~s los dos** both (of them). **~s los tres** all three. **ante ~** above all. **a ~ esto** meanwhile. **con ~** still, however. **del ~** completely. **en ~ el mundo** anywhere. **estar en ~** be on the ball. **es ~ uno** it's all the same. **nosotros ~s** all of us. **sobre ~** above all

toldo *m* sunshade

tolera|ncia *f* tolerance. **~nte** *a* tolerant. **~r** *vt* tolerate

tolondro *m* (*chichón*) lump

toma *f* taking; (*med*) dose; (*de agua*) outlet; (*elec*) socket; (*elec, clavija*) plug. ● *int* well!, fancy that! **~ de corriente** power point. **~dura** *f*. **~dura de pelo** hoax. **~r** *vt* take; catch ‹*autobús, tren etc*›; (*beber*) drink, have; (*comer*) eat, have. ● *vi* take; (*dirigirse*) go. **~rse** *vpr* take; (*beber*) drink, have; (*comer*) eat, have. **~r a bien** take well. **~r a mal** take badly. **~r en serio** take seriously. **~rla con uno** pick on s.o. **~r nota** take note. **~r por** take for. **~ y daca** give and take. **¿qué va a ~r?** what would you like?

tomate *m* tomato

tomavistas *m invar* cine-camera

tómbola *f* tombola

tomillo *m* thyme

tomo *m* volume

ton. **sin ~ ni son** without rhyme or reason

tonada *f*, **tonadilla** *f* tune

tonel *m* barrel. **~ada** *f* ton. **~aje** *m* tonnage

tónic|a *f* tonic water; (*mus*) tonic. **~o** *a* tonic; ‹*sílaba*› stressed. ● *m* tonic

tonificar [7] *vt* invigorate

tono *m* tone; (*mus, modo*) key; (*color*) shade

tont|ería *f* silliness; (*cosa*) silly thing; (*dicho*) silly remark. **~o** *a* silly. ● *m* fool, idiot; (*payaso*) clown. **dejarse de ~erías** stop wasting time. **hacer el ~o** act the fool. **hacerse el ~o** feign ignorance

topacio *m* topaz

topar *vt* ‹*animal*› butt; ‹*persona*› bump into; (*fig*) run into. ● *vi*. ~ **con** run into

tope *a* maximum. ● *m* end; (*de tren*) buffer. **hasta los ~s** crammed full. **ir a ~** go flat out

tópico *a* topical. ● *m* cliché

topo *m* mole

topogr|afía *f* topography. **~áfico** *a* topographical

toque *m* touch; (*sonido*) sound; (*de campana*) peal; (*de reloj*) stroke; (*fig*) crux. ~ **de queda** curfew. **~tear** *vt* keep fingering, fiddle with. **dar el último ~** put the finishing touches

toquilla *f* shawl

tórax *m* thorax

torbellino *m* whirlwind; (*de polvo*) cloud of dust; (*fig*) whirl

torcer [2 & 9] *vt* twist; (*doblar*) bend; wring out ‹*ropa*›. ● *vi* turn. **~se** *vpr* twist; (*fig, desviarse*) go astray; (*fig, frustrarse*) go wrong

tordo *a* dapple grey. ● *m* thrush

tore|ar *vt* fight; (*evitar*) dodge; (*entretener*) put off. ● *vi* fight (bulls). **~o** *m* bullfighting. **~ro** *m* bullfighter

torment|a *f* storm. **~o** *m* torture. **~oso** *a* stormy

tornado *m* tornado

tornar *vt* return

tornasolado *a* irridescent

torneo *m* tournament

tornillo *m* screw

torniquete *m* (*entrada*) turnstile

torno *m* lathe; (*de alfarero*) wheel. **en ~ a** around

toro *m* bull. **~s** *mpl* bullfighting. **ir a los ~s** go to a bullfight

toronja *f* grapefruit

torpe *a* clumsy; (*estúpido*) stupid

torped|ero *m* torpedo-boat. **~o** *m* torpedo

torpeza *f* clumsiness; (*de inteligencia*) slowness

torpor *m* torpor

torrado *m* toasted chick-pea

torre *f* tower; (*en ajedrez*) castle, rook

torrefac|ción *f* roasting. **~to** *a* roasted

torren|cial *a* torrential. **~te** *m* torrent; (*circulatorio*) bloodstream; (*fig*) flood

tórrido *a* torrid

torrija *f* French toast

torsión *f* twisting

torso *m* torso

torta *f* tart; (*bollo, fam*) cake; (*golpe*) slap, punch; (*Mex, bocadillo*) sandwich. **~zo** *m* slap, punch. **no entender ni ~** not understand a word of it. **pegarse un ~zo** have a bad accident

tortícolis *f* stiff neck

tortilla *f* omelette; (*Mex, de maiz*) tortilla, maize cake. ~ **francesa** plain omelette

tórtola *f* turtle-dove

tortuga *f* tortoise; (*de mar*) turtle

tortuoso *a* winding; (*fig*) devious

tortura *f* torture. **~r** *vt* torture

torvo *a* grim

tos *f* cough. ~ **ferina** whooping cough

tosco *a* crude; ‹*persona*› coarse

toser *vi* cough

tósigo *m* poison

tosquedad *f* crudeness; (*de persona*) coarseness

tost|ada *f* toast. **~ado** *a* ‹*pan*› toasted; ‹*café*› roasted; ‹*persona*› tanned; (*marrón*) brown. **~ar** *vt* toast ‹*pan*›; roast ‹*café*›; tan ‹*piel*›. **~ón** *m* (*pan*) crouton; (*lata*) bore

total *a* total. ● *adv* after all. ● *m* total; (*totalidad*) whole. **~idad** *f* whole. **~itario** *a* totalitarian. **~izar** [10] *vt* total. ~ **que** so, to cut a long story short

tóxico *a* toxic

toxicómano *m* drug addict

toxina *f* toxin

tozudo *a* stubborn

traba *f* bond; (*fig, obstáculo*) obstacle. **poner ~s a** hinder

trabaj|ador *a* hard-working. ● *m* worker. **~ar** *vt* work (**de** as); knead ‹*masa*›; (*estudiar*) work at; ‹*actor*› act. ● *vi* work. **~o** *m* work. **~os** *mpl* hardships. **~os forzados** hard labour. **~oso** *a* hard. **costar ~o** be difficult. **¿en qué ~as?** what work do you do?

trabalenguas *m invar* tongue-twister

traba|r *vt* (*sujetar*) fasten; (*unir*) join; (*empezar*) start; (*culin*) thicken. **~rse** *vpr* get tangled up. **trabársele la lengua** get tongue-tied. **~zón** *f* joining; (*fig*) connection

trabucar [7] *vt* mix up

trácala *f* (*Mex*) trick

tracción *f* traction

tractor *m* tractor

tradici|ón *f* tradition. **~onal** *a* traditional. **~onalista** *m* & *f* traditionalist

traduc|ción *f* translation. **~ir** [47] *vt* translate (**al** into). **~tor** *m* translator

traer [41] *vt* bring; (*llevar*) carry; (*atraer*) attract. **traérselas** be difficult

trafica|nte *m* & *f* dealer. **~r** [7] *vi* deal

tráfico *m* traffic; (*com*) trade

traga|deras *fpl* (*fam*) throat. **tener buenas ~deras** (*ser crédulo*) swallow anything; (*ser tolerante*) be easygoing. **~luz** *m* skylight. **~perras** *f invar* slot-machine. **~r** [12] *vt* swallow; (*comer mucho*) devour; (*absorber*) absorb; (*fig*) swallow up. **no (poder) ~r** not be able to stand. **~rse** *vpr* swallow; (*fig*) swallow up

tragedia *f* tragedy

trágico *a* tragic. ● *m* tragedian

trag|o *m* swallow, gulp; (*pequeña porción*) sip; (*fig*, *disgusto*) blow. **~ón** *a* greedy. ● *m* glutton. **echar(se) un ~o** have a drink

trai|ción *f* treachery; (*pol*) treason. **~cionar** *vt* betray. **~cionero** *a* treacherous. **~dor** *a* treacherous. ● *m* traitor

traigo *vb véase* **traer**

traje *m* dress; (*de hombre*) suit. ● *vb véase* **traer**. **~ de baño** swimming-costume. **~ de ceremonia**, **~ de etiqueta**, **~ de noche** evening dress

traj|ín *m* (*transporte*) haulage; (*jaleo*, *fam*) bustle. **~inar** *vt* transport. ● *vi* bustle about

trama *f* weft; (*fig*) link; (*fig*, *argumento*) plot. **~r** *vt* weave; (*fig*) plot

tramitar *vt* negotiate

trámite *m* step. **~s** *mpl* procedure. **en ~** in hand

tramo *m* (*parte*) section; (*de escalera*) flight

tramp|a *f* trap; (*puerta*) trapdoor; (*fig*) trick. **~illa** *f* trapdoor. **hacer ~a** cheat

trampolín *m* trampoline; (*fig*, *de piscina*) springboard

tramposo *a* cheating. ● *m* cheat

tranca *f* stick; (*de puerta*) bar

trance *m* moment; (*hipnótico etc*) trance. **a todo ~** at all costs

tranco *m* stride

tranquil|idad *f* (peace and) quiet; (*de espíritu*) peace of mind. **~izar** [10] *vt* reassure. **~o** *a* quiet; ‹*conciencia*› clear; ‹*mar*› calm; (*despreocupado*) thoughtless. **estáte ~o** don't worry

trans... *pref* (*véase también* **tras...**) trans...

transacción *f* transaction; (*acuerdo*) compromise

transatlántico *a* transatlantic. ● *m* (ocean) liner

transbord|ador *m* ferry. **~ar** *vt* transfer. **~arse** *vpr* change. **~o** *m* transfer. **hacer ~o** change (**en** at)

transcri|bir (*pp* **transcrito**) *vt* transcribe. **~pción** *f* transcription

transcur|rir *vi* pass. **~so** *m* course

transeúnte *a* temporary. ● *m* & *f* passer-by

transfer|encia *f* transfer. **~ir** [4] *vt* transfer

transfigurar *vt* transfigure

transforma|ción *f* transformation. **~dor** *m* transformer. **~r** *vt* transform

transfusión *f* transfusion. **hacer una ~** give a blood transfusion

transgre|dir *vt* transgress. **~sión** *f* transgression

transición *f* transition

transido *a* overcome

transigir [14] *vi* give in, compromise

transistor *m* transistor; (*radio*) radio

transita|ble *a* passable. **~r** *vi* go

transitivo *a* transitive

tránsito *m* transit; (*tráfico*) traffic

transitorio *a* transitory

translúcido *a* translucent

transmi|sión *f* transmission; (*radio*, *TV*) broadcast. **~sor** *m* transmitter. **~sora** *f* broadcasting station. **~tir** *vt* transmit; (*radio*, *TV*) broadcast; (*fig*) pass on

transparen|cia *f* transparency. **~tar** *vt* show. **~te** *a* transparent

transpira|ción *f* perspiration. **~r** *vi* transpire; (*sudar*) sweat

transponer [34] *vt* move. ● *vi* disappear round ‹*esquina etc*›; disappear behind ‹*montaña etc*›. **~se** *vpr* disappear

transport|ar *vt* transport. **~e** *m* transport. **empresa** *f* **de ~es** removals company

transversal *a* transverse; ‹*calle*› side

tranvía *m* tram

trapacería *f* swindle

trapear *vt* (*LAm*) mop

trapecio *m* trapeze; (*math*) trapezium

trapiche *m* (*para azúcar*) mill; (*para aceitunas*) press
trapicheo *m* fiddle
trapisonda *f* (*jaleo, fam*) row; (*enredo, fam*) plot
trapo *m* rag; (*para limpiar*) cloth. **~s** *mpl* (*fam*) clothes. **a todo ~** out of control
tráquea *f* windpipe, trachea
traquete|ar *vt* bang, rattle. **~o** *m* banging, rattle
tras *prep* after; (*detrás*) behind; (*encima de*) as well as
tras... *pref* (*véase también* **trans...**) trans...
trascende|ncia *f* importance. **~ntal** *a* transcendental; (*importante*) important. **~r** [1] *vi* (*oler*) smell (**a** of); (*saberse*) become known; (*extenderse*) spread
trasegar [1 & 12] *vt* move around
trasero *a* back, rear. ● *m* (*anat*) bottom
trasgo *m* goblin
traslad|ar *vt* move; (*aplazar*) postpone; (*traducir*) translate; (*copiar*) copy. **~o** *m* transfer; (*copia*) copy; (*mudanza*) removal. **dar ~o** send a copy
trasl|úcido *a* translucent. **~ucirse** [11] *vpr* be translucent; (*dejarse ver*) show through; (*fig, revelarse*) be revealed. **~uz** *m*. **al ~uz** against the light
trasmano *m*. **a ~** out of reach; (*fig*) out of the way
trasnochar *vt* (*acostarse tarde*) go to bed late; (*no acostarse*) stay up all night; (*no dormir*) be unable to sleep; (*pernoctar*) spend the night
traspas|ar *vt* pierce; (*transferir*) transfer; (*pasar el límite*) go beyond. **~o** *m* transfer. **se ~a** for sale
traspié *m* trip; (*fig*) slip. **dar un ~** stumble; (*fig*) slip up
trasplant|ar *vt* transplant. **~e** *m* transplanting; (*med*) transplant
trastada *f* stupid thing; (*jugada*) dirty trick, practical joke
traste *m* fret. **dar al ~ con** ruin. **ir al ~** fall through
trastero *m* storeroom
trastienda *f* back room; (*fig*) shrewdness
trasto *m* piece of furniture; (*cosa inútil*) piece of junk; (*persona*) useless person, dead loss (*fam*)
trastorn|ado *a* mad. **~ar** *vt* upset; (*volver loco*) drive mad; (*fig, gustar mucho, fam*) delight. **~arse** *vpr* get upset; (*volverse loco*) go mad. **~o** *m* (*incl med*) upset; (*pol*) disturbance; (*fig*) confusion
trastrocar [2 & 7] *vt* change round
trat|able *a* friendly. **~ado** *m* treatise; (*acuerdo*) treaty. **~amiento** *m* treatment; (*título*) title. **~ante** *m & f* dealer. **~ar** *vt* (*incl med*) treat; deal with ‹*asunto etc*›; (*com*) deal; (*manejar*) handle; (*de tú, de Vd*) address (**de** as); (*llamar*) call. ● *vi* deal (with). **~ar con** have to do with; know ‹*persona*›; (*com*) deal in. **~ar de** be about; (*intentar*) try. **~o** *m* treatment; (*acuerdo*) agreement; (*título*) title; (*relación*) relationship. **¡~o hecho!** agreed! **~os** *mpl* dealings. **¿de qué se ~a?** what's it about?
traum|a *m* trauma. **~ático** *a* traumatic
través *m* (*inclinación*) slant. **a ~ de** through; (*de un lado a otro*) across. **de ~** across; (*de lado*) sideways. **mirar de ~** look askance at
travesaño *m* crosspiece
travesía *f* crossing; (*calle*) side-street
trav|esura *f* prank. **~ieso** *a* ‹*niño*› mischievous, naughty
trayecto *m* road; (*tramo*) stretch; (*ruta*) route; (*viaje*) journey. **~ria** *f* trajectory; (*fig*) course
traz|a *f* plan; (*aspecto*) look, appearance; (*habilidad*) skill. **~ado** *a*. **bien ~ado** good-looking. **mal ~ado** unattractive. ● *m* plan. **~ar** [10] *vt* draw; (*bosquejar*) sketch. **~o** *m* line
trébol *m* clover. **~es** *mpl* (*en naipes*) clubs
trece *a & m* thirteen
trecho *m* stretch; (*distancia*) distance; (*tiempo*) while. **a ~s** in places. **de ~ en ~** at intervals
tregua *f* truce; (*fig*) respite
treinta *a & m* thirty
tremendo *a* terrible; (*extraordinario*) terrific
trementina *f* turpentine
tren *m* train; (*equipaje*) luggage. **~ de aterrizaje** landing gear. **~ de vida** lifestyle
tren|cilla *f* braid. **~za** *f* braid; (*de pelo*) plait. **~zar** [10] *vt* plait
trepa|dor *a* climbing. **~r** *vt/i* climb
tres *a & m* three. **~cientos** *a & m* three hundred. **~illo** *m* three-piece suite; (*mus*) triplet
treta *f* trick

tri|angular *a* triangular. **~ángulo** *m* triangle
trib|al *a* tribal. **~u** *f* tribe
tribulación *f* tribulation
tribuna *f* platform; (*de espectadores*) stand
tribunal *m* court; (*de examen etc*) board; (*fig*) tribunal
tribut|ar *vt* pay. **~o** *m* tribute; (*impuesto*) tax
triciclo *m* tricycle
tricolor *a* three-coloured
tricornio *a* three-cornered. ● *m* three-cornered hat
tricotar *vt/i* knit
tridimensional *a* three-dimensional
tridente *m* trident
trigésimo *a* thirtieth
trig|al *m* wheat field. **~o** *m* wheat
trigonometría *f* trigonometry
trigueño *a* olive-skinned; ‹*pelo*› dark blonde
trilogía *f* trilogy
trilla|do *a* (*fig, manoseado*) trite; (*fig, conocido*) well-known. **~r** *vt* thresh
trimestr|al *a* quarterly. **~e** *m* quarter; (*escol, univ*) term
trin|ar *vi* warble. **estar que trina** be furious
trinchar *vt* carve
trinchera *f* ditch; (*mil*) trench; (*rail*) cutting; (*abrigo*) trench coat
trineo *m* sledge
trinidad *f* trinity
Trinidad *f* Trinidad
trino *m* warble
trío *m* trio
tripa *f* intestine; (*culin*) tripe; (*fig, vientre*) tummy, belly. **~s** *fpl* (*de máquina etc*) parts, workings. **me duele la ~** I've got tummy-ache. **revolver las ~s** turn one's stomach
tripicallos *mpl* tripe
tripl|e *a* triple. ● *m*. **el ~e (de)** three times as much (as). **~icado** *a*. **por ~icado** in triplicate. **~icar** [7] *vt* treble
trípode *m* tripod
tríptico *m* triptych
tripula|ción *f* crew. **~nte** *m & f* member of the crew. **~r** *vt* man
triquitraque *m* (*ruido*) clatter
tris *m* crack; (*de papel etc*) ripping noise. **estar en un ~** be on the point of
triste *a* sad; ‹*paisaje, tiempo etc*› gloomy; (*fig, insignificante*) miserable. **~za** *f* sadness
tritón *m* newt
triturar *vt* crush
triunf|al *a* triumphal. **~ante** *a* triumphant. **~ar** *vi* triumph (**de, sobre** over). **~o** *m* triumph
triunvirato *m* triumvirate
trivial *a* trivial
triza *f* piece. **hacer algo ~s** smash sth to pieces
trocar [2 & 7] *vt* (ex)change
trocear *vt* cut up, chop
trocito *m* small piece
trocha *f* narrow path; (*atajo*) short cut
trofeo *m* trophy
tromba *f* waterspout. **~ de agua** heavy downpour
trombón *m* trombone; (*músico*) trombonist
trombosis *f invar* thrombosis
trompa *f* horn; (*de orquesta*) French horn; (*de elefante*) trunk; (*hocico*) snout; (*juguete*) (spinning) top; (*anat*) tube. ● *m* horn player. **coger una ~** (*fam*) get drunk
trompada *f*, **trompazo** *m* bump
trompet|a *f* trumpet; (*músico*) trumpeter, trumpet player; (*clarín*) bugle. **~illa** *f* ear-trumpet
trompicar [7] *vi* trip
trompo *m* (*juguete*) (spinning) top
trona|da *f* thunder storm. **~r** *vt* (*Mex*) shoot. ● *vi* thunder
tronco *m* trunk. **dormir como un ~** sleep like a log
tronchar *vt* bring down; (*fig*) cut short. **~se de risa** laugh a lot
trono *m* throne
trop|a *f* troops. **~el** *m* mob. **ser de ~a** be in the army
tropero *m* (*Arg, vaquero*) cowboy
tropez|ar [1 & 10] *vi* trip; (*fig*) slip up. **~ar con** run into. **~ón** *m* stumble; (*fig*) slip
tropical *a* tropical
trópico *a* tropical. ● *m* tropic
tropiezo *m* slip; (*desgracia*) mishap
trot|ar *vi* trot. **~e** *m* trot; (*fig*) toing and froing. **al ~e** trotting; (*de prisa*) in a rush. **de mucho ~e** hard-wearing
trozo *m* piece, bit. **a ~s** in bits
truco *m* knack; (*ardid*) trick. **coger el ~** get the knack
trucha *f* trout
trueno *m* thunder; (*estampido*) bang
trueque *m* exchange. **aun a ~ de** even at the expense of
trufa *f* truffle. **~r** *vt* stuff with truffles

truhán *m* rogue; (*gracioso*) jester
truncar [7] *vt* truncate; (*fig*) cut short
tu *a* your
tú *pron* you
tuba *f* tuba
tubérculo *m* tuber
tuberculosis *f* tuberculosis
tub|ería *f* pipes; (*oleoducto etc*) pipeline. **~o** *m* tube. **~o de ensayo** test tube. **~o de escape** (*auto*) exhaust (pipe). **~ular** *a* tubular
tuerca *f* nut
tuerto *a* one-eyed, blind in one eye. ● *m* one-eyed person
tuétano *m* marrow; (*fig*) heart. **hasta los ~s** completely
tufo *m* fumes; (*olor*) bad smell
tugurio *m* hovel, slum
tul *m* tulle
tulipán *m* tulip
tulli|do *a* paralysed. **~r** [22] *vt* cripple
tumba *f* grave, tomb
tumb|ar *vt* knock down, knock over; (*fig, en examen, fam*) fail; (*pasmar, fam*) overwhelm. **~arse** *vpr* lie down. **~o** *m* jolt. **dar un ~o** tumble. **~ona** *f* settee; (*sillón*) armchair; (*de lona*) deckchair
tumefacción *f* swelling
tumido *a* swollen
tumor *m* tumour
tumulto *m* turmoil; (*pol*) riot
tuna *f* prickly pear; (*de estudiantes*) student band
tunante *m & f* rogue
túnel *m* tunnel
Túnez *m* (*ciudad*) Tunis; (*país*) Tunisia
túnica *f* tunic
Tunicia *f* Tunisia
tupé *m* toupee; (*fig*) nerve
tupido *a* thick
turba *f* peat; (*muchedumbre*) mob
turba|ción *f* disturbance, upset; (*confusión*) confusion. **~do** *a* upset
turbante *m* turban
turbar *vt* upset; (*molestar*) disturb. **~se** *vpr* be upset
turbina *f* turbine
turbi|o *a* cloudy; ‹*vista*› blurred; ‹*asunto etc*› unclear. **~ón** *m* squall
turbulen|cia *f* turbulence; (*disturbio*) disturbance. **~te** *a* turbulent; ‹*persona*› restless
turco *a* Turkish. ● *m* Turk; (*lengua*) Turkish
tur|ismo *m* tourism; (*coche*) car. **~ista** *m & f* tourist. **~ístico** *a* tourist. **oficina** *f* **de ~ismo** tourist office
turn|arse *vpr* take turns (**para** to). **~o** *m* turn; (*de trabajo*) shift. **por ~o** in turn
turquesa *f* turquoise
Turquía *f* Turkey
turrón *m* nougat
turulato *a* (*fam*) stunned
tutear *vt* address as *tú*. **~se** *vpr* be on familiar terms
tutela *f* (*jurid*) guardianship; (*fig*) protection
tuteo *m* use of the familiar *tú*
tutor *m* guardian; (*escol*) form master
tuve *vb véase* **tener**
tuyo *a & pron* yours. **un amigo ~** a friend of yours

U

u *conj* or
ubicuidad *f* ubiquity
ubre *f* udder
ucraniano *a & m* Ukranian
Ud *abrev* (*Usted*) you
uf *int* phew!; (*de repugnancia*) ugh!
ufan|arse *vpr* be proud (**con, de** of); (*jactarse*) boast (**con, de** about). **~o** *a* proud
ujier *m* usher
úlcera *f* ulcer
ulterior *a* later; ‹*lugar*› further. **~mente** *adv* later, subsequently
últimamente *adv* (*recientemente*) recently; (*al final*) finally; (*en último caso*) as a last resort
ultim|ar *vt* complete. **~átum** *m* ultimatum
último *a* last; (*más reciente*) latest; (*más lejano*) furthest; (*más alto*) top; (*más bajo*) bottom; (*fig, extremo*) extreme. **estar en las últimas** be on one's last legs; (*sin dinero*) be down to one's last penny. **por ~** finally. **ser lo ~** (*muy bueno*) be marvellous; (*muy malo*) be awful. **vestido a la última** dressed in the latest fashion
ultra *a* ultra, extreme
ultraj|ante *a* outrageous. **~e** *m* outrage
ultramar *m* overseas countries. **de ~, en ~** overseas

ultramarino *a* overseas. **~s** *mpl* groceries. **tienda de ~s** grocer's (shop) (*Brit*), grocery store (*Amer*)
ultranza a ~ (*con decisión*) decisively; (*extremo*) extreme
ultra|sónico *a* ultrasonic. **~violeta** *a invar* ultraviolet
ulular *vi* howl; ‹*búho*› hoot
umbilical *a* umbilical
umbral *m* threshold
umbrío *a*, **umbroso** *a* shady
un *art indef m* (*pl* **unos**) a. • *a* one. **~os** *a pl* some
una *art indef f* a. **la ~** one o'clock
un|ánime *a* unanimous. **~animidad** *f* unanimity
undécimo *a* eleventh
ung|ir [14] *vt* anoint. **~üento** *m* ointment
únic|amente *adv* only. **~o** *a* only; (*fig, incomparable*) unique
unicornio *m* unicorn
unid|ad *f* unit; (*cualidad*) unity. **~o** *a* united
unifica|ción *f* unification. **~r** [7] *vt* unite, unify
uniform|ar *vt* standardize; (*poner uniforme a*) put into uniform. **~e** *a* & *m* uniform. **~idad** *f* uniformity
uni|génito *a* only. **~lateral** *a* unilateral
uni|ón *f* union; (*cualidad*) unity; (*tec*) joint. **~r** *vt* join; mix ‹*líquidos*›. **~rse** *vpr* join together
unísono *m* unison. **al ~** in unison
unitario *a* unitary
universal *a* universal
universi|dad *f* university. **U~dad a Distancia** Open University. **~tario** *a* university
universo *m* universe
uno *a* one; (*en plural*) some. • *pron* one; (*alguien*) someone, somebody. • *m* one. **~ a otro** each other. **~ y otro** both. **(los) ~s... (los) otros** some... others
untar *vt* grease; (*med*) rub; (*fig, sobornar, fam*) bribe
uña *f* nail; (*de animal*) claw; (*casco*) hoof
upa *int* up!
uranio *m* uranium
Urano *m* Uranus
urban|idad *f* politeness. **~ismo** *m* town planning. **~ístico** *a* urban. **~ización** *f* development. **~izar** [10] *vt* civilize; develop ‹*terreno*›. **~o** *a* urban
urbe *f* big city
urdimbre *f* warp
urdir *vt* (*fig*) plot
urg|encia *f* urgency; (*emergencia*) emergency; (*necesidad*) urgent need. **~ente** *a* urgent. **~ir** [14] *vi* be urgent. **carta** *f* **~ente** express letter
urinario *m* urinal
urna *f* urn; (*pol*) ballot box
urraca *f* magpie
URSS *abrev* (*historia*) (*Unión de Repúblicas Socialistas Soviéticas*) USSR, Union of Soviet Socialist Republics
Uruguay *m*. **el ~** Uruguay
uruguayo *a* & *m* Uruguayan
us|ado *a* used; ‹*ropa etc*› worn. **~anza** *f* usage, custom. **~ar** *vt* use; (*llevar*) wear. **~o** *m* use; (*costumbre*) usage, custom. **al ~o** (*de moda*) in fashion; (*a la manera de*) in the style of. **de ~o externo** for external use
usted *pron* you
usual *a* usual
usuario *a* user
usur|a *f* usury. **~ero** *m* usurer
usurpar *vt* usurp
usuta *f* (*Arg*) sandal
utensilio *m* tool; (*de cocina*) utensil. **~s** *mpl* equipment
útero *m* womb
útil *a* useful. **~es** *mpl* implements
utili|dad *f* usefulness. **~tario** *a* utilitarian; ‹*coche*› utility. **~zación** *f* use, utilization. **~zar** [10] *vt* use, utilize
uva *f* grape. **~ pasa** raisin. **mala ~** bad mood

V

vaca *f* cow; (*carne*) beef
vacaciones *fpl* holiday(s). **estar de ~** be on holiday. **ir de ~** go on holiday
vaca|nte *a* vacant. • *f* vacancy. **~r** [7] *vi* fall vacant
vaci|ar [20] *vt* empty; (*ahuecar*) hollow out; (*en molde*) cast; (*afilar*) sharpen. **~edad** *f* emptiness; (*tontería*) silly thing, frivolity
vacila|ción *f* hesitation. **~nte** *a* unsteady; (*fig*) hesitant. **~r** *vi* sway; (*dudar*) hesitate; (*fam*) tease
vacío *a* empty; (*vanidoso*) vain. • *m* empty space; (*estado*) emptiness; (*en física*) vacuum; (*fig*) void

vacuidad *f* emptiness; (*tontería*) silly thing, frivolity
vacuna *f* vaccine. **~ción** *f* vaccination. **~r** *vt* vaccinate
vacuno *a* bovine
vacuo *a* empty
vade *m* folder
vad|ear *vt* ford. **~o** *m* ford
vaga|bundear *vi* wander. **~bundo** *a* vagrant; ‹*perro*› stray. ● *m* tramp. **~r** [12] *vi* wander (about)
vagina *f* vagina
vago *a* vague; (*holgazán*) idle; ‹*foto*› blurred. ● *m* idler
vag|ón *m* carriage; (*de mercancías*) truck, wagon. **~ón restaurante** dining-car. **~oneta** *f* truck
vahído *m* dizzy spell
vaho *m* breath; (*vapor*) steam. **~s** *mpl* inhalation
vaina *f* sheath; (*bot*) pod
vainilla *f* vanilla
vaivén *m* swaying; (*de tráfico*) coming and going; (*fig, de suerte*) change. **vaivenes** *mpl* (*fig*) ups and downs
vajilla *f* dishes, crockery. **lavar la ~** wash up
vale *m* voucher; (*pagaré*) IOU. **~dero** *a* valid
valenciano *a* from Valencia
valent|ía *f* courage; (*acción*) brave deed. **~ón** *m* braggart
valer [42] *vt* be worth; (*costar*) cost; (*fig, significar*) mean. ● *vi* be worth; (*costar*) cost; (*servir*) be of use; (*ser valedero*) be valid; (*estar permitido*) be allowed. ● *m* worth. **~ la pena** be worthwhile, be worth it. **¿cuánto vale?** how much is it?. **no ~ para nada** be useless. **¡vale!** all right!, OK! (*fam*). **¿vale?** all right?, OK? (*fam*)
valeroso *a* courageous
valgo *vb véase* **valer**
valía *f* worth
validez *f* validity. **dar ~ a** validate
válido *a* valid
valiente *a* brave; (*valentón*) boastful; (*en sentido irónico*) fine. ● *m* brave person; (*valentón*) braggart
valija *f* case; (*de correos*) mailbag. **~ diplomática** diplomatic bag
val|ioso *a* valuable. **~or** *m* value, worth; (*descaro, fam*) nerve. **~ores** *mpl* securities. **~oración** *f* valuation. **~orar** *vt* value. **conceder ~or a** attach importance to. **objetos** *mpl* **de ~or** valuables. **sin ~or** worthless
vals *m invar* waltz
válvula *f* valve
valla *f* fence; (*fig*) barrier
valle *m* valley
vampiro *m* vampire
vanagloriarse [20 *o regular*] *vpr* boast
vanamente *adv* uselessly, in vain
vandalismo *m* vandalism
vándalo *m* vandal
vanguardia *f* vanguard. **de ~** (*en arte, música etc*) avant-garde
vanid|ad *f* vanity. **~oso** *a* vain
vano *a* vain; (*inútil*) useless. **en ~** in vain
vapor *m* steam; (*gas*) vapour; (*naut*) steamer. **~izador** *m* spray. **~izar** [10] vaporize. **al ~** (*culin*) steamed
vaquer|ía *f* dairy. **~o** *m* cow-herd, cowboy. **~os** *mpl* jeans
vara *f* stick; (*de autoridad*) staff; (*medida*) yard
varar *vi* run aground
varia|ble *a & f* variable. **~ción** *f* variation. **~nte** *f* version. **~ntes** *fpl* hors d'oeuvres. **~r** [20] *vt* change; (*dar variedad a*) vary. ● *vi* vary; (*cambiar*) change
varice *f* varicose vein
varicela *f* chickenpox
varicoso *a* having varicose veins
variedad *f* variety
varilla *f* stick; (*de metal*) rod
vario *a* varied; (*en plural*) several
varita *f* wand
variz *f* varicose vein
var|ón *a* male. ● *m* man; (*niño*) boy. **~onil** *a* manly
vasc|o *a & m* Basque. **~ongado** *a* Basque. **~uence** *a & m* Basque. **las V~ongadas** the Basque provinces
vasectomía *f* vasectomy
vaselina *f* Vaseline (P), petroleum jelly
vasija *f* pot, container
vaso *m* glass; (*anat*) vessel
vástago *m* shoot; (*descendiente*) descendant; (*varilla*) rod
vasto *a* vast
Vaticano *m* Vatican
vaticin|ar *vt* prophesy. **~io** *m* prophesy
vatio *m* watt
vaya *vb véase* **ir**
Vd *abrev* (*Usted*) you
vecin|dad *f* neighbourhood, vicinity; (*vecinos*) neighbours. **~dario** *m*

inhabitants, neighbourhood. **~o** *a* neighbouring; (*de al lado*) next-door. ● *m* neighbour

veda|do *m* preserve. **~do de caza** game preserve. **~r** *vt* prohibit

vega *f* fertile plain

vegeta|ción *f* vegetation. **~l** *a* vegetable. ● *m* plant, vegetable. **~r** *vi* grow; ‹*persona*› vegetate. **~riano** *a & m* vegetarian

vehemente *a* vehement

vehículo *m* vehicle

veinte *a & m* twenty. **~na** *f* score

veinti|cinco *a & m* twenty-five. **~cuatro** *a & m* twenty-four. **~dós** *a & m* twenty-two. **~nueve** *a & m* twenty-nine; **~ocho** *a & m* twenty-eight. **~séis** *a & m* twenty-six. **~siete** *a & m* twenty-seven. **~trés** *a & m* twenty-three. **~ún** *a* twenty-one. **~uno** *a & m* (*delante de nombre masculino* **veintún**) twenty-one

vejar *vt* humiliate; (*molestar*) vex

vejez *f* old age

vejiga *f* bladder; (*med*) blister

vela *f* (*naut*) sail; (*de cera*) candle; (*falta de sueño*) sleeplessness; (*vigilia*) vigil. **pasar la noche en ~** have a sleepless night

velada *f* evening party

vela|do *a* veiled; (*foto*) blurred. **~r** *vt* watch over; (*encubrir*) veil; (*foto*) blur. ● *vi* stay awake, not sleep. **~r por** look after. **~rse** *vpr* (*foto*) blur

velero *m* sailing-ship

veleta *f* weather vane

velo *m* veil

veloc|idad *f* speed; (*auto etc*) gear. **~ímetro** *m* speedometer. **~ista** *m & f* sprinter. **a toda ~idad** at full speed

velódromo *m* cycle-track

veloz *a* fast, quick

vell|o *m* down. **~ón** *m* fleece. **~udo** *a* hairy

vena *f* vein; (*en madera*) grain. **estar de/en ~** be in the mood

venado *m* deer; (*culin*) venison

vencedor *a* winning. ● *m* winner

vencejo *m* (*pájaro*) swift

venc|er [9] *vt* beat; (*superar*) overcome. ● *vi* win; ‹*plazo*› expire. **~erse** *vpr* collapse; ‹*persona*› control o.s. **~ido** *a* beaten; (*com, atrasado*) in arrears. **darse por ~ido** give up. **los ~idos** *mpl* (*en deportes etc*) the losers

venda *f* bandage. **~je** *m* dressing. **~r** *vt* bandage

vendaval *m* gale

vende|dor *a* selling. ● *m* seller, salesman. **~dor ambulante** pedlar. **~r** *vt* sell. **~rse** *vpr* sell. **~rse caro** play hard to get. **se ~** for sale

vendimia *f* grape harvest; (*de vino*) vintage, year

Venecia *f* Venice

veneciano *a* Venetian

veneno *m* poison; (*fig, malevolencia*) spite. **~so** *a* poisonous

venera *f* scallop shell

venera|ble *a* venerable. **~ción** *f* reverence. **~r** *vt* revere

venéreo *a* venereal

venero *m* (*yacimiento*) seam; (*de agua*) spring; (*fig*) source

venezolano *a & m* Venezuelan

Venezuela *f* Venezuela

venga|nza *f* revenge. **~r** [12] *vt* avenge. **~rse** *vpr* take revenge (**de, por** for) (**de, en** on). **~tivo** *a* vindictive

vengo *vb véase* **venir**

venia *f* (*permiso*) permission

venial *a* venial

veni|da *f* arrival; (*vuelta*) return. **~dero** *a* coming. **~r** [53] *vi* come; (*estar, ser*) be. **~r a para** come to. **~r bien** suit. **la semana que viene** next week. **¡venga!** come on!

venta *f* sale; (*posada*) inn. **en ~** for sale

ventaj|a *f* advantage. **~oso** *a* advantageous

ventan|a *f* window; (*de la nariz*) nostril. **~illa** *f* window

ventarrón *m* (*fam*) strong wind

ventear *vt* (*olfatear*) sniff

ventero *m* innkeeper

ventila|ción *f* ventilation. **~dor** *m* fan. **~r** *vt* air

vent|isca *f* blizzard. **~olera** *f* gust of wind. **~osa** *f* sucker. **~osidad** *f* wind, flatulence. **~oso** *a* windy

ventrílocuo *m* ventriloquist

ventrudo *a* pot-bellied

ventur|a *f* happiness; (*suerte*) luck. **~oso** *a* happy, lucky. **a la ~a** at random. **echar la buena ~a a uno** tell s.o.'s fortune. **por ~a** by chance; (*afortunadamente*) fortunately

Venus *f* Venus

ver [43] *vt* see; watch ‹*televisión*›. ● *vi* see. **~se** *vpr* see o.s.; (*encontrarse*) find o.s.; ‹*dos personas*› meet. **a mi (modo de) ~** in my view. **a ~** let's see. **de buen ~** good-looking. **dejarse ~** show. **¡habráse**

visto! did you ever! **no poder ~** not be able to stand. **no tener nada que ~ con** have nothing to do with. **¡para que veas!** so there! **vamos a ~** let's see. **ya lo veo** that's obvious. **ya ~ás** you'll see. **ya ~emos** we'll see
vera *f* edge; (*de río*) bank
veracruzano *a* from Veracruz
veran|eante *m & f* tourist, holiday-maker. **~ear** *vi* spend one's holiday. **~eo** *m* (summer) holiday. **~iego** *a* summer. **~o** *m* summer. **casa** *f* **de ~eo** summer-holiday home. **ir de ~eo** go on holiday. **lugar** *m* **de ~eo** holiday resort
veras *fpl.* **de ~** really
veraz *a* truthful
verbal *a* verbal
verbena *f* (*bot*) verbena; (*fiesta*) fair; (*baile*) dance
verbo *m* verb. **~so** *a* verbose
verdad *f* truth. **¿~?** isn't it?, aren't they?, won't it? etc. **~eramente** *adv* really. **~ero** *a* true; (*fig*) real. **a decir ~** to tell the truth. **de ~** really. **la pura ~** the plain truth. **si bien es ~ que** although
verd|e *a* green; ‹*fruta etc*› unripe; ‹*chiste etc*› dirty, blue. ● *m* green; (*hierba*) grass. **~or** *m* greenness
verdugo *m* executioner; (*fig*) tyrant
verdu|lería *f* greengrocer's (shop). **~lero** *m* greengrocer. **~ra** *f* (green) vegetable(s)
vereda *f* path; (*LAm*, *acera*) pavement (*Brit*), sidewalk (*Amer*)
veredicto *m* verdict
vergel *m* large garden; (*huerto*) orchard
verg|onzoso *a* shameful; (*tímido*) shy. **~üenza** *f* shame; (*timidez*) shyness. **¡es una ~üenza!** it's a disgrace! **me da ~üenza** I'm ashamed; (*tímido*) I'm shy about. **tener ~üenza** be ashamed; (*tímido*) be shy
verídico *a* true
verifica|ción *f* verification. **~r** [7] *vt* check. **~rse** *vpr* take place; (*resultar verdad*) come true
verja *f* grating; (*cerca*) railings; (*puerta*) iron gate
vermú *m*, **vermut** *m* vermouth
vernáculo *a* vernacular
verosímil *a* likely; ‹*relato etc*› credible
verraco *m* boar
verruga *f* wart
versado *a* versed
versar *vi* turn. **~ sobre** be about
versátil *a* versatile; (*fig*) fickle
versión *f* version; (*traducción*) translation
verso *m* verse; (*línea*) line
vértebra *f* vertebra
verte|dero *m* rubbish tip; (*desaguadero*) drain. **~dor** *m* drain. **~r** [1] *vt* pour; (*derramar*) spill. ● *vi* flow
vertical *a & f* vertical
vértice *f* vertex
vertiente *f* slope
vertiginoso *a* dizzy
vértigo *m* dizziness; (*med*) vertigo. **de ~** (*fam*) amazing
vesania *f* rage; (*med*) insanity
vesícula *f* blister. **~ biliar** gall-bladder
vespertino *a* evening
vestíbulo *m* hall; (*de hotel, teatro etc*) foyer
vestido *m* (*de mujer*) dress; (*ropa*) clothes
vestigio *m* trace. **~s** *mpl* remains
vest|imenta *f* clothing. **~ir** [5] *vt* (*ponerse*) put on; (*llevar*) wear; dress ‹*niño etc*›. ● *vi* dress; (*llevar*) wear. **~irse** *vpr* get dressed; (*llevar*) wear. **~uario** *m* wardrobe; (*cuarto*) dressing-room
Vesuvio *m* Vesuvius
vetar *vt* veto
veterano *a* veteran
veterinari|a *f* veterinary science. **~o** *a* veterinary. ● *m* vet (*fam*), veterinary surgeon (*Brit*), veterinarian (*Amer*)
veto *m* veto. **poner el ~ a** veto
vetusto *a* ancient
vez *f* time; (*turno*) turn. **a la ~** at the same time; (*de una vez*) in one go. **alguna que otra ~** from time to time. **alguna ~** sometimes; (*en preguntas*) ever. **algunas veces** sometimes. **a su ~** in (his) turn. **a veces** sometimes. **cada ~ más** more and more. **de una ~** in one go. **de una ~ para siempre** once and for all. **de ~ en cuando** from time to time. **dos veces** twice. **2 veces 4** 2 times 4. **en ~ de** instead of. **érase una ~, había una ~** once upon a time. **muchas veces** often. **otra ~** again. **pocas veces, rara ~** rarely. **repetidas veces** again and again. **tal ~** perhaps. **una ~ (que)** once
vía *f* road; (*rail*) line; (*anat*) tract; (*fig*) way. ● *prep* via. **~ aérea** by air.

~ de comunicación *f* means of communication. **~ férrea** railway (*Brit*), railroad (*Amer*). **~ rápida** fast lane. **estar en ~s de** be in the process of
viab|ilidad *f* viability. **~le** *a* viable
viaducto *m* viaduct
viaj|ante *m & f* commercial traveller. **~ar** *vi* travel. **~e** *m* journey; (*corto*) trip. **~e de novios** honeymoon. **~ero** *m* traveller; (*pasajero*) passenger. **¡buen ~e!** have a good journey!
víbora *f* viper
vibra|ción *f* vibration. **~nte** *a* vibrant. **~r** *vt/i* vibrate
vicario *m* vicar
vice... *pref* vice-...
viceversa *adv* vice versa
vici|ado *a* corrupt; ‹*aire*› stale. **~ar** *vt* corrupt; (*estropear*) spoil. **~o** *m* vice; (*mala costumbre*) bad habit. **~oso** *a* dissolute; ‹*círculo*› vicious
vicisitud *f* vicissitude
víctima *f* victim; (*de un accidente*) casualty
victori|a *f* victory. **~oso** *a* victorious
vid *f* vine
vida *f* life; (*duración*) lifetime. **¡~ mía!** my darling! **de por ~** for life. **en mi ~** never (in my life). **en ~ de** during the lifetime of. **estar en ~** be alive
vídeo *m* video recorder
video|cinta *f* videotape. **~juego** *m* video game
vidriar *vt* glaze
vidri|era *f* stained glass window; (*puerta*) glass door; (*LAm, escaparate*) shop window. **~ería** *f* glass works. **~ero** *m* glazier. **~o** *m* glass. **~oso** *a* glassy
vieira *f* scallop
viejo *a* old. ● *m* old person
Viena *f* Vienna
viene *vb véase* **venir**
viento *m* wind. **hacer ~** be windy
vientre *f* belly; (*matriz*) womb; (*intestino*) bowels. **llevar un niño en el ~** be pregnant
viernes *m* Friday. **V~ Santo** Good Friday
viga *f* beam; (*de metal*) girder
vigen|cia *f* validity. **~te** *a* valid; ‹*ley*› in force. **entrar en ~cia** come into force
vigésimo *a* twentieth
vigía *f* (*torre*) watch-tower; (*persona*) lookout
vigil|ancia *f* vigilance. **~ante** *a* vigilant. ● *m* watchman, supervisor. **~ar** *vt* keep an eye on. ● *vi* be vigilant; ‹*vigía etc*› keep watch. **~ia** *f* vigil; (*relig*) fasting
vigor *m* vigour; (*vigencia*) force. **~oso** *a* vigorous. **entrar en ~** come into force
vil *a* vile. **~eza** *f* vileness; (*acción*) vile deed
vilipendiar *vt* abuse
vilo. en ~ in the air
villa *f* town; (*casa*) villa. **la V~** Madrid
villancico *m* (Christmas) carol
villano *a* rustic; (*grosero*) coarse
vinagre *m* vinegar. **~ra** *f* vinegar bottle. **~ras** *fpl* cruet. **~ta** *f* vinaigrette (sauce)
vincular *vt* bind
vínculo *m* bond
vindicar [7] *vt* avenge; (*justificar*) vindicate
vine *vb véase* **venir**
vinicult|or *m* wine-grower. **~ura** *f* wine growing
vino *m* wine. **~ de Jerez** sherry. **~ de la casa** house wine. **~ de mesa** table wine
viña *f*, **viñedo** *m* vineyard
viola *f* viola; (*músico*) viola player
violación *f* violation; (*de una mujer*) rape
violado *a & m* violet
violar *vt* violate; break ‹*ley*›; rape ‹*mujer*›
violen|cia *f* violence; (*fuerza*) force; (*embarazo*) embarrassment. **~tar** *vt* force; break into ‹*casa etc*›. **~tarse** *vpr* force o.s. **~to** *a* violent; (*fig*) awkward. **hacer ~cia a** force
violeta *a invar & f* violet
viol|ín *m* violin; (*músico*) violinist. **~inista** *m & f* violinist. **~ón** *m* double bass; (*músico*) double-bass player. **~onc(h)elista** *m & f* cellist. **~onc(h)elo** *m* cello
vira|je *m* turn. **~r** *vt* turn. ● *vi* turn; (*fig*) change direction
virg|en *a & f* virgin. **~inal** *a* virginal. **~inidad** *f* virginity
Virgo *m* Virgo
viril *a* virile. **~idad** *f* virility
virtual *a* virtual
virtud *f* virtue; (*capacidad*) ability. **en ~ de** by virtue of
virtuoso *a* virtuous. ● *m* virtuoso

viruela *f* smallpox. **picado de ~s** pock-marked
virulé. a la ~ (*fam*) crooked; (*estropeado*) damaged
virulento *a* virulent
virus *m invar* virus
visa|do *m* visa. **~r** *vt* endorse
vísceras *fpl* entrails
viscos|a *f* viscose. **~o** *a* viscous
visera *f* visor; (*de gorra*) peak
visib|ilidad *f* visibility. **~le** *a* visible
visig|odo *a* Visigothic. ● *m* Visigoth. **~ótico** *a* Visigothic
visillo *m* (*cortina*) net curtain
visi|ón *f* vision; (*vista*) sight. **~onario** *a* & *m* visionary
visita *f* visit; (*persona*) visitor. **~ de cumplido** courtesy call. **~nte** *m* & *f* visitor. **~r** *vt* visit. **tener ~** have visitors
vislumbr|ar *vt* glimpse. **~e** *f* glimpse; (*resplandor*, *fig*) glimmer
viso *m* sheen; (*aspecto*) appearance
visón *m* mink
visor *m* viewfinder
víspera *f* day before, eve
vista *f* sight, vision; (*aspecto, mirada*) look; (*panorama*) view. **apartar la ~** look away; (*fig*) turn a blind eye. **a primera ~**, **a simple ~** at first sight. **clavar la ~ en** stare at. **con ~s a** with a view to. **en ~ de** in view of, considering. **estar a la ~** be obvious. **hacer la ~ gorda** turn a blind eye. **perder de ~** lose sight of. **tener a la ~** have in front of one. **volver la ~ atrás** look back
vistazo *m* glance. **dar/echar un ~ a** glance at
visto *a* seen; (*corriente*) common; (*considerado*) considered. ● *vb véase* **vestir**. **~ bueno** passed. **~ que** since. **bien ~** acceptable. **está ~ que** it's obvious that. **lo nunca ~** an unheard-of thing. **mal ~** unacceptable. **por lo ~** apparently
vistoso *a* colourful, bright
visual *a* visual. ● *f* glance. **echar una ~ a** have a look at
vital *a* vital. **~icio** *a* life. ● *m* (life) annuity. **~idad** *f* vitality
vitamina *f* vitamin
viticult|or *m* wine-grower. **~ura** *f* wine growing
vitorear *vt* cheer
vítreo *a* vitreous
vitrina *f* showcase
vituper|ar *vt* censure. **~io** *m* censure. **~ios** *mpl* abuse
viud|a *f* widow. **~ez** *f* widowhood. **~o** *a* widowed. ● *m* widower
viva *m* cheer
vivacidad *f* liveliness
vivamente *adv* vividly; (*sinceramente*) sincerely
vivaz *a* (*bot*) perennial; (*vivo*) lively
víveres *mpl* supplies
vivero *m* nursery; (*fig*) hotbed
viveza *f* vividness; (*de inteligencia*) sharpness; (*de carácter*) liveliness
vivido *a* true
vívido *a* vivid
vivienda *f* housing; (*casa*) house; (*piso*) flat
viviente *a* living
vivificar [7] *vt* (*animar*) enliven
vivir *vt* live through. ● *vi* live. ● *m* life. **~ de** live on. **de mal ~** dissolute. **¡viva!** hurray! **¡viva el rey!** long live the king!
vivisección *f* vivisection
vivo *a* alive; (*viviente*) living; ‹*color*› bright; (*listo*) clever; (*fig*) lively. **a lo ~**, **al ~** vividly
Vizcaya *f* Biscay
vizconde *m* viscount. **~sa** *f* viscountess
vocab|lo *m* word. **~ulario** *m* vocabulary
vocación *f* vocation
vocal *a* vocal. ● *f* vowel. ● *m* & *f* member. **~ista** *m* & *f* vocalist
voce|ar *vt* call ‹*mercancías*›; (*fig*) proclaim. ● *vi* shout. **~río** *m* shouting
vociferar *vi* shout
vodka *m* & *f* vodka
vola|da *f* flight. **~dor** *a* flying. ● *m* rocket. **~ndas. en ~ndas** in the air; (*fig, rápidamente*) very quickly. **~nte** *a* flying. ● *m* (*auto*) steering-wheel; (*nota*) note; (*rehilete*) shuttlecock; (*tec*) flywheel. **~r** [2] *vt* blow up. ● *vi* fly; (*desaparecer, fam*) disappear
volátil *a* volatile
volcán *m* volcano. **~ico** *a* volcanic
vol|car [2 & 7] *vt* knock over; (*adrede*) empty out. ● *vi* overturn. **~carse** *vpr* fall over; ‹*vehículo*› overturn; (*fig*) do one's utmost. **~carse en** throw o.s. into
vol(e)ibol *m* volleyball
volquete *m* tipper, dump truck
voltaje *m* voltage
volte|ar *vt* turn over; (*en el aire*) toss; ring ‹*campanas*›. **~reta** *f* somersault

voltio *m* volt
voluble *a* (*fig*) fickle
volum|en *m* volume; (*importancia*) importance. **~inoso** *a* voluminous
voluntad *f* will; (*fuerza de voluntad*) will-power; (*deseo*) wish; (*intención*) intention. **buena ~** goodwill. **mala ~** ill will
voluntario *a* voluntary. ● *m* volunteer. **~so** *a* willing; (*obstinado*) wilful
voluptuoso *a* voluptuous
volver [2, *pp* **vuelto**] *vt* turn; (*de arriba a abajo*) turn over; (*devolver*) restore. ● *vi* return; (*fig*) revert. **~se** *vpr* turn round; (*regresar*) return; (*hacerse*) become. **~ a hacer algo** do sth again. **~ en sí** come round
vomit|ar *vt* bring up. ● *vi* be sick, vomit. **~ivo** *m* emetic. ● *a* disgusting
vómito *m* vomit; (*acción*) vomiting
vorágine *f* maelstrom
voraz *a* voracious
vos *pron* (*LAm*) you
vosotros *pron* you; (*reflexivo*) yourselves. **el libro de ~** your book
vot|ación *f* voting; (*voto*) vote. **~ante** *m* & *f* voter. **~ar** *vt* vote for. ● *vi* vote. **~o** *m* vote; (*relig*) vow; (*maldición*) curse. **hacer ~os por** hope for
voy *vb véase* **ir**
voz *f* voice; (*grito*) shout; (*rumor*) rumour; (*palabra*) word. **~ pública** public opinion. **aclarar la ~** clear one's throat. **a media ~** softly. **a una ~** unanimously. **dar voces** shout. **en ~ alta** loudly
vuelco *m* upset. **el corazón me dio un ~** my heart missed a beat
vuelo *m* flight; (*acción*) flying; (*de ropa*) flare. **al ~** in flight; (*fig*) in passing
vuelta *f* turn; (*curva*) bend; (*paseo*) walk; (*revolución*) revolution; (*regreso*) return; (*dinero*) change. **a la ~** on one's return; (*de página*) over the page. **a la ~ de la esquina** round the corner. **dar la ~ al mundo** go round the world. **dar una ~** go for a walk. **estar de ~** be back. **¡hasta la ~!** see you soon!
vuelvo *vb véase* **volver**
vuestro *a* your. ● *pron* yours. **un amigo ~** a friend of yours
vulg|ar *a* vulgar; ‹*persona*› common. **~aridad** *f* ordinariness; (*trivialidad*) triviality; (*grosería*) vulgarity. **~arizar** [10] *vt* popularize. **~o** *m* common people
vulnerab|ilidad *f* vulnerability. **~le** *a* vulnerable

W

wáter *m* toilet
whisky /'wiski/ *m* whisky

xenofobia *f* xenophobia
xilófono *m* xylophone

y *conj* and
ya *adv* already; (*ahora*) now; (*luego*) later; (*en seguida*) immediately; (*pronto*) soon. ● *int* of course! **~ no** no longer. **~ que** since. **¡~, ~!** oh yes!, all right!
yacaré *m* (*LAm*) alligator
yac|er [44] *vi* lie. **~imiento** *m* deposit; (*de petróleo*) oilfield
yanqui *m* & *f* American, Yank(ee)
yate *m* yacht
yegua *f* mare
yeísmo *m* pronunciation of the Spanish *ll* like the Spanish *y*
yelmo *m* helmet
yema *f* (*bot*) bud; (*de huevo*) yolk; (*golosina*) sweet. **~ del dedo** fingertip
yergo *vb véase* **erguir**
yermo *a* uninhabited; (*no cultivable*) barren. ● *m* wasteland
yerno *m* son-in-law
yerro *m* mistake. ● *vb véase* **errar**
yerto *a* stiff
yeso *m* gypsum; (*arquit*) plaster. **~ mate** plaster of Paris
yo *pron* I. ● *m* ego. **~ mismo** I myself. **soy ~** it's me
yodo *m* iodine
yoga *m* yoga
yogur *m* yog(h)urt
York. de ~ ‹*jamón*› cooked
yuca *f* yucca
Yucatán *m* Yucatán
yugo *m* yoke

Yugoslavia *f* Yugoslavia
yugoslavo *a & m* Yugoslav
yunque *m* anvil
yunta *f* yoke
yuxtaponer [34] *vt* juxtapose
yuyo *m* (*Arg*) weed

Z

zafarse *vpr* escape; get out of ‹*obligación etc*›
zafarrancho *m* (*confusión*) mess; (*riña*) quarrel
zafio *a* coarse
zafiro *m* sapphire
zaga *f* rear. **no ir en ~** not be inferior
zaguán *m* hall
zaherir [4] *vt* hurt one's feelings
zahorí *m* clairvoyant; (*de agua*) water diviner
zaino *a* ‹*caballo*› chestnut; ‹*vaca*› black
zalamer|ía *f* flattery. **~o** *a* flattering. ● *m* flatterer
zamarra *f* (*piel*) sheepskin; (*prenda*) sheepskin jacket
zamarrear *vt* shake
zamba *f* (*esp LAm*) South American dance; (*samba*) samba
zambulli|da *f* dive. **~r** [22] *vt* plunge. **~rse** *vpr* dive
zamparse *vpr* fall; (*comer*) gobble up
zanahoria *f* carrot
zancad|a *f* stride. **~illa** *f* trip. **echar la ~illa a uno, poner la ~illa a uno** trip s.o. up
zanc|o *m* stilt. **~udo** *a* long-legged. ● *m* (*LAm*) mosquito
zanganear *vi* idle
zángano *m* drone; (*persona*) idler
zangolotear *vt* fiddle with. ● *vi* rattle; ‹*persona*› fidget
zanja *f* ditch. **~r** *vt* (*fig*) settle
zapapico *m* pickaxe
zapat|ear *vt/i* tap with one's feet. **~ería** *f* shoe shop; (*arte*) shoemaking. **~ero** *m* shoemaker; (*el que remienda zapatos*) cobbler. **~illa** *f* slipper. **~illas deportivas** trainers. **~o** *m* shoe
zaragata *f* turmoil
Zaragoza *f* Saragossa
zarand|a *f* sieve. **~ear** *vt* sieve; (*sacudir*) shake
zarcillo *m* earring
zarpa *f* claw, paw
zarpar *vi* weigh anchor
zarza *f* bramble. **~mora** *f* blackberry
zarzuela *f* musical, operetta
zascandil *m* scatterbrain
zenit *m* zenith
zigzag *m* zigzag. **~uear** *vi* zigzag
zinc *m* zinc
zipizape *m* (*fam*) row
zócalo *m* skirting-board; (*pedestal*) plinth
zodiaco *m*, **zodíaco** *m* zodiac
zona *f* zone; (*área*) area
zoo *m* zoo. **~logía** *f* zoology. **~lógico** *a* zoological
zoólogo *m* zoologist
zopenco *a* stupid. ● *m* idiot
zoquete *m* (*de madera*) block; (*persona*) blockhead
zorr|a *f* fox; (*hembra*) vixen. **~o** *m* fox
zozobra *f* (*fig*) anxiety. **~r** *vi* be shipwrecked; (*fig*) be ruined
zueco *m* clog
zulú *a & m* Zulu
zumb|ar *vt* (*fam*) give ‹*golpe etc*›. ● *vi* buzz. **~ido** *m* buzzing
zumo *m* juice
zurci|do *m* darning. **~r** [9] *vt* darn
zurdo *a* left-handed; ‹*mano*› left
zurrar *vt* (*fig, dar golpes, fam*) beat up
zurriago *m* whip
zutano *m* so-and-so

Spanish in Context

Contents

¿Habla español?

Imagine you are in Spain or any other Spanish-speaking country. Whether you want to travel around, spend a night out, or go shopping, you need to make yourself understood. With what you have learned so far in class or on your own, how well do you think you would cope?

Do you want to find out? The following section is a test-yourself conversation guide, which has been designed precisely to help you practise everyday language. It will help you build up your vocabulary in a fun and relaxed way through role play and model dialogues.

The section includes seven common situations as listed above. For each of them, you will find a variety of role play situations, a reminder of useful structures and vocabulary items, and also a model dialogue.

Jet Set

Role Play

Imagine . . .

1 You want to fly to Madrid/Paris/etc: buy a ticket.
2 You have missed your plane: ask if there is another flight later that day.
3 Your flight is delayed: ask why, and when it will take off.
4 Your flight is cancelled: ask why, and try to find out how you can get to your destination.
5 You are to meet somebody at the airport, but their flight is delayed or cancelled: try to find out the reason for the delay/cancellation and when they are likely to arrive.

Useful vocabulary and structures

el aeropuerto—*the airport*
la terminal—*the terminal*
un billete (Spain), un pasaje (LAm), un boleto (Mex)—*a ticket*
un billete/pasaje de ida y vuelta, un boleto redondo (Mex) —*a return ticket*
un billete/pasaje (sólo) de ida, un boleto sencillo (Mex) —*a one-way/single ticket*
el mostrador de venta de billetes/pasajes/boletos —*the ticket office*
un pasaporte—*a passport*
un visado (Spain),una visa (LAm) —*a visa*
el carné de identidad —*an identity card*
la facturación de equipajes (Spain), el registro de equipaje (Mex)—*check-in*
el mostrador de facturación —*the check-in desk*
ya facturé (Spain) *or* chequeé (LAm) *or* registré (Mex) mi equipaje —*I've checked in my luggage*
¿cuántos bultos tiene? —*how many pieces of luggage do you have?*
el equipaje de mano—*hand luggage*
un asiento junto a la ventanilla —*a window seat*
un asiento junto al pasillo —*an aisle seat*
una tarjeta de embarque —*a boarding card*
un carro *or* un carrito (para el equipaje)—*a luggage trolley*
la sala de espera —*the waiting lounge*

las tiendas libres de impuestos —*duty free shops*
el duty free—*duty free*
salidas internacionales —*international departures*
el vuelo procedente de Londres —*the flight from London*
un vuelo con destino a Roma —*a flight to Rome*
un vuelo de enlace procedente de Berlín —*a connecting flight from Berlin*
el vuelo hace escala en Omán —*the flight stops (over) in Oman*
el vuelo está retrasado *or* atrasado (LAm) —*the flight is delayed*
siempre viaja en clase preferente/clase económica —*he always travels business class/tourist class*
los pasajeros con destino a Praga—*passengers travelling to Prague*
pasamos por la aduana —*we went through customs*
no tengo nada que declarar —*I have nothing to declare*

Model Dialogue

¿En qué vuelo viene?

Una señora = ●
Miembro del personal de tierra = ○

● Buenas tardes, señorita. Perdone que la moleste pero es que he venido a buscar a un amigo que viene de Lisboa y olvidé en casa los datos del vuelo. Sé que llega alrededor de las cuatro pero no sé la hora exacta.

○ Bueno, vamos a ver. ¿Se acuerda del número del vuelo?

● No, pero es el vuelo que hace escala en Lisboa.

○ De acuerdo, pero resulta que hoy hay dos vuelos por la tarde, uno que llega a las cuatro y media y otro a las cinco y veinte.

● ¿De qué líneas aéreas son, por favor?

○ El primero es de Avianca y el segundo de Viasa.

● Tiene que ser el segundo.

○ Por desgracia ese vuelo va a llegar con una hora de retraso, así que tendrá Ud que esperar casi dos horas.

● No importa. Lo bueno es que no se haya ido ya mi amigo. ¿Me dice por favor dónde debo esperar?

○ Claro. La sala de espera para las llegadas internacionales está en la planta baja. Baje por aquella escalera y siga hasta el fondo. Allí verá las puertas automáticas por donde salen los pasajeros.

● Estupendo, pues . . . muchas gracias.

○ No hay de qué. Adiós.

Making Tracks

Role Play

Imagine . . .

1 You want 2 tickets, a return and a single, to Alicante, for the following day.
2 You have just arrived by train and need to leave your suitcases in left luggage. Then you want to have some refreshments and use the toilets before leaving the station: you ask a station guard to direct you.
3 You are meeting a friend off a train from Switzerland but the train is not there: ask the guard if the train is delayed and when it is due to arrive.
4 You are about to get on your train when you hear another passenger talking about changing trains during the journey: check with the guard on the platform if this is true and where and when you have to change.

Useful vocabulary and structures

la estación de tren
—the train station
el andén*—the platform*
el mostrador *or* la ventanilla de venta de billetes (Spain) *or* boletos (LAm)
—the ticket office
un billete (Spain), un boleto (LAm)
—a ticket
un billete sencillo *or* de ida (Spain), un boleto de ida (LAm except Mex), un boleto sencillo (Mex)*—a single ticket*
un billete/boleto de ida y vuelta, un boleto (de viaje) redondo (Mex)
—a return ticket
un billete/boleto de primera clase/segunda (clase)
—a first-class/second-class ticket
un tren expreso/rápido
—an express/a fast train
reservar un asiento
—to reserve a seat
una litera*—a couchette*
el tren para Teruel sale del andén 9
—the train on platform 9 is for Teruel
un carro, un carrito*—a trolley*
un vagón*—a carriage*
un compartimento de fumadores /no fumadores
—a smoking/non-smoking compartment

la rejilla (portaequipajes)
—the luggage rack
¿este tren tiene coche comedor *or* (Mex) carro comedor?
—does this train have a dining car?
el revisor/la revisora
—the ticket collector or *inspector*
voy en tren
—I'm going by train
cambiar de tren, hacer transbordo
—to change trains
¿dónde está la consigna
—where is 'left luggage'?
consigna automática (de equipajes)
—(coin-operated) left-luggage locker

Model dialogue

El tren no la llegado

Un señor = ●
Empleado = ○

● Oiga, por favor, ¿sabe a qué andén llega el expreso que viene de Valencia? Debió haber estado aquí a las once y veinticinco pero no lo encuentro en ningún andén.

○ Lo que pasa es que ese tren viene con retraso.

● ¡Ay no! ¿Se sabe cuándo va a llegar? Es que he venido a buscar a alguien, y si falta mucho puedo salir y volver luego.

○ Bueno, según nos han informado el tren se ha averiado a unos cuantos kilómetros de aquí y se espera que llegará a eso de la una.

● ¿Pero todavía no se sabe a qué hora exactamente?

○ No. De todos modos, seguro que no llegará antes de las doce y media. Si quiere llamar por teléfono más tarde, podrán informarle mejor entonces.

● Pero ya llevo veinte minutos esperando. ¿Por qué no lo anunciaron por el altavoz antes?

○ Acabamos de enterarnos, señor. Se ruega tener paciencia.

● Sí, por supuesto. ¿Me dice por favor dónde está la cafetería?

○ Al otro lado de la estación, en el fondo.

● Gracias, señor. Adiós.

Time Out

Role play

Imagine . . .

1 You want to ask sb out to a concert. You discuss what is on and make arrangements to meet up before.
2 You call the box office of a theatre to book tickets for a play.
3 You call a cinema to find out what is on. Ask if you can book tickets over the phone and check when to pick them up.
4 You call a restaurant to see if they are open on Sundays and book a table for four for that evening.

Useful vocabulary and structures

¿te gustaría ir al cine este fin de semana?
—would you like to go to the cinema at the weekend?

¿qué tienes ganas de hacer *or* qué te apetece hacer (Spain) esta tarde?
—what do you fancy doing this afternoon/evening?

¿estás libre mañana?
—are you free tomorrow?

¿tienes algún compromiso mañana?*—are you doing anything/are you busy tomorrow?*

¿tienes algún plan *or* programa para esta noche?
—do you have any plans for tonight?

salir/ir a tomar una(s) copa(s)
—to go (out) for a drink

vida nocturna*—nightlife*

salir a cenar *or* comer (LAm)
—to go out for dinner/for a meal

¿qué dan *or* qué ponen (Spain) en el Renoir?
—what's on at the Renoir?

¿hay que reservar antes?
—do you have to book in advance?

la taquilla, la boletería (LAm)
—the box office/ticket office

¿a qué hora empieza la película/la función?
—what time does the film/performance start?

ya se agotaron las localidades
—it's sold out

recoger las entradas
—to pick up/collect the tickets

¿qué te parece?
—what do you think (of it)?

¿te gusta?*—are you enjoying it?*

un concierto al aire libre
—an open-air concert

tener una actuación
—*to do* or *play a gig*
reservar una mesa
—*to book* or *reserve a table*
¿me trae la carta ?
—*can I see the menu, please?*
de primer plato *or* de entrada, quisiera las espinacas
—*can I have the spinach to start with, please?*
y de segundo, el salmón
—*and salmon for the main course*
¿me trae una jarra de agua, por favor?
—*could I have a jug of water, please?*
(me trae) la cuenta, por favor
—*can I have the bill, please?*
¿cuánto se deja de propina?
—*how much tip should I leave?*

Model dialogue

¿Qué hacemos el sábado?

Ramón = ●
Eva = ○

● Oye, Eva ¿tienes algún plan para el sábado por la noche?

○ Por el momento no tengo nada planeado, pero iba a llamar a Marichu para quedar con ella*.

● Bueno, pues, vi ayer un cartel de publicidad de 'Los Pachequitos' - van a dar un concierto al aire libre en el Parque del Oeste.

○ ¿Ah, sí? ¿Qué tipo de música tocan?

● ¿No los conoces? Es un grupo cubano que toca son y merengue - seguro que los encontrarás muy buenos. ¿Qué te parece?

○ Me parece bien.

● Estupendo. Entonces iré a sacar las entradas esta tarde. ¿Le compro una a Marichu también?

○ No creo. Que se las arregle como quiera este fin de semana.

● Bueno, entonces nos encontramos en la salida del metro de Moncloa a las ocho, ¿sí? Así podremos ir a tomar una copa antes. De todas maneras te llamo a casa cuando tenga las entradas.

○ De acuerdo. Y primero tengo que hablar con mis padres. Ya sabes como me controlan, ¿eh?

● Claro. Es lo de siempre, ¿no?

○ Bueno, pues nada. Hasta mañana.

● Hasta luego.

* this usage is Peninsular Spanish. In Latin America one would say **'para que nos encontráramos'**.

Hotel Break

Role play

Imagine . . .

1 You call a hotel to book two rooms, a single and a double, for two nights: ask about the price and arrange the dates and arrival time.
2 You arrive at your hotel to find that your room has been double-booked. There is no other hotel nearby and you wish to be given another room also with sea view.
3 When you get up to your room you find that the bathroom is not very clean. You also want something to eat in your room: you call room service.
4 You are going out for the evening and think you may be back quite late: you ask at reception what to do about taking keys with you or ringing the bell.

Useful vocabulary and structures

un hotel de tres/cuatro estrellas —*a three/four star hotel*
alojamiento—*accomodation*
un hostal—*a cheap hotel*
una residencia—*a guesthouse (usually of category between 'hotel' and 'pensión')*
una pensión —*a guesthouse, rooming house*
una reserva, una reservación (LAm)—*a reservation, a booking*
reservar/pagar una habitación —*to book/pay for a room*
¿es necesario dejar un depósito? —*do you require a deposit?*
una habitación sencilla/individual—*a single room*
una habitación doble —*a double room*
una habitación con camas gemelas—*a twin-bedded room*
una cama de matrimonio *or* de dos plazas (LAm) —*a double bed*
una cama individual *or* de una plaza (LAm)—*a single bed*
una habitación con vista al mar —*a (room with) sea view*
registrarse—*to check in*
¿podría ver la habitación, por favor? —*could I see the room, please?*
una caja de seguridad —*a safety deposit box*
¿hay un mozo que nos pueda subir/bajar las maletas? —*is there a porter to take up/bring down our suitcases?*

el servicio a las habitaciones
—room service
no hay agua caliente
—there's no hot water
el televisor/la ducha no funciona
—the television/the shower doesn't work
¿a qué hora se sirve el desayuno?
—what time is breakfast?
¿dónde dejo la llave?
—where should I leave the key?
hay servicio de lavandería?
—is there a laundry service?
¿me podrían despertar mañana por la mañana, por favor?
—could I have a wake-up call in the morning, please?
por favor dejen libre la habitación antes de las once de la mañana
—please vacate your room by 11a.m.
¿a qué hora hay que salir del *or* dejar el hotel?
—what time must I check out by?

Model dialogue

¡Cámbienme de habitación!

Propietaria = ●
Huésped = ○

● Buenos días, señora. ¿Cómo está? ¿Durmió bien?

○ Pues, la verdad es que no muy bien.

● Lo siento, señora. ¿Y a qué se debió eso?

○ Bueno, primero había mucho ruido en la calle. Parece que todas las motos pasaran por allí. Y la nevera hizo un sonido muy raro - una especie de pitido - durante casi toda la noche.

● ¡Uy, lo siento! En seguida llamo a mi marido para que lo arregle.

○ Además la habitación es muy pequeña y está mal ventilada. Con el calor que hacía me costaba respirar. En fin, quiero que me cambie de habitación.

● Mm... Eso podría ser difícil. Es que todas las habitaciones están ocupadas.

○ No me importa. Hice la reserva hace mucho tiempo y esperaba que la habitación fuera cómoda y tranquila. Debo insistir en que me la cambie.

● Vamos a ver. Ah . . . ahora que veo en el libro, resulta que se van hoy los de la 14, de manera que la puede ocupar Ud esta misma mañana.

○ ¿A qué hora estará disponible entonces?

● Cuanto antes, señora. Nada más terminar la limpieza, le aviso.

Money matters and keeping in touch

Role Play

Imagine ...

1 You go to the post office to send a large parcel to your sister in Austria for her birthday.
2 You want stamps for seven postcards to Scotland, two letters to Italy and an airletter to New Zealand.
3 You are at the counter in the bank. You want to cash a cheque and change some English money into Spanish currency.
4 You want to order 80.000 pesetas worth of travellers cheques in US dollars for a trip to Central America.
5 You want to withdraw some cash: talk to the cashier and find out what you have to do.

Useful vocabulary and structures

voy a la oficina de correos *or* voy a Correos (Spain), voy al correo (LAm)
—I'm going to the post office

la ventanilla
—the counter, the window

un sello, una estampilla (LAm), un timbre (Mexico)*—a stamp*

enviar *or* mandar una carta/ una postal/un paquete
—to send a letter/a postcard/ a parcel

mandar algo por avión *or* por vía aérea*—to send sth airmail*

un sobre de avión
—an airmail envelope

un aerograma*—an airletter*

póngalo en la balanza
—put it on the scales

hay que rellenar *or* llenar este formulario *or* este impreso *or* esta forma (Mex)
—you have to fill in or *fill out this form*

abrir una cuenta (bancaria)
—to open a bank account

una cuenta corriente
—a current account

una cuenta de ahorro(s)
—a savings/deposit account

retirar/sacar dinero
—to withdraw/take out money

depositar *or* ingresar (Spain) un cheque/dinero
—to pay in a cheque/some money

en efectivo*—in cash*

un cajero automático
—a cash dispenser

hacer cola—*to queue, stand in line*
pase a (la) caja, por favor
—*please go to the cash desk*
cobrar un cheque *or* talón (Spain)
—*to cash a cheque*
el cheque se lo abonarán *or* cambiarán en (la) caja
—*you can cash the cheque at the cash desk*
¿puede firmar el dorso del cheque?
—*can you sign the back of the cheque?*
quisiera mandar un giro postal
—*I'd like to send a money order*
cheques de viajero
—*traveller's cheques*
una casa de cambio
—*a bureau de change*
divisas, moneda extranjera
—*foreign currency*
la tasa de cambio (Spain), el tipo de cambio
—*the exchange rate*
¿a cuánto está (el cambio de) la libra?
—*what's the rate at the moment for the pound?*
¿me puede dar trescientas libras en cheques de viajero, por favor?
—*can I have three hundred pounds in traveller's cheques, please?*
¿cuánto se paga de comisión?
—*how much commission do you charge?*

Model Dialogue

En la oficina de correos

Cliente = ●
Empleado de la oficina de correos = ○

● Buenos días. Quisiera mandar este paquete a Grecia. ¿Me dice cuánto puede tardar en llegar, por favor?

○ Entre cuatro y cinco días, señor. ¿Quiere enviarlo por correo aéreo?

● Entonces sí. Quiero que llegue antes del sábado.

○ Póngalo en la balanza, por favor. A ver... así son mil cien pesetas.

● Muy bien. Y querría mandar esta carta a Murcia por correo urgente. ¿Me puede dar también nueve sellos para postales, por favor?

○ ¿Para mandar al extranjero o dentro de España?

● Son todas para Estados Unidos.

○ Bueno, aquí tiene.

● Me han dicho que es mejor mandar las postales en sobre, que llegan más rápido así.

○ No hace falta, señor. Hoy día todas llegan igual. ¡Y le aseguro que los carteros no tienen tiempo para leer la correspondencia ajena!

● ¡Por supuesto que no! Gracias. Adiós.

Where to Find What

Role Play

Imagine . . .

1 You want to visit the main museums and galleries in the city: ask what times they are open, whether they are open every day and how much the admission charge is.
2 You would like a map of the city and leaflets about places to visit.
3 You want to find out about good cheap places to eat in the city, and if there are any traditional old cafés to have a drink in.
4 You want some information about visits to the monastery of Montserrat from Barcelona: find out what times the coach trip leaves and returns, how much it costs and what it includes.

Useful vocabulary and structures

un cartel—*a poster*
¿me puedo llevar este folleto?
—*can I take this leaflet/brochure?*
quisiera informarme sobre. . .
—*I'd like to find out about. . .*
alquiler de coches (Spain) *or* carros (LAm)—*car hire, car rental*
alquilar un coche (Spain) *or* un carro (LAm)—*to hire a car, rent a car*
quisiera averiguar a qué horas está abierto el museo
—*I'd like to find out what times the museum is open*
¿a qué hora abre/cierra el museo?
—*what time does the museum open/close?*
¿cuánto cuesta *or* vale la entrada?
—*how much is admission?*

un museo de arte
—*an art gallery (museum)*
el casco viejo—*the old quarter*
¿tiene una lista de hoteles económicos?
—*do you have a list of cheap hotels?*
¿me/nos puede recomendar un buen hotel?
—*can you recommend a good hotel?*
¿me puede/podría decir cómo llegar a . . . ?
—*can you/could you tell me how to get to . . . ?*
¿me puede decir dónde está *or* queda la estación de tren?
—*can you tell me where the train station is?*
un plano de la ciudad/del metro
—*a map of the city/of the underground or subway (US)*

me señaló la ruta en un mapa
—he showed me the route on a map
¿cómo es el horario de los bancos?
—what times are banks open?
¿dónde se puede cambiar dinero?
—where can I change money?
¿hay que pagar por adelantado?
—do you have to pay in advance?
¿nos podría sugerir/recomendar alguna excursión interesante?
—can you suggest/recommend any good day trips?

ir *or* salir a remar
—to go rowing
¿se hacen visitas guiadas del palacio?
—can one do a guided tour of the palace?
¿sabe si hay unos servicios (Spain) *or* baños públicos (LAm) por aquí cerca?
—do you know if there are any public toilets near here?

Model Dialogue

Dónde alojarse y qué conocer

Empleada de la Oficina de Turismo = ●
Turista = ○

● Buenas tardes. ¿En qué les puedo servir?

○ Buenas tardes. Es que acabamos de llegar a la ciudad y necesitamos encontrar alojamiento. ¿Tiene alguna lista de hoteles económicos?

● Por supuesto. Aquí mismo tengo una, en el mostrador. ¿Quieren quedarse en una zona más bien céntrica?

○ Sí, si es posible.

● Entonces les puedo señalar aquí en este plano algunos buenos hostales y hoteles.

○ Gracias. ¿Nos podría dar un folleto o una guía de los lugares de interés más importantes para visitar? Tenemos poco tiempo aquí y no queremos perdernos nada.

● Claro, aquí tiene. Además, les puedo recomendar una visita organizada de tres horas que sale de la plaza central cada dos horas. Aquí verán la información necesaria.

○ Muy bien. Gracias.

● Tal vez también les interese hacer una excursión por los alrededores de la ciudad?

○ Sí, sería genial. Y por último, ¿nos podría decir cómo llegar al consulado australiano?

● Sin problema - ¡está aquí al lado!

Shop Till You Drop

Role Play

Imagine . . .

1 You go to a clothes shop to buy a jacket: ask to try one on.
2 You are at the local food market. You have come to get some ingredients for a big stew of your choice.
3 You bought a jersey yesterday but have found a hole in it. You bring it back to the shop to change it.
4 You buy a rug in a shop. You want to send it back to Dublin: ask if you can pay by credit card and how much it will cost to send.
5 You want to buy a present for your mother to take back home: go to a department store and ask which floor the different items are on (porcelain, fans, perfume, scarves, CDs and tapes).

Useful vocabulary and structures

unos (grandes) almacenes, una tienda de departamentos (Mex)—*a department store*
un centro comercial —*a shopping centre*
una tienda de modas —*a clothes shop*
una tienda de novedades —*a gift shop*
el dependiente/la dependienta (Spain), el vendedor/la vendedora (LAm)—*the shop assistant*
en la planta baja —*on the ground floor*
en la primera planta, en el primer piso—*on the first floor*
a precio de ganga —*at bargain price*
hay que regatear —*you have to bargain*
¿dónde puedo comprar . . .? —*where can I get . . .?*
¿lo/la atienden? —*are you being served?*
estoy mirando solamente —*I'm just looking, thank you*
¿cuánto vale *or* cuesta esto, por favor? —*how much is this, please?*
quisiera una chaqueta —*I'm looking for a jacket*
¿tiene esto en rojo/en otros colores?—*do you have this in red/ in other colours?*
¿me puedo probar esto? —*can I try this on?*

¿me lo/la puedo probar?
—*can I try it on?*

¿dónde está el probador?
—*where is the fitting room?*

¿qué talla tiene *or* usa?
—*what size do you take? [clothes]*

calzo el numero 44
—*I take size 10 shoes*

me llevo éste/ésta, por favor
—*I'll take this (one), please*

quisiera cambiar éste/ésta por otro/otra
—*I'd like to change this for another one*

¿dónde se paga?—*where do I pay?*

¿puedo pagar con cheque?
—*can I pay by cheque?*

¿aceptan tarjetas de crédito?
—*do you take credit cards?*

¿cuánto cobran por el envío (a domicilio)?
—*how much is the delivery charge?*

¿se puede enviar/mandar por barco a . . . ?
—*is it possible to have it shipped to . . . ?*

¿puedo pedir que se me devuelva el impuesto?
—*can I claim tax back on it?*

¿a cuánto están los tomates?
—*how much are the tomatoes?*

¿me da *or* me pone (Spain) medio kilo de queso, por favor?—*can I have half a kilo of cheese, please?*

¿quién es la última?*
—*who's last in the queue* or *line?*

Model dialogue

Compras de última hora

Dependiente = ●

Cliente = ○

● Buenos días, señor. ¿Qué desea?

○ Me da una caja de aspirinas, por favor.

● Sí, señor. ¿Quiere algo más?

○ ¿Tiene repelente para mosquitos?

● ¿Prefiere loción o en espray?

○ En espray, por favor. Y me hace falta también una crema bronceadora. ¿Cuánto vale la de esta marca?

● Pues, cuesta mil trescientas pesetas.

○ ¿Y me puede dar un jarabe para la tos, para niños?

● Muy bien. Aquí tiene. ¿Algo más?

○ Nada más, gracias. ¿Cuánto es, por favor?

● Vamos a ver. Son . . . 2.900 pesetas en total.

○ Tenga Ud.

● Gracias, aquí tiene la vuelta.

ENGLISH-SPANISH
INGLÉS-ESPAÑOL

A

a /ə, eɪ/ *indef art* (*before vowel* **an**) un *m*; una *f*
aback /ə'bæk/ *adv*. **be taken** ~ quedar desconcertado
abacus /'æbəkəs/ *n* ábaco *m*
abandon /ə'bændən/ *vt* abandonar. ● *n* abandono *m*, desenfado *m*. **~ed** *a* abandonado; ‹*behaviour*› perdido. **~ment** *n* abandono *m*
abase /ə'beɪs/ *vt* degradar. **~ment** *n* degradación *f*
abashed /ə'bæʃt/ *a* confuso
abate /ə'beɪt/ *vt* disminuir. ● *vi* disminuir; ‹*storm etc*› calmarse. **~ment** *n* disminución *f*
abattoir /'æbətwɑ:(r)/ *n* matadero *m*
abbess /'æbis/ *n* abadesa *f*
abbey /'æbɪ/ *n* abadía *f*
abbot /'æbət/ *n* abad *m*
abbreviat|e /ə'bri:vɪeɪt/ *vt* abreviar. **~ion** /-'eɪʃn/ *n* abreviatura *f*; (*act*) abreviación *f*
ABC /'eɪbi:'si:/ *n* abecé *m*, abecedario *m*
abdicat|e /'æbdɪkeɪt/ *vt/i* abdicar. **~ion** /-'eɪʃn/ *n* abdicación *f*
abdom|en /'æbdəmən/ *n* abdomen *m*. **~inal** /-'dɒmɪnl/ *a* abdominal
abduct /æb'dʌkt/ *vt* secuestrar. **~ion** /-ʃn/ *n* secuestro *m*. **~or** *n* secuestrador *m*
aberration /æbə'reɪʃn/ *n* aberración *f*
abet /ə'bet/ *vt* (*pt* **abetted**) (*jurid*) ser cómplice de
abeyance /ə'beɪəns/ *n*. **in** ~ en suspenso
abhor /əb'hɔ:(r)/ *vt* (*pt* **abhorred**) aborrecer. **~rence** /-'hɒrəns/ *n* aborrecimiento *m*; (*thing*) abominación *f*. **~rent** /-'hɒrənt/ *a* aborrecible
abide /ə'baɪd/ *vt* (*pt* **abided**) soportar. ● *vi* (*old use*, *pt* **abode**) morar. ~ **by** atenerse a; cumplir ‹*promise*›
abiding /ə'baɪdɪŋ/ *a* duradero, permanente
ability /ə'bɪlətɪ/ *n* capacidad *f*; (*cleverness*) habilidad *f*
abject /'æbdʒekt/ *a* (*wretched*) miserable; (*vile*) abyecto
ablaze /ə'bleɪz/ *a* en llamas
able /'eɪbl/ *a* (**-er**, **-est**) capaz. **be** ~ poder; (*know how to*) saber
ablutions /ə'blu:ʃnz/ *npl* ablución *f*
ably /'eɪblɪ/ *adv* hábilmente
abnormal /æb'nɔ:ml/ *a* anormal. **~ity** /-'mælətɪ/ *n* anormalidad *f*
aboard /ə'bɔ:d/ *adv* a bordo. ● *prep* a bordo de
abode /ə'bəʊd/ *see* **abide**. ● *n* (*old use*) domicilio *m*
abolish /ə'bɒlɪʃ/ *vt* suprimir, abolir
abolition /æbə'lɪʃn/ *n* supresión *f*, abolición *f*
abominable /ə'bɒmɪnəbl/ *a* abominable
abominat|e /ə'bɒmɪneɪt/ *vt* abominar. **~ion** /-'neɪʃn/ *n* abominación *f*
aborigin|al /æbə'rɪdʒənl/ *a* & *n* aborigen (*m* & *f*), indígena (*m* & *f*). **~es** /-i:z/ *npl* aborígenes *mpl*
abort /ə'bɔ:t/ *vt* hacer abortar. ● *vi* abortar. **~ion** /-ʃn/ *n* aborto *m* provocado; (*fig*) aborto *m*. **~ionist** *n* abortista *m* & *f*. **~ive** *a* abortivo; (*fig*) fracasado
abound /ə'baʊnd/ *vi* abundar (**in** de, en)
about /ə'baʊt/ *adv* (*approximately*) alrededor de; (*here and there*) por todas partes; (*in existence*) por aquí. ~ **here** por aquí. **be** ~ **to** estar a punto de. **be up and** ~ estar levantado. ● *prep* sobre; (*around*) alrededor de; (*somewhere in*) en. **talk** ~ hablar de. **~-face** *n* (*fig*) cambio *m* rotundo. **~-turn** *n* (*fig*) cambio *m* rotundo
above /ə'bʌv/ *adv* arriba. ● *prep* encima de; (*more than*) más de. ~ **all** sobre todo. **~-board** *a* honrado.

● *adv* abiertamente. **~-mentioned** *a* susodicho

abrasi|on /ə'breɪʒn/ *n* abrasión *f*. **~ve** /ə'breɪsɪv/ *a & n* abrasivo (*m*); (*fig*) agresivo, brusco

abreast /ə'brest/ *adv* de frente. **keep ~ of** mantenerse al corriente de

abridge /ə'brɪdʒ/ *vt* abreviar. **~ment** *n* abreviación *f*; (*abstract*) resumen *m*

abroad /ə'brɔ:d/ *adv* (*be*) en el extranjero; (*go*) al extranjero; (*far and wide*) por todas partes

abrupt /ə'brʌpt/ *a* brusco. **~ly** *adv* (*suddenly*) repentinamente; (*curtly*) bruscamente. **~ness** *n* brusquedad *f*

abscess /'æbsɪs/ *n* absceso *m*

abscond /əb'skɒnd/ *vi* fugarse

absen|ce /'æbsəns/ *n* ausencia *f*; (*lack*) falta *f*. **~t** /'æbsənt/ *a* ausente. /æb'sent/ *vr*. **~ o.s.** ausentarse. **~tly** *adv* distraídamente. **~t-minded** *a* distraído. **~t-mindedness** *n* distracción *f*, despiste *m*

absentee /æbsən'ti:/ *n* ausente *m & f*. **~ism** *n* absentismo *m*

absinthe /'æbsɪnθ/ *n* ajenjo *m*

absolute /'æbsəlu:t/ *a* absoluto. **~ly** *adv* absolutamente

absolution /æbsə'lu:ʃn/ *n* absolución *f*

absolve /əb'zɒlv/ *vt* (*from sin*) absolver; (*from obligation*) liberar

absor|b /əb'zɔ:b/ *vt* absorber. **~bent** *a* absorbente. **~ption** *n* absorción *f*

abstain /əb'steɪn/ *vi* abstenerse (**from** de)

abstemious /əb'sti:mɪəs/ *a* abstemio

abstention /əb'stenʃn/ *n* abstención *f*

abstinen|ce /'æbstɪnəns/ *n* abstinencia *f*. **~t** *a* abstinente

abstract /'æbstrækt/ *a* abstracto. ● *n* (*quality*) abstracto *m*; (*summary*) resumen *m*. /əb'strækt/ *vt* extraer; (*summarize*) resumir. **~ion** /-ʃn/ *n* abstracción *f*

abstruse /əb'stru:s/ *a* abstruso

absurd /əb'sɜ:d/ *a* absurdo. **~ity** *n* absurdo *m*, disparate *m*

abundan|ce /ə'bʌndəns/ *n* abundancia *f*. **~t** *a* abundante

abuse /ə'bju:z/ *vt* (*misuse*) abusar de; (*ill-treat*) maltratar; (*insult*) insultar. /ə'bju:s/ *n* abuso *m*; (*insults*) insultos *mpl*

abusive /ə'bju:sɪv/ *a* injurioso

abut /ə'bʌt/ *vi* (*pt* **abutted**) confinar (**on** con)

abysmal /ə'bɪzməl/ *a* abismal; (*bad, fam*) pésimo; (*fig*) profundo

abyss /ə'bɪs/ *n* abismo *m*

acacia /ə'keɪʃə/ *n* acacia *f*

academic /ækə'demik/ *a* académico; (*pej*) teórico. ● *n* universitario *m*, catedrático *m*. **~ian** /-də'mɪʃn/ *n* académico *m*

academy /ə'kædəmɪ/ *n* academia *f*. **~ of music** conservatorio *m*

accede /ək'si:d/ *vi*. **~ to** acceder a ‹*request*›; tomar posesión de ‹*office*›. **~ to the throne** subir al trono

accelerat|e /ək'seləreɪt/ *vt* acelerar. **~ion** /-'reɪʃn/ *n* aceleración *f*. **~or** *n* acelerador *m*

accent /'æksənt/ *n* acento *m*. /æk'sent/ *vt* acentuar

accentuate /ək'sentʃʊeɪt/ *vt* acentuar

accept /ək'sept/ *vt* aceptar. **~able** *a* aceptable. **~ance** *n* aceptación *f*; (*approval*) aprobación *f*

access /'ækses/ *n* accceso *m*. **~ibility** /-ɪ'bɪlətɪ/ *n* accesibilidad *f*. **~ible** /ək'sesəbl/ *a* accesible; ‹*person*› tratable

accession /æk'seʃn/ *n* (*to power, throne etc*) ascenso *m*; (*thing added*) adquisición *f*

accessory /ək'sesərɪ/ *a* accesorio. ● *n* accesorio *m*, complemento *m*; (*jurid*) cómplice *m & f*

accident /'æksɪdənt/ *n* accidente *m*; (*chance*) casualidad *f*. **by ~** por accidente, por descuido, sin querer; (*by chance*) por casualidad. **~al** /-'dentl/ *a* accidental, fortuito. **~ally** /-'dentəlɪ/ *adv* por accidente, por descuido, sin querer; (*by chance*) por casualidad

acclaim /ə'kleɪm/ *vt* aclamar. ● *n* aclamación *f*

acclimatiz|ation /əklaɪmətaɪ'zeɪʃn/ *n* aclimatación *f*. **~e** /ə'klaɪmətaɪz/ *vt* aclimatar. ● *vi* aclimatarse

accolade /'ækəleɪd/ *n* (*of knight*) acolada *f*; (*praise*) encomio *m*

accommodat|e /ə'kɒmədeɪt/ *vt* (*give hospitality to*) alojar; (*adapt*) acomodar; (*supply*) proveer; (*oblige*) complacer. **~ing** *a* complaciente. **~ion** /-'deɪʃn/ *n* alojamiento *m*; (*rooms*) habitaciones *fpl*

accompan|iment /ə'kʌmpənɪmənt/ *n* acompañamiento *m*. **~ist** *n* acompañante *m & f*. **~y** /ə'kʌmpənɪ/ *vt* acompañar

accomplice /əˈkʌmplɪs/ *n* cómplice *m & f*
accomplish /əˈkʌmplɪʃ/ *vt* (*complete*) acabar; (*achieve*) realizar; (*carry out*) llevar a cabo. **~ed** *a* consumado. **~ment** *n* realización *f*; (*ability*) talento *m*; (*thing achieved*) triunfo *m*, logro *m*
accord /əˈkɔːd/ *vi* concordar. ● *vt* conceder. ● *n* acuerdo *m*; (*harmony*) armonía *f*. **of one's own ~** espontáneamente. **~ance** *n*. **in ~ance with** de acuerdo con
according /əˈkɔːdɪŋ/ *adv*. **~ to** según. **~ly** *adv* en conformidad; (*therefore*) por consiguiente
accordion /əˈkɔːdɪən/ *n* acordeón *m*
accost /əˈkɒst/ *vt* abordar
account /əˈkaʊnt/ *n* cuenta *f*; (*description*) relato *m*; (*importance*) importancia *f*. **on ~ of** a causa de. **on no ~** de ninguna manera. **on this ~** por eso. **take into ~** tener en cuenta. ● *vt* considerar. **~ for** dar cuenta de, explicar
accountab|ility /əkaʊntəˈbɪlətɪ/ *n* responsabilidad *f*. **~le** *a* responsable (**for** de)
accountan|cy /əˈkaʊntənsɪ/ *n* contabilidad *f*. **~t** *n* contable *m & f*
accoutrements /əˈkuːtrəmənts/ *npl* equipo *m*
accredited /əˈkredɪtɪd/ *a* acreditado; (*authorized*) autorizado
accrue /əˈkruː/ *vi* acumularse
accumulat|e /əˈkjuːmjʊleɪt/ *vt* acumular. ● *vi* acumularse. **~ion** /-ˈleɪʃn/ *n* acumulación *f*. **~or** *n* (*elec*) acumulador *m*
accura|cy /ˈækjərəsɪ/ *n* exactitud *f*, precisión *f*. **~te** *a* exacto, preciso
accus|ation /ækjuːˈzeɪʃn/ *n* acusación *f*. **~e** *vt* acusar
accustom /əˈkʌstəm/ *vt* acostumbrar. **~ed** *a* acostumbrado. **get ~ed (to)** acostumbrarse (a)
ace /eɪs/ *n* as *m*
acetate /ˈæsɪteɪt/ *n* acetato *m*
ache /eɪk/ *n* dolor *m*. ● *vi* doler. **my leg ~s** me duele la pierna
achieve /əˈtʃiːv/ *vt* realizar; lograr ‹*success*›. **~ment** *n* realización *f*; (*feat*) éxito *m*; (*thing achieved*) proeza *f*, logro *m*
acid /ˈæsɪd/ *a & n* ácido (*m*). **~ity** /əˈsɪdətɪ/ *n* acidez *f*
acknowledge /əkˈnɒlɪdʒ/ *vt* reconocer. **~ receipt of** acusar recibo de. **~ment** *n* reconocimiento *m*; (*com*) acuse *m* de recibo
acme /ˈækmɪ/ *n* cima *f*
acne /ˈæknɪ/ *n* acné *m*
acorn /ˈeɪkɔːn/ *n* bellota *f*
acoustic /əˈkuːstɪk/ *a* acústico. **~s** *npl* acústica *f*
acquaint /əˈkweɪnt/ *vt*. **~ s.o. with** poner a uno al corriente de. **be ~ed with** conocer ‹*person*›; saber ‹*fact*›. **~ance** *n* conocimiento *m*; (*person*) conocido *m*
acquiesce /ækwɪˈes/ *vi* consentir (**in** en). **~nce** *n* aquiescencia *f*, consentimiento *m*
acqui|re /əˈkwaɪə(r)/ *vt* adquirir; aprender ‹*language*›. **~re a taste for** tomar gusto a. **~sition** /ækwɪˈzɪʃn/ *n* adquisición *f*. **~sitive** /-ˈkwɪzətɪv/ *a* codicioso
acquit /əˈkwɪt/ *vt* (*pt* **acquitted**) absolver; **~ o.s. well** defenderse bien, tener éxito. **~tal** *n* absolución *f*
acre /ˈeɪkə(r)/ *n* acre *m*. **~age** *n* superficie *f* (en acres)
acrid /ˈækrɪd/ *a* acre
acrimon|ious /ækrɪˈməʊnɪəs/ *a* cáustico, mordaz. **~y** /ˈækrɪmənɪ/ *n* acrimonia *f*, acritud *f*
acrobat /ˈækrəbæt/ *n* acróbata *m & f*. **~ic** /-ˈbætɪk/ *a* acrobático. **~ics** /-ˈbætɪks/ *npl* acrobacia *f*
acronym /ˈækrənɪm/ *n* acrónimo *m*, siglas *fpl*
across /əˈkrɒs/ *adv & prep* (*side to side*) de un lado al otro; (*on other side*) del otro lado de; (*crosswise*) a través. **go** *or* **walk ~** atravesar
act /ækt/ *n* acto *m*; (*action*) acción *f*; (*in variety show*) número *m*; (*decree*) decreto *m*. ● *vt* hacer ‹*part, role*›. ● *vi* actuar; (*pretend*) fingir; (*function*) funcionar. **~ as** actuar de. **~ for** representar. **~ing** *a* interino. ● *n* (*of play*) representación *f*; (*by actor*) interpretación *f*; (*profession*) profesión *f* de actor
action /ˈækʃn/ *n* acción *f*; (*jurid*) demanda *f*; (*plot*) argumento *m*. **out of ~** (*on sign*) no funciona. **put out of ~** inutilizar. **take ~** tomar medidas
activate /ˈæktɪveɪt/ *vt* activar
activ|e /ˈæktɪv/ *a* activo; (*energetic*) enérgico; ‹*volcano*› en actividad. **~ity** /-ˈtɪvətɪ/ *n* actividad *f*
act|or /ˈæktə(r)/ *n* actor *m*. **~ress** *n* actriz *f*

actual /'æktʃʊəl/ *a* verdadero. **~ity** /-'ælətɪ/ *n* realidad *f*. **~ly** *adv* en realidad, efectivamente; (*even*) incluso
actuary /'æktʃʊərɪ/ *n* actuario *m*
actuate /'æktjʊeɪt/ *vt* accionar, impulsar
acumen /'ækjʊmen/ *n* perspicacia *f*
acupunctur|e /'ækjʊpʌŋktʃə(r)/ *n* acupuntura *f*. **~ist** *n* acupunturista *m & f*
acute /ə'kju:t/ *a* agudo. **~ly** *adv* agudamente. **~ness** *n* agudeza *f*
ad /æd/ *n* (*fam*) anuncio *m*
AD /eɪ'di:/ *abbr* (*Anno Domini*) d.J.C.
adamant /'ædəmənt/ *a* inflexible
Adam's apple /'ædəmz'æpl/ *n* nuez *f* (de Adán)
adapt /ə'dæpt/ *vt* adaptar. ● *vi* adaptarse
adaptab|ility /ədæptə'bɪlətɪ/ *n* adaptabilidad *f*. **~le** /ə'dæptəbl/ *a* adaptable
adaptation /ædæp'teɪʃn/ *n* adaptación *f*; (*of book etc*) versión *f*
adaptor /ə'dæptə(r)/ *n* (*elec*) adaptador *m*
add /æd/ *vt* añadir. ● *vi* sumar. **~ up** sumar; (*fig*) tener sentido. **~ up to** equivaler a
adder /'ædə(r)/ *n* víbora *f*
addict /'ædɪkt/ *n* adicto *m*; (*fig*) entusiasta *m & f*. **~ed** /ə'dɪktɪd/ *a*. **~ed to** adicto a; (*fig*) fanático de. **~ion** /-ʃn/ *n* (*med*) dependencia *f*; (*fig*) afición *f*. **~ive** *a* que crea dependencia
adding machine /'ædɪŋməʃi:n/ *n* máquina *f* de sumar, sumadora *f*
addition /ə'dɪʃn/ *n* suma *f*. **in ~** además. **~al** /-ʃənl/ *a* suplementario
additive /'ædɪtɪv/ *a & n* aditivo (*m*)
address /ə'dres/ *n* señas *fpl*, dirección *f*; (*speech*) discurso *m*. ● *vt* poner la dirección; (*speak to*) dirigirse a. **~ee** /ædre'si:/ *n* destinatario *m*
adenoids /'ædɪnɔɪdz/ *npl* vegetaciones *fpl* adenoideas
adept /'ædept/ *a & n* experto (*m*)
adequa|cy /'ædɪkwəsɪ/ *n* suficiencia *f*. **~te** *a* suficiente, adecuado. **~tely** *adv* suficientemente, adecuadamente
adhere /əd'hɪə(r)/ *vi* adherirse (**to** a); observar ‹*rule*›. **~nce** /-rəns/ *n* adhesión *f*; (*to rules*) observancia *f*
adhesion /əd'hi:ʒn/ *n* adherencia *f*
adhesive /əd'hi:sɪv/ *a & n* adhesivo (*m*)
ad infinitum /ædɪnfɪ'naɪtəm/ *adv* hasta el infinito
adjacent /ə'dʒeɪsnt/ *a* contiguo
adjective /'ædʒɪktɪv/ *n* adjetivo *m*
adjoin /ə'dʒɔɪn/ *vt* lindar con. **~ing** *a* contiguo
adjourn /ə'dʒɜ:n/ *vt* aplazar; suspender ‹*meeting etc*›. ● *vi* suspenderse. **~ to** trasladarse a
adjudicate /ə'dʒu:dɪkeɪt/ *vt* juzgar. ● *vi* actuar como juez
adjust /ə'dʒʌst/ *vt* ajustar ‹*machine*›; (*arrange*) arreglar. ● *vi*. **~ (to)** adaptarse (a). **~able** *a* ajustable. **~ment** *n* adaptación *f*; (*tec*) ajuste *m*
ad lib /æd'lɪb/ *a* improvisado. ● *vi* (*pt* **-libbed**) (*fam*) improvisar
administer /əd'mɪnɪstə(r)/ *vt* administrar, dar, proporcionar
administrat|ion /ədmɪnɪ'streɪʃn/ *n* administración *f*. **~or** *n* administrador *m*
admirable /'ædmərəbl/ *a* admirable
admiral /'ædmərəl/ *n* almirante *m*
admiration /ædmə'reɪʃn/ *n* admiración *f*
admire /əd'maɪə(r)/ *vt* admirar. **~r** /-'maɪərə(r)/ *n* admirador *m*; (*suitor*) enamorado *m*
admissible /əd'mɪsəbl/ *a* admisible
admission /əd'mɪʃn/ *n* admisión *f*; (*entry*) entrada *f*
admit /əd'mɪt/ *vt* (*pt* **admitted**) dejar entrar; (*acknowledge*) admitir, reconocer. **~ to** confesar. **be ~ted** (*to hospital etc*) ingresar. **~tance** *n* entrada *f*. **~tedly** *adv* es verdad que
admoni|sh /əd'mɒnɪʃ/ *vt* reprender; (*advise*) aconsejar. **~tion** /-'nɪʃn/ *n* reprensión *f*
ado /ə'du:/ *n* alboroto m; (*trouble*) dificultad *f*. **without more ~** en seguida, sin más
adolescen|ce /ædə'lesns/ *n* adolescencia *f*. **~t** *a & n* adolescente (*m & f*)
adopt /ə'dɒpt/ *vt* adoptar. **~ed** *a* ‹*child*› adoptivo. **~ion** /-ʃn/ *n* adopción *f*. **~ive** *a* adoptivo
ador|able /ədɔ:rəbl/ *a* adorable. **~ation** /ædə'reɪʃn/ *n* adoración *f*. **~e** /ə'dɔ:(r)/ *vt* adorar
adorn /ə'dɔ:n/ *vt* adornar. **~ment** *n* adorno *m*

adrenalin /ə'drenəlɪn/ *n* adrenalina *f*
adrift /ə'drɪft/ *a & adv* a la deriva
adroit /ə'drɔɪt/ *a* diestro
adulation /ædjʊ'leɪʃn/ *n* adulación *f*
adult /'ædʌlt/ *a & n* adulto (*m*)
adulterat|ion /ədʌltə'reɪʃn/ *n* adulteración *f*. **~e** /ə'dʌltəreɪt/ *vt* adulterar
adulter|er /ə'dʌltərə(r)/ *n* adúltero *m*. **~ess** *n* adúltera *f*. **~ous** *a* adúltero. **~y** *n* adulterio *m*
advance /əd'vɑ:ns/ *vt* adelantar. ● *vi* adelantarse. ● *n* adelanto *m*. **in ~** con anticipación, por adelantado. **~d** *a* avanzado; ‹*studies*› superior. **~ment** *n* adelanto *m*; (*in job*) promoción *f*
advantage /əd'vɑ:ntɪdʒ/ *n* ventaja *f*. **take ~ of** aprovecharse de; abusar de ‹*person*›. **~ous** /ædvən'teɪdʒəs/ *a* ventajoso
advent /'ædvənt/ *n* venida *f*. **A~** *n* adviento *m*
adventur|e /əd'ventʃə(r)/ *n* aventura *f*. **~er** *n* aventurero *m*. **~ous** *a* ‹*persona*› aventurero; ‹*cosa*› arriesgado; (*fig*, *bold*) llamativo
adverb /'ædvɜ:b/ *n* adverbio *m*
adversary /'ædvəsərɪ/ *n* adversario *m*
advers|e /'ædvɜ:s/ *a* adverso, contrario, desfavorable. **~ity** /əd'vɜ:sətɪ/ *n* infortunio *m*
advert /'ædvɜ:t/ *n* (*fam*) anuncio *m*. **~ise** /'ædvətaɪz/ *vt* anunciar. ● *vi* hacer publicidad; (*seek*, *sell*) poner un anuncio. **~isement** /əd'vɜ:tɪsmənt/ *n* anuncio *m*. **~iser** /-ə(r)/ *n* anunciante *m & f*
advice /əd'vaɪs/ *n* consejo *m*; (*report*) informe *m*
advis|able /əd'vaɪzəbl/ *a* aconsejable. **~e** *vt* aconsejar; (*inform*) avisar. **~e against** aconsejar en contra de. **~er** *n* consejero *m*; (*consultant*) asesor *m*. **~ory** *a* consultivo
advocate /'ædvəkət/ *n* defensor *m*; (*jurid*) abogado *m*. /'ædvəkeɪt/ *vt* recomendar
aegis /'i:dʒɪs/ *n* égida *f*. **under the ~ of** bajo la tutela de, patrocinado por
aeon /'i:ən/ *n* eternidad *f*
aerial /'eərɪəl/ *a* aéreo. ● *n* antena *f*
aerobatics /eərə'bætɪks/ *npl* acrobacia *f* aérea
aerobics /eə'rɒbɪks/ *npl* aeróbica *f*
aerodrome /'eərədrəʊm/ *n* aeródromo *m*
aerodynamic /eərəʊdaɪ'næmɪk/ *a* aerodinámico
aeroplane /'eərəpleɪn/ *n* avión *m*
aerosol /'eərəsɒl/ *n* aerosol *m*
aesthetic /i:s'θetɪk/ *a* estético
afar /ə'fa:(r)/ *adv* lejos
affable /'æfəbl/ *a* afable
affair /ə'feə(r)/ *n* asunto *m*. **(love) ~** aventura *f*, amorío *m*. **~s** *npl* (*business*) negocios *mpl*
affect /ə'fekt/ *vt* afectar; (*pretend*) fingir
affect|ation /æfek'teɪʃn/ *n* afectación *f*. **~ed** *a* afectado, amanerado
affection /ə'fekʃn/ *n* cariño *m*; (*disease*) afección *f*. **~ate** /-ʃənət/ *a* cariñoso
affiliat|e /ə'fɪlɪeɪt/ *vt* afiliar. **~ion** /-'eɪʃn/ *n* afiliación *f*
affinity /ə'fɪnətɪ/ *n* afinidad *f*
affirm /ə'fɜ:m/ *vt* afirmar. **~ation** /æfə'meɪʃn/ *n* afirmación *f*
affirmative /ə'fɜ:mətɪv/ *a* afirmativo. ● *n* respuesta *f* afirmativa
affix /ə'fɪks/ *vt* sujetar; añadir ‹*signature*›; pegar ‹*stamp*›
afflict /ə'flɪkt/ *vt* afligir. **~ion** /-ʃn/ *n* aflicción *f*, pena *f*
affluen|ce /'æflʊəns/ *n* riqueza *f*. **~t** *a* rico. ● *n* (*geog*) afluente *m*
afford /ə'fɔ:d/ *vt* permitirse; (*provide*) dar
affray /ə'freɪ/ *n* reyerta *f*
affront /ə'frʌnt/ *n* afrenta *f*, ofensa *f*. ● *vt* afrentar, ofender
afield /ə'fi:ld/ *adv*. **far ~** muy lejos
aflame /ə'fleɪm/ *adv & a* en llamas
afloat /ə'fləʊt/ *adv* a flote
afoot /ə'fʊt/ *adv*. **sth is ~** se está tramando algo
aforesaid /ə'fɔ:sed/ *a* susodicho
afraid /ə'freɪd/ *a*. **be ~** tener miedo (**of** a); (*be sorry*) sentir, lamentar
afresh /ə'freʃ/ *adv* de nuevo
Africa /'æfrɪkə/ *n* África *f*. **~n** *a & n* africano (*m*)
after /'ɑ:ftə(r)/ *adv* después; (*behind*) detrás. ● *prep* después de; (*behind*) detrás de. **be ~** (*seek*) buscar, andar en busca de. ● *conj* después de que. ● *a* posterior
afterbirth /'ɑ:ftəbɜ:θ/ *n* placenta *f*
after-effect /'ɑ:ftəɪfekt/ *n* consecuencia *f*, efecto *m* secundario
aftermath /'ɑ:ftəmæθ/ *n* secuelas *fpl*
afternoon /ɑ:ftə'nu:n/ *n* tarde *f*
aftershave /'ɑ:ftəʃeɪv/ *n* loción *f* para después del afeitado

afterthought /'ɑ:ftəθɔ:t/ *n* ocurrencia *f* tardía
afterwards /'ɑ:ftəwədz/ *adv* después
again /ə'gen/ *adv* otra vez; (*besides*) además. **~ and ~** una y otra vez
against /ə'genst/ *prep* contra, en contra de
age /eɪdʒ/ *n* edad *f*. **of ~** mayor de edad. **under ~** menor de edad. ● *vt/i* (*pres p* **ageing**) envejecer. **~d** /'eɪdʒd/ *a* de ... años. **~d 10** de 10 años, que tiene 10 años. **~d** /'eɪdʒɪd/ *a* viejo, anciano. **~less** *a* siempre joven; (*eternal*) eterno, inmemorial. **~s** (*fam*) siglos *mpl*
agency /'eɪdʒənsɪ/ *n* agencia *f*, organismo *m*, oficina *f*; (*means*) mediación *f*
agenda /ə'dʒendə/ *npl* orden *m* del día
agent /'eɪdʒənt/ *n* agente *m* & *f*; (*representative*) representante *m* & *f*
agglomeration /əglɒmə'reɪʃn/ *n* aglomeración *f*
aggravat|e /'ægrəveɪt/ *vt* agravar; (*irritate, fam*) irritar. **~ion** /-'veɪʃn/ *n* agravación *f*; (*irritation, fam*) irritación *f*
aggregate /'ægrɪgət/ *a* total. ● *n* conjunto *m*. /'ægrɪgeɪt/ *vt* agregar. ● *vi* ascender a
aggress|ion /ə'greʃn/ *n* agresión *f*. **~ive** *a* agresivo. **~iveness** *n* agresividad *f*. **~or** *n* agresor *m*
aggrieved /ə'gri:vd/ *a* apenado, ofendido
aghast /ə'gɑ:st/ *a* horrorizado
agil|e /'ædʒaɪl/ *a* ágil. **~ity** /ə'dʒɪlətɪ/ *n* agilidad *f*
agitat|e /'ædʒɪteɪt/ *vt* agitar. **~ion** /-'teɪʃn/ *n* agitación *f*, excitación *f*. **~or** *n* agitador *m*
agnostic /æg'nɒstɪk/ *a* & *n* agnóstico (*m*). **~ism** /-sɪzəm/ *n* agnosticismo *m*
ago /ə'gəʊ/ *adv* hace. **a long time ~** hace mucho tiempo. **3 days ~** hace 3 días
agog /ə'gɒg/ *a* ansioso
agon|ize /'ægənaɪz/ *vi* atormentarse. **~izing** *a* atroz, angustioso, doloroso. **~y** *n* dolor *m* (agudo); (*mental*) angustia *f*
agree /ə'gri:/ *vt* acordar. ● *vi* estar de acuerdo; (*of figures*) concordar; (*get on*) entenderse. **~ with** (*of food etc*) sentar bien a. **~able** /ə'gri:əbl/ *a* agradable. **be ~able** (*willing*) estar de acuerdo. **~d** *a* ‹*time, place*› convenido. **~ment** /ə'gri:mənt/ *n* acuerdo *m*. **in ~ment** de acuerdo
agricultur|al /ægrɪ'kʌltʃərəl/ *a* agrícola. **~e** /'ægrɪkʌltʃə(r)/ *n* agricultura *f*
aground /ə'graʊnd/ *adv*. **run ~** (*of ship*) varar, encallar
ahead /ə'hed/ *adv* delante; (*of time*) antes de. **be ~** ir delante
aid /eɪd/ *vt* ayudar. ● *n* ayuda *f*. **in ~ of** a beneficio de
aide /eɪd/ *n* (*Amer*) ayudante *m* & *f*
AIDS /eɪdz/ *n* (*med*) SIDA *m*
ail /eɪl/ *vt* afligir. **~ing** *a* enfermo. **~ment** *n* enfermedad *f*
aim /eɪm/ *vt* apuntar; (*fig*) dirigir. ● *vi* apuntar; (*fig*) pretender. ● *n* puntería *f*; (*fig*) propósito *m*. **~less** *a*, **~lessly** *adv* sin objeto, sin rumbo
air /eə(r)/ *n* aire *m*. **be on the ~** estar en el aire. **put on ~s** darse aires. ● *vt* airear. ● *a* ‹*base etc*› aéreo. **~borne** *a* en el aire; (*mil*) aerotransportado. **~-conditioned** *a* climatizado, con aire acondicionado. **~craft** /'eəkrɑ:ft/ *n* (*pl invar*) avión *m*. **~field** /'eəfi:ld/ *n* aeródromo *m*. **A~ Force** fuerzas *fpl* aéreas. **~gun** /'eəgʌn/ *n* escopeta *f* de aire comprimido. **~lift** /'eəlɪft/ *n* puente *m* aéreo. **~line** /'eəlaɪn/ *n* línea *f* aérea. **~lock** /'eəlɒk/ *n* (*in pipe*) burbuja *f* de aire; (*chamber*) esclusa *f* de aire. **~ mail** *n* correo *m* aéreo. **~man** /'eəmən/ (*pl* **-men**) *n* aviador *m*. **~port** /'eəpɔ:t/ *n* aeropuerto *m*. **~tight** /'eətaɪt/ *a* hermético. **~worthy** /'eəwɜ:ðɪ/ *a* en condiciones de vuelo. **~y** /'eərɪ/ *a* (**-ier**, **-iest**) aireado; (*manner*) ligero
aisle /aɪl/ *n* nave *f* lateral; (*gangway*) pasillo *m*
ajar /ə'dʒɑ:(r)/ *adv* & *a* entreabierto
akin /ə'kɪn/ *a* semejante (**a** to)
alabaster /'æləbɑ:stə(r)/ *n* alabastro *m*
alacrity /ə'lækrətɪ/ *n* prontitud *f*
alarm /ə'lɑ:m/ *n* alarma *f*; (*clock*) despertador *m*. ● *vt* asustar. **~ist** *n* alarmista *m* & *f*
alas /ə'læs/ *int* ¡ay!, ¡ay de mí!
albatross /'ælbətrɒs/ *n* albatros *m*
albino /æl'bi:nəʊ/ *a* & *n* albino (*m*)
album /'ælbəm/ *n* álbum *m*
alchem|ist /'ælkəmɪst/ *n* alquimista *m* & *f*. **~y** *n* alquimia *f*
alcohol /'ælkəhɒl/ *n* alcohol *m*. **~ic** /-'hɒlɪk/ *a* & *n* alcohólico (*m*). **~ism** *n* alcoholismo *m*

alcove /'ælkəʊv/ *n* nicho *m*
ale /eɪl/ *n* cerveza *f*
alert /ə'lɜ:t/ *a* vivo; (*watchful*) vigilante. ● *n* alerta *f*. **on the ~** alerta. ● *vt* avisar. **~ness** *n* vigilancia *f*
algebra /'ældʒɪbrə/ *n* álgebra *f*
Algeria /æl'dʒɪərɪə/ *n* Argelia *f*. **~n** *a* & *n* argelino (*m*)
alias /'eɪlɪəs/ *n* (*pl* **-ases**) alias *m invar*. ● *adv* alias
alibi /'ælɪbaɪ/ (*pl* **-is**) coartada *f*
alien /'eɪlɪən/ *n* extranjero *m*. ● *a* ajeno
alienat|e /'eɪlɪəneɪt/ *vt* enajenar. **~ion** /-'neɪʃn/ *n* enajenación *f*
alight[1] /ə'laɪt/ *vi* bajar; ‹*bird*› posarse
alight[2] /ə'laɪt/ *a* ardiendo; ‹*light*› encendido
align /ə'laɪn/ *vt* alinear. **~ment** *n* alineación *f*
alike /ə'laɪk/ *a* parecido, semejante. **look** *or* **be ~** parecerse. ● *adv* de la misma manera
alimony /'ælɪmənɪ/ *n* pensión *f* alimenticia
alive /ə'laɪv/ *a* vivo. **~ to** sensible a. **~ with** lleno de
alkali /'ælkəlaɪ/ *n* (*pl* **-is**) álcali *m*. **~ne** *a* alcalino
all /ɔ:l/ *a* & *pron* todo. **~ but one** todos excepto uno. **~ of it** todo. ● *adv* completamente. **~ but** casi. **~ in** (*fam*) rendido. **~ of a sudden** de pronto. **~ over** (*finished*) acabado; (*everywhere*) por todas partes. **~ right!** ¡vale! **be ~ for** estar a favor de. **not at ~** de ninguna manera; (*after thanks!*) ¡no hay de qué!
allay /ə'leɪ/ *vt* aliviar ‹*pain*›; aquietar ‹*fears etc*›
all-clear /ɔ:l'klɪə(r)/ *n* fin *m* de (la) alarma
allegation /ælɪ'geɪʃn/ *n* alegato *m*
allege /ə'ledʒ/ *vt* alegar. **~dly** /-ɪdlɪ/ *adv* según se dice, supuestamente
allegiance /ə'li:dʒəns/ *n* lealtad *f*
allegor|ical /ælɪ'gɒrɪkl/ *a* alegórico. **~y** /'ælɪgərɪ/ *n* alegoría *f*
allerg|ic /ə'lɜ:dʒɪk/ *a* alérgico. **~y** /'ælədʒɪ/ *n* alergia *f*
alleviat|e /ə'li:vɪeɪt/ *vt* aliviar. **~ion** /-'eɪʃn/ *n* alivio *m*
alley /'ælɪ/ (*pl* **-eys**) *n* callejuela *f*; (*for bowling*) bolera *f*
alliance /ə'laɪəns/ *n* alianza *f*
allied /'ælaɪd/ *a* aliado
alligator /'ælɪgeɪtə(r)/ *n* caimán *m*
allocat|e /'æləkeɪt/ *vt* asignar; (*share out*) repartir. **~ion** /-'keɪʃn/ *n* asignación *f*; (*share*) ración *f*; (*distribution*) reparto *m*
allot /ə'lɒt/ *vt* (*pt* **allotted**) asignar. **~ment** *n* asignación *f*; (*share*) ración *f*; (*land*) parcela *f*
all-out /ɔ:l'aʊt/ *a* máximo
allow /ə'laʊ/ *vt* permitir; (*grant*) conceder; (*reckon on*) prever; (*agree*) admitir. **~ for** tener en cuenta. **~ance** /ə'laʊəns/ *n* concesión *f*; (*pension*) pensión *f*; (*com*) rebaja *f*. **make ~ances for** ser indulgente con; (*take into account*) tener en cuenta
alloy /'ælɔɪ/ *n* aleación *f*. /ə'lɔɪ/ *vt* alear
all-round /ɔ:l'raʊnd/ *a* completo
allude /ə'lu:d/ *vi* aludir
allure /ə'lʊə(r)/ *vt* atraer. ● *n* atractivo *m*
allusion /ə'lu:ʒn/ *n* alusión *f*
ally /'ælaɪ/ *n* aliado *m*. /ə'laɪ/ *vt* aliarse
almanac /'ɔ:lmənæk/ *n* almanaque *m*
almighty /ɔ:l'maɪtɪ/ *a* todopoderoso; (*big, fam*) enorme. ● *n*. **the A~** el Todopoderoso *m*
almond /'ɑ:mənd/ *n* almendra *f*; (*tree*) almendro (*m*)
almost /'ɔ:lməʊst/ *adv* casi
alms /ɑ:mz/ *n* limosna *f*
alone /ə'ləʊn/ *a* solo. ● *adv* sólo, solamente
along /ə'lɒŋ/ *prep* por, a lo largo de. ● *adv*. **~ with** junto con. **all ~** todo el tiempo. **come ~** venga
alongside /əlɒŋ'saɪd/ *adv* (*naut*) al costado. ● *prep* al lado de
aloof /ə'lu:f/ *adv* apartado. ● *a* reservado. **~ness** *n* reserva *f*
aloud /ə'laʊd/ *adv* en voz alta
alphabet /'ælfəbet/ *n* alfabeto *m*. **~ical** /-'betɪkl/ *a* alfabético
alpine /'ælpaɪn/ *a* alpino
Alps /ælps/ *npl*. **the ~** los Alpes *mpl*
already /ɔ:l'redɪ/ *adv* ya
Alsatian /æl'seɪʃn/ *n* (*geog*) alsaciano *m*; (*dog*) pastor *m* alemán
also /'ɔ:lsəʊ/ *adv* también; (*moreover*) además
altar /'ɔ:ltə(r)/ *n* altar *m*
alter /'ɔ:ltə(r)/ *vt* cambiar. ● *vi* cambiarse. **~ation** /-'reɪʃn/ *n* modificación *f*; (*to garment*) arreglo *m*
alternate /ɔ:l'tɜ:nət/ *a* alterno. /'ɔ:ltəneɪt/ *vt/i* alternar. **~ly** *adv* alternativamente

alternative /ɔ:l'tɜ:nətɪv/ *a* alternativo. ● *n* alternativa *f*. **~ly** *adv* en cambio, por otra parte
although /ɔ:l'ðəʊ/ *conj* aunque
altitude /'æltɪtju:d/ *n* altitud *f*
altogether /ɔ:ltə'geðə(r)/ *adv* completamente; (*on the whole*) en total
altruis|m /'æltru:ɪzəm/ *n* altruismo *m*. **~t** /'æltru:ɪst/ *n* altruista *m & f*. **~tic** /-'ɪstɪk/ *a* altruista
aluminium /æljʊ'mɪnɪəm/ *n* aluminio *m*
always /'ɔ:lweɪz/ *adv* siempre
am /æm/ *see* **be**
a.m. /'eɪem/ *abbr* (*ante meridiem*) de la mañana
amalgamate /ə'mælgəmeɪt/ *vt* amalgamar. ● *vi* amalgamarse
amass /ə'mæs/ *vt* amontonar
amateur /'æmətə(r)/ *n* aficionado *m*. ● *a* no profesional; (*in sports*) amateur. **~ish** *a* (*pej*) torpe, chapucero
amaz|e /ə'meɪz/ *vt* asombrar. **~ed** *a* asombrado, estupefacto. **be ~ed at** quedarse asombrado de, asombrarse de. **~ement** *n* asombro *m*. **~ingly** *adv* extaordinariamente
ambassador /æm'bæsədə(r)/ *n* embajador *m*
amber /'æmbə(r)/ *n* ámbar *m*; (*auto*) luz *f* amarilla
ambidextrous /æmbɪ'dekstrəs/ *a* ambidextro
ambience /'æmbɪəns/ *n* ambiente *m*
ambigu|ity /æmbɪ'gju:ətɪ/ *n* ambigüedad *f*. **~ous** /æm'bɪgjʊəs/ *a* ambiguo
ambit /'æmbɪt/ *n* ámbito *m*
ambiti|on /æm'bɪʃn/ *n* ambición *f*. **~ous** *a* ambicioso
ambivalen|ce /æm'bɪvələns/ *n* ambivalencia *f*. **~t** *a* ambivalente
amble /'æmbl/ *vi* andar despacio, andar sin prisa
ambulance /'æmbjʊləns/ *n* ambulancia *f*
ambush /'æmbʊʃ/ *n* emboscada *f*. ● *vt* tender una emboscada a
amen /ɑ:'men/ *int* amén
amenable /ə'mi:nəbl/ *a*. **~ to** (*responsive*) sensible a, flexible a
amend /ə'mend/ *vt* enmendar. **~ment** *n* enmienda *f*. **~s** *npl*. **make ~s** reparar
amenities /ə'mi:nətɪz/ *npl* atractivos *mpl*, comodidades *fpl*, instalaciones *fpl*
America /ə'merɪkə/ *n* América; (*North America*) Estados Unidos. **~n** *a & n* americano (*m*); (*North American*) estadounidense (*m & f*). **~nism** *n* americanismo *m*. **~nize** *vt* americanizar
amethyst /'æmɪθɪst/ *n* amatista *f*
amiable /'eɪmɪəbl/ *a* simpático
amicabl|e /'æmɪkəbl/ *a* amistoso. **~y** *adv* amistosamente
amid(st) /ə'mɪd(st)/ *prep* entre, en medio de
amiss /ə'mɪs/ *a* malo. ● *adv* mal. **sth ~** algo que no va bien. **take sth ~** llevar algo a mal
ammonia /ə'məʊnɪə/ *n* amoníaco *m*, amoniaco *m*
ammunition /æmjʊ'nɪʃn/ *n* municiones *fpl*
amnesia /æm'ni:zɪə/ *n* amnesia *f*
amnesty /'æmnəstɪ/ *n* amnistía *f*
amok /ə'mɒk/ *adv*. **run ~** volverse loco
among(st) /ə'mʌŋ(st)/ *prep* entre
amoral /eɪ'mɒrəl/ *a* amoral
amorous /'æmərəs/ *a* amoroso
amorphous /ə'mɔ:fəs/ *a* amorfo
amount /ə'maʊnt/ *n* cantidad *f*; (*total*) total *m*, suma *f*. ● *vi*. **~ to** sumar; (*fig*) equivaler a, significar
amp(ere) /'amp(eə(r))/ *n* amperio *m*
amphibi|an /æm'fɪbɪən/ *n* anfibio *m*. **~ous** *a* anfibio
amphitheatre /'æmfɪθɪətə(r)/ *n* anfiteatro *m*
ampl|e /'æmpl/ *a* (**-er**, **-est**) amplio; (*enough*) suficiente; (*plentiful*) abundante. **~y** *adv* ampliamente, bastante
amplif|ier /'æmplɪfaɪə(r)/ *n* amplificador *m*. **~y** *vt* amplificar
amputat|e /'æmpjʊteɪt/ *vt* amputar. **~ion** /-'teɪʃn/ *n* amputación *f*
amus|e /ə'mju:z/ *vt* divertir. **~ement** *n* diversión *f*. **~ing** *a* divertido
an /ən, æn/ *see* **a**
anachronism /ə'nækrənɪzəm/ *n* anacronismo *m*
anaemi|a /ə'ni:mɪə/ *n* anemia *f*. **~c** *a* anémico
anaesthe|sia /ænɪs'θi:zɪə/ *n* anestesia *f*. **~tic** /ænɪs'θetɪk/ *n* anestésico *m*. **~tist** /ə'ni:sθɪtɪst/ *n* anestesista *m & f*
anagram /'ænəgræm/ *n* anagrama *m*
analogy /ə'nælədʒɪ/ *n* analogía *f*
analys|e /'ænəlaɪz/ *vt* analizar. **~is** /ə'næləsɪs/ *n* (*pl* **-yses** /-si:z/) *n* análisis *m*. **~t** /'ænəlɪst/ *n* analista *m & f*

analytic(al) /ænəˈlɪtɪk(əl)/ *a* analítico
anarch|ist /ˈænəkɪst/ *n* anarquista *m* & *f*. **~y** *n* anarquía *f*
anathema /əˈnæθəmə/ *n* anatema *m*
anatom|ical /ænəˈtɒmɪkl/ *a* anatómico. **~y** /əˈnætəmɪ/ *n* anatomía *f*
ancest|or /ˈænsestə(r)/ *n* antepasado *m*. **~ral** /-ˈsestrəl/ *a* ancestral. **~ry** /ˈænsestrɪ/ *n* ascendencia *f*
anchor /ˈæŋkə(r)/ *n* ancla *f*. ● *vt* anclar; (*fig*) sujetar. ● *vi* anclar
anchovy /ˈæntʃəvɪ/ *n* (*fresh*) boquerón *m*; (*tinned*) anchoa *f*
ancient /ˈeɪnʃənt/ *a* antiguo, viejo
ancillary /ænˈsɪlərɪ/ *a* auxiliar
and /ənd, ænd/ *conj* y; (*before* i- *and* hi-) e. **go ~ see him** vete a verle. **more ~ more** siempre más, cada vez más. **try ~ come** ven si puedes, trata de venir
Andalusia /ændəˈlu:zjə/ *f* Andalucía *f*
anecdote /ˈænɪkdəʊt/ *n* anécdota *f*
anew /əˈnju:/ *adv* de nuevo
angel /ˈeɪndʒl/ *n* ángel *m*. **~ic** /ænˈdʒelɪk/ *a* angélico
anger /ˈæŋgə(r)/ *n* ira *f*. ● *vt* enojar
angle[1] /ˈæŋgl/ *n* ángulo *m*; (*fig*) punto *m* de vista
angle[2] /ˈæŋgl/ *vi* pescar con caña. **~ for** (*fig*) buscar. **~r** /-ə(r)/ *n* pescador *m*
Anglican /ˈæŋglɪkən/ *a* & *n* anglicano (*m*)
Anglo-... /ˈæŋgləʊ/ *pref* anglo...
Anglo-Saxon /ˈæŋgləʊˈsæksn/ *a* & *n* anglosajón (*m*)
angr|ily /ˈæŋgrɪlɪ/ *adv* con enojo. **~y** /ˈæŋgrɪ/ *a* (**-ier**, **-iest**) enojado. **get ~y** enfadarse
anguish /ˈæŋgwɪʃ/ *n* angustia *f*
angular /ˈæŋgjʊlə(r)/ *a* angular; ‹*face*› anguloso
animal /ˈænɪməl/ *a* & *n* animal (*m*)
animat|e /ˈænɪmət/ *a* vivo. /ˈænɪmeɪt/ *vt* animar. **~ion** /-ˈmeɪʃn/ *n* animación *f*
animosity /ænɪˈmɒsətɪ/ *n* animosidad *f*
aniseed /ˈænɪsi:d/ *n* anís *m*
ankle /ˈæŋkl/ *n* tobillo *m*. **~ sock** escarpín *m*, calcetín *m*
annals /ˈænlz/ *npl* anales *mpl*
annex /əˈneks/ *vt* anexionar. **~ation** /ænekˈseɪʃn/ *n* anexión *f*
annexe /ˈæneks/ *n* anexo *m*, dependencia *f*
annihilat|e /əˈnaɪəleɪt/ *vt* aniquilar. **~ion** /-ˈleɪʃn/ *n* aniquilación *f*
anniversary /ænɪˈvɜ:sərɪ/ *n* aniversario *m*
annotat|e /ˈænəteɪt/ *vt* anotar. **~ion** /-ˈteɪʃn/ *n* anotación *f*
announce /əˈnaʊns/ *vt* anunciar, comunicar. **~ment** *n* anuncio *m*, aviso *m*, declaración *f*. **~r** /-e(r)/ *n* (*radio*, *TV*) locutor *m*
annoy /əˈnɔɪ/ *vt* molestar. **~ance** *n* disgusto *m*. **~ed** *a* enfadado. **~ing** *a* molesto
annual /ˈænjʊəl/ *a* anual. ● *n* anuario *m*. **~ly** *adv* cada año
annuity /əˈnju:ətɪ/ *n* anualidad *f*. **life ~** renta *f* vitalicia
annul /əˈnʌl/ *vt* (*pt* **annulled**) anular. **~ment** *n* anulación *f*
anoint /əˈnɔɪnt/ *vt* ungir
anomal|ous /əˈnɒmələs/ *a* anómalo. **~y** *n* anomalía *f*
anon /əˈnɒn/ *adv* (*old use*) dentro de poco
anonymous /əˈnɒnɪməs/ *a* anónimo
anorak /ˈænəræk/ *n* anorac *m*
another /əˈnʌðə(r)/ *a* & *pron* otro (*m*). **~ 10 minutes** 10 minutos más. **in ~ way** de otra manera. **one ~** unos a otros
answer /ˈɑ:nsə(r)/ *n* respuesta *f*; (*solution*) solución *f*. ● *vt* contestar a; escuchar, oír ‹*prayer*›. **~ the door** abrir la puerta. ● *vi* contestar. **~ back** replicar. **~ for** ser responsable de. **~able** *a* responsable. **~ing-machine** *n* contestador *m* automático
ant /ænt/ *n* hormiga *f*
antagoni|sm /ænˈtægənɪzəm/ *n* antagonismo *m*. **~stic** /-ˈnɪstɪk/ *a* antagónico, opuesto. **~ze** /ænˈtægənaɪz/ *vt* provocar la enemistad de
Antarctic /ænˈtɑ:ktɪk/ *a* antártico. ● *n* Antártico *m*
ante-... /ˈæntɪ/ *pref* ante...
antecedent /æntɪˈsi:dnt/ *n* antecedente *m*
antelope /ˈæntɪləʊp/ *n* antílope *m*
antenatal /ˈæntɪneɪtl/ *a* prenatal
antenna /ænˈtenə/ *n* antena *f*
anthem /ˈænθəm/ *n* himno *m*
anthill /ˈænthɪl/ *n* hormiguero *m*
anthology /ænˈθɒlədʒɪ/ *n* antología *f*
anthropolog|ist /ænθrəˈpɒlədʒɪst/ *n* antropológo *m*. **~y** *n* antropología *f*
anti-... /ˈæntɪ/ *pref* anti... **~aircraft** *a* antiaéreo

antibiotic /æntɪbaɪ'ɒtɪk/ *a & n* antibiótico (*m*)
antibody /'æntɪbɒdɪ/ *n* anticuerpo *m*
antic /'æntɪk/ *n* payasada *f*, travesura *f*
anticipat|e /æn'tɪsɪpeɪt/ *vt* anticiparse a; (*foresee*) prever; (*forestall*) prevenir. **~ion** /-'peɪʃn/ *n* anticipación *f*; (*expectation*) esperanza *f*
anticlimax /æntɪ'klaɪmæks/ *n* decepción *f*
anticlockwise /æntɪ'klɒkwaɪz/ *adv & a* en sentido contrario al de las agujas del reloj, hacia la izquierda
anticyclone /æntɪ'saɪkləʊn/ *n* anticiclón *m*
antidote /'æntɪdəʊt/ *m* antídoto *m*
antifreeze /'æntɪfri:z/ *n* anticongelante *m*
antipathy /æn'tɪpəθɪ/ *n* antipatía *f*
antiquarian /æntɪ'kweərɪən/ *a & n* anticuario (*m*)
antiquated /'æntɪkweɪtɪd/ *a* anticuado
antique /æn'ti:k/ *a* antiguo. ● *n* antigüedad *f*. **~ dealer** anticuario *m*. **~ shop** tienda *f* de antigüedades
antiquity /æn'tɪkwətɪ/ *n* antigüedad *f*
anti-Semitic /æntɪsɪ'mɪtɪk/ *a* antisemítico
antiseptic /æntɪ'septɪk/ *a & n* antiséptico (*m*)
antisocial /æntɪ'səʊʃl/ *a* antisocial
antithesis /æn'tɪθəsɪs/ *n* (*pl* **-eses** /-si:z/) antítesis *f*
antler /'æntlər/ *n* cornamenta *f*
anus /'eɪnəs/ *n* ano *m*
anvil /'ænvɪl/ *n* yunque *m*
anxiety /æŋ'zaɪətɪ/ *n* ansiedad *f*; (*worry*) inquietud *f*; (*eagerness*) anhelo *m*
anxious /'æŋkʃəs/ *a* inquieto; (*eager*) deseoso. **~ly** *adv* con inquietud; (*eagerly*) con impaciencia
any /'enɪ/ *a* algún *m*; (*negative*) ningún *m*; (*whatever*) cualquier; (*every*) todo. **at ~ moment** en cualquier momento. **have you ~ wine?** ¿tienes vino? ● *pron* alguno; (*negative*) ninguno. **have we ~?** ¿tenemos algunos? **not ~** ninguno. ● *adv* (*a little*) un poco, algo. **is it ~ better?** ¿está algo mejor? **it isn't ~ good** no sirve para nada
anybody /'enɪbɒdɪ/ *pron* alguien; (*after negative*) nadie. **~ can do it** cualquiera sabe hacerlo, cualquiera puede hacerlo
anyhow /'enɪhaʊ/ *adv* de todas formas; (*in spite of all*) a pesar de todo; (*badly*) de cualquier modo
anyone /'enɪwʌn/ *pron* alguien; (*after negative*) nadie
anything /'enɪθɪŋ/ *pron* algo; (*whatever*) cualquier cosa; (*after negative*) nada. **~ but** todo menos
anyway /'enɪweɪ/ *adv* de todas formas
anywhere /'enɪweə(r)/ *adv* en cualquier parte; (*after negative*) en ningún sitio; (*everywhere*) en todas partes. **~ else** en cualquier otro lugar. **~ you go** dondequiera que vayas
apace /ə'peɪs/ *adv* rápidamente
apart /ə'pɑ:t/ *adv* aparte; (*separated*) apartado, separado. **~ from** aparte de. **come ~** romperse. **take ~** desmontar
apartheid /ə'pɑ:theɪt/ *n* segregación *f* racial, apartheid *m*
apartment /ə'pɑ:tmənt/ *n* (*Amer*) apartamento *m*
apath|etic /æpə'θetɪk/ *a* apático, indiferente. **~y** /'æpəθɪ/ *n* apatía *f*
ape /eɪp/ *n* mono *m*. ● *vt* imitar
aperient /ə'pɪərɪənt/ *a & n* laxante (*m*)
aperitif /ə'perətɪf/ *n* aperitivo *m*
aperture /'æpətʃʊə(r)/ *n* abertura *f*
apex /'eɪpeks/ *n* ápice *m*
aphorism /'æfərɪzəm/ *n* aforismo *m*
aphrodisiac /æfrə'dɪzɪæk/ *a & n* afrodisíaco (*m*), afrodisiaco (*m*)
apiece /ə'pi:s/ *adv* cada uno
aplomb /ə'plɒm/ *n* aplomo *m*
apolog|etic /əpɒlə'dʒetɪk/ *a* lleno de disculpas. **be ~etic** disculparse. **~ize** /ə'pɒlədʒaɪz/ *vi* disculparse (**for** de). **~y** /ə'pɒlədʒɪ/ *n* disculpa *f*; (*poor specimen*) birria *f*
apople|ctic /æpə'plektɪk/ *a* apoplético. **~xy** /'æpəpleksɪ/ *n* apoplejía *f*
apostle /ə'pɒsl/ *n* apóstol *m*
apostrophe /ə'pɒstrəfɪ/ *n* (*punctuation mark*) apóstrofo *m*
appal /ə'pɔ:l/ *vt* (*pt* **appalled**) horrorizar. **~ling** *a* espantoso
apparatus /æpə'reɪtəs/ *n* aparato *m*
apparel /ə'pærəl/ *n* ropa *f*, indumentaria *f*
apparent /ə'pærənt/ *a* aparente; (*clear*) evidente. **~ly** *adv* por lo visto
apparition /æpə'rɪʃn/ *n* aparición *f*

appeal /ə'pi:l/ *vi* apelar; (*attract*) atraer. ● *n* llamamiento *m*; (*attraction*) atractivo *m*; (*jurid*) apelación *f*. **~ing** *a* atrayente

appear /ə'pɪə(r)/ *vi* aparecer; (*arrive*) llegar; (*seem*) parecer; (*on stage*) actuar. **~ance** *n* aparición *f*; (*aspect*) aspecto *m*

appease /ə'pi:z/ *vt* aplacar; (*pacify*) apaciguar

append /ə'pend/ *vt* adjuntar. **~age** /ə'pendɪdʒ/ *n* añadidura *f*

appendicitis /əpendɪ'saɪtɪs/ *n* apendicitis *f*

appendix /ə'pendɪks/ *n* (*pl* **-ices** /-si:z/) (*of book*) apéndice *m*. (*pl* **-ixes**) (*anat*) apéndice *m*

appertain /æpə'teɪn/ *vi* relacionarse (**to** con)

appetite /'æpɪtaɪt/ *n* apetito *m*

appetiz|er /'æpɪtaɪzə(r)/ *n* aperitivo *m*. **~ing** *a* apetitoso

applau|d /ə'plɔ:d/ *vt/i* aplaudir. **~se** *n* aplausos *mpl*

apple /'æpl/ *n* manzana *f*. **~-tree** *n* manzano *m*

appliance /ə'plaɪəns/ *n* aparato *m*. **electrical ~** electrodoméstico *m*

applicable /'æplɪkəbl/ *a* aplicable; (*relevant*) pertinente

applicant /'æplɪkənt/ *n* candidato *m*, solicitante *m & f*

application /æplɪ'keɪʃn/ *n* aplicación *f*; (*request*) solicitud *f*. **~ form** formulario *m* (de solicitud)

appl|ied /ə'plaɪd/ *a* aplicado. **~y** /ə'plaɪ/ *vt* aplicar. ● *vi* aplicarse; (*ask*) dirigirse. **~y for** solicitar ‹*job etc*›

appoint /ə'pɔɪnt/ *vt* nombrar; (*fix*) señalar. **~ment** *n* cita *f*; (*job*) empleo *m*

apportion /ə'pɔ:ʃn/ *vt* repartir

apposite /'æpəzɪt/ *a* apropiado

apprais|al /ə'preɪzl/ *n* evaluación *f*. **~e** *vt* evaluar

appreciable /ə'pri:ʃəbl/ *a* sensible; (*considerable*) considerable

appreciat|e /ə'pri:ʃɪeɪt/ *vt* apreciar; (*understand*) comprender; (*be grateful for*) agradecer. ● *vi* (*increase value*) aumentar en valor. **~ion** /-'eɪʃn/ *n* aprecio *m*; (*gratitude*) agradecimiento *m*. **~ive** /ə'pri:ʃɪətɪv/ *a* (*grateful*) agradecido

apprehen|d /æprɪ'hend/ *vt* detener; (*understand*) comprender. **~sion** /-ʃn/ *n* detención *f*; (*fear*) recelo *m*

apprehensive /æprɪ'hensɪv/ *a* aprensivo

apprentice /ə'prentɪs/ *n* aprendiz *m*. ● *vt* poner de aprendiz. **~ship** *n* aprendizaje *m*

approach /ə'prəʊtʃ/ *vt* acercarse a. ● *vi* acercarse. ● *n* acercamiento *m*; (*to problem*) enfoque *m*; (*access*) acceso *m*. **make ~es to** dirigirse a. **~able** *a* accesible

approbation /æprə'beɪʃn/ *n* aprobación *f*

appropriate /ə'prəʊprɪət/ *a* apropiado. /ə'prəʊprɪeɪt/ *vt* apropiarse de. **~ly** *adv* apropiadamente

approval /ə'pru:vl/ *n* aprobación *f*. **on ~** a prueba

approv|e /ə'pru:v/ *vt/i* aprobar. **~ingly** *adv* con aprobación

approximat|e /ə'prɒksɪmət/ *a* aproximado. /ə'prɒksɪmeɪt/ *vt* aproximarse a. **~ely** *adv* aproximadamente. **~ion** /-'meɪʃn/ *n* aproximación *f*

apricot /'eɪprɪkɒt/ *n* albaricoque *m*, chabacano *m* (*Mex*). **~-tree** *n* albaricoquero *m*, chabacano *m* (*Mex*)

April /'eɪprəl/ *n* abril *m*. **~ fool!** ¡inocentón!

apron /'eɪprən/ *n* delantal *m*

apropos /'æprəpəʊ/ *adv* a propósito

apse /æps/ *n* ábside *m*

apt /æpt/ *a* apropiado; ‹*pupil*› listo. **be ~ to** tener tendencia a

aptitude /'æptɪtju:d/ *n* aptitud *f*

aptly /'æptlɪ/ *adv* acertadamente

aqualung /'ækwəlʌŋ/ *n* pulmón *m* acuático

aquarium /ə'kweərɪəm/ *n* (*pl* **-ums**) acuario *m*

Aquarius /ə'kweərɪəs/ *n* Acuario *m*

aquatic /ə'kwætɪk/ *a* acuático

aqueduct /'ækwɪdʌkt/ *n* acueducto *m*

aquiline /'ækwɪlaɪn/ *a* aquilino

Arab /'ærəb/ *a & n* árabe *m*. **~ian** /ə'reɪbɪən/ *a* árabe. **~ic** /'ærəbɪk/ *a & n* árabe (*m*). **~ic numerals** números *mpl* arábigos

arable /'ærəbl/ *a* cultivable

arbiter /'ɑ:bɪtə(r)/ *n* árbitro *m*

arbitrary /'ɑ:bɪtrərɪ/ *a* arbitrario

arbitrat|e /'ɑ:bɪtreɪt/ *vi* arbitrar. **~ion** /-'treɪʃn/ *n* arbitraje *m*. **~or** *n* árbitro *m*

arc /ɑ:k/ *n* arco *m*

arcade /ɑ:'keɪd/ *n* arcada f; (*around square*) soportales *mpl*; (*shops*)

galería *f*. **amusement ~** galería *f* de atracciones
arcane /ɑ:'keɪn/ *a* misterioso
arch[1] /ɑ:tʃ/ *n* arco *m*. ● *vt* arquear. ● *vi* arquearse
arch[2] /ɑ:tʃ/ *a* malicioso
archaeolog|ical /ɑ:kɪə'lɒdʒɪkl/ *a* arqueológico. **~ist** /ɑ:kɪ'ɒlədʒɪst/ *n* arqueólogo *m*. **~y** /ɑ:kɪ'ɒlədʒɪ/ *n* arqueología *f*
archaic /ɑ:'keɪɪk/ *a* arcaico
archbishop /ɑ:tʃ'bɪʃəp/ *n* arzobispo *m*
arch-enemy /ɑ:tʃ'enəmɪ/ *n* enemigo *m* jurado
archer /'ɑ:tʃə(r)/ *n* arquero *m*. **~y** *n* tiro *m* al arco
archetype /'ɑ:kɪtaɪp/ *n* arquetipo *m*
archipelago /ɑ:kɪ'peləgəʊ/ *n* (*pl* **-os**) archipiélago *m*
architect /'ɑ:kɪtekt/ *n* arquitecto *m*. **~ure** /'ɑ:kɪtektʃə(r)/ *n* arquitectura *f*. **~ural** /-'tektʃərəl/ *a* arquitectónico
archiv|es /'ɑ:kaɪvz/ *npl* archivo *m*. **~ist** /-ɪvɪst/ *n* archivero *m*
archway /'ɑ:tʃweɪ/ *n* arco *m*
Arctic /'ɑ:ktɪk/ *a* ártico. ● *n* Ártico *m*
arctic /'ɑ:ktɪk/ *a* glacial
ardent /'ɑ:dənt/ *a* ardiente, fervoroso, apasionado. **~ly** *adv* ardientemente
ardour /'ɑ:də(r)/ *n* ardor *m*, fervor *m*, pasión *f*
arduous /'ɑ:djʊəs/ *a* arduo
are /ɑ:(r)/ *see* **be**
area /'eərɪə/ *n* (*surface*) superficie *f*; (*region*) zona *f*; (*fig*) campo *m*
arena /ə'ri:nə/ *n* arena *f*; (*in circus*) pista *f*; (*in bullring*) ruedo *m*
aren't /ɑ:nt/ = **are not**
Argentin|a /ɑ:dʒən'ti:nə/ *n* Argentina *f*. **~ian** /-'tɪnɪən/ *a* & *n* argentino (*m*)
arguable /'ɑ:gjʊəbl/ *a* discutible
argue /'ɑ:gju:/ *vi* discutir; (*reason*) razonar
argument /'ɑ:gjʊmənt/ *n* disputa *f*; (*reasoning*) argumento *m*. **~ative** /-'mentətɪv/ *a* discutidor
arid /'ærɪd/ *a* árido
Aries /'eəri:z/ *n* Aries *m*
arise /ə'raɪz/ *vi* (*pt* **arose**, *pp* **arisen**) levantarse; (*fig*) surgir. **~ from** resultar de
aristocra|cy /ærɪ'stɒkrəsɪ/ *n* aristocracia *f*. **~t** /'ærɪstəkræt/ *n* aristócrata *m* & *f*. **~tic** /-'krætɪk/ *a* aristocrático
arithmetic /ə'rɪθmətɪk/ *n* aritmética *f*
ark /ɑ:k/ *n* (*relig*) arca *f*
arm[1] /ɑ:m/ *n* brazo *m*. **~ in ~** cogidos del brazo
arm[2] /ɑ:m/ *n*. **~s** *npl* armas *fpl*. ● *vt* armar
armada /ɑ:'mɑ:də/ *n* armada *f*
armament /'ɑ:məmənt/ *n* armamento *m*
armchair /'ɑ:mtʃeə(r)/ *n* sillón *m*
armed robbery /ɑ:md'rɒbərɪ/ *n* robo *m* a mano armada
armful /'ɑ:mfʊl/ *n* brazada *f*
armistice /'ɑ:mɪstɪs/ *n* armisticio *m*
armlet /'ɑ:mlɪt/ *n* brazalete *m*
armour /'ɑ:mə(r)/ *n* armadura *f*. **~ed** *a* blindado
armoury /'ɑ:mərɪ/ *n* arsenal *m*
armpit /'ɑ:mpɪt/ *n* sobaco *m*, axila *f*
army /'ɑ:mɪ/ *n* ejército *m*
aroma /ə'rəʊmə/ *n* aroma *m*. **~tic** /ærə'mætɪk/ *a* aromático
arose /ə'rəʊz/ *see* **arise**
around /ə'raʊnd/ *adv* alrededor; (*near*) cerca. **all ~** por todas partes. ● *prep* alrededor de; (*with time*) a eso de
arouse /ə'raʊz/ *vt* despertar
arpeggio /ɑ:'pedʒɪəʊ/ *n* arpegio *m*
arrange /ə'reɪndʒ/ *vt* arreglar; (*fix*) fijar. **~ment** *n* arreglo *m*; (*agreement*) acuerdo *m*; (*pl, plans*) preparativos *mpl*
array /ə'reɪ/ *vt* (*dress*) ataviar; (*mil*) formar. ● *n* atavío *m*; (*mil*) orden *m*; (*fig*) colección *f*, conjunto *m*
arrears /ə'rɪəz/ *npl* atrasos *mpl*. **in ~** atrasado en pagos
arrest /ə'rest/ *vt* detener; llamar ‹*attention*›. ● *n* detención *f*. **under ~** detenido
arriv|al /ə'raɪvl/ *n* llegada *f*. **new ~al** recien llegado *m*. **~e** /ə'raɪv/ *vi* llegar
arrogan|ce /'ærəgəns/ *n* arrogancia *f*. **~t** *a* arrogante. **~tly** *adv* con arrogancia
arrow /'ærəʊ/ *n* flecha *f*
arsenal /'ɑ:sənl/ *n* arsenal *m*
arsenic /'ɑ:snɪk/ *n* arsénico *m*
arson /'ɑ:sn/ *n* incendio *m* provocado. **~ist** *n* incendiario *m*
art[1] /ɑ:t/ *n* arte *m*. **A~s** *npl* (*Univ*) Filosofía y Letras *fpl*. **fine ~s** bellas artes *fpl*
art[2] /ɑ:t/ (*old use, with* **thou**) = **are**
artefact /'ɑ:tɪfækt/ *n* artefacto *m*
arterial /ɑ:'tɪərɪəl/ *a* arterial. **~ road** *n* carretera *f* nacional
artery /'ɑ:tərɪ/ *n* arteria *f*

artesian /ɑ:'ti:zjən/ *a*. ~ **well** pozo *m* artesiano

artful /'ɑ:tfʊl/ *a* astuto. ~**ness** *n* astucia *f*

art gallery /'ɑ:tgælərɪ/ *n* museo *m* de pinturas, pinacoteca *f*, galería *f* de arte

arthriti|c /ɑ:'θrɪtɪk/ *a* artrítico. ~**s** /ɑ:'θraɪtɪs/ *n* artritis *f*

artichoke /'ɑ:tɪtʃəʊk/ *n* alcachofa *f*. **Jerusalem** ~ pataca *f*

article /'ɑ:tɪkl/ *n* artículo *m*. ~ **of clothing** prenda *f* de vestir. **leading** ~ artículo de fondo

articulat|e /ɑ:'tɪkjʊlət/ *a* articulado; ‹*person*› elocuente. /ɑ:'tɪkjʊleɪt/ *vt/i* articular. ~**ed lorry** *n* camión *m* con remolque. ~**ion** /-'leɪʃn/ *n* articulación *f*

artifice /'ɑ:tɪfɪs/ *n* artificio *m*

artificial /ɑ:tɪ'fɪʃl/ *a* artificial; ‹*hair etc*› postizo

artillery /ɑ:'tɪlərɪ/ *n* artillería *f*

artisan /ɑ:tɪ'zæn/ *n* artesano *m*

artist /'ɑ:tɪst/ *n* artista *m & f*

artiste /ɑ:'ti:st/ *n* (*in theatre*) artista *m & f*

artist|ic /ɑ:'tɪstɪk/ *a* artístico. ~**ry** *n* arte *m*, habilidad *f*

artless /'ɑ:tlɪs/ *a* ingenuo

arty /'ɑ:tɪ/ *a* (*fam*) que se las da de artista

as /æz, əz/ *adv & conj* como; (*since*) ya que; (*while*) mientras. ~ **big** ~ tan grande como. ~ **far** ~ (*distance*) hasta; (*qualitative*) en cuanto a. ~ **far** ~ **I know** que yo sepa. ~ **if** como si. ~ **long** ~ mientras. ~ **much** ~ tanto como. ~ **soon** ~ tan pronto como. ~ **well** también

asbestos /æz'bestɒs/ *n* amianto *m*, asbesto *m*

ascen|d /ə'send/ *vt/i* subir. ~**t** /ə'sent/ *n* subida *f*

ascertain /æsə'teɪn/ *vt* averiguar

ascetic /ə'setɪk/ *a* ascético. • *n* asceta *m & f*

ascribe /ə'skraɪb/ *vt* atribuir

ash[1] /æʃ/ *n* ceniza *f*

ash[2] /æʃ/ *n*. ~**(-tree)** fresno *m*

ashamed /ə'ʃeɪmd/ *a* avergonzado. **be** ~ avergonzarse

ashen /'æʃn/ *a* ceniciento

ashore /ə'ʃɔ:(r)/ *adv* a tierra. **go** ~ desembarcar

ash: ~**tray** /'æʃtreɪ/ *n* cenicero *m*. **A~ Wednesday** *n* Miércoles *m* de Ceniza

Asia /'eɪʃə/ *n* Asia *f*. ~**n** *a & n* asiático (*m*). ~**tic** /-ɪ'ætɪk/ *a* asiático

aside /ə'saɪd/ *adv* a un lado. • *n* (*in theatre*) aparte *m*

asinine /'æsɪnaɪn/ *a* estúpido

ask /ɑ:sk/ *vt* pedir; preguntar ‹*question*›; (*invite*) invitar. ~ **about** enterarse de. ~ **after** pedir noticias de. ~ **for help** pedir ayuda. ~ **for trouble** buscarse problemas. ~ **s.o. in** invitar a uno a pasar

askance /ə'skæns/ *adv*. **look** ~ **at** mirar de soslayo

askew /ə'skju:/ *adv & a* ladeado

asleep /ə'sli:p/ *adv & a* dormido. **fall** ~ dormirse, quedar dormido

asparagus /ə'spærəgəs/ *n* espárrago *m*

aspect /'æspekt/ *n* aspecto *m*; (*of house etc*) orientación *f*

aspersions /ə'spɜ:ʃnz/ *npl*. **cast** ~ **on** difamar

asphalt /'æsfælt/ *n* asfalto *m*. • *vt* asfaltar

asphyxia /æs'fɪksɪə/ *n* asfixia *f*. ~**te** /əs'fɪksɪeɪt/ *vt* asfixiar. ~**tion** /-'eɪʃn/ *n* asfixia *f*

aspic /'æspɪk/ *n* gelatina *f*

aspir|ation /æspə'reɪʃn/ *n* aspiración *f*. ~**e** /əs'paɪə(r)/ *vi* aspirar

aspirin /'æsprɪn/ *n* aspirina *f*

ass /æs/ *n* asno *m*; (*fig, fam*) imbécil *m*

assail /ə'seɪl/ *vt* asaltar. ~**ant** *n* asaltador *m*

assassin /ə'sæsɪn/ *n* asesino *m*. ~**ate** /ə'sæsɪneɪt/ *vt* asesinar. ~**ation** /-'eɪʃn/ *n* asesinato *m*

assault /ə'sɔ:lt/ *n* (*mil*) ataque *m*; (*jurid*) atentado *m*. • *vt* asaltar

assemblage /ə'semblɪdʒ/ *n* (*of things*) colección *f*; (*of people*) reunión *f*; (*mec*) montaje *m*

assemble /ə'sembl/ *vt* reunir; (*mec*) montar. • *vi* reunirse

assembly /ə'semblɪ/ *n* reunión *f*; (*pol etc*) asamblea *f*. ~ **line** *n* línea *f* de montaje

assent /ə'sent/ *n* asentimiento *m*. • *vi* asentir

assert /ə'sɜ:t/ *vt* afirmar; hacer valer ‹*one's rights*›. ~**ion** /-ʃn/ *n* afirmación *f*. ~**ive** *a* positivo, firme

assess /ə'ses/ *vt* valorar; (*determine*) determinar; fijar ‹*tax etc*›. ~**ment** *n* valoración *f*

asset /'æset/ *n* (*advantage*) ventaja *f*; (*pl, com*) bienes *mpl*

assiduous /ə'sɪdjʊəs/ *a* asiduo

assign /ə'saɪn/ *vt* asignar; (appoint) nombrar

assignation /æsɪg'neɪʃn/ *n* asignación *f*; (*meeting*) cita *f*

assignment /ə'saɪnmənt/ *n* asignación *f*, misión *f*; (*task*) tarea *f*

assimilat|e /ə'sɪmɪleɪt/ *vt* asimilar. ● *vi* asimilarse. **~ion** /-'eɪʃn/ *n* asimilación *f*

assist /ə'sɪst/ *vt/i* ayudar. **~ance** *n* ayuda *f*. **~ant** /ə'sɪstənt/ *n* ayudante *m & f*; (*shop*) dependienta *f*, dependiente *m*. ● *a* auxiliar, adjunto

associat|e /ə'səʊʃɪeɪt/ *vt* asociar. ● *vi* asociarse. /ə'səʊʃɪət/ *a* asociado. ● *n* colega *m & f*; (*com*) socio *m*. **~ion** /-'eɪʃn/ *n* asociación *f*. **A~ion football** *n* fútbol *m*

assort|ed /ə'sɔ:tɪd/ *a* surtido. **~ment** *n* surtido *m*

assume /ə'sju:m/ *vt* suponer; tomar ‹*power, attitude*›; asumir ‹*role, burden*›

assumption /ə'sʌmpʃn/ *n* suposición *f*. **the A~** la Asunción *f*

assur|ance /ə'ʃʊərəns/ *n* seguridad *f*; (*insurance*) seguro *m*. **~e** /ə'ʃʊə(r)/ *vt* asegurar. **~ed** *a* seguro. **~edly** /-rɪdlɪ/ *adv* seguramente

asterisk /'æstərɪsk/ *n* asterisco *m*

astern /ə'stɜ:n/ *adv* a popa

asthma /'æsmə/ *n* asma *f*. **~tic** /-'mætɪk/ *a & n* asmático (*m*)

astonish /ə'stɒnɪʃ/ *vt* asombrar. **~ing** *a* asombroso. **~ment** *n* asombro *m*

astound /ə'staʊnd/ *vt* asombrar

astray /ə'streɪ/ *adv & a*. **go ~** extraviarse. **lead ~** llevar por mal camino

astride /ə'straɪd/ *adv* a horcajadas. ● *prep* a horcajadas sobre

astringent /ə'strɪndʒənt/ *a* astringente; (*fig*) austero. ● *n* astringente *m*

astrolog|er /ə'strɒlədʒə(r)/ *n* astrólogo *m*. **~y** *n* astrología *f*

astronaut /'æstrənɔ:t/ *n* astronauta *m & f*

astronom|er /ə'strɒnəmə(r)/ *n* astrónomo *m*. **~ical** /æstrə'nɒmɪkl/ *a* astronómico. **~y** /ə'strɒnəmɪ/ *n* astronomía *f*

astute /ə'stju:t/ *a* astuto. **~ness** *n* astucia *f*

asunder /ə'sʌndə(r)/ *adv* en pedazos; (*in two*) en dos

asylum /ə'saɪləm/ *n* asilo *m*. **lunatic ~** manicomio *m*

at /ət, æt/ *prep* a. **~ home** en casa. **~ night** por la noche. **~ Robert's** en casa de Roberto. **~ once** en seguida; (*simultaneously*) a la vez. **~ sea** en el mar. **~ the station** en la estación. **~ times** a veces. **not ~ all** nada; (*after thanks*) ¡de nada!

ate /et/ *see* **eat**

atheis|m /'eɪθɪzəm/ *n* ateísmo *m*. **~t** /'eɪθɪɪst/ *n* ateo *m*

athlet|e /'æθli:t/ *n* atleta *m & f*. **~ic** /-'letɪk/ *a* atlético. **~ics** /-'letɪks/ *npl* atletismo *m*

Atlantic /ət'læntɪk/ *a & n* atlántico (*m*). ● *n*. **~ (Ocean)** (Océano *m*) Atlántico *m*

atlas /'ætləs/ *n* atlas *m*

atmospher|e /'ætməsfɪə(r)/ *n* atmósfera *f*; (*fig*) ambiente *m*. **~ic** /-'ferɪk/ *a* atmosférico. **~ics** /-'ferɪks/ *npl* parásitos *mpl*

atom /'ætəm/ *n* átomo *m*. **~ic** /ə'tɒmɪk/ *a* atómico

atomize /'ætəmaɪz/ *vt* atomizar. **~r** /'ætəmaɪzə(r)/ *n* atomizador *m*

atone /ə'təʊn/ *vi*. **~ for** expiar. **~ment** *n* expiación *f*

atroci|ous /ə'trəʊʃəs/ *a* atroz. **~ty** /ə'trɒsətɪ/ *n* atrocidad *f*

atrophy /'ætrəfɪ/ *n* atrofia *f*

attach /ə'tætʃ/ *vt* sujetar; adjuntar ‹*document etc*›. **be ~ed to** (*be fond of*) tener cariño a

attaché /ə'tæʃeɪ/ *n* agregado *m*. **~ case** maletín *m*

attachment /ə'tætʃmənt/ *n* (*affection*) cariño *m*; (*tool*) accesorio *m*

attack /ə'tæk/ *n* ataque *m*. ● *vt/i* atacar. **~er** *n* agresor *m*

attain /ə'teɪn/ *vt* conseguir. **~able** *a* alcanzable. **~ment** *n* logro *m*. **~ments** *npl* conocimientos *mpl*, talento *m*

attempt /ə'tempt/ *vt* intentar. ● *n* tentativa *f*; (*attack*) atentado *m*

attend /ə'tend/ *vt* asistir a; (*escort*) acompañar. ● *vi* prestar atención. **~ to** (*look after*) ocuparse de. **~ance** *n* asistencia *f*; (*people present*) concurrencia *f*. **~ant** /ə'tendənt/ *a* concomitante. ● *n* encargado *m*; (*servant*) sirviente *m*

attention /ə'tenʃn/ *n* atención *f*. **~!** (*mil*) ¡firmes! **pay ~** prestar atención

attentive /ə'tentɪv/ *a* atento. **~ness** *n* atención *f*

attenuate /ə'tenjʊeɪt/ *vt* atenuar

attest /ə'test/ *vt* atestiguar. ● *vi* dar testimonio. **~ation** /æte'steɪʃn/ *n* testimonio *m*

attic /'ætɪk/ *n* desván *m*

attire /ə'taɪə(r)/ *n* atavío *m*. ● *vt* vestir

attitude /'ætɪtju:d/ *n* postura *f*

attorney /ə'tɜ:nɪ/ *n* (*pl* **-eys**) apoderado *m*; (*Amer*) abogado *m*

attract /ə'trækt/ *vt* atraer. **~ion** /-ʃn/ *n* atracción *f*; (*charm*) atractivo *m*

attractive /ə'træktɪv/ *a* atractivo; (*interesting*) atrayente. **~ness** *n* atractivo *m*

attribute /ə'trɪbju:t/ *vt* atribuir. /'ætrɪbju:t/ *n* atributo *m*

attrition /ə'trɪʃn/ *n* desgaste *m*

aubergine /'əʊbəʒi:n/ *n* berenjena *f*

auburn /'ɔ:bən/ *a* castaño

auction /'ɔ:kʃn/ *n* subasta *f*. ● *vt* subastar. **~eer** /-ə'nɪə(r)/ *n* subastador *m*

audaci|ous /ɔ:'deɪʃəs/ *a* audaz. **~ty** /-æsətɪ/ *n* audacia *f*

audible /'ɔ:dəbl/ *a* audible

audience /'ɔ:dɪəns/ *n* (*interview*) audiencia *f*; (*teatro*, *radio*) público *m*

audio-visual /ɔ:dɪəʊ'vɪʒʊəl/ *a* audiovisual

audit /'ɔ:dɪt/ *n* revisión *f* de cuentas. ● *vt* revisar

audition /ɔ:'dɪʃn/ *n* audición *f*. ● *vt* dar audición a

auditor /'ɔ:dɪtə(r)/ *n* interventor *m* de cuentas

auditorium /ɔ:dɪ'tɔ:rɪəm/ *n* sala *f*, auditorio *m*

augment /ɔ:g'ment/ *vt* aumentar

augur /'ɔ:gə(r)/ *vt* augurar. **it ~s well** es de buen agüero

august /ɔ:'gʌst/ *a* augusto

August /'ɔ:gəst/ *n* agosto *m*

aunt /ɑ:nt/ *n* tía *f*

au pair /əʊ'peə(r)/ *n* chica *f* au pair

aura /'ɔ:rə/ *n* atmósfera *f*, halo *m*

auspices /'ɔ:spɪsɪz/ *npl* auspicios *mpl*

auspicious /ɔ:'spɪʃəs/ *a* propicio

auster|e /ɔ:'stɪə(r)/ *a* austero. **~ity** /-erətɪ/ *n* austeridad *f*

Australia /ɒ'streɪlɪə/ *n* Australia *f*. **~n** *a* & *n* australiano (*m*)

Austria /'ɒstrɪə/ *n* Austria *f*. **~n** *a* & *n* austríaco (*m*)

authentic /ɔ:'θentɪk/ *a* auténtico. **~ate** /ɔ:'θentɪkeɪt/ *vt* autenticar. **~ity** /-ən'tɪsətɪ/ *n* autenticidad *f*

author /'ɔ:θə(r)/ *n* autor *m*. **~ess** *n* autora *f*

authoritarian /ɔ:θɒrɪ'teərɪən/ *a* autoritario

authoritative /ɔ:'θɒrɪtətɪv/ *a* autorizado; (*manner*) autoritario

authority /ɔ:'θɒrətɪ/ *n* autoridad *f*; (*permission*) autorización *f*

authoriz|ation /ɔ:θəraɪ'zeɪʃn/ *n* autorización *f*. **~e** /'ɔ:θəraɪz/ *vt* autorizar

authorship /'ɔ:θəʃɪp/ *n* profesión *f* de autor; (*origin*) paternidad *f* literaria

autistic /ɔ:'tɪstɪk/ *a* autista

autobiography /ɔ:təʊbaɪ'ɒgrəfɪ/ *n* autobiografía *f*

autocra|cy /ɔ:'tɒkrəsɪ/ *n* autocracia *f*. **~t** /'ɔ:təkræt/ *n* autócrata *m* & *f*. **~tic** /-'krætɪk/ *a* autocrático

autograph /'ɔ:təgrɑ:f/ *n* autógrafo *m*. ● *vt* firmar

automat|e /'ɔ:təmeɪt/ *vt* automatizar. **~ic** /ɔ:tə'mætɪk/ *a* automático. **~ion** /-'meɪʃn/ *n* automatización *f*. **~on** /ɔ:'tɒmətən/ *n* autómata *m*

automobile /'ɔ:təməbi:l/ *n* (*Amer*) coche *m*, automóvil *m*

autonom|ous /ɔ:'tɒnəməs/ *a* autónomo. **~y** *n* autonomía *f*

autopsy /'ɔ:tɒpsɪ/ *n* autopsia *f*

autumn /'ɔ:təm/ *n* otoño *m*. **~al** /-'tʌmnəl/ *a* de otoño, otoñal

auxiliary /ɔ:g'zɪlɪərɪ/ *a* auxiliar. ● *n* asistente *m*; (*verb*) verbo *m* auxiliar; (*pl*, *troops*) tropas *fpl* auxiliares

avail /ə'veɪl/ *vt*/*i* servir. **~ o.s. of** aprovecharse de. ● *n* ventaja *f*. **to no ~** inútil

availab|ility /əveɪlə'bɪlətɪ/ *n* disponibilidad *f*. **~le** /ə'veɪləbl/ *a* disponible

avalanche /'ævəlɑ:nʃ/ *n* avalancha *f*

avaric|e /'ævərɪs/ *n* avaricia *f*. **~ious** /-'rɪʃəs/ *a* avaro

avenge /ə'vendʒ/ *vt* vengar

avenue /'ævənju:/ *n* avenida *f*; (*fig*) vía *f*

average /'ævərɪdʒ/ *n* promedio *m*. **on ~** por término medio. ● *a* medio. ● *vt* calcular el promedio de. ● *vi* alcanzar un promedio de

avers|e /ə'vɜ:s/ *a* enemigo (**to** de). **be ~e to** sentir repugnancia por, no gustarle. **~ion** /-ʃn/ *n* repugnancia *f*

avert /ə'vɜ:t/ *vt* (*turn away*) apartar; (*ward off*) desviar

aviary /'eɪvɪərɪ/ *n* pajarera *f*

aviation /eɪvɪ'eɪʃn/ *n* aviación *f*

aviator /'eɪvɪeɪtə(r)/ *n* (*old use*) aviador *m*

avid /'ævɪd/ *a* ávido. **~ity** /-'vɪdətɪ/ *n* avidez *f*

avocado /ævə'kɑ:dəʊ/ *n* (*pl* **-os**) aguacate *m*

avoid /ə'vɔɪd/ *vt* evitar. **~able** *a* evitable. **~ance** *n* el evitar *m*

avuncular /ə'vʌŋkjʊlə(r)/ *a* de tío

await /ə'weɪt/ *vt* esperar

awake /ə'weɪk/ *vt/i* (*pt* **awoke,** *pp* **awoken**) despertar. ● *a* despierto. **wide ~** completamente despierto; (*fig*) despabilado. **~n** /ə'weɪkən/ *vt/i* despertar. **~ning** *n* el despertar *m*

award /ə'wɔ:d/ *vt* otorgar; (*jurid*) adjudicar. ● *n* premio *m*; (*jurid*) adjudicación *f*; (*scholarship*) beca *f*

aware /ə'weə(r)/ *a* consciente. **are you ~ that?** ¿te das cuenta de que? **~ness** *n* conciencia *f*

awash /ə'wɒʃ/ *a* inundado

away /ə'weɪ/ *adv* (*absent*) fuera; (*far*) lejos; (*persistently*) sin parar. ● *a* & *n*. **~ (match)** partido *m* fuera de casa

awe /ɔ:/ *n* temor *m*. **~some** *a* imponente. **~struck** *a* atemorizado

awful /'ɔ:fʊl/ *a* terrible, malísimo. **~ly** *adv* terriblemente

awhile /ə'waɪl/ *adv* un rato

awkward /'ɔ:kwəd/ *a* difícil; (*inconvenient*) inoportuno; (*clumsy*) desmañado; (*embarrassed*) incómodo. **~ly** *adv* con dificultad; (*clumsily*) de manera torpe. **~ness** *n* dificultad *f*; (*discomfort*) molestia *f*; (*clumsiness*) torpeza *f*

awning /'ɔ:nɪŋ/ *n* toldo *m*

awoke, awoken /ə'wəʊk, ə'wəʊkən/ *see* **awake**

awry /ə'raɪ/ *adv* & *a* ladeado. **go ~** salir mal

axe /æks/ *n* hacha *f*. ● *vt* (*pres p* **axing**) cortar con hacha; (*fig*) recortar

axiom /'æksɪəm/ *n* axioma *m*

axis /'æksɪs/ *n* (*pl* **axes** /-i:z/) eje *m*

axle /'æksl/ *n* eje *m*

ay(e) /aɪ/ *adv* & *n* sí (*m*)

B

BA *abbr see* **bachelor**

babble /'bæbl/ *vi* balbucir; (*chatter*) parlotear; (*of stream*) murmullar. ● *n* balbuceo *m*; (*chatter*) parloteo *m*; (*of stream*) murmullo *m*

baboon /bə'bu:n/ *n* mandril *m*

baby /'beɪbɪ/ *n* niño *m*, bebé *m*; (*Amer, sl*) chica *f*. **~ish** /'beɪbɪɪʃ/ *a* infantil. **~-sit** *vi* cuidar a los niños, hacer de canguro. **~sitter** *n* persona *f* que cuida a los niños, canguro *m*

bachelor /'bætʃələ(r)/ *n* soltero *m*. **B~ of Arts (BA)** licenciado *m* en filosofía y letras. **B~ of Science (BSc)** licenciado *m* en ciencias

back /bæk/ *n* espalda *f*; (*of car*) parte *f* trasera; (*of chair*) respaldo *m*; (*of cloth*) revés *m*; (*of house*) parte *f* de atrás; (*of animal, book*) lomo *m*; (*of hand, document*) dorso *m*; (*football*) defensa *m* & *f*. **~ of beyond** en el quinto pino. ● *a* trasero; (*taxes*) atrasado. ● *adv* atrás; (*returned*) de vuelta. ● *vt* apoyar; (*betting*) apostar a; dar marcha atrás a ‹*car*›. ● *vi* retroceder; ‹*car*› dar marcha atrás. **~ down** *vi* volverse atrás. **~ out** *vi* retirarse. **~ up** *vi* (*auto*) retroceder. **~ache** /'bækeɪk/ *n* dolor *m* de espalda. **~-bencher** *n* (*pol*) diputado *m* sin poder ministerial. **~biting** /'bækbaɪtɪŋ/ *n* maledicencia *f*. **~bone** /'bækbəʊn/ *n* columna *f* vertebral; (*fig*) pilar *m*. **~chat** /'bæktʃæt/ *n* impertinencias *fpl*. **~date** /bæk'deɪt/ *vt* antedatar. **~er** /'bækə(r)/ *n* partidario *m*; (*com*) financiador *m*. **~fire** /bæk'faɪə(r)/ *vi* (*auto*) petardear; (*fig*) fallar, salir el tiro por la culata. **~gammon** /bæk'gæmən/ *n* backgamon *m*. **~ground** /'bækgraʊnd/ *n* fondo *m*; (*environment*) antecedentes *mpl*. **~hand** /'bækhænd/ *n* (*sport*) revés *m*. **~handed** *a* dado con el dorso de la mano; (*fig*) equívoco, ambiguo. **~hander** *n* (*sport*) revés *m*; (*fig*) ataque *m* indirecto; (*bribe, sl*) soborno *m*. **~ing** /'bækɪŋ/ *n* apoyo *m*. **~lash** /'bæklæʃ/ *n* reacción *f*. **~log** /'bæklɒg/ *n* atrasos *mpl*. **~side** /bæk'saɪd/ *n* (*fam*) trasero *m*. **~stage** /bæk'steɪdʒ/ *a* de bastidores. ● *adv* entre bastidores. **~stroke** /'bækstrəʊk/ *n* (*tennis etc*) revés *m*; (*swimming*) braza *f* de espaldas. **~-up** *n* apoyo *m*. **~ward** /'bækwəd/ *a* ‹*step etc*› hacia atrás;

(*retarded*) atrasado. **~wards** /'bækwədz/ *adv* hacia atrás; (*fall*) de espaldas; (*back to front*) al revés. **go ~wards and forwards** ir de acá para allá. **~water** /'bækwɔ:tə(r)/ *n* agua *f* estancada; (*fig*) lugar *m* apartado

bacon /'beɪkən/ *n* tocino *m*

bacteria /bæk'tɪərɪə/ *npl* bacterias *fpl*. **~l** *a* bacteriano

bad /bæd/ *a* (**worse**, **worst**) malo; (*serious*) grave; (*harmful*) nocivo; ‹*language*› indecente. **feel ~** sentirse mal

bade /beɪd/ *see* **bid**

badge /bædʒ/ *n* distintivo *m*, chapa *f*

badger /'bædʒə(r)/ *n* tejón *m*. ● *vt* acosar

bad: **~ly** *adv* mal. **want ~ly** desear muchísimo. **~ly off** mal de dinero. **~-mannered** *a* mal educado

badminton /'bædmɪntən/ *n* bádminton *m*

bad-tempered /bæd'tempəd/ *a* (*always*) de mal genio; (*temporarily*) de mal humor

baffle /'bæfl/ *vt* desconcertar

bag /bæg/ *n* bolsa *f*; (*handbag*) bolso *m*. ● *vt* (*pt* **bagged**) ensacar; (*take*) coger (*not LAm*), agarrar (*LAm*). **~s** *npl* (*luggage*) equipaje *m*. **~s of** (*fam*) montones de

baggage /'bægɪdʒ/ *n* equipaje *m*

baggy /'bægɪ/ *a* ‹*clothes*› holgado

bagpipes /'bægpaɪps/ *npl* gaita *f*

Bahamas /bə'hɑ:məz/ *npl*. **the ~** las Bahamas *fpl*

bail[1] /beɪl/ *n* caución *f*, fianza *f*. ● *vt* poner en libertad bajo fianza. **~ s.o. out** obtener la libertad de uno bajo fianza

bail[2] /beɪl/ *n* (*cricket*) travesaño *m*

bail[3] /beɪl/ *vt* (*naut*) achicar

bailiff /'beɪlɪf/ *n* alguacil *m*; (*estate*) administrador *m*

bait /beɪt/ *n* cebo *m*. ● *vt* cebar; (*torment*) atormentar

bak|e /beɪk/ *vt* cocer al horno. ● *vi* cocerse. **~er** *n* panadero *m*. **~ery** /'beɪkərɪ/ *n* panadería *f*. **~ing** *n* cocción *f*; (*batch*) hornada *f*. **~ing-powder** *n* levadura *f* en polvo

balance /'bæləns/ *n* equilibrio *m*; (*com*) balance *m*; (*sum*) saldo *m*; (*scales*) balanza *f*; (*remainder*) resto *m*. ● *vt* equilibrar; (*com*) saldar; nivelar ‹*budget*›. ● *vi* equilibrarse; (*com*) saldarse. **~d** *a* equilibrado

balcony /'bælkənɪ/ *n* balcón *m*

bald /bɔ:ld/ *a* (**-er**, **-est**) calvo; ‹*tyre*› desgastado

balderdash /'bɔ:ldədæʃ/ *n* tonterías *fpl*

bald: **~ly** *adv* escuetamente. **~ness** *n* calvicie *f*

bale /beɪl/ *n* bala *f*, fardo *m*. ● *vi*. **~ out** lanzarse en paracaídas

Balearic /bælɪ'ærɪk/ *a*. **~ Islands** Islas *fpl* Baleares

baleful /'beɪlfʊl/ *a* funesto

balk /bɔ:k/ *vt* frustrar. ● *vi*. **~ (at)** resistirse (a)

ball[1] /bɔ:l/ *n* bola *f*; (*tennis etc*) pelota *f*; (*football etc*) balón *m*; (*of yarn*) ovillo *m*

ball[2] /bɔ:l/ (*dance*) baile *m*

ballad /'bæləd/ *n* balada *f*

ballast /'bæləst/ *n* lastre *m*

ball: **~-bearing** *n* cojinete *m* de bolas. **~-cock** *n* llave *f* de bola

ballerina /bælə'ri:nə/ *f* bailarina *f*

ballet /'bæleɪ/ *n* ballet *m*

ballistic /bə'lɪstɪk/ *a* balístico. **~s** *n* balística *f*

balloon /bə'lu:n/ *n* globo *m*

balloonist /bə'lu:nɪst/ *n* aeronauta *m & f*

ballot /'bælət/ *n* votación *f*. **~ (-paper)** *n* papeleta *f*. **~-box** *n* urna *f*

ball-point /'bɔ:lpɔɪnt/ *n*. **~ (pen)** bolígrafo *m*

ballroom /'bɔ:lru:m/ *n* salón *m* de baile

ballyhoo /bælɪ'hu:/ *n* (*publicity*) publicidad *f* sensacionalista; (*uproar*) jaleo *m*

balm /bɑ:m/ *n* bálsamo *m*. **~y** *a* (*mild*) suave; (*sl*) chiflado

baloney /bə'ləʊnɪ/ *n* (*sl*) tonterías *fpl*

balsam /'bɔ:lsəm/ *n* bálsamo *m*

balustrade /bælə'streɪd/ *n* barandilla *f*

bamboo /bæm'bu:/ *n* bambú *m*

bamboozle /bæm'bu:zl/ *vt* engatusar

ban /bæn/ *vt* (*pt* **banned**) prohibir. **~ from** excluir de. ● *n* prohibición *f*

banal /bə'nɑ:l/ *a* banal. **~ity** /-ælətɪ/ *n* banalidad *f*

banana /bə'nɑ:nə/ *n* plátano *m*, banana *f* (*LAm*). **~-tree** plátano *m*, banano *m*

band[1] /bænd/ *n* banda *f*

band[2] /bænd/ *n* (*mus*) orquesta *f*; (*military, brass*) banda *f*. ● *vi*. **~ together** juntarse

bandage /'bændɪdʒ/ *n* venda *f*. ● *vt* vendar

b & b *abbr* (*bed and breakfast*) cama *f* y desayuno

bandit /'bændɪt/ *n* bandido *m*

bandstand /'bændstænd/ *n* quiosco *m* de música

bandwagon /'bændwægən/ *n*. **jump on the ~** (*fig*) subirse al carro

bandy[1] /'bændɪ/ *a* (**-ier**, **-iest**) patizambo

bandy[2] /'bændɪ/ *vt*. **~ about** repetir. **be bandied about** estar en boca de todos

bandy-legged /'bændɪlegd/ *a* patizambo

bane /beɪn/ *n* (*fig*) perdición *f*. **~ful** *a* funesto

bang /bæŋ/ *n* (*noise*) ruido *m*; (*blow*) golpe *m*; (*of gun*) estampido *m*; (*of door*) golpe *m*. ● *vt*/*i* golpear. ● *adv* exactamente. ● *int* ¡pum!

banger /'bæŋə(r)/ *n* petardo *m*; (*culin*, *sl*) salchicha *f*

bangle /'bæŋgl/ *n* brazalete *m*

banish /'bænɪʃ/ *vt* desterrar

banisters /'bænɪstəz/ *npl* barandilla *f*

banjo /'bændʒəʊ/ *n* (*pl* **-os**) banjo *m*

bank[1] /bæŋk/ *n* (*of river*) orilla *f*. ● *vt* cubrir ‹*fire*›. ● *vi* (*aviat*) ladearse

bank[2] /bæŋk/ *n* banco *m*. ● *vt* depositar. **~ on** *vt* contar con. **~ with** tener una cuenta con. **~er** *n* banquero *m*. **~ holiday** *n* día *m* festivo, fiesta *f*. **~ing** *n* (*com*) banca *f*. **~note** /'bæŋknəʊt/ *n* billete *m* de banco

bankrupt /'bæŋkrʌpt/ *a* & *n* quebrado (*m*). ● *vt* hacer quebrar. **~cy** *n* bancarrota *f*, quiebra *f*

banner /'bænə(r)/ *n* bandera *f*; (*in demonstration*) pancarta *f*

banns /bænz/ *npl* amonestaciones *fpl*

banquet /'bæŋkwɪt/ *n* banquete *m*

bantamweight /'bæntəmweɪt/ *n* peso *m* gallo

banter /'bæntə(r)/ *n* chanza *f*. ● *vi* chancearse

bap /bæp/ *n* panecillo *m* blando

baptism /'bæptɪzəm/ *n* bautismo *m*; (*act*) bautizo *m*

Baptist /'bæptɪst/ *n* bautista *m* & *f*

baptize /bæp'taɪz/ *vt* bautizar

bar /bɑ:(r)/ *n* barra *f*; (*on window*) reja *f*; (*of chocolate*) tableta *f*; (*of soap*) pastilla *f*; (*pub*) bar *m*; (*mus*) compás *m*; (*jurid*) abogacía *f*; (*fig*) obstáculo *m*. ● *vt* (*pt* **barred**) atrancar ‹*door*›; (*exclude*) excluir; (*prohibit*) prohibir. ● *prep* excepto

barbar|ian /bɑ:'beərɪən/ *a* & *n* bárbaro (*m*). **~ic** /bɑ:'bærɪk/ *a* bárbaro. **~ity** /-ətɪ/ *n* barbaridad *f*. **~ous** *a* /'bɑ:bərəs/ *a* bárbaro

barbecue /'bɑ:bɪkju:/ *n* barbacoa *f*. ● *vt* asar a la parilla

barbed /bɑ:bd/ *a*. **~ wire** alambre *m* de espinas

barber /'bɑ:bə(r)/ *n* peluquero *m*, barbero *m*

barbiturate /bɑ:'bɪtjʊrət/ *n* barbitúrico *m*

bare /beə(r)/ *a* (**-er**, **est**) desnudo; ‹*room*› con pocos muebles; (*mere*) simple; (*empty*) vacío. ● *vt* desnudar; (*uncover*) descubrir. **~ one's teeth** mostrar los dientes. **~back** /'beəbæk/ *adv* a pelo. **~faced** /'beəfeɪst/ *a* descarado. **~foot** *a* descalzo. **~headed** /'beəhedɪd/ *a* descubierto. **~ly** *adv* apenas. **~ness** *n* desnudez *f*

bargain /'bɑ:gɪn/ *n* (*agreement*) pacto *m*; (*good buy*) ganga *f*. ● *vi* negociar; (*haggle*) regatear. **~ for** esperar, contar con

barge /bɑ:dʒ/ *n* barcaza *f*. ● *vi*. **~ in** irrumpir

baritone /'bærɪtəʊn/ *n* barítono *m*

barium /'beərɪəm/ *n* bario *m*

bark[1] /bɑ:k/ *n* (*of dog*) ladrido *m*. ● *vi* ladrar

bark[2] /bɑ:k/ (*of tree*) corteza *f*

barley /'bɑ:lɪ/ *n* cebada *f*. **~-water** *n* hordiate *m*

bar: ~maid /'bɑ:meɪd/ *n* camarera *f*. **~man** /'bɑ:mən/ *n* (*pl* **-men**) camarero *m*

barmy /'bɑ:mɪ/ *a* (*sl*) chiflado

barn /bɑ:n/ *n* granero *m*

barometer /bə'rɒmɪtə(r)/ *n* barómetro *m*

baron /'bærən/ *n* barón *m*. **~ess** *n* baronesa *f*

baroque /bə'rɒk/ *a* & *n* barroco (*m*)

barracks /'bærəks/ *npl* cuartel *m*

barrage /'bærɑ:ʒ/ *n* (*mil*) barrera *f*; (*dam*) presa *f*; (*of questions*) bombardeo *m*

barrel /'bærəl/ *n* tonel *m*; (*of gun*) cañón *m*. **~-organ** *n* organillo *m*

barren /'bærən/ *a* estéril. **~ness** *n* esterilidad *f*, aridez *f*

barricade /bærɪ'keɪd/ *n* barricada *f*. ● *vt* cerrar con barricadas

barrier /'bærɪə(r)/ *n* barrera *f*

barring /ˈbɑːrɪŋ/ *prep* salvo

barrister /ˈbærɪstə(r)/ *n* abogado *m*

barrow /ˈbærəʊ/ *n* carro *m*; (*wheelbarrow*) carretilla *f*

barter /ˈbɑːtə(r)/ *n* trueque *m*. ● *vt* trocar

base /beɪs/ *n* base *f*. ● *vt* basar. ● *a* vil

baseball /ˈbeɪsbɔːl/ *n* béisbol *m*

baseless /ˈbeɪslɪs/ *a* infundado

basement /ˈbeɪsmənt/ *n* sótano *m*

bash /bæʃ/ *vt* golpear. ● *n* golpe *m*. **have a ~** (*sl*) probar

bashful /ˈbæʃfl/ *a* tímido

basic /ˈbeɪsɪk/ *a* básico, fundamental. **~ally** *adv* fundamentalmente

basil /ˈbæzl/ *n* albahaca *f*

basilica /bəˈzɪlɪkə/ *n* basílica *f*

basin /ˈbeɪsn/ *n* (*for washing*) palangana *f*; (*for food*) cuenco *m*; (*geog*) cuenca *f*

basis /ˈbeɪsɪs/ *n* (*pl* **bases** /-siːz/) base *f*

bask /bɑːsk/ *vi* asolearse; (*fig*) gozar (**in** de)

basket /ˈbɑːskɪt/ *n* cesta *f*; (*big*) cesto *m*. **~ball** /ˈbɑːskɪtbɔːl/ *n* baloncesto *m*

Basque /bɑːsk/ *a* & *n* vasco (*m*). **~ Country** *n* País *m* Vasco. **~ Provinces** *npl* Vascongadas *fpl*

bass[1] /beɪs/ *a* bajo. ● *n* (*mus*) bajo *m*

bass[2] /bæs/ *n* (*marine fish*) róbalo *m*; (*freshwater fish*) perca *f*

bassoon /bəˈsuːn/ *n* fagot *m*

bastard /ˈbɑːstəd/ *a* & *n* bastardo (*m*). **you ~!** (*fam*) ¡cabrón!

baste /beɪst/ *vt* (*sew*) hilvanar; (*culin*) lard(e)ar

bastion /ˈbæstɪən/ *n* baluarte *m*

bat[1] /bæt/ *n* bate *m*; (*for table tennis*) raqueta *f*. **off one's own ~** por sí solo. ● *vt* (*pt* **batted**) golpear. ● *vi* batear

bat[2] /bæt/ *n* (*mammal*) murciélago *m*

bat[3] /bæt/ *vt*. **without ~ting an eyelid** sin pestañear

batch /bætʃ/ *n* (*of people*) grupo *m*; (*of papers*) lío *m*; (*of goods*) remesa *f*; (*of bread*) hornada *f*

bated /ˈbeɪtɪd/ *a*. **with ~ breath** con aliento entrecortado

bath /bɑːθ/ *n* (*pl* **-s** /bɑːðz/) baño *m*; (*tub*) bañera *f*; (*pl, swimming pool*) piscina *f*. ● *vt* bañar. ● *vi* bañarse

bathe /beɪð/ *vt* bañar. ● *vi* bañarse. ● *n* baño *m*. **~r** /-ə(r)/ *n* bañista *m* & *f*

bathing /ˈbeɪðɪŋ/ *n* baños *mpl*. **~-costume** *n* traje *m* de baño

bathroom /ˈbɑːθrʊm/ *n* cuarto *m* de baño

batman /ˈbætmən/ *n* (*pl* **-men**) (*mil*) ordenanza *f*

baton /ˈbætən/ *n* (*mil*) bastón *m*; (*mus*) batuta *f*

batsman /ˈbætsmən/ *n* (*pl* **-men**) bateador *m*

battalion /bəˈtælɪən/ *n* batallón *m*

batter[1] /ˈbætə(r)/ *vt* apalear

batter[2] /ˈbætə(r)/ *n* batido *m* para rebozar, albardilla *f*

batter: **~ed** *a* ‹*car etc*› estropeado; ‹*wife etc*› golpeado. **~ing** *n* (*fam*) bombardeo *m*

battery /ˈbætərɪ/ *n* (*mil, auto*) batería *f*; (*of torch, radio*) pila *f*

battle /ˈbætl/ *n* batalla *f*; (*fig*) lucha *f*. ● *vi* luchar. **~axe** /ˈbætlæks/ *n* (*woman, fam*) arpía *f*. **~field** /ˈbætlfiːld/ *n* campo *m* de batalla. **~ments** /ˈbætlmənts/ *npl* almenas *fpl*. **~ship** /ˈbætlʃɪp/ *n* acorazado *m*

batty /ˈbætɪ/ *a* (*sl*) chiflado

baulk /bɔːlk/ *vt* frustrar. ● *vi*. **~ (at)** resistirse (a)

bawd|iness /ˈbɔːdɪnəs/ *n* obscenidad *f*. **~y** /ˈbɔːdɪ/ *a* (**-ier, -iest**) obsceno, verde

bawl /bɔːl/ *vt/i* gritar

bay[1] /beɪ/ *n* (*geog*) bahía *f*

bay[2] /beɪ/ *n* (*bot*) laurel *m*

bay[3] /beɪ/ *n* (*of dog*) ladrido *m*. **keep at ~** mantener a raya. ● *vi* ladrar

bayonet /ˈbeɪənet/ *n* bayoneta *f*

bay window /beɪˈwɪndəʊ/ *n* ventana *f* salediza

bazaar /bəˈzɑː(r)/ *n* bazar *m*

BC /biːˈsiː/ *abbr* (*before Christ*) a. de C., antes de Cristo

be /biː/ *vi* (*pres* **am, are, is**; *pt* **was, were**; *pp* **been**) (*position or temporary*) estar; (*permanent*) ser. **~ cold/hot, etc** tener frío/calor, etc. **~ reading/singing, etc** (*aux*) leer/cantar, etc. **~ that as it may** sea como fuere. **he is 30** (*age*) tiene 30 años. **he is to come** (*must*) tiene que venir. **how are you?** ¿cómo estás? **how much is it?** ¿cuánto vale?, ¿cuánto es? **have been to** haber estado en. **it is cold/hot, etc** (*weather*) hace frío/calor, etc

beach /biːtʃ/ *n* playa *f*

beachcomber /ˈbiːtʃkəʊmə(r)/ *n* raquero *m*

beacon /ˈbiːkən/ *n* faro *m*

bead /bi:d/ *n* cuenta *f*; (*of glass*) abalorio *m*
beak /bi:k/ *n* pico *m*
beaker /'bi:kə(r)/ *n* jarra *f*, vaso *m*
beam /bi:m/ *n* viga *f*; (*of light*) rayo *m*; (*naut*) bao *m*. ● *vt* emitir. ● *vi* irradiar; (*smile*) sonreír. **~-ends** *npl*. **be on one's ~-ends** no tener más dinero. **~ing** *a* radiante
bean /bi:n/ *n* judía; (*broad bean*) haba *f*; (*of coffee*) grano *m*
beano /'bi:nəʊ/ *n* (*pl* **-os**) (*fam*) juerga *f*
bear[1] /beə(r)/ *vt* (*pt* **bore**, *pp* **borne**) llevar; parir ‹*niño*›; (*endure*) soportar. **~ right** torcer a la derecha. **~ in mind** tener en cuenta. **~ with** tener paciencia con
bear[2] /beə(r)/ *n* oso *m*
bearable /'beərəbl/ *a* soportable
beard /bɪəd/ *n* barba *f*. **~ed** *a* barbudo
bearer /'beərə(r)/ *n* portador *m*; (*of passport*) poseedor *m*
bearing /'beərɪŋ/ *n* comportamiento *m*; (*relevance*) relación *f*; (*mec*) cojinete *m*. **get one's ~s** orientarse
beast /bi:st/ *n* bestia *f*; (*person*) bruto *m*. **~ly** /'bi:stlɪ/ *a* (**-ier**, **-iest**) bestial; (*fam*) horrible
beat /bi:t/ *vt* (*pt* **beat**, *pp* **beaten**) golpear; (*culin*) batir; (*defeat*) derrotar; (*better*) sobrepasar; (*baffle*) dejar perplejo. **~ a retreat** (*mil*) batirse en retirada. **~ it** (*sl*) largarse. ● *vi* ‹*heart*› latir. ● *n* latido *m*; (*mus*) ritmo *m*; (*of policeman*) ronda *f*. **~ up** dar una paliza a; (*culin*) batir. **~er** *n* batidor *m*. **~ing** *n* paliza *f*
beautician /bju:'tɪʃn/ *n* esteticista *m* & *f*
beautiful /'bju:tɪfl/ *a* hermoso. **~ly** *adv* maravillosamente
beautify /'bju:tɪfaɪ/ *vt* embellecer
beauty /'bju:tɪ/ *n* belleza *f*. **~ parlour** *n* salón *m* de belleza. **~ spot** (*on face*) lunar *m*; (*site*) lugar *m* pintoresco
beaver /'bi:və(r)/ *n* castor *m*
became /bɪ'keɪm/ *see* **become**
because /bɪ'kɒz/ *conj* porque. ● *adv*. **~ of** a causa de
beck /bek/ *n*. **be at the ~ and call of** estar a disposición de
beckon /'bekən/ *vt*/*i*. **~ (to)** hacer señas (a)
become /bɪ'kʌm/ *vt* (*pt* **became**, *pp* **become**) ‹*clothes*› sentar bien. ● *vi* hacerse, llegar a ser, volverse, convertirse en. **what has ~ of her?** ¿qué es de ella?
becoming /bɪ'kʌmɪŋ/ *a* ‹*clothes*› favorecedor
bed /bed/ *n* cama *f*; (*layer*) estrato *m*; (*of sea, river*) fondo *m*; (*of flowers*) macizo *m*. ● *vi* (*pt* **bedded**). **~ down** acostarse. **~ and breakfast (b & b)** cama y desayuno. **~bug** /'bedbʌg/ *n* chinche *f*. **~clothes** /'bedkləʊðz/ *npl*, **~ding** *n* ropa *f* de cama
bedevil /bɪ'devl/ *vt* (*pt* **bedevilled**) (*torment*) atormentar
bedlam /'bedləm/ *n* confusión *f*, manicomio *m*
bed: **~pan** /'bedpæn/ *n* orinal *m* de cama. **~post** /'bedpəʊst/ *n* columna *f* de la cama
bedraggled /bɪ'drægld/ *a* sucio
bed: **~ridden** /'bedrɪdn/ *a* encamado. **~room** /'bedrʊm/ *n* dormitorio *m*, habitación *f*. **~side** /'bedsaɪd/ *n* cabecera *f*. **~-sitting-room** /bed'sɪtɪŋru:m/ *n* salón *m* con cama, estudio *m*. **~spread** /'bedspred/ *n* colcha *f*. **~time** /'bedtaɪm/ *n* hora *f* de acostarse
bee /bi:/ *n* abeja *f*. **make a ~-line for** ir en línea recta hacia
beech /bi:tʃ/ *n* haya *f*
beef /bi:f/ *n* carne *f* de vaca, carne *f* de res (*LAm*). ● *vi* (*sl*) quejarse. **~burger** /'bi:fbɜ:gə(r)/ *n* hamburguesa *f*
beefeater /'bi:fi:tə(r)/ *n* alabardero *m* de la torre de Londres
beefsteak /bi:f'steɪk/ *n* filete *m*, bistec *m*, bife *m* (*Arg*)
beefy /'bi:fɪ/ *a* (**-ier**, **-iest**) musculoso
beehive /'bi:haɪv/ *n* colmena *f*
been /bi:n/ *see* **be**
beer /bɪə(r)/ *n* cerveza *f*
beet /bi:t/ *n* remolacha *f*
beetle /'bi:tl/ *n* escarabajo *m*
beetroot /'bi:tru:t/ *n* *invar* remolacha *f*
befall /bɪ'fɔ:l/ *vt* (*pt* **befell**, *pp* **befallen**) acontecer a. ● *vi* acontecer
befit /bɪ'fɪt/ *vt* (*pt* **befitted**) convenir a
before /bɪ'fɔ:(r)/ *prep* (*time*) antes de; (*place*) delante de. **~ leaving** antes de marcharse. ● *adv* (*place*) delante; (*time*) antes. **a week ~** una semana antes. **the week ~** la semana anterior. ● *conj* (*time*) antes de que. **~ he leaves** antes de que se

vaya. **~hand** /bɪ'fɔ:hænd/ *adv* de antemano
befriend /bɪ'frend/ *vt* ofrecer amistad a
beg /beg/ *vt/i* (*pt* **begged**) mendigar; (*entreat*) suplicar; (*ask*) pedir. **~ s.o.'s pardon** pedir perdón a uno. **I ~ your pardon!** ¡perdone Vd! **I ~ your pardon?** ¿cómo? **it's going ~ging** no lo quiere nadie
began /bɪ'gæn/ *see* **begin**
beget /bɪ'get/ *vt* (*pt* **begot**, *pp* **begotten**, *pres p* **begetting**) engendrar
beggar /'begə(r)/ *n* mendigo *m*; (*sl*) individuo *m*, tío *m* (*fam*)
begin /bɪ'gɪn/ *vt/i* (*pt* **began**, *pp* **begun**, *pres p* **beginning**) comenzar, empezar. **~ner** *n* principiante *m & f*. **~ning** *n* principio *m*
begot, begotten /bɪ'gɒt, bɪ'gɒtn/ *see* **beget**
begrudge /bɪ'grʌdʒ/ *vt* envidiar; (*give*) dar de mala gana
beguile /bɪ'gaɪl/ *vt* engañar, seducir; (*entertain*) entretener
begun /bɪ'gʌn/ *see* **begin**
behalf /bɪ'hɑ:f/ *n*. **on ~ of** de parte de, en nombre de
behav|e /bɪ'heɪv/ *vi* comportarse, portarse. **~ (o.s.)** portarse bien. **~iour** /bɪ'heɪvjə(r)/ *n* comportamiento *m*
behead /bɪ'hed/ *vt* decapitar
beheld /bɪ'held/ *see* **behold**
behind /bɪ'haɪnd/ *prep* detrás de. ● *adv* detrás; (*late*) atrasado. ● *n* (*fam*) trasero *m*
behold /bɪ'həʊld/ *vt* (*pt* **beheld**) (*old use*) mirar, contemplar
beholden /bɪ'həʊldən/ *a* agradecido
being /'bi:ɪŋ/ *n* ser *m*. **come into ~** nacer
belated /bɪ'leɪtɪd/ *a* tardío
belch /beltʃ/ *vi* eructar. ● *vt*. **~ out** arrojar ‹*smoke*›
belfry /'belfrɪ/ *n* campanario *m*
Belgi|an /'beldʒən/ *a & n* belga (*m & f*). **~um** /'beldʒəm/ *n* Bélgica *f*
belie /bɪ'laɪ/ *vt* desmentir
belie|f /bɪ'li:f/ *n* (*trust*) fe *f*; (*opinion*) creencia *f*. **~ve** /bɪ'li:v/ *vt/i* creer. **make ~ve** fingir. **~ver** /-ə(r)/ *n* creyente *m & f*; (*supporter*) partidario *m*
belittle /bɪ'lɪtl/ *vt* empequeñecer; (*fig*) despreciar
bell /bel/ *n* campana *f*; (*on door*) timbre *m*
belligerent /bɪ'lɪdʒərənt/ *a & n* beligerante (*m & f*)
bellow /'beləʊ/ *vt* gritar. ● *vi* bramar
bellows /'beləʊz/ *npl* fuelle *m*
belly /'belɪ/ *n* vientre *m*. **~ful** /'belɪfʊl/ *n* panzada *f*. **have a ~ful of** (*sl*) estar harto de
belong /bɪ'lɒŋ/ *vi* pertenecer; (*club*) ser socio (**to** de)
belongings /bɪ'lɒŋɪŋz/ *npl* pertenencias *fpl*. **personal ~** efectos *mpl* personales
beloved /bɪ'lʌvɪd/ *a & n* querido (*m*)
below /bɪ'ləʊ/ *prep* debajo de; (*fig*) inferior a. ● *adv* abajo
belt /belt/ *n* cinturón *m*; (*area*) zona *f*. ● *vt* (*fig*) rodear; (*sl*) pegar
bemused /bɪ'mju:zd/ *a* perplejo
bench /bentʃ/ *n* banco *m*. **the B~** (*jurid*) la magistratura *f*
bend /bend/ *vt* (*pt & pp* **bent**) doblar; torcer ‹*arm, leg*›. ● *vi* doblarse; ‹*road*› torcerse. ● *n* curva *f*. **~ down/over** inclinarse
beneath /bɪ'ni:θ/ *prep* debajo de; (*fig*) inferior a. ● *adv* abajo
benediction /benɪ'dɪkʃn/ *n* bendición *f*
benefactor /'benɪfæktə(r)/ *n* bienhechor *m*, benefactor *m*
beneficial /benɪ'fɪʃl/ *a* provechoso
beneficiary /benɪ'fɪʃərɪ/ *a & n* beneficiario (*m*)
benefit /'benɪfɪt/ *n* provecho *m*, ventaja *f*; (*allowance*) subsidio *m*; (*financial gain*) beneficio *m*. ● *vt* (*pt* **benefited**, *pres p* **benefiting**) aprovechar. ● *vi* aprovecharse
benevolen|ce /bɪ'nevələns/ *n* benevolencia *f*. **~t** *a* benévolo
benign /bɪ'naɪn/ *a* benigno
bent /bent/ *see* **bend**. ● *n* inclinación *f*. ● *a* encorvado; (*sl*) corrompido
bequeath /bɪ'kwi:ð/ *vt* legar
bequest /bɪ'kwest/ *n* legado *m*
bereave|d /bɪ'ri:vd/ *n*. **the ~d** la familia *f* del difunto. **~ment** *n* pérdida *f*; (*mourning*) luto *m*
bereft /bɪ'reft/ *a*. **~ of** privado de
beret /'bereɪ/ *n* boina *f*
Bermuda /bə'mju:də/ *n* Islas *fpl* Bermudas
berry /'berɪ/ *n* baya *f*
berserk /bə'sɜ:k/ *a*. **go ~** volverse loco, perder los estribos
berth /bɜ:θ/ *n* litera *f*; (*anchorage*) amarradero *m*. **give a wide ~ to** evitar. ● *vi* atracar

beseech /bɪ'siːtʃ/ *vt* (*pt* **besought**) suplicar

beset /bɪ'set/ *vt* (*pt* **beset**, *pres p* **besetting**) acosar

beside /bɪ'saɪd/ *prep* al lado de. **be ~ o.s.** estar fuera de sí

besides /bɪ'saɪdz/ *prep* además de; (*except*) excepto. ● *adv* además

besiege /bɪ'siːdʒ/ *vt* asediar; (*fig*) acosar

besought /bɪ'sɔːt/ *see* **beseech**

bespoke /bɪ'spəʊk/ *a* ‹*tailor*› que confecciona a la medida

best /best/ *a* (el) mejor. **the ~ thing is to...** lo mejor es... ● *adv* (lo) mejor. **like ~** preferir. ● *n* lo mejor. **at ~** a lo más. **do one's ~** hacer todo lo posible. **make the ~ of** contentarse con. **~ man** *n* padrino *m* (de boda)

bestow /bɪ'stəʊ/ *vt* conceder

bestseller /best'selə(r)/ *n* éxito *m* de librería, bestseller *m*

bet /bet/ *n* apuesta *f*. ● *vt/i* (*pt* **bet** *or* **betted**) apostar

betray /bɪ'treɪ/ *vt* traicionar. **~al** *n* traición *f*

betroth|al /bɪ'trəʊðəl/ *n* esponsales *mpl*. **~ed** *a* prometido

better /'betə(r)/ *a* & *adv* mejor. **~ off** en mejores condiciones; (*richer*) más rico. **get ~** mejorar. **all the ~** tanto mejor. **I'd ~** más vale que. **the ~ part of** la mayor parte de. **the sooner the ~** cuanto antes mejor. ● *vt* mejorar; (*beat*) sobrepasar. ● *n* superior *m*. **get the ~ of** vencer a. **one's ~s** sus superiores *mpl*

between /bɪ'twiːn/ *prep* entre. ● *adv* en medio

beverage /'bevərɪdʒ/ *n* bebida *f*

bevy /bevɪ/ *n* grupo *m*

beware /bɪ'weə(r)/ *vi* tener cuidado. ● *int* ¡cuidado!

bewilder /bɪ'wɪldə(r)/ *vt* desconcertar. **~ment** *n* aturdimiento *m*

bewitch /bɪ'wɪtʃ/ *vt* hechizar

beyond /bɪ'jɒnd/ *prep* más allá de; (*fig*) fuera de. **~ doubt** sin lugar a duda. **~ reason** irrazonable. ● *adv* más allá

bias /'baɪəs/ *n* predisposición *f*; (*prejudice*) prejuicio *m*; (*sewing*) sesgo *m*. ● *vt* (*pt* **biased**) influir en. **~ed** *a* parcial

bib /bɪb/ *n* babero *m*

Bible /'baɪbl/ *n* Biblia *f*

biblical /'bɪblɪkl/ *a* bíblico

bibliography /bɪblɪ'ɒgrəfɪ/ *n* bibliografía *f*

biceps /'baɪseps/ *n* bíceps *m*

bicker /'bɪkə(r)/ *vi* altercar

bicycle /'baɪsɪkl/ *n* bicicleta *f*. ● *vi* ir en bicicleta

bid /bɪd/ *n* (*offer*) oferta *f*; (*attempt*) tentativa *f*. ● *vi* hacer una oferta. ● *vt* (*pt* **bid**, *pres p* **bidding**) ofrecer; (*pt* **bid**, *pp* **bidden**, *pres p* **bidding**) mandar; dar ‹*welcome, good-day etc*›. **~der** *n* postor *m*. **~ding** *n* (*at auction*) ofertas *fpl*; (*order*) mandato *m*

bide /baɪd/ *vt*. **~ one's time** esperar el momento oportuno

biennial /baɪ'enɪəl/ *a* bienal. ● *n* (*event*) bienal *f*; (*bot*) planta *f* bienal

bifocals /baɪ'fəʊklz/ *npl* gafas *fpl* bifocales, anteojos *mpl* bifocales (*LAm*)

big /bɪg/ *a* (**bigger**, **biggest**) grande; (*generous, sl*) generoso. ● *adv*. **talk ~** fanfarronear

bigam|ist /'bɪgəmɪst/ *n* bígamo *m*. **~ous** *a* bígamo. **~y** *n* bigamia *f*

big-headed /bɪg'hedɪd/ *a* engreído

bigot /'bɪgət/ *n* fanático *m*. **~ed** *a* fanático. **~ry** *n* fanatismo *m*

bigwig /'bɪgwɪg/ *n* (*fam*) pez *m* gordo

bike /baɪk/ *n* (*fam*) bicicleta *f*, bici *f* (*fam*)

bikini /bɪ'kiːnɪ/ *n* (*pl* **-is**) biquini *m*, bikini *m*

bilberry /'bɪlbərɪ/ *n* arándano *m*

bile /baɪl/ *n* bilis *f*

bilingual /baɪ'lɪŋgwəl/ *a* bilingüe

bilious /'bɪlɪəs/ *a* (*med*) bilioso

bill[1] /bɪl/ *n* cuenta *f*; (*invoice*) factura *f*; (*notice*) cartel *m*; (*Amer, banknote*) billete *m*; (*pol*) proyecto *m* de ley. ● *vt* pasar la factura; (*in theatre*) anunciar

bill[2] /bɪl/ *n* (*of bird*) pico *m*

billet /'bɪlɪt/ *n* (*mil*) alojamiento *m*. ● *vt* alojar

billiards /'bɪlɪədz/ *n* billar *m*

billion /'bɪlɪən/ *n* billón *m*; (*Amer*) mil millones *mpl*

billy-goat /'bɪlɪgəʊt/ *n* macho *m* cabrío

bin /bɪn/ *n* recipiente *m*; (*for rubbish*) cubo *m*; (*for waste paper*) papelera *f*

bind /baɪnd/ *vt* (*pt* **bound**) atar; encuadernar ‹*book*›; (*jurid*) obligar. ● *n* (*sl*) lata *f*. **~ing**

/ˈbaɪndɪŋ/ *n* (*of books*) encuadernación *f*; (*braid*) ribete *m*
binge /bɪndʒ/ *n* (*sl*) (*of food*) comilona *f*; (*of drink*) borrachera *f*. **go on a ~** ir de juerga
bingo /ˈbɪŋgəʊ/ *n* bingo *m*
binoculars /bɪˈnɒkjʊləz/ *npl* prismáticos *mpl*
biochemistry /baɪəʊˈkemɪstrɪ/ *n* bioquímica *f*
biograph|er /baɪˈɒgrəfə(r)/ *n* biógrafo m. **~y** *n* biografía *f*
biolog|ical /baɪəˈlɒdʒɪkl/ *a* biológico. **~ist** *n* biólogo *m*. **~** /baɪˈɒlədʒɪ/ *n* biología *f*
biped /ˈbaɪped/ *n* bípedo *m*
birch /bɜ:tʃ/ *n* (*tree*) abedul *m*; (*whip*) férula *f*
bird /bɜ:d/ *n* ave *f*; (*small*) pájaro *m*; (*fam*) tipo *m*; (*girl*, *sl*) chica *f*
Biro /ˈbaɪərəʊ/ *n* (*pl* **-os**) (*P*) bolígrafo *m*, biromen *m* (*Arg*)
birth /bɜ:θ/ *n* nacimiento *m*. **~certificate** *n* partida *f* de nacimiento. **~-control** *n* control *m* de la natalidad. **~day** /ˈbɜ:θdeɪ/ *n* cumpleaños *m invar*. **~mark** /ˈbɜ:θmɑ:k/ *n* marca *f* de nacimiento. **~-rate** *n* natalidad *f*. **~right** /ˈbɜ:θraɪt/ *n* derechos *mpl* de nacimiento
biscuit /ˈbɪskɪt/ *n* galleta *f*
bisect /baɪˈsekt/ *vt* bisecar
bishop /ˈbɪʃəp/ *n* obispo *m*
bit[1] /bɪt/ *n* trozo *m*; (*quantity*) poco *m*
bit[2] /bɪt/ *see* **bite**
bit[3] /bɪt/ *n* (*of horse*) bocado *m*; (*mec*) broca *f*
bitch /bɪtʃ/ *n* perra *f*; (*woman*, *fam*) mujer *f* maligna, bruja *f* (*fam*). ● *vi* (*fam*) quejarse (**about** de). **~y** *a* malintencionado
bit|e /baɪt/ *vt*/*i* (*pt* **bit**, *pp* **bitten**) morder. **~e one's nails** morderse las uñas. ● *n* mordisco *m*; (*mouthful*) bocado *m*; (*of insect etc*) picadura *f*. **~ing** /ˈbaɪtɪŋ/ *a* mordaz
bitter /ˈbɪtə(r)/ *a* amargo; (*of weather*) glacial. **to the ~ end** hasta el final. ● *n* cerveza *f* amarga. **~ly** *adv* amargamente. **it's ~ly cold** hace un frío glacial. **~ness** *n* amargor *m*; (*resentment*) amargura *f*
bizarre /bɪˈzɑ:(r)/ *a* extraño
blab /blæb/ *vi* (*pt* **blabbed**) chismear
black /blæk/ *a* (**-er**, **-est**) negro. **~ and blue** amoratado. ● *n* negro *m*. ● *vt* ennegrecer; limpiar ‹*shoes*›. **~ out** desmayarse; (*make dark*) apagar las luces de
blackball /ˈblækbɔ:l/ *vt* votar en contra de
blackberry /ˈblækbərɪ/ *n* zarzamora *f*
blackbird /ˈblækbɜ:d/ *n* mirlo *m*
blackboard /ˈblækbɔ:d/ *n* pizarra *f*
blackcurrant /blækˈkʌrənt/ *n* casis *f*
blacken /ˈblækən/ *vt* ennegrecer. ● *vi* ennegrecerse
blackguard /ˈblægɑ:d/ *n* canalla *m*
blackleg /ˈblækleg/ *n* esquirol *m*
blacklist /ˈblæklɪst/ *vt* poner en la lista negra
blackmail /ˈblækmeɪl/ *n* chantaje *m*. ● *vt* chantajear. **~er** *n* chantajista *m & f*
black-out /ˈblækaʊt/ *n* apagón *m*; (*med*) desmayo *m*; (*of news*) censura *f*
blacksmith /ˈblæksmɪθ/ *n* herrero *m*
bladder /ˈblædə(r)/ *n* vejiga *f*
blade /bleɪd/ *n* hoja *f*; (*razor-blade*) cuchilla *f*. **~ of grass** brizna *f* de hierba
blame /bleɪm/ *vt* echar la culpa a. **be to ~** tener la culpa. ● *n* culpa *f*. **~less** *a* inocente
bland /blænd/ *a* (**-er**, **-est**) suave
blandishments /ˈblændɪʃmənts/ *npl* halagos *mpl*
blank /blæŋk/ *a* en blanco; ‹*cartridge*› sin bala; (*fig*) vacío. **~ verse** *n* verso *m* suelto. ● *n* blanco *m*
blanket /ˈblæŋkɪt/ *n* manta *f*; (*fig*) capa *f*. ● *vt* (*pt* **blanketed**) (*fig*) cubrir (**in**, **with** de)
blare /bleə(r)/ *vi* sonar muy fuerte. ● *n* estrépito *m*
blarney /ˈblɑ:nɪ/ *n* coba *f*. ● *vt* dar coba
blasé /ˈblɑ:zeɪ/ *a* hastiado
blasphem|e /blæsˈfi:m/ *vt*/*i* blasfemar. **~er** *n* blasfemador *m*. **~ous** /ˈblæsfəməs/ *a* blasfemo. **~y** /ˈblæsfəmɪ/ *n* blasfemia *f*
blast /blɑ:st/ *n* explosión *f*; (*gust*) ráfaga *f*; (*sound*) toque *m*. ● *vt* volar. **~ed** *a* maldito. **~-furnace** *n* alto horno *m*. **~-off** *n* (*of missile*) despegue *m*
blatant /ˈbleɪtnt/ *a* patente; (*shameless*) descarado
blaze /ˈbleɪz/ *n* llamarada *f*; (*of light*) resplandor *m*; (*fig*) arranque *m*. ● *vi* arder en llamas; (*fig*) brillar. **~ a trail** abrir un camino
blazer /ˈbleɪzə(r)/ *n* chaqueta *f*

bleach /bli:tʃ/ *n* lejía *f*; (*for hair*) decolorante *m*. ● *vt* blanquear; decolorar ‹*hair*›. ● *vi* blanquearse

bleak /bli:k/ *a* (**-er**, **-est**) desolado; (*fig*) sombrío

bleary /ˈblɪərɪ/ *a* ‹*eyes*› nublado; (*indistinct*) indistinto

bleat /bli:t/ *n* balido *m*. ● *vi* balar

bleed /bli:d/ *vt/i* (*pt* **bled**) sangrar

bleep /bli:p/ *n* pitido *m*. **~er** *n* busca *m*, buscapersonas *m*

blemish /ˈblemɪʃ/ *n* tacha *f*

blend /blend/ *n* mezcla *f*. ● *vt* mezclar. ● *vi* combinarse

bless /bles/ *vt* bendecir. **~ you!** (*on sneezing*) ¡Jesús! **~ed** *a* bendito. **be ~ed with** estar dotado de. **~ing** *n* bendición *f*; (*advantage*) ventaja *f*

blew /blu:/ *see* **blow**[1]

blight /blaɪt/ *n* añublo *m*, tizón *m*; (*fig*) plaga *f*. ● *vt* añublar, atizonar; (*fig*) destrozar

blighter /ˈblaɪtə(r)/ *n* (*sl*) tío *m* (*fam*), sinvergüenza *m*

blind /blaɪnd/ *a* ciego. **~ alley** *n* callejón *m* sin salida. ● *n* persiana *f*; (*fig*) pretexto *m*. ● *vt* cegar. **~fold** /ˈblaɪndfəʊld/ *a* & *adv* con los ojos vendados. ● *n* venda *f*. ● *vt* vendar los ojos. **~ly** *adv* a ciegas. **~ness** *n* ceguera *f*

blink /blɪŋk/ *vi* parpadear; (*of light*) centellear

blinkers /ˈblɪŋkəz/ *npl* anteojeras *fpl*; (*auto*) intermitente *m*

bliss /blɪs/ *n* felicidad *f*. **~ful** *a* feliz. **~fully** *adv* felizmente; (*completely*) completamente

blister /ˈblɪstə(r)/ *n* ampolla *f*. ● *vi* formarse ampollas

blithe /blaɪð/ *a* alegre

blitz /blɪts/ *n* bombardeo *m* aéreo. ● *vt* bombardear

blizzard /ˈblɪzəd/ *n* ventisca *f*

bloated /ˈbləʊtɪd/ *a* hinchado (**with** de)

bloater /ˈbləʊtə(r)/ *n* arenque *m* ahumado

blob /blɒb/ *n* gota *f*; (*stain*) mancha *f*

bloc /blɒk/ *n* (*pol*) bloque *m*

block /blɒk/ *n* bloque *m*; (*of wood*) zoquete *m*; (*of buildings*) manzana *f*, cuadra *f* (*LAm*); (*in pipe*) obstrucción *f*. **in ~ letters** en letra de imprenta. **traffic ~** embotellamiento *m*. ● *vt* obstruir. **~ade** /blɒˈkeɪd/ *n* bloqueo *m*. ● *vt* bloquear. **~age** *n* obstrucción *f*

blockhead /ˈblɒkhed/ *n* (*fam*) zopenco *m*

bloke /bləʊk/ *n* (*fam*) tío *m* (*fam*), tipo *m*

blond /blɒnd/ *a* & *n* rubio (*m*). **~e** *a* & *n* rubia (*f*)

blood /blʌd/ *n* sangre *f*. **~ count** *n* recuento *m* sanguíneo. **~-curdling** *a* horripilante

bloodhound /ˈblʌdhaʊnd/ *n* sabueso *m*

blood: **~ pressure** *n* tensión *f* arterial. **high ~ pressure** hipertensión *f*. **~shed** /ˈblʌdʃed/ *n* efusión *f* de sangre, derramamiento *m* de sangre, matanza *f*. **~shot** /ˈblʌdʃɒt/ *a* sanguinolento; ‹*eye*› inyectado de sangre. **~stream** /ˈblʌdstri:m/ *n* sangre *f*

bloodthirsty /ˈblʌdθɜ:stɪ/ *a* sanguinario

bloody /ˈblʌdɪ/ *a* (**-ier**, **-iest**) sangriento; (*stained*) ensangrentado; (*sl*) maldito. **~y-minded** *a* (*fam*) terco

bloom /blu:m/ *n* flor *f*. ● *vi* florecer

bloomer /ˈblu:mə(r)/ *n* (*sl*) metedura *f* de pata

blooming *a* floreciente; (*fam*) maldito

blossom /ˈblɒsəm/ *n* flor *f*. ● *vi* florecer. **~ out (into)** (*fig*) llegar a ser

blot /blɒt/ *n* borrón *m*. ● *vt* (*pt* **blotted**) manchar; (*dry*) secar. **~ out** oscurecer

blotch /blɒtʃ/ *n* mancha *f*. **~y** *a* lleno de manchas

blotter /ˈblɒtə(r)/ *n*, **blotting-paper** /ˈblɒtɪŋpeɪpə(r)/ *n* papel *m* secante

blouse /blaʊz/ *n* blusa *f*

blow[1] /bləʊ/ *vt* (*pt* **blew**, *pp* **blown**) soplar; fundir ‹*fuse*›; tocar ‹*trumpet*›. ● *vi* soplar; ‹*fuse*› fundirse; (*sound*) sonar. ● *n* (*puff*) soplo *m*. **~ down** *vt* derribar. **~ out** apagar ‹*candle*›. **~ over** pasar. **~ up** *vt* inflar; (*explode*) volar; (*photo*) ampliar. ● *vi* (*explode*) estallar; (*burst*) reventar

blow[2] /bləʊ/ *n* (*incl fig*) golpe *m*

blow-dry /ˈbləʊdraɪ/ *vt* secar con secador

blowlamp /ˈbləʊlæmp/ *n* soplete *m*

blow: **~-out** *n* (*of tyre*) reventón *m*. **~-up** *n* (*photo*) ampliación *f*

blowzy /ˈblaʊzɪ/ *a* desaliñado

blubber /ˈblʌbə(r)/ *n* grasa *f* de ballena

bludgeon /ˈblʌdʒən/ *n* cachiporra *f*. ● *vt* aporrear

blue /bluː/ *a* (**-er**, **-est**) azul; ‹*joke*› verde. ● *n* azul *m*. **out of the ~** totalmente inesperado. **~s** *npl*. **have the ~s** tener tristeza

bluebell /ˈbluːbel/ *n* campanilla *f*

bluebottle /ˈbluːbɒtl/ *n* moscarda *f*

blueprint /ˈbluːprɪnt/ *n* ferroprusiato *m*; (*fig*, *plan*) anteproyecto *m*

bluff /blʌf/ *a* ‹*person*› brusco. ● *n* (*poker*) farol *m*. ● *vt* engañar. ● *vi* (*poker*) tirarse un farol

blunder /ˈblʌndə(r)/ *vi* cometer un error. ● *n* metedura *f* de pata

blunt /blʌnt/ *a* desafilado; ‹*person*› directo, abrupto. ● *vt* desafilar. **~ly** *adv* francamente. **~ness** *n* embotadura *f*; (*fig*) franqueza *f*, brusquedad *f*

blur /blɜː(r)/ *n* impresión *f* indistinta. ● *vt* (*pt* **blurred**) hacer borroso

blurb /blɜːb/ *n* resumen *m* publicitario

blurt /blɜːt/ *vt*. **~ out** dejar escapar

blush /blʌʃ/ *vi* ruborizarse. ● *n* sonrojo *m*

bluster /ˈblʌstə(r)/ *vi* ‹*weather*› bramar; ‹*person*› fanfarronear. **~y** *a* tempestuoso

boar /bɔː(r)/ *n* verraco *m*

board /bɔːd/ *n* tabla *f*, tablero *m*; (*for notices*) tablón *m*; (*food*) pensión *f*; (*admin*) junta *f*. **~ and lodging** casa y comida. **above ~** correcto. **full ~** pensión *f* completa. **go by the ~** ser abandonado. ● *vt* alojar; (*naut*) embarcar en. ● *vi* alojarse (**with** en casa de); (*at school*) ser interno. **~er** *n* huésped *m*; (*schol*) interno *m*. **~ing-house** *n* casa *f* de huéspedes, pensión *f*. **~ing-school** *n* internado *m*

boast /bəʊst/ *vt* enorgullecerse de. ● *vi* jactarse. ● *n* jactancia *f*. **~er** *n* jactancioso *m*. **~ful** *a* jactancioso

boat /bəʊt/ *n* barco *m*; (*large*) navío *m*; (*small*) barca *f*

boater /ˈbəʊtə(r)/ *n* (*hat*) canotié *m*

boatswain /ˈbəʊsn/ *n* contramaestre *m*

bob[1] /bɒb/ *vi* (*pt* **bobbed**) menearse, subir y bajar. **~ up** presentarse súbitamente

bob[2] /bɒb/ *n invar* (*sl*) chelín *m*

bobbin /ˈbɒbɪn/ *n* carrete *m*; (*in sewing machine*) canilla *f*

bobby /ˈbɒbɪ/ *n* (*fam*) policía *m*, poli *m* (*fam*)

bobsleigh /ˈbɒbsleɪ/ *n* bob(sleigh) *m*

bode /bəʊd/ *vi* presagiar. **~ well/ill** ser de buen/mal agüero

bodice /ˈbɒdɪs/ *n* corpiño *m*

bodily /ˈbɒdɪlɪ/ *a* físico, corporal. ● *adv* físicamente; (*in person*) en persona

body /ˈbɒdɪ/ *n* cuerpo *m*. **~guard** /ˈbɒdɪgɑːd/ *n* guardaespaldas *m invar*. **~work** *n* carrocería *f*

boffin /ˈbɒfɪn/ *n* (*sl*) científico *m*

bog /bɒg/ *n* ciénaga *f*. ● *vt* (*pt* **bogged**). **get ~ged down** empantanarse

bogey /ˈbəʊgɪ/ *n* duende *m*; (*nuisance*) pesadilla *f*

boggle /ˈbɒgl/ *vi* sobresaltarse. **the mind ~s** ¡no es posible!

bogus /ˈbəʊgəs/ *a* falso

bogy /ˈbəʊgɪ/ *n* duende *m*; (*nuisance*) pesadilla *f*

boil[1] /bɔɪl/ *vt*/*i* hervir. **be ~ing hot** estar ardiendo; ‹*weather*› hacer mucho calor. **~ away** evaporarse. **~ down to** reducirse a. **~ over** rebosar

boil[2] /bɔɪl/ *n* furúnculo *m*

boiled /ˈbɔɪld/ *a* hervido; ‹*egg*› pasado por agua

boiler /ˈbɔɪlə(r)/ *n* caldera *f*. **~ suit** *n* mono *m*

boisterous /ˈbɔɪstərəs/ *a* ruidoso, bullicioso

bold /bəʊld/ *a* (**-er**, **-est**) audaz. **~ness** *n* audacia *f*

Bolivia /bəˈlɪvɪə/ *n* Bolivia *f*. **~n** *a* & *n* boliviano (*m*)

bollard /ˈbɒləd/ *n* (*naut*) noray *m*; (*Brit*, *auto*) poste *m*

bolster /ˈbəʊlstə(r)/ *n* cabezal *m*. ● *vt*. **~ up** sostener

bolt /bəʊlt/ *n* cerrojo *m*; (*for nut*) perno *m*; (*lightning*) rayo *m*; (*leap*) fuga *f*. ● *vt* echar el cerrojo a ‹*door*›; engullir ‹*food*›. ● *vi* fugarse. ● *adv*. **~ upright** rígido

bomb /bɒm/ *n* bomba *f*. ● *vt* bombardear. **~ard** /bɒmˈbɑːd/ *vt* bombardear

bombastic /bɒmˈbæstɪk/ *a* ampuloso

bomb: **~er** /ˈbɒmə(r)/ *n* bombardero *m*. **~ing** *n* bombardeo *m*. **~shell** *n* bomba *f*

bonanza /bəˈnænzə/ *n* bonanza *f*

bond /bɒnd/ *n* (*agreement*) obligación *f*; (*link*) lazo *m*; (*com*) bono *m*

bondage /ˈbɒndɪdʒ/ *n* esclavitud *f*

bone /bəʊn/ *n* hueso *m*; (*of fish*) espina *f*. ● *vt* deshuesar. **~-dry** *a* completamente seco. **~ idle** *a* holgazán

bonfire /ˈbɒnfaɪə(r)/ *n* hoguera *f*

bonnet /ˈbɒnɪt/ *n* gorra *f*; (*auto*) capó *m*, tapa *f* del motor (*Mex*)

bonny /ˈbɒnɪ/ *a* (**-ier**, **-iest**) bonito

bonus /ˈbəʊnəs/ *n* prima *f*; (*fig*) plus *m*

bony /ˈbəʊnɪ/ *a* (**-ier**, **-iest**) huesudo; ‹*fish*› lleno de espinas

boo /bu:/ *int* ¡bu! ● *vt/i* abuchear

boob /bu:b/ *n* (*mistake, sl*) metedura *f* de pata. ● *vi* (*sl*) meter la pata

booby /ˈbu:bɪ/ *n* bobo *m*. **~ trap** trampa *f*; (*mil*) trampa *f* explosiva

book /bʊk/ *n* libro *m*; (*of cheques etc*) talonario *m*; (*notebook*) libreta *f*; (*exercise book*) cuaderno *m*; (*pl, com*) cuentas *fpl*. ● *vt* (*enter*) registrar; (*reserve*) reservar. ● *vi* reservar. **~able** *a* que se puede reservar. **~case** /ˈbʊkkeɪs/ *n* estantería *f*, librería *f*. **~ing-office** (*in theatre*) taquilla *f*; (*rail*) despacho *m* de billetes. **~let** /ˈbʊklɪt/ *n* folleto *m*

bookkeeping /ˈbʊkki:pɪŋ/ *n* contabilidad *f*

bookmaker /ˈbʊkmeɪkə(r)/ *n* corredor *m* de apuestas

book: **~mark** /ˈbʊkmɑ:(r)k/ *n* señal *f*. **~seller** /ˈbʊkselə(r)/ *n* librero *m*. **~shop** /ˈbʊkʃɒp/ *n* librería *f*. **~stall** /ˈbʊkstɔ:l/ *n* quiosco *m* de libros. **~worm** /ˈbʊkwɜ:m/ *n* (*fig*) ratón *m* de biblioteca

boom /bu:m/ *vi* retumbar; (*fig*) prosperar. ● *n* estampido *m*; (*com*) auge *m*

boon /bu:n/ *n* beneficio *m*

boor /bʊə(r)/ *n* patán *m*. **~ish** *a* grosero

boost /bu:st/ *vt* estimular; reforzar ‹*morale*›; aumentar ‹*price*›; (*publicize*) hacer publicidad por. ● *n* empuje *m*. **~er** *n* (*med*) revacunación *f*

boot /bu:t/ *n* bota *f*; (*auto*) maletero *m*, baúl *m* (*LAm*). **get the ~** (*sl*) ser despedido

booth /bu:ð/ *n* cabina *f*; (*at fair*) puesto *m*

booty /ˈbu:tɪ/ *n* botín *m*

booze /bu:z/ *vi* (*fam*) beber mucho. ● *n* (*fam*) alcohol *m*; (*spree*) borrachera *f*

border /ˈbɔ:də(r)/ *n* borde *m*; (*frontier*) frontera *f*; (*in garden*) arriate *m*. ● *vi*. **~ on** lindar con

borderline /ˈbɔ:dəlaɪn/ *n* línea *f* divisoria. **~ case** *n* caso *m* dudoso

bore[1] /bɔ:(r)/ *vt* (*tec*) taladrar. ● *vi* taladrar

bore[2] /bɔ:(r)/ *vt* (*annoy*) aburrir. ● *n* (*person*) pelmazo *m*; (*thing*) lata *f*

bore[3] /bɔ:(r)/ *see* **bear**[1]

boredom /ˈbɔ:dəm/ *n* aburrimiento *m*

boring /ˈbɔ:rɪŋ/ *a* aburrido, pesado

born /bɔ:n/ *a* nato. **be ~** nacer

borne /bɔ:n/ *see* **bear**[1]

borough /ˈbʌrə/ *n* municipio *m*

borrow /ˈbɒrəʊ/ *vt* pedir prestado

Borstal /ˈbɔ:stl/ *n* reformatorio *m*

bosh /bɒʃ/ *int* & *n* (*sl*) tonterías (*fpl*)

bosom /ˈbʊzəm/ *n* seno *m*. **~ friend** *n* amigo *m* íntimo

boss /bɒs/ *n* (*fam*) jefe *m*. ● *vt*. **~ (about)** (*fam*) dar órdenes a. **~y** /ˈbɒsɪ/ *a* mandón

botan|ical /bəˈtænɪkl/ *a* botánico. **~ist** /ˈbɒtənɪst/ *n* botánico *m*. **~y** /ˈbɒtənɪ/ *n* botánica *f*

botch /bɒtʃ/ *vt* chapucear. ● *n* chapuza *f*

both /bəʊθ/ *a* & *pron* ambos (*mpl*), los dos (*mpl*). ● *adv* al mismo tiempo, a la vez

bother /ˈbɒðə(r)/ *vt* molestar; (*worry*) preocupar. **~ it!** *int* ¡caramba! ● *vi* molestarse. **~ about** preocuparse de. **~ doing** tenerse la molestia de hacer. ● *n* molestia *f*

bottle /ˈbɒtl/ *n* botella; (*for baby*) biberón *m*. ● *vt* embotellar. **~ up** (*fig*) reprimir. **~neck** /ˈbɒtlnek/ *n* (*traffic jam*) embotellamiento *m*. **~-opener** *n* destapador *m*, abrebotellas *m invar*; (*corkscrew*) sacacorchos *m invar*

bottom /ˈbɒtəm/ *n* fondo *m*; (*of hill*) pie *m*; (*buttocks*) trasero *m*. ● *a* último, inferior. **~less** *a* sin fondo

bough /baʊ/ *n* rama *f*

bought /bɔ:t/ *see* **buy**

boulder /ˈbəʊldə(r)/ *n* canto *m*

boulevard /ˈbu:ləvɑ:d/ *n* bulevar *m*

bounc|e /baʊns/ *vt* hacer rebotar. ● *vi* rebotar; ‹*person*› saltar; ‹*cheque, sl*› ser rechazado. ● *n* rebote *m*. **~ing** /ˈbaʊnsɪŋ/ *a* robusto

bound[1] /baʊnd/ *vi* saltar. ● *n* salto *m*

bound[2] /baʊnd/ *n*. **out of ~s** zona *f* prohibida

bound[3] /baʊnd/ *a*. **be ~ for** dirigirse a

bound[4] /baʊnd/ *see* **bind**. **~ to** obligado a; (*certain*) seguro de

boundary /ˈbaʊndərɪ/ *n* límite *m*

boundless /ˈbaʊndləs/ *a* ilimitado

bountiful /ˈbaʊntɪfl/ *a* abundante

bouquet /bʊˈkeɪ/ *n* ramo *m*; (*perfume*) aroma *m*; (*of wine*) buqué *m*, nariz *f*

bout /baʊt/ *n* período *m*; (*med*) ataque *m*; (*sport*) encuentro *m*

bow[1] /bəʊ/ *n* (*weapon, mus*) arco *m*; (*knot*) lazo *m*

bow[2] /baʊ/ *n* reverencia *f*. ● *vi* inclinarse. ● *vt* inclinar

bow[3] /baʊ/ *n* (*naut*) proa *f*

bowels /ˈbaʊəlz/ *npl* intestinos *mpl*; (*fig*) entrañas *fpl*

bowl[1] /bəʊl/ *n* cuenco *m*; (*for washing*) palangana *f*; (*of pipe*) cazoleta *f*

bowl[2] /bəʊl/ *n* (*ball*) bola *f*. ● *vt* (*cricket*) arrojar. ● *vi* (*cricket*) arrojar la pelota. **~ over** derribar

bow-legged /bəʊˈlegɪd/ *a* estevado

bowler[1] /ˈbəʊlə(r)/ *n* (*cricket*) lanzador *m*

bowler[2] /ˈbəʊlə(r)/ *n*. **~ (hat)** hongo *m*, bombín *m*

bowling /ˈbəʊlɪŋ/ *n* bolos *mpl*

bow-tie /bəʊˈtaɪ/ *n* corbata *f* de lazo, pajarita *f*

box[1] /bɒks/ *n* caja *f*; (*for jewels etc*) estuche *m*; (*in theatre*) palco *m*

box[2] /bɒks/ *vt* boxear contra. **~ s.o.'s ears** dar una manotada a uno. ● *vi* boxear. **~er** *n* boxeador *m*. **~ing** *n* boxeo *m*

box: **B~ing Day** *n* el 26 de diciembre. **~-office** *n* taquilla *f*. **~-room** *n* trastero *m*

boy /bɔɪ/ *n* chico *m*, muchacho *m*; (*young*) niño *m*

boycott /ˈbɔɪkɒt/ *vt* boicotear. ● *n* boicoteo *m*

boy: **~-friend** *n* novio *m*. **~hood** *n* niñez *f*. **~ish** *a* de muchacho; (*childish*) infantil

bra /brɑː/ *n* sostén *m*, sujetador *m*

brace /breɪs/ *n* abrazadera *f*; (*dental*) aparato *m*. ● *vt* asegurar. **~ o.s.** prepararse. **~s** *npl* tirantes *mpl*

bracelet /ˈbreɪslɪt/ *n* pulsera *f*

bracing /ˈbreɪsɪŋ/ *a* vigorizante

bracken /ˈbrækən/ *n* helecho *m*

bracket /ˈbrækɪt/ *n* soporte *m*; (*group*) categoría *f*; (*typ*) paréntesis *m invar*. **square ~s** corchetes *mpl*. ● *vt* poner entre paréntesis; (*join together*) agrupar

brag /bræg/ *vi* (*pt* **bragged**) jactarse (**about** de)

braid /breɪd/ *n* galón *m*; (*of hair*) trenza *f*

brain /breɪn/ *n* cerebro *m*. ● *vt* romper la cabeza

brain-child /ˈbreɪntʃaɪld/ *n* invento *m*

brain: **~ drain** (*fam*) fuga *f* de cerebros. **~less** *a* estúpido. **~s** *npl* (*fig*) inteligencia *f*

brainstorm /ˈbreɪnstɔːm/ *n* ataque *m* de locura; (*Amer, brainwave*) idea *f* genial

brainwash /ˈbreɪnwɒʃ/ *vt* lavar el cerebro

brainwave /ˈbreɪnweɪv/ *n* idea *f* genial

brainy /ˈbreɪnɪ/ *a* (**-ier**, **-iest**) inteligente

braise /breɪz/ *vt* cocer a fuego lento

brake /breɪk/ *n* freno *m*. **disc ~** freno de disco. **hand ~** freno de mano. ● *vt/i* frenar. **~ fluid** *n* líquido *m* de freno. **~ lining** *n* forro *m* del freno. **~ shoe** *n* zapata *f* del freno

bramble /bræmbl/ *n* zarza *f*

bran /bræn/ *n* salvado *m*

branch /brɑːntʃ/ *n* rama *f*; (*of road*) bifurcación *f*; (*com*) sucursal *m*; (*fig*) ramo *m*. ● *vi*. **~ off** bifurcarse. **~ out** ramificarse

brand /brænd/ *n* marca *f*; (*iron*) hierro *m*. ● *vt* marcar; (*reputation*) tildar de

brandish /ˈbrændɪʃ/ *vt* blandir

brand-new /brændˈnjuː/ *a* flamante

brandy /ˈbrændɪ/ *n* coñac *m*

brash /bræʃ/ *a* descarado

brass /brɑːs/ *n* latón *m*. **get down to ~ tacks** (*fig*) ir al grano. **top ~** (*sl*) peces *mpl* gordos. **~y** *a* (**-ier**, **-iest**) descarado

brassière /ˈbræsjeə(r)/ *n* sostén *m*, sujetador *m*

brat /bræt/ *n* (*pej*) mocoso *m*

bravado /brəˈvɑːdəʊ/ *n* bravata *f*

brave /breɪv/ *a* (**-er**, **-est**) valiente. ● *n* (*Red Indian*) guerrero *m* indio. ● *vt* afrontar. **~ry** /-ərɪ/ *n* valentía *f*, valor *m*

brawl /brɔːl/ *n* alboroto *m*. ● *vi* pelearse

brawn /brɔːn/ *n* músculo *m*; (*strength*) fuerza *f* muscular. **~y** *a* musculoso

bray /breɪ/ *n* rebuzno *m*. ● *vi* rebuznar
brazen /ˈbreɪzn/ *a* descarado
brazier /ˈbreɪzɪə(r)/ *n* brasero *m*
Brazil /brəˈzɪl/ *n* el Brasil *m*. **~ian** *a* & *n* brasileño (*m*)
breach /briːtʃ/ *n* violación *f*; (*of contract*) incumplimiento *m*; (*gap*) brecha *f*. ● *vt* abrir una brecha en
bread /bred/ *n* pan *m*. **loaf of ~** pan. **~crumbs** /ˈbredkrʌmz/ *npl* migajas *fpl*; (*culin*) pan rallado. **~line** *n*. **on the ~line** en la miseria
breadth /bredθ/ *n* anchura *f*
bread-winner /ˈbredwɪnə(r)/ *n* sostén *m* de la familia, cabeza *f* de familia
break /breɪk/ *vt* (*pt* **broke**, *pp* **broken**) romper; quebrantar ‹*law*›; batir ‹*record*›; comunicar ‹*news*›; interrumpir ‹*journey*›. ● *vi* romperse; ‹*news*› divulgarse. ● *n* ruptura *f*; (*interval*) intervalo *m*; (*chance*, *fam*) oportunidad *f*; (*in weather*) cambio *m*. **~ away** escapar. **~ down** *vt* derribar; analizar ‹*figures*›. ● *vi* estropearse; (*auto*) averiarse; (*med*) sufrir un colapso; (*cry*) deshacerse en lágrimas. **~ into** forzar ‹*house etc*›; (*start doing*) ponerse a. **~ off** interrumpirse. **~ out** ‹*war*, *disease*› estallar; (*run away*) escaparse. **~ up** romperse; ‹*schools*› terminar. **~able** *a* frágil. **~age** *n* rotura *f*
breakdown /ˈbreɪkdaʊn/ *n* (*tec*) falla *f*; (*med*) colapso *m*, crisis *f* nerviosa; (*of figures*) análisis *f*
breaker /ˈbreɪkə(r)/ *n* (*wave*) cachón *m*
breakfast /ˈbrekfəst/ *n* desayuno *m*
breakthrough /ˈbreɪkθruː/ *n* adelanto *m*
breakwater /ˈbreɪkwɔːtə(r)/ *n* rompeolas *m invar*
breast /brest/ *n* pecho *m*; (*of chicken etc*) pechuga *f*. **~-stroke** *n* braza *f* de pecho
breath /breθ/ *n* aliento *m*, respiración *f*. **out of ~** sin aliento. **under one's ~** a media voz. **~alyser** /ˈbreθəlaɪzə(r)/ *n* alcoholímetro *m*
breath|e /briːð/ *vt/i* respirar. **~er** /ˈbriːðə(r)/ *n* descanso *m*, pausa *f*. **~ing** *n* respiración *f*
breathtaking /ˈbreθteɪkɪŋ/ *a* impresionante
bred /bred/ *see* **breed**
breeches /ˈbrɪtʃɪz/ *npl* calzones *mpl*
breed /briːd/ *vt/i* (*pt* **bred**) reproducirse; (*fig*) engendrar. ● *n* raza *f*. **~er** *n* criador *m*. **~ing** *n* cría *f*; (*manners*) educación *f*
breez|e /briːz/ *n* brisa *f*. **~y** *a* de mucho viento; ‹*person*› despreocupado. **it is ~y** hace viento
Breton /ˈbretən/ *a* & *n* bretón (*m*)
brew /bruː/ *vt* hacer. ● *vi* fermentar; ‹*tea*› reposar; (*fig*) prepararse. ● *n* infusión *f*. **~er** *n* cervecero *m*. **~ery** *n* fábrica *f* de cerveza, cervecería *f*
bribe /braɪb/ *n* soborno *m*. ● *vt* sobornar. **~ry** /-ərɪ/ *n* soborno *m*
brick /brɪk/ *n* ladrillo *m*. ● *vt*. **~ up** tapar con ladrillos. **~layer** /ˈbrɪkleɪə(r)/ *n* albañil *m*
bridal /braɪdl/ *a* nupcial
bride /braɪd/ *m* novia *f*. **~groom** /ˈbraɪdgrʊm/ *n* novio *m*. **~smaid** /ˈbraɪdzmeɪd/ *n* dama *f* de honor
bridge[1] /brɪdʒ/ *n* puente *m*; (*of nose*) caballete *m*. ● *vt* tender un puente sobre. **~ a gap** llenar un vacío
bridge[2] /brɪdʒ/ *n* (*cards*) bridge *m*
bridle /ˈbraɪdl/ *n* brida *f*. ● *vt* embridar. **~-path** *n* camino *m* de herradura
brief /briːf/ *a* (**-er**, **-est**) breve. ● *n* (*jurid*) escrito *m*. ● *vt* dar instrucciones a. **~case** /ˈbriːfkeɪs/ *n* maletín *m*. **~ly** *adv* brevemente. **~s** *npl* (*man's*) calzoncillos *mpl*; (*woman's*) bragas *fpl*
brigad|e /brɪˈgeɪd/ *n* brigada *f*. **~ier** /-əˈdɪə(r)/ *n* general *m* de brigada
bright /braɪt/ *a* (**-er**, **-est**) brillante, claro; (*clever*) listo; (*cheerful*) alegre. **~en** /ˈbraɪtn/ *vt* aclarar; hacer más alegre ‹*house etc*›. ● *vi* ‹*weather*› aclararse; ‹*face*› animarse. **~ly** *adv* brillantemente. **~ness** *n* claridad *f*
brillian|ce /ˈbrɪljəns/ *n* brillantez *f*, brillo *m*. **~t** *a* brillante
brim /brɪm/ *n* borde *m*; (*of hat*) ala *f*. ● *vi* (*pt* **brimmed**). **~ over** desbordarse
brine /braɪn/ *n* salmuera *f*
bring /brɪŋ/ *vt* (*pt* **brought**) traer ‹*thing*›; conducir ‹*person*, *vehicle*›. **~ about** causar. **~ back** devolver. **~ down** derribar; rebajar ‹*price*›. **~ off** lograr. **~ on** causar. **~ out** sacar; lanzar ‹*product*›; publicar ‹*book*›. **~ round/to** hacer volver en sí ‹*unconscious person*›. **~ up** (*med*) vomitar; educar ‹*children*›; plantear ‹*question*›

brink /brɪŋk/ *n* borde *m*
brisk /brɪsk/ *a* (**-er**, **-est**) enérgico, vivo. **~ness** *n* energía *f*
bristl|e /ˈbrɪsl/ *n* cerda *f*. ● *vi* erizarse. **~ing with** erizado de
Brit|ain /ˈbrɪtən/ *n* Gran Bretaña *f*. **~ish** /ˈbrɪtɪʃ/ *a* británico. **the ~ish** los británicos. **~on** /ˈbrɪtən/ *n* británico *m*
Brittany /ˈbrɪtənɪ/ *n* Bretaña *f*
brittle /ˈbrɪtl/ *a* frágil, quebradizo
broach /brəʊtʃ/ *vt* abordar ‹*subject*›; espitar ‹*cask*›
broad /brɔ:d/ *a* (**-er**, **-est**) ancho. **in ~ daylight** en pleno día. **~ bean** *n* haba *f*
broadcast /ˈbrɔ:dkɑ:st/ *n* emisión *f*. ● *vt* (*pt* **broadcast**) emitir. ● *vi* hablar por la radio. **~ing** *a* de radiodifusión. ● *n* radio-difusión *f*
broad: ~en /ˈbrɔ:dn/ *vt* ensanchar. ● *vi* ensancharse. **~ly** *adv* en general. **~-minded** *a* de miras amplias, tolerante, liberal
brocade /brəˈkeɪd/ *n* brocado *m*
broccoli /ˈbrɒkəlɪ/ *n invar* brécol *m*
brochure /ˈbrəʊʃə(r)/ *n* folleto *m*
brogue /brəʊg/ *n* abarca *f*; (*accent*) acento *m* regional
broke /brəʊk/ *see* **break**. ● *a* (*sl*) sin blanca
broken /ˈbrəʊkən/ *see* **break**. ● *a*. **~ English** inglés *m* chapurreado. **~-hearted** *a* con el corazón destrozado
broker /ˈbrəʊkə(r)/ *n* corredor *m*
brolly /ˈbrɒlɪ/ *n* (*fam*) paraguas *m invar*
bronchitis /brɒŋˈkaɪtɪs/ *n* bronquitis *f*
bronze /brɒnz/ *n* bronce *m*. ● *vt* broncear. ● *vi* broncearse
brooch /brəʊtʃ/ *n* broche *m*
brood /bru:d/ *n* cría *f*; (*joc*) prole *m*. ● *vi* empollar; (*fig*) meditar. **~y** *a* contemplativo
brook[1] /brʊk/ *n* arroyo *m*
brook[2] /brʊk/ *vt* soportar
broom /bru:m/ *n* hiniesta *f*; (*brush*) escoba *f*. **~stick** /ˈbru:mstɪk/ *n* palo *m* de escoba
broth /brɒθ/ *n* caldo *m*
brothel /ˈbrɒθl/ *n* burdel *m*
brother /ˈbrʌðə(r)/ *n* hermano *m*. **~hood** *n* fraternidad *f*, (*relig*) hermandad *f*. **~-in-law** *n* cuñado *m*. **~ly** *a* fraternal
brought /brɔ:t/ *see* **bring**
brow /braʊ/ *n* frente *f*; (*of hill*) cima *f*
browbeat /ˈbraʊbi:t/ *vt* (*pt* **-beat**, *pp* **-beaten**) intimidar
brown /braʊn/ *a* (**-er**, **-est**) marrón; ‹*skin*› moreno; ‹*hair*› castaño. ● *n* marrón *m*. ● *vt* poner moreno; (*culin*) dorar. ● *vi* ponerse moreno; (*culin*) dorarse. **be ~ed off** (*sl*) estar hasta la coronilla
Brownie /ˈbraʊnɪ/ *n* niña *f* exploradora
browse /braʊz/ *vi* (*in a shop*) curiosear; ‹*animal*› pacer
bruise /bru:z/ *n* magulladura *f*. ● *vt* magullar; machucar ‹*fruit*›. ● *vi* magullarse; ‹*fruit*› machacarse
brunch /brʌntʃ/ *n* (*fam*) desayuno *m* tardío
brunette /bru:ˈnet/ *n* morena *f*
brunt /brʌnt/ *n*. **the ~ of** lo más fuerte de
brush /brʌʃ/ *n* cepillo *m*; (*large*) escoba; (*for decorating*) brocha *f*; (*artist's*) pincel; (*skirmish*) escaramuza *f*. ● *vt* cepillar. **~ against** rozar. **~ aside** rechazar. **~ off** (*rebuff*) desairar. **~ up (on)** refrescar
brusque /bru:sk/ *a* brusco. **~ly** *adv* bruscamente
Brussels /ˈbrʌslz/ *n* Bruselas *f*. **~ sprout** col *m* de Bruselas
brutal /ˈbru:tl/ *a* brutal. **~ity** /-ˈtælətɪ/ *n* brutalidad *f*
brute /bru:t/ *n* bestia *f*. **~ force** fuerza *f* bruta
BSc *abbr see* **bachelor**
bubble /ˈbʌbl/ *n* burbuja *f*. ● *vi* burbujear. **~ over** desbordarse
bubbly /ˈbʌblɪ/ *a* burbujeante. ● *n* (*fam*) champaña *m*, champán *m* (*fam*)
buck[1] /bʌk/ *a* macho. ● *n* (*deer*) ciervo *m*. ● *vi* (*of horse*) corcovear. **~ up** (*hurry, sl*) darse prisa; (*cheer up, sl*) animarse
buck[2] /bʌk/ (*Amer, sl*) dólar *m*
buck[3] /bʌk/ *n*. **pass the ~ to s.o.** echarle a uno el muerto
bucket /ˈbʌkɪt/ *n* cubo *m*
buckle /ˈbʌkl/ *n* hebilla *f*. ● *vt* abrochar. ● *vi* torcerse. **~ down to** dedicarse con empeño a
bud /bʌd/ *n* brote *m*. ● *vi* (*pt* **budded**) brotar.
Buddhis|m /ˈbʊdɪzəm/ *n* budismo *m*. **~t** /ˈbʊdɪst/ *a* & *n* budista (*m* & *f*)
budding /ˈbʌdɪŋ/ *a* (*fig*) en ciernes

buddy /ˈbʌdɪ/ *n* (*fam*) compañero *m*, amigote *m* (*fam*)
budge /bʌdʒ/ *vt* mover. ● *vi* moverse
budgerigar /ˈbʌdʒərɪgɑ:(r)/ *n* periquito *m*
budget /ˈbʌdʒɪt/ *n* presupuesto *m*. ● *vi* (*pt* **budgeted**) presupuestar
buff /bʌf/ *n* (*colour*) color *m* de ante; (*fam*) aficionado *m*. ● *vt* pulir
buffalo /ˈbʌfələʊ/ *n* (*pl* **-oes** *or* **-o**) búfalo *m*
buffer /ˈbʌfə(r)/ *n* parachoques *m invar*. ~ **state** *n* estado *m* tapón
buffet /ˈbʊfeɪ/ *n* (*meal, counter*) bufé *m*. /ˈbʌfɪt/ *n* golpe *m*; (*slap*) bofetada *f*. ● *vt* (*pt* **buffeted**) golpear
buffoon /bəˈfu:n/ *n* payaso *m*, bufón *m*
bug /bʌg/ *n* bicho *m*; (*germ, sl*) microbio *m*; (*device, sl*) micrófono *m* oculto. ● *vt* (*pt* **bugged**) ocultar un micrófono en; intervenir ‹*telephone*›; (*Amer, sl*) molestar
bugbear /ˈbʌgbeə(r)/ *n* pesadilla *f*
buggy /ˈbʌgɪ/ *n*. **baby** ~ (*esp Amer*) cochecito *m* de niño
bugle /ˈbju:gl/ *n* corneta *f*
build /bɪld/ *vt*/*i* (*pt* **built**) construir. ~ **up** *vt* urbanizar; (*increase*) aumentar. ● *n* (*of person*) figura *f*, tipo *m*. ~**er** *n* constructor *m*. ~**-up** *n* aumento *m*; (*of gas etc*) acumulación *f*; (*fig*) propaganda *f*
built /bɪlt/ *see* **build**. ~**-in** *a* empotrado. ~**-up area** *n* zona *f* urbanizada
bulb /bʌlb/ *n* bulbo *m*; (*elec*) bombilla *f*. ~**ous** *a* bulboso
Bulgaria /bʌlˈgeərɪə/ *n* Bulgaria *f*. ~**n** *a* & *n* búlgaro (*m*)
bulg|e /bʌldʒ/ *n* protuberancia *f*. ● *vi* pandearse; (*jut out*) sobresalir. ~**ing** *a* abultado; ‹*eyes*› saltón
bulk /bʌlk/ *n* bulto *m*, volumen *m*. **in** ~ a granel; (*loose*) suelto. **the** ~ **of** la mayor parte de. ~**y** *a* voluminoso
bull /bʊl/ *n* toro *m*
bulldog /ˈbʊldɒg/ *n* buldog *m*
bulldozer /ˈbʊldəʊzə(r)/ *n* oruga *f* aplanadora, bulldozer *m*
bullet /ˈbʊlɪt/ *n* bala *f*
bulletin /ˈbʊlətɪn/ *n* anuncio *m*; (*journal*) boletín *m*
bullet-proof /ˈbʊlɪtpru:f/ *a* a prueba de balas
bullfight /ˈbʊlfaɪt/ *n* corrida *f* (de toros). ~**er** *n* torero *m*
bullion /ˈbʊljən/ *n* (*gold*) oro *m* en barras; (*silver*) plata *f* en barras
bull: ~**ring** /ˈbʊlrɪŋ/ *n* plaza *f* de toros. ~**'s-eye** *n* centro *m* del blanco, diana *f*
bully /ˈbʊlɪ/ *n* matón *m*. ● *vt* intimidar. ~**ing** *n* intimidación *f*
bum[1] /bʌm/ *n* (*bottom, sl*) trasero *m*
bum[2] /bʌm/ *n* (*Amer, sl*) holgazán *m*
bumble-bee /ˈbʌmblbi:/ *n* abejorro *m*
bump /bʌmp/ *vt* chocar contra. ● *vi* dar sacudidas. ● *n* choque *m*; (*swelling*) chichón *m*. ~ **into** chocar contra; (*meet*) encontrar
bumper /ˈbʌmpə(r)/ *n* parachoques *m invar*. ● *a* abundante. ~ **edition** *n* edición *f* especial
bumpkin /ˈbʌmpkɪn/ *n* patán *m*, paleto *m* (*fam*)
bumptious /ˈbʌmpʃəs/ *a* presuntuoso
bun /bʌn/ *n* bollo *m*; (*hair*) moño *m*
bunch /bʌntʃ/ *n* manojo *m*; (*of people*) grupo *m*; (*of bananas, grapes*) racimo *m*, (*of flowers*) ramo *m*
bundle /ˈbʌndl/ *n* bulto *m*; (*of papers*) legajo *m*; (*of nerves*) manojo *m*. ● *vt*. ~ **up** atar
bung /bʌŋ/ *n* tapón *m*. ● *vt* tapar; (*sl*) tirar
bungalow /ˈbʌŋgələʊ/ *n* casa *f* de un solo piso, chalé *m*, bungalow *m*
bungle /ˈbʌŋgl/ *vt* chapucear
bunion /ˈbʌnjən/ *n* juanete *m*
bunk /bʌŋk/ *n* litera *f*
bunker /ˈbʌŋkə(r)/ *n* carbonera *f*; (*golf*) obstáculo *m*; (*mil*) refugio *m*, búnker *m*
bunkum /ˈbʌŋkəm/ *n* tonterías *fpl*
bunny /ˈbʌnɪ/ *n* conejito *m*
buoy /bɔɪ/ *n* boya *f*. ● *vt*. ~ **up** hacer flotar; (*fig*) animar
buoyan|cy /ˈbɔɪənsɪ/ *n* flotabilidad *f*; (*fig*) optimismo *m*. ~**t** /ˈbɔɪənt/ *a* boyante; (*fig*) alegre
burden /ˈbɜ:dn/ *n* carga *f*. ● *vt* cargar (**with** de). ~**some** *a* pesado
bureau /ˈbjʊərəʊ/ *n* (*pl* **-eaux** /-əʊz/) escritorio *m*; (*office*) oficina *f*
bureaucra|cy /bjʊəˈrɒkrəsɪ/ *n* burocracia *f*. ~**t** /ˈbjʊərəkræt/ *n* burócrata *m* & *f*. ~**tic** /-ˈkrætɪk/ *a* burocrático
burgeon /ˈbɜ:dʒən/ *vi* brotar; (*fig*) crecer
burgl|ar /ˈbɜ:glə(r)/ *n* ladrón *m*. ~**ary** *n* robo *m* con allanamiento de

morada. **~e** /'bɜ:gl/ *vt* robar con allanamiento
Burgundy /'bɜ:gəndɪ/ *n* Borgoña *f*; (*wine*) vino *m* de Borgoña
burial /'berɪəl/ *n* entierro *m*
burlesque /bɜ:'lesk/ *n* burlesco *m*
burly /'bɜ:lɪ/ *a* (**-ier**, **-iest**) corpulento
Burm|a /'bɜ:mə/ Birmania *f*. **~ese** /-'mi:z/ *a* & *n* birmano (*m*)
burn /bɜ:n/ *vt* (*pt* **burned** *or* **burnt**) quemar. ● *vi* quemarse. **~ down** *vt* destruir con fuego. ● *n* quemadura *f*. **~er** *n* quemador *m*. **~ing** *a* ardiente; ‹*food*› que quema; ‹*question*› candente
burnish /'bɜ:nɪʃ/ *vt* lustrar, pulir
burnt /bɜ:nt/ *see* **burn**
burp /bɜ:p/ *n* (*fam*) eructo *m*. ● *vi* (*fam*) eructar
burr /bɜ:(r)/ *n* (*bot*) erizo *m*
burrow /'bʌrəʊ/ *n* madriguera *f*. ● *vt* excavar
bursar /'bɜ:sə(r)/ *n* tesorero *m*. **~y** /'bɜ:sərɪ/ *n* beca *f*
burst /bɜ:st/ *vt* (*pt* **burst**) reventar. ● *vi* reventarse; ‹*tyre*› pincharse. ● *n* reventón *m*; (*mil*) ráfaga *f*; (*fig*) explosión *f*. **~ of laughter** carcajada *f*
bury /'berɪ/ *vt* enterrar; (*hide*) ocultar
bus /bʌs/ *n* (*pl* **buses**) autobús *m*, camión *m* (*Mex*). ● *vi* (*pt* **bussed**) ir en autobús
bush /bʊʃ/ *n* arbusto *m*; (*land*) monte *m*. **~y** *a* espeso
busily /'bɪzɪlɪ/ *adv* afanosamente
business /'bɪznɪs/ *n* negocio *m*; (*com*) negocios *mpl*; (*profession*) ocupación *f*; (*fig*) asunto *m*. **mind one's own ~** ocuparse de sus propios asuntos. **~-like** *a* práctico, serio. **~man** *n* hombre *m* de negocios
busker /'bʌskə(r)/ *n* músico *m* ambulante
bus-stop /'bʌsstɒp/ *n* parada *f* de autobús
bust[1] /bʌst/ *n* busto *m*; (*chest*) pecho *m*
bust[2] /bʌst/ *vt* (*pt* **busted** *or* **bust**) (*sl*) romper. ● *vi* romperse. ● *a* roto. **go ~** (*sl*) quebrar
bustle /'bʌsl/ *vi* apresurarse. ● *n* bullicio *m*
bust-up /'bʌstʌp/ *n* (*sl*) riña *f*
busy /'bɪzɪ/ *a* (**-ier**, **-iest**) ocupado; ‹*street*› concurrido. ● *vt*. **~ o.s. with** ocuparse de
busybody /'bɪzɪbɒdɪ/ *n* entrometido *m*
but /bʌt/ *conj* pero; (*after negative*) sino. ● *prep* menos. **~ for** si no fuera por. **last ~ one** penúltimo. ● *adv* solamente
butane /'bju:teɪn/ *n* butano *m*
butcher /'bʊtʃə(r)/ *n* carnicero *m*. ● *vt* matar; (*fig*) hacer una carnicería con. **~y** *n* carnicería *f*, matanza *f*
butler /'bʌtlə(r)/ *n* mayordomo *m*
butt /bʌt/ *n* (*of gun*) culata *f*; (*of cigarette*) colilla *f*; (*target*) blanco *m*. ● *vi* topar. **~ in** interrumpir
butter /'bʌtə(r)/ *n* mantequilla *f*. ● *vt* untar con mantequilla. **~ up** *vt* (*fam*) lisonjear, dar jabón a. **~-bean** *n* judía *f*
buttercup /'bʌtəkʌp/ *n* ranúnculo *m*
butter-fingers /'bʌtəfɪŋgəz/ *n* manazas *m invar*, torpe *m*
butterfly /'bʌtəflaɪ/ *n* mariposa *f*
buttock /'bʌtək/ *n* nalga *f*
button /'bʌtn/ *n* botón *m*. ● *vt* abotonar. ● *vi* abotonarse. **~hole** /'bʌtnhəʊl/ *n* ojal *m*. ● *vt* (*fig*) detener
buttress /'bʌtrɪs/ *n* contrafuerte *m*. ● *vt* apoyar
buxom /'bʌksəm/ *a* ‹*woman*› rollizo
buy /baɪ/ *vt* (*pt* **bought**) comprar. ● *n* compra *f*. **~er** *n* comprador *m*
buzz /bʌz/ *n* zumbido *m*; (*phone call, fam*) llamada *f*. ● *vi* zumbar. **~ off** (*sl*) largarse. **~er** *n* timbre *m*
by /baɪ/ *prep* por; (*near*) cerca de; (*before*) antes de; (*according to*) según. **~ and large** en conjunto, en general. **~ car** en coche. **~ oneself** por sí solo
bye-bye /'baɪbaɪ/ *int* (*fam*) ¡adiós!
by-election /'baɪɪlekʃn/ *n* elección *f* parcial
bygone /'baɪgɒn/ *a* pasado
by-law /'baɪlɔ:/ *n* reglamento *m* (local)
bypass /'baɪpɑ:s/ *n* carretera *f* de circunvalación. ● *vt* evitar
by-product /'baɪprɒdʌkt/ *n* subproducto *m*
bystander /'baɪstændə(r)/ *n* espectador *m*
byword /'baɪwɜ:d/ *n* sinónimo *m*. **be a ~ for** ser conocido por

C

cab /kæb/ *n* taxi *m*; (*of lorry, train*) cabina *f*
cabaret /ˈkæbəreɪ/ *n* espectáculo *m*
cabbage /ˈkæbɪdʒ/ *n* col *m*, repollo *m*
cabin /ˈkæbɪn/ *n* cabaña *f*; (*in ship*) camarote *m*; (*in plane*) cabina *f*
cabinet /ˈkæbɪnɪt/ *n* (*cupboard*) armario *m*; (*for display*) vitrina *f*. **C~** (*pol*) gabinete *m*. **~-maker** *n* ebanista *m & f*
cable /ˈkeɪbl/ *n* cable *m*. ● *vt* cablegrafiar. **~ railway** *n* funicular *m*
cache /kæʃ/ *n* (*place*) escondrijo *m*; (*things*) reservas *fpl* escondidas. ● *vt* ocultar
cackle /ˈkækl/ *n* (*of hen*) cacareo *m*; (*laugh*) risotada *f*. ● *vi* cacarear; (*laugh*) reírse a carcajadas
cacophon|ous /kəˈkɒfənəs/ *a* cacofónico. **~y** *n* cacofonía *f*
cactus /ˈkæktəs/ *n* (*pl* **-ti** /-taɪ/) cacto *m*
cad /kæd/ *n* sinvergüenza m. **~dish** *a* desvergonzado
caddie /ˈkædɪ/ *n* (*golf*) portador *m* de palos
caddy /ˈkædɪ/ *n* cajita *f*
cadence /ˈkeɪdəns/ *n* cadencia *f*
cadet /kəˈdet/ *n* cadete *m*
cadge /kædʒ/ *vt/i* gorronear. **~r** /-ə(r)/ *n* gorrón *m*
Caesarean /sɪˈzeərɪən/ *a* cesáreo. **~ section** *n* cesárea *f*
café /ˈkæfeɪ/ *n* cafetería *f*
cafeteria /kæfɪˈtɪərɪə/ *n* autoservicio *m*
caffeine /ˈkæfiːn/ *n* cafeína *f*
cage /keɪdʒ/ *n* jaula *f*. ● *vt* enjaular
cagey /ˈkeɪdʒɪ/ *a* (*fam*) evasivo
Cairo /ˈkaɪərəʊ/ *n* el Cairo *m*
cajole /kəˈdʒəʊl/ *vt* engatusar. **~ry** *n* engatusamiento *m*
cake /keɪk/ *n* pastel *m*, tarta *f*; (*sponge*) bizcocho *m*. **~ of soap** pastilla *f* de jabón. **~d** *a* incrustado
calamit|ous /kəˈlæmɪtəs/ *a* desastroso. **~y** /kəˈlæmətɪ/ *n* calamidad *f*
calcium /ˈkælsɪəm/ *n* calcio *m*
calculat|e /ˈkælkjʊleɪt/ *vt/i* calcular; (*Amer*) suponer. **~ing** *a* calculador. **~ion** /-ˈleɪʃn/ *n* cálculo *m*. **~or** *n* calculadora *f*
calculus /ˈkælkjʊləs/ *n* (*pl* **-li**) cálculo *m*
calendar /ˈkælɪndə(r)/ *n* calendario *m*
calf[1] /kɑːf/ *n* (*pl* **calves**) ternero *m*
calf[2] /kɑːf/ *n* (*pl* **calves**) (*of leg*) pantorrilla *f*
calibre /ˈkælɪbə(r)/ *n* calibre *m*
calico /ˈkælɪkəʊ/ *n* calicó *m*
call /kɔːl/ *vt/i* llamar. ● *n* llamada *f*; (*shout*) grito *m*; (*visit*) visita *f*. **be on ~** estar de guardia. **long distance ~** conferencia *f*. **~ back** *vt* hacer volver; (*on phone*) volver a llamar. ● *vi* volver; (*on phone*) volver a llamar. **~ for** pedir; (*fetch*) ir a buscar. **~ off** cancelar. **~ on** visitar. **~ out** dar voces. **~ together** convocar. **~ up** (*mil*) llamar al servicio militar; (*phone*) llamar. **~-box** *n* cabina *f* telefónica. **~er** *n* visita *f*; (*phone*) el que llama *m*. **~ing** *n* vocación *f*
callous /ˈkæləs/ *a* insensible, cruel. **~ness** *n* crueldad *f*
callow /ˈkæləʊ/ *a* (**-er, -est**) inexperto
calm /kɑːm/ *a* (**-er, -est**) tranquilo; ‹*weather*› calmoso. ● *n* tranquilidad *f*, calma *f*. ● *vt* calmar. ● *vi* calmarse. **~ness** *n* tranquilidad *f*, calma *f*
calorie /ˈkælərɪ/ *n* caloría *f*
camber /ˈkæmbə(r)/ *n* curvatura *f*
came /keɪm/ *see* **come**
camel /ˈkæml/ *n* camello *m*
camellia /kəˈmiːljə/ *n* camelia *f*
cameo /ˈkæmɪəʊ/ *n* (*pl* **-os**) camafeo *m*
camera /ˈkæmərə/ *n* máquina *f* (fotográfica); (*TV*) cámara *f*. **~man** *n* (*pl* **-men**) operador *m*, cámara *m*
camouflage /ˈkæməflɑːʒ/ *n* camuflaje *m*. ● *vt* encubrir; (*mil*) camuflar
camp[1] /kæmp/ *n* campamento *m*. ● *vi* acamparse
camp[2] /kæmp/ *a* (*affected*) amanerado
campaign /kæmˈpeɪn/ *n* campaña *f*. ● *vi* hacer campaña
camp: **~-bed** *n* catre *m* de tijera. **~er** *n* campista *m & f*; (*vehicle*) caravana *f*. **~ing** *n* camping *m*. **go ~ing** hacer camping. **~site** /ˈkæmpsaɪt/ *n* camping *m*
campus /ˈkæmpəs/ *n* (*pl* **-puses**) ciudad *f* universitaria
can[1] /kæn/ *v aux* (*pt* **could**) (*be able to*) poder; (*know how to*) saber. **~not** (*neg*), **~'t** (*neg, fam*). **I ~not/~'t go** no puedo ir

can[2] /kæn/ *n* lata *f*. ● *vt* (*pt* **canned**) enlatar. **~ned music** música *f* grabada

Canad|a /ˈkænədə/ *n* el Canadá *m*. **~ian** /kəˈneɪdɪən/ *a* & *n* canadiense (*m* & *f*)

canal /kəˈnæl/ *n* canal *m*

canary /kəˈneərɪ/ *n* canario *m*

cancel /ˈkænsl/ *vt/i* (*pt* **cancelled**) anular; cancelar ‹*contract etc*›; suspender ‹*appointment etc*›; (*delete*) tachar. **~lation** /-ˈleɪʃn/ *n* cancelación *f*

cancer /ˈkænsə(r)/ *n* cáncer *m*. **C~** *n* (*Astr*) Cáncer *m*. **~ous** *a* canceroso

candid /ˈkændɪd/ *a* franco

candida|cy /ˈcændɪdəsɪ/ *n* candidatura *f*. **~te** /ˈkændɪdeɪt/ *n* candidato *m*

candle /ˈkændl/ *n* vela *f*. **~stick** /ˈkændlstɪk/ *n* candelero *m*

candour /ˈkændə(r)/ *n* franqueza *f*

candy /ˈkændɪ/ *n* (*Amer*) caramelo *m*. **~-floss** *n* algodón *m* de azúcar

cane /keɪn/ *n* caña *f*; (*for baskets*) mimbre *m*; (*stick*) bastón *m*. ● *vt* (*strike*) castigar con palmeta

canine /ˈkeɪnaɪn/ *a* canino

canister /ˈkænɪstə(r)/ *n* bote *m*

cannabis /ˈkænəbɪs/ *n* cáñamo *m* índico, hachís *m*, mariguana *f*

cannibal /ˈkænɪbl/ *n* caníbal *m*. **~ism** *n* canibalismo *m*

cannon /ˈkænən/ *n invar* cañón *m*. **~ shot** cañonazo *m*

cannot /ˈkænət/ *see* **can**[1]

canny /ˈkænɪ/ *a* astuto

canoe /kəˈnuː/ *n* canoa *f*, piragua *f*. ● *vi* ir en canoa. **~ist** *n* piragüista *m* & *f*

canon /ˈkænən/ *n* canon *m*; (*person*) canónigo *m*. **~ize** /ˈkænənaɪz/ *vt* canonizar

can-opener /ˈkænəʊpnə(r)/ *n* abrelatas *m invar*

canopy /ˈkænəpɪ/ *n* dosel *m*; (*of parachute*) casquete *m*

cant /kænt/ *n* jerga *f*

can't /kɑːnt/ *see* **can**[1]

cantankerous /kænˈtæŋkərəs/ *a* malhumorado

canteen /kænˈtiːn/ *n* cantina *f*; (*of cutlery*) juego *m*; (*flask*) cantimplora *f*

canter /ˈkæntə(r)/ *n* medio galope *m*. ● *vi* ir a medio galope

canvas /ˈkænvəs/ *n* lona *f*; (*artist's*) lienzo *m*

canvass /ˈkænvəs/ *vi* hacer campaña, solicitar votos. **~ing** *n* solicitación *f* (de votos)

canyon /ˈkænjən/ *n* cañón *m*

cap /kæp/ *n* gorra *f*; (*lid*) tapa *f*; (*of cartridge*) cápsula *f*; (*academic*) birrete *m*; (*of pen*) capuchón *m*; (*mec*) casquete *m*. ● *vt* (*pt* **capped**) tapar, poner cápsula a; (*outdo*) superar

capab|ility /keɪpəˈbɪlətɪ/ *n* capacidad *f*. **~le** /ˈkeɪpəbl/ *a* capaz. **~ly** *adv* competentemente

capacity /kəˈpæsətɪ/ *n* capacidad *f*; (*function*) calidad *f*

cape[1] /keɪp/ *n* (*cloak*) capa *f*

cape[2] /keɪp/ *n* (*geog*) cabo *m*

caper[1] /ˈkeɪpə(r)/ *vi* brincar. ● *n* salto *m*; (*fig*) travesura *f*

caper[2] /ˈkeɪpə(r)/ *n* (*culin*) alcaparra *f*

capital /ˈkæpɪtl/ *a* capital. **~ letter** *n* mayúscula *f*. ● *n* (*town*) capital *f*; (*money*) capital *m*

capitalis|m /ˈkæpɪtəlɪzəm/ *n* capitalismo *m*. **~t** *a* & *n* capitalista (*m* & *f*)

capitalize /ˈkæpɪtəlaɪz/ *vt* capitalizar; (*typ*) escribir con mayúsculas. **~ on** aprovechar

capitulat|e /kəˈpɪtʃʊleɪt/ *vi* capitular. **~ion** /-ˈleɪʃn/ *n* capitulación *f*

capon /ˈkeɪpən/ *n* capón *m*

capricious /kəˈprɪʃəs/ *a* caprichoso

Capricorn /ˈkæprɪkɔːn/ *n* Capricornio *m*

capsicum /ˈkæpsɪkəm/ *n* pimiento *m*

capsize /kæpˈsaɪz/ *vt* hacer zozobrar. ● *vi* zozobrar

capsule /ˈkæpsjuːl/ *n* cápsula *f*

captain /ˈkæptɪn/ *n* capitán *m*. ● *vt* capitanear

caption /ˈkæpʃn/ *n* (*heading*) título *m*; (*of cartoon etc*) leyenda *f*

captivate /ˈkæptɪveɪt/ *vt* encantar

captiv|e /ˈkæptɪv/ *a* & *n* cautivo (*m*). **~ity** /-ˈtɪvətɪ/ *n* cautiverio *m*, cautividad *f*

capture /ˈkæptʃə(r)/ *vt* prender; llamar ‹*attention*›; (*mil*) tomar. ● *n* apresamiento *m*; (*mil*) toma *f*

car /kɑː(r)/ *n* coche *m*, carro *m* (*LAm*)

carafe /kəˈræf/ *n* jarro *m*, garrafa *f*

caramel /ˈkærəmel/ *n* azúcar *m* quemado; (*sweet*) caramelo *m*

carat /ˈkærət/ *n* quilate *m*

caravan /ˈkærəvæn/ *n* caravana *f*

carbohydrate /kɑːbəʊˈhaɪdreɪt/ *n* hidrato *m* de carbono

carbon /'kɑ:bən/ *n* carbono *m*; (*paper*) carbón *m*. ~ **copy** copia *f* al carbón

carburettor /kɑ:bjʊ'retə(r)/ *n* carburador *m*

carcass /'kɑ:kəs/ *n* cadáver *m*, esqueleto *m*

card /kɑ:d/ *n* tarjeta *f*; (*for games*) carta *f*; (*membership*) carnet *m*; (*records*) ficha *f*

cardboard /'kɑ:dbɔ:d/ *n* cartón *m*

cardiac /'kɑ:dɪæk/ *a* cardíaco

cardigan /'kɑ:dɪgən/ *n* chaqueta *f* de punto, rebeca *f*

cardinal /'kɑ:dɪnəl/ *a* cardinal. ● *n* cardenal *m*

card-index /'kɑ:dɪndeks/ *n* fichero *m*

care /keə(r)/ *n* cuidado *m*; (*worry*) preocupación *f*; (*protection*) cargo *m*. ~ **of** a cuidado de, en casa de. **take** ~ **of** cuidar de ‹*person*›; ocuparse de ‹*matter*›. ● *vi* interesarse. **I don't** ~ me es igual. ~ **about** interesarse por. ~ **for** cuidar de; (*like*) querer

career /kə'rɪə(r)/ *n* carrera *f*. ● *vi* correr a toda velocidad

carefree /'keəfri:/ *a* despreocupado

careful /'keəfʊl/ *a* cuidadoso; (*cautious*) prudente. ~**ly** *adv* con cuidado

careless /'keəlɪs/ *a* negligente; (*not worried*) indiferente. ~**ly** *adv* descuidadamente. ~**ness** *n* descuido *m*

caress /kə'res/ *n* caricia *f*. ● *vt* acariciar

caretaker /'keəteɪkə(r)/ *n* vigilante *m*; (*of flats etc*) portero *m*

car-ferry /'kɑ:ferɪ/ *n* transbordador *m* de coches

cargo /'kɑ:gəʊ/ *n* (*pl* **-oes**) carga *f*

Caribbean /kærɪ'bi:ən/ *a* caribe. ~ **Sea** *n* mar *m* Caribe

caricature /'kærɪkətʃʊə(r)/ *n* caricatura *f*. ● *vt* caricaturizar

carnage /'kɑ:nɪdʒ/ *n* carnicería *f*, matanza *f*

carnal /'kɑ:nl/ *a* carnal

carnation /kɑ:'neɪʃn/ *n* clavel *m*

carnival /'kɑ:nɪvl/ *n* carnaval *m*

carol /'kærəl/ *n* villancico *m*

carouse /kə'raʊz/ *vi* correrse una juerga

carousel /kærə'sel/ *n* tiovivo *m*

carp[1] /kɑ:p/ *n invar* carpa *f*

carp[2] /kɑ:p/ *vi*. ~ **at** quejarse de

car park /'kɑ:pɑ:k/ *n* aparcamiento *m*

carpent|er /'kɑ:pɪntə(r)/ *n* carpintero *m*. ~**ry** *n* carpintería *f*

carpet /'kɑ:pɪt/ *n* alfombra *f*. **be on the** ~ (*fam*) recibir un rapapolvo; (*under consideration*) estar sobre el tapete. ● *vt* alfombrar. ~**-sweeper** *n* escoba *f* mecánica

carriage /'kærɪdʒ/ *n* coche *m*; (*mec*) carro *m*; (*transport*) transporte *m*; (*cost, bearing*) porte *m*

carriageway /'kærɪdʒweɪ/ *n* calzada *f*, carretera *f*

carrier /'kærɪə(r)/ *n* transportista *m* & *f*; (*company*) empresa *f* de transportes; (*med*) portador *m*. ~**-bag** bolsa *f*

carrot /'kærət/ *n* zanahoria *f*

carry /'kærɪ/ *vt* llevar; transportar ‹*goods*›; (*involve*) llevar consigo, implicar. ● *vi* ‹*sounds*› llegar, oírse. ~ **off** llevarse. ~ **on** continuar; (*complain, fam*) quejarse. ~ **out** realizar; cumplir ‹*promise, threat*›. ~**-cot** *n* capazo *m*

cart /kɑ:t/ *n* carro *m*. ● *vt* acarrear; (*carry, fam*) llevar

cartilage /'kɑ:tɪlɪdʒ/ *n* cartílago *m*

carton /'kɑ:tən/ *n* caja *f* (de cartón)

cartoon /kɑ:'tu:n/ *n* caricatura *f*, chiste *m*; (*strip*) historieta *f*; (*film*) dibujos *mpl* animados. ~**ist** *n* caricaturista *m* & *f*

cartridge /'kɑ:trɪdʒ/ *n* cartucho *m*

carve /kɑ:v/ *vt* tallar; trinchar ‹*meat*›

cascade /kæs'keɪd/ *n* cascada *f*. ● *vi* caer en cascadas

case /keɪs/ *n* caso *m*; (*jurid*) proceso *m*; (*crate*) cajón *m*; (*box*) caja *f*; (*suitcase*) maleta *f*. **in any** ~ en todo caso. **in** ~ **he comes** por si viene. **in** ~ **of** en caso de. **lower** ~ caja *f* baja, minúscula *f*. **upper** ~ caja *f* alta, mayúscula *f*

cash /kæʃ/ *n* dinero *m* efectivo. **pay (in)** ~ pagar al contado. ● *vt* cobrar. ~ **in (on)** aprovecharse de. ~ **desk** *n* caja *f*

cashew /'kæʃu:/ *n* anacardo *m*

cashier /kæ'ʃɪə(r)/ *n* cajero *m*

cashmere /kæʃ'mɪə(r)/ *n* casimir *m*, cachemir *m*

casino /kə'si:nəʊ/ *n* (*pl* **-os**) casino *m*

cask /kɑ:sk/ *n* barril *m*

casket /'kɑ:skɪt/ *n* cajita *f*

casserole /'kæsərəʊl/ *n* cacerola *f*; (*stew*) cazuela *f*

cassette /kə'set/ *n* casete *m*

cast /kɑ:st/ *vt* (*pt* **cast**) arrojar; fundir ‹*metal*›; dar ‹*vote*›; (*in theatre*)

repartir. ● *n* lanzamiento *m*; (*in play*) reparto *m*; (*mould*) molde *m*
castanets /kæstəˈnets/ *npl* castañuelas *fpl*
castaway /ˈkɑːstəweɪ/ *n* náufrago *m*
caste /kɑːst/ *n* casta *f*
cast: ~ **iron** *n* hierro *m* fundido. **~-iron** *a* de hierro fundido; (*fig*) sólido
castle /ˈkɑːsl/ *n* castillo *m*; (*chess*) torre *f*
cast-offs /ˈkɑːstɒfs/ *npl* desechos *mpl*
castor /ˈkɑːstə(r)/ *n* ruedecilla *f*
castor oil /kɑːstərˈɔɪl/ *n* aceite *m* de ricino
castor sugar /ˈkɑːstəʃʊgə(r)/ *n* azúcar *m* extrafino
castrat|e /kæˈstreɪt/ *vt* castrar. **~ion** /-ʃn/ *n* castración *f*
casual /ˈkæʒʊəl/ *a* casual; ‹*meeting*› fortuito; ‹*work*› ocasional; ‹*attitude*› despreocupado; ‹*clothes*› informal, de sport. **~ly** *adv* de paso
casualt|y /ˈkæʒʊəltɪ/ *n* accidente *m*; (*injured*) víctima *f*, herido *m*; (*dead*) víctima *f*, muerto *m*. **~ies** *npl* (*mil*) bajas *fpl*
cat /kæt/ *n* gato *m*
cataclysm /ˈkætəklɪzəm/ *n* cataclismo *m*
catacomb /ˈkætəkuːm/ *n* catacumba *f*
catalogue /ˈkætəlɒg/ *n* catálogo *m*. ● *vt* catalogar
catalyst /ˈkætəlɪst/ *n* catalizador *m*
catamaran /kætəməˈræn/ *n* catamarán *m*
catapult /ˈkætəpʌlt/ *n* catapulta *f*; (*child's*) tirador *m*, tirachinos *m invar*
cataract /ˈkætərækt/ *n* catarata *f*
catarrh /kəˈtɑː(r)/ *n* catarro *m*
catastroph|e /kəˈtæstrəfɪ/ *n* catástrofe *m*. **~ic** /kætəˈstrɒfɪk/ *a* catastrófico
catch /kætʃ/ *vt* (*pt* **caught**) coger (*not LAm*), agarrar; (*grab*) asir; tomar ‹*train, bus*›; (*unawares*) sorprender; (*understand*) comprender; contraer ‹*disease*›. ~ **a cold** resfriarse. ~ **sight of** avistar. ● *vi* (*get stuck*) engancharse; ‹*fire*› prenderse. ● *n* cogida *f*; (*of fish*) pesca *f*; (*on door*) pestillo *m*; (*on window*) cerradura *f*. ~ **on** (*fam*) hacerse popular. ~ **up** poner al día. ~ **up with** alcanzar; ponerse al corriente de ‹*news etc*›
catching /ˈkætʃɪŋ/ *a* contagioso
catchment /ˈkætʃmənt/ *n*. ~ **area** *n* zona *f* de captación
catch-phrase /ˈkætʃfreɪz/ *n* eslogan *m*
catchword /ˈkætʃwɜːd/ *n* eslogan *m*, consigna *f*
catchy /ˈkætʃɪ/ *a* pegadizo
catechism /ˈkætɪkɪzəm/ *n* catecismo *m*
categorical /kætɪˈgɒrɪkl/ *a* categórico
category /ˈkætɪgərɪ/ *n* categoría *f*
cater /ˈkeɪtə(r)/ *vi* proveer comida a. ~ **for** proveer a ‹*needs*›. **~er** *n* proveedor *m*
caterpillar /ˈkætəpɪlə(r)/ *n* oruga *f*
cathedral /kəˈθiːdrəl/ *n* catedral *f*
catholic /ˈkæθəlɪk/ *a* universal. **C~** *a* & *n* católico (*m*). **C~ism** /kəˈθɒlɪsɪzəm/ *n* catolicismo *m*
catnap /ˈkætnæp/ *n* sueñecito *m*
cat's eyes /ˈkætsaɪz/ *npl* catafotos *mpl*
cattle /ˈkætl/ *npl* ganado *m* (vacuno)
cat|ty /ˈkætɪ/ *a* malicioso. **~walk** /ˈkætwɔːk/ *n* pasarela *f*
caucus /ˈkɔːkəs/ *n* comité *m* electoral
caught /kɔːt/ *see* **catch**
cauldron /ˈkɔːldrən/ *n* caldera *f*
cauliflower /ˈkɒlɪflaʊə(r)/ *n* coliflor *f*
cause /kɔːz/ *n* causa *f*, motivo *m*. ● *vt* causar
causeway /ˈkɔːzweɪ/ *n* calzada *f* elevada, carretera *f* elevada
caustic /ˈkɔːstɪk/ *a* & *n* cáustico (*m*)
cauterize /ˈkɔːtəraɪz/ *vt* cauterizar
caution /ˈkɔːʃn/ *n* cautela *f*; (*warning*) advertencia *f*. ● *vt* advertir; (*jurid*) amonestar
cautious /ˈkɔːʃəs/ *a* cauteloso, prudente. **~ly** *adv* con precaución, cautelosamente
cavalcade /kævəlˈkeɪd/ *n* cabalgata *f*
cavalier /kævəˈlɪə(r)/ *a* arrogante
cavalry /ˈkævəlrɪ/ *n* caballería *f*
cave /keɪv/ *n* cueva *f*. ● *vi*. ~ **in** hundirse. **~-man** *n* (*pl* **-men**) troglodita *m*
cavern /ˈkævən/ *n* caverna *f*, cueva *f*
caviare /ˈkævɪɑː(r)/ *n* caviar *m*
caving /ˈkeɪvɪŋ/ *n* espeleología *f*
cavity /ˈkævətɪ/ *n* cavidad *f*; (*in tooth*) caries *f*
cavort /kəˈvɔːt/ *vi* brincar
cease /siːs/ *vt/i* cesar. ● *n*. **without** ~ sin cesar. **~-fire** *n* tregua *f*, alto *m* el fuego. **~less** *a* incesante
cedar /ˈsiːdə(r)/ *n* cedro *m*

cede /si:d/ *vt* ceder
cedilla /sɪ'dɪlə/ *n* cedilla *f*
ceiling /'si:lɪŋ/ *n* techo *m*
celebrat|e /'selɪbreɪt/ *vt* celebrar. • *vi* divertirse. **~ed** /'selɪbreɪtɪd/ *a* célebre. **~ion** /-'breɪʃn/ *n* celebración *f*; (*party*) fiesta *f*
celebrity /sɪ'lebrətɪ/ *n* celebridad *f*
celery /'selərɪ/ *n* apio *m*
celestial /sɪ'lestjəl/ *a* celestial
celiba|cy /'selɪbəsɪ/ *n* celibato *m*. **~te** /'selɪbət/ *a* & *n* célibe (*m* & *f*)
cell /sel/ *n* celda *f*; (*biol*) célula *f*; (*elec*) pila *f*
cellar /'selə(r)/ *n* sótano *m*; (*for wine*) bodega *f*
cell|ist /'tʃelɪst/ *n* violonc(h)elo *m* & *f*, violonc(h)elista *m* & *f*. **~o** /'tʃeləʊ/ *n* (*pl* **-os**) violonc(h)elo *m*
Cellophane /'seləfeɪn/ *n* (*P*) celofán *m* (*P*)
cellular /'seljʊlə(r)/ *a* celular
celluloid /'seljʊlɔɪd/ *n* celuloide *m*
cellulose /'seljʊləʊs/ *n* celulosa *f*
Celt /kelt/ *n* celta *m* & *f*. **~ic** *a* céltico
cement /sɪ'ment/ *n* cemento *m*. • *vt* cementar; (*fig*) consolidar
cemetery /'semətrɪ/ *n* cementerio *m*
cenotaph /'senətɑ:f/ *n* cenotafio *m*
censor /'sensə(r)/ *n* censor *m*. • *vt* censurar. **~ship** *n* censura *f*
censure /'senʃə(r)/ *n* censura *f*. • *vt* censurar
census /'sensəs/ *n* censo *m*
cent /sent/ *n* centavo *m*
centenary /sen'ti:nərɪ/ *n* centenario *m*
centigrade /'sentɪgreɪd/ *a* centígrado
centilitre /'sentɪli:tə(r)/ *n* centilitro *m*
centimetre /'sentɪmi:tə(r)/ *n* centímetro *m*
centipede /'sentɪpi:d/ *n* ciempiés *m invar*
central /'sentrəl/ *a* central; (*of town*) céntrico. **~ heating** *n* calefacción *f* central. **~ize** *vt* centralizar. **~ly** *adv* (*situated*) en el centro
centre /'sentə(r)/ *n* centro *m*. • *vt* (*pt* **centred**) *vi* concentrarse
centrifugal /sen'trɪfjʊgəl/ *a* centrífugo
century /'sentʃərɪ/ *n* siglo *m*
ceramic /sɪ'ræmɪk/ *a* cerámico. **~s** *npl* cerámica *f*
cereal /'sɪərɪəl/ *n* cereal *m*
cerebral /'serɪbrəl/ *a* cerebral
ceremon|ial /serɪ'məʊnɪəl/ *a* & *n* ceremonial (*m*). **~ious** /-'məʊnɪəs/ *a* ceremonioso. **~y** /'serɪmənɪ/ *n* ceremonia *f*
certain /'sɜ:tn/ *a* cierto. **for ~** seguro. **make ~ of** asegurarse de. **~ly** *adv* desde luego. **~ty** *n* certeza *f*
certificate /sə'tɪfɪkət/ *n* certificado *m*; (*of birth, death etc*) partida *f*
certify /'sɜ:tɪfaɪ/ *vt* certificar
cessation /se'seɪʃən/ *n* cesación *f*
cesspit /'sespɪt/ *n*, **cesspool** /'sespu:l/ *n* pozo *m* negro; (*fam*) sentina *f*
chafe /tʃeɪf/ *vt* rozar. • *vi* rozarse; (*fig*) irritarse
chaff /tʃæf/ *vt* zumbarse de
chaffinch /'tʃæfɪntʃ/ *n* pinzón *m*
chagrin /'ʃægrɪn/ *n* disgusto *m*
chain /tʃeɪn/ *n* cadena *f*. • *vt* encadenar. **~ reaction** *n* reacción *f* en cadena. **~-smoker** *n* fumador *m* que siempre tiene un cigarillo encendido. **~ store** *n* sucursal *m*
chair /tʃeə(r)/ *n* silla *f*; (*univ*) cátedra *f*. • *vt* presidir. **~-lift** *n* telesilla *m*
chairman /'tʃeəmən/ *n* (*pl* **-men**) presidente *m*
chalet /'ʃæleɪ/ *n* chalé *m*
chalice /'tʃælɪs/ *n* cáliz *m*
chalk /tʃɔ:k/ *n* creta *f*; (*stick*) tiza *f*. **~y** *a* cretáceo
challeng|e /'tʃælɪndʒ/ *n* desafío *m*; (*fig*) reto *m*. • *vt* desafiar; (*question*) poner en duda. **~ing** *a* estimulante
chamber /'tʃeɪmbə(r)/ *n* (*old use*) cámara *f*. **~maid** /'tʃeɪmbəmeɪd/ *n* camarera *f*. **~-pot** *n* orinal *m*. **~s** *npl* despacho *m*, bufete *m*
chameleon /kə'mi:ljən/ *n* camaleón *m*
chamois /'ʃæmɪ/ *n* gamuza *f*
champagne /ʃæm'peɪn/ *n* champaña *m*, champán *m* (*fam*)
champion /'tʃæmpɪən/ *n* campeón *m*. • *vt* defender. **~ship** *n* campeonato *m*
chance /tʃɑ:ns/ *n* casualidad *f*; (*likelihood*) probabilidad *f*; (*opportunity*) oportunidad *f*; (*risk*) riesgo *m*. **by ~** por casualidad. • *a* fortuito. • *vt* arriesgar. • *vi* suceder. **~ upon** tropezar con
chancellor /'tʃɑ:nsələ(r)/ *n* canciller *m*; (*univ*) rector *m*. **C~ of the Exchequer** Ministro *m* de Hacienda
chancy /'tʃɑ:nsɪ/ *a* arriesgado; (*uncertain*) incierto

chandelier /ʃændəˈlɪə(r)/ *n* araña *f* (de luces)

change /tʃeɪndʒ/ *vt* cambiar; (*substitute*) reemplazar. **~ one's mind** cambiar de idea. ● *vi* cambiarse. ● *n* cambio *m*; (*small coins*) suelto *m*. **~ of life** menopausia *f*. **~able** *a* cambiable; ‹*weather*› variable. **~-over** *n* cambio *m*

channel /ˈtʃænl/ *n* canal *m*; (*fig*) medio *m*. **the C~ Islands** *npl* las islas *fpl* Anglonormandas. **the (English) C~** el canal de la Mancha. ● *vt* (*pt* **channelled**) acanalar; (*fig*) encauzar

chant /tʃɑːnt/ *n* canto *m*. ● *vt/i* cantar; (*fig*) salmodiar

chao|s /ˈkeɪɒs/ *n* caos *m*, desorden *m*. **~tic** /-ˈɒtɪk/ *a* caótico, desordenado

chap[1] /tʃæp/ *n* (*crack*) grieta *f*. ● *vt* (*pt* **chapped**) agrietar. ● *vi* agrietarse

chap[2] /tʃæp/ *n* (*fam*) hombre *m*, tío *m* (*fam*)

chapel /ˈtʃæpl/ *n* capilla *f*

chaperon /ˈʃæpərəʊn/ *n* acompañanta *f*. ● *vt* acompañar

chaplain /ˈtʃæplɪn/ *n* capellán *m*

chapter /ˈtʃæptə(r)/ *n* capítulo *m*

char[1] /tʃɑː(r)/ *vt* (*pt* **charred**) carbonizar

char[2] /tʃɑː(r)/ *n* asistenta *f*

character /ˈkærəktə(r)/ *n* carácter *m*; (*in play*) personaje *m*. **in ~** característico

characteristic /kærəktəˈrɪstɪk/ *a* característico. **~ally** *adv* típicamente

characterize /ˈkærəktəraɪz/ *vt* caracterizar

charade /ʃəˈrɑːd/ *n* charada *f*, farsa *f*

charcoal /ˈtʃɑːkəʊl/ *n* carbón *m* vegetal; (*for drawing*) carboncillo *m*

charge /tʃɑːdʒ/ *n* precio *m*; (*elec, mil*) carga *f*; (*jurid*) acusación *f*; (*task, custody*) encargo *m*; (*responsibility*) responsabilidad *f*. **in ~ of** responsable de, encargado de. **take ~ of** encargarse de. ● *vt* pedir; (*elec, mil*) cargar; (*jurid*) acusar; (*entrust*) encargar. ● *vi* cargar; (*money*) cobrar. **~able** *a* a cargo (de)

chariot /ˈtʃærɪət/ *n* carro *m*

charisma /kəˈrɪzmə/ *n* carisma *m*. **~tic** /-ˈmætɪk/ *a* carismático

charitable /ˈtʃærɪtəbl/ *a* caritativo

charity /ˈtʃærɪtɪ/ *n* caridad *f*; (*society*) institución *f* benéfica

charlatan /ˈʃɑːlətən/ *n* charlatán *m*

charm /tʃɑːm/ *n* encanto *m*; (*spell*) hechizo *m*; (*on bracelet*) dije *m*, amuleto *m*. ● *vt* encantar. **~ing** *a* encantador

chart /tʃɑːt/ *n* (*naut*) carta *f* de marear; (*table*) tabla *f*. ● *vt* poner en una carta de marear

charter /ˈtʃɑːtə(r)/ *n* carta *f*. ● *vt* conceder carta a, estatuir; alquilar ‹*bus, train*›; fletar ‹*plane, ship*›. **~ed accountant** *n* contador *m* titulado. **~ flight** *n* vuelo *m* charter

charwoman /ˈtʃɑːwʊmən/ *n* (*pl* **-women**) asistenta *f*

chary /ˈtʃeərɪ/ *a* cauteloso

chase /tʃeɪs/ *vt* perseguir. ● *vi* correr. ● *n* persecución *f*. **~ away**, **~ off** ahuyentar

chasm /ˈkæzəm/ *n* abismo *m*

chassis /ˈʃæsɪ/ *n* chasis *m*

chaste /tʃeɪst/ *a* casto

chastise /tʃæsˈtaɪz/ *vt* castigar

chastity /ˈtʃæstətɪ/ *n* castidad *f*

chat /tʃæt/ *n* charla *f*. **have a ~** charlar. ● *vi* (*pt* **chatted**) charlar

chattels /ˈtʃætlz/ *n* bienes *mpl* muebles

chatter /ˈtʃætə(r)/ *n* charla *f*. ● *vi* charlar. **his teeth are ~ing** le castañetean los dientes. **~box** /ˈtʃætəbɒks/ *n* parlanchín *m*

chatty *a* hablador; ‹*style*› familiar

chauffeur /ˈʃəʊfə(r)/ *n* chófer *m*

chauvinis|m /ˈʃəʊvɪnɪzəm/ *n* patriotería *f*; (*male*) machismo *m*. **~t** /ˈʃəʊvɪnɪst/ *n* patriotero *m*; (*male*) machista *m & f*

cheap /tʃiːp/ *a* (**-er**, **-est**) barato; (*poor quality*) de baja calidad; ‹*rate*› económico. **~en** /ˈtʃiːpən/ *vt* abaratar. **~(ly)** *adv* barato, a bajo precio. **~ness** *n* baratura *f*

cheat /ˈtʃiːt/ *vt* defraudar; (*deceive*) engañar. ● *vi* (*at cards*) hacer trampas. ● *n* trampa *f*; (*person*) tramposo *m*

check[1] /tʃek/ *vt* comprobar; (*examine*) inspeccionar; (*curb*) detener; (*chess*) dar jaque a. ● *vi* comprobar. ● *n* comprobación *f*; (*of tickets*) control *m*; (*curb*) freno *m*; (*chess*) jaque *m*; (*bill, Amer*) cuenta *f*. **~ in** registrarse; (*at airport*) facturar el equipaje. **~ out** pagar la cuenta y marcharse. **~ up** comprobar. **~ up on** investigar

check[2] /tʃek/ *n* (*pattern*) cuadro *m*. **~ed** *a* a cuadros

checkmate /'tʃekmeɪt/ *n* jaque *m* mate. ● *vt* dar mate a
check-up /'tʃekʌp/ *n* examen *m*
cheek /tʃi:k/ *n* mejilla *f*; (*fig*) descaro *m*. **~-bone** *n* pómulo *m*. **~y** *a* descarado
cheep /tʃi:p/ *vi* piar
cheer /tʃɪə(r)/ *n* alegría *f*; (*applause*) viva *m*. ● *vt* alegrar; (*applaud*) aplaudir. ● *vi* alegrarse; (*applaud*) aplaudir. **~ up!** ¡anímate! **~ful** *a* alegre. **~fulness** *n* alegría *f*
cheerio /tʃɪərɪ'əʊ/ *int* (*fam*) ¡adiós!, ¡hasta luego!
cheer: **~less** /'tʃɪəlɪs/ *a* triste. **~s!** ¡salud!
cheese /tʃi:z/ *n* queso *m*
cheetah /'tʃi:tə/ *n* guepardo *m*
chef /ʃef/ *n* cocinero *m*
chemical /'kemɪkl/ *a* químico. ● *n* producto *m* químico
chemist /'kemɪst/ *n* farmacéutico *m*; (*scientist*) químico *m*. **~ry** *n* química *f*. **~'s (shop)** *n* farmacia *f*
cheque /tʃek/ *n* cheque *m*, talón *m*. **~-book** *n* talonario *m*
chequered /'tʃekəd/ *a* a cuadros; (*fig*) con altibajos
cherish /'tʃerɪʃ/ *vt* cuidar; (*love*) querer; abrigar ‹*hope*›
cherry /'tʃerɪ/ *n* cereza *f*. **~-tree** *n* cerezo *m*
cherub /'tʃerəb/ *n* (*pl* **-im**) (*angel*) querubín *m*
chess /tʃes/ *n* ajedrez *m*. **~-board** *n* tablero *m* de ajedrez
chest /tʃest/ *n* pecho *m*; (*box*) cofre *m*, cajón *m*. **~ of drawers** *n* cómoda *f*
chestnut /'tʃesnʌt/ *n* castaña *f*. **~-tree** *n* castaño *m*
chew /tʃu:/ *vt* masticar; (*fig*) rumiar. **~ing-gum** *n* chicle *m*
chic /ʃi:k/ *a* elegante. ● *n* elegancia *f*
chick /tʃɪk/ *n* polluelo *m*. **~en** /'tʃɪkɪn/ *n* pollo *m*. ● *a* (*sl*) cobarde. ● *vi*. **~en out** (*sl*) retirarse. **~en-pox** *n* varicela *f*
chicory /'tʃɪkərɪ/ *n* (*in coffee*) achicoria *f*; (*in salad*) escarola *f*
chide /tʃaɪd/ *vt* (*pt* **chided**) reprender
chief /tʃi:f/ *n* jefe *m*. ● *a* principal. **~ly** *adv* principalmente
chilblain /'tʃɪlbleɪn/ *n* sabañón *m*
child /tʃaɪld/ *n* (*pl* **children** /'tʃɪldrən/) niño *m*; (*offspring*) hijo *m*. **~birth** /'tʃaɪldbɜ:θ/ *n* parto *m*. **~hood** *n* niñez *f*. **~ish** *a* infantil. **~less** *a* sin hijos. **~like** *a* inocente, infantil
Chile /'tʃɪlɪ/ *n* Chile *m*. **~an** *a* & *n* chileno (*m*)
chill /tʃɪl/ *n* frío *m*; (*illness*) resfriado *m*. ● *a* frío. ● *vt* enfriar; refrigerar ‹*food*›
chilli /'tʃɪlɪ/ *n* (*pl* **-ies**) chile *m*
chilly /'tʃɪlɪ/ *a* frío
chime /tʃaɪm/ *n* carillón *m*. ● *vt* tocar ‹*bells*›; dar ‹*hours*›. ● *vi* repicar
chimney /'tʃɪmnɪ/ *n* (*pl* **-eys**) chimenea *f*. **~-pot** *n* cañón *m* de chimenea. **~-sweep** *n* deshollinador *m*
chimpanzee /tʃɪmpæn'zi:/ *n* chimpancé *m*
chin /tʃɪn/ *n* barbilla *f*
china /'tʃaɪnə/ *n* porcelana *f*
Chin|a /'tʃaɪnə/ *n* China *f*. **~ese** /-'ni:z/ *a* & *n* chino (*m*)
chink[1] /tʃɪŋk/ *n* (*crack*) grieta *f*
chink[2] /tʃɪŋk/ *n* (*sound*) tintín *m*. ● *vt* hacer tintinear. ● *vi* tintinear
chip /tʃɪp/ *n* pedacito *m*; (*splinter*) astilla *f*; (*culin*) patata *f* frita; (*gambling*) ficha *f*. **have a ~ on one's shoulder** guardar rencor. ● *vt* (*pt* **chipped**) desportillar. ● *vi* desportillarse. **~ in** (*fam*) interrumpir; (*with money*) contribuir
chiropodist /kɪ'rɒpədɪst/ *n* callista *m* & *f*
chirp /tʃɜ:p/ *n* pío *m*. ● *vi* piar
chirpy /'tʃɜ:pɪ/ *a* alegre
chisel /'tʃɪzl/ *n* formón *m*. ● *vt* (*pt* **chiselled**) cincelar
chit /tʃɪt/ *n* vale *m*, nota *f*
chit-chat /'tʃɪttʃæt/ *n* cháchara *f*
chivalr|ous *a* /'ʃɪvəlrəs/ *a* caballeroso. **~y** /'ʃɪvəlrɪ/ *n* caballerosidad *f*
chive /tʃaɪv/ *n* cebollino *m*
chlorine /'klɔ:ri:n/ *n* cloro *m*
chock /tʃɒk/ *n* calzo *m*. **~-a-block** *a*, **~-full** *a* atestado
chocolate /'tʃɒklɪt/ *n* chocolate *m*; (*individual sweet*) bombón *m*
choice /tʃɔɪs/ *n* elección *f*; (*preference*) preferencia *f*. ● *a* escogido
choir /'kwaɪə(r)/ *n* coro *m*. **~boy** /'kwaɪəbɔɪ/ *n* niño *m* de coro
choke /tʃəʊk/ *vt* sofocar. ● *vi* sofocarse. ● *n* (*auto*) estrangulador *m*, estárter *m*
cholera /'kɒlərə/ *n* cólera *m*
cholesterol /kə'lestərɒl/ *n* colesterol *m*

choose /tʃu:z/ *vt/i* (*pt* **chose,** *pp* **chosen**) elegir. **~y** /'tʃu:zɪ/ *a* (*fam*) exigente

chop /tʃɒp/ *vt* (*pt* **chopped**) cortar. ● *n* (*culin*) chuleta *f*. **~ down** talar. **~ off** cortar. **~per** *n* hacha *f*; (*butcher's*) cuchilla *f*; (*sl*) helicóptero *m*

choppy /'tʃɒpɪ/ *a* picado

chopstick /'tʃɒpstɪk/ *n* palillo *m* (chino)

choral /'kɔ:rəl/ *a* coral

chord /kɔ:d/ *n* cuerda *f*; (*mus*) acorde *m*

chore /tʃɔ:(r)/ *n* tarea *f*, faena *f*. **household ~s** *npl* faenas *fpl* domésticas

choreographer /kɒrɪ'ɒgrəfə(r)/ *n* coreógrafo *m*

chorister /'kɒrɪstə(r)/ *n* (*singer*) corista *m & f*

chortle /'tʃɔ:tl/ *n* risita *f* alegre. ● *vi* reírse alegremente

chorus /'kɔ:rəs/ *n* coro *m*; (*of song*) estribillo *m*

chose, chosen /tʃəʊz, 'tʃəʊzn/ *see* **choose**

Christ /kraɪst/ *n* Cristo *m*

christen /'krɪsn/ *vt* bautizar. **~ing** *n* bautizo *m*

Christian /'krɪstjən/ *a & n* cristiano (*m*). **~ name** *n* nombre *m* de pila

Christmas /'krɪsməs/ *n* Navidad *f*; (*period*) Navidades *fpl*. ● *a* de Navidad, navideño. **~-box** *n* aguinaldo *m*. **~ day** *n* día *m* de Navidad. **~ Eve** *n* Nochebuena *f*. **Father ~** *n* Papá *m* Noel. **Happy ~!** ¡Felices Pascuas!

chrom|e /krəʊm/ *n* cromo *m*. **~ium** /'krəʊmɪəm/ *n* cromo *m*. **~ium plating** *n* cromado *m*

chromosome /'krəʊməsəʊm/ *n* cromosoma *m*

chronic /'krɒnɪk/ *a* crónico; (*bad, fam*) terrible

chronicle /'krɒnɪkl/ *n* crónica *f*. ● *vt* historiar

chronolog|ical /krɒnə'lɒdʒɪkl/ *a* cronológico. **~y** /krə'nɒlədʒɪ/ *n* cronología *f*

chrysanthemum /krɪ'sænθəməm/ *n* crisantemo *m*

chubby /'tʃʌbɪ/ *a* (**-ier, -iest**) regordete; ‹*face*› mofletudo

chuck /tʃʌk/ *vt* (*fam*) arrojar. **~ out** tirar

chuckle /'tʃʌkl/ *n* risa *f* ahogada. ● *vi* reírse entre dientes

chuffed /tʃʌft/ *a* (*sl*) contento

chug /tʃʌg/ *vi* (*pt* **chugged**) (*of motor*) traquetear

chum /tʃʌm/ *n* amigo *m*, compinche *m*. **~my** *a*. **be ~my** ‹*2 people*› ser muy amigos. **be ~my with** ser muy amigo de

chump /tʃʌmp/ *n* (*sl*) tonto *m*. **~ chop** *n* chuleta *f*

chunk /tʃʌŋk/ *n* trozo *m* grueso. **~y** /tʃʌŋkɪ/ *a* macizo

church /'tʃɜ:tʃ/ *n* iglesia *f*. **~yard** /'tʃɜ:tʃjɑ:d/ *n* cementerio *m*

churlish /'tʃɜ:lɪʃ/ *a* grosero

churn /'tʃɜ:n/ *n* (*for milk*) lechera *f*, cántara *f*; (*for butter*) mantequera *f*. ● *vt* agitar. **~ out** producir en profusión

chute /ʃu:t/ *n* tobogán *m*

chutney /'tʃʌtnɪ/ *n* (*pl* **-eys**) condimento *m* agridulce

cider /'saɪdə(r)/ *n* sidra *f*

cigar /sɪ'gɑ:(r)/ *n* puro *m*

cigarette /sɪgə'ret/ *n* cigarillo *m*. **~-holder** *n* boquilla *f*

cine-camera /'sɪnɪkæmərə/ *n* cámara *f*, tomavistas *m invar*

cinema /'sɪnəmə/ *n* cine *m*

cinnamon /'sɪnəmən/ *n* canela *f*

cipher /'saɪfə(r)/ *n* (*math, fig*) cero *m*; (*secret system*) cifra *f*

circle /'sɜ:kl/ *n* círculo *m*; (*in theatre*) anfiteatro *m*. ● *vt* girar alrededor de. ● *vi* dar vueltas

circuit /'sɜ:kɪt/ *n* circuito *m*; (*chain*) cadena *f*

circuitous /sɜ:'kju:ɪtəs/ *a* indirecto

circular /'sɜ:kjʊlə(r)/ *a & n* circular (*f*)

circularize /'sɜ:kjʊləraɪz/ *vt* enviar circulares a

circulat|e /'sɜ:kjʊleɪt/ *vt* hacer circular. ● *vi* circular. **~ion** /-'leɪʃn/ *n* circulación *f*; (*of journals*) tirada *f*

circumcis|e /'sɜ:kəmsaɪz/ *vt* circuncidar. **~ion** /-'sɪʒn/ *n* circuncisión *f*

circumference /sə'kʌmfərəns/ *n* circunferencia *f*

circumflex /'sɜ:kəmfleks/ *a & n* circunflejo (*m*)

circumspect /'sɜ:kəmspekt/ *a* circunspecto

circumstance /'sɜ:kəmstəns/ *n* circunstancia *f*. **~s** (*means*) *npl* situación *f* económica

circus /'sɜ:kəs/ *n* circo *m*

cistern /'sɪstən/ *n* depósito *m*; (*of WC*) cisterna *f*

citadel /'sɪtədl/ *n* ciudadela *f*

citation /saɪ'teɪʃn/ *n* citación *f*
cite /saɪt/ *vt* citar
citizen /'sɪtɪzn/ *n* ciudadano *m*; (*inhabitant*) habitante *m* & *f*. **~ship** *n* ciudadanía *f*
citrus /'sɪtrəs/ *n*. **~ fruits** cítricos *mpl*
city /'sɪtɪ/ *n* ciudad *f*; **the C~** el centro *m* financiero de Londres
civic /'sɪvɪk/ *a* cívico. **~s** *npl* cívica *f*
civil /'sɪvl/ *a* civil, cortés
civilian /sɪ'vɪlɪən/ *a* & *n* civil (*m* & *f*). **~ clothes** *npl* traje *m* de paisano
civility /sɪ'vɪlətɪ/ *n* cortesía *f*
civiliz|ation /sɪvɪlaɪ'zeɪʃn/ *n* civilización *f*. **~e** /'sɪvəlaɪz/ *vt* civilizar.
civil: **~ servant** *n* funcionario *m*. **~ service** *n* administración *f* pública
civvies /'sɪvɪz/ *npl*. **in ~** (*sl*) en traje *m* de paisano
clad /klæd/ *see* **clothe**
claim /kleɪm/ *vt* reclamar; (*assert*) pretender. • *n* reclamación *f*; (*right*) derecho *m*; (*jurid*) demanda *f*. **~ant** *n* demandante *m* & *f*; (*to throne*) pretendiente *m*
clairvoyant /kleə'vɔɪənt/ *n* clarividente *m* & *f*
clam /klæm/ *n* almeja *f*
clamber /'klæmbə(r)/ *vi* trepar a gatas
clammy /'klæmɪ/ *a* (**-ier**, **-iest**) húmedo
clamour /'klæmə(r)/ *n* clamor *m*. • *vi*. **~ for** pedir a voces
clamp /klæmp/ *n* abrazadera *f*; (*auto*) cepo *m*. • *vt* sujetar con abrazadera. **~ down on** reprimir
clan /klæn/ *n* clan *m*
clandestine /klæn'destɪn/ *a* clandestino
clang /klæŋ/ *n* sonido *m* metálico
clanger /'klæŋə(r)/ *n* (*sl*) metedura *f* de pata
clap /klæp/ *vt* (*pt* **clapped**) aplaudir; batir ‹*hands*›. • *vi* aplaudir. • *n* palmada *f*; (*of thunder*) trueno *m*
claptrap /'klæptræp/ *n* charlatanería *f*, tonterías *fpl*
claret /'klærət/ *n* clarete *m*
clarif|ication /klærɪfɪ'keɪʃn/ *n* aclaración *f*. **~y** /'klærɪfaɪ/ *vt* aclarar. • *vi* aclararse
clarinet /klærɪ'net/ *n* clarinete *m*
clarity /'klærətɪ/ *n* claridad *f*
clash /klæʃ/ *n* choque *m*; (*noise*) estruendo *m*; (*contrast*) contraste *m*; (*fig*) conflicto *m*. • *vt* golpear. • *vi* encontrarse; ‹*dates*› coincidir; ‹*opinions*› estar en desacuerdo; ‹*colours*› desentonar
clasp /klɑ:sp/ *n* cierre *m*. • *vt* agarrar; apretar ‹*hand*›; (*fasten*) abrochar
class /klɑ:s/ *n* clase *f*. **evening ~** *n* clase nocturna. • *vt* clasificar
classic /'klæsɪk/ *a* & *n* clásico (*m*). **~al** *a* clásico. **~s** *npl* estudios *mpl* clásicos
classif|ication /klæsɪfɪ'keɪʃn/ *n* clasificación *f*. **~y** /'klæsɪfaɪ/ *vt* clasificar
classroom /'klɑ:sru:m/ *n* aula *f*
classy /'klɑ:sɪ/ *a* (*sl*) elegante
clatter /'klætə(r)/ *n* estrépito *m*. • *vi* hacer ruido
clause /klɔ:z/ *n* cláusula *f*; (*gram*) oración *f*
claustrophobia /klɔ:strə'fəʊbɪə/ *n* claustrofobia *f*
claw /klɔ:/ *n* garra *f*; (*of cat*) uña *f*; (*of crab*) pinza *f*; (*device*) garfio *m*. • *vt* arañar
clay /kleɪ/ *n* arcilla *f*
clean /kli:n/ *a* (**-er**, **-est**) limpio; ‹*stroke*› neto. • *adv* completamente. • *vt* limpiar. • *vi* hacer la limpieza. **~ up** hacer la limpieza. **~-cut** *a* bien definido. **~er** *n* mujer *f* de la limpieza. **~liness** /'klenlɪnɪs/ *n* limpieza *f*
cleans|e /klenz/ *vt* limpiar; (*fig*) purificar. **~ing cream** *n* crema *f* desmaquilladora
clear /klɪə(r)/ *a* (**-er**, **-est**) claro; (*transparent*) transparente; (*without obstacles*) libre; ‹*profit*› neto; ‹*sky*› despejado. **keep ~ of** evitar. • *adv* claramente. • *vt* despejar; liquidar ‹*goods*›; (*jurid*) absolver; (*jump over*) saltar por encima de; quitar ‹*table*›. • *vi* ‹*weather*› despejarse; ‹*fog*› disolverse. **~ off** *vi* (*sl*), **~ out** *vi* (*sl*) largarse. **~ up** *vt* (*tidy*) poner en orden; aclarar ‹*mystery*›; • *vi* ‹*weather*› despejarse
clearance /'klɪərəns/ *n* espacio *m* libre; (*removal of obstructions*) despeje *m*; (*authorization*) permiso *m*; (*by customs*) despacho *m*; (*by security*) acreditación *f*. **~ sale** *n* liquidación *f*
clearing /'klɪərɪŋ/ *n* claro *m*
clearly /'klɪəlɪ/ *adv* evidentemente
clearway /'klɪəweɪ/ *n* carretera *f* en la que no se permite parar
cleavage /'kli:vɪdʒ/ *n* escote *m*; (*fig*) división *f*
cleave /kli:v/ *vt* (*pt* **cleaved, clove** *or* **cleft**; *pp* **cloven** *or* **cleft**) hender. • *vi* henderse

clef /klef/ *n* (*mus*) clave *f*
cleft /kleft/ *see* **cleave**
clemen|cy /'klemənsɪ/ *n* clemencia *f*. **~t** *a* clemente
clench /klentʃ/ *vt* apretar
clergy /'klɜ:dʒɪ/ *n* clero *m*. **~man** *n* (*pl* **-men**) clérigo *m*
cleric /'klerɪk/ *n* clérigo *m*. **~al** *a* clerical; (*of clerks*) de oficina
clerk /klɑ:k/ *n* empleado *m*; (*jurid*) escribano *m*
clever /'klevə(r)/ *a* (**-er**, **-est**) listo; (*skilful*) hábil. **~ly** *adv* inteligentemente; (*with skill*) hábilmente. **~ness** *n* inteligencia *f*
cliché /'kli:ʃeɪ/ *n* tópico *m*, frase *f* hecha
click /klɪk/ *n* golpecito *m*. ● *vi* chascar; (*sl*) llevarse bien
client /'klaɪənt/ *n* cliente *m* & *f*
clientele /kli:ən'tel/ *n* clientela *f*
cliff /klɪf/ *n* acantilado *m*
climat|e /'klaɪmɪt/ *n* clima *m*. **~ic** /-'mætɪk/ *a* climático
climax /'klaɪmæks/ *n* punto *m* culminante
climb /klaɪm/ *vt* subir ‹*stairs*›; trepar ‹*tree*›; escalar ‹*mountain*›. ● *vi* subir. ● *n* subida *f*. **~ down** bajar; (*fig*) volverse atrás, rajarse. **~er** *n* (*sport*) alpinista *m* & *f*; (*plant*) trepadora *f*
clinch /klɪntʃ/ *vt* cerrar ‹*deal*›
cling /klɪŋ/ *vi* (*pt* **clung**) agarrarse; (*stick*) pegarse
clinic /'klɪnɪk/ *n* clínica *f*. **~al** /'klɪnɪkl/ *a* clínico
clink /klɪŋk/ *n* sonido *m* metálico. ● *vt* hacer tintinear. ● *vi* tintinear
clinker /'klɪŋkə(r)/ *n* escoria *f*
clip[1] /klɪp/ *n* (*for paper*) sujetapapeles *m invar*; (*for hair*) horquilla *f*. ● *vt* (*pt* **clipped**) (*join*) sujetar
clip[2] /klɪp/ *n* (*with scissors*) tijeretada *f*; (*blow, fam*) golpe *m*. ● *vt* (*pt* **clipped**) (*cut*) cortar; (*fam*) golpear. **~pers** /'klɪpəz/ *npl* (*for hair*) maquinilla *f* para cortar el pelo; (*for nails*) cortauñas *m invar*. **~ping** *n* recorte *m*
clique /kli:k/ *n* pandilla *f*
cloak /kləʊk/ *n* capa *f*. **~room** /'kləʊkru:m/ *n* guardarropa *m*; (*toilet*) servicios *mpl*
clobber /'klɒbə(r)/ *n* (*sl*) trastos *mpl*. ● *vt* (*sl*) dar una paliza a
clock /klɒk/ *n* reloj *m*. **grandfather ~** reloj de caja. ● *vi*. **~ in** fichar, registrar la llegada. **~wise** /'klɒkwaɪz/ *a* & *adv* en el sentido de las agujas del reloj, a la derecha. **~work** /'klɒkwɜ:k/ *n* mecanismo *m* de relojería. **like ~work** con precisión
clod /klɒd/ *n* terrón *m*
clog /klɒg/ *n* zueco *m*. ● *vt* (*pt* **clogged**) atascar. ● *vi* atascarse
cloister /'klɔɪstə(r)/ *n* claustro *m*
close[1] /kləʊs/ *a* (**-er**, **-est**) cercano; (*together*) apretado; ‹*friend*› íntimo; ‹*weather*› bochornoso; ‹*link etc*› estrecho; ‹*game, battle*› reñido. **have a ~ shave** (*fig*) escaparse de milagro. ● *adv* cerca. ● *n* recinto *m*
close[2] /kləʊz/ *vt* cerrar. ● *vi* cerrarse; (*end*) terminar. ● *n* fin *m*. **~d shop** *n* empresa *f* que emplea solamente a miembros del sindicato
close: **~ly** *adv* de cerca; (*with attention*) atentamente; (*exactly*) exactamente. **~ness** *n* proximidad *f*; (*togetherness*) intimidad *f*
closet /'klɒzɪt/ *n* (*Amer*) armario *m*
close-up /'kləʊsʌp/ *n* (*cinema etc*) primer plano *m*
closure /'kləʊʒə(r)/ *n* cierre *m*
clot /klɒt/ *n* (*culin*) grumo *m*; (*med*) coágulo *m*; (*sl*) tonto *m*. ● *vi* (*pt* **clotted**) cuajarse
cloth /klɒθ/ *n* tela *f*; (*duster*) trapo *m*; (*table-cloth*) mantel *m*
cloth|e /kləʊð/ *vt* (*pt* **clothed** *or* **clad**) vestir. **~es** /kləʊðz/ *npl*, **~ing** *n* ropa *f*
cloud /klaʊd/ *n* nube *f*. ● *vi* nublarse. **~burst** /'klaʊdbɜ:st/ *n* chaparrón *m*. **~y** *a* (**-ier**, **-iest**) nublado; ‹*liquid*› turbio
clout /klaʊt/ *n* bofetada *f*. ● *vt* abofetear
clove[1] /kləʊv/ *n* clavo *m*
clove[2] /kləʊv/ *n*. **~ of garlic** *n* diente *m* de ajo
clove[3] /kləʊv/ *see* **cleave**
clover /'kləʊvə(r)/ *n* trébol *m*
clown /klaʊn/ *n* payaso *m*. ● *vi* hacer el payaso
cloy /klɔɪ/ *vt* empalagar
club /klʌb/ *n* club *m*; (*weapon*) porra *f*; (*at cards*) trébol *m*. ● *vt* (*pt* **clubbed**) aporrear. ● *vi*. **~ together** reunirse, pagar a escote
cluck /klʌk/ *vi* cloquear
clue /klu:/ *n* pista *f*; (*in crosswords*) indicación *f*. **not to have a ~** no tener la menor idea

clump /klʌmp/ *n* grupo *m*. ● *vt* agrupar. ● *vi* pisar fuertemente
clums|iness /'klʌmzɪnɪs/ *n* torpeza *f*. **~y** /'klʌmzɪ/ *a* (**-ier**, **-iest**) torpe
clung /klʌŋ/ *see* **cling**
cluster /'klʌstə(r)/ *n* grupo *m*. ● *vi* agruparse
clutch /klʌtʃ/ *vt* agarrar. ● *n* (*auto*) embrague *m*
clutter /'klʌtə(r)/ *n* desorden *m*. ● *vt* llenar desordenadamente
coach /kəʊtʃ/ *n* autocar *m*; (*of train*) vagón *m*; (*horse-drawn*) coche *m*; (*sport*) entrenador *m*. ● *vt* dar clases particulares; (*sport*) entrenar
coagulate /kəʊ'ægjʊleɪt/ *vt* coagular. ● *vi* coagularse
coal /kəʊl/ *n* carbón *m*. **~field** /'kəʊlfi:ld/ *n* yacimiento *m* de carbón
coalition /kəʊə'lɪʃn/ *n* coalición *f*
coarse /kɔ:s/ *a* (**-er**, **-est**) grosero; ‹*material*› basto. **~ness** *n* grosería *f*; (*texture*) basteza *f*
coast /kəʊst/ *n* costa *f*. ● *vi* (*with cycle*) deslizarse cuesta abajo; (*with car*) ir en punto muerto. **~al** *a* costero. **~er** /'kəʊstə(r)/ *n* (*ship*) barco *m* de cabotaje; (*for glass*) posavasos *m invar*. **~guard** /'kəʊstgɑ:d/ *n* guardacostas *m invar*. **~line** /'kəʊstlaɪn/ *n* litoral *m*
coat /kəʊt/ *n* abrigo *m*; (*jacket*) chaqueta *f*; (*of animal*) pelo *m*; (*of paint*) mano *f*. ● *vt* cubrir, revestir. **~ing** *n* capa *f*. **~ of arms** *n* escudo *m* de armas
coax /kəʊks/ *vt* engatusar
cob /kɒb/ *n* (*of corn*) mazorca *f*
cobble[1] /'kɒbl/ *n* guijarro *m*, adoquín *m*. ● *vt* empedrar con guijarros, adoquinar
cobble[2] /'kɒbl/ *vt* (*mend*) remendar. **~r** /'kɒblə(r)/ *n* (*old use*) remendón *m*
cobweb /'kɒbweb/ *n* telaraña *f*
cocaine /kə'keɪn/ *n* cocaína *f*
cock /kɒk/ *n* gallo *m*; (*mec*) grifo m; (*of gun*) martillo *m*. ● *vt* amartillar ‹*gun*›; aguzar ‹*ears*›. **~-and-bull story** *n* patraña *f*. **~erel** /'kɒkərəl/ *n* gallo *m*. **~-eyed** *a* (*sl*) torcido
cockle /'kɒkl/ *n* berberecho *m*
cockney /'kɒknɪ/ *a* & *n* (*pl* **-eys**) londinense (*m* & *f*) (del este de Londres)
cockpit /'kɒkpɪt/ *n* (*in aircraft*) cabina *f* del piloto
cockroach /'kɒkrəʊtʃ/ *n* cucaracha *f*
cocksure /kɒk'ʃʊə(r)/ *a* presuntuoso
cocktail /'kɒkteɪl/ *n* cóctel *m*. **fruit ~** macedonia *f* de frutas
cock-up /'kɒkʌp/ *n* (*sl*) lío *m*
cocky /'kɒkɪ/ *a* (**-ier**, **-iest**) engreído
cocoa /'kəʊkəʊ/ *n* cacao *m*; (*drink*) chocolate *m*
coconut /'kəʊkənʌt/ *n* coco *m*
cocoon /kə'ku:n/ *n* capullo *m*
cod /kɒd/ *n* (*pl* **cod**) bacalao *m*, abadejo *m*
coddle /'kɒdl/ *vt* mimar; (*culin*) cocer a fuego lento
code /kəʊd/ *n* código *m*; (*secret*) cifra *f*
codify /'kəʊdɪfaɪ/ *vt* codificar
cod-liver oil /'kɒdlɪvə(r)ɒɪl/ *n* aceite *m* de hígado de bacalao
coeducational /kəʊedʒʊ'keɪʃənl/ *a* mixto
coerc|e /kəʊ'ɜ:s/ *vt* obligar. **~ion** /-ʃn/ *n* coacción *f*
coexist /kəʊɪg'zɪst/ *vi* coexistir. **~ence** *n* coexistencia *f*
coffee /'kɒfɪ/ *n* café *m*. **~-mill** *n* molinillo *m* de café. **~-pot** *n* cafetera *f*
coffer /'kɒfə(r)/ *n* cofre *m*
coffin /'kɒfɪn/ *n* ataúd *m*
cog /kɒg/ *n* diente *m*; (*fig*) pieza *f*
cogent /'kəʊdʒənt/ *a* convincente
cohabit /kəʊ'hæbɪt/ *vi* cohabitar
coherent /kəʊ'hɪərənt/ *a* coherente
coil /kɔɪl/ *vt* enrollar. ● *n* rollo *m*; (*one ring*) vuelta *f*
coin /kɔɪn/ *n* moneda *f*. ● *vt* acuñar. **~age** *n* sistema *m* monetario
coincide /kəʊɪn'saɪd/ *vi* coincidir
coinciden|ce /kəʊ'ɪnsɪdəns/ *n* casualidad *f*. **~tal** /-'dentl/ *a* casual; (*coinciding*) coincidente
coke /kəʊk/ *n* (*coal*) coque *m*
colander /'kʌləndə(r)/ *n* colador *m*
cold /kəʊld/ *a* (**-er**, **-est**) frío. **be ~** tener frío. **it is ~** hace frío. ● *n* frío *m*; (*med*) resfriado *m*. **have a ~** estar constipado. **~-blooded** *a* insensible. **~ cream** *n* crema *f*. **~ feet** (*fig*) mieditis *f*. **~ness** *n* frialdad *f*. **~-shoulder** *vt* tratar con frialdad. **~ sore** *n* herpes *m* labial. **~ storage** *n* conservación *f* en frigorífico
coleslaw /'kəʊlslɔ:/ *n* ensalada *f* de col
colic /'kɒlɪk/ *n* cólico *m*
collaborat|e /kə'læbəreɪt/ *vi* colaborar. **~ion** /-'reɪʃn/ *n* colaboración *f*. **~or** *n* colaborador *m*

collage /'kɒlɑ:ʒ/ *n* collage *m*
collaps|e /kə'læps/ *vi* derrumbarse; (*med*) sufrir un colapso. ● *n* derrumbamiento *m*; (*med*) colapso *m*. **~ible** /kə'læpsəbl/ *a* plegable
collar /'kɒlə(r)/ *n* cuello *m*; (*for animals*) collar *m*. ● *vt* (*fam*) hurtar. **~-bone** *n* clavícula *f*
colleague /'kɒli:g/ *n* colega *m & f*
collect /kə'lekt/ *vt* reunir; (*hobby*) coleccionar; (*pick up*) recoger; recaudar ‹*rent*›. ● *vi* ‹*people*› reunirse; ‹*things*› acumularse. **~ed** /kə'lektɪd/ *a* reunido; ‹*person*› tranquilo. **~ion** /-ʃn/ *n* colección *f*; (*in church*) colecta *f*; (*of post*) recogida *f*. **~ive** /kə'lektɪv/ *a* colectivo. **~or** *n* coleccionista *m & f*; (*of taxes*) recaudador *m*
college /'kɒlɪdʒ/ *n* colegio *m*; (*of art, music etc*) escuela *f*; (*univ*) colegio *m* mayor
collide /kə'laɪd/ *vi* chocar
colliery /'kɒlɪərɪ/ *n* mina *f* de carbón
collision /kə'lɪʒn/ *n* choque *m*
colloquial /kə'ləʊkwɪəl/ *a* familiar. **~ism** *n* expresión *f* familiar
collusion /kə'lu:ʒn/ *n* connivencia *f*
colon /'kəʊlən/ *n* (*gram*) dos puntos *mpl*; (*med*) colon *m*
colonel /'kɜ:nl/ *n* coronel *m*
colon|ial /kə'ləʊnɪəl/ *a* colonial. **~ize** /'kɒlənaɪz/ *vt* colonizar. **~y** /'kɒlənɪ/ *n* colonia *f*
colossal /kə'lɒsl/ *a* colosal
colour /'kʌlə(r)/ *n* color *m*. **off ~** (*fig*) indispuesto. ● *a* de color(es), en color(es). ● *vt* colorar; (*dye*) teñir. ● *vi* (*blush*) sonrojarse. **~ bar** *n* barrera *f* racial. **~-blind** *a* daltoniano. **~ed** /'kʌləd/ *a* de color. **~ful** *a* lleno de color; (*fig*) pintoresco. **~less** *a* incoloro. **~s** *npl* (*flag*) bandera *f*
colt /kəʊlt/ *n* potro *m*
column /'kɒləm/ *n* columna *f*. **~ist** /'kɒləmnɪst/ *n* columnista *m & f*
coma /'kəʊmə/ *n* coma *m*
comb /kəʊm/ *n* peine *m*. ● *vt* peinar; (*search*) registrar
combat /'kɒmbæt/ *n* combate *m*. ● *vt* (*pt* **combated**) combatir. **~ant** /-ətənt/ *n* combatiente *m & f*
combination /kɒmbɪ'neɪʃn/ *n* combinación *f*
combine /kəm'baɪn/ *vt* combinar. ● *vi* combinarse. /'kɒmbaɪn/ *n* asociación *f*. **~harvester** *n* cosechadora *f*
combustion /kəm'bʌstʃən/ *n* combustión *f*
come /kʌm/ *vi* (*pt* **came**, *pp* **come**) venir; (*occur*) pasar. **~ about** ocurrir. **~ across** encontrarse con ‹*person*›; encontrar ‹*object*›. **~ apart** deshacerse. **~ away** marcharse. **~ back** volver. **~ by** obtener; (*pass*) pasar. **~ down** bajar. **~ in** entrar. **~ in for** recibir. **~ into** heredar ‹*money*›. **~ off** desprenderse; (*succeed*) tener éxito. **~ off it!** (*fam*) ¡no me vengas con eso! **~ out** salir; (*result*) resultar. **~ round** (*after fainting*) volver en sí; (*be converted*) cambiar de idea. **~ to** llegar a ‹*decision etc*›. **~ up** subir; (*fig*) salir. **~ up with** proponer ‹*idea*›
comeback /'kʌmbæk/ *n* retorno *m*; (*retort*) réplica *f*
comedian /kə'mi:dɪən/ *n* cómico *m*
comedown /'kʌmdaʊn/ *n* revés *m*
comedy /'kɒmədɪ/ *n* comedia *f*
comely /'kʌmlɪ/ *a* (**-ier**, **-iest**) (*old use*) bonito
comet /'kɒmɪt/ *n* cometa *m*
comeuppance /kʌm'ʌpəns/ *n* (*Amer*) merecido *m*
comf|ort /'kʌmfət/ *n* bienestar *m*; (*consolation*) consuelo *m*. ● *vt* consolar. **~ortable** *a* cómodo; (*wealthy*) holgado. **~y** /'kʌmfɪ/ *a* (*fam*) cómodo
comic /'kɒmɪk/ *a* cómico. ● *n* cómico *m*; (*periodical*) tebeo *m*. **~al** *a* cómico. **~ strip** *n* historieta *f*
coming /'kʌmɪŋ/ *n* llegada *f*. ● *a* próximo; ‹*week, month etc*› que viene. **~ and going** ir y venir
comma /'kɒmə/ *n* coma *f*
command /kə'mɑ:nd/ *n* orden *f*; (*mastery*) dominio *m*. ● *vt* mandar; (*deserve*) merecer
commandeer /kɒmən'dɪə(r)/ *vt* requisar
commander /kə'mɑ:ndə(r)/ *n* comandante *m*
commanding /kə'mɑ:ndɪŋ/ *a* imponente
commandment /kə'mɑ:ndmənt/ *n* mandamiento *m*
commando /kə'mɑ:ndəʊ/ *n* (*pl* **-os**) comando *m*
commemorat|e /kə'meməreɪt/ *vt* conmemorar. **~ion** /-'reɪʃn/ *n* conmemoración *f*. **~ive** /-ətɪv/ *a* conmemorativo
commence /kə'mens/ *vt/i* empezar. **~ment** *n* principio *m*

commend /kə'mend/ *vt* alabar; (*entrust*) encomendar. **~able** *a* loable. **~ation** /kɒmen'deɪʃn/ *n* elogio *m*

commensurate /kə'menʃərət/ *a* proporcionado

comment /'kɒment/ *n* observación *f*. ● *vi* hacer observaciones

commentary /'kɒməntrɪ/ *n* comentario *m*; (*radio, TV*) reportaje *m*

commentat|e /'kɒmənteɪt/ *vi* narrar. **~or** *n* (*radio, TV*) locutor *m*

commerc|e /'kɒmɜ:s/ *n* comercio *m*. **~ial** /kə'mɜ:ʃl/ *a* comercial. ● *n* anuncio *m*. **~ialize** *vt* comercializar

commiserat|e /kə'mɪzəreɪt/ *vt* compadecer. ● *vi* compadecerse (**with** de). **~ion** /-'reɪʃn/ *n* conmiseración *f*

commission /kə'mɪʃn/ *n* comisión *f*. **out of ~** fuera de servicio. ● *vt* encargar; (*mil*) nombrar

commissionaire /kəmɪʃə'neə(r)/ *n* portero *m*

commissioner /kə'mɪʃənə(r)/ *n* comisario *m*; (*of police*) jefe *m*

commit /kə'mɪt/ *vt* (*pt* **committed**) cometer; (*entrust*) confiar. **~ o.s.** comprometerse. **~ to memory** aprender de memoria. **~ment** *n* compromiso *m*

committee /kə'mɪtɪ/ *n* comité *m*

commodity /kə'mɒdətɪ/ *n* producto *m*, artículo *m*

common /'kɒmən/ *a* (**-er, -est**) común; (*usual*) corriente; (*vulgar*) ordinario. ● *n* ejido *m*

commoner /'kɒmənə(r)/ *n* plebeyo *m*

common: **~ law** *n* derecho *m* consuetudinario. **~ly** *adv* comúnmente. **C~ Market** *n* Mercado *m* Común

commonplace /'kɒmənpleɪs/ *a* banal. ● *n* banalidad *f*

common: **~-room** *n* sala *f* común, salón *m* común. **~ sense** *n* sentido *m* común

Commonwealth /'kɒmənwelθ/ *n*. **the ~** la Mancomunidad *f* Británica

commotion /kə'məʊʃn/ *n* confusión *f*

communal /'kɒmjʊnl/ *a* comunal

commune[1] /'kɒmju:n/ *n* comuna *f*

commune[2] /kə'mju:n/ *vi* comunicarse

communicat|e /kə'mju:nɪkeɪt/ *vt* comunicar. ● *vi* comunicarse. **~ion** /-'keɪʃn/ *n* comunicación *f*. **~ive** /-ətɪv/ *a* comunicativo

communion /kə'mju:nɪən/ *n* comunión *f*

communiqué /kə'mju:nɪkeɪ/ *n* comunicado *m*

communis|m /'kɒmjʊnɪsəm/ *n* comunismo *m*. **~t** /'kɒmjʊnɪst/ *n* comunista *m & f*

community /kə'mju:nətɪ/ *n* comunidad *f*. **~ centre** *n* centro *m* social

commute /kə'mju:t/ *vi* viajar diariamente. ● *vt* (*jurid*) conmutar. **~r** /-ə(r)/ *n* viajero *m* diario

compact /kəm'pækt/ *a* compacto. /'kɒmpækt/ *n* (*for powder*) polvera *f*. **~ disc** /'kɒm-/ *n* disco *m* compacto

companion /kəm'pænɪən/ *n* compañero *m*. **~ship** *n* compañerismo *m*

company /'kʌmpənɪ/ *n* compañía *f*; (*guests, fam*) visita *f*; (*com*) sociedad *f*

compar|able /'kɒmpərəbl/ *a* comparable. **~ative** /kəm'pærətɪv/ *a* comparativo; (*fig*) relativo. ● *n* (*gram*) comparativo *m*. **~e** /kəm'peə(r)/ *vt* comparar. ● *vi* poderse comparar. **~ison** /kəm'pærɪsn/ *n* comparación *f*

compartment /kəm'pɑ:tmənt/ *n* compartimiento *m*; (*on train*) departamento *m*

compass /'kʌmpəs/ *n* brújula *f*. **~es** *npl* compás *m*

compassion /kəm'pæʃn/ *n* compasión *f*. **~ate** *a* compasivo

compatib|ility /kəmpætə'bɪlətɪ/ *n* compatibilidad *f*. **~le** /kəm'pætəbl/ *a* compatible

compatriot /kəm'pætrɪət/ *n* compatriota *m & f*

compel /kəm'pel/ *vt* (*pt* **compelled**) obligar. **~ling** *a* irresistible

compendium /kəm'pendɪəm/ *n* compendio *m*

compensat|e /'kɒmpənseɪt/ *vt* compensar; (*for loss*) indemnizar. ● *vi* compensar. **~ion** /-'seɪʃn/ *n* compensación *f*; (*financial*) indemnización *f*

compère /'kɒmpeə(r)/ *n* presentador *m*. ● *vt* presentar

compete /kəm'pi:t/ *vi* competir

competen|ce /'kɒmpətəns/ *n* competencia *f*, aptitud *f*. **~t** /'kɒmpɪtənt/ *a* competente, capaz

competit|ion /kɒmpə'tɪʃn/ *n* (*contest*) concurso *m*; (*com*) competencia *f*. **~ive** /kəm'petətɪv/ *a* competidor; ‹*price*› competitivo. **~or** /kəm'petɪtə(r)/ *n* competidor *m*; (*in contest*) concursante *m & f*

compile /kəm'paɪl/ *vt* compilar. **~r** /-ə(r)/ *n* recopilador *m*, compilador *m*

complacen|cy /kəm'pleɪsənsɪ/ *n* satisfacción *f* de sí mismo. **~t** /kəm'pleɪsnt/ *a* satisfecho de sí mismo

complain /kəm'pleɪn/ *vi*. **~ (about)** quejarse (de). **~ of** (*med*) sufrir de. **~t** /kəm'pleɪnt/ *n* queja *f*; (*med*) enfermedad *f*

complement /'kɒmplɪmənt/ *n* complemento *m*. ● *vt* complementar. **~ary** /-'mentrɪ/ *a* complementario

complet|e /kəm'pli:t/ *a* completo; (*finished*) acabado; (*downright*) total. ● *vt* acabar; llenar ‹*a form*›. **~ely** *adv* completamente. **~ion** /-ʃn/ *n* conclusión *f*

complex /'kɒmpleks/ *a* complejo. ● *n* complejo *m*

complexion /kəm'plekʃn/ *n* tez *f*; (*fig*) aspecto *m*

complexity /kəm'pleksətɪ/ *n* complejidad *f*

complian|ce /kəm'plaɪəns/ *n* sumisión *f*. **in ~ce with** de acuerdo con. **~t** *a* sumiso

complicat|e /'kɒmplɪkeɪt/ *vt* complicar. **~ed** *a* complicado. **~ion** /-'keɪʃn/ *n* complicación *f*

complicity /kəm'plɪsətɪ/ *n* complicidad *f*

compliment /'kɒmplɪmənt/ *n* cumplido *m*; (*amorous*) piropo *m*. ● *vt* felicitar. **~ary** /-'mentrɪ/ *a* halagador; (*given free*) de favor. **~s** *npl* saludos *mpl*

comply /kəm'plaɪ/ *vi*. **~ with** conformarse con

component /kəm'pəʊnənt/ *a & n* componente (*m*)

compose /kəm'pəʊz/ *vt* componer. **~ o.s.** tranquilizarse. **~d** *a* sereno

compos|er /kəm'pəʊzə(r)/ *n* compositor *m*. **~ition** /kɒmpə'zɪʃn/ *n* composición *f*

compost /'kɒmpɒst/ *n* abono *m*

composure /kəm'pəʊʒə(r)/ *n* serenidad *f*

compound[1] /'kɒmpaʊnd/ *n* compuesto *m*. ● *a* compuesto; ‹*fracture*› complicado. /kəm'paʊnd/ *vt* componer; agravar ‹*problem etc*›. ● *vi* (*settle*) arreglarse

compound[2] /'kɒmpaʊnd/ *n* (*enclosure*) recinto *m*

comprehen|d /kɒmprɪ'hend/ *vt* comprender. **~sion** /kɒmprɪ'henʃn/ *n* comprensión *f*

comprehensive /kɒmprɪ'hensɪv/ *a* extenso; ‹*insurance*› a todo riesgo. **~ school** *n* instituto *m*

compress /'kɒmpres/ *n* (*med*) compresa *f*. /kəm'pres/ *vt* comprimir; (*fig*) condensar. **~ion** /-ʃn/ *n* compresión *f*

comprise /kəm'praɪz/ *vt* comprender

compromise /'kɒmprəmaɪz/ *n* acuerdo *m*, acomodo *m*, arreglo *m*. ● *vt* comprometer. ● *vi* llegar a un acuerdo

compuls|ion /kəm'pʌlʃn/ *n* obligación *f*, impulso *m*. **~ive** /kəm'pʌlsɪv/ *a* compulsivo. **~ory** /kəm'pʌlsərɪ/ *a* obligatorio

compunction /kəm'pʌŋkʃn/ *n* remordimiento *m*

computer /kəm'pju:tə(r)/ *n* ordenador *m*. **~ize** *vt* instalar ordenadores en. **be ~ized** tener ordenador

comrade /'kɒmreɪd/ *n* camarada *m & f*. **~ship** *n* camaradería *f*

con[1] /kɒn/ *vt* (*pt* **conned**) (*fam*) estafar. ● *n* (*fam*) estafa *f*

con[2] /kɒn/ *see* **pro and con**

concave /'kɒŋkeɪv/ *a* cóncavo

conceal /kən'si:l/ *vt* ocultar. **~ment** *n* encubrimiento *m*

concede /kən'si:d/ *vt* conceder

conceit /kən'si:t/ *n* vanidad *f*. **~ed** *a* engreído

conceiv|able /kən'si:vəbl/ *a* concebible. **~ably** *adv*. **may ~ably** es concebible que. **~e** /kən'si:v/ *vt/i* concebir

concentrat|e /'kɒnsəntreɪt/ *vt* concentrar. ● *vi* concentrarse. **~ion** /-'treɪʃn/ *n* concentración *f*. **~ion camp** *n* campo *m* de concentración

concept /'kɒnsept/ *n* concepto *m*

conception /kən'sepʃn/ *n* concepción *f*

conceptual /kən'septʃʊəl/ *a* conceptual

concern /kən'sɜ:n/ *n* asunto *m*; (*worry*) preocupación *f*; (*com*) empresa *f*. ● *vt* tener que ver con; (*deal with*) tratar de. **as far as I'm ~ed** en cuanto a mí. **be ~ed about** preocuparse por. **~ing** *prep* acerca de

concert /'kɒnsət/ *n* concierto *m*. **in ~** de común acuerdo. **~ed** /kən'sɜ:tɪd/ *a* concertado
concertina /kɒnsə'ti:nə/ *n* concertina *f*
concerto /kən'tʃɜ:təʊ/ *n* (*pl* **-os**) concierto *m*
concession /kən'seʃn/ *n* concesión *f*
conciliat|e /kən'sɪlɪeɪt/ *vt* conciliar. **~ion** /-'eɪʃn/ *n* conciliación *f*
concise /kən'saɪs/ *a* conciso. **~ly** *adv* concisamente. **~ness** *n* concisión *f*
conclu|de /kən'klu:d/ *vt* concluir. ● *vi* concluirse. **~ding** *a* final. **~sion** *n* conclusión *f*
conclusive /kən'klu:sɪv/ *a* decisivo. **~ly** *adv* concluyentemente
concoct /kən'kɒkt/ *vt* confeccionar; (*fig*) inventar. **~ion** /-ʃn/ *n* mezcla *f*; (*drink*) brebaje *m*
concourse /'kɒŋkɔ:s/ *n* (*rail*) vestíbulo *m*
concrete /'kɒŋkri:t/ *n* hormigón *m*. ● *a* concreto. ● *vt* cubrir con hormigón
concur /kən'kɜ:(r)/ *vi* (*pt* **concurred**) estar de acuerdo
concussion /kən'kʌʃn/ *n* conmoción *f* cerebral
condemn /kən'dem/ *vt* condenar. **~ation** /kɒndem'neɪʃn/ *n* condenación *f*, condena *f*; (*censure*) censura *f*
condens|ation /kɒnden'seɪʃn/ *n* condensación *f*. **~e** /kən'dens/ *vt* condensar. ● *vi* condensarse
condescend /kɒndɪ'send/ *vi* dignarse (**to** a). **~ing** *a* superior
condiment /'kɒndɪmənt/ *n* condimento *m*
condition /kən'dɪʃn/ *n* condición *f*. **on ~ that** a condición de que. ● *vt* condicionar. **~al** *a* condicional. **~er** *n* acondicionador *m*; (*for hair*) suavizante *m*
condolences /kən'dəʊlənsɪz/ *npl* pésame *m*
condom /'kɒndɒm/ *n* condón *m*
condone /kən'dəʊn/ *vt* condonar
conducive /kən'dju:sɪv/ *a*. **be ~ to** ser favorable a
conduct /kən'dʌkt/ *vt* conducir; dirigir ‹*orchestra*›. /'kɒndʌkt/ *n* conducta *f*. **~or** /kən'dʌktə(r)/ *n* director *m*; (*of bus*) cobrador *m*. **~ress** *n* cobradora *f*
cone /kəʊn/ *n* cono *m*; (*for ice-cream*) cucurucho *m*
confectioner /kən'fekʃənə(r)/ *n* pastelero *m*. **~y** *n* dulces *mpl*, golosinas *fpl*
confederation /kənfedə'reɪʃn/ *n* confederación *f*
confer /kən'fɜ:(r)/ *vt* (*pt* **conferred**) conferir. ● *vi* consultar
conference /'kɒnfərəns/ *n* congreso *m*
confess /kən'fes/ *vt* confesar. ● *vi* confesarse. **~ion** /-ʃn/ *n* confesión *f*. **~ional** *n* confes(i)onario *m*. **~or** *n* confesor *m*
confetti /kən'fetɪ/ *n* confeti *m*, confetis *mpl*
confide /kən'faɪd/ *vt/i* confiar
confiden|ce /'kɒnfɪdəns/ *n* confianza *f*; (*secret*) confidencia *f*. **~ce trick** *n* estafa *f*, timo *m*. **~t** /'kɒnfɪdənt/ *a* seguro
confidential /kɒnfɪ'denʃl/ *a* confidencial
confine /kən'faɪn/ *vt* confinar; (*limit*) limitar. **~ment** *n* (*imprisonment*) prisión *f*; (*med*) parto *m*
confines /'kɒnfaɪnz/ *npl* confines *mpl*
confirm /kən'fɜ:m/ *vt* confirmar. **~ation** /kɒnfə'meɪʃn/ *n* confirmación *f*. **~ed** *a* inveterado
confiscat|e /'kɒnfɪskeɪt/ *vt* confiscar. **~ion** /-'keɪʃn/ *n* confiscación *f*
conflagration /kɒnflə'greɪʃn/ *n* conflagración *f*
conflict /'kɒnflɪkt/ *n* conflicto *m*. /kən'flɪkt/ *vi* chocar. **~ing** /kən-/ *a* contradictorio
conform /kən'fɔ:m/ *vt* conformar. ● *vi* conformarse. **~ist** *n* conformista *m & f*
confound /kən'faʊnd/ *vt* confundir. **~ed** *a* (*fam*) maldito
confront /kən'frʌnt/ *vt* hacer frente a; (*face*) enfrentarse con. **~ation** /kɒnfrʌn'teɪʃn/ *n* confrontación *f*
confus|e /kən'fju:z/ *vt* confundir. **~ing** *a* desconcertante. **~ion** /-ʒn/ *n* confusión *f*
congeal /kən'dʒi:l/ *vt* coagular. ● *vi* coagularse
congenial /kən'dʒi:nɪəl/ *a* simpático
congenital /kən'dʒenɪtl/ *a* congénito
congest|ed /kən'dʒestɪd/ *a* congestionado. **~ion** /-tʃən/ *n* congestión *f*
congratulat|e /kən'grætjʊleɪt/ *vt* felicitar. **~ions** /-'leɪʃnz/ *npl* felicitaciones *fpl*

congregat|e /'kɒŋgrɪgeɪt/ *vi* congregarse. **~ion** /-'geɪʃn/ *n* asamblea *f*; (*relig*) fieles *mpl*, feligreses *mpl*
congress /'kɒŋgres/ *n* congreso *m*. **C~** (*Amer*) el Congreso
conic(al) /'kɒnɪk(l)/ *a* cónico
conifer /'kɒnɪfə(r)/ *n* conífera *f*
conjecture /kən'dʒektʃə(r)/ *n* conjetura *f*. ● *vt* conjeturar. ● *vi* hacer conjeturas
conjugal /'kɒndʒʊgl/ *a* conyugal
conjugat|e /'kɒndʒʊgeɪt/ *vt* conjugar. **~ion** /-'geɪʃn/ *n* conjugación *f*
conjunction /kən'dʒʌŋkʃn/ *n* conjunción *f*
conjur|e /'kʌndʒə(r)/ *vi* hacer juegos de manos. ● *vt*. **~e up** evocar. **~or** *n* prestidigitador *m*
conk /kɒŋk/ *vi*. **~ out** (*sl*) fallar; ‹*person*› desmayarse
conker /'kɒŋkə(r)/ *n* (*fam*) castaña *f* de Indias
conman /'kɒnmæn/ *n* (*fam*) estafador *m*, timador *m*
connect /kə'nekt/ *vt* juntar; (*elec*) conectar. ● *vi* unirse; (*elec*) conectarse. **~ with** ‹*train*› enlazar con. **~ed** *a* unido; (*related*) relacionado. **be ~ed with** tener que ver con, estar emparentado con
connection /kə'nekʃn/ *n* unión *f*; (*rail*) enlace *m*; (*elec*, *mec*) conexión *f*; (*fig*) relación *f*. **in ~ with** a propósito de, con respecto a. **~s** *npl* relaciones *fpl*
conniv|ance /kə'naɪvəns/ *n* connivencia *f*. **~e** /kə'naɪv/ *vi*. **~e at** hacer la vista gorda a
connoisseur /kɒnə'sɜ:(r)/ *n* experto *m*
connot|ation /kɒnə'teɪʃn/ *n* connotación *f*. **~e** /kə'nəʊt/ *vt* connotar; (*imply*) implicar
conquer /'kɒŋkə(r)/ *vt* conquistar; (*fig*) vencer. **~or** *n* conquistador *m*
conquest /'kɒŋkwest/ *n* conquista *f*
conscience /'kɒnʃəns/ *n* conciencia *f*
conscientious /kɒnʃɪ'enʃəs/ *a* concienzudo
conscious /'kɒnʃəs/ *a* consciente; (*deliberate*) intencional. **~ly** *adv* a sabiendas. **~ness** *n* consciencia *f*; (*med*) conocimiento *m*
conscript /'kɒnskrɪpt/ *n* recluta *m*. /kən'skrɪpt/ *vt* reclutar. **~ion** /kən'skrɪpʃn/ *n* reclutamiento *m*
consecrat|e /'kɒnsɪkreɪt/ *vt* consagrar. **~ion** /-'kreɪʃn/ *n* consagración *f*
consecutive /kən'sekjʊtɪv/ *a* sucesivo
consensus /kən'sensəs/ *n* consenso *m*
consent /kən'sent/ *vi* consentir. ● *n* consentimiento *m*
consequen|ce /'kɒnsɪkwəns/ *n* consecuencia *f*. **~t** /'kɒnsɪkwənt/ *a* consiguiente. **~tly** *adv* por consiguiente
conservation /kɒnsə'veɪʃn/ *n* conservación *f*, preservación *f*. **~ist** /kɒnsə'veɪʃənɪst/ *n* conservacionista *m & f*
conservative /kən'sɜ:vətɪv/ *a* conservador; (*modest*) prudente, moderado. **C~** *a & n* conservador (*m*)
conservatory /kən'sɜ:vətrɪ/ *n* (*greenhouse*) invernadero *m*
conserve /kən'sɜ:v/ *vt* conservar
consider /kən'sɪdə(r)/ *vt* considerar; (*take into account*) tomar en cuenta. **~able** /kən'sɪdərəbl/ *a* considerable. **~ably** *adv* considerablemente
considerat|e /kən'sɪdərət/ *a* considerado. **~ion** /-'reɪʃn/ *n* consideración *f*
considering /kən'sɪdərɪŋ/ *prep* en vista de
consign /kən'saɪn/ *vt* consignar; (*send*) enviar. **~ment** *n* envío *m*
consist /kən'sɪst/ *vi*. **~ of** consistir en
consistency /kən'sɪstənsɪ/ *n* consistencia *f*; (*fig*) coherencia *f*
consistent /kən'sɪstənt/ *a* coherente; (*unchanging*) constante. **~ with** compatible con. **~ly** *adv* constantemente
consolation /kɒnsə'leɪʃn/ *n* consuelo *m*
console /kən'səʊl/ *vt* consolar
consolidat|e /kən'sɒlɪdeɪt/ *vt* consolidar. ● *vi* consolidarse. **~ion** /-'deɪʃn/ *n* consolidación *f*
consonant /'kɒnsənənt/ *n* consonante *f*
consort /'kɒnsɔ:t/ *n* consorte *m & f*. /kən'sɔ:t/ *vi*. **~ with** asociarse con
consortium /kən'sɔ:tɪəm/ *n* (*pl* **-tia**) consorcio *m*
conspicuous /kən'spɪkjʊəs/ *a* (*easily seen*) visible; (*showy*) llamativo; (*noteworthy*) notable
conspir|acy /kən'spɪrəsɪ/ *n* complot *m*, conspiración *f*. **~e** /kən'spaɪə(r)/ *vi* conspirar

constab|le /'kʌnstəbl/ *n* policía *m*, guardia *m*. **~ulary** /kən'stæbjʊlərɪ/ *n* policía *f*

constant /'kɒnstənt/ *a* constante. **~ly** *adv* constantemente

constellation /kɒnstə'leɪʃn/ *n* constelación *f*

consternation /kɒnstə'neɪʃn/ *n* consternación *f*

constipat|ed /'kɒnstɪpeɪtɪd/ *a* estreñido. **~ion** /-'peɪʃn/ *n* estreñimiento *m*

constituen|cy /kən'stɪtjʊənsɪ/ *n* distrito *m* electoral. **~t** /kən'stɪtjʊənt/ *n* componente *m*; (*pol*) elector *m*

constitut|e /'kɒnstɪtju:t/ *vt* constituir. **~ion** /-'tju:ʃn/ *n* constitución *f*. **~ional** /-'tju:ʃənl/ *a* constitucional. ● *n* paseo *m*

constrain /kən'streɪn/ *vt* forzar, obligar, constreñir. **~t** /kən'streɪnt/ *n* fuerza *f*

constrict /kən'strɪkt/ *vt* apretar. **~ion** /-ʃn/ *n* constricción *f*

construct /kən'strʌkt/ *vt* construir. **~ion** /-ʃn/ *n* construcción *f*. **~ive** /kən'strʌktɪv/ *a* constructivo

construe /kən'stru:/ *vt* interpretar; (*gram*) construir

consul /'kɒnsl/ *n* cónsul *m*. **~ar** /-jʊlə(r)/ *a* consular. **~ate** /-ət/ *n* consulado *m*

consult /kən'sʌlt/ *vt/i* consultar. **~ant** /kən'sʌltənt/ *n* asesor *m*; (*med*) especialista *m* & *f*; (*tec*) consejero *m* técnico. **~ation** /kɒnsəl'teɪʃn/ *n* consulta *f*

consume /kən'sju:m/ *vt* consumir; (*eat*) comer; (*drink*) beber. **~r** /-ə(r)/ *n* consumidor *m*. ● *a* de consumo. **~rism** /kən'sju:mərɪzəm/ *n* protección *f* del consumidor, consumismo *m*

consummat|e /'kɒnsəmeɪt/ *vt* consumar. **~ion** /-'meɪʃn/ *n* consumación *f*

consumption /kən'sʌmpʃn/ *n* consumo *m*; (*med*) tisis *f*

contact /'kɒntækt/ *n* contacto *m*. ● *vt* ponerse en contacto con

contagious /kən'teɪdʒəs/ *a* contagioso

contain /kən'teɪn/ *vt* contener. **~ o.s.** contenerse. **~er** *n* recipiente *m*; (*com*) contenedor *m*

contaminat|e /kən'tæmɪneɪt/ *vt* contaminar. **~ion** /-'neɪʃn/ *n* contaminación *f*

contemplat|e /'kɒntəmpleɪt/ *vt* contemplar; (*consider*) considerar. **~ion** /-'pleɪʃn/ *n* contemplación *f*

contemporary /kən'tempərərɪ/ *a* & *n* contemporáneo (*m*)

contempt /kən'tempt/ *n* desprecio *m*. **~ible** *a* despreciable. **~uous** /-tjʊəs/ *a* desdeñoso

contend /kən'tend/ *vt* sostener. ● *vi* contender. **~er** *n* contendiente *m* & *f*

content[1] /kən'tent/ *a* satisfecho. ● *vt* contentar

content[2] /'kɒntent/ *n* contenido *m*

contented /kən'tentɪd/ *a* satisfecho

contention /kən'tenʃn/ *n* contienda *f*; (*opinion*) opinión *f*, argumento *m*

contentment /kən'tentmənt/ *n* contento *m*

contest /'kɒntest/ *n* (*competition*) concurso *m*; (*fight*) contienda *f*. /kən'test/ *vt* disputar. **~ant** *n* contendiente *m* & *f*, concursante *m* & *f*

context /'kɒntekst/ *n* contexto *m*

continent /'kɒntɪnənt/ *n* continente *m*. **the C~** Europa *f*. **~al** /-'nentl/ *a* continental

contingency /kən'tɪndʒənsɪ/ *n* contingencia *f*

contingent /kən'tɪndʒənt/ *a* & *n* contingente (*m*)

continu|al /kən'tɪnjʊəl/ *a* continuo. **~ance** /kən'tɪnjʊəns/ *n* continuación *f*. **~ation** /-ʊ'eɪʃn/ *n* continuación *f*. **~e** /kən'tɪnju:/ *vt/i* continuar; (*resume*) seguir. **~ed** *a* continuo. **~ity** /kɒntɪ'nju:ətɪ/ *n* continuidad *f*. **~ity girl** (*cinema*, *TV*) secretaria *f* de rodaje. **~ous** /kən'tɪnjʊəs/ *a* continuo. **~ously** *adv* continuamente

contort /kən'tɔ:t/ *vt* retorcer. **~ion** /-ʃn/ *n* contorsión *f*. **~ionist** /-ʃənɪst/ *n* contorsionista *m* & *f*

contour /'kɒntʊə(r)/ *n* contorno *m*. **~ line** *n* curva *f* de nivel

contraband /'kɒntrəbænd/ *n* contrabando *m*

contracepti|on /kɒntrə'sepʃn/ *n* contracepción *f*. **~ve** /kɒntrə'septɪv/ *a* & *n* anticonceptivo (*m*)

contract /'kɒntrækt/ *n* contrato *m*. /kən'trækt/ *vt* contraer. ● *vi* contraerse. **~ion** /kən'trækʃn/ *n* contracción *f*. **~or** /kən'træktə(r)/ *n* contratista *m* & *f*

contradict /kɒntrə'dɪkt/ *vt* contradecir. **~ion** /-ʃn/ *n* contradicción *f*. **~ory** *a* contradictorio

contraption /kən'træpʃn/ *n* (*fam*) artilugio *m*
contrary /'kɒntrərɪ/ *a & n* contrario (*m*). **on the ~** al contrario. ● *adv*. **~ to** contrariamente a. /kən'treərɪ/ *a* terco
contrast /'kɒntrɑ:st/ *n* contraste *m*. /kən'trɑ:st/ *vt* poner en contraste. ● *vi* contrastar. **~ing** *a* contrastante
contraven|e /kɒntrə'vi:n/ *vt* contravenir. **~tion** /-'venʃn/ *n* contravención *f*
contribut|e /kən'trɪbju:t/ *vt/i* contribuir. **~e to** escribir para ‹*newspaper*›. **~ion** /kɒntrɪ'bju:ʃn/ *n* contribución *f*; (*from salary*) cotización *f*. **~or** *n* contribuyente *m & f*; (*to newspaper*) colaborador *m*
contrite /'kɒntraɪt/ *a* arrepentido, pesaroso
contriv|ance /kən'traɪvəns/ *n* invención *f*. **~e** /kən'traɪv/ *vt* idear. **~e to** conseguir
control /kən'trəʊl/ *vt* (*pt* **controlled**) controlar. ● *n* control *m*. **~s** *npl* (*mec*) mandos *mpl*
controvers|ial /kɒntrə'vɜ:ʃl/ *a* polémico, discutible. **~y** /'kɒntrəvɜ:sɪ/ *n* controversia *f*
conundrum /kə'nʌndrəm/ *n* adivinanza *f*; (*problem*) enigma *m*
conurbation /kɒnɜ:'beɪʃn/ *n* conurbación *f*
convalesce /kɒnvə'les/ *vi* convalecer. **~nce** *n* convalecencia *f*. **~nt** *a & n* convaleciente (*m & f*). **~nt home** *n* casa *f* de convalecencia
convector /kən'vektə(r)/ *n* estufa *f* de convección
convene /kən'vi:n/ *vt* convocar. ● *vi* reunirse
convenien|ce /kən'vi:nɪəns/ *n* conveniencia *f*, comodidad *f*. **all modern ~ces** todas las comodidades. **at your ~ce** según le convenga. **~ces** *npl* servicios *mpl*. **~t** /kən'vi:nɪənt/ *a* cómodo; ‹*place*› bien situado; ‹*time*› oportuno. **be ~t** convenir. **~tly** *adv* convenientemente
convent /'kɒnvənt/ *n* convento *m*
convention /kən'venʃn/ *n* convención *f*; (*meeting*) congreso *m*. **~al** *a* convencional
converge /kən'vɜ:dʒ/ *vi* convergir
conversant /kən'vɜ:sənt/ *a*. **~ with** versado en
conversation /kɒnvə'seɪʃn/ *n* conversación *f*. **~al** *a* de la conversación. **~alist** *n* hábil conversador *m*
converse[1] /kən'vɜ:s/ *vi* conversar
converse[2] /'kɒnvɜ:s/ *a* inverso. ● *n* lo contrario. **~ly** *adv* a la inversa
conver|sion /kən'vɜ:ʃn/ *n* conversión *f*. **~t** /kən'vɜ:t/ *vt* convertir. /'kɒnvɜ:t/ *n* converso *m*. **~tible** /kən'vɜ:tɪbl/ *a* convertible. ● *n* (*auto*) descapotable *m*
convex /'kɒnveks/ *a* convexo
convey /kən'veɪ/ *vt* llevar; transportar ‹*goods*›; comunicar ‹*idea, feeling*›. **~ance** *n* transporte *m*. **~or belt** *n* cinta *f* transportadora
convict /kən'vɪkt/ *vt* condenar. /'kɒnvɪkt/ *n* presidiario *m*. **~ion** /kən'vikʃn/ *n* condena *f*; (*belief*) creencia *f*
convinc|e /kən'vɪns/ *vt* convencer. **~ing** *a* convincente
convivial /kən'vɪvɪəl/ *a* alegre
convoke /kən'vəʊk/ *vt* convocar
convoluted /'kɒnvəlu:tɪd/ *a* enrollado; ‹*argument*› complicado
convoy /'kɒnvɔɪ/ *n* convoy *m*
convuls|e /kən'vʌls/ *vt* convulsionar. **be ~ed with laughter** desternillarse de risa. **~ion** /-ʃn/ *n* convulsión *f*
coo /ku:/ *vi* arrullar
cook /kʊk/ *vt* cocinar; (*alter, fam*) falsificar. **~ up** (*fam*) inventar. ● *n* cocinero *m*
cooker /'kʊkə(r)/ *n* cocina *f*
cookery /'kʊkərɪ/ *n* cocina *f*
cookie /'kʊkɪ/ *n* (*Amer*) galleta *f*
cool /ku:l/ *a* (**-er**, **-est**) fresco; (*calm*) tranquilo; (*unfriendly*) frío. ● *n* fresco *m*; (*sl*) calma *f*. ● *vt* enfriar. ● *vi* enfriarse. **~ down** ‹*person*› calmarse. **~ly** *adv* tranquilamente. **~ness** *n* frescura *f*
coop /ku:p/ *n* gallinero *m*. ● *vt*. **~ up** encerrar
co-operat|e /kəʊ'ɒpəreɪt/ *vi* cooperar. **~ion** /-'reɪʃn/ *n* cooperación *f*
cooperative /kəʊ'ɒpərətɪv/ *a* cooperativo. ● *n* cooperativa *f*
co-opt /kəʊ'ɒpt/ *vt* cooptar
co-ordinat|e /kəʊ'ɔ:dɪneɪt/ *vt* coordinar. **~ion** /-'neɪʃn/ *n* coordinación *f*
cop /kɒp/ *vt* (*pt* **copped**) (*sl*) prender. ● *n* (*sl*) policía *m*

cope /kəʊp/ *vi* (*fam*) arreglárselas. **~ with** enfrentarse con
copious /ˈkəʊpɪəs/ *a* abundante
copper[1] /ˈkɒpə(r)/ *n* cobre *m*; (*coin*) perra *f*. ● *a* de cobre
copper[2] /ˈkɒpə(r)/ *n* (*sl*) policía *m*
coppice /ˈkɒpɪs/ *n*, **copse** /kɒps/ *n* bosquecillo *m*
Coptic /ˈkɒptɪk/ *a* copto
copulat|e /ˈkɒpjʊleɪt/ *vi* copular. **~ion** /-ˈleɪʃn/ *n* cópula *f*
copy /ˈkɒpɪ/ *n* copia *f*; (*typ*) material *m*. ● *vt* copiar
copyright /ˈkɒpɪraɪt/ *n* derechos *mpl* de autor
copy-writer /ˈkɒpɪraɪtə(r)/ *n* redactor *m* de textos publicitarios
coral /ˈkɒrəl/ *n* coral *m*
cord /kɔ:d/ *n* cuerda *f*; (*fabric*) pana *f*. **~s** *npl* pantalones *mpl* de pana
cordial /ˈkɔ:dɪəl/ *a* & *n* cordial (*m*)
cordon /ˈkɔ:dn/ *n* cordón *m*. ● *vt*. **~ off** acordonar
corduroy /ˈkɔ:dərɔɪ/ *n* pana *f*
core /kɔ:(r)/ *n* (*of apple*) corazón *m*; (*fig*) meollo *m*
cork /kɔ:k/ *n* corcho *m*. ● *vt* taponar. **~screw** /ˈkɔ:kskru:/ *n* sacacorchos *m invar*
corn[1] /kɔ:n/ *n* (*wheat*) trigo *m*; (*Amer*) maíz *m*; (*seed*) grano *m*
corn[2] /kɔ:n/ *n* (*hard skin*) callo *m*
corned /kɔ:nd/ *a*. **~ beef** *n* carne *f* de vaca en lata
corner /ˈkɔ:nə(r)/ *n* ángulo *m*; (*inside*) rincón *m*; (*outside*) esquina *f*; (*football*) saque *m* de esquina. ● *vt* arrinconar; (*com*) acaparar. **~-stone** *n* piedra *f* angular
cornet /ˈkɔ:nɪt/ *n* (*mus*) corneta *f*; (*for ice-cream*) cucurucho *m*
cornflakes /ˈkɔ:nfleɪks/ *npl* copos *mpl* de maíz
cornflour /ˈkɔ:nflaʊə(r)/ *n* harina *f* de maíz
cornice /ˈkɔ:nɪs/ *n* cornisa *f*
cornucopia /kɔ:njʊˈkəʊpɪə/ *n* cuerno *m* de la abundancia
Corn|ish /ˈkɔ:nɪʃ/ *a* de Cornualles. **~wall** /ˈkɔ:nwəl/ *n* Cornualles *m*
corny /ˈkɔ:nɪ/ *a* (*trite, fam*) gastado; (*mawkish*) sentimental, sensiblero
corollary /kəˈrɒlərɪ/ *n* corolario *m*
coronary /ˈkɒrənərɪ/ *n* trombosis *f* coronaria
coronation /kɒrəˈneɪʃn/ *n* coronación *f*
coroner /ˈkɒrənə(r)/ *n* juez *m* de primera instancia
corporal[1] /ˈkɔ:pərəl/ *n* cabo *m*
corporal[2] /ˈkɔ:pərəl/ *a* corporal
corporate /ˈkɔ:pərət/ *a* corporativo
corporation /kɔ:pəˈreɪʃn/ *n* corporación *f*; (*of town*) ayuntamiento *m*
corps /kɔ:(r)/ *n* (*pl* **corps** /kɔ:z/) cuerpo *m*
corpse /kɔ:ps/ *n* cadáver *m*
corpulent /ˈkɔ:pjʊlənt/ *a* gordo, corpulento
corpuscle /ˈkɔ:pʌsl/ *n* glóbulo *m*
corral /kəˈrɑ:l/ *n* (*Amer*) corral *m*
correct /kəˈrekt/ *a* correcto; (*time*) exacto. ● *vt* corregir. **~ion** /-ʃn/ *n* corrección *f*
correlat|e /ˈkɒrəleɪt/ *vt* poner en correlación. **~ion** /-ˈleɪʃn/ *n* correlación *f*
correspond /kɒrɪˈspɒnd/ *vi* corresponder; (*write*) escribirse. **~ence** *n* correspondencia *f*. **~ent** *n* corresponsal *m* & *f*
corridor /ˈkɒrɪdɔ:(r)/ *n* pasillo *m*
corroborate /kəˈrɒbəreɪt/ *vt* corroborar
corro|de /kəˈrəʊd/ *vt* corroer. ● *vi* corroerse. **~sion** *n* corrosión *f*
corrugated /ˈkɒrəgeɪtɪd/ *a* ondulado. **~ iron** *n* hierro *m* ondulado
corrupt /kəˈrʌpt/ *a* corrompido. ● *vt* corromper. **~ion** /-ʃn/ *n* corrupción *f*
corset /ˈkɔ:sɪt/ *n* corsé *m*
Corsica /ˈkɔ:sɪkə/ *n* Córcega *f*. **~n** *a* & *n* corso (*m*)
cortège /ˈkɔ:teɪʒ/ *n* cortejo *m*
cos /kɒs/ *n* lechuga *f* romana
cosh /kɒʃ/ *n* cachiporra *f*. ● *vt* aporrear
cosiness /ˈkəʊzɪnɪs/ *n* comodidad *f*
cosmetic /kɒzˈmetɪk/ *a* & *n* cosmético (*m*)
cosmic /ˈkɒzmɪk/ *a* cósmico
cosmonaut /ˈkɒzmənɔ:t/ *n* cosmonauta *m* & *f*
cosmopolitan /kɒzməˈpɒlɪtən/ *a* & *n* cosmopolita (*m* & *f*)
cosmos /ˈkɒzmɒs/ *n* cosmos *m*
Cossack /ˈkɒsæk/ *a* & *n* cosaco (*m*)
cosset /ˈkɒsɪt/ *vt* (*pt* **cosseted**) mimar
cost /kɒst/ *vi* (*pt* **cost**) costar, valer. ● *vt* (*pt* **costed**) calcular el coste de. ● *n* precio *m*. **at all ~s** cueste lo que cueste. **to one's ~** a sus expensas. **~s** *npl* (*jurid*) costas *fpl*
Costa Rica /kɒstəˈri:kə/ *n* Costa *f* Rica. **~n** *a* & *n* costarricense (*m* & *f*), costarriqueño (*m*)

costly /ˈkɒstlɪ/ *a* (**-ier**, **-iest**) caro, costoso
costume /ˈkɒstju:m/ *n* traje *m*
cosy /ˈkəʊzɪ/ *a* (**-ier**, **-iest**) cómodo; ‹*place*› acogedor. ● *n* cubierta *f* (de tetera)
cot /kɒt/ *n* cuna *f*
cottage /ˈkɒtɪdʒ/ *n* casita *f* de campo. ~ **cheese** *n* requesón *m*. ~ **industry** *n* industria *f* casera. ~ **pie** *n* carne *f* picada con puré de patatas
cotton /ˈkɒtn/ *n* algodón *m*. ● *vi*. ~ **on** (*sl*) comprender. ~ **wool** *n* algodón hidrófilo
couch /kaʊtʃ/ *n* sofá *m*. ● *vt* expresar
couchette /ku:ˈʃet/ *n* litera *f*
cough /kɒf/ *vi* toser. ● *n* tos *f*. ~ **up** (*sl*) pagar. ~ **mixture** *n* jarabe *m* para la tos
could /kʊd, kəd/ *pt of* **can**
couldn't /ˈkʊdnt/ = **could not**
council /ˈkaʊnsl/ *n* consejo *m*; (*of town*) ayuntamiento *m*. ~ **house** *n* vivienda *f* protegida. ~**lor** /ˈkaʊnsələ(r)/ *n* concejal *m*
counsel /ˈkaʊnsl/ *n* consejo *m*; (*pl invar*) (*jurid*) abogado *m*. ~**lor** *n* consejero *m*
count[1] /kaʊnt/ *n* recuento *m*. ● *vt/i* contar
count[2] /kaʊnt/ *n* (*nobleman*) conde *m*
countdown /ˈkaʊntdaʊn/ *n* cuenta *f* atrás
countenance /ˈkaʊntɪnəns/ *n* semblante *m*. ● *vt* aprobar
counter /ˈkaʊntə(r)/ *n* (*in shop etc*) mostrador *m*; (*token*) ficha *f*. ● *adv*. ~ **to** en contra de. ● *a* opuesto. ● *vt* oponerse a; parar ‹*blow*›. ● *vi* contraatacar
counter... /ˈkaʊntə(r)/ *pref* contra...
counteract /kaʊntərˈækt/ *vt* contrarrestar
counter-attack /ˈkaʊntərətæk/ *n* contraataque *m*. ● *vt/i* contraatacar
counterbalance /ˈkaʊntəbæləns/ *n* contrapeso *m*. ● *vt/i* contrapesar
counterfeit /ˈkaʊntəfɪt/ *a* falsificado. ● *n* falsificación *f*. ● *vt* falsificar
counterfoil /ˈkaʊntəfɔɪl/ *n* talón *m*
counterpart /ˈkaʊntəpɑ:t/ *n* equivalente *m*; (*person*) homólogo *m*
counter-productive /ˈkaʊntəprəˈdʌktɪv/ *a* contraproducente
countersign /ˈkaʊntəsaɪn/ *vt* refrendar
countess /ˈkaʊntɪs/ *n* condesa *f*
countless /ˈkaʊntlɪs/ *a* innumerable
countrified /ˈkʌntrɪfaɪd/ *a* rústico
country /ˈkʌntrɪ/ *n* (*native land*) país *m*; (*countryside*) campo *m*. ~ **folk** *n* gente *f* del campo. **go to the** ~ ir al campo; (*pol*) convocar elecciones generales
countryman /ˈkʌntrɪmən/ *n* (*pl* **-men**) campesino *m*; (*of one's own country*) compatriota *m*
countryside /ˈkʌntrɪsaɪd/ *n* campo *m*
county /ˈkaʊntɪ/ *n* condado *m*, provincia *f*
coup /ku:/ *n* golpe *m*
coupé /ˈku:peɪ/ *n* cupé *m*
couple /ˈkʌpl/ *n* (*of things*) par *m*; (*of people*) pareja *f*; (*married*) matrimonio *m*. **a** ~ **of** un par de. ● *vt* unir; (*tec*) acoplar. ● *vi* copularse
coupon /ˈku:pɒn/ *n* cupón *m*
courage /ˈkʌrɪdʒ/ *n* valor *m*. ~**ous** /kəˈreɪdʒəs/ *a* valiente. ~**ously** *adv* valientemente
courgette /kʊəˈʒet/ *n* calabacín *m*
courier /ˈkʊrɪə(r)/ *n* mensajero *m*; (*for tourists*) guía *m* & *f*
course /kɔ:s/ *n* curso *m*; (*behaviour*) conducta *f*; (*aviat, naut*) rumbo *m*; (*culin*) plato *m*; (*for golf*) campo *m*. **in due** ~ a su debido tiempo. **in the** ~ **of** en el transcurso de, durante. **of** ~ desde luego, por supuesto
court /kɔ:t/ *n* corte *f*; (*tennis*) pista *f*; (*jurid*) tribunal *m*. ● *vt* cortejar; buscar ‹*danger*›
courteous /ˈkɜ:tɪəs/ *a* cortés
courtesan /kɔ:tɪˈzæn/ *n* (*old use*) cortesana *f*
courtesy /ˈkɜ:təsɪ/ *n* cortesía *f*
court: ~**ier** /ˈkɔ:tɪə(r)/ *n* (*old use*) cortesano *m*. ~ **martial** *n* (*pl* **courts martial**) consejo *m* de guerra. ~**-martial** *vt* (*pt* ~**-martialled**) juzgar en consejo de guerra. ~**ship** /ˈkɔ:tʃɪp/ *n* cortejo *m*
courtyard /ˈkɔ:tjɑ:d/ *n* patio *m*
cousin /ˈkʌzn/ *n* primo *m*. **first** ~ primo carnal. **second** ~ primo segundo
cove /kəʊv/ *n* cala *f*
covenant /ˈkʌvənənt/ *n* acuerdo *m*
Coventry /ˈkɒvntrɪ/ *n*. **send to** ~ hacer el vacío
cover /ˈkʌvə(r)/ *vt* cubrir; (*journalism*) hacer un reportaje sobre. ~

up cubrir; (*fig*) ocultar. ● *n* cubierta *f*; (*shelter*) abrigo *m*; (*lid*) tapa *f*; (*for furniture*) funda *f*; (*pretext*) pretexto *m*; (*of magazine*) portada *f*. **~age** /'kʌvərɪdʒ/ *n* reportaje *m*. **~ charge** *n* precio *m* del cubierto. **~ing** *n* cubierta *f*. **~ing letter** *n* carta *f* explicatoria, carta *f* adjunta

covet /'kʌvɪt/ *vt* codiciar

cow /kaʊ/ *n* vaca *f*

coward /'kaʊəd/ *n* cobarde *m*. **~ly** *a* cobarde. **~ice** /'kaʊədɪs/ *n* cobardía *f*

cowboy /'kaʊbɔɪ/ *n* vaquero *m*

cower /'kaʊə(r)/ *vi* encogerse, acobardarse

cowl /kaʊl/ *n* capucha *f*; (*of chimney*) sombrerete *m*

cowshed /'kaʊʃed/ *n* establo *m*

coxswain /'kɒksn/ *n* timonel *m*

coy /kɔɪ/ *a* (**-er**, **-est**) (falsamente) tímido, remilgado

crab[1] /kræb/ *n* cangrejo *m*

crab[2] /kræb/ *vi* (*pt* **crabbed**) quejarse

crab-apple /'kræbæpl/ *n* manzana *f* silvestre

crack /kræk/ *n* grieta *f*; (*noise*) crujido *m*; (*of whip*) chasquido *m*; (*joke, sl*) chiste *m*. ● *a* (*fam*) de primera. ● *vt* agrietar; chasquear ‹*whip, fingers*›; cascar ‹*nut*›; gastar ‹*joke*›; resolver ‹*problem*›. ● *vi* agrietarse. **get ~ing** (*fam*) darse prisa. **~ down on** (*fam*) tomar medidas enérgicas contra. **~ up** *vi* fallar; ‹*person*› volverse loco. **~ed** /krækt/ *a* (*sl*) chiflado

cracker /'krækə(r)/ *n* petardo *m*; (*culin*) galleta *f* (soso); (*culin, Amer*) galleta *f*

crackers /'krækəz/ *a* (*sl*) chiflado

crackl|e /'krækl/ *vi* crepitar. ● *n* crepitación *f*, crujido *m*. **~ing** /'kræklɪŋ/ *n* crepitación *f*, crujido *m*; (*of pork*) chicharrón *m*

crackpot /'krækpɒt/ *n* (*sl*) chiflado *m*

cradle /'kreɪdl/ *n* cuna *f*. ● *vt* acunar

craft /krɑ:ft/ *n* destreza *f*; (*technique*) arte *f*; (*cunning*) astucia *f*. ● *n invar* (*boat*) barco *m*

craftsman /'krɑ:ftsmən/ *n* (*pl* **-men**) artesano *m*. **~ship** *n* artesanía *f*

crafty /'krɑ:ftɪ/ *a* (**-ier**, **-iest**) astuto

crag /kræg/ *n* despeñadero *m*. **~gy** *a* peñascoso

cram /kræm/ *vt* (*pt* **crammed**) rellenar. **~ with** llenar de. ● *vi* (*for exams*) empollar. **~-full** *a* atestado

cramp /kræmp/ *n* calambre *m*

cramped /kræmpt/ *a* apretado

cranberry /'krænbərɪ/ *n* arándano *m*

crane /kreɪn/ *n* grúa *f*; (*bird*) grulla *f*. ● *vt* estirar ‹*neck*›

crank[1] /kræŋk/ *n* manivela *f*

crank[2] /kræŋk/ *n* (*person*) excéntrico *m*. **~y** *a* excéntrico

cranny /'krænɪ/ *n* grieta *f*

crash /kræʃ/ *n* accidente *m*; (*noise*) estruendo *m*; (*collision*) choque *m*; (*com*) quiebra *f*. ● *vt* estrellar. ● *vi* quebrar con estrépito; (*have accident*) tener un accidente; ‹*car etc*› chocar; (*fail*) fracasar. **~ course** *n* curso *m* intensivo. **~-helmet** *n* casco *m* protector. **~-land** *vi* hacer un aterrizaje de emergencia, hacer un aterrizaje forzoso

crass /kræs/ *a* craso, burdo

crate /kreɪt/ *n* cajón *m*. ● *vt* embalar

crater /'kreɪtə(r)/ *n* cráter *m*

cravat /krə'væt/ *n* corbata *f*, fular *m*

crav|e /kreɪv/ *vi*. **~e for** anhelar. **~ing** *n* ansia *f*

crawl /krɔ:l/ *vi* andar a gatas; (*move slowly*) avanzar lentamente; (*drag o.s.*) arrastrarse. ● *n* (*swimming*) crol *m*. **at a ~** a paso lento. **~ to** humillarse ante. **~ with** hervir de

crayon /'kreɪən/ *n* lápiz *m* de color

craze /kreɪz/ *n* manía *f*

craz|iness /'kreɪzɪnɪs/ *n* locura *f*. **~y** /'kreɪzɪ/ *a* (**-ier**, **-iest**) loco. **be ~y about** andar loco por. **~y paving** *n* enlosado *m* irregular

creak /kri:k/ *n* crujido *m*; (*of hinge*) chirrido *m*. ● *vi* crujir; ‹*hinge*› chirriar

cream /kri:m/ *n* crema *f*; (*fresh*) nata *f*. ● *a* (*colour*) color de crema. ● *vt* (*remove*) desnatar; (*beat*) batir. **~ cheese** *n* queso *m* de nata. **~y** *a* cremoso

crease /kri:s/ *n* pliegue *m*; (*crumple*) arruga *f*. ● *vt* plegar; (*wrinkle*) arrugar. ● *vi* arrugarse

creat|e /kri:'eɪt/ *vt* crear. **~ion** /-ʃn/ *n* creación *f*. **~ive** *a* creativo. **~or** *n* creador *m*

creature /'kri:tʃə(r)/ *n* criatura *f*, bicho *m*, animal *m*

crèche /kreɪʃ/ *n* guardería *f* infantil

credence /'kri:dns/ *n* creencia *f*, fe *f*

credentials /krɪ'denʃlz/ *npl* credenciales *mpl*

credib|ility /kredə'bɪlətɪ/ *n* credibilidad *f*. **~le** /'kredəbl/ *a* creíble

credit /'kredɪt/ *n* crédito *m*; (*honour*) honor *m*. **take the ~ for** atribuirse

el mérito de. ● *vt* (*pt* **credited**) acreditar; (*believe*) creer. ~ **s.o. with** atribuir a uno. ~**able** *a* loable. ~ **card** *n* tarjeta *f* de crédito. ~**or** *n* acreedor *m*
credulous /ˈkrədjʊləs/ *a* crédulo
creed /kri:d/ *n* credo *m*
creek /kri:k/ *n* ensenada *f*. **up the** ~ (*sl*) en apuros
creep /kri:p/ *vi* (*pt* **crept**) arrastrarse; ‹*plant*› trepar. ● *n* (*sl*) persona *f* desagradable. ~**er** *n* enredadera *f*. ~**s** /kri:ps/ *npl*. **give s.o. the** ~**s** dar repugnancia a uno
cremat|e /krɪˈmeɪt/ *vt* incinerar. ~**ion** /-ʃn/ *n* cremación *f*. ~**orium** /kreməˈtɔ:rɪəm/ *n* (*pl* **-ia**) crematorio *m*
Creole /ˈkrɪəʊl/ *a* & *n* criollo (*m*)
crêpe /kreɪp/ *n* crespón *m*
crept /krept/ *see* **creep**
crescendo /krɪˈʃendəʊ/ *n* (*pl* **-os**) crescendo *m*
crescent /ˈkresnt/ *n* media luna *f*; (*street*) calle *f* en forma de media luna
cress /kres/ *n* berro *m*
crest /krest/ *n* cresta *f*; (*coat of arms*) blasón *m*
Crete /kri:t/ *n* Creta *f*
cretin /ˈkretɪn/ *n* cretino *m*
crevasse /krɪˈvæs/ *n* grieta *f*
crevice /ˈkrevɪs/ *n* grieta *f*
crew[1] /kru:/ *n* tripulación *f*; (*gang*) pandilla *f*
crew[2] /kru:/ *see* **crow**[2]
crew: ~ **cut** *n* corte *m* al rape. ~ **neck** *n* cuello *m* redondo
crib /krɪb/ *n* cuna *f*; (*relig*) belén *m*; (*plagiarism*) plagio *m*. ● *vt/i* (*pt* **cribbed**) plagiar
crick /krɪk/ *n* calambre *m*; (*in neck*) tortícolis *f*
cricket[1] /ˈkrɪkɪt/ *n* criquet *m*
cricket[2] /ˈkrɪkɪt/ *n* (*insect*) grillo *m*
cricketer /ˈkrɪkɪtə(r)/ *n* jugador *m* de criquet
crim|e /kraɪm/ *n* crimen *m*; (*acts*) criminalidad *f*. ~**inal** /ˈkrɪmɪnl/ *a* & *n* criminal (*m*)
crimp /krɪmp/ *vt* rizar
crimson /ˈkrɪmzn/ *a* & *n* carmesí (*m*)
cringe /krɪndʒ/ *vi* encogerse; (*fig*) humillarse
crinkle /ˈkrɪŋkl/ *vt* arrugar. ● *vi* arrugarse. ● *n* arruga *f*
crinoline /ˈkrɪnəlɪn/ *n* miriñaque *m*
cripple /ˈkrɪpl/ *n* lisiado *m*, mutilado *m*. ● *vt* lisiar; (*fig*) paralizar
crisis /ˈkraɪsɪs/ *n* (*pl* **crises** /ˈkraɪsi:z/) crisis *f*
crisp /krɪsp/ *a* (**-er**, **-est**) (*culin*) crujiente; ‹*air*› vigorizador. ~**s** *npl* patatas *fpl* fritas a la inglesa
criss-cross /ˈkrɪskrɒs/ *a* entrecruzado. ● *vt* entrecruzar. ● *vi* entrecruzarse
criterion /kraɪˈtɪərɪən/ *n* (*pl* **-ia**) criterio *m*
critic /ˈkrɪtɪk/ *n* crítico *m*
critical /ˈkrɪtɪkl/ *a* crítico. ~**ly** *adv* críticamente; (*ill*) gravemente
critici|sm /ˈkrɪtɪsɪzəm/ *n* crítica *f*. ~**ze** /ˈkrɪtɪsaɪz/ *vt/i* criticar
croak /krəʊk/ *n* (*of person*) gruñido *m*; (*of frog*) canto *m*. ● *vi* gruñir; ‹*frog*› croar
crochet /ˈkrəʊʃeɪ/ *n* croché *m*, ganchillo *m*. ● *vt* hacer ganchillo
crock[1] /krɒk/ *n* (*person*, *fam*) vejancón *m*; (*old car*) cacharro *m*
crock[2] /krɒk/ *n* vasija *f* de loza
crockery /ˈkrɒkərɪ/ *n* loza *f*
crocodile /ˈkrɒkədaɪl/ *n* cocodrilo *m*. ~ **tears** *npl* lágrimas *fpl* de cocodrilo
crocus /ˈkrəʊkəs/ *n* (*pl* **-es**) azafrán *m*
crony /ˈkrəʊnɪ/ *n* amigote *m*
crook /krʊk/ *n* (*fam*) maleante *m* & *f*, estafador *m*, criminal *m*; (*stick*) cayado *m*; (*of arm*) pliegue *m*
crooked /ˈkrʊkɪd/ *a* torcido; (*winding*) tortuoso; (*dishonest*) poco honrado
croon /kru:n/ *vt/i* canturrear
crop /krɒp/ *n* cosecha *f*; (*fig*) montón *m*. ● *vt* (*pt* **cropped**) *vi* cortar. ~ **up** surgir
cropper /ˈkrɒpər/ *n*. **come a** ~ (*fall*, *fam*) caer; (*fail*, *fam*) fracasar
croquet /ˈkrəʊkeɪ/ *n* croquet *m*
croquette /krəˈket/ *n* croqueta *f*
cross /krɒs/ *n* cruz *f*; (*of animals*) cruce *m*. ● *vt/i* cruzar; (*oppose*) contrariar. ~ **off** tachar. ~ **o.s.** santiguarse. ~ **out** tachar. ~ **s.o.'s mind** ocurrírsele a uno. ● *a* enfadado. **talk at** ~ **purposes** hablar sin entenderse
crossbar /ˈkrɒsbɑ:(r)/ *n* travesaño *m*
cross-examine /krɒsɪgˈzæmɪn/ *vt* interrogar
cross-eyed /ˈkrɒsaɪd/ *a* bizco
crossfire /ˈkrɒsfaɪə(r)/ *n* fuego *m* cruzado
crossing /ˈkrɒsɪŋ/ *n* (*by boat*) travesía *f*; (*on road*) paso *m* para peatones

crossly /'krɒslɪ/ *adv* con enfado
cross-reference /krɒs'refrəns/ *n* referencia *f*
crossroads /'krɒsrəʊdz/ *n* cruce *m* (de carreteras)
cross-section /krɒs'sekʃn/ *n* sección *f* transversal; (*fig*) muestra *f* representativa
crosswise /'krɒswaɪz/ *adv* al través
crossword /'krɒswɜ:d/ *n* crucigrama *m*
crotch /krɒtʃ/ *n* entrepiernas *fpl*
crotchety /'krɒtʃɪtɪ/ *a* de mal genio
crouch /kraʊtʃ/ *vi* agacharse
crow[1] /krəʊ/ *n* cuervo *m*. **as the ~ flies** en línea recta
crow[2] /krəʊ/ *vi* (*pt* **crew**) cacarear
crowbar /'krəʊbɑ:(r)/ *n* palanca *f*
crowd /kraʊd/ *n* muchedumbre *f*. ● *vt* amontonar; (*fill*) llenar. ● *vi* amontonarse; (*gather*) reunirse. **~ed** *a* atestado
crown /kraʊn/ *n* corona *f*; (*of hill*) cumbre *f*; (*of head*) coronilla *f*. ● *vt* coronar; poner una corona a ‹*tooth*›. **C~ Court** *n* tribunal *m* regional. **C~ prince** *n* príncipe *m* heredero
crucial /'kru:ʃl/ *a* crucial
crucifix /'kru:sɪfɪks/ *n* crucifijo *m*. **~ion** /-'fɪkʃn/ *n* crucifixión *f*
crucify /'kru:sɪfaɪ/ *vt* crucificar
crude /kru:d/ *a* (**-er**, **-est**) (*raw*) crudo; (*rough*) tosco; (*vulgar*) ordinario
cruel /krʊəl/ *a* (**crueller**, **cruellest**) cruel. **~ty** *n* crueldad *f*
cruet /'kru:ɪt/ *n* vinagreras *fpl*
cruise /kru:z/ *n* crucero *m*. ● *vi* hacer un crucero; (*of car*) circular lentamente. **~r** *n* crucero *m*
crumb /krʌm/ *n* migaja *f*
crumble /'krʌmbl/ *vt* desmenuzar. ● *vi* desmenuzarse; (*collapse*) derrumbarse
crummy /'krʌmɪ/ *a* (**-ier**, **-iest**) (*sl*) miserable
crumpet /'krʌmpɪt/ *n* bollo *m* blando
crumple /'krʌmpl/ *vt* arrugar; estrujar ‹*paper*›. ● *vi* arrugarse
crunch /krʌntʃ/ *vt* hacer crujir; (*bite*) ronzar, morder, masticar. ● *n* crujido *m*; (*fig*) momento *m* decisivo
crusade /kru:'seɪd/ *n* cruzada *f*. **~r** /-ə(r)/ *n* cruzado *m*
crush /krʌʃ/ *vt* aplastar; arrugar ‹*clothes*›; estrujar ‹*paper*›. ● *n* (*crowd*) aglomeración *f*. **have a ~ on** (*sl*) estar perdido por. **orange ~** *n* naranjada *f*
crust /krʌst/ *n* corteza *f*. **~y** *a* ‹*bread*› de corteza dura; ‹*person*› malhumorado
crutch /krʌtʃ/ *n* muleta *f*; (*anat*) entrepiernas *fpl*
crux /krʌks/ *n* (*pl* **cruxes**) punto *m* más importante, quid *m*, busilis *m*
cry /kraɪ/ *n* grito *m*. **be a far ~ from** (*fig*) distar mucho de. ● *vi* llorar; (*call out*) gritar. **~ off** rajarse. **~-baby** *n* llorón *m*
crypt /krɪpt/ *n* cripta *f*
cryptic /'krɪptɪk/ *a* enigmático
crystal /'krɪstl/ *n* cristal *m*. **~lize** *vt* cristalizar. ● *vi* cristalizarse
cub /kʌb/ *n* cachorro *m*. **C~ (Scout)** *n* niño *m* explorador
Cuba /'kju:bə/ *n* Cuba *f*. **~n** *a* & *n* cubano (*m*)
cubby-hole /'kʌbɪhəʊl/ *n* casilla *f*; (*room*) chiribitil *m*, cuchitril *m*
cub|e /kju:b/ *n* cubo *m*. **~ic** *a* cúbico
cubicle /'kju:bɪkl/ *n* cubículo *m*; (*changing room*) caseta *f*
cubis|m /'kju:bɪzm/ *n* cubismo *m*. **~t** *a* & *n* cubista (*m* & *f*)
cuckold /'kʌkəʊld/ *n* cornudo *m*
cuckoo /'kʊku:/ *n* cuco *m*, cuclillo *m*
cucumber /'kju:kʌmbə(r)/ *n* pepino *m*
cuddl|e /'kʌdl/ *vt* abrazar. ● *vi* abrazarse. ● *n* abrazo *m*. **~y** *a* mimoso
cudgel /'kʌdʒl/ *n* porra *f*. ● *vt* (*pt* **cudgelled**) aporrear
cue[1] /kju:/ *n* indicación *f*; (*in theatre*) pie *m*
cue[2] /kju:/ *n* (*in billiards*) taco *m*
cuff /kʌf/ *n* puño *m*; (*blow*) bofetada *f*. **speak off the ~** hablar de improviso. ● *vt* abofetear. **~-link** *n* gemelo *m*
cul-de-sac /'kʌldəsæk/ *n* callejón *m* sin salida
culinary /'kʌlɪnərɪ/ *a* culinario
cull /kʌl/ *vt* coger ‹*flowers*›; entresacar ‹*animals*›
culminat|e /'kʌlmɪneɪt/ *vi* culminar. **~ion** /-'neɪʃn/ *n* culminación *f*
culottes /kʊ'lɒts/ *npl* falda *f* pantalón
culprit /'kʌlprɪt/ *n* culpable *m*
cult /kʌlt/ *n* culto *m*
cultivat|e /'kʌltɪveɪt/ *vt* cultivar. **~ion** /-'veɪʃn/ *n* cultivo *m*; (*fig*) cultura *f*

cultur|al /ˈkʌltʃərəl/ *a* cultural. **~e** /ˈkʌltʃə(r)/ *n* cultura *f*; (*bot etc*) cultivo *m*. **~ed** *a* cultivado; ‹*person*› culto
cumbersome /ˈkʌmbəsəm/ *a* incómodo; (*heavy*) pesado
cumulative /ˈkjuːmjʊlətɪv/ *a* cumulativo
cunning /ˈkʌnɪŋ/ *a* astuto. ● *n* astucia *f*
cup /kʌp/ *n* taza *f*; (*prize*) copa *f*
cupboard /ˈkʌbəd/ *n* armario *m*
Cup Final /kʌpˈfaɪnl/ *n* final *f* del campeonato
cupful /ˈkʌpfʊl/ *n* taza *f*
cupidity /kjuːˈpɪdɪtɪ/ *n* codicia *f*
curable /ˈkjʊərəbl/ *a* curable
curate /ˈkjʊərət/ *n* coadjutor *m*
curator /kjʊəˈreɪtə(r)/ *n* (*of museum*) conservador *m*
curb /kɜːb/ *n* freno *m*. ● *vt* refrenar
curdle /ˈkɜːdl/ *vt* cuajar. ● *vi* cuajarse; ‹*milk*› cortarse
curds /kɜːdz/ *npl* cuajada *f*, requesón *m*
cure /kjʊə(r)/ *vt* curar. ● *n* cura *f*
curfew /ˈkɜːfjuː/ *n* queda *f*; (*signal*) toque *m* de queda
curio /ˈkjʊərɪəʊ/ *n* (*pl* **-os**) curiosidad *f*
curio|us /ˈkjʊərɪəs/ *a* curioso. **~sity** /-ˈɒsətɪ/ *n* curiosidad *f*
curl /kɜːl/ *vt* rizar ‹*hair*›. **~ o.s. up** acurrucarse. ● *vi* ‹*hair*› rizarse; ‹*paper*› arrollarse. ● *n* rizo *m*. **~er** /ˈkɜːlə(r)/ *n* bigudí *m*, rulo *m*. **~y** /ˈkɜːlɪ/ *a* (**-ier**, **-iest**) rizado
currant /ˈkʌrənt/ *n* pasa *f* de Corinto
currency /ˈkʌrənsɪ/ *n* moneda *f*; (*acceptance*) uso *m* (corriente)
current /ˈkʌrənt/ *a & n* corriente (*f*). **~ events** asuntos *mpl* de actualidad. **~ly** *adv* actualmente
curriculum /kəˈrɪkjʊləm/ *n* (*pl* **-la**) programa *m* de estudios. **~ vitae** *n* curriculum *m* vitae
curry[1] /ˈkʌrɪ/ *n* curry *m*
curry[2] /ˈkʌrɪ/ *vt*. **~ favour with** congraciarse con
curse /kɜːs/ *n* maldición *f*; (*oath*) palabrota *f*. ● *vt* maldecir. ● *vi* decir palabrotas
cursory /ˈkɜːsərɪ/ *a* superficial
curt /kɜːt/ *a* brusco
curtail /kɜːˈteɪl/ *vt* abreviar; reducir ‹*expenses*›
curtain /ˈkɜːtn/ *n* cortina *f*; (*in theatre*) telón *m*
curtsy /ˈkɜːtsɪ/ *n* reverencia *f*. ● *vi* hacer una reverencia
curve /kɜːv/ *n* curva *f*. ● *vt* encurvar. ● *vi* encorvarse; ‹*road*› torcerse
cushion /ˈkʊʃn/ *n* cojín *m*. ● *vt* amortiguar ‹*a blow*›; (*fig*) proteger
cushy /ˈkʊʃɪ/ *a* (**-ier**, **-iest**) (*fam*) fácil
custard /ˈkʌstəd/ *n* natillas *fpl*
custodian /kʌˈstəʊdɪən/ *n* custodio *m*
custody /ˈkʌstədɪ/ *n* custodia *f*. **be in ~** (*jurid*) estar detenido
custom /ˈkʌstəm/ *n* costumbre *f*; (*com*) clientela *f*
customary /ˈkʌstəmərɪ/ *a* acostumbrado
customer /ˈkʌstəmə(r)/ *n* cliente *m*
customs /ˈkʌstəmz/ *npl* aduana *f*. **~ officer** *n* aduanero *m*
cut /kʌt/ *vt/i* (*pt* **cut**, *pres p* **cutting**) cortar; reducir ‹*prices*›. ● *n* corte *m*; (*reduction*) reducción *f*. **~ across** atravesar. **~ back**, **~ down** reducir. **~ in** interrumpir. **~ off** cortar; (*phone*) desconectar; (*fig*) aislar. **~ out** recortar; (*omit*) suprimir. **~ through** atravesar. **~ up** cortar en pedazos. **be ~ up about** (*fig*) afligirse por
cute /kjuːt/ *a* (**-er**, **-est**) (*fam*) listo; (*Amer*) mono
cuticle /ˈkjuːtɪkl/ *n* cutícula *f*
cutlery /ˈkʌtlərɪ/ *n* cubiertos *mpl*
cutlet /ˈkʌtlɪt/ *n* chuleta *f*
cut-price /ˈkʌtpraɪs/ *a* a precio reducido
cut-throat /ˈkʌtθrəʊt/ *a* despiadado
cutting /ˈkʌtɪŋ/ *a* cortante; ‹*remark*› mordaz. ● *n* (*from newspaper*) recorte *m*; (*of plant*) esqueje *m*
cyanide /ˈsaɪənaɪd/ *n* cianuro *m*
cybernetics /saɪbəˈnetɪks/ *n* cibernética *f*
cyclamen /ˈsɪkləmən/ *n* ciclamen *m*
cycle /ˈsaɪkl/ *n* ciclo *m*; (*bicycle*) bicicleta *f*. ● *vi* ir en bicicleta
cyclic(al) /ˈsaɪklɪk(l)/ *a* cíclico
cycli|ng /ˈsaɪklɪŋ/ *n* ciclismo *m*. **~st** *n* ciclista *m & f*
cyclone /ˈsaɪkləʊn/ *n* ciclón *m*
cylind|er /ˈsɪlɪndə(r)/ *n* cilindro *m*. **~er head** (*auto*) *n* culata *f*. **~rical** /-ˈlɪndrɪkl/ *a* cilíndrico
cymbal /ˈsɪmbl/ *n* címbalo *m*
cynic /ˈsɪnɪk/ *n* cínico *m*. **~al** *a* cínico. **~ism** /-sɪzəm/ *n* cinismo *m*
cypress /ˈsaɪprəs/ *n* ciprés *m*

Cypr|iot /ˈsɪprɪət/ *a & n* chipriota (*m & f*). **~us** /ˈsaɪprəs/ *n* Chipre *f*
cyst /sɪst/ *n* quiste *m*
czar /zɑ:(r)/ *n* zar *m*
Czech /tʃek/ *a & n* checo (*m*). **the ~ Republic** *n* la república *f* Checa
Czechoslovak /tʃekəʊˈsləʊvæk/ *a & n* (*history*) checoslovaco (*m*). **~ia** /-əˈvækɪə/ *n* (*history*) Checoslovaquia *f*

D

dab /dæb/ *vt* (*pt* **dabbed**) tocar ligeramente. ● *n* toque *m* suave. **a ~ of** un poquito de
dabble /ˈdæbl/ *vi.* **~ in** meterse (superficialmente) en. **~r** /ə(r)/ *n* aficionado *m*
dad /dæd/ *n* (*fam*) papá *m*. **~dy** *n* (*children's use*) papá *m*. **~dy-long-legs** *n* típula *f*
daffodil /ˈdæfədɪl/ *n* narciso *m*
daft /dɑ:ft/ *a* (**-er**, **-est**) tonto
dagger /ˈdægə(r)/ *n* puñal *m*
dahlia /ˈdeɪlɪə/ *n* dalia *f*
daily /ˈdeɪlɪ/ *a* diario. ● *adv* diariamente, cada día. ● *n* diario *m*; (*cleaner, fam*) asistenta *f*
dainty /ˈdeɪntɪ/ *a* (**-ier**, **-iest**) delicado
dairy /ˈdeərɪ/ *n* vaquería *f*; (*shop*) lechería *f*. ● *a* lechero
dais /dei:s/ *n* estrado *m*
daisy /ˈdeɪzɪ/ *n* margarita *f*
dale /deɪl/ *n* valle *m*
dally /ˈdælɪ/ *vi* tardar; (*waste time*) perder el tiempo
dam /dæm/ *n* presa *f*. ● *vt* (*pt* **dammed**) embalsar
damag|e /ˈdæmɪdʒ/ *n* daño *m*; (*pl, jurid*) daños *mpl* y perjuicios *mpl*. ● *vt* (*fig*) dañar, estropear. **~ing** *a* perjudicial
damask /ˈdæməsk/ *n* damasco *m*
dame /deɪm/ *n* (*old use*) dama *f*; (*Amer, sl*) chica *f*
damn /dæm/ *vt* condenar; (*curse*) maldecir. ● *int* ¡córcholis! ● *a* maldito. ● *n*. **I don't care a ~** (no) me importa un comino. **~ation** /-ˈneɪʃn/ *n* condenación *f*, perdición *f*
damp /dæmp/ *n* humedad *f*. ● *a* (**-er**, **-est**) húmedo. ● *vt* mojar; (*fig*) ahogar. **~er** /ˈdæmpə(r)/ *n* apagador *m*, sordina *f*; (*fig*) aguafiestas *m invar*. **~ness** *n* humedad *f*
damsel /ˈdæmzl/ *n* (*old use*) doncella *f*
dance /dɑ:ns/ *vt/i* bailar. ● *n* baile *m*. **~-hall** *n* salón *m* de baile. **~r** /-ə(r)/ *n* bailador *m*; (*professional*) bailarín *m*
dandelion /ˈdændɪlaɪən/ *n* diente *m* de león
dandruff /ˈdændrʌf/ *n* caspa *f*
dandy /ˈdændɪ/ *n* petimetre *m*
Dane /deɪn/ *n* danés *m*
danger /ˈdeɪndʒə(r)/ *n* peligro *m*; (*risk*) riesgo *m*. **~ous** *a* peligroso
dangle /ˈdæŋgl/ *vt* balancear. ● *vi* suspender, colgar
Danish /ˈdeɪnɪʃ/ *a* danés. ● *m* (*lang*) danés *m*
dank /dæŋk/ *a* (**-er**, **-est**) húmedo, malsano
dare /deə(r)/ *vt* desafiar. ● *vi* atreverse a. **I ~ say** probablemente. ● *n* desafío *m*
daredevil /ˈdeədevl/ *n* atrevido *m*
daring /ˈdeərɪŋ/ *a* atrevido
dark /dɑ:k/ *a* (**-er**, **-est**) oscuro; (*gloomy*) sombrío; ‹*skin, hair*› moreno. ● *n* oscuridad *f*; (*nightfall*) atardecer. **in the ~** a oscuras. **~en** /ˈdɑ:kən/ *vt* oscurecer. ● *vi* oscurecerse. **~ horse** *n* persona *f* de talentos desconocidos. **~ness** *n* oscuridad *f*. **~-room** *n* cámara *f* oscura
darling /ˈdɑ:lɪŋ/ *a* querido. ● *n* querido *m*
darn /dɑ:n/ *vt* zurcir
dart /dɑ:t/ *n* dardo *m*. ● *vi* lanzarse; (*run*) precipitarse. **~board** /ˈdɑ:tbɔ:d/ *n* blanco *m*. **~s** *npl* los dardos *mpl*
dash /dæʃ/ *vi* precipitarse. **~ off** marcharse apresuradamente. **~ out** salir corriendo. ● *vt* lanzar; (*break*) romper; defraudar ‹*hopes*›. ● *n* carrera *f*; (*small amount*) poquito *m*; (*stroke*) raya *f*. **cut a ~** causar sensación
dashboard /ˈdæʃbɔ:d/ *n* tablero *m* de mandos
dashing /ˈdæʃɪŋ/ *a* vivo; (*showy*) vistoso
data /ˈdeɪtə/ *npl* datos *mpl*. **~ processing** *n* proceso *m* de datos
date[1] /deɪt/ *n* fecha *f*; (*fam*) cita *f*. **to ~** hasta la fecha. ● *vt* fechar; (*go out with, fam*) salir con. ● *vi* datar; (*be old-fashioned*) quedar anticuado
date[2] /deɪt/ *n* (*fruit*) dátil *m*
dated /ˈdeɪtɪd/ *a* pasado de moda

daub /dɔ:b/ *vt* embadurnar

daughter /ˈdɔ:tə(r)/ *n* hija *f*. **~-in-law** *n* nuera *f*

daunt /dɔ:nt/ *vt* intimidar

dauntless /ˈdɔ:ntlɪs/ *a* intrépido

dawdle /ˈdɔ:dl/ *vi* andar despacio; (*waste time*) perder el tiempo. **~r** /-ə(r)/ *n* rezagado *m*

dawn /dɔ:n/ *n* amanecer *m*. ● *vi* amanecer; (*fig*) nacer. **it ~ed on me that** caí en la cuenta de que, comprendí que

day /deɪ/ *n* día *m*; (*whole day*) jornada *f*; (*period*) época *f*. **~-break** *n* amanecer *m*. **~-dream** *n* ensueño *m*. ● *vi* soñar despierto. **~light** /ˈdeɪlaɪt/ *n* luz *f* del día. **~time** /ˈdeɪtaɪm/ *n* día *m*

daze /deɪz/ *vt* aturdir. ● *n* aturdimiento *m*. **in a ~** aturdido

dazzle /ˈdæzl/ *vt* deslumbrar

deacon /ˈdi:kən/ *n* diácono *m*

dead /ded/ *a* muerto; (*numb*) entumecido. **~ centre** justo en medio. ● *adv* completamente. **~ beat** rendido. **~ on time** justo a tiempo. **~ slow** muy lento. **stop ~** parar en seco. ● *n* muertos *mpl*. **in the ~ of night** en plena noche. **the ~** los muertos *mpl*. **~en** /ˈdedn/ *vt* amortiguar ‹*sound, blow*›; calmar ‹*pain*›. **~ end** *n* callejón *m* sin salida. **~ heat** *n* empate *m*

deadline /ˈdedlaɪn/ *n* fecha *f* tope, fin *m* de plazo

deadlock /ˈdedlɒk/ *n* punto *m* muerto

deadly /ˈdedlɪ/ *a* (**-ier**, **-iest**) mortal; (*harmful*) nocivo; (*dreary*) aburrido

deadpan/ˈdedpæn/ *a* impasible

deaf /def/ *a* (**-er**, **-est**) sordo. **~-aid** *n* audífono *m*. **~en** /ˈdefn/ *vt* ensordecer. **~ening** *a* ensordecedor. **~ mute** *n* sordomudo *m*. **~ness** *n* sordera *f*

deal /di:l/ *n* (*transaction*) negocio *m*; (*agreement*) pacto *m*; (*of cards*) reparto m; (*treatment*) trato *m*; (*amount*) cantidad *f*. **a great ~** muchísimo. ● *vt* (*pt* **dealt**) distribuir; dar ‹*a blow, cards*›. ● *vi*. **~ in** comerciar en. **~ with** tratar con ‹*person*›; tratar de ‹*subject etc*›; ocuparse de ‹*problem etc*›. **~er** *n* comerciante *m*. **~ings** /ˈdi:lɪŋz/ *npl* trato *m*

dean /di:n/ *n* deán *m*; (*univ*) decano *m*

dear /dɪə(r)/ *a* (**-er**, **-est**) querido; (*expensive*) caro. ● *n* querido *m*; (*child*) pequeño *m*. ● *adv* caro. ● *int* ¡Dios mío! **~ me!** ¡Dios mío! **~ly** *adv* tiernamente; (*pay*) caro; (*very much*) muchísimo

dearth /dɜ:θ/ *n* escasez *f*

death /deθ/ *n* muerte *f*. **~ duty** *n* derechos *mpl* reales. **~ly** *a* mortal; ‹*silence*› profundo. ● *adv* como la muerte. **~'s head** *n* calavera *f*. **~-trap** *n* lugar *m* peligroso.

débâcle /deɪˈbɑ:kl/ *n* fracaso *m*, desastre *m*

debar /dɪˈbɑ:(r)/ *vt* (*pt* **debarred**) excluir

debase /dɪˈbeɪs/ *vt* degradar

debat|able /dɪˈbeɪtəbl/ *a* discutible. **~e** /dɪˈbeɪt/ *n* debate *m*. ● *vt* debatir, discutir. ● *vi* discutir; (*consider*) considerar

debauch /dɪˈbɔ:tʃ/ *vt* corromper. **~ery** *n* libertinaje *m*

debilit|ate /dɪˈbɪlɪteɪt/ *vt* debilitar. **~y** /dɪˈbɪlətɪ/ *n* debilidad *f*

debit /ˈdebɪt/ *n* debe *m*. ● *vt*. **~ s.o.'s account** cargar en cuenta a uno

debonair /debəˈneə(r)/ *a* alegre

debris /ˈdebri:/ *n* escombros *mpl*

debt /det/ *n* deuda *f*. **be in ~** tener deudas. **~or** *n* deudor *m*

debutante /ˈdebju:tɑ:nt/ *n* (*old use*) debutante *f*

decade /ˈdekeɪd/ *n* década *f*

decaden|ce /ˈdekədəns/ *n* decadencia *f*. **~t** /ˈdekədənt/ *a* decadente

decant /dɪˈkænt/ *vt* decantar. **~er** /ə(r)/ *n* garrafa *f*

decapitate /dɪˈkæpɪteɪt/ *vt* decapitar

decay /dɪˈkeɪ/ *vi* decaer; ‹*tooth*› cariarse. ● *n* decadencia *f*; (*of tooth*) caries *f*

deceased /dɪˈsi:st/ *a* difunto

deceit /dɪˈsi:t/ *n* engaño *m*. **~ful** *a* falso. **~fully** *adv* falsamente

deceive /dɪˈsi:v/ *vt* engañar

December /dɪˈsembə(r)/ *n* diciembre *m*

decen|cy /ˈdi:sənsɪ/ *n* decencia *f*. **~t** /ˈdi:snt/ *a* decente; (*good, fam*) bueno; (*kind, fam*) amable. **~tly** *adv* decentemente

decentralize /di:ˈsentrəlaɪz/ *vt* descentralizar

decepti|on /dɪˈsepʃn/ *n* engaño *m*. **~ve** /dɪˈseptɪv/ *a* engañoso

decibel /ˈdesɪbel/ *n* decibel(io) *m*

decide /dɪ'saɪd/ *vt/i* decidir. **~d** /-ɪd/ *a* resuelto; (*unquestionable*) indudable. **~dly** /-ɪdlɪ/ *adv* decididamente; (*unquestionably*) indudablemente

decimal /'desɪml/ *a & n* decimal (*f*). **~ point** *n* coma *f* (decimal)

decimate /'desɪmeɪt/ *vt* diezmar

decipher /dɪ'saɪfə(r)/ *vt* descifrar

decision /dɪ'sɪʒn/ *n* decisión *f*

decisive /dɪ'saɪsɪv/ *a* decisivo; ‹*manner*› decidido. **~ly** *adv* de manera decisiva

deck /dek/ *n* cubierta *f*; (*of cards, Amer*) baraja *f*. **top ~** (*of bus*) imperial *m*. ● *vt* adornar. **~-chair** *n* tumbona *f*

declaim /dɪ'kleɪm/ *vt* declamar

declar|ation /deklə'reɪʃn/ *n* declaración *f*. **~e** /dɪ'kleə(r)/ *vt* declarar

decline /dɪ'klaɪn/ *vt* rehusar; (*gram*) declinar. ● *vi* disminuir; (*deteriorate*) deteriorarse; (*fall*) bajar. ● *n* decadencia *f*; (*decrease*) disminución *f*; (*fall*) baja *f*

decode /di:'kəʊd/ *vt* descifrar

decompos|e /di:kəm'pəʊz/ *vt* descomponer. ● *vi* descomponerse. **~ition** /-ɒmpə'zɪʃn/ *n* descomposición *f*

décor /'deɪkɔ:(r)/ *n* decoración *f*

decorat|e /'dekəreɪt/ *vt* decorar; empapelar y pintar ‹*room*›. **~ion** /-'reɪʃn/ *n* (*act*) decoración *f*; (*ornament*) adorno *m*. **~ive** /-ətɪv/ *a* decorativo. **~or** /'dekəreɪtə(r)/ *n* pintor *m* decorador. **interior ~or** decorador *m* de interiores

decorum /dɪ'kɔ:rəm/ *n* decoro *m*

decoy /'di:kɔɪ/ *n* señuelo *m*. /dɪ'kɔɪ/ *vt* atraer con señuelo

decrease /dɪ'kri:s/ *vt* disminuir. ● *vi* disminuirse. /'di:kri:s/ *n* disminución *f*

decree /dɪ'kri:/ *n* decreto *m*; (*jurid*) sentencia *f*. ● *vt* (*pt* **decreed**) decretar

decrepit /dɪ'krepɪt/ *a* decrépito

decry /dɪ'kraɪ/ *vt* denigrar

dedicat|e /'dedɪkeɪt/ *vt* dedicar. **~ion** /-'keɪʃn/ *n* dedicación *f*; (*in book*) dedicatoria *f*

deduce /dɪ'dju:s/ *vt* deducir

deduct /dɪ'dʌkt/ *vt* deducir. **~ion** /-ʃn/ *n* deducción *f*

deed /di:d/ *n* hecho *m*; (*jurid*) escritura *f*

deem /di:m/ *vt* juzgar, considerar

deep /di:p/ *a* (**-er**, **est**) *adv* profundo. **get into ~ waters** meterse en honduras. **go off the ~ end** enfadarse. ● *adv* profundamente. **be ~ in thought** estar absorto en sus pensamientos. **~en** /'di:pən/ *vt* profundizar. ● *vi* hacerse más profundo. **~-freeze** *n* congelador *m*. **~ly** *adv* profundamente

deer /dɪə(r)/ *n invar* ciervo *m*

deface /dɪ'feɪs/ *vt* desfigurar

defamation /defə'meɪʃn/ *n* difamación *f*

default /dɪ'fɔ:lt/ *vi* faltar. ● *n*. **by ~** en rebeldía. **in ~ of** en ausencia de

defeat /dɪ'fi:t/ *vt* vencer; (*frustrate*) frustrar. ● *n* derrota *f*; (*of plan etc*) fracaso *m*. **~ism** /dɪ'fi:tɪzm/ *n* derrotismo *m*. **~ist** /dɪ'fi:tɪst/ *n* derrotista *m & f*

defect /'di:fekt/ *n* defecto *m*. /dɪ'fekt/ *vi* desertar. **~ to** pasar a. **~ion** /dɪ'fekʃn/ *n* deserción *f*. **~ive** /dɪ'fektɪv/ *a* defectuoso

defence /dɪ'fens/ *n* defensa *f*. **~less** *a* indefenso

defend /dɪ'fend/ *vt* defender. **~ant** *n* (*jurid*) acusado *m*

defensive /dɪ'fensɪv/ *a* defensivo. ● *n* defensiva *f*

defer /dɪ'fɜ:(r)/ *vt* (*pt* **deferred**) aplazar

deferen|ce /'defərəns/ *n* deferencia *f*. **~tial** /-'renʃl/ *a* deferente

defian|ce /dɪ'faɪəns/ *n* desafío *m*. **in ~ce of** a despecho de. **~t** *a* desafiante. **~tly** *adv* con tono retador

deficien|cy /dɪ'fɪʃənsɪ/ *n* falta *f*. **~t** /dɪ'fɪʃnt/ *a* deficiente. **be ~t in** carecer de

deficit /'defɪsɪt/ *n* déficit *m*

defile /dɪ'faɪl/ *vt* ensuciar; (*fig*) deshonrar

define /dɪ'faɪn/ *vt* definir

definite /'defɪnɪt/ *a* determinado; (*clear*) claro; (*firm*) categórico. **~ly** *adv* claramente; (*certainly*) seguramente

definition /defɪ'nɪʃn/ *n* definición *f*

definitive /dɪ'fɪnətɪv/ *a* definitivo

deflat|e /dɪ'fleɪt/ *vt* desinflar. ● *vi* desinflarse. **~ion** /-ʃn/ *n* (*com*) deflación *f*

deflect /dɪ'flekt/ *vt* desviar. ● *vi* desviarse

deform /dɪ'fɔ:m/ *vt* deformar. **~ed** *a* deforme. **~ity** *n* deformidad *f*

defraud /dɪ'frɔ:d/ *vt* defraudar

defray /dɪ'freɪ/ *vt* pagar

defrost /diːˈfrɒst/ *vt* descongelar
deft /deft/ *a* (**-er**, **-est**) hábil. **~ness** *n* destreza *f*
defunct /dɪˈfʌŋkt/ *a* difunto
defuse /diːˈfjuːz/ *vt* desactivar ‹*bomb*›; (*fig*) calmar
defy /dɪˈfaɪ/ *vt* desafiar; (*resist*) resistir
degenerate /dɪˈdʒenəreɪt/ *vi* degenerar. /dɪˈdʒenərət/ *a* & *n* degenerado (*m*)
degrad|ation /degrəˈdeɪʃn/ *n* degradación *f*. **~e** /dɪˈgreɪd/ *vt* degradar
degree /dɪˈgriː/ *n* grado *m*; (*univ*) licenciatura *f*; (*rank*) rango *m*. **to a certain ~** hasta cierto punto. **to a ~** (*fam*) sumamente
dehydrate /diːˈhaɪdreɪt/ *vt* deshidratar
de-ice /diːˈaɪs/ *vt* descongelar
deign /deɪn/ *vi*. **~ to** dignarse
deity /ˈdiːɪtɪ/ *n* deidad *f*
deject|ed /dɪˈdʒektɪd/ *a* desanimado. **~ion** /-ʃn/ *n* abatimiento *m*
delay /dɪˈleɪ/ *vt* retardar; (*postpone*) aplazar. ● *vi* demorarse. ● *n* demora *f*
delectable /dɪˈlektəbl/ *a* deleitable
delegat|e /ˈdelɪgeɪt/ *vt* delegar. /ˈdelɪgət/ *n* delegado *m*. **~ion** /-ˈgeɪʃn/ *n* delegación *f*
delet|e /dɪˈliːt/ *vt* tachar. **~ion** /-ʃn/ *n* tachadura *f*
deliberat|e /dɪˈlɪbəreɪt/ *vt/i* deliberar. /dɪˈlɪbərət/ *a* intencionado; ‹*steps etc*› pausado. **~ely** *adv* a propósito. **~ion** /-ˈreɪʃn/ *n* deliberación *f*
delica|cy /ˈdelɪkəsɪ/ *n* delicadeza *f*; (*food*) manjar *m*; (*sweet food*) golosina *f*. **~te** /ˈdelɪkət/ *a* delicado
delicatessen /delɪkəˈtesn/ *n* charcutería *f* fina
delicious /dɪˈlɪʃəs/ *a* delicioso
delight /dɪˈlaɪt/ *n* placer *m*. ● *vt* encantar. ● *vi* deleitarse. **~ed** *a* encantado. **~ful** *a* delicioso
delineat|e /dɪˈlɪnɪeɪt/ *vt* delinear. **~ion** /-ˈeɪʃn/ *n* delineación *f*
delinquen|cy /dɪˈlɪŋkwənsɪ/ *n* delincuencia *f*. **~t** /dɪˈlɪŋkwənt/ *a* & *n* delincuente (*m* & *f*)
deliri|ous /dɪˈlɪrɪəs/ *a* delirante. **~um** *n* delirio *m*
deliver /dɪˈlɪvə(r)/ *vt* entregar; (*utter*) pronunciar; (*aim*) lanzar; (*set free*) librar; (*med*) asistir al parto de. **~ance** *n* liberación *f*. **~y** *n* entrega *f*; (*of post*) reparto *m*; (*med*) parto *m*
delta /ˈdeltə/ *n* (*geog*) delta *m*
delude /dɪˈluːd/ *vt* engañar. **~ o.s.** engañarse
deluge /ˈdeljuːdʒ/ *n* diluvio *m*
delusion /dɪˈluːʒn/ *n* ilusión *f*
de luxe /dɪˈlʌks/ *a* de lujo
delve /delv/ *vi* cavar. **~ into** (*investigate*) investigar
demagogue /ˈdeməgɒg/ *n* demagogo *m*
demand /dɪˈmɑːnd/ *vt* exigir. ● *n* petición *f*; (*claim*) reclamación *f*; (*com*) demanda *f*. **in ~** muy popular, muy solicitado. **on ~** a solicitud. **~ing** *a* exigente. **~s** *npl* exigencias *fpl*
demarcation /diːmɑːˈkeɪʃn/ *n* demarcación *f*
demean /dɪˈmiːn/ *vt*. **~ o.s.** degradarse. **~our** /dɪˈmiːnə(r)/ *n* conducta *f*
demented /dɪˈmentɪd/ *a* demente
demerara /deməˈreərə/ *n*. **~ (sugar)** *n* azúcar *m* moreno
demise /dɪˈmaɪz/ *n* fallecimiento *m*
demo /ˈdeməʊ/ *n* (*pl* **-os**) (*fam*) manifestación *f*
demobilize /diːˈməʊbəlaɪz/ *vt* desmovilizar
democra|cy /dɪˈmɒkrəsɪ/ *n* democracia *f*. **~t** /ˈdeməkræt/ *n* demócrata *m* & *f*. **~tic** /-ˈkrætɪk/ *a* democrático
demoli|sh /dɪˈmɒlɪʃ/ *vt* derribar. **~tion** /deməˈlɪʃn/ *n* demolición *f*
demon /ˈdiːmən/ *n* demonio *m*
demonstrat|e /ˈdemənstreɪt/ *vt* demostrar. ● *vi* manifestarse, hacer una manifestación. **~ion** /-ˈstreɪʃn/ *n* demostración *f*; (*pol etc*) manifestación *f*
demonstrative /dɪˈmɒnstrətɪv/ *a* demostrativo
demonstrator /ˈdemənstreɪtə(r)/ *n* demostrador *m*: (*pol etc*) manifestante *m* & *f*
demoralize /dɪˈmɒrəlaɪz/ *vt* desmoralizar
demote /dɪˈməʊt/ *vt* degradar
demure /dɪˈmjʊə(r)/ *a* recatado
den /den/ *n* (*of animal*) guarida *f*, madriguera *f*
denial /dɪˈnaɪəl/ *n* denegación *f*; (*statement*) desmentimiento *m*
denigrate /ˈdenɪgreɪt/ *vt* denigrar
denim /ˈdenɪm/ *n* dril *m* (de algodón azul grueso). **~s** *npl* pantalón *m* vaquero

Denmark /'denmɑ:k/ *n* Dinamarca *f*
denomination /dɪnɒmɪ'neɪʃn/ *n* denominación *f*; (*relig*) secta *f*
denote /dɪ'nəʊt/ *vt* denotar
denounce /dɪ'naʊns/ *vt* denunciar
dens|e /dens/ *a* (**-er**, **-est**) espeso; ‹*person*› torpe. **~ely** *adv* densamente. **~ity** *n* densidad *f*
dent /dent/ *n* abolladura *f*. ● *vt* abollar
dental /'dentl/ *a* dental. **~ surgeon** *n* dentista *m & f*
dentist /'dentɪst/ *n* dentista *m & f*. **~ry** *n* odontología *f*
denture /'dentʃə(r)/ *n* dentadura *f* postiza
denude /dɪ'nju:d/ *vt* desnudar; (*fig*) despojar
denunciation /dɪnʌnsɪ'eɪʃn/ *n* denuncia *f*
deny /dɪ'naɪ/ *vt* negar; desmentir ‹*rumour*›; (*disown*) renegar
deodorant /di'əʊdərənt/ *a & n* desodorante (*m*)
depart /dɪ'pɑ:t/ *vi* marcharse; ‹*train etc*› salir. **~ from** apartarse de
department /dɪ'pɑ:tment/ *n* departamento *m*; (*com*) sección *f*. **~ store** *n* grandes almacenes *mpl*
departure /dɪ'pɑ:tʃə(r)/ *n* partida *f*; (*of train etc*) salida *f*. **~ from** (*fig*) desviación *f*
depend /dɪ'pend/ *vi* depender. **~ on** depender de; (*rely*) contar con. **~able** *a* seguro. **~ant** /dɪ'pendənt/ *n* familiar *m & f* dependiente. **~ence** *n* dependencia *f*. **~ent** *a* dependiente. **be ~ent on** depender de
depict /dɪ'pɪkt/ *vt* pintar; (*in words*) describir
deplete /dɪ'pli:t/ *vt* agotar
deplor|able /dɪ'plɔ:rəbl/ *a* lamentable. **~e** /dɪ'plɔ:(r)/ *vt* lamentar
deploy /dɪ'plɔɪ/ *vt* desplegar. ● *vi* desplegarse
depopulate /di:'pɒpjʊleɪt/ *vt* despoblar
deport /dɪ'pɔ:t/ *vt* deportar. **~ation** /di:pɔ:'teɪʃn/ *n* deportación *f*
depose /dɪ'pəʊz/ *vt* deponer
deposit /dɪ'pɒzɪt/ *vt* (*pt* **deposited**) depositar. ● *n* depósito *m*. **~or** *n* depositante *m & f*
depot /'depəʊ/ *n* depósito *m*; (*Amer*) estación *f*
deprav|e /dɪ'preɪv/ *vt* depravar. **~ity** /-'prævətɪ/ *n* depravación *f*
deprecate /'deprɪkeɪt/ *vt* desaprobar
depreciat|e /dɪ'pri:ʃɪeɪt/ *vt* depreciar. ● *vi* depreciarse. **~ion** /-'eɪʃn/ *n* depreciación *f*
depress /dɪ'pres/ *vt* deprimir; (*press down*) apretar. **~ion** /-ʃn/ *n* depresión *f*
depriv|ation /deprɪ'veɪʃn/ *n* privación *f*. **~e** /dɪ'praɪv/ *vt*. **~ of** privar de
depth /depθ/ *n* profundidad *f*. **be out of one's ~** perder pie; (*fig*) meterse en honduras. **in the ~s of** en lo más hondo de
deputation /depjʊ'teɪʃn/ *n* diputación *f*
deputize /'depjʊtaɪz/ *vi*. **~ for** sustituir a
deputy /'depjʊtɪ/ *n* sustituto *m*. **~ chairman** *n* vicepresidente *m*
derail /dɪ'reɪl/ *vt* hacer descarrilar. **~ment** *n* descarrilamiento *m*
deranged /dɪ'reɪndʒd/ *a* ‹*mind*› trastornado
derelict /'derəlɪkt/ *a* abandonado
deri|de /dɪ'raɪd/ *vt* mofarse de. **~sion** /-'rɪʒn/ *n* mofa *f*. **~sive** *a* burlón. **~sory** /dɪ'raɪsərɪ/ *a* mofador; ‹*offer etc*› irrisorio
deriv|ation /derɪ'veɪʃn/ *n* derivación *f*. **~ative** /dɪ'rɪvətɪv/ *a & n* derivado (*m*). **~e** /dɪ'raɪv/ *vt/i* derivar
derogatory /dɪ'rɒgətrɪ/ *a* despectivo
derv /dɜ:v/ *n* gasóleo *m*
descen|d /dɪ'send/ *vt/i* descender, bajar. **~dant** *n* descendiente *m & f*. **~t** /dɪ'sent/ *n* descenso *m*; (*lineage*) descendencia *f*
descri|be /dɪs'kraɪb/ *vt* describir. **~ption** /-'krɪpʃn/ *n* descripción *f*. **~ptive** /-'krɪptɪv/ *a* descriptivo
desecrat|e /'desɪkreɪt/ *vt* profanar. **~ion** /-'kreɪʃn/ *n* profanación *f*
desert[1] /dɪ'zɜ:t/ *vt* abandonar. ● *vi* (*mil*) desertar
desert[2] /'dezət/ *a & n* desierto (*m*)
deserter /dɪ'zɜ:tə(r)/ *n* desertor *m*
deserts /dɪ'zɜ:ts/ *npl* lo merecido. **get one's ~** llevarse su merecido
deserv|e /dɪ'zɜ:v/ *vt* merecer. **~edly** *adv* merecidamente. **~ing** *a* ‹*person*› digno de; ‹*action*› meritorio
design /dɪ'zaɪn/ *n* diseño *m*; (*plan*) proyecto *m*; (*pattern*) modelo *m*; (*aim*) propósito *m*. **have ~s on**

poner la mira en. ● *vt* diseñar; (*plan*) proyectar

designat|e /ˈdezɪgneɪt/ *vt* designar; (*appoint*) nombrar. **~ion** /-ˈneɪʃn/ *n* denominación *f*; (*appointment*) nombramiento *m*

designer /dɪˈzaɪnə(r)/ *n* diseñador *m*; (*of clothing*) modisto *m*; (*in theatre*) escenógrafo *m*

desirab|ility /dɪzaɪərəˈbɪlətɪ/ *n* conveniencia *f*. **~le** /dɪˈzaɪrəbl/ *a* deseable

desire /dɪˈzaɪə(r)/ *n* deseo *m*. ● *vt* desear

desist /dɪˈzɪst/ *vi* desistir

desk /desk/ *n* escritorio *m*; (*at school*) pupitre *m*; (*in hotel*) recepción *f*; (*com*) caja *f*

desolat|e /ˈdesələt/ *a* desolado; (*uninhabited*) deshabitado. **~ion** /-ˈleɪʃn/ *n* desolación *f*

despair /dɪˈspeə(r)/ *n* desesperación *f*. ● *vi*. **~ of** desesperarse de

desperat|e /ˈdespərət/ *a* desesperado; (*dangerous*) peligroso. **~ely** *adv* desesperadamente. **~ion** /-ˈreɪʃn/ *n* desesperación *f*

despicable /dɪˈspɪkəbl/ *a* despreciable

despise /dɪˈspaɪz/ *vt* despreciar

despite /dɪˈspaɪt/ *prep* a pesar de

desponden|cy /dɪˈspɒndənsɪ/ *n* abatimiento *m*. **~t** /dɪˈspɒndənt/ *a* desanimado

despot /ˈdespɒt/ *n* déspota *m*

dessert /dɪˈzɜ:t/ *n* postre *m*. **~spoon** *n* cuchara *f* de postre

destination /destɪˈneɪʃn/ *n* destino *m*

destine /ˈdestɪn/ *vt* destinar

destiny /ˈdestɪnɪ/ *n* destino *m*

destitute /ˈdestɪtju:t/ *a* indigente. **~ of** desprovisto de

destroy /dɪˈstrɔɪ/ *vt* destruir

destroyer /dɪˈstrɔɪə(r)/ *n* (*naut*) destructor *m*

destructi|on /dɪˈstrʌkʃn/ *n* destrucción *f*. **~ve** *a* destructivo

desultory /ˈdesəltrɪ/ *a* irregular

detach /dɪˈtætʃ/ *vt* separar. **~able** *a* separable. **~ed** *a* separado. **~ed house** *n* chalet *m*. **~ment** /dɪˈtætʃmənt/ *n* separación *f*; (*mil*) destacamento *m*; (*fig*) indiferencia *f*

detail /ˈdi:teɪl/ *n* detalle *m*. ● *vt* detallar; (*mil*) destacar. **~ed** *a* detallado

detain /dɪˈteɪn/ *vt* detener; (*delay*) retener. **~ee** /di:teɪˈni:/ *n* detenido *m*

detect /dɪˈtekt/ *vt* percibir; (*discover*) descubrir. **~ion** /-ʃn/ *n* descubrimiento *m*, detección *f*. **~or** *n* detector *m*

detective /dɪˈtektɪv/ *n* detective *m*. **~ story** *n* novela *f* policíaca

detention /dɪˈtenʃn/ *n* detención *f*

deter /dɪˈtɜ:(r)/ *vt* (*pt* **deterred**) disuadir; (*prevent*) impedir

detergent /dɪˈtɜ:dʒənt/ *a* & *n* detergente (*m*)

deteriorat|e /dɪˈtɪərɪəreɪt/ *vi* deteriorarse. **~ion** /-ˈreɪʃn/ *n* deterioro *m*

determination /dɪtɜ:mɪˈneɪʃn/ *n* determinación *f*

determine /dɪˈtɜ:mɪn/ *vt* determinar; (*decide*) decidir. **~d** *a* determinado; (*resolute*) resuelto

deterrent /dɪˈterənt/ *n* fuerza *f* de disuasión

detest /dɪˈtest/ *vt* aborrecer. **~able** *a* odioso

detonat|e /ˈdetəneɪt/ *vt* hacer detonar. ● *vi* detonar. **~ion** /-ˈneɪʃn/ *n* detonación *f*. **~or** *n* detonador *m*

detour /ˈdi:tʊə(r)/ *n* desviación *f*

detract /dɪˈtrækt/ *vi*. **~ from** (*lessen*) disminuir

detriment /ˈdetrɪmənt/ *n* perjuicio *m*. **~al** /-ˈmentl/ *a* perjudicial

devalu|ation /di:væljuːˈeɪʃn/ *n* desvalorización *f*. **~e** /di:ˈvælju:/ *vt* desvalorizar

devastat|e /ˈdevəsteɪt/ *vt* devastar. **~ing** *a* devastador; (*fig*) arrollador

develop /dɪˈveləp/ *vt* desarrollar; contraer ‹*illness*›; urbanizar ‹*land*›. ● *vi* desarrollarse; (*show*) aparecerse. **~er** *n* (*foto*) revelador *m*. **~ing country** *n* país *m* en vías de desarrollo. **~ment** *n* desarrollo *m*. **(new) ~ment** novedad *f*

deviant /ˈdi:vɪənt/ *a* desviado

deviat|e /ˈdi:vɪeɪt/ *vi* desviarse. **~ion** /-ˈeɪʃn/ *n* desviación *f*

device /dɪˈvaɪs/ *n* dispositivo *m*; (*scheme*) estratagema *f*

devil /ˈdevl/ *n* diablo *m*. **~ish** *a* diabólico

devious /ˈdi:vɪəs/ *a* tortuoso

devise /dɪˈvaɪz/ *vt* idear

devoid /dɪˈvɔɪd/ *a*. **~ of** desprovisto de

devolution /di:vəˈlu:ʃn/ *n* descentralización *f*; (*of power*) delegación *f*

devot|e /dɪˈvəʊt/ *vt* dedicar. **~ed** *a* leal. **~edly** *adv* con devoción *f*. **~ee**

/devə'ti:/ *n* partidario *m*. **~ion** /-ʃn/ *n* dedicación *f*. **~ions** *npl* (*relig*) oraciones *fpl*

devour /dɪ'vaʊə(r)/ *vt* devorar

devout /dɪ'vaʊt/ *a* devoto

dew /dju:/ *n* rocío *m*

dext|erity /dek'sterətɪ/ *n* destreza *f*. **~(e)rous** /'dekstrəs/ *a* diestro

diabet|es /daɪə'bi:ti:z/ *n* diabetes *f*. **~ic** /-'betɪk/ *a* & *n* diabético (*m*)

diabolical /daɪə'bɒlɪkl/ *a* diabólico

diadem /'daɪədem/ *n* diadema *f*

diagnos|e /'daɪəgnəʊz/ *vt* diagnosticar. **~is** /daɪəg'nəʊsɪs/ *n* (*pl* **-oses** /-si:z/) diagnóstico *m*

diagonal /daɪ'ægənl/ *a* & *n* diagonal (*f*)

diagram /'daɪəgræm/ *n* diagrama *m*

dial /'daɪəl/ *n* cuadrante *m*; (*on phone*) disco *m*. ● *vt* (*pt* **dialled**) marcar

dialect /'daɪəlekt/ *n* dialecto *m*

dial: **~ling code** *n* prefijo *m*. **~ling tone** *n* señal *f* para marcar

dialogue /'daɪəlɒg/ *n* diálogo *m*

diameter /daɪ'æmɪtə(r)/ *n* diámetro *m*

diamond /'daɪəmənd/ *n* diamante *m*; (*shape*) rombo *m*. **~s** *npl* (*cards*) diamantes *mpl*

diaper /'daɪəpə(r)/ *n* (*Amer*) pañal *m*

diaphanous /daɪ'æfənəs/ *a* diáfano

diaphragm /'daɪəfræm/ *n* diafragma *m*

diarrhoea /daɪə'rɪə/ *n* diarrea *f*

diary /'daɪərɪ/ *n* diario *m*; (*book*) agenda *f*

diatribe /'daɪətraɪb/ *n* diatriba *f*

dice /daɪs/ *n* *invar* dado *m*. ● *vt* (*culin*) cortar en cubitos

dicey /'daɪsɪ/ *a* (*sl*) arriesgado

dictat|e /dɪk'teɪt/ *vt*/*i* dictar. **~es** /'dɪkteɪts/ *npl* dictados *mpl*. **~ion** /dɪk'teɪʃn/ *n* dictado *m*

dictator /dɪk'teɪtə(r)/ *n* dictador *m*. **~ship** *n* dictadura *f*

diction /'dɪkʃn/ *n* dicción *f*

dictionary /'dɪkʃənərɪ/ *n* diccionario *m*

did /dɪd/ *see* **do**

didactic /daɪ'dæktɪk/ *a* didáctico

diddle /'dɪdl/ *vt* (*sl*) estafar

didn't /'dɪdnt/ = **did not**

die[1] /daɪ/ *vi* (*pres p* **dying**) morir. **be dying to** morirse por. **~ down** disminuir. **~ out** extinguirse

die[2] /daɪ/ *n* (*tec*) cuño *m*

die-hard /'daɪhɑ:d/ *n* intransigente *m* & *f*

diesel /'di:zl/ *n* (*fuel*) gasóleo *m*. **~ engine** *n* motor *m* diesel

diet /'daɪət/ *n* alimentación *f*; (*restricted*) régimen *m*. ● *vi* estar a régimen. **~etic** /daɪə'tetɪk/ *a* dietético. **~itian** *n* dietético *m*

differ /'dɪfə(r)/ *vi* ser distinto; (*disagree*) no estar de acuerdo. **~ence** /'dɪfrəns/ *n* diferencia *f*; (*disagreement*) desacuerdo *m*. **~ent** /'dɪfrənt/ *a* distinto, diferente

differentia|l /dɪfə'renʃl/ *a* & *n* diferencial (*f*). **~te** /dɪfə'renʃɪeɪt/ *vt* diferenciar. ● *vi* diferenciarse

differently /'dɪfrəntlɪ/ *adv* de otra manera

difficult /'dɪfɪkəlt/ *a* difícil. **~y** *n* dificultad *f*

diffiden|ce /'dɪfɪdəns/ *n* falta *f* de confianza. **~t** /'dɪfɪdənt/ *a* que falta confianza

diffus|e /dɪ'fju:s/ *a* difuso. /dɪ'fju:z/ *vt* difundir. ● *vi* difundirse. **~ion** /-ʒn/ *n* difusión *f*

dig /dɪg/ *n* (*poke*) empujón *m*; (*poke with elbow*) codazo *m*; (*remark*) indirecta *f*; (*archaeol*) excavación *f*. ● *vt* (*pt* **dug**, *pres p* **digging**) cavar; (*thrust*) empujar. ● *vi* cavar. **~ out** extraer. **~ up** desenterrar. **~s** *npl* (*fam*) alojamiento *m*

digest /'daɪdʒest/ *n* resumen *m*. ● *vt* digerir. **~ible** *a* digerible. **~ion** /-ʃn/ *n* digestión *f*. **~ive** *a* digestivo

digger /'dɪgə(r)/ *n* (*mec*) excavadora *f*

digit /'dɪdʒɪt/ *n* cifra *f*; (*finger*) dedo *m*. **~al** /'dɪdʒɪtl/ *a* digital

dignif|ied /'dɪgnɪfaɪd/ *a* solemne. **~y** /'dɪgnɪfaɪ/ *vt* dignificar

dignitary /'dɪgnɪtərɪ/ *n* dignatario *m*

dignity /'dɪgnətɪ/ *n* dignidad *f*

digress /daɪ'gres/ *vi* divagar. **~ from** apartarse de. **~ion** /-ʃn/ *n* digresión *f*

dike /daɪk/ *n* dique *m*

dilapidated /dɪ'læpɪdeɪtɪd/ *a* ruinoso

dilat|e /daɪ'leɪt/ *vt* dilatar. ● *vi* dilatarse. **~ion** /-ʃn/ *n* dilatación *f*

dilatory /'dɪlətərɪ/ *a* dilatorio, lento

dilemma /daɪ'lemə/ *n* dilema *m*

diligen|ce /'dɪlɪdʒəns/ *n* diligencia *f*. **~t** /'dɪlɪdʒənt/ *a* diligente

dilly-dally /'dɪlɪdælɪ/ *vi* (*fam*) perder el tiempo

dilute /daɪ'lju:t/ *vt* diluir

dim /dɪm/ *a* (**dimmer**, **dimmest**) (*weak*) débil; (*dark*) oscuro; (*stupid*,

fam) torpe. ● *vt* (*pt* **dimmed**) amortiguar. ● *vi* apagarse. ~ **the headlights** bajar los faros
dime /daɪm/ *n* (*Amer*) moneda *f* de diez centavos
dimension /daɪ'menʃn/ *n* dimensión *f*
diminish /dɪ'mɪnɪʃ/ *vt/i* disminuir
diminutive /dɪ'mɪnjʊtɪv/ *a* diminuto. ● *n* diminutivo *m*
dimness /'dɪmnɪs/ *n* debilidad *f*; (*of room etc*) oscuridad *f*
dimple /'dɪmpl/ *n* hoyuelo *m*
din /dɪn/ *n* jaleo *m*
dine /daɪn/ *vi* cenar. ~**r** /-ə(r)/ *n* comensal *m* & *f*; (*rail*) coche *m* restaurante
dinghy /'dɪŋgɪ/ *n* (*inflatable*) bote *m* neumático
ding|iness /'dɪndʒɪnɪs/ *n* suciedad *f*. ~**y** /'dɪndʒɪ/ *a* (**-ier**, **-iest**) miserable, sucio
dining-room /'daɪnɪŋru:m/ *n* comedor *m*
dinner /'dɪnə(r)/ *n* cena *f*. ~**-jacket** *n* esmoquin *m*. ~ **party** *n* cena *f*
dinosaur /'daɪnəsɔ:(r)/ *n* dinosaurio *m*
dint /dɪnt/ *n*. **by** ~ **of** a fuerza de
diocese /'daɪəsɪs/ *n* diócesis *f*
dip /dɪp/ *vt* (*pt* **dipped**) sumergir. ● *vi* bajar. ~ **into** hojear ‹*book*›. ● *n* (*slope*) inclinación *f*; (*in sea*) baño *m*
diphtheria /dɪf'θɪərɪə/ *n* difteria *f*
diphthong /'dɪfθɒŋ/ *n* diptongo *m*
diploma /dɪ'pləʊmə/ *n* diploma *m*
diplomacy /dɪ'pləʊməsɪ/ *n* diplomacia *f*
diplomat /'dɪpləmæt/ *n* diplomático *m*. ~**ic** /-'mætɪk/ *a* diplomático
dipstick /'dɪpstɪk/ *n* (*auto*) varilla *f* del nivel de aceite
dire /daɪə(r)/ *a* (**-er**, **-est**) terrible; ‹*need, poverty*› extremo
direct /dɪ'rekt/ *a* directo. ● *adv* directamente. ● *vt* dirigir; (*show the way*) indicar
direction /dɪ'rekʃn/ *n* dirección *f*. ~**s** *npl* instrucciones *fpl*
directly /dɪ'rektlɪ/ *adv* directamente; (*at once*) en seguida. ● *conj* (*fam*) en cuanto
director /dɪ'rektə(r)/ *n* director *m*
directory /dɪ'rektərɪ/ *n* guía *f*
dirge /dɜ:dʒ/ *n* canto *m* fúnebre
dirt /dɜ:t/ *n* suciedad *f*. ~**-track** *n* (*sport*) pista *f* de ceniza. ~**y** /'dɜ:tɪ/ *a* (**-ier**, **-iest**) sucio. ~**y trick** *n* mala jugada *f*. ~**y word** *n* palabrota *f*. ● *vt* ensuciar
disability /dɪsə'bɪlətɪ/ *n* invalidez *f*
disable /dɪs'eɪbl/ *vt* incapacitar. ~**d** *a* minusválido
disabuse /dɪsə'bju:z/ *vt* desengañar
disadvantage /dɪsəd'vɑ:ntɪdʒ/ *n* desventaja *f*. ~**d** *a* desventajado
disagree /dɪsə'gri:/ *vi* no estar de acuerdo; ‹*food, climate*› sentar mal a. ~**able** /dɪsə'gri:əbl/ *a* desagradable. ~**ment** *n* desacuerdo *m*; (*quarrel*) riña *f*
disappear /dɪsə'pɪə(r)/ *vi* desaparecer. ~**ance** *n* desaparición *f*
disappoint /dɪsə'pɔɪnt/ *vt* desilusionar, decepcionar. ~**ment** *n* desilusión *f*, decepción *f*
disapprov|al /dɪsə'pru:vl/ *n* desaprobación *f*. ~**e** /dɪsə'pru:v/ *vi*. ~ **of** desaprobar
disarm /dɪs'ɑ:m/ *vt/i* desarmar. ~**ament** *n* desarme *m*
disarray /dɪsə'reɪ/ *n* desorden *m*
disast|er /dɪ'zɑ:stə(r)/ *n* desastre *m*. ~**rous** *a* catastrófico
disband /dɪs'bænd/ *vt* disolver. ● *vi* disolverse
disbelief /dɪsbɪ'li:f/ *n* incredulidad *f*
disc /dɪsk/ *n* disco *m*
discard /dɪs'kɑ:d/ *vt* descartar; abandonar ‹*beliefs etc*›
discern /dɪ'sɜ:n/ *vt* percibir. ~**ible** *a* perceptible. ~**ing** *a* perspicaz
discharge /dɪs'tʃɑ:dʒ/ *vt* descargar; cumplir ‹*duty*›; (*dismiss*) despedir; poner en libertad ‹*prisoner*›; (*mil*) licenciar. /'dɪstʃɑ:dʒ/ *n* descarga *f*; (*med*) secreción *f*; (*mil*) licenciamiento *m*; (*dismissal*) despedida *f*
disciple /dɪ'saɪpl/ *n* discípulo *m*
disciplin|arian /dɪsəplɪ'neərɪən/ *n* ordenancista *m* & *f*. ~**ary** *a* disciplinario. ~**e** /'dɪsɪplɪn/ *n* disciplina *f*. ● *vt* disciplinar; (*punish*) castigar
disc jockey /'dɪskdʒɒkɪ/ *n* (*on radio*) pinchadiscos *m & f invar*
disclaim /dɪs'kleɪm/ *vt* desconocer. ~**er** *n* renuncia *f*
disclos|e /dɪs'kləʊz/ *vt* revelar. ~**ure** /-ʒə(r)/ *n* revelación *f*
disco /'dɪskəʊ/ *n* (*pl* **-os**) (*fam*) discoteca *f*
discolo|ur /dɪs'kʌlə(r)/ *vt* decolorar. ● *vi* decolorarse. ~**ration** /-'reɪʃn/ *n* decoloración *f*

discomfort /dɪs'kʌmfət/ *n* malestar *m*; (*lack of comfort*) incomodidad *f*

disconcert /dɪskən'sɜ:t/ *vt* desconcertar

disconnect /dɪskə'nekt/ *vt* separar; (*elec*) desconectar

disconsolate /dɪs'kɒnsələt/ *a* desconsolado

discontent /dɪskən'tent/ *n* descontento *m*. **~ed** *a* descontento

discontinue /dɪskən'tɪnju:/ *vt* interrumpir

discord /'dɪskɔ:d/ *n* discordia *f*; (*mus*) disonancia *f*. **~ant** /-'skɔ:dənt/ *a* discorde; (*mus*) disonante

discothèque /'dɪskətek/ *n* discoteca *f*

discount /'dɪskaʊnt/ *n* descuento *m*. /dɪs'kaʊnt/ *vt* hacer caso omiso de; (*com*) descontar

discourage /dɪs'kʌrɪdʒ/ *vt* desanimar; (*dissuade*) disuadir

discourse /'dɪskɔ:s/ *n* discurso *m*

discourteous /dɪs'kɜ:tɪəs/ *a* descortés

discover /dɪs'kʌvə(r)/ *vt* descubrir. **~y** *n* descubrimiento *m*

discredit /dɪs'kredɪt/ *vt* (*pt* **discredited**) desacreditar. ● *n* descrédito *m*

discreet /dɪs'kri:t/ *a* discreto. **~ly** *adv* discretamente

discrepancy /dɪ'skrepənsɪ/ *n* discrepancia *f*

discretion /dɪ'skreʃn/ *n* discreción *f*

discriminat|e /dɪs'krɪmɪneɪt/ *vt/i* discriminar. **~e between** distinguir entre. **~ing** *a* perspicaz. **~ion** /-'neɪʃn/ *n* discernimiento *m*; (*bias*) discriminación *f*

discus /'dɪskəs/ *n* disco *m*

discuss /dɪ'skʌs/ *vt* discutir. **~ion** /-ʃn/ *n* discusión *f*

disdain /dɪs'deɪn/ *n* desdén *m*. ● *vt* desdeñar. **~ful** *a* desdeñoso

disease /dɪ'zi:z/ *n* enfermedad *f*. **~d** *a* enfermo

disembark /dɪsɪm'bɑ:k/ *vt/i* desembarcar

disembodied /dɪsɪm'bɒdɪd/ *a* incorpóreo

disenchant /dɪsɪn'tʃɑ:nt/ *vt* desencantar. **~ment** *n* desencanto *m*

disengage /dɪsɪn'geɪdʒ/ *vt* soltar. **~ the clutch** desembragar. **~ment** *n* soltura *f*

disentangle /dɪsɪn'tæŋgl/ *vt* desenredar

disfavour /dɪs'feɪvə(r)/ *n* desaprobación *f*. **fall into ~** ‹*person*› caer en desgracia; ‹*custom, word*› caer en desuso

disfigure /dɪs'fɪgə(r)/ *vt* desfigurar

disgorge /dɪs'gɔ:dʒ/ *vt* arrojar; ‹*river*› descargar; (*fig*) restituir

disgrace /dɪs'greɪs/ *n* deshonra *f*; (*disfavour*) desgracia *f*. ● *vt* deshonrar. **~ful** *a* vergonzoso

disgruntled /dɪs'grʌntld/ *a* descontento

disguise /dɪs'gaɪz/ *vt* disfrazar. ● *n* disfraz *m*. **in ~** disfrazado

disgust /dɪs'gʌst/ *n* repugnancia *f*, asco *m*. ● *vt* repugnar, dar asco. **~ing** *a* repugnante, asqueroso

dish /dɪʃ/ *n* plato *m*. ● *vt*. **~ out** (*fam*) distribuir. **~ up** servir. **~cloth** /'dɪʃklɒθ/ *n* bayeta *f*

dishearten /dɪs'hɑ:tn/ *vt* desanimar

dishevelled /dɪ'ʃevld/ *a* desaliñado; ‹*hair*› despeinado

dishonest /dɪs'ɒnɪst/ *a* ‹*person*› poco honrado; ‹*means*› fraudulento. **~y** *n* falta *f* de honradez

dishonour /dɪs'ɒnə(r)/ *n* deshonra *f*. ● *vt* deshonrar. **~able** *a* deshonroso. **~ably** *adv* deshonrosamente

dishwasher /'dɪʃwɒʃə(r)/ *n* lavaplatos *m & f*

disillusion /dɪsɪ'lu:ʒn/ *vt* desilusionar. **~ment** *n* desilusión

disincentive /dɪsɪn'sentɪv/ *n* freno *m*

disinclined /dɪsɪn'klaɪnd/ *a* poco dispuesto

disinfect /dɪsɪn'fekt/ *vt* desinfectar. **~ant** *n* desinfectante *m*

disinherit /dɪsɪn'herɪt/ *vt* desheredar

disintegrate /dɪs'ɪntɪgreɪt/ *vt* desintegrar. ● *vi* desintegrarse

disinterested /dɪs'ɪntrəstɪd/ *a* desinteresado

disjointed /dɪs'dʒɔɪntɪd/ *a* inconexo

disk /dɪsk/ *n* disco *m*

dislike /dɪs'laɪk/ *n* aversión *f*. ● *vt* tener aversión a

dislocat|e /'dɪsləkeɪt/ *vt* dislocar(se) ‹*limb*›. **~ion** /-'keɪʃn/ *n* dislocación *f*

dislodge /dɪs'lɒdʒ/ *vt* sacar; (*oust*) desalojar

disloyal /dɪs'lɔɪəl/ *a* desleal. **~ty** *n* deslealtad *f*

dismal /'dɪzməl/ *a* triste; (*bad*) fatal

dismantle /dɪs'mæntl/ *vt* desarmar

dismay /dɪsˈmeɪ/ *n* consternación *f*. ● *vt* consternar
dismiss /dɪsˈmɪs/ *vt* despedir; (*reject*) rechazar. **~al** *n* despedida *f*; (*of idea*) abandono *m*
dismount /dɪsˈmaʊnt/ *vi* apearse
disobedien|ce /dɪsəˈbiːdɪəns/ *n* desobediencia *f*. **~t** /dɪsəˈbiːdɪənt/ *a* desobediente
disobey /dɪsəˈbeɪ/ *vt/i* desobedecer
disorder /dɪsˈɔːdə(r)/ *n* desorden *m*; (*ailment*) trastorno *m*. **~ly** *a* desordenado
disorganize /dɪsˈɔːgənaɪz/ *vt* desorganizar
disorientate /dɪsˈɔːrɪənteɪt/ *vt* desorientar
disown /dɪsˈəʊn/ *vt* repudiar
disparaging /dɪsˈpærɪdʒɪŋ/ *a* despreciativo. **~ly** *adv* con desprecio
disparity /dɪsˈpærətɪ/ *n* disparidad *f*
dispassionate /dɪsˈpæʃənət/ *a* desapasionado
dispatch /dɪsˈpætʃ/ *vt* enviar. ● *n* envío *m*; (*report*) despacho *m*. **~-rider** *n* correo *m*
dispel /dɪsˈpel/ *vt* (*pt* **dispelled**) disipar
dispensable /dɪsˈpensəbl/ *a* prescindible
dispensary /dɪsˈpensərɪ/ *n* farmacia *f*
dispensation /dɪspenˈseɪʃn/ *n* distribución *f*; (*relig*) dispensa *f*
dispense /dɪsˈpens/ *vt* distribuir; (*med*) preparar; (*relig*) dispensar; administrar ‹*justice*›. **~ with** prescindir de. **~r** /-ə(r)/ *n* (*mec*) distribuidor *m* automático; (*med*) farmacéutico *m*
dispers|al /dɪˈspɜːsl/ *n* dispersión *f*. **~e** /dɪˈspɜːs/ *vt* dispersar. ● *vi* dispersarse
dispirited /dɪsˈpɪrɪtɪd/ *a* desanimado
displace /dɪsˈpleɪs/ *vt* desplazar
display /dɪsˈpleɪ/ *vt* mostrar; exhibir ‹*goods*›; manifestar ‹*feelings*›. ● *n* exposición *f*; (*of feelings*) manifestación *f*; (*pej*) ostentación *f*
displeas|e /dɪsˈpliːz/ *vt* desagradar. **be ~ed with** estar disgustado con. **~ure** /-ˈpleʒə(r)/ *n* desagrado *m*
dispos|able /dɪsˈpəʊzəbl/ *a* desechable. **~al** *n* (*of waste*) eliminación *f*. **at s.o.'s ~al** a la disposición de uno. **~e** /dɪsˈpəʊz/ *vt* disponer. **be well ~ed towards** estar bien dispuesto hacia. ● *vi*. **~e of** deshacerse de
disposition /dɪspəˈzɪʃn/ *n* disposición *f*
disproportionate /dɪsprəˈpɔːʃənət/ *a* desproporcionado
disprove /dɪsˈpruːv/ *vt* refutar
dispute /dɪsˈpjuːt/ *vt* disputar. ● *n* disputa *f*. **in ~** disputado
disqualif|ication /dɪskwɒlɪfɪˈkeɪʃn/ *n* descalificación *f*. **~y** /dɪsˈkwɒlɪfaɪ/ *vt* incapacitar; (*sport*) descalificar
disquiet /dɪsˈkwaɪət/ *n* inquietud *f*
disregard /dɪsrɪˈgɑːd/ *vt* no hacer caso de. ● *n* indiferencia *f* (**for** a)
disrepair /dɪsrɪˈpeə(r)/ *n* mal estado *m*
disreputable /dɪsˈrepjʊtəbl/ *a* de mala fama
disrepute /dɪsrɪˈpjuːt/ *n* discrédito *m*
disrespect /dɪsrɪsˈpekt/ *n* falta *f* de respeto
disrobe /dɪsˈrəʊb/ *vt* desvestir. ● *vi* desvestirse
disrupt /dɪsˈrʌpt/ *vt* interrumpir; trastornar ‹*plans*›. **~ion** /-ʃn/ *n* interrupción *f*; (*disorder*) desorganización *f*. **~ive** *a* desbaratador
dissatisfaction /dɪsætɪsˈfækʃn/ *n* descontento *m*
dissatisfied /dɪˈsætɪsfaɪd/ *a* descontento
dissect /dɪˈsekt/ *vt* disecar. **~ion** /-ʃn/ *n* disección *f*
disseminat|e /dɪˈsemɪneɪt/ *vt* diseminar. **~ion** /-ˈneɪʃn/ *n* diseminación *f*
dissent /dɪˈsent/ *vi* disentir. ● *n* disentimiento *m*
dissertation /dɪsəˈteɪʃn/ *n* disertación *f*; (*univ*) tesis *f*
disservice /dɪsˈsɜːvɪs/ *n* mal servicio *m*
dissident /ˈdɪsɪdənt/ *a & n* disidente (*m & f*)
dissimilar /dɪˈsɪmɪlə(r)/ *a* distinto
dissipate /ˈdɪsɪpeɪt/ *vt* disipar; (*fig*) desvanecer. **~d** *a* disoluto
dissociate /dɪˈsəʊʃɪeɪt/ *vt* disociar
dissolut|e /ˈdɪsəluːt/ *a* disoluto. **~ion** /dɪsəˈluːʃn/ *n* disolución *f*
dissolve /dɪˈzɒlv/ *vt* disolver. ● *vi* disolverse
dissuade /dɪˈsweɪd/ *vt* disuadir
distan|ce /ˈdɪstəns/ *n* distancia *f*. **from a ~ce** desde lejos. **in the ~ce** a

lo lejos. **~t** /'dɪstənt/ *a* lejano; (*aloof*) frío
distaste /dɪs'teɪst/ *n* aversión *f*. **~ful** *a* desagradable
distemper[1] /dɪ'stempə(r)/ *n* (*paint*) temple *m*. ● *vt* pintar al temple
distemper[2] /dɪ'stempə(r)/ *n* (*of dogs*) moquillo *m*
distend /dɪs'tend/ *vt* dilatar. ● *vi* dilatarse
distil /dɪs'tɪl/ *vt* (*pt* **distilled**) destilar. **~lation** /-'leɪʃn/ *n* destilación *f*. **~lery** /dɪs'tɪlərɪ/ *n* destilería *f*
distinct /dɪs'tɪŋkt/ *a* distinto; (*clear*) claro; (*marked*) marcado. **~ion** /-ʃn/ *n* distinción *f*; (*in exam*) sobresaliente *m*. **~ive** *a* distintivo. **~ly** *adv* claramente
distinguish /dɪs'tɪŋgwɪʃ/ *vt/i* distinguir. **~ed** *a* distinguido
distort /dɪs'tɔ:t/ *vt* torcer. **~ion** /-ʃn/ *n* deformación *f*
distract /dɪs'trækt/ *vt* distraer. **~ed** *a* aturdido. **~ing** *a* molesto. **~ion** /-ʃn/ *n* distracción *f*; (*confusion*) aturdimiento *m*
distraught /dɪs'trɔ:t/ *a* aturdido
distress /dɪs'tres/ *n* angustia *f*; (*poverty*) miseria *f*; (*danger*) peligro *m*. ● *vt* afligir. **~ing** *a* penoso
distribut|e /dɪs'trɪbju:t/ *vt* distribuir. **~ion** /-'bju:ʃn/ *n* distribución *f*. **~or** *n* distribuidor *m*; (*auto*) distribuidor *m* de encendido
district /'dɪstrɪkt/ *n* districto *m*; (*of town*) barrio *m*
distrust /dɪs'trʌst/ *n* desconfianza *f*. ● *vt* desconfiar de
disturb /dɪs'tɜ:b/ *vt* molestar; (*perturb*) inquietar; (*move*) desordenar; (*interrupt*) interrumpir. **~ance** *n* disturbio *m*; (*tumult*) alboroto *m*. **~ed** *a* trastornado. **~ing** *a* inquietante
disused /dɪs'ju:zd/ *a* fuera de uso
ditch /dɪtʃ/ *n* zanja *f*; (*for irrigation*) acequia *f*. ● *vt* (*sl*) abandonar
dither /'dɪðə(r)/ *vi* vacilar
ditto /'dɪtəʊ/ *adv* ídem
divan /dɪ'væn/ *n* diván *m*
dive /daɪv/ *vi* tirarse de cabeza; (*rush*) meterse (precipitadamente); (*underwater*) bucear. ● *n* salto *m*; (*of plane*) picado *m*; (*place, fam*) taberna *f*. **~r** *n* saltador *m*; (*underwater*) buzo *m*
diverge /daɪ'vɜ:dʒ/ *vi* divergir. **~nt** /daɪ'vɜ:dʒənt/ *a* divergente
divers|e /daɪ'vɜ:s/ *a* diverso. **~ify** /daɪ'vɜ:sɪfaɪ/ *vt* diversificar. **~ity** /daɪ'vɜ:sətɪ/ *n* diversidad *f*
diver|sion /daɪ'vɜ:ʃn/ *n* desvío *m*; (*distraction*) diversión *f*. **~t** /daɪ'vɜ:t/ *vt* desviar; (*entertain*) divertir
divest /daɪ'vest/ *vt*. **~ of** despojar de
divide /dɪ'vaɪd/ *vt* dividir. ● *vi* dividirse
dividend /'dɪvɪdend/ *n* dividendo *m*
divine /dɪ'vaɪn/ *a* divino
diving-board /'daɪvɪŋbɔ:d/ *n* trampolín *m*
diving-suit /'daɪvɪŋsu:t/ *n* escafandra *f*
divinity /dɪ'vɪnɪtɪ/ *n* divinidad *f*
division /dɪ'vɪʒn/ *n* división *f*
divorce /dɪ'vɔ:s/ *n* divorcio *m*. ● *vt* divorciarse de; ‹*judge*› divorciar. ● *vi* divorciarse. **~e** /dɪvɔ:'si:/ *n* divorciado *m*
divulge /daɪ'vʌldʒ/ *vt* divulgar
DIY *abbr see* **do-it-yourself**
dizz|iness /'dɪzɪnɪs/ *n* vértigo *m*. **~y** /'dɪzɪ/ *a* (**-ier, -iest**) mareado; ‹*speed*› vertiginoso. **be** *or* **feel ~y** marearse
do /du:/ *vt* (*3 sing pres* **does,** *pt* **did**, *pp* **done**) hacer; (*swindle, sl*) engañar. ● *vi* hacer; (*fare*) ir; (*be suitable*) convenir; (*be enough*) bastar. ● *n* (*pl* **dos** *or* **do's**) (*fam*) fiesta *f*. ● *v aux*. **~ you speak Spanish? Yes I ~** ¿habla Vd español? Sí. **doesn't he?**, **don't you?** ¿verdad? **~ come in!** (*emphatic*) ¡pase Vd! **~ away with** abolir. **~ in** (*exhaust, fam*) agotar; (*kill, sl*) matar. **~ out** (*clean*) limpiar. **~ up** abotonar ‹*coat etc*›; renovar ‹*house*›. **~ with** tener que ver con; (*need*) necesitar. **~ without** prescindir de. **~ne for** (*fam*) arruinado. **~ne in** (*fam*) agotado. **well ~ne** (*culin*) bien hecho. **well ~ne!** ¡muy bien!
docile /'dəʊsaɪl/ *a* dócil
dock[1] /dɒk/ *n* dique *m*. ● *vt* poner en dique. ● *vi* atracar al muelle
dock[2] /dɒk/ *n* (*jurid*) banquillo *m* de los acusados
dock: **~er** *n* estibador *m*. **~yard** /'dɒkjɑ:d/ *n* astillero *m*
doctor /'dɒktə(r)/ *n* médico *m*, doctor *m*; (*univ*) doctor *m*. ● *vt* castrar ‹*cat*›; (*fig*) adulterar
doctorate /'dɒktərət/ *n* doctorado *m*
doctrine /'dɒktrɪn/ *n* doctrina *f*
document /'dɒkjʊmənt/ *n* documento *m*. **~ary** /-'mentrɪ/ *a & n* documental (*m*)

doddering /'dɒdərɪŋ/ *a* chocho
dodge /dɒdʒ/ *vt* esquivar. ● *vi* esquivarse. ● *n* regate *m*; (*fam*) truco *m*
dodgems /'dɒdʒəmz/ *npl* autos *mpl* de choque
dodgy /'dɒdʒɪ/ *a* (**-ier, -iest**) (*awkward*) difícil
does /dʌz/ *see* **do**
doesn't /'dʌznt/ = **does not**
dog /dɒg/ *n* perro *m*. ● *vt* (*pt* **dogged**) perseguir. **~-collar** *n* (*relig, fam*) alzacuello *m*. **~-eared** *a* ‹*book*› sobado
dogged /'dɒgɪd/ *a* obstinado
doghouse /'dɒghaʊs/ *n* (*Amer*) perrera *f*. **in the ~** (*sl*) en desgracia
dogma /'dɒgmə/ *n* dogma *m*. **~tic** /-'mætɪk/ *a* dogmático
dogsbody /'dɒgzbɒdɪ/ *n* (*fam*) burro *m* de carga
doh /dəʊ/ *n* (*mus, first note of any musical scale*) do *m*
doily /'dɔɪlɪ/ *n* tapete *m*
doings /'du:ɪŋz/ *npl* (*fam*) actividades *fpl*
do-it-yourself /du:ɪtjɔ:'self/ (*abbr* **DIY**) *n* bricolaje *m*. **~ enthusiast** *n* manitas *m*
doldrums /'dɒldrəmz/ *npl*. **be in the ~** estar abatido
dole /dəʊl/ *vt*. **~ out** distribuir. ● *n* (*fam*) subsidio *m* de paro. **on the ~** (*fam*) parado
doleful /'dəʊlfl/ *a* triste
doll /dɒl/ *n* muñeca *f*. ● *vt*. **~ up** (*fam*) emperejilar
dollar /'dɒlə(r)/ *n* dólar *m*
dollop /'dɒləp/ *n* (*fam*) masa *f*
dolphin /'dɒlfɪn/ *n* delfín *m*
domain /dəʊ'meɪn/ *n* dominio *m*; (*fig*) campo *m*
dome /dəʊm/ *n* cúpula *f*. **~d** *a* abovedado
domestic /də'mestɪk/ *a* doméstico; ‹*trade, flights, etc*› nacional
domesticated *a* ‹*animal*› domesticado
domesticity /dɒme'stɪsətɪ/ *n* domesticidad *f*
domestic: **~ science** *n* economía *f* doméstica. **~ servant** *n* doméstico *m*
dominant /'dɒmɪnənt/ *a* dominante
dominat|e /'dɒmɪneɪt/ *vt/i* dominar. **~ion** /-'neɪʃn/ *n* dominación *f*
domineer /dɒmɪ'nɪə(r)/ *vi* tiranizar
Dominican Republic /dəmɪnɪkən rɪ'pʌblɪk/ *n* República *f* Dominicana
dominion /də'mɪnjən/ *n* dominio *m*
domino /'dɒmɪnəʊ/ *n* (*pl* **~es**) ficha *f* de dominó. **~es** *npl* (*game*) dominó *m*
don[1] /dɒn/ *n* profesor *m*
don[2] /dɒn/ *vt* (*pt* **donned**) ponerse
donat|e /dəʊ'neɪt/ *vt* donar. **~ion** /-ʃn/ *n* donativo *m*
done /dʌn/ *see* **do**
donkey /'dɒŋkɪ/ *n* burro *m*. **~-work** *n* trabajo *m* penoso
donor /'dəʊnə(r)/ *n* donante *m* & *f*
don't /dəʊnt/ = **do not**
doodle /'du:dl/ *vi* garrapatear
doom /du:m/ *n* destino *m*; (*death*) muerte *f*. ● *vt*. **be ~ed to** ser condenado a
doomsday /'du:mzdeɪ/ *n* día *m* del juicio final
door /dɔ:(r)/ *n* puerta *f*. **~man** /'dɔ:mən/ *n* (*pl* **-men**) portero *m*. **~mat** /'dɔ:mæt/ *n* felpudo *m*. **~step** /'dɔ:step/ *n* peldaño *m*. **~way** /'dɔ:weɪ/ *n* entrada *f*
dope /dəʊp/ *n* (*fam*) droga *f*; (*idiot, sl*) imbécil *m*. ● *vt* (*fam*) drogar. **~y** *a* (*sl*) torpe
dormant /'dɔ:mənt/ *a* inactivo
dormer /'dɔ:mə(r)/ *n*. **~ (window)** buhardilla *f*
dormitory /'dɔ:mɪtrɪ/ *n* dormitorio *m*
dormouse /'dɔ:maʊs/ *n* (*pl* **-mice**) lirón *m*
dos|age /'dəʊsɪdʒ/ *n* dosis *f*. **~e** /dəʊs/ *n* dosis *f*
doss /dɒs/ *vi* (*sl*) dormir. **~-house** *n* refugio *m*
dot /dɒt/ *n* punto *m*. **on the ~** en punto. ● *vt* (*pt* **dotted**) salpicar. **be ~ted with** estar salpicado de
dote /dəʊt/ *vi*. **~ on** adorar
dotted line /dɒtɪd'laɪn/ *n* línea *f* de puntos
dotty /'dɒtɪ/ *a* (**-ier, -iest**) (*fam*) chiflado
double /'dʌbl/ *a* doble. ● *adv* doble, dos veces. ● *n* doble *m*; (*person*) doble *m* & *f*. **at the ~** corriendo. ● *vt* doblar; redoblar ‹*efforts etc*›. ● *vi* doblarse. **~-bass** *n* contrabajo *m*. **~ bed** *n* cama *f* de matrimonio. **~-breasted** *a* cruzado. **~ chin** *n* papada *f*. **~-cross** *vt* traicionar. **~-dealing** *n* doblez *m* & *f*. **~-decker** *n* autobús *m* de dos pisos. **~ Dutch** *n* galimatías

m. **~-jointed** *a* con articulaciones dobles. **~s** *npl* (*tennis*) doble *m*

doubt /daʊt/ *n* duda *f.* ● *vt* dudar; (*distrust*) dudar de, desconfiar de. **~ful** *a* dudoso. **~less** *adv* sin duda

doubly /ˈdʌblɪ/ *adv* doblemente

dough /dəʊ/ *n* masa *f*; (*money, sl*) dinero *m*, pasta *f* (*sl*)

doughnut /ˈdəʊnʌt/ *n* buñuelo *m*

douse /daʊs/ *vt* mojar; apagar ‹*fire*›

dove /dʌv/ *n* paloma *f*

dowager /ˈdaʊədʒə(r)/ *n* viuda *f* (con bienes o título del marido)

dowdy /ˈdaʊdɪ/ *a* (**-ier, -iest**) poco atractivo

down[1] /daʊn/ *adv* abajo. **~ with** abajo. **come ~** bajar. **go ~** bajar; ‹*sun*› ponerse. ● *prep* abajo. ● *a* (*sad*) triste. ● *vt* derribar; (*drink, fam*) beber

down[2] /daʊn/ *n* (*feathers*) plumón *m*

down-and-out /ˈdaʊnəndˈaʊt/ *n* vagabundo *m*

downcast /ˈdaʊnkɑ:st/ *a* abatido

downfall /ˈdaʊnfɔ:l/ *n* caída *f*; (*fig*) perdición *f*

downgrade /daʊnˈgreɪd/ *vt* degradar

down-hearted /daʊnˈhɑ:tɪd/ *a* abatido

downhill /daʊnˈhɪl/ *adv* cuesta abajo

down payment /ˈdaʊnpeɪmənt/ *n* depósito *m*

downpour /ˈdaʊnpɔ:(r)/ *n* aguacero *m*

downright /ˈdaʊnraɪt/ *a* completo; (*honest*) franco. ● *adv* completamente

downs /daʊnz/ *npl* colinas *fpl*

downstairs /daʊnˈsteəz/ *adv* abajo. /ˈdaʊnsteəz/ *a* de abajo

downstream /ˈdaʊnstri:m/ *adv* río abajo

down-to-earth /daʊntʊˈɜ:θ/ *a* práctico

downtrodden /ˈdaʊntrɒdn/ *a* oprimido

down: **~ under** en las antípodas; (*in Australia*) en Australia. **~ward** /ˈdaʊnwəd/ *a & adv*, **~wards** *adv* hacia abajo

dowry /ˈdaʊərɪ/ *n* dote *f*

doze /dəʊz/ *vi* dormitar. **~ off** dormirse, dar una cabezada. ● *n* sueño *m* ligero

dozen /ˈdʌzn/ *n* docena *f.* **~s of** (*fam*) miles de, muchos

Dr *abbr* (*Doctor*) Dr, Doctor *m.* **~ Broadley** (el) Doctor Broadley

drab /dræb/ *a* monótono

draft /drɑ:ft/ *n* borrador *m*; (*outline*) bosquejo *m*; (*com*) letra *f* de cambio; (*Amer, mil*) reclutamiento *m*; (*Amer, of air*) corriente *f* de aire. ● *vt* bosquejar; (*mil*) destacar; (*Amer, conscript*) reclutar

drag /dræg/ *vt* (*pt* **dragged**) arrastrar; rastrear ‹*river*›. ● *vi* arrastrarse por el suelo. ● *n* (*fam*) lata *f.* **in ~** (*man, sl*) vestido de mujer

dragon /ˈdrægən/ *n* dragón *m*

dragon-fly /ˈdrægənflaɪ/ *n* libélula *f*

drain /dreɪn/ *vt* desaguar; apurar ‹*tank, glass*›; (*fig*) agotar. ● *vi* escurrirse. ● *n* desaguadero *m.* **be a ~ on** agotar. **~ing-board** *n* escurridero *m*

drama /ˈdrɑ:mə/ *n* drama *m*; (*art*) arte *m* teatral. **~tic** /drəˈmætɪk/ *a* dramático. **~tist** /ˈdræmətɪst/ *n* dramaturgo *m.* **~tize** /ˈdræmətaɪz/ *vt* adaptar al teatro; (*fig*) dramatizar

drank /dræŋk/ *see* **drink**

drape /dreɪp/ *vt* cubrir; (*hang*) colgar. **~s** *npl* (*Amer*) cortinas *fpl*

drastic /ˈdræstɪk/ *a* drástico

draught /drɑ:ft/ *n* corriente *f* de aire. **~ beer** *n* cerveza *f* de barril. **~s** *n pl* (*game*) juego *m* de damas

draughtsman /ˈdrɑ:ftsmən/ *n* (*pl* **-men**) diseñador *m*

draughty /ˈdrɑ:ftɪ/ *a* lleno de corrientes de aire

draw /drɔ:/ *vt* (*pt* **drew**, *pp* **drawn**) tirar; (*attract*) atraer; dibujar ‹*picture*›; trazar ‹*line*›; retirar ‹*money*›. **~ the line at** trazar el límite. ● *vi* (*sport*) empatar; dibujar ‹*pictures*›; (*in lottery*) sortear. ● *n* (*sport*) empate *m*; (*in lottery*) sorteo *m.* **~ in** ‹*days*› acortarse. **~ out** sacar ‹*money*›. **~ up** pararse; redactar ‹*document*›; acercar ‹*chair*›

drawback /ˈdrɔ:bæk/ *n* desventaja *f*

drawbridge /ˈdrɔ:brɪdʒ/ *n* puente *m* levadizo

drawer /drɔ:(r)/ *n* cajón *m.* **~s** /drɔ:z/ *npl* calzoncillos *mpl*; (*women's*) bragas *fpl*

drawing /ˈdrɔ:ɪŋ/ *n* dibujo *m.* **~-pin** *n* chinche *m*, chincheta *f*

drawing-room /ˈdrɔ:ɪŋru:m/ *n* salón *m*

drawl /drɔ:l/ *n* habla *f* lenta

drawn /drɔ:n/ *see* **draw**. ● *a* ‹*face*› ojeroso

dread /dred/ *n* terror *m*. ● *vt* temer. **~ful** /'dredfl/ *a* terrible. **~fully** *adv* terriblemente

dream /dri:m/ *n* sueño *m*. ● *vt/i* (*pt* **dreamed** *or* **dreamt**) soñar. ● *a* ideal. **~ up** idear. **~er** *n* soñador *m*. **~y** *a* soñador

drear|iness /'drɪərɪnɪs/ *n* tristeza *f*; (*monotony*) monotonía *f*. **~y** /'drɪərɪ/ *a* (**-ier, -iest**) triste; (*boring*) monótono

dredge[1] /dredʒ/ *n* draga *f*. ● *vt* dragar

dredge[2] /dredʒ/ *n* (*culin*) espolvorear

dredger[1] /'dredʒə(r)/ *n* draga *f*

dredger[2] /'dredʒə(r)/ *n* (*for sugar*) espolvoreador *m*

dregs /dregz/ *npl* heces *fpl*; (*fig*) hez *f*

drench /drentʃ/ *vt* empapar

dress /dres/ *n* vestido *m*; (*clothing*) ropa *f*. ● *vt* vestir; (*decorate*) adornar; (*med*) vendar; (*culin*) aderezar, aliñar. ● *vi* vestirse. **~ circle** *n* primer palco *m*

dresser[1] /'dresə(r)/ *n* (*furniture*) aparador *m*

dresser[2] /'dresə(r)/ *n* (*in theatre*) camarero *m*

dressing /'dresɪŋ/ *n* (*sauce*) aliño *m*; (*bandage*) vendaje *m*. **~-case** *n* neceser *m*. **~-down** *n* rapapolvo *m*, reprensión *f*. **~-gown** *n* bata *f*. **~-room** *n* tocador *m*; (*in theatre*) camarín *m*. **~-table** *n* tocador *m*

dressmak|er /'dresmeɪkə(r)/ *n* modista *m & f*. **~ing** *n* costura *f*

dress rehearsal /'dresrɪhɜ:sl/ *n* ensayo *m* general

dressy /'dresɪ/ *a* (**-ier, -iest**) elegante

drew /dru:/ *see* **draw**

dribble /'drɪbl/ *vi* gotear; ‹*baby*› babear; (*in football*) regatear

dribs and drabs /drɪbzn'dræbz/ *npl*. **in ~** poco a poco, en cantidades pequeñas

drie|d /draɪd/ *a* ‹*food*› seco; ‹*fruit*› paso. **~r** /'draɪə(r)/ *n* secador *m*

drift /drɪft/ *vi* ir a la deriva; ‹*snow*› amontonarse. ● *n* (*movement*) dirección *f*; (*of snow*) montón *m*; (*meaning*) significado *m*. **~er** *n* persona *f* sin rumbo. **~wood** /'drɪftwʊd/ *n* madera *f* flotante

drill /drɪl/ *n* (*tool*) taladro *m*; (*training*) ejercicio *m*; (*fig*) lo normal. ● *vt* taladrar, perforar; (*train*) entrenar. ● *vi* entrenarse

drily /'draɪlɪ/ *adv* secamente

drink /drɪŋk/ *vt/i* (*pt* **drank**, *pp* **drunk**) beber. ● *n* bebida *f*. **~able** *a* bebible; ‹*water*› potable. **~er** *n* bebedor *m*. **~ing-water** *n* agua *f* potable

drip /drɪp/ *vi* (*pt* **dripped**) gotear. ● *n* gota *f*; (*med*) goteo *m* intravenoso; (*person, sl*) mentecato *m*. **~-dry** *a* que no necesita plancharse

dripping /'drɪpɪŋ/ *n* (*culin*) pringue *m*

drive /draɪv/ *vt* (*pt* **drove,** *pp* **driven**) empujar; conducir, manejar (*LAm*) ‹*car etc*›. **~ in** clavar ‹*nail*›. **~ s.o. mad** volver loco a uno. ● *vi* conducir. **~ in** (*in car*) entrar en coche. ● *n* paseo *m*; (*road*) calle *f*; (*private road*) camino *m* de entrada; (*fig*) energía *f*; (*pol*) campaña *f*. **~ at** querer decir. **~r** /'draɪvə(r)/ *n* conductor *m*, chófer *m* (*LAm*)

drivel /'drɪvl/ *n* tonterías *fpl*

driving /'draɪvɪŋ/ *n* conducción *f*. **~-licence** *n* carné *m* de conducir. **~ school** *n* autoescuela *f*

drizzl|e /'drɪzl/ *n* llovizna *f*. ● *vi* lloviznar. **~y** *a* lloviznoso

dromedary /'drɒmədərɪ/ *n* dromedario *m*

drone /drəʊn/ *n* (*noise*) zumbido *m*; (*bee*) zángano *m*. ● *vi* zumbar; (*fig*) hablar en voz monótona; (*idle, fam*) holgazanear

drool /dru:l/ *vi* babear

droop /dru:p/ *vt* inclinar. ● *vi* inclinarse; ‹*flowers*› marchitarse

drop /drɒp/ *n* gota *f*; (*fall*) caída *f*; (*decrease*) baja *f*; (*of cliff*) precipicio *m*. ● *vt* (*pt* **dropped**) dejar caer; (*lower*) bajar. ● *vi* caer. **~ in on** pasar por casa de. **~ off** (*sleep*) dormirse. **~ out** retirarse; ‹*student*› abandonar los estudios. **~-out** *n* marginado *m*

droppings /'drɒpɪŋz/ *npl* excremento *m*

dross /drɒs/ *n* escoria *f*

drought /draʊt/ *n* sequía *f*

drove[1] /drəʊv/ *see* **drive**

drove[2] /drəʊv/ *n* manada *f*

drown /draʊn/ *vt* ahogar. ● *vi* ahogarse

drowsy /'draʊzɪ/ *a* soñoliento

drudge /drʌdʒ/ *n* esclavo *m* del trabajo. **~ry** /-ərɪ/ *n* trabajo *m* pesado

drug /drʌg/ *n* droga *f*; (*med*) medicamento *m*. ● *vt* (*pt* **drugged**) drogar. **~ addict** *n* toxicómano *m*

drugstore /ˈdrʌgstɔ:(r)/ *n* (*Amer*) farmacia *f* (que vende otros artículos también)
drum /drʌm/ *n* tambor *m*; (*for oil*) bidón *m*. ● *vi* (*pt* **drummed**) tocar el tambor. ● *vt*. ~ **into s.o.** inculcar en la mente de uno. ~**mer** *n* tambor *m*; (*in group*) batería *f*. ~**s** *npl* batería *f*. ~**stick** /ˈdrʌmstɪk/ *n* baqueta *f*; (*culin*) pierna *f* (de pollo)
drunk /drʌŋk/ *see* **drink**. ● *a* borracho. **get** ~ emborracharse. ~**ard** *n* borracho *m*. ~**en** *a* borracho. ~**enness** *n* embriaguez *f*
dry /draɪ/ *a* (**drier, driest**) seco. ● *vt* secar. ● *vi* secarse. ~ **up** (*fam*) secar los platos. ~**-clean** *vt* limpiar en seco. ~**-cleaner** *n* tintorero *m*. ~**cleaner's** (*shop*) tintorería *f*. ~**ness** *n* sequedad *f*
dual /ˈdju:əl/ *a* doble. ~ **carriageway** *n* autovía *f*, carretera *f* de doble calzada. ~**-purpose** *a* de doble uso
dub /dʌb/ *vt* (*pt* **dubbed**) doblar ‹*film*›; (*nickname*) apodar
dubious /ˈdju:bɪəs/ *a* dudoso; ‹*person*› sospechoso
duchess /ˈdʌtʃɪs/ *n* duquesa *f*
duck[1] /dʌk/ *n* pato *m*
duck[2] /dʌk/ *vt* sumergir; bajar ‹*head etc*›. ● *vi* agacharse
duckling /ˈdʌklɪŋ/ *n* patito *m*
duct /dʌkt/ *n* conducto *m*
dud /dʌd/ *a* inútil; ‹*cheque*› sin fondos; ‹*coin*› falso
due /dju:/ *a* debido; (*expected*) esperado. ~ **to** debido a. ● *adv*. ~ **north** *n* derecho hacia el norte. ~**s** *npl* derechos *mpl*
duel /ˈdju:əl/ *n* duelo *m*
duet /dju:ˈet/ *n* dúo *m*
duffle /ˈdʌfl/ *a*. ~ **bag** *n* bolsa *f* de lona. ~**-coat** *n* trenca *f*
dug /dʌg/ *see* **dig**
duke /dju:k/ *n* duque *m*
dull /dʌl/ *a* (**-er, -est**) ‹*weather*› gris; ‹*colour*› apagado; ‹*person, play, etc*› pesado; ‹*sound*› sordo; (*stupid*) torpe. ● *vt* aliviar ‹*pain*›; entorpecer ‹*mind*›
duly /ˈdju:lɪ/ *adv* debidamente
dumb /dʌm/ *a* (**-er, -est**) mudo; (*fam*) estúpido
dumbfound /dʌmˈfaʊnd/ *vt* pasmar
dummy /ˈdʌmɪ/ *n* muñeco *m*; (*of tailor*) maniquí *m*; (*of baby*) chupete *m*. ● *a* falso. ~ **run** *n* prueba *f*
dump /dʌmp/ *vt* descargar; (*fam*) deshacerse de. ● *n* vertedero *m*; (*mil*) depósito *m*; (*fam*) lugar *m* desagradable. **be down in the** ~**s** estar deprimido
dumpling /ˈdʌmplɪŋ/ *n* bola *f* de masa hervida
dumpy /ˈdʌmpɪ/ *a* (**-ier, -iest**) regordete
dunce /dʌns/ *n* burro *m*
dung /dʌŋ/ *n* excremento *m*; (*manure*) estiércol *m*
dungarees /dʌŋgəˈri:z/ *npl* mono *m*, peto *m*
dungeon /ˈdʌndʒən/ *n* calabozo *m*
dunk /dʌŋk/ *vt* remojar
duo /ˈdju:əʊ/ *n* dúo *m*
dupe /dju:p/ *vt* engañar. ● *n* inocentón *m*
duplicat|e /ˈdju:plɪkət/ *a* & *n* duplicado (*m*). /ˈdju:plɪkeɪt/ *vt* duplicar; (*on machine*) reproducir. ~**or** *n* multicopista *f*
duplicity /dju:ˈplɪsətɪ/ *n* doblez *f*
durable /ˈdjʊərəbl/ *a* resistente; (*enduring*) duradero
duration /djʊˈreɪʃn/ *n* duración *f*
duress /djʊˈres/ *n* coacción *f*
during /ˈdjʊərɪŋ/ *prep* durante
dusk /dʌsk/ *n* crepúsculo *m*
dusky /ˈdʌskɪ/ *a* (**-ier, -iest**) oscuro
dust /dʌst/ *n* polvo *m*. ● *vt* quitar el polvo a; (*sprinkle*) espolvorear
dustbin /ˈdʌstbɪn/ *n* cubo *m* de la basura
dust-cover /ˈdʌstkʌvə(r)/ *n* sobrecubierta *f*
duster /ˈdʌstə(r)/ *n* trapo *m*
dust-jacket /ˈdʌstdʒækɪt/ *n* sobrecubierta *f*
dustman /ˈdʌstmən/ *n* (*pl* **-men**) basurero *m*
dustpan /ˈdʌstpæn/ *n* recogedor *m*
dusty /ˈdʌstɪ/ *a* (**-ier, -iest**) polvoriento
Dutch /dʌtʃ/ *a* & *n* holandés (*m*). **go** ~ pagar a escote. ~**man** *m* holandés *m*. ~**woman** *n* holandesa *f*
dutiful /ˈdju:tɪfl/ *a* obediente
duty /ˈdju:tɪ/ *n* deber *m*; (*tax*) derechos *mpl* de aduana. **on** ~ de servicio. ~**-free** *a* libre de impuestos
duvet /ˈdju:veɪ/ *n* edredón *m*
dwarf /dwɔ:f/ *n* (*pl* **-s**) enano *m*. ● *vt* empequeñecer
dwell /dwel/ *vi* (*pt* **dwelt**) morar. ~ **on** dilatarse. ~**er** *n* habitante *m* & *f*. ~**ing** *n* morada *f*
dwindle /ˈdwɪndl/ *vi* disminuir

dye /daɪ/ *vt* (*pres p* **dyeing**) teñir. ● *n* tinte *m*
dying /'daɪɪŋ/ *see* **die**
dynamic /daɪ'næmɪk/ *a* dinámico. **~s** *npl* dinámica *f*
dynamite /'daɪnəmaɪt/ *n* dinamita *f*. ● *vt* dinamitar
dynamo /'daɪnəməʊ/ *n* dinamo *f*, dínamo *f*
dynasty /'dɪnəstɪ/ *n* dinastía *f*
dysentery /'dɪsəntrɪ/ *n* disentería *f*
dyslexia /dɪs'leksɪə/ *n* dislexia *f*

E

each /i:tʃ/ *a* cada. ● *pron* cada uno. **~ one** cada uno. **~ other** uno a otro, el uno al otro. **they love ~ other** se aman
eager /'i:gə(r)/ *a* impaciente; (*enthusiastic*) ávido. **~ly** *adv* con impaciencia. **~ness** *n* impaciencia *f*, ansia *f*
eagle /'i:gl/ *n* águila *f*
ear[1] /ɪə(r)/ *n* oído *m*; (*outer*) oreja *f*
ear[2] /ɪə(r)/ *n* (*of corn*) espiga *f*
ear: **~ache** /'ɪəreɪk/ *n* dolor *m* de oído. **~-drum** *n* tímpano *m*
earl /ɜ:l/ *n* conde *m*
early /'ɜ:lɪ/ *a* (**-ier, -iest**) temprano; (*before expected time*) prematuro. **in the ~ spring** a principios de la primavera. ● *adv* temprano; (*ahead of time*) con anticipación
earmark /'ɪəmɑ:k/ *vt*. **~ for** destinar a
earn /ɜ:n/ *vt* ganar; (*deserve*) merecer
earnest /'ɜ:nɪst/ *a* serio. **in ~** en serio
earnings /'ɜ:nɪŋz/ *npl* ingresos *mpl*; (*com*) ganacias *fpl*
ear: **~phones** /'ɪəfəʊnz/ *npl* auricular *m*. **~-ring** *n* pendiente *m*
earshot /'ɪəʃɒt/ *n*. **within ~** al alcance del oído
earth /ɜ:θ/ *n* tierra *f*. ● *vt* (*elec*) conectar a tierra. **~ly** *a* terrenal
earthenware /'ɜ:θnweə(r)/ *n* loza *f* de barro
earthquake /'ɜ:θkweɪk/ *n* terremoto *m*
earthy /'ɜ:θɪ/ *a* terroso; (*coarse*) grosero
earwig /'ɪəwɪg/ *n* tijereta *f*
ease /i:z/ *n* facilidad *f*; (*comfort*) tranquilidad *f*. **at ~** a gusto; (*mil*) en posición de descanso. **ill at ~** molesto. **with ~** fácilmente. ● *vt* calmar; aliviar ‹*pain*›; tranquilizar ‹*mind*›; (*loosen*) aflojar. ● *vi* calmarse; (*lessen*) disminuir
easel /'i:zl/ *n* caballete *m*
east /i:st/ *n* este *m*, oriente *m*. ● *a* del este, oriental. ● *adv* hacia el este.
Easter /'i:stə(r)/ *n* Semana *f* Santa; (*relig*) Pascua *f* de Resurrección. **~ egg** *n* huevo *m* de Pascua
east: **~erly** *a* este; ‹*wind*› del este. **~ern** *a* del este, oriental. **~ward** *adv*, **~wards** *adv* hacia el este
easy /'i:zɪ/ *a* (**-ier, -iest**) fácil; (*relaxed*) tranquilo. **go ~ on** (*fam*) tener cuidado con. **take it ~** no preocuparse. ● *int* ¡despacio! **~ chair** *n* sillón *m*. **~-going** *a* acomodadizo
eat /i:t/ *vt/i* (*pt* **ate**, *pp* **eaten**) comer. **~ into** corroer. **~able** *a* comestible. **~er** *n* comedor *m*
eau-de-Cologne /əʊdəkə'ləʊn/ *n* agua *f* de colonia
eaves /i:vz/ *npl* alero *m*
eavesdrop /'i:vzdrɒp/ *vi* (*pt* **-dropped**) escuchar a escondidas
ebb /eb/ *n* reflujo *m*. ● *vi* bajar; (*fig*) decaer
ebony /'ebənɪ/ *n* ébano *m*
ebullient /ɪ'bʌlɪənt/ *a* exuberante
EC /i:'si:/ *abbr* (*European Community*) CE (Comunidad *f* Europea)
eccentric /ɪk'sentrɪk/ *a* & *n* excéntrico (*m*). **~ity** /eksen'trɪsətɪ/ *n* excentricidad *f*
ecclesiastical /ɪkli:zɪ'æstɪkl/ *a* eclesiástico
echelon /'eʃəlɒn/ *n* escalón *m*
echo /'ekəʊ/ *n* (*pl* **-oes**) eco *m*. ● *vt* (*pt* **echoed**, *pres p* **echoing**) repetir; (*imitate*) imitar. ● *vi* hacer eco
eclectic /ɪk'lektɪk/ *a* & *n* ecléctico (*m*)
eclipse /ɪ'klɪps/ *n* eclipse *m*. ● *vt* eclipsar
ecology /ɪ'kɒlədʒɪ/ *n* ecología *f*
econom|ic /i:kə'nɒmɪk/ *a* económico. **~ical** *a* económico. **~ics** *n* economía *f*. **~ist** /ɪ'kɒnəmɪst/ *n* economista *m* & *f*. **~ize** /ɪ'kɒnəmaɪz/ *vi* economizar. **~y** /ɪ'kɒnəmɪ/ *n* economía *f*
ecsta|sy /'ekstəsɪ/ *n* éxtasis *f*. **~tic** /ɪk'stætɪk/ *a* extático. **~tically** *adv* con éxtasis
Ecuador /'ekwədɔ:(r)/ *n* el Ecuador *m*

ecumenical /i:kju:'menɪkl/ *a* ecuménico
eddy /'edɪ/ *n* remolino *m*
edge /edʒ/ *n* borde *m*, margen *m*; (*of knife*) filo *m*; (*of town*) afueras *fpl*. **have the ~ on** (*fam*) llevar la ventaja a. **on ~** nervioso. ● *vt* ribetear; (*move*) mover poco a poco. ● *vi* avanzar cautelosamente. **~ways** *adv* de lado
edging /'edʒɪŋ/ *n* borde *m*; (*sewing*) ribete *m*
edgy /'edʒɪ/ *a* nervioso
edible /'edɪbl/ *a* comestible
edict /'i:dɪkt/ *n* edicto *m*
edifice /'edɪfɪs/ *n* edificio *m*
edify /'edɪfaɪ/ *vt* edificar
edit /'edɪt/ *vt* dirigir ‹*newspaper*›; preparar una edición de ‹*text*›; (*write*) redactar; montar ‹*film*›. **~ed by** a cargo de. **~ion** /ɪ'dɪʃn/ *n* edición *f*. **~or** /'edɪtə(r)/ *n* (*of newspaper*) director *m*; (*of text*) redactor *m*. **~orial** /edɪ'tɔ:rɪəl/ *a* editorial. ● *n* artículo *m* de fondo. **~or in chief** *n* jefe *m* de redacción
educat|e /'edʒʊkeɪt/ *vt* instruir, educar. **~ed** *a* culto. **~ion** /-'keɪʃn/ *n* enseñanza *f*; (*culture*) cultura *f*; (*upbringing*) educación *f*. **~ional** /-'keɪʃənl/ *a* instructivo
EEC /i:i:'si:/ *abbr* (*European Economic Community*) CEE (Comunidad *f* Económica Europea)
eel /i:l/ *n* anguila *f*
eerie /'ɪərɪ/ *a* (**-ier, -iest**) misterioso
efface /ɪ'feɪs/ *vt* borrar
effect /ɪ'fekt/ *n* efecto *m*. **in ~** efectivamente. **take ~** entrar en vigor. ● *vt* efectuar
effective /ɪ'fektɪv/ *a* eficaz; (*striking*) impresionante; (*mil*) efectivo. **~ly** *adv* eficazmente. **~ness** *n* eficacia *f*
effeminate /ɪ'femɪnət/ *a* afeminado
effervescent /efə'vesnt/ *a* efervescente
effete /ɪ'fi:t/ *a* agotado
efficien|cy /ɪ'fɪʃənsɪ/ *n* eficiencia *f*; (*mec*) rendimiento *m*. **~t** /ɪ'fɪʃnt/ *a* eficiente. **~tly** *adv* eficientemente
effigy /'efɪdʒɪ/ *n* efigie *f*
effort /'efət/ *n* esfuerzo *m*. **~less** *a* fácil
effrontery /ɪ'frʌntərɪ/ *n* descaro *m*
effusive /ɪ'fju:sɪv/ *a* efusivo
e.g. /i:'dʒi:/ *abbr* (*exempli gratia*) p.ej., por ejemplo
egalitarian /ɪgælɪ'teərɪən/ *a* & *n* igualitario (*m*)
egg[1] /eg/ *n* huevo *m*
egg[2] /eg/ *vt*. **~ on** (*fam*) incitar
egg-cup /'egkʌp/ *n* huevera *f*
egg-plant /'egplɑ:nt/ *n* berenjena *f*
eggshell /'egʃel/ *n* cáscara *f* de huevo
ego /'i:gəʊ/ *n* (*pl* **-os**) yo *m*. **~ism** *n* egoísmo *m*. **~ist** *n* egoísta *m* & *f*. **~centric** /i:gəʊ'sentrɪk/ *a* egocéntrico. **~tism** *n* egotismo *m*. **~tist** *n* egotista *m* & *f*
Egypt /'i:dʒɪpt/ *n* Egipto *m*. **~ian** /ɪ'dʒɪpʃn/ *a* & *n* egipcio (*m*)
eh /eɪ/ *int* (*fam*) ¡eh!
eiderdown /'aɪdədaʊn/ *n* edredón *m*
eight /eɪt/ *a* & *n* ocho (*m*)
eighteen /eɪ'ti:n/ *a* & *n* dieciocho (*m*). **~th** *a* & *n* decimoctavo (*m*)
eighth /eɪtθ/ *a* & *n* octavo (*m*)
eight|ieth /'eɪtɪəθ/ *a* & *n* ochenta (*m*), octogésimo (*m*). **~y** /'eɪtɪ/ *a* & *n* ochenta (*m*)
either /'aɪðə(r)/ *a* cualquiera de los dos; (*negative*) ninguno de los dos; (*each*) cada. ● *pron* uno u otro; (*with negative*) ni uno ni otro. ● *adv* (*negative*) tampoco. ● *conj* o. **~ he or** o él o; (*with negative*) ni él ni
ejaculate /ɪ'dʒækjʊleɪt/ *vt/i* (*exclaim*) exclamar
eject /ɪ'dʒekt/ *vt* expulsar, echar
eke /i:k/ *vt*. **~ out** hacer bastar; (*increase*) complementar
elaborate /ɪ'læbərət/ *a* complicado. /ɪ'læbəreɪt/ *vt* elaborar. ● *vi* explicarse
elapse /ɪ'læps/ *vi* (*of time*) transcurrir
elastic /ɪ'læstɪk/ *a* & *n* elástico (*m*). **~ band** *n* goma *f* (elástica)
elasticity /ɪlæ'stɪsətɪ/ *n* elasticidad *f*
elat|ed /ɪ'leɪtɪd/ *a* regocijado. **~ion** /-ʃn/ *n* regocijo *m*
elbow /'elbəʊ/ *n* codo *m*
elder[1] /'eldə(r)/ *a* & *n* mayor (*m*)
elder[2] /'eldə(r)/ *n* (*tree*) saúco *m*
elderly /'eldəlɪ/ *a* mayor, anciano
eldest /'eldɪst/ *a* & *n* el mayor (*m*)
elect /ɪ'lekt/ *vt* elegir. **~ to do** decidir hacer. ● *a* electo. **~ion** /-ʃn/ *n* elección *f*
elector /ɪ'lektə(r)/ *n* elector *m*. **~al** *a* electoral. **~ate** *n* electorado *m*
electric /ɪ'lektrɪk/ *a* eléctrico. **~al** *a* eléctrico. **~ blanket** *n* manta *f* eléctrica. **~ian** /ɪlek'trɪʃn/ *n* electricista

m & f. **~ity** /ɪlek'trɪsətɪ/ *n* electricidad *f*
electrify /ɪ'lektrɪfaɪ/ *vt* electrificar; (*fig*) electrizar
electrocute /ɪ'lektrəkju:t/ *vt* electrocutar
electrolysis /ɪlek'trɒlɪsɪs/ *n* electrólisis *f*
electron /ɪ'lektrɒn/ *n* electrón *m*
electronic /ɪlek'trɒnɪk/ *a* electrónico. **~s** *n* electrónica *f*
elegan|ce /'elɪgəns/ *n* elegancia *f.* **~t** /'elɪgənt/ *a* elegante. **~tly** *adv* elegantemente
element /'elɪmənt/ *n* elemento *m.* **~ary** /-'mentrɪ/ *a* elemental
elephant /'elɪfənt/ *n* elefante *m*
elevat|e /'elɪveɪt/ *vt* elevar. **~ion** /-'veɪʃn/ *n* elevación *f.* **~or** /'elɪveɪtə(r)/ *n* (*Amer*) ascensor *m*
eleven /ɪ'levn/ *a & n* once (*m*). **~th** *a & n* undécimo (*m*)
elf /elf/ *n* (*pl* **elves**) duende *m*
elicit /ɪ'lɪsɪt/ *vt* sacar
eligible /'elɪdʒəbl/ *a* elegible. **be ~ for** tener derecho a
eliminat|e /ɪ'lɪmɪneɪt/ *vt* eliminar. **~ion** /-'neɪʃn/ *n* eliminación *f*
élite /eɪ'li:t/ *n* elite *f*, élite *m*
elixir /ɪ'lɪksɪə(r)/ *n* elixir *m*
ellip|se /ɪ'lɪps/ *n* elipse *f.* **~tical** *a* elíptico
elm /elm/ *n* olmo *m*
elocution /elə'kju:ʃn/ *n* elocución *f*
elongate /'i:lɒŋgeɪt/ *vt* alargar
elope /ɪ'ləʊp/ *vi* fugarse con el amante. **~ment** *n* fuga *f*
eloquen|ce /'eləkwəns/ *n* elocuencia *f.* **~t** /'eləkwənt/ *a* elocuente. **~tly** *adv* con elocuencia
El Salvador /el'sælvədɔ:(r)/ *n* El Salvador *m*
else /els/ *adv* más. **everybody ~** todos los demás. **nobody ~** ningún otro, nadie más. **nothing ~** nada más. **or ~** o bien. **somewhere ~** en otra parte
elsewhere /els'weə(r)/ *adv* en otra parte
elucidate /ɪ'lu:sɪdeɪt/ *vt* aclarar
elude /ɪ'lu:d/ *vt* eludir
elusive /ɪ'lu:sɪv/ *a* esquivo
emaciated /ɪ'meɪʃɪeɪtɪd/ *a* esquelético
emanate /'eməneɪt/ *vi* emanar
emancipat|e /ɪ'mænsɪpeɪt/ *vt* emancipar. **~ion** /-'peɪʃn/ *n* emancipación *f*
embalm /ɪm'bɑ:m/ *vt* embalsamar
embankment /ɪm'bæŋkmənt/ *n* terraplén *m*; (*of river*) dique *m*
embargo /ɪm'bɑ:gəʊ/ *n* (*pl* **-oes**) prohibición *f*
embark /ɪm'bɑ:k/ *vt* embarcar. ● *vi* embarcarse. **~ on** (*fig*) emprender. **~ation** /embɑ:'keɪʃn/ *n* (*of people*) embarco *m*; (*of goods*) embarque *m*
embarrass /ɪm'bærəs/ *vt* desconcertar; (*shame*) dar vergüenza. **~ment** *n* desconcierto *m*; (*shame*) vergüenza *f*
embassy /'embəsɪ/ *n* embajada *f*
embed /ɪm'bed/ *vt* (*pt* **embedded**) embutir; (*fig*) fijar
embellish /ɪm'belɪʃ/ *vt* embellecer. **~ment** *n* embellecimiento *m*
embers /'embəz/ *npl* ascua *f*
embezzle /ɪm'bezl/ *vt* desfalcar. **~ment** *n* desfalco *m*
embitter /ɪm'bɪtə(r)/ *vt* amargar
emblem /'embləm/ *n* emblema *m*
embod|iment /ɪm'bɒdɪmənt/ *n* encarnación *f.* **~y** /ɪm'bɒdɪ/ *vt* encarnar; (*include*) incluir
emboss /ɪm'bɒs/ *vt* grabar en relieve, repujar. **~ed** *a* en relieve, repujado
embrace /ɪm'breɪs/ *vt* abrazar; (*fig*) abarcar. ● *vi* abrazarse. ● *n* abrazo *m*
embroider /ɪm'brɔɪdə(r)/ *vt* bordar. **~y** *n* bordado *m*
embroil /ɪm'brɔɪl/ *vt* enredar
embryo /'embrɪəʊ/ *n* (*pl* **-os**) embrión *m.* **~nic** /-'ɒnɪk/ *a* embrionario
emend /ɪ'mend/ *vt* enmendar
emerald /'emərəld/ *n* esmeralda *f*
emerge /ɪ'mɜ:dʒ/ *vi* salir. **~nce** /-əns/ *n* aparición *f*
emergency /ɪ'mɜ:dʒənsɪ/ *n* emergencia *f.* **in an ~** en caso de emergencia. **~ exit** *n* salida *f* de emergencia
emery /'emərɪ/ *n* esmeril *m.* **~-board** *n* lima *f* de uñas
emigrant /'emɪgrənt/ *n* emigrante *m & f*
emigrat|e /'emɪgreɪt/ *vi* emigrar. **~ion** /-'greɪʃn/ *n* emigración *f*
eminen|ce /'emɪnəns/ *n* eminencia *f.* **~t** /'emɪnənt/ *a* eminente. **~tly** *adv* eminentemente
emissary /'emɪsərɪ/ *n* emisario *m*
emission /ɪ'mɪʃn/ *n* emisión *f*
emit /ɪ'mɪt/ *vt* (*pt* **emitted**) emitir
emollient /ɪ'mɒlɪənt/ *a & n* emoliente (*m*)

emoti|on /ɪ'məʊʃn/ *n* emoción *f*. **~onal** *a* emocional; ‹*person*› emotivo; (*moving*) conmovedor. **~ve** /ɪ'məʊtɪv/ *a* emotivo

empathy /'empəθɪ/ *n* empatía *f*

emperor /'empərə(r)/ *n* emperador *m*

emphasi|s /'emfəsɪs/ *n* (*pl* **~ses** /-si:z/) énfasis *m*. **~ze** /'emfəsaɪz/ *vt* subrayar; (*single out*) destacar

emphatic /ɪm'fætɪk/ *a* categórico; (*resolute*) decidido

empire /'empaɪə(r)/ *n* imperio *m*

empirical /ɪm'pɪrɪkl/ *a* empírico

employ /ɪm'plɔɪ/ *vt* emplear. **~ee** /emplɔɪ'i:/ *n* empleado *m*. **~er** *n* patrón *m*. **~ment** *n* empleo *m*. **~ment agency** *n* agencia *f* de colocaciones

empower /ɪm'paʊə(r)/ *vt* autorizar (**to do** a hacer)

empress /'emprɪs/ *n* emperatriz *f*

empt|ies /'emptɪz/ *npl* envases *mpl*. **~iness** *n* vacío *m*. **~y** /'emptɪ/ *a* vacío; ‹*promise*› vano. **on an ~y stomach** con el estómago vacío. ● *vt* vaciar. ● *vi* vaciarse

emulate /'emjʊleɪt/ *vt* emular

emulsion /ɪ'mʌlʃn/ *n* emulsión *f*

enable /ɪ'neɪbl/ *vt*. **~ s.o. to** permitir a uno

enact /ɪ'nækt/ *vt* (*jurid*) decretar; (*in theatre*) representar

enamel /ɪ'næml/ *n* esmalte *m*. ● *vt* (*pt* **enamelled**) esmaltar

enamoured /ɪ'næməd/ *a*. **be ~ of** estar enamorado de

encampment /ɪn'kæmpmənt/ *n* campamento *m*

encase /ɪn'keɪs/ *vt* encerrar

enchant /ɪn'tʃɑ:nt/ *vt* encantar. **~ing** *a* encantador. **~ment** *n* encanto *m*

encircle /ɪn'sɜ:kl/ *vt* rodear

enclave /'enkleɪv/ *n* enclave *m*

enclos|e /ɪn'kləʊz/ *vt* cercar ‹*land*›; (*with letter*) adjuntar; (*in receptacle*) encerrar. **~ed** *a* ‹*space*› encerrado; (*com*) adjunto. **~ure** /ɪn'kləʊʒə(r)/ *n* cercamiento *m*; (*area*) recinto *m*; (*com*) documento *m* adjunto

encompass /ɪn'kʌmpəs/ *vt* cercar; (*include*) incluir, abarcar

encore /'ɒŋkɔ:(r)/ *int* ¡bis! ● *n* bis *m*, repetición *f*

encounter /ɪn'kaʊntə(r)/ *vt* encontrar. ● *n* encuentro *m*

encourage /ɪn'kʌrɪdʒ/ *vt* animar; (*stimulate*) estimular. **~ment** *n* estímulo *m*

encroach /ɪn'krəʊtʃ/ *vi*. **~ on** invadir ‹*land*›; quitar ‹*time*›. **~ment** *n* usurpación *f*

encumb|er /ɪn'kʌmbə(r)/ *vt* (*hamper*) estorbar; (*burden*) cargar. **be ~ered with** estar cargado de. **~rance** *n* estorbo *m*; (*burden*) carga *f*

encyclical /ɪn'sɪklɪkl/ *n* encíclica *f*

encyclopaedi|a /ɪnsaɪklə'pi:dɪə/ *n* enciclopedia *f*. **~c** *a* enciclopédico

end /end/ *n* fin *m*; (*furthest point*) extremo *m*. **in the ~** por fin. **make ~s meet** poder llegar a fin de mes. **no ~** (*fam*) muy. **no ~ of** muchísimos. **on ~** de pie; (*consecutive*) seguido. ● *vt/i* terminar, acabar

endanger /ɪn'deɪndʒə(r)/ *vt* arriesgar

endear|ing /ɪn'dɪərɪŋ/ *a* simpático. **~ment** *n* palabra *f* cariñosa

endeavour /ɪn'devə(r)/ *n* tentativa *f*. ● *vi*. **~ to** esforzarse por

ending /'endɪŋ/ *n* fin *m*

endive /'endɪv/ *n* escarola *f*, endibia *f*

endless /'endlɪs/ *a* interminable; ‹*patience*› infinito

endorse /ɪn'dɔ:s/ *vt* endosar; (*fig*) aprobar. **~ment** *n* endoso *m*; (*fig*) aprobación *f*; (*auto*) nota *f* de inhabilitación

endow /ɪn'daʊ/ *vt* dotar

endur|able /ɪn'djʊərəbl/ *a* aguantable. **~ance** *n* resistencia *f*. **~e** /ɪn'djʊə(r)/ *vt* aguantar. ● *vi* durar. **~ing** *a* perdurable

enemy /'enəmɪ/ *n* & *a* enemigo (*m*)

energ|etic /enə'dʒetɪk/ *a* enérgico. **~y** /'enədʒɪ/ *n* energía *f*

enervat|e /'enə:veɪt/ *vt* debilitar. **~ing** *a* debilitante

enfold /ɪn'fəʊld/ *vt* envolver; (*in arms*) abrazar

enforce /ɪn'fɔ:s/ *vt* aplicar; (*impose*) imponer; hacer cumplir ‹*law*›. **~d** *a* forzado

engage /ɪn'geɪdʒ/ *vt* emplear ‹*staff*›; (*reserve*) reservar; ocupar ‹*attention*›; (*mec*) hacer engranar. ● *vi* (*mec*) engranar. **~d** *a* prometido; (*busy*) ocupado. **get ~d** prometerse. **~ment** *n* compromiso *m*; (*undertaking*) obligación *f*

engaging /ɪn'geɪdʒɪŋ/ *a* atractivo

engender /ɪn'dʒendə(r)/ *vt* engendrar

engine /'endʒɪn/ *n* motor *m*; (*of train*) locomotora *f*. **~-driver** *n* maquinista *m*

engineer /endʒɪ'nɪə(r)/ *n* ingeniero *m*; (*mechanic*) mecánico *m*. ● *vt* (*contrive, fam*) lograr. **~ing** *n* ingeniería *f*

England /'ɪŋglənd/ *n* Inglaterra *f*

English /'ɪŋglɪʃ/ *a* inglés. ● *n* (*lang*) inglés *m*; (*people*) ingleses *mpl*. **~man** *n* inglés *m*. **~woman** *n* inglesa *f*. **the ~ Channel** *n* el canal *m* de la Mancha

engrav|e /ɪn'greɪv/ *vt* grabar. **~ing** *n* grabado *m*

engrossed /ɪn'grəʊst/ *a* absorto

engulf /ɪn'gʌlf/ *vt* tragar(se)

enhance /ɪn'hɑ:ns/ *vt* aumentar

enigma /ɪ'nɪgmə/ *n* enigma *m*. **~tic** /enɪg'mætɪk/ *a* enigmático

enjoy /ɪn'dʒɔɪ/ *vt* gozar de. **~ o.s.** divertirse. **I ~ reading** me gusta la lectura. **~able** *a* agradable. **~ment** *n* placer *m*

enlarge /ɪn'lɑ:dʒ/ *vt* agrandar; (*foto*) ampliar. ● *vi* agrandarse. **~ upon** extenderse sobre. **~ment** *n* (*foto*) ampliación *f*

enlighten /ɪn'laɪtn/ *vt* aclarar; (*inform*) informar. **~ment** *n* aclaración *f*. **the E~ment** el siglo *m* de la luces

enlist /ɪn'lɪst/ *vt* alistar; (*fig*) conseguir. ● *vi* alistarse

enliven /ɪn'laɪvn/ *vt* animar

enmity /'enmətɪ/ *n* enemistad *f*

ennoble /ɪ'nəʊbl/ *vt* ennoblecer

enorm|ity /ɪ'nɔ:mətɪ/ *n* enormidad *f*. **~ous** /ɪ'nɔ:məs/ *a* enorme

enough /ɪ'nʌf/ *a & adv* bastante. ● *n* bastante *m*, suficiente *m*. ● *int* ¡basta!

enquir|e /ɪn'kwaɪə(r)/ *vt/i* preguntar. **~e about** informarse de. **~y** *n* pregunta *f*; (*investigation*) investigación *f*

enrage /ɪn'reɪdʒ/ *vt* enfurecer

enrapture /ɪn'ræptʃə(r)/ *vt* extasiar

enrich /ɪn'rɪtʃ/ *vt* enriquecer

enrol /ɪn'rəʊl/ *vt* (*pt* **enrolled**) inscribir; matricular ‹*student*›. ● *vi* inscribirse; ‹*student*› matricularse. **~ment** *n* inscripción *f*; (*of student*) matrícula *f*

ensconce /ɪn'skɒns/ *vt*. **~ o.s.** arrellanarse

ensemble /ɒn'sɒmbl/ *n* conjunto *m*

enshrine /ɪn'ʃraɪn/ *vt* encerrar

ensign /'ensaɪn/ *n* enseña *f*

enslave /ɪn'sleɪv/ *vt* esclavizar

ensue /ɪn'sju:/ *vi* resultar, seguirse

ensure /ɪn'ʃʊə(r)/ *vt* asegurar

entail /ɪn'teɪl/ *vt* suponer; acarrear ‹*trouble etc*›

entangle /ɪn'tæŋgl/ *vt* enredar. **~ment** *n* enredo *m*; (*mil*) alambrada *f*

enter /'entə(r)/ *vt* entrar en; (*write*) escribir; matricular ‹*school etc*›; hacerse socio de ‹*club*›. ● *vi* entrar

enterprise /'entəpraɪz/ *n* empresa *f*; (*fig*) iniciativa *f*

enterprising /'entəpraɪzɪŋ/ *a* emprendedor

entertain /entə'teɪn/ *vt* divertir; recibir ‹*guests*›; abrigar ‹*ideas, hopes*›; (*consider*) considerar. **~ment** *n* diversión *f*; (*performance*) espectáculo *m*; (*reception*) recepción *f*

enthral /ɪn'θrɔ:l/ *vt* (*pt* **enthralled**) cautivar

enthuse /ɪn'θju:z/ *vi*. **~ over** entusiasmarse por

enthusias|m /ɪn'θju:zɪæzəm/ *n* entusiasmo *m*. **~tic** /-'æstɪk/ *a* entusiasta; ‹*thing*› entusiástico. **~tically** /-'æstɪklɪ/ *adv* con entusiasmo. **~t** /ɪn'θju:zɪæst/ *n* entusiasta *m & f*

entice /ɪn'taɪs/ *vt* atraer. **~ment** *n* atracción *f*

entire /ɪn'taɪə(r)/ *a* entero. **~ly** *adv* completamente. **~ty** /ɪn'taɪərətɪ/ *n*. **in its ~ty** en su totalidad

entitle /ɪn'taɪtl/ *vt* titular; (*give a right*) dar derecho a. **be ~d to** tener derecho a. **~ment** *n* derecho *m*

entity /'entətɪ/ *n* entidad *f*

entomb /ɪn'tu:m/ *vt* sepultar

entrails /'entreɪlz/ *npl* entrañas *fpl*

entrance[1] /'entrəns/ *n* entrada *f*; (*right to enter*) admisión *f*

entrance[2] /ɪn'trɑ:ns/ *vt* encantar

entrant /'entrənt/ *n* participante *m & f*; (*in exam*) candidato *m*

entreat /ɪn'tri:t/ *vt* suplicar. **~y** *n* súplica *f*

entrench /ɪn'trentʃ/ *vt* atrincherar

entrust /ɪn'trʌst/ *vt* confiar

entry /'entrɪ/ *n* entrada *f*; (*of street*) bocacalle *f*; (*note*) apunte *m*

entwine /ɪn'twaɪn/ *vt* entrelazar

enumerate /ɪ'nju:məreɪt/ *vt* enumerar

enunciate /ɪ'nʌnsɪeɪt/ *vt* pronunciar; (*state*) enunciar

envelop /ɪn'veləp/ *vt* (*pt* **enveloped**) envolver

envelope /'envələʊp/ *n* sobre *m*

enviable /'envɪəbl/ *a* envidiable

envious /'envɪəs/ *a* envidioso. **~ly** *adv* con envidia

environment /ɪn'vaɪərənmənt/ *n* medio *m* ambiente. **~al** /-'mentl/ *a* ambiental

envisage /ɪn'vɪzɪdʒ/ *vt* prever; (*imagine*) imaginar

envoy /'envɔɪ/ *n* enviado *m*

envy /'envɪ/ *n* envidia *f*. ● *vt* envidiar

enzyme /'enzaɪm/ *n* enzima *f*

epaulette /'epəʊlet/ *n* charretera *f*

ephemeral /ɪ'femərəl/ *a* efímero

epic /'epɪk/ *n* épica *f*. ● *a* épico

epicentre /'epɪsentə(r)/ *n* epicentro *m*

epicure /'epɪkjʊə(r)/ *n* sibarita *m* & *f*; (*gourmet*) gastrónomo *m*

epidemic /epɪ'demɪk/ *n* epidemia *f*. ● *a* epidémico

epilep|sy /'epɪlepsɪ/ *n* epilepsia *f*. **~tic** /-'leptɪk/ *a* & *n* epiléptico (*m*)

epilogue /'epɪlɒg/ *n* epílogo *m*

episode /'epɪsəʊd/ *n* episodio *m*

epistle /ɪ'pɪsl/ *n* epístola *f*

epitaph /'epɪtɑ:f/ *n* epitafio *m*

epithet /'epɪθet/ *n* epíteto *m*

epitom|e /ɪ'pɪtəmɪ/ *n* epítome *m*, personificación *f*. **~ize** *vt* epitomar, personificar, ser la personificación de

epoch /'i:pɒk/ *n* época *f*. **~-making** *a* que hace época

equal /'i:kwəl/ *a* & *n* igual (*m* & *f*). **~ to** (*a task*) a la altura de. ● *vt* (*pt* **equalled**) ser igual a; (*math*) ser. **~ity** /ɪ'kwɒlətɪ/ *n* igualdad *f*. **~ize** /'i:kwəlaɪz/ *vt/i* igualar. **~izer** /-ə(r)/ *n* (*sport*) tanto *m* de empate. **~ly** *adv* igualmente

equanimity /ekwə'nɪmətɪ/ *n* ecuanimidad *f*

equate /ɪ'kweɪt/ *vt* igualar

equation /ɪ'kweɪʒn/ *n* ecuación *f*

equator /ɪ'kweɪtə(r)/ *n* ecuador *m*. **~ial** /ekwə'tɔ:rɪəl/ *a* ecuatorial

equestrian /ɪ'kwestrɪən/ *a* ecuestre

equilateral /i:kwɪ'lætərl/ *a* equilátero

equilibrium /i:kwɪ'lɪbrɪəm/ *n* equilibrio *m*

equinox /'i:kwɪnɒks/ *n* equinoccio *m*

equip /ɪ'kwɪp/ *vt* (*pt* **equipped**) equipar. **~ment** *n* equipo *m*

equitable /'ekwɪtəbl/ *a* equitativo

equity /'ekwətɪ/ *n* equidad *f*; (*pl*, *com*) acciones *fpl* ordinarias

equivalen|ce /ɪ'kwɪvələns/ *n* equivalencia *f*. **~t** /ɪ'kwɪvələnt/ *a* & *n* equivalente (*m*)

equivocal /ɪ'kwɪvəkl/ *a* equívoco

era /'ɪərə/ *n* era *f*

eradicate /ɪ'rædɪkeɪt/ *vt* extirpar

erase /ɪ'reɪz/ *vt* borrar. **~r** /-ə(r)/ *n* borrador *m*

erect /ɪ'rekt/ *a* erguido. ● *vt* levantar. **~ion** /-ʃn/ *n* erección *f*, montaje *m*

ermine /'ɜ:mɪn/ *n* armiño *m*

ero|de /ɪ'rəʊd/ *vt* desgastar. **~sion** /-ʒn/ *n* desgaste *m*

erotic /ɪ'rɒtɪk/ *a* erótico. **~ism** /-sɪzəm/ *n* erotismo *m*

err /ɜ:(r)/ *vi* errar; (*sin*) pecar

errand /'erənd/ *n* recado *m*

erratic /ɪ'rætɪk/ *a* irregular; ‹*person*› voluble

erroneous /ɪ'rəʊnɪəs/ *a* erróneo

error /'erə(r)/ *n* error *m*

erudit|e /'eru:daɪt/ *a* erudito. **~ion** /-'dɪʃn/ *n* erudición *f*

erupt /ɪ'rʌpt/ *vi* estar en erupción; (*fig*) estallar. **~ion** /-ʃn/ *n* erupción *f*

escalat|e /'eskəleɪt/ *vt* intensificar. ● *vi* intensificarse. **~ion** /-'leɪʃn/ *n* intensificación *f*

escalator /'eskəleɪtə(r)/ *n* escalera *f* mecánica

escapade /eskə'peɪd/ *n* aventura *f*

escap|e /ɪ'skeɪp/ *vi* escaparse. ● *vt* evitar. ● *n* fuga *f*; (*avoidance*) evasión *f*. **have a narrow ~e** escapar por un pelo. **~ism** /ɪ'skeɪpɪzəm/ *n* escapismo *m*

escarpment /ɪs'kɑ:pmənt/ *n* escarpa *f*

escort /'eskɔ:t/ *n* acompañante *m*; (*mil*) escolta *f*. /ɪ'skɔ:t/ *vt* acompañar; (*mil*) escoltar

Eskimo /'eskɪməʊ/ *n* (*pl* **-os**, **-o**) esquimal (*m* & *f*)

especial /ɪ'speʃl/ *a* especial. **~ly** *adv* especialmente

espionage /'espɪənɑ:ʒ/ *n* espionaje *m*

esplanade /esplə'neɪd/ *n* paseo *m* marítimo

Esq. /ɪ'skwaɪə(r)/ *abbr* (*Esquire*) (*in address*). **E. Ashton, ~** Sr. D. E. Ashton

essay /'eseɪ/ *n* ensayo *m*; (*at school*) composición *f*

essence /'esns/ *n* esencia *f*. **in ~** esencialmente

essential /ɪ'senʃl/ *a* esencial. ● *n* lo esencial. **~ly** *adv* esencialmente

establish /ɪ'stæblɪʃ/ *vt* establecer; (*prove*) probar. **~ment** *n* establecimiento *m*. **the E~ment** los que mandan, el sistema *m*
estate /ɪ'steɪt/ *n* finca *f*; (*possessions*) bienes *mpl*. **~ agent** *n* agente *m* inmobiliario. **~ car** *n* furgoneta *f*
esteem /ɪ'sti:m/ *vt* estimar. ● *n* estimación *f*, estima *f*
estimat|e /'estɪmət/ *n* cálculo *m*; (*com*) presupuesto *m*. /'estɪmeɪt/ *vt* calcular. **~ion** /-'meɪʃn/ *n* estima *f*, estimación *f*; (*opinion*) opinión *f*
estranged /ɪs'treɪndʒd/ *a* alejado
estuary /'estʃʊərɪ/ *n* estuario *m*
etc. /et'setrə/ *abbr* (*et cetera*) etc., etcétera
etching /'etʃɪŋ/ *n* aguafuerte *m*
eternal /ɪ'tɜ:nl/ *a* eterno
eternity /ɪ'tɜ:nətɪ/ *n* eternidad *f*
ether /'i:θə(r)/ *n* éter *m*
ethereal /ɪ'θɪərɪəl/ *a* etéreo
ethic /'eθɪk/ *n* ética *f*. **~s** *npl* ética *f*. **~al** *a* ético
ethnic /'eθnɪk/ *a* étnico
ethos /'i:θɒs/ *n* carácter *m* distintivo
etiquette /'etɪket/ *n* etiqueta *f*
etymology /etɪ'mɒlədʒɪ/ *n* etimología *f*
eucalyptus /ju:kə'lɪptəs/ *n* (*pl* **-tuses**) eucalipto *m*
eulogy /'ju:lədʒɪ/ *n* encomio *m*
euphemism /'ju:fəmɪzəm/ *n* eufemismo *m*
euphoria /ju:'fɔ:rɪə/ *n* euforia *f*
Europe /'jʊərəp/ *n* Europa *f*. **~an** /-'pɪən/ *a* & *n* europeo (*m*)
euthanasia /ju:θə'neɪzɪə/ *n* eutanasia *f*
evacuat|e /ɪ'vækjʊeɪt/ *vt* evacuar; desocupar ‹*building*›. **~ion** /-'eɪʃn/ *n* evacuación *f*
evade /ɪ'veɪd/ *vt* evadir
evaluate /ɪ'væljʊeɪt/ *vt* evaluar
evangeli|cal /i:væn'dʒelɪkl/ *a* evangélico. **~st** /ɪ'vændʒəlɪst/ *n* evangelista *m & f*
evaporat|e /ɪ'væpəreɪt/ *vi* evaporarse. **~ion** /-'reɪʃn/ *n* evaporación *f*
evasion /ɪ'veɪʒn/ *n* evasión *f*
evasive /ɪ'veɪsɪv/ *a* evasivo
eve /i:v/ *n* víspera *f*
even /'i:vn/ *a* regular; (*flat*) llano; ‹*surface*› liso; ‹*amount*› igual; ‹*number*› par. **get ~ with** desquitarse con. ● *vt* nivelar. **~ up** igualar. ● *adv* aun, hasta, incluso. **~ if** aunque. **~ so** aun así. **not ~** ni siquiera
evening /'i:vnɪŋ/ *n* tarde *f*; (*after dark*) noche *f*. **~ class** *n* clase *f* nocturna. **~ dress** *n* (*man's*) traje *m* de etiqueta; (*woman's*) traje *m* de noche
evensong /'i:vənsɒŋ/ *n* vísperas *fpl*
event /ɪ'vent/ *n* acontecimiento *m*; (*sport*) prueba *f*. **in the ~ of** en caso de. **~ful** *a* lleno de acontecimientos
eventual /ɪ'ventʃʊəl/ *a* final, definitivo. **~ity** /-'ælətɪ/ *n* eventualidad *f*. **~ly** *adv* finalmente
ever /'evə(r)/ *adv* jamás, nunca; (*at all times*) siempre. **~ after** desde entonces. **~ since** desde entonces. ● *conj* después de que. **~ so** (*fam*) muy. **for ~** para siempre. **hardly ~** casi nunca
evergreen /'evəgri:n/ *a* de hoja perenne. ● *n* árbol *m* de hoja perenne
everlasting /'evəlɑ:stɪŋ/ *a* eterno
every /'evrɪ/ *a* cada, todo. **~ child** todos los niños. **~ one** cada uno. **~ other day** cada dos días
everybody /'evrɪbɒdɪ/ *pron* todo el mundo
everyday /'evrɪdeɪ/ *a* todos los días
everyone /'evrɪwʌn/ *pron* todo el mundo. **~ else** todos los demás
everything /'evrɪθɪŋ/ *pron* todo
everywhere /'evrɪweə(r)/ *adv* en todas partes
evict /ɪ'vɪkt/ *vt* desahuciar. **~ion** /-ʃn/ *n* desahucio *m*
eviden|ce /'evɪdəns/ *n* evidencia *f*; (*proof*) pruebas *fpl*; (*jurid*) testimonio *m*. **~ce of** señales de. **in ~ce** visible. **~t** /'evɪdənt/ *a* evidente. **~tly** *adv* evidentemente
evil /'i:vl/ *a* malo. ● *n* mal *m*, maldad *f*
evocative /ɪ'vɒkətɪv/ *a* evocador
evoke /ɪ'vəʊk/ *vt* evocar
evolution /i:və'lu:ʃn/ *n* evolución *f*
evolve /ɪ'vɒlv/ *vt* desarrollar. ● *vi* desarrollarse, evolucionar
ewe /ju:/ *n* oveja *f*
ex... /eks/ *pref* ex...
exacerbate /ɪg'zæsəbeɪt/ *vt* exacerbar
exact /ɪg'zækt/ *a* exacto. ● *vt* exigir (**from** a). **~ing** *a* exigente. **~itude** *n* exactitud *f*. **~ly** *adv* exactamente
exaggerat|e /ɪg'zædʒəreɪt/ *vt* exagerar. **~ion** /-'reɪʃn/ *n* exageración *f*
exalt /ɪg'zɔ:lt/ *vt* exaltar

exam /ɪg'zæm/ *n* (*fam*) examen *m*. **~ination** /ɪgzæmɪ'neɪʃn/ *n* examen *m*. **~ine** /ɪg'zæmɪn/ *vt* examinar; interrogar ‹*witness*›. **~iner** /-ə(r)/ *n* examinador *m*

example /ɪg'zɑ:mpl/ *n* ejemplo *m*. **make an ~ of** infligir castigo ejemplar a

exasperat|e /ɪg'zæspəreɪt/ *vt* exasperar. **~ion** /-'reɪʃn/ *n* exasperación *f*

excavat|e /'ekskəveɪt/ *vt* excavar. **~ion** /-'veɪʃn/ *n* excavación *f*

exceed /ɪk'si:d/ *vt* exceder. **~ingly** *adv* extremadamente

excel /ɪk'sel/ *vi* (*pt* **excelled**) sobresalir. ● *vt* superar

excellen|ce /'eksələns/ *n* excelencia *f*. **~t** /'eksələnt/ *a* excelente. **~tly** *adv* excelentemente

except /ɪk'sept/ *prep* excepto, con excepción de. **~ for** con excepción de. ● *vt* exceptuar. **~ing** *prep* con excepción de

exception /ɪk'sepʃən/ *n* excepción *f*. **take ~ to** ofenderse por. **~al** /ɪk'sepʃənl/ *a* excepcional. **~ally** *adv* excepcionalmente

excerpt /'eksɜ:pt/ *n* extracto *m*

excess /ɪk'ses/ *n* exceso *m*. /'ekses/ *a* excedente. **~ fare** *n* suplemento *m*. **~ luggage** *n* exceso *m* de equipaje

excessive /ɪk'sesɪv/ *a* excesivo. **~ly** *adv* excesivamente

exchange /ɪk'stʃeɪndʒ/ *vt* cambiar. ● *n* cambio *m*. **(telephone) ~** central *f* telefónica

exchequer /ɪks'tʃekə(r)/ *n* (*pol*) erario *m*, hacienda *f*

excise[1] /'eksaɪz/ *n* impuestos *mpl* indirectos

excise[2] /ek'saɪz/ *vt* quitar

excit|able /ɪk'saɪtəbl/ *a* excitable. **~e** /ɪk'saɪt/ *vt* emocionar; (*stimulate*) excitar. **~ed** *a* entusiasmado. **~ement** *n* emoción *f*; (*enthusiasm*) entusiasmo *m*. **~ing** *a* emocionante

excla|im /ɪk'skleɪm/ *vi* exclamar. **~mation** /eksklə'meɪʃn/ *n* exclamación *f*. **~mation mark** *n* signo *m* de admiración *f*, punto *m* de exclamación

exclu|de /ɪk'sklu:d/ *vt* excluir. **~sion** /-ʒən/ *n* exclusión *f*

exclusive /ɪk'sklu:sɪv/ *a* exclusivo; ‹*club*› selecto. **~ of** excluyendo. **~ly** *adv* exclusivamente

excomunicate /ekskə'mju:nɪkeɪt/ *vt* excomulgar

excrement /'ekskrɪmənt/ *n* excremento *m*

excruciating /ɪk'skru:ʃɪeɪtɪŋ/ *a* atroz, insoportable

excursion /ɪk'skɜ:ʃn/ *n* excursión *f*

excus|able *a* /ɪk'skju:zəbl/ *a* perdonable. **~e** /ɪk'skju:z/ *vt* perdonar. **~e from** dispensar de. **~e me!** ¡perdón! /ɪk'skju:s/ *n* excusa *f*

ex-directory /eksdɪ'rektərɪ/ *a* que no está en la guía telefónica

execrable /'eksɪkrəbl/ *a* execrable

execut|e /'eksɪkju:t/ *vt* ejecutar. **~ion** /eksɪ'kju:ʃn/ *n* ejecución *f*. **~ioner** *n* verdugo *m*

executive /ɪg'zekjʊtɪv/ *a & n* ejecutivo (*m*)

executor /ɪg'zekjʊtə(r)/ *n* (*jurid*) testamentario *m*

exemplary /ɪg'zemplərɪ/ *a* ejemplar

exemplify /ɪg'zemplɪfaɪ/ *vt* ilustrar

exempt /ɪg'zempt/ *a* exento. ● *vt* dispensar. **~ion** /-ʃn/ *n* exención *f*

exercise /'eksəsaɪz/ *n* ejercicio *m*. ● *vt* ejercer. ● *vi* hacer ejercicios. **~ book** *n* cuaderno *m*

exert /ɪg'zɜ:t/ *vt* ejercer. **~ o.s.** esforzarse. **~ion** /-ʃn/ *n* esfuerzo *m*

exhal|ation /ekshə'leɪʃn/ *n* exhalación *f*. **~e** /eks'heɪl/ *vt/i* exhalar

exhaust /ɪg'zɔ:st/ *vt* agotar. ● *n* (*auto*) tubo *m* de escape. **~ed** *a* agotado. **~ion** /-stʃən/ *n* agotamiento *m*. **~ive** /ɪg'zɔ:stɪv/ *a* exhaustivo

exhibit /ɪg'zɪbɪt/ *vt* exponer; (*jurid*) exhibir; (*fig*) mostrar. ● *n* objeto *m* expuesto; (*jurid*) documento *m*

exhibition /eksɪ'bɪʃn/ *n* exposición *f*; (*act of showing*) demostración *f*; (*univ*) beca *f*. **~ist** *n* exhibicionista *m & f*

exhibitor /ɪg'zɪbɪtə(r)/ *n* expositor *m*

exhilarat|e /ɪg'zɪləreɪt/ *vt* alegrar. **~ion** /-'reɪʃn/ *n* regocijo *m*

exhort /ɪg'zɔ:t/ *vt* exhortar

exile /'eksaɪl/ *n* exilio *m*; (*person*) exiliado *m*. ● *vt* desterrar

exist /ɪg'zɪst/ *vi* existir. **~ence** *n* existencia *f*. **in ~ence** existente

existentialism /egzɪs'tenʃəlɪzəm/ *n* existencialismo *m*

exit /'eksɪt/ *n* salida *f*

exodus /'eksədəs/ *n* éxodo *m*

exonerate /ɪg'zɒnəreɪt/ *vt* disculpar

exorbitant /ɪg'zɔ:bɪtənt/ *a* exorbitante

exorcis|e /'eksɔ:saɪz/ *vt* exorcizar. **~m** /-sɪzəm/ *n* exorcismo *m*
exotic /ɪg'zɒtɪk/ *a* exótico
expand /ɪk'spænd/ *vt* extender; dilatar ‹*metal*›; (*develop*) desarrollar. ● *vi* extenderse; (*develop*) desarrollarse; ‹*metal*› dilatarse
expanse /ɪk'spæns/ *n* extensión *f*
expansion /ɪk'spænʃn/ *n* extensión *f*; (*of metal*) dilatación *f*
expansive /ɪk'spænsɪv/ *a* expansivo
expatriate /eks'pætrɪət/ *a* & *n* expatriado (*m*)
expect /ɪk'spekt/ *vt* esperar; (*suppose*) suponer; (*demand*) contar con. **I ~ so** supongo que sí
expectan|cy /ɪk'spektənsɪ/ *n* esperanza *f*. **life ~cy** esperanza *f* de vida. **~t** /ɪk'spektənt/ *a* expectante. **~t mother** *n* futura madre *f*
expectation /ekspek'teɪʃn/ *n* esperanza *f*
expedien|cy /ɪk'spi:dɪənsɪ/ *n* conveniencia *f*. **~t** /ɪk'spi:dɪənt/ *a* conveniente
expedite /'ekspɪdaɪt/ *vt* acelerar
expedition /ekspɪ'dɪʃn/ *n* expedición *f*. **~ary** *a* expedicionario
expel /ɪk'spel/ *vt* (*pt* **expelled**) expulsar
expend /ɪk'spend/ *vt* gastar. **~able** *a* prescindible
expenditure /ɪk'spendɪtʃə(r)/ *n* gastos *mpl*
expens|e /ɪk'spens/ *n* gasto *m*; (*fig*) costa *f*. **at s.o.'s ~e** a costa de uno. **~ive** /ɪk'spensɪv/ *a* caro. **~ively** *adv* costosamente
experience /ɪk'spɪərɪəns/ *n* experiencia. ● *vt* experimentar. **~d** *a* experto
experiment /ɪk'sperɪmənt/ *n* experimento *m*. ● *vi* experimentar. **~al** /-'mentl/ *a* experimental
expert /'ekspɜ:t/ *a* & *n* experto (*m*). **~ise** /ekspɜ:'ti:z/ *n* pericia *f*. **~ly** *adv* hábilmente
expir|e /ɪk'spaɪə(r)/ *vi* expirar. **~y** *n* expiración *f*
expla|in /ɪk'spleɪn/ *vt* explicar. **~nation** /eksplə'neɪʃn/ *n* explicación *f*. **~natory** /ɪks'plænətərɪ/ *a* explicativo
expletive /ɪk'spli:tɪv/ *n* palabrota *f*
explicit /ɪk'splɪsɪt/ *a* explícito
explode /ɪk'spləʊd/ *vt* hacer explotar; (*tec*) explosionar. ● *vi* estallar
exploit /'eksplɔɪt/ *n* hazaña *f*. /ɪk'splɔɪt/ *vt* explotar. **~ation** /eksplɔɪ'teɪʃn/ *n* explotación *f*
explor|ation /eksplə'reɪʃn/ *n* exploración *f*. **~atory** /ɪk'splɒrətrɪ/ *a* exploratorio. **~e** /ɪk'splɔ:(r)/ *vt* explorar. **~er** *n* explorador *m*
explosi|on /ɪk'spləʊʒn/ *n* explosión *f*. **~ve** *a* & *n* explosivo (*m*)
exponent /ɪk'spəʊnənt/ *n* exponente *m*
export /ɪk'spɔ:t/ *vt* exportar. /'ekspɔ:t/ *n* exportación *f*. **~er** /ɪks'pɔ:tə(r)/ exportador *m*
expos|e /ɪk'spəʊz/ *vt* exponer; (*reveal*) descubrir. **~ure** /-ʒə(r)/ *n* exposición *f*. **die of ~ure** morir de frío
expound /ɪk'spaʊnd/ *vt* exponer
express[1] /ɪk'spres/ *vt* expresar
express[2] /ɪk'spres/ *a* expreso; ‹*letter*› urgente. ● *adv* (*by express post*) por correo urgente. ● *n* (*train*) rápido *m*, expreso *m*
expression /ɪk'spreʃn/ *n* expresión *f*
expressive /ɪk'spresɪv/ *a* expresivo
expressly /ɪk'spreslɪ/ *adv* expresamente
expulsion /ɪk'spʌlʃn/ *n* expulsión *f*
expurgate /'ekspəgeɪt/ *vt* expurgar
exquisite /'ekskwɪzɪt/ *a* exquisito. **~ly** *adv* primorosamente
ex-serviceman /eks'sɜ:vɪsmən/ *n* (*pl* **-men**) excombatiente *m*
extant /ek'stænt/ *a* existente
extempore /ek'stempərɪ/ *a* improvisado. ● *adv* de improviso
exten|d /ɪk'stend/ *vt* extender; (*prolong*) prolongar; ensanchar ‹*house*›. ● *vi* extenderse. **~sion** *n* extensión *f*; (*of road, time*) prolongación *f*; (*building*) anejo *m*; (*com*) prórroga *f*
extensive /ɪk'stensɪv/ *a* extenso. **~ly** *adv* extensamente
extent /ɪk'stent/ *n* extensión *f*; (*fig*) alcance. **to a certain ~** hasta cierto punto
extenuate /ɪk'stenjʊeɪt/ *vt* atenuar
exterior /ɪk'stɪərɪə(r)/ *a* & *n* exterior (*m*)
exterminat|e /ɪk'stɜ:mɪneɪt/ *vt* exterminar. **~ion** /-'neɪʃn/ *n* exterminio *m*
external /ɪk'stɜ:nl/ *a* externo. **~ly** *adv* externamente
extinct /ɪk'stɪŋkt/ *a* extinto. **~ion** /-ʃn/ *n* extinción *f*
extinguish /ɪk'stɪŋgwɪʃ/ *vt* extinguir. **~er** *n* extintor *m*
extol /ɪk'stəʊl/ *vt* (*pt* **extolled**) alabar

extort /ɪk'stɔ:t/ *vt* sacar por la fuerza. **~ion** /-ʃn/ *n* exacción *f*. **~ionate** /ɪk'stɔ:ʃənət/ *a* exorbitante

extra /'ekstrə/ *a* suplementario. ● *adv* extraordinariamente. ● *n* suplemento *m*; (*cinema*) extra *m* & *f*

extract /'ekstrækt/ *n* extracto *m*. /ɪk'strækt/ *vt* extraer; (*fig*) arrancar. **~ion** /-ʃn/ *n* extracción *f*; (*lineage*) origen *m*

extradit|e /'ekstrədaɪt/ *vt* extraditar. **~ion** /-'dɪʃn/ *n* extradición *f*

extramarital /ekstrə'mærɪtl/ *a* fuera del matrimonio

extramural /ekstrə'mjʊərəl/ *a* fuera del recinto universitario; (*for external students*) para estudiantes externos

extraordinary /ɪk'strɔ:dnrɪ/ *a* extraordinario

extra-sensory /ekstrə'sensərɪ/ *a* extrasensorial

extravagan|ce /ɪk'strævəgəns/ *n* prodigalidad *f*, extravagancia *f*. **~t** /ɪk'strævəgənt/ *a* pródigo, extravagante

extrem|e /ɪk'stri:m/ *a* & *n* extremo (*m*). **~ely** *adv* extremadamente. **~ist** *n* extremista *m* & *f*. **~ity** /ɪk'streməti/ *n* extremidad *f*

extricate /'ekstrɪkeɪt/ *vt* desenredar, librar

extrovert /'ekstrəvɜ:t/ *n* extrovertido *m*

exuberan|ce /ɪg'zju:bərəns/ *n* exuberancia *f*. **~t** /ɪg'zju:bərənt/ *a* exuberante

exude /ɪg'zju:d/ *vt* rezumar

exult /ɪg'zʌlt/ *vi* exultar

eye /aɪ/ *n* ojo *m*. **keep an ~ on** no perder de vista. **see ~ to ~** estar de acuerdo con. ● *vt* (*pt* **eyed,** *pres p* **eyeing**) mirar. **~ball** /'aɪbɔ:l/ *n* globo *m* del ojo. **~brow** /'aɪbraʊ/ *n* ceja *f*. **~ful** /'aɪfʊl/ *n* (*fam*) espectáculo *m* sorprendente. **~lash** /'aɪlæʃ/ *n* pestaña *f*. **~let** /'aɪlɪt/ *n* ojete *m*. **~lid** /'aɪlɪd/ *n* párpado *m*. **~-opener** *n* (*fam*) revelación *f*. **~-shadow** *n* sombra *f* de ojos, sombreador *m*. **~sight** /'aɪsaɪt/ *n* vista *f*. **~sore** /'aɪsɔ:(r)/ *n* (*fig, fam*) monstruosidad *f*, horror *m*. **~witness** /'aɪwɪtnɪs/ *n* testigo *m* ocular

F

fable /'feɪbl/ *n* fábula *f*

fabric /'fæbrɪk/ *n* tejido *m*, tela *f*

fabrication /fæbrɪ'keɪʃn/ *n* invención *f*

fabulous /'fæbjʊləs/ *a* fabuloso

façade /fə'sɑ:d/ *n* fachada *f*

face /feɪs/ *n* cara *f*, rostro *m*; (*of watch*) esfera *f*; (*aspect*) aspecto *m*. **~ down(wards)** boca abajo. **~ up(wards)** boca arriba. **in the ~ of** frente a. **lose ~** quedar mal. **pull ~s** hacer muecas. ● *vt* mirar hacia; ‹*house*› dar a; (*confront*) enfrentarse con. ● *vi* volverse. **~ up to** enfrentarse con. **~ flannel** *n* paño *m* (para lavarse la cara). **~less** *a* anónimo. **~-lift** *n* cirugía *f* estética en la cara

facet /'fæsɪt/ *n* faceta *f*

facetious /fə'si:ʃəs/ *a* chistoso, gracioso

facial /'feɪʃl/ *a* facial. ● *n* masaje *m* facial

facile /'fæsaɪl/ *a* fácil

facilitate /fə'sɪlɪteɪt/ *vt* facilitar

facility /fə'sɪlɪtɪ/ *n* facilidad *f*

facing /'feɪsɪŋ/ *n* revestimiento *m*. **~s** *npl* (*on clothes*) vueltas *fpl*

facsimile /fæk'sɪmɪlɪ/ *n* facsímile *m*

fact /fækt/ *n* hecho *m*. **as a matter of ~, in ~** en realidad, a decir verdad

faction /'fækʃn/ *n* facción *f*

factor /'fæktə(r)/ *n* factor *m*

factory /'fæktərɪ/ *n* fábrica *f*

factual /'fæktʃʊəl/ *a* basado en hechos, factual

faculty /'fækəltɪ/ *n* facultad *f*

fad /fæd/ *n* manía *f*, capricho *m*

fade /feɪd/ *vi* ‹*colour*› descolorarse; ‹*flowers*› marchitarse; ‹*light*› apagarse; ‹*memory, sound*› desvanecerse

faeces /'fi:si:z/ *npl* excrementos *mpl*

fag[1] /fæg/ *n* (*chore, fam*) faena *f*; (*cigarette, sl*) cigarillo *m*, pitillo *m*

fag[2] /fæg/ *n* (*homosexual, Amer, sl*) marica *m*

fagged /fægd/ *a*. **~ (out)** rendido

fah /fɑ/ *n* (*mus, fourth note of any musical scale*) fa *m*

fail /feɪl/ *vi* fallar; (*run short*) acabarse. **he ~ed to arrive** no llegó. ● *vt* no aprobar ‹*exam*›; suspender ‹*candidate*›; (*disappoint*) fallar. **~ s.o.** ‹*words etc*› faltarle a uno. ● *n*. **without ~** sin falta

failing /'feɪlɪŋ/ *n* defecto *m*. ● *prep* a falta de

failure /ˈfeɪljə(r)/ *n* fracaso *m*; (*person*) fracasado *m*; (*med*) ataque *m*; (*mec*) fallo *m*. **~ to do** dejar *m* de hacer

faint /feɪnt/ *a* (**-er, -est**) (*weak*) débil; (*indistinct*) indistinto. **feel ~** estar mareado. **the ~est idea** la más remota idea. ● *vi* desmayarse. ● *n* desmayo *m*. **~-hearted** *a* pusilánime, cobarde. **~ly** *adv* (*weakly*) débilmente; (*indistinctly*) indistintamente. **~ness** *n* debilidad *f*

fair[1] /feə(r)/ *a* (**-er, -est**) (*just*) justo; ‹*weather*› bueno; ‹*amount*› razonable; ‹*hair*› rubio; ‹*skin*› blanco. **~ play** *n* juego *m* limpio. ● *adv* limpio

fair[2] /feə(r)/ *n* feria *f*

fair: **~ly** *adv* (*justly*) justamente; (*rather*) bastante. **~ness** *n* justicia *f*

fairy /ˈfeərɪ/ *n* hada *f*. **~land** *n* país *m* de las hadas. **~ story**, **~-tale** cuento *m* de hadas

fait accompli /feɪtəˈkɒmpli:/ *n* hecho *m* consumado

faith /feɪθ/ *n* (*trust*) confianza *f*; (*relig*) fe *f*. **~ful** *a* fiel. **~fully** *adv* fielmente. **~fulness** *n* fidelidad *f*. **~-healing** *n* curación *f* por la fe

fake /feɪk/ *n* falsificación *f*; (*person*) impostor *m*. ● *a* falso. ● *vt* falsificar; (*pretend*) fingir

fakir /ˈfeɪkɪə(r)/ *n* faquir *m*

falcon /ˈfɔ:lkən/ *n* halcón *m*

Falkland /ˈfɔ:lklənd/ *n*. **the ~ Islands** *npl* las islas *fpl* Malvinas

fall /fɔ:l/ *vi* (*pt* **fell**, *pp* **fallen**) caer. ● *n* caída *f*; (*autumn, Amer*) otoño *m*; (*in price*) baja *f*. **~ back on** recurrir a. **~ down** (*fall*) caer; (*be unsuccessful*) fracasar. **~ for** (*fam*) enamorarse de ‹*person*›; (*fam*) dejarse engañar por ‹*trick*›. **~ in** (*mil*) formar filas. **~ off** (*diminish*) disminuir. **~ out** (*quarrel*) reñir (**with** con); (*drop out*) caer. **~ over** caer(se). **~ over sth** tropezar con algo. **~ short** ser insuficiente. **~ through** fracasar

fallacy /ˈfæləsɪ/ *n* error *m*

fallible /ˈfælɪbl/ *a* falible

fallout /ˈfɔ:laʊt/ *n* lluvia *f* radiactiva

fallow /ˈfæləʊ/ *a* en barbecho

false /fɔ:ls/ *a* falso. **~hood** *n* mentira *f*. **~ly** *adv* falsamente. **~ness** *n* falsedad *f*

falsetto /fɔ:lˈsetəʊ/ *n* (*pl* **-os**) falsete *m*

falsify /ˈfɔ:lsɪfaɪ/ *vt* falsificar

falter /ˈfɔ:ltə(r)/ *vi* vacilar

fame /feɪm/ *n* fama *f*. **~d** *a* famoso

familiar /fəˈmɪlɪə(r)/ *a* familiar. **be ~ with** conocer. **~ity** /-ˈærətɪ/ *n* familiaridad *f*. **~ize** *vt* familiarizar

family /ˈfæməlɪ/ *n* familia *f*. ● *a* de (la) familia, familiar

famine /ˈfæmɪn/ *n* hambre *f*, hambruna *f* (*Amer*)

famished /ˈfæmɪʃt/ *a* hambriento

famous /ˈfeɪməs/ *a* famoso. **~ly** *adv* (*fam*) a las mil maravillas

fan[1] /fæn/ *n* abanico *m*; (*mec*) ventilador *m*. ● *vt* (*pt* **fanned**) abanicar; soplar ‹*fire*›. ● *vi*. **~ out** desparramarse en forma de abanico

fan[2] /fæn/ *n* (*of person*) admirador *m*; (*enthusiast*) aficionado *m*, entusiasta *m & f*

fanatic /fəˈnætɪk/ *n* fanático *m*. **~al** *a* fanático. **~ism** /-sɪzəm/ *n* fanatismo *m*

fan belt /ˈfænbelt/ *n* correa *f* de ventilador

fancier /ˈfænsɪə(r)/ *n* aficionado *m*

fanciful /ˈfænsɪfl/ *a* (*imaginative*) imaginativo; (*unreal*) imaginario

fancy /ˈfænsɪ/ *n* fantasía *f*; (*liking*) gusto *m*. **take a ~ to** tomar cariño a ‹*person*›; aficionarse a ‹*thing*›. ● *a* de lujo; (*extravagant*) excesivo. ● *vt* (*imagine*) imaginar; (*believe*) creer; (*want, fam*) apetecer a. **~ dress** *n* disfraz *m*

fanfare /ˈfænfeə(r)/ *n* fanfarria *f*

fang /fæŋ/ *n* (*of animal*) colmillo *m*; (*of snake*) diente *m*

fanlight /ˈfænlaɪt/ *n* montante *m*

fantasize /ˈfæntəsaɪz/ *vi* fantasear

fantastic /fænˈtæstɪk/ *a* fantástico

fantasy /ˈfæntəsɪ/ *n* fantasía *f*

far /fɑ:(r)/ *adv* lejos; (*much*) mucho. **as ~ as** hasta. **as ~ as I know** que yo sepa. **by ~** con mucho. ● *a* (**further, furthest** *or* **farther, farthest**) lejano

far-away /ˈfɑ:rəweɪ/ *a* lejano

farc|e /fɑ:s/ *n* farsa *f*. **~ical** *a* ridículo

fare /feə(r)/ *n* (*for transport*) tarifa *f*; (*food*) comida *f*. ● *vi* irle. **how did you ~?** ¿qué tal te fue?

Far East /fɑ:(r)ˈi:st/ *n* Extremo/Lejano Oriente *m*

farewell /feəˈwel/ *int & n* adiós (*m*)

far-fetched /fɑ:ˈfetʃt/ *a* improbable

farm /fɑ:m/ *n* granja *f*. ● *vt* cultivar. **~ out** arrendar. ● *vi* ser agricultor. **~er** *n* agricultor *m*. **~house** *n* granja *f*. **~ing** *n* agricultura *f*. **~yard** *n* corral *m*

far: ~**-off** *a* lejano. ~**-reaching** *a* trascendental. ~**-seeing** *a* clarividente. ~**-sighted** *a* hipermétrope; (*fig*) clarividente

farther, farthest /'fɑ:ðə(r),'fɑ:ðəst/ *see* **far**

fascinat|e /'fæsɪneɪt/ *vt* fascinar. ~**ion** /-'neɪʃn/ *n* fascinación *f*

fascis|m /'fæʃɪzəm/ *n* fascismo *m*. ~**t** /'fæʃɪst/ *a* & *n* fascista (*m* & *f*)

fashion /'fæʃn/ *n* (*manner*) manera *f*; (*vogue*) moda *f*. ~**able** *a* de moda

fast[1] /fɑ:st/ *a* (**-er, -est**) rápido; ‹*clock*› adelantado; (*secure*) fijo; ‹*colours*› sólido. ● *adv* rápidamente; (*securely*) firmemente. ~ **asleep** profundamente dormido

fast[2] /fɑ:st/ *vi* ayunar. ● *n* ayuno *m*

fasten /'fɑ:sn/ *vt/i* sujetar; cerrar ‹*windows, doors*›; abrochar ‹*belt etc*›. ~**er** *n*, ~**ing** *n* (*on box, window*) cierre *m*; (*on door*) cerrojo *m*

fastidious /fə'stɪdɪəs/ *a* exigente, minucioso

fat /fæt/ *n* grasa *f*. ● *a* (**fatter, fattest**) gordo; ‹*meat*› que tiene mucha grasa; (*thick*) grueso. **a** ~ **lot of** (*sl*) muy poco

fatal /'feɪtl/ *a* mortal; (*fateful*) fatídico

fatalis|m /'feɪtəlɪzəm/ *n* fatalismo *m*. ~**t** *n* fatalista *m* & *f*

fatality /fə'tæləti/ *n* calamidad *f*; (*death*) muerte *f*

fatally /'feɪtəlɪ/ *adv* mortalmente; (*by fate*) fatalmente

fate /feɪt/ *n* destino *m*; (*one's lot*) suerte *f*. ~**d** *a* predestinado. ~**ful** *a* fatídico

fat-head /'fæthed/ *n* imbécil *m*

father /'fɑ:ðə(r)/ *n* padre *m*. ~**hood** *m* paternidad *f*. ~**-in-law** *m* (*pl* **fathers-in-law**) *m* suegro *m*. ~**ly** *a* paternal

fathom /'fæðəm/ *n* braza *f*. ● *vt*. ~ **(out)** comprender

fatigue /fə'ti:g/ *n* fatiga *f*. ● *vt* fatigar

fat: ~**ness** *n* gordura *f*. ~**ten** *vt/i* engordar. ~**tening** *a* que engorda. ~**ty** *a* graso. ● *n* (*fam*) gordinflón *m*

fatuous /'fætjʊəs/ *a* fatuo

faucet /'fɔ:sɪt/ *n* (*Amer*) grifo *m*

fault /fɔ:lt/ *n* defecto *m*; (*blame*) culpa *f*; (*tennis*) falta *f*; (*geol*) falla *f*. **at** ~ culpable. ● *vt* criticar. ~**less** *a* impecable. ~**y** *a* defectuoso

fauna /'fɔ:nə/ *n* fauna *f*

faux pas /fəʊ'pɑ:/ (*pl* **faux pas** /fəʊ'pɑ:/) *n* metedura *f* de pata, paso *m* en falso

favour /'feɪvə(r)/ *n* favor *m*. ● *vt* favorecer; (*support*) estar a favor de; (*prefer*) preferir. ~**able** *a* favorable. ~**ably** *adv* favorablemente

favourit|e /'feɪvərɪt/ *a* & *n* preferido (*m*). ~**ism** *n* favoritismo *m*

fawn[1] /fɔ:n/ *n* cervato *m*. ● *a* color de cervato, beige, beis

fawn[2] /fɔ:n/ *vi*. ~ **on** adular

fax /fæks/ *n* telefacsímil *m*, fax *m*

fear /fɪə(r)/ *n* miedo *m*. ● *vt* temer. ~**ful** *a* (*frightening*) espantoso; (*frightened*) temeroso. ~**less** *a* intrépido. ~**lessness** *n* intrepidez *f*. ~**some** *a* espantoso

feasib|ility /fi:zə'bɪlətɪ/ *n* viabilidad *f*. ~**le** /'fi:zəbl/ *a* factible; (*likely*) posible

feast /fi:st/ *n* (*relig*) fiesta *f*; (*meal*) banquete *m*, comilona *f*. ● *vt* banquetear, festejar. ~ **on** regalarse con

feat /fi:t/ *n* hazaña *f*

feather /'feðə(r)/ *n* pluma *f*. ● *vt*. ~ **one's nest** hacer su agosto. ~**-brained** *a* tonto. ~**weight** *n* peso *m* pluma

feature /'fi:tʃə(r)/ *n* (*on face*) facción *f*; (*characteristic*) característica *f*; (*in newspaper*) artículo *m*; ~ **(film)** película *f* principal, largometraje *m*. ● *vt* presentar; (*give prominence to*) destacar. ● *vi* figurar

February /'februərɪ/ *n* febrero *m*

feckless /'feklɪs/ *a* inepto; (*irresponsible*) irreflexivo

fed /fed/ *see* **feed**. ● *a*. ~ **up** (*sl*) harto (**with** de)

federal /'fedərəl/ *a* federal

federation /fedə'reɪʃn/ *n* federación *f*

fee /fi:/ *n* (*professional*) honorarios *mpl*; (*enrolment*) derechos *mpl*; (*club*) cuota *f*

feeble /'fi:bl/ *a* (**-er, -est**) débil. ~**-minded** *a* imbécil

feed /fi:d/ *vt* (*pt* **fed**) dar de comer a; (*supply*) alimentar. ● *vi* comer. ● *n* (*for animals*) pienso *m*; (*for babies*) comida *f*. ~**back** *n* reacciones *fpl*, comentarios *mpl*

feel /fi:l/ *vt* (*pt* **felt**) sentir; (*touch*) tocar; (*think*) parecerle. **do you** ~ **it's a good idea?** te parece buena idea? **I** ~ **it is necessary** me parece necesario. ~ **as if** tener la impresión de que. ~ **hot/hungry** tener calor/hambre. ~ **like** (*want, fam*)

tener ganas de. ~ **up to** sentirse capaz de
feeler /'fi:lə(r)/ *n* (*of insects*) antena *f*. **put out a** ~ (*fig*) hacer un sondeo
feeling /'fi:lɪŋ/ *n* sentimiento *m*; (*physical*) sensación *f*
feet /fi:t/ *see* **foot**
feign /feɪn/ *vt* fingir
feint /feɪnt/ *n* finta *f*
felicitous /fə'lɪsɪtəs/ *a* feliz, oportuno
feline /'fi:laɪn/ *a* felino
fell[1] /fel/ *see* **fall**
fell[2] /fel/ *vt* derribar
fellow /'feləʊ/ *n* (*fam*) tipo *m*; (*comrade*) compañero *m*; (*society*) socio *m*. ~**-countryman** *n* compatriota *m* & *f*. ~ **passenger/traveller** *n* compañero *m* de viaje. ~**ship** *n* compañerismo *m*; (*group*) asociación *f*
felony /'felənɪ/ *n* crimen *m*
felt[1] /felt/ *n* fieltro *m*
felt[2] /felt/ *see* **feel**
female /'fi:meɪl/ *a* hembra; ‹*voice, sex etc*› femenino. ● *n* mujer *f*; (*animal*) hembra *f*
femini|ne /'femənɪn/ *a* & *n* femenino (*m*). ~**nity** /-'nɪnətɪ/ *n* feminidad *f*. ~**st** *n* feminista *m* & *f*
fenc|e /fens/ *n* cerca *f*; (*person, sl*) perista *m* & *f* (*fam*). ● *vt*. ~**e (in)** encerrar, cercar. ● *vi* (*sport*) practicar la esgrima. ~**er** *n* esgrimidor *m*. ~**ing** *n* (*sport*) esgrima *f*
fend /fend/ *vi*. ~ **for o.s.** valerse por sí mismo. ● *vt*. ~ **off** defenderse de
fender /'fendə(r)/ *n* guardafuego *m*; (*mudguard, Amer*) guardabarros *m invar*; (*naut*) defensa *f*
fennel /'fenl/ *n* hinojo *m*
ferment /'fɜ:ment/ *n* fermento *m*; (*fig*) agitación *f*. /fə'ment/ *vt*/*i* fermentar. ~**ation** /-'teɪʃn/ *n* fermentación *f*
fern /fɜ:n/ *n* helecho *m*
feroci|ous /fə'rəʊʃəs/ *a* feroz. ~**ty** /fə'rɒsətɪ/ *n* ferocidad *f*
ferret /'ferɪt/ *n* hurón *m*. ● *vi* (*pt* **ferreted**) huronear. ● *vt*. ~ **out** descubrir
ferry /'ferɪ/ *n* ferry *m*. ● *vt* transportar
fertil|e /'fɜ:taɪl/ *a* fértil; (*biol*) fecundo. ~**ity** /-'tɪlətɪ/ *n* fertilidad *f*; (*biol*) fecundidad *f*
fertilize /'fɜ:təlaɪz/ *vt* abonar; (*biol*) fecundar. ~**r** *n* abono *m*
fervent /'fɜ:vənt/ *a* ferviente
fervour /'fɜ:və(r)/ *n* fervor *m*
fester /'festə(r)/ *vi* enconarse
festival /'festəvl/ *n* fiesta *f*; (*of arts*) festival *m*
festive /'festɪv/ *a* festivo. ~ **season** *n* temporada *f* de fiestas
festivity /fe'stɪvətɪ/ *n* festividad *f*
festoon /fe'stu:n/ *vi*. ~ **with** adornar de
fetch /fetʃ/ *vt* (*go for*) ir a buscar; (*bring*) traer; (*be sold for*) venderse por
fetching /'fetʃɪŋ/ *a* atractivo
fête /feɪt/ *n* fiesta *f*. ● *vt* festejar
fetid /'fetɪd/ *a* fétido
fetish /'fetɪʃ/ *n* fetiche *m*; (*psych*) obsesión *f*
fetter /'fetə(r)/ *vt* encadenar. ~**s** *npl* grilletes *mpl*
fettle /'fetl/ *n* condición *f*
feud /fju:d/ *n* enemistad *f* (inveterada)
feudal /fjú:dl/ *a* feudal. ~**ism** *n* feudalismo m
fever /'fi:və(r)/ *n* fiebre *f*. ~**ish** *a* febril
few /fju:/ *a* pocos. ● *n* pocos *mpl*. **a** ~ unos (pocos). **a good** ~, **quite a** ~ (*fam*) muchos. ~**er** *a* & *n* menos. ~**est** *a* & *n* el menor número de
fiancé /fɪ'ɒnseɪ/ *n* novio *m*. ~**e** /fɪ'ɒnseɪ/ *n* novia *f*
fiasco /fɪ'æskəʊ/ *n* (*pl* **-os**) fiasco *m*
fib /fɪb/ *n* mentirijilla *f*. ~**ber** *n* mentiroso *m*
fibre /'faɪbə(r)/ *n* fibra *f*. ~**glass** *n* fibra *f* de vidrio
fickle /'fɪkl/ *a* inconstante
fiction /'fɪkʃn/ *n* ficción *f*. **(works of)** ~ novelas *fpl*. ~**al** *a* novelesco
fictitious /fɪk'tɪʃəs/ *a* ficticio
fiddle /'fɪdl/ *n* (*fam*) violín *m*; (*swindle, sl*) trampa *f*. ● *vt* (*sl*) falsificar. ~ **with** juguetear con, toquetear, manosear. ~**r** *n* (*fam*) violinista *m* & *f*; (*cheat, sl*) tramposo *m*
fidelity /fɪ'delətɪ/ *n* fidelidad *f*
fidget /'fɪdʒɪt/ *vi* (*pt* **fidgeted**) moverse, ponerse nervioso. ~ **with** juguetear con. ● *n* azogado *m*. ~**y** *a* azogado
field /fi:ld/ *n* campo *m*. ~ **day** *n* gran ocasión *f*. ~ **glasses** *npl* gemelos *mpl*. **F**~ **Marshal** *n* mariscal *m* de campo, capitán *m* general. ~**work** *n* investigaciones *fpl* en el terreno
fiend /fi:nd/ *n* demonio *m*. ~**ish** *a* diabólico

fierce /fɪəs/ *a* (**-er, -est**) feroz; ‹*attack*› violento. **~ness** *n* ferocidad *f*, violencia *f*
fiery /ˈfaɪərɪ/ *a* (**-ier, -iest**) ardiente
fifteen /fɪfˈtiːn/ *a* & *n* quince (*m*). **~th** *a* & *n* quince (*m*), decimoquinto (*m*). ● *n* (*fraction*) quinzavo *m*
fifth /fɪfθ/ *a* & *n* quinto (*m*). **~ column** *n* quinta columna *f*
fift|ieth /ˈfɪftɪəθ/ *a* & *n* cincuenta (*m*). **~y** *a* & *n* cincuenta (*m*). **~y-~y** mitad y mitad, a medias. **a ~y-~y chance** una posibilidad *f* de cada dos
fig /fɪg/ *n* higo *m*
fight /faɪt/ *vt*/*i* (*pt* **fought**) luchar; (*quarrel*) disputar. **~ shy of** evitar. ● *n* lucha *f*; (*quarrel*) disputa *f*; (*mil*) combate *m*. **~ back** defenderse. **~ off** rechazar ‹*attack*›; luchar contra ‹*illness*›. **~er** *n* luchador *m*; (*mil*) combatiente *m* & *f*; (*aircraft*) avión *m* de caza. **~ing** *n* luchas *fpl*
figment /ˈfɪgmənt/ *n* invención *f*
figurative /ˈfɪgjʊrətɪv/ *a* figurado
figure /ˈfɪgə(r)/ *n* (*number*) cifra *f*; (*diagram*) figura *f*; (*shape*) forma *f*; (*of woman*) tipo *m*. ● *vt* imaginar. ● *vi* figurar. **that ~s** (*Amer*, *fam*) es lógico. **~ out** explicarse. **~head** *n* testaferro *m*, mascarón *m* de proa. **~ of speech** *n* tropo *m*, figura *f*. **~s** *npl* (*arithmetic*) aritmética *f*
filament /ˈfɪləmənt/ *n* filamento *m*
filch /fɪltʃ/ *vt* hurtar
file[1] /faɪl/ *n* carpeta *f*; (*set of papers*) expediente *m*. ● *vt* archivar ‹*papers*›
file[2] /faɪl/ *n* (*row*) fila *f*. ● *vi*. **~ in** entrar en fila. **~ past** desfilar ante
file[3] /faɪl/ *n* (*tool*) lima *f*. ● *vt* limar
filings /ˈfaɪlɪŋz/ *npl* limaduras *fpl*
fill /fɪl/ *vt* llenar. ● *vi* llenarse. **~ in** rellenar ‹*form*›. **~ out** (*get fatter*) engordar. **~ up** (*auto*) llenar, repostar. ● *n*. **eat one's ~** hartarse de comer. **have had one's ~ of** estar harto de
fillet /ˈfɪlɪt/ *n* filete *m*. ● *vt* (*pt* **filleted**) cortar en filetes
filling /ˈfɪlɪŋ/ *n* (*in tooth*) empaste *m*. **~ station** *n* estación *f* de servicio
film /fɪlm/ *n* película *f*. ● *vt* filmar. **~ star** *n* estrella *f* de cine. **~-strip** *n* tira *f* de película
filter /ˈfɪltə(r)/ *n* filtro *m*. ● *vt* filtrar. ● *vi* filtrarse. **~-tipped** *a* con filtro
filth /fɪlθ/ *n* inmundicia *f*. **~iness** *n* inmundicia *f*. **~y** *a* inmundo
fin /fɪn/ *n* aleta *f*
final /ˈfaɪnl/ *a* último; (*conclusive*) decisivo. ● *n* (*sport*) final *f*. **~s** *npl* (*schol*) exámenes *mpl* de fin de curso
finale /fɪˈnɑːlɪ/ *n* final *m*
final: **~ist** *n* finalista *m* & *f*. **~ize** *vt* concluir. **~ly** *adv* (*lastly*) finalmente, por fin; (*once and for all*) definitivamente
financ|e /ˈfaɪnæns/ *n* finanzas *fpl*. ● *vt* financiar. **~ial** /faɪˈnænʃl/ *a* financiero. **~ially** *adv* económicamente. **~ier** /faɪˈnænsɪə(r)/ *n* financiero *m*
finch /fɪntʃ/ *n* pinzón *m*
find /faɪnd/ *vt* (*pt* **found**) encontrar. **~ out** enterarse de. **~er** *n* el *m* que encuentra, descubridor *m*. **~ings** *npl* resultados *mpl*
fine[1] /faɪn/ *a* (**-er, -est**) fino; (*excellent*) excelente. ● *adv* muy bien; (*small*) en trozos pequeños
fine[2] /faɪn/ *n* multa *f*. ● *vt* multar
fine: **~ arts** *npl* bellas artes *fpl*. **~ly** *adv* (*admirably*) espléndidamente; (*cut*) en trozos pequeños. **~ry** /ˈfaɪnərɪ/ *n* galas *fpl*
finesse /fɪˈnes/ *n* tino *m*
finger /ˈfɪŋgə(r)/ *n* dedo *m*. ● *vt* tocar. **~-nail** *n* uña *f*. **~print** *n* huella *f* dactilar. **~-stall** *n* dedil *m*. **~tip** *n* punta *f* del dedo
finicking /ˈfɪnɪkɪŋ/ *a*, **finicky** /ˈfɪnɪkɪ/ *a* melindroso
finish /ˈfɪnɪʃ/ *vt*/*i* terminar. **~ doing** terminar de hacer. **~ up doing** terminar por hacer. ● *n* fin *m*; (*of race*) llegada *f*, meta *f*; (*appearance*) acabado *m*
finite /ˈfaɪnaɪt/ *a* finito
Fin|land /ˈfɪnlənd/ *n* Finlandia *f*. **~n** *n* finlandés *m*. **~nish** *a* & *n* finlandés (*m*)
fiord /fjɔːd/ *n* fiordo *m*
fir /fɜː(r)/ *n* abeto *m*
fire /faɪə(r)/ *n* fuego *m*; (*conflagration*) incendio *m*. ● *vt* disparar ‹*bullet etc*›; (*dismiss*) despedir; (*fig*) excitar, enardecer, inflamar. ● *vi* tirar. **~arm** *n* arma *f* de fuego. **~ brigade** *n* cuerpo *m* de bomberos. **~cracker** *n* (*Amer*) petardo *m*. **~ department** *n* (*Amer*) cuerpo *m* de bomberos. **~-engine** *n* coche *m* de bomberos. **~-escape** *n* escalera *f* de incendios. **~light** *n*

lumbre *f*. **~man** *n* bombero *m*. **~place** *n* chimenea *f*. **~side** *n* hogar *m*. **~ station** *n* parque *m* de bomberos. **~wood** *n* leña *f*. **~work** *n* fuego *m* artificial

firing-squad /'faɪərɪŋskwɒd/ *n* pelotón *m* de ejecución

firm[1] /fɜ:m/ *n* empresa *f*

firm[2] /fɜ:m/ *a* (**-er, -est**) firme. **~ly** *adv* firmemente. **~ness** *n* firmeza *f*

first /fɜ:st/ *a* primero. **at ~ hand** directamente. **at ~ sight** a primera vista. ● *n* primero *m*. ● *adv* primero; (*first time*) por primera vez. **~ of all** ante todo. **~ aid** *n* primeros auxilios *mpl*. **~-born** *a* primogénito. **~-class** *a* de primera clase. **~ floor** *n* primer piso *m*; (*Amer*) planta *f* baja. **F~ Lady** *n* (*Amer*) Primera Dama *f*. **~ly** *adv* en primer lugar. **~ name** *n* nombre *m* de pila. **~-rate** *a* excelente

fiscal /'fɪskl/ *a* fiscal

fish /fɪʃ/ *n* (*usually invar*) (*alive in water*) pez *m*; (*food*) pescado *m*. ● *vi* pescar. **~ for** pescar. **~ out** (*take out, fam*) sacar. **go ~ing** ir de pesca. **~erman** /'fɪʃəmən/ *n* pescador *m*. **~ing** *n* pesca *f*. **~ing-rod** *n* caña *f* de pesca. **~monger** *n* pescadero *m*. **~-shop** *n* pescadería *f*. **~y** *a* ‹*smell*› a pescado; (*questionable, fam*) sospechoso

fission /'fɪʃn/ *n* fisión *f*

fist /fɪst/ *n* puño *m*

fit[1] /fɪt/ *a* (**fitter, fittest**) conveniente; (*healthy*) sano; (*good enough*) adecuado; (*able*) capaz. ● *n* (*of clothes*) corte *m*. ● *vt* (*pt* **fitted**) (*adapt*) adaptar; (*be the right size for*) sentar bien a; (*install*) colocar. ● *vi* encajar; (*in certain space*) caber; ‹*clothes*› sentar. **~ out** equipar. **~ up** equipar

fit[2] /fɪt/ *n* ataque *m*

fitful /'fɪtfl/ *a* irregular

fitment /'fɪtmənt/ *n* mueble *m*

fitness /'fɪtnɪs/ *n* (*buena*) salud *f*; (*of remark*) conveniencia *f*

fitting /'fɪtɪŋ/ *a* apropiado. ● *n* (*of clothes*) prueba *f*. **~s** /'fɪtɪŋz/ *npl* (*in house*) accesorios *mpl*

five /faɪv/ *a* & *n* cinco (*m*). **~r** /'faɪvə(r)/ *n* (*fam*) billete *m* de cinco libras

fix /fɪks/ *vt* (*make firm, attach, decide*) fijar; (*mend, deal with*) arreglar. ● *n*. **in a ~** en un aprieto. **~ation** /-eɪʃn/ *n* fijación *f*. **~ed** *a* fijo

fixture /'fɪkstʃə(r) *n* (*sport*) partido *m*. **~s** (*in house*) accesorios *mpl*

fizz /fɪz/ *vi* burbujear. ● *n* efervescencia *f*. **~le** /fɪzl/ *vi* burbujear. **~le out** fracasar. **~y** *a* efervescente; ‹*water*› con gas

flab /flæb/ *n* (*fam*) flaccidez *f*

flabbergast /'flæbəgɑ:st/ *vt* pasmar

flabby /'flaæbɪ/ *a* flojo

flag /flæg/ *n* bandera *f*. ● *vt* (*pt* **flagged**). **~ down** hacer señales de parada a. ● *vi* (*pt* **flagged**) (*weaken*) flaquear; ‹*interest*› decaer; ‹*conversation*› languidecer

flagon /'flægən/ *n* botella *f* grande, jarro *m*

flag-pole /'flægpəʊl/ *n* asta *f* de bandera

flagrant /'fleɪgrənt/ *a* (*glaring*) flagrante; (*scandalous*) escandaloso

flagstone /'flægstəʊn/ *n* losa *f*

flair /fleə(r)/ *n* don *m* (**for** de)

flak|e /fleɪk/ *n* copo *m*; (*of paint, metal*) escama *f*. ● *vi* desconcharse. **~e out** (*fam*) caer rendido. **~y** *a* escamoso

flamboyant /flæm'bɔɪənt/ *a* ‹*clothes*› vistoso; ‹*manner*› extravagante

flame /fleɪm/ *n* llama *f*. ● *vi* llamear

flamingo /flə'mɪŋgəʊ/ *n* (*pl* **-o(e)s**) flamenco *m*

flammable /'flæməbl/ *a* inflamable

flan /flæn/ *n* tartaleta *f*, tarteleta *f*

flank /flænk/ *n* (*of animal*) ijada *f*, flanco *m*; (*of person*) costado *m*; (*of mountain*) falda *f*; (*mil*) flanco *m*

flannel /'flænl/ *n* franela *f* (de lana); (*for face*) paño *m* (para lavarse la cara). **~ette** *n* franela *f* (de algodón), muletón *m*

flap /flæp/ *vi* (*pt* **flapped**) ondear; ‹*wings*› aletear; (*become agitated, fam*) ponerse nervioso. ● *vt* sacudir; batir ‹*wings*›. ● *n* (*of pocket*) cartera *f*; (*of table*) ala *f*. **get into a ~** ponerse nervioso

flare /fleə(r)/ ● *n* llamarada *f*; (*mil*) bengala *f*; (*in skirt*) vuelo *m*. ● *vi*. **~ up** llamear; ‹*fighting*› estallar; ‹*person*› encolerizarse. **~d** *a* ‹*skirt*› acampanado

flash /flæʃ/ ● *vi* brillar; (*on and off*) destellar. ● *vt* despedir; (*aim torch*) dirigir; (*flaunt*) hacer ostentación de. **~ past** pasar como un rayo. ● *n* relámpago *m*; (*of news, camera*) flash *m*. **~back** *n* escena *f* retrospectiva. **~light** *n* (*torch*) linterna *f*

flashy /'flæʃɪ/ *a* ostentoso

flask /flɑ:sk/ *n* frasco *m*; (*vacuum flask*) termo *m*

flat[1] /flæt/ *a* (**flatter**, **flattest**) llano; ‹*tyre*› desinflado; ‹*refusal*› categórico; ‹*fare, rate*› fijo; (*mus*) desafinado. ● *adv*. **~ out** (*at top speed*) a toda velocidad

flat[2] /flæt/ *n* (*rooms*) piso *m*, apartamento *m*; ‹*tyre*› (*fam*) pinchazo *m*; (*mus*) bemol *m*

flat: **~ly** *adv* categóricamente. **~ness** *n* llanura *f*. **~ten** /'flætn/ *vt* allanar, aplanar. ● *vi* allanarse, aplanarse

flatter /flætə(r)/ *vt* adular. **~er** *n* adulador *m*. **~ing** *a* ‹*person*› lisonjero; ‹*clothes*› favorecedor. **~y** *n* adulación *f*

flatulence /'flætjʊləns/ *n* flatulencia *f*

flaunt /flɔ:nt/ *vt* hacer ostentación de

flautist /'flɔ:tɪst/ *n* flautista *m & f*

flavour /'fleɪvə(r)/ *n* sabor *m*. ● *vt* condimentar. **~ing** *n* condimento *m*

flaw /flɔ:/ *n* defecto *m*. **~less** *a* perfecto

flax /flæks/ *n* lino *m*. **~en** *a* de lino; ‹*hair*› rubio

flea /fli:/ *n* pulga *f*

fleck /flek/ *n* mancha *f*, pinta *f*

fled /fled/ *see* **flee**

fledged /fledʒd/ *a*. **fully ~** ‹*doctor etc*› hecho y derecho; ‹*member*› de pleno derecho

fledg(e)ling /'fledʒlɪŋ/ *n* pájaro *m* volantón

flee /fli:/ *vi* (*pt* **fled**) huir. ● *vt* huir de

fleece /fli:s/ *n* vellón *m*. ● *vt* (*rob*) desplumar

fleet /fli:t/ *n* (*naut, aviat*) flota *f*; (*of cars*) parque *m*

fleeting /'fli:tɪŋ/ *a* fugaz

Flemish /'flemɪʃ/ *a & n* flamenco (*m*)

flesh /fleʃ/ *n* carne *f*. **in the ~** en persona. **one's own ~ and blood** los de su sangre. **~y** *a* ‹*fruit*› carnoso

flew /flu:/ *see* **fly**[1]

flex /fleks/ *vt* doblar; flexionar ‹*muscle*›. ● *n* (*elec*) cable *m*, flexible *m*

flexib|ility /fleksə'bɪlətɪ/ *n* flexibilidad *f*. **~le** /'fleksəbl/ *a* flexible

flexitime /fleksɪ'taɪm/ *n* horario *m* flexible

flick /flɪk/ *n* golpecito *m*. ● *vt* dar un golpecito a. **~ through** hojear

flicker /'flɪkə(r)/ *vi* temblar; ‹*light*› parpadear. ● *n* temblor *m*; (*of hope*) resquicio *m*; (*of light*) parpadeo *m*

flick: **~-knife** *n* navaja *f* de muelle. **~s** *npl* cine *m*

flier /'flaɪə(r)/ *n* aviador *m*; (*circular, Amer*) prospecto *m*, folleto *m*

flies /flaɪz/ *npl* (*on trousers, fam*) bragueta *f*

flight /flaɪt/ *n* vuelo *m*; (*fleeing*) huida *f*, fuga *f*. **~ of stairs** tramo *m* de escalera *f*. **put to ~** poner en fuga. **take (to) ~** darse a la fuga. **~-deck** *n* cubierta *f* de vuelo

flighty /'flaɪtɪ/ *a* (**-ier, -iest**) frívolo

flimsy /'flɪmzɪ/ *a* (**-ier, -iest**) flojo, débil, poco substancioso

flinch /flɪntʃ/ *vi* (*draw back*) retroceder (**from** ante). **without ~ing** (*without wincing*) sin pestañear

fling /flɪŋ/ *vt* (*pt* **flung**) arrojar. ● *n*. **have a ~** echar una cana al aire

flint /flɪnt/ *n* pedernal *m*; (*for lighter*) piedra *f*

flip /flɪp/ *vt* (*pt* **flipped**) dar un golpecito a. **~ through** hojear. ● *n* golpecito *m*. **~ side** *n* otra cara *f*

flippant /'flɪpənt/ *a* poco serio; (*disrespectful*) irrespetuoso

flipper /'flɪpə(r)/ *n* aleta *f*

flirt /flɜ:t/ *vi* coquetear. ● *n* (*woman*) coqueta *f*; (*man*) mariposón *m*, coqueto *m*. **~ation** /-'teɪʃn/ *n* coqueteo *m*

flit /flɪt/ *vi* (*pt* **flitted**) revolotear

float /fləʊt/ *vi* flotar. ● *vt* hacer flotar. ● *n* flotador *m*; (*on fishing line*) corcho *m*; (*cart*) carroza *f*

flock /flɒk/ *n* (*of birds*) bandada *f*; (*of sheep*) rebaño *m*; (*of people*) muchedumbre *f*, multitud *f*. ● *vi* congregarse

flog /flɒg/ *vt* (*pt* **flogged**) (*beat*) azotar; (*sell, sl*) vender

flood /flʌd/ *n* inundación *f*; (*fig*) torrente *m*. ● *vt* inundar. ● *vi* ‹*building etc*› inundarse; ‹*river*› desbordar

floodlight /'flʌdlaɪt/ *n* foco *m*. ● *vt* (*pt* **floodlit**) iluminar (con focos)

floor /flɔ:(r)/ *n* suelo *m*; (*storey*) piso *m*; (*for dancing*) pista *f*. ● *vt* (*knock down*) derribar; (*baffle*) confundir

flop /flɒp/ *vi* (*pt* **flopped**) dejarse caer pesadamente; (*fail, sl*) fracasar. ● *n* (*sl*) fracaso *m*. **~py** *a* flojo

flora /ˈflɔ:rə/ *n* flora *f*

floral /ˈflɔ:rəl/ *a* floral

florid /ˈflɒrɪd/ *a* florido

florist /ˈflɒrɪst/ *n* florista *m & f*

flounce /flaʊns/ *n* volante *m*

flounder[1] /ˈflaʊndə(r)/ *vi* avanzar con dificultad, no saber qué hacer

flounder[2] /ˈflaʊndə(r)/ *n* (*fish*) platija *f*

flour /flaʊə(r)/ *n* harina *f*

flourish /ˈflʌrɪʃ/ *vi* prosperar. ● *vt* blandir. ● *n* ademán *m* elegante; (*in handwriting*) rasgo *m*. **~ing** *a* próspero

floury /ˈflaʊərɪ/ *a* harinoso

flout /flaʊt/ *vt* burlarse de

flow /fləʊ/ *vi* correr; (*hang loosely*) caer. **~ into** ‹*river*› desembocar en. ● *n* flujo *m*; (*jet*) chorro *m*; (*stream*) corriente *f*; (*of words, tears*) torrente *m*. **~ chart** *n* organigrama *m*

flower /ˈflaʊə(r)/ *n* flor *f*. **~-bed** *n* macizo *m* de flores. **~ed** *a* floreado, de flores. **~y** *a* florido

flown /fləʊn/ *see* **fly**[1]

flu /flu:/ *n* (*fam*) gripe *f*

fluctuat|e /ˈflʌktjʊeɪt/ *vi* fluctuar. **~ion** /-eɪʃn/ *n* fluctuación *f*

flue /flu:/ *n* humero *m*

fluen|cy /ˈflu:ənsɪ/ *n* facilidad *f*. **~t** *a* ‹*style*› fluido; ‹*speaker*› elocuente. **be ~t (in a language)** hablar (un idioma) con soltura. **~tly** *adv* con fluidez; (*lang*) con soltura

fluff /flʌf/ *n* pelusa *f*. **~y** *a* (**-ier, -iest**) velloso

fluid /ˈflu:id/ *a & n* fluido (*m*)

fluke /flu:k/ *n* (*stroke of luck*) chiripa *f*

flung /flʌŋ/ *see* **fling**

flunk /flʌŋk/ *vt* (*Amer, fam*) ser suspendido en ‹*exam*›; suspender ‹*person*›. ● *vi* (*fam*) ser suspendido

fluorescent /flʊəˈresnt/ *a* fluorescente

fluoride /flʊəraɪd/ *n* fluoruro *m*

flurry /ˈflʌrɪ/ *n* (*squall*) ráfaga *f*; (*fig*) agitación *f*

flush[1] /flʌʃ/ *vi* ruborizarse. ● *vt* limpiar con agua. **~ the toilet** tirar de la cadena. ● *n* (*blush*) rubor *m*; (*fig*) emoción *f*

flush[2] /flʌʃ/ *a*. **~ (with)** a nivel (con)

flush[3] /flʌʃ/ *vt/i*. **~ out** (*drive out*) echar fuera

fluster /ˈflʌstə(r)/ *vt* poner nervioso

flute /flu:t/ *n* flauta *f*

flutter /ˈflʌtə(r)/ *vi* ondear; ‹*bird*› revolotear. ● *n* (*of wings*) revoloteo *m*; (*fig*) agitación *f*

flux /flʌks/ *n* flujo *m*. **be in a state of ~** estar siempre cambiando

fly[1] /flaɪ/ *vi* (*pt* **flew**, *pp* **flown**) volar; ‹*passenger*› ir en avión; ‹*flag*› flotar; (*rush*) correr. ● *vt* pilotar ‹*aircraft*›; transportar en avión ‹*passengers, goods*›; izar ‹*flag*›. ● *n* (*of trousers*) bragueta *f*

fly[2] /flaɪ/ *n* mosca *f*

flyer /ˈflaɪə(r)/ *n* aviador *m*; (*circular, Amer*) prospecto *m*, folleto *m*

flying /ˈflaɪɪŋ/ *a* volante; (*hasty*) relámpago *invar*. ● *n* (*activity*) aviación *f*. **~ visit** *n* visita *f* relámpago

fly: **~leaf** *n* guarda *f*. **~over** *n* paso *m* elevado. **~weight** *n* peso *m* mosca

foal /fəʊl/ *n* potro *m*

foam /fəʊm/ *n* espuma *f*. **~(rubber)** *n* goma *f* espuma. ● *vi* espumar

fob /fɒb/ *vt* (*pt* **fobbed**). **~ off on s.o.** (*palm off*) encajar a uno

focal /ˈfəʊkl/ *a* focal

focus /ˈfəʊkəs/ *n* (*pl* **-cuses** *or* **-ci** /-saɪ/) foco *m*; (*fig*) centro *m*. **in ~** enfocado. **out of ~** desenfocado. ● *vt/i* (*pt* **focused**) enfocar(se); (*fig*) concentrar

fodder /ˈfɒdə(r)/ *n* forraje *m*

foe /fəʊ/ *n* enemigo *m*

foetus /ˈfi:təs/ *n* (*pl* **-tuses**) feto *m*

fog /fɒg/ *n* niebla *f*. ● *vt* (*pt* **fogged**) envolver en niebla; (*photo*) velar. ● *vi*. **~ (up)** empañarse; (*photo*) velarse

fog(e)y /fəʊgɪ/ *n*. **be an old ~** estar chapado a la antigua

foggy /ˈfɒgɪ/ *a* (**-ier, -iest**) nebuloso. **it is ~** hay niebla

foghorn /ˈfɒghɔ:n/ *n* sirena *f* de niebla

foible /ˈfɔɪbl/ *n* punto *m* débil

foil[1] /fɔɪl/ *vt* (*thwart*) frustrar

foil[2] /fɔɪl/ *n* papel *m* de plata; (*fig*) contraste *m*

foist /fɔɪst/ *vt* encajar (on a)

fold[1] /fəʊld/ *vt* doblar; cruzar ‹*arms*›. ● *vi* doblarse; (*fail*) fracasar. ● *n* pliegue *m*

fold[2] /fəʊld/ *n* (*for sheep*) redil *m*

folder /ˈfəʊldə(r)/ *n* (*file*) carpeta *f*; (*leaflet*) folleto *m*

folding /ˈfəʊldɪŋ/ *a* plegable

foliage /ˈfəʊlɪɪdʒ/ *n* follaje *m*

folk /fəʊk/ *n* gente *f*. ● *a* popular. **~lore** *n* folklore *m*. **~s** *npl* (*one's relatives*) familia *f*
follow /ˈfɒləʊ/ *vt/i* seguir. **~ up** seguir; (*investigate further*) investigar. **~er** *n* seguidor *m*. **~ing** *n* partidarios *mpl*. ● *a* siguiente. ● *prep* después de
folly /ˈfɒlɪ/ *n* locura *f*
foment /fəˈment/ *vt* fomentar
fond /fɒnd/ *a* (**-er**, **-est**) (*loving*) cariñoso; ‹*hope*› vivo. **be ~ of s.o.** tener(le) cariño a uno. **be ~ of sth** ser aficionado a algo
fondle /ˈfɒndl/ *vt* acariciar
fondness /ˈfɒndnɪs/ *n* cariño *m*; (*for things*) afición *f*
font /fɒnt/ *n* pila *f* bautismal
food /fu:d/ *n* alimento *m*, comida *f*. **~ processor** *n* robot *m* de cocina, batidora *f*
fool /fu:l/ *n* tonto *m*. ● *vt* engañar. ● *vi* hacer el tonto
foolhardy /ˈfu:lhɑ:dɪ/ *a* temerario
foolish /ˈfu:lɪʃ/ *a* tonto. **~ly** *adv* tontamente. **~ness** *n* tontería *f*
foolproof /ˈfu:lpru:f/ *a* infalible, a toda prueba, a prueba de tontos
foot /fʊt/ *n* (*pl* **feet**) pie *m*; (*measure*) pie *m* (=*30,48 cm*); (*of animal, furniture*) pata *f*. **get under s.o.'s feet** estorbar a uno. **on ~** a pie. **on/to one's feet** de pie. **put one's ~ in it** meter la pata. ● *vt* pagar ‹*bill*›. **~ it** ir andando
footage /ˈfʊtɪdʒ/ *n* (*of film*) secuencia *f*
football /ˈfʊtbɔ:l/ *n* (*ball*) balón *m*; (*game*) fútbol *m*. **~er** *n* futbolista *m & f*
footbridge /ˈfʊtbrɪdʒ/ *n* puente *m* para peatones
foothills /ˈfʊthɪlz/ *npl* estribaciones *fpl*
foothold /ˈfʊthəʊld/ *n* punto *m* de apoyo *m*
footing /ˈfʊtɪŋ/ *n* pie *m*
footlights /ˈfʊtlaɪts/ *npl* candilejas *fpl*
footloose /ˈfʊtlu:s/ *a* libre
footman /ˈfʊtmən/ *n* lacayo *m*
footnote /ˈfʊtnəʊt/ *n* nota *f* (al pie de la página)
foot: **~path** *n* (*in country*) senda *f*; (*in town*) acera *f*, vereda *f* (*Arg*), banqueta *f* (*Mex*). **~print** *n* huella *f*. **~sore** *a*. **be ~sore** tener los pies doloridos. **~step** *n* paso *m*. **~stool** *n* escabel *m*. **~wear** *n* calzado *m*

for /fɔ:(r)/, *unstressed* /fə(r)/ *prep* (*expressing purpose*) para; (*on behalf of*) por; (*in spite of*) a pesar de; (*during*) durante; (*in favour of*) a favor de. **he has been in Madrid ~ two months** hace dos meses que está en Madrid. ● *conj* ya que
forage /ˈfɒrɪdʒ/ *vi* forrajear. ● *n* forraje *m*
foray /ˈfɒreɪ/ *n* incursión *f*
forbade /fəˈbæd/ *see* **forbid**
forbear /fɔ:ˈbeər/ *vt/i* (*pt* **forbore**, *pp* **forborne**) contenerse. **~ance** *n* paciencia *f*
forbid /fəˈbɪd/ *vt* (*pt* **forbade**, *pp* **forbidden**) prohibir (**s.o. to do** a uno hacer). **~ s.o. sth** prohibir algo a uno
forbidding /fəˈbɪdɪŋ/ *a* imponente
force /fɔ:s/ *n* fuerza *f*. **come into ~** entrar en vigor. **the ~s** las fuerzas *fpl* armadas. ● *vt* forzar. **~ on** imponer a. **~d** *a* forzado. **~-feed** *vt* alimentar a la fuerza. **~ful** /ˈfɔ:sfʊl/ *a* enérgico
forceps /ˈfɔ:seps/ *n invar* tenazas *fpl*; (*for obstetric use*) fórceps *m invar*; (*for dental use*) gatillo *m*
forcibl|e /ˈfɔ:səbl/ *a* a la fuerza. **~y** *adv* a la fuerza
ford /fɔ:d/ *n* vado *m*, botadero *m* (*Mex*). ● *vt* vadear
fore /fɔ:(r)/ *a* anterior. ● *n*. **come to the ~** hacerse evidente
forearm /ˈfɔ:rɑ:m/ *n* antebrazo *m*
foreboding /fɔ:ˈbəʊdɪŋ/ *n* presentimiento *m*
forecast /ˈfɔ:kɑ:st/ *vt* (*pt* **forecast**) pronosticar. ● *n* pronóstico *m*
forecourt /ˈfɔ:kɔ:t/ *n* patio *m*
forefathers /ˈfɔ:fɑ:ðəz/ *npl* antepasados *mpl*
forefinger /ˈfɔ:fɪŋgə(r)/ *n* (dedo *m*) índice *m*
forefront /ˈfɔ:frʌnt/ *n* vanguardia *f*. **in the ~** a/en vanguardia, en primer plano
foregone /ˈfɔ:gɒn/ *a*. **~ conclusion** resultado *m* previsto
foreground /ˈfɔ:graʊnd/ *n* primer plano *m*
forehead /ˈfɒrɪd/ *n* frente *f*
foreign /ˈfɒrən/ *a* extranjero; ‹*trade*› exterior; ‹*travel*› al extranjero, en el extranjero. **~er** *n* extranjero *m*. **F~ Secretary** *n* ministro *m* de Asuntos Exteriores
foreman /ˈfɔ:mən/ *n* capataz *m*, caporal *m*

foremost /ˈfɔːməʊst/ *a* primero. ● *adv*. **first and ~** ante todo
forensic /fəˈrensɪk/ *a* forense
forerunner /ˈfɔːrʌnə(r)/ *n* precursor *m*
foresee /fɔːˈsiː/ *vt* (*pt* **-saw**, *pp* **-seen**) prever. **~able** *a* previsible
foreshadow /fɔːˈʃædəʊ/ *vt* presagiar
foresight /ˈfɔːsaɪt/ *n* previsión *f*
forest /ˈfɒrɪst/ *n* bosque *m*
forestall /fɔːˈstɔːl/ *vt* anticiparse a
forestry /ˈfɒrɪstrɪ/ *n* silvicultura *f*
foretaste /ˈfɔːteɪst/ *n* anticipación *f*
foretell /fɔːˈtel/ *vt* (*pt* **foretold**) predecir
forever /fəˈrevə(r)/ *adv* para siempre
forewarn /fɔːˈwɔːn/ *vt* prevenir
foreword /ˈfɔːwɜːd/ *n* prefacio *m*
forfeit /ˈfɔːfɪt/ *n* (*penalty*) pena *f*; (*in game*) prenda *f*; (*fine*) multa *f*. ● *vt* perder
forgave /fəˈgeɪv/ *see* **forgive**
forge[1] /fɔːdʒ/ *n* fragua *f*. ● *vt* fraguar; (*copy*) falsificar
forge[2] /fɔːdʒ/ *vi* avanzar. **~ahead** adelantarse rápidamente
forge: **~r** /ˈfɔːdʒə(r)/ *n* falsificador *m*. **~ry** *n* falsificación *f*
forget /fəˈget/ *vt* (*pt* **forgot**, *pp* **forgotten**) olvidar. **~ o.s.** propasarse, extralimitarse. ● *vi* olvidar(se). **I forgot** se me olvidó. **~ful** *a* olvidadizo. **~ful of** olvidando. **~-me-not** *n* nomeolvides *f invar*
forgive /fəˈgɪv/ *vt* (*pt* **forgave**, *pp* **forgiven**) perdonar. **~ness** *n* perdón *m*
forgo /fɔːˈgəʊ/ *vt* (*pt* **forwent**, *pp* **forgone**) renunciar a
fork /fɔːk/ *n* tenedor *m*; (*for digging*) horca *f*; (*in road*) bifurcación *f*. ● *vi* ‹*road*› bifurcarse. **~ out** (*sl*) aflojar la bolsa (*fam*), pagar. **~ed** *a* ahorquillado; ‹*road*› bifurcado. **~-lift truck** *n* carretilla *f* elevadora
forlorn /fəˈlɔːn/ *a* (*hopeless*) desesperado; (*abandoned*) abandonado. **~ hope** *n* empresa *f* desesperada
form /fɔːm/ *n* forma *f*; (*document*) impreso *m*, formulario *m*; (*schol*) clase *f*. ● *vt* formar. ● *vi* formarse
formal /ˈfɔːml/ *a* formal; ‹*person*› formalista; ‹*dress*› de etiqueta. **~ity** /-ˈmælətɪ/ *n* formalidad *f*. **~ly** *adv* oficialmente
format /ˈfɔːmæt/ *n* formato *m*
formation /fɔːˈmeɪʃn/ *n* formación *f*
formative /ˈfɔːmətɪv/ *a* formativo
former /ˈfɔːmə(r)/ *a* anterior; (*first of two*) primero. **~ly** *adv* antes
formidable /ˈfɔːmɪdəbl/ *a* formidable
formless /ˈfɔːmlɪs/ *a* informe
formula /ˈfɔːmjʊlə/ *n* (*pl* **-ae** /-iː/ *or* **-as**) fórmula *f*
formulate /ˈfɔːmjʊleɪt/ *vt* formular
fornicat|e /ˈfɔːnɪkeɪt/ *vi* fornicar. **~ion** /-ˈkeɪʃn/ *n* fornicación *f*
forsake /fəˈseɪk/ *vt* (*pt* **forsook**, *pp* **forsaken**) abandonar
fort /fɔːt/ *n* (*mil*) fuerte *m*
forte /ˈfɔːteɪ/ *n* (*talent*) fuerte *m*
forth /fɔːθ/ *adv* en adelante. **and so ~** y así sucesivamente. **go back and ~** ir y venir
forthcoming /fɔːθˈkʌmɪŋ/ *a* próximo, venidero; (*sociable*, *fam*) comunicativo
forthright /ˈfɔːθraɪt/ *a* directo
forthwith /fɔːθˈwɪθ/ *adv* inmediatamente
fortieth /ˈfɔːtɪɪθ/ *a* cuarenta, cuadragésimo. ● *n* cuadragésima parte *f*
fortif|ication /fɔːtɪfɪˈkeɪʃn/ *n* fortificación *f*. **~y** /ˈfɔːtɪfaɪ/ *vt* fortificar
fortitude /ˈfɔːtɪtjuːd/ *n* valor *m*
fortnight /ˈfɔːtnaɪt/ *n* quince días *mpl*, quincena *f*. **~ly** *a* bimensual. ● *adv* cada quince días
fortress /ˈfɔːtrɪs/ *n* fortaleza *f*
fortuitous /fɔːˈtjuːɪtəs/ *a* fortuito
fortunate /ˈfɔːtʃənət/ *a* afortunado. **be ~** tener suerte. **~ly** *adv* afortunadamente
fortune /ˈfɔːtʃuːn/ *n* fortuna *f*. **have the good ~ to** tener la suerte de. **~-teller** *n* adivino *m*
forty /ˈfɔːtɪ/ *a* & *n* cuarenta (*m*). **~ winks** un sueñecito *m*
forum /ˈfɔːrəm/ *n* foro *m*
forward /ˈfɔːwəd/ *a* delantero; (*advanced*) precoz; (*pert*) impertinente. ● *n* (*sport*) delantero *m*. ● *adv* adelante. **come ~** presentarse. **go ~** avanzar. ● *vt* hacer seguir ‹*letter*›; enviar ‹*goods*›; (*fig*) favorecer. **~ness** *n* precocidad *f*
forwards /ˈfɔːwədz/ *adv* adelante
fossil /ˈfɒsl/ *a* & *n* fósil (*m*)
foster /ˈfɒstə(r)/ *vt* (*promote*) fomentar; criar ‹*child*›. **~-child** *n* hijo *m* adoptivo. **~-mother** *n* madre *f* adoptiva
fought /fɔːt/ *see* **fight**

foul /faʊl/ *a* (**-er, -est**) ‹*smell, weather*› asqueroso; (*dirty*) sucio; ‹*language*› obsceno; ‹*air*› viciado. **~ play** *n* jugada *f* sucia; (*crime*) delito *m*. ● *n* (*sport*) falta *f*. ● *vt* ensuciar; manchar ‹*reputation*›. **~-mouthed** *a* obsceno

found[1] /faʊnd/ *see* **find**

found[2] /faʊnd/ *vt* fundar

found[3] /faʊnd/ *vt* (*tec*) fundir

foundation /faʊn'deɪʃn/ *n* fundación *f*; (*basis*) fundamento. **~s** *npl* (*archit*) cimientos *mpl*

founder[1] /'faʊndə(r)/ *n* fundador *m*

founder[2] /'faʊndə(r)/ *vi* ‹*ship*› hundirse

foundry /'faʊndrɪ/ *n* fundición *f*

fountain /'faʊntɪn/ *n* fuente *f*. **~-pen** *n* estilográfica *f*

four /fɔ:(r)/ *a* & *n* cuatro (*m*). **~fold** *a* cuádruple. ● *adv* cuatro veces. **~-poster** *n* cama *f* con cuatro columnas

foursome /'fɔ:səm/ *n* grupo *m* de cuatro personas

fourteen /'fɔ:ti:n/ *a* & *n* catorce (*m*). **~th** *a* & *n* catorce (*m*), decimocuarto (*m*). ● *n* (*fraction*) catorceavo *m*

fourth /fɔ:θ/ *a* & *n* cuarto (*m*)

fowl /faʊl/ *n* ave *f*

fox /fɒks/ *n* zorro *m*, zorra *f*. ● *vt* (*baffle*) dejar perplejo; (*deceive*) engañar

foyer /'fɔɪeɪ/ *n* (*hall*) vestíbulo *m*

fraction /'frækʃn/ *n* fracción *f*

fractious /'frækʃəs/ *a* díscolo

fracture /'fræktʃə(r)/ *n* fractura *f*. ● *vt* fracturar. ● *vi* fracturarse

fragile /'frædʒaɪl/ *a* frágil

fragment /'frægmənt/ *n* fragmento *m*. **~ary** *a* fragmentario

fragran|ce /'freɪgrəns/ *n* fragancia *f*. **~t** *a* fragante

frail /freɪl/ *a* (**-er, -est**) frágil

frame /freɪm/ *n* (*of picture, door, window*) marco *m*; (*of spectacles*) montura *f*; (*fig, structure*) estructura *f*; (*temporary state*) estado *m*. **~ of mind** estado *m* de ánimo. ● *vt* enmarcar; (*fig*) formular; (*jurid, sl*) incriminar falsamente. **~-up** *n* (*sl*) complot *m*

framework /'freɪmwɜ:k/ *n* estructura *f*; (*context*) marco *m*

France /frɑ:ns/ *n* Francia *f*

franchise /'fræntʃaɪz/ *n* (*pol*) derecho *m* a votar; (*com*) concesión *f*

Franco... /'fræŋkəʊ/ *pref* franco...

frank /fræŋk/ *a* sincero. ● *vt* franquear. **~ly** *adv* sinceramente. **~ness** *n* sinceridad *f*

frantic /'fræntɪk/ *a* frenético. **~ with** loco de

fraternal /frə'tɜ:nl/ *a* fraternal

fraternity /frə'tɜ:nɪtɪ/ *n* fraternidad *f*; (*club*) asociación *f*

fraternize /'frætənaɪz/ *vi* fraternizar

fraud /frɔ:d/ *n* (*deception*) fraude *m*; (*person*) impostor *m*. **~ulent** *a* fraudulento

fraught /frɔ:t/ *a* (*tense*) tenso. **~ with** cargado de

fray[1] /freɪ/ *vt* desgastar. ● *vi* deshilacharse

fray[2] /freɪ/ *n* riña *f*

freak /fri:k/ *n* (*caprice*) capricho *m*; (*monster*) monstruo *m*; (*person*) chalado *m*. ● *a* anormal. **~ish** *a* anormal

freckle /'frekl/ *n* peca *f*. **~d** *a* pecoso

free /fri:/ *a* (**freer** /'fri:ə(r)/, **freest** /'fri:ɪst/) libre; (*gratis*) gratis; (*lavish*) generoso. **~ kick** *n* golpe *m* franco. **~ of charge** gratis. **~ speech** *n* libertad *f* de expresión. **give a ~ hand** dar carta blanca. ● *vt* (*pt* **freed**) (*set at liberty*) poner en libertad; (*relieve from*) liberar (**from/of** de); (*untangle*) desenredar; (*loosen*) soltar

freedom /'fri:dəm/ *n* libertad *f*

freehold /'fri:həʊld/ *n* propiedad *f* absoluta

freelance /'fri:lɑ:ns/ *a* independiente

freely /'fri:lɪ/ *adv* libremente

Freemason /'fri:meɪsn/ *n* masón *m*. **~ry** *n* masonería *f*

free-range /'fri:reɪndʒ/ *a* ‹*eggs*› de granja

freesia /'fri:zjə/ *n* fresia *f*

freeway /'fri:weɪ/ *n* (*Amer*) autopista *f*

freez|e /'fri:z/ *vt* (*pt* **froze**, *pp* **frozen**) helar; congelar ‹*food, wages*›. ● *vi* helarse, congelarse; (*become motionless*) quedarse inmóvil. ● *n* helada *f*; (*of wages, prices*) congelación *f*. **~er** *n* congelador *m*. **~ing** *a* glacial. ● *n* congelación *f*. **below ~ing** bajo cero

freight /freɪt/ *n* (*goods*) mercancías *fpl*; (*hire of ship etc*) flete *m*. **~er** *n* (*ship*) buque *m* de carga

French /frentʃ/ *a* francés. ● *n* (*lang*) francés *m*. **~man** *n* francés *m*. **~-speaking** *a* francófono. **~ window** *n* puertaventana *f*. **~woman** *f* francesa *f*

frenz|ied /ˈfrenzɪd/ *a* frenético. **~y** *n* frenesí *m*

frequency /ˈfriːkwənsɪ/ *n* frecuencia *f*

frequent /frɪˈkwent/ *vt* frecuentar. /ˈfriːkwənt/ *a* frecuente. **~ly** *adv* frecuentemente

fresco /ˈfreskəʊ/ *n* (*pl* **-o(e)s**) fresco *m*

fresh /freʃ/ *a* (**-er**, **-est**) fresco; (*different*, *additional*) nuevo; (*cheeky*) fresco, descarado; ‹*water*› dulce. **~en** *vi* refrescar. **~en up** ‹*person*› refrescarse. **~ly** *adv* recientemente. **~man** *n* estudiante *m* de primer año. **~ness** *n* frescura *f*

fret /fret/ *vi* (*pt* **fretted**) inquietarse. **~ful** *a* (*discontented*) quejoso; (*irritable*) irritable

Freudian /ˈfrɔɪdjən/ *a* freudiano

friar /ˈfraɪə(r)/ *n* fraile *m*

friction /ˈfrɪkʃn/ *n* fricción *f*

Friday /ˈfraɪdeɪ/ *n* viernes *m*. **Good ~** Viernes Santo

fridge /frɪdʒ/ *n* (*fam*) nevera *f*, refrigerador *m*, refrigeradora *f*

fried /fraɪd/ *see* **fry**. ● *a* frito

friend /frend/ *n* amigo *m*. **~liness** /ˈfrendlɪnɪs/ *n* simpatía *f*. **~ly** *a* (**-ier**, **-iest**) simpático. **F~ly Society** *n* mutualidad *f*. **~ship** /ˈfrendʃɪp/ *n* amistad *f*

frieze /friːz/ *n* friso *m*

frigate /ˈfrɪgət/ *n* fragata *f*

fright /fraɪt/ *n* susto *m*; (*person*) espantajo *m*; (*thing*) horror *m*

frighten /ˈfraɪtn/ *vt* asustar. **~ off** ahuyentar. **~ed** *a* asustado. **be ~ed** tener miedo (**of** de)

frightful /ˈfraɪtfl/ *a* espantoso, horrible. **~ly** *adv* terriblemente

frigid /ˈfrɪdʒɪd/ *a* frío; (*psych*) frígido. **~ity** /-ˈdʒɪdətɪ/ *n* frigidez *f*

frill /frɪl/ *n* volante *m*. **~s** *npl* (*fig*) adornos *mpl*. **with no ~s** sencillo

fringe /frɪndʒ/ *n* (*sewing*) fleco *m*; (*ornamental border*) franja *f*; (*of hair*) flequillo *m*; (*of area*) periferia *f*; (*of society*) margen *m*. **~ benefits** *npl* beneficios *mpl* suplementarios. **~ theatre** *n* teatro *m* de vanguardia

frisk /frɪsk/ *vt* (*search*) cachear

frisky /ˈfrɪskɪ/ *a* (**-ier**, **-iest**) retozón; ‹*horse*› fogoso

fritter[1] /ˈfrɪtə(r)/ *vt*. **~ away** desperdiciar

fritter[2] /ˈfrɪtə(r)/ *n* buñuelo *m*

frivol|ity /frɪˈvɒlətɪ/ *n* frivolidad *f*. **~ous** /ˈfrɪvələs/ *a* frívolo

frizzy /ˈfrɪzɪ/ *a* crespo

fro /frəʊ/ *see* **to and fro**

frock /frɒk/ *n* vestido *m*; (*of monk*) hábito *m*

frog /frɒg/ *n* rana *f*. **have a ~ in one's throat** tener carraspera

frogman /ˈfrɒgmən/ *n* hombre *m* rana

frolic /ˈfrɒlɪk/ *vi* (*pt* **frolicked**) retozar. ● *n* broma *f*

from /frɒm/, *unstressed* /frəm/ *prep* de; (*with time*, *prices*, *etc*) a partir de; (*habit*, *conviction*) por; (*according to*) según. **take ~** (*away from*) quitar a

front /frʌnt/ *n* parte *f* delantera; (*of building*) fachada *f*; (*of clothes*) delantera *f*; (*mil*, *pol*) frente *f*; (*of book*) principio *m*; (*fig*, *appearance*) apariencia *f*; (*sea front*) paseo *m* marítimo. **in ~ of** delante de. **put a bold ~ on** hacer de tripas corazón, mostrar firmeza. ● *a* delantero; (*first*) primero. **~age** *n* fachada *f*. **~al** *a* frontal; ‹*attack*› de frente. **~ door** *n* puerta *f* principal. **~ page** *n* (*of newspaper*) primera plana *f*

frontier /ˈfrʌntɪə(r)/ *n* frontera *f*

frost /frɒst/ *n* (*freezing*) helada *f*; (*frozen dew*) escarcha *f*. **~-bite** *n* congelación *f*. **~-bitten** *a* congelado. **~ed** *a* ‹*glass*› esmerilado

frosting /ˈfrɒstɪŋ/ *n* (*icing*, *Amer*) azúcar *m* glaseado

frosty *a* ‹*weather*› de helada; ‹*window*› escarchado; (*fig*) glacial

froth /frɒθ/ *n* espuma *f*. ● *vi* espumar. **~y** *a* espumoso

frown /fraʊn/ *vi* fruncir el entrecejo. **~ on** desaprobar. ● *n* ceño *m*

froze /frəʊz/, **frozen** /ˈfrəʊzn/ *see* **freeze**

frugal /ˈfruːgl/ *a* frugal. **~ly** *adv* frugalmente

fruit /fruːt/ *n* (*bot*, *on tree*, *fig*) fruto *m*; (*as food*) fruta *f*. **~erer** *n* frutero *m*. **~ful** /ˈfruːtfl/ *a* fértil; (*fig*) fructífero. **~less** *a* infructuoso. **~ machine** *n* (máquina *f*) tragaperras *m*. **~ salad** *n* macedonia *f* de frutas. **~y** /ˈfruːtɪ/ *a* ‹*taste*› que sabe a fruta

fruition /fruːˈɪʃn/ *n*. **come to ~** realizarse

frump /frʌmp/ *n* espantajo *m*

frustrat|e /frʌˈstreɪt/ *vt* frustrar. **~ion** /-ʃn/ *n* frustración *f*; (*disappointment*) decepción *f*

fry[1] /fraɪ/ *vt* (*pt* **fried**) freír. ● *vi* freírse

fry[2] /fraɪ/ *n* (*pl* **fry**). **small ~** gente *f* de poca monta
frying-pan /ˈfraɪɪŋpæn/ *n* sartén *f*
fuchsia /ˈfjuːʃə/ *n* fucsia *f*
fuddy-duddy /ˈfʌdɪdʌdɪ/ *n*. **be a ~** (*sl*) estar chapado a la antigua
fudge /fʌdʒ/ *n* dulce *m* de azúcar
fuel /ˈfjuːəl/ *n* combustible *m*; (*for car engine*) carburante *m*; (*fig*) pábulo *m*. ● *vt* (*pt* **fuelled**) alimentar de combustible
fugitive /ˈfjuːdʒɪtɪv/ *a* & *n* fugitivo (*m*)
fugue /fjuːg/ *n* (*mus*) fuga *f*
fulfil /fʊlˈfɪl/ *vt* (*pt* **fulfilled**) cumplir (con) ‹*promise, obligation*›; satisfacer ‹*condition*›; realizar ‹*hopes, plans*›; llevar a cabo ‹*task*›. **~ment** *n* (*of promise, obligation*) cumplimiento *m*; (*of conditions*) satisfacción *f*; (*of hopes, plans*) realización *f*; (*of task*) ejecución *f*
full /fʊl/ *a* (**-er**, **-est**) lleno; ‹*bus, hotel*› completo; ‹*skirt*› amplio; ‹*account*› detallado. **at ~ speed** a máxima velocidad. **be ~ (up)** (*with food*) no poder más. **in ~ swing** en plena marcha. ● *n*. **in ~** sin quitar nada. **to the ~** completamente. **write in ~** escribir con todas las letras. **~ back** *n* (*sport*) defensa *m* & *f*. **~-blooded** *a* vigoroso. **~ moon** *n* plenilunio *m*. **~-scale** *a* ‹*drawing*› de tamaño natural; (*fig*) amplio. **~ stop** *n* punto *m*; (*at end of paragraph, fig*) punto *m* final. **~ time** *a* de jornada completa. **~y** *adv* completamente
fulsome /ˈfʊlsəm/ *a* excesivo
fumble /ˈfʌmbl/ *vi* buscar (torpemente)
fume /fjuːm/ *vi* humear; (*fig, be furious*) estar furioso. **~s** *npl* humo *m*
fumigate /ˈfjuːmɪgeɪt/ *vt* fumigar
fun /fʌn/ *n* (*amusement*) diversión *f*; (*merriment*) alegría *f*. **for ~** en broma. **have ~** divertirse. **make ~ of** burlarse de
function /ˈfʌŋkʃn/ *n* (*purpose, duty*) función *f*; (*reception*) recepción *f*. ● *vi* funcionar. **~al** *a* funcional
fund /fʌnd/ *n* fondo *m*. ● *vt* proveer fondos para
fundamental /fʌndəˈmentl/ *a* fundamental
funeral /ˈfjuːnərəl/ *n* funeral *m*, funerales *mpl*. ● *a* fúnebre
fun-fair /ˈfʌnfeə(r)/ *n* parque *m* de atracciones
fungus /ˈfʌŋgəs/ *n* (*pl* **-gi** /-gaɪ/) hongo *m*
funicular /fjuːˈnɪkjʊlə(r)/ *n* funicular *m*
funk /fʌŋk/ *m* (*fear, sl*) miedo *m*; (*state of depression, Amer, sl*) depresión *f*. **be in a (blue) ~** tener (mucho) miedo; (*Amer*) estar (muy) deprimido. ● *vi* rajarse
funnel /ˈfʌnl/ *n* (*for pouring*) embudo *m*; (*of ship*) chimenea *f*
funn|ily /ˈfʌnɪlɪ/ *adv* graciosamente; (*oddly*) curiosamente. **~y** *a* (**-ier**, **-iest**) divertido, gracioso; (*odd*) curioso, raro. **~y-bone** *n* cóndilo *m* del húmero. **~y business** *n* engaño *m*
fur /fɜː(r)/ *n* pelo *m*; (*pelt*) piel *f*; (*in kettle*) sarro *m*
furbish /ˈfɜːbɪʃ/ *vt* pulir; (*renovate*) renovar
furious /ˈfjʊərɪəs/ *a* furioso. **~ly** *adv* furiosamente
furnace /ˈfɜːnɪs/ *n* horno *m*
furnish /ˈfɜːnɪʃ/ *vt* (*with furniture*) amueblar; (*supply*) proveer. **~ings** *npl* muebles *mpl*, mobiliario *m*
furniture /ˈfɜːnɪtʃə(r)/ *n* muebles *mpl*, mobiliario *m*
furrier /ˈfʌrɪə(r)/ *n* peletero *m*
furrow /ˈfʌrəʊ/ *n* surco *m*
furry /ˈfɜːrɪ/ *a* peludo
furthe|r /ˈfɜːðə(r)/ *a* más lejano; (*additional*) nuevo. ● *adv* más lejos; (*more*) además. ● *vt* fomentar. **~rmore** *adv* además. **~rmost** *a* más lejano. **~st** *a* más lejano. ● *adv* más lejos
furtive /ˈfɜːtɪv/ *a* furtivo
fury /ˈfjʊərɪ/ *n* furia *f*
fuse[1] /fjuːz/ *vt* (*melt*) fundir; (*fig, unite*) fusionar. **~ the lights** fundir los plomos. ● *vi* fundirse; (*fig*) fusionarse. ● *n* fusible *m*, plomo *m*
fuse[2] /fjuːz/ *n* (*of bomb*) mecha *f*
fuse-box /ˈfjuːzbɒks/ *n* caja *f* de fusibles
fuselage /ˈfjuːzəlɑːʒ/ *n* fuselaje *m*
fusion /ˈfjuːʒn/ *n* fusión *f*
fuss /fʌs/ *n* (*commotion*) jaleo *m*. **kick up a ~** armar un lío, armar una bronca, protestar. **make a ~ of** tratar con mucha atención. **~y** *a* (**-ier**, **-iest**) (*finicky*) remilgado; (*demanding*) exigente; (*ornate*) recargado
fusty /ˈfʌstɪ/ *a* (**-ier**, **-iest**) que huele a cerrado

futile /ˈfjuːtaɪl/ *a* inútil, vano
future /ˈfjuːtʃə(r)/ *a* futuro. ● *n* futuro *m*, porvenir *m*; (*gram*) futuro *m*. **in ~** en lo sucesivo, de ahora en adelante
futuristic /fjuːtʃəˈrɪstɪk/ *a* futurista
fuzz /fʌz/ *n* (*fluff*) pelusa *f*; (*police, sl*) policía *f*, poli *f* (*fam*)
fuzzy /ˈfʌzɪ/ *a* ‹*hair*› crespo; ‹*photograph*› borroso

G

gab /gæb/ *n* charla *f*. **have the gift of the ~** tener un pico de oro
gabardine /gæbəˈdiːn/ *n* gabardina *f*
gabble /ˈgæbl/ *vt* decir atropelladamente. ● *vi* hablar atropelladamente. ● *n* torrente *m* de palabras
gable /ˈgeɪbl/ *n* aguilón *m*
gad /gæd/ *vi* (*pt* **gadded**). **~ about** callejear
gadget /ˈgædʒɪt/ *n* chisme *m*
Gaelic /ˈgeɪlɪk/ *a* & *n* gaélico (*m*)
gaffe /gæf/ *n* plancha *f*, metedura *f* de pata
gag /gæg/ *n* mordaza *f*; (*joke*) chiste *m*. ● *vt* (*pt* **gagged**) amordazar
gaga /ˈgɑːgɑː/ *a* (*sl*) chocho
gaiety /ˈgeɪətɪ/ *n* alegría *f*
gaily /ˈgeɪlɪ/ *adv* alegremente
gain /geɪn/ *vt* ganar; (*acquire*) adquirir; (*obtain*) conseguir. ● *vi* ‹*clock*› adelantar. ● *n* ganancia *f*; (*increase*) aumento *m*. **~ful** *a* lucrativo
gainsay /geɪnˈseɪ/ *vt* (*pt* **gainsaid**) (*formal*) negar
gait /geɪt/ *n* modo *m* de andar
gala /ˈgɑːlə/ *n* fiesta *f*; (*sport*) competición *f*
galaxy /ˈgæləksɪ/ *n* galaxia *f*
gale /geɪl/ *n* vendaval *m*; (*storm*) tempestad *f*
gall /gɔːl/ *n* bilis *f*; (*fig*) hiel *f*; (*impudence*) descaro *m*
gallant /ˈgælənt/ *a* (*brave*) valiente; (*chivalrous*) galante. **~ry** *n* valor *m*
gall-bladder /ˈgɔːlblædə(r)/ *n* vesícula *f* biliar
galleon /ˈgælɪən/ *n* galeón *m*
gallery /ˈgælərɪ/ *n* galería *f*
galley /ˈgælɪ/ *n* (*ship*) galera *f*; (*ship's kitchen*) cocina *f*. **~ (proof)** *n* (*typ*) galerada *f*
Gallic /ˈgælɪk/ *a* gálico. **~ism** *n* galicismo *m*
gallivant /ˈgælɪvænt/ *vi* (*fam*) callejear
gallon /ˈgælən/ *n* galón *m* (*imperial = 4,546l; Amer = 3,785l*)
gallop /ˈgæləp/ *n* galope *m*. ● *vi* (*pt* **galloped**) galopar
gallows /ˈgæləʊz/ *n* horca *f*
galore /gəˈlɔː(r)/ *adv* en abundancia
galosh /gəˈlɒʃ/ *n* chanclo *m*
galvanize /ˈgælvənaɪz/ *vt* galvanizar
gambit /ˈgæmbɪt/ *n* (*in chess*) gambito *m*; (*fig*) táctica *f*
gambl|e /ˈgæmbl/ *vt/i* jugar. **~e on** contar con. ● *n* (*venture*) empresa *f* arriesgada; (*bet*) jugada *f*; (*risk*) riesgo *m*. **~er** *n* jugador *m*. **~ing** *n* juego *m*
game[1] /geɪm/ *n* juego *m*; (*match*) partido *m*; (*animals, birds*) caza *f*. ● *a* valiente. **~ for** listo para
game[2] /geɪm/ *a* (*lame*) cojo
gamekeeper /ˈgeɪmkiːpə(r)/ *n* guardabosque *m*
gammon /ˈgæmən/ *n* jamón *m* ahumado
gamut /ˈgæmət/ *n* gama *f*
gamy /ˈgeɪmɪ/ *a* manido
gander /ˈgændə(r)/ *n* ganso *m*
gang /gæŋ/ *n* pandilla *f*; (*of workmen*) equipo *m*. ● *vi*. **~ up** unirse (**on** contra)
gangling /ˈgæŋglɪŋ/ *a* larguirucho
gangrene /ˈgæŋgriːn/ *n* gangrena *f*
gangster /ˈgæŋstə(r)/ *n* bandido *m*, gangster *m*
gangway /ˈgæŋweɪ/ *n* pasillo *m*; (*of ship*) pasarela *f*
gaol /dʒeɪl/ *n* cárcel *f*. **~bird** *n* criminal *m* empedernido. **~er** *n* carcelero *m*
gap /gæp/ *n* vacío *m*; (*breach*) brecha *f*; (*in time*) intervalo *m*; (*deficiency*) laguna *f*; (*difference*) diferencia *f*
gap|e /ˈgeɪp/ *vi* quedarse boquiabierto; (*be wide open*) estar muy abierto. **~ing** *a* abierto; (*person*) boquiabierto
garage /ˈgærɑːʒ/ *n* garaje *m*; (*petrol station*) gasolinera *f*; (*for repairs*) taller *m*. ● *vt* dejar en (el) garaje
garb /gɑːb/ *n* vestido *m*
garbage /ˈgɑːbɪdʒ/ *n* basura *f*
garble /ˈgɑːbl/ *vt* mutilar
garden /ˈgɑːdn/ *n* (*of flowers*) jardín *m*; (*of vegetables/fruit*) huerto *m*. ● *vi* trabajar en el jardín/huerto. **~er** *n* jardinero/hortelano *m*. **~ing** *n* jardinería/horticultura *f*

gargantuan /gɑː'gæntjʊən/ *a* gigantesco
gargle /'gɑːgl/ *vi* hacer gárgaras. *n* gargarismo *m*
gargoyle /'gɑːgɔɪl/ *n* gárgola *f*
garish /'geərɪʃ/ *a* chillón
garland /'gɑːlənd/ *n* guirnalda *f*
garlic /'gɑːlɪk/ *n* ajo *m*
garment /'gɑːmənt/ *n* prenda *f* (de vestir)
garnet /'gɑːnɪt/ *n* granate *m*
garnish /'gɑːnɪʃ/ *vt* aderezar. ● *n* aderezo *m*
garret /'gærət/ *n* guardilla *f*, buhardilla *f*
garrison /'gærɪsn/ *n* guarnición *f*
garrulous /'gærələs/ *a* hablador
garter /'gɑːtə(r)/ *n* liga *f*
gas /gæs/ *n* (*pl* **gases**) gas *m*; (*med*) anestésico *m*; (*petrol*, *Amer*, *fam*) gasolina *f*. ● *vt* (*pt* **gassed**) asfixiar con gas. ● *vi* (*fam*) charlar. ~ **fire** *n* estufa *f* de gas
gash /gæʃ/ *n* cuchillada *f*. ● *vt* acuchillar
gasket /'gæskɪt/ *n* junta *f*
gas: ~ **mask** *n* careta *f* antigás *a invar*. ~ **meter** *n* contador *m* de gas
gasoline /'gæsəliːn/ *n* (*petrol*, *Amer*) gasolina *f*
gasometer /gæ'sɒmɪtə(r)/ *n* gasómetro *m*
gasp /gɑːsp/ *vi* jadear; (*with surprise*) quedarse boquiabierto. ● *n* jadeo *m*
gas: ~ **ring** *n* hornillo *m* de gas. ~ **station** *n* (*Amer*) gasolinera *f*
gastric /'gæstrɪk/ *a* gástrico
gastronomy /gæ'strɒnəmɪ/ *n* gastronomía *f*
gate /geɪt/ *n* puerta *f*; (*of metal*) verja *f*; (*barrier*) barrera *f*
gateau /'gætəʊ/ *n* (*pl* **gateaux**) tarta *f*
gate: ~**crasher** *n* intruso *m* (que ha entrado sin ser invitado o sin pagar). ~**way** *n* puerta f
gather /'gæðə(r)/ *vt* reunir ‹*people, things*›; (*accumulate*) acumular; (*pick up*) recoger; recoger ‹*flowers*›; (*fig*, *infer*) deducir; (*sewing*) fruncir. ~ **speed** acelerar. ● *vi* ‹*people*› reunirse; ‹*things*› acumularse. ~**ing** *n* reunión *f*
gauche /gəʊʃ/ *a* torpe
gaudy /'gɔːdɪ/ *a* (**-ier**, **-iest**) chillón
gauge /geɪdʒ/ *n* (*measurement*) medida *f*; (*rail*) entrevía *f*; (*instrument*) indicador *m*. ● *vt* medir; (*fig*) estimar
gaunt /gɔːnt/ *a* macilento; (*grim*) lúgubre
gauntlet /'gɔːntlɪt/ *n*. **run the ~ of** estar sometido a
gauze /gɔːz/ *n* gasa *f*
gave /geɪv/ *see* **give**
gawk /gɔːk/ *vi*. ~ **at** mirar como un tonto
gawky /'gɔːkɪ/ *a* (**-ier**, **-iest**) torpe
gawp /gɔːp/ *vi*. ~ **at** mirar como un tonto
gay /geɪ/ *a* (**-er**, **-est**) (*joyful*) alegre; (*homosexual*, *fam*) homosexual, gay (*fam*)
gaze /geɪz/ *vi*. ~ **(at)** mirar (fijamente). ● *n* mirada *f* (fija)
gazelle /gə'zel/ *n* gacela *f*
gazette /gə'zet/ *n* boletín *m* oficial, gaceta *f*
gazump /gə'zʌmp/ *vt* aceptar un precio más elevado de otro comprador
GB *abbr see* **Great Britain**
gear /gɪə(r)/ *n* equipo *m*; (*tec*) engranaje *m*; (*auto*) marcha *f*. **in ~** engranado. **out of ~** desengranado. ● *vt* adaptar. ~**box** *n* (*auto*) caja *f* de cambios
geese /giːs/ *see* **goose**
geezer /'giːzə(r)/ *n* (*sl*) tipo *m*
gelatine /'dʒelətiːn/ *n* gelatina *f*
gelignite /'dʒelɪgnaɪt/ *n* gelignita *f*
gem /dʒem/ *n* piedra *f* preciosa
Gemini /'dʒemɪnaɪ/ *n* (*astr*) Gemelos *mpl*, Géminis *mpl*
gen /dʒen/ *n* (*sl*) información *f*
gender /'dʒendə(r)/ *n* género *m*
gene /dʒiːn/ *n* gene *m*
genealogy /dʒiːnɪ'ælədzɪ/ *n* genealogía *f*
general /'dʒenərəl/ *a* general. ● *n* general *m*. **in ~** generalmente. ~ **election** *n* elecciones *fpl* generales
generaliz|ation /dʒenərəlaɪ'zeɪʃn/ *n* generalización *f*. ~**e** *vt/i* generalizar
generally /'dʒenərəlɪ/ *adv* generalmente
general practitioner /'dʒenərəl præk'tɪʃənə(r)/ *n* médico *m* de cabecera
generate /'dʒenəreɪt/ *vt* producir; (*elec*) generar
generation /dzenə'reɪʃn/ *n* generación *f*
generator /'dʒenəreɪtə(r)/ *n* (*elec*) generador *m*
genero|sity /dʒenə'rɒsətɪ/ *n* generosidad *f*. ~**us** /'dʒenərəs/ *a* generoso; (*plentiful*) abundante

genetic /dʒɪ'netɪk/ *a* genético. **~s** *n* genética *f*

Geneva /dʒɪ'ni:və/ *n* Ginebra *f*

genial /'dʒi:nɪəl/ *a* simpático, afable; ‹*climate*› suave, templado

genital /'dʒenɪtl/ *a* genital. **~s** *npl* genitales *mpl*

genitive /'dʒenɪtɪv/ *a* & *n* genitivo (*m*)

genius /'dʒi:nɪəs/ *n* (*pl* **-uses**) genio *m*

genocide /'dʒenəsaɪd/ *n* genocidio *m*

genre /ʒɑ:ŋr/ *n* género *m*

gent /dʒent/ *n* (*sl*) señor *m*. **~s** *n* aseo *m* de caballeros

genteel /dʒen'ti:l/ *a* distinguido; (*excessively refined*) cursi

gentle /'dʒentl/ *a* (**-er**, **-est**) (*mild, kind*) amable, dulce; (*slight*) ligero; ‹*hint*› discreto

gentlefolk /'dʒentlfəʊk/ *npl* gente *f* de buena familia

gentleman /'dʒentlmən/ *n* señor *m*; (*well-bred*) caballero *m*

gentleness /'dʒentlnɪs/ *n* amabilidad *f*

gentlewoman /'dʒentlwʊmən/ *n* señora *f* (de buena familia)

gently /'dʒentlɪ/ *adv* amablemente; (*slowly*) despacio

gentry /'dʒentrɪ/ *npl* pequeña aristocracia *f*

genuflect /'dʒenju:flekt/ *vi* doblar la rodilla

genuine /'dʒenjʊɪn/ *a* verdadero; ‹*person*› sincero

geograph|er /dʒɪ'ɒgrəfə(r)/ *n* geógrafo *m*. **~ical** /dʒɪə'græfɪkl/ *a* geográfico. **~y** /dʒɪ'ɒgrəfɪ/ *n* geografía *f*

geolog|ical /dʒɪə'lɒdʒɪkl/ *a* geológico. **~ist** *n* geólogo *m*. **~y** /dʒɪ'ɒlədʒɪ/ *n* geología *f*

geometr|ic(al) /dʒɪə'metrɪk(l)/ *a* geométrico. **~y** /dʒɪ'ɒmətrɪ/ *n* geometría *f*

geranium /dʒə'reɪnɪəm/ *n* geranio *m*

geriatrics /dʒerɪ'ætrɪks/ *n* geriatría *f*

germ /dʒɜ:m/ *n* (*rudiment, seed*) germen *m*; (*med*) microbio *m*

German /'dʒɜ:mən/ *a* & *n* alemán (*m*). **~ic** /dʒə'mænɪk/ *a* germánico. **~ measles** *n* rubéola *f*. **~ shepherd (dog)** *n* (perro *m*) pastor *m* alemán. **~y** *n* Alemania *f*

germicide /'dʒɜ:mɪsaɪd/ *n* germicida *m*

germinate /'dʒɜ:mɪneɪt/ *vi* germinar. ● *vt* hacer germinar

gerrymander /'dʒerɪmændə(r)/ *n* falsificación *f* electoral

gestation /dʒe'steɪʃn/ *n* gestación *f*

gesticulate /dʒe'stɪkjʊleɪt/ *vi* hacer ademanes, gesticular

gesture /'dʒestʃə(r)/ *n* ademán *m*; (*fig*) gesto *m*

get /get/ *vt* (*pt* & *pp* **got**, *pp Amer* **gotten**, *pres p* **getting**) obtener, tener; (*catch*) coger (*not LAm*), agarrar (*esp LAm*); (*buy*) comprar; (*find*) encontrar; (*fetch*) buscar, traer; (*understand, sl*) comprender, caer (*fam*). **~ s.o. to do sth** conseguir que uno haga algo. ● *vi* (*go*) ir; (*become*) hacerse; (*start to*) empezar a; (*manage*) conseguir. **~ married** casarse. **~ ready** prepararse. **~ about** ‹*person*› salir mucho; (*after illness*) levantarse. **~ along** (*manage*) ir tirando; (*progress*) hacer progresos. **~ along with** llevarse bien con. **~ at** (*reach*) llegar a; (*imply*) querer decir. **~ away** salir; (*escape*) escaparse. **~ back** *vi* volver. ● *vt* (*recover*) recobrar. **~ by** (*manage*) ir tirando; (*pass*) pasar. **~ down** bajar; (*depress*) deprimir. **~ in** entrar; subir ‹*vehicle*›; (*arrive*) llegar. **~ off** bajar de ‹*train, car etc*›; (*leave*) irse; (*jurid*) salir absuelto. **~ on** (*progress*) hacer progresos; (*succeed*) tener éxito. **~ on with** (*be on good terms with*) llevarse bien con; (*continue*) seguir. **~ out** ‹*person*› salir; (*take out*) sacar. **~ out of** (*fig*) librarse de. **~ over** reponerse de ‹*illness*›. **~ round** soslayar ‹*difficulty etc*›; engatusar ‹*person*›. **~ through** (*pass*) pasar; (*finish*) terminar; (*on phone*) comunicar con. **~ up** levantarse; (*climb*) subir; (*organize*) preparar. **~away** *n* huida *f*. **~-up** *n* traje *m*

geyser /'gi:zə(r)/ *n* calentador *m* de agua; (*geog*) géiser *m*

Ghana /'gɑ:nə/ *n* Ghana *f*

ghastly /'gɑ:stlɪ/ *a* (**-ier**, **-iest**) horrible; (*pale*) pálido

gherkin /'gɜ:kɪn/ *n* pepinillo *m*

ghetto /'getəʊ/ *n* (*pl* **-os**) (*Jewish quarter*) judería *f*; (*ethnic settlement*) barrio *m* pobre habitado por un grupo étnico

ghost /gəʊst/ *n* fantasma *m*. **~ly** *a* espectral

ghoulish /'gu:lɪʃ/ *a* macabro

giant /'dʒaɪənt/ *n* gigante *m*. ● *a* gigantesco

gibberish /ˈdʒibərɪʃ/ *n* jerigonza *f*
gibe /dʒaɪb/ *n* mofa *f*
giblets /ˈdʒɪblɪts/ *npl* menudillos *mpl*
Gibraltar /dʒɪˈbrɔːltə(r)/ *n* Gibraltar *m*
gidd|iness /ˈgɪdɪnɪs/ *n* vértigo *m*. **~y** *a* (**-ier**, **-iest**) mareado; ‹*speed*› vertiginoso. **be/feel ~y** estar/sentirse mareado
gift /gɪft/ *n* regalo *m*; (*ability*) don *m*. **~ed** *a* dotado de talento. **~-wrap** *vt* envolver para regalo
gig /gɪg/ *n* (*fam*) concierto *m*
gigantic /dʒaɪˈgæntɪk/ *a* gigantesco
giggle /ˈgɪgl/ *vi* reírse tontamente. ● *n* risita *f*. **the ~s** la risa *f* tonta
gild /gɪld/ *vt* dorar
gills /gɪlz/ *npl* agallas *fpl*
gilt /gɪlt/ *a* dorado. **~-edged** *a* (*com*) de máxima garantía
gimmick /ˈgɪmɪk/ *n* truco *m*
gin / dʒɪn/ *n* ginebra *f*
ginger /ˈdʒɪndʒə(r)/ *n* jengibre *m*. ● *a* rojizo. ● *vt*. **~ up** animar. **~ ale** *n*, **~ beer** *n* cerveza *f* de jengibre. **~bread** *n* pan *m* de jengibre
gingerly /ˈdʒɪndʒəlɪ/ *adv* cautelosamente
gingham /ˈgɪŋəm/ *n* guinga *f*
gipsy /ˈdʒɪpsɪ/ *n* gitano *m*
giraffe /dʒɪˈrɑːf/ *n* jirafa *f*
girder /ˈgɜːdə(r)/ *n* viga *f*
girdle /ˈgɜːdl/ *n* (*belt*) cinturón *m*; (*corset*) corsé *m*
girl /gɜːl/ *n* chica *f*, muchacha *f*; (*child*) niña *f*. **~-friend** *n* amiga *f*; (*of boy*) novia *f*. **~hood** *n* (*up to adolescence*) niñez *f*; (*adolescence*) juventud *f*. **~ish** *a* de niña; ‹*boy*› afeminado
giro /ˈdʒaɪrəʊ/ *n* (*pl* **-os**) giro *m* (bancario)
girth /gɜːθ/ *n* circunferencia *f*
gist /dʒɪst/ *n* lo esencial *invar*
give /gɪv/ *vt* (*pt* **gave**, *pp* **given**) dar; (*deliver*) entregar; regalar ‹*present*›; prestar ‹*aid, attention*›; (*grant*) conceder; (*yield*) ceder; (*devote*) dedicar. **~ o.s. to** darse a. ● *vi* dar; (*yield*) ceder; (*stretch*) estirarse. ● *n* elasticidad *f*. **~ away** regalar; descubrir ‹*secret*›. **~ back** devolver. **~ in** (*yield*) rendirse. **~ off** emitir. **~ o.s. up** entregarse (a). **~ out** distribuir; (*announce*) anunciar; (*become used up*) agotarse. **~ over** (*devote*) dedicar; (*stop, fam*) dejar (de). **~ up** (*renounce*) renunciar a; (*yield*) ceder
given /ˈgɪvn/ *see* **give**. ● *a* dado. **~ name** *n* nombre *m* de pila
glacier /ˈglæsɪə(r)/ *n* glaciar *m*
glad /glæd/ *a* contento. **~den** *vt* alegrar
glade /gleɪd/ *n* claro *m*
gladiator /ˈglædɪeɪtə(r)/ *n* gladiador *m*
gladiolus /glædɪˈəʊləs/ *n* (*pl* **-li** /-laɪ/) estoque *m*, gladiolo *m*, gladíolo *m*
gladly /ˈglædlɪ/ *adv* alegremente; (*willingly*) con mucho gusto
glamo|rize /ˈglæməraɪz/ *vt* embellecer. **~rous** *a* atractivo. **~ur** *n* encanto *m*
glance /glɑːns/ *n* ojeada *f*. ● *vi*. **~ at** dar un vistazo a
gland /glænd/ *n* glándula *f*
glar|e /gleə(r)/ *vi* deslumbrar; (*stare angrily*) mirar airadamente. ● *n* deslumbramiento *m*; (*stare, fig*) mirada *f* airada. **~ing** *a* deslumbrador; (*obvious*) manifiesto
glass /glɑːs/ *n* (*material*) vidrio *m*; (*without stem or for wine*) vaso *m*; (*with stem*) copa *f*; (*for beer*) caña *f*; (*mirror*) espejo *m*. **~es** *npl* (*spectacles*) gafas *fpl*, anteojos (*LAm*) *mpl*. **~y** *a* vítreo
glaze /gleɪz/ *vt* poner cristales a ‹*windows, doors*›; vidriar ‹*pottery*›. ● *n* barniz *m*; (*for pottery*) esmalte *m*. **~d** *a* ‹*object*› vidriado; ‹*eye*› vidrioso
gleam /gliːm/ *n* destello *m*. ● *vi* destellar
glean /gliːn/ *vt* espigar
glee /gliː/ *n* regocijo *m*. **~ club** *n* orfeón *m*. **~ful** *a* regocijado
glen /glen/ *n* cañada *f*
glib /glɪb/ *a* de mucha labia; ‹*reply*› fácil. **~ly** *adv* con poca sinceridad
glid|e /glaɪd/ *vi* deslizarse; ‹*plane*› planear. **~er** *n* planeador *m*. **~ing** *n* planeo *m*
glimmer /ˈglɪmə(r)/ *n* destello *m*. ● *vi* destellar
glimpse /glɪmps/ *n* vislumbre *f*. **catch a ~ of** vislumbrar. ● *vt* vislumbrar
glint /glɪnt/ *n* destello *m*. ● *vi* destellar
glisten /ˈglɪsn/ *vi* brillar
glitter /ˈglɪtə(r)/ *vi* brillar. ● *n* brillo *m*
gloat /gləʊt/ *vi*. **~ on/over** regodearse

global /gləʊbl/ *a* (*world-wide*) mundial; (*all-embracing*) global
globe /gləʊb/ *n* globo *m*
globule /'glɒbju:l/ *n* glóbulo *m*
gloom /glu:m/ *n* oscuridad *f*; (*sadness, fig*) tristeza *f*. **~y** *a* (**-ier**, **-iest**) triste; (*pessimistic*) pesimista
glorify /'glɔ:rɪfaɪ/ *vt* glorificar
glorious /'glɔ:rɪəs/ *a* espléndido; ‹*deed, hero etc*› glorioso
glory /'glɔ:rɪ/ *n* gloria *f*; (*beauty*) esplendor *m*. ● *vi*. **~ in** enorgullecerse de. **~-hole** *n* (*untidy room*) leonera *f*
gloss /glɒs/ *n* lustre *m*. ● *a* brillante. ● *vi*. **~ over** (*make light of*) minimizar; (*cover up*) encubrir
glossary /'glɒsərɪ/ *n* glosario *m*
glossy /'glɒsɪ/ *a* brillante
glove /glʌv/ *n* guante *m*. **~ compartment** *n* (*auto*) guantera *f*, gaveta *f*. **~d** *a* enguantado
glow /gləʊ/ *vi* brillar; (*with health*) rebosar de; (*with passion*) enardecerse. ● *n* incandescencia *f*; (*of cheeks*) rubor *m*
glower /'glaʊə(r)/ *vi*. **~ (at)** mirar airadamente
glowing /'gləʊɪŋ/ *a* incandescente; ‹*account*› entusiasta; ‹*complexion*› rojo; (*with health*) rebosante de
glucose /'glu:kəʊs/ *n* glucosa *f*
glue /glu:/ *n* cola *f*. ● *vt* (*pres p* **gluing**) pegar
glum /glʌm/ *a* (**glummer**, **glummest**) triste
glut /glʌt/ *n* superabundancia *f*
glutton /'glʌtn/ *n* glotón *m*. **~ous** *a* glotón. **~y** *n* glotonería *f*
glycerine /'glɪsəri:n/ *n* glicerina *f*
gnarled /nɑ:ld/ *a* nudoso
gnash /næʃ/ *vt*. **~ one's teeth** rechinar los dientes
gnat /næt/ *n* mosquito *m*
gnaw /nɔ:/ *vt/i* roer
gnome /nəʊm/ *n* gnomo *m*
go /gəʊ/ *vi* (*pt* **went**, *pp* **gone**) ir; (*leave*) irse; (*work*) funcionar; (*become*) hacerse; (*be sold*) venderse; (*vanish*) desaparecer. **~ ahead!** ¡adelante! **~ bad** pasarse. **~ riding** montar a caballo. **~ shopping** ir de compras. **be ~ing to do** ir a hacer. ● *n* (*pl* **goes**) (*energy*) energía *f*. **be on the ~** trabajar sin cesar. **have a ~** intentar. **it's your ~** te toca a ti. **make a ~ of** tener éxito en. **~ across** cruzar. **~ away** irse. **~ back** volver. **~ back on** faltar a ‹*promise etc*›. **~ by** pasar. **~ down** bajar; ‹*sun*› ponerse. **~ for** buscar, traer; (*like*) gustar; (*attack, sl*) atacar. **~ in** entrar. **~ in for** presentarse para ‹*exam*›. **~ off** (*leave*) irse; (*go bad*) pasarse; (*explode*) estallar. **~ on** seguir; (*happen*) pasar. **~ out** salir; ‹*light, fire*› apagarse. **~ over** (*check*) examinar. **~ round** (*be enough*) ser bastante. **~ through** (*suffer*) sufrir; (*check*) examinar. **~ under** hundirse. **~ up** subir. **~ without** pasarse sin
goad /gəʊd/ *vt* aguijonear
go-ahead /'gəʊəhed/ *n* luz *f* verde. ● *a* dinámico
goal /gəʊl/ *n* fin *m*, objeto *m*; (*sport*) gol *m*. **~ie** *n* (*fam*) portero *m*. **~keeper** *n* portero *m*. **~-post** *n* poste *m* (de la portería)
goat /gəʊt/ *n* cabra *f*
goatee /gəʊ'ti:/ *n* perilla *f*, barbas *fpl* de chivo
gobble /'gɒbl/ *vt* engullir
go-between /'gəʊbɪtwi:n/ *n* intermediario *m*
goblet /'gɒblɪt/ *n* copa *f*
goblin /'gɒblɪn/ *n* duende *m*
God /gɒd/ *n* Dios *m*. **~-forsaken** *a* olvidado de Dios
god /gɒd/ *n* dios *m*. **~child** *n* ahijado *m*. **~-daughter** *n* ahijada *f*. **~dess** /'gɒdɪs/ *n* diosa *f*. **~father** *n* padrino *m*. **~ly** *a* devoto. **~mother** *n* madrina *f*. **~send** *n* beneficio *m* inesperado. **~son** *n* ahijado *m*
go-getter /gəʊ'getə(r)/ *n* persona *f* ambiciosa
goggle /'gɒgl/ *vi*. **~ (at)** mirar con los ojos desmesuradamente abiertos
goggles /'gɒglz/ *npl* gafas *fpl* protectoras
going /'gəʊɪŋ/ *n* camino *m*; (*racing*) (estado *m* del) terreno *m*. **it is slow/hard ~** es lento/difícil. ● *a* ‹*price*› actual; ‹*concern*› en funcionamiento. **~s-on** *npl* actividades *fpl* anormales, tejemaneje *m*
gold /gəʊld/ *n* oro *m*. ● *a* de oro. **~en** /'gəʊldən/ *a* de oro; (*in colour*) dorado; ‹*opportunity*› único. **~en wedding** *n* bodas *fpl* de oro. **~fish** *n invar* pez *m* de colores, carpa *f* dorada. **~-mine** *n* mina *f* de oro; (*fig*) fuente *f* de gran riqueza. **~-plated** *a* chapado en oro. **~smith** *n* orfebre *m*

golf /gɒlf/ *n* golf *m*. **~-course** *n* campo *m* de golf. **~er** *n* jugador *m* de golf

golly /ˈgɒlɪ/ *int* ¡caramba!

golosh /gəˈlɒʃ/ *n* chanclo *m*

gondol|a /ˈgɒndələ/ *n* góndola *f*. **~ier** /gɒndəˈlɪə(r)/ *n* gondolero *m*

gone /gɒn/ *see* **go**. ● *a* pasado. **~ six o'clock** después de las seis

gong /gɒŋ/ *n* gong(o) *m*

good /gʊd/ *a* (**better**, **best**) bueno, (*before masculine singular noun*) buen. **~ afternoon!** ¡buenas tardes! **~ evening!** (*before dark*) ¡buenas tardes!; (*after dark*) ¡buenas noches! **G~ Friday** *n* Viernes *m* Santo. **~ morning!** ¡buenos días! **~ name** *n* (buena) reputación *f*. **~ night!** ¡buenas noches! **a ~ deal** bastante. **as ~ as** (*almost*) casi. **be ~ with** entender. **do ~** hacer bien. **feel ~** sentirse bien. **have a ~ time** divertirse. **it is ~ for you** le sentará bien. ● *n* bien *m*. **for ~** para siempre. **it is no ~ shouting/etc** es inútil gritar/etc.

goodbye /gʊdˈbaɪ/ *int* ¡adiós! ● *n* adiós *m*. **say ~ to** despedirse de

good: **~-for-nothing** *a* & *n* inútil (*m*). **~-looking** *a* guapo

goodness /ˈgʊdnɪs/ *n* bondad *f*. **~!**, **~ gracious!**, **~ me!**, **my ~!** ¡Dios mío!

goods /gʊdz/ *npl* (*merchandise*) mercancías *fpl*

goodwill /gʊdˈwɪl/ *n* buena voluntad *f*

goody /ˈgʊdɪ/ *n* (*culin*, *fam*) golosina *f*; (*in film*) bueno *m*. **~-goody** *n* mojigato *m*

gooey /ˈguːɪ/ *a* (**gooier**, **gooiest**) (*sl*) pegajoso; (*fig*) sentimental

goof /guːf/ *vi* (*Amer*, *blunder*) cometer una pifia. **~y** *a* (*sl*) necio

goose /guːs/ *n* (*pl* **geese**) oca *f*

gooseberry /ˈgʊzbərɪ/ *n* uva *f* espina, grosella *f*

goose-flesh /ˈguːsfleʃ/ *n*, **goose-pimples** /ˈguːspɪmplz/ *n* carne *f* de gallina

gore /gɔː(r)/ *n* sangre *f*. ● *vt* cornear

gorge /gɔːdʒ/ *n* (*geog*) garganta *f*. ● *vt*. **~ o.s.** hartarse (**on** de)

gorgeous /ˈgɔːdʒəs/ *a* magnífico

gorilla /gəˈrɪlə/ *n* gorila *m*

gormless /ˈgɔːmlɪs/ *a* (*sl*) idiota

gorse /gɔːs/ *n* aulaga *f*

gory /ˈgɔːrɪ/ *a* (**-ier**, **-iest**) (*covered in blood*) ensangrentado; (*horrific*, *fig*) horrible

gosh /gɒʃ/ *int* ¡caramba!

go-slow /gəʊˈsləʊ/ *n* huelga *f* de celo

gospel /ˈgɒspl/ *n* evangelio *m*

gossip /ˈgɒsɪp/ *n* (*idle chatter*) charla *f*; (*tittle-tattle*) comadreo *m*; (*person*) chismoso *m*. ● *vi* (*pt* **gossiped**) (*chatter*) charlar; (*repeat scandal*) comadrear. **~y** *a* chismoso

got /gɒt/ *see* **get**. **have ~** tener. **have ~ to do** tener que hacer

Gothic /ˈgɒθɪk/ *a* (*archit*) gótico; ‹*people*› godo

gouge /gaʊdʒ/ *vt*. **~ out** arrancar

gourmet /ˈgʊəmeɪ/ *n* gastrónomo *m*

gout /gaʊt/ *n* (*med*) gota *f*

govern /ˈgʌvn/ *vt*/*i* gobernar

governess /ˈgʌvənɪs/ *n* institutriz *f*

government /ˈgʌvənmənt/ *n* gobierno *m*. **~al** /gʌvənˈmentl/ *a* gubernamental

governor /ˈgʌvənə(r)/ *n* gobernador *m*

gown /gaʊn/ *n* vestido *m*; (*of judge, teacher*) toga *f*

GP *abbr see* **general practitioner**

grab /græb/ *vt* (*pt* **grabbed**) agarrar

grace /greɪs/ *n* gracia *f*. **~ful** *a* elegante

gracious /ˈgreɪʃəs/ *a* (*kind*) amable; (*elegant*) elegante

gradation /grəˈdeɪʃn/ *n* gradación *f*

grade /greɪd/ *n* clase *f*, categoría *f*; (*of goods*) clase *f*, calidad *f*; (*on scale*) grado *m*; (*school mark*) nota *f*; (*class, Amer*) curso *m*. **~ school** *n* (*Amer*) escuela *f* primaria. ● *vt* clasificar; (*schol*) calificar

gradient /ˈgreɪdɪənt/ *n* (*slope*) pendiente *f*

gradual /ˈgrædʒʊəl/ *a* gradual. **~ly** *adv* gradualmente

graduat|e /ˈgrædjʊət/ *n* (*univ*) licenciado. ● *vi* /ˈgrædjʊeɪt/ licenciarse. ● *vt* graduar. **~ion** /-ˈeɪʃn/ *n* entrega *f* de títulos

graffiti /grəˈfiːtɪ/ *npl* pintada *f*

graft[1] /grɑːft/ *n* (*med*, *bot*) injerto *m*. ● *vt* injertar

graft[2] /grɑːft/ *n* (*bribery*, *fam*) corrupción *f*

grain /greɪn/ *n* grano *m*

gram /græm/ *n* gramo *m*

gramma|r /ˈgræmə(r)/ *n* gramática *f*. **~tical** /grəˈmætɪkl/ *a* gramatical

gramophone /ˈgræməfəʊn/ *n* tocadiscos *m invar*

grand /grænd/ *a* (**-er**, **-est**) magnífico; (*excellent*, *fam*) estupendo. **~child** *n* nieto *m*. **~daughter** *n* nieta *f*

grandeur /'grændʒə(r)/ *n* grandiosidad *f*
grandfather /'grændfɑ:ðə(r)/ *n* abuelo *m*
grandiose /'grændɪəʊs/ *a* grandioso
grand: **~mother** *n* abuela *f*. **~parents** *npl* abuelos *mpl*. **~ piano** *n* piano *m* de cola. **~son** *n* nieto *m*
grandstand /'grænstænd/ *n* tribuna *f*
granite /'grænɪt/ *n* granito *m*
granny /'grænɪ/ *n* (*fam*) abuela *f*, nana *f* (*fam*)
grant /grɑ:nt/ *vt* conceder; (*give*) donar; (*admit*) admitir (**that** que). **take for ~ed** dar por sentado. ● *n* concesión *f*; (*univ*) beca *f*
granulated /'grænjʊleɪtɪd/ *a*. **~ sugar** *n* azúcar *m* granulado
granule /'grænu:l/ *n* gránulo *m*
grape /greɪp/ *n* uva *f*
grapefruit /'greɪpfru:t/ *n invar* toronja *f*, pomelo *m*
graph /grɑ:f/ *n* gráfica *f*
graphic /'græfɪk/ *a* gráfico
grapple /'græpl/ *vi*. **~ with** intentar vencer
grasp /grɑ:sp/ *vt* agarrar. ● *n* (*hold*) agarro *m*; (*strength of hand*) apretón *m*; (*reach*) alcance *m*; (*fig*) comprensión *f*
grasping /'grɑ:spɪŋ/ *a* avaro
grass /grɑ:s/ *n* hierba *f*. **~hopper** *n* saltamontes *m invar*. **~land** *n* pradera *f*. **~ roots** *npl* base *f* popular. ● *a* popular. **~y** *a* cubierto de hierba
grate /greɪt/ *n* (*fireplace*) parrilla *f*. ● *vt* rallar. **~ one's teeth** hacer rechinar los dientes. ● *vi* rechinar
grateful /'greɪtfl/ *a* agradecido. **~ly** *adv* con gratitud
grater /'greɪtə(r)/ *n* rallador *m*
gratif|ied /'grætɪfaɪd/ *a* contento. **~y** *vt* satisfacer; (*please*) agradar a. **~ying** *a* agradable
grating /'greɪtɪŋ/ *n* reja *f*
gratis /'grɑ:tɪs/ *a & adv* gratis (*a invar*)
gratitude /'grætɪtju:d/ *n* gratitud *f*
gratuitous /grə'tju:ɪtəs/ *a* gratuito
gratuity /grə'tju:ətɪ/ *n* (*tip*) propina *f*; (*gift of money*) gratificación *f*
grave[1] /greɪv/ *n* sepultura *f*
grave[2] /greɪv/ *a* (**-er, -est**) (*serious*) serio. /grɑ:v/ *a*. **~ accent** *n* acento *m* grave
grave-digger /'greɪvdɪgə(r)/ *n* sepulturero *m*
gravel /'grævl/ *n* grava *f*
gravely /'greɪvlɪ/ *a* (*seriously*) seriamente
grave: **~stone** *n* lápida *f*. **~yard** *n* cementerio *m*
gravitat|e /'grævɪteɪt/ *vi* gravitar. **~ion** /-'teɪʃn/ *n* gravitación *f*
gravity /'grævətɪ/ *n* gravedad *f*
gravy /'greɪvɪ/ *n* salsa *f*
graze[1] /greɪz/ *vt/i* (*eat*) pacer
graze[2] /greɪz/ *vt* (*touch*) rozar; (*scrape*) raspar. ● *n* rozadura *f*
greas|e /gri:s/ *n* grasa *f*. ● *vt* engrasar. **~e-paint** *n* maquillaje *m*. **~e-proof paper** *n* papel *m* a prueba de grasa, apergaminado *m*. **~y** *a* grasiento
great /greɪt/ *a* (**-er, -est**) grande, (*before singular noun*) gran; (*very good, fam*) estupendo. **G~ Britain** *n* Gran Bretaña *f*. **~-grandfather** *n* bisabuelo *m*. **~-grandmother** *n* bisabuela *f*. **~ly** /'greɪtlɪ/ *adv* (*very*) muy; (*much*) mucho. **~ness** *n* grandeza *f*
Greece /gri:s/ *n* Grecia *f*
greed /gri:d/ *n* avaricia *f*; (*for food*) glotonería *f*. **~y** *a* avaro; (*for food*) glotón
Greek /gri:k/ *a & n* griego (*m*)
green /gri:n/ *a* (**-er, -est**) verde; (*fig*) crédulo. ● *n* verde *m*; (*grass*) césped *m*. **~ belt** *n* zona *f* verde. **~ery** *n* verdor *m*. **~ fingers** *npl* habilidad *f* con las plantas
greengage /'gri:ngeɪdʒ/ *n* (*plum*) claudia *f*
greengrocer /'gri:ngrəʊsə(r)/ *n* verdulero *m*
greenhouse /'gri:nhaʊs/ *n* invernadero *m*
green: **~ light** *n* luz *f* verde. **~s** *npl* verduras *fpl*
Greenwich Mean Time /grenɪtʃ'mi:ntaɪm/ *n* hora *f* media de Greenwich
greet /gri:t/ *vt* saludar; (*receive*) recibir. **~ing** *n* saludo *m*. **~ings** *npl* (*in letter*) recuerdos *mpl*
gregarious /grɪ'geərɪəs/ *a* gregario
grenade /grɪ'neɪd/ *n* granada *f*
grew /gru:/ *see* **grow**
grey /greɪ/ *a & n* (**-er, -est**) gris (*m*). ● *vi* ‹*hair*› encanecer
greyhound /'greɪhaʊnd/ *n* galgo *m*
grid /grɪd/ *n* reja *f*; (*network, elec*) red *f*; (*culin*) parrilla *f*; (*on map*) cuadrícula *f*

grief /gri:f/ *n* dolor *m*. **come to ~** ‹*person*› sufrir un accidente; (*fail*) fracasar

grievance /'gri:vns/ *n* queja *f*

grieve /gri:v/ *vt* afligir. ● *vi* afligirse. **~ for** llorar

grievous /'gri:vəs/ *a* doloroso; (*serious*) grave

grill /grɪl/ *n* (*cooking device*) parrilla *f*; (*food*) parrillada *f*, asado *m*, asada *f*. ● *vt* asar a la parrilla; (*interrogate*) interrogar

grille /grɪl/ *n* rejilla *f*

grim /grɪm/ *a* (**grimmer**, **grimmest**) severo

grimace /'grɪməs/ *n* mueca *f*. ● *vi* hacer muecas

grim|e /graɪm/ *n* mugre *f*. **~y** *a* mugriento

grin /grɪn/ *vt* (*pt* **grinned**) sonreír. ● *n* sonrisa *f* (abierta)

grind /graɪnd/ *vt* (*pt* **ground**) moler ‹*coffee, corn etc*›; (*pulverize*) pulverizar; (*sharpen*) afilar. **~ one's teeth** hacer rechinar los dientes. ● *n* faena *f*

grip /grɪp/ *vt* (*pt* **gripped**) agarrar; (*interest*) captar la atención de. ● *n* (*hold*) agarro *m*; (*strength of hand*) apretón *m*. **come to ~s** encararse (**with** a/con)

gripe /graɪp/ *n*. **~s** *npl* (*med*) cólico *m*

grisly /'grɪzlɪ/ *a* (**-ier**, **-iest**) horrible

gristle /'grɪsl/ *n* cartílago *m*

grit /grɪt/ *n* arena *f*; (*fig*) valor *m*, aguante *m*. ● *vt* (*pt* **gritted**) echar arena en ‹*road*›. **~ one's teeth** (*fig*) acorazarse

grizzle /'grɪzl/ *vi* lloriquear

groan /grəʊn/ *vi* gemir. ● *n* gemido *m*

grocer /'grəʊsə(r)/ *n* tendero *m*. **~ies** *npl* comestibles *mpl*. **~y** *n* tienda *f* de comestibles

grog /grɒg/ *n* grog *m*

groggy /'grɒgɪ/ *a* (*weak*) débil; (*unsteady*) inseguro; (*ill*) malucho

groin /grɔɪn/ *n* ingle *f*

groom /gru:m/ *n* mozo *m* de caballos; (*bridegroom*) novio *m*. ● *vt* almohazar ‹*horses*›; (*fig*) preparar. **well-~ed** *a* bien arreglado

groove /gru:v/ *n* ranura *f*; (*in record*) surco *m*

grope /grəʊp/ *vi* (*find one's way*) moverse a tientas. **~ for** buscar a tientas

gross /grəʊs/ *a* (**-er**, **-est**) (*coarse*) grosero; (*com*) bruto; (*fat*) grueso; (*flagrant*) grave. ● *n invar* gruesa *f*. **~ly** *adv* groseramente; (*very*) enormemente

grotesque /grəʊ'tesk/ *a* grotesco

grotto /'grɒtəʊ/ *n* (*pl* **-oes**) gruta *f*

grotty /'grɒtɪ/ *a* (*sl*) desagradable; (*dirty*) sucio

grouch /graʊtʃ/ *vi* (*grumble, fam*) rezongar

ground[1] /graʊnd/ *n* suelo *m*; (*area*) terreno *m*; (*reason*) razón *f*; (*elec, Amer*) toma *f* de tierra. ● *vt* varar ‹*ship*›; prohibir despegar ‹*aircraft*›. **~s** *npl* jardines *mpl*; (*sediment*) poso *m*

ground[2] /graʊnd/ *see* **grind**

ground: **~ floor** *n* planta *f* baja. **~ rent** *n* alquiler *m* del terreno

grounding /'graʊndɪŋ/ *n* base *f*, conocimientos *mpl* (**in** de)

groundless /'graʊndlɪs/ *a* infundado

ground: **~sheet** *n* tela *f* impermeable. **~swell** *n* mar *m* de fondo. **~work** *n* trabajo *m* preparatorio

group /gru:p/ *n* grupo *m*. ● *vt* agrupar. ● *vi* agruparse

grouse[1] /graʊs/ *n invar* (*bird*) urogallo *m*. **red ~** lagópodo *m* escocés

grouse[2] /graʊs/ *vi* (*grumble, fam*) rezongar

grove /grəʊv/ *n* arboleda *f*. **lemon ~** *n* limonar *m*. **olive ~** *n* olivar *m*. **orange ~** *n* naranjal *m*. **pine ~** *n* pinar *m*

grovel /'grɒvl/ *vi* (*pt* **grovelled**) arrastrarse, humillarse. **~ling** *a* servil

grow /grəʊ/ *vi* (*pt* **grew**, *pp* **grown**) crecer; ‹*cultivated plant*› cultivarse; (*become*) volverse, ponerse. ● *vt* cultivar. **~ up** hacerse mayor. **~er** *n* cultivador *m*

growl /graʊl/ *vi* gruñir. ● *n* gruñido *m*

grown /grəʊn/ *see* **grow**. ● *a* adulto. **~-up** *a* & *n* adulto (*m*)

growth /grəʊθ/ *n* crecimiento *m*; (*increase*) aumento *m*; (*development*) desarrollo *m*; (*med*) tumor *m*

grub /grʌb/ *n* (*larva*) larva *f*; (*food, sl*) comida *f*

grubby /'grʌbɪ/ *a* (**-ier**, **-iest**) mugriento

grudg|e /grʌdʒ/ *vt* dar de mala gana; (*envy*) envidiar. **~e doing** molestarle hacer. **he ~ed paying** le molestó pagar. ● *n* rencor *m*. **bear/have a ~e against s.o.** guardar rencor a alguien. **~ingly** *adv* de mala gana

gruelling /'gru:əlɪŋ/ *a* agotador

gruesome /'gru:səm/ *a* horrible

gruff /grʌf/ *a* (**-er**, **-est**) ‹*manners*› brusco; ‹*voice*› ronco

grumble /'grʌmbl/ *vi* rezongar

grumpy /'grʌmpɪ/ *a* (**-ier**, **-iest**) malhumorado

grunt /grʌnt/ *vi* gruñir. ● *n* gruñido *m*

guarant|ee /gærən'ti:/ *n* garantía *f*. ● *vt* garantizar. **~or** *n* garante *m* & *f*

guard /gɑ:d/ *vt* proteger; (*watch*) vigilar. ● *vi*. **~ against** guardar de. ● *n* (*vigilance, mil group*) guardia *f*; (*person*) guardia *m*; (*on train*) jefe *m* de tren

guarded /'gɑ:dɪd/ *a* cauteloso

guardian /'gɑ:dɪən/ *n* guardián *m*; (*of orphan*) tutor *m*

guer(r)illa /gə'rɪlə/ *n* guerrillero *m*. **~ warfare** *n* guerra *f* de guerrillas

guess /ges/ *vt/i* adivinar; (*suppose, Amer*) creer. ● *n* conjetura *f*. **~work** *n* conjetura(s) *f(pl)*

guest /gest/ *n* invitado *m*; (*in hotel*) huésped *m*. **~house** *n* casa *f* de huéspedes

guffaw /gʌ'fɔ:/ *n* carcajada *f*. ● *vi* reírse a carcajadas

guidance /'gaɪdəns/ *n* (*advice*) consejos *mpl*; (*information*) información *f*

guide /gaɪd/ *n* (*person*) guía *m* & *f*; (*book*) guía *f*. **Girl G~** exploradora *f*, guía *f* (*fam*). ● *vt* guiar. **~book** *n* guía *f*. **~d missile** *n* proyectil *m* teledirigido. **~-lines** *npl* pauta *f*

guild /gɪld/ *n* gremio *m*

guile /gaɪl/ *n* astucia *f*

guillotine /'gɪləti:n/ *n* guillotina *f*

guilt /gɪlt/ *n* culpabilidad *f*. **~y** *a* culpable

guinea-pig /'gɪnɪpɪg/ *n* (*including fig*) cobaya *f*

guise /gaɪz/ *n* (*external appearance*) apariencia *f*; (*style*) manera *f*

guitar /gɪ'ta:(r)/ *n* guitarra *f*. **~ist** *n* guitarrista *m* & *f*

gulf /gʌlf/ *n* (*part of sea*) golfo *m*; (*hollow*) abismo *m*

gull /gʌl/ *n* gaviota *f*

gullet /'gʌlɪt/ *n* esófago *m*

gullible /'gʌləbl/ *a* crédulo

gully /'gʌlɪ/ *n* (*ravine*) barranco *m*

gulp /gʌlp/ *vt*. **~ down** tragarse de prisa. ● *vi* tragar; (*from fear etc*) sentir dificultad para tragar. ● *n* trago *m*

gum[1] /gʌm/ *n* goma *f*; (*for chewing*) chicle *m*. ● *vt* (*pt* **gummed**) engomar

gum[2] /gʌm/ *n* (*anat*) encía *f*. **~boil** /'gʌmbɔɪl/ *n* flemón *m*

gumboot /'gʌmbu:t/ *n* bota *f* de agua

gumption /'gʌmpʃn/ *n* (*fam*) iniciativa *f*; (*common sense*) sentido *m* común

gun /gʌn/ *n* (*pistol*) pistola *f*; (*rifle*) fusil *m*; (*large*) cañón *m*. ● *vt* (*pt* **gunned**). **~ down** abatir a tiros. **~fire** *n* tiros *mpl*

gunge /gʌndʒ/ *n* (*sl*) materia *f* sucia (y pegajosa)

gun: **~man** /'gʌnmən/ *n* pistolero *m*. **~ner** /'gʌnə(r)/ *n* artillero *m*. **~powder** *n* pólvora *f*. **~shot** *n* disparo *m*

gurgle /'gɜ:gl/ *n* (*of liquid*) gorgoteo *m*; (*of baby*) gorjeo *m*. ● *vi* ‹*liquid*› gorgotear; ‹*baby*› gorjear

guru /'gʊru:/ *n* (*pl* **-us**) mentor *m*

gush /gʌʃ/ *vi*. **~ (out)** salir a borbotones. ● *n* (*of liquid*) chorro *m*; (*fig*) torrente *m*. **~ing** *a* efusivo

gusset /'gʌsɪt/ *n* escudete *m*

gust /gʌst/ *n* ráfaga *f*; (*of smoke*) bocanada *f*

gusto /'gʌstəʊ/ *n* entusiasmo *m*

gusty /'gʌstɪ/ *a* borrascoso

gut /gʌt/ *n* tripa *f*, intestino *m*. ● *vt* (*pt* **gutted**) destripar; ‹*fire*› destruir. **~s** *npl* tripas *fpl*; (*courage, fam*) valor *m*

gutter /'gʌtə(r)/ *n* (*on roof*) canalón *m*; (*in street*) cuneta *f*; (*slum, fig*) arroyo *m*. **~snipe** *n* golfillo *m*

guttural /'gʌtərəl/ *a* gutural

guy /gaɪ/ *n* (*man, fam*) hombre *m*, tío *m* (*fam*)

guzzle /'gʌzl/ *vt/i* soplarse, tragarse

gym /dʒɪm/ *n* (*gymnasium, fam*) gimnasio *m*; (*gymnastics, fam*) gimnasia *f*

gymkhana /dʒɪmkɑ:nə/ *n* gincana *f*, gymkhana *f*

gymnasium /dʒɪm'neɪzɪəm/ *n* gimnasio *m*

gymnast /'dʒɪmnæst/ *n* gimnasta *m* & *f*. **~ics** *npl* gimnasia *f*

gym-slip /'dʒɪmslɪp/ *n* túnica *f* (de gimnasia)

gynaecolog|ist /gaɪnɪ'kɒlədʒɪst/ *n* ginecólogo *m*. **~y** *n* ginecología *f*

gypsy /'dʒɪpsɪ/ *n* gitano *m*

gyrate /dʒaɪə'reɪt/ *vi* girar

gyroscope /'dʒaɪərəskəʊp/ *n* giroscopio *m*

H

haberdashery /hæbə'dæʃərɪ/ *n* mercería *f*

habit /'hæbɪt/ *n* costumbre *f*; (*costume, relig*) hábito *m*. **be in the ~ of** (+ *gerund*) tener la costumbre de (+ *infintive*), soler (+ *infinitive*). **get into the ~ of** (+ *gerund*) acostumbrarse a (+ *infinitive*)

habitable /'hæbɪtəbl/ *a* habitable

habitat /'hæbɪtæt/ *n* hábitat *m*

habitation /hæbɪ'teɪʃn/ *n* habitación *f*

habitual /hə'bɪtjʊəl/ *a* habitual; ‹*smoker, liar*› inveterado. **~ly** *adv* de costumbre

hack /hæk/ *n* (*old horse*) jamelgo *m*; (*writer*) escritorzuelo *m*. ● *vt* cortar. **~ to pieces** cortar en pedazos

hackney /'hæknɪ/ *a*. **~ carriage** *n* coche *m* de alquiler, taxi *m*

hackneyed /'hæknɪd/ *a* manido

had /hæd/ *see* **have**

haddock /'hædək/ *n invar* eglefino *m*. **smoked ~** *n* eglefino *m* ahumado

haemorrhage /'hemərɪdʒ/ *n* hemorragia *f*

haemorrhoids /'hemərɔɪdz/ *npl* hemorroides *fpl*, almorranas *fpl*

hag /hæg/ *n* bruja *f*

haggard /'hægəd/ *a* ojeroso

haggle /'hægl/ *vi* regatear

Hague /heɪg/ *n*. **The ~** La Haya *f*

hail[1] /heɪl/ *n* granizo *m*. ● *vi* granizar

hail[2] /heɪl/ *vt* (*greet*) saludar; llamar ‹*taxi*›. ● *vi*. **~ from** venir de

hailstone /'heɪlstəʊn/ *n* grano *m* de granizo

hair /heə(r)/ *n* pelo *m*. **~brush** *n* cepillo *m* para el pelo. **~cut** *n* corte *m* de pelo. **have a ~cut** cortarse el pelo. **~do** *n* (*fam*) peinado *m*. **~dresser** *n* peluquero *m*. **~dresser's (shop)** *n* peluquería *f*. **~-dryer** *n* secador *m*. **~pin** *n* horquilla *f*. **~pin bend** *n* curva *f* cerrada. **~-raising** *a* espeluznante. **~-style** *n* peinado *m*

hairy /'heərɪ/ *a* (**-ier**, **-iest**) peludo; (*terrifying, sl*) espeluznante

hake /heɪk/ *n invar* merluza *f*

halcyon /'hælsɪən/ *a* sereno. **~ days** *npl* época *f* feliz

hale /heɪl/ *a* robusto

half /hɑ:f/ *n* (*pl* **halves**) mitad *f*. ● *a* medio. **~ a dozen** media docena *f*. **~ an hour** media hora *f*. ● *adv* medio, a medias. **~-back** *n* (*sport*) medio *m*. **~-caste** *a & n* **mestizo** (*m*). **~-hearted** *a* poco entusiasta. **~-term** *n* vacaciones *fpl* de medio trimestre. **~-time** *n* (*sport*) descanso *m*. **~way** *a* medio. ● *adv* a medio camino. **~-wit** *n* imbécil *m & f*. **at ~-mast** a media asta

halibut /'hælɪbət/ *n invar* hipogloso *m*, halibut *m*

hall /hɔ:l/ *n* (*room*) sala *f*; (*mansion*) casa *f* solariega; (*entrance*) vestíbulo *m*. **~ of residence** *n* colegio *m* mayor

hallelujah /hælɪ'lu:jə/ *int & n* aleluya (*f*)

hallmark /'hɔ:lmɑ:k/ *n* (*on gold etc*) contraste *m*; (*fig*) sello *m* (distintivo)

hallo /hə'ləʊ/ *int* = **hello**

hallow /'hæləʊ/ *vt* santificar. **H~e'en** *n* víspera *f* de Todos los Santos

hallucination /həlu:sɪ'neɪʃn/ *n* alucinación *f*

halo /'heɪləʊ/ *n* (*pl* **-oes**) aureola *f*

halt /hɔ:lt/ *n* alto *m*. ● *vt* parar. ● *vi* pararse

halve /hɑ:v/ *vt* dividir por mitad

ham /hæm/ *n* jamón *m*; (*theatre, sl*) racionista *m & f*

hamburger /'hæmbɜ:gə(r)/ *n* hamburguesa *f*

hamlet /'hæmlɪt/ *n* aldea *f*, caserío *m*

hammer /'hæmə(r)/ *n* martillo *m*. ● *vt* martill(e)ar; (*defeat, fam*) machacar

hammock /'hæmək/ *n* hamaca *f*

hamper[1] /'hæmpə(r)/ *n* cesta *f*

hamper[2] /'hæmpə(r)/ *vt* estorbar, poner trabas

hamster /'hæmstə(r)/ *n* hámster *m*

hand /'hænd/ *n* (*including cards*) mano *f*; (*of clock*) manecilla *f*; (*writing*) escritura *f*, letra *f*; (*worker*) obrero *m*. **at ~** a mano. **by ~** a mano. **lend a ~** echar una mano. **on ~** a mano. **on one's ~s** (*fig*) en (las) manos de uno. **on the one ~... on the other ~** por un lado... por otro.

out of ~ fuera de control. **to ~** a mano. ● *vt* dar. **~ down** pasar. **~ in** entregar. **~ over** entregar. **~ out** distribuir. **~bag** *n* bolso *m*, cartera *f* (*LAm*). **~book** *n* (*manual*) manual *m*; (*guidebook*) guía *f*. **~cuffs** *npl* esposas *fpl*. **~ful** /'hændfʊl/ *n* puñado *m*; (*person, fam*) persona *f* difícil. **~-luggage** *n* equipaje *m* de mano. **~-out** *n* folleto *m*; (*money*) limosna *f*

handicap /'hændɪkæp/ *n* desventaja *f*; (*sport*) handicap *m*. ● *vt* (*pt* **handicapped**) imponer impedimentos a

handicraft /'hændɪkrɑ:ft/ *n* artesanía *f*

handiwork /'hændɪwɜ:k/ *n* obra *f*, trabajo *m* manual

handkerchief /'hæŋkətʃɪf/ *n* (*pl* **-fs**) pañuelo *m*

handle /'hændl/ *n* (*of door etc*) tirador *m*; (*of implement*) mango *m*; (*of cup, bag, basket etc*) asa *f*. ● *vt* manejar; (*touch*) tocar; (*control*) controlar

handlebar /'hændlbɑ:(r)/ *n* (*on bicycle*) manillar *m*

handshake /'hændʃeɪk/ *n* apretón *m* de manos

handsome /'hænsəm/ *a* (*good-looking*) guapo; (*generous*) generoso; (*large*) considerable

handwriting /'hændraɪtɪŋ/ *n* escritura *f*, letra *f*

handy /'hændɪ/ *a* (**-ier**, **-iest**) (*useful*) cómodo; ‹*person*› diestro; (*near*) a mano. **~man** *n* hombre *m* habilidoso

hang /hæŋ/ *vt* (*pt* **hung**) colgar; (*pt* **hanged**) (*capital punishment*) ahorcar. ● *vi* colgar; ‹*hair*› caer. ● *n*. **get the ~ of sth** coger el truco de algo. **~ about** holgazanear. **~ on** (*hold out*) resistir; (*wait, sl*) esperar. **~ out** *vi* tender; (*live, sl*) vivir. **~ up** (*telephone*) colgar

hangar /'hæŋə(r)/ *n* hangar *m*

hanger /'hæŋə(r)/ *n* (*for clothes*) percha *f*. **~-on** *n* parásito *m*, pegote *m*

hang-gliding /'hæŋglaɪdɪŋ/ *n* vuelo *m* libre

hangman /'hæŋmən/ *n* verdugo *m*

hangover /'hæŋəʊvə(r)/ *n* (*after drinking*) resaca *f*

hang-up /'hæŋʌp/ *n* (*sl*) complejo *m*

hanker /'hæŋkə(r)/ *vi*. **~ after** anhelar. **~ing** *n* anhelo *m*

hanky-panky /'hæŋkɪpæŋkɪ/ *n* (*trickery, sl*) trucos *mpl*

haphazard /hæp'hæzəd/ *a* fortuito. **~ly** *adv* al azar

hapless /'hæplɪs/ *a* desafortunado

happen /'hæpən/ *vi* pasar, suceder, ocurrir. **if he ~s to come** si acaso viene. **~ing** *n* acontecimiento *m*

happ|ily /'hæpɪlɪ/ *adv* felizmente; (*fortunately*) afortunadamente. **~iness** *n* felicidad *f*. **~y** *a* (**-ier**, **-iest**) feliz. **~y-go-lucky** *a* despreocupado. **~y medium** *n* término *m* medio

harangue /hə'ræŋ/ *n* arenga *f*. ● *vt* arengar

harass /'hærəs/ *vt* acosar. **~ment** *n* tormento *m*

harbour /'hɑ:bə(r)/ *n* puerto *m*. ● *vt* encubrir ‹*criminal*›; abrigar ‹*feelings*›

hard /hɑ:d/ *a* (**-er**, **-est**) duro; (*difficult*) difícil. **~ of hearing** duro de oído. ● *adv* mucho; (*pull*) fuerte. **~ by** (muy) cerca. **~ done by** tratado injustamente. **~ up** (*fam*) sin un cuarto. **~board** *n* chapa *f* de madera, tabla *f*. **~-boiled egg** *n* huevo *m* duro. **~en** /'hɑ:dn/ *vt* endurecer. ● *vi* endurecerse. **~-headed** *a* realista

hardly /'hɑ:dlɪ/ *adv* apenas. **~ ever** casi nunca

hardness /'hɑ:dnɪs/ *n* dureza *f*

hardship /'hɑ:dʃɪp/ *n* apuro *m*

hard: **~ shoulder** *n* arcén *m*. **~ware** *n* ferretería *f*; ‹*computer*› hardware *m*. **~-working** *a* trabajador

hardy /'hɑ:dɪ/ *a* (**-ier**, **-iest**) (*bold*) audaz; (*robust*) robusto; (*bot*) resistente

hare /heə(r)/ *n* liebre *f*. **~-brained** *a* aturdido

harem /'hɑ:ri:m/ *n* harén *m*

haricot /'hærɪkəʊ/ *n*. **~ bean** alubia *f*, judía *f*

hark /hɑ:k/ *vi* escuchar. **~ back to** volver a

harlot /'hɑ:lət/ *n* prostituta *f*

harm /hɑ:m/ *n* daño *m*. **there is no ~ in** (+ *gerund*) no hay ningún mal en (+ *infinitive*). ● *vt* hacer daño a ‹*person*›; dañar ‹*thing*›; perjudicar ‹*interests*›. **~ful** *a* perjudical. **~less** *a* inofensivo

harmonica /hɑ:'mɒnɪkə/ *n* armónica *f*

harmon|ious /hɑ:'məʊnɪəs/ *a* armonioso. **~ize** *vt*/*i* armonizar. **~y** *n* armonía *f*

harness /ˈhɑːnɪs/ *n* (*for horses*) guarniciones *fpl*; (*for children*) andadores *mpl*. ● *vt* poner guarniciones a ‹*horse*›; (*fig*) aprovechar

harp /hɑːp/ *n* arpa *f*. ● *vi*. **~ on (about)** machacar. **~ist** /ˈhɑːpɪst/ *n* arpista *m & f*

harpoon /hɑːˈpuːn/ *n* arpón *m*

harpsichord /ˈhɑːpsɪkɔːd/ *n* clavicémbalo *m*, clave *m*

harrowing /ˈhærəʊɪŋ/ *a* desgarrador

harsh /hɑːʃ/ *a* (**-er**, **-est**) duro, severo; ‹*taste, sound*› áspero. **~ly** *adv* severamente. **~ness** *n* severidad *f*

harvest /ˈhɑːvɪst/ *n* cosecha *f*. ● *vt* cosechar. **~er** *n* (*person*) segador; (*machine*) cosechadora *f*

has /hæz/ *see* **have**

hash /hæʃ/ *n* picadillo *m*. **make a ~ of sth** hacer algo con los pies, estropear algo

hashish /ˈhæʃiːʃ/ *n* hachís *m*

hassle /ˈhæsl/ *n* (*quarrel*) pelea *f*; (*difficulty*) problema *m*, dificultad *f*; (*bother, fam*) pena *f*, follón *m*, lío *m*. ● *vt* (*harass*) acosar, dar la lata

haste /heɪst/ *n* prisa *f*. **in ~** de prisa. **make ~** darse prisa

hasten /ˈheɪsn/ *vt* apresurar. ● *vi* apresurarse, darse prisa

hast|ily /ˈheɪstɪlɪ/ *adv* de prisa. **~y** *a* (**-ier**, **-iest**) precipitado; (*rash*) irreflexivo

hat /hæt/ *n* sombrero *m*. **a ~ trick** *n* tres victorias *fpl* consecutivas

hatch[1] /hætʃ/ *n* (*for food*) ventanilla *f*; (*naut*) escotilla *f*

hatch[2] /hætʃ/ *vt* empollar ‹*eggs*›; tramar ‹*plot*›. ● *vi* salir del cascarón

hatchback /ˈhætʃbæk/ *n* (coche *m*) cincopuertas *m invar*, coche *m* con puerta trasera

hatchet /ˈhætʃɪt/ *n* hacha *f*

hate /heɪt/ *n* odio *m*. ● *vt* odiar. **~ful** *a* odioso

hatred /ˈheɪtrɪd/ *n* odio *m*

haughty /ˈhɔːtɪ/ *a* (**-ier**, **-iest**) altivo

haul /hɔːl/ *vt* arrastrar; transportar ‹*goods*›. ● *n* (*catch*) redada *f*; (*stolen goods*) botín *m*; (*journey*) recorrido *m*. **~age** *n* transporte *m*. **~ier** *n* transportista *m & f*

haunch /hɔːntʃ/ *n* anca *f*

haunt /hɔːnt/ *vt* frecuentar. ● *n* sitio *m* preferido. **~ed house** *n* casa *f* frecuentada por fantasmas

Havana /həˈvænə/ *n* La Habana *f*

have /hæv/ *vt* (*3 sing pres tense* **has**, *pt* **had**) tener; (*eat, drink*) tomar. **~ it out with** resolver el asunto. **~ sth done** hacer hacer algo. **~ to do** tener que hacer. ● *v aux* haber. **~ just done** acabar de hacer. ● *n*. **the ~s and ~-nots** los ricos *mpl* y los pobres *mpl*

haven /ˈheɪvn/ *n* puerto *m*; (*refuge*) refugio *m*

haversack /ˈhævəsæk/ *n* mochila *f*

havoc /ˈhævək/ *n* estragos *mpl*

haw /hɔː/ *see* **hum**

hawk[1] /hɔːk/ *n* halcón *m*

hawk[2] /hɔːk/ *vt* vender por las calles. **~er** *n* vendedor *m* ambulante

hawthorn /ˈhɔːθɔːn/ *n* espino *m* (blanco)

hay /heɪ/ *n* heno *m*. **~ fever** *n* fiebre *f* del heno. **~stack** *n* almiar *m*

haywire /ˈheɪwaɪə(r)/ *a*. **go ~** ‹*plans*› desorganizarse; ‹*machine*› estropearse

hazard /ˈhæzəd/ *n* riesgo *m*. ● *vt* arriesgar; aventurar ‹*guess*›. **~ous** *a* arriesgado

haze /heɪz/ *n* neblina *f*

hazel /ˈheɪzl/ *n* avellano *m*. **~-nut** *n* avellana *f*

hazy /ˈheɪzɪ/ *a* (**-ier**, **-iest**) nebuloso

he /hiː/ *pron* él. ● *n* (*animal*) macho *m*; (*man*) varón *m*

head /hed/ *n* cabeza *f*; (*leader*) jefe *m*; (*of beer*) espuma *f*. **~s or tails** cara o cruz. ● *a* principal. **~ waiter** *n* jefe *m* de comedor. ● *vt* encabezar. **~ the ball** dar un cabezazo. **~ for** dirigirse a. **~ache** *n* dolor *m* de cabeza. **~-dress** *n* tocado *m*. **~er** *n* (*football*) cabezazo *m*. **~ first** *adv* de cabeza. **~gear** *n* tocado *m*

heading /ˈhedɪŋ/ *n* título *m*, encabezamiento *m*

headlamp /ˈhedlæmp/ *n* faro *m*

headland /ˈhedlənd/ *n* promontorio *m*

headlight /ˈhedlaɪt/ *n* faro *m*

headline /ˈhedlaɪn/ *n* titular *m*

headlong /ˈhedlɒŋ/ *adv* de cabeza; (*precipitately*) precipitadamente

head: **~master** *n* director *m*. **~mistress** *n* directora *f*. **~-on** *a & adv* de frente. **~phone** *n* auricular *m*, audífono *m* (*LAm*)

headquarters /hedˈkwɔːtəz/ *n* (*of organization*) sede *f*; (*of business*) oficina *f* central; (*mil*) cuartel *m* general

headstrong /'hedstrɒŋ/ *a* testarudo
headway /'hedweɪ/ *n* progreso *m*. **make ~** hacer progresos
heady /'hedɪ/ *a* (**-ier**, **-iest**) (*impetuous*) impetuoso; (*intoxicating*) embriagador
heal /hi:l/ *vt* curar. ● *vi* ‹*wound*› cicatrizarse; (*fig*) curarse
health /helθ/ *n* salud *f*. **~y** *a* sano
heap /hi:p/ *n* montón *m*. ● *vt* amontonar. **~s of** (*fam*) montones de, muchísimos
hear /hɪə(r)/ *vt/i* (*pt* **heard** /hɜ:d/) oír. **~, ~!** ¡bravo! **not ~ of** (*refuse to allow*) no querer oír. **~ about** oir hablar de. **~ from** recibir noticias de. **~ of** oir hablar de
hearing /'hɪərɪŋ/ *n* oído *m*; (*of witness*) audición *f*. **~-aid** *n* audífono *m*
hearsay /'hɪəseɪ/ *n* rumores *mpl*. **from ~** según los rumores
hearse /hɜ:s/ *n* coche *m* fúnebre
heart / hɑ:t/ *n* corazón *m*. **at ~** en el fondo. **by ~** de memoria. **lose ~** descorozonarse. **~ache** *n* pena *f*. **~ attack** *n* ataque *m* al corazón. **~-break** *n* pena *f*. **~-breaking** *a* desgarrador. **~-broken** *a*. **be ~-broken** partírsele el corazón
heartburn /'hɑ:tbɜ:n/ *n* acedía *f*
hearten /'hɑ:tn/ *vt* animar
heartfelt /'hɑ:tfelt/ *a* sincero
hearth /hɑ:θ/ *n* hogar *m*
heartily /'hɑ:tɪlɪ/ *adv* de buena gana; (*sincerely*) sinceramente
heart: **~less** *a* cruel. **~-searching** *n* examen *m* de conciencia. **~-to-~** *a* abierto
hearty /'hɑ:tɪ/ *a* (*sincere*) sincero; ‹*meal*› abundante
heat /hi:t/ *n* calor *m*; (*contest*) eliminatoria *f*. ● *vt* calentar. ● *vi* calentarse. **~ed** *a* (*fig*) acalorado. **~er** /'hi:tə(r)/ *n* calentador *m*
heath /hi:θ/ *n* brezal *m*, descampado *m*, terreno *m* baldío
heathen /'hi:ðn/ *n* & *a* pagano (*m*)
heather /'heðə(r)/ *n* brezo *m*
heat: **~ing** *n* calefacción *f*. **~-stroke** *n* insolación *f*. **~wave** *n* ola *f* de calor
heave /hi:v/ *vt* (*lift*) levantar; exhalar ‹*sigh*›; (*throw, fam*) lanzar. ● *vi* (*retch*) sentir náuseas
heaven /'hevn/ *n* cielo *m*. **~ly** *a* celestial; (*astronomy*) celeste; (*excellent, fam*) divino
heav|ily /'hevɪlɪ/ *adv* pesadamente; (*smoke, drink*) mucho. **~y** *a* (**-ier**, **-iest**) pesado; ‹*sea*› grueso; ‹*traffic*› denso; ‹*work*› duro. **~yweight** *n* peso *m* pesado
Hebrew /'hi:bru:/ *a* & *n* hebreo (*m*)
heckle /'hekl/ *vt* interrumpir ‹*speaker*›
hectic /'hektɪk/ *a* febril
hedge /hedʒ/ *n* seto *m* vivo. ● *vt* rodear con seto vivo. ● *vi* escaparse por la tangente
hedgehog /'hedʒhɒg/ *n* erizo *m*
heed /hi:d/ *vt* hacer caso de. ● *n* atención *f*. **pay ~ to** hacer caso de. **~less** *a* desatento
heel /hi:l/ *n* talón *m*; (*of shoe*) tacón *m*. **down at ~**, **down at the ~s** (*Amer*) desharrapado
hefty /'heftɪ/ *a* (**-ier**, **-iest**) (*sturdy*) fuerte; (*heavy*) pesado
heifer /'hefə(r)/ *n* novilla *f*
height /haɪt/ *n* altura *f*; (*of person*) estatura *f*; (*of fame, glory*) cumbre *f*; (*of joy, folly, pain*) colmo *m*
heighten /'haɪtn/ *vt* (*raise*) elevar; (*fig*) aumentar
heinous /'heɪnəs/ *a* atroz
heir /eə(r)/ *n* heredero *m*. **~ess** *n* heredera *f*. **~loom** /'eəlu:m/ *n* reliquia *f* heredada
held /held/ *see* **hold**[1]
helicopter /'helɪkɒptə(r)/ *n* helicóptero *m*
heliport /'helɪpɔ:t/ *n* helipuerto *m*
hell /hel/ *n* infierno *m*. **~-bent** *a* resuelto. **~ish** *a* infernal
hello /hə'ləʊ/ *int* ¡hola!; (*telephone, caller*) ¡oiga!, ¡bueno! (*Mex*), ¡hola! (*Arg*); (*telephone, person answering*) ¡diga!, ¡bueno! (*Mex*), ¡hola! (*Arg*); (*surprise*) ¡vaya! **say ~ to** saludar
helm /helm/ *n* (*of ship*) timón *m*
helmet /'helmɪt/ *n* casco *m*
help /help/ *vt/i* ayudar. **he cannot ~ laughing** no puede menos de reír. **~ o.s. to** servirse. **it cannot be ~ed** no hay más remedio. ● *n* ayuda *f*; (*charwoman*) asistenta *f*. **~er** *n* ayudante *m*. **~ful** *a* útil; ‹*person*› amable
helping /'helpɪŋ/ *n* porción *f*
helpless /'helplɪs/ *a* (*unable to manage*) incapaz; (*powerless*) impotente
helter-skelter /heltə'skeltə(r)/ *n* tobogán *m*. ● *adv* atropelladamente
hem /hem/ *n* dobladillo *m*. ● *vt* (*pt* **hemmed**) hacer un dobladillo. **~ in** encerrar

hemisphere /'hemɪsfɪə(r)/ *n* hemisferio *m*
hemp /hemp/ *n* (*plant*) cáñamo *m*; (*hashish*) hachís *m*
hen /hen/ *n* gallina *f*
hence /hens/ *adv* de aquí. **~forth** *adv* de ahora en adelante
henchman /'hentʃmən/ *n* secuaz *m*
henna /'henə/ *n* alheña *f*
hen-party /'henpɑ:tɪ/ *n* (*fam*) reunión *f* de mujeres
henpecked /'henpekt/ *a* dominado por su mujer
her /hɜ:(r)/ *pron* (*accusative*) la; (*dative*) le; (*after prep*) ella. **I know ~** la conozco. ● *a* su, sus *pl*
herald /'herəld/ *vt* anunciar
heraldry /'herəldrɪ/ *n* heráldica *f*
herb /hɜ:b/ *n* hierba *f*. **~s** *npl* hierbas *fpl* finas
herbaceous /hɜ:'beɪʃəs/ *a* herbáceo
herbalist /'hɜ:bəlɪst/ *n* herbolario *m*
herculean /hɜ:kjʊ'li:ən/ *a* hercúleo
herd /hɜ:d/ *n* rebaño *m*. ● *vt*. **~ together** reunir
here /hɪə(r)/ *adv* aquí. **~!** (*take this*) ¡tenga! **~abouts** *adv* por aquí. **~after** *adv* en el futuro. **~by** *adv* por este medio; (*in letter*) por la presente
heredit|ary /hɪ'redɪtərɪ/ *a* hereditario. **~y** /hɪ'redətɪ/ *n* herencia *f*
here|sy /'herəsɪ/ *n* herejía *f*. **~tic** *n* hereje *m & f*
herewith /hɪə'wɪð/ *adv* adjunto
heritage /'herɪtɪdʒ/ *n* herencia *f*; (*fig*) patrimonio *m*
hermetic /hɜ:'metɪk/ *a* hermético
hermit /'hɜ:mɪt/ *n* ermitaño *m*
hernia /'hɜ:nɪə/ *n* hernia *f*
hero /'hɪərəʊ/ *n* (*pl* **-oes**) héroe *m*. **~ic** *a* heroico
heroin /'herəʊɪn/ *n* heroína *f*
hero: **~ine** /'herəʊɪn/ *n* heroína *f*. **~ism** /'herəʊɪzm/ *n* heroismo *m*
heron /'herən/ *n* garza *f* real
herring /'herɪŋ/ *n* arenque *m*
hers /hɜ:z/ *poss pron* suyo *m*, suya *f*, suyos *mpl*, suyas *fpl*, de ella
herself /hɜ:'self/ *pron* ella misma; (*reflexive*) se; (*after prep*) sí
hesitant /'hezɪtənt/ *a* vacilante
hesitat|e /'hezɪteɪt/ *vi* vacilar. **~ion** /-'teɪʃn/ *n* vacilación *f*
hessian /'hesɪən/ *n* arpillera *f*
het /het/ *a*. **~ up** (*sl*) nervioso
heterogeneous /hetərəʊ'dʒi:nɪəs/ *a* heterogéneo
heterosexual /hetərəʊ'seksjʊəl/ *a* heterosexual
hew /hju:/ *vt* (*pp* **hewn**) cortar; (*cut into shape*) tallar
hexagon /'heksəgən/ *n* hexágono *m*. **~al** /-'ægənl/ *a* hexagonal
hey /heɪ/ *int* ¡eh!
heyday /heɪdeɪ/ *n* apogeo *m*
hi /haɪ/ *int* (*fam*) ¡hola!
hiatus /haɪ'eɪtəs/ *n* (*pl* **-tuses**) hiato *m*
hibernat|e /'haɪbəneɪt/ *vi* hibernar. **~ion** *n* hibernación *f*
hibiscus /hɪ'bɪskəs/ *n* hibisco *m*
hiccup /'hɪkʌp/ *n* hipo *m*. **have (the) ~s** tener hipo. ● *vi* tener hipo
hide[1] /haɪd/ *vt* (*pt* **hid**, *pp* **hidden**) esconder. ● *vi* esconderse
hide[2] /haɪd/ *n* piel *f*, cuero *m*
hideous /'hɪdɪəs/ *a* (*dreadful*) horrible; (*ugly*) feo
hide-out /'haɪdaʊt/ *n* escondrijo *m*
hiding[1] /'haɪdɪŋ/ *n* (*thrashing*) paliza *f*
hiding[2] /'haɪdɪŋ/ *n*. **go into ~** esconderse
hierarchy /'haɪərɑ:kɪ/ *n* jerarquía *f*
hieroglyph /'haɪərəglɪf/ *n* jeroglífico *m*
hi-fi /'haɪfaɪ/ *a* de alta fidelidad. ● *n* (equipo *m* de) alta fidelidad (*f*)
higgledy-piggledy /hɪgldɪ'pɪgldɪ/ *adv* en desorden
high /haɪ/ *a* (**-er**, **-est**) alto; ‹*price*› elevado; ‹*number*, *speed*› grande; ‹*wind*› fuerte; (*intoxicated*, *fam*) ebrio; ‹*voice*› agudo; ‹*meat*› manido. **in the ~ season** en plena temporada. ● *n* alto nivel *m*. **a (new) ~** un récord *m*. ● *adv* alto
highbrow /'haɪbraʊ/ *a & n* intelectual (*m & f*)
higher education /haɪər edʒʊ'keɪʃn/ *n* enseñanza *f* superior
high-falutin /haɪfə'lu:tɪn/ *a* pomposo
high-handed /haɪ'hændɪd/ *a* despótico
high jump /'haɪdʒʌmp/ *n* salto *m* de altura
highlight /'haɪlaɪt/ *n* punto *m* culminante. ● *vt* destacar
highly /'haɪlɪ/ *adv* muy; ‹*paid*› muy bien. **~ strung** *a* nervioso
highness /'haɪnɪs/ *n* (*title*) alteza *f*
high: **~-rise building** *n* rascacielos *m*. **~ school** *n* instituto *m*. **~-speed** *a* de gran velocidad. **~ spot** *n* (*fam*) punto *m* culminante. **~ street** *n*

calle *f* mayor. **~-strung** *a* (*Amer*) nervioso. **~ tea** *n* merienda *f* substanciosa
highway /ˈhaɪweɪ/ *n* carretera *f*. **~man** *n* salteador *m* de caminos
hijack /ˈhaɪdʒæk/ *vt* secuestrar. ● *n* secuestro *m*. **~er** *n* secuestrador
hike /haɪk/ *n* caminata *f*. ● *vi* darse la caminata. **~r** *n* excursionista *m* & *f*
hilarious /hɪˈleərɪəs/ *a* (*funny*) muy divertido
hill /hɪl/ *n* colina *f*; (*slope*) cuesta *f*. **~-billy** *n* rústico *m*. **~side** *n* ladera *f*. **~y** *a* montuoso
hilt /hɪlt/ *n* (*of sword*) puño *m*. **to the ~** totalmente
him /hɪm/ *pron* le, lo; (*after prep*) él. **I know ~** le/lo conozco
himself /hɪmˈself/ *pron* él mismo; (*reflexive*) se
hind /haɪnd/ *a* trasero
hinder /ˈhɪndə(r)/ *vt* estorbar; (*prevent*) impedir
hindrance /ˈhɪndrəns/ *n* obstáculo *m*
hindsight /ˈhaɪnsaɪt/ *n*. **with ~** retrospectivamente
Hindu /hɪnˈduː/ *n* & *a* hindú (*m* & *f*). **~ism** *n* hinduismo *m*
hinge /hɪndʒ/ *n* bisagra *f*. ● *vi*. **~ on** (*depend on*) depender de
hint /hɪnt/ *n* indirecta *f*; (*advice*) consejo *m*. ● *vt* dar a entender. ● *vi* soltar una indirecta. **~ at** hacer alusión a
hinterland /ˈhɪntəlænd/ *n* interior *m*
hip /hɪp/ *n* cadera *f*
hippie /ˈhɪpɪ/ *n* hippie *m* & *f*
hippopotamus /hɪpəˈpɒtəməs/ *n* (*pl* **-muses** *or* **-mi**) hipopótamo *m*
hire /haɪə(r)/ *vt* alquilar ‹*thing*›; contratar ‹*person*›. ● *n* alquiler *m*. **~-purchase** *n* compra *f* a plazos
hirsute /ˈhɜːsjuːt/ *a* hirsuto
his /hɪz/ *a* su, sus *pl*. ● *poss pron* el suyo *m*, la suya *f*, los suyos *mpl*, las suyas *fpl*
Hispan|ic /hɪˈspænɪk/ *a* hispánico. **~ist** /ˈhɪspənɪst/ *n* hispanista *m* & *f*. **~o...** *pref* hispano...
hiss /hɪs/ *n* silbido. ● *vt/i* silbar
histor|ian /hɪˈstɔːrɪən/ *n* historiador *m*. **~ic(al)** /hɪˈstɒrɪkl/ *a* histórico. **~y** /ˈhɪstərɪ/ *n* historia *f*. **make ~y** pasar a la historia
histrionic /hɪstrɪˈɒnɪk/ *a* histriónico
hit /hɪt/ *vt* (*pt* **hit**, *pres p* **hitting**) golpear; (*collide with*) chocar con; (*find*) dar con; (*affect*) afectar. **~ it off with** hacer buenas migas con. ● *n* (*blow*) golpe *m*; (*fig*) éxito *m*. **~ on** *vi* encontrar, dar con
hitch /hɪtʃ/ *vt* (*fasten*) atar. ● *n* (*snag*) problema *m*. **~ a lift**, **~-hike** *vi* hacer autostop, hacer dedo (*Arg*), pedir aventón (*Mex*). **~-hiker** *n* autostopista *m* & *f*
hither /ˈhɪðə(r)/ *adv* acá. **~ and thither** acá y allá
hitherto /ˈhɪðətuː/ *adv* hasta ahora
hit-or-miss /ˈhɪtɔːˈmɪs/ *a* (*fam*) a la buena de Dios, a ojo
hive /haɪv/ *n* colmena *f*. ● *vt*. **~off** separar; (*industry*) desnacionalizar
hoard /hɔːd/ *vt* acumular. ● *n* provisión *f*; (*of money*) tesoro *m*
hoarding /ˈhɔːdɪŋ/ *n* cartelera *f*, valla *f* publicitaria
hoar-frost /ˈhɔːfrɒst/ *n* escarcha *f*
hoarse /hɔːs/ *a* (**-er**, **-est**) ronco. **~ness** *n* (*of voice*) ronquera *f*; (*of sound*) ronquedad *f*
hoax /həʊks/ *n* engaño *m*. ● *vt* engañar
hob /hɒb/ *n* repisa *f*; (*of cooker*) fogón *m*
hobble /ˈhɒbl/ *vi* cojear
hobby /ˈhɒbɪ/ *n* pasatiempo *m*
hobby-horse /ˈhɒbɪhɔːs/ *n* (*toy*) caballito *m* (de niño); (*fixation*) caballo *m* de batalla
hobnail /ˈhɒbneɪl/ *n* clavo *m*
hob-nob /ˈhɒbnɒb/ *vi* (*pt* **hob-nobbed**). **~ with** codearse con
hock[1] /hɒk/ *n* vino *m* del Rin
hock[2] /hɒk/ *vt* (*pawn*, *sl*) empeñar
hockey /ˈhɒkɪ/ *n* hockey *m*
hodgepodge /ˈhɒdʒpɒdʒ/ *n* mezcolanza *f*
hoe /həʊ/ *n* azada *f*. ● *vt* (*pres p* **hoeing**) azadonar
hog / hɒg/ *n* cerdo *m*. ● *vt* (*pt* **hogged**) (*fam*) acaparar
hoist /hɔɪst/ *vt* levantar; izar ‹*flag*›. ● *n* montacargas *m invar*
hold[1] /həʊld/ *vt* (*pt* **held**) tener; (*grasp*) coger (*not LAm*), agarrar; (*contain*) contener; mantener ‹*interest*›; (*believe*) creer; contener ‹*breath*›. **~ one's tongue** callarse. ● *vi* mantenerse. ● *n* asidero *m*; (*influence*) influencia *f*. **get ~ of** agarrar; (*fig*, *acquire*) adquirir. **~ back** (*contain*) contener; (*conceal*) ocultar. **~ on** (*stand firm*) resistir; (*wait*) esperar. **~ on to** (*keep*) guardar; (*cling to*) agarrarse a. **~**

out *vt* (*offer*) ofrecer. ● *vi* (*resist*) resistir. **~ over** aplazar. **~ up** (*support*) sostener; (*delay*) retrasar; (*rob*) atracar. **~ with** aprobar

hold² /həʊld/ *n* (*of ship*) bodega *f*

holdall /'həʊldɔ:l/ *n* bolsa *f* (de viaje)

holder /'həʊldə(r)/ *n* tenedor *m*; (*of post*) titular *m*; (*for object*) soporte *m*

holding /'həʊldɪŋ/ *n* (*land*) propiedad *f*

hold-up /'həʊldʌp/ *n* atraco *m*

hole /həʊl/ *n* agujero *m*; (*in ground*) hoyo *m*; (*in road*) bache *m*. ● *vt* agujerear

holiday /'hɒlɪdeɪ/ *n* vacaciones *fpl*; (*public*) fiesta *f*. ● *vi* pasar las vacaciones. **~-maker** *n* veraneante *m*

holiness /'həʊlɪnɪs/ *n* santidad *f*

Holland /'hɒlənd/ *n* Holanda *f*

hollow /'hɒləʊ/ *a* & *n* hueco (*m*). ● *vt* ahuecar

holly /'hɒlɪ/ *n* acebo *m*. **~hock** *n* malva *f* real

holocaust /'hɒləkɔ:st/ *n* holocausto *m*

holster /'həʊlstə(r)/ *n* pistolera *f*

holy /'həʊlɪ/ *a* (**-ier**, **-iest**) santo, sagrado. **H~ Ghost** *n*, **H~ Spirit** *n* Espíritu *m* Santo. **~ water** *n* agua *f* bendita

homage /'hɒmɪdʒ/ *n* homenaje *m*

home /həʊm/ *n* casa *f*; (*institution*) asilo *m*; (*for soldiers*) hogar *m*; (*native land*) patria *f*. **feel at ~ with** sentirse como en su casa. ● *a* casera, de casa; (*of family*) de familia; (*pol*) interior; ‹*match*› de casa. ● *adv*. **(at) ~** en casa. **H~ Counties** *npl* región *f* alrededor de Londres. **~land** *n* patria *f*. **~less** *a* sin hogar. **~ly** /'həʊmlɪ/ *a* (**-ier**, **-iest**) casero; (*ugly*) feo. **H~ Office** *n* Ministerio *m* del Interior. **H~ Secretary** *n* Ministro *m* del Interior. **~sick** *a*. **be ~sick** tener morriña. **~ town** *n* ciudad *f* natal. **~ truths** *npl* las verdades *fpl* del barquero, las cuatro verdades *fpl*. **~ward** /'həʊmwəd/ *a* ‹*journey*› de vuelta. ● *adv* hacia casa. **~work** *n* deberes *mpl*

homicide /'hɒmɪsaɪd/ *n* homicidio *m*

homoeopath|ic /həʊmɪəʊ'pæθɪk/ *a* homeopático. **~y** /-'ɒpəθɪ/ *n* homeopatía *f*

homogeneous /həʊməʊ'dʒi:nɪəs/ *a* homogéneo

homosexual /həʊməʊ'seksjʊəl/ *a* & *n* homosexual (*m*)

hone /həʊn/ *vt* afilar

honest /'ɒnɪst/ *a* honrado; (*frank*) sincero. **~ly** *adv* honradamente. **~y** *n* honradez *f*

honey /'hʌnɪ/ *n* miel *f*; (*person, fam*) cielo *m*, cariño *m*. **~comb** /'hʌnɪkəʊm/ *n* panal *m*

honeymoon /'hʌnɪmu:n/ *n* luna *f* de miel

honeysuckle /'hʌnɪsʌkl/ *n* madreselva *f*

honk /hɒŋk/ *vi* tocar la bocina

honorary /'ɒnərərɪ/ *a* honorario

honour /'ɒnə(r)/ *n* honor *m*. ● *vt* honrar. **~able** *a* honorable

hood /hʊd/ *n* capucha *f*; (*car roof*) capota *f*; (*car bonnet*) capó *m*

hoodlum /'hu:dləm/ *n* gamberro *m*, matón *m*

hoodwink /'hʊdwɪŋk/ *vt* engañar

hoof /hu:f/ *n* (*pl* **hoofs** *or* **hooves**) casco *m*

hook /hʊk/ *n* gancho *m*; (*on garment*) corchete *m*; (*for fishing*) anzuelo *m*. **by ~ or by crook** por fas o por nefas, por las buenas o por las malas. **get s.o. off the ~** sacar a uno de un apuro. **off the ~** ‹*telephone*› descolgado. ● *vt* enganchar. ● *vi* engancharse

hooked /hʊkt/ *a* ganchudo. **~ on** (*sl*) adicto a

hooker /'hʊkə(r)/ *n* (*rugby*) talonador *m*; (*Amer*, *sl*) prostituta *f*

hookey /'hʊkɪ/ *n*. **play ~** (*Amer*, *sl*) hacer novillos

hooligan /'hu:lɪgən/ *n* gamberro *m*

hoop /hu:p/ *n* aro *m*

hooray /hʊ'reɪ/ *int* & *n* ¡viva! (*m*)

hoot /hu:t/ *n* (*of horn*) bocinazo *m*; (*of owl*) ululato *m*. ● *vi* tocar la bocina; ‹*owl*› ulular

hooter /'hu:tə(r)/ *n* (*of car*) bocina *f*; (*of factory*) sirena *f*

Hoover /'hu:və(r)/ *n* (*P*) aspiradora *f*. ● *vt* pasar la aspiradora

hop¹ /hɒp/ *vi* (*pt* **hopped**) saltar a la pata coja. **~ in** (*fam*) subir. **~ it** (*sl*) largarse. **~ out** (*fam*) bajar. ● *n* salto *m*; (*flight*) etapa *f*

hop² /hɒp/ *n*. **~(s)** lúpulo *m*

hope /həʊp/ *n* esperanza *f*. ● *vt/i* esperar. **~ for** esperar. **~ful** *a* esperanzador. **~fully** *adv* con optimismo; (*it is hoped*) se espera. **~less** *a* desesperado. **~lessly** *adv* sin esperanza

hopscotch /'hɒpskɒtʃ/ *n* tejo *m*

horde /hɔ:d/ *n* horda *f*

horizon /hə'raɪzn/ *n* horizonte *m*
horizontal /hɒrɪ'zɒntl/ *a* horizontal. **~ly** *adv* horizontalmente
hormone /'hɔ:məʊn/ *n* hormona *f*
horn /hɔ:n/ *n* cuerno *m*; (*of car*) bocina *f*; (*mus*) trompa *f*. ● *vt*. **~ in** (*sl*) entrometerse. **~ed** *a* con cuernos
hornet /'hɔ:nɪt/ *n* avispón *m*
horny /'hɔ:nɪ/ *a* ‹*hands*› calloso
horoscope /'hɒrəskəʊp/ *n* horóscopo *m*
horri|ble /'hɒrəbl/ *a* horrible. **~d** /'hɒrɪd/ *a* horrible
horrif|ic /hə'rɪfɪk/ *a* horroroso. **~y** /'hɒrɪfaɪ/ *vt* horrorizar
horror /'hɒrə(r)/ *n* horror *m*. **~ film** *n* película *f* de miedo
hors-d'oevre /ɔ:'dɜ:vr/ *n* entremés *m*
horse /hɔ:s/ *n* caballo *m*. **~back** *n*. **on ~back** a caballo
horse chestnut /hɔ:s'tʃesnʌt/ *n* castaña *f* de Indias
horse: **~man** *n* jinete *m*. **~play** *n* payasadas *fpl*. **~power** *n* (*unit*) caballo *m* (de fuerza). **~-racing** *n* carreras *fpl* de caballos
horseradish /'hɔ:srædɪʃ/ *n* rábano *m* picante
horse: **~ sense** *n* (*fam*) sentido *m* común. **~shoe** /'hɔ:sʃu:/ *n* herradura *f*
horsy /'hɔ:sɪ/ *a* ‹*face etc*› caballuno
horticultur|al /hɔ:tɪ'kʌltʃərəl/ *a* hortícola. **~e** /'hɔ:tɪkʌltʃə(r)/ *n* horticultura *f*
hose /həʊz/ *n* (*tube*) manga *f*. ● *vt* (*water*) regar con una manga; (*clean*) limpiar con una manga. **~-pipe** *n* manga *f*
hosiery /'həʊzɪərɪ/ *n* calcetería *f*
hospice /'hɒspɪs/ *n* hospicio *m*
hospitabl|e /hɒ'spɪtəbl/ *a* hospitalario. **~y** *adv* con hospitalidad
hospital /'hɒspɪtl/ *n* hospital *m*
hospitality /hɒspɪ'tælətɪ/ *n* hospitalidad *f*
host[1] /həʊst/ *n*. **a ~ of** un montón de
host[2] /həʊst/ *n* (*master of house*) huésped *m*, anfitrión *m*
host[3] /həʊst/ *n* (*relig*) hostia *f*
hostage /'hɒstɪdʒ/ *n* rehén *m*
hostel /'hɒstl/ *n* (*for students*) residencia *f*. **youth ~** albergue *m* juvenil
hostess /'həʊstɪs/ *n* huéspeda *f*, anfitriona *f*
hostil|e /'hɒstaɪl/ *a* hostil. **~ity** *n* hostilidad *f*
hot /hɒt/ *a* (**hotter**, **hottest**) caliente; (*culin*) picante; ‹*news*› de última hora. **be/feel ~** tener calor. **in ~ water** (*fam*) en un apuro. **it is ~** hace calor. ● *vt*/*i*. **~ up** (*fam*) calentarse
hotbed /'hɒtbed/ *n* (*fig*) semillero *m*
hotchpotch /'hɒtʃpɒtʃ/ *n* mezcolanza *f*
hot dog /hɒt'dɒg/ *n* perrito *m* caliente
hotel /həʊ'tel/ *n* hotel *m*. **~ier** *n* hotelero *m*
hot: **~head** *n* impetuoso *m*. **~-headed** *a* impetuoso. **~-house** *n* invernadero *m*. **~ line** *n* teléfono *m* rojo. **~plate** *n* calentador *m*. **~-water bottle** *n* bolsa *f* de agua caliente
hound /haʊnd/ *n* perro *m* de caza. ● *vt* perseguir
hour /aʊə(r)/ *n* hora *f*. **~ly** *a* & *adv* cada hora. **~ly pay** *n* sueldo *m* por hora. **paid ~ly** pagado por hora
house /haʊs/ *n* (*pl* **-s** /'haʊzɪz/) casa *f*; (*theatre building*) sala *f*; (*theatre audience*) público *m*; (*pol*) cámara *f*. /haʊz/ *vt* alojar; (*keep*) guardar. **~boat** *n* casa *f* flotante. **~breaking** *n* robo *m* de casa. **~hold** /'haʊshəʊld/ *n* casa *f*, familia *f*. **~holder** *n* dueño *m* de una casa; (*head of household*) cabeza *f* de familia. **~keeper** *n* ama *f* de llaves. **~keeping** *n* gobierno *m* de la casa. **~maid** *n* criada *f*, mucama *f* (*LAm*). **H~ of Commons** *n* Cámara *f* de los Comunes. **~-proud** *a* meticuloso. **~-warming** *n* inauguración *f* de una casa. **~wife** /'haʊswaɪf/ *n* ama *f* de casa. **~work** *n* quehaceres *mpl* domésticos
housing /'haʊzɪŋ/ *n* alojamiento *m*. **~ estate** *n* urbanización *f*
hovel /'hɒvl/ *n* casucha *f*
hover /'hɒvə(r)/ *vi* ‹*bird, threat etc*› cernerse; (*loiter*) rondar. **~craft** *n* aerodeslizador *m*
how /haʊ/ *adv* cómo. **~ about a walk?** ¿qué le parece si damos un paseo? **~ are you?** ¿cómo está Vd? **~ do you do?** (*in introduction*) mucho gusto. **~ long?** ¿cuánto tiempo? **~ many?** ¿cuántos? **~ much?** ¿cuánto? **~ often?** ¿cuántas veces? **and ~!** ¡y cómo!
however /haʊ'evə(r)/ *adv* (*with verb*) de cualquier manera que (+ *subjunctive*); (*with adjective or adverb*) por... que (+ *subjunctive*);

(*nevertheless*) no obstante, sin embargo. ~ **much it rains** por mucho que llueva

howl /haʊl/ *n* aullido. ● *vi* aullar

howler /'haʊlə(r)/ *n* (*fam*) plancha *f*

HP *abbr see* **hire-purchase**

hp *abbr see* **horsepower**

hub /hʌb/ *n* (*of wheel*) cubo *m*; (*fig*) centro *m*

hubbub /'hʌbʌb/ *n* barahúnda *f*

hub-cap /'hʌbkæp/ *n* tapacubos *m invar*

huddle /'hʌdl/ *vi* apiñarse

hue[1] /hju:/ *n* (*colour*) color *m*

hue[2] /hju:/ *n*. ~ **and cry** clamor *m*

huff /hʌf/ *n*. **in a** ~ enojado

hug /hʌg/ *vt* (*pt* **hugged**) abrazar; (*keep close to*) no apartarse de. ● *n* abrazo *m*

huge /hju:dʒ/ *a* enorme. ~**ly** *adv* enormemente

hulk /hʌlk/ *n* (*of ship*) barco *m* viejo; (*person*) armatoste *m*

hull /hʌl/ *n* (*of ship*) casco *m*

hullabaloo /hʌləbə'lu:/ *n* tumulto *m*

hullo /hə'ləʊ/ *int* = **hello**

hum /hʌm/ *vt/i* (*pt* **hummed**) ‹*person*› canturrear; ‹*insect, engine*› zumbar. ● *n* zumbido *m*. ~ **(or hem) and haw (or ha)** vacilar

human /'hju:mən/ *a* & *n* humano (*m*). ~ **being** *n* ser *m* humano

humane /hju:'meɪn/ *a* humano

humanism /'hju:mənɪzəm/ *n* humanismo *m*

humanitarian /hju:mænɪ'teərɪən/ *a* humanitario

humanity /hju:'mænətɪ/ *n* humanidad *f*

humbl|e /'hʌmbl/ *a* (**-er**, **-est**) humilde. ● *vt* humillar. ~**y** *adv* humildemente

humbug /'hʌmbʌg/ *n* (*false talk*) charlatanería *f*; (*person*) charlatán *m*; (*sweet*) caramelo *m* de menta

humdrum /'hʌmdrʌm/ *a* monótono

humid /'hju:mɪd/ *a* húmedo. ~**ifier** *n* humedecedor *m*. ~**ity** /hju:'mɪdətɪ/ *n* humedad *f*

humiliat|e /hju:'mɪlɪeɪt/ *vt* humillar. ~**ion** /-'eɪʃn/ *n* humillación *f*

humility /hju:'mɪlətɪ/ *n* humildad *f*

humorist /'hju:mərɪst/ *n* humorista *m* & *f*

humo|rous /'hju:mərəs/ *a* divertido. ~**rously** *adv* con gracia. ~**ur** *n* humorismo *m*; (*mood*) humor *m*. **sense of** ~**ur** *n* sentido *m* del humor

hump /hʌmp/ *n* montecillo *m*; (*of the spine*) joroba *f*. **the** ~ (*sl*) malhumor *m*. ● *vt* encorvarse; (*hoist up*) llevar al hombro

hunch /hʌntʃ/ *vt* encorvar. ~**ed up** encorvado. ● *n* presentimiento *m*; (*lump*) joroba *f*. ~**back** /'hʌntʃbæk/ *n* jorobado *m*

hundred /'hʌndrəd/ *a* ciento, (*before noun*) cien. ● *n* ciento *m*. ~**fold** *a* céntuplo. ● *adv* cien veces. ~**s of** centenares de. ~**th** *a* centésimo. ● *n* centésimo *m*, centésima parte *f*

hundredweight /'hʌndrədweɪt/ *n* 50,8kg; (*Amer*) 45,36kg

hung /hʌŋ/ *see* **hang**

Hungar|ian /hʌŋ'geərɪən/ *a* & *n* húngaro (*m*). ~**y** /'hʌŋgərɪ/ *n* Hungría *f*

hunger /'hʌŋgə(r)/ *n* hambre *f*. ● *vi*. ~ **for** tener hambre de. ~**-strike** *n* huelga *f* de hambre

hungr|ily /'hʌŋgrəlɪ/ *adv* ávidamente. ~**y** *a* (**-ier**, **-iest**) hambriento. **be** ~**y** tener hambre

hunk /hʌŋk/ *n* (buen) pedazo *m*

hunt /hʌnt/ *vt/i* cazar. ~ **for** buscar. ● *n* caza *f*. ~**er** *n* cazador *m*. ~**ing** *n* caza *f*

hurdle /'hɜ:dl/ *n* (*sport*) valla *f*; (*fig*) obstáculo *m*

hurdy-gurdy /'hɜ:dɪgɜ:dɪ/ *n* organillo *m*

hurl /hɜ:l/ *vt* lanzar

hurly-burly /'hɜ:lɪbɜ:lɪ/ *n* tumulto *m*

hurrah /hʊ'rɑ:/, **hurray** /hʊ'reɪ/ *int* & *n* ¡viva! (*m*)

hurricane /'hʌrɪkən/ *n* huracán *m*

hurried /'hʌrɪd/ *a* apresurado. ~**ly** *adv* apresuradamente

hurry /'hʌrɪ/ *vi* apresurarse, darse prisa. ● *vt* apresurar, dar prisa a. ● *n* prisa *f*. **be in a** ~ tener prisa

hurt /hɜ:t/ *vt/i* (*pt* **hurt**) herir. ● *n* (*injury*) herida *f*; (*harm*) daño *m*. ~**ful** *a* hiriente; (*harmful*) dañoso

hurtle /'hɜ:tl/ *vt* lanzar. ● *vi*. ~ **along** mover rápidamente

husband /'hʌzbənd/ *n* marido *m*

hush /hʌʃ/ *vt* acallar. ● *n* silencio *m*. ~ **up** ocultar ‹*affair*›. ~-~ *a* (*fam*) muy secreto

husk /hʌsk/ *n* cáscara *f*

husky /'hʌskɪ/ *a* (**-ier**, **-iest**) (*hoarse*) ronco; (*burly*) fornido

hussy /'hʌsɪ/ *n* desvergonzada *f*

hustle /'hʌsl/ *vt* (*jostle*) empujar. ● *vi* (*hurry*) darse prisa. ● *n* empuje *m*. ~ **and bustle** *n* bullicio *m*

hut /hʌt/ *n* cabaña *f*

hutch /hʌtʃ/ *n* conejera *f*
hyacinth /'haɪəsɪnθ/ *n* jacinto *m*
hybrid /'haɪbrɪd/ *a* & *n* híbrido (*m*)
hydrangea /haɪ'dreɪndʒə/ *n* hortensia *f*
hydrant /'haɪdrənt/ *n*. **(fire)** ~ *n* boca *f* de riego
hydraulic /haɪ'drɔ:lɪk/ *a* hidráulico
hydroelectric /haɪdrəʊɪ'lektrɪk/ *a* hidroeléctrico
hydrofoil /'haɪdrəfɔɪl/ *n* aerodeslizador *m*
hydrogen /'haɪdrədʒən/ *n* hidrógeno *m*. ~ **bomb** *n* bomba *f* de hidrógeno. ~ **peroxide** *n* peróxido *m* de hidrógeno
hyena /haɪ'i:nə/ *n* hiena *f*
hygien|e /'haɪdʒi:n/ *n* higiene *f*. ~**ic** *a* higiénico
hymn /hɪm/ *n* himno *m*
hyper... /'haɪpə(r)/ *pref* hiper...
hypermarket /'haɪpəmɑ:kɪt/ *n* hipermercado *m*
hyphen /'haɪfn/ *n* guión *m*. ~**ate** *vt* escribir con guión
hypno|sis /hɪp'nəʊsɪs/ *n* hipnosis *f*. ~**tic** /-'nɒtɪk/ *a* hipnótico. ~**tism** /hɪpnə'tɪzəm/ *n* hipnotismo *m*. ~**tist** *n* hipnotista *m* & *f*. ~**tize** *vt* hipnotizar
hypochondriac /haɪpə'kɒndrɪæk/ *n* hipocondríaco *m*
hypocrisy /hɪ'pɒkrəsɪ/ *n* hipocresía *f*
hypocrit|e /'hɪpəkrɪt/ *n* hipócrita *m* & *f*. ~**ical** *a* hipócrita
hypodermic /haɪpə'dɜ:mɪk/ *a* hipodérmico. • *n* jeringa *f* hipodérmica
hypothe|sis /haɪ'pɒθəsɪs/ *n* (*pl* **-theses** /-si:z/) hipótesis *f*. ~**tical** /-ə'θetɪkl/ *a* hipotético
hysteri|a /hɪ'stɪərɪə/ *n* histerismo *m*. ~**cal** /-'terɪkl/ *a* histérico. ~**cs** /hɪ'sterɪks/ *npl* histerismo *m*. **have** ~**cs** ponerse histérico; (*laugh*) morir de risa

I

I /aɪ/ *pron* yo
ice /aɪs/ *n* hielo *m*. • *vt* helar; glasear ‹*cake*›. • *vi*. ~ **(up)** helarse. ~**berg** *n* iceberg *m*, témpano *m*. ~**-cream** *n* helado *m*. ~**-cube** *n* cubito *m* de hielo. ~ **hockey** *n* hockey *m* sobre hielo
Iceland /'aɪslənd/ *n* Islandia *f*. ~**er** *n* islandés *m*. ~**ic** /-'lændɪk/ *a* islandés
ice lolly /aɪs'lɒlɪ/ polo *m*, paleta *f* (*LAm*)
icicle /'aɪsɪkl/ *n* carámbano *m*
icing /'aɪsɪŋ/ *n* (*sugar*) azúcar *m* glaseado
icon /'aɪkɒn/ *n* icono *m*
icy /'aɪsɪ/ *a* (**-ier**, **-iest**) glacial
idea /ai'dɪə/ *n* idea *f*
ideal /ai'dɪəl/ *a* ideal. • *n* ideal *m*. ~**ism** *n* idealismo *m*. ~**ist** *n* idealista *m* & *f*. ~**istic** /-'lɪstɪk/ *a* idealista. ~**ize** *vt* idealizar. ~**ly** *adv* idealmente
identical /ai'dentɪkl/ *a* idéntico
identif|ication /aɪdentɪfɪ'keɪʃn/ *n* identificación *f*. ~**y** /aɪ'dentɪfaɪ/ *vt* identificar. • *vi*. ~**y with** identificarse con
identikit /aɪ'dentɪkɪt/ *n* retrato-robot *m*
identity /aɪ'dentɪtɪ/ *n* identidad *f*
ideolog|ical /aɪdɪə'lɒdʒɪkl/ *a* ideológico. ~**y** /aɪdɪ'ɒlədʒɪ/ *n* ideología *f*
idiocy /'ɪdɪəsɪ/ *n* idiotez *f*
idiom /'ɪdɪəm/ *n* locución *f*. ~**atic** /-'mætɪk/ *a* idiomático
idiosyncrasy /ɪdɪəʊ'sɪnkrəsɪ/ *n* idiosincrasia *f*
idiot /'ɪdɪət/ *n* idiota *m* & *f*. ~**ic** /-'ɒtɪk/ *a* idiota
idle /'aɪdl/ *a* (**-er**, **-est**) ocioso; (*lazy*) holgazán; (*out of work*) desocupado; ‹*machine*› parado. • *vi* ‹*engine*› marchar en vacío. • *vt*. ~ **away** perder. ~**ness** *n* ociosidad *f*. ~**r** /-ə(r)/ *n* ocioso *m*
idol /'aɪdl/ *n* ídolo *m*. ~**ize** *vt* idolatrar
idyllic /ɪ'dɪlɪk/ *a* idílico
i.e. /aɪ'i:/ *abbr* (*id est*) es decir
if /ɪf/ *conj* si
igloo /'iglu:/ *n* iglú *m*
ignite /ɪg'naɪt/ *vt* encender. • *vi* encenderse
ignition /ɪg'nɪʃn/ *n* ignición *f*; (*auto*) encendido *m*. ~ **(switch)** *n* contacto *m*
ignoramus /ɪgnə'reɪməs/ *n* (*pl* **-muses**) ignorante
ignoran|ce /'ɪgnərəns/ *n* ignorancia *f*. ~**t** *a* ignorante. ~**tly** *adv* por ignorancia
ignore /ɪg'nɔ:(r)/ *vt* no hacer caso de
ilk /ɪlk/ *n* ralea *f*
ill /ɪl/ *a* enfermo; (*bad*) malo. ~ **will** *n* mala voluntad *f*. • *adv* mal. ~ **at**

ease inquieto. ● *n* mal *m*. **~-advised** *a* imprudente. **~-bred** *a* mal educado

illegal /ɪ'li:gl/ *a* ilegal

illegible /ɪ'ledʒəbl/ *a* ilegible

illegitima|cy /ɪlɪ'dʒɪtɪməsɪ/ *n* ilegitimidad *f*. **~te** *a* ilegítimo

ill: **~-fated** *a* malogrado. **~-gotten** *a* mal adquirido

illitera|cy /ɪ'lɪtərəsɪ/ *n* analfabetismo *m*. **~te** *a* & *n* analfabeto (*m*)

ill: **~-natured** *a* poco afable. **~ness** *n* enfermedad *f*

illogical /ɪ'lɒdʒɪkl/ *a* ilógico

ill: **~-starred** *a* malogrado. **~-treat** *vt* maltratar

illuminat|e /ɪ'lu:mɪneɪt/ *vt* iluminar. **~ion** /-'neɪʃn/ *n* iluminación *f*

illus|ion /ɪ'lu:ʒn/ *n* ilusión *f*. **~sory** *a* ilusorio

illustrat|e /'ɪləstreɪt/ *vt* ilustrar. **~ion** *n* (*example*) ejemplo *m*; (*picture in book*) grabado *m*, lámina *f*. **~ive** *a* ilustrativo

illustrious /ɪ'lʌstrɪəs/ *a* ilustre

image /'ɪmɪdʒ/ *n* imagen *f*. **~ry** *n* imágenes *fpl*

imagin|able /ɪ'mædʒɪnəbl/ *a* imaginable. **~ary** *a* imaginario. **~ation** /-'neɪʃn/ *n* imaginación *f*. **~ative** *a* imaginativo. **~e** *vt* imaginar(se)

imbalance /ɪm'bæləns/ *n* desequilibrio *m*

imbecil|e /'ɪmbəsi:l/ *a* & *n* imbécil (*m* & *f*). **~ity** /-'sɪlətɪ/ *n* imbecilidad *f*

imbibe /ɪm'baɪb/ *vt* embeber; (*drink*) beber

imbue /ɪm'bju:/ *vt* empapar (**with** de)

imitat|e /'ɪmɪteɪt/ *vt* imitar. **~ion** /-'teɪʃn/ *n* imitación *f*. **~or** *n* imitador *m*

immaculate /ɪ'mækjʊlət/ *a* inmaculado

immaterial /ɪmə'tɪərɪəl/ *a* inmaterial; (*unimportant*) insignificante

immature /ɪmə'tjʊə(r)/ *a* inmaduro

immediate /ɪ'mi:dɪət/ *a* inmediato. **~ly** *adv* inmediatamente. **~ly you hear me** en cuanto me oigas. ● *conj* en cuanto (+ *subj*)

immens|e /ɪ'mens/ *a* inmenso. **~ely** *adv* inmensamente; (*very much*, *fam*) muchísimo. **~ity** *n* inmensidad *f*

immers|e /ɪ'mɜ:s/ *vt* sumergir. **~ion** /ɪ'mɜ:ʃn/ *n* inmersión *f*. **~ion heater** *n* calentador *m* de inmersión

immigra|nt /'ɪmɪgrənt/ *a* & *n* inmigrante (*m* & *f*). **~te** *vi* inmigrar. **~tion** /-'greɪʃn/ *n* inmigración *f*

imminen|ce /'ɪmɪnəns/ *n* inminencia *f*. **~t** *a* inminente

immobil|e /ɪ'məʊbaɪl/ *a* inmóvil. **~ize** /-bɪlaɪz/ *vt* inmovilizar

immoderate /ɪ'mɒdərət/ *a* inmoderado

immodest /ɪ'mɒdɪst/ *a* inmodesto

immoral /ɪ'mɒrəl/ *a* inmoral. **~ity** /ɪmə'rælətɪ/ *n* inmoralidad *f*

immortal /ɪ'mɔ:tl/ *a* inmortal. **~ity** /-'tælətɪ/ *n* inmortalidad *f*. **~ize** *vt* inmortalizar

immun|e /ɪ'mju:n/ *a* inmune (**from, to** a, contra). **~ity** *n* inmunidad *f*. **~ization** /ɪmjʊnaɪ'zeɪʃn/ *n* inmunización *f*. **~ize** *vt* inmunizar

imp /ɪmp/ *n* diablillo *m*

impact /'ɪmpækt/ *n* impacto *m*

impair /ɪm'peə(r)/ *vt* perjudicar

impale /ɪm'peɪl/ *vt* empalar

impart /ɪm'pɑ:t/ *vt* comunicar

impartial /ɪm'pɑ:ʃl/ *a* imparcial. **~ity** /-ɪ'ælətɪ/ *n* imparcialidad *f*

impassable /ɪm'pɑ:səbl/ *a* ‹*barrier etc*› infranqueable; ‹*road*› impracticable

impasse /æm'pɑ:s/ *n* callejón *m* sin salida

impassioned /ɪm'pæʃnd/ *a* apasionado

impassive /ɪm'pæsɪv/ *a* impasible

impatien|ce /ɪm'peɪʃəns/ *n* impaciencia *f*. **~t** *a* impaciente. **~tly** *adv* con impaciencia

impeach /ɪm'pi:tʃ/ *vt* acusar

impeccable /ɪm'pekəbl/ *a* impecable

impede /ɪm'pi:d/ *vt* estorbar

impediment /ɪm'pedɪmənt/ *n* obstáculo *m*. **(speech) ~** *n* defecto *m* del habla

impel /ɪm'pel/ *vt* (*pt* **impelled**) impeler

impending /ɪm'pendɪŋ/ *a* inminente

impenetrable /ɪm'penɪtrəbl/ *a* impenetrable

imperative /ɪm'perətɪv/ *a* imprescindible. ● *n* (*gram*) imperativo *m*

imperceptible /ɪmpə'septəbl/ *a* imperceptible

imperfect /ɪm'pɜ:fɪkt/ *a* imperfecto. **~ion** /ə-'fekʃn/ *n* imperfección *f*

imperial /ɪm'pɪərɪəl/ *a* imperial. **~ism** *n* imperialismo *m*

imperil /ɪm'perəl/ *vt* (*pt* **imperilled**) poner en peligro

imperious /ɪm'pɪərɪəs/ *a* imperioso

impersonal /ɪm'pɜ:sənl/ *a* impersonal

impersonat|e /ɪm'pɜ:səneɪt/ *vt* hacerse pasar por; (*mimic*) imitar. **~ion** /-'neɪʃn/ *n* imitación *f*. **~or** *n* imitador *m*

impertinen|ce /ɪm'pɜ:tɪnəns/ *n* impertinencia *f*. **~t** *a* impertinente. **~tly** *adv* impertinentemente

impervious /ɪm'pɜ:vɪəs/ *a*. **~ to** impermeable a; (*fig*) insensible a

impetuous /ɪm'petjʊəs/ *a* impetuoso

impetus /'ɪmpɪtəs/ *n* ímpetu *m*

impinge /ɪm'pɪndʒ/ *vi*. **~ on** afectar a

impish /'ɪmpɪʃ/ *a* travieso

implacable /ɪm'plækəbl/ *a* implacable

implant /ɪm'plɑ:nt/ *vt* implantar

implement /'ɪmplɪmənt/ *n* herramienta *f*. /'ɪmplɪment/ *vt* realizar

implicat|e /'ɪmplɪkeɪt/ *vt* implicar. **~ion** /-'keɪʃn/ *n* implicación *f*

implicit /ɪm'plɪsɪt/ *a* (*implied*) implícito; (*unquestioning*) absoluto

implied /ɪm'plaɪd/ *a* implícito

implore /ɪm'plɔ:(r)/ *vt* implorar

imply /ɪm'plaɪ/ *vt* implicar; (*mean*) querer decir; (*insinuate*) dar a entender

impolite /ɪmpə'laɪt/ *a* mal educado

imponderable /ɪm'pɒndərəbl/ *a & n* imponderable (*m*)

import /ɪm'pɔ:t/ *vt* importar. /'ɪmpɔ:t/ *n* (*article*) importación *f*; (*meaning*) significación *f*

importan|ce /ɪm'pɔ:təns/ *n* importancia *f*. **~t** *a* importante

importation /ɪmpɔ:'teɪʃn/ *n* importación *f*

importer /ɪm'pɔ:tə(r)/ *n* importador *m*

impose /ɪm'pəʊz/ *vt* imponer. ● *vi*. **~ on** abusar de la amabilidad de

imposing /ɪm'pəʊzɪŋ/ *a* imponente

imposition /ɪmpə'zɪʃn/ *n* imposición *f*; (*fig*) molestia *f*

impossib|ility /ɪmpɒsə'bɪlətɪ/ *n* imposibilidad *f*. **~le** *a* imposible

impostor /ɪm'pɒstə(r)/ *n* impostor *m*

impoten|ce /'ɪmpətəns/ *n* impotencia *f*. **~t** *a* impotente

impound /ɪm'paʊnd/ *vt* confiscar

impoverish /ɪm'pɒvərɪʃ/ *vt* empobrecer

impracticable /ɪm'præktɪkəbl/ *a* impracticable

impractical /ɪm'præktɪkl/ *a* poco práctico

imprecise /ɪmprɪ'saɪs/ *a* impreciso

impregnable /ɪm'pregnəbl/ *a* inexpugnable

impregnate /'ɪmpregneɪt/ *vt* impregnar (**with** de)

impresario /ɪmprɪ'sɑ:rɪəʊ/ *n* (*pl* **-os**) empresario *m*

impress /ɪm'pres/ *vt* impresionar; (*imprint*) imprimir. **~ on s.o.** hacer entender a uno

impression /ɪm'preʃn/ *n* impresión *f*. **~able** *a* impresionable

impressive /ɪm'presɪv/ *a* impresionante

imprint /'ɪmprɪnt/ *n* impresión *f*. /ɪm'prɪnt/ *vt* imprimir

imprison /ɪm'prɪzn/ *vt* encarcelar. **~ment** *n* encarcelamiento *m*

improbab|ility /ɪmprɒbə'bɪlətɪ/ *n* improbabilidad *f*. **~le** *a* improbable

impromptu /ɪm'prɒmptju:/ *a* improvisado. ● *adv* de improviso

improper /ɪm'prɒpə(r)/ *a* impropio; (*incorrect*) incorrecto

impropriety /ɪmprə'praɪətɪ/ *n* inconveniencia *f*

improve /ɪm'pru:v/ *vt* mejorar. ● *vi* mejorar(se). **~ment** *n* mejora *f*

improvis|ation /ɪmprəvaɪ'zeɪʃn/ *n* improvisación *f*. **~e** *vt/i* improvisar

imprudent /ɪm'pru:dənt/ *a* imprudente

impuden|ce /'ɪmpjʊdəns/ *n* insolencia *f*. **~t** *a* insolente

impulse /'ɪmpʌls/ *n* impulso *m*. **on ~** sin reflexionar

impulsive /ɪm'pʌlsɪv/ *a* irreflexivo. **~ly** *adv* sin reflexionar

impunity /ɪm'pju:nətɪ/ *n* impunidad *f*. **with ~** impunemente

impur|e /ɪm'pjʊə(r)/ *a* impuro. **~ity** *n* impureza *f*

impute /ɪm'pu:t/ *vt* imputar

in /ɪn/ *prep* en, dentro de. **~ a firm manner** de una manera terminante. **~ an hour('s time)** dentro de una hora. **~ doing** al hacer. **~ so far as** en cuanto que. **~ the evening** por la tarde. **~ the main** por la mayor parte. **~ the rain** bajo la lluvia. **~ the sun** al sol. **one ~ ten** uno de cada diez. **the best ~** el mejor de. ● *adv* (*inside*) dentro; (*at home*) en casa; (*in fashion*) de moda. ● *n*. **the ~s and outs of** los detalles *mpl* de

inability /ɪnə'bɪlətɪ/ *n* incapacidad *f*

inaccessible /ɪnæk'sesəbl/ *a* inaccesible

inaccura|cy /ɪn'ækjʊrəsɪ/ *n* inexactitud *f*. **~te** *a* inexacto

inaction /ɪn'ækʃn/ *n* inacción *f*

inactiv|e /ɪn'æktɪv/ *a* inactivo. **~ity** /-'tɪvətɪ/ *n* inactividad *f*

inadequa|cy /ɪn'ædɪkwəsɪ/ *a* insuficiencia *f*. **~te** *a* insuficiente

inadmissible /ɪnəd'mɪsəbl/ *a* inadmisible

inadvertently /ɪnəd'vɜ:təntlɪ/ *adv* por descuido

inadvisable /ɪnəd'vaɪzəbl/ *a* no aconsejable

inane /ɪ'neɪn/ *a* estúpido

inanimate /ɪn'ænɪmət/ *a* inanimado

inappropriate /ɪnə'prəʊpriət/ *a* inoportuno

inarticulate /ɪnɑ:'tɪkjʊlət/ *a* incapaz de expresarse claramente

inasmuch as /ɪnəz'mʌtʃəz/ *adv* ya que

inattentive /ɪnə'tentɪv/ *a* desatento

inaudible /ɪn'ɔ:dəbl/ *a* inaudible

inaugural /ɪ'nɔ:gjʊrəl/ *a* inaugural

inaugurat|e /ɪ'nɔ:gjʊreɪt/ *vt* inaugurar. **~ion** /-'reɪʃn/ *n* inauguración *f*

inauspicious /ɪnɔ:'spɪʃəs/ *a* poco propicio

inborn /'ɪnbɔ:n/ *a* innato

inbred /ɪn'bred/ *a* (*inborn*) innato

incalculable /ɪn'kælkjʊləbl/ *a* incalculable

incapab|ility /ɪnkeɪpə'bɪlətɪ/ *n* incapacidad *f*. **~le** *a* incapaz

incapacit|ate /ɪnkə'pæsɪteɪt/ *vt* incapacitar. **~y** *n* incapacidad *f*

incarcerat|e /ɪn'kɑ:səreɪt/ *vt* encarcelar. **~ion** /-'reɪʃn/ *n* encarcelamiento *m*

incarnat|e /ɪn'kɑ:nət/ *a* encarnado. **~ion** /-'neɪʃn/ *n* encarnación *f*

incautious /ɪn'kɔ:ʃəs/ *a* incauto. **~ly** *adv* incautamente

incendiary /ɪn'sendɪərɪ/ *a* incendiario. ● *n* (*person*) incendiario *m*; (*bomb*) bomba *f* incendiaria

incense[1] /'ɪnsens/ *n* incienso *m*

incense[2] /ɪn'sens/ *vt* enfurecer

incentive /ɪn'sentɪv/ *n* incentivo *m*; (*payment*) prima *f* de incentivo

inception /ɪn'sepʃn/ *n* principio *m*

incertitude /ɪn'sɜ:tɪtju:d/ *n* incertidumbre *f*

incessant /ɪn'sesnt/ *a* incesante. **~ly** *adv* sin cesar

incest /'ɪnsest/ *n* incesto *m*. **~uous** /-'sestjʊəs/ *a* incestuoso

inch /ɪntʃ/ *n* pulgada *f* (= *2,54cm*). ● *vi* avanzar palmo a palmo

incidence /'ɪnsɪdəns/ *n* frecuencia *f*

incident /'ɪnsɪdənt/ *n* incidente *m*

incidental /ɪnsɪ'dentl/ *a* fortuito. **~ly** *adv* incidentemente; (*by the way*) a propósito

incinerat|e /ɪn'sɪnəreɪt/ *vt* incinerar. **~or** *n* incinerador *m*

incipient /ɪn'sɪpɪənt/ *a* incipiente

incision /ɪn'sɪʒn/ *n* incisión *f*

incisive /ɪn'saɪsɪv/ *a* incisivo

incite /ɪn'saɪt/ *vt* incitar. **~ment** *n* incitación *f*

inclement /ɪn'klemənt/ *a* inclemente

inclination /ɪnklɪ'neɪʃn/ *n* inclinación *f*

incline[1] /ɪn'klaɪn/ *vt* inclinar. ● *vi* inclinarse. **be ~d to** tener tendencia a

incline[2] /'ɪnklaɪn/ *n* cuesta *f*

inclu|de /ɪn'klu:d/ *vt* incluir. **~ding** *prep* incluso. **~sion** /-ʒn/ *n* inclusión *f*

inclusive /ɪn'klu:sɪv/ *a* inclusivo. **be ~ of** incluir. ● *adv* inclusive

incognito /ɪnkɒg'ni:təʊ/ *adv* de incógnito

incoherent /ɪnkəʊ'hɪərənt/ *a* incoherente

income /'ɪnkʌm/ *n* ingresos *mpl*. **~ tax** *n* impuesto *m* sobre la renta

incoming /'ɪnkʌmɪŋ/ *a* ‹*tide*› ascendente; ‹*tenant etc*› nuevo

incomparable /ɪn'kɒmpərəbl/ *a* incomparable

incompatible /ɪnkəm'pætəbl/ *a* incompatible

incompeten|ce /ɪn'kɒmpɪtəns/ *n* incompetencia *f*. **~t** *a* incompetente

incomplete /ɪnkəm'pli:t/ *a* incompleto

incomprehensible /ɪnkɒmprɪ'hensəbl/ *a* incomprensible

inconceivable /ɪnkən'si:vəbl/ *a* inconcebible

inconclusive /ɪnkən'klu:sɪv/ *a* poco concluyente

incongruous /ɪn'kɒŋgrʊəs/ *a* incongruente

inconsequential /ɪnkɒnsɪ'kwenʃl/ *a* sin importancia

inconsiderate /ɪnkən'sɪdərət/ *a* desconsiderado

inconsisten|cy /ɪnkən'sɪstənsɪ/ *n* inconsecuencia *f*. **~t** *a* inconsecuente. **be ~t with** no concordar con

inconspicuous /ɪnkən'spɪkjʊəs/ *a* que no llama la atención. **~ly** *adv* sin llamar la atención

incontinen|ce /ɪn'kɒntɪnəns/ *a* incontinencia *f*. **~t** *a* incontinente

inconvenien|ce /ɪnkən'vi:nɪəns/ *a* incomodidad *f*; (*drawback*) inconveniente m. **~t** *a* incómodo; ‹*time*› inoportuno

incorporat|e /ɪn'kɔ:pəreɪt/ *vt* incorporar; (*include*) incluir. **~ion** /-'reɪʃn/ *n* incorporación *f*

incorrect /ɪnkə'rekt/ *a* incorrecto

incorrigible /ɪn'kɒrɪdʒəbl/ *a* incorregible

incorruptible /ɪnkə'rʌptəbl/ *a* incorruptible

increase /'ɪnkri:s/ *n* aumento *m* (**in**, **of** de). /ɪn'kri:s/ *vt/i* aumentar

increasing /ɪn'kri:sɪŋ/ *a* creciente. **~ly** *adv* cada vez más

incredible /ɪn'kredəbl/ *a* increíble

incredulous /ɪn'kredjʊləs/ *a* incrédulo

increment /'ɪnkrɪmənt/ *n* aumento *m*

incriminat|e /ɪn'krɪmɪneɪt/ *vt* acriminar. **~ing** *a* acriminador

incubat|e /'ɪŋkjʊbeɪt/ *vt* incubar. **~ion** /-'beɪʃn/ *n* incubación *f*. **~or** *n* incubadora *f*

inculcate /'ɪnkʌlkeɪt/ *vt* inculcar

incumbent /ɪn'kʌmbənt/ *n* titular. ● *a*. **be ~ on** incumbir a

incur /ɪn'kɜ:(r)/ *vt* (*pt* **incurred**) incurrir en; contraer ‹*debts*›

incurable /ɪn'kjʊərəbl/ *a* incurable

incursion /ɪn'kɜ:ʃn/ *n* incursión *f*

indebted /ɪn'detɪd/ *a*. **~ to s.o.** estar en deuda con uno

indecen|cy /ɪn'di:snsɪ/ *n* indecencia *f*. **~t** *a* indecente

indecisi|on /ɪndɪ'sɪʒn/ *n* indecisión *f*. **~ve** /ɪndɪ'saɪsɪv/ *a* indeciso

indeed /ɪn'dɪ:d/ *adv* en efecto; (*really?*) ¿de veras?

indefatigable /ɪndɪ'fætɪgəbl/ *a* incansable

indefinable /ɪndɪ'faɪnəbl/ *a* indefinible

indefinite /ɪn'defɪnət/ *a* indefinido. **~ly** *adv* indefinidamente

indelible /ɪn'delɪbl/ *a* indeleble

indemni|fy /ɪn'demnɪfaɪ/ *vt* indemnizar. **~ty** /-ətɪ/ *n* indemnización *f*

indent /ɪn'dent/ *vt* endentar ‹*text*›. **~ation** /-'teɪʃn/ *n* mella *f*

independen|ce /ɪndɪ'pendəns/ *n* independencia *f*. **~t** *a* independiente. **~tly** *adv* independientemente. **~tly of** independientemente de

indescribable /ɪndɪ'skraɪbəbl/ *a* indescriptible

indestructible /ɪndɪ'strʌktəbl/ *a* indestructible

indeterminate /ɪndɪ'tɜ:mɪnət/ *a* indeterminado

index /'ɪndeks/ *n* (*pl* **indexes**) índice *m*. ● *vt* poner índice a; (*enter in the/an index*) poner en el/un índice. **~ finger** *n* (dedo *m*) índice *m*. **~-linked** *a* indexado

India /'ɪndɪə/ *n* la India *f*. **~n** *a & n* indio (*m*). **~n summer** *n* veranillo *m* de San Martín

indicat|e /'ɪndɪkeɪt/ *vt* indicar. **~ion** /-'keɪʃn/ *n* indicación *f*. **~ive** /ɪn'dɪkətɪv/ *a & n* indicativo (*m*). **~or** /'ɪndɪkeɪtə(r)/ *n* indicador *m*

indict /ɪn'daɪt/ *vt* acusar. **~ment** *n* acusación *f*

indifferen|ce /ɪn'dɪfrəns/ *n* indiferencia *f*. **~t** *a* indiferente; (*not good*) mediocre

indigenous /ɪn'dɪdʒɪnəs/ *a* indígena

indigesti|ble /ɪndɪ'dʒestəbl/ *a* indigesto. **~on** /-tʃən/ *n* indigestión *f*

indigna|nt /ɪn'dɪgnənt/ *a* indignado. **~tion** /-'neɪʃn/ *n* indignación *f*

indignity /ɪn'dɪgnətɪ/ *n* indignidad *f*

indigo /'ɪndɪgəʊ/ *n* añil (*m*)

indirect /ɪndɪ'rekt/ *a* indirecto. **~ly** *adv* indirectamente

indiscre|et /ɪndɪ'skri:t/ *a* indiscreto. **~tion** /-'kreʃn/ *n* indiscreción *f*

indiscriminate /ɪndɪ'skrɪmɪnət/ *a* indistinto. **~ly** *adv* indistintamente

indispensable /ɪndɪ'spensəbl/ *a* imprescindible

indispos|ed /ɪndɪ'spəʊzd/ *a* indispuesto. **~ition** /-ə'zɪʃn/ *n* indisposición *f*

indisputable /ɪndɪ'spju:təbl/ *a* indiscutible

indissoluble /ɪndɪ'sɒljʊbl/ *a* indisoluble

indistinct /ɪndɪ'stɪŋkt/ *a* indistinto

indistinguishable /ɪndɪ'stɪŋgwɪʃəbl/ *a* indistinguible

individual /ɪndɪ'vɪdjʊəl/ *a* individual. ● *n* individuo *m*. **~ist** *n* individualista *m & f*. **~ity** *n*

individualidad *f.* **~ly** *adv* individualmente
indivisible /ɪndɪ'vɪzəbl/ *a* indivisible
Indo-China /ɪndəʊ'tʃaɪnə/ *n* Indochina *f*
indoctrinat|e /ɪn'dɒktrɪneɪt/ *vt* adoctrinar. **~ion** /-'neɪʃn/ *n* adoctrinamiento *m*
indolen|ce /'ɪndələns/ *n* indolencia *f.* **~t** *a* indolente
indomitable /ɪn'dɒmɪtəbl/ *a* indomable
Indonesia /ɪndəʊ'ni:zɪə/ *n* Indonesia *f.* **~n** *a* & *n* indonesio (*m*)
indoor /'ɪndɔ:(r)/ *a* interior; ‹*clothes etc*› de casa; (*covered*) cubierto. **~s** *adv* dentro; (*at home*) en casa
induce /ɪn'dju:s/ *vt* inducir; (*cause*) provocar. **~ment** *n* incentivo *m*
induct /ɪn'dʌkt/ *vt* instalar; (*mil, Amer*) incorporar
indulge /ɪn'dʌldʒ/ *vt* satisfacer ‹*desires*›; complacer ‹*person*›. ● *vi.* **~ in** entregarse a. **~nce** /ɪn'dʌldʒəns/ *n* (*of desires*) satisfacción *f*; (*relig*) indulgencia *f.* **~nt** *a* indulgente
industrial /ɪn'dʌstrɪəl/ *a* industrial; ‹*unrest*› laboral. **~ist** *n* industrial *m* & *f.* **~ized** *a* industrializado
industrious /ɪn'dʌstrɪəs/ *a* trabajador
industry /'ɪndəstrɪ/ *n* industria *f*; (*zeal*) aplicación *f*
inebriated /ɪ'ni:brɪeɪtɪd/ *a* borracho
inedible /ɪn'edɪbl/ *a* incomible
ineffable /ɪn'efəbl/ *a* inefable
ineffective /ɪnɪ'fektɪv/ *a* ineficaz; ‹*person*› incapaz
ineffectual /ɪnɪ'fektjʊəl/ *a* ineficaz
inefficien|cy /ɪnɪ'fɪʃnsɪ/ *n* ineficacia *f*; (of *person*) incompetencia *f.* **~t** *a* ineficaz; ‹*person*› incompetente
ineligible /ɪn'elɪdʒəbl/ *a* inelegible. **be ~ for** no tener derecho a
inept /ɪ'nept/ *a* inepto
inequality /ɪnɪ'kwɒlətɪ/ *n* desigualdad *f*
inert /ɪ'nɜ:t/ *a* inerte
inertia /ɪ'nɜ:ʃə/ *n* inercia *f*
inescapable /ɪnɪ'skeɪpəbl/ *a* ineludible
inestimable /ɪn'estɪməbl/ *a* inestimable
inevitabl|e /ɪn'evɪtəbl/ *a* inevitable. **~ly** *adv* inevitablemente
inexact /ɪnɪg'zækt/ *a* inexacto
inexcusable /ɪnɪk'skju:səbl/ *a* imperdonable
inexhaustible /ɪnɪg'zɔ:stəbl/ *a* inagotable
inexorable /ɪn'eksərəbl/ *a* inexorable
inexpensive /ɪnɪk'spensɪv/ *a* económico, barato
inexperience /ɪnɪk'spɪərɪəns/ *n* falta *f* de experiencia. **~d** *a* inexperto
inexplicable /ɪnɪk'splɪkəbl/ *a* inexplicable
inextricable /ɪnɪk'strɪkəbl/ *a* inextricable
infallib|ility /ɪn'fæləbɪlətɪ/ *n* infalibilidad *f.* **~le** *a* infalible
infam|ous /'ɪnfəməs/ *a* infame. **~y** *n* infamia *f*
infan|cy /'ɪnfənsɪ/ *n* infancia *f.* **~t** *n* niño *m.* **~tile** /'ɪnfəntaɪl/ *a* infantil
infantry /'ɪnfəntrɪ/ *n* infantería *f*
infatuat|ed /ɪn'fætjʊeɪtɪd/ *a.* **be ~ed with** encapricharse por. **~ion** /-'eɪʃn/ *n* encaprichamiento *m*
infect /ɪn'fekt/ *vt* infectar; (*fig*) contagiar. **~ s.o. with** contagiar a uno. **~ion** /-'fekʃn/ *n* infección *f*; (*fig*) contagio *m.* **~ious** /ɪn'fekʃəs/ *a* contagioso
infer /ɪn'fɜ:(r)/ *vt* (*pt* **inferred**) deducir. **~ence** /'ɪnfərəns/ *n* deducción *f*
inferior /ɪn'fɪərɪə(r)/ *a* inferior. ● *n* inferior *m* & *f.* **~ity** /-'ɒrətɪ/ *n* inferioridad *f*
infernal /ɪn'fɜ:nl/ *a* infernal. **~ly** *adv* (*fam*) atrozmente
inferno /ɪn'fɜ:nəʊ/ *n* (*pl* **-os**) infierno *m*
infertil|e /ɪn'fɜ:taɪl/ *a* estéril. **~ity** /-'tɪlətɪ/ *n* esterilidad *f*
infest /ɪn'fest/ *vt* infestar. **~ation** /-'steɪʃn/ *n* infestación *f*
infidelity /ɪnfɪ'delətɪ/ *n* infidelidad *f*
infighting /'ɪnfaɪtɪŋ/ *n* lucha *f* cuerpo a cuerpo; (*fig*) riñas *fpl* (internas)
infiltrat|e /ɪnfɪl'treɪt/ *vt* infiltrar. ● *vi* infiltrarse. **~ion** /-'treɪʃn/ *n* infiltración *f*
infinite /'ɪnfɪnət/ *a* infinito. **~ly** *adv* infinitamente
infinitesimal /ɪnfɪnɪ'tesɪml/ *a* infinitesimal
infinitive /ɪn'fɪnətɪv/ *n* infinitivo *m*
infinity /ɪn'fɪnətɪ/ *n* (*infinite distance*) infinito *m*; (*infinite quantity*) infinidad *f*
infirm /ɪn'fɜ:m/ *a* enfermizo
infirmary /ɪn'fɜ:mərɪ/ *n* hospital *m*; (*sick bay*) enfermería *f*

infirmity /ɪn'fɜ:mətɪ/ *n* enfermedad *f*; (*weakness*) debilidad *f*
inflam|e /ɪn'fleɪm/ *vt* inflamar. **~mable** /ɪn'flæməbl/ *a* inflamable. **~mation** /-ə'meɪʃn/ *n* inflamación *f*. **~matory** /ɪn'flæmətərɪ/ *a* inflamatorio
inflate /ɪn'fleɪt/ *vt* inflar
inflation /ɪn'fleɪʃn/ *n* inflación *f*. **~ary** *a* inflacionario
inflection /ɪn'flekʃn/ *n* inflexión *f*
inflexible /ɪn'fleksəbl/ *a* inflexible
inflict /ɪn'flɪkt/ *vt* infligir (**on** a)
inflow /'ɪnfləʊ/ *n* afluencia *f*
influence /'ɪnflʊəns/ *n* influencia *f*. **under the ~** (*drunk, fam*) borracho. ● *vt* influir, influenciar (*esp LAm*)
influential /ɪnflʊ'enʃl/ *a* influyente
influenza /ɪnflʊ'enzə/ *n* gripe *f*
influx /'ɪnflʌks/ *n* afluencia *f*
inform /ɪn'fɔ:m/ *vt* informar. **keep ~ed** tener al corriente
informal /ɪn'fɔ:ml/ *a* (*simple*) sencillo, sin ceremonia; (*unofficial*) oficioso. **~ity** /'mælətɪ/ *n* falta *f* de ceremonia. **~ly** *adv* sin ceremonia
inform|ant /ɪn'fɔ:mənt/ *n* informador *m*. **~ation** /ɪnfə'meɪʃn/ *n* información *f*. **~ative** /ɪn'fɔ:mətɪv/ *a* informativo. **~er** /ɪn'fɔ:mə(r)/ *n* denunciante *m*
infra-red /ɪnfrə'red/ *a* infrarrojo
infrequent /ɪn'fri:kwənt/ *a* poco frecuente. **~ly** *adv* raramente
infringe /ɪn'frɪndʒ/ *vt* infringir. **~ on** usurpar. **~ment** *n* infracción *f*
infuriate /ɪn'fjʊərɪeɪt/ *vt* enfurecer
infus|e /ɪn'fju:z/ *vt* infundir. **~ion** /-ʒn/ *n* infusión *f*
ingen|ious /ɪn'dʒi:nɪəs/ *a* ingenioso. **~uity** /ɪndʒɪ'nju:ətɪ/ *n* ingeniosidad *f*
ingenuous /ɪn'dʒenjʊəs/ *a* ingenuo
ingest /ɪn'dʒest/ *vt* ingerir
ingot /'ɪŋgət/ *n* lingote *m*
ingrained /ɪn'greɪnd/ *a* arraigado
ingratiate /ɪn'greɪʃɪeɪt/ *vt*. **~ o.s. with** congraciarse con
ingratitude /ɪn'grætɪtju:d/ *n* ingratitud *f*
ingredient /ɪn'gri:dɪənt/ *n* ingrediente *m*
ingrowing /'ɪngrəʊɪŋ/ *a*. **~ nail** *n* uñero *m*, uña *f* encarnada
inhabit /ɪn'hæbɪt/ *vt* habitar. **~able** *a* habitable. **~ant** *n* habitante *m*
inhale /ɪn'heɪl/ *vt* aspirar. ● *vi* (*tobacco*) aspirar el humo
inherent /ɪn'hɪərənt/ *a* inherente. **~ly** *adv* intrínsecamente
inherit /ɪn'herɪt/ *vt* heredar. **~ance** *n* herencia *f*
inhibit /ɪn'hɪbɪt/ *vt* inhibir. **be ~ed** tener inhibiciones. **~ion** /-'bɪʃn/ *n* inhibición *f*
inhospitable /ɪnhə'spɪtəbl/ *a* ‹*place*› inhóspito; ‹*person*› inhospitalario
inhuman /ɪn'hju:mən/ *a* inhumano. **~e** /ɪnhju:'meɪn/ *a* inhumano. **~ity** /ɪnhju:'mænətɪ/ *n* inhumanidad *f*
inimical /ɪ'nɪmɪkl/ *a* hostil
inimitable /ɪ'nɪmɪtəbl/ *a* inimitable
iniquit|ous /ɪ'nɪkwɪtəs/ *a* inicuo. **~y** /-ətɪ/ *n* iniquidad *f*
initial /ɪ'nɪʃl/ *n* inicial *f*. ● *vt* (*pt* **initialled**) firmar con iniciales. **he ~led the document** firmó el documento con sus iniciales. ● *a* inicial. **~ly** *adv* al principio
initiat|e /ɪ'nɪʃɪeɪt/ *vt* iniciar; promover ‹*scheme etc*›. **~ion** /-'eɪʃn/ *n* iniciación *f*
initiative /ɪ'nɪʃətɪv/ *n* iniciativa *f*
inject /ɪn'dʒekt/ *vt* inyectar; (*fig*) injertar ‹*new element*›. **~ion** /-ʃn/ *n* inyección *f*
injunction /ɪn'dʒʌŋkʃn/ *n* (*court order*) entredicho *m*
injur|e /'ɪndʒə(r)/ *vt* (*wound*) herir; (*fig, damage*) perjudicar. **~y** /'ɪndʒərɪ/ *n* herida *f*; (*damage*) perjuicio *m*
injustice /ɪn'dʒʌstɪs/ *n* injusticia *f*
ink /ɪŋk/ *n* tinta *f*
inkling /'ɪŋklɪŋ/ *n* atisbo *m*
ink: **~-well** *n* tintero *m*. **~y** *a* manchado de tinta
inland /'ɪnlənd/ *a* interior. ● *adv* tierra adentro. **I~ Revenue** *n* Hacienda *f*
in-laws /'ɪnlɔ:z/ *npl* parientes *mpl* políticos
inlay /ɪn'leɪ/ *vt* (*pt* **inlaid**) taracear, incrustar. /'ɪnleɪ/ *n* taracea *f*, incrustación *f*
inlet /'ɪnlet/ *n* ensenada *f*; (*tec*) entrada *f*
inmate /'ɪnmeɪt/ *n* (*of asylum*) internado *m*; (*of prison*) preso *m*
inn /ɪn/ *n* posada *f*
innards /'ɪnədz/ *npl* tripas *fpl*
innate /ɪ'neɪt/ *a* innato
inner /'ɪnə(r)/ *a* interior; (*fig*) íntimo. **~most** *a* más íntimo. **~ tube** *n* cámara *f* de aire, llanta *f* (*LAm*)
innings /'ɪnɪŋz/ *n invar* turno *m*
innkeeper /'ɪnki:pə(r)/ *n* posadero *m*

innocen|ce /ˈɪnəsns/ *n* inocencia *f*. **~t** *a* & *n* inocente (*m* & *f*)

innocuous /ɪˈnɒkjʊəs/ *a* inocuo

innovat|e /ˈɪnəveɪt/ *vi* innovar. **~ion** /-ˈveɪʃn/ *n* innovación *f*. **~or** *n* innovador *m*

innuendo /ɪnjuːˈendəʊ/ *n* (*pl* **-oes**) insinuación *f*

innumerable /ɪˈnjuːmərəbl/ *a* innumerable

inoculat|e /ɪˈnɒkjʊleɪt/ *vt* inocular. **~ion** /-ˈleɪʃn/ *n* inoculación *f*

inoffensive /ɪnəˈfensɪv/ *a* inofensivo

inoperative /ɪnˈɒpərətɪv/ *a* inoperante

inopportune /ɪnˈɒpətjuːn/ *a* inoportuno

inordinate /ɪˈnɔːdɪnət/ *a* excesivo. **~ly** *adv* excesivamente

in-patient /ˈɪnpeɪʃnt/ *n* paciente *m* interno

input /ˈɪnpʊt/ *n* (*data*) datos *mpl*; (*comput process*) entrada *f*, input *m*; (*elec*) energía *f*

inquest /ˈɪnkwest/ *n* investigación *f* judicial

inquir|e /ɪnˈkwaɪə(r)/ *vi* preguntar. **~y** *n* (*question*) pregunta *f*; (*investigation*) investigación *f*

inquisition /ɪnkwɪˈzɪʃn/ *n* inquisición *f*

inquisitive /ɪnˈkwɪzətɪv/ *a* inquisitivo

inroad /ˈɪnrəʊd/ *n* incursión *f*

inrush /ˈɪnrʌʃ/ *n* irrupción *f*

insan|e /ɪnˈseɪn/ *a* loco. **~ity** /-ˈsænətɪ/ *n* locura *f*

insanitary /ɪnˈsænɪtərɪ/ *a* insalubre

insatiable /ɪnˈseɪʃəbl/ *a* insaciable

inscri|be /ɪnˈskraɪb/ *vt* inscribir; dedicar ‹*book*›. **~ption** /-ɪpʃn/ *n* inscripción *f*; (*in book*) dedicatoria *f*

inscrutable /ɪnˈskruːtəbl/ *a* inescrutable

insect /ˈɪnsekt/ *n* insecto *m*. **~icide** /ɪnˈsektɪsaɪd/ *n* insecticida *f*

insecur|e /ɪnsɪˈkjʊə(r)/ *a* inseguro. **~ity** *n* inseguridad *f*

insemination /ɪnsemɪˈneɪʃn/ *n* inseminación *f*

insensible /ɪnˈsensəbl/ *a* insensible; (*unconscious*) sin conocimiento

insensitive /ɪnˈsensətɪv/ *a* insensible

inseparable /ɪnˈsepərəbl/ *a* inseparable

insert /ˈɪnsɜːt/ *n* materia *f* insertada. /ɪnˈsɜːt/ *vt* insertar. **~ion** /-ʃn/ *n* inserción *f*

inshore /ɪnˈʃɔː(r)/ *a* costero

inside /ɪnˈsaɪd/ *n* interior *m*. **~ out** al revés; (*thoroughly*) a fondo. ● *a* interior. ● *adv* dentro. ● *prep* dentro de. **~s** *npl* tripas *fpl*

insidious /ɪnˈsɪdɪəs/ *a* insidioso

insight /ˈɪnsaɪt/ *n* (*perception*) penetración *f*, revelación *f*

insignia /ɪnˈsɪgnɪə/ *npl* insignias *fpl*

insignificant /ɪnsɪgˈnɪfɪkənt/ *a* insignificante

insincer|e /ɪnsɪnˈsɪə(r)/ *a* poco sincero. **~ity** /-ˈserətɪ/ *n* falta *f* de sinceridad *f*

insinuat|e /ɪnˈsɪnjʊeɪt/ *vt* insinuar. **~ion** /-ˈeɪʃn/ *n* insinuación *f*

insipid /ɪnˈsɪpɪd/ *a* insípido

insist /ɪnˈsɪst/ *vt/i* insistir. **~ on** insistir en; (*demand*) exigir

insisten|ce /ɪnˈsɪstəns/ *n* insistencia *f*. **~t** *a* insistente. **~tly** *adv* con insistencia

insolen|ce /ˈɪnsələns/ *n* insolencia *f*. **~t** *a* insolente

insoluble /ɪnˈsɒljʊbl/ *a* insoluble

insolvent /ɪnˈsɒlvənt/ *a* insolvente

insomnia /ɪnˈsɒmnɪə/ *n* insomnio *m*. **~c** /-ɪæk/ *n* insomne *m* & *f*

inspect /ɪnˈspekt/ *vt* inspeccionar; revisar ‹*ticket*›. **~ion** /-ʃn/ *n* inspección *f*. **~or** *n* inspector *m*; (*on train, bus*) revisor *m*

inspir|ation /ɪnspəˈreɪʃn/ *n* inspiración *f*. **~e** /ɪnˈspaɪə(r)/ *vt* inspirar

instability /ɪnstəˈbɪlətɪ/ *n* inestabilidad *f*

install /ɪnˈstɔːl/ *vt* instalar. **~ation** /-əˈleɪʃn/ *n* instalación *f*

instalment /ɪnˈstɔːlmənt/ *n* (*payment*) plazo *m*; (*of serial*) entrega *f*

instance /ˈɪnstəns/ *n* ejemplo *m*; (*case*) caso *m*. **for ~** por ejemplo. **in the first ~** en primer lugar

instant /ˈɪnstənt/ *a* inmediato; ‹*food*› instantáneo. ● *n* instante *m*. **~aneous** /ɪnstənˈteɪnɪəs/ *a* instantáneo. **~ly** /ˈɪnstəntlɪ/ *adv* inmediatamente

instead /ɪnˈsted/ *adv* en cambio. **~ of doing** en vez de hacer. **~ of s.o.** en lugar de uno

instep /ˈɪnstep/ *n* empeine *m*

instigat|e /ˈɪnstɪgeɪt/ *vt* instigar. **~ion** /-ˈgeɪʃn/ *n* instigación *f*. **~or** *n* instigador *m*

instil /ɪnˈstɪl/ *vt* (*pt* **instilled**) infundir

instinct /ˈɪnstɪŋkt/ *n* instinto *m*. **~ive** /ɪnˈstɪŋktɪv/ *a* instintivo

institut|e /'ɪnstɪtju:t/ *n* instituto *m*. ● *vt* instituir; iniciar ‹*enquiry etc*›. **~ion** /-'tju:ʃn/ *n* institución *f*

instruct /ɪn'strʌkt/ *vt* instruir; (*order*) mandar. **~ s.o. in sth** enseñar algo a uno. **~ion** /-ʃn/ *n* instrucción *f*. **~ions** /-ʃnz/ *npl* (*for use*) modo *m* de empleo. **~ive** *a* instructivo

instrument /'ɪnstrəmənt/ *n* instrumento *m*. **~al** /ɪnstrə'mentl/ *a* instrumental. **be ~al in** contribuir a. **~alist** *n* instrumentalista *m & f*

insubordinat|e /ɪnsə'bɔ:dɪnət/ *a* insubordinado. **~ion** /-'neɪʃn/ *n* insubordinación *f*

insufferable /ɪn'sʌfərəbl/ *a* insufrible, insoportable

insufficient /ɪnsə'fɪʃnt/ *a* insuficiente. **~ly** *adv* insuficientemente

insular /'ɪnsjʊlə(r)/ *a* insular; (*narrow-minded*) de miras estrechas

insulat|e /'ɪnsjʊleɪt/ *vt* aislar. **~ing tape** *n* cinta *f* aisladora/aislante. **~ion** /-'leɪʃn/ *n* aislamiento *m*

insulin /'ɪnsjʊlɪn/ *n* insulina *f*

insult /ɪn'sʌlt/ *vt* insultar. /'ɪnsʌlt/ *n* insulto *m*

insuperable /ɪn'sju:pərəbl/ *a* insuperable

insur|ance /ɪn'ʃʊərəns/ *n* seguro *m*. **~e** *vt* asegurar. **~e that** asegurarse de que

insurgent /ɪn'sɜ:dʒənt/ *a & n* insurrecto (*m*)

insurmountable /ɪnsə'maʊntəbl/ *a* insuperable

insurrection /ɪnsə'rekʃn/ *n* insurrección *f*

intact /ɪn'tækt/ *a* intacto

intake /'ɪnteɪk/ *n* (*quantity*) número *m*; (*mec*) admisión *f*; (*of food*) consumo *m*

intangible /ɪn'tændʒəbl/ *a* intangible

integral /'ɪntɪgrəl/ *a* íntegro. **be an ~ part of** ser parte integrante de

integrat|e /'ɪntɪgreɪt/ *vt* integrar. ● *vi* integrarse. **~ion** /-'greɪʃn/ *n* integración *f*

integrity /ɪn'tegrətɪ/ *n* integridad *f*

intellect /'ɪntəlekt/ *n* intelecto *m*. **~ual** *a & n* intelectual (*m*)

intelligen|ce /ɪn'telɪdʒəns/ *n* inteligencia *f*; (*information*) información *f*. **~t** *a* inteligente. **~tly** *adv* inteligentemente. **~tsia** /ɪntelɪ'dʒentsɪə/ *n* intelectualidad *f*

intelligible /ɪn'telɪdʒəbl/ *a* inteligible

intemperance /ɪn'tempərəns/ *n* inmoderación *f*

intend /ɪn'tend/ *vt* destinar. **~ to do** tener la intención de hacer. **~ed** *a* intencionado. ● *n* (*future spouse*) novio *m*

intense /ɪn'tens/ *a* intenso; ‹*person*› apasionado. **~ly** *adv* intensamente; (*very*) sumamente

intensif|ication /ɪntensɪfɪ'keɪʃn/ *n* intensificación *f*. **~y** /-faɪ/ *vt* intensificar

intensity /ɪn'tensətɪ/ *n* intensidad *f*

intensive /ɪn'tensɪv/ *a* intensivo. **~ care** *n* asistencia *f* intensiva, cuidados *mpl* intensivos

intent /ɪn'tent/ *n* propósito *m*. ● *a* atento. **~ on** absorto en. **~ on doing** resuelto a hacer

intention /ɪn'tenʃn/ *n* intención *f*. **~al** *a* intencional

intently /ɪn'tentlɪ/ *adv* atentamente

inter /ɪn'tɜ:(r)/ *vt* (*pt* **interred**) enterrar

inter... /'ɪntə(r)/ *pref* inter..., entre...

interact /ɪntər'ækt/ *vi* obrar recíprocamente. **~ion** /-ʃn/ *n* interacción *f*

intercede /ɪntə'si:d/ *vi* interceder

intercept /ɪntə'sept/ *vt* interceptar. **~ion** /-ʃn/ *n* interceptación *f*; (*in geometry*) intersección *f*

interchange /'ɪntətʃeɪndʒ/ *n* (*road junction*) cruce *m*. **~able** /-'tʃeɪndʒəbl/ *a* intercambiable

intercom /'ɪntəkɒm/ *n* intercomunicador *m*

interconnected /ɪntəkə'nektɪd/ *a* relacionado

intercourse /'ɪntəkɔ:s/ *n* trato *m*; (*sexual*) trato *m* sexual

interest /'ɪntrest/ *n* interés *m*; (*advantage*) ventaja *f*. ● *vt* interesar. **~ed** *a* interesado. **be ~ed in** interesarse por. **~ing** *a* interesante

interfere /ɪntə'fɪə(r)/ *vi* entrometerse. **~ in** entrometerse en. **~ with** entrometerse en, interferir en; interferir ‹*radio*›. **~nce** *n* interferencia *f*

interim *a* provisional. ● *n*. **in the ~** entre tanto

interior /ɪn'tɪərɪə(r)/ *a & n* interior (*m*)

interjection /ɪntəˈdʒekʃn/ *n* interjección *f*
interlock /ɪntəˈlɒk/ *vt/i* (*tec*) engranar
interloper /ˈɪntələʊpə(r)/ *n* intruso *m*
interlude /ˈɪntəlu:d/ *n* intervalo *m*; (*theatre, music*) interludio *m*
intermarr|iage /ɪntəˈmærɪdʒ/ *n* matrimonio *m* entre personas de distintas razas. **~y** *vi* casarse (con personas de distintas razas)
intermediary /ɪntəˈmi:dɪərɪ/ *a & n* intermediario (*m*)
intermediate /ɪntəˈmi:dɪət/ *a* intermedio
interminable /ɪnˈtɜ:mɪnəbl/ *a* interminable
intermission /ɪntəˈmɪʃn/ *n* pausa *f*; (*theatre*) descanso *m*
intermittent /ɪntəˈmɪtnt/ *a* intermitente. **~ly** *adv* con discontinuidad
intern /ɪnˈtɜ:n/ *vt* internar. /ˈɪntɜ:n/ *n* (*doctor, Amer*) interno *m*
internal /ɪnˈtɜ:nl/ *a* interior. **~ly** *adv* interiormente
international /ɪntəˈnæʃənl/ *a & n* internacional (*m*)
internee /ˌɪntɜ:ˈni:/ *n* internado *m*
internment /ɪnˈtɜ:nmənt/ *n* internamiento *m*
interplay /ˈɪntəpleɪ/ *n* interacción *f*
interpolate /ɪnˈtɜ:pəleɪt/ *vt* interpolar
interpret /ɪnˈtɜ:prɪt/ *vt/i* interpretar. **~ation** /-ˈteɪʃn/ *n* interpretación *f*. **~er** *n* intérprete *m & f*
interrelated /ɪntərɪˈleɪtɪd/ *a* interrelacionado
interrogat|e /ɪnˈterəgeɪt/ *vt* interrogar. **~ion** /-ˈgeɪʃn/ *n* interrogación *f*; (*session of questions*) interrogatorio *m*
interrogative /ɪntəˈrɒgətɪv/ *a & n* interrogativo (*m*)
interrupt /ɪntəˈrʌpt/ *vt* interrumpir. **~ion** /-ʃn/ *n* interrupción *f*
intersect /ɪntəˈsekt/ *vt* cruzar. • *vi* ‹*roads*› cruzarse; (*geometry*) intersecarse. **~ion** /-ʃn/ *n* (*roads*) cruce *m*; (*geometry*) intersección *f*
interspersed /ɪntəˈspɜ:st/ *a* disperso. **~ with** salpicado de
intertwine /ɪntəˈtwaɪn/ *vt* entrelazar. • *vi* entrelazarse
interval /ˈɪntəvl/ *n* intervalo *m*; (*theatre*) descanso *m*. **at ~s** a intervalos
interven|e /ɪntəˈvi:n/ *vi* intervenir. **~tion** /-ˈvenʃn/ *n* intervención *f*
interview /ˈɪntəvju:/ *n* entrevista *f*. • *vt* entrevistarse con. **~er** *n* entrevistador *m*
intestin|al /ɪnteˈstaɪnl/ *a* intestinal. **~e** /ɪnˈtestɪn/ *n* intestino *m*
intimacy /ˈɪntɪməsɪ/ *n* intimidad *f*
intimate[1] /ˈɪntɪmət/ *a* íntimo
intimate[2] /ˈɪntɪmeɪt/ *vt* (*state*) anunciar; (*imply*) dar a entender
intimately /ˈɪntɪmətlɪ/ *adv* íntimamente
intimidat|e /ɪnˈtɪmɪdeɪt/ *vt* intimidar. **~ion** /-ˈdeɪʃn/ *n* intimidación *f*
into /ˈɪntu:/, *unstressed* /ˈɪntə/ *prep* en; (*translate*) a
intolerable /ɪnˈtɒlərəbl/ *a* intolerable
intoleran|ce /ɪnˈtɒlərəns/ *n* intolerancia *f*. **~t** *a* intolerante
intonation /ɪntəˈneɪʃn/ *n* entonación *f*
intoxicat|e /ɪnˈtɒksɪkeɪt/ *vt* embriagar; (*med*) intoxicar. **~ed** *a* ebrio. **~ion** /-ˈkeɪʃn/ *n* embriaguez *f*; (*med*) intoxicación *f*
intra... /ˈɪntrə/ *pref* intra...
intractable /ɪnˈtræktəbl/ *a* ‹*person*› intratable; ‹*thing*› muy difícil
intransigent /ɪnˈtrænsɪdʒənt/ *a* intransigente
intransitive /ɪnˈtrænsɪtɪv/ *a* intransitivo
intravenous /ɪntrəˈvi:nəs/ *a* intravenoso
intrepid /ɪnˈtrepɪd/ *a* intrépido
intrica|cy /ˈɪntrɪkəsɪ/ *n* complejidad *f*. **~te** *a* complejo
intrigu|e /ɪnˈtri:g/ *vt/i* intrigar. • *n* intriga *f*. **~ing** *a* intrigante
intrinsic /ɪnˈtrɪnsɪk/ *a* intrínseco. **~ally** *adv* intrínsecamente
introduc|e /ɪntrəˈdju:s/ *vt* introducir; presentar ‹*person*›. **~tion** /ɪntrəˈdʌkʃn/ *n* introducción *f*; (*to person*) presentación *f*. **~tory** /-tərɪ/ *a* preliminar
introspective /ɪntrəˈspektɪv/ *a* introspectivo
introvert /ˈɪntrəvɜ:t/ *n* introvertido *m*
intru|de /ɪnˈtru:d/ *vi* entrometerse; (*disturb*) molestar. **~der** *n* intruso *m*. **~sion** *n* intrusión *f*
intuiti|on /ɪntju:ˈɪʃn/ *n* intuición *f*. **~ve** /ɪnˈtju:ɪtɪv/ *a* intuitivo

inundat|e /ˈɪnʌndeɪt/ *vt* inundar. **~ion** /-ˈdeɪʃn/ *n* inundación *f*
invade /ɪnˈveɪd/ *vt* invadir. **~r** /-ə(r)/ *n* invasor *m*
invalid[1] /ˈɪnvəlɪd/ *n* enfermo *m*, inválido *m*
invalid[2] /ɪnˈvælɪd/ *a* nulo. **~ate** *vt* invalidar
invaluable /ɪnˈvæljʊəbl/ *a* inestimable
invariabl|e /ɪnˈveərɪəbl/ *a* invariable. **~y** *adv* invariablemente
invasion /ɪnˈveɪʒn/ *n* invasión *f*
invective /ɪnˈvektɪv/ *n* invectiva *f*
inveigh /ɪnˈveɪ/ *vi* dirigir invectivas (**against** contra)
inveigle /ɪnˈveɪgl/ *vt* engatusar, persuadir
invent /ɪnˈvent/ *vt* inventar. **~ion** /-ˈvenʃn/ *n* invención *f*. **~ive** *a* inventivo. **~or** *n* inventor *m*
inventory /ˈɪnvəntərɪ/ *n* inventario *m*
invers|e /ɪnˈvɜːs/ *a* & *n* inverso (*m*). **~ely** *adv* inversamente. **~ion** /ɪnˈvɜːʃn/ *n* inversión *f*
invert /ɪnˈvɜːt/ *vt* invertir. **~ed commas** *npl* comillas *fpl*
invest /ɪnˈvest/ *vt* invertir. ● *vi*. **~ in** hacer una inversión *f*
investigat|e /ɪnˈvestɪgeɪt/ *vt* investigar. **~ion** /-ˈgeɪʃn/ *n* investigación *f*. **under ~ion** sometido a examen. **~or** *n* investigador *m*
inveterate /ɪnˈvetərət/ *a* inveterado
invidious /ɪnˈvɪdɪəs/ *a* (*hateful*) odioso; (*unfair*) injusto
invigilat|e /ɪnˈvɪdʒɪleɪt/ *vi* vigilar. **~or** *n* celador *m*
invigorate /ɪnˈvɪgəreɪt/ *vt* vigorizar; (*stimulate*) estimular
invincible /ɪnˈvɪnsɪbl/ *a* invencible
invisible /ɪnˈvɪzəbl/ *a* invisible
invit|ation /ɪnvɪˈteɪʃn/ *n* invitación *f*. **~e** /ɪnˈvaɪt/ *vt* invitar; (*ask for*) pedir. **~ing** *a* atrayente
invoice /ˈɪnvɔɪs/ *n* factura *f*. ● *vt* facturar
invoke /ɪnˈvəʊk/ *vt* invocar
involuntary /ɪnˈvɒləntərɪ/ *a* involuntario
involve /ɪnˈvɒlv/ *vt* enredar. **~d** *a* (*complex*) complicado. **~d in** embrollado en. **~ment** *n* enredo *m*
invulnerable /ɪnˈvʌlnərəbl/ *a* invulnerable
inward /ˈɪnwəd/ *a* interior. ● *adv* interiormente. **~s** *adv* hacia/para dentro
iodine /ˈaɪədiːn/ *n* yodo *m*
iota /aɪˈəʊtə/ *n* (*amount*) pizca *f*
IOU /aɪəʊˈjuː/ *abbr* (*I owe you*) pagaré *m*
IQ /aɪˈkjuː/ *abbr* (*intelligence quotient*) cociente *m* intelectual
Iran /ɪˈrɑːn/ *n* Irán *m*. **~ian** /ɪˈreɪnɪən/ *a* & *n* iraní (*m*)
Iraq /ɪˈrɑːk/ *n* Irak *m*. **~i** *a* & *n* iraquí (*m*)
irascible /ɪˈræsəbl/ *a* irascible
irate /aɪˈreɪt/ *a* colérico
ire /aɪə(r)/ *n* ira *f*
Ireland /ˈaɪələnd/ *n* Irlanda *f*
iris /ˈaɪərɪs/ *n* (*anat*) iris *m*; (*bot*) lirio *m*
Irish /ˈaɪərɪʃ/ *a* irlandés. ● *n* (*lang*) irlandés *m*. **~man** *n* irlandés *m*. **~woman** *n* irlandesa *f*
irk /ɜːk/ *vt* fastidiar. **~some** *a* fastidioso
iron /ˈaɪən/ *n* hierro *m*; (*appliance*) plancha *f*. ● *a* de hierro. ● *vt* planchar. **~ out** allanar. **I~ Curtain** *n* telón *m* de acero
ironic(al) /aɪˈrɒnɪk(l)/ *a* irónico
ironing-board /ˈaɪənɪŋˈbɔːd/ *n* tabla *f* de planchar
ironmonger /ˈaɪənmʌŋgə(r)/ *n* ferretero *m*. **~y** *n* ferretería *f*
ironwork /ˈaɪənwɜːk/ *n* herraje *m*
irony /ˈaɪərənɪ/ *n* ironía *f*
irrational /ɪˈræʃənl/ *a* irracional
irreconcilable /ɪrekənˈsaɪləbl/ *a* irreconciliable
irrefutable /ɪrɪˈfjuːtəbl/ *a* irrefutable
irregular /ɪˈregjʊlə(r)/ *a* irregular. **~ity** /-ˈlærətɪ/ *n* irregularidad *f*
irrelevan|ce /ɪˈreləvəns/ *n* inoportunidad *f*, impertinencia *f*. **~t** *a* no pertinente
irreparable /ɪˈrepərəbl/ *a* irreparable
irreplaceable /ɪrɪˈpleɪsəbl/ *a* irreemplazable
irrepressible /ɪrɪˈpresəbl/ *a* irreprimible
irresistible /ɪrɪˈzɪstəbl/ *a* irresistible
irresolute /ɪˈrezəluːt/ *a* irresoluto, indeciso
irrespective /ɪrɪˈspektɪv/ *a*. **~ of** sin tomar en cuenta
irresponsible /ɪrɪˈspɒnsəbl/ *a* irresponsable
irretrievable /ɪrɪˈtriːvəbl/ *a* irrecuperable
irreverent /ɪˈrevərənt/ *a* irreverente

irreversible /ɪrɪ'vɜ:səbl/ *a* irreversible; ‹*decision*› irrevocable
irrevocable /ɪ'revəkəbl/ *a* irrevocable
irrigat|e /'ɪrɪgeɪt/ *vt* regar; (*med*) irrigar. **~ion** /-'geɪʃn/ *n* riego *m*; (*med*) irrigación *f*
irritable /'ɪrɪtəbl/ *a* irritable
irritat|e /'ɪrɪteɪt/ *vt* irritar. **~ion** /-'teɪʃn/ *n* irritación *f*
is /ɪz/ *see* **be**
Islam /'ɪzlɑ:m/ *n* Islam *m*. **~ic** /ɪz'læmɪk/ *a* islámico
island /'aɪlənd/ *n* isla *f*. **traffic ~** *n* refugio *m* (en la calle). **~er** *n* isleño *m*
isle /aɪl/ *n* isla *f*
isolat|e /'aɪsəleɪt/ *vt* aislar. **~ion** /-'leɪʃn/ *n* aislamiento *m*
isotope /'aɪsətəʊp/ *n* isotopo *m*
Israel /'ɪzreɪl/ *n* Israel *m*. **~i** /ɪz'reɪlɪ/ *a & n* israelí (*m*)
issue /'ɪʃu:/ *n* asunto *m*; (*outcome*) resultado *m*; (*of magazine etc*) número *m*; (*of stamps*) emisión *f*; (*offspring*) descendencia *f*. **at ~** en cuestión. **take ~ with** oponerse a. ● *vt* distribuir; emitir ‹*stamps etc*›; publicar ‹*book*›. ● *vi*. **~ from** salir de
isthmus /'ɪsməs/ *n* istmo *m*
it /ɪt/ *pron* (*subject*) el, ella, ello; (*direct object*) lo, la; (*indirect object*) le; (*after preposition*) él, ella, ello. **~ is hot** hace calor. **~ is me** soy yo. **far from ~** ni mucho menos. **that's ~** eso es. **who is ~?** ¿quién es?
italic /ɪ'tælɪk/ *a* bastardillo *m*. **~s** *npl* (letra *f*) bastardilla *f*
ital|ian /ɪ'tæljən/ *a & n* italiano (*m*). **I~y** /'ɪtəlɪ/ *n* Italia f
itch /ɪtʃ/ *n* picazón *f*. ● *vi* picar. **I'm ~ing to** rabio por. **my arm ~es** me pica el brazo. **~y** *a* que pica
item /'aɪtəm/ *n* artículo *m*; (*on agenda*) asunto *m*. **news ~** *n* noticia *f*. **~ize** *vt* detallar
itinerant /aɪ'tɪnərənt/ *a* ambulante
itinerary /aɪ'tɪnərərɪ/ *n* itinerario *m*
its /ɪts/ *a* su, sus (*pl*). ● *pron* (el) suyo *m*, (la) suya *f*, (los) suyos *mpl*, (las) suyas *fpl*
it's /ɪts/ = **it is, it has**
itself /ɪt'self/ *pron* él mismo, ella misma, ello mismo; (*reflexive*) se; (*after prep*) sí mismo, sí misma
ivory /'aɪvərɪ/ *n* marfil *m*. **~ tower** *n* torre *f* de marfil
ivy /'aɪvɪ/ *n* hiedra *f*

J

jab /dʒæb/ *vt* (*pt* **jabbed**) pinchar; (*thrust*) hurgonear. ● *n* pinchazo *m*
jabber /'dʒæbə(r)/ *vi* barbullar. ● *n* farfulla *f*
jack /dʒæk/ *n* (*mec*) gato *m*; (*cards*) sota *f*. ● *vt*. **~ up** alzar con gato
jackal /'dʒækl/ *n* chacal *m*
jackass /'dʒækæs/ *n* burro *m*
jackdaw /'dʒækdɔ:/ *n* grajilla *f*
jacket /'dʒækɪt/ *n* chaqueta *f*, saco *m* (*LAm*); (*of book*) sobrecubierta *f*, camisa *f*
jack-knife /'dʒæknaɪf/ *n* navaja *f*
jackpot /'dʒækpɒt/ *n* premio *m* gordo. **hit the ~** sacar el premio gordo
jade /dʒeɪd/ *n* (*stone*) jade *m*
jaded /'dʒeɪdɪd/ *a* cansado
jagged /'dʒægɪd/ *a* dentado
jaguar /'dʒægjʊə(r)/ *n* jaguar *m*
jail /dʒeɪl/ *n* cárcel *m*. **~bird** *n* criminal *m* empedernido. **~er** *n* carcelero *m*
jalopy /dʒə'lɒpɪ/ *n* cacharro *m*
jam[1] /dʒæm/ *vt* (*pt* **jammed**) interferir con ‹*radio*›; ‹*traffic*› embotellar; ‹*people*› agolparse en. ● *vi* obstruirse; ‹*mechanism etc*› atascarse. ● *n* (*of people*) agolpamiento *m*; (*of traffic*) embotellamiento *m*; (*situation, fam*) apuro *m*
jam[2] /dʒæm/ *n* mermelada *f*
Jamaica /dʒə'meɪkə/ *n* Jamaica *f*
jamboree /dʒæmbə'ri:/ *n* reunión *f*
jam-packed /'dʒæm'pækt/ *a* atestado
jangle /'dʒæŋgl/ *n* sonido *m* metálico (y áspero). ● *vt/i* sonar discordemente
janitor /'dʒænɪtə(r)/ *n* portero *m*
January /'dʒænjʊərɪ/ *n* enero *m*
Japan /dʒə'pæn/ *n* el Japón *m*. **~ese** /dʒæpə'ni:z/ *a & n* japonés (*m*)
jar[1] /dʒɑ:(r)/ *n* tarro *m*, frasco *m*
jar[2] /dʒɑ:(r)/ *vi* (*pt* **jarred**) ‹*sound*› sonar mal; ‹*colours*› chillar. ● *vt* sacudir
jar[3] /dʒɑ:(r)/ *n*. **on the ~** (*ajar*) entreabierto
jargon /'dʒɑ:gən/ *n* jerga *f*
jarring /'dʒɑ:rɪŋ/ *a* discorde
jasmine /'dʒæsmɪn/ *n* jazmín *m*
jaundice /'dʒɔ:ndɪs/ *n* ictericia *f*. **~d** *a* (*envious*) envidioso; (*bitter*) amargado
jaunt /dʒɔ:nt/ *n* excursión *f*

jaunty /ˈdʒɔːntɪ/ *a* (**-ier**, **-iest**) garboso

javelin /ˈdʒævəlɪn/ *n* jabalina *f*

jaw /dʒɔː/ *n* mandíbula *f*. ● *vi* (*talk lengthily, sl*) hablar por los codos

jay /dʒeɪ/ *n* arrendajo *m*. **~-walk** *vi* cruzar la calle descuidadamente

jazz /dʒæz/ *n* jazz *m*. ● *vt*. **~ up** animar. **~y** *a* chillón

jealous /dʒeləs/ *a* celoso. **~y** *n* celos *mpl*

jeans /dʒiːnz/ *npl* (pantalones *mpl*) vaqueros *mpl*

jeep /dʒiːp/ *n* jeep *m*

jeer /dʒɪə(r)/ *vt/i*. **~ at** mofarse de, befar; (*boo*) abuchear. ● *n* mofa *f*; (*boo*) abucheo *m*

jell /dʒel/ *vi* cuajar. **~ied** *a* en gelatina

jelly /dʒelɪ/ *n* jalea *f*. **~fish** *n* medusa *f*

jeopard|ize /ˈdʒepədaɪz/ *vt* arriesgar. **~y** *n* peligro *m*

jerk /dʒɜːk/ *n* sacudida *f*; (*fool, sl*) idiota *m & f*. ● *vt* sacudir. **~ily** *adv* a sacudidads. **~y** *a* espasmódico

jersey /ˈdʒɜːzɪ/ *n* (*pl* **-eys**) jersey *m*

jest /dʒest/ *n* broma *f*. ● *vi* bromear. **~er** *n* bufón *m*

Jesus /ˈdʒiːzəs/ *n* Jesús *m*

jet[1] /dʒet/ *n* (*stream*) chorro *m*; (*plane*) yet *m*, avión *m* de propulsión por reacción

jet[2] /dʒet/ *n* (*mineral*) azabache *m*. **~-black** *a* de azabache, como el azabache

jet: **~ lag** *n* cansancio *m* retardado después de un vuelo largo. **have ~ lag** estar desfasado. **~-propelled** *a* (de propulsión) a reacción

jettison /ˈdʒetɪsn/ *vt* echar al mar; (*fig, discard*) deshacerse de

jetty /ˈdʒetɪ/ *n* muelle *m*

Jew /dʒuː/ *n* judío *m*

jewel /ˈdʒuːəl/ *n* joya *f*. **~led** *a* enjoyado. **~ler** *n* joyero *m*. **~lery** *n* joyas *fpl*

Jew: **~ess** *n* judía *f*. **~ish** *a* judío. **~ry** /ˈdʒʊərɪ/ *n* los judíos *mpl*

jib[1] /dʒɪb/ *n* (*sail*) foque *m*

jib[2] /dʒɪb/ *vi* (*pt* **jibbed**) rehusar. **~ at** oponerse a.

jiffy /ˈdʒɪfɪ/ *n* momentito *m*. **do sth in a ~** hacer algo en un santiamén

jig /dʒɪg/ *n* (*dance*) giga *f*

jiggle /ˈdʒɪgl/ *vt* zangolotear

jigsaw /ˈdʒɪgsɔː/ *n* rompecabezas *m invar*

jilt /dʒɪlt/ *vt* plantar, dejar plantado

jingle /ˈdʒɪŋgl/ *vt* hacer sonar. ● *vi* tintinear. ● *n* tintineo *m*; (*advert*) anuncio *m* cantado

jinx /dʒɪŋks/ *n* (*person*) gafe *m*; (*spell*) maleficio *m*

jitter|s /ˈdʒɪtəz/ *npl*. **have the ~s** estar nervioso. **~y** /-ərɪ/ *a* nervioso. **be ~y** estar nervioso

job /dʒɒb/ *n* trabajo *m*; (*post*) empleo *m*, puesto *m*. **have a ~ doing** costar trabajo hacer. **it is a good ~ that** menos mal que. **~centre** *n* bolsa *f* de trabajo. **~less** *a* sin trabajo.

jockey /ˈdʒɒkɪ/ *n* jockey *m*. ● *vi* (*manoevre*) maniobrar (**for** para)

jocular /ˈdʒɒkjʊlə(r)/ *a* jocoso

jog /dʒɒg/ *vt* (*pt* **jogged**) empujar; refrescar ‹*memory*›. ● *vi* hacer footing. **~ging** *n* jogging *m*

join /dʒɔɪn/ *vt* unir, juntar; hacerse socio de ‹*club*›; hacerse miembro de ‹*political group*›; alistarse en ‹*army*›; reunirse con ‹*another person*›. ● *vi* ‹*roads etc*› empalmar; ‹*rivers*› confluir. **~ in** participar (en). **~ up** (*mil*) alistarse. ● *n* juntura

joiner /ˈdʒɔɪnə(r)/ *n* carpintero *m*

joint /dʒɔɪnt/ *a* común. **~ author** *n* coautor *m*. ● *n* (*join*) juntura *f*; (*anat*) articulación *f*; (*culin*) asado *m*; (*place, sl*) garito *m*; (*marijuana, sl*) cigarillo *m* de marijuana. **out of ~** descoyuntado. **~ly** *adv* conjuntamente

joist /dʒɔɪst/ *n* viga *f*

jok|e /dʒəʊk/ *n* broma *f*; (*funny story*) chiste *m*. ● *vi* bromear. **~er** *n* bromista *m & f*; (*cards*) comodín *m*. **~ingly** *adv* en broma

joll|ification /dʒɒlɪfɪˈkeɪʃn/ *n* jolgorio *m*. **~ity** *n* jolgorio *m*. **~y** *a* (**-ier**, **-iest**) alegre. ● *adv* (*fam*) muy

jolt /dʒɒlt/ *vt* sacudir. ● *vt* ‹*vehicle*› traquetear. ● *n* sacudida *f*

Jordan /ˈdʒɔːdən/ *n* Jordania *f*. **~ian** *a & n* /-ˈdeɪnɪən/ jordano (*m*)

jostle /ˈdʒɒsl/ *vt/i* empujar(se)

jot /dʒɒt/ *n* pizca *f*. ● *vt* (*pt* **jotted**) apuntar. **~ter** *n* bloc *m*

journal /ˈdʒɜːnl/ *n* (*diary*) diario *m*; (*newspaper*) periódico *m*; (*magazine*) revista *f*. **~ese** /dʒɜːnəˈliːz/ *n* jerga *f* periodística. **~ism** *n* periodismo *m*. **~ist** *n* periodista *m & f*

journey /ˈdʒɜːnɪ/ *n* viaje *m*. ● *vi* viajar

jovial /ˈdʒəʊvɪəl/ *a* jovial

jowl /dʒaʊl/ *n* (*jaw*) quijada *f*; (*cheek*) mejilla *f*. **cheek by ~** muy cerca

joy /dʒɔɪ/ *n* alegría *f*. **~ful** *a* alegre. **~ride** *n* paseo *m* en coche sin permiso del dueño. **~ous** *a* alegre

jubila|nt /'dʒu:bɪlənt/ *a* jubiloso. **~tion** /-'leɪʃn/ *n* júbilo *m*

jubilee /'dʒu:bɪli:/ *n* aniversario *m* especial

Judaism /'dʒu:deɪɪzəm/ *n* judaísmo *m*

judder /'dʒʌdə(r)/ *vi* vibrar. ● *n* vibración *f*

judge /dʒʌdʒ/ *n* juez *m*. ● *vt* juzgar. **~ment** *n* juicio *m*

judicia|l /dʒu:'dɪʃl/ *a* judicial. **~ry** *n* magistratura *f*

judicious /dʒu:'dɪʃəs/ *a* juicioso

judo /'dʒu:dəʊ/ *n* judo *m*

jug /dʒʌg/ *n* jarra *f*

juggernaut /'dʒʌgənɔ:t/ *n* (*lorry*) camión *m* grande

juggle /'dʒʌgl/ *vt/i* hacer juegos malabares (con). **~r** *n* malabarista *m & f*

juic|e /dʒu:s/ *n* jugo *m*, zumo *m*. **~y** *a* jugoso, zumoso; ‹*story etc*› (*fam*) picante

juke-box /'dʒu:kbɒks/ *n* tocadiscos *m invar* tragaperras

July /dʒu:'laɪ/ *n* julio

jumble /'dʒʌmbl/ *vt* mezclar. ● *n* (*muddle*) revoltijo *m*. **~ sale** *n* venta *f* de objetos usados, mercadillo *m*

jumbo /'dʒʌmbəʊ/ *a*. **~ jet** *n* jumbo *m*

jump /dʒʌmp/ *vt/i* saltar. **~ the gun** obrar prematuramente. **~ the queue** colarse. ● *vi* saltar; (*start*) asustarse; ‹*prices*› alzarse. **~ at** apresurarse a aprovechar. ● *n* salto *m*; (*start*) susto *m*; (*increase*) aumento *m*

jumper /'dʒʌmpə(r)/ *n* jersey *m*; (*dress, Amer*) mandil *m*, falda *f* con peto

jumpy /'dʒʌmpɪ/ *a* nervioso

junction /'dʒʌŋkʃn/ *n* juntura *f*; (*of roads*) cruce *m*, entronque *m* (*LAm*); (*rail*) empalme *m*, entronque *m* (*LAm*)

juncture /'dʒʌŋktʃə(r)/ *n* momento *m*; (*state of affairs*) coyuntura *f*

June /dʒu:n/ *n* junio *m*

jungle /'dʒʌŋgl/ *n* selva *f*

junior /'dʒu:nɪə(r)/ *a* (*in age*) más joven (**to** que); (*in rank*) subalterno. ● *n* menor *m*. **~ school** *n* escuela *f*

junk /dʒʌŋk/ *n* trastos *mpl* viejos. ● *vt* (*fam*) tirar

junkie /'dʒʌŋkɪ/ *n* (*sl*) drogadicto *m*

junk shop /'dʒʌŋkʃɒp/ *n* tienda *f* de trastos viejos

junta /'dʒʌntə/ *n* junta *f*

jurisdiction /dʒʊərɪs'dɪkʃn/ *n* jurisdicción *f*

jurisprudence /dʒʊərɪs'pru:dəns/ *n* jurisprudencia *f*

juror /'dʒʊərə(r)/ *n* jurado *m*

jury /'dʒʊərɪ/ *n* jurado *m*

just /dʒʌst/ *a* (*fair*) justo. ● *adv* exactamente; (*slightly*) apenas; (*only*) sólo, solamente. **~ as tall** tan alto (**as** como). **~ listen!** ¡escucha! **he has ~ left** acaba de marcharse

justice /'dʒʌstɪs/ *n* justicia *f*. **J~ of the Peace** juez *m* de paz

justif|iable /dʒʌstɪ'faɪəbl/ *a* justificable. **~iably** *adv* con razón. **~ication** /dʒʌstɪfɪ'keɪʃn/ *n* justificación *f*. **~y** /'dʒʌstɪfaɪ/ *vt* justificar

justly /'dʒʌstlɪ/ *adv* con justicia

jut /dʒʌt/ *vi* (*pt* **jutted**). **~ out** sobresalir

juvenile /'dʒu:vənaɪl/ *a* juvenil; (*childish*) infantil. ● *n* joven *m & f*. **~ court** *n* tribunal *m* de menores

juxtapose /dʒʌkstə'pəʊz/ *vt* yuxtaponer

K

kaleidoscope /kə'laɪdəskəʊp/ *n* calidoscopio *m*

kangaroo /kæŋgə'ru:/ *n* canguro *m*

kapok /'keɪpɒk/ *n* miraguano *m*

karate /kə'rɑ:tɪ/ *n* karate *m*

kebab /kɪ'bæb/ *n* broqueta *f*

keel /ki:l/ *n* (*of ship*) quilla *f*. ● *vi*. **~ over** volcarse

keen /ki:n/ *a* (**-er, -est**) ‹*interest, feeling*› vivo; ‹*wind, mind, analysis*› penetrante; ‹*edge*› afilado; ‹*appetite*› bueno; ‹*eyesight*› agudo; (*eager*) entusiasta. **be ~ on** gustarle a uno. **he's ~ on Shostakovich** le gusta Shostakovich. **~ly** *adv* vivamente; (*enthusiastically*) con entusiasmo. **~ness** *n* intensidad *f*; (*enthusiasm*) entusiasmo *m*.

keep /ki:p/ *vt* (*pt* **kept**) guardar; cumplir ‹*promise*›; tener ‹*shop, animals*›; mantener ‹*family*›; observar ‹*rule*›; (*celebrate*) celebrar; (*delay*) detener; (*prevent*) impedir. ● *vi* ‹*food*› conservarse; (*remain*) quedarse. ● *n* subsistencia *f*; (*of castle*) torreón *m*. **for ~s** (*fam*) para siempre. **~ back** *vt* retener. ● *vi* no acercarse. **~ in** no dejar salir. **~ in with** mantenerse en buenas relaciones con. **~ out** no dejar entrar. **~ up** mantener. **~ up (with)** estar al día (en). **~er** *n* guarda *m*

keeping /'ki:pɪŋ/ *n* cuidado *m*. **in ~ with** de acuerdo con

keepsake /'ki:pseɪk/ *n* recuerdo *m*

keg /keg/ *n* barrilete *m*

kennel /'kenl/ *n* perrera *f*

Kenya /'kenjə/ *n* Kenia *f*

kept /kept/ *see* **keep**

kerb /kɜ:b/ *n* bordillo *m*

kerfuffle /kə'fʌfl/ *n* (*fuss, fam*) lío *m*

kernel /'kɜ:nl/ *n* almendra *f*; (*fig*) meollo *m*

kerosene /'kerəsi:n/ *n* queroseno *m*

ketchup /'ketʃʌp/ *n* salsa *f* de tomate

kettle /'ketl/ *n* hervidor *m*

key /ki:/ *n* llave *f*; (*of typewriter, piano etc*) tecla *f*. ● *a* clave. ● *vt*. **~ up** excitar. **~board** *n* teclado *m*. **~hole** *n* ojo *m* de la cerradura. **~note** *n* (*mus*) tónica *f*; (*speech*) idea *f* fundamental. **~-ring** *n* llavero *m*. **~stone** *n* piedra *f* clave

khaki /'kɑ:kɪ/ *a* caqui

kibbutz /kɪ'bʊts/ *n* (*pl* **-im** /-i:m/ *or* **-es**) kibbutz *m*

kick /kɪk/ *vt* dar una patada a; ‹*animals*› tirar una coz a. ● *vi* dar patadas; ‹*firearm*› dar culatazo. ● *n* patada *f*; (*of animal*) coz *f*; (*of firearm*) culatazo *m*; (*thrill, fam*) placer *m*. **~ out** (*fam*) echar a patadas. **~ up** armar ‹*fuss etc*›. **~back** *n* culatazo *m*; (*payment*) soborno *m*. **~-off** *n* (*sport*) saque *m* inicial

kid /kɪd/ *n* (*young goat*) cabrito *m*; (*leather*) cabritilla *f*; (*child, sl*) chaval *m*. ● *vt* (*pt* **kidded**) tomar el pelo a. ● *vi* bromear

kidnap /'kɪdnæp/ *vt* (*pt* **kidnapped**) secuestrar. **~ping** *n* secuestro *m*

kidney /'kɪdnɪ/ *n* riñón *m*. ● *a* renal

kill /kɪl/ *vt* matar; (*fig*) acabar con. ● *n* matanza *f*; (*in hunt*) pieza(s) *f(pl)*. **~er** *n* matador *m*; (*murderer*) asesino *m*. **~ing** *n* matanza *f*; (*murder*) asesinato *m*. ● *a* (*funny, fam*) para morirse de risa; (*tiring, fam*) agotador. **~joy** *n* aguafiestas *m & f invar*

kiln /kɪln/ *n* horno *m*

kilo /'ki:ləʊ/ *n* (*pl* **-os**) kilo *m*

kilogram(me) /'kɪləgræm/ *n* kilogramo *m*

kilohertz /'kɪləhɜ:ts/ *n* kilohercio *m*

kilometre /'kɪləmi:tə(r)/ *n* kilómetro *m*

kilowatt /'kɪləwɒt/ *n* kilovatio *m*

kilt /kɪlt/ *n* falda *f* escocesa

kin /kɪn/ *n* parientes *mpl*. **next of ~** pariente *m* más próximo, parientes *mpl* más próximos

kind[1] /kaɪnd/ *n* clase *f*. **~ of** (*somewhat, fam*) un poco. **in ~** en especie. **be two of a ~** ser tal para cual

kind[2] /kaɪnd/ *a* amable

kindergarten /'kɪndəgɑ:tn/ *n* escuela *f* de párvulos

kind-hearted /kaɪnd'hɑ:tɪd/ *a* bondadoso

kindle /'kɪndl/ *vt/i* encender(se)

kind: ~liness *n* bondad *f*. **~ly** *a* (**-ier, -iest**) bondadoso. ● *adv* bondadosamente; (*please*) haga el favor de. **~ness** *n* bondad *f*

kindred /'kɪndrɪd/ *a* emparentado. **~ spirits** *npl* almas *fpl* afines

kinetic /kɪ'netɪk/ *a* cinético

king /kɪŋ/ *n* rey *m*

kingdom /'kɪŋdəm/ *n* reino *m*

kingpin /'kɪŋpɪn/ *n* (*person*) persona *f* clave; (*thing*) piedra *f* angular

king-size(d) /'kɪŋsaɪz(d)/ *a* extraordinariamente grande

kink /kɪŋk/ *n* (*in rope*) retorcimiento *m*; (*fig*) manía *f*. **~y** *a* (*fam*) pervertido

kiosk /'ki:ɒsk/ *n* quiosco *m*. **telephone ~** cabina *f* telefónica

kip /kɪp/ *n* (*sl*) sueño *m*. ● *vi* (*pt* **kipped**) dormir

kipper /'kɪpə(r)/ *n* arenque *m* ahumado

kiss /kɪs/ *n* beso *m*. ● *vt/i* besar(se)

kit /kɪt/ *n* avíos *mpl*; (*tools*) herramientos *mpl*. ● *vt* (*pt* **kitted**). **~ out** equipar de. **~bag** *n* mochila *f*

kitchen /'kɪtʃɪn/ *n* cocina *f*. **~ette** /kɪtʃɪ'net/ *n* cocina *f* pequeña. **~ garden** *n* huerto *m*

kite /kaɪt/ *n* (*toy*) cometa *f*

kith /kɪθ/ *n*. **~ and kin** amigos *mpl* y parientes *mpl*

kitten /'kɪtn/ *n* gatito *m*

kitty /'kɪtɪ/ *n* (*fund*) fondo *m* común

kleptomaniac /kleptəʊ'meɪnɪæk/ *n* cleptómano *m*

knack /næk/ *n* truco *m*

knapsack /'næpsæk/ *n* mochila *f*

knave /neɪv/ *n* (*cards*) sota *f*

knead /ni:d/ *vt* amasar

knee /ni:/ *n* rodilla *f*. **~cap** *n* rótula *f*

kneel /ni:l/ *vi* (*pt* **knelt**). ~ **(down)** arrodillarse
knees-up /'ni:zʌp/ *n* (*fam*) baile *m*
knell /nel/ *n* toque *m* de difuntos
knelt /nelt/ *see* **kneel**
knew /nju:/ *see* **know**
knickerbockers /'nɪkəbɒkəz/ *npl* pantalón *m* bombacho
knickers /'nɪkəz/ *npl* bragas *fpl*
knick-knack /'nɪknæk/ *n* chuchería *f*
knife /naɪf/ *n* (*pl* **knives**) cuchillo *m*. ● *vt* acuchillar
knight /naɪt/ *n* caballero *m*; (*chess*) caballo *m*. ● *vt* conceder el título de Sir a. **~hood** *n* título *m* de Sir
knit /nɪt/ *vt* (*pt* **knitted** *or* **knit**) tejer. ● *vi* hacer punto. ~ **one's brow** fruncir el ceño. **~ting** *n* labor *f* de punto. **~wear** *n* artículos *mpl* de punto
knob /nɒb/ *n* botón *m*; (*of door, drawer etc*) tirador *m*. **~bly** *a* nudoso
knock /nɒk/ *vt* golpear; (*criticize*) criticar. ● *vi* golpear; (*at door*) llamar. ● *n* golpe *m*. ~ **about** *vt* maltratar. ● *vi* rodar. ~ **down** derribar; atropellar ‹*person*›; rebajar ‹*prices*›. ~ **off** *vt* hacer caer; (*complete quickly, fam*) despachar; (*steal, sl*) birlar. ● *vi* (*finish work, fam*) terminar, salir del trabajo. ~ **out** (*by blow*) dejar sin conocimiento; (*eliminate*) eliminar; (*tire*) agotar. ~ **over** tirar; atropellar ‹*person*›. ~ **up** preparar de prisa ‹*meal etc*›. **~-down** *a* ‹*price*› de saldo. **~er** *n* aldaba *f*. **~-kneed** *a* patizambo. **~out** *n* (*boxing*) knockout *m*
knot /nɒt/ *n* nudo *m*. ● *vt* (*pt* **knotted**) anudar. **~ty** /'nɒtɪ/ *a* nudoso
know /nəʊ/ *vt* (*pt* **knew**) saber; (*be acquainted with*) conocer. ● *vi* saber. ● *n*. **be in the** ~ estar al tanto. ~ **about** entender de ‹*cars etc*›. ~ **of** saber de. **~-all** *n*, **~-it-all** (*Amer*) *n* sabelotodo *m* & *f*. **~-how** *n* habilidad *f*. **~ingly** *adv* deliberadamente
knowledge /'nɒlɪdʒ/ *n* conocimiento *m*; (*learning*) conocimientos *mpl*. **~able** *a* informado
known /nəʊn/ *see* **know**. ● *a* conocido
knuckle /'nʌkl/ *n* nudillo *m*. ● *vi*. ~ **under** someterse
Koran /kə'rɑ:n/ *n* Corán *m*, Alcorán *m*
Korea /kə'rɪə/ *n* Corea *f*
kosher /'kəʊʃə(r)/ *a* preparado según la ley judía
kowtow /kaʊ'taʊ/ *vi* humillarse (**to** ante)
kudos /'kju:dɒs/ *n* prestigio *m*

L

lab /læb/ *n* (*fam*) laboratorio *m*
label /'leɪbl/ *n* etiqueta *f*. ● *vt* (*pt* **labelled**) poner etiqueta a; (*fig, describe as*) describir como
laboratory /lə'bɒrətərɪ/ *n* laboratorio *m*
laborious /lə'bɔ:rɪəs/ *a* penoso
labour /'leɪbə(r)/ *n* trabajo *m*; (*workers*) mano *f* de obra. **in** ~ de parto. ● *vi* trabajar. ● *vt* insistir en
Labour /'leɪbə(r)/ *n* el partido *m* laborista. ● *a* laborista
laboured /'leɪbəd/ *a* penoso
labourer /'leɪbərə(r)/ *n* obrero *m*; (*on farm*) labriego *m*
labyrinth /'læbərɪnθ/ *n* laberinto *m*
lace /leɪs/ *n* encaje *m*; (*of shoe*) cordón *m*, agujeta *f* (*Mex*). ● *vt* (*fasten*) atar. ~ **with** echar a ‹*a drink*›
lacerate /'læsəreɪt/ *vt* lacerar
lack /læk/ *n* falta *f*. **for** ~ **of** por falta de. ● *vt* faltarle a uno. **he ~s money** carece de dinero. **be ~ing** faltar
lackadaisical /lækə'deɪzɪkl/ *a* indolente, apático
lackey /'lækɪ/ *n* lacayo *m*
laconic /lə'kɒnɪk/ *a* lacónico
lacquer /'lækə(r)/ *n* laca *f*
lad /læd/ *n* muchacho *m*
ladder /'lædə(r)/ *n* escalera *f* (de mano); (*in stocking*) carrera *f*. ● *vt* hacer una carrera en. ● *vi* hacerse una carrera
laden /'leɪdn/ *a* cargado (**with** de)
ladle /'leɪdl/ *n* cucharón *m*
lady /'leɪdɪ/ *n* señora *f*. **young** ~ señorita *f*. **~bird** *n*, **~bug** *n* (*Amer*) mariquita *f*. ~ **friend** *n* amiga *f*. **~-in-waiting** *n* dama *f* de honor. **~like** *a* distinguido. **~ship** *n* Señora *f*
lag[1] /læg/ *vi* (*pt* **lagged**). ~ **(behind)** retrasarse. ● *n* (*interval*) intervalo *m*
lag[2] /læg/ *vt* (*pt* **lagged**) revestir ‹*pipes*›
lager /'lɑ:gə(r)/ *n* cerveza *f* dorada
laggard /'lægəd/ *n* holgazán *m*

lagging /ˈlægɪŋ/ *n* revestimiento *m* calorífugo
lagoon /ləˈgu:n/ *n* laguna *f*
lah /lɑ:/ *n* (*mus, sixth note of any musical scale*) la *m*
laid /leɪd/ *see* **lay**[1]
lain /leɪn/ *see* **lie**[1]
lair /leə(r)/ *n* guarida *f*
laity /ˈleɪətɪ/ *n* laicado *m*
lake /leɪk/ *n* lago *m*
lamb /læm/ *n* cordero *m*. **~swool** *n* lana *f* de cordero
lame /leɪm/ *a* (**-er**, **-est**) cojo; ‹*excuse*› poco convincente. **~ly** *adv* (*argue*) con poca convicción *f*
lament /ləˈment/ *n* lamento *m*. ● *vt/i* lamentarse (de). **~able** /ˈlæməntəbl/ *a* lamentable
laminated /ˈlæmɪneɪtɪd/ *a* laminado
lamp /læmp/ *n* lámpara *f*. **~post** *n* farol *m*. **~shade** *n* pantalla *f*
lance /lɑ:ns/ *n* lanza *f*. ● *vt* (*med*) abrir con lanceta. **~-corporal** *n* cabo *m* interino
lancet /ˈlɑ:nsɪt/ *n* lanceta *f*
land /lænd/ *n* tierra *f*; (*country*) país *m*; (*plot*) terreno *m*. ● *a* terrestre; ‹*breeze*› de tierra; ‹*policy, reform*› agrario. ● *vt* desembarcar; (*obtain*) conseguir; dar ‹*blow*›; (*put*) meter. ● *vi* (*from ship*) desembarcar; ‹*aircraft*› aterrizar; (*fall*) caer. **~ up** ir a parar
landed /ˈlændɪd/ *a* hacendado
landing /ˈlændɪŋ/ *n* desembarque *m*; (*aviat*) aterrizaje *m*; (*top of stairs*) descanso *m*. **~-stage** *n* desembarcadero *m*
landlady /ˈlændleɪdɪ/ *n* propietaria *f*; (*of inn*) patrona *f*
land-locked /ˈlændlɒkt/ *a* rodeado de tierra
landlord /ˈlændlɔ:d/ *n* propietario *m*; (*of inn*) patrón *m*
land: **~mark** *n* punto *m* destacado. **~scape** /ˈlændskeɪp/ *n* paisaje *m*. ● *vt* ajardinar. **~slide** *n* desprendimiento *m* de tierras; (*pol*) victoria *f* arrolladora
lane /leɪn/ *n* (*path, road*) camino *m*; (*strip of road*) carril *m*; (*aviat*) ruta *f*
language /ˈlæŋgwɪdʒ/ *n* idioma *m*; (*speech, style*) lenguaje *m*
langu|id /ˈlæŋgwɪd/ *a* lánguido. **~ish** /ˈlæŋgwɪʃ/ *vi* languidecer. **~or** /ˈlæŋgə(r)/ *n* languidez *f*
lank /læŋk/ *a* larguirucho; ‹*hair*› lacio. **~y** /ˈlæŋkɪ/ *a* (**-ier**, **-iest**) larguirucho
lantern /ˈlæntən/ *n* linterna *f*
lap[1] /læp/ *n* regazo *m*
lap[2] /læp/ *n* (*sport*) vuelta *f*. ● *vt/i* (*pt* **lapped**). **~ over** traslapar(se)
lap[3] /læp/ *vt* (*pt* **lapped**). **~ up** beber a lengüetazos; (*fig*) aceptar con entusiasmo. ● *vi* ‹*waves*› chapotear
lapel /ləˈpel/ *n* solapa *f*
lapse /læps/ *vi* (*decline*) degradarse; (*expire*) caducar; ‹*time*› transcurrir. **~ into** recaer en. ● *n* error *m*; (*of time*) intervalo *m*
larceny /ˈlɑ:sənɪ/ *n* robo *m*
lard /lɑ:d/ *n* manteca *f* de cerdo
larder /ˈlɑ:də(r)/ *n* despensa *f*
large /lɑ:dʒ/ *a* (**-er**, **-est**) grande, (*before singular noun*) gran. ● *n*. **at ~** en libertad. **~ly** *adv* en gran parte. **~ness** *n* (gran) tamaño *m*
largesse /lɑ:ˈʒes/ *n* generosidad *f*
lark[1] /lɑ:k/ *n* alondra *f*
lark[2] /lɑ:k/ *n* broma *f*; (*bit of fun*) travesura *f*. ● *vi* andar de juerga
larva /ˈlɑ:və/ *n* (*pl* **-vae** /-vi:/) larva *f*
laryn|gitis /lærɪnˈdʒaɪtɪs/ *n* laringitis *f*. **~x** /ˈlærɪŋks/ *n* laringe *f*
lascivious /ləˈsɪvɪəs/ *a* lascivo
laser /ˈleɪzə(r)/ *n* láser *m*
lash /læʃ/ *vt* azotar. **~ out** (*spend*) gastar. **~ out against** atacar. ● *n* latigazo *m*; (*eyelash*) pestaña *f*
lashings /ˈlæʃɪŋz/ *npl*. **~ of** (*cream etc, sl*) montones de
lass /læs/ *n* muchacha *f*
lassitude /ˈlæsɪtju:d/ *n* lasitud *f*
lasso /læˈsu:/ *n* (*pl* **-os**) lazo *m*
last[1] /lɑ:st/ *a* último; ‹*week etc*› pasado. **~ Monday** *n* el lunes pasado. **have the ~ word** decir la última palabra. **the ~ straw** *n* el colmo *m*. ● *adv* por último; (*most recently*) la última vez. **he came ~** llegó el último. ● *n* último *m*; (*remainder*) lo que queda. **~ but one** penúltimo. **at (long) ~** en fin.
last[2] /lɑ:st/ *vi* durar. **~ out** sobrevivir
last[3] /lɑ:st/ *n* horma *f*
lasting /ˈlɑ:stɪŋ/ *a* duradero
last: **~ly** *adv* por último. **~ night** *n* anoche *m*
latch /lætʃ/ *n* picaporte *m*
late /leɪt/ *a* (**-er**, **-est**) (*not on time*) tarde; (*recent*) reciente; (*former*) antiguo, ex; ‹*fruit*› tardío; ‹*hour*› avanzado; (*deceased*) difunto. **in ~ July** a fines de julio. **the ~ Dr Phillips** el difunto Dr. Phillips. ● *adv* tarde. **of ~** últimamente. **~ly** *adv* últimamente. **~ness** *n* (*delay*) retraso *m*; (*of hour*) lo avanzado

latent /ˈleɪtnt/ *a* latente
lateral /ˈlætərəl/ *a* lateral
latest /ˈleɪtɪst/ *a* último. **at the ~** a más tardar
lathe /leɪð/ *n* torno *m*
lather /ˈlɑːðə(r)/ *n* espuma *f*. ● *vt* enjabonar. ● *vi* hacer espuma
Latin /ˈlætɪn/ *n* (*lang*) latín *m*. ● *a* latino
latitude /ˈlætɪtjuːd/ *n* latitud *m*
latrine /ləˈtriːn/ *n* letrina *f*
latter /ˈlætə(r)/ *a* último; (*of two*) segundo. ● *n*. **the ~** éste *m*, ésta *f*, éstos *mpl*, éstas *fpl*. **~-day** *a* moderno. **~ly** *adv* últimamente
lattice /ˈlætɪs/ *n* enrejado *m*
laudable /ˈlɔːdəbl/ *a* laudable
laugh /lɑːf/ *vi* reír(se) (**at** de). ● *n* risa *f*. **~able** *a* ridículo. **~ing-stock** *n* hazmerreír *m invar*. **~ter** /ˈlɑːftə(r)/ *n* (*act*) risa *f*; (*sound of laughs*) risas *fpl*
launch[1] /lɔːntʃ/ *vt* lanzar. ● *n* lanzamiento *m*. **~ (out) into** lanzarse a
launch[2] /lɔːntʃ/ *n* (*boat*) lancha *f*
launching pad /ˈlɔːntʃɪŋpæd/ *n* plataforma *f* de lanzamiento
laund|er /ˈlɔːndə(r)/ *vt* lavar (y planchar). **~erette** *n* lavandería *f* automática. **~ress** *n* lavandera *f*. **~ry** /ˈlɔːndrɪ/ *n* (*place*) lavandería *f*; (*dirty clothes*) ropa *f* sucia; (*clean clothes*) colada *f*
laurel /ˈlɒrəl/ *n* laurel *m*
lava /ˈlɑːvə/ *n* lava *f*
lavatory /ˈlævətərɪ/ *n* retrete *m*. **public ~** servicios *mpl*
lavender /ˈlævəndə(r)/ *n* lavanda *f*
lavish /ˈlævɪʃ/ *a* ‹*person*› pródigo; (*plentiful*) abundante; (*lush*) suntuoso. ● *vt* prodigar. **~ly** *adv* profusamente
law /lɔː/ *n* ley *f*; (*profession, subject of study*) derecho *m*. **~-abiding** *a* observante de la ley. **~ and order** *n* orden *m* público. **~ court** *n* tribunal *m*. **~ful** *a* (*permitted by law*) lícito; (*recognized by law*) legítimo. **~fully** *adv* legalmente. **~less** *a* sin leyes
lawn /lɔːn/ *n* césped *m*. **~-mower** *n* cortacésped *f*. **~ tennis** *n* tenis *m* (sobre hierba)
lawsuit /ˈlɔːsuːt/ *n* pleito *m*
lawyer /ˈlɔɪə(r)/ *n* abogado *m*
lax /læks/ *a* descuidado; ‹*morals etc*› laxo
laxative /ˈlæksətɪv/ *n* laxante *m*
laxity /ˈlæksətɪ/ *n* descuido *m*
lay[1] /leɪ/ *vt* (*pt* **laid**) poner ‹*incl table, eggs*›; tender ‹*trap*›; formar ‹*plan*›. **~ hands on** echar mano a. **~ hold of** agarrar. **~ waste** asolar. **~ aside** dejar a un lado. **~ down** dejar a un lado; imponer ‹*condition*›. **~ into** (*sl*) dar una paliza a. **~ off** *vt* despedir ‹*worker*›; ● *vi* (*fam*) terminar. **~ on** (*provide*) proveer. **~ out** (*design*) disponer; (*display*) exponer; desembolsar ‹*money*›. **~ up** (*store*) guardar; obligar a guardar cama ‹*person*›
lay[2] /leɪ/ *a* (*non-clerical*) laico; ‹*opinion etc*› profano
lay[3] /leɪ/ *see* **lie**
layabout /ˈleɪəbaʊt/ *n* holgazán *m*
lay-by /ˈleɪbaɪ/ *n* apartadero *m*
layer /ˈleɪə(r)/ *n* capa *f*
layette /leɪˈet/ *n* canastilla *f*
layman /ˈleɪmən/ *n* lego *m*
lay-off /ˈleɪɒf/ *n* paro *m* forzoso
layout /ˈleɪaʊt/ *n* disposición *f*
laze /leɪz/ *vi* holgazanear; (*relax*) descansar
laz|iness /ˈleɪzɪnɪs/ *n* pereza *f*. **~y** *a* perezoso. **~y-bones** *n* holgazán *m*
lb. *abbr* (*pound*) libra *f*
lead[1] /liːd/ *vt* (*pt* **led**) conducir; dirigir ‹*team*›; llevar ‹*life*›; (*induce*) inducir a. ● *vi* (*go first*) ir delante; ‹*road*› ir, conducir; (*in cards*) salir. ● *n* mando *m*; (*clue*) pista *f*; (*leash*) correa *f*; (*in theatre*) primer papel *m*; (*wire*) cable *m*; (*example*) ejemplo *m*. **in the ~** en cabeza. **~ away** llevar. **~ up to** preparar el terreno para
lead[2] /led/ *n* plomo *m*; (*of pencil*) mina *f*. **~en** /ˈledn/ *a* de plomo
leader /ˈliːdə(r)/ *n* jefe *m*; (*leading article*) editorial *m*. **~ship** *n* dirección *f*
leading /ˈliːdɪŋ/ *a* principal; (*in front*) delantero. **~ article** *n* editorial *m*
leaf /liːf/ *n* (*pl* **leaves**) hoja *f*. ● *vi*. **~ through** hojear
leaflet /ˈliːflɪt/ *n* folleto *m*
leafy /ˈliːfɪ/ *a* frondoso
league /liːg/ *n* liga *f*. **be in ~ with** conchabarse con
leak /liːk/ *n* (*hole*) agujero *m*; (*of gas, liquid*) escape *m*; (*of information*) filtración *f*; (*in roof*) gotera *f*; (*in boat*) vía *f* de agua. ● *vi* ‹*receptacle, gas, liquid*› salirse; ‹*information*› filtrarse; (*drip*) gotear; ‹*boat*› hacer agua. ● *vt* dejar escapar; filtrar ‹*in-*

formation›. ~**age** *n* = **leak**. ~**y** *a* ‹*receptacle*› agujereado; ‹*roof*› que tiene goteras; ‹*boat*› que hace agua

lean[1] /li:n/ *vt* (*pt* **leaned** *or* **leant** /lent/) apoyar. ● *vi* inclinarse. ~ **against** apoyarse en. ~ **on** apoyarse en. ~**out** asomarse (**of** a). ~ **over** inclinarse

lean[2] /li:n/ *a* (**-er**, **-est**) magro. ● *n* carne *f* magra

leaning /'li:nɪŋ/ *a* inclinado. ● *n* inclinación *f*

leanness /'li:nnɪs/ *n* (*of meat*) magrez *f*; (*of person*) flaqueza *f*

lean-to /'li:ntu:/ *n* colgadizo *m*

leap /li:p/ *vi* (*pt* **leaped** *or* **leapt** /lept/) saltar. ● *n* salto *m*. ~**-frog** *n* salto *m*, saltacabrilla *f*. ● *vi* (*pt* **-frogged**) jugar a saltacabrilla. ~ **year** *n* año *m* bisiesto

learn /lɜ:n/ *vt/i* (*pt* **learned** *or* **learnt**) aprender (**to do** a hacer). ~**ed** /'lɜ:nɪd/ *a* culto. ~**er** /'lɜ:nə(r)/ *n* principiante *m*; (*apprentice*) aprendiz *m*; (*student*) estudiante *m* & *f*. ~**ing** *n* saber *m*

lease /li:s/ *n* arriendo *m*. ● *vt* arrendar

leash /li:ʃ/ *n* correa *f*

least /li:st/ *a*. **the** ~ (*smallest amount of*) mínimo; (*slightest*) menor; (*smallest*) más pequeño. ● *n* lo menos. **at** ~ por lo menos. **not in the** ~ en absoluto. ● *adv* menos

leather /'leðə(r)/ *n* piel *f*, cuero *m*

leave /li:v/ *vt* (*pt* **left**) dejar; (*depart from*) marcharse de. ~ **alone** dejar de tocar ‹*thing*›; dejar en paz ‹*person*›. **be left (over)** quedar. ● *vi* marcharse; ‹*train*› salir. ● *n* permiso *m*. **on** ~ (*mil*) de permiso. **take one's** ~ **of** despedirse de. ~ **out** omitir

leavings /'li:vɪŋz/ *npl* restos *mpl*

Leban|on /'lebənən/ *n* el Líbano *m*. ~**ese** /-'ni:z/ *a* & *n* libanés (*m*)

lecher /'letʃə(r)/ *n* libertino *m*. ~**ous** *a* lascivo. ~**y** *n* lascivia *f*

lectern /'lektɜ:n/ *n* atril *m*; (*in church*) facistol *m*

lecture /'lektʃə(r)/ *n* conferencia *f*; (*univ*) clase *f*; (*rebuke*) sermón *m*. ● *vt/i* dar una conferencia (a); (*univ*) dar clases (a); (*rebuke*) sermonear. ~**r** *n* conferenciante *m*; (*univ*) profesor *m*

led /led/ *see* **lead**[1]

ledge /ledʒ/ *n* repisa *f*; (*of window*) antepecho *m*

ledger /'ledʒə(r)/ *n* libro *m* mayor

lee /li:/ *n* sotavento *m*; (*fig*) abrigo *m*

leech /li:tʃ/ *n* sanguijuela *f*

leek /li:k/ *n* puerro *m*

leer /'lɪə(r)/ *vi*. ~ **(at)** mirar impúdicamente. ● *n* mirada *f* impúdica

leeway /'li:weɪ/ *n* deriva *f*; (*fig, freedom of action*) libertad *f* de acción. **make up** ~ recuperar los atrasos

left[1] /left/ *a* izquierdo. ● *adv* a la izquierda. ● *n* izquierda *f*

left[2] /left/ *see* **leave**

left: ~**-hand** *a* izquierdo. ~**-handed** *a* zurdo. ~**ist** *n* izquierdista *m* & *f*. ~ **luggage** *n* consigna *f*. ~**-overs** *npl* restos *mpl*

left-wing /left'wɪŋ/ *a* izquierdista

leg /leg/ *n* pierna *f*; (*of animal, furniture*) pata *f*; (*of pork*) pernil *m*; (*of lamb*) pierna *f*; (*of journey*) etapa *f*. **on its last** ~**s** en las últimas

legacy /'legəsɪ/ *n* herencia *f*

legal /'li:gl/ *a* (*permitted by law*) lícito; (*recognized by law*) legítimo; ‹*affairs etc*› jurídico. ~ **aid** *n* abogacía *f* de pobres. ~**ity** /-'gælətɪ/ *n* legalidad *f*. ~**ize** *vt* legalizar. ~**ly** *adv* legalmente

legation /lɪ'geɪʃn/ *n* legación *f*

legend /'ledʒənd/ *n* leyenda *f*. ~**ary** *a* legendario

leggings /'legɪŋz/ *npl* polainas *fpl*

legib|ility /'ledʒəbɪlətɪ/ *n* legibilidad *f*. ~**le** *a* legible. ~**ly** *a* legiblemente

legion /'li:dʒən/ *n* legión *f*

legislat|e /'ledʒɪsleɪt/ *vi* legislar. ~**ion** /-'leɪʃn/ *n* legislación *f*. ~**ive** *a* legislativo. ~**ure** /-eɪtʃə(r)/ *n* cuerpo *m* legislativo

legitima|cy /lɪ'dʒɪtɪməsɪ/ *f* legitimidad *f*. ~**te** *a* legítimo

leisure /'leʒə(r)/ *n* ocio *m*. **at one's** ~ cuando tenga tiempo. ~**ly** *adv* sin prisa

lemon /'lemən/ *n* limón *m*. ~**ade** /lemə'neɪd/ *n* (*fizzy*) gaseosa *f* (de limón); (*still*) limonada *f*

lend /lend/ *vt* (*pt* **lent**) prestar. ~ **itself to** prestarse a. ~**er** *n* prestador *m*; (*moneylender*) prestamista *m* & *f*. ~**ing** *n* préstamo *m*. ~**ing library** *n* biblioteca *f* de préstamo

length /leŋθ/ *n* largo *m*; (*in time*) duración *f*; (*of cloth*) largo *m*; (*of road*) tramo *m*. **at** ~ (*at last*) por fin. **at (great)** ~ detalladamente. ~**en**

/ˈleŋθən/ *vt* alargar. ● *vi* alargarse. **~ways** *adv* a lo largo. **~y** *a* largo
lenien|cy /ˈliːnɪənsɪ/ *n* indulgencia *f*. **~t** *a* indulgente. **~tly** *adv* con indulgencia
lens /lens/ *n* lente *f*. **contact ~es** *npl* lentillas *fpl*
lent /lent/ *see* **lend**
Lent /lent/ *n* cuaresma *f*
lentil /ˈlentl/ *n* (*bean*) lenteja *f*
Leo /ˈliːəʊ/ *n* (*astr*) Leo *m*
leopard /ˈlepəd/ *n* leopardo *m*
leotard /ˈliːətɑːd/ *n* leotardo *m*
lep|er /ˈlepə(r)/ *n* leproso *m*. **~rosy** /ˈleprəsɪ/ *n* lepra *f*
lesbian /ˈlezbɪən/ *n* lesbiana *f*. ● *a* lesbiano
lesion /ˈliːʒn/ *n* lesión *f*
less /les/ *a* (*in quantity*) menos; (*in size*) menor. ● *adv & prep* menos. **~ than** menos que; (*with numbers*) menos de. ● *n* menor *m*. **~ and ~** cada vez menos. **none the ~** sin embargo. **~en** /ˈlesn/ *vt/i* disminuir. **~er** /ˈlesə(r)/ *a* menor
lesson /ˈlesn/ *n* clase *f*
lest /lest/ *conj* por miedo de que
let /let/ *vt* (*pt* **let**, *pres p* **letting**) dejar; (*lease*) alquilar. **~ me do it** déjame hacerlo. ● *v aux*. **~'s go!** ¡vamos!, ¡vámonos! **~'s see** (vamos) a ver. **~'s talk/drink** hablemos/bebamos. ● *n* alquiler *m*. **~ down** bajar; (*deflate*) desinflar; (*fig*) defraudar. **~ go** soltar. **~ in** dejar entrar. **~ off** disparar ‹*gun*›; (*cause to explode*) hacer explotar; hacer estallar ‹*firework*›; (*excuse*) perdonar. **~ off steam** (*fig*) desfogarse. **~ on** (*sl*) revelar. **~ o.s. in for** meterse en. **~ out** dejar salir. **~ through** dejar pasar. **~ up** disminuir. **~-down** *n* desilusión *f*
lethal /ˈliːθl/ *a* ‹*dose, wound*› mortal; ‹*weapon*› mortífero
letharg|ic /lɪˈtɑːdʒɪk/ *a* letárgico. **~y** /ˈleθədʒɪ/ *n* letargo *m*
letter /ˈletə(r)/ *n* (*of alphabet*) letra *f*; (*written message*) carta *f*. **~-bomb** *n* carta *f* explosiva. **~-box** *n* buzón *m*. **~-head** *n* membrete *m*. **~ing** *n* letras *fpl*
lettuce /ˈletɪs/ *n* lechuga *f*
let-up /ˈletʌp/ *n* (*fam*) descanso *m*
leukaemia /luːˈkiːmɪə/ *n* leucemia *f*
level /ˈlevl/ *a* (*flat*) llano; (*on surface*) horizontal; (*in height*) a nivel; (*in score*) igual; ‹*spoonful*› raso. ● *n* nivel *m*. **be on the ~** (*fam*) ser honrado. ● *vt* (*pt* **levelled**) nivelar; (*aim*) apuntar. **~ crossing** *n* paso *m* a nivel. **~-headed** *a* juicioso
lever /ˈliːvə(r)/ *n* palanca *f*. ● *vt* apalancar. **~age** /ˈliːvərɪdʒ/ *n* apalancamiento *m*
levity /ˈlevətɪ/ *n* ligereza *f*
levy /ˈlevɪ/ *vt* exigir ‹*tax*›. ● *n* impuesto *m*
lewd /luːd/ *a* (**-er**, **-est**) lascivo
lexicography /leksɪˈkɒgrəfɪ/ *n* lexicografía *f*
lexicon /ˈleksɪkən/ *n* léxico *m*
liable /ˈlaɪəbl/ *a*. **be ~ to do** tener tendencia a hacer. **~ for** responsable de. **~ to** susceptible de; expuesto a ‹*fine*›
liability /laɪəˈbɪlətɪ/ *n* responsabilidad *f*; (*disadvantage, fam*) inconveniente *m*. **liabilities** *npl* (*debts*) deudas *fpl*
liais|e /lɪˈeɪz/ *vi* hacer un enlace, enlazar. **~on** /lɪˈeɪzɒn/ *n* enlace *m*; (*love affair*) lío *m*
liar /ˈlaɪə(r)/ *n* mentiroso *m*
libel /ˈlaɪbl/ *n* libelo *m*. ● *vt* (*pt* **libelled**) difamar (por escrito)
Liberal /ˈlɪbərəl/ *a & n* liberal (*m & f*)
liberal /ˈlɪbərəl/ *a* liberal; (*generous*) generoso; (*tolerant*) tolerante. **~ly** *adv* liberalmente; (*generously*) generosamente; (*tolerantly*) tolerantemente
liberat|e /ˈlɪbəreɪt/ *vt* liberar. **~ion** /-ˈreɪʃn/ *n* liberación *f*
libertine /ˈlɪbətiːn/ *n* libertino *m*
liberty /ˈlɪbətɪ/ *n* libertad *f*. **be at ~ to** estar autorizado para. **take liberties** tomarse libertades. **take the ~ of** tomarse la libertad de
libido /lɪˈbiːdəʊ/ *n* (*pl* **-os**) libido *m*
Libra /ˈliːbrə/ *n* (*astr*) Libra *f*
librar|ian /laɪˈbreərɪən/ *n* bibliotecario *m*. **~y** /ˈlaɪbrərɪ/ *n* biblioteca *f*
libretto /lɪˈbretəʊ/ *n* (*pl* **-os**) libreto *m*
Libya /ˈlɪbɪə/ *n* Libia *f*. **~n** *a & n* libio (*m*)
lice /laɪs/ *see* **louse**
licence /ˈlaɪsns/ *n* licencia *f*, permiso *m*; (*fig, liberty*) libertad *f*. **~ plate** *n* (placa *f* de) matrícula *f*. **driving ~** carné *m* de conducir
license /ˈlaɪsns/ *vt* autorizar
licentious /laɪˈsenʃəs/ *a* licencioso
lichen /ˈlaɪkən/ *n* liquen *m*

lick /lɪk/ *vt* lamer; (*defeat, sl*) dar una paliza a. ~ **one's chops** relamerse. ● *n* lametón *m*
licorice /'lɪkərɪs/ *n* (*Amer*) regaliz *m*
lid /lɪd/ *n* tapa *f*; (*of pan*) cobertera *f*
lido /'li:dəʊ/ *n* (*pl* **-os**) piscina *f*
lie[1] /laɪ/ *vi* (*pt* **lay**, *pp* **lain**, *pres p* **lying**) echarse; (*state*) estar echado; (*remain*) quedarse; (*be*) estar, encontrarse; (*in grave*) yacer. **be lying** estar echado. ~ **down** acostarse. ~ **low** quedarse escondido
lie[2] /laɪ/ *n* mentira *f*. ● *vi* (*pt* **lied**, *pres p* **lying**) mentir. **give the ~ to** desmentir
lie-in /laɪ'ɪn/ *n*. **have a ~-in** quedarse en la cama
lieu /lju:/ *n*. **in ~ of** en lugar de
lieutenant /lef'tenənt/ *n* (*mil*) teniente *m*
life /laɪf/ *n* (*pl* **lives**) vida *f*. **~belt** *n* cinturón *m* salvavidas. **~boat** *n* lancha *f* de salvamento; (*on ship*) bote *m* salvavidas. **~buoy** *n* boya *f* salvavidas. ~ **cycle** *n* ciclo *m* vital. **~-guard** *n* bañero *m*. **~-jacket** *n* chaleco *m* salvavidas. **~less** *a* sin vida. **~like** *a* natural. **~line** *n* cuerda *f* salvavidas; (*fig*) cordón *m* umbilical. **~long** *a* de toda la vida. **~-size(d)** *a* de tamaño natural. **~time** *n* vida *f*
lift /lɪft/ *vt* levantar; (*steal, fam*) robar. ● *vi* ‹*fog*› disiparse. ● *n* ascensor *m*, elevador *m* (*LAm*). **give a ~ to s.o.** llevar a uno en su coche, dar aventón a uno (*LAm*). **~-off** *n* (*aviat*) despegue *m*
ligament /'lɪgəmənt/ *n* ligamento *m*
light[1] /laɪt/ *n* luz *f*; (*lamp*) lámpara *f*, luz *f*; (*flame*) fuego *m*; (*headlight*) faro *m*. **bring to ~** sacar a luz. **come to ~** salir a luz. **have you got a ~?** ¿tienes fuego? **the ~s** *npl* (*auto, traffic signals*) el semáforo *m*. ● *a* claro. ● *vt* (*pt* **lit** *or* **lighted**) encender; (*illuminate*) alumbrar. ~ **up** *vt/i* iluminar(se)
light[2] /laɪt/ *a* (**-er**, **-est**) (*not heavy*) ligero
lighten[1] /'laɪtn/ *vt* (*make less heavy*) aligerar
lighten[2] /'laɪtn/ *vt* (*give light to*) iluminar; (*make brighter*) aclarar
lighter /'laɪtə(r)/ *n* (*for cigarettes*) mechero *m*
light-fingered /laɪt'fɪŋgəd/ *a* largo de uñas
light-headed /laɪt'hedɪd/ *a* (*dizzy*) mareado; (*frivolous*) casquivano
light-hearted /laɪt'hɑ:tɪd/ *a* alegre
lighthouse /'laɪthaʊs/ *n* faro *m*
lighting /'laɪtɪŋ/ *n* (*system*) alumbrado *m*; (*act*) iluminación *f*
light: **~ly** *adv* ligeramente. **~ness** *n* ligereza *f*
lightning /'laɪtnɪŋ/ *n* relámpago *m*. ● *a* relámpago
lightweight /'laɪtweɪt/ *a* ligero. ● *n* (*boxing*) peso *m* ligero
light-year /'laɪtjɪə(r)/ *n* año *m* luz
like[1] /laɪk/ *a* parecido. ● *prep* como. ● *conj* (*fam*) como. ● *n* igual. **the ~s of you** la gente como tú
like[2] /laɪk/ *vt* gustarle (a uno). **I ~ chocolate** me gusta el chocolate. **I should ~** quisiera. **they ~ swimming** (a ellos) les gusta nadar. **would you ~?** ¿quieres? **~able** *a* simpático. **~s** *npl* gustos *mpl*
likelihood /'laɪklɪhʊd/ *n* probabilidad *f*
likely *a* (**-ier**, **-iest**) probable. **he is ~ to come** es probable que venga. ● *adv* probablemente. **not ~!** ¡ni hablar!
like-minded /laɪk'maɪndɪd/ *a*. **be ~** tener las mismas opiniones
liken /'laɪkən/ *vt* comparar
likeness /'laɪknɪs/ *n* parecido *m*. **be a good ~** parecerse mucho
likewise /'laɪkwaɪz/ *adv* (*also*) también; (*the same way*) lo mismo
liking /'laɪkɪŋ/ *n* (*for thing*) afición *f*; (*for person*) simpatía *f*
lilac /'laɪlək/ *n* lila *f*. ● *a* color de lila
lilt /lɪlt/ *n* ritmo *m*
lily /'lɪlɪ/ *n* lirio *m*. **~ of the valley** lirio *m* de los valles
limb /lɪm/ *n* miembro *m*. **out on a ~** aislado
limber /'lɪmbə(r)/ *vi*. ~ **up** hacer ejercicios preliminares
limbo /'lɪmbəʊ/ *n* limbo *m*. **be in ~** (*forgotten*) estar olvidado
lime[1] /laɪm/ *n* (*white substance*) cal *f*
lime[2] /laɪm/ *n* (*fruit*) lima *f*
lime[3] /laɪm/ *n*. **~(-tree)** (*linden tree*) tilo *m*
limelight /'laɪmlaɪt/ *n*. **be in the ~** estar muy a la vista
limerick /'lɪmərɪk/ *n* quintilla *f* humorística
limestone /'laɪmstəʊn/ *n* caliza *f*
limit /'lɪmɪt/ *n* límite *m*. ● *vt* limitar. **~ation** /-'teɪʃn/ *n* limitación *f*. **~ed**

a limitado. **~ed company** *n* sociedad *f* anónima
limousine /ˈlɪməzi:n/ *n* limusina *f*
limp[1] /lɪmp/ *vi* cojear. ● *n* cojera *f*. **have a ~** cojear
limp[2] /lɪmp/ *a* (**-er, -est**) flojo
limpid /ˈlɪmpɪd/ *a* límpido
linctus /ˈlɪŋktəs/ *n* jarabe *m* (para la tos)
line[1] /laɪn/ *n* línea *f*; (*track*) vía *f*; (*wrinkle*) arruga *f*; (*row*) fila *f*; (*of poem*) verso *m*; (*rope*) cuerda *f*; (*of goods*) surtido *m*; (*queue, Amer*) cola *f*. **in ~ with** de acuerdo con. ● *vt* (*on paper etc*) rayar; bordear ‹*streets etc*›. **~ up** alinearse; (*in queue*) hacer cola
line[2] /laɪn/ *vt* forrar; (*fill*) llenar
lineage /ˈlɪnɪɪdʒ/ *n* linaje *m*
linear /ˈlɪnɪə(r)/ *a* lineal
linen /ˈlɪnɪn/ *n* (*sheets etc*) ropa *f* blanca; (*material*) lino *m*
liner /ˈlaɪnə(r)/ *n* transatlántico *m*
linesman /ˈlaɪnzmən/ *n* (*football*) juez *m* de línea
linger /ˈlɪŋgə(r)/ *vi* tarder en marcharse; ‹*smells etc*› persistir. **~ over** dilatarse en
lingerie /ˈlænʒərɪ/ *n* ropa *f* interior, lencería *f*
lingo /ˈlɪŋgəʊ/ *n* (*pl* **-os**) idioma *m*; (*specialized vocabulary*) jerga *f*
linguist /ˈlɪŋgwɪst/ *n* (*specialist in languages*) polígIota *m* & *f*; (*specialist in linguistics*) lingüista *m* & *f*. **~ic** /lɪŋˈgwɪstɪk/ *a* lingüístico. **~ics** *n* lingüística *f*
lining /ˈlaɪnɪŋ/ *n* forro *m*; (*auto, of brakes*) guarnición *f*
link /lɪŋk/ *n* (*of chain*) eslabón *m*; (*fig*) lazo *m*. ● *vt* eslabonar; (*fig*) enlazar. **~ up with** reunirse con. **~age** *n* enlace *m*
links /lɪŋks/ *n invar* campo *m* de golf
lino /ˈlaɪnəʊ/ *n* (*pl* **-os**) linóleo *m*. **~leum** /lɪˈnəʊlɪəm/ *n* linóleo *m*
lint /lɪnt/ *n* (*med*) hilas *fpl*; (*fluff*) pelusa *f*
lion /ˈlaɪən/ *n* león *m*. **the ~'s share** la parte *f* del león. **~ess** *n* leona *f*
lionize /ˈlaɪənaɪz/ *vt* tratar como una celebridad
lip /lɪp/ *n* labio *m*; (*edge*) borde *m*. **pay ~ service to** aprobar de boquilla. **stiff upper ~** *n* imperturbabilidad *f*. **~-read** *vt/i* leer en los labios. **~salve** *n* crema *f* para los labios. **~stick** *n* lápiz *m* de labios.
liquefy /ˈlɪkwɪfaɪ/ *vt/i* licuar(se)
liqueur /lɪˈkjʊə(r)/ *n* licor *m*
liquid /ˈlɪkwɪd/ *a* & *n* líquido (*m*)
liquidat|e /ˈlɪkwɪdeɪt/ *vt* liquidar. **~ion** /-ˈdeɪʃn/ *n* liquidación *f*
liquidize /ˈlikwɪdaɪz/ *vt* licuar. **~r** *n* licuadora *f*
liquor /ˈlɪkə(r)/ *n* bebida *f* alcohólica
liquorice /ˈlɪkərɪs/ *n* regaliz *m*
lira /ˈlɪərə/ *n* (*pl* **lire** /ˈlɪəreɪ/ *or* **liras**) lira *f*
lisle /laɪl/ *n* hilo *m* de Escocia
lisp /lɪsp/ *n* ceceo *m*. **speak with a ~** cecear. ● *vi* cecear
lissom /ˈlɪsəm/ *a* flexible, ágil
list[1] /lɪst/ *n* lista *f*. ● *vt* hacer una lista de; (*enter in a list*) inscribir
list[2] /lɪst/ *vi* ‹*ship*› escorar
listen /ˈlɪsn/ *vi* escuchar. **~ in (to)** escuchar. **~ to** escuchar. **~er** *n* oyente *m* & *f*
listless /ˈlɪstlɪs/ *a* apático
lit /lɪt/ *see* **light**[1]
litany /ˈlɪtənɪ/ *n* letanía *f*
literacy /ˈlɪtərəsɪ/ *n* capacidad *f* de leer y escribir
literal /ˈlɪtərəl/ *a* literal; (*fig*) prosaico. **~ly** *adv* al pie de la letra, literalmente
literary /ˈlɪtərərɪ/ *a* literario
literate /ˈlɪtərət/ *a* que sabe leer y escribir
literature /ˈlɪtərətʃə(r)/ *n* literatura *f*; (*fig*) impresos *mpl*
lithe /laɪð/ *a* ágil
lithograph /ˈlɪθəgrɑ:f/ *n* litografía *f*
litigation /lɪtɪˈgeɪʃn/ *n* litigio *m*
litre /ˈli:tə(r)/ *n* litro *m*
litter /ˈlɪtə(r)/ *n* basura *f*; (*of animals*) camada *f*. ● *vt* ensuciar; (*scatter*) esparcir. **~ed with** lleno de. **~-bin** *n* papelera *f*
little /ˈlɪtl/ *a* pequeño; (*not much*) poco de. ● *n* poco *m*. **a ~** un poco. **a ~ water** un poco de agua. ● *adv* poco. **~ by ~** poco a poco. **~ finger** *n* meñique *m*
liturgy /ˈlɪtədʒɪ/ *n* liturgia *f*
live[1] /lɪv/ *vt/i* vivir. **~ down** lograr borrar. **~ it up** echar una cana al aire. **~ on** (*feed o.s. on*) vivir de; (*continue*) perdurar. **~ up to** vivir de acuerdo con; cumplir ‹*a promise*›
live[2] /laɪv/ *a* vivo; ‹*wire*› con corriente; ‹*broadcast*› en directo. **be a ~ wire** ser una persona enérgica
livelihood /ˈlaɪvlɪhʊd/ *n* sustento *m*
livel|iness /ˈlaɪvlɪnɪs/ *n* vivacidad *f*. **~y** *a* (**-ier, -iest**) vivo

liven /'laɪvn/ *vt/i.* ~ **up** animar(se); (*cheer up*) alegrar(se)
liver /'lɪvə(r)/ *n* hígado *m*
livery /'lɪvərɪ/ *n* librea *f*
livestock /'laɪvstɒk/ *n* ganado *m*
livid /'lɪvɪd/ *a* lívido; (*angry*, *fam*) furioso
living /'lɪvɪŋ/ *a* vivo. ● *n* vida *f.* **~-room** *n* cuarto *m* de estar, cuarto *m* de estancia (*LAm*)
lizard /'lɪzəd/ *n* lagartija *f*; (*big*) lagarto *m*
llama /'lɑ:mə/ *n* llama *f*
load /ləʊd/ *n* (*incl elec*) carga *f*; (*quantity*) cantidad *f*; (*weight*, *strain*) peso *m.* ● *vt* cargar. **~ed** *a* ‹*incl dice*› cargado; (*wealthy*, *sl*) muy rico. **~s of** (*fam*) montones de
loaf[1] /ləʊf/ *n* (*pl* **loaves**) pan *m*; (*stick of bread*) barra *f*
loaf[2] /ləʊf/ *vi.* ~ **(about)** holgazanear. **~er** *n* holgazán *m*
loam /ləʊm/ *n* marga *f*
loan /ləʊn/ *n* préstamo *m.* **on** ~ prestado. ● *vt* prestar
loath /ləʊθ/ *a* poco dispuesto (**to** a)
loath|e /ləʊð/ *vt* odiar. **~ing** *n* odio *m* (**of** a). **~some** *a* odioso
lobby /'lɒbɪ/ *n* vestíbulo *m*; (*pol*) grupo *m* de presión. ● *vt* hacer presión sobre
lobe /ləʊb/ *n* lóbulo *m*
lobster /'lɒbstə(r)/ *n* langosta *f*
local /'ləʊkl/ *a* local. ● *n* (*pub*, *fam*) bar *m.* **the ~s** los vecinos *mpl*
locale /ləʊ'kɑ:l/ *n* escenario *m*
local government /ləʊkl'gʌvənmənt/ *n* gobierno *m* municipal
locality /ləʊ'kælətɪ/ *n* localidad *f*
localized /'ləʊkəlaɪzd/ *a* localizado
locally /'ləʊkəlɪ/ *adv* localmente; (*nearby*) en la localidad
locate /ləʊ'keɪt/ *vt* (*situate*) situar; (*find*) encontrar
location /ləʊ'keɪʃn/ *n* colocación *f*; (*place*) situación *f.* **on** ~ fuera del estudio. **to film on ~ in Andalusia** rodar en Andalucía
lock[1] /lɒk/ *n* (*of door etc*) cerradura *f*; (*on canal*) esclusa *f.* ● *vt/i* cerrar(se) con llave. ~ **in** encerrar. ~ **out** cerrar la puerta a. ~ **up** encerrar
lock[2] /lɒk/ *n* (*of hair*) mechón *m.* **~s** *npl* pelo *m*
locker /'lɒkə(r)/ *n* armario *m*
locket /'lɒkɪt/ *n* medallón *m*
lock-out /'lɒkaʊt/ *n* lock-out *m*
locksmith /'lɒksmɪθ/ *n* cerrajero *m*
locomotion /ləʊkə'məʊʃn/ *n* locomoción *f*
locomotive /ləʊkə'məʊtɪv/ *n* locomotora *f*
locum /'ləʊkəm/ *n* interino *m*
locust /'ləʊkəst/ *n* langosta *f*
lodge /lɒdʒ/ *n* (*in park*) casa *f* del guarda; (*of porter*) portería *f.* ● *vt* alojar; presentar ‹*complaint*›; depositar ‹*money*›. ● *vi* alojarse. **~r** /-ə(r)/ *n* huésped *m*
lodgings /'lɒdʒɪŋz/ *n* alojamiento *m*; (*room*) habitación *f*
loft /lɒft/ *n* desván *m*
lofty /'lɒftɪ/ *a* (**-ier**, **-iest**) elevado; (*haughty*) altanero
log /lɒg/ *n* (*of wood*) leño *m*; (*naut*) cuaderno *m* de bitácora. **sleep like a ~** dormir como un lirón. ● *vt* (*pt* **logged**) apuntar; (*travel*) recorrer
logarithm /'lɒgərɪðəm/ *n* logaritmo *m*
log-book /'lɒgbʊk/ *n* cuaderno *m* de bitácora; (*aviat*) diario *m* de vuelo
loggerheads /'lɒgəhedz/ *npl.* **be at ~ with** estar a matar con
logic /'lɒdʒɪk/ *a* lógica *f.* **~al** *a* lógico. **~ally** *adv* lógicamente
logistics /lə'dʒɪstɪks/ *n* logística *f*
logo /'ləʊgəʊ/ *n* (*pl* **-os**) logotipo *m*
loin /lɔɪn/ *n* (*culin*) solomillo *m.* **~s** *npl* ijadas *fpl*
loiter /'lɔɪtə(r)/ *vi* holgazanear
loll /lɒl/ *vi* repantigarse
loll|ipop /'lɒlɪpɒp/ *n* (*boiled sweet*) pirulí *m.* **~y** *n* (*iced*) polo *m*; (*money*, *sl*) dinero *m*
London /'lʌndən/ *n* Londres *m.* ● *a* londinense. **~er** *n* londinense *m* & *f*
lone /ləʊn/ *a* solitario. **~ly** /'ləʊnlɪ/ *a* (**-ier**, **-iest**) solitario. **feel ~ly** sentirse muy solo. **~r** /'ləʊnə(r)/ *n* solitario *m.* **~some** *a* solitario
long[1] /lɒŋ/ *a* (**-er**, **-est**) largo. **a ~ time** mucho tiempo. **how ~ is it?** ¿cuánto tiene de largo? **in the ~ run** a la larga. ● *adv* largo/mucho tiempo. **as ~ as** (*while*) mientras; (*provided that*) con tal que (+ *subjunctive*). **before ~** dentro de poco. **so ~!** ¡hasta luego! **so ~ as** (*provided that*) con tal que (+ *subjunctive*)
long[2] /lɒŋ/ *vi.* ~ **for** anhelar
long-distance /lɒŋ'dɪstəns/ *a* de larga distancia. ~ **(tele)phone call** *n* conferencia *f*
longer /'lɒŋgə(r)/ *adv.* **no ~er** ya no

longevity /lɒn'dʒevətɪ/ *n* longevidad *f*

long: ~ **face** *n* cara *f* triste. ~**hand** *n* escritura *f* a mano. ~ **johns** *npl* (*fam*) calzoncillos *mpl* largos. ~ **jump** *n* salto *m* de longitud

longing /'lɒŋɪŋ/ *n* anhelo *m*, ansia *f*

longitude /'lɒŋgɪtju:d/ *n* longitud *f*

long: ~**-playing record** *n* elepé *m*. ~**-range** *a* de gran alcance. ~**-sighted** *a* présbita. ~**-standing** *a* de mucho tiempo. ~**-suffering** *a* sufrido. ~**-term** *a* a largo plazo. ~ **wave** *n* onda *f* larga. ~**-winded** *a* ‹*speaker etc*› prolijo

loo /lu:/ *n* (*fam*) servicios *mpl*

look /lʊk/ *vt* mirar; (*seem*) parecer; representar ‹*age*›. ● *vi* mirar; (*seem*) parecer; (*search*) buscar. ● *n* mirada *f*; (*appearance*) aspecto *m*. ~ **after** ocuparse de; cuidar ‹*person*›. ~ **at** mirar. ~ **down on** despreciar. ~ **for** buscar. ~ **forward to** esperar con ansia. ~ **in on** pasar por casa de. ~ **into** investigar. ~ **like** (*resemble*) parecerse a. ~ **on to** ‹*room, window*› dar a. ~ **out** tener cuidado. ~ **out for** buscar; (*watch*) tener cuidado con. ~ **round** volver la cabeza. ~ **through** hojear. ~ **up** buscar ‹*word*›; (*visit*) ir a ver. ~ **up to** respetar. ~**er-on** *n* espectador *m*. ~**ing-glass** *n* espejo *m*. ~**-out** *n* (*mil*) atalaya *f*; (*person*) vigía *m*. ~**s** *npl* belleza *f*. **good** ~**s** *mpl* belleza *f*

loom[1] /lu:m/ *n* telar *m*

loom[2] /lu:m/ *vi* aparecerse

loony /'lu:nɪ/ *a* & *n* (*sl*) chiflado (*m*) (*fam*), loco (*m*). ~ **bin** *n* (*sl*) manicomio *m*

loop /lu:p/ *n* lazo *m*. ● *vt* hacer presilla con

loophole /'lu:phəʊl/ *n* (*in rule*) escapatoria *f*

loose /lu:s/ *a* (**-er**, **-est**) (*untied*) suelto; (*not tight*) flojo; (*inexact*) vago; (*immoral*) inmoral; (*not packed*) suelto. **be at a** ~ **end**, **be at** ~ **ends** (*Amer*) no tener nada que hacer. ~**ly** *adv* sueltamente; (*roughly*) aproximadamente. ~**n** /'lu:sn/ *vt* (*slacken*) aflojar; (*untie*) desatar

loot /lu:t/ *n* botín *m*. ● *vt* saquear. ~**er** *n* saqueador *m*. ~**ing** *n* saqueo *m*

lop /lɒp/ *vt* (*pt* **lopped**). ~ **off** cortar

lop-sided /lɒp'saɪdɪd/ *a* ladeado

loquacious /ləʊ'kweɪʃəs/ *a* locuaz

lord /lɔ:d/ *n* señor *m*; (*British title*) lord *m*. **(good) L~!** ¡Dios mío! **the L~** el Señor *m*. **the (House of) L~s** la Cámara *f* de los Lores. ~**ly** señorial; (*haughty*) altivo. ~**ship** *n* señoría *f*

lore /lɔ:(r)/ *n* tradiciones *fpl*

lorgnette /lɔ:'njet/ *n* impertinentes *mpl*

lorry /'lɒrɪ/ *n* camión *m*

lose /lu:z/ *vt/i* (*pt* **lost**) perder. ~**r** *n* perdedor *m*

loss /lɒs/ *n* pérdida *f*. **be at a** ~ estar perplejo. **be at a** ~ **for words** no encontrar palabras. **be at a** ~ **to** no saber cómo

lost /lɒst/ *see* **lose**. ● *a* perdido. ~ **property** *n*, ~ **and found** (*Amer*) *n* oficina *f* de objetos perdidos. **get** ~ perderse

lot /lɒt/ *n* (*fate*) suerte *f*; (*at auction*) lote *m*; (*land*) solar *m*. **a** ~ **(of)** muchos. **quite a** ~ **of** (*fam*) bastante. ~**s (of)** (*fam*) muchos. **the** ~ todos *mpl*

lotion /'ləʊʃn/ *n* loción *f*

lottery /'lɒtərɪ/ *n* lotería *f*

lotto /'lɒtəʊ/ *n* lotería *f*

lotus /'ləʊtəs/ *n* (*pl* **-uses**) loto *m*

loud /laʊd/ *a* (**-er**, **-est**) fuerte; (*noisy*) ruidoso; (*gaudy*) chillón. **out** ~ en voz alta. ~ **hailer** *n* megáfono *m*. ~**ly** *adv* (*speak etc*) en voz alta; (*noisily*) ruidosamente. ~**speaker** *n* altavoz *m*

lounge /laʊndʒ/ *vi* repantigarse. ● *n* salón *m*. ~ **suit** *n* traje *m* de calle

louse /laʊs/ *n* (*pl* **lice**) piojo *m*

lousy /'laʊzɪ/ *a* (**-ier**, **-iest**) piojoso; (*bad, sl*) malísimo

lout /laʊt/ *n* patán *m*

lovable /'lʌvəbl/ *a* adorable

love /lʌv/ *n* amor *m*; (*tennis*) cero *m*. **be in** ~ **with** estar enamorado de. **fall in** ~ **with** enamorarse de. ● *vt* querer ‹*person*›; gustarle mucho a uno, encantarle a uno ‹*things*›. **I** ~ **milk** me encanta la leche. ~ **affair** *n* amores *mpl*

lovely /'lʌvlɪ/ *a* (**-ier**, **-iest**) hermoso; (*delightful, fam*) precioso. **have a** ~ **time** divertirse

lover /'lʌvə(r)/ *n* amante *m & f*

lovesick /'lʌvsɪk/ *a* atortolado

loving /'lʌvɪŋ/ *a* cariñoso

low[1] /ləʊ/ *a* & *adv* (**-er**, **-est**) bajo. ● *n* (*low pressure*) área *f* de baja presión

low[2] /ləʊ/ *vi* mugir

lowbrow /'ləʊbraʊ/ *a* poco culto
low-cut /'ləʊkʌt/ *a* escotado
low-down /'ləʊdaʊn/ *a* bajo. ● *n* (*sl*) informes *mpl*
lower /'ləʊə(r)/ *a & adv see* **low**². ● *vt* bajar. ~ **o.s.** envilecerse
low-key /'ləʊ'ki:/ *a* moderado
lowlands /'ləʊləndz/ *npl* tierra *f* baja
lowly /'ləʊlɪ/ *a* (**-ier, -iest**) humilde
loyal /'lɔɪəl/ *a* leal. ~**ly** *adv* lealmente. ~**ty** *n* lealtad *f*
lozenge /'lɒzɪŋdʒ/ *n* (*shape*) rombo *m*; (*tablet*) pastilla *f*
LP /el'pi:/ *abbr* (*long-playing record*) elepé *m*
Ltd /'lɪmɪtɪd/ *abbr* (*Limited*) S.A., Sociedad Anónima
lubrica|nt /'lu:brɪkənt/ *n* lubricante *m*. ~**te** /-'keɪt/ *vt* lubricar. ~**tion** /-'keɪʃn/ *n* lubricación *f*
lucid /'lu:sɪd/ *a* lúcido. ~**ity** /-'sɪdətɪ/ *n* lucidez *f*
luck /lʌk/ *n* suerte *f*. **bad** ~ *n* mala suerte *f*. ~**ily** /'lʌkɪlɪ/ *adv* afortunadamente. ~**y** *a* (**-ier, -iest**) afortunado
lucrative /'lu:krətɪv/ *a* lucrativo
lucre /'lu:kə(r)/ *n* (*pej*) dinero *m*. **filthy** ~ vil metal *m*
ludicrous /'lu:dɪkrəs/ *a* ridículo
lug /lʌg/ *vt* (*pt* **lugged**) arrastrar
luggage /'lʌgɪdʒ/ *n* equipaje *m*. ~**-rack** *n* rejilla *f*. ~**-van** *n* furgón *m*
lugubrious /lu:'gu:brɪəs/ *a* lúgubre
lukewarm /'lu:kwɔ:m/ *a* tibio
lull /lʌl/ *vt* (*soothe, send to sleep*) adormecer; (*calm*) calmar. ● *n* periodo *m* de calma
lullaby /'lʌləbaɪ/ *n* canción *f* de cuna
lumbago /lʌm'beɪgəʊ/ *n* lumbago *m*
lumber /'lʌmbə(r)/ *n* trastos *mpl* viejos; (*wood*) maderos *mpl*. ● *vt*. ~ **s.o. with** hacer que uno cargue con. ~**jack** *n* leñador *m*
luminous /'lu:mɪnəs/ *a* luminoso
lump¹ /'lʌmp/ *n* protuberancia *f*; (*in liquid*) grumo *m*; (*of sugar*) terrón *m*; (*in throat*) nudo *m*. ● *vt*. ~ **together** agrupar
lump² /lʌmp/ *vt*. ~ **it** (*fam*) aguantarlo
lump: ~ **sum** *n* suma *f* global. ~**y** *a* ‹*sauce*› grumoso; (*bumpy*) cubierto de protuberancias
lunacy /'lu:nəsɪ/ *n* locura *f*
lunar /'lu:nə(r)/ *a* lunar
lunatic /'lu:nətɪk/ *n* loco *m*
lunch /lʌntʃ/ *n* comida *f*, almuerzo *m*. ● *vi* comer
luncheon /'lʌntʃən/ *n* comida *f*, almuerzo *m*. ~ **meat** *n* carne *f* en lata. ~ **voucher** *n* vale *m* de comida
lung /lʌŋ/ *n* pulmón *m*
lunge /lʌndʒ/ *n* arremetida *f*
lurch¹ /lɜ:tʃ/ *vi* tambalearse
lurch² /lɜ:tʃ/ *n*. **leave in the** ~ dejar en la estacada
lure /ljʊə(r)/ *vt* atraer. ● *n* (*attraction*) atractivo *m*
lurid /'ljʊərɪd/ *a* chillón; (*shocking*) espeluznante
lurk /lɜ:k/ *vi* esconderse; (*in ambush*) estar al acecho; (*prowl*) rondar
luscious /'lʌʃəs/ *a* delicioso
lush /lʌʃ/ *a* exuberante. ● *n* (*Amer, sl*) borracho *m*
lust /lʌst/ *n* lujuria *f*; (*fig*) ansia *f*. ● *vi*. ~ **after** codiciar. ~**ful** *a* lujurioso
lustre /'lʌstə(r)/ *n* lustre *m*
lusty /'lʌstɪ/ *a* (**-ier, -iest**) fuerte
lute /lu:t/ *n* laúd *m*
Luxemburg /'lʌksəmbɜ:g/ *n* Luxemburgo *m*
luxuriant /lʌg'zjʊərɪənt/ *a* exuberante
luxur|ious /lʌg'zjʊərɪəs/ *a* lujoso. ~**y** /'lʌkʃərɪ/ *n* lujo *m*. ● *a* de lujo
lye /laɪ/ *n* lejía *f*
lying /'laɪɪŋ/ *see* **lie**¹, **lie**². ● *n* mentiras *fpl*
lynch /lɪntʃ/ *vt* linchar
lynx /lɪŋks/ *n* lince *m*
lyre /'laɪə(r)/ *n* lira *f*
lyric /'lɪrɪk/ *a* lírico. ~**al** *a* lírico. ~**ism** /-sɪzəm/ *n* lirismo *m*. ~**s** *npl* letra *f*

M

MA *abbr* (*Master of Arts*) Master *m*, grado *m* universitario entre el de licenciado y doctor
mac /mæk/ *n* (*fam*) impermeable *m*
macabre /mə'kɑ:brə/ *a* macabro
macaroni /mækə'rəʊnɪ/ *n* macarrones *mpl*
macaroon /mækə'ru:n/ *n* mostachón *m*
mace¹ /meɪs/ *n* (*staff*) maza *f*
mace² /meɪs/ *n* (*spice*) macis *f*
Mach /mɑ:k/ *n*. ~ **(number)** *n* (número *m* de) Mach (*m*)
machiavellian /mækɪə'velɪən/ *a* maquiavélico

machinations /mækɪ'neɪʃnz/ *npl* maquinaciones *fpl*
machine /mə'ʃi:n/ *n* máquina *f*. ● *vt* (*sew*) coser a máquina; (*tec*) trabajar a máquina. **~-gun** *n* ametralladora *f*. **~ry** /mə'ʃi:nərɪ/ *n* maquinaria *f*; (*working parts, fig*) mecanismo *m*. **~ tool** *n* máquina *f* herramienta
machinist /mə'ʃi:nɪst/ *n* maquinista *m & f*
mach|ismo /mæ'tʃɪzməʊ/ *n* machismo *m*. **~o** *a* macho
mackerel /'mækrəl/ *n invar* (*fish*) caballa *f*
mackintosh /'mækɪntɒʃ/ *n* impermeable *m*
macrobiotic /mækrəʊbaɪ'ɒtɪk/ *a* macrobiótico
mad /mæd/ *a* (**madder, maddest**) loco; (*foolish*) insensato; ‹*dog*› rabioso; (*angry, fam*) furioso. **be ~ about** estar loco por. **like ~** como un loco; (*a lot*) muchísimo
Madagascar /mædə'gæskə(r)/ *n* Madagascar *m*
madam /'mædəm/ *n* señora *f*; (*unmarried*) señorita *f*
madcap /'mædkæp/ *a* atolondrado. ● *n* locuelo *m*
madden /'mædn/ *vt* (*make mad*) enloquecer; (*make angry*) enfurecer
made /meɪd/ *see* **make**. **~ to measure** hecho a la medida
Madeira /mə'dɪərə/ *n* (*wine*) vino *m* de Madera
mad: **~house** *n* manicomio *m*. **~ly** *adv* (*interested, in love etc*) locamente; (*frantically*) como un loco. **~man** *n* loco *m*. **~ness** *n* locura *f*
madonna /mə'dɒnə/ *n* Virgen *f* María
madrigal /'mædrɪgl/ *n* madrigal *m*
maelstrom /'meɪlstrəm/ *n* remolino *m*
maestro /'maɪstrəʊ/ *n* (*pl* **maestri** /-stri:/ *or* **os**) maestro *m*
Mafia /'mæfɪə/ *n* mafia *f*
magazine /mægə'zi:n/ *n* revista *f*; (*of gun*) recámara *f*
magenta /mə'dʒentə/ *a* rojo purpúreo
maggot /'mægət/ *n* gusano *m*. **~y** *a* agusanado
Magi /'meɪdʒaɪ/ *npl*. **the ~** los Reyes *mpl* Magos
magic /'mædʒɪk/ *n* magia *f*. ● *a* mágico. **~al** *a* mágico. **~ian** /mə'dʒɪʃn/ *n* mago *m*
magisterial /mædʒɪ'stɪərɪəl/ *a* magistral; (*imperious*) autoritario
magistrate /'mædʒɪstreɪt/ *n* magistrado *m*, juez *m*
magnanim|ity /mægnə'nɪmətɪ/ *n* magnanimidad *f*. **~ous** /-'nænɪməs/ *a* magnánimo
magnate /'mægneɪt/ *n* magnate *m*
magnesia /mæg'ni:ʒə/ *n* magnesia *f*
magnet /'mægnɪt/ *n* imán *m*. **~ic** /-'netɪk/ *a* magnético. **~ism** *n* magnetismo *m*. **~ize** *vt* magnetizar
magnificen|ce /mæg'nɪfɪsns/ *a* magnificencia *f*. **~t** *a* magnífico
magnif|ication /mægnɪfɪ'keɪʃn/ *n* aumento *m*. **~ier** /-'faɪə(r)/ *n* lupa *f*, lente *f* de aumento. **~y** /-'faɪ/ *vt* aumentar. **~ying-glass** *n* lupa *f*, lente *f* de aumento
magnitude /'mægnɪtju:d/ *n* magnitud *f*
magnolia /mæg'nəʊlɪə/ *n* magnolia *f*
magnum /'mægnəm/ *n* botella *f* de litro y medio
magpie /'mægpaɪ/ *n* urraca *f*
mahogany /mə'hɒgənɪ/ *n* caoba *f*
maid /meɪd/ *n* (*servant*) criada *f*; (*girl, old use*) doncella *f*. **old ~** solterona *f*
maiden /'meɪdn/ *n* doncella *f*. ● *a* ‹*aunt*› soltera; ‹*voyage*› inaugural. **~hood** *n* doncellez *f*, virginidad *f*, soltería *f*. **~ly** *adv* virginal. **~ name** *n* apellido *m* de soltera
mail[1] /meɪl/ *n* correo *m*; (*letters*) cartas *fpl*. ● *a* postal, de correos. ● *vt* (*post*) echar al correo; (*send*) enviar por correo
mail[2] /meɪl/ *n* (*armour*) (cota *f* de) malla *f*
mail: **~ing list** *n* lista *f* de direcciones. **~man** *n* (*Amer*) cartero *m*. **~ order** *n* venta *f* por correo
maim /meɪm/ *vt* mutilar
main /meɪn/ *n*. **(waterugas) ~** cañería *f* principal. **in the ~** en su mayor parte. **the ~s** *npl* (*elec*) la red *f* eléctrica. ● *a* principal. **a ~ road** *n* una carretera *f*. **~land** *n* continente *m*. **~ly** *adv* principalmente. **~spring** *n* muelle *m* real; (*fig, motive*) móvil *m* principal. **~stay** *n* sostén *m*. **~stream** *n* corriente *f* principal. **~ street** *n* calle *f* principal
maintain /meɪn'teɪn/ *vt* mantener
maintenance /'meɪntənəns/ *n* mantenimiento *m*; (*allowance*) pensión *f* alimenticia

maisonette /meɪzə'net/ *n* (*small house*) casita *f*; (*part of house*) dúplex *m*
maize /meɪz/ *n* maíz *m*
majestic /mə'dʒestɪk/ *a* majestuoso
majesty /'mædʒəstɪ/ *n* majestad *f*
major /'meɪdʒə(r)/ *a* mayor. **a ~ road** una calle *f* prioritaria. ● *n* comandante *m*. ● *vi*. **~ in** (*univ, Amer*) especializarse en
Majorca /mə'jɔ:kə/ *n* Mallorca *f*
majority /mə'dʒɒrətɪ/ *n* mayoría *f*. **the ~ of people** la mayoría *f* de la gente. ● *a* mayoritario
make /meɪk/ *vt*/*i* (*pt* **made**) hacer; (*manufacture*) fabricar; ganar ‹*money*›; tomar ‹*decision*›; llegar a ‹*destination*›. **~ s.o. do sth** obligar a uno a hacer algo. **be made of** estar hecho de. **I cannot ~ anything of it** no me lo explico. **I ~ it two o'clock** yo tengo las dos. ● *n* fabricación *f*; (*brand*) marca *f*. **~ as if to** estar a punto de. **~ believe** fingir. **~ do** (*manage*) arreglarse. **~ do with** (*content o.s.*) contentarse con. **~ for** dirigirse a. **~ good** *vi* tener éxito. ● *vt* compensar; (*repair*) reparar. **~ it** llegar; (*succeed*) tener éxito. **~ it up** (*become reconciled*) hacer las paces. **~ much of** dar mucha importancia a. **~ off** escaparse (**with** con). **~ out** *vt* distinguir; (*understand*) entender; (*draw up*) extender; (*assert*) dar a entender. ● *vi* arreglárselas. **~ over** ceder (**to** a). **~ up** formar; (*prepare*) preparar; inventar ‹*story*›; (*compensate*) compensar. ● *vi* hacer las paces. **~ up (one's face)** maquillarse. **~ up for** compensar; recuperar ‹*time*›. **~ up to** congraciarse con. **~-believe** *a* fingido, simulado. ● *n* ficción *f*
maker /'meɪkə(r)/ *n* fabricante *m* & *f*. **the M~** el Hacedor *m*, el Creador *m*
makeshift /'meɪkʃɪft/ *n* expediente *m*. ● *a* (*temporary*) provisional; (*improvised*) improvisado
make-up /'meɪkʌp/ *n* maquillaje *m*
makeweight /'meɪkweɪt/ *n* complemento *m*
making /'meɪkɪŋ/ *n*. **be the ~ of** ser la causa del éxito de. **he has the ~s of** tiene madera de. **in the ~** en vías de formación
maladjust|ed /mælə'dʒʌstɪd/ *a* inadaptado. **~ment** *n* inadaptación *f*
maladministration /mæləd mɪnɪ'streɪʃn/ *n* mala administración *f*
malady /'mælədɪ/ *n* enfermedad *f*
malaise /mæ'leɪz/ *n* malestar *m*
malaria /mə'leərɪə/ *n* paludismo *m*
Malay /mə'leɪ/ *a* & *n* malayo (*m*). **~sia** *n* Malasia *f*
male /meɪl/ *a* masculino; (*bot, tec*) macho. ● *n* macho *m*; (*man*) varón *m*
malefactor /'mælɪfæktə(r)/ *n* malhechor *m*
malevolen|ce /mə'levəlns/ *n* malevolencia *f*. **~t** *a* malévolo
malform|ation /mælfɔ:'meɪʃn/ *n* malformación *f*. **~ed** *a* deforme
malfunction /mæl'fʌŋkʃn/ *n* funcionamiento *m* defectuoso. ● *vi* funcionar mal
malic|e /'mælɪs/ *n* rencor *m*. **bear s.o. ~e** guardar rencor a uno. **~ious** /mə'lɪʃəs/ *a* malévolo. **~iously** *adv* con malevolencia
malign /mə'laɪn/ *a* maligno. ● *vt* calumniar
malignan|cy /mə'lɪgnənsɪ/ *n* malignidad *f*. **~t** *a* maligno
malinger /mə'lɪŋgə(r)/ *vi* fingirse enfermo. **~er** *n* enfermo *m* fingido
malleable /'mælɪəbl/ *a* maleable
mallet /'mælɪt/ *n* mazo *m*
malnutrition /mælnju:'trɪʃn/ *n* desnutrición *f*
malpractice /mæl'præktɪs/ *n* falta *f* profesional
malt /mɔ:lt/ *n* malta *f*
Malt|a /'mɔ:ltə/ *n* Malta *f*. **~ese** /-'ti:z/ *a* & *n* maltés (*m*)
maltreat /mæl'tri:t/ *vt* maltratar. **~ment** *n* maltrato *m*
malt whisky /mɔ:lt'wɪskɪ/ *n* gúisqui *m* de malta
mammal /'mæml/ *n* mamífero *m*
mammoth /'mæməθ/ *n* mamut *m*. ● *a* gigantesco
man /mæn/ *n* (*pl* **men**) hombre *m*; (*in sports team*) jugador *m*; (*chess*) pieza *f*. **~ in the street** hombre *m* de la calle. **~ to ~** de hombre a hombre. ● *vt* (*pt* **manned**) guarnecer (de hombres); tripular ‹*ship*›; servir ‹*guns*›
manacle /'mænəkl/ *n* manilla *f*. ● *vt* poner esposas a
manage /'mænɪdʒ/ *vt* dirigir; llevar ‹*shop, affairs*›; (*handle*) manejar. ● *vi* arreglárselas. **~ to do** lograr

hacer. **~able** *a* manejable. **~ment** *n* dirección *f*

manager /ˈmænɪdʒə(r)/ *n* director *m*; (*of actor*) empresario *m*. **~ess** /-ˈres/ *n* directora *f*. **~ial** /-ˈdʒɪərɪəl/ *a* directivo. **~ial staff** *n* personal *m* dirigente

managing director /mænɪdʒɪŋ daɪˈrektə(r)/ *n* director *m* gerente

mandarin /ˈmændərɪn/ *n* mandarín *m*; (*orange*) mandarina *f*

mandate /ˈmændeɪt/ *n* mandato *m*

mandatory /ˈmændətərɪ/ *a* obligatorio

mane /meɪn/ *n* (*of horse*) crin *f*; (*of lion*) melena *f*

manful /ˈmænfl/ *a* valiente

manganese /ˈmæŋgəniːz/ *n* manganeso *m*

manger /ˈmeɪndʒə(r)/ *n* pesebre *m*

mangle[1] /ˈmæŋgl/ *n* (*for wringing*) exprimidor *m*; (*for smoothing*) máquina *f* de planchar

mangle[2] /ˈmæŋgl/ *vt* destrozar

mango /ˈmæŋgəʊ/ *n* (*pl* **-oes**) mango *m*

mangy /ˈmeɪndʒɪ/ *a* sarnoso

man: **~handle** *vt* maltratar. **~hole** *n* registro *m*. **~hole cover** *n* tapa *f* de registro. **~hood** *n* edad *f* viril; (*quality*) virilidad *f*. **~-hour** *n* hora-hombre *f*. **~-hunt** *n* persecución *f*

mania /ˈmeɪnɪə/ *n* manía *f*. **~c** /-ɪæk/ *n* maníaco *m*

manicur|e /ˈmænɪkjʊə(r)/ *n* manicura *f*. ● *vt* hacer la manicura a ‹*person*›. **~ist** *n* manicuro *m*

manifest /ˈmænɪfest/ *a* manifiesto. ● *vt* mostrar. **~ation** /-ˈsteɪʃn/ *n* manifestación *f*

manifesto /mænɪˈfestəʊ/ *n* (*pl* **-os**) manifiesto *m*

manifold /ˈmænɪfəʊld/ *a* múltiple

manipulat|e /məˈnɪpjʊleɪt/ *vt* manipular. **~ion** /-ˈleɪʃn/ *n* manipulación *f*

mankind /mænˈkaɪnd/ *n* la humanidad *f*

man: **~ly** *adv* viril. **~-made** *a* artificial

mannequin /ˈmænɪkɪn/ *n* maniquí *m*

manner /ˈmænə(r)/ *n* manera *f*; (*behaviour*) comportamiento *m*; (*kind*) clase *f*. **~ed** *a* amanerado. **bad-~ed** *a* mal educado. **~s** *npl* (*social behaviour*) educación *f*. **have no ~s** no tener educación

mannerism /ˈmænərɪzəm/ *n* peculiaridad *f*

mannish /ˈmænɪʃ/ *a* ‹*woman*› hombruna

manoeuvre /məˈnuːvə(r)/ *n* maniobra *f*. ● *vt/i* maniobrar

man-of-war /mænəvˈwɔː(r)/ *n* buque *m* de guerra

manor /ˈmænə(r)/ *n* casa *f* solariega

manpower /ˈmænpaʊə(r)/ *n* mano *f* de obra

manservant /ˈmænsɜːvənt/ *n* criado *m*

mansion /ˈmænʃn/ *n* mansión *f*

man: **~-size(d)** *a* grande. **~slaughter** *n* homicidio *m* impremeditado

mantelpiece /ˈmæntlpiːs/ *n* repisa *f* de chimenea

mantilla /mænˈtɪlə/ *n* mantilla *f*

mantle /ˈmæntl/ *n* manto *m*

manual /ˈmænjʊəl/ *a* manual. ● *n* (*handbook*) manual *m*

manufacture /mænjʊˈfæktʃə(r)/ *vt* fabricar. ● *n* fabricación *f*. **~r** /-ə(r)/ *n* fabricante *m*

manure /məˈnjʊə(r)/ *n* estiércol *m*

manuscript /ˈmænjʊskrɪpt/ *n* manuscrito *m*

many /ˈmenɪ/ *a* & *n* muchos (*mpl*). **~ people** mucha gente *f*. **~ a time** muchas veces. **a great/good ~** muchísimos

map /mæp/ *n* mapa *m*; (*of streets etc*) plano *m*. ● *vt* (*pt* **mapped**) levantar un mapa de. **~ out** organizar

maple /ˈmeɪpl/ *n* arce *m*

mar /mɑː/ *vt* (*pt* **marred**) estropear; aguar ‹*enjoyment*›

marathon /ˈmærəθən/ *n* maratón *m*

maraud|er /məˈrɔːdə(r)/ *n* merodeador *m*. **~ing** *a* merodeador

marble /ˈmɑːbl/ *n* mármol *m*; (*for game*) canica *f*

March /mɑːtʃ/ *n* marzo *m*

march /mɑːtʃ/ *vi* (*mil*) marchar. **~ off** irse. ● *vt*. **~ off** (*lead away*) llevarse. ● *n* marcha *f*

marchioness /mɑːʃəˈnes/ *n* marquesa *f*

march-past /ˈmɑːtʃpɑːst/ *n* desfile *m*

mare /meə(r)/ *n* yegua *f*

margarine /mɑːdʒəˈriːn/ *n* margarina *f*

margin /ˈmɑːdʒɪn/ *n* margen *f*. **~al** *a* marginal. **~al seat** *n* (*pol*) escaño *m* inseguro. **~ally** *adv* muy poco

marguerite /mɑːgəˈriːt/ *n* margarita *f*

marigold /ˈmærɪgəʊld/ *n* caléndula *f*

marijuana /mærɪ'hwɑ:nə/ *n* marihuana *f*
marina /mə'ri:nə/ *n* puerto *m* deportivo
marina|de /mærɪ'neɪd/ *n* escabeche *m*. **~te** /'mærɪneɪt/ *vt* escabechar
marine /mə'ri:n/ *a* marino. ● *n* (*sailor*) soldado *m* de infantería de marina; (*shipping*) marina *f*
marionette /mærɪə'net/ *n* marioneta *f*
marital /'mærɪtl/ *a* marital, matrimonial. **~ status** *n* estado *m* civil
maritime /'mærɪtaɪm/ *a* marítimo
marjoram /'mɑ:dʒərəm/ *n* mejorana *f*
mark[1] /mɑ:k/ *n* marca *f*; (*trace*) huella *f*; (*schol*) nota *f*; (*target*) blanco *m*. ● *vt* marcar; poner nota a ‹*exam*›. **~ time** marcar el paso. **~ out** trazar; escoger ‹*person*›
mark[2] /mɑ:k/ *n* (*currency*) marco *m*
marked /mɑ:kt/ *a* marcado. **~ly** /-kɪdlɪ/ *adv* marcadamente
marker /'mɑ:kə(r)/ *n* marcador *m*; (*for book*) registro *m*
market /'mɑ:kɪt/ *n* mercado *m*. **on the ~** en venta. ● *vt* (*sell*) vender; (*launch*) comercializar. **~ garden** *n* huerto *m*. **~ing** *n* marketing *m*
marking /'mɑ:kɪŋ/ *n* (*marks*) marcas *fpl*
marksman /'mɑ:ksmən/ *n* tirador *m*. **~ship** *n* puntería *f*
marmalade /'mɑ:məleɪd/ *n* mermelada *f* de naranja
marmot /'mɑ:mət/ *n* marmota *f*
maroon /mə'ru:n/ *n* granate *m*. ● *a* de color granate
marooned /mə'ru:nd/ *a* abandonado; (*snow-bound etc*) aislado
marquee /mɑ:ki:/ *n* tienda de campaña *f* grande; (*awning*, *Amer*) marquesina *f*
marquetry /'mɑ:kɪtrɪ/ *n* marquetería *f*
marquis /'mɑ:kwɪs/ *n* marqués *m*
marriage /'mærɪdʒ/ *n* matrimonio *m*; (*wedding*) boda *f*. **~able** *a* casadero
married /'mærɪd/ *a* casado; ‹*life*› conjugal
marrow /'mærəʊ/ *n* (*of bone*) tuétano *m*; (*vegetable*) calabacín *m*
marry /'mærɪ/ *vt* casarse con; (*give or unite in marriage*) casar. ● *vi* casarse. **get married** casarse
marsh /mɑ:ʃ/ *n* pantano *m*
marshal /'mɑ:ʃl/ *n* (*mil*) mariscal *m*; (*master of ceremonies*) maestro *m* de ceremonias; (*at sports events*) oficial *m*. ● *vt* (*pt* **marshalled**) ordenar; formar ‹*troops*›
marsh mallow /mɑ:ʃ'mæləʊ/ *n* (*plant*) malvavisco *m*
marshmallow /mɑ:ʃ'mæləʊ/ *n* (*sweet*) caramelo *m* blando
marshy /'mɑ:ʃɪ/ *a* pantanoso
martial /'mɑ:ʃl/ *a* marcial. **~ law** *n* ley *f* marcial
Martian /'mɑ:ʃn/ *a* & *n* marciano (*m*)
martinet /mɑ:tɪ'net/ *n* ordenancista *m* & *f*
martyr /'mɑ:tə(r)/ *n* mártir *m* & *f*. ● *vt* martirizar. **~dom** *n* martirio *m*
marvel /'mɑ:vl/ *n* maravilla *f*. ● *vi* (*pt* **marvelled**) maravillarse (**at** con, de). **~lous** /'mɑ:vələs/ *a* maravilloso
Marxis|m /'mɑ:ksɪzəm/ *n* marxismo *m*. **~t** *a* & *n* marxista (*m* & *f*)
marzipan /'mɑ:zɪpæn/ *n* mazapán *m*
mascara /mæ'skɑ:rə/ *n* rimel *m*
mascot /'mæskɒt/ *n* mascota *f*
masculin|e /'mæskjʊlɪn/ *a* & *n* masculino (*m*). **~ity** /-'lɪnətɪ/ *n* masculinidad *f*
mash /mæʃ/ *n* mezcla *f*; (*potatoes*, *fam*) puré *m* de patatas. ● *vt* (*crush*) machacar; (*mix*) mezclar. **~ed potatoes** *n* puré *m* de patatas
mask /mɑ:sk/ *n* máscara *f*. ● *vt* enmascarar
masochis|m /'mæsəkɪzəm/ *n* masoquismo *m*. **~t** *n* masoquista *m* & *f*
mason /'meɪsn/ *n* (*builder*) albañil *m*
Mason /'meɪsn/ *n*. **~** masón *m*. **~ic** /mə'sɒnɪk/ *a* masónico
masonry /'meɪsnrɪ/ *n* albañilería *f*
masquerade /mɑ:skə'reɪd/ *n* mascarada *f*. ● *vi*. **~ as** hacerse pasar por
mass[1] /mæs/ *n* masa *f*; (*large quantity*) montón *m*. **the ~es** *npl* las masas *fpl*. ● *vt/i* agrupar(se)
mass[2] /mæs/ *n* (*relig*) misa *f*. **high ~** misa *f* mayor
massacre /'mæsəkə(r)/ *n* masacre *f*, matanza *f*. ● *vt* masacrar
massage /'mæsɑ:ʒ/ *n* masaje *m*. ● *vt* dar masaje a
masseu|r /mæ'sɜ:(r)/ *n* masajista *m*. **~se** /mæ'sɜ:z/ *n* masajista *f*
massive /'mæsɪv/ *a* masivo; (*heavy*) macizo; (*huge*) enorme
mass: **~ media** *n* medios *mpl* de comunicación. **~-produce** *vt* fabricar en serie

mast /mɑ:st/ *n* mástil *m*; (*for radio, TV*) torre *f*
master /'mɑ:stə(r)/ *n* maestro *m*; (*in secondary school*) profesor *m*; (*of ship*) capitán *m*. ● *vt* dominar. **~-key** *n* llave *f* maestra. **~ly** *a* magistral. **~mind** *n* cerebro *m*. ● *vt* dirigir. **M~ of Arts** master *m*, grado *m* universitario entre el de licenciado y el de doctor
masterpiece /'mɑ:stəpi:s/ *n* obra *f* maestra
master-stroke /'mɑ:stəstrəʊk/ *n* golpe *m* maestro
mastery /'mɑ:stərɪ/ *n* dominio *m*; (*skill*) maestría *f*
masturbat|e /'mæstəbeɪt/ *vi* masturbarse. **~ion** /-'beɪʃn/ *n* masturbación *f*
mat /mæt/ *n* estera *f*; (*at door*) felpudo *m*
match[1] /mætʃ/ *n* (*sport*) partido *m*; (*equal*) igual *m*; (*marriage*) matrimonio *m*; (*s.o. to marry*) partido *m*. ● *vt* emparejar; (*equal*) igualar; ‹*clothes, colours*› hacer juego con. ● *vi* hacer juego
match[2] /mætʃ/ *n* (*of wood*) fósforo *m*; (*of wax*) cerilla *f*. **~box** /'mætʃbɒks/ *n* (*for wooden matches*) caja *f* de fósforos; (*for wax matches*) caja *f* de cerillas
matching /'mætʃɪŋ/ *a* que hace juego
mate[1] /meɪt/ *n* compañero *m*; (*of animals*) macho *m*, hembra *f*; (*assistant*) ayudante *m*. ● *vt/i* acoplar(se)
mate[2] /meɪt/ *n* (*chess*) mate *m*
material /mə'tɪərɪəl/ *n* material *m*; (*cloth*) tela *f*. ● *a* material; (*fig*) importante. **~istic** /-'lɪstɪk/ *a* materialista. **~s** *npl* materiales *mpl*. **raw ~s** *npl* materias *fpl* primas
materialize /mə'tɪərɪəlaɪz/ *vi* materializarse
maternal /mə'tɜ:nl/ *a* maternal; ‹*relation*› materno
maternity /mə'tɜ:nɪtɪ/ *n* maternidad *f*. ● *a* de maternidad. **~ clothes** *npl* vestido *m* pre-mamá. **~ hospital** *n* maternidad *f*
matey /'meɪtɪ/ *a* (*fam*) simpático
mathematic|ian /mæθəmə'tɪʃn/ *n* matemático *m*. **~al** /-'mætɪkl/ *a* matemático. **~s** /-'mætɪks/ *n & npl* matemáticas *fpl*
maths /mæθs/, **math** (*Amer*) *n & npl* matemáticas *fpl*

matinée /'mætɪneɪ/ *n* función *f* de tarde
matriculat|e /mə'trɪkjʊleɪt/ *vt/i* matricular(se). **~ion** /-'leɪʃn/ *n* matriculación *f*
matrimon|ial /mætrɪ'məʊnɪəl/ *a* matrimonial. **~y** /'mætrɪmənɪ/ *n* matrimonio *m*
matrix /'meɪtrɪks/ *n* (*pl* **matrices** /-si:z/) matriz *f*
matron /'meɪtrən/ *n* (*married, elderly*) matrona *f*; (*in school*) ama *f* de llaves; (*former use, in hospital*) enfermera *f* jefe. **~ly** *a* matronil
matt /mæt/ *a* mate
matted /'mætɪd/ *a* enmarañado
matter /'mætə(r)/ *n* (*substance*) materia *f*; (*affair*) asunto *m*; (*pus*) pus *m*. **as a ~ of fact** en realidad. **no ~** no importa. **what is the ~?** ¿qué pasa? ● *vi* importar. **it does not ~** no importa. **~-of-fact** *a* realista
matting /'mætɪŋ/ *n* estera *f*
mattress /'mætrɪs/ *n* colchón *m*
matur|e /mə'tjʊə(r)/ *a* maduro. ● *vt/i* madurar. **~ity** *n* madurez *f*
maul /mɔ:l/ *vt* maltratar
Mauritius /mə'rɪʃəs/ *n* Mauricio *m*
mausoleum /mɔ:sə'lɪəm/ *n* mausoleo *m*
mauve /məʊv/ *a & n* color (*m*) de malva
mawkish /'mɔ:kɪʃ/ *a* empalagoso
maxim /'mæksɪm/ *n* máxima *f*
maxim|ize /'mæksɪmaɪz/ *vt* llevar al máximo. **~um** *a & n* (*pl* **-ima**) máximo (*m*)
may /meɪ/ *v aux* (*pt* **might**) poder. **~ I smoke?** ¿se permite fumar? **~ he be happy** ¡que sea feliz! **he ~/might come** puede que venga. **I ~/might as well stay** más vale quedarme. **it ~/might be true** puede ser verdad
May /meɪ/ *n* mayo *m*. **~ Day** *n* el primero *m* de mayo
maybe /'meɪbɪ/ *adv* quizá(s)
mayhem /'meɪhem/ *n* (*havoc*) alboroto *m*
mayonnaise /meɪə'neɪz/ *n* mayonesa *f*
mayor /meə(r)/ *n* alcalde *m*, alcaldesa *f*. **~ess** *n* alcaldesa *f*
maze /meɪz/ *n* laberinto *m*
me[1] /mi:/ *pron* me; (*after prep*) mí. **he knows ~** me conoce. **it's ~** soy yo
me[2] /mi:/ *n* (*mus, third note of any musical scale*) mi *m*
meadow /'medəʊ/ *n* prado *m*

meagre /'mi:gə(r)/ *a* escaso
meal[1] /mi:l/ *n* comida *f*
meal[2] /mi:l/ *n* (*grain*) harina *f*
mealy-mouthed /mi:lɪ'maʊðd/ *a* hipócrita
mean[1] /mi:n/ *vt* (*pt* **meant**) (*intend*) tener la intención de, querer; (*signify*) querer decir, significar. ~ **to do** tener la intención de hacer. ~ **well** tener buenas intenciones. **be meant for** estar destinado a
mean[2] /mi:n/ *a* (**-er**, **-est**) (*miserly*) tacaño; (*unkind*) malo; (*poor*) pobre
mean[3] /mi:n/ *a* medio. ● *n* medio *m*; (*average*) promedio *m*
meander /mɪ'ændə(r)/ *vi* ‹*river*› serpentear; ‹*person*› vagar
meaning /'mi:nɪŋ/ *n* sentido *m*. **~ful** *a* significativo. **~less** *a* sin sentido
meanness /'mi:nnɪs/ *n* (*miserliness*) tacañería *f*; (*unkindness*) maldad *f*
means /mi:nz/ *n* medio *m*. **by all** ~ por supuesto. **by no** ~ de ninguna manera. ● *npl* (*wealth*) recursos *mpl*. ~ **test** *n* investigación *f* financial
meant /ment/ *see* **mean**[1]
meantime /'mi:natɪm/ *adv* entretanto. **in the** ~ entretanto
meanwhile /'mi:nwaɪl/ *adv* entretanto
measles /'mi:zlz/ *n* sarampión *m*
measly /'mi:zlɪ/ *a* (*sl*) miserable
measurable /'meʒərəbl/ *a* mensurable
measure /'meʒə(r)/ *n* medida *f*; (*ruler*) regla *f*. ● *vt*/*i* medir. ~ **up to** estar a la altura de. **~d** *a* (*rhythmical*) acompasado; (*carefully considered*) prudente. **~ment** *n* medida *f*
meat /mi:t/ *n* carne *f*. **~y** *a* carnoso; (*fig*) sustancioso
mechanic /mɪ'kænɪk/ *n* mecánico *m*. **~al** /mɪ'kænɪkl/ *a* mecánico. **~s** *n* mecánica *f*
mechani|sm /'mekənɪzəm/ *n* mecanismo *m*. **~ze** *vt* mecanizar
medal /'medl/ *n* medalla *f*
medallion /mɪ'dælɪən/ *n* medallón *m*
medallist /'medəlɪst/ *n* ganador *m* de una medalla. **be a gold** ~ ganar una medalla de oro
meddle /'medl/ *vi* entrometerse (**in** en); (*tinker*) tocar. ~ **with** (*tinker*) tocar. **~some** *a* entrometido
media /'mi:dɪə/ *see* **medium**. ● *npl*. **the** ~ *npl* los medios *mpl* de comunicación
mediat|e /'mi:dɪeɪt/ *vi* mediar. **~ion** /-eɪʃn/ *n* mediación *f*. **~or** *n* mediador *m*
medical /'medɪkl/ *a* médico; ‹*student*› de medicina. ● *n* (*fam*) reconocimiento *m* médico
medicat|ed /'medɪkeɪtɪd/ *a* medicinal. **~ion** /-'keɪʃn/ *n* medicación *f*
medicin|e /'medsɪn/ *n* medicina *f*. **~al** /mɪ'dɪsɪnl/ *a* medicinal
medieval /medɪ'i:vl/ *a* medieval
mediocr|e /mi:dɪ'əʊkə(r)/ *a* mediocre. **~ity** /-'ɒkrətɪ/ *n* mediocridad *f*
meditat|e /'medɪteɪt/ *vt*/*i* meditar. **~ion** /-'teɪʃn/ *n* meditación *f*
Mediterranean /medɪtə'reɪnɪən/ *a* mediterráneo. ● *n*. **the** ~ el Mediterráneo *m*
medium /'mi:dɪəm/ *n* (*pl* **media**) medio *m*; (*pl* **mediums**) (*person*) médium *m*. ● *a* mediano
medley /'medlɪ/ *n* popurrí *m*
meek /mi:k/ *a* (**-er**, **-est**) manso
meet /mi:t/ *vt* (*pt* **met**) encontrar; (*bump into s.o.*) encontrarse con; (*see again*) ver; (*fetch*) ir a buscar; (*get to know, be introduced to*) conocer. ~ **the bill** pagar la cuenta. ● *vi* encontrarse; (*get to know*) conocerse; (*in session*) reunirse. ~ **with** tropezar con ‹*obstacles*›
meeting /'mi:tɪŋ/ *n* reunión *f*; (*accidental between two people*) encuentro *m*; (*arranged between two people*) cita *f*
megalomania /megələʊ'meɪnɪə/ *n* megalomanía *f*
megaphone /'megəfəʊn/ *n* megáfono *m*
melanchol|ic /melən'kɒlɪk/ *a* melancólico. **~y** /'melənkɒlɪ/ *n* melancolía *f*. ● *a* melancólico
mêlée /me'leɪ/ *n* pelea *f* confusa
mellow /'meləʊ/ *a* (**-er**, **-est**) ‹*fruit, person*› maduro; ‹*sound, colour*› dulce. ● *vt*/*i* madurar(se)
melodi|c /mɪ'lɒdɪk/ *a* melódico. **~ous** /mɪ'ləʊdɪəs/ *a* melodioso
melodrama /'melədrɑ:mə/ *n* melodrama *m*. **~tic** /-ə'mætɪk/ *a* melodramático
melody /'melədɪ/ *n* melodía *f*
melon /'melən/ *n* melón *m*
melt /melt/ *vt* (*make liquid*) derretir; fundir ‹*metals*›. ● *vi* (*become liquid*) derretirse; ‹*metals*› fundirse. **~ing-pot** *n* crisol *m*
member /'membə(r)/ *n* miembro *m*. **M~ of Parliament** *n* diputado *m*.

~ship *n* calidad *f* de miembro; (*members*) miembros *mpl*
membrane /'membreɪn/ *n* membrana *f*
memento /mɪ'mentəʊ/ *n* (*pl* **-oes**) recuerdo *m*
memo /'meməʊ/ *n* (*pl* **-os**) (*fam*) nota *f*
memoir /'memwɑ:(r)/ *n* memoria *f*
memorable /'memərəbl/ *a* memorable
memorandum /memə'rændəm/ *n* (*pl* **-ums**) nota *f*
memorial /mɪ'mɔ:rɪəl/ *n* monumento *m*. ● *a* conmemorativo
memorize /'meməraɪz/ *vt* aprender de memoria
memory /'memərɪ/ *n* (*faculty*) memoria *f*; (*thing remembered*) recuerdo *m*. **from ~** de memoria. **in ~ of** en memoria de
men /men/ *see* **man**
menac|e /'menəs/ *n* amenaza *f*; (*nuisance*) pesado *m*. ● *vt* amenazar. **~ingly** *adv* de manera amenazadora
menagerie /mɪ'nædʒərɪ/ *n* casa *f* de fieras
mend /mend/ *vt* reparar; (*darn*) zurcir. **~ one's ways** enmendarse. ● *n* remiendo *m*. **be on the ~** ir mejorando
menfolk /'menfəʊk/ *n* hombres *mpl*
menial /'mi:nɪəl/ *a* servil
meningitis /menɪn'dʒaɪtɪs/ *n* meningitis *f*
menopause /'menəpɔ:z/ *n* menopausia *f*
menstruat|e /'menstrʊeɪt/ *vi* menstruar. **~ion** /-eɪʃn/ *n* menstruación *f*
mental /'mentl/ *a* mental; ‹*hospital*› psiquiátrico
mentality /men'tælətɪ/ *n* mentalidad *f*
menthol /'menθɒl/ *n* mentol *m*. **~ated** *a* mentolado
mention /'menʃn/ *vt* mencionar. **don't ~ it!** ¡no hay de qué! ● *n* mención *f*
mentor /'mentɔ:(r)/ *n* mentor *m*
menu /'menju:/ *n* (*set meal*) menú *m*; (*a la carte*) lista *f* (de platos)
mercantile /'mɜ:kəntaɪl/ *a* mercantil
mercenary /'mɜ:sɪnərɪ/ *a* & n mercenario (*m*)
merchandise /'mɜ:tʃəndaɪz/ *n* mercancías *fpl*
merchant /'mɜ:tʃənt/ *n* comerciante *m*. ● *a* ‹*ship, navy*› mercante. **~ bank** *n* banco *m* mercantil
merci|ful /'mɜ:sɪfl/ *a* misericordioso. **~fully** *adv* (*fortunately, fam*) gracias a Dios. **~less** /'mɜ:sɪlɪs/ *a* despiadado
mercur|ial /mɜ:'kjʊərɪəl/ *a* mercurial; (*fig, active*) vivo. **~y** /'mɜ:kjʊrɪ/ *n* mercurio *m*
mercy /'mɜ:sɪ/ *n* compasión *f*. **at the ~ of** a merced de
mere /mɪə(r)/ *a* simple. **~ly** *adv* simplemente
merest /'mɪərɪst/ *a* mínimo
merge /mɜ:dʒ/ *vt* unir; fusionar ‹*companies*›. ● *vi* unirse; ‹*companies*› fusionarse. **~r** /-ə(r)/ *n* fusión *f*
meridian /mə'rɪdɪən/ *n* meridiano *m*
meringue /mə'ræŋ/ *n* merengue *m*
merit /'merɪt/ *n* mérito *m*. ● *vt* (*pt* **merited**) merecer. **~orious** /-'tɔ:rɪəs/ *a* meritorio
mermaid /'mɜ:meɪd/ *n* sirena *f*
merr|ily /'merəlɪ/ *adv* alegremente. **~iment** /'merɪmənt/ *n* alegría *f*. **~y** /'merɪ/ *a* (**-ier**, **-iest**) alegre. **make ~** divertirse. **~y-go-round** *n* tiovivo *m*. **~y-making** *n* holgorio *m*
mesh /meʃ/ *n* malla *f*; (*network*) red *f*
mesmerize /'mezməraɪz/ *vt* hipnotizar
mess /mes/ *n* desorden *m*; (*dirt*) suciedad *f*; (*mil*) rancho *m*. **make a ~ of** chapucear, estropear. ● *vt*. **~ up** desordenar; (*dirty*) ensuciar. ● *vi*. **~ about** entretenerse. **~ with** (*tinker with*) manosear
message /'mesɪdʒ/ *n* recado *m*
messenger /'mesɪndʒə(r)/ *n* mensajero *m*
Messiah /mɪ'saɪə/ *n* Mesías *m*
Messrs /'mesəz/ *npl*. **~ Smith** los señores *mpl or* Sres. Smith
messy /'mesɪ/ *a* (**-ier**, **-iest**) en desorden; (*dirty*) sucio
met /met/ *see* **meet**
metabolism /mɪ'tæbəlɪzəm/ *n* metabolismo *m*
metal /'metl/ *n* metal. ● *a* de metal. **~lic** /mɪ'tælɪk/ *a* metálico
metallurgy /mɪ'tælədʒɪ/ *n* metalurgia *f*
metamorphosis /metə'mɔ:fəsɪs/ *n* (*pl* **-phoses** /-sɪ:z/) metamorfosis *f*
metaphor /'metəfə(r)/ *n* metáfora *f*. **~ical** /-'fɔrɪkl/ *a* metafórico

mete /mi:t/ *vt*. ~ **out** repartir; dar ‹*punishment*›
meteor /'mi:tɪə(r)/ *n* meteoro *m*
meteorite /'mi:tɪəraɪt/ *n* meteorito *m*
meteorolog|ical /mi:tɪərə'lɒdʒɪkl/ *a* meteorológico. ~**y** /-'rɒlədʒɪ/ *n* meteorología *f*
meter[1] /'mi:tə(r)/ *n* contador *m*
meter[2] /'mi:tə(r)/ *n* (*Amer*) = **metre**
method /'meθəd/ *n* método *m*
methodical /mɪ'θɒdɪkl/ *a* metódico
Methodist /'meθədɪst/ *a* & *n* metodista (*m* & *f*)
methylated /'meθɪleɪtɪd/ *a*. ~ **spirit** *n* alcohol *m* desnaturalizado
meticulous /mɪ'tɪkjʊləs/ *a* meticuloso
metre /'mi:tə(r)/ *n* metro *m*
metric /'metrɪk/ *a* métrico. ~**ation** /-'keɪʃn/ *n* cambio *m* al sistema métrico
metropolis /mɪ'trɒpəlɪs/ *n* metrópoli *f*
metropolitan /metrə'pɒlɪtən/ *a* metropolitano
mettle /'metl/ *n* valor *m*
mew /mju:/ *n* maullido *m*. • *vi* maullar
mews /mju:z/ *npl* casas *fpl* pequeñas (que antes eran caballerizas)
Mexic|an /'meksɪkən/ *a* & *n* mejicano (*m*); (*in Mexico*) mexicano (*m*). ~**o** /-kəʊ/ *n* Méjico *m*; (*in Mexico*) México *m*
mezzanine /'metsəni:n/ *n* entresuelo *m*
mi /mi:/ *n* (*mus, third note of any musical scale*) mi *m*
miaow /mi:'aʊ/ *n* & *vi* = **mew**
mice /maɪs/ *see* **mouse**
mickey /'mɪkɪ/ *n*. **take the ~ out of** (*sl*) tomar el pelo a
micro... /'maɪkrəʊ/ *pref* micro...
microbe /'maɪkrəʊb/ *n* microbio *m*
microchip /'maɪkrəʊtʃɪp/ *n* pastilla *f*
microfilm /'maɪkrəʊfɪlm/ *n* microfilme *m*
microphone /'maɪkrəfəʊn/ *n* micrófono *m*
microprocessor /maɪkrəʊ'prəʊsesə(r)/ *n* microprocesador *m*
microscop|e /'maɪkrəskəʊp/ *n* microscopio *m*. ~**ic** /-'skɒpɪk/ *a* microscópico
microwave /'maɪkrəʊweɪv/ *n* microonda *f*. ~ **oven** *n* horno *m* de microondas
mid /mɪd/ *a*. **in ~ air** en pleno aire. **in ~ March** a mediados de marzo. **in ~ ocean** en medio del océano
midday /mɪd'deɪ/ *n* mediodía *m*
middle /'mɪdl/ *a* de en medio; ‹*quality*› mediano. • *n* medio *m*. **in the ~ of** en medio de. ~**-aged** *a* de mediana edad. **M~ Ages** *npl* Edad *f* Media. ~ **class** *n* clase *f* media. ~**-class** *a* de la clase media. **M~ East** *n* Oriente *m* Medio. ~**man** *n* intermediario *m*
middling /'mɪdlɪŋ/ *a* regular
midge /mɪdʒ/ *n* mosquito *m*
midget /'mɪdʒɪt/ *n* enano *m*. • *a* minúsculo
Midlands /'mɪdləndz/ *npl* región *f* central de Inglaterra
midnight /'mɪdnaɪt/ *n* medianoche *f*
midriff /'mɪdrɪf/ *n* diafragma *m*; (*fam*) vientre *m*
midst /mɪdst/ *n*. **in our ~** entre nosotros. **in the ~ of** en medio de
midsummer /mɪd'sʌmə(r)/ *n* pleno verano *m*; (*solstice*) solsticio *m* de verano
midway /mɪd'weɪ/ *adv* a medio camino
midwife /'mɪdwaɪf/ *n* comadrona *f*
midwinter /mɪd'wɪntə(r)/ *n* pleno invierno *m*
might[1] /maɪt/ *see* **may**
might[2] /maɪt/ *n* (*strength*) fuerza *f*; (*power*) poder *m*. ~**y** *a* (*strong*) fuerte; (*powerful*) poderoso; (*very great, fam*) enorme. • *adv* (*fam*) muy
migraine /'mi:greɪn/ *n* jaqueca *f*
migrant /'maɪgrənt/ *a* migratorio. • *n* (*person*) emigrante *m* & *f*
migrat|e /maɪ'greɪt/ *vi* emigrar. ~**ion** /-ʃn/ *n* migración *f*
mike /maɪk/ *n* (*fam*) micrófono *m*
mild /maɪld/ *a* (**-er**, **-est**) ‹*person*› apacible; ‹*climate*› templado; (*slight*) ligero; ‹*taste*› suave; ‹*illness*› benigno
mildew /'mɪldju:/ *n* moho *m*
mild: ~**ly** *adv* (*slightly*) ligeramente. ~**ness** *n* (*of person*) apacibilidad *f*; (*of climate, illness*) benignidad *f*; (*of taste*) suavidad *f*
mile /maɪl/ *n* milla *f*. ~**s better** (*fam*) mucho mejor. ~**s too big** (*fam*) demasiado grande. ~**age** *n* (*loosely*) kilometraje *m*. ~**stone** *n* mojón *m*; (*event, stage, fig*) hito *m*
milieu /mɪ'ljɜ:/ *n* ambiente *m*

militant /'mɪlɪtənt/ *a & n* militante (*m & f*)
military /'mɪlɪtərɪ/ *a* militar
militate /'mɪlɪteɪt/ *vi* militar (**against** contra)
militia /mɪ'lɪʃə/ *n* milicia *f*
milk /mɪlk/ *n* leche *f*. ● *a* ‹*product*› lácteo; ‹*chocolate*› con leche. ● *vt* ordeñar ‹*cow*›; (*exploit*) chupar. **~man** *n* repartidor *m* de leche. **~shake** *n* batido *m* de leche. **~y** *a* lechoso. **M~y Way** *n* Vía *f* Láctea
mill /mɪl/ *n* molino *m*; (*for coffee, pepper*) molinillo *m*; (*factory*) fábrica *f*. ● *vt* moler. ● *vi*. **~ about/around** apiñarse, circular
millennium /mɪ'lenɪəm/ *n* (*pl* **-ia** *or* **-iums**) milenio *m*
miller /'mɪlə(r)/ *n* molinero *m*
millet /'mɪlɪt/ *n* mijo *m*
milli... /'mɪlɪ/ *pref* mili...
milligram(me) /'mɪlɪgræm/ *n* miligramo *m*
millimetre /'mɪlɪmi:tə(r)/ *n* milímetro *m*
milliner /'mɪlɪnə(r)/ *n* sombrerero *m*
million /'mɪlɪən/ *n* millón *m*. **a ~ pounds** un millón *m* de libras. **~aire** *n* millonario *m*
millstone /'mɪlstəʊn/ *n* muela *f* (de molino); (*fig, burden*) losa *f*
mime /maɪm/ *n* pantomima *f*. ● *vt* hacer en pantomima. ● *vi* actuar de mimo
mimic /'mɪmɪk/ *vt* (*pt* **mimicked**) imitar. ● *n* imitador *m*. **~ry** *n* imitación *f*
mimosa /mɪ'məʊzə/ *n* mimosa *f*
minaret /mɪnə'ret/ *n* alminar *m*
mince /mɪns/ *vt* desmenuzar; picar ‹*meat*›. **not to ~ matters/words** no tener pelos en la lengua. ● *n* carne *f* picada. **~meat** *n* conserva *f* de fruta picada. **make ~meat of s.o.** hacer trizas a uno. **~ pie** *n* pastel *m* con frutas picadas. **~r** *n* máquina *f* de picar carne
mind /maɪnd/ *n* mente *f*; (*sanity*) juicio *m*; (*opinion*) parecer *m*; (*intention*) intención *f*. **be on one's ~** preocuparle a uno. ● *vt* (*look after*) cuidar; (*heed*) hacer caso de. **I don't ~** me da igual. **I don't ~ the noise** no me molesta el ruido. **never ~** no te preocupes, no se preocupe. **~er** *n* cuidador *m*. **~ful** *a* atento (**of** a). **~less** *a* estúpido
mine[1] /maɪn/ *poss pron* (el) mío *m*, (la) mía *f*, (los) míos *mpl*, (las) mías *fpl*. **it is ~** es mío
mine[2] /maɪn/ *n* mina *f*. ● *vt* extraer. **~field** *n* campo *m* de minas. **~r** *n* minero *m*
mineral /'mɪnərəl/ *a & n* mineral (*m*). **~ (water)** *n* (*fizzy soft drink*) gaseosa *f*. **~ water** *n* (*natural*) agua *f* mineral
minesweeper /'maɪnswi:pə(r)/ *n* (*ship*) dragaminas *m invar*
mingle /'mɪŋgl/ *vt/i* mezclar(se)
mingy /'mɪndʒɪ/ *a* tacaño
mini... /'mɪnɪ/ *pref* mini...
miniature /'mɪnɪtʃə(r)/ *a & n* miniatura (*f*)
mini: **~bus** *n* microbús *m*. **~cab** *n* taxi *m*
minim /'mɪnɪm/ *n* (*mus*) blanca *f*
minim|al /'mɪnɪml/ *a* mínimo. **~ize** *vt* minimizar. **~um** *a & n* (*pl* **-ima**) mínimo (*m*)
mining /'maɪnɪŋ/ *n* explotación *f*. ● *a* minero
miniskirt /'mɪnɪskɜ:t/ *n* minifalda *f*
minist|er /'mɪnɪstə(r)/ *n* ministro *m*; (*relig*) pastor *m*. **~erial** /-'stɪərɪəl/ *a* ministerial. **~ry** *n* ministerio *m*
mink /mɪŋk/ *n* visón *m*
minor /'maɪnə(r)/ *a* (*incl mus*) menor; (*of little importance*) sin importancia. ● *n* menor *m & f* de edad
minority /maɪ'nɒrətɪ/ *n* minoría *f*. ● *a* minoritario
minster /'mɪnstə(r)/ *n* catedral *f*
minstrel /'mɪnstrəl/ *n* juglar *m*
mint[1] /mɪnt/ *n* (*plant*) menta *f*; (*sweet*) caramelo *m* de menta
mint[2] /mɪnt/ *n*. **the M~** *n* casa *f* de la moneda. **a ~** un dineral *m*. ● *vt* acuñar. **in ~ condition** como nuevo
minuet /mɪnjʊ'et/ *n* minué *m*
minus /'maɪnəs/ *prep* menos; (*without, fam*) sin. ● *n* (*sign*) menos *m*. **~ sign** *n* menos *m*
minuscule /'mɪnəskju:l/ *a* minúsculo
minute[1] /'mɪnɪt/ *n* minuto *m*. **~s** *npl* (*of meeting*) actas *fpl*
minute[2] /maɪ'nju:t/ *a* minúsculo; (*detailed*) minucioso
minx /mɪŋks/ *n* chica *f* descarada
mirac|le /'mɪrəkl/ *n* milagro *m*. **~ulous** /mɪ'rækjʊləs/ *a* milagroso
mirage /'mɪrɑ:ʒ/ *n* espejismo *m*
mire /'maɪə(r)/ *n* fango *m*
mirror /'mɪrə(r)/ *n* espejo *m*. ● *vt* reflejar
mirth /mɜ:θ/ *n* (*merriment*) alegría *f*; (*laughter*) risas *fpl*

misadventure /mɪsəd'ventʃə(r)/ *n* desgracia *f*
misanthropist /mɪ'zænθrəpɪst/ *n* misántropo *m*
misapprehension /mɪsæpri'henʃŋ/ *n* malentendido *m*
misbehav|e /mɪsbɪ'heɪv/ *vi* portarse mal. **~iour** *n* mala conducta *f*
miscalculat|e /mɪs'kælkjʊleɪt/ *vt/i* calcular mal. **~ion** /-'leɪʃn/ *n* desacierto *m*
miscarr|iage /'mɪskærɪdʒ/ *n* aborto *m*. **~iage of justice** *n* error *m* judicial. **~y** *vi* abortar
miscellaneous /mɪsə'leɪnɪəs/ *a* vario
mischief /'mɪstʃɪf/ *n* (*foolish conduct*) travesura *f*; (*harm*) daño *m*. **get into ~** cometer travesuras. **make ~** armar un lío
mischievous /'mɪstʃɪvəs/ *a* travieso; (*malicious*) perjudicial
misconception /mɪskən'sepʃn/ *n* equivocación *f*
misconduct /mɪs'kɒndʌkt/ *n* mala conducta *f*
misconstrue /mɪskən'stru:/ *vt* interpretar mal
misdeed /mɪs'di:d/ *n* fechoría *f*
misdemeanour /mɪsdɪ'mi:nə(r)/ *n* fechoría *f*
misdirect /mɪsdɪ'rekt/ *vt* dirigir mal ‹*person*›
miser /'maɪzə(r)/ *n* avaro *m*
miserable /'mɪzərəbl/ *a* (*sad*) triste; (*wretched*) miserable; ‹*weather*› malo
miserly /'maɪzəlɪ/ *a* avariento
misery /'mɪzərɪ/ *n* (*unhappiness*) tristeza *f*; (*pain*) sufrimiento *m*; (*poverty*) pobreza *f*; (*person, fam*) aguafiestas *m & f*
misfire /mɪs'faɪə(r)/ *vi* fallar
misfit /'mɪsfɪt/ *n* (*person*) inadaptado *m*; (*thing*) cosa *f* mal ajustada
misfortune /mɪs'fɔ:tʃu:n/ *n* desgracia *f*
misgiving /mɪs'gɪvɪŋ/ *n* (*doubt*) duda *f*; (*apprehension*) presentimiento *m*
misguided /mɪs'gaɪdɪd/ *a* equivocado. **be ~** equivocarse
mishap /'mɪshæp/ *n* desgracia *f*
misinform /mɪsɪn'fɔ:m/ *vt* informar mal
misinterpret /mɪsɪn'tɜ:prɪt/ *vt* interpretar mal
misjudge /mɪs'dʒʌdʒ/ *vt* juzgar mal
mislay /mɪs'leɪ/ *vt* (*pt* **mislaid**) extraviar
mislead /mɪs'li:d/ *vt* (*pt* **misled**) engañar. **~ing** *a* engañoso
mismanage /mɪs'mænɪdʒ/ *vt* administrar mal. **~ment** *n* mala administración *f*
misnomer /mɪs'nəʊmə(r)/ *n* nombre *m* equivocado
misplace /mɪs'pleɪs/ *vt* colocar mal; (*lose*) extraviar
misprint /'mɪsprɪnt/ *n* errata *f*
misquote /mɪs'kwəʊt/ *vt* citar mal
misrepresent /mɪsreprɪ'zent/ *vt* describir engañosamente
miss[1] /mɪs/ *vt* (*fail to hit*) errar; (*notice absence of*) echar de menos; perder ‹*train*›. **~ the point** no comprender. ● *n* fallo *m*. **~ out** omitir
miss[2] /mɪs/ *n* (*pl* **misses**) señorita *f*
misshapen /mɪs'ʃeɪpən/ *a* deforme
missile /'mɪsaɪl/ *n* proyectil *m*
missing /'mɪsɪŋ/ *a* ‹*person*› (*absent*) ausente; ‹*person*› (*after disaster*) desaparecido; (*lost*) perdido. **be ~** faltar
mission /'mɪʃn/ *n* misión *f*. **~ary** /'mɪʃənərɪ/ *n* misionero *m*
missive /'mɪsɪv/ *n* misiva *f*
misspell /mɪs'spel/ *vt* (*pt* **misspelt** *or* **misspelled**) escribir mal
mist /mɪst/ *n* neblina *f*; (*at sea*) bruma *f*. ● *vt/i* empañar(se)
mistake /mɪ'steɪk/ *n* error *m*. ● *vt* (*pt* **mistook**, *pp* **mistaken**) equivocarse de; (*misunderstand*) entender mal. **~ for** tomar por. **~n** /-ən/ *a* equivocado. **be ~n** equivocarse. **~nly** *adv* equivocadamente
mistletoe /'mɪsltəʊ/ *n* muérdago *m*
mistreat /mɪs'tri:t/ *vt* maltratar
mistress /'mɪstrɪs/ *n* (*of house*) señora *f*; (*primary school teacher*) maestra *f*; (*secondary school teacher*) profesora *f*; (*lover*) amante *f*
mistrust /mɪs'trʌst/ *vt* desconfiar de. ● *n* desconfianza *f*
misty /'mɪstɪ/ *a* (**-ier**, **-iest**) nebuloso; ‹*day*› de niebla; ‹*glass*› empañado. **it is ~** hay neblina
misunderstand /mɪsʌndə'stænd/ *vt* (*pt* **-stood**) entender mal. **~ing** *n* malentendido *m*
misuse /mɪs'ju:z/ *vt* emplear mal; abusar de ‹*power etc*›. /mɪs'ju:s/ *n* mal uso *m*; (*unfair use*) abuso *m*
mite /maɪt/ *n* (*insect*) ácaro *m*, garrapata *f*; (*child*) niño *m* pequeño
mitigate /'mɪtɪgeɪt/ *vt* mitigar

mitre /ˈmaɪtə(r)/ *n* (*head-dress*) mitra *f*

mitten /ˈmɪtn/ *n* manopla *f*; (*leaving fingers exposed*) mitón *m*

mix /mɪks/ *vt/i* mezclar(se). **~ up** mezclar; (*confuse*) confundir. **~ with** frecuentar ‹*people*›. ● *n* mezcla *f*

mixed /mɪkst/ *a* ‹*school etc*› mixto; (*assorted*) variado. **be ~ up** estar confuso

mixer /ˈmɪksə(r)/ *n* (*culin*) batidora *f*. **be a good ~** tener don de gentes

mixture /ˈmɪkstʃə(r)/ *n* mezcla *f*

mix-up /ˈmɪksʌp/ *n* lío *m*

moan /məʊn/ *n* gemido *m*. ● *vi* gemir; (*complain*) quejarse (**about** de). **~er** *n* refunfuñador *m*

moat /məʊt/ *n* foso *m*

mob /mɒb/ *n* (*crowd*) muchedumbre *f*; (*gang*) pandilla *f*; (*masses*) populacho *m*. ● *vt* (*pt* **mobbed**) acosar

mobil|e /ˈməʊbaɪl/ *a* móvil. **~e home** *n* caravana *f*. ● *n* móvil *m*. **~ity** /məˈbɪlətɪ/ *n* movilidad *f*

mobiliz|ation /məʊbɪlaɪˈzeɪʃn/ *n* movilización *f*. **~e** /ˈməʊbɪlaɪz/ *vt/i* movilizar

moccasin /ˈmɒkəsɪn/ *n* mocasín *m*

mocha /ˈmɒkə/ *n* moca *m*

mock /mɒk/ *vt* burlarse de. ● *vi* burlarse. ● *a* fingido

mockery /ˈmɒkərɪ/ *n* burla *f*. **a ~ of** una parodia *f* de

mock-up /ˈmɒkʌp/ *n* maqueta *f*

mode /məʊd/ *n* (*way, method*) modo *m*; (*fashion*) moda *f*

model /ˈmɒdl/ *n* modelo *m*; (*mock-up*) maqueta *f*; (*for fashion*) maniquí *m*. ● *a* (*exemplary*) ejemplar; ‹*car etc*› en miniatura. ● *vt* (*pt* **modelled**) modelar; presentar ‹*clothes*›. ● *vi* ser maniquí; (*pose*) posar. **~ling** *n* profesión *f* de maniquí

moderate /ˈmɒdərət/ *a* & *n* moderado (*m*). /ˈmɒdəreɪt/ *vt/i* moderar(se). **~ly** /ˈmɒdərətlɪ/ *adv* (*in moderation*) moderadamente; (*fairly*) medianamente

moderation /mɒdəˈreɪʃn/ *n* moderación *f*. **in ~** con moderación

modern /ˈmɒdn/ *a* moderno. **~ize** *vt* modernizar

modest /ˈmɒdɪst/ *a* modesto. **~y** *n* modestia *f*

modicum /ˈmɒdɪkəm/ *n*. **a ~ of** un poquito *m* de

modif|ication /mɒdɪfɪˈkeɪʃn/ *n* modificación *f*. **~y** /-faɪ/ *vt/i* modificar(se)

modulat|e /ˈmɒdjʊleɪt/ *vt/i* modular. **~ion** /-ˈleɪʃn/ *n* modulación *f*

module /ˈmɒdju:l/ *n* módulo *m*

mogul /ˈməʊgəl/ *n* (*fam*) magnate *m*

mohair /ˈməʊheə(r)/ *n* mohair *m*

moist /mɔɪst/ *a* (**-er**, **-est**) húmedo. **~en** /ˈmɔɪsn/ *vt* humedecer

moistur|e /ˈmɔɪstʃə(r)/ *n* humedad *f*. **~ize** /ˈmɔɪstʃəraɪz/ *vt* humedecer. **~izer** *n* crema *f* hidratante

molar /ˈməʊlə(r)/ *n* muela *f*

molasses /məˈlæsɪz/ *n* melaza *f*

mold /məʊld/ (*Amer*) = **mould**

mole[1] /məʊl/ *n* (*animal*) topo *m*

mole[2] /məʊl/ *n* (*on skin*) lunar *m*

mole[3] /məʊl/ *n* (*breakwater*) malecón *m*

molecule /ˈmɒlɪkju:l/ *n* molécula *f*

molehill /ˈməʊlhɪl/ *n* topera *f*

molest /məˈlest/ *vt* importunar

mollify /ˈmɒlɪfaɪ/ *vt* apaciguar

mollusc /ˈmɒləsk/ *n* molusco *m*

mollycoddle /ˈmɒlɪkɒdl/ *vt* mimar

molten /ˈməʊltən/ *a* fundido

mom /mɒm/ *n* (*Amer*) mamá *f*

moment /ˈməʊmənt/ *n* momento *m*. **~arily** /ˈməʊməntərɪlɪ/ *adv* momentáneamente. **~ary** *a* momentáneo

momentous /məˈmentəs/ *a* importante

momentum /məˈmentəm/ *n* momento *m*; (*speed*) velocidad *f*; (*fig*) ímpetu *m*

Monaco /ˈmɒnəkəʊ/ *n* Mónaco *m*

monarch /ˈmɒnək/ *n* monarca *m*. **~ist** *n* monárquico *m*. **~y** *n* monarquía *f*

monast|ery /ˈmɒnəstərɪ/ *n* monasterio *m*. **~ic** /məˈnæstɪk/ *a* monástico

Monday /ˈmʌndeɪ/ *n* lunes *m*

monetar|ist /ˈmʌnɪtərɪst/ *n* monetarista *m* & *f*. **~y** *a* monetario

money /ˈmʌnɪ/ *n* dinero *m*. **~-box** *n* hucha *f*. **~ed** *a* adinerado. **~-lender** *n* prestamista *m* & *f*. **~ order** *n* giro *m* postal. **~s** *npl* cantidades *fpl* de dinero. **~spinner** *n* mina *f* de dinero

mongol /ˈmɒŋgl/ *n* & *a* (*med*) mongólico (*m*)

mongrel /ˈmʌŋgrəl/ *n* perro *m* mestizo

monitor /ˈmɒnɪtə(r)/ *n* (*pupil*) monitor *m* & *f*; (*tec*) monitor *m*. ● *vt* controlar; escuchar ‹*a broadcast*›

monk /mʌŋk/ *n* monje *m*
monkey /'mʌŋkɪ/ *n* mono *m*. **~-nut** *n* cacahuete *m*, maní *m* (*LAm*). **~-wrench** *n* llave *f* inglesa
mono /'mɒnəʊ/ *a* monofónico
monocle /'mɒnəkl/ *n* monóculo *m*
monogram /'mɒnəgræm/ *n* monograma *m*
monologue /'mɒnəlɒg/ *n* monólogo *m*
monopol|ize /mə'nɒpəlaɪz/ *vt* monopolizar. **~y** *n* monopolio *m*
monosyllab|ic /mɒnəsɪ'læbɪk/ *a* monosilábico. **~le** /-'sɪləbl/ *n* monosílabo *m*
monotone /'mɒnətəʊn/ *n* monotonía *f*. **speak in a ~** hablar con una voz monótona
monoton|ous /mə'nɒtənəs/ *a* monótono. **~y** *n* monotonía *f*
monsoon /mɒn'suːn/ *n* monzón *m*
monster /'mɒnstə(r)/ *n* monstruo *m*
monstrosity /mɒn'strɒsətɪ/ *n* monstruosidad *f*
monstrous /'mɒnstrəs/ *a* monstruoso
montage /mɒn'tɑːʒ/ *n* montaje *m*
month /mʌnθ/ *n* mes *m*. **~ly** /'mʌnθlɪ/ *a* mensual. ● *adv* mensualmente. ● *n* (*periodical*) revista *f* mensual
monument /'mɒnjʊmənt/ *n* monumento *m*. **~al** /-'mentl/ *a* monumental
moo /muː/ *n* mugido *m*. ● *vi* mugir
mooch /muːtʃ/ *vi* (*sl*) haraganear. ● *vt* (*Amer, sl*) birlar
mood /muːd/ *n* humor *m*. **be in the ~ for** tener ganas de. **in a good/bad ~** de buen/mal humor. **~y** *a* (**-ier, -iest**) de humor cambiadizo; (*bad-tempered*) malhumorado
moon /muːn/ *n* luna *f*. **~light** *n* luz *f* de la luna. **~lighting** *n* (*fam*) pluriempleo *m*. **~lit** *a* iluminado por la luna; ‹*night*› de luna
moor[1] /mʊə(r)/ *n* (*open land*) páramo *m*
moor[2] /mʊə(r)/ *vt* amarrar. **~ings** *npl* (*ropes*) amarras *fpl*; (*place*) amarradero *m*
Moor /mʊə(r)/ *n* moro *m*
moose /muːs/ *n invar* alce *m*
moot /muːt/ *a* discutible. ● *vt* proponer ‹*question*›
mop /mɒp/ *n* fregona *f*. **~ of hair** pelambrera *f*. ● *vt* (*pt* **mopped**) fregar. **~ (up)** limpiar
mope /məʊp/ *vi* estar abatido
moped /'məʊped/ *n* ciclomotor *m*
moral /'mɒrəl/ *a* moral. ● *n* moraleja *f*. **~s** *npl* moralidad *f*
morale /mə'rɑːl/ *n* moral *f*
moral|ist /'mɒrəlɪst/ *n* moralista *m* & *f*. **~ity** /mə'rælətɪ/ *n* moralidad *f*. **~ize** *vi* moralizar. **~ly** *adv* moralmente
morass /mə'ræs/ *n* (*marsh*) pantano *m*; (*fig, entanglement*) embrollo *m*
morbid /'mɔːbɪd/ *a* morboso
more /mɔː(r)/ *a* & *n* & *adv* más. **~ and ~** cada vez más. **~ or less** más o menos. **once ~** una vez más. **some ~** más
moreover /mɔː'rəʊvə(r)/ *adv* además
morgue /mɔːg/ *n* depósito *m* de cadáveres
moribund /'mɒrɪbʌnd/ *a* moribundo
morning /'mɔːnɪŋ/ *n* mañana *f*; (*early hours*) madrugada *f*. **at 11 o'clock in the ~** a las once de la mañana. **in the ~** por la mañana
Morocc|an /mə'rɒkən/ *a* & n marroquí (*m* & *f*). **~o** /-kəʊ/ *n* Marruecos *mpl*
moron /'mɔːrɒn/ *n* imbécil *m* & *f*
morose /mə'rəʊs/ *a* malhumorado
morphine /'mɔːfiːn/ *n* morfina *f*
Morse /mɔːs/ *n* Morse *m*. **~ (code)** *n* alfabeto *m* Morse
morsel /'mɔːsl/ *n* pedazo *m*; (*mouthful*) bocado *m*
mortal /'mɔːtl/ *a* & *n* mortal (*m*). **~ity** /-'tælətɪ/ *n* mortalidad *f*
mortar /'mɔːtə(r)/ *n* (*all senses*) mortero *m*
mortgage /'mɔːgɪdʒ/ *n* hipoteca *f*. ● *vt* hipotecar
mortify /'mɔːtɪfaɪ/ *vt* mortificar
mortuary /'mɔːtjʊərɪ/ *n* depósito *m* de cadáveres
mosaic /məʊ'zeɪk/ *n* mosaico *m*
Moscow /'mɒskəʊ/ *n* Moscú *m*
Moses /'məʊzɪz/ *a*. **~ basket** *n* moisés *m*
mosque /mɒsk/ *n* mezquita *f*
mosquito /mɒs'kiːtəʊ/ *n* (*pl* **-oes**) mosquito *m*
moss /mɒs/ *n* musgo *m*. **~y** *a* musgoso
most /məʊst/ *a* más. **for the ~ part** en su mayor parte. ● *n* la mayoría *f*. **~ of** la mayor parte de. **at ~** a lo más. **make the ~ of** aprovechar al máximo. ● *adv* más; (*very*) muy. **~ly** *adv* principalmente

MOT *abbr* (*Ministry of Transport*). ~ **(test)** ITV, inspección *f* técnica de vehículos

motel /məʊ'tel/ *n* motel *m*

moth /mɒθ/ *n* mariposa *f* (nocturna); (*in clothes*) polilla *f*. ~**-ball** *n* bola *f* de naftalina. ~**-eaten** *a* apolillado

mother /'mʌðə(r)/ *n* madre *f*. ● *vt* cuidar como a un hijo. ~**hood** *n* maternidad *f*. ~**-in-law** *n* (*pl* ~**s-in-law**) suegra *f*. ~**land** *n* patria *f*. ~**ly** *adv* maternalmente. ~**-of-pearl** *n* nácar *m*. **M~'s Day** *n* el día *m* de la Madre. ~**-to-be** *n* futura madre *f*. ~ **tongue** *n* lengua *f* materna

motif /məʊ'ti:f/ *n* motivo *m*

motion /'məʊʃn/ *n* movimiento *m*; (*proposal*) moción *f*. ● *vt/i*. ~ **(to) s.o. to** hacer señas a uno para que. ~**less** *a* inmóvil

motivat|e /'məʊtɪveɪt/ *vt* motivar. ~**ion** /-'veɪʃn/ *n* motivación *f*

motive /'məʊtɪv/ *n* motivo *m*

motley /'mɒtlɪ/ *a* abigarrado

motor /'məʊtə(r)/ *n* motor *m*; (*car*) coche *m*. ● *a* motor; (*fem*) motora, motriz. ● *vi* ir en coche. ~ **bike** *n* (*fam*) motocicleta *f*, moto *f* (*fam*). ~ **boat** *n* lancha *f* motora. ~**cade** /'məʊtəkeɪd/ *n* (*Amer*) desfile *m* de automóviles. ~ **car** *n* coche *m*, automóvil *m*. ~ **cycle** *n* motocicleta *f*. ~**cyclist** *n* motociclista *m* & *f*. ~**ing** *n* automovilismo *m*. ~**ist** *n* automovilista *m* & *f*. ~**ize** *vt* motorizar. ~**way** *n* autopista *f*

mottled /'mɒtld/ *a* abigarrado

motto /'mɒtəʊ/ *n* (*pl* **-oes**) lema *m*

mould[1] /məʊld/ *n* molde *m*. ● *vt* moldear

mould[2] /məʊld/ *n* (*fungus, rot*) moho *m*

moulding /'məʊldɪŋ/ *n* (*on wall etc*) moldura *f*

mouldy /'məʊldɪ/ *a* mohoso

moult /məʊlt/ *vi* mudar

mound /maʊnd/ *n* montículo *m*; (*pile, fig*) montón *m*

mount[1] /maʊnt/ *vt/i* subir. ● *n* montura *f*. ~ **up** aumentar

mount[2] /maʊnt/ *n* (*hill*) monte *m*

mountain /'maʊntɪn/ *n* montaña *f*. ~**eer** /maʊntɪ'nɪə(r)/ *n* alpinista *m* & *f*. ~**eering** *n* alpinismo *m*. ~**ous** /'maʊntɪnəs/ *a* montañoso

mourn /mɔ:n/ *vt* llorar. ● *vi* lamentarse. ~ **for** llorar la muerte de. ~**er** *n* persona *f* que acompaña el cortejo fúnebre. ~**ful** *a* triste. ~**ing** *n* luto *m*

mouse /maʊs/ *n* (*pl* **mice**) ratón *m*. ~**trap** *n* ratonera *f*

mousse /mu:s/ *n* (*dish*) crema *f* batida

moustache /mə'stɑ:ʃ/ *n* bigote *m*

mousy /'maʊsɪ/ *a* ‹*hair*› pardusco; (*fig*) tímido

mouth /maʊð/ *vt* formar con los labios. /maʊθ/ *n* boca *f*. ~**ful** *n* bocado *m*. ~**-organ** *n* armónica *f*. ~**piece** *n* (*mus*) boquilla *f*; (*fig, person*) portavoz *f*, vocero *m* (*LAm*). ~**wash** *n* enjuague *m*

movable /'mu:vəbl/ *a* móvil, movible

move /mu:v/ *vt* mover; mudarse de ‹*house*›; (*with emotion*) conmover; (*propose*) proponer. ● *vi* moverse; (*be in motion*) estar en movimiento; (*progress*) hacer progresos; (*take action*) tomar medidas; (*depart*) irse. ~ **(out)** irse. ● *n* movimiento *m*; (*in game*) jugada *f*; (*player's turn*) turno *m*; (*removal*) mudanza *f*. **on the** ~ en movimiento. ~ **along** (hacer) circular. ~ **away** alejarse. ~ **back** (hacer) retroceder. ~ **forward** (hacer) avanzar. ~ **in** instalarse. ~ **on** (hacer) circular. ~ **over** apartarse. ~**ment** /'mu:vmənt/ *n* movimiento *m*

movie /'mu:vɪ/ *n* (*Amer*) película *f*. **the** ~**s** *npl* el cine *m*

moving /'mu:vɪŋ/ *a* en movimiento; (*touching*) conmovedor

mow /məʊ/ *vt* (*pt* **mowed** *or* **mown**) segar. ~ **down** derribar. ~**er** *n* (*for lawn*) cortacésped *m inv*

MP *abbr see* **Member of Parliament**

Mr /'mɪstə(r)/ *abbr* (*pl* **Messrs**) (*Mister*) señor *m*. ~ **Coldbeck** (el) Sr. Coldbeck

Mrs /'mɪsɪz/ *abbr* (*pl* **Mrs**) (*Missis*) señora *f*. ~ **Andrews** (la) Sra. Andrews. **the** ~ **Andrews** (las) Sras. Andrews

Ms /mɪz/ *abbr* (*title of married or unmarried woman*) señora *f*, señorita. **Ms Lawton** (la) Sra. Lawton

much /mʌtʃ/ *a* & *n* mucho (*m*). ● *adv* mucho; (*before pp*) muy. ~ **as** por mucho que. ~ **the same** más o menos lo mismo. **so** ~ tanto. **too** ~ demasiado

muck /mʌk/ *n* estiércol *m*; (*dirt, fam*) suciedad *f*. ● *vi*. ~ **about** (*sl*) perder el tiempo. ~ **about with** (*sl*) juguetear con. ● *vt.* ~ **up** (*sl*) echar a perder. ~ **in** (*sl*) participar. ~**y** *a* sucio

mucus /'mju:kəs/ *n* moco *m*

mud /mʌd/ *n* lodo *m*, barro *m*

muddle /'mʌdl/ *vt* embrollar. ● *vi.* ~ **through** salir del paso. ● *n* desorden *m*; (*mix-up*) lío *m*

muddy /'mʌdɪ/ *a* lodoso; ‹*hands etc*› cubierto de lodo

mudguard /'mʌdgɑ:d/ *n* guardabarros *m invar*

muff /mʌf/ *n* manguito *m*

muffin /'mʌfɪn/ *n* mollete *m*

muffle /'mʌfl/ *vt* tapar; amortiguar ‹*a sound*›. ~**r** *n* (*scarf*) bufanda *f*

mug /mʌg/ *n* tazón *m*; (*for beer*) jarra *f*; (*face*, *sl*) cara *f*, jeta *f* (*sl*); (*fool*, *sl*) primo *m*. ● *vt* (*pt* **mugged**) asaltar. ~**ger** *n* asaltador *m*. ~**ging** *n* asalto *m*

muggy /'mʌgɪ/ *a* bochornoso

Muhammadan /mə'hæmɪdən/ *a* & *n* mahometano (*m*)

mule[1] /mju:l/ *n* mula *f*, mulo *m*

mule[2] /mju:l/ *n* (*slipper*) babucha *f*

mull[1] /mʌl/ *vt.* ~ **over** reflexionar sobre

mull[2] /mʌl/ *vt* calentar con especias ‹*wine*›

multi... /'mʌltɪ/ *pref* multi...

multicoloured /mʌltɪ'kʌləd/ *a* multicolor

multifarious /mʌltɪ'feərɪəs/ *a* múltiple

multinational /mʌltɪ'næʃənl/ *a* & *n* multinacional (*f*)

multipl|e /'mʌltɪpl/ *a* & *n* múltiplo (*m*). ~**ication** /mʌltɪplɪ'keɪʃn/ *n* multiplicación *f*. ~**y** /'mʌltɪplaɪ/ *vt*/*i* multiplicar(se)

multitude /'mʌltɪtju:d/ *n* multitud *f*

mum[1] /mʌm/ *n* (*fam*) mamá *f* (*fam*)

mum[2] /mʌm/ *a.* **keep** ~ (*fam*) guardar silencio

mumble /'mʌmbl/ *vt* decir entre dientes. ● *vi* hablar entre dientes

mummify /'mʌmɪfaɪ/ *vt*/*i* momificar(se)

mummy[1] /'mʌmɪ/ *n* (*mother*, *fam*) mamá *f* (*fam*)

mummy[2] /'mʌmɪ/ *n* momia *f*

mumps /mʌmps/ *n* paperas *fpl*

munch /mʌntʃ/ *vt*/*i* mascar

mundane /mʌn'deɪn/ *a* mundano

municipal /mju:'nɪsɪpl/ *a* municipal. ~**ity** /-'pælətɪ/ *n* municipio *m*

munificent /mju:'nɪfɪsənt/ *a* munífico

munitions /mju:'nɪʃnz/ *npl* municiones *fpl*

mural /'mjʊərəl/ *a* & *n* mural (*f*)

murder /'mɜ:də(r)/ *n* asesinato *m*. ● *vt* asesinar. ~**er** *n* asesino *m*. ~**ess** *n* asesina *f*. ~**ous** *a* homicida

murky /'mɜ:kɪ/ *a* (**-ier**, **-iest**) oscuro

murmur /'mɜ:mə(r)/ *n* murmullo *m*. ● *vt*/*i* murmurar

muscle /'mʌsl/ *n* músculo *m*. ● *vi.* ~ **in** (*Amer*, *sl*) meterse por fuerza en

muscular /'mʌskjʊlə(r)/ *a* muscular; (*having well-developed muscles*) musculoso

muse /mju:z/ *vi* meditar

museum /mju:'zɪəm/ *n* museo *m*

mush /mʌʃ/ *n* pulpa *f*

mushrom /'mʌʃrʊm/ *n* champiñón *m*; (*bot*) seta *f*. ● *vi* (*appear in large numbers*) crecer como hongos

mushy /'mʌʃɪ/ *a* pulposo

music /'mju:zɪk/ *n* música *f*. ~**al** *a* musical; ‹*instrument*› de música; (*talented*) que tiene don de música. ● *n* comedia *f* musical. ~ **hall** *n* teatro *m* de variedades. ~**ian** /mju:'zɪʃn/ *n* músico *m*

musk /mʌsk/ *n* almizcle *m*

Muslim /'mʊzlɪm/ *a* & *n* musulmán (*m*)

muslin /'mʌzlɪn/ *n* muselina *f*

musquash /'mʌskwɒʃ/ *n* ratón *m* almizclero

mussel /mʌsl/ *n* mejillón *m*

must /mʌst/ *v aux* deber, tener que. **he** ~ **be old** debe ser viejo. **I** ~ **have done it** debo haberlo hecho. **you** ~ **go** debes marcharte. ● *n.* **be a** ~ ser imprescindible

mustard /'mʌstəd/ *n* mostaza *f*

muster /'mʌstə(r)/ *vt*/*i* reunir(se)

musty /'mʌstɪ/ *a* (**-ier**, **-iest**) que huele a cerrado

mutation /mju:'teɪʃn/ *n* mutación *f*

mute /mju:t/ *a* & *n* mudo (*m*). ~**d** *a* ‹*sound*› sordo; ‹*criticism*› callado

mutilat|e /'mju:tɪleɪt/ *vt* mutilar. ~**ion** /-'leɪʃn/ *n* mutilación *f*

mutin|ous /'mju:tɪnəs/ *a* ‹*sailor etc*› amotinado; (*fig*) rebelde. ~**y** *n* motín *m*. ● *vi* amotinarse

mutter /'mʌtə(r)/ *vt*/*i* murmurar

mutton /'mʌtn/ *n* cordero *m*

mutual /'mju:tʃʊəl/ *a* mutuo; (*common*, *fam*) común. ~**ly** *adv* mutuamente

muzzle /mʌzl/ *n* (*snout*) hocico *m*; (*device*) bozal *m*; (*of gun*) boca *f*. ● *vt* poner el bozal a

my /maɪ/ *a* mi, mis *pl*
myopic /maɪˈɒpɪk/ *a* miope
myriad /ˈmɪrɪəd/ *n* miríada *f*
myself /maɪˈself/ *pron* yo mismo *m*, yo misma *f*; (*reflexive*) me; (*after prep*) mí (mismo) *m*, mí (misma) *f*
myster|ious /mɪˈstɪərɪəs/ *a* misterioso. **~y** /ˈmɪstərɪ/ *n* misterio *m*
mystic /ˈmɪstɪk/ *a* & *n* místico (*m*). **~al** *a* místico. **~ism** /-sɪzəm/ *n* misticismo *m*
mystif|ication /mɪstɪfɪˈkeɪʃn/ *n* confusión *f*. **~y** /-faɪ/ *vt* dejar perplejo
mystique /mɪˈsti:k/ *n* mística *f*
myth /mɪθ/ *n* mito *m*. **~ical** *a* mítico. **~ology** /mɪˈθɒlədʒɪ/ *n* mitología *f*

N

N *abbr* (*north*) norte *m*
nab /næb/ *vt* (*pt* **nabbed**) (*arrest, sl*) coger (*not LAm*), agarrar (*esp LAm*)
nag /næg/ *vt* (*pt* **nagged**) fastidiar; (*scold*) regañar. ● *vi* criticar
nagging /ˈnægɪŋ/ *a* persistente, regañón
nail /neɪl/ *n* clavo *m*; (*of finger, toe*) uña *f*. **pay on the ~** pagar a tocateja. ● *vt* clavar. **~ polish** *n* esmalte *m* para las uñas
naïve /naɪˈi:v/ *a* ingenuo
naked /ˈneɪkɪd/ *a* desnudo. **to the ~ eye** a simple vista. **~ly** *adv* desnudamente. **~ness** *n* desnudez *f*
namby-pamby /næmbɪˈpæmbɪ/ *a* & *n* ñoño (*m*)
name /neɪm/ *n* nombre *m*; (*fig*) fama *f*. ● *vt* nombrar; (*fix*) fijar. **be ~d after** llevar el nombre de. **~less** *a* anónimo. **~ly** /ˈneɪmlɪ/ *adv* a saber. **~sake** /ˈneɪmseɪk/ *n* (*person*) tocayo *m*
nanny /ˈnænɪ/ *n* niñera *f*. **~-goat** *n* cabra *f*
nap[1] /næp/ *n* (*sleep*) sueñecito *m*; (*after lunch*) siesta *f*. ● *vi* (*pt* **napped**) echarse un sueño. **catch s.o. ~ping** coger a uno desprevenido
nap[2] /næp/ *n* (*fibres*) lanilla *f*
nape /neɪp/ *n* nuca *f*
napkin /ˈnæpkɪn/ *n* (*at meals*) servilleta *f*; (*for baby*) pañal *m*
nappy /ˈnæpɪ/ *n* pañal *m*
narcotic /nɑ:ˈkɒtɪk/ *a* & *n* narcótico (*m*)
narrat|e /nəˈreɪt/ *vt* contar. **~ion** /-ʃn/ *n* narración *f*. **~ive** /ˈnærətɪv/ *n* relato *m*. **~or** /nəˈreɪtə(r)/ *n* narrador *m*
narrow /ˈnærəʊ/ *a* (**-er**, **-est**) estrecho. **have a ~ escape** escaparse por los pelos. ● *vt* estrechar; (*limit*) limitar. ● *vi* estrecharse. **~ly** *adv* estrechamente; (*just*) por poco. **~-minded** *a* de miras estrechas. **~ness** *n* estrechez *f*
nasal /ˈneɪzl/ *a* nasal
nast|ily /ˈnɑ:stɪlɪ/ *adv* desagradablemente; (*maliciously*) con malevolencia. **~iness** *n* (*malice*) malevolencia *f*. **~y** *a* /ˈnɑ:stɪ/ (**-ier**, **-iest**) desagradable; (*malicious*) malévolo; ‹*weather*› malo; ‹*taste, smell*› asqueroso; ‹*wound*› grave; ‹*person*› antipático
natal /ˈneɪtl *a* natal
nation /ˈneɪʃn/ *n* nación *f*
national /ˈnæʃənl/ *a* nacional. ● *n* súbdito *m*. **~ anthem** *n* himno *m* nacional. **~ism** *n* nacionalismo *m*. **~ity** /næʃəˈnælətɪ/ *n* nacionalidad *f*. **~ize** *vt* nacionalizar. **~ly** *adv* a nivel nacional
nationwide /ˈneɪʃnwaɪd/ *a* nacional
native /ˈneɪtɪv/ *n* natural *m* & *f*. **be a ~ of** ser natural de. ● *a* nativo; ‹*country, town*› natal; (*inborn*) innato. **~ speaker of Spanish** hispanohablante *m* & *f*. **~ language** *n* lengua *f* materna
Nativity /nəˈtɪvətɪ/ *n*. **the ~** la Natividad *f*
NATO /ˈneɪtəʊ/ *abbr* (*North Atlantic Treaty Organization*) OTAN *f*, Organización *f* del Tratado del Atlántico Norte
natter /ˈnætə(r)/ *vi* (*fam*) charlar. ● *n* (*fam*) charla *f*
natural /ˈnætʃərəl/ *a* natural. **~ history** *n* historia *f* natural. **~ist** *n* naturalista *m* & *f*
naturaliz|ation /nætʃərəlaɪˈzeɪʃn/ *n* naturalización *f*. **~e** *vt* naturalizar
naturally /ˈnætʃərəlɪ/ *adv* (*of course*) naturalmente; (*by nature*) por naturaleza
nature /ˈneɪtʃə(r)/ *n* naturaleza *f*; (*kind*) género *m*; (*of person*) carácter *m*
naught /nɔ:t/ *n* (*old use*) nada *f*; (*maths*) cero *m*
naught|ily /ˈnɔ:tɪlɪ/ *adv* mal. **~y** *a* (**-ier**, **-iest**) malo; ‹*child*› travieso; ‹*joke*› verde

nause|a /'nɔ:zɪə/ *n* náusea *f*. **~ate** *vt* dar náuseas a. **~ous** *a* nauseabundo
nautical /'nɔ:tɪkl/ *a* náutico. **~ mile** *n* milla *f* marina
naval /'neɪvl/ *a* naval; ‹*officer*› de marina
Navarre /nə'vɑ:(r)/ *n* Navarra *f*. **~se** *a* navarro
nave /neɪv/ *n* (*of church*) nave *f*
navel /'neɪvl/ *n* ombligo *m*
navigable /'nævɪgəbl/ *a* navegable
navigat|e /'nævɪgeɪt/ *vt* navegar por ‹*sea etc*›; gobernar ‹*ship*›. ● *vi* navegar. **~ion** *n* navegación *f*. **~or** *n* navegante *m*
navvy /'nævɪ/ *n* peón *m* caminero
navy /'neɪvɪ/ *n* marina *f*. **~ (blue)** azul *m* marino
NE *abbr* (*north-east*) noreste *m*
near /'nɪə(r)/ *adv* cerca. **~ at hand** muy cerca. **~ by** *adv* cerca. **draw ~** acercarse. ● *prep*. **~ (to)** cerca de. ● *a* cercano. ● *vt* acercarse a. **~by** *a* cercano. **N~ East** *n* Oriente *m* Próximo. **~ly** /'nɪəlɪ/ *adv* casi. **not ~ly as pretty as** no es ni con mucho tan guapa como. **~ness** /'nɪənɪs/ *n* proximidad *f*
neat /ni:t/ *a* (**-er**, **-est**) pulcro; ‹*room etc*› bien arreglado; (*clever*) diestro; (*ingenious*) hábil; ‹*whisky, brandy etc*› solo. **~ly** *adv* pulcramente. **~ness** *n* pulcritud *f*
nebulous /'nebjʊləs/ *a* nebuloso
necessar|ies /'nesəsərɪz/ *npl* lo indispensable. **~ily** /nesə'serɪlɪ/ *adv* necesariamente. **~y** *a* necesario, imprescindible
necessit|ate /nə'sesɪteɪt/ *vt* necesitar. **~y** /nɪ'sesətɪ/ *n* necesidad *f*; (*thing*) cosa *f* indispensable
neck /nek/ *n* (*of person, bottle, dress*) cuello *m*; (*of animal*) pescuezo *m*. **~ and ~** parejos. **~lace** /'nekləs/ *n* collar *m*. **~line** *n* escote *m*. **~tie** n corbata *f*
nectar /'nektə(r)/ *n* néctar *m*
nectarine /nektə'ri:n/ *n* nectarina *f*
née /neɪ/ *a* de soltera
need /ni:d/ *n* necesidad *f*. ● *vt* necesitar; (*demand*) exigir. **you ~ not speak** no tienes que hablar
needle /'ni:dl/ *n* aguja *f*. ● *vt* (*annoy, fam*) pinchar
needless /'ni:dlɪs/ *a* innecesario. **~ly** *adv* innecesariamente
needlework /'ni:dlwɜ:k/ *n* costura *f*; (*embroidery*) bordado *m*
needy /'ni:dɪ/ *a* (**-ier**, **-iest**) necesitado
negation /nɪ'geɪʃn/ *n* negación *f*
negative /'negətɪv/ *a* negativo. ● *n* (*of photograph*) negativo *m*; (*word, gram*) negativa *f*. **~ly** *adv* negativamente
neglect /nɪ'glekt/ *vt* descuidar; no cumplir con ‹*duty*›. **~ to do** dejar de hacer. ● *n* descuido *m*, negligencia *f*. **(state of) ~** abandono *m*. **~ful** *a* descuidado
négligé /'neglɪʒeɪ/ *n* bata *f*, salto *m* de cama
negligen|ce /'neglidʒəns/ *n* negligencia *f*, descuido *m*. **~t** *a* descuidado
negligible /'neglɪdʒəbl/ *a* insignificante
negotiable /nɪ'gəʊʃəbl/ *a* negociable
negotiat|e /nɪ'gəʊʃɪeɪt/ *vt/i* negociar. **~ion** /-'eɪʃn/ *n* negociación *f*. **~or** *n* negociador *m*
Negr|ess /'ni:grɪs/ *n* negra *f*. **~o** *n* (*pl* **-oes**) negro *m*. ● *a* negro
neigh /neɪ/ *n* relincho *m*. ● *vi* relinchar
neighbour /'neɪbə(r)/ *n* vecino *m*. **~hood** *n* vecindad *f*, barrio *m*. **in the ~hood of** alrededor de. **~ing** *a* vecino. **~ly** /'neɪbəlɪ/ *a* amable
neither /'naɪðə(r)/ *a & pron* ninguno *m* de los dos, ni el uno *m* ni el otro *m*. ● *adv* ni. **~ big nor small** ni grande ni pequeño. **~ shall I come** no voy yo tampoco. ● *conj* tampoco
neon /'ni:ɒn/ *n* neón *m*. ● *a* ‹*lamp etc*› de neón
nephew /'nevju:/ *n* sobrino *m*
nepotism /'nepətɪzəm/ *m* nepotismo *m*
nerve /nɜ:v/ *n* nervio *m*; (*courage*) valor *m*; (*calm*) sangre *f* fría; (*impudence, fam*) descaro *m*. **~-racking** *a* exasperante. **~s** *npl* (*before exams etc*) nervios *mpl*
nervous /'nɜ:vəs/ *a* nervioso. **be/feel ~** (*afraid*) tener miedo **(of** a). **~ly** *adv* (*tensely*) nerviosamente; (*timidly*) tímidamente. **~ness** *n* nerviosidad *f*; (*fear*) miedo *m*
nervy /'nɜ:vɪ/ *a see* **nervous**; (*Amer, fam*) descarado
nest /nest/ *n* nido *m*. ● *vi* anidar. **~-egg** *n* (*money*) ahorros *mpl*
nestle /'nesl/ *vi* acomodarse. **~ up to** arrimarse a

net /net/ *n* red *f*. ● *vt* (*pt* **netted**) coger (*not LAm*), agarrar (*esp LAm*). ● *a* (*weight etc*) neto
netball /'netbɔ:l/ *n* baloncesto *m*
Netherlands /'neðələndz/ *npl*. **the ~** los Países *mpl* Bajos
netting /'netɪŋ/ *n* (*nets*) redes *fpl*; (*wire*) malla *f*; (*fabric*) tul *m*
nettle /'netl/ *n* ortiga *f*
network /'netwɜ:k/ *n* red *f*
neuralgia /njʊə'rældʒɪə/ *n* neuralgia *f*
neuro|sis /njʊə'rəʊsɪs/ *n* (*pl* **-oses** /-si:z/) neurosis *f*. **~tic** *a* & *n* neurótico (*m*)
neuter /'nju:tə(r)/ *a* & *n* neutro (*m*). ● *vt* castrar ‹*animals*›
neutral /'nju:trəl/ *a* neutral; ‹*colour*› neutro; (*elec*) neutro. **~ (gear)** (*auto*) punto *m* muerto. **~ity** /-'trælətɪ/ *n* neutralidad *f*
neutron /'nju:trɒn/ *n* neutrón *m*. **~ bomb** *n* bomba *f* de neutrones
never /nevə(r)/ *adv* nunca, jamás; (*not, fam*) no. **~ again** nunca más. **~ mind** (*don't worry*) no te preocupes, no se preocupe; (*it doesn't matter*) no importa. **he ~ smiles** no sonríe nunca. **I ~ saw him** (*fam*) no le vi. **~-ending** *a* interminable
nevertheless /nevəðə'les/ *adv* sin embargo, no obstante
new /nju:/ *a* (**-er, -est**) (*new to owner*) nuevo (*placed before noun*); (*brand new*) nuevo (*placed after noun*). **~-born** *a* recién nacido. **~comer** *n* recién llegado *m*. **~fangled** *a* (*pej*) moderno. **~-laid egg** *n* huevo *m* fresco. **~ly** *adv* nuevamente; (*recently*) recién. **~ly-weds** *npl* recién casados *mpl*. **~ moon** *n* luna *f* nueva. **~ness** *n* novedad *f*
news /nju:z/ *n* noticias *fpl*; (*broadcasting, press*) informaciones *fpl*; (*on TV*) telediario *m*; (*on radio*) diario *m* hablado. **~agent** *n* vendedor *m* de periódicos. **~caster** *n* locutor *m*. **~letter** *n* boletín *m*. **~paper** *n* periódico *m*. **~-reader** *n* locutor *m*. **~reel** *n* noticiario *m*, nodo *m* (*in Spain*)
newt /nju:t/ *n* tritón *m*
new year /nju:'jɪə(r)/ *n* año *m* nuevo. **N~'s Day** *n* día *m* de Año Nuevo. **N~'s Eve** *n* noche *f* vieja
New Zealand /nju:'zi:lənd/ *n* Nueva Zelanda *f*. **~er** *n* neozelandés *m*
next /nekst/ *a* próximo; ‹*week, month etc*› que viene, próximo; (*adjoining*) vecino; (*following*) siguiente. ● *adv* la próxima vez; (*afterwards*) después. ● *n* siguiente *m*. **~ to** junto a. **~ to nothing** casi nada. **~ door** al lado (**to** de). **~-door** de al lado. **~-best** mejor alternativa *f*. **~ of kin** *n* pariente *m* más próximo, parientes *mpl* más próximos
nib /nɪb/ *n* (*of pen*) plumilla *f*
nibble /'nɪbl/ *vt/i* mordisquear. ● *n* mordisco *m*
nice /naɪs/ *a* (**-er, -est**) agradable; (*likeable*) simpático; (*kind*) amable; (*pretty*) bonito; ‹*weather*› bueno; (*subtle*) sutil. **~ly** *adv* agradablemente; (*kindly*) amablemente; (*well*) bien
nicety /'naɪsətɪ/ *n* (*precision*) precisión *f*; (*detail*) detalle. **to a ~** exactamente
niche /nɪtʃ, ni:ʃ/ *n* (*recess*) nicho *m*; (*fig*) buena posición *f*
nick /nɪk/ *n* corte *m* pequeño; (*prison, sl*) cárcel *f*. **in the ~ of time** justo a tiempo. ● *vt* (*steal, arrest, sl*) birlar
nickel /'nɪkl/ *n* níquel *m*; (*Amer*) moneda *f* de cinco centavos
nickname /'nɪkneɪm/ *n* apodo *m*; (*short form*) diminutivo *m*. ● *vt* apodar
nicotine /'nɪkəti:n/ *n* nicotina *f*
niece /ni:s/ *n* sobrina *f*
nifty /'nɪftɪ/ *a* (*sl*) (*smart*) elegante
Nigeria /naɪ'dʒɪərɪə/ *n* Nigeria *f*. **~n** *a* & *n* nigeriano (*m*)
niggardly /'nɪgədlɪ/ *a* ‹*person*› tacaño; ‹*thing*› miserable
niggling /'nɪglɪŋ/ *a* molesto
night /naɪt/ *n* noche *f*; (*evening*) tarde *f*. ● *a* nocturno, de noche. **~cap** *n* (*hat*) gorro *m* de dormir; (*drink*) bebida *f* (tomada antes de acostarse). **~-club** *n* sala *f* de fiestas, boîte *f*. **~-dress** *n* camisón *m*. **~fall** *n* anochecer *m*. **~-gown** *n* camisón *m*
nightingale /'naɪtɪŋgeɪl/ *n* ruiseñor *m*
night: **~-life** *n* vida *f* nocturna. **~ly** *adv* todas las noches. **~mare** *n* pesadilla *f*. **~-school** *n* escuela *f* nocturna. **~-time** *n* noche *f*. **~-watchman** *n* sereno *m*
nil /nɪl/ *n* nada *f*; (*sport*) cero *m*
nimble /'nɪmbl/ *a* (**-er, -est**) ágil
nine /naɪn/ *a* & *n* nueve (*m*)
nineteen /naɪn'ti:n/ *a* & *n* diecinueve (*m*). **~th** *a* & *n* diecinueve (*m*), decimonoveno (*m*)

ninet|ieth /'naɪntɪəθ/ *a* noventa, nonagésimo. **~y** *a & n* noventa (*m*)

ninth /'naɪnθ/ *a & n* noveno (*m*)

nip[1] /nɪp/ *vt* (*pt* **nipped**) (*pinch*) pellizcar; (*bite*) mordisquear. ● *vi* (*rush, sl*) correr. ● *n* (*pinch*) pellizco *m*; (*cold*) frío *m*

nip[2] /nɪp/ *n* (*of drink*) trago *m*

nipper /'nɪpə(r)/ *n* (*sl*) chaval *m*

nipple /'nɪpl/ *n* pezón *m*; (*of baby's bottle*) tetilla *f*

nippy /'nɪpɪ/ *a* (**-ier**, **-iest**) (*nimble, fam*) ágil; (*quick, fam*) rápido; (*chilly, fam*) fresquito

nitrogen /'naɪtrədʒən/ *n* nitrógeno *m*

nitwit /'nɪtwɪt/ *n* (*fam*) imbécil *m & f*

no /nəʊ/ *a* ninguno. **~ entry** prohibido el paso. **~ man's land** *n* tierra *f* de nadie. **~ smoking** se prohibe fumar. **~ way!** (*Amer, fam*) ¡ni hablar! ● *adv* no. ● *n* (*pl* **noes**) no *m*

nobility /nəʊ'bɪlətɪ/ *n* nobleza *f*

noble /nəʊbl/ *a* (**-er**, **-est**) noble. **~man** *n* noble *m*

nobody /'nəʊbədɪ/ *pron* nadie *m*. ● *n* nadie *m*. **~ is there** no hay nadie. **he knows ~** no conoce a nadie

nocturnal /nɒk'tɜːnl/ *a* nocturno

nod /nɒd/ *vt* (*pt* **nodded**). **~ one's head** asentir con la cabeza. ● *vi* (*in agreement*) asentir con la cabeza; (*in greeting*) saludar; (*be drowsy*) dar cabezadas. ● *n* inclinación *f* de cabeza

nodule /'nɒdjuːl/ *n* nódulo *m*

nois|e /nɔɪz/ *n* ruido *m*. **~eless** *a* silencioso. **~ily** /'nɔɪzɪlɪ/ *adv* ruidosamente. **~y** *a* (**-ier**, **-iest**) ruidoso

nomad /'nəʊmæd/ *n* nómada *m & f*. **~ic** /-'mædɪk/ *a* nómada

nominal /'nɒmɪnl/ *a* nominal

nominat|e /'nɒmɪneɪt/ *vt* nombrar; (*put forward*) proponer. **~ion** /-'neɪʃn/ *n* nombramiento *m*

non-... /nɒn/ pref *no ...*

nonagenarian /nəʊnədʒɪ'neərɪən/ *a & n* nonagenario (*m*), noventón (*m*)

nonchalant /'nɒnʃələnt/ *a* imperturbable

non-commissioned /nɒnkə'mɪʃnd/ *a*. **~ officer** *n* suboficial *m*

non-comittal /nɒnkə'mɪtl/ *a* evasivo

nondescript /'nɒndɪskript/ *a* inclasificable, anodino

none /nʌn/ *pron* (*person*) nadie, ninguno; (*thing*) ninguno, nada. **~ of** nada de. **~ of us** ninguno de nosotros. **I have ~** no tengo nada. ● *adv* no, de ninguna manera. **he is ~ the happier** no está más contento

nonentity /nɒ'nentətɪ/ *n* nulidad *f*

non-existent /nɒnɪg'zɪstənt/ *a* inexistente

nonplussed /nɒn'plʌst/ *a* perplejo

nonsens|e /'nɒnsns/ *n* tonterías *fpl*, disparates *mpl*. **~ical** /-'sensɪkl/ *a* absurdo

non-smoker /nɒn'sməʊkə(r)/ *n* persona *f* que no fuma; (*rail*) departamento *m* de no fumadores

non-starter /nɒn'stɑːtə(r)/ *n* (*fam*) proyecto *m* imposible

non-stop /nɒn'stɒp/ *a* ‹*train*› directo; ‹*flight*› sin escalas. ● *adv* sin parar; (*by train*) directamente; (*by air*) sin escalas

noodles /'nuːdlz/ *npl* fideos *mpl*

nook /nʊk/ *n* rincón *m*

noon /nuːn/ *n* mediodía *m*

no-one /'nəʊwʌn/ *pron* nadie. *see* **nobody**

noose /nuːs/ *n* nudo *m* corredizo

nor /nɔː(r)/ *conj* ni, tampoco. **neither blue ~ red** ni azul ni rojo. **he doesn't play the piano, ~ do I** no sabe tocar el piano, ni yo tampoco

Nordic /'nɔːdɪk/ *a* nórdico

norm /nɔːm/ *n* norma *f*; (*normal*) lo normal

normal /'nɔːml/ *a* normal. **~cy** *n* (*Amer*) normalidad *f*. **~ity** /-'mælətɪ/ *n* normalidad *f*. **~ly** *adv* normalmente

Norman /'nɔːmən/ *a & n* normando (*m*)

Normandy /'nɔːməndɪ/ *n* Normandia *f*

north /nɔːθ/ *n* norte *m*. ● *a* del norte, norteño. ● *adv* hacia el norte. **N~ America** *n* América *f* del Norte, Norteamérica *f*. **N~ American** *a & n* norteamericano (*m*). **~-east** *n* nordeste *m*. **~erly** /'nɔːðəlɪ/ *a* del norte. **~ern** /'nɔːðən/ *a* del norte. **~erner** *n* norteño *m*. **N~ Sea** *n* mar *m* del Norte. **~ward** *a* hacia el norte. **~wards** *adv* hacia el norte. **~west** *n* noroeste *m*

Norw|ay /'nɔːweɪ/ *n* Noruega *f*. **~egian** *a & n* noruego (*m*)

nose /nəʊz/ *n* nariz *f*. ● *vi*. **~ about** curiosear. **~bleed** *n* hemorragia *f* nasal. **~dive** *n* picado *m*

nostalgi|a /nɒ'stældʒə/ *n* nostalgia *f*. **~c** *a* nostálgico

nostril /ˈnɒstrɪl/ *n* nariz *f*; (*of horse*) ollar *m*

nosy /ˈnəʊzɪ/ *a* (**-ier**, **-iest**) (*fam*) entrometido

not /nɒt/ *adv* no. **~ at all** no... nada; (*after thank you*) de nada. **~ yet** aún no. **I do ~ know** no sé. **I suppose ~** supongo que no

notabl|e /ˈnəʊtəbl/ *a* notable. ● *n* (*person*) notabilidad *f*. **~y** /ˈnəʊtəblɪ/ *adv* notablemente

notary /ˈnəʊtərɪ/ *n* notario *m*

notation /nəʊˈteɪʃn/ *n* notación *f*

notch /nɒtʃ/ *n* muesca *f*. ● *vt*. **~ up** apuntar ‹*score etc*›

note /nəʊt/ *n* nota *f*; (*banknote*) billete *m*. **take ~s** tomar apuntes. ● *vt* notar. **~book** *n* libreta *f*. **~d** *a* célebre. **~paper** *n* papel *m* de escribir. **~worthy** *a* notable

nothing /ˈnʌθɪŋ/ *pron* nada. **he eats ~** no come nada. **for ~** (*free*) gratis; (*in vain*) inútilmente. ● *n* nada *f*; (*person*) nulidad *f*; (*thing of no importance*) frusleria *f*; (*zero*) cero *m*. ● *adv* de ninguna manera. **~ big** nada grande. **~ else** nada más. **~ much** poca cosa

notice /ˈnəʊtɪs/ *n* (*attention*) atención *f*; (*advert*) anuncio *m*; (*sign*) letrero *m*; (*poster*) cartel *m*; (*termination of employment*) despido *m*; (*warning*) aviso *m*. **(advance) ~** previo aviso *m*. **~ (of dismissal)** despido *m*. **take ~ of** prestar atención a, hacer caso a ‹*person*›; hacer caso de ‹*thing*›. ● *vt* notar. **~able** *a* evidente. **~ably** *adv* visiblemente. **~-board** *n* tablón *m* de anuncios

notif|ication /nəʊtɪfɪˈkeɪʃn/ *n* aviso *m*, notificación *f*. **~y** *vt* avisar

notion /ˈnəʊʃn/ *n* (*concept*) concepto *m*; (*idea*) idea *f*. **~s** *npl* (*sewing goods etc, Amer*) artículos *mpl* de mercería

notori|ety /nəʊtəˈraɪətɪ/ *n* notoriedad *f*; (*pej*) mala fama *f*. **~ous** /nəʊˈtɔːrɪəs/ *a* notorio. **~ously** *adv* notoriamente

notwithstanding /nɒtwɪθˈstændɪŋ/ *prep* a pesar de. ● *adv* sin embargo

nougat /ˈnuːgɑː/ *n* turrón *m*

nought /nɔːt/ *n* cero *m*

noun /naʊn/ *n* sustantivo *m*, nombre *m*

nourish /ˈnʌrɪʃ/ *vt* alimentar; (*incl fig*) nutrir. **~ment** *n* alimento *m*

novel /ˈnɒvl/ *n* novela *f*. ● *a* nuevo. **~ist** *n* novelista *m* & *f*. **~ty** *n* novedad *f*

November /nəʊˈvembə(r)/ *n* noviembre *m*

novice /ˈnɒvɪs/ *n* principiante *m* & *f*

now /naʊ/ *adv* ahora. **~ and again**, **~ and then** de vez en cuando. **just ~** ahora mismo; (*a moment ago*) hace poco. ● *conj* ahora que

nowadays /ˈnaʊədeɪz/ *adv* hoy (en) día

nowhere /ˈnəʊweə(r)/ *adv* en/por ninguna parte; (*after motion towards*) a ninguna parte

noxious /ˈnɒkʃəs/ *a* nocivo

nozzle /ˈnɒzl/ *n* boquilla *f*; (*tec*) tobera *f*

nuance /ˈnjʊɑːns/ *n* matiz *m*

nuclear /ˈnjuːklɪə(r)/ *a* nuclear

nucleus /ˈnjuːklɪəs/ *n* (*pl* **-lei** /-lɪaɪ/) núcleo *m*

nude /njuːd/ *a* & *n* desnudo (*m*). **in the ~** desnudo

nudge /nʌdʒ/ *vt* dar un codazo a. ● *n* codazo *m*

nudi|sm /ˈnjuːdɪzəm/ *n* desnudismo *m*. **~st** *n* nudista *m* & *f*. **~ty** /ˈnjuːdətɪ/ *n* desnudez *f*

nuisance /ˈnjuːsns/ *n* (*thing, event*) fastidio *m*; (*person*) pesado *m*. **be a ~** dar la lata

null /nʌl/ *a* nulo. **~ify** *vt* anular

numb /nʌm/ *a* entumecido. ● *vt* entumecer

number /ˈnʌmbə(r)/ *n* número *m*. ● *vt* numerar; (*count, include*) contar. **~-plate** *n* matrícula *f*

numeracy /ˈnjuːmərəsɪ/ *n* conocimientos *mpl* de matemáticas

numeral /ˈnjuːmərəl/ *n* número *m*

numerate /ˈnjuːmərət/ *a* que tiene buenos conocimientos de matemáticas

numerical /njuːˈmerɪkl/ *a* numérico

numerous /ˈnjuːmərəs/ *a* numeroso

nun /nʌn/ *n* monja *f*

nurse /nɜːs/ *n* enfermera *f*, enfermero *m*; (*nanny*) niñera *f*. **wet ~** *n* nodriza *f*. ● *vt* cuidar; abrigar ‹*hope etc*›. **~maid** *n* niñera *f*

nursery /ˈnɜːsərɪ/ *n* cuarto *m* de los niños; (*for plants*) vivero *m*. **(day) ~** *n* guardería *f* infantil. **~ rhyme** *n* canción *f* infantil. **~ school** *n* escuela *f* de párvulos

nursing home /ˈnɜːsɪŋhəʊm/ *n* (*for old people*) asilo *m* de ancianos

nurture /ˈnɜːtʃə(r)/ *vt* alimentar

nut /nʌt/ *n* (*walnut, Brazil nut etc*) nuez *f*; (*hazlenut*) avellana *f*; (*peanut*) cacahuete *m*; (*tec*) tuerca *f*;

(*crazy person, sl*) chiflado *m*. **~crackers** *npl* cascanueces *m invar*
nutmeg /'nʌtmeg/ *n* nuez *f* moscada
nutrient /'nju:trɪənt/ *n* alimento *m*
nutrit|ion /nju:'trɪʃn/ *n* nutrición *f*. **~ious** *a* nutritivo
nuts /nʌtz/ *a* (*crazy, sl*) chiflado
nutshell /'nʌtʃel/ *n* cáscara *f* de nuez. **in a ~** en pocas palabras
nuzzle /'nʌzl/ *vt* acariciar con el hocico
NW *abbr* (*north-west*) noroeste *m*
nylon /'naɪlɒn/ *n* nailon *m*. **~s** *npl* medias *fpl* de nailon
nymph /nɪmf/ *n* ninfa *f*

O

oaf /əʊf/ *n* (*pl* **oafs**) zoquete *m*
oak /əʊk/ *n* roble *m*
OAP /əʊeɪ'pi:/ *abbr* (*old-age pensioner*) *n* pensionista *m & f*
oar /ɔ:(r)/ *n* remo *m*. **~sman** /'ɔ:zmən/ *n* (*pl* **-men**) remero *m*
oasis /əʊ'eɪsɪs/ *n* (*pl* **oases** /-si:z/) oasis *m invar*
oath /əʊθ/ *n* juramento *m*; (*swear-word*) palabrota *f*
oat|meal /'əʊtmi:l/ *n* harina *f* de avena. **~s** /əʊts/ *npl* avena *f*
obedien|ce /əʊ'bi:dɪəns/ *n* obediencia *f*. **~t** /əʊ'bi:dɪənt/ *a* obediente. **~tly** *adv* obedientemente
obelisk /'ɒbəlɪsk/ *n* obelisco *m*
obes|e /əʊ'bi:s/ *a* obeso. **~ity** *n* obesidad *f*
obey /əʊ'beɪ/ *vt* obedecer; cumplir ‹*instructions etc*›
obituary /ə'bɪtʃʊərɪ/ *n* necrología *f*
object /'ɒbdʒɪkt/ *n* objeto *m*. /əb'dʒekt/ *vi* oponerse
objection /əb'dʒekʃn/ *n* objeción *f*. **~able** /əb'dʒekʃnəbl/ *a* censurable; (*unpleasant*) desagradable
objective /əb'dʒektɪv/ *a & n* objetivo (*m*). **~ively** *adv* objetivamente
objector /əb'dʒektə(r)/ *n* objetante *m & f*
oblig|ation /ɒblɪ'geɪʃn/ *n* obligación *f*. **be under an ~ation to** tener obligación de. **~atory** /ə'blɪgətrɪ/ *a* obligatorio. **~e** /ə'blaɪdʒ/ *vt* obligar; (*do a small service*) hacer un favor a. **~ed** *a* agradecido. **much ~ed!** ¡muchas gracias! **~ing** *a* atento
oblique /ə'bli:k/ *a* oblicuo
obliterat|e /ə'blɪtəreɪt/ *vt* borrar. **~ion** /-'reɪʃn/ *n* borradura *f*
oblivio|n /ə'blɪvɪən/ *n* olvido *m*. **~us** /ə'blɪvɪəs/ *a* (*unaware*) inconsciente (**to, of** de)
oblong /'ɒblɒŋ/ *a & n* oblongo (*m*)
obnoxious /əb'nɒkʃəs/ *a* odioso
oboe /'əʊbəʊ/ *n* oboe *m*
obscen|e /əb'si:n/ *a* obsceno. **~ity** /-enətɪ/ *n* obscenidad *f*
obscur|e /əb'skjʊə(r)/ *a* oscuro. • *vt* oscurecer; (*conceal*) esconder; (*confuse*) confundir. **~ity** *n* oscuridad *f*
obsequious /əb'si:kwɪəs/ *a* obsequioso
observan|ce /əb'zɜ:vəns/ *n* observancia *f*. **~t** /əb'zɜ:vənt/ *a* observador
observation /ɒbzə'veɪʃn/ *n* observación *f*
observatory /əb'zɜ:vətrɪ/ *n* observatorio *m*
observe /əb'zɜ:v/ *vt* observar. **~r** *n* observador *m*
obsess /əb'ses/ *vt* obsesionar. **~ion** /-ʃn/ *n* obsesión *f*. **~ive** *a* obsesivo
obsolete /'ɒbsəli:t/ *a* desusado
obstacle /'ɒbstəkl/ *n* obstáculo *m*
obstetrics /əb'stetrɪks/ *n* obstetricia *f*
obstina|cy /'ɒbstɪnəsɪ/ *n* obstinación *f*. **~te** /'ɒbstɪnət/ *a* obstinado. **~tely** *adv* obstinadamente
obstreperous /ɒb'strepərəs/ *a* turbulento, ruidoso, protestón
obstruct /əb'strʌkt/ *vt* obstruir. **~ion** /-ʃn/ *n* obstrucción *f*
obtain /əb'teɪn/ *vt* obtener. • *vi* prevalecer. **~able** *a* asequible
obtrusive /əb'tru:sɪv/ *a* importuno
obtuse /əb'tju:s/ *a* obtuso
obviate /'ɒbvɪeɪt/ *vt* evitar
obvious /'ɒbvɪəs/ *a* obvio. **~ly** obviamente
occasion /ə'keɪʒn/ *n* ocasión *f*, oportunidad *f*. **on ~** de vez en cuando. • *vt* ocasionar. **~al** /ə'keɪʒənl/ *a* poco frecuente. **~ally** *adv* de vez en cuando
occult /ɒ'kʌlt/ *a* oculto
occup|ant /'ɒkjʊpənt/ *n* ocupante *m & f*. **~ation** /ɒkjʊ'peɪʃn/ *n* ocupación *f*; (*job*) trabajo *m*, profesión *f*. **~ational** *a* profesional. **~ier** *n* ocupante *m & f*. **~y** /'ɒkjʊpaɪ/ *vt* ocupar
occur /ə'kɜ:(r)/ *vi* (*pt* **occurred**) ocurrir, suceder; (*exist*) encontrarse. **it ~red to me that** se me ocurrió que.

~rence /ə'kʌrəns/ *n* suceso *m*, acontecimiento *m*
ocean /'əʊʃn/ *n* océano *m*
o'clock /ə'klɒk/ *adv*. **it is 7 ~** son las siete
octagon /'ɒktəgən/ *n* octágono *m*
octane /'ɒkteɪn/ *n* octano *m*
octave /'ɒktɪv/ *n* octava *f*
October /ɒk'təʊbə(r)/ *n* octubre *m*
octopus /'ɒktəpəs/ *n* (*pl* **-puses**) pulpo *m*
oculist /'ɒkjʊlɪst/ *n* oculista *m & f*
odd /ɒd/ *a* (**-er**, **-est**) extraño, raro; ‹*number*› impar; (*one of pair*) sin pareja; (*occasional*) poco frecuente; (*left over*) sobrante. **fifty-~** unos cincuenta, cincuenta y pico. **the ~ one out** la excepción *f*. **~ity** *n* (*thing*) curiosidad *f*; (*person*) excéntrico *m*. **~ly** *adv* extrañamente. **~ly enough** por extraño que parezca. **~ment** /'ɒdmənt/ *n* retazo *m*. **~s** /ɒdz/ *npl* probabilidades *fpl*; (*in betting*) apuesta *f*. **~s and ends** retazos *mpl*. **at ~s** de punta, de malas
ode /əʊd/ *n* oda *f*
odious /'əʊdɪəs/ *a* odioso
odour /'əʊdə(r)/ *n* olor *m*. **~less** *a* inodoro
of /əv, ɒv/ *prep* de. **a friend ~ mine** un amigo mío. **how kind ~ you** es Vd muy amable
off /ɒf/ *adv* lejos; ‹*light etc*› apagado; ‹*tap*› cerrado; ‹*food*› pasado. ● *prep* de, desde; (*away from*) fuera de; (*distant from*) lejos de. **be better ~** estar mejor. **be ~** marcharse. **day ~** *n* día *m* de asueto, día *m* libre
offal /'ɒfl/ *n* menudos *mpl*, asaduras *fpl*
off: **~-beat** *a* insólito. **~ chance** *n* posibilidad *f* remota. **~ colour** *a* indispuesto
offen|ce /ə'fens/ *n* ofensa *f*; (*illegal act*) delito *m*. **take ~ce** ofenderse. **~d** /ə'fend/ *vt* ofender. **~der** *n* delincuente *m & f*. **~sive** /ə'fensɪv/ *a* ofensivo; (*disgusting*) repugnante. ● *n* ofensiva *f*
offer /'ɒfə(r)/ *vt* ofrecer. ● *n* oferta *f*. **on ~** en oferta
offhand /ɒf'hænd/ *a* (*casual*) desenvuelto; (*brusque*) descortés. ● *adv* de improviso
office /'ɒfɪs/ *n* oficina *f*; (*post*) cargo *m*
officer /'ɒfɪsə(r)/ *n* oficial *m*; (*policeman*) policía *f*, guardia *m*; (*of organization*) director *m*
official /ə'fɪʃl/ *a & n* oficial (*m*). **~ly** *adv* oficialmente
officiate /ə'fɪʃɪeɪt/ *vi* oficiar. **~ as** desempeñar las funciones de
officious /ə'fɪʃəs/ *a* oficioso
offing /'ɒfɪŋ/ *n*. **in the ~** en perspectiva
off: **~-licence** *n* tienda *f* de bebidas alcohólicas. **~-load** *vt* descargar. **~-putting** *a* (*disconcerting*, *fam*) desconcertante; (*repellent*) repugnante. **~set** /'ɒfset/ *vt* (*pt* **-set**, *pres p* **-setting**) contrapesar. **~shoot** /'ɒfʃu:t/ *n* retoño *m*; (*fig*) ramificación *f*. **~side** /ɒf'saɪd/ *a* (*sport*) fuera de juego. **~spring** /'ɒfsprɪŋ/ *n invar* progenie *f*. **~-stage** *a* entre bastidores. **~-white** *a* blancuzco, color hueso
often /'ɒfn/ *adv* muchas veces, con frecuencia, a menudo. **how ~?** ¿cuántas veces?
ogle /'əʊgl/ *vt* comerse con los ojos
ogre /'əʊgə(r)/ *n* ogro *m*
oh /əʊ/ *int* ¡oh!, ¡ay!
oil /ɔɪl/ *n* aceite *m*; (*petroleum*) petróleo *m*. ● *vt* lubricar. **~field** /'ɔɪlfi:ld/ *n* yacimiento *m* petrolífero. **~-painting** *n* pintura *f* al óleo. **~-rig** /'ɔɪlrɪg/ *n* plataforma *f* de perforación. **~skins** /'ɔɪlskɪnz/ *npl* chubasquero *m*. **~y** *a* aceitoso; ‹*food*› grasiento
ointment /'ɔɪntmənt/ *n* ungüento *m*
OK /əʊ'keɪ/ *int* ¡vale!, ¡de acuerdo! ● *a* bien; (*satisfactory*) satisfactorio. ● *adv* muy bien
old /əʊld/ *a* (**-er**, **-est**) viejo; (*not modern*) anticuado; (*former*) antiguo. **how ~ is she?** ¿cuántos años tiene? **she is ten years ~** tiene diez años. **of ~** de antaño. **~ age** *n* vejez *f*. **~-fashioned** *a* anticuado. **~ maid** *n* solterona *f*. **~-world** *a* antiguo
oleander /əʊlɪ'ændə(r)/ *n* adelfa *f*
olive /'ɒlɪv/ *n* (*fruit*) aceituna *f*; (*tree*) olivo *m*. ● *a* de oliva; (*colour*) aceitunado
Olympic /ə'lɪmpɪk/ *a* olímpico. **~s** *npl*, **~ Games** *npl* Juegos *mpl* Olímpicos
omelette /'ɒmlɪt/ *n* tortilla *f*, tortilla *f* de huevos (*Mex*)
om|en /'əʊmen/ *n* agüero *m*. **~inous** /'ɒmɪnəs/ *a* siniestro
omi|ssion /ə'mɪʃn/ *n* omisión *f*. **~t** /ə'mɪt/ *vt* (*pt* **omitted**) omitir
omnipotent /ɒm'nɪpətənt/ *a* omnipotente

on /ɒn/ *prep* en, sobre. **~ foot** a pie. **~ Monday** el lunes. **~ Mondays** los lunes. **~ seeing** al ver. **~ the way** de camino. ● *adv* (*light etc*) encendido; (*put on*) puesto, poco natural; (*machine*) en marcha; (*tap*) abierto. **~ and off** de vez en cuando. **~ and ~** sin cesar. **and so ~** y así sucesivamente. **be ~ at** (*fam*) criticar. **go ~** continuar. **later ~** más tarde

once /wʌns/ *adv* una vez; (*formerly*) antes. ● *conj* una vez que. **at ~** en seguida. **~-over** *n* (*fam*) ojeada *f*

oncoming /'ɒnkʌmɪŋ/ *a* que se acerca; ‹*traffic*› que viene en sentido contrario, de frente

one /wʌn/ *a* & *n* uno (*m*). ● *pron* uno. **~ another** el uno al otro. **~ by ~** uno a uno. **~ never knows** nunca se sabe. **the blue ~** el azul. **this ~** éste. **~-off** *a* (*fam*) único

onerous /'ɒnərəs/ *a* oneroso

one: **~self** /wʌn'self/ *pron* (*subject*) uno mismo; (*object*) se; (*after prep*) sí (mismo). **by ~self** solo. **~-sided** *a* unilateral. **~-way** *a* ‹*street*› de dirección única; ‹*ticket*› de ida

onion /'ʌnɪən/ *n* cebolla *f*

onlooker /'ɒnlʊkə(r)/ *n* espectador *m*

only /'əʊnlɪ/ *a* único. **~ son** *n* hijo *m* único. ● *adv* sólo, solamente. **~ just** apenas. **~ too** de veras. ● *conj* pero, sólo que

onset /'ɒnset/ *n* principio *m*; (*attack*) ataque *m*

onslaught /'ɒnslɔ:t/ *n* ataque *m* violento

onus /'əʊnəs/ *n* responsabilidad *f*

onward(s) /'ɒnwəd(z)/ *a* & *adv* hacia adelante

onyx /'ɒnɪks/ *n* ónice *f*

ooze /u:z/ *vt*/*i* rezumar

opal /'əʊpl/ *n* ópalo *m*

opaque /əʊ'peɪk/ *a* opaco

open /'əʊpən/ *a* abierto; (*free to all*) público; (*undisguised*) manifiesto; ‹*question*› discutible; ‹*view*› despejado. **~ sea** *n* alta mar *f*. **~ secret** *n* secreto *m* a voces. **O~ University** *n* Universidad *f* a Distancia. **half-~** *a* medio abierto. **in the ~** *n* al aire libre. ● *vt*/*i* abrir. **~-ended** *a* abierto. **~er** /'əʊpənə(r)/ *n* (*for tins*) abrelatas *m invar*; (*for bottles with caps*) abrebotellas *m invar*; (*corkscrew*) sacacorchos *m invar*. **eye-~er** *n* (*fam*) revelación *f*. **~ing** /'əʊpənɪŋ/ *n* abertura *f*; (*beginning*) principio *m*; (*job*) vacante *m*. **~ly** /'əʊpənlɪ/ *adv* abiertamente. **~-minded** *a* imparcial

opera /'ɒprə/ *n* ópera *f*. **~-glasses** *npl* gemelos *mpl* de teatro

operate /'ɒpəreɪt/ *vt* hacer funcionar. ● *vi* funcionar; ‹*medicine etc*› operar. **~ on** (*med*) operar a

operatic /ɒpə'rætɪk/ *a* operístico

operation /ɒpə'reɪʃn/ *n* operación *f*; (*mec*) funcionamiento *m*. **in ~** en vigor. **~al** /ɒpə'reɪʃnl/ *a* operacional

operative /'ɒpərətɪv/ *a* operativo; ‹*law etc*› en vigor

operator *n* operario *m*; (*telephonist*) telefonista *m* & *f*

operetta /ɒpə'retə/ *n* opereta *f*

opinion /ə'pɪnɪən/ *n* opinión *f*. **in my ~** a mi parecer. **~ated** *a* dogmático

opium /'əʊpɪəm/ *n* opio *m*

opponent /ə'pəʊnənt/ *n* adversario *m*

opportun|e /'ɒpətju:n/ *a* oportuno. **~ist** /ɒpə'tju:nɪst/ *n* oportunista *m* & *f*. **~ity** /ɒpə'tju:nətɪ/ *n* oportunidad *f*

oppos|e /ə'pəʊz/ *vt* oponerse a. **~ed to** en contra de. **be ~ed to** oponerse a. **~ing** *a* opuesto

opposite /'ɒpəzɪt/ *a* opuesto; (*facing*) de enfrente. ● *n* contrario *m*. ● *adv* enfrente. ● *prep* enfrente de. **~ number** *n* homólogo *m*

opposition /ɒpə'zɪʃn/ *n* oposición *f*; (*resistence*) resistencia *f*

oppress /ə'pres/ *vt* oprimir. **~ion** /-ʃn/ *n* opresión *f*. **~ive** *a* (*cruel*) opresivo; ‹*heat*› sofocante. **~or** *n* opresor *m*

opt /ɒpt/ *vi*. **~ for** elegir. **~ out** negarse a participar

optic|al /'ɒptɪkl/ *a* óptico. **~ian** /ɒp'tɪʃn/ *n* óptico *m*

optimis|m /'ɒptɪmɪzəm/ *n* optimismo *m*. **~t** /'ɒptɪmɪst/ *n* optimista *m* & *f*. **~tic** /-'mɪstɪk/ *a* optimista

optimum /'ɒptɪməm/ *n* lo óptimo, lo mejor

option /'ɒpʃn/ *n* opción *f*. **~al** /'ɒpʃənl/ *a* facultativo

opulen|ce /'ɒpjʊləns/ *n* opulencia *f*. **~t** /'ɒpjʊlənt/ *a* opulento

or /ɔ:(r)/ *conj* o; (*before Spanish* o- *and* ho-) u; (*after negative*) ni. **~ else** si no, o bien

oracle /'ɒrəkl/ *n* oráculo *m*

oral /'ɔ:rəl/ *a* oral. ● *n* (*fam*) examen *m* oral
orange /'ɒrɪndʒ/ *n* naranja *f*; (*tree*) naranjo *m*; (*colour*) color *m* naranja. ● *a* de color naranja. **~ade** *n* naranjada *f*
orator /'ɒrətə(r)/ *n* orador *m*
oratorio /ɒrə'tɔ:rɪəʊ/ *n* (*pl* **-os**) oratorio *m*
oratory /'ɒrətrɪ/ *n* oratoria *f*
orb /ɔ:b/ *n* orbe *m*
orbit /'ɔ:bɪt/ *n* órbita *f*. ● *vt* orbitar
orchard /'ɔ:tʃəd/ *n* huerto *m*
orchestra /'ɔ:kɪstrə/ *n* orquesta *f*. **~l** /-'kestrəl/ *a* orquestal. **~te** /'ɔ:kɪstreɪt/ *vt* orquestar
orchid /'ɔ:kɪd/ *n* orquídea *f*
ordain /ɔ:'deɪn/ *vt* ordenar
ordeal /ɔ:'di:l/ *n* prueba *f* dura
order /'ɔ:də(r)/ *n* orden *m*; (*com*) pedido *m*. **in ~ that** para que. **in ~ to** para. ● *vt* (*command*) mandar; (*com*) pedir
orderly /'ɔ:dəlɪ/ *a* ordenado. ● *n* asistente *m* & *f*
ordinary /'ɔ:dɪnrɪ/ *a* corriente; (*average*) medio; (*mediocre*) ordinario
ordination /ɔ:dɪ'neɪʃn/ *n* ordenación *f*
ore /ɔ:(r)/ *n* mineral *m*
organ /'ɔ:gən/ *n* órgano *m*
organic /ɔ:'gænɪk/ *a* orgánico
organism /'ɔ:gənɪzəm/ *n* organismo *m*
organist /'ɔ:gənɪst/ *n* organista *m* & *f*
organiz|ation /ɔ:gənaɪ'zeɪʃn/ *n* organización *f*. **~e** /'ɔ:gənaɪz/ *vt* organizar. **~er** *n* organizador *m*
orgasm /'ɔ:gæzəm/ *n* orgasmo *m*
orgy /'ɔ:dʒɪ/ *n* orgía *f*
Orient /'ɔ:rɪənt/ *n* Oriente *m*. **~al** /-'entl/ *a* & *n* oriental (*m* & *f*)
orientat|e /'ɔ:rɪənteɪt/ *vt* orientar. **~ion** /-'teɪʃn/ *n* orientación *f*
orifice /'ɒrɪfɪs/ *n* orificio *m*
origin /'ɒrɪdʒɪn/ *n* origen *m*. **~al** /ə'rɪdʒənl/ *a* original. **~ality** /-'nælətɪ/ *n* originalidad *f*. **~ally** *adv* originalmente. **~ate** /ə'rɪdʒɪneɪt/ *vi*. **~ate from** provenir de. **~ator** *n* autor *m*
ormolu /'ɔ:məlu:/ *n* similor *m*
ornament /'ɔ:nəmənt/ *n* adorno *m*. **~al** /-'mentl/ *a* de adorno. **~ation** /-en'teɪʃn/ *n* ornamentación *f*
ornate /ɔ:'neɪt/ *a* adornado; ‹*style*› florido
ornithology /ɔ:nɪ'θɒlədʒɪ/ *n* ornitología *f*
orphan /'ɔ:fn/ *n* huérfano *m*. ● *vt* dejar huérfano. **~age** *n* orfanato *m*
orthodox /'ɔ:θədɒks/ *a* ortodoxo. **~y** *n* ortodoxia *f*
orthopaedic /ɔ:θə'pi:dɪk/ *a* ortopédico. **~s** *n* ortopedia *f*
oscillate /'ɒsɪleɪt/ *vi* oscilar
ossify /'ɒsɪfaɪ/ *vt* osificar. ● *vi* osificarse
ostensibl|e /ɒs'tensɪbl/ *a* aparente. **~y** *adv* aparentemente
ostentat|ion /ɒsten'teɪʃn/ *n* ostentación *f*. **~ious** *a* ostentoso
osteopath /'ɒstɪəpæθ/ *n* osteópata *m* & *f*. **~y** /-'ɒpəθɪ/ *n* osteopatía *f*
ostracize /'ɒstrəsaɪz/ *vt* excluir
ostrich /'ɒstrɪtʃ/ *n* avestruz *m*
other /'ʌðə(r)/ *a* & *n* & *pron* otro (*m*). **~ than** de otra manera que. **the ~ one** el otro. **~wise** /'ʌðəwaɪz/ *adv* de otra manera; (*or*) si no
otter /'ɒtə(r)/ *n* nutria *f*
ouch /aʊtʃ/ *int* ¡ay!
ought /ɔ:t/ *v aux* deber. **I ~ to see it** debería verlo. **he ~ to have done it** debería haberlo hecho
ounce /aʊns/ *n* onza *f* (= *28.35 gr.*)
our /'aʊə(r)/ *a* nuestro. **~s** /'aʊəz/ *poss pron* el nuestro, la nuestra, los nuestros, las nuestras. **~selves** /aʊə'selvz/ *pron* (*subject*) nosotros mismos, nosotras mismas; (*reflexive*) nos; (*after prep*) nosotros (mismos), nosotras (mismas)
oust /aʊst/ *vt* expulsar, desalojar
out /aʊt/ *adv* fuera; ‹*light*› apagado; (*in blossom*) en flor; (*in error*) equivocado. **~-and-~** *a* cien por cien. **~ of date** anticuado; (*not valid*) caducado. **~ of doors** fuera. **~ of order** estropeado; (*sign*) no funciona. **~ of pity** por compasión. **~ of place** fuera de lugar; (*fig*) inoportuno. **~ of print** agotado. **~ of sorts** indispuesto. **~ of stock** agotado. **~ of tune** desafinado. **~ of work** parado, desempleado. **be ~** equivocarse. **be ~ of** quedarse sin. **be ~ to** estar resuelto a. **five ~ of six** cinco de cada seis. **made ~ of** hecho de
outbid /aʊt'bɪd/ *vt* (*pt* **-bid**, *pres p* **-bidding**) ofrecer más que
outboard /'aʊtbɔ:d/ *a* fuera borda
outbreak /'aʊtbreɪk/ *n* (*of anger*) arranque *m*; (*of war*) comienzo *m*; (*of disease*) epidemia *f*

outbuilding /'aʊtbɪldɪŋ/ *n* dependencia *f*
outburst /'aʊtbɜ:st/ *n* explosión *f*
outcast /'aʊtkɑ:st/ *n* paria *m & f*
outcome /'aʊtkʌm/ *n* resultado *m*
outcry /'aʊtkraɪ/ *n* protesta *f*
outdated /aʊt'deɪtɪd/ *a* anticuado
outdo /aʊt'du:/ *vt* (*pt* **-did**, *pp* **-done**) superar
outdoor /'aʊtdɔ:(r)/ *a* al aire libre. **~s** /-'dɔ:z/ *adv* al aire libre
outer /'aʊtə(r)/ *a* exterior
outfit /'aʊtfɪt/ *n* equipo *m*; (*clothes*) traje *m*. **~ter** *n* camisero *m*
outgoing /'aʊtgəʊɪŋ/ *a* ‹*minister etc*› saliente; (*sociable*) abierto. **~s** *npl* gastos *mpl*
outgrow /æʊt'grəʊ/ *vt* (*pt* **-grew**, *pp* **-grown**) crecer más que ‹*person*›; hacerse demasiado grande para ‹*clothes*›. **he's ~n his trousers** le quedan pequeños los pantalones
outhouse /'aʊthaʊs/ *n* dependencia *f*
outing /'aʊtɪŋ/ *n* excursión *f*
outlandish /aʊt'lændɪʃ/ *a* extravagante
outlaw /'aʊtlɔ:/ *n* proscrito *m*. ● *vt* proscribir
outlay /'aʊtleɪ/ *n* gastos *mpl*
outlet /'aʊtlet/ *n* salida *f*
outline /'aʊtlaɪn/ *n* contorno *m*; (*summary*) resumen *m*. ● *vt* trazar; (*describe*) dar un resumen de
outlive /aʊt'lɪv/ *vt* sobrevivir a
outlook /'aʊtlʊk/ *n* perspectiva *f*
outlying /'aʊtlaɪɪŋ/ *a* remoto
outmoded /aʊt'məʊdɪd/ *a* anticuado
outnumber /aʊt'nʌmbə(r)/ *vt* sobrepasar en número
outpatient /aʊt'peɪʃnt/ *n* paciente *m* externo
outpost /'aʊtpəʊst/ *n* avanzada *f*
output /'aʊtpʊt/ *n* producción *f*
outrage /'aʊtreɪdʒ/ *n* ultraje *m*. ● *vt* ultrajar. **~ous** /aʊt'reɪdʒəs/ *a* escandaloso, atroz
outright /'aʊtraɪt/ *adv* completamente; (*at once*) inmediatamente; (*frankly*) francamente. ● *a* completo; ‹*refusal*› rotundo
outset /'aʊtset/ *n* principio *m*
outside /'aʊtsaɪd/ *a & n* exterior (*m*). /aʊt'saɪd/ *adv* fuera. ● *prep* fuera de. **~r** /aʊt'saɪdə(r)/ *n* forastero *m*; (*in race*) caballo *m* no favorito
outsize /'aʊtsaɪz/ *a* de tamaño extraordinario
outskirts /'aʊtskɜ:ts/ *npl* afueras *fpl*
outspoken /aʊt'spəʊkn/ *a* franco. **be ~** no tener pelos en la lengua
outstanding /aʊt'stændɪŋ/ *a* excepcional; (*not settled*) pendiente; (*conspicuous*) sobresaliente
outstretched /aʊt'stretʃt/ *a* extendido
outstrip /aʊt'strɪp/ *vt* (*pt* **-stripped**) superar
outward /'aʊtwəd/ *a* externo; ‹*journey*› de ida. **~ly** *adv* por fuera, exteriormente. **~(s)** *adv* hacia fuera
outweigh /aʊt'weɪ/ *vt* pesar más que; (*fig*) valer más que
outwit /aʊt'wɪt/ *vt* (*pt* **-witted**) ser más listo que
oval /'əʊvl/ *a* oval(ado). ● *n* óvalo *m*
ovary /'əʊvərɪ/ *n* ovario *m*
ovation /əʊ'veɪʃn/ *n* ovación *f*
oven /'ʌvn/ *n* horno *m*
over /'əʊvə(r)/ *prep* por encima de; (*across*) al otro lado de; (*during*) durante; (*more than*) más de. **~ and above** por encima de. ● *adv* por encima; (*ended*) terminado; (*more*) más; (*in excess*) de sobra. **~ again** otra vez. **~ and ~** una y otra vez. **~ here** por aquí. **~ there** por allí. **all ~** por todas partes
over... /'əʊvə(r)/ *pref* sobre..., super...
overall /əʊvər'ɔ:l/ *a* global; ‹*length, cost*› total. ● *adv* en conjunto. /'əʊvərɔ:l/ *n*, **~s** *npl* mono *m*
overawe /əʊvər'ɔ:/ *vt* intimidar
overbalance /əʊvə'bæləns/ *vt* hacer perder el equilibrio. ● *vi* perder el equilibrio
overbearing /əʊvə'beərɪŋ/ *a* dominante
overboard /'əʊvəbɔ:d/ *adv* al agua
overbook /əʊvə'bʊk/ *vt* aceptar demasiadas reservaciones para
overcast /əʊvə'kɑ:st/ *a* nublado
overcharge /əʊvə'tʃɑ:dʒ/ *vt* (*fill too much*) sobrecargar; (*charge too much*) cobrar demasiado
overcoat /'əʊvəkəʊt/ *n* abrigo *m*
overcome /əʊvə'kʌm/ *vt* (*pt* **-came**, *pp* **-come**) superar, vencer. **be ~ by** estar abrumado de
overcrowded /əʊvə'kraʊdɪd/ *a* atestado (de gente)
overdo /əʊvə'du:/ *vt* (*pt* **-did**, *pp* **-done**) exagerar; (*culin*) cocer demasiado

overdose /'əʊvədəʊs/ *n* sobredosis *f*

overdraft /'əʊvədrɑ:ft/ *n* giro *m* en descubierto

overdraw /əʊvə'drɔ:/ *vt* (*pt* **-drew**, *pp* **-drawn**) girar en descubierto. **be ~n** tener un saldo deudor

overdue /əʊvə'dju:/ *a* retrasado; (*belated*) tardío; ‹*bill*› vencido y no pagado

overestimate /əʊvər'estɪmeɪt/ *vt* sobrestimar

overflow /əʊvə'fləʊ/ *vi* desbordarse. /'əʊvəfləʊ/ *n* (*excess*) exceso *m*; (*outlet*) rebosadero *m*

overgrown /əʊvə'grəʊn/ *a* demasiado grande; ‹*garden*› cubierto de hierbas

overhang /əʊvə'hæŋ/ *vt* (*pt* **-hung**) sobresalir por encima de; (*fig*) amenazar. ● *vi* sobresalir. /'əʊvəhæŋ/ *n* saliente *f*

overhaul /əʊvə'hɔ:l/ *vt* revisar. /'əʊvəhɔ:l/ *n* revisión *f*

overhead /əʊvə'hed/ *adv* por encima. /'əʊvəhed/ *a* de arriba. **~s** *npl* gastos *mpl* generales

overhear /əʊvə'hɪə(r)/ *vt* (*pt* **-heard**) oír por casualidad

overjoyed /əʊvə'dʒɔɪd/ *a* muy contento. **he was ~** rebosaba de alegría

overland /'əʊvəlænd/ *a* terrestre. ● *adv* por tierra

overlap /əʊvə'læp/ *vt* (*pt* **-lapped**) traslapar. ● *vi* traslaparse

overleaf /əʊvə'li:f/ *adv* a la vuelta. **see ~** véase al dorso

overload /əʊvə'ləʊd/ *vt* sobrecargar

overlook /əʊvə'lʊk/ *vt* dominar; ‹*building*› dar a; (*forget*) olvidar; (*oversee*) inspeccionar; (*forgive*) perdonar

overnight /əʊvə'naɪt/ *adv* por la noche, durante la noche; (*fig, instantly*) de la noche a la mañana. **stay ~** pasar la noche. ● *a* de noche

overpass /'əʊvəpɑ:s/ *n* paso *m* a desnivel, paso *m* elevado

overpay /əʊvə'peɪ/ *vt* (*pt* **-paid**) pagar demasiado

overpower /əʊvə'paʊə(r)/ *vt* subyugar; dominar ‹*opponent*›; (*fig*) abrumar. **~ing** *a* abrumador

overpriced /əʊvə'praɪst/ *a* demasiado caro

overrate /əʊvə'reɪt/ *vt* supervalorar

overreach /əʊvə'ri:tʃ/ *vr*. **~ o.s.** extralimitarse

overreact /əʊvərɪ'ækt/ *vi* reaccionar excesivamente

overrid|e /əʊvə'raɪd/ *vt* (*pt* **-rode**, *pp* **-ridden**) pasar por encima de. **~ing** *a* dominante

overripe /'əʊvəraɪp/ *a* pasado, demasiado maduro

overrule /əʊvə'ru:l/ *vt* anular; denegar ‹*claim*›

overrun /əʊvə'rʌn/ *vt* (*pt* **-ran**, *pp* **-run**, *pres p* **-running**) invadir; exceder ‹*limit*›

overseas /əʊvə'si:z/ *a* de ultramar. ● *adv* al extranjero, en ultramar

oversee /əʊvə'si:/ *vt* (*pt* **-saw**, *pp* **-seen**) vigilar. **~r** /'əʊvəsɪə(r)/ *n* supervisor *m*

overshadow /əʊvə'ʃædəʊ/ *vt* (*darken*) sombrear; (*fig*) eclipsar

overshoot /əʊvə'ʃu:t/ *vt* (*pt* **-shot**) excederse. **~ the mark** pasarse de la raya

oversight /'əʊvəsaɪt/ *n* descuido *m*

oversleep /əʊvə'sli:p/ *vi* (*pt* **-slept**) despertarse tarde. **I overslept** se me pegaron las sábanas

overstep /əʊvə'step/ *vt* (*pt* **-stepped**) pasar de. **~ the mark** pasarse de la raya

overt /'əʊvɜ:t/ *a* manifiesto

overtak|e /əʊvə'teɪk/ *vt/i* (*pt* **-took**, *pp* **-taken**) sobrepasar; (*auto*) adelantar. **~ing** *n* adelantamiento *m*

overtax /əʊvə'tæks/ *vt* exigir demasiado

overthrow /əʊvə'θrəʊ/ *vt* (*pt* **-threw**, *pp* **-thrown**) derrocar. /'əʊvəθrəʊ/ *n* derrocamiento *m*

overtime /'əʊvətaɪm/ *n* horas *fpl* extra

overtone /'əʊvətəʊn/ *n* (*fig*) matiz *m*

overture /'əʊvətjʊə(r)/ *n* obertura *f*. **~s** *npl* (*fig*) propuestas *fpl*

overturn /əʊvə'tɜ:n/ *vt/i* volcar

overweight /əʊvə'weɪt/ *a* demasiado pesado. **be ~** pesar demasiado, ser gordo

overwhelm /əʊvə'welm/ *vt* aplastar; (*with emotion*) abrumar. **~ing** *a* aplastante; (*fig*) abrumador

overwork /əʊvə'wɜ:k/ *vt* hacer trabajar demasiado. ● *vi* trabajar demasiado. ● *n* trabajo *m* excesivo

overwrought /əʊvə'rɔ:t/ *a* agotado, muy nervioso

ovulation /ɒvjʊ'leɪʃn/ *n* ovulación *f*

ow|e /əʊ/ *vt* deber. **~ing** *a* debido. **~ing to** a causa de

owl /aʊl/ *n* lechuza *f*, búho *m*

own /əʊn/ *a* propio. **get one's ~ back** (*fam*) vengarse. **hold one's ~**

mantenerse firme, saber defenderse. **on one's ~** por su cuenta. ● *vt* poseer, tener. ● *vi.* **~ up (to)** (*fam*) confesar. **~er** *n* propietario *m*, dueño *m*. **~ership** *n* posesión *f*; (*right*) propiedad *f*

ox /ɒks/ *n* (*pl* **oxen**) buey *m*

oxide /'ɒksaɪd/ *n* óxido *m*

oxygen /'ɒksɪdʒən/ *n* oxígeno *m*

oyster /'ɔɪstə(r)/ *n* ostra *f*

P

p /pi:/ *abbr* (*pence, penny*) penique(s) (*m*(*pl*))

pace /peɪs/ *n* paso *m*. ● *vi.* **~ up and down** pasearse de aquí para allá. **~-maker** *n* (*runner*) el que marca el paso; (*med*) marcapasos *m invar*. **keep ~ with** andar al mismo paso que

Pacific /pə'sɪfɪk/ *a* pacífico. ● *n.* **~ (Ocean)** (Océano *m*) Pacífico *m*

pacif|ist /'pæsɪfɪst/ *n* pacifista *m & f*. **~y** /'pæsɪfaɪ/ *vt* apaciguar

pack /pæk/ *n* fardo *m*; (*of cards*) baraja *f*; (*of hounds*) jauría *f*; (*of wolves*) manada *f*; (*large amount*) montón *m*. ● *vt* empaquetar; hacer ‹*suitcase*›; (*press down*) apretar. ● *vi* hacer la maleta. **~age** /'pækɪdʒ/ *n* paquete *m*. ● *vt* empaquetar. **~age deal** *n* acuerdo *m* global. **~age tour** *n* viaje *m* organizado. **~ed lunch** *n* almuerzo *m* frío. **~ed out** (*fam*) de bote en bote. **~et** /'pækɪt/ *n* paquete *m*. **send ~ing** echar a paseo

pact /pækt/ *n* pacto *m*, acuerdo *m*

pad /pæd/ *n* almohadilla *f*; (*for writing*) bloc *m*; (*for ink*) tampón *m*; (*flat, fam*) piso *m*. ● *vt* (*pt* **padded**) rellenar. **~ding** *n* relleno *m*. ● *vi* andar a pasos quedos. **launching ~** plataforma *f* de lanzamiento

paddle[1] /'pædl/ *n* canalete *m*

paddle[2] /'pædl/ *vi* mojarse los pies

paddle-steamer /'pædlsti:mə(r)/ *n* vapor *m* de ruedas

paddock /'pædək/ *n* recinto *m*; (*field*) prado *m*

paddy /'pædɪ/ *n* arroz *m* con cáscara. **~-field** *n* arrozal *m*

padlock /'pædlɒk/ *n* candado *m*. ● *vt* cerrar con candado

paediatrician /pi:dɪə'trɪʃn/ *n* pediatra *m & f*

pagan /'peɪgən/ *a & n* pagano (*m*)

page[1] /peɪdʒ/ *n* página *f*. ● *vt* paginar

page[2] /peɪdʒ/ (*in hotel*) botones *m invar*. ● *vt* llamar

pageant /'pædʒənt/ *n* espectáculo *m* (histórico). **~ry** *n* boato *m*

pagoda /pə'gəʊdə/ *n* pagoda *f*

paid /peɪd/ *see* **pay**. ● *a.* **put ~ to** (*fam*) acabar con

pail /peɪl/ *n* cubo *m*

pain /peɪn/ *n* dolor *m*. **~ in the neck** (*fam*) ‹*persona*› pesado *m*; ‹*thing*› lata *f*. **be in ~** tener dolores. **~s** *npl* (*effort*) esfuerzos *mpl*. **be at ~s** esmerarse. ● *vt* doler. **~ful** /'peɪnfl/ *a* doloroso; (*laborious*) penoso. **~-killer** *n* calmante *m*. **~less** *a* indoloro. **~staking** /'peɪnzteɪkɪŋ/ *a* esmerado

paint /peɪnt/ *n* pintura *f*. ● *vt/i* pintar. **~er** *n* pintor *m*. **~ing** *n* pintura *f*

pair /peə(r)/ *n* par *m*; (*of people*) pareja *f*. **~ of trousers** pantalón *m*, pantalones *mpl*. ● *vi* emparejarse. **~ off** emparejarse

pajamas /pə'dʒɑ:məz/ *npl* pijama *m*

Pakistan /pɑ:kɪ'stɑ:n/ *n* el Pakistán *m*. **~i** *a & n* paquistaní (*m & f*)

pal /pæl/ *n* (*fam*) amigo *m*

palace /'pælɪs/ *n* palacio *m*

palat|able /'pælətəbl/ *a* sabroso; (*fig*) aceptable. **~e** /'pælət/ *n* paladar *m*

palatial /pə'leɪʃl/ *a* suntuoso

palaver /pə'lɑ:və(r)/ *n* (*fam*) lío *m*

pale[1] /peɪl/ *a* (**-er, -est**) pálido; ‹*colour*› claro. ● *vi* palidecer

pale[2] /peɪl/ *n* estaca *n*

paleness /'peɪlnɪs/ *n* palidez *f*

Palestin|e /'pælɪstaɪn/ *n* Palestina *f*. **~ian** /-'stɪnɪən/ *a & n* palestino (*m*)

palette /'pælɪt/ *n* paleta *f*. **~-knife** *n* espátula *f*

pall[1] /pɔ:l/ *n* paño *m* mortuorio; (*fig*) capa *f*

pall[2] /pɔ:l/ *vi.* **~ (on)** perder su sabor (para)

pallid /'pælɪd/ *a* pálido

palm /pɑ:m/ *n* palma *f*. ● *vt.* **~ off** encajar (**on** a). **~ist** /'pɑ:mɪst/ *n* quiromántico *m*. **P~ Sunday** *n* Domingo *m* de Ramos

palpable /'pælpəbl/ *a* palpable

palpitat|e /'pælpɪteɪt/ *vi* palpitar. **~ion** /-'teɪʃn/ *n* palpitación *f*

paltry /'pɔ:ltrɪ/ *a* (**-ier, -iest**) insignificante

pamper /'pæmpə(r)/ *vt* mimar

pamphlet /'pæmflɪt/ *n* folleto *m*
pan /pæn/ *n* cacerola *f*; (*for frying*) sartén *f*; (*of scales*) platillo *m*; (*of lavatory*) taza *f*
panacea /pænə'sɪə/ *n* panacea *f*
panache /pæ'næʃ/ *n* brío *m*
pancake /'pænkeɪk/ *n* hojuela *f*, crêpe *f*
panda /'pændə/ *n* panda *m*. ~ **car** *n* coche *m* de la policía
pandemonium /pændɪ'məʊnɪəm/ *n* pandemonio *m*
pander /'pændə(r)/ *vi*. ~ **to** complacer
pane /peɪn/ *n* (*of glass*) vidrio *m*
panel /'pænl/ *n* panel *m*; (*group of people*) jurado *m*. ~**ling** *n* paneles *mpl*
pang /pæŋ/ *n* punzada *f*
panic /'pænɪk/ *n* pánico *m*. ● *vi* (*pt* **panicked**) ser preso de pánico. ~**-stricken** *a* preso de pánico
panoram|a /pænə'rɑːmə/ *n* panorama *m*. ~**ic** /-'ræmɪk/ *a* panorámico
pansy /'pænzɪ/ *n* pensamiento *m*; (*effeminate man, fam*) maricón *m*
pant /pænt/ *vi* jadear
pantechnicon /pæn'teknɪkən/ *n* camión *m* de mudanzas
panther /'pænθə(r)/ *n* pantera *f*
panties /'pæntɪz/ *npl* bragas *fpl*
pantomime /'pæntəmaɪm/ *n* pantomima *f*
pantry /'pæntrɪ/ *n* despensa *f*
pants /pænts/ *npl* (*man's underwear, fam*) calzoncillos *mpl*; (*woman's underwear, fam*) bragas *fpl*; (*trousers, fam*) pantalones *mpl*
papa|cy /'peɪpəsɪ/ *n* papado *m*. ~**l** *a* papal
paper /'peɪpə(r)/ *n* papel *m*; (*newspaper*) periódico *m*; (*exam*) examen *m*; (*document*) documento *m*. **on** ~ en teoría. ● *vt* empapelar, tapizar (*LAm*). ~**back** /'peɪpəbæk/ *a* en rústica. ● *n* libro *m* en rústica. ~**-clip** *n* sujetapapeles *m invar*, clip *m*. ~**weight** /'peɪpəweɪt/ *n* pisapapeles *m invar*. ~**work** *n* papeleo *m*, trabajo *m* de oficina
papier mâché /pæpɪeɪ'mæʃeɪ/ *n* cartón *m* piedra
par /pɑː(r)/ *n* par *f*; (*golf*) par *m*. **feel below** ~ no estar en forma. **on a** ~ **with** a la par con
parable /'pærəbl/ *n* parábola *f*
parachut|e /'pærəʃuːt/ *n* paracaídas *m invar*. ● *vi* lanzarse en paracaídas. ~**ist** *n* paracaidista *m & f*
parade /pə'reɪd/ *n* desfile *m*; (*street*) paseo *m*; (*display*) alarde *m*. ● *vi* desfilar. ● *vt* hacer alarde de
paradise /'pærədaɪs/ *n* paraíso *m*
paradox /'pærədɒks/ *n* paradoja *f*. ~**ical** /-'dɒksɪkl/ *a* paradójico
paraffin /'pærəfɪn/ *n* queroseno *m*
paragon /'pærəgən/ *n* dechado *m*
paragraph /'pærəgrɑːf/ *n* párrafo *m*
parallel /'pærəlel/ *a* paralelo. ● *n* paralelo *m*; (*line*) paralela *f*. ● *vt* ser paralelo a
paraly|se /'pærəlaɪz/ *vt* paralizar. ~**sis** /pə'ræləsɪs/ *n* (*pl* **-ses** /-siːz/) parálisis *f*. ~**tic** /pærə'lɪtɪk/ *a & n* paralítico (*m*)
parameter /pə'ræmɪtə(r)/ *n* parámetro *m*
paramount /'pærəmaʊnt/ *a* supremo
paranoia /pærə'nɔɪə/ *n* paranoia *f*
parapet /'pærəpɪt/ *n* parapeto *m*
paraphernalia /pærəfə'neɪlɪə/ *n* trastos *mpl*
paraphrase /'pærəfreɪz/ *n* paráfrasis *f*. ● *vt* parafrasear
paraplegic /pærə'pliːdʒɪk/ *n* parapléjico *m*
parasite /'pærəsaɪt/ *n* parásito *m*
parasol /'pærəsɒl/ *n* sombrilla *f*
paratrooper /'pærətruːpə(r)/ *n* paracaidista *m*
parcel /'pɑːsl/ *n* paquete *m*
parch /pɑːtʃ/ *vt* resecar. **be** ~**ed** tener mucha sed
parchment /'pɑːtʃmənt/ *n* pergamino *m*
pardon /'pɑːdn/ *n* perdón *m*; (*jurid*) indulto *m*. **I beg your** ~! ¡perdone Vd! **I beg your** ~? ¿cómo?, ¿mande? (*Mex*). ● *vt* perdonar
pare /peə(r)/ *vt* cortar ‹*nails*›; (*peel*) pelar, mondar
parent /'peərənt/ *n* (*father*) padre *m*; (*mother*) madre *f*; (*source*) origen *m*. ~**s** *npl* padres *mpl*. ~**al** /pə'rentl/ *a* de los padres
parenthesis /pə'renθəsɪs/ *n* (*pl* **-theses** /-siːz/) paréntesis *m invar*
parenthood /'peərənthʊd/ *n* paternidad *f*, maternidad *f*
Paris /'pærɪs/ *n* París *m*
parish /'pærɪʃ/ *n* parroquia *f*; (*municipal*) municipio *m*. ~**ioner** /pə'rɪʃənə(r)/ *n* feligrés *m*
Parisian /pə'rɪzɪən/ *a & n* parisino (*m*)
parity /'pærətɪ/ *n* igualdad *f*

park /pɑ:k/ *n* parque *m*. ● *vt/i* aparcar. ~ **oneself** *vr* (*fam*) instalarse

parka /'pɑ:kə/ *n* anorak *m*

parking-meter /'pɑ:kɪŋmi:tə(r)/ *n* parquímetro *m*

parliament /'pɑ:ləmənt/ *n* parlamento *m*. ~**ary** /-'mentrɪ/ *a* parlamentario

parlour /'pɑ:lə(r)/ *n* salón *m*

parochial /pə'rəʊkɪəl/ *a* parroquial; (*fig*) pueblerino

parody /'pærədɪ/ *n* parodia *f*. ● *vt* parodiar

parole /pə'rəʊl/ *n* libertad *f* bajo palabra, libertad *f* provisional. **on** ~ libre bajo palabra. ● *vt* liberar bajo palabra

paroxysm /'pærəksɪzəm/ *n* paroxismo *m*

parquet /'pɑ:keɪ/ *n*. ~ **floor** *n* parqué *m*

parrot /'pærət/ *n* papagayo *m*

parry /'pærɪ/ *vt* parar; (*avoid*) esquivar. ● *n* parada *f*

parsimonious /pɑ:sɪ'məʊnɪəs/ *a* parsimonioso

parsley /'pɑ:slɪ/ *n* perejil *m*

parsnip /'pɑ:snɪp/ *n* pastinaca *f*

parson /'pɑ:sn/ *n* cura *m*, párroco *m*

part /pɑ:t/ *n* parte *f*; (*of machine*) pieza *f*; (*of serial*) entrega *f*; (*in play*) papel *m*; (*side in dispute*) partido *m*. **on the** ~ **of** por parte de. ● *adv* en parte. ● *vt* separar. ~ **with** *vt* separarse de. ● *vi* separarse

partake /pɑ:'teɪk/ *vt* (*pt* **-took**, *pp* **-taken**) participar. ~ **of** compartir

partial /'pɑ:ʃl/ *a* parcial. **be** ~ **to** ser aficionado a. ~**ity** /-ɪ'ælətɪ/ *n* parcialidad *f*. ~**ly** *adv* parcialmente

participa|nt /pɑ:'tɪsɪpənt/ *n* participante *m & f*. ~**te** /pɑ:'tɪsɪpeɪt/ *vi* participar. ~**tion** /-'peɪʃn/ *n* participación *f*

participle /'pɑ:tɪsɪpl/ *n* participio *m*

particle /'pɑ:tɪkl/ *n* partícula *f*

particular /pə'tɪkjʊlə(r)/ *a* particular; (*precise*) meticuloso; (*fastidious*) quisquilloso. ● *n*. **in** ~ especialmente. ~**ly** *adv* especialmente. ~**s** *npl* detalles *mpl*

parting /'pɑ:tɪŋ/ *n* separación *f*; (*in hair*) raya *f*. ● *a* de despedida

partisan /pɑ:tɪ'zæn/ *n* partidario *m*

partition /pɑ:'tɪʃn/ *n* partición *f*; (*wall*) tabique *m*. ● *vt* dividir

partly /'pɑ:tlɪ/ *adv* en parte

partner /'pɑ:tnə(r)/ *n* socio *m*; (*sport*) pareja *f*. ~**ship** *n* asociación *f*; (*com*) sociedad *f*

partridge /'pɑ:trɪdʒ/ *n* perdiz *f*

part-time /pɑ:t'taɪm/ *a & adv* a tiempo parcial

party /'pɑ:tɪ/ *n* reunión *f*, fiesta *f*; (*group*) grupo *m*; (*pol*) partido *m*; (*jurid*) parte *f*. ~ **line** *n* (*telephone*) línea *f* colectiva

pass /pɑ:s/ *vt* pasar; (*in front of*) pasar por delante de; (*overtake*) adelantar; (*approve*) aprobar ‹*exam, bill, law*›; hacer ‹*remark*›; pronunciar ‹*judgement*›. ~ **down** transmitir. ~ **over** pasar por alto de. ~ **round** distribuir. ~ **through** pasar por; (*cross*) atravesar. ~ **up** (*fam*) dejar pasar. ● *vi* pasar; (*in exam*) aprobar. ~ **away** morir. ~ **out** (*fam*) desmayarse. ● *n* (*permit*) permiso *m*; (*in mountains*) puerto *m*, desfiladero *m*; (*sport*) pase *m*; (*in exam*) aprobado *m*. **make a** ~ **at** (*fam*) hacer proposiciones amorosas a. ~**able** /'pɑ:səbl/ *a* pasable; ‹*road*› transitable

passage /'pæsɪdʒ/ *n* paso *m*; (*voyage*) travesía *f*; (*corridor*) pasillo *m*; (*in book*) pasaje *m*

passenger /'pæsɪndʒə(r)/ *n* pasajero *m*

passer-by /pɑ:sə'baɪ/ *n* (*pl* **passers-by**) transeúnte *m & f*

passion /'pæʃn/ *n* pasión *f*. ~**ate** *a* apasionado. ~**ately** *adv* apasionadamente

passive /'pæsɪv/ *a* pasivo. ~**ness** *n* pasividad *f*

passmark /'pɑ:smɑ:k/ *n* aprobado *m*

Passover /'pɑ:səʊvə(r)/ *n* Pascua *f* de los hebreos

passport /'pɑ:spɔ:t/ *n* pasaporte *m*

password /'pɑ:swɜ:d/ *n* contraseña *f*

past /pɑ:st/ *a & n* pasado (*m*). **in times** ~ en tiempos pasados. **the** ~ **week** *n* la semana *f* pasada. ● *prep* por delante de; (*beyond*) más allá de. ● *adv* por delante. **drive** ~ pasar en coche. **go** ~ pasar

paste /peɪst/ *n* pasta *f*; (*adhesive*) engrudo *m*. ● *vt* (*fasten*) pegar; (*cover*) engrudar. ~**board** /'peɪstbɔ:d/ *n* cartón *m*. ~ **jewellery** *n* joyas *fpl* de imitación

pastel /'pæstl/ *a & n* pastel (*m*)

pasteurize /'pæstʃəraɪz/ *vt* pasteurizar

pastiche /pæ'sti:ʃ/ *n* pastiche *m*

pastille /'pæstɪl/ *n* pastilla *f*
pastime /'pɑ:staɪm/ *n* pasatiempo *m*
pastoral /'pɑ:stərəl/ *a* pastoral
pastr|ies *npl* pasteles *mpl*, pastas *fpl*. **~y** /'peɪstrɪ/ *n* pasta *f*
pasture /'pɑ:stʃə(r)/ *n* pasto *m*
pasty[1] /'pæstɪ/ *n* empanada *f*
pasty[2] /'peɪstɪ/ *a* pastoso; (*pale*) pálido
pat[1] /pæt/ *vt* (*pt* **patted**) dar palmaditas en; acariciar ‹*dog etc*›. ● *n* palmadita *f*; (*of butter*) porción *f*
pat[2] /pæt/ *adv* en el momento oportuno
patch /pætʃ/ *n* pedazo *m*; (*period*) período *m*; (*repair*) remiendo *m*; (*piece of ground*) terreno *m*. **not a ~ on** (*fam*) muy inferior a. ● *vt* remendar. **~ up** arreglar. **~work** *n* labor *m* de retazos; (*fig*) mosaico *m*. **~y** *a* desigual
pâté /'pæteɪ/ *n* pasta *f*, paté *m*
patent /'peɪtnt/ *a* patente. ● *n* patente *f*. ● *vt* patentar. **~ leather** *n* charol *m*. **~ly** *adv* evidentemente
patern|al /pə'tɜ:nl/ *a* paterno. **~ity** /pə'tɜ:nətɪ/ *n* paternidad *f*
path /pɑ:θ/ *n* (*pl* **-s** /pɑ:ðz/) sendero *m*; (*sport*) pista *f*; (*of rocket*) trayectoria *f*; (*fig*) camino *m*
pathetic /pə'θetɪk/ *a* patético, lastimoso
pathology /pə'θɒlədʒɪ/ *n* patología *f*
pathos /'peɪθɒs/ *n* patetismo *m*
patien|ce /'peɪʃns/ *n* paciencia *f*. **~t** /'peɪʃnt/ *a* & *n* paciente (*m* & *f*). **~tly** *adv* con paciencia
patio /'pætɪəʊ/ *n* (*pl* **-os**) patio *m*
patriarch /'peɪtrɪɑ:k/ *n* patriarca *m*
patrician /pə'trɪʃn/ *a* & *n* patricio (*m*)
patriot /'pætrɪət/ *n* patriota *m* & *f*. **~ic** /-'ɒtɪk/ *a* patriótico. **~ism** *n* patriotismo *m*
patrol /pə'trəʊl/ *n* patrulla *f*. ● *vt/i* patrullar
patron /'peɪtrən/ *n* (*of the arts etc*) mecenas *m* & *f*; (*customer*) cliente *m* & *f*; (*of charity*) patrocinador *m*. **~age** /'pætrənɪdʒ/ *n* patrocinio *m*; (*of shop etc*) clientela *f*. **~ize** *vt* ser cliente de; (*fig*) tratar con condescendencia
patter[1] /'pætə(r)/ *n* (*of steps*) golpeteo *m*; (*of rain*) tamborileo *m*. ● *vi* correr con pasos ligeros; ‹*rain*› tamborilear
patter[2] /'pætə(r)/ (*speech*) jerga *f*; (*chatter*) parloteo *m*
pattern /'pætn/ *n* diseño *m*; (*model*) modelo *m*; (*sample*) muestra *f*; (*manner*) modo *m*; (*in dressmaking*) patrón *m*
paunch /pɔ:ntʃ/ *n* panza *f*
pauper /'pɔ:pə(r)/ *n* indigente *m* & *f*, pobre *m* & *f*
pause /pɔ:z/ *n* pausa *f*. ● *vi* hacer una pausa
pave /peɪv/ *vt* pavimentar. **~ the way for** preparar el terreno para
pavement /'peɪvmənt/ *n* pavimento *m*; (*at side of road*) acera *f*
pavilion /pə'vɪlɪən/ *n* pabellón *m*
paving-stone /'peɪvɪŋstəʊn/ *n* losa *f*
paw /pɔ:/ *n* pata *f*; (*of cat*) garra *f*. ● *vi* tocar con la pata; ‹*person*› manosear
pawn[1] /pɔ:n/ *n* (*chess*) peón *m*; (*fig*) instrumento *m*
pawn[2] /pɔ:n/ *vt* empeñar. ● *n*. **in ~** en prenda. **~broker** /'pɔ:nbrəʊkə(r)/ *n* prestamista *m* & *f*. **~shop** *n* monte *m* de piedad
pawpaw /'pɔ:pɔ:/ *n* papaya *f*
pay /peɪ/ *vt* (*pt* **paid**) pagar; prestar ‹*attention*›; hacer ‹*compliment, visit*›. **~ back** devolver. **~ cash** pagar al contado. **~ in** ingresar. **~ off** pagar. **~ out** pagar. ● *vi* pagar; (*be profitable*) rendir. ● *n* paga *f*. **in the ~ of** al servicio de. **~able** /'peɪəbl/ *a* pagadero. **~ment** /'peɪmənt/ *n* pago *m*. **~-off** *n* (*sl*) liquidación *f*; (*fig*) ajuste *m* de cuentas. **~roll** /'peɪrəʊl/ *n* nómina *f*. **~ up** pagar
pea /pi:/ *n* guisante *m*
peace /pi:s/ *n* paz *f*. **~ of mind** tranquilidad *f*. **~able** *a* pacífico. **~ful** /'pi:sfl/ *a* tranquilo. **~maker** /'pi:smeɪkə(r)/ *n* pacificador *m*
peach /pi:tʃ/ *n* melocotón *m*, durazno *m* (*LAm*); (*tree*) melocotonero *m*, duraznero *m* (*LAm*)
peacock /'pi:kɒk/ *n* pavo *m* real
peak /pi:k/ *n* cumbre *f*; (*maximum*) máximo *m*. **~ hours** *npl* horas *fpl* punta. **~ed cap** *n* gorra *f* de visera
peaky /'pi:kɪ/ *a* pálido
peal /pi:l/ *n* repique *m*. **~s of laughter** risotadas *fpl*
peanut /'pi:nʌt/ *n* cacahuete *m*, maní *m* (*Mex*). **~s** (*sl*) una bagatela *f*
pear /peə(r)/ *n* pera *f*; (*tree*) peral *m*
pearl /pɜ:l/ *n* perla *f*. **~y** *a* nacarado
peasant /'peznt/ *n* campesino *m*
peat /pi:t/ *n* turba *f*
pebble /'pebl/ *n* guijarro *m*

peck /pek/ *vt* picotear; (*kiss, fam*) dar un besito a. ● *n* picotazo *m*; (*kiss*) besito *m*. **~ish** /'pekɪʃ/ *a*. **be ~ish** (*fam*) tener hambre, tener gazuza (*fam*)

peculiar /pɪ'kju:lɪə(r)/ *a* raro; (*special*) especial. **~ity** /-'ærətɪ/ *n* rareza *f*; (*feature*) particularidad *f*

pedal /'pedl/ *n* pedal *m*. ● *vi* pedalear

pedantic /pɪ'dæntɪk/ *a* pedante

peddle /'pedl/ *vt* vender por las calles

pedestal /'pedɪstl/ *n* pedestal *m*

pedestrian /pɪ'destrɪən/ *n* peatón *m*. ● *a* de peatones; (*dull*) prosaico. **~ crossing** *n* paso *m* de peatones

pedigree /'pedɪgri:/ linaje *m*; (*of animal*) pedigrí *m*. ● *a* ‹*animal*› de raza

pedlar /'pedlə(r)/ *n* buhonero *m*, vendedor *m* ambulante

peek /pi:k/ *vi* mirar a hurtadillas

peel /pi:l/ *n* cáscara *f*. ● *vt* pelar ‹*fruit, vegetables*›. ● *vi* pelarse. **~ings** *npl* peladuras *fpl*, monda *f*

peep[1] /pi:p/ *vi* mirar a hurtadillas. ● *n* mirada *f* furtiva

peep[2] /pi:p/ ‹*bird*› piar. ● *n* pío *m*

peep-hole /'pi:phəʊl/ *n* mirilla *f*

peer[1] /pɪə(r)/ *vi* mirar. **~ at** escudriñar

peer[2] /pɪə(r)/ *n* par *m*, compañero *m*. **~age** *n* pares *mpl*

peev|ed /pi:vd/ *a* (*sl*) irritado. **~ish** /'pi:vɪʃ/ *a* picajoso

peg /peg/ *n* clavija *f*; (*for washing*) pinza *f*; (*hook*) gancho *m*; (*for tent*) estaca *f*. **off the ~** de percha. ● *vt* (*pt* **pegged**) fijar ‹*precios*›. **~ away at** afanarse por

pejorative /pɪ'dʒɒrətɪv/ *a* peyorativo, despectivo

pelican /'pelɪkən/ *n* pelícano *m*. **~ crossing** *n* paso *m* de peatones (con semáforo)

pellet /'pelɪt/ *n* pelotilla *f*; (*for gun*) perdigón *m*

pelt[1] /pelt/ *n* pellejo *m*

pelt[2] /pelt/ *vt* tirar. ● *vi* llover a cántaros

pelvis /'pelvɪs/ *n* pelvis *f*

pen[1] /pen/ *n* (*enclosure*) recinto *m*

pen[2] /pen/ (*for writing*) pluma *f*, estilográfica *f*; (*ball-point*) bolígrafo *m*

penal /'pi:nl/ *a* penal. **~ize** *vt* castigar. **~ty** /'penltɪ/ *n* castigo *m*; (*fine*) multa *f*. **~ty kick** *n* (*football*) penalty *m*

penance /'penəns/ *n* penitencia *f*

pence /pens/ *see* **penny**

pencil /'pensl/ *n* lápiz *m*. ● *vt* (*pt* **pencilled**) escribir con lápiz. **~-sharpener** *n* sacapuntas *m invar*

pendant /'pendənt/ *n* dije *m*, medallón *m*

pending /'pendɪŋ/ *a* pendiente. ● *prep* hasta

pendulum /'pendjʊləm/ *n* péndulo *m*

penetrat|e /'penɪtreɪt/ *vt/i* penetrar. **~ing** *a* penetrante. **~ion** /-'treɪʃn/ *n* penetración *f*

penguin /'peŋgwɪn/ *n* pingüino *m*

penicillin /penɪ'sɪlɪn/ *n* penicilina *f*

peninsula /pə'nɪnsjʊlə/ *n* península *f*

penis /'pi:nɪs/ *n* pene *m*

peniten|ce /'penɪtəns/ *n* penitencia *f*. **~t** /'penɪtənt/ *a* & *n* penitente (*m* & *f*). **~tiary** /penɪ'tenʃərɪ/ *n* (*Amer*) cárcel *m*

pen: **~knife** /'pennaɪf/ *n* (*pl* **penknives**) navaja *f*; (*small*) cortaplumas *m invar*. **~-name** *n* seudónimo *m*

pennant /'penənt/ *n* banderín *m*

penn|iless /'penɪlɪs/ *a* sin un céntimo. **~y** /'penɪ/ *n* (*pl* **pennies** *or* **pence**) penique *m*

pension /'penʃn/ *n* pensión *f*; (*for retirement*) jubilación *f*. ● *vt* pensionar. **~able** *a* con derecho a pensión; ‹*age*› de la jubilación. **~er** *n* jubilado *m*. **~ off** jubilar

pensive /'pensɪv/ *a* pensativo

pent-up /pent'ʌp/ *a* reprimido; (*confined*) encerrado

pentagon /'pentəgən/ *n* pentágono *m*

Pentecost /'pentɪkɒst/ *n* Pentecostés *m*

penthouse /'penthaʊs/ *n* ático *m*

penultimate /pen'ʌltɪmət/ *a* penúltimo

penury /'penjʊərɪ/ *n* penuria *f*

peony /'pi:ənɪ/ *n* peonía *f*

people /'pi:pl/ *npl* gente *f*; (*citizens*) pueblo *m*. **~ say** se dice. **English ~** los ingleses *mpl*. **my ~** (*fam*) mi familia *f*. ● *vt* poblar

pep /pep/ *n* vigor *m*. ● *vt*. **~ up** animar

pepper /'pepə(r)/ *n* pimienta *f*; (*vegetable*) pimiento *m*. ● *vt* sazonar con pimienta. **~y** *a* picante. **~corn**

/'pepəkɔ:n/ *n* grano *m* de pimienta. **~corn rent** *n* alquiler *m* nominal
peppermint /'pepəmɪnt/ *n* menta *f*; (*sweet*) pastilla *f* de menta
pep talk /'peptɔ:k/ *n* palabras *fpl* animadoras
per /pɜ:(r)/ *prep* por. **~ annum** al año. **~ cent** por ciento. **~ head** por cabeza, por persona. **ten miles ~ hour** diez millas por hora
perceive /pə'si:v/ *vt* percibir; (*notice*) darse cuenta de
percentage /pə'sentɪdʒ/ *n* porcentaje *m*
percepti|ble /pə'septəbl/ *a* perceptible. **~on** /pə'sepʃn/ *n* percepción *f*. **~ve** *a* perspicaz
perch[1] /pɜ:tʃ/ *n* (*of bird*) percha *f*. ● *vi* posarse
perch[2] /pɜ:tʃ/ (*fish*) perca *f*
percolat|e /'pɜ:kəleɪt/ *vt* filtrar. ● *vi* filtrarse. **~or** *n* cafetera *f*
percussion /pə'kʌʃn/ *n* percusión *f*
peremptory /pə'remptərɪ/ *a* perentorio
perennial /pə'renɪəl/ *a* & *n* perenne (*m*)
perfect /'pɜ:fɪkt/ *a* perfecto. /pə'fekt/ *vt* perfeccionar. **~ion** /pə'fekʃn/ *n* perfección *f*. **to ~ion** a la perfección. **~ionist** *n* perfeccionista *m* & *f*. **~ly** /'pɜ:fɪktlɪ/ *adv* perfectamente
perforat|e /'pɜ:fəreɪt/ *vt* perforar. **~ion** /-'reɪʃn/ *n* perforación *f*
perform /pə'fɔ:m/ *vt* hacer, realizar; representar ‹*play*›; desempeñar ‹*role*›; (*mus*) interpretar. **~ an operation** (*med*) operar. **~ance** *n* ejecución *f*; (*of play*) representación *f*; (*of car*) rendimiento *m*; (*fuss, fam*) jaleo *m*. **~er** *n* artista *m* & *f*
perfume /'pɜ:fju:m/ *n* perfume *m*
perfunctory /pə'fʌŋktərɪ/ *a* superficial
perhaps /pə'hæps/ *adv* quizá(s), tal vez
peril /'perəl/ *n* peligro *m*. **~ous** *a* arriesgado, peligroso
perimeter /pə'rɪmɪtə(r)/ *n* perímetro *m*
period /'pɪərɪəd/ *n* período *m*; (*lesson*) clase *f*; (*gram*) punto *m*. ● *a* de (la) época. **~ic** /-'ɒdɪk/ *a* periódico. **~ical** /pɪərɪ'ɒdɪkl/ *n* revista *f*. **~ically** /-'ɒdɪklɪ/ *adv* periódico
peripher|al /pə'rɪfərəl/ *a* periférico. **~y** /pə'rɪfərɪ/ *n* periferia *f*
periscope /'perɪskəʊp/ *n* periscopio *m*
perish /'perɪʃ/ *vi* perecer; (*rot*) estropearse. **~able** *a* perecedero. **~ing** *a* (*fam*) glacial
perjur|e /'pɜ:dʒə(r)/ *vr*. **~e o.s.** perjurarse. **~y** *n* perjurio *m*
perk[1] /pɜ:k/ *n* gaje *m*
perk[2] /pɜ:k/ *vt/i*. **~ up** *vt* reanimar. ● *vi* reanimarse. **~y** *a* alegre
perm /pɜ:m/ *n* permanente *f*. ● *vt* hacer una permanente a
permanen|ce /'pɜ:mənəns/ *n* permanencia *f*. **~t** /'pɜ:mənənt/ *a* permanente. **~tly** *adv* permanentemente
permea|ble /'pɜ:mɪəbl/ *a* permeable. **~te** /'pɜ:mɪeɪt/ *vt* penetrar; (*soak*) empapar
permissible /pə'mɪsəbl/ *a* permisible
permission /pə'mɪʃn/ *n* permiso *m*
permissive /pə'mɪsɪv/ *a* indulgente. **~ness** *n* tolerancia *f*. **~ society** *n* sociedad *f* permisiva
permit /pə'mɪt/ *vt* (*pt* **permitted**) permitir. /'pɜ:mɪt/ *n* permiso *m*
permutation /pɜ:mju:'teɪʃn/ *n* permutación *f*
pernicious /pə'nɪʃəs/ *a* pernicioso
peroxide /pə'rɒksaɪd/ *n* peróxido *m*
perpendicular /pɜ:pən'dɪkjʊlə(r)/ *a* & *n* perpendicular (*f*)
perpetrat|e /'pɜ:pɪtreɪt/ *vt* cometer. **~or** *n* autor *m*
perpetua|l /pə'petʃʊəl/ *a* perpetuo. **~te** /pə'petʃʊeɪt/ *vt* perpetuar. **~tion** /-'eɪʃn/ *n* perpetuación *f*
perplex /pə'pleks/ *vt* dejar perplejo. **~ed** *a* perplejo. **~ing** *a* desconcertante. **~ity** *n* perplejidad *f*
persecut|e /'pɜ:sɪkju:t/ *vt* perseguir. **~ion** /-'kju:ʃn/ *n* persecución *f*
persever|ance /pɜ:sɪ'vɪərəns/ *n* perseverancia *f*. **~e** /pɜ:sɪ'vɪə(r)/ *vi* perseverar, persistir
Persian /'pɜ:ʃn/ *a* persa. **the ~ Gulf** *n* el golfo *m* Pérsico. ● *n* persa (*m* & *f*); (*lang*) persa *m*
persist /pə'sɪst/ *vi* persistir. **~ence** *n* persistencia *f*. **~ent** *a* persistente; (*continual*) continuo. **~ently** *adv* persistentemente
person /'pɜ:sn/ *n* persona *f*
personal /'pɜ:sənl/ *a* personal
personality /pɜ:sə'nælətɪ/ *n* personalidad *f*; (*on TV*) personaje *m*
personally /'pɜ:sənəlɪ/ *adv* personalmente; (*in person*) en persona

personify /pə'sɒnɪfaɪ/ *vt* personificar
personnel /pɜ:sə'nel/ *n* personal *m*
perspective /pə'spektɪv/ *n* perspectiva *f*
perspicacious /pɜ:spɪ'keɪʃəs/ *a* perspicaz
perspir|ation /pɜ:spə'reɪʃn/ *n* sudor *m*. **~e** /pəs'paɪə(r)/ *vi* sudar
persua|de /pə'sweɪd/ *vt* persuadir. **~sion** *n* persuasión *f*. **~sive** /pə'sweɪsɪv/ *a* persuasivo. **~sively** *adv* de manera persuasiva
pert /pɜ:t/ *a* (*saucy*) impertinente; (*lively*) animado
pertain /pə'teɪn/ *vi*. **~ to** relacionarse con
pertinent /'pɜ:tɪnənt/ *a* pertinente. **~ly** *adv* pertinentemente
pertly /'pɜ:tlɪ/ *adv* impertinentemente
perturb /pə'tɜ:b/ *vt* perturbar
Peru /pə'ru:/ *n* el Perú *m*
perus|al /pə'ru:zl/ *n* lectura *f* cuidadosa. **~e** /pə'ru:z/ *vt* leer cuidadosamente
Peruvian /pə'ru:vɪan/ *a* & *n* peruano (*m*)
perva|de /pə'veɪd/ *vt* difundirse por. **~sive** *a* penetrante
perver|se /pə'vɜ:s/ *a* (*stubborn*) terco; (*wicked*) perverso. **~sity** *n* terquedad *f*; (*wickedness*) perversidad *f*. **~sion** *n* perversión *f*. **~t** /pə'vɜ:t/ *vt* pervertir. /'pɜ:vɜ:t/ *n* pervertido *m*
pessimis|m /'pesɪmɪzəm/ *n* pesimismo *m*. **~t** /'pesɪmɪst/ *n* pesimista *m* & *f*. **~tic** /-'mɪstɪk/ *a* pesimista
pest /pest/ *n* insecto *m* nocivo, plaga *f*; (*person*) pelma *m*; (*thing*) lata *f*
pester /'pestə(r)/ *vt* importunar
pesticide /'pestɪsaɪd/ *n* pesticida *f*
pet /pet/ *n* animal *m* doméstico; (*favourite*) favorito *m*. • *a* preferido. • *vt* (*pt* **petted**) acariciar
petal /'petl/ *n* pétalo *m*
peter /'pi:tə(r)/ *vi*. **~ out** ‹*supplies*› agotarse; (*disappear*) desparecer
petite /pə'ti:t/ *a* (*of woman*) chiquita
petition /pɪ'tɪʃn/ *n* petición *f*. • *vt* dirigir una petición a
pet name /'petneɪm/ *n* apodo *m* cariñoso
petrify /'petrɪfaɪ/ *vt* petrificar. • *vi* petrificarse
petrol /'petrəl/ *n* gasolina *f*. **~eum** /pɪ'trəʊlɪəm/ *n* petróleo *m*. **~ gauge** *n* indicador *m* de nivel de gasolina. **~ pump** *n* (*in car*) bomba *f* de gasolina; (*at garage*) surtidor *m* de gasolina. **~ station** *n* gasolinera *f*. **~ tank** *n* depósito *m* de gasolina
petticoat /'petɪkəʊt/ *n* enaguas *fpl*
pett|iness /'petɪnɪs/ *n* mezquindad *f*. **~y** /'petɪ/ *a* (**-ier**, **-iest**) insignificante; (*mean*) mezquino. **~y cash** *n* dinero *m* para gastos menores. **~y officer** *n* suboficial *m* de marina
petulan|ce /'petjʊləns/ *n* irritabilidad *f*. **~t** /'petjʊlənt/ *a* irritable
pew /pju:/ *n* banco *m* (de iglesia)
pewter /'pju:tə(r)/ *n* peltre *m*
phallic /'fælɪk/ *a* fálico
phantom /'fæntəm/ *n* fantasma *m*
pharmaceutical /fɑ:mə'sju:tɪkl/ *a* farmacéutico
pharmac|ist /'fɑ:məsɪst/ *n* farmacéutico *m*. **~y** /'fɑ:məsɪ/ *n* farmacia *f*
pharyngitis /færɪn'dʒaɪtɪs/ *n* faringitis *f*
phase /feɪz/ *n* etapa *f*. • *vt*. **~ in** introducir progresivamente. **~ out** retirar progresivamente
PhD *abbr* (*Doctor of Philosophy*) *n* Doctor *m* en Filosofía
pheasant /'feznt/ *n* faisán *m*
phenomenal /fɪ'nɒmɪnl/ *a* fenomenal
phenomenon /fɪ'nɒmɪnən/ *n* (*pl* **-ena**) fenómeno *m*
phew /fju:/ *int* ¡uy!
phial /'faɪəl/ *n* frasco *m*
philanderer /fɪ'lændərə(r)/ *n* mariposón *m*
philanthrop|ic /fɪlən'θrɒpɪk/ *a* filantrópico. **~ist** /fɪ'lænθrəpɪst/ *n* filántropo *m*
philatel|ist /fɪ'lætəlɪst/ *n* filatelista *m* & *f*. **~y** /fɪ'lætəlɪ/ *n* filatelia *f*
philharmonic /fɪlhɑ:'mɒnɪk/ *a* filarmónico
Philippines /'fɪlɪpi:nz/ *npl* Filipinas *fpl*
philistine /'fɪlɪstaɪn/ *a* & *n* filisteo (*m*)
philosoph|er /fɪ'lɒsəfə(r)/ *n* filósofo *m*. **~ical** /-ə'sɒfɪkl/ *a* filosófico. **~y** /fɪ'lɒsəfɪ/ *n* filosofía *f*
phlegm /flem/ *n* flema *f*. **~atic** /fleg'mætɪk/ *a* flemático
phobia /'fəʊbɪə/ *n* fobia *f*
phone /fəʊn/ *n* (*fam*) teléfono *m*. • *vt*/*i* llamar por teléfono. **~ back**

‹*caller*› volver a llamar; ‹*person called*› llamar. ~ **box** *n* cabina *f* telefónica
phonetic /fəˈnetɪk/ *a* fonético. ~**s** *n* fonética *f*
phoney /ˈfəʊnɪ/ *a* (**-ier**, **-iest**) (*sl*) falso. ● *n* (*sl*) farsante *m & f*
phosphate /ˈfɒsfeɪt/ *n* fosfato *m*
phosphorus /ˈfɒsfərəs/ *n* fósforo *m*
photo /ˈfəʊtəʊ/ *n* (*pl* **-os**) (*fam*) fotografía *f*, foto *f* (*fam*)
photocopy /ˈfəʊtəʊkɒpɪ/ *n* fotocopia *f*. ● *vt* fotocopiar
photogenic /fəʊtəʊˈdʒenɪk/ *a* fotogénico
photograph /ˈfəʊtəgrɑːf/ *n* fotografía *f*. ● *vt* hacer una fotografía de, sacar fotos de. ~**er** /fəˈtɒgrəfə(r)/ *n* fotógrafo *m*. ~**ic** /-ˈgræfɪk/ *a* fotográfico ~**y** /fəˈtɒgrəfɪ/ *n* fotografía *f*
phrase /freɪz/ *n* frase *f*, locución *f*, expresión *f*. ● *vt* expresar. ~**-book** *n* libro *m* de frases
physical /ˈfɪzɪkl/ *a* físico
physician /fɪˈzɪʃn/ *n* médico *m*
physic|ist /ˈfɪzɪsɪst/ *n* físico *m*. ~**s** /ˈfɪzɪks/ *n* física *f*
physiology /fɪzɪˈɒlədʒɪ/ *n* fisiología *f*
physiotherap|ist /fɪzɪəʊˈθerəpɪst/ *n* fisioterapeuta *m & f*. ~**y** /fɪzɪəʊˈθerəpɪ/ *n* fisioterapia *f*
physique /fɪˈziːk/ *n* constitución *f*; (*appearance*) físico *m*
pian|ist /ˈpɪənɪst/ *n* pianista *m & f*. ~**o** /pɪˈænəʊ/ *n* (*pl* **-os**) piano *m*
piccolo /ˈpɪkələʊ/ *n* flautín *m*, píccolo *m*
pick[1] /pɪk/ (*tool*) pico *m*
pick[2] /pɪk/ *vt* escoger; recoger ‹*flowers etc*›; forzar ‹*a lock*›; (*dig*) picar. ~ **a quarrel** buscar camorra. ~ **holes in** criticar. ● *n* (*choice*) selección *f*; (*the best*) lo mejor. ~ **on** *vt* (*nag*) meterse con. ~ **out** *vt* escoger; (*identify*) identificar; destacar ‹*colour*›. ~ **up** *vt* recoger; (*lift*) levantar; (*learn*) aprender; adquirir ‹*habit, etc*›; obtener ‹*information*›; contagiarse de ‹*illness*›. ● *vi* mejorar; (*med*) reponerse
pickaxe /ˈpɪkæks/ *n* pico *m*
picket /ˈpɪkɪt/ *n* (*striker*) huelguista *m & f*; (*group of strikers*) piquete *m*; (*stake*) estaca *f*. ~ **line** *n* piquete *m*. ● *vt* vigilar por piquetes. ● *vi* estar de guardia
pickle /ˈpɪkl/ *n* (*in vinegar*) encurtido *m*; (*in brine*) salmuera *f*. **in a** ~ (*fam*) en un apuro. ● *vt* encurtir. ~**s** *npl* encurtido *m*
pick: ~**pocket** /ˈpɪkpɒkɪt/ *n* ratero *m*. ~**-up** *n* (*sl*) ligue *m*; (*truck*) camioneta *f*; (*stylus-holder*) fonocaptor *m*, brazo *m*
picnic /ˈpɪknɪk/ *n* comida *f* campestre. ● *vi* (*pt* **picnicked**) merendar en el campo
pictorial /pɪkˈtɔːrɪəl/ *a* ilustrado
picture /ˈpɪktʃə(r)/ *n* (*painting*) cuadro *m*; (*photo*) fotografía *f*; (*drawing*) dibujo *m*; (*beautiful thing*) preciosidad *f*; (*film*) película *f*; (*fig*) descripción *f*. **the** ~**s** *npl* el cine *m*. ● *vt* imaginarse; (*describe*) describir
picturesque /pɪktʃəˈresk/ *a* pintoresco
piddling /ˈpɪdlɪŋ/ *a* (*fam*) insignificante
pidgin /ˈpɪdʒɪn/ *a*. ~ **English** *n* inglés *m* corrompido
pie /paɪ/ *n* empanada *f*; (*sweet*) pastel *m*, tarta *f*
piebald /ˈpaɪbɔːld/ *a* pío
piece /piːs/ *n* pedazo *m*; (*coin*) moneda *f*; (*in game*) pieza *f*. **a** ~ **of advice** un consejo *m*. **a** ~ **of news** una noticia *f*. **take to** ~**s** desmontar. ● *vt*. ~ **together** juntar. ~**meal** /ˈpiːsmiːl/ *a* gradual; (*unsystematic*) poco sistemático. —adv poco a poco. ~**-work** *n* trabajo *m* a destajo
pier /pɪə(r)/ *n* muelle *m*
pierc|e /pɪəs/ *vt* perforar. ~**ing** *a* penetrante
piety /ˈpaɪətɪ/ *n* piedad *f*
piffl|e /ˈpɪfl/ *n* (*sl*) tonterías *fpl*. ~**ing** *a* (*sl*) insignificante
pig /pɪg/ *n* cerdo *m*
pigeon /ˈpɪdʒɪn/ *n* paloma *f*; (*culin*) pichón *m*. ~**-hole** *n* casilla *f*
pig: ~**gy** /ˈpɪgɪ/ *a* (*greedy, fam*) glotón. ~**gy-back** *adv* a cuestas. ~**gy bank** *n* hucha *f*. ~**-headed** *a* terco
pigment /ˈpɪgmənt/ *n* pigmento *m*. ~**ation** /-ˈteɪʃn/ *n* pigmentación *f*
pig: ~**skin** /ˈpɪgskɪn/ *n* piel *m* de cerdo. ~**sty** /ˈpɪgstaɪ/ *n* pocilga *f*
pigtail /ˈpɪgteɪl/ *n* (*plait*) trenza *f*
pike /paɪk/ *n invar* (*fish*) lucio *m*
pilchard /ˈpɪltʃəd/ *n* sardina *f*
pile[1] /paɪl/ *n* (*heap*) montón *m*. ● *vt* amontonar. ~ **it on** exagerar. ● *vi* amontonarse. ~ **up** *vt* amontonar. ● *vi* amontonarse. ~**s** /paɪlz/ *npl* (*med*) almorranas *fpl*
pile[2] /paɪl/ *n* (*of fabric*) pelo *m*

pile-up /'paɪlʌp/ *n* accidente *m* múltiple
pilfer /'pɪlfə(r)/ *vt/i* hurtar. **~age** *n*, **~ing** *n* hurto *m*
pilgrim /'pɪlgrɪm/ *n* peregrino. **~age** *n* peregrinación *f*
pill /pɪl/ *n* píldora *f*
pillage /'pɪlɪdʒ/ *n* saqueo *m*. ● *vt* saquear
pillar /'pɪlə(r)/ *n* columna *f*. **~-box** *n* buzón *m*
pillion /'pɪlɪən/ *n* asiento *m* trasero. **ride ~** ir en el asiento trasero
pillory /'pɪlərɪ/ *n* picota *f*
pillow /'pɪləʊ/ *n* almohada *f*. **~case** /'pɪləʊkeɪs/ *n* funda *f* de almohada
pilot /'paɪlət/ *n* piloto *m*. ● *vt* pilotar. **~-light** *n* fuego *m* piloto
pimp /pɪmp/ *n* alcahuete *m*
pimple /'pɪmpl/ *n* grano *m*
pin /pɪn/ *n* alfiler *m*; (*mec*) perno *m*. **~s and needles** hormigueo *m*. ● *vt* (*pt* **pinned**) prender con alfileres; (*hold down*) enclavijar; (*fix*) sujetar. **~ s.o. down** obligar a uno a que se decida. **~ up** fijar
pinafore /'pɪnəfɔ:(r)/ *n* delantal *m*. **~ dress** *n* mandil *m*
pincers /'pɪnsəz/ *npl* tenazas *fpl*
pinch /pɪntʃ/ *vt* pellizcar; (*steal*, *sl*) hurtar. ● *vi* ‹*shoe*› apretar. ● *n* pellizco *m*; (*small amount*) pizca *f*. **at a ~** en caso de necesidad
pincushion /'pɪnkʊʃn/ *n* acerico *m*
pine[1] /paɪn/ *n* pino *m*
pine[2] /paɪn/ *vi*. **~ away** consumirse. **~ for** suspirar por
pineapple /'paɪnæpl/ *n* piña *f*, ananás *m*
ping /pɪŋ/ *n* sonido *m* agudo. **~-pong** /'pɪŋpɒŋ/ *n* pimpón *m*, ping-pong *m*
pinion /'pɪnjən/ *vt* maniatar
pink /pɪŋk/ *a* & *n* color (*m*) de rosa
pinnacle /'pɪnəkl/ *n* pináculo *m*
pin: **~point** *vt* determinar con precisión *f*. **~stripe** /'pɪnstraɪp/ *n* raya *f* fina
pint /paɪnt/ *n* pinta *f* (= *0.57 litre*)
pin-up /'pɪnʌp/ *n* (*fam*) fotografía *f* de mujer
pioneer /paɪə'nɪə(r)/ *n* pionero *m*. ● *vt* ser el primero, promotor de, promover
pious /'paɪəs/ *a* piadoso
pip[1] /pɪp/ *n* (*seed*) pepita *f*
pip[2] /pɪp/ (*time signal*) señal *f*
pip[3] /pɪp/ (*on uniform*) estrella *f*
pipe /paɪp/ *n* tubo *m*; (*mus*) caramillo *m*; (*for smoking*) pipa *f*. ● *vt* conducir por tuberías. **~down** (*fam*) bajar la voz, callarse. **~-cleaner** *n* limpiapipas *m invar*. **~-dream** *n* ilusión *f*. **~line** /'paɪplaɪn/ *n* tubería *f*; (*for oil*) oleoducto *m*. **in the ~line** en preparación *f*. **~r** *n* flautista *m* & *f*
piping /'paɪpɪŋ/ *n* tubería *f*. **~ hot** muy caliente, hirviendo
piquant /'pi:kənt/ *a* picante
pique /pi:k/ *n* resentimiento *m*
pira|cy /'paɪərəsɪ/ *n* piratería *f*. **~te** /'paɪərət/ *n* pirata *m*
pirouette /pɪrʊ'et/ *n* pirueta *f*. ● *vi* piruetear
Pisces /'paɪsi:z/ *n* (*astr*) Piscis *m*
pistol /'pɪstl/ *n* pistola *f*
piston /'pɪstən/ *n* pistón *m*
pit /pɪt/ *n* foso *m*; (*mine*) mina *f*; (*of stomach*) boca *f*. ● *vt* (*pt* **pitted**) marcar con hoyos; (*fig*) oponer. **~ o.s. against** medirse con
pitch[1] /pɪtʃ/ *n* brea *f*
pitch[2] /pɪtʃ/ (*degree*) grado *m*; (*mus*) tono *m*; (*sport*) campo *m*. ● *vt* lanzar; armar ‹*tent*›. **~ into** (*fam*) atacar. ● *vi* caerse; ‹*ship*› cabecear. **~ in** (*fam*) contribuir. **~ed battle** *n* batalla *f* campal
pitch-black /pɪtʃ'blæk/ *a* oscuro como boca de lobo
pitcher /'pɪtʃə(r)/ *n* jarro *m*
pitchfork /'pɪtʃfɔ:k/ *n* horca *f*
piteous /'pɪtɪəs/ *a* lastimoso
pitfall /'pɪtfɔ:l/ *n* trampa *f*
pith /pɪθ/ *n* (*of orange*, *lemon*) médula *f*; (*fig*) meollo *m*
pithy /'pɪθɪ/ *a* (**-ier**, **-iest**) conciso
piti|ful /'pɪtɪfl/ *a* lastimoso. **~less** *a* despiadado
pittance /'pɪtns/ *n* sueldo *m* irrisorio
pity /'pɪtɪ/ *n* piedad *f*; (*regret*) lástima *f*. ● *vt* compadecerse de
pivot /'pɪvət/ *n* pivote *m*. ● *vt* montonar sobre un pivote. ● *vi* girar sobre un pivote; (*fig*) depender (**on** de)
pixie /'pɪksɪ/ *n* duende *m*
placard /'plækɑ:d/ *n* pancarta *f*; (*poster*) cartel *m*
placate /plə'keɪt/ *vt* apaciguar
place /pleɪs/ *n* lugar *m*; (*seat*) asiento *m*; (*post*) puesto *m*; (*house*, *fam*) casa *f*. **take ~** tener lugar. ● *vt* poner, colocar; (*remember*) recordar; (*identify*) identificar. **be**

~d (*in race*) colocarse. **~-mat** *n* salvamanteles *m invar*. **~ment** /'pleɪsmənt/ *n* colocación *f*
placid /'plæsɪd/ *a* plácido
plagiari|sm /'pleɪdʒərɪzm/ *n* plagio *m*. **~ze** /'pleɪdʒəraɪz/ *vt* plagiar
plague /pleɪg/ *n* peste *f*; (*fig*) plaga *f*. ● *vt* atormentar
plaice /pleɪs/ *n invar* platija *f*
plaid /plæd/ *n* tartán *m*
plain /pleɪn/ *a* (**-er**, **-est**) claro; (*simple*) sencillo; (*candid*) franco; (*ugly*) feo. **in ~ clothes** en traje de paisano. ● *adv* claramente. ● *n* llanura *f*. **~ly** *adv* claramente; (*frankly*) francamente; (*simply*) sencillamente. **~ness** *n* claridad *f*; (*simplicity*) sencillez *f*
plaintiff /'pleɪntɪf/ *n* demandante *m* & *f*
plait /plæt/ *vt* trenzar. ● *n* trenza *f*
plan /plæn/ *n* proyecto *m*; (*map*) plano *m*. ● *vt* (*pt* **planned**) planear, proyectar; (*intend*) proponerse
plane[1] /pleɪn/ *n* (*tree*) plátano *m*
plane[2] /pleɪn/ (*level*) nivel *m*; (*aviat*) avión *m*. ● *a* plano
plane[3] /pleɪn/ (*tool*) cepillo *m*. ● *vt* cepillar
planet /'plænɪt/ *n* planeta *m*. **~ary** *a* planetario
plank /plæŋk/ *n* tabla *f*
planning /'plænɪŋ/ *n* planificación *f*. **family ~** *n* planificación familiar. **town ~** *n* urbanismo *m*
plant /plɑ:nt/ *n* planta *f*; (*mec*) maquinaria *f*; (*factory*) fábrica *f*. ● *vt* plantar; (*place in position*) colocar. **~ation** /plæn'teɪʃn/ *n* plantación *f*
plaque /plæk/ *n* placa *f*
plasma /'plæzmə/ *n* plasma *m*
plaster /'plɑ:stə(r)/ *n* yeso *m*; (*adhesive*) esparadrapo *m*; (*for setting bones*) escayola *f*. **~ of Paris** *n* yeso *m* mate. ● *vt* enyesar; (*med*) escayolar ‹*broken bone*›; (*cover*) cubrir (**with** de). **~ed** *a* (*fam*) borracho
plastic /'plæstɪk/ *a* & *n* plástico (*m*)
Plasticine /'plæstɪsi:n/ *n* (*P*) pasta *f* de modelar, plastilina *f* (*P*)
plastic surgery /plæstɪk'sɜ:dʒərɪ/ *n* cirugía *f* estética
plate /pleɪt/ *n* plato *m*; (*of metal*) chapa *f*; (*silverware*) vajilla *f* de plata; (*in book*) lámina *f*. ● *vt* (*cover with metal*) chapear
plateau /'plætəʊ/ *n* (*pl* **plateaux**) *n* meseta *f*
plateful /'pleɪtfl/ *n* (*pl* **-fuls**) plato *m*
platform /'plætfɔ:m/ *n* plataforma *f*; (*rail*) andén *m*
platinum /'plætɪnəm/ *n* platino *m*
platitude /'plætɪtju:d/ *n* tópico *m*, perogrullada *f*, lugar *m* común
platonic /plə'tɒnɪk/ *a* platónico
platoon /plə'tu:n/ *n* pelotón *m*
platter /'plætə(r)/ *n* fuente *f*, plato *m* grande
plausible /'plɔ:zəbl/ *a* plausible; ‹*person*› convincente
play /pleɪ/ *vt* jugar; (*act role*) desempeñar el papel de; tocar ‹*instrument*›. **~ safe** no arriesgarse. **~ up to** halagar. ● *vi* jugar. **~ed out** agotado. ● *n* juego *m*; (*drama*) obra *f* de teatro. **~ on words** *n* juego *m* de palabras. **~ down** *vt* minimizar. **~ on** *vt* aprovecharse de. **~ up** *vi* (*fam*) causar problemas. **~-act** *vi* hacer la comedia. **~boy** /'pleɪbɔɪ/ *n* calavera *m*. **~er** *n* jugador *m*; (*mus*) músico *m*. **~ful** /'pleɪfl/ *a* juguetón. **~fully** *adv* jugando; (*jokingly*) en broma. **~ground** /'pleɪgraʊnd/ *n* parque *m* de juegos infantiles; (*in school*) campo *m* de recreo. **~-group** *n* jardín *m* de la infancia. **~ing** /'pleɪɪŋ/ *n* juego *m*. **~ing-card** *n* naipe *m*. **~ing-field** *n* campo *m* de deportes. **~mate** /'pleɪmeɪt/ *n* compañero *m* (de juego). **~-pen** *n* corralito *m*. **~thing** *n* juguete *m*. **~wright** /'pleɪraɪt/ *n* dramaturgo *m*
plc /pi:el'si:/ *abbr* (*public limited company*) S.A., sociedad *f* anónima
plea /pli:/ *n* súplica *f*; (*excuse*) excusa *f*; (*jurid*) defensa *f*
plead /pli:d/ *vt* (*jurid*) alegar; (*as excuse*) pretextar. ● *vi* suplicar; (*jurid*) abogar. **~ with** suplicar
pleasant /'pleznt/ *a* agradable
pleas|e /pli:z/ *int* por favor. ● *vt* agradar, dar gusto a. ● *vi* agradar; (*wish*) querer. **~e o.s.** hacer lo que quiera. **do as you ~e** haz lo que quieras. **~ed** *a* contento. **~ed with** satisfecho de. **~ing** *a* agradable
pleasur|e /'pleʒə(r)/ *n* placer *m*. **~able** *a* agradable
pleat /pli:t/ *n* pliegue *m*. ● *vt* hacer pliegues en
plebiscite /'plebɪsɪt/ *n* plebiscito *m*
plectrum /'plektrəm/ *n* plectro *m*
pledge /pledʒ/ *n* prenda *f*; (*promise*) promesa *f*. ● *vt* empeñar; (*promise*) prometer
plent|iful /'plentɪfl/ *a* abundante. **~y** /'plentɪ/ *n* abundancia *f*. **~y (of)** muchos (de)

pleurisy /ˈplʊərəsɪ/ *n* pleuresía *f*
pliable /ˈplaɪəbl/ *a* flexible
pliers /ˈplaɪəz/ *npl* alicates *mpl*
plight /plaɪt/ *n* situación *f* (difícil)
plimsolls /ˈplɪmsəlz/ *npl* zapatillas *fpl* de lona
plinth /plɪnθ/ *n* plinto *m*
plod /plɒd/ *vi* (*pt* **plodded**) caminar con paso pesado; (*work hard*) trabajar laboriosamente. **~der** *n* empollón *m*
plonk /plɒŋk/ *n* (*sl*) vino *m* peleón
plop /plɒp/ *n* paf *m*. ● *vi* (*pt* **plopped**) caerse con un paf
plot /plɒt/ *n* complot *m*; (*of novel etc*) argumento *m*; (*piece of land*) parcela *f*. ● *vt* (*pt* **plotted**) tramar; (*mark out*) trazar. ● *vi* conspirar
plough /plaʊ/ *n* arado *m*. ● *vt/i* arar. **~ through** avanzar laboriosamente por
ploy /plɔɪ/ *n* (*fam*) estratagema *f*, truco *m*
pluck /plʌk/ *vt* arrancar; depilarse ‹*eyebrows*›; desplumar ‹*bird*›; recoger ‹*flowers*›. **~ up courage** hacer de tripas corazón. ● *n* valor *m*. **~y** *a* (**-ier**, **-iest**) valiente
plug /plʌg/ *n* tapón *m*; (*elec*) enchufe *m*; (*auto*) bujía *f*. ● *vt* (*pt* **plugged**) tapar; (*advertise*, *fam*) dar publicidad a. **~ in** (*elec*) enchufar
plum /plʌm/ *n* ciruela *f*; (*tree*) ciruelo *m*
plumage /ˈplu:mɪdʒ/ *n* plumaje *m*
plumb /plʌm/ *a* vertical. ● *n* plomada *f*. ● *adv* verticalmente; (*exactly*) exactamente. ● *vt* sondar
plumb|er /ˈplʌmə(r)/ *n* fontanero *m*. **~ing** *n* instalación *f* sanitaria, instalación *f* de cañerías
plume /plu:m/ *n* pluma *f*
plum job /plʌmˈdʒɒb/ *n* (*fam*) puesto *m* estupendo
plummet /ˈplʌmɪt/ *n* plomada *f*. ● *vi* caer a plomo, caer en picado
plump /plʌmp/ *a* (**-er**, **-est**) rechoncho. ● *vt*. **~ for** elegir. **~ness** *n* gordura *f*
plum pudding /plʌmˈpʊdɪŋ/ *n* budín *m* de pasas
plunder /ˈplʌndə(r)/ *n* (*act*) saqueo *m*; (*goods*) botín *m*. ● *vt* saquear
plung|e /plʌndʒ/ *vt* hundir; (*in water*) sumergir. ● *vi* zambullirse; (*fall*) caer. ● *n* salto *m*. **~er** *n* (*for sink*) desatascador *m*; (*mec*) émbolo *m*. **~ing** *a* ‹*neckline*› bajo, escotado
plural /ˈplʊərəl/ *a* & *n* plural (*m*)
plus /plʌs/ *prep* más. ● *a* positivo. ● *n* signo *m* más; (*fig*) ventaja *f*. **five ~** más de cinco
plush /plʌʃ/ *n* felpa *f*. ● *a* de felpa, afelpado; (*fig*) lujoso. **~y** *a* lujoso
plutocrat /ˈplu:təkræt/ *n* plutócrata *m* & *f*
plutonium /plu:ˈtəʊnjəm/ *n* plutonio *m*
ply /plaɪ/ *vt* manejar ‹*tool*›; ejercer ‹*trade*›. **~ s.o. with drink** dar continuamente de beber a uno. **~wood** *n* contrachapado *m*
p.m. /pi:ˈem/ *abbr* (*post meridiem*) de la tarde
pneumatic /nju:ˈmætɪk/ *a* neumático
pneumonia /nju:ˈməʊnjə/ *n* pulmonía *f*
PO /pi:ˈəʊ/ *abbr* (*Post Office*) oficina *f* de correos
poach /pəʊtʃ/ *vt* escalfar ‹*egg*›; cocer ‹*fish etc*›; (*steal*) cazar en vedado. **~er** *n* cazador *m* furtivo
pocket /ˈpɒkɪt/ *n* bolsillo *m*; (*of air*, *resistance*) bolsa *f*. **be in ~** salir ganado. **be out of ~** salir perdiendo. ● *vt* poner en el bolsillo. **~-book** *n* (*notebook*) libro *m* de bolsillo; (*purse*, *Amer*) cartera *f*; (*handbag*, *Amer*) bolso *m*. **~-money** *n* dinero *m* para los gastos personales
pock-marked /ˈpɒkmɑ:kt/ *a* ‹*face*› picado de viruelas
pod /pɒd/ *n* vaina *f*
podgy /ˈpɒdʒɪ/ *a* (**-ier**, **-iest**) rechoncho
poem /ˈpəʊɪm/ *n* poesía *f*
poet /ˈpəʊɪt/ *n* poeta *m*. **~ess** *n* poetisa *f*. **~ic** /-ˈetɪk/ *a*, **~ical** / -ˈetɪkl/ *a* poético. **P~ Laureate** *n* poeta laureado. **~ry** /ˈpəʊɪtrɪ/ *n* poesía *f*
poignant /ˈpɔɪnjənt/ *a* conmovedor
point /pɔɪnt/ *n* punto *m*; (*sharp end*) punta *f*; (*significance*) lo importante; (*elec*) toma *f* de corriente. **good ~s** cualidades *fpl*. **to the ~** pertinente. **up to a ~** hasta cierto punto. **what is the ~?** ¿para qué?, ¿a qué fin? ● *vt* (*aim*) apuntar; (*show*) indicar. **~ out** señalar. ● *vi* señalar. **~-blank** *a* & *adv* a boca de jarro, a quemarropa. **~ed** /ˈpɔɪntɪd/ *a* puntiagudo; (*fig*) mordaz. **~er** /ˈpɔɪntə(r)/ *n* indicador *m*; (*dog*) perro *m* de muestra; (*clue*, *fam*) indicación *f*. **~less** /ˈpɔɪntlɪs/ *a* inútil

poise /pɔɪz/ *n* equilibrio *m*; (*elegance*) elegancia *f*; (*fig*) aplomo *m*. **~d** *a* en equilibrio. **~d for** listo para

poison /'pɔɪzn/ *n* veneno *m*. ● *vt* envenenar. **~ous** *a* venenoso; ‹*chemical etc*› tóxico

poke /pəʊk/ *vt* empujar; atizar ‹*fire*›. **~ fun at** burlarse de. **~ out** asomar ‹*head*›. ● *vi* hurgar; (*pry*) meterse. **~ about** fisgonear. ● *n* empuje *m*

poker[1] /'pəʊkə(r)/ *n* atizador *m*

poker[2] /'pəʊkə(r)/ (*cards*) póquer *m*. **~-face** *n* cara *f* inmutable

poky /'pəʊkɪ/ *a* (**-ier, -iest**) estrecho

Poland /'pəʊlənd/ *n* Polonia *f*

polar /'pəʊlə(r)/ *a* polar. **~ bear** *n* oso *m* blanco

polarize /'pəʊləraɪz/ *vt* polarizar

Pole /pəʊl/ polaco *n*

pole[1] /pəʊl/ *n* palo *m*; (*for flag*) asta *f*

pole[2] /pəʊl/ (*geog*) polo *m*. **~-star** *n* estrella *f* polar

polemic /pə'lemɪk/ *a* polémico. ● *n* polémica *f*

police /pə'li:s/ *n* policía *f*. ● *vt* vigilar. **~man** /pə'li:smən/ *n* (*pl* **-men**) policía *m*, guardia *m*. **~ record** *n* antecedentes *mpl* penales. **~ state** *n* estado *m* policíaco. **~ station** *n* comisaría *f*. **~woman** /-wʊmən/ *n* (*pl* **-women**) mujer *m* policía

policy[1] /'pɒlɪsɪ/ *n* política *f*

policy[2] /'pɒlɪsɪ/ (*insurance*) póliza *f* (de seguros)

polio(myelitis) /'pəʊlɪəʊ(maɪə'laɪtɪs)/ *n* polio(mielitis) *f*

polish /'pɒlɪʃ/ *n* (*for shoes*) betún *m*; (*for floor*) cera *f*; (*for nails*) esmalte *m* de uñas; (*shine*) brillo *m*; (*fig*) finura *f*. **nail ~** esmalte *m* de uñas. ● *vt* pulir; limpiar ‹*shoes*›; encerar ‹*floor*›. **~ off** despachar. **~ed** *a* pulido; ‹*manner*› refinado. **~er** *n* pulidor *m*; (*machine*) pulidora *f*

Polish /'pəʊlɪʃ/ *a* & *n* polaco (*m*)

polite /pə'laɪt/ *a* cortés. **~ly** *adv* cortésmente. **~ness** *n* cortesía *f*

politic|al /pə'lɪtɪkl/ *a* político. **~ian** /pɒlɪ'tɪʃn/ *n* político *m*. **~s** /'pɒlətɪks/ *n* política *f*

polka /'pɒlkə/ *n* polca *f*. **~ dots** *npl* diseño *m* de puntos

poll /pəʊl/ *n* elección *f*; (*survey*) encuesta *f*. ● *vt* obtener ‹*votes*›

pollen /'pɒlən/ *n* polen *m*

polling-booth /'pəulɪŋbu:ð/ *n* cabina *f* de votar

pollut|e /pə'lu:t/ *vt* contaminar. **~ion** /-ʃn/ *n* contaminación *f*

polo /'pəʊləʊ/ *n* polo *m*. **~-neck** *n* cuello *m* vuelto

poltergeist /'pɒltəgaɪst/ *n* duende *m*

polyester /pɒlɪ'estə(r)/ *n* poliéster *m*

polygam|ist /pə'lɪgəmɪst/ *n* polígamo *m*. **~ous** *a* polígamo. **~y** /pə'lɪgəmɪ/ *n* poligamia *f*

polyglot /'pɒlɪglɒt/ *a* & *n* políglota (*m* & *f*)

polygon /'pɒlɪgən/ *n* polígono *m*

polyp /'pɒlɪp/ *n* pólipo *m*

polystyrene /pɒlɪ'staɪri:n/ *n* poliestireno *m*

polytechnic /pɒlɪ'teknɪk/ *n* escuela *f* politécnica

polythene /'pɒlɪθi:n/ *n* polietileno *m*. **~ bag** *n* bolsa *f* de plástico

pomegranate /'pɒmɪgrænɪt/ *n* (*fruit*) granada *f*

pommel /'pʌml/ *n* pomo *m*

pomp /pɒmp/ *n* pompa *f*

pompon /'pɒmpɒn/ *n* pompón *m*

pompo|sity /pɒm'pɒsətɪ/ *n* pomposidad *f*. **~us** /'pɒmpəs/ *a* pomposo

poncho /'pɒntʃəʊ/ *n* (*pl* **-os**) poncho *m*

pond /pɒnd/ *n* charca *f*; (*artificial*) estanque *m*

ponder /'pɒndə(r)/ *vt* considerar. ● *vi* reflexionar. **~ous** /'pɒndərəs/ *a* pesado

pong /pɒŋ/ *n* (*sl*) hedor *m*. ● *vi* (*sl*) apestar

pontif|f /'pɒntɪf/ *n* pontífice *m*. **~ical** /-'tɪfɪkl/ *a* pontifical; (*fig*) dogmático. **~icate** /pɒn'tɪfɪkeɪt/ *vi* pontificar

pontoon /pɒn'tu:n/ *n* pontón *m*. **~ bridge** *n* puente *m* de pontones

pony /'pəʊnɪ/ *n* poni *m*. **~-tail** *n* cola *f* de caballo. **~-trekking** *n* excursionismo *m* en poni

poodle /'pu:dl/ *n* perro *m* de lanas, caniche *m*

pool[1] /pu:l/ *n* charca *f*; (*artificial*) estanque *m*. **(swimming-)~** *n* piscina *f*

pool[2] /pu:l/ (*common fund*) fondos *mpl* comunes; (*snooker*) billar *m* americano. ● *vt* aunar. **~s** *npl* quinielas *fpl*

poor /pʊə(r)/ *a* (**-er, -est**) pobre; (*not good*) malo. **be in ~ health** estar mal de salud. **~ly** *a* (*fam*) indispuesto. ● *adv* pobremente; (*badly*) mal

pop[1] /pɒp/ *n* ruido *m* seco; (*of bottle*) taponazo *m*. ● *vt* (*pt* **popped**) hacer reventar; (*put*) poner. ~ **in** *vi* entrar; (*visit*) pasar por. ~ **out** *vi* saltar; ‹*person*› salir un rato. ~ **up** *vi* surgir, aparecer
pop[2] /pɒp/ *a* (*popular*) pop *invar*. ● *n* (*fam*) música *f* pop. ~ **art** *n* arte *m* pop
popcorn /'pɒpkɔ:n/ *n* palomitas *fpl*
pope /pəʊp/ *n* papa *m*
popgun /'pɒpgʌn/ *n* pistola *f* de aire comprimido
poplar /'pɒplə(r)/ *n* chopo *m*
poplin /'pɒplɪn/ *n* popelina *f*
poppy /'pɒpɪ/ *n* amapola *f*
popular /'pɒpjʊlə(r)/ *a* popular. ~**ity** /-'lærətɪ/ *n* popularidad *f*. ~**ize** *vt* popularizar
populat|e /'pɒpjʊleɪt/ *vt* poblar. ~**ion** /-'leɪʃn/ *n* población *f*; (*number of inhabitants*) habitantes *mpl*
porcelain /'pɔ:səlɪn/ *n* porcelana *f*
porch /pɔ:tʃ/ *n* porche *m*
porcupine /'pɔ:kjʊpaɪn/ *n* puerco *m* espín
pore[1] /pɔ:(r)/ *n* poro *m*
pore[2] /pɔ:(r)/ *vi*. ~ **over** estudiar detenidamente
pork /pɔ:k/ *n* cerdo *m*
porn /pɔ:n/ *n* (*fam*) pornografía *f*. ~**ographic** /-ə'græfɪk/ *a* pornográfico. ~**ography** /pɔ:'nɒgrəfɪ/ *n* pornografía *f*
porous /'pɔ:rəs/ *a* poroso
porpoise /'pɔ:pəs/ *n* marsopa *f*
porridge /'pɒrɪdʒ/ *n* gachas *fpl* de avena
port[1] /pɔ:t/ *n* puerto *m*; (*porthole*) portilla *f*. ~ **of call** puerto de escala
port[2] /pɔ:t/ (*naut, left*) babor *m*. ● *a* de babor
port[3] /pɔ:t/ (*wine*) oporto *m*
portable /'pɔ:təbl/ *a* portátil
portal /'pɔ:tl/ *n* portal *m*
portent /'pɔ:tent/ *n* presagio *m*
porter /'pɔ:tə(r)/ *n* portero *m*; (*for luggage*) mozo *m*. ~**age** *n* porte *m*
portfolio /pɔ:t'fəʊljəʊ/ *n* (*pl* **-os**) cartera *f*
porthole /'pɔ:thəʊl/ *n* portilla *f*
portico /'pɔ:tɪkəʊ/ *n* (*pl* **-oes**) pórtico *m*
portion /'pɔ:ʃn/ *n* porción *f*. ● *vt* repartir
portly /'pɔ:tlɪ/ *a* (**-ier**, **-iest**) corpulento
portrait /'pɔ:trɪt/ *n* retrato *m*
portray /pɔ:'treɪ/ *vt* retratar; (*represent*) representar. ~**al** *n* retrato *m*
Portug|al /'pɔ:tjʊgl/ *n* Portugal *m*. ~**uese** /-'gi:z/ *a* & *n* portugués (*m*)
pose /pəʊz/ *n* postura *f*. ● *vt* colocar; hacer ‹*question*›; plantear ‹*problem*›. ● *vi* posar. ~ **as** hacerse pasar por. ~**r** /'pəʊzə(r)/ *n* pregunta *f* difícil
posh /pɒʃ/ *a* (*sl*) elegante
position /pə'zɪʃn/ *n* posición *f*; (*job*) puesto *m*; (*status*) rango *m*. ● *vt* colocar
positive /'pɒzətɪv/ *a* positivo; (*real*) verdadero; (*certain*) seguro. ● *n* (*foto*) positiva *f*. ~**ly** *adv* positivamente
possess /pə'zes/ *vt* poseer. ~**ion** /pə'zeʃn/ *n* posesión *f*. **take** ~**ion of** tomar posesión de. ~**ions** *npl* posesiones *fpl*; (*jurid*) bienes *mpl*. ~**ive** /pə'zesɪv/ *a* posesivo. ~**or** *n* poseedor *m*
possib|ility /pɒsə'bɪlətɪ/ *n* posibilidad *f*. ~**le** /'pɒsəbl/ *a* posible. ~**ly** *adv* posiblemente
post[1] /pəʊst/ *n* (*pole*) poste *m*. ● *vt* fijar ‹*notice*›
post[2] /pəʊst/ (*place*) puesto *m*
post[3] /pəʊst/ (*mail*) correo *m*. ● *vt* echar ‹*letter*›. **keep s.o.** ~**ed** tener a uno al corriente
post... /pəʊst/ *pref* post
post: ~**age** /'pəʊstɪdʒ/ *n* franqueo *m*. ~**al** /'pəʊstl/ *a* postal. ~**al order** *n* giro *m* postal. ~**-box** *n* buzón *m*. ~**card** /'pəʊstkɑ:d/ *n* (tarjeta *f*) postal *f*. ~**-code** *n* código *m* postal
post-date /pəʊst'deɪt/ *vt* poner fecha posterior a
poster /'pəʊstə(r)/ *n* cartel *m*
poste restante /pəʊst'restɑ:nt/ *n* lista *f* de correos
posteri|or /pɒ'stɪərɪə(r)/ *a* posterior. ● *n* trasero *m*. ~**ty** /pɒs'terətɪ/ *n* posteridad *f*
posthumous /'pɒstjʊməs/ *a* póstumo. ~**ly** *adv* después de la muerte
post: ~**man** /'pəʊstmən/ *n* (*pl* **-men**) cartero *m*. ~**mark** /'pəʊstmɑ:k/ *n* matasellos *m invar*. ~**master** /'pəʊstmɑ:stə(r) *n* administrador *m* de correos. ~**mistress** /'pəʊstmɪstrɪs/ *n* administradora *f* de correos
post-mortem /'pəʊstmɔ:təm/ *n* autopsia *f*
Post Office /'pəʊstɒfɪs/ *n* oficina *f* de correos, correos *mpl*

postpone /pəʊst'pəʊn/ *vt* aplazar. **~ment** *n* aplazamiento *m*
postscript /'pəʊstskrɪpt/ *n* posdata *f*
postulant /'pɒstjʊlənt/ *n* postulante *m & f*
postulate /'pɒstjʊleɪt/ *vt* postular
posture /'pɒstʃə(r)/ *n* postura *f*. ● *vi* adoptar una postura
posy /'pəʊzɪ/ *n* ramillete *m*
pot /pɒt/ *n* (*for cooking*) olla *f*; (*for flowers*) tiesto *m*; (*marijuana, sl*) mariguana *f*. **go to ~** (*sl*) echarse a perder. ● *vt* (*pt* **potted**) poner en tiesto
potassium /pə'tæsjəm/ *n* potasio *m*
potato /pə'teɪtəʊ/ *n* (*pl* **-oes**) patata *f*, papa *f* (*LAm*)
pot: **~-belly** *n* barriga *f*. **~-boiler** *n* obra *f* literaria escrita sólo para ganar dinero
poten|cy /'pəʊtənsɪ/ *n* potencia *f*. **~t** /'pəʊtnt/ *a* potente; ‹*drink*› fuerte
potentate /'pəʊtənteɪt/ *n* potentado *m*
potential /pəʊ'tenʃl/ *a & n* potencial (*m*). **~ity** /-ʃɪ'ælətɪ/ *n* potencialidad *f*. **~ly** *adv* potencialmente
pot-hole /'pɒthəʊl/ *n* caverna *f*; (*in road*) bache *m*. **~r** *n* espeleólogo *m*
potion /'pəʊʃn/ *n* poción *f*
pot: **~ luck** *n* lo que haya. **~-shot** *n* tiro *m* al azar. **~ted** /'pɒtɪd/ *see* **pot**. ● *a* ‹*food*› en conserva
potter[1] /'pɒtə(r)/ *n* alfarero *m*
potter[2] /'pɒtə(r)/ *vi* hacer pequeños trabajos agradables, no hacer nada de particular
pottery /'pɒtərɪ/ *n* cerámica *f*
potty /'pɒtɪ/ *a* (**-ier**, **-iest**) (*sl*) chiflado. ● *n* orinal *m*
pouch /paʊtʃ/ *n* bolsa *f* pequeña
pouffe /pu:f/ *n* (*stool*) taburete *m*
poulterer /'pəʊltərə(r)/ *n* pollero *m*
poultice /'pəʊltɪs/ *n* cataplasma *f*
poultry /'pəʊltrɪ/ *n* aves *fpl* de corral
pounce /paʊns/ *vi* saltar, atacar de repente. ● *n* salto *m*, ataque *m* repentino
pound[1] /paʊnd/ *n* (*weight*) libra *f* (= *454g*); (*money*) libra *f* (esterlina)
pound[2] /paʊnd/ *n* (*for cars*) depósito *m*
pound[3] /paʊnd/ *vt* (*crush*) machacar; (*bombard*) bombardear. ● *vi* golpear; ‹*heart*› palpitar; (*walk*) ir con pasos pesados
pour /pɔ:(r)/ *vt* verter. **~ out** servir ‹*drink*›. ● *vi* fluir; (*rain*) llover a cántaros. **~ in** ‹*people*› entrar en tropel. **~ing rain** *n* lluvia *f* torrencial. **~ out** ‹*people*› salir en tropel
pout /paʊt/ *vi* hacer pucheros. ● *n* puchero *m*, mala cara *f*
poverty /'pɒvətɪ/ *n* pobreza *f*
powder /'paʊdə(r)/ *n* polvo *m*; (*cosmetic*) polvos *mpl*. ● *vt* polvorear; (*pulverize*) pulverizar. **~ one's face** ponerse polvos en la cara. **~ed** *a* en polvo. **~y** *a* polvoriento
power /'paʊə(r)/ *n* poder *m*; (*elec*) corriente *f*; (*energy*) energía *f*; (*nation*) potencia *f*. **~ cut** *n* apagón *m*. **~ed** *a* con motor. **~ed by** impulsado por. **~ful** *a* poderoso. **~less** *a* impotente. **~-station** *n* central *f* eléctrica
practicable /'præktɪkəbl/ *a* practicable
practical /'præktɪkl/ *a* práctico. **~ joke** *n* broma *f* pesada. **~ly** *adv* prácticamente
practi|ce /'præktɪs/ *n* práctica *f*; (*custom*) costumbre *f*; (*exercise*) ejercicio *m*; (*sport*) entrenamiento *m*; (*clients*) clientela *f*. **be in ~ce** ‹*doctor, lawyer*› ejercer. **be out of ~ce** no estar en forma. **in ~ce** (*in fact*) en la práctica; (*on form*) en forma. **~se** /'præktɪs/ *vt* hacer ejercicios en; (*put into practice*) poner en práctica; (*sport*) entrenarse en; ejercer ‹*profession*›. ● *vi* ejercitarse; ‹*professional*› ejercer. —**~sed** *a* experto
practitioner /præk'tɪʃənə(r)/ *n* profesional *m & f*. **general ~** médico *m* de cabecera. **medical ~** médico *m*
pragmatic /præg'mætɪk/ *a* pragmático
prairie /'preərɪ/ *n* pradera *f*
praise /preɪz/ *vt* alabar. ● *n* alabanza *f*. **~worthy** *a* loable
pram /præm/ *n* cochecito *m* de niño
prance /prɑ:ns/ *vi* ‹*horse*› hacer cabriolas; ‹*person*› pavonearse
prank /præŋk/ *n* travesura *f*
prattle /prætl/ *vi* parlotear. ● *n* parloteo *m*
prawn /prɔ:n/ *n* gamba *f*
pray /preɪ/ *vi* rezar. **~er** /preə(r)/ *n* oración *f*. **~ for** rogar
pre.. /pri:/ *pref* pre...
preach /pri:tʃ/ *vt/i* predicar. **~er** *n* predicador *m*
preamble /pri:'æmbl/ *n* preámbulo *m*

pre-arrange /pri:ə'reɪndʒ/ *vt* arreglar de antemano. **~ment** *n* arreglo *m* previo

precarious /prɪ'keərɪəs/ *a* precario. **~ly** *adv* precariamente

precaution /prɪ'kɔ:ʃn/ *n* precaución *f*. **~ary** *a* de precaución; (*preventive*) preventivo

precede /prɪ'si:d/ *vt* preceder

preceden|ce /'presɪdəns/ *n* precedencia *f*. **~t** /'presɪdənt/ *n* precedente *m*

preceding /prɪ'si:dɪŋ/ *a* precedente

precept /'pri:sept/ *n* precepto *m*

precinct /'pri:sɪŋkt/ *n* recinto *m*. **pedestrian ~** zona *f* peatonal. **~s** *npl* contornos *mpl*

precious /'preʃəs/ *a* precioso. ● *adv* (*fam*) muy

precipice /'presɪpɪs/ *n* precipicio *m*

precipitat|e /prɪ'sɪpɪteɪt/ *vt* precipitar. /prɪ'sɪpɪtət/ *n* precipitado *m*. ● *a* precipitado. **~ion** /-'teɪʃn/ *n* precipitación *f*

precipitous /prɪ'sɪpɪtəs/ *a* escarpado

précis /'preɪsi:/ *n* (*pl* **précis** /-si:z/) resumen *m*

precis|e /prɪ'saɪs/ *a* preciso; (*careful*) meticuloso. **~ely** *adv* precisamente. **~ion** /-'sɪʒn/ *n* precisión *f*

preclude /prɪ'klu:d/ *vt* (*prevent*) impedir; (*exclude*) excluir

precocious /prɪ'kəʊʃəs/ *a* precoz. **~ly** *adv* precozmente

preconce|ived /pri:kən'si:vd/ *a* preconcebido. **~ption** /-'sepʃn/ *n* preconcepción *f*

precursor /pri:'kɜ:sə(r)/ *n* precursor *m*

predator /'predətə(r)/ *n* animal *m* de rapiña. **~y** *a* de rapiña

predecessor /'pri:dɪsesə(r)/ *n* predecesor *m*, antecesor *m*

predestin|ation /prɪdestɪ'neɪʃn/ *n* predestinación *f*. **~e** /pri:'destɪn/ *vt* predestinar

predicament /prɪ'dɪkəmənt/ *n* apuro *m*

predicat|e /'predɪkət/ *n* predicado *m*. **~ive** /prɪ'dɪkətɪv/ *a* predicativo

predict /prɪ'dɪkt/ *vt* predecir. **~ion** /-ʃn/ *n* predicción *f*

predilection /pri:dɪ'lekʃn/ *n* predilección *f*

predispose /pri:dɪ'spəʊz/ *vt* predisponer

predomina|nt /prɪ'dɒmɪnənt/ *a* predominante. **~te** /prɪ'dɒmɪneɪt/ *vi* predominar

pre-eminent /pri:'emɪnənt/ *a* preeminente

pre-empt /pri:'empt/ *vt* adquirir por adelantado, adelantarse a

preen /pri:n/ *vt* limpiar, arreglar. **~ o.s.** atildarse

prefab /'pri:fæb/ *n* (*fam*) casa *f* prefabricada. **~ricated** /-'fæbrɪkeɪtɪd/ *a* prefabricado

preface /'prefəs/ *n* prólogo *m*

prefect /'pri:fekt/ *n* monitor *m*; (*official*) prefecto *m*

prefer /prɪ'fɜ:(r)/ *vt* (*pt* **preferred**) preferir. **~able** /'prefrəbl/ *a* preferible. **~ence** /'prefrəns/ *n* preferencia *f*. **~ential** /-ə'renʃl/ *a* preferente

prefix /'pri:fɪks/ *n* (*pl* **-ixes**) prefijo *m*

pregnan|cy /'pregnənsɪ/ *n* embarazo *m*. **~t** /'pregnənt/ *a* embarazada

prehistoric /pri:hɪ'stɒrɪk/ *a* prehistórico

prejudge /pri:'dʒʌdʒ/ *vt* prejuzgar

prejudice /'predʒʊdɪs/ *n* prejuicio *m*; (*harm*) perjuicio *m*. ● *vt* predisponer; (*harm*) perjudicar. **~d** *a* parcial

prelate /'prelət/ *n* prelado *m*

preliminar|ies /prɪ'lɪmɪnərɪz/ *npl* preliminares *mpl*. **~y** /prɪ'lɪmɪnərɪ/ *a* preliminar

prelude /'prelju:d/ *n* preludio *m*

pre-marital /pri:'mærɪtl/ *a* prematrimonial

premature /'premətjʊə(r)/ *a* prematuro

premeditated /pri:'medɪteɪtɪd/ *a* premeditado

premier /'premɪə(r)/ *a* primero. ● *n* (*pol*) primer ministro

première /'premɪə(r)/ *n* estreno *m*

premises /'premɪsɪz/ *npl* local *m*. **on the ~** en el local

premiss /'premɪs/ *n* premisa *f*

premium /'pri:mɪəm/ *n* premio *m*. **at a ~** muy solicitado

premonition /pri:mə'nɪʃn/ *n* presentimiento *m*

preoccup|ation /pri:ɒkjʊ'peɪʃn/ *n* preocupación *f*. **~ied** /-'ɒkjʊpaɪd/ *a* preocupado

prep /prep/ *n* deberes *mpl*

preparation /prepə'reɪʃn/ *n* preparación *f*. **~s** *npl* preparativos *mpl*

preparatory /prɪ'pærətrɪ/ *a* preparatorio. ~ **school** *n* escuela *f* primaria privada

prepare /prɪ'peə(r)/ *vt* preparar. ● *vi* prepararse. ~**d to** dispuesto a

prepay /pri:'peɪ/ *vt* (*pt* **-paid**) pagar por adelantado

preponderance /prɪ'pɒndərəns/ *n* preponderancia *f*

preposition /prepə'zɪʃn/ *n* preposición *f*

prepossessing /pri:pə'zesɪŋ/ *a* atractivo

preposterous /prɪ'pɒstərəs/ *a* absurdo

prep school /'prepsku:l/ *n* escuela *f* primaria privada

prerequisite /pri:'rekwɪzɪt/ *n* requisito *m* previo

prerogative /prɪ'rɒgətɪv/ *n* prerrogativa *f*

Presbyterian /prezbɪ'tɪərɪən/ *a* & *n* presbiteriano (*m*)

prescri|be /prɪ'skraɪb/ *vt* prescribir; (*med*) recetar. ~**ption** /-'ɪpʃn/ *n* prescripción *f*; (*med*) receta *f*

presence /'prezns/ *n* presencia *f*; (*attendance*) asistencia *f*. ~ **of mind** presencia *f* de ánimo

present[1] /'preznt/ *a* & *n* presente (*m* & *f*). **at** ~ actualmente. **for the** ~ por ahora

present[2] /'preznt/ *n* (*gift*) regalo *m*

present[3] /prɪ'zent/ *vt* presentar; (*give*) obsequiar. ~ **s.o. with** obsequiar a uno con. ~**able** *a* presentable. ~**ation** /prezn'teɪʃn/ *n* presentación *f*; (*ceremony*) ceremonia *f* de entrega

presently /'prezntlɪ/ *adv* dentro de poco

preserv|ation /prezə'veɪʃn/ *n* conservación *f*. ~**ative** /prɪ'zɜ:vətɪv/ *n* preservativo *m*. ~**e** /prɪ'zɜ:v/ *vt* conservar; (*maintain*) mantener; (*culin*) poner en conserva. ● *n* coto *m*; (*jam*) confitura *f*

preside /prɪ'zaɪd/ *vi* presidir. ~ **over** presidir

presiden|cy /'prezɪdənsɪ/ *n* presidencia *f*. ~**t** /'prezɪdənt/ *n* presidente *m*. ~**tial** /-'denʃl/ *a* presidencial

press /pres/ *vt* apretar; exprimir ‹*fruit etc*›; (*insist on*) insistir en; (*iron*) planchar. **be** ~**ed for** tener poco. ● *vi* apretar; ‹*time*› apremiar; (*fig*) urgir. ~ **on** seguir adelante. ● *n* presión *f*; (*mec, newspapers*) prensa *f*; (*printing*) imprenta *f*. ~ **conference** *n* rueda *f* de prensa. ~ **cutting** *n* recorte *m* de periódico. ~**ing** /'presɪŋ/ *a* urgente. ~**-stud** *n* automático *m*. ~**-up** *n* plancha *f*

pressure /'preʃə(r)/ *n* presión *f*. ● *vt* hacer presión sobre. ~**-cooker** *n* olla *f* a presión. ~ **group** *n* grupo *m* de presión

pressurize /'preʃəraɪz/ *vt* hacer presión sobre

prestig|e /pre'sti:ʒ/ *n* prestigio *m*. ~**ious** /pre'stɪdʒəs/ *a* prestigioso

presum|ably /prɪ'zju:məblɪ/ *adv* presumiblemente, probablemente. ~**e** /prɪ'zju:m/ *vt* presumir. ~**e (up)on** *vi* abusar de. ~**ption** /-'zʌmpʃn/ *n* presunción *f*. ~**ptuous** /prɪ'zʌmptʃʊəs/ *a* presuntuoso

presuppose /pri:sə'pəʊz/ *vt* presuponer

preten|ce /prɪ'tens/ *n* fingimiento *m*; (*claim*) pretensión *f*; (*pretext*) pretexto *m*. ~**d** /prɪ'tend/ *vt*/*i* fingir. ~**d to** (*lay claim*) pretender

pretentious /prɪ'tenʃəs/ *a* pretencioso

pretext /'pri:tekst/ *n* pretexto *m*

pretty /'prɪtɪ/ *a* (**-ier**, **-iest**) *adv* bonito, lindo (*esp LAm*); ‹*person*› guapo

prevail /prɪ'veɪl/ *vi* predominar; (*win*) prevalecer. ~ **on** persuadir

prevalen|ce /'prevələns/ *n* costumbre *f*. ~**t** /'prevələnt/ *a* extendido

prevaricate /prɪ'værɪkeɪt/ *vi* despistar

prevent /prɪ'vent/ *vt* impedir. ~**able** *a* evitable. ~**ion** /-ʃn/ *n* prevención *f*. ~**ive** *a* preventivo

preview /'pri:vju:/ *n* preestreno *m*, avance *m*

previous /'pri:vɪəs/ *a* anterior. ~ **to** antes de. ~**ly** *adv* anteriormente, antes

pre-war /pri:'wɔ:(r)/ *a* de antes de la guerra

prey /preɪ/ *n* presa *f*; (*fig*) víctima *f*. **bird of** ~ *n* ave *f* de rapiña. ● *vi*. ~ **on** alimentarse de; (*worry*) atormentar

price /praɪs/ *n* precio *m*. ● *vt* fijar el precio de. ~**less** *a* inapreciable; (*amusing, fam*) muy divertido. ~**y** *a* (*fam*) caro

prick /prɪk/ *vt*/*i* pinchar. ~ **up one's ears** aguzar las orejas. ● *n* pinchazo *m*

prickl|e /ˈprɪkl/ *n* (*bot*) espina f; (*of animal*) púa *f*; (*sensation*) picor *m*. **~y** *a* espinoso; ‹*animal*› lleno de púas; ‹*person*› quisquilloso

pride /praɪd/ *n* orgullo *m*. **~ of place** *n* puesto *m* de honor. ● *vr*. **~ o.s. on** enorgullecerse de

priest /priːst/ *n* sacerdote *m*. **~hood** *n* sacerdocio *m*. **~ly** *a* sacerdotal

prig /prɪg/ *n* mojigato *m*. **~gish** *a* mojigato

prim /prɪm/ *a* (**primmer**, **primmest**) estirado; (*prudish*) gazmoño

primarily /ˈpraɪmərɪlɪ/ *adv* en primer lugar

primary /ˈpraɪmərɪ/ *a* primario; (*chief*) principal. **~ school** *n* escuela *f* primaria

prime[1] /praɪm/ *vt* cebar ‹*gun*›; (*prepare*) preparar; aprestar ‹*surface*›

prime[2] /praɪm/ *a* principal; (*first rate*) excelente. **~ minister** *n* primer ministro *m*. ● *n*. **be in one's ~** estar en la flor de la vida

primer[1] /ˈpraɪmə(r)/ *n* (*of paint*) primera mano *f*

primer[2] /ˈpraɪmə(r)/ (*book*) silabario *m*

primeval /praɪˈmiːvl/ *a* primitivo

primitive /ˈprɪmɪtɪv/ *a* primitivo

primrose /ˈprɪmrəʊz/ *n* primavera *f*

prince /prɪns/ *n* príncipe *m*. **~ly** *a* principesco. **~ss** /prɪnˈses/ *n* princesa *f*

principal /ˈprɪnsəpl/ *a* principal. ● *n* (*of school etc*) director *m*

principality /prɪnsɪˈpælətɪ/ *n* principado *m*

principally /ˈprɪnsɪpəlɪ/ *adv* principalmente

principle /ˈprɪnsəpl/ *n* principio *m*. **in ~** en principio. **on ~** por principio

print /prɪnt/ *vt* imprimir; (*write in capitals*) escribir con letras de molde. ● *n* (*of finger, foot*) huella *f*; (*letters*) caracteres *mpl*; (*of design*) estampado *m*; (*picture*) grabado *m*; (*photo*) copia *f*. **in ~** ‹*book*› disponible. **out of ~** agotado. **~ed matter** *n* impresos *mpl*. **~er** /ˈprɪntə(r)/ *n* impresor *m*; (*machine*) impresora *f*. **~ing** *n* tipografía *f*. **~-out** *n* listado *m*

prior /ˈpraɪə(r)/ *n* prior *m*. ● *a* anterior. **~ to** antes de

priority /praɪˈɒrətɪ/ *n* prioridad *f*

priory /ˈpraɪərɪ/ *n* priorato *m*

prise /praɪz/ *vt* apalancar. **~ open** abrir por fuerza

prism /ˈprɪzəm/ *n* prisma *m*

prison /ˈprɪzn/ *n* cárcel *m*. **~er** *n* prisionero *m*; (*in prison*) preso *m*; (*under arrest*) detenido *m*. **~ officer** *n* carcelero *m*

pristine /ˈprɪstiːn/ *a* prístino

privacy /ˈprɪvəsɪ/ *n* intimidad *f*; (*private life*) vida *f* privada. **in ~** en la intimidad

private /ˈpraɪvət/ *a* privado; (*confidential*) personal; ‹*lessons, house*› particular; ‹*ceremony*› en la intimidad. ● *n* soldado *m* raso. **in ~** en privado; (*secretly*) en secreto. **~ eye** *n* (*fam*) detective *m* privado. **~ly** *adv* en privado; (*inwardly*) interiormente

privation /praɪˈveɪʃn/ *n* privación *f*

privet /ˈprɪvɪt/ *n* alheña *f*

privilege /ˈprɪvəlɪdʒ/ *n* privilegio *m*. **~d** *a* privilegiado

privy /ˈprɪvɪ/ *a*. **~ to** al corriente de

prize /praɪz/ *n* premio *m*. ● *a* ‹*idiot etc*› de remate. ● *vt* estimar. **~-fighter** *n* boxeador *m* profesional. **~-giving** *n* reparto *m* de premios. **~winner** *n* premiado *m*

pro /prəʊ/ *n*. **~s and cons** el pro *m* y el contra *m*

probab|ility /prɒbəˈbɪlətɪ/ *n* probabilidad *f*. **~le** /ˈprɒbəbl/ *a* probable. **~ly** *adv* probablemente

probation /prəˈbeɪʃn/ *n* prueba *f*; (*jurid*) libertad *f* condicional. **~ary** *a* de prueba

probe /prəʊb/ *n* sonda *f*; (*fig*) encuesta *f*. ● *vt* sondar. ● *vi*. **~ into** investigar

problem /ˈprɒbləm/ *n* problema *m*. ● *a* difícil. **~atic** /-ˈmætɪk/ *a* problemático

procedure /prəˈsiːdʒə(r)/ *n* procedimiento *m*

proceed /prəˈsiːd/ *vi* proceder. **~ing** *n* procedimiento *m*. **~ings** /prəˈsiːdɪŋz/ *npl* (*report*) actas *fpl*; (*jurid*) proceso *m*

proceeds /ˈprəʊsiːdz/ *npl* ganancias *fpl*

process /ˈprəʊses/ *n* proceso *m*. **in ~ of** en vías de. **in the ~ of time** con el tiempo. ● *vt* tratar; revelar ‹*photo*›. **~ion** /prəˈseʃn/ *n* desfile *m*

procla|im /prəˈkleɪm/ *vt* proclamar. **~mation** /prɒkləˈmeɪʃn/ *n* proclamación *f*

procrastinate /prəʊ'kræstɪneɪt/ *vi* aplazar, demorar, diferir
procreation /prəʊkrɪ'eɪʃn/ *n* procreación *f*
procure /prə'kjʊə(r)/ *vt* obtener
prod /prɒd/ *vt* (*pt* **prodded**) empujar; (*with elbow*) dar un codazo a. ● *vi* dar con el dedo. ● *n* empuje *m*; (*with elbow*) codazo *m*
prodigal /'prɒdɪgl/ *a* pródigo
prodigious /prə'dɪdʒəs/ *a* prodigioso
prodigy /'prɒdɪdʒɪ/ *n* prodigio *m*
produce /prə'dju:s/ *vt* (*show*) presentar; (*bring out*) sacar; poner en escena ‹*play*›; (*cause*) causar; (*manufacture*) producir. /'prɒdju:s/ *n* productos *mpl*. **~er** /prə'dju:sə(r)/ *n* productor *m*; (*in theatre*) director *m*
product /'prɒdʌkt/ *n* producto *m*. **~ion** /prə'dʌkʃn/ *n* producción *f*; (*of play*) representación *f*
productiv|e /prə'dʌktɪv/ *a* productivo. **~ity** /prɒdʌk'tɪvətɪ/ *n* productividad *f*
profan|e /prə'feɪn/ *a* profano; (*blasphemous*) blasfemo. **~ity** /-'fænətɪ/ *n* profanidad *f*
profess /prə'fes/ *vt* profesar; (*pretend*) pretender
profession /prə'feʃn/ *n* profesión *f*. **~al** *a* & *n* profesional (*m* & *f*)
professor /prə'fesə(r)/ *n* catedrático *m*; (*Amer*) profesor *m*
proffer /'prɒfə(r)/ *vt* ofrecer
proficien|cy /prə'fɪʃənsɪ/ *n* competencia *f*. **~t** /prə'fɪʃnt/ *a* competente
profile /'prəʊfaɪl/ *n* perfil *m*
profit /'prɒfɪt/ *n* (*com*) ganancia *f*; (*fig*) provecho *m*. ● *vi*. **~ from** sacar provecho de. **~able** *a* provechoso
profound /prə'faʊnd/ *a* profundo. **~ly** *adv* profundamente
profus|e /prə'fju:s/ *a* profuso. **~ely** *adv* profusamente. **~ion** /-ʒn/ *n* profusión *f*
progeny /'prɒdʒənɪ/ *n* progenie *f*
prognosis /prɒ'gnəʊsɪs/ *n* (*pl* **-oses**) pronóstico *m*
program(|me) /'prəʊgræm/ *n* programa *m*. ● *vt* (*pt* **programmed**) programar. **~mer** *n* programador *m*
progress /'prəʊgres/ *n* progreso *m*, progresos *mpl*; (*development*) desarrollo *m*. **in ~** en curso. /prə'gres/ *vi* hacer progresos; (*develop*) desarrollarse. **~ion** /prə'greʃn/ *n* progresión *f*
progressive /prə'gresɪv/ *a* progresivo; (*reforming*) progresista. **~ly** *adv* progresivamente
prohibit /prə'hɪbɪt/ *vt* prohibir. **~ive** /-bətɪv/ *a* prohibitivo
project /prə'dʒekt/ *vt* proyectar. ● *vi* (*stick out*) sobresalir. /'prɒdʒekt/ *n* proyecto *m*
projectile /prə'dʒektaɪl/ *n* proyectil *m*
projector /prə'dʒektə(r)/ *n* proyector *m*
proletari|an /prəʊlɪ'teərɪən/ *a* & *n* proletario (*m*). **~at** /prəʊlɪ'teərɪət/ *n* proletariado *m*
prolif|erate /prə'lɪfəreɪt/ *vi* proliferar. **~eration** /-'reɪʃn/ *n* proliferación *f*. **~ic** /prə'lɪfɪk/ *a* prolífico
prologue /'prəʊlɒg/ *n* prólogo *m*
prolong /prə'lɒŋ/ *vt* prolongar
promenade /prɒmə'nɑ:d/ *n* paseo *m*; (*along beach*) paseo *m* marítimo. ● *vt* pasear. ● *vi* pasearse. **~ concert** *n* concierto *m* (que forma parte de un festival de música clásica en Londres, en que no todo el público tiene asientos)
prominen|ce /'prɒmɪnəns/ *n* prominencia *f*; (*fig*) importancia *f*. **~t** /'prɒmɪnənt/ *a* prominente; (*important*) importante; (*conspicuous*) conspicuo
promiscu|ity /prɒmɪ'skju:ətɪ/ *n* libertinaje *m*. **~ous** /prə'mɪskjʊəs/ *a* libertino
promis|e /'prɒmɪs/ *n* promesa *f*. ● *vt*/*i* prometer. **~ing** *a* prometedor; ‹*person*› que promete
promontory /'prɒməntrɪ/ *n* promontorio *m*
promot|e /prə'məʊt/ *vt* promover. **~ion** /-'məʊʃn/ *n* promoción *f*
prompt /prɒmpt/ *a* pronto; (*punctual*) puntual. ● *adv* en punto. ● *vt* incitar; apuntar ‹*actor*›. **~er** *n* apuntador *m*. **~ly** *adv* puntualmente. **~ness** *n* prontitud *f*
promulgate /'prɒməlgeɪt/ *vt* promulgar
prone /prəʊn/ *a* echado boca abajo. **~ to** propenso a
prong /prɒŋ/ *n* (*of fork*) diente *m*
pronoun /'prəʊnaʊn/ *n* pronombre *m*
pronounc|e /prə'naʊns/ *vt* pronunciar; (*declare*) declarar. **~ement** *n* declaración *f*. **~ed**

/prə'naʊnst/ *a* pronunciado; (*noticeable*) marcado
pronunciation /prənʌnsɪ'eɪʃn/ *n* pronunciación *f*
proof /pru:f/ *n* prueba *f*; (*of alcohol*) graduación *f* normal. ● *a*. ~ **against** a prueba de. **~-reading** *n* corrección *f* de pruebas
prop[1] /prɒp/ *n* puntal *m*; (*fig*) apoyo *m*. ● *vt* (*pt* **propped**) apoyar. ~ **against** (*lean*) apoyar en
prop[2] /prɒp/ (*in theatre, fam*) accesorio *m*
propaganda /prɒpə'gændə/ *n* propaganda *f*
propagat|e /'prɒpəgeɪt/ *vt* propagar. ● *vi* propagarse. **~ion** /-'geɪʃn/ *n* propagación *f*
propel /prə'pel/ *vt* (*pt* **propelled**) propulsar. **~ler** /prə'pelə(r)/ *n* hélice *f*
propensity /prə'pensətɪ/ *n* propensión *f*
proper /'prɒpə(r)/ *a* correcto; (*suitable*) apropiado; (*gram*) propio; (*real, fam*) verdadero. **~ly** *adv* correctamente
property /'prɒpətɪ/ *n* propiedad *f*; (*things owned*) bienes *mpl*. ● *a* inmobiliario
prophe|cy /'prɒfəsɪ/ *n* profecía *f*. **~sy** /'prɒfɪsaɪ/ *vt/i* profetizar. **~t** /'prɒfɪt/ *n* profeta *m*. **~tic** /prə'fetɪk/ *a* profético
propitious /prə'pɪʃəs/ *a* propicio
proportion /prə'pɔ:ʃn/ *n* proporción *f*. **~al** *a*, **~ate** *a* proporcional
propos|al /prə'pəʊzl/ *n* propuesta *f*. **~al of marriage** oferta *f* de matrimonio. **~e** /prə'pəʊz/ *vt* proponer. ● *vi* hacer una oferta de matrimonio
proposition /prɒpə'zɪʃn/ *n* proposición *f*; (*project, fam*) asunto *m*
propound /prə'paʊnd/ *vt* proponer
proprietor /prə'praɪətə(r)/ *n* propietario *m*
propriety /prə'praɪətɪ/ *n* decoro *m*
propulsion /prə'pʌlʃn/ *n* propulsión *f*
prosaic /prə'zeɪk/ *a* prosaico
proscribe /prə'skraɪb/ *vt* proscribir
prose /prəʊz/ *n* prosa *f*
prosecut|e /'prɒsɪkju:t/ *vt* procesar; (*carry on*) proseguir. **~ion** /-'kju:ʃn/ *n* proceso *m*. **~or** *n* acusador *m*. **Public P~or** fiscal *m*
prospect /'prɒspekt/ *n* vista *f*; (*expectation*) perspectiva *f*. /prə'spekt/ *vi* prospectar
prospective /prə'spektɪv/ *a* probable; (*future*) futuro
prospector /prə'spektə(r)/ *n* prospector *m*, explorador *m*
prospectus /prə'spektəs/ *n* prospecto *m*
prosper /'prɒspə(r)/ *vi* prosperar. **~ity** /-'sperətɪ/ *n* prosperidad *f*. **~ous** /'prɒspərəs/ *a* próspero
prostitut|e /'prɒstɪtju:t/ *n* prostituta *f*. **~ion** /-'tju:ʃn/ *n* prostitución *f*
prostrate /'prɒstreɪt/ *a* echado boca abajo; (*fig*) postrado
protagonist /prə'tægənɪst/ *n* protagonista *m & f*
protect /prə'tekt/ *vt* proteger. **~ion** /-ʃn/ *n* protección *f*. **~ive** /prə'tektɪv/ *a* protector. **~or** *n* protector *m*
protégé /'prɒtɪʒeɪ/ *n* protegido *m*. **~e** *n* protegida *f*
protein /'prəʊti:n/ *n* proteína *f*
protest /'prəʊtest/ *n* protesta *f*. **under ~** bajo protesta. /prə'test/ *vt/i* protestar. **~er** *n* (*demonstrator*) manifestante *m & f*
Protestant /'prɒtɪstənt/ *a & n* protestante (*m & f*)
protocol /'prəʊtəkɒl/ *n* protocolo *m*
prototype /'prəʊtətaɪp/ *n* prototipo *m*
protract /prə'trækt/ *vt* prolongar
protractor /prə'træktə(r)/ *n* transportador *m*
protrude /prə'tru:d/ *vi* sobresalir
protuberance /prə'tju:bərəns/ *n* protuberancia *f*
proud /praʊd/ *a* orgulloso. **~ly** *adv* orgullosamente
prove /pru:v/ *vt* probar. ● *vi* resultar. **~n** *a* probado
provenance /'prɒvənəns/ *n* procedencia *f*
proverb /'prɒvɜ:b/ *n* proverbio *m*. **~ial** /prə'vɜ:bɪəl/ *a* proverbial
provide /prə'vaɪd/ *vt* proveer. ● *vi*. ~ **against** precaverse de. ~ **for** (*allow for*) prever; mantener ‹*person*›. **~d** /prə'vaɪdɪd/ *conj*. ~ **(that)** con tal que
providen|ce /'prɒvɪdəns/ *n* providencia *f*. **~t** *a* providente. **~tial** /prɒvɪ'denʃl/ *a* providencial
providing /prə'vaɪdɪŋ/ *conj*. ~ **that** con tal que
provinc|e /'prɒvɪns/ *n* provincia *f*; (*fig*) competencia *f*. **~ial** /prə'vɪnʃl/ *a* provincial

provision /prə'vɪʒn/ *n* provisión *f*; (*supply*) suministro *m*; (*stipulation*) condición *f*. **~s** *npl* comestibles *mpl*

provisional /prə'vɪʒənl/ *a* provisional. **~ly** *adv* provisionalmente

proviso /prə'vaɪzəʊ/ *n* (*pl* **-os**) condición *f*

provo|cation /prɒvə'keɪʃn/ *n* provocación *f*. **~cative** /-'vɒkətɪv/ *a* provocador. **~ke** /prə'vəʊk/ *vt* provocar

prow /praʊ/ *n* proa *f*

prowess /'praʊɪs/ *n* habilidad *f*; (*valour*) valor *m*

prowl /praʊl/ *vi* merodear. ● *n* ronda *f*. **be on the ~** merodear. **~er** *n* merodeador *m*

proximity /prɒk'sɪmətɪ/ *n* proximidad *f*

proxy /'prɒksɪ/ *n* poder *m*. **by ~** por poder

prude /pru:d/ *n* mojigato *m*

pruden|ce /'pru:dəns/ *n* prudencia *f*. **~t** /'pru:dənt/ *a* prudente. **~tly** *adv* prudentemente

prudish /'pru:dɪʃ/ *a* mojigato

prune[1] /pru:n/ *n* ciruela *f* pasa

prune[2] /pru:n/ *vt* podar

pry /praɪ/ *vi* entrometerse

psalm /sɑ:m/ *n* salmo *m*

pseudo... /'sju:dəʊ/ *pref* seudo...

pseudonym /'sju:dənɪm/ *n* seudónimo *m*

psychiatr|ic /saɪkɪ'ætrɪk/ *a* psiquiátrico. **~ist** /saɪ'kaɪətrɪst/ *n* psiquiatra *m & f*. **~y** /saɪ'kaɪətrɪ/ *n* psiquiatría *f*

physic /'saɪkɪk/ *a* psíquico

psycho-analys|e /saɪkəʊ'ænəlaɪz/ *vt* psicoanalizar. **~is** /saɪkəʊə'næləsɪs/ *n* psicoanálisis *m*. **~t** /-ɪst/ *n* psicoanalista *m & f*

psycholog|ical /saɪkə'lɒdʒɪkl/ *a* psicológico. **~ist** /saɪ'kɒlədʒɪst/ *n* psicólogo *m*. **~y** /saɪ'kɒlədʒɪ/ *n* psicología *f*

psychopath /'saɪkəpæθ/ *n* psicópata *m & f*

pub /pʌb/ *n* bar *m*

puberty /'pju:bətɪ/ *n* pubertad *f*

pubic /'pju:bɪk/ *a* pubiano, púbico

public /'pʌblɪk/ *a* público

publican /'pʌblɪkən/ *n* tabernero *m*

publication /pʌblɪ'keɪʃn/ *n* publicación *f*

public house /pʌblɪk'haʊs/ *n* bar *m*

publicity /pʌb'lɪsətɪ/ *n* publicidad *f*

publicize /'pʌblɪsaɪz/ *vt* publicar, anunciar

publicly /'pʌblɪklɪ/ *adv* públicamente

public school /pʌblɪk'sku:l/ *n* colegio *m* privado; (*Amer*) instituto *m*

public-spirited /pʌblɪk'spɪrɪtɪd/ *a* cívico

publish /'pʌblɪʃ/ *vt* publicar. **~er** *n* editor *m*. **~ing** *n* publicación *f*

puck /pʌk/ *n* (*ice hockey*) disco *m*

pucker /'pʌkə(r)/ *vt* arrugar. ● *vi* arrugarse

pudding /'pʊdɪŋ/ *n* postre *m*; (*steamed*) budín *m*

puddle /'pʌdl/ *n* charco *m*

pudgy /'pʌdʒɪ/ *a* (**-ier**, **-iest**) rechoncho

puerile /'pjʊəraɪl/ *a* pueril

puff /pʌf/ *n* soplo *m*; (*for powder*) borla *f*. ● *vt/i* soplar. **~ at** chupar ‹*pipe*›. **~ out** apagar ‹*candle*›; (*swell up*) hinchar. **~ed** *a* (*out of breath*) sin aliento. **~ pastry** *n* hojaldre *m*. **~y** /'pʌfɪ/ *a* hinchado

pugnacious /pʌg'neɪʃəs/ *a* belicoso

pug-nosed /'pʌgnəʊzd/ *a* chato

pull /pʊl/ *vt* tirar de; sacar ‹*tooth*›; torcer ‹*muscle*›. **~ a face** hacer una mueca. **~ a fast one** hacer una mala jugada. **~ down** derribar ‹*building*›. **~ off** quitarse; (*fig*) lograr. **~ one's weight** poner de su parte. **~ out** sacar. **~ s.o.'s leg** tomarle el pelo a uno. **~ up** (*uproot*) desarraigar; (*reprimand*) reprender. ● *vi* tirar (**at** de). **~ away** (*auto*) alejarse. **~ back** retirarse. **~ in** (*enter*) entrar; (*auto*) parar. **~ o.s. together** tranquilizarse. **~ out** (*auto*) salirse. **~ through** recobrar la salud. **~ up** (*auto*) parar. ● *n* tirón *m*; (*fig*) atracción *f*; (*influence*) influencia *f*. **give a ~** tirar

pulley /'pʊlɪ/ *n* polea *f*

pullover /'pʊləʊvə(r)/ *n* jersey *m*

pulp /pʌlp/ *n* pulpa *f*; (*for paper*) pasta *f*

pulpit /'pʊlpɪt/ *n* púlpito *m*

pulsate /'pʌlseɪt/ *vi* pulsar

pulse /pʌls/ *n* (*med*) pulso *m*

pulverize /'pʌlvəraɪz/ *vt* pulverizar

pumice /'pʌmɪs/ *n* piedra *f* pómez

pummel /'pʌml/ *vt* (*pt* **pummelled**) aporrear

pump[1] /pʌmp/ *n* bomba *f*; ● *vt* sacar con una bomba; (*fig*) sonsacar. **~ up** inflar

pump[2] /pʌmp/ (*plimsoll*) zapatilla *f* de lona; (*dancing shoe*) escarpín *m*

pumpkin /'pʌmpkɪn/ *n* calabaza *f*

pun /pʌn/ *n* juego *m* de palabras
punch[1] /pʌntʃ/ *vt* dar un puñetazo a; (*perforate*) perforar; hacer ‹*hole*›. ● *n* puñetazo *m*; (*vigour*, *sl*) empuje *m*; (*device*) punzón *m*
punch[2] /pʌntʃ/ (*drink*) ponche *m*
punch: **~-drunk** *a* aturdido a golpes. **~ line** *n* gracia *f*. **~-up** *n* riña *f*
punctilious /pʌŋk'tɪlɪəs/ *a* meticuloso
punctual /'pʌŋktʃʊəl/ *a* puntual. **~ity** /-'ælətɪ/ *n* puntualidad *f*. **~ly** *adv* puntualmente
punctuat|e /'pʌŋkʃʊeɪt/ *vt* puntuar. **~ion** /-'eɪʃn/ *n* puntuación *f*
puncture /'pʌŋktʃə(r)/ *n* (*in tyre*) pinchazo *m*. ● *vt* pinchar. ● *vi* pincharse
pundit /'pʌndɪt/ *n* experto *m*
pungen|cy /'pʌndʒənsɪ/ *n* acritud *f*; (*fig*) mordacidad *f*. **~t** /'pʌndʒənt/ *a* acre; ‹*remark*› mordaz
punish /'pʌnɪʃ/ *vt* castigar. **~able** *a* castigable. **~ment** *n* castigo *m*
punitive /'pju:nɪtɪv/ *a* punitivo
punk /pʌŋk/ *a* ‹*music, person*› punk
punnet /'pʌnɪt/ *n* canastilla *f*
punt[1] /pʌnt/ *n* (*boat*) batea *f*
punt[2] /pʌnt/ *vi* apostar. **~er** *n* apostante *m & f*
puny /'pju:nɪ/ *a* (**-ier**, **-iest**) diminuto; (*weak*) débil; (*petty*) insignificante
pup /pʌp/ *n* cachorro *m*
pupil[1] /'pju:pl/ *n* alumno *m*
pupil[2] /'pju:pl/ (*of eye*) pupila *f*
puppet /'pʌpɪt/ *n* títere *m*
puppy /'pʌpɪ/ *n* cachorro *m*
purchase /'pɜ:tʃəs/ *vt* comprar. ● *n* compra *f*. **~r** *n* comprador *m*
pur|e /'pjʊə(r)/ *a* (**-er**, **-est**) puro. **~ely** *adv* puramente. **~ity** *n* pureza *f*
purée /'pjʊəreɪ/ *n* puré *m*
purgatory /'pɜ:gətrɪ/ *n* purgatorio *m*
purge /pɜ:dʒ/ *vt* purgar. ● *n* purga *f*
purif|ication /pjʊərɪfɪ'keɪʃn/ *n* purificación *f*. **~y** /'pjʊərɪfaɪ/ *vt* purificar
purist /'pjʊərɪst/ *n* purista *m & f*
puritan /'pjʊərɪtən/ *n* puritano *m*. **~ical** /-'tænɪkl/ *a* puritano
purl /pɜ:l/ *n* (*knitting*) punto *m* del revés
purple /'pɜ:pl/ *a* purpúreo, morado. ● *n* púrpura *f*
purport /pə'pɔ:t/ *vt*. **~ to be** pretender ser
purpose /'pɜ:pəs/ *n* propósito *m*; (*determination*) resolución *f*. **on ~** a propósito. **to no ~** en vano. **~-built** *a* construido especialmente. **~ful** *a* (*resolute*) resuelto. **~ly** *adv* a propósito
purr /pɜ:(r)/ *vi* ronronear
purse /pɜ:s/ *n* monedero *m*; (*Amer*) bolso *m*, cartera *f* (*LAm*). ● *vt* fruncir
pursu|e /pə'sju:/ *vt* perseguir, seguir. **~er** *n* perseguidor *m*. **~it** /pə'sju:t/ *n* persecución *f*; (*fig*) ocupación *f*
purveyor /pə'veɪə(r)/ *n* proveedor *m*
pus /pʌs/ *n* pus *m*
push /pʊʃ/ *vt* empujar; apretar ‹*button*›. ● *vi* empujar. ● *n* empuje *m*; (*effort*) esfuerzo *m*; (*drive*) dinamismo *m*. **at a ~** en caso de necesidad. **get the ~** (*sl*) ser despedido. **~ aside** *vt* apartar. **~ back** *vt* hacer retroceder. **~ off** *vi* (*sl*) marcharse. **~ on** *vi* seguir adelante. **~ up** *vt* levantar. **~-button telephone** *n* teléfono *m* de teclas. **~-chair** *n* sillita *f* con ruedas. **~ing** /'pʊʃɪŋ/ *a* ambicioso. **~-over** *n* (*fam*) cosa *f* muy fácil, pan comido. **~y** *a* (*pej*) ambicioso
puss /pʊs/ *n* minino *m*
put /pʊt/ *vt* (*pt* **put**, *pres p* **putting**) poner; (*express*) expresar; (*say*) decir; (*estimate*) estimar; hacer ‹*question*›. **~ across** comunicar; (*deceive*) engañar. **~ aside** poner aparte. **~ away** guardar. **~ back** devolver; retrasar ‹*clock*›. **~ by** guardar; ahorrar ‹*money*›. **~ down** depositar; (*suppress*) suprimir; (*write*) apuntar; (*kill*) sacrificar. **~ forward** avanzar. **~ in** introducir; (*submit*) presentar. **~ in for** pedir. **~ off** aplazar; (*disconcert*) desconcertar. **~ on** (*wear*) ponerse; cobrar ‹*speed*›; encender ‹*light*›. **~ one's foot down** mantenerse firme. **~ out** (*extinguish*) apagar; (*inconvenience*) incomodar; extender ‹*hand*›; (*disconcert*) desconcertar. **~ to sea** hacerse a la mar. **~ through** (*phone*) poner. **~ up** levantar; subir ‹*price*›; alojar ‹*guest*›. **~ up with** soportar. **stay ~** (*fam*) no moverse
putrefy /'pju:trɪfaɪ/ *vi* pudrirse
putt /pʌt/ *n* (*golf*) golpe *m* suave
putty /'pʌtɪ/ *n* masilla *f*
put-up /'pʊtʌp/ *a*. **~ job** *n* confabulación *f*

puzzl|e /ˈpʌzl/ *n* enigma *m*; (*game*) rompecabezas *m invar*. ● *vt* dejar perplejo. ● *vi* calentarse los sesos. **~ing** *a* incomprensible; (*odd*) curioso
pygmy /ˈpɪɡmɪ/ *n* pigmeo *m*
pyjamas /pəˈdʒɑːməz/ *npl* pijama *m*
pylon /ˈpaɪlɒn/ *n* pilón *m*
pyramid /ˈpɪrəmɪd/ *n* pirámide *f*
python /ˈpaɪθn/ *n* pitón *m*

Q

quack[1] /kwæk/ *n* (*of duck*) graznido *m*
quack[2] /kwæk/ (*person*) charlatán *m*. **~ doctor** *n* curandero *m*
quadrangle /ˈkwɒdræŋɡl/ *n* cuadrilátero *m*; (*court*) patio *m*
quadruped /ˈkwɒdrʊped/ *n* cuadrúpedo *m*
quadruple /ˈkwɒdrʊpl/ *a* & *n* cuádruplo (*m*). ● *vt* cuadruplicar. **~t** /-plət/ *n* cuatrillizo *m*
quagmire /ˈkwæɡmaɪə(r)/ *n* ciénaga *f*; (*fig*) atolladero *m*
quail /kweɪl/ *n* codorniz *f*
quaint /kweɪnt/ *a* (**-er**, **-est**) pintoresco; (*odd*) curioso
quake /kweɪk/ *vi* temblar. ● *n* (*fam*) terremoto *m*
Quaker /ˈkweɪkə(r)/ *n* cuáquero (*m*)
qualification /kwɒlɪfɪˈkeɪʃn/ *n* título *m*; (*requirement*) requisito *m*; (*ability*) capacidad *f*; (*fig*) reserva *f*
qualif|ied /ˈkwɒlɪfaɪd/ *a* cualificado; (*limited*) limitado; (*with degree, diploma*) titulado. **~y** /ˈkwɒlɪfaɪ/ *vt* calificar; (*limit*) limitar. ● *vi* sacar el título; (*sport*) clasificarse; (*fig*) llenar los requisitos
qualitative /ˈkwɒlɪtətɪv/ *a* cualitativo
quality /ˈkwɒlɪtɪ/ *n* calidad *f*; (*attribute*) cualidad *f*
qualm /kwɑːm/ *n* escrúpulo *m*
quandary /ˈkwɒndrɪ/ *n*. **in a ~** en un dilema
quantitative /ˈkwɒntɪtətɪv/ *a* cuantitativo
quantity /ˈkwɒntɪtɪ/ *n* cantidad *f*
quarantine /ˈkwɒrəntiːn/ *n* cuarentena *f*
quarrel /ˈkwɒrəl/ *n* riña *f*. ● *vi* (*pt* **quarrelled**) reñir. **~some** *a* pendenciero
quarry[1] /ˈkwɒrɪ/ *n* (*excavation*) cantera *f*
quarry[2] /ˈkwɒrɪ/ *n* (*animal*) presa *f*
quart /kwɔːt/ *n* (poco más de un) litro *m*
quarter /ˈkwɔːtə(r)/ *n* cuarto *m*; (*of year*) trimestre *m*; (*district*) barrio *m*. **from all ~s** de todas partes. ● *vt* dividir en cuartos; (*mil*) acuartelar. **~s** *npl* alojamiento *m*
quartermaster /ˈkwɔːtəmɑːstə(r)/ *n* intendente *m*
quarter: **~-final** *n* cuarto *m* de final. **~ly** *a* trimestral. ● *adv* cada tres meses
quartet /kwɔːˈtet/ *n* cuarteto *m*
quartz /kwɔːts/ *n* cuarzo *m*. ● *a* ‹*watch etc*› de cuarzo
quash /kwɒʃ/ *vt* anular
quasi.. /ˈkweɪsaɪ/ *pref* cuasi...
quaver /ˈkweɪvə(r)/ *vi* temblar. ● *n* (*mus*) corchea *f*
quay /kiː/ *n* muelle *m*
queasy /ˈkwiːzɪ/ *a* ‹*stomach*› delicado
queen /kwiːn/ *n* reina *f*. **~ mother** *n* reina *f* madre
queer /kwɪə(r)/ *a* (**-er**, **-est**) extraño; (*dubious*) sospechoso; (*ill*) indispuesto. ● *n* (*sl*) homosexual *m*
quell /kwel/ *vt* reprimir
quench /kwentʃ/ *vt* apagar; sofocar ‹*desire*›
querulous /ˈkwerʊləs/ *a* quejumbroso
query /ˈkwɪərɪ/ *n* pregunta *f*. ● *vt* preguntar; (*doubt*) poner en duda
quest /kwest/ *n* busca *f*
question /ˈkwestʃən/ *n* pregunta *f*; (*for discussion*) cuestión *f*. **in ~** en cuestión. **out of the ~** imposible. **without ~** sin duda. ● *vt* preguntar; ‹*police etc*› interrogar; (*doubt*) poner en duda. **~able** /ˈkwestʃənəbl/ *a* discutible. **~ mark** *n* signo *m* de interrogación. **~naire** /kwestʃəˈneə(r)/ *n* cuestionario *m*
queue /kjuː/ *n* cola *f*. ● *vi* (*pres p* **queuing**) hacer cola
quibble /ˈkwɪbl/ *vi* discutir; (*split hairs*) sutilizar
quick /kwɪk/ *a* (**-er**, **-est**) rápido. **be ~!** ¡date prisa! ● *adv* rápidamente. ● *n*. **to the ~** en lo vivo. **~en** /ˈkwɪkən/ *vt* acelerar. ● *vi* acelerarse. **~ly** *adv* rápidamente. **~sand** /ˈkwɪksænd/ *n* arena *f* movediza. **~-tempered** *a* irascible

quid /kwɪd/ *n invar* (*sl*) libra *f* (esterlina)

quiet /'kwaɪət/ *a* (**-er**, **-est**) tranquilo; (*silent*) callado; (*discreet*) discreto. ● *n* tranquilidad *f*. **on the ~** a escondidas. **~en** /'kwaɪətn/ *vt* calmar. ● *vi* calmarse. **~ly** *adv* tranquilamente; (*silently*) silenciosamente; (*discreetly*) discretamente. **~ness** *n* tranquilidad *f*

quill /kwɪl/ *n* pluma *f*

quilt /kwɪlt/ *n* edredón *m*. ● *vt* acolchar

quince /kwɪns/ *n* membrillo *m*

quinine /kwɪ'ni:n/ *n* quinina *f*

quintessence /kwɪn'tesns/ *n* quintaesencia *f*

quintet /kwɪn'tet/ *n* quinteto *m*

quintuplet /'kwɪntju:plət/ *n* quintillizo *m*

quip /kwɪp/ *n* ocurrencia *f*

quirk /kwɜ:k/ *n* peculiaridad *f*

quit /kwɪt/ *vt* (*pt* **quitted**) dejar. ● *vi* abandonar; (*leave*) marcharse; (*resign*) dimitir. **~ doing** (*cease*, *Amer*) dejar de hacer

quite /kwaɪt/ *adv* bastante; (*completely*) totalmente; (*really*) verdaderamente. **~ (so)!** ¡claro! **~ a few** bastante

quits /kwɪts/ *a* a la par. **call it ~** darlo por terminado

quiver /'kwɪvə(r)/ *vi* temblar

quixotic /kwɪk'sɒtɪk/ *a* quijotesco

quiz /kwɪz/ *n* (*pl* **quizzes**) serie *f* de preguntas; (*game*) concurso *m*. ● *vt* (*pt* **quizzed**) interrogar. **~zical** /'kwɪzɪkl/ *a* burlón

quorum /'kwɔ:rəm/ *n* quórum *m*

quota /'kwəʊtə/ *n* cuota *f*

quot|ation /kwəʊ'teɪʃn/ *n* cita *f*; (*price*) presupuesto *m*. **~ation marks** *npl* comillas *fpl*. **~e** /kwəʊt/ *vt* citar; (*com*) cotizar. ● *n* (*fam*) cita *f*; (*price*) presupuesto *m*. **in ~es** *npl* entre comillas

quotient /'kwəʊʃnt/ *n* cociente *m*

R

rabbi /'ræbaɪ/ *n* rabino *m*

rabbit /'ræbɪt/ *n* conejo *m*

rabble /'ræbl/ *n* gentío *m*. **the ~** (*pej*) el populacho *m*

rabi|d /'ræbɪd/ *a* feroz; ‹*dog*› rabioso. **~es** /'reɪbi:z/ *n* rabia *f*

race[1] /reɪs/ *n* carrera *f*. ● *vt* hacer correr ‹*horse*›; acelerar ‹*engine*›. ● *vi* (*run*) correr, ir corriendo; (*rush*) ir de prisa

race[2] /reɪs/ (*group*) raza *f*

race: **~course** /'reɪskɔ:s/ *n* hipódromo *m*. **~horse** /'reɪshɔ:s/ *n* caballo *m* de carreras. **~riots** /'reɪsraɪəts/ *npl* disturbios *mpl* raciales. **~-track** /'reɪstræk/ *n* hipódromo *m*

racial /'reɪʃl/ *a* racial. **~ism** /-ɪzəm/ *n* racismo *m*

racing /'reɪsɪŋ/ *n* carreras *fpl*. **~ car** *n* coche *m* de carreras

racis|m /'reɪsɪzəm/ *n* racismo *m*. **~t** /'reɪsɪst/ *a* & *n* racista (*m* & *f*)

rack[1] /ræk/ *n* (*shelf*) estante *m*; (*for luggage*) rejilla *f*; (*for plates*) escurreplatos *m invar*. ● *vt*. **~ one's brains** devanarse los sesos

rack[2] /ræk/ *n*. **go to ~ and ruin** quedarse en la ruina

racket[1] /'rækɪt/ *n* (*for sports*) raqueta

racket[2] /'rækɪt/ (*din*) alboroto *m*; (*swindle*) estafa *f*. **~eer** /-ə'tɪə(r)/ *n* estafador *m*

raconteur /rækɒn'tɜ:/ *n* anecdotista *m* & *f*

racy /'reɪsɪ/ *a* (**-ier**, **-iest**) vivo

radar /'reɪdɑ:(r)/ *n* radar *m*

radian|ce /'reɪdɪəns/ *n* resplandor *m*. **~t** /'reɪdɪənt/ *a* radiante. **~tly** *adv* con resplandor

radiat|e /'reɪdɪeɪt/ *vt* irradiar. ● *vi* divergir. **~ion** /-'eɪʃn/ *n* radiación *f*. **~or** /'reɪdɪeɪtə(r)/ *n* radiador *m*

radical /'rædɪkl/ *a* & *n* radical (*m*)

radio /'reɪdɪəʊ/ *n* (*pl* **-os**) radio *f*. ● *vt* transmitir por radio

radioactiv|e /reɪdɪəʊ'æktɪv/ *a* radiactivo. **~ity** /-'tɪvətɪ/ *n* radiactividad *f*

radiograph|er /reɪdɪ'ɒgrəfə(r)/ *n* radiógrafo *m*. **~y** *n* radiografía *f*

radish /'rædɪʃ/ *n* rábano *m*

radius /'reɪdɪəs/ *n* (*pl* **-dii** /-dɪaɪ/) radio *m*

raffish /'ræfɪʃ/ *a* disoluto

raffle /ræfl/ *n* rifa *f*

raft /rɑ:ft/ *n* balsa *f*

rafter /'rɑ:ftə(r)/ *n* cabrio *m*

rag[1] /ræg/ *n* andrajo *m*; (*for wiping*) trapo *m*; (*newspaper*) periodicucho *m*. **in ~s** ‹*person*› andrajoso; ‹*clothes*› hecho jirones

rag[2] /ræg/ *n* (*univ*) festival *m* estudiantil; (*prank*, *fam*) broma *f*

pesada. ● *vt* (*pt* **ragged**) (*sl*) tomar el pelo a
ragamuffin /'rægəmʌfɪn/ *n* granuja *m*, golfo *m*
rage /reɪdʒ/ *n* rabia *f*; (*fashion*) moda *f*. ● *vi* estar furioso; ‹*storm*› bramar
ragged /'rægɪd/ *a* ‹person› andrajoso; ‹*clothes*› hecho jirones; ‹*edge*› mellado
raid /reɪd/ *n* (*mil*) incursión *f*; (*by police, etc*) redada *f*; (*by thieves*) asalto *m*. ● *vt* (*mil*) atacar; ‹*police*› hacer una redada en; ‹*thieves*› asaltar. **~er** *n* invasor *m*; (*thief*) ladrón *m*
rail[1] /reɪl/ *n* barandilla *f*; (*for train*) riel *m*; (*rod*) barra *f*. **by ~** por ferrocarril
rail[2] /reɪl/ *vi*. **~ against**, **~ at** insultar
railing /'reɪlɪŋ/ *n* barandilla *f*; (*fence*) verja *f*
rail|road /'reɪlrəʊd/ *n* (*Amer*), **~way** /'reɪlweɪ/ *n* ferrocarril *m*. **~wayman** *n* (*pl* **-men**) ferroviario *m*. **~way station** *n* estación *f* de ferrocarril
rain /reɪn/ *n* lluvia *f*. ● *vi* llover. **~bow** /'reɪnbəʊ/ *n* arco *m* iris. **~coat** /'reɪnkəʊt/ *n* impermeable *m*. **~fall** /'reɪnfɔ:l/ *n* precipitación *f*. **~-water** *n* agua *f* de lluvia. **~y** /'reɪnɪ/ *a* (**-ier**, **-iest**) lluvioso
raise /reɪz/ *vt* levantar; (*breed*) criar; obtener ‹*money etc*›; hacer ‹*question*›; plantear ‹*problem*›; subir ‹*price*›. **~ one's glass to** brindar por. **~ one's hat** descubrirse. ● *n* (*Amer*) aumento *m*
raisin /'reɪzn/ *n* (uva *f*) pasa *f*
rake[1] /reɪk/ *n* rastrillo *m*. ● *vt* rastrillar; (*search*) buscar en. **~ up** remover
rake[2] /reɪk/ *n* (*man*) calavera *m*
rake-off /'reɪkɒf/ *n* (*fam*) comisión *f*
rally /'rælɪ/ *vt* reunir; (*revive*) reanimar. ● *vi* reunirse; (*in sickness*) recuperarse. ● *n* reunión *f*; (*recovery*) recuperación *f*; (*auto*) rallye *m*
ram /ræm/ *n* carnero *m*. ● *vt* (*pt* **rammed**) (*thrust*) meter por la fuerza; (*crash into*) chocar con
rambl|e /'ræmbl/ *n* excursión *f* a pie. ● *vi* ir de paseo; (*in speech*) divagar. **~e on** divagar. **~er** *n* excursionista *m* & *f*. **~ing** *a* ‹*speech*› divagador
ramification /ræmɪfɪ'keɪʃn/ *n* ramificación *f*
ramp /ræmp/ *n* rampa *f*
rampage /ræm'peɪdʒ/ *vi* alborotarse. /'ræmpeɪdʒ/ *n*. **go on the ~** alborotarse
rampant /'ræmpənt/ *a*. **be ~** ‹*disease etc*› estar extendido
rampart /'ræmpɑ:t/ *n* muralla *f*
ramshackle /'ræmʃækl/ *a* desvencijado
ran /ræn/ *see* **run**
ranch /rɑ:ntʃ/ *n* hacienda *f*
rancid /'rænsɪd/ *a* rancio
rancour /'ræŋkə(r)/ *n* rencor *m*
random /'rændəm/ *a* hecho al azar; (*chance*) fortuito. ● *n*. **at ~** al azar
randy /'rændɪ/ *a* (**-ier**, **-iest**) lujurioso, cachondo (*fam*)
rang /ræŋ/ *see* **ring**[2]
range /reɪndʒ/ *n* alcance *m*; (*distance*) distancia *f*; (*series*) serie *f*; (*of mountains*) cordillera *f*; (*extent*) extensión *f*; (*com*) surtido *m*; (*open area*) dehesa *f*; (*stove*) cocina *f* económica. ● *vi* extenderse; (*vary*) variar
ranger /'reɪndʒə(r)/ *n* guardabosque *m*
rank[1] /ræŋk/ *n* posición *f*, categoría *f*; (*row*) fila *f*; (*for taxis*) parada *f*. **the ~ and file** la masa *f*. ● *vt* clasificar. ● *vi* clasificarse. **~s** *npl* soldados *mpl* rasos
rank[2] /ræŋk/ *a* (**-er**, **-est**) exuberante; (*smell*) fétido; (*fig*) completo
rankle /'ræŋkl/ *vi* (*fig*) causar rencor
ransack /'rænsæk/ *vt* registrar; (*pillage*) saquear
ransom /'rænsəm/ *n* rescate *m*. **hold s.o. to ~** exigir rescate por uno; (*fig*) hacer chantaje a uno. ● *vt* rescatar; (*redeem*) redimir
rant /rænt/ *vi* vociferar
rap /ræp/ *n* golpe *m* seco. ● *vt/i* (*pt* **rapped**) golpear
rapacious /rə'peɪʃəs/ *a* rapaz
rape /reɪp/ *vt* violar. ● *n* violación *f*
rapid /'ræpɪd/ *a* rápido. **~ity** /rə'pɪdətɪ/ *n* rapidez *f*. **~s** /'ræpɪdz/ *npl* rápido *m*
rapist /'reɪpɪst/ *n* violador *m*
rapport /ræ'pɔ:(r)/ *n* armonía *f*, relación *f*
rapt /ræpt/ *a* ‹*attention*› profundo. **~ in** absorto en
raptur|e /'ræptʃə(r)/ *n* éxtasis *m*. **~ous** *a* extático
rare[1] /reə(r)/ *a* (**-er**, **-est**) raro
rare[2] /reə(r)/ *a* (*culin*) poco hecho
rarefied /'reərɪfaɪd/ *a* enrarecido

rarely /'reəlɪ/ *adv* raramente
rarity /'reərətɪ/ *n* rareza *f*
raring /'reərɪŋ/ *a* (*fam*). **~ to** impaciente por
rascal /'rɑ:skl/ *n* tunante *m & f*
rash[1] /ræʃ/ *a* (**-er**, **-est**) imprudente, precipitado
rash[2] /ræʃ/ *n* erupción *f*
rasher /'ræʃə(r)/ *n* loncha *f*
rash|ly /'ræʃlɪ/ *adv* imprudentemente, a la ligera. **~ness** *n* imprudencia *f*
rasp /rɑ:sp/ *n* (*file*) escofina *f*
raspberry /'rɑ:zbrɪ/ *n* frambuesa *f*
rasping /'rɑ:spɪŋ/ *a* áspero
rat /ræt/ *n* rata *f*. ● *vi* (*pt* **ratted**). **~ on** (*desert*) desertar; (*inform on*) denunciar, chivarse
rate /reɪt/ *n* (*ratio*) proporción *f*; (*speed*) velocidad *f*; (*price*) precio *m*; (*of interest*) tipo *m*. **at any ~** de todas formas. **at the ~ of** (*on the basis of*) a razón de. **at this ~** así. ● *vt* valorar; (*consider*) considerar; (*deserve*, *Amer*) merecer. ● *vi* ser considerado. **~able value** *n* valor *m* imponible. **~payer** /'reɪtpeɪə(r)/ *n* contribuyente *m & f*. **~s** *npl* (*taxes*) impuestos *mpl* municipales
rather /'rɑ:ðə(r)/ *adv* mejor dicho; (*fairly*) bastante; (*a little*) un poco. ● *int* claro. **I would ~ not** prefiero no
ratif|ication /rætɪfɪ'keɪʃn/ *n* ratificación *f*. **~y** /'rætɪfaɪ/ *vt* ratificar
rating /'reɪtɪŋ/ *n* clasificación *f*; (*sailor*) marinero *m*; (*number*, *TV*) índice *m*
ratio /'reɪʃɪəʊ/ *n* (*pl* **-os**) proporción *f*
ration /'ræʃn/ *n* ración *f*. ● *vt* racionar
rational /'ræʃənəl/ *a* racional. **~ize** /'ræʃənəlaɪz/ *vt* racionalizar
rat race /'rætreɪs/ *n* lucha *f* incesante para triunfar
rattle /rætl/ *vi* traquetear. ● *vt* (*shake*) agitar; (*sl*) desconcertar. ● *n* traqueteo *m*; (*toy*) sonajero *m*. **~ off** (*fig*) decir de corrida
rattlesnake /'rætlsneɪk/ *n* serpiente *f* de cascabel
ratty /'rætɪ/ *a* (**-ier**, **-iest**) (*sl*) irritable
raucous /'rɔ:kəs/ *a* estridente
ravage /'rævɪdʒ/ *vt* estragar. **~s** /'rævɪdʒɪz/ *npl* estragos *mpl*
rave /reɪv/ *vi* delirar; (*in anger*) enfurecerse. **~ about** entusiasmarse por
raven /'reɪvn/ *n* cuervo *m*. ● *a* ‹*hair*› negro
ravenous /'rævənəs/ *a* voraz; ‹*person*› hambriento. **be ~** morirse de hambre
ravine /rə'vi:n/ *n* barranco *m*
raving /'reɪvɪŋ/ *a*. **~ mad** loco de atar. **~s** *npl* divagaciones *fpl*
ravish /'rævɪʃ/ *vt* (*rape*) violar. **~ing** *a* (*enchanting*) encantador
raw /rɔ:/ *a* (**-er**, **-est**) crudo; (*not processed*) bruto; ‹*wound*› en carne viva; (*inexperienced*) inexperto; ‹*weather*› crudo. **~ deal** *n* tratamiento *m* injusto, injusticia *f*. **~ materials** *npl* materias *fpl* primas
ray /reɪ/ *n* rayo *m*
raze /reɪz/ *vt* arrasar
razor /'reɪzə(r)/ *n* navaja *f* de afeitar; (*electric*) maquinilla *f* de afeitar
Rd *abbr* (*Road*) C/, Calle *f*
re[1] /ri:/ *prep* con referencia a. ● *pref* re...
re[2] /reɪ/ *n* (*mus*, *second note of any musical scale*) re *m*
reach /ri:tʃ/ *vt* alcanzar; (*extend*) extender; (*arrive at*) llegar a; (*achieve*) lograr; (*hand over*) pasar, dar. ● *vi* extenderse. ● *n* alcance *m*; (*of river*) tramo *m* recto. **within ~ of** al alcance de; (*close to*) a corta distancia de
react /rɪ'ækt/ *vi* reaccionar. **~ion** /rɪ'ækʃn/ *n* reacción *f*. **~ionary** *a & n* reaccionario (*m*)
reactor /rɪ'æktə(r)/ *n* reactor *m*
read /ri:d/ *vt* (*pt* **read** /red/) leer; (*study*) estudiar; (*interpret*) interpretar. ● *vi* leer; ‹*instrument*› indicar. ● *n* (*fam*) lectura *f*. **~ out** *vt* leer en voz alta. **~able** *a* interesante, agradable; (*clear*) legible. **~er** /'ri:də(r)/ *n* lector *m*. **~ership** *n* lectores *m*
readi|ly /'redɪlɪ/ *adv* (*willingly*) de buena gana; (*easily*) fácilmente. **~ness** /'redɪnɪs/ *n* prontitud *f*. **in ~ness** preparado, listo
reading /'ri:dɪŋ/ *n* lectura *f*
readjust /ri:ə'dʒʌst/ *vt* reajustar. ● *vi* readaptarse (**to** a)
ready /'redɪ/ *a* (**-ier**, **-iest**) listo, preparado; (*quick*) pronto. **~-made** *a* confeccionado. **~ money** *n* dinero *m* contante. **~ reckoner** *n* baremo *m*. **get ~** preparase
real /rɪəl/ *a* verdadero. ● *adv* (*Amer*, *fam*) verdaderamente. **~ estate** *n* bienes *mpl* raíces

realis|m /'rɪəlɪzəm/ *n* realismo *m*. **~t** /'rɪəlɪst/ *n* realista *m & f*. **~tic** /-'lɪstɪk/ *a* realista. **~tically** /-'lɪstɪklɪ/ *adv* de manera realista
reality /rɪ'ælətɪ/ *n* realidad *f*
realiz|ation /rɪəlaɪ'zeɪʃn/ *n* comprensión *f*; (*com*) realización *f*. **~e** /'rɪəlaɪz/ *vt* darse cuenta de; (*fulfil, com*) realizar
really /'rɪəlɪ/ *adv* verdaderamente
realm /relm/ *n* reino *m*
ream /ri:m/ *n* resma *f*
reap /ri:p/ *vt* segar; (*fig*) cosechar
re: ~appear /ri:ə'pɪə(r)/ *vi* reaparecer. **~appraisal** /ri:ə'preɪzl/ *n* revaluación *f*
rear[1] /rɪə(r)/ *n* parte *f* de atrás. ● *a* posterior, trasero
rear[2] /rɪə(r)/ *vt* (*bring up, breed*) criar. **~ one's head** levantar la cabeza. ● *vi* ‹*horse*› encabritarse. **~ up** ‹*horse*› encabritarse
rear: ~admiral *n* contraalmirante *m*. **~guard** /'rɪəgɑ:d/ *n* retaguardia *f*
re: ~arm /ri:'ɑ:m/ *vt* rearmar. ● *vi* rearmarse. **~arrange** /ri:ə'reɪndʒ/ *vt* arreglar de otra manera
reason /'ri:zn/ *n* razón *f*, motivo *m*. **within ~** dentro de lo razonable. ● *vi* razonar
reasonable /'ri:zənəbl/ *a* razonable
reasoning /'ri:znɪŋ/ *n* razonamiento *m*
reassur|ance /ri:ə'ʃʊərəns/ *n* promesa *f* tranquilizadora; (*guarantee*) garantía *f*. **~e** /ri:ə'ʃʊə(r)/ *vt* tranquilizar
rebate /'ri:beɪt/ *n* reembolso *m*; (*discount*) rebaja *f*
rebel /'rebl/ *n* rebelde *m & f*. /rɪ'bel/ *vi* (*pt* **rebelled**) rebelarse. **~lion** *n* rebelión *f*. **~lious** *a* rebelde
rebound /rɪ'baʊnd/ *vi* rebotar; (*fig*) recaer. /'ri:baʊnd/ *n* rebote *m*. **on the ~** (*fig*) por reacción
rebuff /rɪ'bʌf/ *vt* rechazar. ● *n* desaire *m*
rebuild /ri:'bɪld/ *vt* (*pt* **rebuilt**) reconstruir
rebuke /rɪ'bju:k/ *vt* reprender. ● *n* reprensión *f*
rebuttal /rɪ'bʌtl/ *n* refutación *f*
recall /rɪ'kɔ:l/ *vt* (*call s.o. back*) llamar; (*remember*) recordar. ● *n* llamada *f*
recant /rɪ'kænt/ *vi* retractarse
recap /'ri:kæp/ *vt/i* (*pt* **recapped**) (*fam*) resumir. ● *n* (*fam*) resumen *m*
recapitulat|e /ri:kə'pɪtʃʊleɪt/ *vt/i* resumir. **~ion** /-'leɪʃn/ *n* resumen *m*
recapture /ri:'kæptʃə(r)/ *vt* recobrar; (*recall*) hacer revivir
reced|e /rɪ'si:d/ *vi* retroceder. **~ing** *a* ‹*forehead*› huidizo
receipt /rɪ'si:t/ *n* recibo *m*. **~s** *npl* (*com*) ingresos *mpl*
receive /rɪ'si:v/ *vt* recibir. **~r** /-ə(r)/ *n* (*of stolen goods*) perista *m & f*; (*of phone*) auricular *m*
recent /'ri:snt/ *a* reciente. **~ly** *adv* recientemente
receptacle /rɪ'septəkl/ *n* recipiente *m*
reception /rɪ'sepʃn/ *n* recepción *f*; (*welcome*) acogida *f*. **~ist** *n* recepcionista *m & f*
receptive /rɪ'septɪv/ *a* receptivo
recess /rɪ'ses/ *n* hueco *m*; (*holiday*) vacaciones *fpl*; (*fig*) parte *f* recóndita
recession /rɪ'seʃn/ *n* recesión *f*
recharge /ri:'tʃɑ:dʒ/ *vt* cargar de nuevo, recargar
recipe /'resəpɪ/ *n* receta *f*
recipient /rɪ'sɪpɪənt/ *n* recipiente *m & f*; (*of letter*) destinatario *m*
reciprocal /rɪ'sɪprəkl/ *a* recíproco
reciprocate /rɪ'sɪprəkeɪt/ *vt* corresponder a
recital /rɪ'saɪtl/ *n* (*mus*) recital *m*
recite /rɪ'saɪt/ *vt* recitar; (*list*) enumerar
reckless /'reklɪs/ *a* imprudente. **~ly** *adv* imprudentemente. **~ness** *n* imprudencia *f*
reckon /'rekən/ *vt/i* calcular; (*consider*) considerar; (*think*) pensar. **~ on** (*rely*) contar con. **~ing** *n* cálculo *m*
reclaim /rɪ'kleɪm/ *vt* reclamar; recuperar ‹*land*›
reclin|e /rɪ'klaɪn/ *vi* recostarse. **~ing** *a* acostado; ‹*seat*› reclinable
recluse /rɪ'klu:s/ *n* solitario *m*
recogni|tion /rekəg'nɪʃn/ *n* reconocimiento *m*. **beyond ~tion** irreconocible. **~ze** /'rekəgnaɪz/ *vt* reconocer
recoil /rɪ'kɔɪl/ *vi* retroceder. ● *n* (*of gun*) culatazo *m*
recollect /rekə'lekt/ *vt* recordar. **~ion** /-ʃn/ *n* recuerdo *m*
recommend /rekə'mend/ *vt* recomendar. **~ation** /-'deɪʃn/ *n* recomendación *f*

recompense /'rekəmpens/ *vt* recompensar. • *n* recompensa *f*

reconcil|e /'rekənsaɪl/ *vt* reconciliar ‹*people*›; conciliar ‹*facts*›. **~e o.s.** resignarse **(to** a). **~iation** /-sɪlɪ'eɪʃn/ *n* reconciliación *f*

recondition /ri:kən'dɪʃn/ *vt* reacondicionar, arreglar

reconnaissance /rɪ'kɒnɪsns/ *n* reconocimiento *m*

reconnoitre /rekə'nɔɪtə(r)/ *vt* (*pres p* **-tring**) (*mil*) reconocer. • *vi* hacer un reconocimiento

re: **~consider** /ri:kən'sɪdə(r)/ *vt* volver a considerar. **~construct** /ri:kən'strʌkt/ *vt* reconstruir. **~construction** /-ʃn/ *n* reconstrucción *f*

record /rɪ'kɔ:d/ *vt* (*in register*) registrar; (*in diary*) apuntar; (*mus*) grabar. /'rekɔ:d/ *n* (*file*) documentación *f*, expediente *m*; (*mus*) disco *m*; (*sport*) récord *m*. **off the ~** en confianza. **~er** /rɪ'kɔ:də(r)/ *n* registrador *m*; (*mus*) flauta *f* dulce. **~ing** *n* grabación *f*. **~-player** *n* tocadiscos *m invar*

recount /rɪ'kaʊnt/ *vt* contar, relatar, referir

re-count /ri:'kaʊnt/ *vt* recontar. /'ri:kaʊnt/ *n* (*pol*) recuento *m*

recoup /rɪ'ku:p/ *vt* recuperar

recourse /rɪ'kɔ:s/ *n* recurso *m*. **have ~ to** recurrir a

recover /rɪ'kʌvə(r)/ *vt* recuperar. • *vi* reponerse. **~y** *n* recuperación *f*

recreation /rekrɪ'eɪʃn/ *n* recreo *m*. **~al** *a* de recreo

recrimination /rɪkrɪmɪ'neɪʃn/ *n* recriminación *f*

recruit /rɪ'kru:t/ *n* recluta *m*. • *vt* reclutar. **~ment** *n* reclutamiento *m*

rectang|le /'rektæŋgl/ *n* rectángulo *m*. **~ular** /-'tæŋgjʊlə(r)/ *a* rectangular

rectif|ication /rektɪfɪ'keɪʃn/ *n* rectificación *f*. **~y** /'rektɪfaɪ/ *vt* rectificar

rector /'rektə(r)/ *n* párroco *m*; (*of college*) rector *m*. **~y** *n* rectoría *f*

recumbent /rɪ'kʌmbənt/ *a* recostado

recuperat|e /rɪ'ku:pəreɪt/ *vt* recuperar. • *vi* reponerse. **~ion** /-'reɪʃn/ *n* recuperación *f*

recur /rɪ'kɜ:(r)/ *vi* (*pt* **recurred**) repetirse. **~rence** /rɪ'kʌrns/ *n* repetición *f*. **~rent** /rɪ'kʌrənt/ *a* repetido

recycle /ri:'saɪkl/ *vt* reciclar

red /red/ *a* **(redder, reddest)** rojo. • *n* rojo. **in the ~** ‹*account*› en descubierto. **~breast** /'redbrest/ *n* petirrojo *m*. **~brick** /'redbrɪk/ *a* ‹*univ*› de reciente fundación. **~den** /'redn/ *vt* enrojecer. • *vi* enrojecerse. **~dish** *a* rojizo

redecorate /ri:'dekəreɪt/ *vt* pintar de nuevo

rede|em /rɪ'di:m/ *vt* redimir. **~eming quality** *n* cualidad *f* compensadora. **~mption** /-'dempʃn/ *n* redención *f*

redeploy /ri:dɪ'plɔɪ/ *vt* disponer de otra manera; (*mil*) cambiar de frente

red: **~-handed** *a* en flagrante. **~ herring** *n* (*fig*) pista *f* falsa. **~-hot** *a* al rojo; ‹*news*› de última hora

Red Indian /red'ɪndjən/ *n* piel *m & f* roja

redirect /ri:daɪ'rekt/ *vt* reexpedir

red: **~-letter day** *n* día *m* señalado, día *m* memorable. **~ light** *n* luz *f* roja. **~ness** *n* rojez *f*

redo /ri:'du:/ *vt* (*pt* **redid**, *pp* **redone**) rehacer

redouble /rɪ'dʌbl/ *vt* redoblar

redress /rɪ'dres/ *vt* reparar. • *n* reparación *f*

red tape /red'teɪp/ *n* (*fig*) papeleo *m*

reduc|e /rɪ'dju:s/ *vt* reducir. • *vi* reducirse; (*slim*) adelgazar. **~tion** /'dʌkʃn/ *n* reducción *f*

redundan|cy /rɪ'dʌndənsɪ/ *n* superfluidad *f*; (*unemployment*) desempleo *m*. **~t** /rɪ'dʌndənt/ superfluo. **be made ~t** perder su empleo

reed /ri:d/ *n* caña *f*; (*mus*) lengüeta *f*

reef /ri:f/ *n* arrecife *m*

reek /ri:k/ *n* mal olor *m*. • *vi*. **~ (of)** apestar a

reel /ri:l/ *n* carrete *m*. • *vi* dar vueltas; (*stagger*) tambalearse. • *vt*. **~ off** (*fig*) enumerar

refectory /rɪ'fektərɪ/ *n* refectorio *m*

refer /rɪ'fɜ:(r)/ *vt* (*pt* **referred**) remitir. • *vi* referirse. **~ to** referirse a; (*consult*) consultar

referee /refə'ri:/ *n* árbitro *m*; (*for job*) referencia *f*. • *vi* (*pt* **refereed**) arbitrar

reference /'refrəns/ *n* referencia *f*. **~ book** *n* libro *m* de consulta. **in ~ to, with ~ to** en cuanto a; (*com*) respecto a

referendum /refə'rendəm/ *n* (*pl* **-ums**) referéndum *m*

refill /ri:'fɪl/ *vt* rellenar. /'ri:fɪl/ *n* recambio *m*

refine /rɪ'faɪn/ *vt* refinar. **~d** *a* refinado. **~ment** *n* refinamiento *m*; (*tec*) refinación *f*. **~ry** /-ərɪ/ *n* refinería *f*

reflect /rɪ'flekt/ *vt* reflejar. ● *vi* reflejar; (*think*) reflexionar. **~ upon** perjudicar. **~ion** /-ʃn/ *n* reflexión *f*; (*image*) reflejo *m*. **~ive** /rɪ'flektɪv/ *a* reflector; (*thoughtful*) pensativo. **~or** *n* reflector *m*

reflex /'ri:fleks/ *a* & *n* reflejo (*m*)

reflexive /rɪ'fleksɪv/ *a* (*gram*) reflexivo

reform /rɪ'fɔ:m/ *vt* reformar. ● *vi* reformarse. ● *n* reforma *f*. **~er** *n* reformador *m*

refract /rɪ'frækt/ *vt* refractar

refrain[1] /rɪ'freɪn/ *n* estribillo *m*

refrain[2] /rɪ'freɪn/ *vi* abstenerse (**from** de)

refresh /rɪ'freʃ/ *vt* refrescar. **~er** /rɪ'freʃə(r)/ *a* ‹*course*› de repaso. **~ing** *a* refrescante. **~ments** *npl* (*food and drink*) refrigerio *m*

refrigerat|e /rɪ'frɪdʒəreɪt/ *vt* refrigerar. **~or** *n* nevera *f*, refrigeradora *f* (*LAm*)

refuel /ri:'fju:əl/ *vt*/*i* (*pt* **refuelled**) repostar

refuge /'refju:dʒ/ *n* refugio *m*. **take ~** refugiarse. **~e** /refjʊ'dʒi:/ *n* refugiado *m*

refund /rɪ'fʌnd/ *vt* reembolsar. /'ri:fʌnd/ *n* reembolso *m*

refurbish /ri:'fɜ:bɪʃ/ *vt* renovar

refusal /rɪ'fju:zl/ *n* negativa *f*

refuse[1] /rɪ'fju:z/ *vt* rehusar. ● *vi* negarse

refuse[2] /'refju:s/ *n* basura *f*

refute /rɪ'fju:t/ *vt* refutar

regain /rɪ'geɪn/ *vt* recobrar

regal /'ri:gl/ *a* real

regale /rɪ'geɪl/ *vt* festejar

regalia /rɪ'geɪlɪə/ *npl* insignias *fpl*

regard /rɪ'gɑ:d/ *vt* mirar; (*consider*) considerar. **as ~s** en cuanto a. ● *n* mirada *f*; (*care*) atención *f*; (*esteem*) respeto *m*. **~ing** *prep* en cuanto a. **~less** /rɪ'gɑ:dlɪs/ *adv* a pesar de todo. **~less of** sin tener en cuenta. **~s** *npl* saludos *mpl*. **kind ~s** *npl* recuerdos *mpl*

regatta /rɪ'gætə/ *n* regata *f*

regency /'ri:dʒənsɪ/ *n* regencia *f*

regenerate /rɪ'dʒenəreɪt/ *vt* regenerar

regent /'ri:dʒənt/ *n* regente *m* & *f*

regime /reɪ'ʒi:m/ *n* régimen *m*

regiment /'redʒɪmənt/ *n* regimiento *m*. **~al** /-'mentl/ *a* del regimiento. **~ation** /-en'teɪʃn/ *n* reglamentación *f* rígida

region /'ri:dʒən/ *n* región *f*. **in the ~ of** alrededor de. **~al** *a* regional

register /'redʒɪstə(r)/ *n* registro *m*. ● *vt* registrar; matricular ‹*vehicle*›; declarar ‹*birth*›; certificar ‹*letter*›; facturar ‹*luggage*›; (*indicate*) indicar; (*express*) expresar. ● *vi* (*enrol*) inscribirse; (*fig*) producir impresión. **~ office** *n* registro *m* civil

registrar /redʒɪ'strɑ:(r)/ *n* secretario *m* del registro civil; (*univ*) secretario *m* general

registration /redʒɪ'streɪʃn/ *n* registración *f*; (*in register*) inscripción *f*; (*of vehicle*) matrícula *f*

registry /'redʒɪstrɪ/ *n*. **~ office** *n* registro *m* civil

regression /rɪ'greʃn/ *n* regresión *f*

regret /rɪ'gret/ *n* pesar *m*. ● *vt* (*pt* **regretted**) lamentar. **I ~ that** siento (que). **~fully** *adv* con pesar. **~table** *a* lamentable. **~tably** *adv* lamentablemente

regular /'regjʊlə(r)/ *a* regular; (*usual*) habitual. ● *n* (*fam*) cliente *m* habitual. **~ity** /-'lærətɪ/ *n* regularidad *f*. **~ly** *adv* regularmente

regulat|e /'regjʊleɪt/ *vt* regular. **~ion** /-'leɪʃn/ *n* arreglo *m*; (*rule*) regla *f*

rehabilitat|e /ri:hə'bɪlɪteɪt/ *vt* rehabilitar. **~ion** /-'teɪʃn/ *n* rehabilitación *f*

rehash /ri:'hæʃ/ *vt* volver a presentar. /'ri:hæʃ/ *n* refrito *m*

rehears|al /rɪ'hɜ:sl/ *n* ensayo *m*. **~e** /rɪ'hɜ:s/ *vt* ensayar

reign /reɪn/ *n* reinado *m*. ● *vi* reinar

reimburse /ri:ɪm'bɜ:s/ *vt* reembolsar

reins /reɪnz/ *npl* riendas *fpl*

reindeer /'reɪndɪə(r)/ *n invar* reno *m*

reinforce /ri:ɪn'fɔ:s/ *vt* reforzar. **~ment** *n* refuerzo *m*

reinstate /ri:ɪn'steɪt/ *vt* reintegrar

reiterate /ri:'ɪtəreɪt/ *vt* reiterar

reject /rɪ'dʒekt/ *vt* rechazar. /'ri:dʒekt/ *n* producto *m* defectuoso. **~ion** /'dʒekʃn/ *n* rechazamiento *m*, rechazo *m*

rejoic|e /rɪ'dʒɔɪs/ *vi* regocijarse. **~ing** *n* regocijo *m*

rejoin /rɪ'dʒɔɪn/ *vt* reunirse con; (*answer*) replicar. **~der** /rɪ'dʒɔɪndə(r)/ *n* réplica *f*

rejuvenate /rɪ'dʒu:vəneɪt/ *vt* rejuvenecer
rekindle /ri:'kɪndl/ *vt* reavivar
relapse /rɪ'læps/ *n* recaída *f*. • *vi* recaer; (*into crime*) reincidir
relate /rɪ'leɪt/ *vt* contar; (*connect*) relacionar. • *vi* relacionarse (**to** con). **~d** *a* emparentado; ‹*ideas etc*› relacionado
relation /rɪ'leɪʃn/ *n* relación *f*; (*person*) pariente *m* & *f*. **~ship** *n* relación *f*; (*blood tie*) parentesco *m*; (*affair*) relaciones *fpl*
relative /'relətɪv/ *n* pariente *m* & *f*. • *a* relativo. **~ly** *adv* relativamente
relax /rɪ'læks/ *vt* relajar. • *vi* relajarse. **~ation** /ri:læk'seɪʃn/ *n* relajación *f*; (*rest*) descanso *m*; (*recreation*) recreo *m*. **~ing** *a* relajante
relay /'ri:leɪ/ *n* relevo *m*. **~ (race)** *n* carrera *f* de relevos. /rɪ'leɪ/ *vt* retransmitir
release /rɪ'li:s/ *vt* soltar; poner en libertad ‹*prisoner*›; lanzar ‹*bomb*›; estrenar ‹*film*›; (*mec*) desenganchar; publicar ‹*news*›; emitir ‹*smoke*›. • *n* liberación *f*; (*of film*) estreno *m*; (*record*) disco *m* nuevo
relegate /'relɪgeɪt/ *vt* relegar
relent /rɪ'lent/ *vi* ceder. **~less** *a* implacable; (*continuous*) incesante
relevan|ce /'reləvəns/ *n* pertinencia *f*. **~t** /'reləvənt/ *a* pertinente
reliab|ility /rɪlaɪə'bɪlətɪ/ *n* fiabilidad *f*. **~le** /rɪ'laɪəbl/ *a* seguro; ‹*person*› de fiar; (*com*) serio
relian|ce /rɪ'laɪəns/ *n* dependencia *f*; (*trust*) confianza *f*. **~t** *a* confiado
relic /'relɪk/ *n* reliquia *f*. **~s** *npl* restos *mpl*
relie|f /rɪ'li:f/ *n* alivio *m*; (*assistance*) socorro *m*; (*outline*) relieve *m*. **~ve** /rɪ'li:v/ *vt* aliviar; (*take over from*) relevar
religio|n /rɪ'lɪdʒən/ *n* religión *f*. **~us** /rɪ'lɪdʒəs/ *a* religioso
relinquish /rɪ'lɪnkwɪʃ/ *vt* abandonar, renunciar
relish /'relɪʃ/ *n* gusto *m*; (*culin*) salsa *f*. • *vt* saborear. **I don't ~ the idea** no me gusta la idea
relocate /ri:ləʊ'keɪt/ *vt* colocar de nuevo
reluctan|ce /rɪ'lʌktəns/ *n* desgana *f*. **~t** /rɪ'lʌktənt/ *a* mal dispuesto. **be ~t to** no tener ganas de. **~tly** *adv* de mala gana
rely /rɪ'laɪ/ *vi*. **~ on** contar con; (*trust*) fiarse de; (*depend*) depender
remain /rɪ'meɪn/ *vi* quedar. **~der** /rɪ'meɪndə(r)/ *n* resto *m*. **~s** *npl* restos *mpl*; (*left-overs*) sobras *fpl*
remand /rɪ'mɑ:nd/ *vt*. **~ in custody** mantener bajo custodia. • *n*. **on ~** bajo custodia
remark /rɪ'mɑ:k/ *n* observación *f*. • *vt* observar. **~able** *a* notable
remarry /ri:'mærɪ/ *vi* volver a casarse
remedial /rɪ'mi:dɪəl/ *a* remediador
remedy /'remədɪ/ *n* remedio *m*. • *vt* remediar
rememb|er /rɪ'membə(r)/ *vt* acordarse de. • *vi* acordarse. **~rance** *n* recuerdo *m*
remind /rɪ'maɪnd/ *vt* recordar. **~er** *n* recordatorio *m*; (*letter*) notificación *f*
reminisce /remɪ'nɪs/ *vi* recordar el pasado. **~nces** *npl* recuerdos *mpl*. **~nt** /remɪ'nɪsnt/ *a*. **be ~nt of** recordar
remiss /rɪ'mɪs/ *a* negligente
remission /rɪ'mɪʃn/ *n* remisión *f*; (*of sentence*) reducción *f* de condena
remit /rɪ'mɪt/ *vt* (*pt* **remitted**) perdonar; enviar ‹*money*›. • *vi* moderarse. **~tance** *n* remesa *f*
remnant /'remnənt/ *n* resto *m*; (*of cloth*) retazo *m*; (*trace*) vestigio *m*
remonstrate /'remənstreɪt/ *vi* protestar
remorse /rɪ'mɔ:s/ *n* remordimiento *m*. **~ful** *a* lleno de remordimiento. **~less** *a* implacable
remote /rɪ'məʊt/ *a* remoto; (*slight*) leve; ‹*person*› distante. **~ control** *n* mando *m* a distancia. **~ly** *adv* remotamente. **~ness** *n* lejanía *f*; (*isolation*) aislamiento *m*, alejamiento *m*; (*fig*) improbabilidad *f*
remov|able /rɪ'mu:vəbl/ *a* movible; (*detachable*) de quita y pon, separable. **~al** *n* eliminación *f*; (*from house*) mudanza *f*. **~e** /rɪ'mu:v/ *vt* quitar; (*dismiss*) despedir; (*get rid of*) eliminar; (*do away with*) suprimir
remunerat|e /rɪ'mju:nəreɪt/ *vt* remunerar. **~ion** /-'reɪʃn/ *n* remuneración *f*. **~ive** *a* remunerador
Renaissance /rə'neɪsəns/ *n* Renacimiento *m*
rend /rend/ *vt* (*pt* **rent**) rasgar

render /'rendə(r)/ *vt* rendir; (*com*) presentar; (*mus*) interpretar; prestar ‹*help etc*›. **~ing** *n* (*mus*) interpretación *f*

rendezvous /'rɒndɪvu:/ *n* (*pl* **-vous** /-vu:z/) cita *f*

renegade /'renɪgeɪd/ *n* renegado

renew /rɪ'nju:/ *vt* renovar; (*resume*) reanudar. **~able** *a* renovable. **~al** *n* renovación *f*

renounce /rɪ'naʊns/ *vt* renunciar a; (*disown*) repudiar

renovat|e /'renəveɪt/ *vt* renovar. **~ion** /-'veɪʃn/ *n* renovación *f*

renown /rɪ'naʊn/ *n* fama *f*. **~ed** *a* célebre

rent[1] /rent/ *n* alquiler *m*. ● *vt* alquilar

rent[2] /rent/ *see* **rend**

rental /rentl/ *n* alquiler *m*

renunciation /rɪnʌnsɪ'eɪʃn/ *n* renuncia *f*

reopen /ri:'əʊpən/ *vt* reabrir. ● *vi* reabrirse. **~ing** *n* reapertura *f*

reorganize /ri:'ɔ:gənaɪz/ *vt* reorganizar

rep[1] /rep/ *n* (*com, fam*) representante *m & f*

rep[2] /rep/ (*theatre, fam*) teatro *m* de repertorio

repair /rɪ'peə(r)/ *vt* reparar; remendar ‹*clothes, shoes*›. ● *n* reparación *f*; (*patch*) remiendo *m*. **in good ~** en buen estado

repartee /repɑ:'ti:/ *n* ocurrencias *fpl*

repatriat|e /ri:'pætrɪeɪt/ *vt* repatriar. **~ion** /-'eɪʃn/ *n* repatriación *f*

repay /ri:'peɪ/ *vt* (*pt* **repaid**) reembolsar; pagar ‹*debt*›; (*reward*) recompensar. **~ment** *n* reembolso *m*, pago *m*

repeal /rɪ'pi:l/ *vt* abrogar. ● *n* abrogación *f*

repeat /rɪ'pi:t/ *vt* repetir. ● *vi* repetir(se). ● *n* repetición *f*. **~edly** /rɪ'pi:tɪdlɪ/ *adv* repetidas veces

repel /rɪ'pel/ *vt* (*pt* **repelled**) repeler. **~lent** *a* repelente

repent /rɪ'pent/ *vi* arrepentirse. **~ance** *n* arrepentimiento *m*. **~ant** *a* arrepentido

repercussion /ri:pə'kʌʃn/ *n* repercusión *f*

reperto|ire /'repətwɑ:(r)/ *n* repertorio *m*. **~ry** /'repətrɪ/ *n* repertorio *m*. **~ry (theatre)** *n* teatro *m* de repertorio

repetit|ion /repɪ'tɪʃn/ *n* repetición *f*. **~ious** /-'tɪʃəs/ *a*, **~ive** /rɪ'petətɪv/ *a* que se repite; (*dull*) monótono

replace /rɪ'pleɪs/ *vt* reponer; (*take the place of*) sustituir. **~ment** *n* sustitución *f*; (*person*) sustituto *m*. **~ment part** *n* recambio *m*

replay /'ri:pleɪ/ *n* (*sport*) repetición *f* del partido; (*recording*) repetición *f* inmediata

replenish /rɪ'plenɪʃ/ *vt* reponer; (*refill*) rellenar

replete /rɪ'pli:t/ *a* repleto

replica /'replɪkə/ *n* copia *f*

reply /rɪ'plaɪ/ *vt/i* contestar. ● *n* respuesta *f*

report /rɪ'pɔ:t/ *vt* anunciar; (*denounce*) denunciar. ● *vi* presentar un informe; (*present o.s.*) presentarse. ● *n* informe *m*; (*schol*) boletín *m*; (*rumour*) rumor *m*; (*newspaper*) reportaje *m*; (*sound*) estallido *m*. **~age** /repɔ:'tɑ:ʒ/ *n* reportaje *m*. **~edly** *adv* según se dice. **~er** /rɪ'pɔ:tə(r)/ *n* reportero *m*, informador *m*

repose /rɪ'pəʊz/ *n* reposo *m*

repository /rɪ'pɒzɪtrɪ/ *n* depósito *m*

repossess /ri:pə'zes/ *vt* recuperar

reprehen|d /reprɪ'hend/ *vt* reprender. **~sible** /-səbl/ *a* reprensible

represent /reprɪ'zent/ *vt* representar. **~ation** /-'teɪʃn/ *n* representación *f*. **~ative** /reprɪ'zentətɪv/ *a* representativo. ● *n* representante *m & f*

repress /rɪ'pres/ *vt* reprimir. **~ion** /-ʃn/ *n* represión *f*. **~ive** *a* represivo

reprieve /rɪ'pri:v/ *n* indulto *m*; (*fig*) respiro *m*. ● *vt* indultar; (*fig*) aliviar

reprimand /'reprɪmɑ:nd/ *vt* reprender. ● *n* reprensión *f*

reprint /'ri:prɪnt/ *n* reimpresión *f*; (*offprint*) tirada *f* aparte. /ri:'prɪnt/ *vt* reimprimir

reprisal /rɪ'praɪzl/ *n* represalia *f*

reproach /rɪ'prəʊtʃ/ *vt* reprochar. ● *n* reproche *m*. **~ful** *a* de reproche, reprobador. **~fully** *adv* con reproche

reprobate /'reprəbeɪt/ *n* malvado *m*; (*relig*) réprobo *m*

reproduc|e /ri:prə'dju:s/ *vt* reproducir. ● *vi* reproducirse. **~tion** /-'dʌkʃn/ *n* reproducción *f*. **~tive** /-'dʌktɪv/ *a* reproductor

reprove /rɪ'pru:v/ *vt* reprender

reptile /'reptaɪl/ *n* reptil *m*

republic /rɪ'pʌblɪk/ *n* república *f*. **~an** *a & n* republicano (*m*)
repudiate /rɪ'pju:dɪeɪt/ *vt* repudiar; (*refuse to recognize*) negarse a reconocer
repugnan|ce /rɪ'pʌgnəns/ *n* repugnancia *f*. **~t** /rɪ'pʌgnənt/ *a* repugnante
repuls|e /rɪ'pʌls/ *vt* rechazar, repulsar. **~ion** /-ʃn/ *n* repulsión *f*. **~ive** *a* repulsivo
reputable /'repjʊtəbl/ *a* acreditado, de confianza, honroso
reputation /repjʊ'teɪʃn/ *n* reputación *f*
repute /rɪ'pju:t/ *n* reputación *f*. **~d** /-ɪd/ *a* supuesto. **~dly** *adv* según se dice
request /rɪ'kwest/ *n* petición *f*. ● *vt* pedir. **~ stop** *n* parada *f* discrecional
require /rɪ'kwaɪə(r)/ *vt* requerir; (*need*) necesitar; (*demand*) exigir. **~d** *a* necesario. **~ment** *n* requisito *m*
requisite /'rekwɪzɪt/ *a* necesario. ● *n* requisito *m*
requisition /rekwɪ'zɪʃn/ *n* requisición *f*. ● *vt* requisar
resale /'ri:seɪl/ *n* reventa *f*
rescind /rɪ'sɪnd/ *vt* rescindir
rescue /'reskju:/ *vt* salvar. ● *n* salvamento *m*. **~r** /-ə(r)/ *n* salvador *m*
research /rɪ'sɜ:tʃ/ *n* investigación *f*. ● *vt* investigar. **~er** *n* investigador *m*
resembl|ance /rɪ'zembləns/ *n* parecido *m*. **~e** /rɪ'zembl/ *vt* parecerse a
resent /rɪ'zent/ *vt* resentirse por. **~ful** *a* resentido. **~ment** *n* resentimiento *m*
reservation /rezə'veɪʃn/ *n* reserva *f*; (*booking*) reservación *f*
reserve /rɪ'zɜ:v/ *vt* reservar. ● *n* reserva *f*; (*in sports*) suplente *m & f*. **~d** *a* reservado
reservist /rɪ'zɜ:vɪst/ *n* reservista *m & f*
reservoir /'rezəvwɑ:(r)/ *n* embalse *m*; (*tank*) depósito *m*
reshape /ri:'ʃeɪp/ *vt* formar de nuevo, reorganizar
reshuffle /ri:'ʃʌfl/ *vt* (*pol*) reorganizar. ● *n* (*pol*) reorganización *f*
reside /rɪ'zaɪd/ *vi* residir
residen|ce /'rezɪdəns/ *n* residencia *f*. **~ce permit** *n* permiso *m* de residencia. **be in ~ce** ‹*doctor etc*› interno. **~t** /'rezɪdənt/ *a & n* residente (*m & f*). **~tial** /rezɪ'denʃl/ *a* residencial
residue /'rezɪdju:/ *n* residuo *m*
resign /rɪ'zaɪn/ *vt/i* dimitir. **~ o.s. to** resignarse a. **~ation** /rezɪg'neɪʃn/ *n* resignación *f*; (*from job*) dimisión *f*. **~ed** *a* resignado
resilien|ce /rɪ'zɪlɪəns/ *n* elasticidad *f*; (*of person*) resistencia *f*. **~t** /rɪ'zɪlɪənt/ *a* elástico; ‹*person*› resistente
resin /'rezɪn/ *n* resina *f*
resist /rɪ'zɪst/ *vt* resistir. ● *vi* resistirse. **~ance** *n* resistencia *f*. **~ant** *a* resistente
resolut|e /'rezəlu:t/ *a* resuelto. **~ion** /-'lu:ʃn/ *n* resolución *f*
resolve /rɪ'zɒlv/ *vt* resolver. **~ to do** resolverse a hacer. ● *n* resolución *f*. **~d** *a* resuelto
resonan|ce /'rezənəns/ *n* resonancia *f*. **~t** /'rezənənt/ *a* resonante
resort /rɪ'zɔ:t/ *vi*. **~ to** recurrir a. ● *n* recurso *m*; (*place*) lugar *m* turístico. **in the last ~** como último recurso
resound /rɪ'zaʊnd/ *vi* resonar. **~ing** *a* resonante
resource /rɪ'sɔ:s/ *n* recurso *m*. **~ful** *a* ingenioso. **~fulness** *n* ingeniosidad *f*
respect /rɪ'spekt/ *n* (*esteem*) respeto *m*; (*aspect*) respecto *m*. **with ~ to** con respecto a. ● *vt* respetar
respectab|ility /rɪspektə'bɪlətɪ/ *n* respetabilidad *f*. **~le** /rɪ'spektəbl/ *a* respetable. **~ly** *adv* respetablemente
respectful /rɪ'spektfl/ *a* respetuoso
respective /rɪ'spektɪv/ *a* respectivo. **~ly** *adv* respectivamente
respiration /respə'reɪʃn/ *n* respiración *f*
respite /'respaɪt/ *n* respiro *m*, tregua *f*
resplendent /rɪ'splendənt/ *a* resplandeciente
respon|d /rɪ'spɒnd/ *vi* responder. **~se** /rɪ'spɒns/ *n* respuesta *f*; (*reaction*) reacción *f*
responsib|ility /rɪspɒnsə'bɪlətɪ/ *n* responsabilidad *f*. **~le** /rɪ'spɒnsəbl/ *a* responsable; ‹*job*› de responsabilidad. **~ly** *adv* con formalidad
responsive /rɪ'spɒnsɪv/ *a* que reacciona bien. **~ to** sensible a
rest[1] /rest/ *vt* descansar; (*lean*) apoyar; (*place*) poner, colocar. ● *vi*

descansar; (*lean*) apoyarse. ● *n* descanso *m*; (*mus*) pausa *f*

rest² /rest/ *n* (*remainder*) resto *m*, lo demás; (*people*) los demás, los otros *mpl*. ● *vi* (*remain*) quedar

restaurant /'restərɒnt/ *n* restaurante *m*

restful /'restfl/ *a* sosegado

restitution /restɪ'tju:ʃn/ *n* restitución *f*

restive /'restɪv/ *a* inquieto

restless /'restlɪs/ *a* inquieto. **~ly** *adv* inquietamente. **~ness** *n* inquietud *f*

restor|ation /restə'reɪʃn/ *n* restauración *f*. **~e** /rɪ'stɔ:(r)/ *vt* restablecer; restaurar ‹*building*›; (*put back in position*) reponer; (*return*) devolver

restrain /rɪ'streɪn/ *vt* contener. **~ o.s.** contenerse. **~ed** *a* (*moderate*) moderado; (*in control of self*) comedido. **~t** *n* restricción *f*; (*moderation*) moderación *f*

restrict /rɪ'strɪkt/ *vt* restringir. **~ion** /-ʃn/ *n* restricción *f*. **~ive** /rɪ'strɪktɪv/ *a* restrictivo

result /rɪ'zʌlt/ *n* resultado *m*. ● *vi*. **~ from** resultar de. **~ in** dar como resultado

resume /rɪ'zju:m/ *vt* reanudar. ● *vi* continuar

résumé /'rezjʊmeɪ/ *n* resumen *m*

resumption /rɪ'zʌmpʃn/ *n* continuación *f*

resurgence /rɪ'sɜ:dʒəns/ *n* resurgimiento *m*

resurrect /rezə'rekt/ *vt* resucitar. **~ion** /-ʃn/ *n* resurrección *f*

resuscitat|e /rɪ'sʌsɪteɪt/ *vt* resucitar. **~ion** /-'teɪʃn/ *n* resucitación *f*

retail /'ri:teɪl/ *n* venta *f* al por menor. ● *a & adv* al por menor. ● *vt* vender al por menor. ● *vi* venderse al por menor. **~er** *n* minorista *m & f*

retain /rɪ'teɪn/ *vt* retener; (*keep*) conservar

retainer /rɪ'teɪnə(r)/ *n* (*fee*) anticipo *m*

retaliat|e /rɪ'tælɪeɪt/ *vi* desquitarse. **~ion** /-'eɪʃn/ *n* represalias *fpl*

retarded /rɪ'tɑ:dɪd/ *a* retrasado

retentive /rɪ'tentɪv/ *a* ‹*memory*› bueno

rethink /ri:'θɪŋk/ *vt* (*pt* **rethought**) considerar de nuevo

reticen|ce /'retɪsns/ *n* reserva *f*. **~t** /'retɪsnt/ *a* reservado, callado

retina /'retɪnə/ *n* retina *f*

retinue /'retɪnju:/ *n* séquito *m*

retir|e /rɪ'taɪə(r)/ *vi* (*from work*) jubilarse; (*withdraw*) retirarse; (*go to bed*) acostarse. ● *vt* jubilar. **~ed** *a* jubilado. **~ement** *n* jubilación *f*. **~ing** /rɪ'taɪərɪŋ/ *a* reservado

retort /rɪ'tɔ:t/ *vt/i* replicar. ● *n* réplica *f*

retrace /ri:'treɪs/ *vt* repasar. **~ one's steps** volver sobre sus pasos

retract /rɪ'trækt/ *vt* retirar. ● *vi* retractarse

retrain /ri:'treɪn/ *vt* reciclar, reeducar

retreat /rɪ'tri:t/ *vi* retirarse. ● *n* retirada *f*; (*place*) refugio *m*

retrial /ri:'traɪəl/ *n* nuevo proceso *m*

retribution /retrɪ'bju:ʃn/ *n* justo *m* castigo

retriev|al /rɪ'tri:vl/ *n* recuperación *f*. **~e** /rɪ'tri:v/ *vt* (*recover*) recuperar; (*save*) salvar; (*put right*) reparar. **~er** *n* (*dog*) perro *m* cobrador

retrograde /'retrəgreɪd/ *a* retrógrado

retrospect /'retrəspekt/ *n* retrospección *f*. **in ~** retrospectivamente. **~ive** /-'spektɪv/ *a* retrospectivo

return /rɪ'tɜ:n/ *vi* volver; (*reappear*) reaparecer. ● *vt* devolver; (*com*) declarar; (*pol*) elegir. ● *n* vuelta *f*; (*com*) ganancia *f*; (*restitution*) devolución *f*. **~ of income** *n* declaración *f* de ingresos. **in ~ for** a cambio de. **many happy ~s!** ¡feliz cumpleaños! **~ing** /rɪ'tɜ:nɪŋ/ *a*. **~ing officer** *n* escrutador *m*. **~ match** *n* partido *m* de desquite. **~ ticket** *n* billete *m* de ida y vuelta. **~s** *npl* (*com*) ingresos *mpl*

reunion /ri:'ju:nɪən/ *n* reunión *f*

reunite /ri:ju:'naɪt/ *vt* reunir

rev /rev/ *n* (*auto, fam*) revolución *f*. ● *vt/i*. **~ (up)** (*pt* **revved**) (*auto, fam*) acelerar(se)

revamp /ri:'væmp/ *vt* renovar

reveal /rɪ'vi:l/ *vt* revelar. **~ing** *a* revelador

revel /'revl/ *vi* (*pt* **revelled**) jaranear. **~ in** deleitarse en. **~ry** *n* juerga *f*

revelation /revə'leɪʃn/ *n* revelación *f*

revenge /rɪ'vendʒ/ *n* venganza *f*; (*sport*) desquite *m*. **take ~** vengarse. ● *vt* vengar. **~ful** *a* vindicativo, vengativo

revenue /'revənju:/ *n* ingresos *mpl*

reverberate /rɪ'vɜ:bəreɪt/ *vi* ‹*light*› reverberar; ‹*sound*› resonar
revere /rɪ'vɪə(r)/ *vt* venerar
reverence /'revərəns/ *n* reverencia *f*
reverend /'revərənd/ *a* reverendo
reverent /'revərənt/ *a* reverente
reverie /'revərɪ/ *n* ensueño *m*
revers /rɪ'vɪə/ *n* (*pl* **revers** /rɪ'vɪəz/) *n* solapa *f*
revers|al /rɪ'vɜ:sl/ *n* inversión *f*. **~e** /rɪ'vɜ:s/ *a* inverso. ● *n* contrario *m*; (*back*) revés *m*; (*auto*) marcha *f* atrás. ● *vt* invertir; anular ‹*decision*›; (*auto*) dar marcha atrás a. ● *vi* (*auto*) dar marcha atrás
revert /rɪ'vɜ:t/ *vi*. **~ to** volver a
review /rɪ'vju:/ *n* repaso *m*; (*mil*) revista *f*; (*of book, play, etc*) crítica *f*. ● *vt* analizar ‹*situation*›; reseñar ‹*book, play, etc*›. **~er** *n* crítico *m*
revile /rɪ'vaɪl/ *vt* injuriar
revis|e /rɪ'vaɪz/ *vt* revisar; (*schol*) repasar. **~ion** /-ɪʒn/ *n* revisión *f*; (*schol*) repaso *m*
reviv|al /rɪ'vaɪvl/ *n* restablecimiento *m*; (*of faith*) despertar *m*; (*of play*) reestreno *m*. **~e** /rɪ'vaɪv/ *vt* restablecer; resucitar ‹*person*›. ● *vi* restablecerse; ‹*person*› volver en sí
revoke /rɪ'vəʊk/ *vt* revocar
revolt /rɪ'vəʊlt/ *vi* sublevarse. ● *vt* dar asco a. ● *n* sublevación *f*
revolting /rɪ'vəʊltɪŋ/ *a* asqueroso
revolution /revə'lu:ʃn/ *n* revolución *f*. **~ary** *a & n* revolucionario (*m*). **~ize** *vt* revolucionar
revolve /rɪ'vɒlv/ *vi* girar
revolver /rɪ'vɒlvə(r)/ *n* revólver *m*
revolving /rɪ'vɒlvɪŋ/ *a* giratorio
revue /rɪ'vju:/ *n* revista *f*
revulsion /rɪ'vʌlʃn/ *n* asco *m*
reward /rɪ'wɔ:d/ *n* recompensa *f*. ● *vt* recompensar. **~ing** *a* remunerador; (*worthwhile*) que vale la pena
rewrite /ri:'raɪt/ *vt* (*pt* **rewrote,** *pp* **rewritten**) escribir de nuevo; (*change*) redactar de nuevo
rhapsody /'ræpsədɪ/ *n* rapsodia *f*
rhetoric /'retərɪk/ *n* retórica *f*. **~al** /rɪ'tɒrɪkl/ *a* retórico
rheumati|c /ru:'mætɪk/ *a* reumático. **~sm** /'ru:mətɪzəm/ *n* reumatismo *m*
rhinoceros /raɪ'nɒsərəs/ *n* (*pl* **-oses**) rinoceronte *m*
rhubarb /'ru:bɑ:b/ *n* ruibarbo *m*
rhyme /raɪm/ *n* rima *f*; (*poem*) poesía *f*. ● *vt/i* rimar
rhythm /'rɪðəm/ *n* ritmo *m*. **~ic(al)** /'rɪðmɪk(l)/ *a* rítmico
rib /rɪb/ *n* costilla *f*. —**vt** (*pt* **ribbed**) (*fam*) tomar el pelo a
ribald /'rɪbld/ *a* obsceno, verde
ribbon /'rɪbən/ *n* cinta *f*
rice /raɪs/ *n* arroz *m*. **~ pudding** *n* arroz con leche
rich /rɪtʃ/ *a* (**-er, -est**) rico. ● *n* ricos *mpl*. **~es** *npl* riquezas *fpl*. **~ly** *adv* ricamente. **~ness** *n* riqueza *f*
rickety /'rɪkətɪ/ *a* (*shaky*) cojo, desvencijado
ricochet /'rɪkəʃeɪ/ *n* rebote *m*. ● *vi* rebotar
rid /rɪd/ *vt* (*pt* **rid,** *pres p* **ridding**) librar (**of** de). **get ~ of** deshacerse de. **~dance** /'rɪdns/ *n*. **good ~dance!** ¡qué alivio!
ridden /'rɪdn/ *see* **ride**. ● *a* (*infested*) infestado. **~ by** (*oppressed*) agobiado de
riddle[1] /'rɪdl/ *n* acertijo *m*
riddle[2] /'ridl/ *vt* acribillar. **be ~d with** estar lleno de
ride /raɪd/ *vi* (*pt* **rode,** *pp* **ridden**) (*on horseback*) montar; (*go*) ir (en bicicleta, a caballo etc). **take s.o. for a ~** (*fam*) engañarle a uno. ● *vt* montar a ‹*horse*›; ir en ‹*bicycle*›; recorrer ‹*distance*›. ● *n* (*on horse*) cabalgata *f*; (*in car*) paseo *m* en coche. **~r** /-ə(r)/ *n* (*on horse*) jinete *m*; (*cyclist*) ciclista *m & f*; (*in document*) cláusula *f* adicional
ridge /rɪdʒ/ *n* línea *f*, arruga *f*; (*of mountain*) cresta *f*; (*of roof*) caballete *m*
ridicul|e /'rɪdɪkju:l/ *n* irrisión *f*. ● *vt* ridiculizar. **~ous** /rɪ'dɪkjʊləs/ *a* ridículo
riding /'raɪdɪŋ/ *n* equitación *f*
rife /raɪf/ *a* difundido. **~ with** lleno de
riff-raff /'rɪfræf/ *n* gentuza *f*
rifle[1] /'raɪfl/ *n* fusil *m*
rifle[2] /'raɪfl/ *vt* saquear
rifle-range /'raɪflreɪndʒ/ *n* campo *m* de tiro
rift /rɪft/ *n* grieta *f*; (*fig*) ruptura *f*
rig[1] /rɪg/ *vt* (*pt* **rigged**) aparejar. ● *n* (*at sea*) plataforma *f* de perforación. **~ up** *vt* improvisar
rig[2] /rɪg/ *vt* (*pej*) amañar
right /raɪt/ *a* (*correct, fair*) exacto, justo; (*morally*) bueno; (*not left*) derecho; (*suitable*) adecuado. ● *n* (*entitlement*) derecho *m*; (*not left*) derecha *f*; (*not evil*) bien *m*. **~ of**

way *n* (*auto*) prioridad *f*. **be in the ~** tener razón. **on the ~** a la derecha. **put ~** rectificar. ● *vt* enderezar; (*fig*) corregir. ● *adv* a la derecha; (*directly*) derecho; (*completely*) completamente; (*well*) bien. **~ away** *adv* inmediatamente. **~ angle** *n* ángulo *m* recto

righteous /'raɪtʃəs/ *a* recto; ‹*cause*› justo

right: **~ful** /'raɪtfl/ *a* legítimo. **~fully** *adv* legítimamente. **~-hand man** *n* brazo *m* derecho. **~ly** *adv* justamente. **~ wing** *a* (*pol*) *n* derechista

rigid /'rɪdʒɪd/ *a* rígido. **~ity** /-'dʒɪdətɪ/ *n* rigidez *f*

rigmarole /'rɪgmərəʊl/ *n* galimatías *m invar*

rig|orous /'rɪgərəs/ *a* riguroso. **~our** /'rɪgə(r)/ *n* rigor *m*

rig-out /'rɪgaʊt/ *n* (*fam*) atavío *m*

rile /raɪl/ *vt* (*fam*) irritar

rim /rɪm/ *n* borde *m*; (*of wheel*) llanta *f*; (*of glasses*) montura *f*. **~med** *a* bordeado

rind /raɪnd/ *n* corteza *f*; (*of fruit*) cáscara *f*

ring[1] /rɪŋ/ *n* (*circle*) círculo *m*; (*circle of metal etc*) aro *m*; (*on finger*) anillo *m*; (*on finger with stone*) sortija *f*; (*boxing*) cuadrilátero *m*; (*bullring*) ruedo *m*, redondel *m*, plaza *f*; (*for circus*) pista *f*; ● *vt* rodear

ring[2] /rɪŋ/ *n* (*of bell*) toque *m*; (*tinkle*) tintineo *m*; (*telephone call*) llamada *f*. ● *vt* (*pt* **rang**, *pp* **rung**) hacer sonar; (*telephone*) llamar por teléfono. **~ the bell** tocar el timbre. ● *v* sonar. **~ back** *vt/i* volver a llamar. **~ off** *vi* colgar. **~ up** *vt* llamar por teléfono

ring: **~leader** /'rɪŋli:də(r)/ *n* cabecilla *f*. **~ road** *n* carretera *f* de circunvalación

rink /rɪŋk/ *n* pista *f*

rinse /rɪns/ *vt* enjuagar. ● *n* aclarado *m*; (*of dishes*) enjuague *m*; (*for hair*) reflejo *m*

riot /'raɪət/ *n* disturbio *m*; (*of colours*) profusión *f*. **run ~** desenfrenarse. ● *vi* amotinarse. **~er** *n* amotinador *m*. **~ous** *a* tumultuoso

rip /rɪp/ *vt* (*pt* **ripped**) rasgar. ● *vi* rasgarse. **let ~** (*fig*) soltar. ● *n* rasgadura *f*. **~ off** *vt* (*sl*) robar. **~-cord** *n* (*of parachute*) cuerda *f* de abertura

ripe /raɪp/ *a* (**-er**, **-est**) maduro. **~n** /'raɪpən/ *vt/i* madurar. **~ness** *n* madurez *f*

rip-off /'rɪpɒf/ *n* (*sl*) timo *m*

ripple /'rɪpl/ *n* rizo *m*; (*sound*) murmullo *m*. ● *vt* rizar. ● *vi* rizarse

rise /raɪz/ *vi* (*pt* **rose**, *pp* **risen**) levantarse; (*rebel*) sublevarse; ‹*river*› crecer; ‹*prices*› subir. ● *n* subida *f*; (*land*) altura *f*; (*increase*) aumento *m*; (*to power*) ascenso *m*. **give ~ to** ocasionar. **~r** /-ə(r)/ *n*. **early ~r** *n* madrugador *m*

rising /'raɪzɪŋ/ *n* (*revolt*) sublevación *f*. ● *a* ‹*sun*› naciente. **~ generation** *n* nueva generación *f*

risk /rɪsk/ *n* riesgo *m*. ● *vt* arriesgar. **~y** *a* (**-ier**, **-iest**) arriesgado

risqué /'ri:skeɪ/ *a* subido de color

rissole /'rɪsəʊl/ *n* croqueta *f*

rite /raɪt/ *n* rito *m*

ritual /'rɪtʃʊəl/ *a* & *n* ritual (*m*)

rival /'raɪvl/ *a* & *n* rival (*m*). ● *vt* (*pt* **rivalled**) rivalizar con. **~ry** *n* rivalidad *f*

river /'rɪvə(r)/ *n* río *m*

rivet /'rɪvɪt/ *n* remache *m*. ● *vt* remachar. **~ing** *a* fascinante

Riviera /rɪvɪ'erə/ *n*. **the (French) ~** la Costa *f* Azul. **the (Italian) ~** la Riviera *f* (Italiana)

rivulet /'rɪvjʊlɪt/ *n* riachuelo *m*

road /rəʊd/ *n* (*in town*) calle *f*; (*between towns*) carretera *f*; (*way*) camino *m*. **on the ~** en camino. **~-hog** *n* conductor *m* descortés. **~-house** *n* albergue *m*. **~-map** *n* mapa *m* de carreteras. **~side** /'rəʊdsaɪd/ *n* borde *m* de la carretera. **~ sign** *n* señal *f* de tráfico. **~way** /'rəʊdweɪ/ *n* calzada *f*. **~-works** *npl* obras *fpl*. **~worthy** /'rəʊdwɜ:ðɪ/ *a* ‹*vehicle*› seguro

roam /rəʊm/ *vi* vagar

roar /rɔ:(r)/ *n* rugido *m*; (*laughter*) carcajada *f*. ● *vt/i* rugir. **~ past** ‹*vehicles*› pasar con estruendo. **~ with laughter** reírse a carcajadas. **~ing** /'rɔ:rɪŋ/ *a* ‹*trade etc*› activo

roast /rəʊst/ *vt* asar; tostar ‹*coffee*›. ● *vi* asarse; ‹*person*, *coffee*› tostarse. ● *a* & *n* asado (*m*). **~ beef** *n* rosbif *m*

rob /rɒb/ *vt* (*pt* **robbed**) robar; asaltar ‹*bank*›. **~ of** privar de. **~ber** *n* ladrón *m*; (*of bank*) atracador *m*. **~bery** *n* robo *m*

robe /rəʊb/ *n* manto *m*; (*univ etc*) toga *f*. **bath-~** *n* albornoz *m*

robin /'rɒbɪn/ *n* petirrojo *m*

robot /'rəʊbɒt/ *n* robot *m*, autómata *m*

robust /rəʊ'bʌst/ *a* robusto

rock[1] /rɒk/ *n* roca *f*; (*boulder*) peñasco *m*; (*sweet*) caramelo *m* en forma de barra; (*of Gibraltar*) peñón *m*. **on the ~s** ‹*drink*› con hielo; (*fig*) arruinado. **be on the ~s** ‹*marriage etc*› andar mal

rock[2] /rɒk/ *vt* mecer; (*shake*) sacudir. ● *vi* mecerse; (*shake*) sacudirse. ● *n* (*mus*) música *f* rock

rock: **~-bottom** *a* (*fam*) bajísimo. **~ery** /'rɒkərɪ/ *n* cuadro *m* alpino, rocalla *f*

rocket /'rɒkɪt/ *n* cohete *m*

rock: **~ing-chair** *n* mecedora *f*. **~ing-horse** *n* caballo *m* de balancín. **~y** /'rɒkɪ/ *a* (**-ier**, **-iest**) rocoso; (*fig*, *shaky*) bamboleante

rod /rɒd/ *n* vara *f*; (*for fishing*) caña *f*; (*metal*) barra *f*

rode /rəʊd/ *see* **ride**

rodent /'rəʊdnt/ *n* roedor *m*

rodeo /rə'deɪəʊ/ *n* (*pl* **-os**) rodeo *m*

roe[1] /rəʊ/ *n* (*fish eggs*) hueva *f*

roe[2] /rəʊ/ (*pl* **roe**, *or* **roes**) (*deer*) corzo *m*

rogu|e /rəʊg/ *n* pícaro *m*. **~ish** *a* picaresco

role /rəʊl/ *n* papel *m*

roll /rəʊl/ *vt* hacer rodar; (*roll up*) enrollar; (*flatten lawn*) allanar; aplanar ‹*pastry*›. ● *vi* rodar; ‹*ship*› balancearse; (*on floor*) revolcarse. **be ~ing (in money)** (*fam*) nadar (en dinero). ● *n* rollo *m*; (*of ship*) balanceo *m*; (*of drum*) redoble *m*; (*of thunder*) retumbo *m*; (*bread*) panecillo *m*; (*list*) lista *f*. **~ over** *vi* (*turn over*) dar una vuelta. **~ up** *vt* enrollar; arremangar ‹*sleeve*›. ● *vi* (*fam*) llegar. **~-call** *n* lista *f*

roller /'rəʊlə(r)/ *n* rodillo *m*; (*wheel*) rueda *f*; (*for hair*) rulo *m*, bigudí *m*. **~-coaster** *n* montaña *f* rusa. **~-skate** *n* patín *m* de ruedas

rollicking /'rɒlɪkɪŋ/ *a* alegre

rolling /'rəʊlɪŋ/ *a* ondulado. **~-pin** *n* rodillo *m*

Roman /'rəʊmən/ *a* & *n* romano (*m*). **~ Catholic** *a* & *n* católico (*m*) (romano)

romance /rəʊ'mæns/ *n* novela *f* romántica; (*love*) amor *m*; (*affair*) aventura *f*

Romania /rəʊ'meɪnɪə/ *n* Rumania *f*. **~n** *a* & *n* rumano (*m*)

romantic /rəʊ'mæntɪk/ *a* romántico. **~ism** *n* romanticismo *m*

Rome /'rəʊm/ *n* Roma *f*

romp /rɒmp/ *vi* retozar. ● *n* retozo *m*

rompers /'rɒmpəz/ *npl* pelele *m*

roof /ru:f/ *n* techo *m*, tejado *m*; (*of mouth*) paladar *m*. ● *vt* techar. **~-garden** *n* jardín *m* en la azotea. **~-rack** *n* baca *f*. **~-top** *n* tejado *m*

rook[1] /rʊk/ *n* grajo *m*

rook[2] /rʊk/ (*in chess*) torre *f*

room /ru:m/ *n* cuarto *m*, habitación *f*; (*bedroom*) dormitorio *m*; (*space*) sitio *m*; (*large hall*) sala *f*. **~y** *a* espacioso; ‹*clothes*› holgado

roost /ru:st/ *n* percha *f*. ● *vi* descansar. **~er** *n* gallo *m*

root[1] /ru:t/ *n* raíz *f*. **take ~** echar raíces. ● *vt* hacer arraigar. ● *vi* echar raíces, arraigarse

root[2] /ru:t/ *vt*/*i*. **~ about** *vi* hurgar. **~ for** *vi* (*Amer*, *sl*) alentar. **~ out** *vt* extirpar

rootless /'ru:tlɪs/ *a* desarraigado

rope /rəʊp/ *n* cuerda *f*. **know the ~s** estar al corriente. ● *vt* atar. **~ in** *vt* agarrar

rosary /'rəʊzərɪ/ *n* (*relig*) rosario *m*

rose[1] /rəʊz/ *n* rosa *f*; (*nozzle*) roseta *f*

rose[2] /rəʊz/ *see* **rise**

rosé /'rəʊzeɪ/ *n* (vino *m*) rosado *m*

rosette /rəʊ'zet/ *n* escarapela *f*

roster /'rɒstə(r)/ *n* lista *f*

rostrum /'rɒstrəm/ *n* tribuna *f*

rosy /'rəʊzɪ/ *a* (**-ier**, **-iest**) rosado; ‹*skin*› sonrosado

rot /rɒt/ *vt* (*pt* **rotted**) pudrir. ● *vi* pudrirse. ● *n* putrefacción *f*; (*sl*) tonterías *fpl*

rota /'rəʊtə/ *n* lista *f*

rotary /'rəʊtərɪ/ *a* giratorio, rotativo

rotat|e /rəʊ'teɪt/ *vt* girar; (*change round*) alternar. ● *vi* girar; (*change round*) alternarse. **~ion** /-ʃn/ *n* rotación *f*

rote /rəʊt/ *n*. **by ~** maquinalmente, de memoria

rotten /'rɒtn/ *a* podrido; (*fam*) desagradable

rotund /rəʊ'tʌnd/ *a* redondo; ‹*person*› regordete

rouge /ru:ʒ/ *n* colorete *m*

rough /rʌf/ *a* (**-er**, **-est**) áspero; ‹*person*› tosco; (*bad*) malo; ‹*ground*› accidentado; (*violent*) brutal; (*aproximate*) aproximado; ‹*diamond*› bruto. ● *adv* duro. **~ copy** *n*, **~ draft** *n* borrador *m*. ● *n* (*ruffian*) matón *m*. ● *vt*. ~ **it** vivir sin comodidades. ~ **out** *vt* esbozar

roughage /ˈrʌfɪdʒ/ *n* alimento *m* indigesto, afrecho *m*; (*for animals*) forraje *m*

rough: **~-and-ready** *a* improvisado. **~-and-tumble** *n* riña *f*. **~ly** *adv* toscamente; (*more or less*) más o menos. **~ness** *n* aspereza *f*; (*lack of manners*) incultura *f*; (*crudeness*) tosquedad *f*

roulette /ru:ˈlet/ *n* ruleta *f*

round /raʊnd/ *a* (**-er**, **-est**) redondo. ● *n* círculo *m*; (*slice*) tajada *f*; (*of visits, drinks*) ronda *f*; (*of competition*) vuelta *f*; (*boxing*) asalto *m*.● *prep* alrededor de. ● *adv* alrededor. ~ **about** (*approximately*) aproximadamente. **come ~ to**, **go ~ to** (*a friend etc*) pasar por casa de. ● *vt* redondear; doblar ‹*corner*›. ~ **off** *vt* terminar. ~ **up** *vt* reunir; redondear ‹*price*›

roundabout /ˈraʊndəbaʊt/ *n* tiovivo *m*; (*for traffic*) glorieta *f*. ● *a* indirecto

rounders /ˈraʊndəz/ *n* juego *m* parecido al béisbol

round: **~ly** *adv* (*bluntly*) francamente. ~ **trip** *n* viaje *m* de ida y vuelta. **~-up** *n* reunión *f*; (*of suspects*) redada *f*

rous|e /raʊz/ *vt* despertar. **~ing** *a* excitante

rout /raʊt/ *n* derrota *f*. ● *vt* derrotar

route /ru:t/ *n* ruta *f*; (*naut, aviat*) rumbo *m*; (*of bus*) línea *f*

routine /ru:ˈti:n/ *n* rutina *f*. ● *a* rutinario

rov|e /rəʊv/ *vt/i* vagar (por). **~ing** *a* errante

row[1] /rəʊ/ *n* fila *f*

row[2] /rəʊ/ *n* (*in boat*) paseo *m* en bote (de remos). ● *vi* remar

row[3] /raʊ/ *n* (*noise, fam*) ruido *m*; (*quarrel*) pelea *f*. ● *vi* (*fam*) pelearse

rowdy /ˈraʊdɪ/ *a* (**-ier**, **-iest**) *n* ruidoso

rowing /ˈrəʊɪŋ/ *n* remo *m*. **~-boat** *n* bote *m* de remos

royal /ˈrɔɪəl/ *a* real. **~ist** *a* & *n* monárquico (*m*). **~ly** *adv* magníficamente. **~ty** /ˈrɔɪəltɪ/ *n* familia *f* real; (*payment*) derechos *mpl* de autor

rub /rʌb/ *vt* (*pt* **rubbed**) frotar. ~ **it in** insistir en algo. ● *n* frotamiento *m*. ~ **off on s.o.** *vi* pegársele a uno. ~ **out** *vt* borrar

rubber /ˈrʌbə(r)/ *n* goma *f*. ~ **band** *n* goma *f* (elástica). ~ **stamp** *n* sello *m* de goma. **~-stamp** *vt* (*fig*) aprobar maquinalmente. **~y** *a* parecido al caucho

rubbish /ˈrʌbɪʃ/ *n* basura *f*; (*junk*) trastos *mpl*; (*fig*) tonterías *fpl*. **~y** *a* sin valor

rubble /ˈrʌbl/ *n* escombros; (*small*) cascajo *m*

ruby /ˈru:bɪ/ *n* rubí *m*

rucksack /ˈrʌksæk/ *n* mochila *f*

rudder /ˈrʌdə(r)/ *n* timón *m*

ruddy /ˈrʌdɪ/ *a* (**-ier**, **-iest**) rubicundo; (*sl*) maldito

rude /ru:d/ *a* (**-er**, **-est**) descortés, mal educado; (*improper*) indecente; (*brusque*) brusco. **~ly** *adv* con descortesía. **~ness** *n* descortesía *f*

rudiment /ˈru:dɪmənt/ *n* rudimento *m*. **~ary** /-ˈmentrɪ/ *a* rudimentario

rueful /ˈru:fl/ *a* triste

ruffian /ˈrʌfɪən/ *n* rufián *m*

ruffle /ˈrʌfl/ *vt* despeinar ‹*hair*›; arrugar ‹*clothes*›. ● *n* (*frill*) volante *m*, fruncido *m*

rug /rʌg/ *n* tapete *m*; (*blanket*) manta *f*

Rugby /ˈrʌgbɪ/ *n*. ~ **(football)** *n* rugby *m*

rugged /ˈrʌgɪd/ *a* desigual; (*landscape*) accidentado; (*fig*) duro

ruin /ˈru:ɪn/ *n* ruina *f*. ● *vt* arruinar. **~ous** *a* ruinoso

rule /ru:l/ *n* regla *f*; (*custom*) costumbre *f*; (*pol*) dominio *m*. **as a ~** por regla general. ● *vt* gobernar; (*master*) dominar; (*jurid*) decretar; (*decide*) decidir. ~ **out** *vt* descartar. **~d paper** *n* papel *m* rayado

ruler /ˈru:lə(r)/ *n* (*sovereign*) soberano *m*; (*leader*) gobernante *m* & *f*; (*measure*) regla *f*

ruling /ˈru:lɪŋ/ *a* ‹*class*› dirigente. ● *n* decisión *f*

rum /rʌm/ *n* ron *m*

rumble /rʌmbl/ *vi* retumbar; ‹*stomach*› hacer ruidos. ● *n* retumbo *m*; (*of stomach*) ruido *m*

ruminant /ˈru:mɪnənt/ *a* & *n* rumiante (*m*)

rummage /ˈrʌmɪdʒ/ *vi* hurgar

rumour /ˈru:mə(r)/ *n* rumor *m*. ● *vt*. **it is ~ed that** se dice que

rump /rʌmp/ *n* (*of horse*) grupa *f*; (*of fowl*) rabadilla *f*. ~ **steak** *n* filete *m*

rumpus /ˈrʌmpəs/ *n* (*fam*) jaleo *m*

run /rʌn/ *vi* (*pt* **ran**, *pp* **run**, *pres p* **running**) correr; (*flow*) fluir; (*pass*) pasar; (*function*) funcionar; (*melt*) derretirse; ‹*bus etc*› circular; ‹*play*› representarse (continuamente); ‹*colours*› correrse; (*in election*) presentarse. ● *vt* tener ‹*house*›; (*control*) dirigir; correr ‹*risk*›; (*drive*) conducir; (*pass*) pasar; (*present*) presentar; forzar ‹*blockade*›. **~ a temperature** tener fiebre. ● *n* corrida *f*, carrera *f*; (*journey*) viaje *m*; (*outing*) paseo *m*, excursión *f*; (*distance travelled*) recorrido *m*; (*ladder*) carrera *f*; (*ski*) pista *f*; (*series*) serie *f*. **at a ~** corriendo. **have the ~ of** tener a su disposición. **in the long ~** a la larga. **on the ~** de fuga. **~ across** *vt* toparse con ‹*friend*›. **~ away** *vi* escaparse. **~ down** *vi* bajar corriendo; ‹*clock*› quedarse sin cuerda. ● *vt* (*auto*) atropellar; (*belittle*) denigrar. **~ in** *vt* rodar ‹*vehicle*›. ● *vi* entrar corriendo. **~ into** *vt* toparse con ‹*friend*›; (*hit*) chocar con. **~ off** *vt* tirar ‹*copies etc*›. **~ out** *vi* salir corriendo; ‹*liquid*› salirse; (*fig*) agotarse. **~ out of** quedar sin. **~ over** *vt* (*auto*) atropellar. **~ through** *vt* traspasar; (*revise*) repasar. **~ up** *vt* hacerse ‹*bill*›. ● *vi* subir corriendo. **~ up against** tropezar con ‹*difficulties*›. **~away** /ˈrʌnəweɪ/ *a* fugitivo; ‹*success*› decisivo; ‹*inflation*› galopante. ● *n* fugitivo *m*. **~ down** *a* ‹*person*› agotado. **~-down** *n* informe *m* detallado

rung[1] /rʌŋ/ *n* (*of ladder*) peldaño *m*

rung[2] /rʌŋ/ *see* **ring**

run: **~ner** /ˈrʌnə(r)/ *n* corredor *m*; (*on sledge*) patín *m*. **~ner bean** *n* judía *f* escarlata. **~ner-up** *n* subcampeón *m*, segundo *m*. **~ning** /ˈrʌnɪŋ/ *n* (*race*) carrera *f*. **be in the ~ning** tener posibilidades de ganar. ● *a* en marcha; ‹*water*› corriente; ‹*commentary*› en directo. **four times ~ning** cuatro veces seguidas. **~ny** /ˈrʌnɪ/ *a* líquido; ‹*nose*› que moquea. **~-of-the-mill** *a* ordinario. **~-up** *n* período *m* que precede. **~way** /ˈrʌnweɪ/ *n* pista *f*

rupture /ˈrʌptʃə(r)/ *n* ruptura *f*; (*med*) hernia *f*. ● *vt/i* quebrarse

rural /ˈrʊərəl/ *a* rural

ruse /ru:z/ *n* ardid *m*

rush[1] /rʌʃ/ *n* (*haste*) prisa *f*; (*crush*) bullicio *m*. ● *vi* precipitarse. ● *vt* apresurar; (*mil*) asaltar

rush[2] /rʌʃ/ *n* (*plant*) junco *m*

rush-hour /ˈrʌʃaʊə(r)/ *n* hora *f* punta

rusk /rʌsk/ *n* galleta *f*, tostada *f*

russet /ˈrʌsɪt/ *a* rojizo. ● *n* (*apple*) manzana *f* rojiza

Russia /ˈrʌʃə/ *n* Rusia *f*. **~n** *a & n* ruso (*m*)

rust /rʌst/ *n* orín *m*. ● *vt* oxidar. ● *vi* oxidarse

rustic /ˈrʌstɪk/ *a* rústico

rustle /ˈrʌsl/ *vt* hacer susurrar; (*Amer*) robar. **~ up** (*fam*) preparar. ● *vi* susurrar

rust: **~-proof** *a* inoxidable. **~y** *a* (**-ier**, **-iest**) oxidado

rut /rʌt/ *n* surco *m*. **in a ~** en la rutina de siempre

ruthless /ˈru:θlɪs/ *a* despiadado. **~ness** *n* crueldad *f*

rye /raɪ/ *n* centeno *m*

S

S *abbr* (*south*) sur *m*

sabbath /ˈsæbəθ/ *n* día *m* de descanso; (*Christian*) domingo *m*; (*Jewish*) sábado *m*

sabbatical /səˈbætɪkl/ *a* sabático

sabot|age /ˈsæbətɑ:ʒ/ *n* sabotaje *m*. ● *vt* sabotear. **~eur** /-ˈtɜ:(r)/ *n* saboteador *m*

saccharin /ˈsækərɪn/ *n* sacarina *f*

sachet /ˈsæʃeɪ/ *n* bolsita *f*

sack[1] /sæk/ *n* saco *m*. **get the ~** (*fam*) ser despedido. ● *vt* (*fam*) despedir. **~ing** *n* arpillera *f*; (*fam*) despido *m*

sack[2] /sæk/ *vt* (*plunder*) saquear

sacrament /ˈsækrəmənt/ *n* sacramento *m*

sacred /ˈseɪkrɪd/ *a* sagrado

sacrifice /ˈsækrɪfaɪs/ *n* sacrificio *m*. ● *vt* sacrificar

sacrileg|e /ˈsækrɪlɪdʒ/ *n* sacrilegio *m*. **~ious** /-ˈlɪdʒəs/ *a* sacrílego

sacrosanct /ˈsækrəʊsæŋkt/ *a* sacrosanto

sad /sæd/ *a* (**sadder**, **saddest**) triste. **~den** /ˈsædn/ *vt* entristecer

saddle /ˈsædl/ *n* silla *f*. **be in the ~** (*fig*) tener las riendas. ● *vt* ensillar ‹*horse*›. **~ s.o. with** (*fig*) cargar a uno con. **~-bag** *n* alforja *f*

sad: **~ly** *adv* tristemente; (*fig*) desgraciadamente. **~ness** *n* tristeza *f*

sadis|m /ˈseɪdɪzəm/ *n* sadismo *m*. **~t** /ˈseɪdɪst/ *n* sádico *m*. **~tic** /səˈdɪstɪk/ *a* sádico

safari /sə'fɑ:rɪ/ *n* safari *m*

safe /seɪf/ *a* (**-er**, **-est**) seguro; (*out of danger*) salvo; (*cautious*) prudente. **~ and sound** sano y salvo. ● *n* caja *f* fuerte. **~ deposit** *n* caja *f* de seguridad. **~guard** /'seɪfgɑ:d/ *n* salvaguardia *f*. ● *vt* salvaguardar. **~ly** *adv* sin peligro; (*in safe place*) en lugar seguro. **~ty** /'seɪftɪ/ *n* seguridad *f*. **~ty belt** *n* cinturón *m* de seguridad. **~ty-pin** *n* imperdible *m*. **~ty-valve** *n* válvula *f* de seguridad

saffron /'sæfrən/ *n* azafrán *m*

sag /sæg/ *vi* (*pt* **sagged**) hundirse; (*give*) aflojarse

saga /'sɑ:gə/ *n* saga *f*

sage[1] /seɪdʒ/ *n* (*wise person*) sabio *m*. ● *a* sabio

sage[2] /seɪdʒ/ *n* (*herb*) salvia *f*

sagging /'sægɪŋ/ *a* hundido; (*fig*) decaído

Sagittarius /sædʒɪ'teərɪəs/ *n* (*astr*) Sagitario *m*

sago /'seɪgəʊ/ *n* sagú *m*

said /sed/ *see* **say**

sail /seɪl/ *n* vela *f*; (*trip*) paseo *m* (en barco). ● *vi* navegar; (*leave*) partir; (*sport*) practicar la vela; (*fig*) deslizarse. ● *vt* manejar ‹*boat*›. **~ing** *n* (*sport*) vela *f*. **~ing-boat** *n*, **~ing-ship** *n* barco *m* de vela. **~or** /'seɪlə(r)/ *n* marinero *m*

saint /seɪnt, *before name* sənt/ *n* santo *m*. **~ly** *a* santo

sake /seɪk/ *n*. **for the ~ of** por, por el amor de

salacious /sə'leɪʃəs/ *a* salaz

salad /'sæləd/ *n* ensalada *f*. **~ bowl** *n* ensaladera *f*. **~ cream** *n* mayonesa *f*. **~-dressing** *n* aliño *m*

salar|ied /'sælərɪd/ *a* asalariado. **~y** /'sælərɪ/ *n* sueldo *m*

sale /seɪl/ *n* venta *f*; (*at reduced prices*) liquidación *f*. **for ~** (*sign*) se vende. **on ~** en venta. **~able** /'seɪləbl/ *a* vendible. **~sman** /'seɪlzmən/ *n* (*pl* **-men**) vendedor *m*; (*in shop*) dependiente *m*; (*traveller*) viajante *m*. **~swoman** *n* (*pl* **-women**) vendedora *f*; (*in shop*) dependienta *f*

salient /'seɪlɪənt/ *a* saliente, destacado

saliva /sə'laɪvə/ *n* saliva *f*

sallow /'sæləʊ/ *a* (**-er**, **-est**) amarillento

salmon /'sæmən/ *n invar* salmón *m*. **~ trout** *n* trucha *f* salmonada

salon /'sælɒn/ *n* salón *m*

saloon /sə'lu:n/ *n* (*on ship*) salón *m*; (*Amer*, *bar*) bar *m*; (*auto*) turismo *m*

salt /sɔ:lt/ *n* sal *f*. ● *a* salado. ● *vt* salar. **~-cellar** *n* salero *m*. **~y** *a* salado

salutary /'sæljʊtrɪ/ *a* saludable

salute /sə'lu:t/ *n* saludo *m*. ● *vt* saludar. ● *vi* hacer un saludo

salvage /'sælvɪdʒ/ *n* salvamento *m*; (*goods*) objetos *mpl* salvados. ● *vt* salvar

salvation /sæl'veɪʃn/ *n* salvación *f*

salve /sælv/ *n* ungüento *m*

salver /'sælvə(r)/ *n* bandeja *f*

salvo /'sælvəʊ/ *n* (*pl* **-os**) salva *f*

same /seɪm/ *a* igual (**as** que); (*before noun*) mismo (**as** que). **at the ~ time** al mismo tiempo. ● *pron*. **the ~** el mismo, la misma, los mismos, las mismas. **do the ~ as** hacer como. ● *adv*. **the ~** de la misma manera. **all the ~** de todas formas

sample /'sɑ:mpl/ *n* muestra *f*. ● *vt* probar ‹*food*›

sanatorium /sænə'tɔ:rɪəm/ *n* (*pl* **-ums**) sanatorio *m*

sanctify /'sæŋktɪfaɪ/ *vt* santificar

sanctimonious /sæŋktɪ'məʊnɪəs/ *a* beato

sanction /'sæŋkʃn/ *n* sanción *f*. ● *vt* sancionar

sanctity /'sæŋktətɪ/ *n* santidad *f*

sanctuary /'sæŋktʃʊərɪ/ *n* (*relig*) santuario *m*; (*for wildlife*) reserva *f*; (*refuge*) asilo *m*

sand /sænd/ *n* arena *f*. ● *vt* enarenar. **~s** *npl* (*beach*) playa *f*

sandal /'sændl/ *n* sandalia *f*

sand: **~castle** *n* castillo *m* de arena. **~paper** /'sændpeɪpə(r)/ *n* papel *m* de lija. ● *vt* lijar. **~storm** /'sændstɔ:m/ *n* tempestad *f* de arena

sandwich /'sænwɪdʒ/ *n* bocadillo *m*, sandwich *m*. ● *vt*. **~ed between** intercalado

sandy /'sændɪ/ *a* arenoso

sane /seɪn/ *a* (**-er**, **-est**) ‹*person*› cuerdo; ‹*judgement*, *policy*› razonable. **~ly** *adv* sensatamente

sang /sæŋ/ *see* **sing**

sanitary /'sænɪtrɪ/ *a* higiénico; ‹*system etc*› sanitario. **~ towel** *n*, **~ napkin** *n* (*Amer*) compresa *f* (higiénica)

sanitation /sænɪ'teɪʃn/ *n* higiene *f*; (*drainage*) sistema *m* sanitario

sanity /'sænɪtɪ/ *n* cordura *f*; (*fig*) sensatez *f*

sank /sæŋk/ *see* **sink**

Santa Claus /'sæntəklɔːz/ *n* Papá *m* Noel

sap /sæp/ *n* (*in plants*) savia *f*. ● *vt* (*pt* **sapped**) agotar

sapling /'sæplɪŋ/ *n* árbol *m* joven

sapphire /'sæfaɪə(r)/ *n* zafiro *m*

sarcas|m /'sɑːkæzəm/ *n* sarcasmo *m*. **~tic** /-'kæstɪk/ *a* sarcástico

sardine /sɑː'diːn/ *n* sardina *f*

Sardinia /sɑː'dɪnɪə/ *n* Cerdeña *f*. **~n** *a* & *n* sardo (*m*)

sardonic /sɑː'dɒnɪk/ *a* sardónico

sash /sæʃ/ *n* (*over shoulder*) banda *f*; (*round waist*) fajín *m*. **~-window** *n* ventana *f* de guillotina

sat /sæt/ *see* **sit**

satanic /sə'tænɪk/ *a* satánico

satchel /'sætʃl/ *n* cartera *f*

satellite /'sætəlaɪt/ *n* & *a* satélite (*m*)

satiate /'seɪʃɪeɪt/ *vt* saciar

satin /'sætɪn/ *n* raso *m*. ● *a* de raso; (*like satin*) satinado

satir|e /'sætaɪə(r)/ *n* sátira *f*. **~ical** /sə'tɪrɪkl/ *a* satírico. **~ist** /'sætərɪst/ *n* satírico *m*. **~ize** /'sætəraɪz/ *vt* satirizar

satisfaction /sætɪs'fækʃn/ *n* satisfacción *f*

satisfactor|ily /sætɪs'fæktərɪlɪ/ *adv* satisfactoriamente. **~y** /sætɪs 'fæktərɪ/ *a* satisfactorio

satisfy /'sætɪsfaɪ/ *vt* satisfacer; (*convince*) convencer. **~ing** *a* satisfactorio

satsuma /sæt'suːmə/ *n* mandarina *f*

saturat|e /'sætʃəreɪt/ *vt* saturar, empapar. **~ed** *a* saturado, empapado. **~ion** /-'reɪʃn/ *n* saturación *f*

Saturday /'sætədeɪ/ *n* sábado *m*

sauce /sɔːs/ *n* salsa *f*; (*cheek*) descaro *m*. **~pan** /'sɔːspən/ *n* cazo *m*

saucer /'sɔːsə(r)/ *n* platillo *m*

saucy /'sɔːsɪ/ *a* (**-ier**, **-iest**) descarado

Saudi Arabia /saʊdɪə'reɪbɪə/ *n* Arabia *f* Saudí

sauna /'sɔːnə/ *n* sauna *f*

saunter /'sɔːntə(r)/ *vi* deambular, pasearse

sausage /'sɒsɪdʒ/ *n* salchicha *f*

savage /'sævɪdʒ/ *a* salvaje; (*fierce*) feroz; (*furious*, *fam*) rabioso. ● *n* salvaje *m* & *f*. ● *vt* atacar. **~ry** *n* ferocidad *f*

sav|e /seɪv/ *vt* salvar; ahorrar ‹*money*, *time*›; (*prevent*) evitar. ● *n* (*football*) parada *f*. ● *prep* salvo, con excepción de. **~er** *n* ahorrador *m*. **~ing** *n* ahorro *m*. **~ings** *npl* ahorros *mpl*

saviour /'seɪvɪə(r)/ *n* salvador *m*

savour /'seɪvə(r)/ *n* sabor *m*. ● *vt* saborear. **~y** *a* (*appetizing*) sabroso; (*not sweet*) no dulce. ● *n* aperitivo *m* (no dulce)

saw[1] /sɔː/ *see* **see**[1]

saw[2] /sɔː/ *n* sierra *f*. ● *vt* (*pt* **sawed**, *pp* **sawn**) serrar. **~dust** /'sɔːdʌst/ *n* serrín *m*. **~n** /sɔːn/ *see* **saw**

saxophone /'sæksəfəʊn/ *n* saxófono *m*

say /seɪ/ *vt/i* (*pt* **said** /sed/) decir; rezar ‹*prayer*›. **I ~!** ¡no me digas! ● *n*. **have a ~** expresar una opinión; (*in decision*) tener voz en capítulo. **have no ~** no tener ni voz ni voto. **~ing** /'seɪɪŋ/ *n* refrán *m*

scab /skæb/ *n* costra *f*; (*blackleg*, *fam*) esquirol *m*

scaffold /'skæfəʊld/ *n* (*gallows*) cadalso *m*, patíbulo *m*. **~ing** /'skæfəldɪŋ/ *n* (*for workmen*) andamio *m*

scald /skɔːld/ *vt* escaldar; calentar ‹*milk etc*›. ● *n* escaldadura *f*

scale[1] /skeɪl/ *n* escala *f*

scale[2] /skeɪl/ *n* (*of fish*) escama *f*

scale[3] /skeɪl/ *vt* (*climb*) escalar. **~ down** *vt* reducir (proporcionalmente)

scales /skeɪlz/ *npl* (*for weighing*) balanza *f*, peso *m*

scallop /'skɒləp/ *n* venera *f*; (*on dress*) festón *m*

scalp /skælp/ *n* cuero *m* cabelludo. ● *vt* quitar el cuero cabelludo a

scalpel /'skælpəl/ *n* escalpelo *m*

scamp /skæmp/ *n* bribón *m*

scamper /'skæmpə(r)/ *vi*. **~ away** marcharse corriendo

scampi /'skæmpɪ/ *npl* gambas *fpl* grandes

scan /skæn/ *vt* (*pt* **scanned**) escudriñar; (*quickly*) echar un vistazo a; ‹*radar*› explorar. ● *vi* ‹*poetry*› estar bien medido

scandal /'skændl/ *n* escándalo *m*; (*gossip*) chismorreo *m*. **~ize** /'skændəlaɪz/ *vt* escandalizar. **~ous** *a* escandaloso

Scandinavia /skændɪ'neɪvɪə/ *n* Escandinavia *f*. **~n** *a* & *n* escandinavo (*m*)

scant /skænt/ *a* escaso. **~ily** *adv* insuficientemente. **~y** /'skæntɪ/ *a* (**-ier**, **-iest**) escaso

scapegoat /'skeɪpgəʊt/ *n* cabeza *f* de turco

scar /skɑ:(r)/ *n* cicatriz *f*. ● *vt* (*pt* **scarred**) dejar una cicatriz en. ● *vi* cicatrizarse

scarc|e /skeəs/ *a* (**-er**, **-est**) escaso. **make o.s. ~e** (*fam*) mantenerse lejos. **~ely** /'skeəslɪ/ *adv* apenas. **~ity** *n* escasez *f*

scare /'skeə(r)/ *vt* asustar. **be ~d** tener miedo. ● *n* susto *m*. **~crow** /'skeəkrəʊ/ *n* espantapájaros *m invar*. **~monger** /'skeəmʌŋgə(r)/ *n* alarmista *m & f*

scarf /skɑ:f/ *n* (*pl* **scarves**) bufanda *f*; (*over head*) pañuelo *m*

scarlet /'skɑ:lət/ *a* escarlata *f*. **~ fever** *n* escarlatina *f*

scary /'skeərɪ/ *a* (**-ier**, **-iest**) que da miedo

scathing /'skeɪðɪŋ/ *a* mordaz

scatter /'skætə(r)/ *vt* (*throw*) esparcir; (*disperse*) dispersar. ● *vi* dispersarse. **~-brained** *a* atolondrado. **~ed** *a* disperso; (*occasional*) esporádico

scatty /'skætɪ/ *a* (**-ier**, **-iest**) (*sl*) atolondrado

scavenge /'skævɪndʒ/ *vi* buscar (en la basura). **~r** /-ə(r)/ *n* (*vagrant*) persona *f* que busca objetos en la basura

scenario /sɪ'nɑ:rɪəʊ/ *n* (*pl* **-os**) argumento; (*of film*) guión *m*

scen|e /si:n/ *n* escena *f*; (*sight*) vista *f*; (*fuss*) lío *m*. **behind the ~es** entre bastidores. **~ery** /'si:nərɪ/ *n* paisaje *m*; (*in theatre*) decorado *m*. **~ic** /'si:nɪk/ *a* pintoresco

scent /sent/ *n* olor *m*; (*perfume*) perfume *m*; (*trail*) pista *f*. ● *vt* presentir; (*make fragrant*) perfumar

sceptic /'skeptɪk/ *n* escéptico *m*. **~al** *a* escéptico. **~ism** /-sɪzəm/ *n* escepticismo *m*

sceptre /'septə(r)/ *n* cetro *m*

schedule /'ʃedju:l, 'skedju:l/ *n* programa *f*; (*timetable*) horario *m*. **behind ~** con retraso. **on ~** sin retraso. ● *vt* proyectar. **~d flight** *n* vuelo *m* regular

scheme /ski:m/ *n* proyecto *m*; (*plot*) intriga *f*. ● *vi* hacer proyectos; (*pej*) intrigar. **~r** *n* intrigante *m & f*

schism /'sɪzəm/ *n* cisma *m*

schizophrenic /skɪtsə'frenɪk/ *a & n* esquizofrénico (*m*)

scholar /'skɒlə(r)/ *n* erudito *m*. **~ly** *a* erudito. **~ship** *n* erudición *f*; (*grant*) beca *f*

scholastic /skə'læstɪk/ *a* escolar

school /sku:l/ *n* escuela *f*; (*of univ*) facultad *f*. ● *a* ‹*age, holidays, year*› escolar. ● *vt* enseñar; (*discipline*) disciplinar. **~boy** /'sku:lbɔɪ/ *n* colegial *m*. **~girl** /-gɜ:l/ *n* colegiala *f*. **~ing** *n* instrucción *f*. **~master** /'sku:lmɑ:stə(r)/ *n* (*primary*) maestro *m*; (*secondary*) profesor *m*. **~-mistress** *n* (*primary*) maestra *f*; (*secondary*) profesora *f*. **~-teacher** *n* (*primary*) maestro *m*; (*secondary*) profesor *m*

schooner /'sku:nə(r)/ *n* goleta *f*; (*glass*) vaso *m* grande

sciatica /saɪ'ætɪkə/ *n* ciática *f*

scien|ce /'saɪəns/ *n* ciencia *f*. **~ce fiction** *n* ciencia *f* ficción. **~tific** /-'tɪfɪk/ *a* científico. **~tist** /'saɪəntɪst/ *n* científico *m*

scintillate /'sɪntɪleɪt/ *vi* centellear

scissors /'sɪsəz/ *npl* tijeras *fpl*

sclerosis /sklə'rəʊsɪs/ *n* esclerosis *f*

scoff /skɒf/ *vt* (*sl*) zamparse. ● *vi*. **~ at** mofarse de

scold /skəʊld/ *vt* regañar. **~ing** *n* regaño *m*

scone /skɒn/ *n* (tipo *m* de) bollo *m*

scoop /sku:p/ *n* paleta *f*; (*news*) noticia *f* exclusiva. ● *vt*. **~ out** excavar. **~ up** recoger

scoot /sku:t/ *vi* (*fam*) largarse corriendo. **~er** /'sku:tə(r)/ *n* escúter *m*; (*for child*) patinete *m*

scope /skəʊp/ *n* alcance *m*; (*opportunity*) oportunidad *f*

scorch /skɔ:tʃ/ *vt* chamuscar. **~er** *n* (*fam*) día *m* de mucho calor. **~ing** *a* (*fam*) de mucho calor

score /skɔ:(r)/ *n* tanteo *m*; (*mus*) partitura *f*; (*twenty*) veintena *f*; (*reason*) motivo *m*. **on that ~** en cuanto a eso. ● *vt* marcar; (*slash*) rayar; (*mus*) instrumentar; conseguir ‹*success*›. ● *vi* marcar un tanto; (*keep score*) tantear. **~ over s.o.** aventajar a. **~r** /-ə(r)/ *n* tanteador *m*

scorn /skɔ:n/ *n* desdén m. ● *vt* desdeñar. **~ful** *a* desdeñoso. **~fully** *adv* desdeñosamente

Scorpio /'skɔ:pɪəʊ/ *n* (*astr*) Escorpión *m*

scorpion /'skɔ:pɪən/ *n* escorpión *m*

Scot /skɒt/ *n* escocés *m*. **~ch** /skɒtʃ/ *a* escocés. ● *n* güisqui *m*

scotch /skɒtʃ/ *vt* frustrar; (*suppress*) suprimir

scot-free /skɒt'fri:/ *a* impune; (*gratis*) sin pagar

Scot: **~land** /'skɒtlənd/ *n* Escocia *f*. **~s** *a* escocés. **~sman** *n* escocés *m*. **~swoman** *n* escocesa *f*. **~tish** *a* escocés

scoundrel /'skaʊndrəl/ *n* canalla *f*

scour /'skaʊə(r)/ *vt* estregar; (*search*) registrar. **~er** *n* estropajo *m*

scourge /skɜ:dʒ/ *n* azote *m*

scout /skaʊt/ *n* explorador *m*. **Boy S~** explorador *m*. • *vi*. **~ (for)** buscar

scowl /skaʊl/ *n* ceño *m*. • *vi* fruncir el entrecejo

scraggy /'skrægɪ/ *a* (**-ier**, **-iest**) descarnado

scram /skræm/ *vi* (*sl*) largarse

scramble /'skræmbl/ *vi* (*clamber*) gatear. **~ for** pelearse para obtener. • *vt* revolver ‹*eggs*›. • *n* (*difficult climb*) subida *f* difícil; (*struggle*) lucha *f*

scrap /skræp/ *n* pedacito *m*; (*fight, fam*) pelea *f*. • *vt* (*pt* **scrapped**) desechar. **~-book** *n* álbum *m* de recortes. **~s** *npl* sobras *fpl*

scrape /skreɪp/ *n* raspadura *f*; (*fig*) apuro *m*. • *vt* raspar; (*graze*) arañar; (*rub*) frotar. • *vi*. **~ through** lograr pasar; aprobar por los pelos ‹*exam*›. **~ together** reunir. **~r** /-ə(r)/ *n* raspador *m*

scrap: **~ heap** *n* montón *m* de deshechos. **~-iron** *n* chatarra *f*

scrappy /'skræpɪ/ *a* fragmentario, pobre, de mala calidad

scratch /skrætʃ/ *vt* rayar; (*with nail etc*) arañar; rascar ‹*itch*›. • *vi* arañar. • *n* raya *f*; (*from nail etc*) arañazo *m*. **start from ~** empezar sin nada, empezar desde el principio. **up to ~** al nivel requerido

scrawl /skrɔ:l/ *n* garrapato *m*. • *vt*/*i* garrapatear

scrawny /'skrɔ:nɪ/ *a* (**-ier**, **-iest**) descarnado

scream /skri:m/ *vt*/*i* gritar. • *n* grito *m*

screech /skri:tʃ/ *vi* gritar; ‹*brakes etc*› chirriar. • *n* grito *m*; (*of brakes etc*) chirrido *m*

screen /skri:n/ *n* pantalla *f*; (*folding*) biombo *m*. • *vt* (*hide*) ocultar; (*protect*) proteger; proyectar ‹*film*›; seleccionar ‹*candidates*›

screw /skru:/ *n* tornillo *m*. • *vt* atornillar. **~driver** /'skru:draɪvə(r)/ *n* destornillador *m*. **~ up** atornillar; entornar ‹*eyes*›; torcer ‹*face*›; (*ruin, sl*) arruinar. **~y** /'skru:ɪ/ *a* (**-ier**, **-iest**) (*sl*) chiflado

scribble /'skrɪbl/ *vt*/*i* garrapatear. • *n* garrapato *m*

scribe /skraɪb/ *n* copista *m & f*

script /skrɪpt/ *n* escritura *f*; (*of film etc*) guión *m*

Scriptures /'skrɪptʃəz/ *npl* Sagradas Escrituras *fpl*

script-writer /'skrɪptraɪtə(r)/ *n* guionista *m & f*

scroll /skrəʊl/ *n* rollo *m* (de pergamino)

scrounge /skraʊndʒ/ *vt*/*i* obtener de gorra; (*steal*) birlar. **~r** /-ə(r)/ *n* gorrón *m*

scrub /skrʌb/ *n* (*land*) maleza *f*; (*clean*) fregado *m*. • *vt*/*i* (*pt* **scrubbed**) fregar

scruff /skrʌf/ *n*. **the ~ of the neck** el cogote *m*

scruffy /'skrʌfɪ/ *a* (**-ier**, **-iest**) desaliñado

scrum /skrʌm/ *n*, **scrummage** /'skrʌmɪdʒ/ *n* (*Rugby*) melée *f*

scrup|le /'skru:pl/ *n* escrúpulo *m*. **~ulous** /'skru:pjʊləs/ *a* escrupuloso. **~ulously** *adv* escrupulosamente

scrutin|ize /'skru:tɪnaɪz/ *vt* escudriñar. **~y** /'skru:tɪnɪ/ *n* examen *m* minucioso

scuff /skʌf/ *vt* arañar ‹*shoes*›

scuffle /'skʌfl/ *n* pelea *f*

scullery /'skʌlərɪ/ *n* trascocina *f*

sculpt /skʌlpt/ *vt*/*i* esculpir. **~or** *n* escultor *m*. **~ure** /-tʃə(r)/ *n* escultura *f*. • *vt*/*i* esculpir

scum /skʌm/ *n* espuma *f*; (*people, pej*) escoria *f*

scurf /skɜ:f/ *n* caspa *f*

scurrilous /'skʌrɪləs/ *a* grosero

scurry /'skʌrɪ/ *vi* correr

scurvy /'skɜ:vɪ/ *n* escorbuto *m*

scuttle[1] /'skʌtl/ *n* cubo *m* del carbón

scuttle[2] /'skʌtl/ *vt* barrenar ‹*ship*›

scuttle[3] /'skʌtl/ *vi*. **~ away** correr, irse de prisa

scythe /saɪð/ *n* guadaña *f*

SE *abbr* (*south-east*) sudeste *m*

sea /si:/ *n* mar *m*. **at ~** en el mar; (*fig*) confuso. **by ~** por mar. **~board** /'si:bɔ:d/ *n* litoral *m*. **~farer** /'si:feərə(r)/ *n* marinero *m*. **~food** /'si:fu:d/ *n* mariscos *mpl*. **~gull** /'si:gʌl/ *n* gaviota *f*. **~-horse** *n* caballito *m* de mar, hipocampo *m*

seal[1] /si:l/ *n* sello *m*. • *vt* sellar. **~ off** acordonar ‹*area*›

seal[2] /si:l/ (*animal*) foca *f*
sea level /'si:levl/ *n* nivel *m* del mar
sealing-wax /'si:lɪŋwæks/ *n* lacre *m*
sea lion /'si:laɪən/ *n* león *m* marino
seam /si:m/ *n* costura *f*; (*of coal*) veta *f*
seaman /'si:mən/ *n* (*pl* **-men**) marinero *m*
seamy /'si:mɪ/ *a*. **the ~ side** *n* el lado *m* sórdido, el revés *m*
seance /'seɪɑ:ns/ *n* sesión *f* de espiritismo
sea: **~plane** /'si:pleɪn/ *n* hidroavión *f*. **~port** /'si:pɔ:t/ *n* puerto *m* de mar
search /sɜ:tʃ/ *vt* registrar; (*examine*) examinar. ● *vi* buscar. ● *n* (*for sth*) búsqueda *f*; (*of sth*) registro *m*. **in ~ of** en busca de. **~ for** buscar. **~ing** *a* penetrante. **~-party** *n* equipo *m* de salvamento. **~light** /'sɜ:tʃlaɪt/ *n* reflector *m*
sea: **~scape** /'si:skeɪp/ *n* marina *f*. **~shore** *n* orilla *f* del mar. **~sick** /'si:sɪk/ *a* mareado. **be ~sick** marearse. **~side** /'si:saɪd/ *n* playa *f*
season /'si:zn/ *n* estación *f*; (*period*) temporada *f*. ● *vt* (*culin*) sazonar; secar ‹*wood*›. **~able** *a* propio de la estación. **~al** *a* estacional. **~ed** /'si:znd/ *a* (*fig*) experto. **~ing** *n* condimento *m*. **~-ticket** *n* billete *m* de abono
seat /si:t/ *n* asiento *m*; (*place*) lugar *m*; (*of trousers*) fondillos *mpl*; (*bottom*) trasero *m*. **take a ~** sentarse. ● *vt* sentar; (*have seats for*) tener asientos para. **~-belt** *n* cinturón *m* de seguridad
sea: **~-urchin** *n* erizo *m* de mar. **~weed** /'si:wi:d/ *n* alga *f*. **~worthy** /'si:wɜ:ðɪ/ *a* en estado de navegar
secateurs /'sekətɜ:z/ *npl* tijeras *fpl* de podar
sece|de /sɪ'si:d/ *vi* separase. **~ssion** /-eʃn/ *n* secesión *f*
seclu|de /sɪ'klu:d/ *vt* aislar. **~ded** *a* aislado. **~sion** /-ʒn/ *n* aislamiento *m*
second[1] /'sekənd/ *a* & *n* segundo (*m*). **on ~ thoughts** pensándolo bien. ● *adv* (*in race etc*) en segundo lugar. ● *vt* apoyar. **~s** *npl* (*goods*) artículos *mpl* de segunda calidad; (*more food, fam*) otra porción *f*
second[2] /sɪ'kɒnd/ *vt* (*transfer*) trasladar temporalmente
secondary /'sekəndrɪ/ *a* secundario. **~ school** *n* instituto *m*
second: **~-best** *a* segundo. **~-class** *a* de segunda clase. **~-hand** *a* de segunda mano. **~ly** *adv* en segundo lugar. **~-rate** *a* mediocre
secre|cy /'si:krəsɪ/ *n* secreto *m*. **~t** /'si:krɪt/ *a* & *n* secreto (*m*). **in ~t** en secreto
secretar|ial /sekrə'teərɪəl/ *a* de secretario. **~iat** /sekrə'teərɪət/ *n* secretaría *f*. **~y** /'sekrətrɪ/ *n* secretario *m*. **S~y of State** ministro *m*: (*Amer*) Ministro *m* de Asuntos Exteriores
secret|e /sɪ'kri:t/ *vt* (*med*) secretar. **~ion** /-ʃn/ *n* secreción *f*
secretive /'si:krɪtɪv/ *a* reservado
secretly /'si:krɪtlɪ/ *adv* en secreto
sect /sekt/ *n* secta *f*. **~arian** /-'teərɪən/ *a* sectario
section /'sekʃn/ *n* sección *f*; (*part*) parte *f*
sector /'sektə(r)/ *n* sector *m*
secular /'sekjʊlə(r)/ *a* seglar
secur|e /sɪ'kjʊə(r)/ *a* seguro; (*fixed*) fijo. ● *vt* asegurar; (*obtain*) obtener. **~ely** *adv* seguramente. **~ity** /sɪ'kjʊərətɪ/ *n* seguridad *f*; (*for loan*) garantía *f*, fianza *f*
sedate /sɪ'deɪt/ *a* sosegado
sedat|ion /sɪ'deɪʃn/ *n* sedación *f*. **~ive** /'sedətɪv/ *a* & *n* sedante (*m*)
sedentary /'sedəntrɪ/ *a* sedentario
sediment /'sedɪmənt/ *n* sedimento *m*
seduc|e /sɪ'dju:s/ *vt* seducir. **~er** /-ə(r)/ *n* seductor *m*. **~tion** /sɪ'dʌkʃn/ *n* seducción *f*. **~tive** /-tɪv/ *a* seductor
see[1] /si:/ ● *vt* (*pt* **saw**, *pp* **seen**) ver; (*understand*) comprender; (*notice*) notar; (*escort*) acompañar. **~ing that** visto que. **~ you later!** ¡hasta luego! ● *vi* ver; (*understand*) comprender. **~ about** ocuparse de. **~ off** despedirse de. **~ through** llevar a cabo; descubrir el juego de ‹*person*›. **~ to** ocuparse de
see[2] /si:/ *n* diócesis *f*
seed /si:d/ *n* semilla *f*; (*fig*) germen *m*; (*tennis*) preseleccionado *m*. **~ling** *n* plantón *m*. **go to ~** granar; (*fig*) echarse a perder. **~y** /'si:dɪ/ *a* (**-ier**, **-iest**) sórdido
seek /si:k/ *vt* (*pt* **sought**) buscar. **~ out** buscar
seem /si:m/ *vi* parecer. **~ingly** *adv* aparentemente
seemly /'si:mlɪ/ *a* (**-ier**, **-iest**) correcto

seen /si:n/ *see* **see**[1]
seep /si:p/ *vi* filtrarse. **~age** *n* filtración *f*
see-saw /'si:sɔ:/ *n* balancín *m*
seethe /si:ð/ *vi* (*fig*) hervir. **be seething with anger** estar furioso
see-through /'si:θru:/ *a* transparente
segment /'segmənt/ *n* segmento *m*; (*of orange*) gajo *m*
segregat|e /'segrɪgeɪt/ *vt* segregar. **~ion** /-'geɪʃn/ *n* segregación *f*
seiz|e /si:z/ *vt* agarrar; (*jurid*) incautarse de. **~e on** *vi* valerse de. **~e up** *vi* (*tec*) agarrotarse. **~ure** /'si:ʒə(r)/ *n* incautación *f*; (*med*) ataque *m*
seldom /'seldəm/ *adv* raramente
select /sɪ'lekt/ *vt* escoger; (*sport*) seleccionar. ● *a* selecto; (*exclusive*) exclusivo. **~ion** /-ʃn/ *n* selección *f*. **~ive** *a* selectivo
self /self/ *n* (*pl* **selves**) sí mismo. **~-addressed** *a* con su propia dirección. **~-assurance** *n* confianza *f* en sí mismo. **~-assured** *a* seguro de sí mismo. **~-catering** *a* con facilidades para cocinar. **~-centred** *a* egocéntrico. **~-confidence** *n* confianza *f* en sí mismo. **~-confident** *a* seguro de sí mismo. **~-conscious** *a* cohibido. **~-contained** *a* independiente. **~-control** *n* dominio *m* de sí mismo. **~-defence** *n* defensa *f* propia. **~-denial** *n* abnegación *f*. **~-employed** *a* que trabaja por cuenta propia. **~-esteem** *n* amor *m* propio. **~-evident** *a* evidente. **~-government** *n* autonomía *f*. **~-important** *a* presumido. **~-indulgent** *a* inmoderado. **~-interest** *n* interés *m* propio. **~ish** /'selfɪʃ/ *a* egoísta. **~ishness** *n* egoísmo *m*. **~less** /'selflɪs/ *a* desinteresado. **~-made** *a* rico por su propio esfuerzo. **~-opinionated** *a* intransigente; (*arrogant*) engreído. **~-pity** *n* compasión *f* de sí mismo. **~-portrait** *n* autorretrato *m*. **~-possessed** *a* dueño de sí mismo. **~-reliant** *a* independiente. **~-respect** *n* amor *m* propio. **~-righteous** *a* santurrón. **~-sacrifice** *n* abnegación *f*. **~-satisfied** *a* satisfecho de sí mismo. **~-seeking** *a* egoísta. **~-service** *a* & *n* autoservicio (*m*). **~-styled** *a* sedicente, llamado. **~-sufficient** *a* independiente. **~-willed** *a* terco
sell /sel/ *vt* (*pt* **sold**) vender. **be sold on** (*fam*) entusiasmarse por. **be sold out** estar agotado. ● *vi* venderse. **~-by date** *n* fecha *f* de caducidad. **~ off** *vt* liquidar. **~ up** *vt* vender todo. **~er** *n* vendedor *m*
Sellotape /'seləteɪp/ *n* (*P*) (papel *m*) celo *m*, cinta *f* adhesiva
sell-out /'selaʊt/ *n* (*betrayal*, *fam*) traición *f*
semantic /sɪ'mæntɪk/ *a* semántico. **~s** *n* semántica *f*
semaphore /'seməfɔ:(r)/ *n* semáforo *m*
semblance /'sembləns/ *n* apariencia *f*
semen /'si:mən/ *n* semen *m*
semester /sɪ'mestə(r)/ *n* (*Amer*) semestre *m*
semi... /'semɪ/ *pref* semi...
semi|breve /'semɪbri:v/ *n* semibreve *f*, redonda *f*. **~circle** /'semɪsɜ:kl/ *n* semicírculo *m*. **~circular** /-'sɜ:kjʊlə(r)/ *a* semicircular. **~colon** /semɪ'kəʊlən/ *n* punto *m* y coma. **~-detached** /semɪdɪ'tætʃt/ *a* ‹*house*› adosado. **~final** /semɪ'faɪnl/ *n* semifinal *f*
seminar /'semɪnɑ:(r)/ *n* seminario *m*
seminary /'semɪnərɪ/ *n* (*college*) seminario *m*
semiquaver /'semɪkweɪvə(r)/ *n* (*mus*) semicorchea *f*
Semit|e /'si:maɪt/ *n* semita *m* & *f*. **~ic** /sɪ'mɪtɪk/ *a* semítico
semolina /semə'li:nə/ *n* sémola *f*
senat|e /'senɪt/ *n* senado *m*. **~or** /-ətə(r)/ *n* senador *m*
send /send/ *vt/i* (*pt* **sent**) enviar. **~ away** despedir. **~ away for** pedir (por correo). **~ for** enviar a buscar. **~ off for** pedir (por correo). **~ up** (*fam*) parodiar. **~er** *n* remitente *m*. **~-off** *n* despedida *f*
senil|e /'si:naɪl/ *a* senil. **~ity** /sɪ'nɪlətɪ/ *n* senilidad *f*
senior /'si:nɪə(r)/ *a* mayor; (*in rank*) superior; ‹*partner etc*› principal. ● *n* mayor *m* & *f*. **~ citizen** *n* jubilado *m*. **~ity** /-'ɒrətɪ/ *n* antigüedad *f*
sensation /sen'seɪʃn/ *n* sensación *f*. **~al** *a* sensacional
sense /sens/ *n* sentido *m*; (*common sense*) juicio *m*; (*feeling*) sensación *f*. **make ~** *vt* tener sentido. **make ~ of** comprender. **~less** *a* insensato; (*med*) sin sentido
sensibilities /sensɪ'bɪlətɪz/ *npl* susceptibilidad *f*. **~ibility** /sensɪ'bɪlətɪ/ *n* sensibilidad *f*

sensible /'sensəbl/ *a* sensato; ‹*clothing*› práctico

sensitiv|e /'sensɪtɪv/ *a* sensible; (*touchy*) susceptible. **~ity** /-'tɪvətɪ/ *n* sensibilidad *f*

sensory /'sensərɪ/ *a* sensorio

sensual /'senʃʊəl/ *a* sensual. **~ity** /-'ælətɪ/ *n* sensualidad *f*

sensuous /'sensʊəs/ *a* sensual

sent /sent/ *see* **send**

sentence /'sentəns/ *n* frase *f*; (*jurid*) sentencia *f*; (*punishment*) condena *f*. ● *vt*. **~ to** condenar a

sentiment /'sentɪmənt/ *n* sentimiento *m*; (*opinion*) opinión *f*. **~al** /sentɪ'mentl/ *a* sentimental. **~ality** /-'tælətɪ/ *n* sentimentalismo *m*

sentry /'sentrɪ/ *n* centinela *f*

separable /'sepərəbl/ *a* separable

separate[1] /'sepərət/ *a* separado; (*independent*) independiente. **~ly** *adv* por separado. **~s** *npl* coordinados *mpl*

separat|e[2] /'sepəreɪt/ *vt* separar. ● *vi* separarse. **~ion** /-'reɪʃn/ *n* separación *f*. **~ist** /'sepərətɪst/ *n* separatista *m & f*

September /sep'tembə(r)/ *n* se(p)tiembre *m*

septic /'septɪk/ *a* séptico. **~ tank** *n* fosa *f* séptica

sequel /'si:kwəl/ *n* continuación *f*; (*consequence*) consecuencia *f*

sequence /'si:kwəns/ *n* sucesión *f*; (*of film*) secuencia *f*

sequin /'si:kwɪn/ *n* lentejuela *f*

serenade /serə'neɪd/ *n* serenata *f*. ● *vt* dar serenata a

seren|e /sɪ'ri:n/ *a* sereno. **~ity** /-enətɪ/ *n* serenidad *f*

sergeant /'sɑ:dʒənt/ *n* sargento *m*

serial /'sɪərɪəl/ *n* serial *m*. ● *a* de serie. **~ize** *vt* publicar por entregas

series /'sɪəri:z/ *n* serie *f*

serious /'sɪərɪəs/ *a* serio. **~ly** *adv* seriamente; (*ill*) gravemente. **take ~ly** tomar en serio. **~ness** *n* seriedad *f*

sermon /'sɜ:mən/ *n* sermón *m*

serpent /'sɜ:pənt/ *n* serpiente *f*

serrated /sɪ'reɪtɪd/ *a* serrado

serum /'sɪərəm/ *n* (*pl* **-a**) suero *m*

servant /'sɜ:vənt/ *n* criado *m*; (*fig*) servidor *m*

serve /sɜ:v/ *vt* servir; (*in the army etc*) prestar servicio; cumplir ‹*sentence*›. **~ as** servir de. **~ its purpose** servir para el caso. **it ~s you right** ¡bien te lo mereces! ¡te está bien merecido! ● *vi* servir. ● *n* (*in tennis*) saque *m*

service /'sɜ:vɪs/ *n* servicio *m*; (*maintenance*) revisión *f*. **of ~ to** útil a. ● *vt* revisar ‹*car etc*›. **~able** /'sɜ:vɪsəbl/ *a* práctico; (*durable*) duradero. **~ charge** *n* servicio *m*. **~man** /'sɜ:vɪsmən/ *n* (*pl* **-men**) militar *m*. **~s** *npl* (*mil*) fuerzas *fpl* armadas. **~ station** *n* estación *f* de servicio

serviette /sɜ:vɪ'et/ *n* servilleta *f*

servile /'sɜ:vaɪl/ *a* servil

session /'seʃn/ *n* sesión *f*; (*univ*) curso *m*

set /set/ *vt* (*pt* **set**, *pres p* **setting**) poner; poner en hora ‹*clock etc*›; fijar ‹*limit etc*›; (*typ*) componer. **~ fire to** pegar fuego a. **~ free** *vt* poner en libertad. ● *vi* ‹*sun*› ponerse; ‹*jelly*› cuajarse. ● *n* serie *f*; (*of cutlery etc*) juego *m*; (*tennis*) set *m*; (*TV, radio*) aparato *m*; (*of hair*) marcado *m*; (*in theatre*) decorado *m*; (*of people*) círculo *m*. ● *a* fijo. **be ~ on** estar resuelto a. **~ about** *vi* empezar a. **~ back** *vt* (*delay*) retardar; (*cost, sl*) costar. **~ off** *vi* salir. ● *vt* (*make start*) poner en marcha; hacer estallar ‹*bomb*›. **~ out** *vi* (*declare*) declarar; (*leave*) salir. **~ sail** salir. **~ the table** poner la mesa. **~ up** *vt* establecer. **~-back** *n* revés *m*. **~ square** *n* escuadra *f* de dibujar

settee /se'ti:/ *n* sofá *m*

setting /'setɪŋ/ *n* (*of sun*) puesta *f*; (*of jewel*) engaste *m*; (*in theatre*) escenario *m*; (*typ*) composición *f*. **~-lotion** *n* fijador *m*

settle /'setl/ *vt* (*arrange*) arreglar; (*pay*) pagar; fijar ‹*date*›; calmar ‹*nerves*›. ● *vi* (*come to rest*) posarse; (*live*) instalarse. **~ down** calmarse; (*become orderly*) sentar la cabeza. **~ for** aceptar. **~ up** ajustar cuentas. **~ment** /'setlmənt/ *n* establecimiento *m*; (*agreement*) acuerdo *m*; (*com*) liquidación *f*; (*place*) colonia *f*. **~r** /-ə(r)/ *n* colonizador *m*

set: **~-to** *n* pelea *f*. **~-up** *n* (*fam*) sistema *m*

seven /'sevn/ *a & n* siete (*m*). **~teen** /sevn'ti:n/ *a & n* diecisiete (*m*). **~teenth** *a & n* decimoséptimo (*m*). **~th** *a & n* séptimo (*m*). **~tieth** *a & n* setenta (*m*), septuagésimo (*m*). **~ty** /'sevntɪ/ *a & n* setenta (*m*)

sever /'sevə(r)/ *vt* cortar; (*fig*) romper

several /'sevrəl/ *a & pron* varios

severance /'sevərəns/ *n* (*breaking off*) ruptura *f*

sever|e /sɪ'vɪə(r)/ *a* (**-er**, **-est**) severo; (*violent*) violento; (*serious*) grave; ‹*weather*› riguroso. **~ely** *adv* severamente; (*seriously*) gravemente. **~ity** /-'verətɪ/ *n* severidad *f*; (*violence*) violencia *f*; (*seriousness*) gravedad *f*

sew /səʊ/ *vt/i* (*pt* **sewed**, *pp* **sewn**, *or* **sewed**) coser

sew|age /'su:ɪdʒ/ *n* aguas *fpl* residuales. **~er** /'su:ə(r)/ *n* cloaca *f*

sewing /'səʊɪŋ/ *n* costura *f*. **~-machine** *n* máquina *f* de coser

sewn /səʊn/ *see* **sew**

sex /seks/ *n* sexo *m*. **have ~** tener relaciones sexuales. ● *a* sexual. **~ist** /'seksɪst/ *a & n* sexista (*m & f*)

sextet /seks'tet/ *n* sexteto *m*

sexual /'sekʃʊəl/ *a* sexual. **~ intercourse** *n* relaciones *fpl* sexuales. **~ity** /-'ælətɪ/ *n* sexualidad *f*

sexy /'seksɪ/ *a* (**-ier**, **-iest**) excitante, sexy, provocativo

shabb|ily /'ʃæbɪlɪ/ *adv* pobremente; (*act*) mezquinamente. **~iness** *n* pobreza *f*; (*meanness*) mezquindad *f*. **~y** /'ʃæbɪ/ *a* (**-ier**, **-iest**) ‹*clothes*› gastado; ‹*person*› pobremente vestido; (*mean*) mezquino

shack /ʃæk/ *n* choza *f*

shackles /'ʃæklz/ *npl* grillos *mpl*, grilletes *mpl*

shade /ʃeɪd/ *n* sombra *f*; (*of colour*) matiz *m*; (*for lamp*) pantalla *f*. **a ~ better** un poquito mejor. ● *vt* dar sombra a

shadow /'ʃædəʊ/ *n* sombra *f*. **S~ Cabinet** *n* gobierno *m* en la sombra. ● *vt* (*follow*) seguir. **~y** *a* (*fig*) vago

shady /'ʃeɪdɪ/ *a* (**-ier**, **-iest**) sombreado; (*fig*) dudoso

shaft /ʃɑ:ft/ *n* (*of arrow*) astil *m*; (*mec*) eje *m*; (*of light*) rayo *m*; (*of lift, mine*) pozo *m*

shaggy /'ʃægɪ/ *a* (**-ier**, **-iest**) peludo

shak|e /ʃeɪk/ *vt* (*pt* **shook**, *pp* **shaken**) sacudir; agitar ‹*bottle*›; (*shock*) desconcertar. **~e hands with** estrechar la mano a. ● *vi* temblar. **~e off** *vi* deshacerse de. ● *n* sacudida *f*. **~e-up** *n* reorganización *f*. **~y** /'ʃeɪkɪ/ *a* (**-ier**, **-iest**) tembloroso; ‹*table etc*› inestable; (*unreliable*) incierto

shall /ʃæl/ *v, aux* (*first person in future tense*). **I ~ go** iré. **we ~ see** veremos

shallot /ʃə'lɒt/ *n* chalote *m*

shallow /'ʃæləʊ/ *a* (**-er**, **-est**) poco profundo; (*fig*) superficial

sham /ʃæm/ *n* farsa *f*; (*person*) impostor *m*. ● *a* falso; (*affected*) fingido. ● *vt* (*pt* **shammed**) fingir

shambles /'ʃæmblz/ *npl* (*mess, fam*) desorden *m* total

shame /ʃeɪm/ *n* vergüenza *f*. **what a ~!** ¡qué lástima! ● *vt* avergonzar. **~faced** /'ʃeɪmfeɪst/ *a* avergonzado. **~ful** *a* vergonzoso. **~fully** *adv* vergonzosamente. **~less** *a* desvergonzado

shampoo /ʃæm'pu:/ *n* champú *m*. ● *vt* lavar

shamrock /'ʃæmrɒk/ *n* trébol *m*

shandy /'ʃændɪ/ *n* cerveza *f* con gaseosa, clara *f*

shan't /ʃɑ:nt/ = **shall not**

shanty /'ʃæntɪ/ *n* chabola *f*. **~ town** *n* chabolas *fpl*

shape /ʃeɪp/ *n* forma *f*. ● *vt* formar; determinar ‹*future*›. ● *vi* formarse. **~ up** prometer. **~less** *a* informe. **~ly** /'ʃeɪplɪ/ *a* (**-ier**, **-iest**) bien proporcionado

share /ʃeə(r)/ *n* porción *f*; (*com*) acción *f*. **go ~s** compartir. ● *vt* compartir; (*divide*) dividir. ● *vi* participar. **~ in** participar en. **~holder** /'ʃeəhəʊldə(r)/ *n* accionista *m & f*. **~-out** *n* reparto *m*

shark /ʃɑ:k/ *n* tiburón *m*; (*fig*) estafador *m*

sharp /ʃɑ:p/ *a* (**-er**, **-est**) ‹*knife etc*› afilado; ‹*pin etc*› puntiagudo; ‹*pain, sound*› agudo; ‹*taste*› acre; (*sudden, harsh*) brusco; (*well defined*) marcado; (*dishonest*) poco escrupuloso; (*clever*) listo. ● *adv* en punto. **at seven o'clock ~** a las siete en punto. ● *n* (*mus*) sostenido *m*. **~en** /'ʃɑ:pn/ *vt* afilar; sacar punta a ‹*pencil*›. **~ener** *n* (*mec*) afilador *m*; (*for pencils*) sacapuntas *m invar*. **~ly** *adv* bruscamente

shatter /'ʃætə(r)/ *vt* hacer añicos; (*upset*) perturbar. ● *vi* hacerse añicos. **~ed** *a* (*exhausted*) agotado

shav|e /ʃeɪv/ *vt* afeitar. ● *vi* afeitarse. ● *n* afeitado *m*. **have a ~e** afeitarse. **~en** *a* ‹*face*› afeitado; ‹*head*› rapado. **~er** *n* maquinilla *f* (de afeitar). **~ing-brush** *n* brocha *f* de afietar. **~ing-cream** *n* crema *f* de afeitar

shawl /ʃɔ:l/ *n* chal *m*

she /ʃi:/ *pron* ella. ● *n* hembra *f*

sheaf /ʃi:f/ *n* (*pl* **sheaves**) gavilla *f*

shear /ʃɪə(r)/ *vt* (*pp* **shorn**, *or* **sheared**) esquilar. **~s** /ʃɪəz/ *npl* tijeras *fpl* grandes

sheath /ʃi:θ/ *n* (*pl* **-s** /ʃi:ðz/) vaina *f*; (*contraceptive*) condón *m*. **~e** /ʃi:ð/ *vt* envainar

shed[1] /ʃed/ *n* cobertizo *m*

shed[2] /ʃed/ *vt* (*pt* **shed**, *pres p* **shedding**) perder; derramar ‹*tears*›; despojarse de ‹*clothes*›. **~ light on** aclarar

sheen /ʃi:n/ *n* lustre *m*

sheep /ʃi:p/ *n invar* oveja *f*. **~-dog** *n* perro *m* pastor. **~ish** /'ʃi:pɪʃ/ *a* vergonzoso. **~ishly** *adv* tímidamente. **~skin** /'ʃi:pskɪn/ *n* piel *f* de carnero, zamarra *f*

sheer /ʃɪə(r)/ *a* puro; (*steep*) perpendicular; ‹*fabric*› muy fino. ● *adv* a pico

sheet /ʃi:t/ *n* sábana *f*; (*of paper*) hoja *f*; (*of glass*) lámina *f*; (*of ice*) capa *f*

sheikh /ʃeɪk/ *n* jeque *m*

shelf /ʃelf/ *n* (*pl* **shelves**) estante *m*. **be on the ~** quedarse para vestir santos

shell /ʃel/ *n* concha *f*; (*of egg*) cáscara *f*; (*of building*) casco *m*; (*explosive*) proyectil *m*. ● *vt* desgranar ‹*peas etc*›; (*mil*) bombardear. **~fish** /'ʃelfɪʃ/ *n invar* (*crustacean*) crustáceo *m*; (*mollusc*) marisco *m*

shelter /'ʃeltə(r)/ *n* refugio *m*, abrigo *m*. ● *vt* abrigar; (*protect*) proteger; (*give lodging to*) dar asilo a. ● *vi* abrigarse. **~ed** *a* ‹*spot*› abrigado; ‹*life etc*› protegido

shelv|e /ʃelv/ *vt* (*fig*) dar carpetazo a. **~ing** /'ʃelvɪŋ/ *n* estantería *f*

shepherd /'ʃepəd/ *n* pastor *m*. ● *vt* guiar. **~ess** /-'des/ *n* pastora *f*. **~'s pie** *n* carne *f* picada con puré de patatas

sherbet /'ʃɜ:bət/ *n* (*Amer*, *water-ice*) sorbete *m*

sheriff /'ʃerɪf/ *n* alguacil *m*, sheriff *m*

sherry /'ʃerɪ/ *n* (vino *m* de) jerez *m*

shield /ʃi:ld/ *n* escudo *m*. ● *vt* proteger

shift /ʃɪft/ *vt* cambiar; cambiar de sitio ‹*furniture etc*›; echar ‹*blame etc*›. ● *n* cambio *m*; (*work*) turno *m*; (*workers*) tanda *f*. **make ~** arreglárselas. **~less** /'ʃɪftlɪs/ *a* holgazán

shifty /'ʃɪftɪ/ *a* (**-ier**, **-iest**) taimado

shilling /'ʃɪlɪŋ/ *n* chelín *m*

shilly-shally /'ʃɪlɪʃælɪ/ *vi* titubear

shimmer /'ʃɪmə(r)/ *vi* rielar, relucir. ● *n* luz *f* trémula

shin /ʃɪn/ *n* espinilla *f*

shine /ʃaɪn/ *vi* (*pt* **shone**) brillar. ● *vt* sacar brillo a. **~ on** dirigir ‹*torch*›. ● *n* brillo *m*

shingle /'ʃɪŋgl/ *n* (*pebbles*) guijarros *mpl*

shingles /'ʃɪŋglz/ *npl* (*med*) herpes *mpl* & *fpl*

shiny /'ʃaɪnɪ/ *a* (**-ier**, **-iest**) brillante

ship /ʃɪp/ *n* buque *m*, barco *m*. ● *vt* (*pt* **shipped**) transportar; (*send*) enviar; (*load*) embarcar. **~building** /'ʃɪpbɪldɪŋ/ *n* construcción *f* naval. **~ment** *n* envío *m*. **~per** *n* expedidor *m*. **~ping** *n* envío *m*; (*ships*) barcos *mpl*. **~shape** /'ʃɪpʃeɪp/ *adv* & *a* en buen orden, en regla. **~wreck** /'ʃɪprek/ *n* naufragio *m*. **~wrecked** *a* naufragado. **be ~wrecked** naufragar. **~yard** /'ʃɪpjɑ:d/ *n* astillero *m*

shirk /ʃɜ:k/ *vt* esquivar. **~er** *n* gandul *m*

shirt /ʃɜ:t/ *n* camisa *f*. **in ~-sleeves** en mangas de camisa. **~y** /'ʃɜ:tɪ/ *a* (*sl*) enfadado

shiver /'ʃɪvə(r)/ *vi* temblar. ● *n* escalofrío *m*

shoal /ʃəʊl/ *n* banco *m*

shock /ʃɒk/ *n* sacudida *f*; (*fig*) susto *m*; (*elec*) descarga *f*; (*med*) choque *m*. ● *vt* escandalizar. **~ing** *a* escandaloso; (*fam*) espantoso. **~ingly** *adv* terriblemente

shod /ʃɒd/ *see* **shoe**

shodd|ily /'ʃɒdɪlɪ/ *adv* mal. **~y** /'ʃɒdɪ/ *a* (**-ier**, **-iest**) mal hecho, de pacotilla

shoe /ʃu:/ *n* zapato *m*; (*of horse*) herradura *f*. ● *vt* (*pt* **shod**, *pres p* **shoeing**) herrar ‹*horse*›. **be well shod** estar bien calzado. **~horn** /'ʃu:hɔ:n/ *n* calzador *m*. **~-lace** *n* cordón *m* de zapato. **~maker** /'ʃu:meɪkə(r)/ *n* zapatero *m*. **~ polish** *n* betún *m*. **~-string** *n*. **on a ~-string** con poco dinero. **~-tree** *n* horma *f*

shone /ʃɒn/ *see* **shine**

shoo /ʃu:/ *vt* ahuyentar

shook /ʃʊk/ *see* **shake**

shoot /ʃu:t/ *vt* (*pt* **shot**) disparar; rodar ‹*film*›. ● *vi* (*hunt*) cazar. ● *n* (*bot*) retoño *m*; (*hunt*) cacería *f*. **~ down** *vt* derribar. **~ out** *vi* (*rush*) salir disparado. **~ up** ‹*prices*› subir de repente; (*grow*) crecer. **~ing-range** *n* campo *m* de tiro

shop /ʃɒp/ *n* tienda *f*; (*work-shop*) taller *m*. **talk ~** hablar de su trabajo. ● *vi* (*pt* **shopping**) hacer compras. **~ around** buscar el mejor precio. **go ~ping** ir de compras. **~ assistant** *n* dependiente *m*. **~keeper** /'ʃɒpki:pə(r)/ *n* tendero *m*. **~-lifter** *n* ratero *m* (de tiendas). **~-lifting** *n* ratería *f* (de tiendas). **~per** *n* comprador *m*. **~ping** /'ʃɒpɪŋ/ *n* compras *fpl*. **~ping bag** *n* bolsa *f* de la compra. **~ping centre** *n* centro *m* comercial. **~ steward** *n* enlace *m* sindical. **~-window** *n* escaparate *m*

shore /ʃɔ:(r)/ *n* orilla *f*

shorn /ʃɔ:n/ *see* **shear**

short /ʃɔ:t/ *a* (**-er**, **-est**) corto; (*not lasting*) breve; ‹*person*› bajo; (*curt*) brusco. **a ~ time ago** hace poco. **be ~ of** necesitar. **Mick is ~ for Michael** Mick es el diminutivo de Michael. ● *adv* (*stop*) en seco. **~ of doing** a menos que no hagamos. ● *n*. **in ~** en resumen. **~age** /'ʃɔ:tɪdʒ/ *n* escasez *f*. **~bread** /'ʃɔ:tbred/ *n* galleta *f* de mantequilla. **~-change** *vt* estafar, engañar. **~ circuit** *n* cortocircuito *m*. **~coming** /'ʃɔ:tkʌmɪŋ/ *n* deficiencia *f*. **~ cut** *n* atajo *m*. **~en** /'ʃɔ:tn/ *vt* acortar. **~hand** /'ʃɔ:thænd/ *n* taquigrafía *f*. **~hand typist** *n* taquimecanógrafo *m*, taquimeca *f* (*fam*). **~-lived** *a* efímero. **~ly** /'ʃɔ:tlɪ/ *adv* dentro de poco. **~s** *npl* pantalón *m* corto. **~-sighted** *a* miope. **~tempered** *a* de mal genio

shot /ʃɒt/ *see* **shoot**. ● *n* tiro *m*; (*person*) tirador *m*; (*photo*) foto *f*; (*injection*) inyección *f*. **like a ~** como una bala; (*willingly*) de buena gana. **~-gun** *n* escopeta *f*

should /ʃʊd, ʃəd/ *v*, *aux*. **I ~ go** debería ir. **I ~ have seen him** debiera haberlo visto. **I ~ like** me gustaría. **if he ~ come** si viniese

shoulder /'ʃəʊldə(r)/ *n* hombro *m*. ● *vt* cargar con ‹*responsibility*›; llevar a hombros ‹*burden*›. **~-blade** *n* omóplato *m*. **~-strap** *n* correa *f* del hombro; (*of bra etc*) tirante *m*

shout /ʃaʊt/ *n* grito *m*. ● *vt*/*i* gritar. **~ at s.o.** gritarle a uno. **~ down** hacer callar a gritos

shove /ʃʌv/ *n* empujón *m*. ● *vt* empujar; (*put*, *fam*) poner. ● *vi* empujar. **~ off** *vi* (*fam*) largarse

shovel /'ʃʌvl/ *n* pala *f*. ● *vt* (*pt* **shovelled**) mover con la pala

show /ʃəʊ/ *vt* (*pt* **showed**, *pp* **shown**) mostrar; (*put on display*) exponer; poner ‹*film*›. ● *vi* (*be visible*) verse. ● *n* demostración *f*; (*exhibition*) exposición *f*; (*ostentation*) pompa *f*; (*in theatre*) espectáculo *m*; (*in cinema*) sesión *f*. **on ~** expuesto. **~ off** *vt* lucir; (*pej*) ostentar. ● *vi* presumir. **~ up** *vi* destacar; (*be present*) presentarse. ● *vt* (*unmask*) desenmascarar. **~-case** *n* vitrina *f*. **~-down** *n* confrontación *f*

shower /'ʃaʊə(r)/ *n* chaparrón *m*; (*of blows etc*) lluvia *f*; (*for washing*) ducha *f*. **have a ~** ducharse. ● *vi* ducharse. ● *vt*. **~ with** colmar de. **~proof** /'ʃaʊəpru:f/ *a* impermeable. **~y** *a* lluvioso

show: **~-jumping** *n* concurso *m* hípico. **~manship** /'ʃəʊmənʃɪp/ *n* teatralidad *f*, arte *f* de presentar espectáculos

shown /ʃəʊn/ *see* **show**

show: **~-off** *n* fanfarrón *m*. **~-place** *n* lugar *m* de interés turístico. **~room** /'ʃəʊru:m/ *n* sala *f* de exposición *f*

showy /'ʃəʊɪ/ *a* (**-ier**, **-iest**) llamativo; ‹*person*› ostentoso

shrank /ʃræŋk/ *see* **shrink**

shrapnel /'ʃræpnəl/ *n* metralla *f*

shred /ʃred/ *n* pedazo *m*; (*fig*) pizca *f*. ● *vt* (*pt* **shredded**) hacer tiras; (*culin*) cortar en tiras. **~der** *n* desfibradora *f*, trituradora *f*

shrew /ʃru:/ *n* musaraña *f*; (*woman*) arpía *f*

shrewd /ʃru:d/ *a* (**-er**, **-est**) astuto. **~ness** *n* astucia *f*

shriek /ʃri:k/ *n* chillido *m*. ● *vt*/*i* chillar

shrift /ʃrɪft/ *n*. **give s.o. short ~** despachar a uno con brusquedad

shrill /ʃrɪl/ *a* agudo

shrimp /ʃrɪmp/ *n* camarón *m*

shrine /ʃraɪn/ *n* (*place*) lugar *m* santo; (*tomb*) sepulcro *m*

shrink /ʃrɪŋk/ *vt* (*pt* **shrank**, *pp* **shrunk**) encoger. ● *vi* encogerse; (*draw back*) retirarse; (*lessen*) disminuir. **~age** *n* encogimiento *m*

shrivel /'ʃrɪvl/ *vi* (*pt* **shrivelled**) (*dry up*) secarse; (*become wrinkled*) arrugarse

shroud /ʃraʊd/ *n* sudario *m*; (*fig*) velo *m*. ● *vt* (*veil*) velar
Shrove /ʃrəʊv/ *n*. ~ **Tuesday** *n* martes *m* de carnaval
shrub /ʃrʌb/ *n* arbusto *m*
shrug /ʃrʌg/ *vt* (*pt* **shrugged**) encogerse de hombros. ● *n* encogimiento *m* de hombros
shrunk /ʃrʌŋk/ *see* **shrink**
shrunken /ˈʃrʌnkən/ *a* encogido
shudder /ˈʃʌdə(r)/ *vi* estremecerse. ● *n* estremecimiento *m*
shuffle /ˈʃʌfl/ *vi* arrastrar los pies. ● *vt* barajar ‹*cards*›. ● *n* arrastramiento *m* de los pies; (*of cards*) barajadura *f*
shun /ʃʌn/ *vt* (*pt* **shunned**) evitar
shunt /ʃʌnt/ *vt* apartar, desviar
shush /ʃʊʃ/ *int* ¡chitón!
shut /ʃʌt/ *vt* (*pt* **shut**, *pres p* **shutting**) cerrar. ● *vi* cerrarse. ~ **down** cerrar. ~ **up** *vt* cerrar; (*fam*) hacer callar. ● *vi* callarse. **~-down** *n* cierre *m*. **~ter** /ˈʃʌtə(r)/ *n* contraventana *f*; (*photo*) obturador *m*
shuttle /ˈʃʌtl/ *n* lanzadera *f*; (*train*) tren *m* de enlace. ● *vt* transportar. ● *vi* ir y venir. **~cock** /ˈʃʌtlkɒk/ *n* volante *m*. ~ **service** *n* servicio *m* de enlace
shy /ʃaɪ/ *a* (**-er**, **-est**) tímido. ● *vi* (*pt* **shied**) asustarse. ~ **away from** huir. **~ness** *n* timidez *f*
Siamese /saɪəˈmi:z/ *a* siamés
sibling /ˈsɪblɪŋ/ *n* hermano *m*, hermana *f*
Sicil|ian /sɪˈsɪljən/ *a* & *n* siciliano (*m*). **~y** /ˈsɪsɪlɪ/ *n* Sicilia *f*
sick /sɪk/ *a* enfermo; ‹*humour*› negro; (*fed up*, *fam*) harto. **be** ~ (*vomit*) vomitar. **be** ~ **of** (*fig*) estar harto de. **feel** ~ sentir náuseas. **~en** /ˈsɪkən/ *vt* dar asco. ● *vi* caer enfermo. **be ~ening for** incubar
sickle /ˈsɪkl/ *n* hoz *f*
sick: **~ly** /ˈsɪklɪ/ *a* (**-ier**, **-iest**) enfermizo; ‹*taste, smell etc*› nauseabundo. **~ness** /ˈsɪknɪs/ *n* enfermedad *f*. **~-room** *n* cuarto *m* del enfermo
side /saɪd/ *n* lado *m*; (*of river*) orilla *f*; (*of hill*) ladera *f*; (*team*) equipo *m*; (*fig*) parte *f*. ~ **by** ~ uno al lado del otro. **on the** ~ (*sideline*) como actividad secundaria; (*secretly*) a escondidas. ● *a* lateral. ● *vi*. ~ **with** tomar el partido de. **~board** /ˈsaɪdbɔ:d/ *n* aparador *m*. **~boards** *npl*, **~burns** *npl* (*sl*) patillas *fpl*. **~-car** *n* sidecar *m*. **~-effect** *n* efecto *m* secundario. **~light** /ˈsaɪdlaɪt/ *n* luz *f* de posición. **~line** /ˈsaɪdlaɪn/ *n* actividad *f* secundaria. **~long** /-lɒŋ/ *a* & *adv* de soslayo. **~-road** *n* calle *f* secundaria. **~-saddle** *n* silla *f* de mujer. **ride ~-saddle** *adv* a mujeriegas. **~-show** *n* atracción *f* secundaria. **~-step** *vt* evitar. **~-track** *vt* desviar del asunto. **~walk** /ˈsaɪdwɔ:k/ *n* (*Amer*) acera *f*, vereda *f* (*LAm*). **~ways** /ˈsaɪdweɪz/ *a* & *adv* de lado. **~-whiskers** *npl* patillas *fpl*
siding /ˈsaɪdɪŋ/ *n* apartadero *m*
sidle /ˈsaɪdl/ *vi* avanzar furtivamente. ~ **up to** acercarse furtivamente
siege /si:dʒ/ *n* sitio *m*, cerco *m*
siesta /sɪˈestə/ *n* siesta *f*
sieve /sɪv/ *n* cernedor *m*. ● *vt* cerner
sift /sɪft/ *vt* cerner. ● *vi*. ~ **through** examinar
sigh /saɪ/ *n* suspiro. ● *vi* suspirar
sight /saɪt/ *n* vista *f*; (*spectacle*) espectáculo *m*; (*on gun*) mira *f*. **at (first)** ~ a primera vista. **catch** ~ **of** vislumbrar. **lose** ~ **of** perder de vista. **on** ~ a primera vista. **within** ~ **of** (*near*) cerca de. ● *vt* ver, divisar. **~-seeing** /ˈsaɪtsi:ɪŋ/ *n* visita *f* turística. **~seer** /-ə(r)/ *n* turista *m* & *f*
sign /saɪn/ *n* señal *f*. ● *vt* firmar. ~ **on**, ~ **up** *vt* inscribir. ● *vi* inscribirse
signal /ˈsɪgnəl/ *n* señal *f*. ● *vt* (*pt* **signalled**) comunicar; hacer señas a ‹*person*›. **~-box** *n* casilla *f* del guardavía. **~man** /ˈsɪgnəlmən/ *n* (*pl* **-men**) guardavía *f*
signatory /ˈsɪgnətrɪ/ *n* firmante *m* & *f*
signature /ˈsɪgnətʃə(r)/ *n* firma *f*. ~ **tune** *n* sintonía *f*
signet-ring /ˈsɪgnɪtrɪŋ/ *n* anillo *m* de sello
significan|ce /sɪgˈnɪfɪkəns/ *n* significado *m*. **~t** /sɪgˈnɪfɪkənt/ *a* significativo; (*important*) importante. **~tly** *adv* significativamente
signify /ˈsɪgnɪfaɪ/ *vt* significar. ● *vi* (*matter*) importar, tener importancia
signpost /ˈsaɪnpəʊst/ *n* poste *m* indicador
silen|ce /ˈsaɪləns/ *n* silencio *m*. ● *vt* hacer callar. **~cer** /-ə(r)/ *n* silenciador *m*. **~t** /ˈsaɪlənt/ *a* silencioso;

‹*film*› mudo. **~tly** *adv* silenciosamente

silhouette /sɪlu:ˈet/ *n* silueta *f*. ● *vt*. **be ~d** perfilarse, destacarse (**against** contra)

silicon /ˈsɪlɪkən/ *n* silicio *m*. **~ chip** *n* pastilla *f* de silicio

silk /sɪlk/ *n* seda *f*. **~en** *a*, **~y** *a* (*of silk*) de seda; (*like silk*) sedoso. **~worm** *n* gusano *m* de seda

sill /sɪl/ *n* antepecho *m*; (*of window*) alféizar *m*; (*of door*) umbral *m*

silly /ˈsɪlɪ/ *a* (**-ier**, **-iest**) tonto. ● *n*. **~-billy** (*fam*) tonto *m*

silo /ˈsaɪləʊ/ *n* (*pl* **-os**) silo *m*

silt /sɪlt/ *n* sedimento *m*

silver /ˈsɪlvə(r)/ *n* plata *f*. ● *a* de plata. **~ plated** *a* bañado en plata, plateado. **~side** /ˈsɪlvəsaɪd/ *n* (*culin*) contra *f*. **~smith** /ˈsɪlvəsmɪθ/ *n* platero *m*. **~ware** /ˈsɪlvəweə(r)/ *n* plata *f*. **~ wedding** *n* bodas *fpl* de plata. **~y** *a* plateado; ‹*sound*› argentino

simil|ar /ˈsɪmɪlə(r)/ *a* parecido. **~arity** /-ɪˈlærətɪ/ *n* parecido *m*. **~arly** *adv* de igual manera

simile /ˈsɪmɪlɪ/ *n* símil *m*

simmer /ˈsɪmə(r)/ *vt/i* hervir a fuego lento; (*fig*) hervir. **~ down** calmarse

simpl|e /ˈsɪmpl/ *a* (**-er**, **-est**) sencillo; ‹*person*› ingenuo. **~e-minded** *a* ingenuo. **~eton** /ˈsɪmpltən/ *n* simplón *m*. **~icity** /-ˈplɪsetɪ/ *n* secillez *f*. **~ification** /-ɪˈkeɪʃn/ *n* simplificación *f*. **~ify** /ˈsɪmplɪfaɪ/ *vt* simplificar. **~y** *adv* sencillamente; (*absolutely*) absolutamente

simulat|e /ˈsɪmjʊleɪt/ *vt* simular. **~ion** /-ˈleɪʃn/ *n* simulación *f*

simultaneous /sɪmlˈteɪnɪəs/ *a* simultáneo. **~ly** *adv* simultáneamente

sin /sɪn/ *n* pecado *m*. ● *vi* (*pt* **sinned**) pecar

since /sɪns/ *prep* desde. ● *adv* desde entonces. ● *conj* desde que; (*because*) ya que

sincer|e /sɪnˈsɪə(r)/ *a* sincero. **~ely** *adv* sinceramente. **~ity** /-ˈserətɪ/ *n* sinceridad *f*

sinew /ˈsɪnju:/ *n* tendón *m*. **~s** *npl* músculos *mpl*

sinful /ˈsɪnfl/ *a* pecaminoso; (*shocking*) escandaloso

sing /sɪŋ/ *vt/i* (*pt* **sang**, *pp* **sung**) cantar

singe /sɪndʒ/ *vt* (*pres p* **singeing**) chamuscar

singer /ˈsɪŋə(r)/ *n* cantante *m & f*

singl|e /ˈsɪŋgl/ *a* único; (*not double*) sencillo; (*unmarried*) soltero; ‹*bed, room*› individual. ● *n* (*tennis*) juego *m* individual; (*ticket*) billete *m* sencillo. ● *vt*. **~e out** escoger; (*distinguish*) distinguir. **~e-handed** *a & adv* sin ayuda. **~e-minded** *a* resuelto

singlet /ˈsɪŋglɪt/ *n* camiseta *f*

singly /ˈsɪŋglɪ/ *adv* uno a uno

singsong /ˈsɪŋsɒŋ/ *a* monótono. ● *n*. **have a ~** cantar juntos

singular /ˈsɪŋgjʊlə(r)/ *n* singular *f*. ● *a* singular; (*uncommon*) raro; ‹*noun*› en singular. **~ly** *adv* singularmente

sinister /ˈsɪnɪstə(r)/ *a* siniestro

sink /sɪŋk/ *vt* (*pt* **sank**, *pp* **sunk**) hundir; perforar ‹*well*›; invertir ‹*money*›. ● *vi* hundirse; ‹*patient*› debilitarse. ● *n* fregadero *m*. **~ in** *vi* penetrar

sinner /ˈsɪnə(r)/ *n* pecador *m*

sinuous /ˈsɪnjʊəs/ *a* sinuoso

sinus /ˈsaɪnəs/ *n* (*pl* **-uses**) seno *m*

sip /sɪp/ *n* sorbo *m*. ● *vt* (*pt* **sipped**) sorber

siphon /ˈsaɪfən/ *n* sifón *m*. *vt*. **~ out** sacar con sifón

sir /sɜ:(r)/ *n* señor *m*. **S~** *n* (*title*) sir *m*

siren /ˈsaɪərən/ *n* sirena *f*

sirloin /ˈsɜ:lɔɪn/ *n* solomillo *m*, lomo *m* bajo

sirocco /sɪˈrɒkəʊ/ *n* siroco *m*

sissy /ˈsɪsɪ/ *n* hombre *m* afeminado, marica *m*, mariquita *m*; (*coward*) gallina *m & f*

sister /ˈsɪstə(r)/ *n* hermana *f*; (*nurse*) enfermera *f* jefe. **S~ Mary** Sor María. **~-in-law** *n* (*pl* **~s-in-law**) cuñada *f*. **~ly** *a* de hermana; (*like sister*) como hermana

sit /sɪt/ *vt* (*pt* **sat**, *pres p* **sitting**) sentar. ● *vi* sentarse; ‹*committee etc*› reunirse. **be ~ting** estar sentado. **~ back** *vi* (*fig*) relajarse. **~ down** *vi* sentarse. **~ for** *vi* presentarse a ‹*exam*›; posar para ‹*portrait*›. **~ up** *vi* enderezarse; (*stay awake*) velar. **~-in** *n* ocupación *f*

site /saɪt/ *n* sitio *m*. **building ~** *n* solar *m*. ● *vt* situar

sit: ~ting *n* sesión *f*; (*in restaurant*) turno *m*. **~ting-room** *n* cuarto *m* de estar

situat|e /ˈsɪtjʊeɪt/ *vt* situar. **~ed** *a* situado. **~ion** /-ˈeɪʃn/ *n* situación *f*; (*job*) puesto *m*

six /sɪks/ *a* & *n* seis (*m*). **~teen** /sɪkˈsti:n/ *a* & *n* dieciséis (*m*). **~teenth** *a* & *n* decimosexto (*m*). **~th** *a* & *n* sexto (*m*). **~tieth** *a* & *n* sesenta (*m*), sexagésimo (*m*). **~ty** /ˈsɪkstɪ/ *a* & *n* sesenta (*m*)

size /saɪz/ *n* tamaño *m*; (*of clothes*) talla *f*; (*of shoes*) número *m*; (*extent*) magnitud *f*. ● *vt*. **~ up** (*fam*) juzgar. **~able** *a* bastante grande

sizzle /ˈsɪzl/ *vi* crepitar

skate[1] /skeɪt/ *n* patín *m*. ● *vi* patinar. **~board** /ˈskeɪtbɔ:d/ *n* monopatín *m*. **~r** *n* patinador *m*

skate[2] /skeɪt/ *n invar* (*fish*) raya *f*

skating /ˈskeɪtɪŋ/ *n* patinaje *m*. **~-rink** *n* pista *f* de patinaje

skein /skeɪn/ *n* madeja *f*

skelet|al /ˈskelɪtl/ *a* esquelético. **~on** /ˈskelɪtn/ *n* esqueleto *m*. **~on staff** *n* personal *m* reducido

sketch /sketʃ/ *n* esbozo *m*; (*drawing*) dibujo *m*; (*in theatre*) pieza *f* corta y divertida. ● *vt* esbozar. ● *vi* dibujar. **~y** /ˈsketʃɪ/ *a* (**-ier**, **-iest**) incompleto

skew /skju:/ *n*. **on the ~** sesgado

skewer /ˈskju:ə(r)/ *n* broqueta *f*

ski /ski:/ *n* (*pl* **skis**) esquí *m*. ● *vi* (*pt* **skied**, *pres p* **skiing**) esquiar. **go ~ing** ir a esquiar

skid /skɪd/ *vi* (*pt* **skidded**) patinar. ● *n* patinazo *m*

ski: **~er** *n* esquiador *m*. **~ing** *n* esquí *m*

skilful /ˈskɪlfl/ *a* diestro

ski-lift /ˈski:lɪft/ *n* telesquí *m*

skill /skɪl/ *n* destreza *f*, habilidad *f*. **~ed** *a* hábil; ‹*worker*› cualificado

skim /skɪm/ *vt* (*pt* **skimmed**) espumar; desnatar ‹*milk*›; (*glide over*) rozar. **~ over** *vt* rasar. **~ through** *vi* hojear

skimp /skɪmp/ *vt* escatimar. **~y** /ˈskɪmpɪ/ *a* (**-ier**, **-iest**) insuficiente; ‹*skirt, dress*› corto

skin /skɪn/ *n* piel *f*. ● *vt* (*pt* **skinned**) despellejar; pelar (*fruit*). **~-deep** *a* superficial. **~-diving** *n* natación *f* submarina. **~flint** /ˈskɪnflɪnt/ *n* tacaño *m*. **~ny** /ˈskɪnɪ/ *a* (**-ier**, **-iest**) flaco

skint /skɪnt/ *a* (*sl*) sin una perra

skip[1] /skɪp/ *vi* (*pt* **skipped**) *vi* saltar; (*with rope*) saltar a la comba. ● *vt* saltarse. ● *n* salto *m*

skip[2] /skɪp/ *n* (*container*) cuba *f*

skipper /ˈskɪpə(r)/ *n* capitán *m*

skipping-rope /ˈskɪpɪŋrəʊp/ *n* comba *f*

skirmish /ˈskɜ:mɪʃ/ *n* escaramuza *f*

skirt /skɜ:t/ *n* falda *f*. ● *vt* rodear; (*go round*) ladear

skirting-board /ˈskɜ:tɪŋbɔ:d/ *n* rodapié *m*, zócalo *m*

skit /skɪt/ *n* pieza *f* satírica

skittish /ˈskɪtɪʃ/ *a* juguetón; ‹*horse*› nervioso

skittle /ˈskɪtl/ *n* bolo *m*

skive /skaɪv/ *vi* (*sl*) gandulear

skivvy /ˈskɪvɪ/ *n* (*fam*) criada *f*

skulk /skʌlk/ *vi* avanzar furtivamente; (*hide*) esconderse

skull /skʌl/ *n* cráneo *m*; (*remains*) calavera *f*. **~-cap** *n* casquete *m*

skunk /skʌŋk/ *n* mofeta *f*; (*person*) canalla *f*

sky /skaɪ/ *n* cielo *m*. **~-blue** *a* & *n* azul (*m*) celeste. **~jack** /ˈskaɪdʒæk/ *vt* secuestrar. **~jacker** *n* secuestrador *m*. **~light** /ˈskaɪlaɪt/ *n* tragaluz *m*. **~scraper** /ˈskaɪ skreɪpə(r)/ *n* rascacielos *m invar*

slab /slæb/ *n* bloque *m*; (*of stone*) losa *f*; (*of chocolate*) tableta *f*

slack /slæk/ *a* (**-er**, **-est**) flojo; ‹*person*› negligente; ‹*period*› de poca actividad. ● *n* (*of rope*) parte *f* floja. ● *vt* aflojar. ● *vi* aflojarse; ‹*person*› descansar. **~en** /ˈslækən/ *vt* aflojar. ● *vi* aflojarse; ‹*person*› descansar. **~en (off)** *vt* aflojar. **~ off** (*fam*) aflojar

slacks /slæks/ *npl* pantalones *mpl*

slag /slæg/ *n* escoria *f*

slain /sleɪn/ *see* **slay**

slake /sleɪk/ *vt* apagar

slam /slæm/ *vt* (*pt* **slammed**) golpear; (*throw*) arrojar; (*criticize, sl*) criticar. **~ the door** dar un portazo. ● *vi* cerrarse de golpe. ● *n* golpe *m*; (*of door*) portazo *m*

slander /ˈslɑ:ndə(r)/ *n* calumnia *f*. ● *vt* difamar. **~ous** *a* calumnioso

slang /slæŋ/ *n* jerga *f*, argot *m*. **~y** *a* vulgar

slant /slɑ:nt/ *vt* inclinar; presentar con parcialidad ‹*news*›. ● *n* inclinación *f*; (*point of view*) punto *m* de vista

slap /slæp/ *vt* (*pt* **slapped**) abofetear; (*on the back*) dar una palmada; (*put*) arrojar. ● *n* bofetada *f*; (*on back*) palmada *f*. ● *adv* de lleno. **~dash**

/'slæpdæʃ/ *a* descuidado. **~-happy** *a* (*fam*) despreocupado; (*dazed, fam*) aturdido. **~stick** /'slæpstɪk/ *n* payasada *f*. **~-up** *a* (*sl*) de primera categoría

slash /slæʃ/ *vt* acuchillar; (*fig*) reducir radicalmente. ● *n* cuchillada *f*

slat /slæt/ *n* tablilla *f*

slate /sleɪt/ *n* pizarra *f*. ● *vt* (*fam*) criticar

slaughter /'slɔ:tə(r)/ *vt* masacrar; matar ‹*animal*›. ● *n* carnicería *f*; (*of animals*) matanza *f*. **~house** /'slɔ:təhaʊs/ *n* matadero *m*

Slav /slɑ:v/ *a* & *n* eslavo (*m*)

slav|e /sleɪv/ *n* esclavo *m*. ● *vi* trabajar como un negro. **~e-driver** *n* negrero *m*. **~ery** /-ərɪ/ *n* esclavitud *f*. **~ish** /'sleɪvɪʃ/ *a* servil

Slavonic /slə'vɒnɪk/ *a* eslavo

slay /sleɪ/ *vt* (*pt* **slew**, *pp* **slain**) matar

sleazy /'sli:zɪ/ *a* (**-ier**, **-iest**) (*fam*) sórdido

sledge /sledʒ/ *n* trineo *m*. **~-hammer** *n* almádena *f*

sleek /sli:k/ *a* (**-er**, **-est**) liso, brillante; (*elegant*) elegante

sleep /sli:p/ *n* sueño *m*. **go to ~** dormirse. ● *vi* (*pt* **slept**) dormir. ● *vt* poder alojar. **~er** *n* durmiente *m* & *f*; (*on track*) traviesa *f*; (*berth*) coche-cama *m*. **~ily** *adv* soñolientamente. **~ing-bag** *n* saco *m* de dormir. **~ing-pill** *n* somnífero *m*. **~less** *a* insomne. **~lessness** *n* insomnio *m*. **~-walker** *n* sonámbulo *m*. **~y** /'sli:pɪ/ *a* (**-ier**, **-iest**) soñoliento. **be ~y** tener sueño

sleet /sli:t/ *n* aguanieve *f*. ● *vi* caer aguanieve

sleeve /sli:v/ *n* manga *f*; (*for record*) funda *f*. **up one's ~** en reserva. **~less** *a* sin mangas

sleigh /sleɪ/ *n* trineo *m*

sleight /slaɪt/ *n*. **~ of hand** prestidigitación *f*

slender /'slendə(r)/ *a* delgado; (*fig*) escaso

slept /slept/ *see* **sleep**

sleuth /slu:θ/ *n* investigador *m*

slew[1] /slu:/ *see* **slay**

slew[2] /slu:/ *vi* (*turn*) girar

slice /slaɪs/ *n* lonja *f*; (*of bread*) rebanada *f*; (*of sth round*) rodaja *f*; (*implement*) paleta *f*. ● *vt* cortar; rebanar ‹*bread*›

slick /slɪk/ *a* liso; (*cunning*) astuto. ● *n*. **(oil)-~** capa *f* de aceite

slid|e /slaɪd/ *vt* (*pt* **slid**) deslizar. ● *vi* resbalar. **~e over** pasar por alto de. ● *n* resbalón *m*; (*in playground*) tobogán *m*; (*for hair*) pasador *m*; (*photo*) diapositiva *f*; (*fig, fall*) baja *f*. **~e-rule** *n* regla *f* de cálculo. **~ing** *a* corredizo. **~ing scale** *n* escala *f* móvil

slight /slaɪt/ *a* (**-er**, **-est**) ligero; (*slender*) delgado. ● *vt* ofender. ● *n* desaire *m*. **~est** *a* mínimo. **not in the ~est** en absoluto. **~ly** *adv* un poco

slim /slɪm/ *a* (**slimmer**, **slimmest**) delgado. ● *vi* (*pt* **slimmed**) adelgazar

slime /slaɪm/ *n* légamo *m*, lodo *m*, fango *m*

slimness /'slɪmnɪs/ *n* delgadez *f*

slimy /'slaɪmɪ/ *a* legamoso, fangoso, viscoso; (*fig*) rastrero

sling /slɪŋ/ *n* honda *f*; (*toy*) tirador; (*med*) cabestrillo *m*. ● *vt* (*pt* **slung**) lanzar

slip /slɪp/ *vt* (*pt* **slipped**) deslizar. **~ s.o.'s mind** olvidársele a uno. ● *vi* deslizarse. ● *n* resbalón *m*; (*mistake*) error *m*; (*petticoat*) combinación *f*; (*paper*) trozo *m*. **~ of the tongue** *n* lapsus *m* linguae. **give the ~ to** zafarse de, dar esquinazo a. **~ away** *vi* escabullirse. **~ into** *vi* ponerse ‹*clothes*›. **~ up** *vi* (*fam*) equivocarse

slipper /'slɪpə(r)/ *n* zapatilla *f*

slippery /'slɪpərɪ/ *a* resbaladizo

slip: **~-road** *n* rampa *f* de acceso. **~shod** /'slɪpʃɒd/ *a* descuidado. **~-up** *n* (*fam*) error *m*

slit /slɪt/ *n* raja *f*; (*cut*) corte *m*. ● *vt* (*pt* **slit**, *pres p* **slitting**) rajar; (*cut*) cortar

slither /'slɪðə(r)/ *vi* deslizarse

sliver /'slɪvə(r)/ *n* trocito *m*; (*splinter*) astilla *f*

slobber /'slɒbə(r)/ *vi* babear

slog /slɒg/ *vt* (*pt* **slogged**) golpear. ● *vi* trabajar como un negro. ● *n* golpetazo *m*; (*hard work*) trabajo *m* penoso

slogan /'sləʊgən/ *n* eslogan *m*

slop /slɒp/ *vt* (*pt* **slopped**) derramar. ● *vi* derramarse. **~s** *npl* (*fam*) agua *f* sucia

slop|e /sləʊp/ *vi* inclinarse. ● *vt* inclinar. ● *n* declive *m*, pendiente *m*. **~ing** *a* inclinado

sloppy /'slɒpɪ/ *a* (**-ier**, **-iest**) (*wet*) mojado; ‹*food*› líquido; ‹*work*› descuidado; ‹*person*› desaliñado; (*fig*) sentimental

slosh /slɒʃ/ *vi* (*fam*) chapotear. ● *vt* (*hit, sl*) pegar

slot /slɒt/ *n* ranura *f*. ● *vt* (*pt* **slotted**) encajar

sloth /sləʊθ/ *n* pereza *f*

slot-machine /ˈslɒtməʃi:n/ *n* distribuidor *m* automático; (*for gambling*) máquina *f* tragaperras

slouch /slaʊtʃ/ *vi* andar cargado de espaldas; (*in chair*) repanchigarse

Slovak /ˈsləʊvæk/ *a* & *n* eslovaco (*m*). **~ia** /sləʊˈvækɪə/ *n* Eslovaquia *f*

slovenl|iness /ˈslʌvnlɪnɪs/ *n* despreocupación *f*. **~y** /ˈslʌvnlɪ/ *a* descuidado

slow /sləʊ/ *a* (**-er**, **-est**) lento. **be ~** ‹*clock*› estar atrasado. **in ~ motion** a cámara lenta. ● *adv* despacio. ● *vt* retardar. ● *vi* ir más despacio. **~ down**, **~ up** *vt* retardar. ● *vi* ir más despacio. **~coach** /ˈsləʊkəʊtʃ/ *n* tardón *m*. **~ly** *adv* despacio. **~ness** *n* lentitud *f*

sludge /slʌdʒ/ *n* fango *m*; (*sediment*) sedimento *m*

slug /slʌg/ *n* babosa *f*; (*bullet*) posta *f*. **~gish** /ˈslʌgɪʃ/ *a* lento

sluice /slu:s/ *n* (*gate*) compuerta *f*; (*channel*) canal *m*

slum /slʌm/ *n* tugurio *m*

slumber /ˈslʌmbə(r)/ *n* sueño *m*. ● *vi* dormir

slump /slʌmp/ *n* baja *f* repentina; (*in business*) depresión *f*. ● *vi* bajar repentinamente; (*flop down*) dejarse caer pesadamente; (*collapse*) desplomarse

slung /slʌŋ/ *see* **sling**

slur /slɜ:(r)/ *vt*/*i* (*pt* **slurred**) articular mal. ● *n* dicción *f* defectuosa; (*discredit*) calumnia *f*

slush /slʌʃ/ *n* nieve *f* medio derretida; (*fig*) sentimentalismo *m*. **~ fund** *n* fondos *mpl* secretos para fines deshonestos. **~y** *a* ‹*road*› cubierto de nieve medio derretida

slut /slʌt/ *n* mujer *f* desaseada

sly /slaɪ/ *a* (**slyer, slyest**) (*crafty*) astuto; (*secretive*) furtivo. ● *n*. **on the ~** a escondidas. **~ly** *adv* astutamente

smack[1] /smæk/ *n* golpe *m*; (*on face*) bofetada *f*. ● *adv* (*fam*) de lleno. ● *vt* pegar

smack[2] /smæk/ *vi*. **~ of** saber a; (*fig*) oler a

small /smɔ:l/ *a* (**-er**, **-est**) pequeño. ● *n*. **the ~ of the back** la región *f* lumbar. **~ ads** *npl* anuncios *mpl* por palabras. **~ change** *n* cambio *m*. **~holding** /ˈsmɔ:lhəʊldɪŋ/ *n* parcela *f*. **~pox** /ˈsmɔ:lpɒks/ *n* viruela *f*. **~ talk** *n* charla *f*. **~-time** *a* (*fam*) de poca monta

smarmy /ˈsmɑ:mɪ/ *a* (**-ier**, **-iest**) (*fam*) zalamero

smart /smɑ:t/ *a* (**-er**, **-est**) elegante; (*clever*) inteligente; (*brisk*) rápido. ● *vi* escocer. **~en** /ˈsmɑ:tn/ *vt* arreglar. ● *vi* arreglarse. **~en up** *vi* arreglarse. **~ly** *adv* elegantemente; (*quickly*) rápidamente. **~ness** *n* elegancia *f*

smash /smæʃ/ *vt* romper; (*into little pieces*) hacer pedazos; batir ‹*record*›. ● *vi* romperse; (*collide*) chocar (**into** con). ● *n* (*noise*) estruendo *m*; (*collision*) choque *m*; (*com*) quiebra *f*. **~ing** /ˈsmæʃɪŋ/ *a* (*fam*) estupendo

smattering /ˈsmætərɪŋ/ *n* conocimientos *mpl* superficiales

smear /smɪə(r)/ *vt* untar (**with** de); (*stain*) manchar (**with** de); (*fig*) difamar. ● *n* mancha *f*; (*med*) frotis *m*

smell /smel/ *n* olor *m*; (*sense*) olfato *m*. ● *vt*/*i* (*pt* **smelt**) oler. **~y** *a* maloliente

smelt[1] /smelt/ *see* **smell**

smelt[2] /smelt/ *vt* fundir

smile /smaɪl/ *n* sonrisa *f*. ● *vi* sonreír(se)

smirk /smɜ:k/ *n* sonrisa *f* afectada

smite /smaɪt/ *vt* (*pt* **smote,** *pp* **smitten**) golpear

smith /smɪθ/ *n* herrero *m*

smithereens /smɪðəˈri:nz/ *npl* añicos *mpl*. **smash to ~** hacer añicos

smitten /ˈsmɪtn/ *see* **smite**. ● *a* encaprichado (**with** por)

smock /smɒk/ *n* blusa *f*, bata *f*

smog /smɒg/ *n* niebla *f* con humo

smok|e /sməʊk/ *n* humo *m*. ● *vt*/*i* fumar. **~eless** *a* sin humo. **~er** /-ə(r)/ *n* fumador *m*. **~e-screen** *n* cortina *f* de humo. **~y** *a* ‹*room*› lleno de humo

smooth /smu:ð/ *a* (**-er**, **-est**) liso; ‹*sound, movement*› suave; ‹*sea*› tranquilo; ‹*manners*› zalamero. ● *vt* alisar; (*fig*) allanar. **~ly** *adv* suavemente

smote /sməʊt/ *see* **smite**

smother /ˈsmʌðə(r)/ *vt* sofocar; (*cover*) cubrir
smoulder /ˈsməʊldə(r)/ *vi* arder sin llama; (*fig*) arder
smudge /smʌdʒ/ *n* borrón *m*, mancha *f*. ● *vt* tiznar. ● *vi* tiznarse
smug /smʌg/ *a* (**smugger**, **smuggest**) satisfecho de sí mismo
smuggl|e /ˈsmʌgl/ *vt* pasar de contrabando. **~er** *n* contrabandista *m* & *f*. **~ing** *n* contrabando *m*
smug: **~ly** *adv* con suficiencia. **~ness** *n* suficiencia *f*
smut /smʌt/ *n* tizne *m*; (*mark*) tiznajo *m*. **~ty** *a* (**-ier**, **-iest**) tiznado; (*fig*) obsceno
snack /snæk/ *n* tentempié *m*. **~-bar** *n* cafetería *f*
snag /snæg/ *n* problema *m*; (*in cloth*) rasgón *m*
snail /sneɪl/ *n* caracol *m*. **~'s pace** *n* paso *m* de tortuga
snake /sneɪk/ *n* serpiente *f*
snap /snæp/ *vt* (*pt* **snapped**) (*break*) romper; castañetear ‹*fingers*›. ● *vi* romperse; ‹*dog*› intentar morder; (*say*) contestar bruscamente; ‹*whip*› chasquear. **~ at** ‹*dog*› intentar morder; (*say*) contestar bruscamente. ● *n* chasquido *m*; (*photo*) foto *f*. ● *a* instantáneo. **~ up** *vt* agarrar. **~py** /ˈsnæpɪ/ *a* (**-ier**, **-iest**) (*fam*) rápido. **make it ~py!** (*fam*) ¡date prisa! **~shot** /ˈsnæpʃɒt/ *n* foto *f*
snare /sneə(r)/ *n* trampa *f*
snarl /snɑ:l/ *vi* gruñir. ● *n* gruñido *m*
snatch /snætʃ/ *vt* agarrar; (*steal*) robar. ● *n* arrebatamiento *m*; (*short part*) trocito *m*; (*theft*) robo *m*
sneak /sni:k/ ● *n* soplón *m*. ● *vi*. **~ in** entrar furtivamente. **~ out** salir furtivamente
sneakers /ˈsni:kəz/ *npl* zapatillas *fpl* de lona
sneak|ing /ˈsni:kɪŋ/ *a* furtivo. **~y** *a* furtivo
sneer /snɪə(r)/ *n* sonrisa *f* de desprecio. ● *vi* sonreír con desprecio. **~ at** hablar con desprecio a
sneeze /sni:z/ *n* estornudo *m*. ● *vi* estornudar
snide /snaɪd/ *a* (*fam*) despreciativo
sniff /snɪf/ *vt* oler. ● *vi* aspirar por la nariz. ● *n* aspiración *f*
snigger /ˈsnɪgə(r)/ *n* risa *f* disimulada. ● *vi* reír disimuladamente
snip /snɪp/ *vt* (*pt* **snipped**) tijeretear. ● *n* tijeretada *f*; (*bargain*, *sl*) ganga *f*
snipe /snaɪp/ *vi* disparar desde un escondite. **~r** /ə(r)/ *n* tirador *m* emboscado, francotirador *m*
snippet /ˈsnɪpɪt/ *n* retazo *m*
snivel /ˈsnɪvl/ *vi* (*pt* **snivelled**) lloriquear. **~ling** *a* llorón
snob /snɒb/ *n* esnob *m*. **~bery** *n* esnobismo *m*. **~bish** *a* esnob
snooker /ˈsnu:kə(r)/ *n* billar *m*
snoop /snu:p/ *vi* (*fam*) curiosear
snooty /ˈsnu:tɪ/ *a* (*fam*) desdeñoso
snooze /snu:z/ *n* sueñecito *m*. ● *vi* echarse un sueñecito
snore /snɔ:(r)/ *n* ronquido *m*. ● *vi* roncar
snorkel /ˈsnɔ:kl/ *n* tubo *m* respiratorio
snort /snɔ:t/ *n* bufido *m*. ● *vi* bufar
snout /snaʊt/ *n* hocico *m*
snow /snəʊ/ *n* nieve *f*. ● *vi* nevar. **be ~ed under with** estar inundado por. **~ball** /ˈsnəʊbɔ:l/ *n* bola *f* de nieve. **~-drift** *n* nieve amontonada. **~drop** /ˈsnəʊdrɒp/ *n* campanilla *f* de invierno. **~fall** /ˈsnəʊfɔ:l/ *n* nevada *f*. **~flake** /ˈsnəʊfleɪk/ *n* copo *m* de nieve. **~man** /ˈsnəʊmæn/ *n* (*pl* **-men**) muñeco *m* de nieve. **~-plough** *n* quitanieves *m invar*. **~storm** /ˈsnəʊstɔ:m/ *n* nevasca *f*. **~y** *a* ‹*place*› de nieves abundantes; ‹*weather*› con nevadas seguidas
snub /snʌb/ *vt* (*pt* **snubbed**) desairar. ● *n* desaire *m*. **~-nosed** /ˈsnʌbnəʊzd/ *a* chato
snuff /snʌf/ *n* rapé *m*. ● *vt* despabilar ‹*candle*›. **~ out** apagar ‹*candle*›
snuffle /ˈsnʌfl/ *vi* respirar ruidosamente
snug /snʌg/ *a* (**snugger**, **snuggest**) cómodo; (*tight*) ajustado
snuggle /ˈsnʌgl/ *vi* acomodarse
so /səʊ/ *adv* (*before a or adv*) tan; (*thus*) así. ● *conj* así que. **~ am I** yo tambien. **~ as to** para. **~ far** *adv* (*time*) hasta ahora; (*place*) hasta aquí. **~ far as I know** que yo sepa. **~ long!** (*fam*) ¡hasta luego! **~ much** tanto. **~ that** *conj* para que. **and ~ forth**, **and ~ on** y así sucesivamente. **if ~** si es así. **I think ~** creo que sí. **or ~** más o menos
soak /səʊk/ *vt* remojar. ● *vi* remojarse. **~ in** penetrar. **~ up** absorber. **~ing** *a* empapado. ● *n* remojón *m*
so-and-so /ˈsəʊənsəʊ/ *n* fulano *m*

soap /səʊp/ *n* jabón *m*. ● *vt* enjabonar. ~ **powder** *n* jabón en polvo. ~**y** *a* jabonoso
soar /sɔ:(r)/ *vi* elevarse; ‹*price etc*› ponerse por las nubes
sob /sɒb/ *n* sollozo *m*. ● *vi* (*pt* **sobbed**) sollozar
sober /'səʊbə(r)/ *a* sobrio; ‹*colour*› discreto
so-called /'səʊkɔ:ld/ *a* llamado, supuesto
soccer /'sɒkə(r)/ *n* (*fam*) fútbol *m*
sociable /'səʊʃəbl/ *a* sociable
social /'səʊʃl/ *a* social; (*sociable*) sociable. ● *n* reunión *f*. ~**ism** /-zəm/ *n* socialismo *m*. ~**ist** /'səʊʃəlɪst/ *a* & *n* socialista *m* & *f*. ~**ize** /'səʊʃəlaɪz/ *vt* socializar. ~**ly** *adv* socialmente. ~ **security** *n* seguridad *f* social. ~ **worker** *n* asistente *m* social
society /sə'saɪətɪ/ *n* sociedad *f*
sociolog|ical /səʊsɪə'lɒdʒɪkl/ *a* sociológico. ~**ist** *n* sociólogo *m*. ~**y** /səʊsɪ'ɒlədʒɪ/ *n* sociología *f*
sock[1] /sɒk/ *n* calcetín *m*
sock[2] /sɒk/ *vt* (*sl*) pegar
socket /'sɒkɪt/ *n* hueco *m*; (*of eye*) cuenca *f*; (*wall plug*) enchufe *m*; (*for bulb*) portalámparas *m invar*, casquillo *m*
soda /'səʊdə/ *n* sosa *f*; (*water*) soda *f*. ~**-water** *n* soda *f*
sodden /'sɒdn/ *a* empapado
sodium /'səʊdɪəm/ *n* sodio *m*
sofa /'səʊfə/ *n* sofá *m*
soft /sɒft/ *a* (**-er**, **-est**) blando; ‹*sound, colour*› suave; (*gentle*) dulce, tierno; (*silly*) estúpido. ~ **drink** *n* bebida *f* no alcohólica. ~ **spot** *n* debilidad *f*. ~**en** /'sɒfn/ *vt* ablandar; (*fig*) suavizar. ● *vi* ablandarse; (*fig*) suavizarse. ~**ly** *adv* dulcemente. ~**ness** *n* blandura *f*; (*fig*) dulzura *f*. ~**ware** /'sɒftweə(r)/ *n* programación *f*, software *m*
soggy /'sɒgɪ/ *a* (**-ier**, **-iest**) empapado
soh /səʊ/ *n* (*mus, fifth note of any musical scale*) sol *m*
soil[1] /sɔɪl/ *n* suelo *m*
soil[2] /sɔɪl/ *vt* ensuciar. ● *vi* ensuciarse
solace /'sɒləs/ *n* consuelo *m*
solar /'səʊlə(r)/ *a* solar. ~**ium** /sə'leərɪəm/ *n* (*pl* **-a**) solario *m*
sold /səʊld/ *see* **sell**
solder /'sɒldə(r)/ *n* soldadura *f*. ● *vt* soldar
soldier /'səʊldʒə(r)/ *n* soldado *m*. ● *vi*. ~ **on** (*fam*) perseverar
sole[1] /səʊl/ *n* (*of foot*) planta *f*; (*of shoe*) suela *f*
sole[2] /səʊl/ (*fish*) lenguado *m*
sole[3] /səʊl/ *a* único, solo. ~**ly** *adv* únicamente
solemn /'sɒləm/ *a* solemne. ~**ity** /sə'lemnətɪ/ *n* solemnidad *f*. ~**ly** *adv* solemnemente
solicit /sə'lɪsɪt/ *vt* solicitar. ● *vi* importunar
solicitor /sə'lɪsɪtə(r)/ *n* abogado *m*; (*notary*) notario *m*
solicitous /sə'lɪsɪtəs/ *a* solícito
solid /'sɒlɪd/ *a* sólido; ‹*gold etc*› macizo; (*unanimous*) unánime; ‹*meal*› sustancioso. ● *n* sólido *m*. ~**arity** /sɒlɪ'dærətɪ/ *n* solidaridad *f*. ~**ify** /sə'lɪdɪfaɪ/ *vt* solidificar. ● *vi* solidificarse. ~**ity** /sə'lɪdətɪ/ *n* solidez *f*. ~**ly** *adv* sólidamente. ~**s** *npl* alimentos *mpl* sólidos
soliloquy /sə'lɪləkwɪ/ *n* soliloquio *m*
solitaire /sɒlɪ'teə(r)/ *n* solitario *m*
solitary /'sɒlɪtrɪ/ *a* solitario
solitude /'sɒlɪtju:d/ *n* soledad *f*
solo /'səʊləʊ/ *n* (*pl* **-os**) (*mus*) solo *m*. ~**ist** *n* solista *m* & *f*
solstice /'sɒlstɪs/ *n* solsticio *m*
soluble /'sɒljʊbl/ *a* soluble
solution /sə'lu:ʃn/ *n* solución *f*
solvable *a* soluble
solve /sɒlv/ *vt* resolver
solvent /'sɒlvənt/ *a* & *n* solvente (*m*)
sombre /'sɒmbə(r)/ *a* sombrío
some /sʌm/ *a* alguno; (*a little*) un poco de. ~ **day** algún día. ~ **two hours** unas dos horas. **will you have** ~ **wine?** ¿quieres vino? ● *pron* algunos; (*a little*) un poco. ~ **of us** algunos de nosotros. **I want** ~ quiero un poco. ● *adv* (*approximately*) unos. ~**body** /'sʌmbədɪ/ *pron* alguien. ● *n* personaje *m*. ~**how** /'sʌmhaʊ/ *adv* de algún modo. ~**how or other** de una manera u otra. ~**one** /'sʌmwʌn/ *pron* alguien. ● *n* personaje *m*
somersault /'sʌməsɔ:lt/ *n* salto *m* mortal. ● *vi* dar un salto mortal
some: ~**thing** /'sʌmθɪŋ/ *pron* algo *m*. ~**thing like** algo como; (*approximately*) cerca de. ~**time** /'sʌmtaɪm/ *a* ex. ● *adv* algún día; (*in past*) durante. ~**time last summer** *a* (durante) el verano pasado. ~**times** /'sʌmtaɪmz/ *adv* de vez en cuando, a veces. ~**what** /'sʌmwɒt/ *adv* algo, un poco. ~**where** /'sʌmweə(r)/ *adv* en alguna parte

son /sʌn/ *n* hijo *m*
sonata /səˈnɑːtə/ *n* sonata *f*
song /sɒŋ/ *n* canción *f*. **sell for a ~** vender muy barato. **~-book** *n* cancionero *m*
sonic /ˈsɒnɪk/ *a* sónico
son-in-law /ˈsʌnɪnlɔː/ *n* (*pl* **sons-in-law**) yerno *m*
sonnet /ˈsɒnɪt/ *n* soneto *m*
sonny /ˈsʌnɪ/ *n* (*fam*) hijo *m*
soon /suːn/ *adv* (**-er**, **-est**) pronto; (*in a short time*) dentro de poco; (*early*) temprano. **~ after** poco después. **~er or later** tarde o temprano. **as ~ as** en cuanto; **as ~ as possible** lo antes posible. **I would ~er not go** prefiero no ir
soot /sʊt/ *n* hollín *m*
sooth|e /suːð/ *vt* calmar. **~ing** *a* calmante
sooty /ˈsʊtɪ/ *a* cubierto de hollín
sophisticated /səˈfɪstɪkeɪtɪd/ *a* sofisticado; (*complex*) complejo
soporific /sɒpəˈrɪfɪk/ *a* soporífero
sopping /ˈsɒpɪŋ/ *a*. **~ (wet)** empapado
soppy /ˈsɒpɪ/ *a* (**-ier**, **-iest**) (*fam*) sentimental; (*silly*, *fam*) tonto
soprano /səˈprɑːnəʊ/ *n* (*pl* **-os**) (*voice*) soprano *m*; (*singer*) soprano *f*
sorcerer /ˈsɔːsərə(r)/ *n* hechicero *m*
sordid /ˈsɔːdɪd/ *a* sórdido
sore /ˈsɔː(r)/ *a* (**-er**, **-est**) que duele, dolorido; (*distressed*) penoso; (*vexed*) enojado. ● *n* llaga *f*. **~ly** /ˈsɔːlɪ/ *adv* gravemente. **~ throat** *n* dolor *m* de garganta. **I've got a ~ throat** me duele la garganta
sorrow /ˈsɒrəʊ/ *n* pena *f*, tristeza *f*. **~ful** *a* triste
sorry /ˈsɒrɪ/ *a* (**-ier**, **-ier**) arrepentido; (*wretched*) lamentable; (*sad*) triste. **be ~** sentirlo; (*repent*) arrepentirse. **be ~ for s.o.** (*pity*) compadecerse de uno. **~!** ¡perdón!, ¡perdone!
sort /sɔːt/ *n* clase *f*; (*person*, *fam*) tipo *m*. **be out of ~s** estar indispuesto; (*irritable*) estar de mal humor. ● *vt* clasificar. **~ out** (*choose*) escoger; (*separate*) separar; resolver ‹*problem*›
so-so /ˈsəʊsəʊ/ *a & adv* regular
soufflé /ˈsuːfleɪ/ *n* suflé *m*
sought /sɔːt/ *see* **seek**
soul /səʊl/ *n* alma *f*. **~ful** /ˈsəʊlfl/ *a* sentimental
sound[1] /saʊnd/ *n* sonido *m*; ruido *m*. ● *vt* sonar; (*test*) sondar. ● *vi* sonar; (*seem*) parecer (**as if** que)
sound[2] /saʊnd/ *a* (**-er**, **-est**) sano; ‹*argument etc*› lógico; (*secure*) seguro. **~ asleep** profundamente dormido
sound[3] /saʊnd/ (*strait*) estrecho *m*
sound barrier /ˈsaʊndbærɪə(r)/ *n* barrera *f* del sonido
soundly /ˈsaʊndlɪ/ *adv* sólidamente; (*asleep*) profundamente
sound: **~-proof** *a* insonorizado. **~-track** *n* banda *f* sonora
soup /suːp/ *n* sopa *f*. **in the ~** (*sl*) en apuros
sour /ˈsaʊə(r)/ *a* (**-er**, **-est**) agrio; ‹*cream*, *milk*› cortado. ● *vt* agriar. ● *vi* agriarse
source /sɔːs/ *n* fuente *f*
south /saʊθ/ *n* sur *m*. ● *a* del sur. ● *adv* hacia el sur. **S~ Africa** *n* Africa *f* del Sur. **S~ America** *n* América *f* (del Sur), Sudamérica *f*. **S~ American** *a & n* sudamericano (*m*). **~-east** *n* sudeste *m*. **~erly** /ˈsʌðəlɪ/ *a* sur; ‹*wind*› del sur. **~ern** /ˈsʌðən/ *a* del sur, meridional. **~erner** *n* meridional *m*. **~ward** *a* sur; ● *adv* hacia el sur. **~wards** *adv* hacia el sur. **~-west** *n* sudoeste *m*
souvenir /suːvəˈnɪə(r)/ *n* recuerdo *m*
sovereign /ˈsɒvrɪn/ *n & a* soberano (*m*). **~ty** *n* soberanía *f*
Soviet /ˈsəʊvɪət/ *a* (*history*) soviético. **the ~ Union** *n* la Unión *f* Soviética
sow[1] /səʊ/ *vt* (*pt* **sowed**, *pp* **sowed** *or* **sown**) sembrar
sow[2] /saʊ/ *n* cerda *f*
soya /ˈsɔɪə/ *n*. **~ bean** *n* soja *f*
spa /spɑː/ *n* balneario *m*
space /speɪs/ *n* espacio *m*; (*room*) sitio *m*; (*period*) período *m*. ● *a* ‹*research etc*› espacial. ● *vt* espaciar. **~ out** espaciar. **~craft** /ˈspeɪskrɑːft/ *n*, **~ship** *n* nave *f* espacial. **~suit** *n* traje *m* espacial
spacious /ˈspeɪʃəs/ *a* espacioso
spade /speɪd/ *n* pala *f*. **~s** *npl* (*cards*) picos *mpl*, picas *fpl*; (*in Spanish pack*) espadas *fpl*. **~work** /ˈspeɪdwɜːk/ *n* trabajo *m* preparatorio
spaghetti /spəˈgetɪ/ *n* espaguetis *mpl*
Spain /speɪn/ *n* España *f*
span[1] /spæn/ *n* (*of arch*) luz *f*; (*of time*) espacio *m*; (*of wings*) envergadura *f*. ● *vt* (*pt* **spanned**) extenderse sobre
span[2] /spæn/ *see* **spick**
Spaniard /ˈspænjəd/ *n* español *m*

spaniel /ˈspænjəl/ *n* perro *m* de aguas

Spanish /ˈspænɪʃ/ *a & n* español (*m*)

spank /spæŋk/ *vt* dar un azote a. **~ing** *n* azote *m*

spanner /ˈspænə(r)/ *n* llave *f*

spar /spɑ:(r)/ *vi* (*pt* **sparred**) entrenarse en el boxeo; (*argue*) disputar

spare /speə(r)/ *vt* salvar; (*do without*) prescindir de; (*afford to give*) dar; (*use with restraint*) escatimar. ● *a* de reserva; (*surplus*) sobrante; ‹*person*› enjuto; ‹*meal etc*› frugal. **~ (part)** *n* repuesto *m*. **~ time** *n* tiempo *m* libre. **~ tyre** *n* neumático *m* de repuesto

sparing /ˈspeərɪŋ/ *a* frugal. **~ly** *adv* frugalmente

spark /spɑ:k/ *n* chispa *f*. ● *vt*. **~ off** (*initiate*) provocar. **~ing-plug** *n* (*auto*) bujía *f*

sparkl|e /ˈspɑ:kl/ *vi* centellear. ● *n* centelleo *m*. **~ing** *a* centelleante; ‹*wine*› espumoso

sparrow /ˈspærəʊ/ *n* gorrión *m*

sparse /spɑ:s/ *a* escaso; ‹*population*› poco denso. **~ly** *adv* escasamente

spartan /ˈspɑ:tn/ *a* espartano

spasm /ˈspæzəm/ *n* espasmo *m*; (*of cough*) acceso *m*. **~odic** /spæzˈmɒdɪk/ *a* espasmódico

spastic /ˈspæstɪk/ *n* víctima *f* de parálisis cerebral

spat /spæt/ *see* **spit**

spate /speɪt/ *n* avalancha *f*

spatial /ˈspeɪʃl/ *a* espacial

spatter /ˈspætə(r)/ *vt* salpicar (**with** de)

spatula /ˈspætjʊlə/ *n* espátula *f*

spawn /spɔ:n/ *n* hueva *f*. ● *vt* engendrar. ● *vi* desovar

speak /spi:k/ *vt/i* (*pt* **spoke,** *pp* **spoken**) hablar. **~ for** *vi* hablar en nombre de. **~ up** *vi* hablar más fuerte. **~er** /ˈspi:kə(r)/ *n* (*in public*) orador *m*; (*loudspeaker*) altavoz *m*. **be a Spanish ~er** hablar español

spear /spɪə(r)/ *n* lanza *f*. **~head** /ˈspɪəhed/ *n* punta *f* de lanza. ● *vt* (*lead*) encabezar. **~mint** /ˈspɪəmɪnt/ *n* menta *f* verde

spec /spek/ *n*. **on ~** (*fam*) por si acaso

special /ˈspeʃl/ *a* especial. **~ist** /ˈspeʃəlɪst/ *n* especialista *m & f*. **~ity** /-ɪˈælətɪ/ *n* especialidad *f*. **~ization** /-ˈzeɪʃn/ *n* especialización *f*. **~ize** /ˈspeʃəlaɪz/ *vi* especializarse. **~ized** *a* especializado. **~ty** *n* especialidad *f*. **~ly** *adv* especialmente

species /ˈspi:ʃi:z/ *n* especie *f*

specif|ic /spəˈsɪfɪk/ *a* específico. **~ically** *adv* específicamente. **~ication** /-ɪˈkeɪʃn/ *n* especificación *f*; (*details*) descripción *f*. **~y** /ˈspesɪfaɪ/ *vt* especificar

specimen /ˈspesɪmɪn/ *n* muestra *f*

speck /spek/ *n* manchita *f*; (*particle*) partícula *f*

speckled /ˈspekld/ *a* moteado

specs /speks/ *npl* (*fam*) gafas *fpl*, anteojos *mpl* (*LAm*)

spectac|le /ˈspektəkl/ *n* espectáculo *m*. **~les** *npl* gafas *fpl*, anteojos *mpl* (*LAm*). **~ular** /spekˈtækjʊlə(r)/ *a* espectacular

spectator /spekˈteɪtə(r)/ *n* espectador *m*

spectre /ˈspektə(r)/ *n* espectro *m*

spectrum /ˈspektrəm/ *n* (*pl* **-tra**) espectro *m*; (*of ideas*) gama *f*

speculat|e /ˈspekjʊleɪt/ *vi* especular. **~ion** /-ˈleɪʃn/ *n* especulación *f*. **~ive** /-lətɪv/ *a* especulativo. **~or** *n* especulador *m*

sped /sped/ *see* **speed**

speech /spi:tʃ/ *n* (*faculty*) habla *f*; (*address*) discurso *m*. **~less** *a* mudo

speed /spi:d/ *n* velocidad *f*; (*rapidity*) rapidez *f*; (*haste*) prisa *f*. ● *vi* (*pt* **sped**) apresurarse. (*pt* **speeded**) (*drive too fast*) ir a una velocidad excesiva. **~ up** *vt* acelerar. ● *vi* acelerarse. **~boat** /ˈspi:dbəʊt/ *n* lancha *f* motora. **~ily** *adv* rápidamente. **~ing** *n* exceso *m* de velocidad. **~ometer** /spi:ˈdɒmɪtə(r)/ *n* velocímetro *m*. **~way** /ˈspi:dweɪ/ *n* pista *f*; (*Amer*) autopista *f*. **~y** /ˈspi:dɪ/ *a* (**-ier, -iest**) rápido

spell[1] /spel/ *n* (*magic*) hechizo *m*

spell[2] /spel/ *vt/i* (*pt* **spelled** *or* **spelt**) escribir; (*mean*) significar. **~ out** *vt* deletrear; (*fig*) explicar. **~ing** *n* ortografía *f*

spell[3] /spel/ (*period*) período *m*

spellbound /ˈspelbaʊnd/ *a* hechizado

spelt /spelt/ *see* **spell**[2]

spend /spend/ *vt* (*pt* **spent**) gastar; pasar ‹*time etc*›; dedicar ‹*care etc*›. ● *vi* gastar dinero. **~thrift** /ˈspendθrɪft/ *n* derrochador *m*

spent /spent/ *see* **spend**

sperm /spɜ:m/ *n* (*pl* **sperms** *or* **sperm**) esperma *f*

spew /spju:/ *vt/i* vomitar

spher|e /sfɪə(r)/ *n* esfera *f*. **~ical** /ˈsferɪkl/ *a* esférico
sphinx /sfɪŋks/ *n* esfinge *f*
spice /spaɪs/ *n* especia *f*; (*fig*) sabor *m*
spick /spɪk/ *a*. **~ and span** impecable
spicy /ˈspaɪsɪ/ *a* picante
spider /ˈspaɪdə(r)/ *n* araña *f*
spik|e /spaɪk/ *n* (*of metal etc*) punta *f*. **~y** *a* puntiagudo; ‹*person*› quisquilloso
spill /spɪl/ *vt* (*pt* **spilled** *or* **spilt**) derramar. ● *vi* derramarse. **~ over** desbordarse
spin /spɪn/ *vt* (*pt* **spun**, *pres p* **spinning**) hacer girar; hilar ‹*wool etc*›. ● *vi* girar. ● *n* vuelta *f*; (*short drive*) paseo *m*
spinach /ˈspɪnɪdʒ/ *n* espinacas *fpl*
spinal /ˈspaɪnl/ *a* espinal. **~ cord** *n* médula *f* espinal
spindl|e /ˈspɪndl/ *n* (*for spinning*) huso *m*. **~y** *a* larguirucho
spin-drier /spɪnˈdraɪə(r)/ *n* secador *m* centrífugo
spine /spaɪn/ *n* columna *f* vertebral; (*of book*) lomo *m*. **~less** *a* (*fig*) sin carácter
spinning /ˈspɪnɪŋ/ *n* hilado *m*. **~-top** *n* trompa *f*, peonza *f*. **~-wheel** *n* rueca *f*
spin-off /ˈspɪnɒf/ *n* beneficio *m* incidental; (*by-product*) subproducto *m*
spinster /ˈspɪnstə(r)/ *n* soltera *f*; (*old maid, fam*) solterona *f*
spiral /ˈspaɪərəl/ *a* espiral, helicoidal. ● *n* hélice *f*. ● *vi* (*pt* **spiralled**) moverse en espiral. **~ staircase** *n* escalera *f* de caracol
spire /ˈspaɪə(r)/ *n* (*archit*) aguja *f*
spirit /ˈspɪrɪt/ *n* espíritu *m*; (*boldness*) valor *m*. **in low ~s** abatido. ● *vt*. **~ away** hacer desaparecer. **~ed** /ˈspɪrɪtɪd/ *a* animado, fogoso. **~-lamp** *n* lamparilla *f* de alcohol. **~-level** *n* nivel *m* de aire. **~s** *npl* (*drinks*) bebidas *fpl* alcohólicas
spiritual /ˈspɪrɪtjʊəl/ *a* espiritual. ● *n* canción *f* religiosa de los negros. **~ualism** /-zəm/ *n* espiritismo *m*. **~ualist** /ˈspɪrɪtjʊəlɪst/ *n* espiritista *m & f*
spit[1] /spɪt/ *vt* (*pt* **spat** *or* **spit**, *pres p* **spitting**) escupir. ● *vi* escupir; (*rain*) lloviznar. ● *n* esputo *m*; (*spittle*) saliva *f*
spit[2] /spɪt/ (*for roasting*) asador *m*
spite /spaɪt/ *n* rencor *m*. **in ~ of** a pesar de. ● *vt* fastidiar. **~ful** *a* rencoroso. **~fully** *adv* con rencor
spitting image /spɪtɪŋˈɪmɪdʒ/ *n* vivo retrato *m*
spittle /spɪtl/ *n* saliva *f*
splash /splæʃ/ *vt* salpicar. ● *vi* esparcirse; ‹*person*› chapotear. ● *n* salpicadura *f*; (*sound*) chapoteo *m*; (*of colour*) mancha *f*; (*drop, fam*) gota *f*. **~ about** *vi* chapotear. **~ down** *vi* ‹*spacecraft*› amerizar
spleen /spli:n/ *n* bazo *m*; (*fig*) esplín *m*
splendid /ˈsplendɪd/ *a* espléndido
splendour /ˈsplendə(r)/ *n* esplendor *m*
splint /splɪnt/ *n* tablilla *f*
splinter /ˈsplɪntə(r)/ *n* astilla *f*. ● *vi* astillarse. **~ group** *n* grupo *m* disidente
split /splɪt/ *vt* (*pt* **split**, *pres p* **splitting**) hender, rajar; (*tear*) rajar; (*divide*) dividir; (*share*) repartir. **~ one's sides** caerse de risa. ● *vi* partirse; (*divide*) dividirse. **~ on s.o.** (*sl*) traicionar. ● *n* hendidura *f*; (*tear*) desgarrón *m*; (*quarrel*) ruptura *f*; (*pol*) escisión *f*. **~ up** *vi* separarse. **~ second** *n* fracción *f* de segundo
splurge /splɜ:dʒ/ *vi* (*fam*) derrochar
splutter /ˈsplʌtə(r)/ *vi* chisporrotear; ‹*person*› farfullar. ● *n* chisporroteo *m*; (*speech*) farfulla *f*
spoil /spɔɪl/ *vt* (*pt* **spoilt** *or* **spoiled**) estropear, echar a perder; (*ruin*) arruinar; (*indulge*) mimar. ● *n* botín *m*. **~s** *npl* botín *m*. **~-sport** *n* aguafiestas *m invar*
spoke[1] /spəʊk/ *see* **speak**
spoke[2] /spəʊk/ *n* (*of wheel*) radio *m*
spoken /spəʊkən/ *see* **speak**
spokesman /ˈspəʊksmən/ *n* (*pl* **-men**) portavoz *m*
spong|e /spʌndʒ/ *n* esponja *f*. ● *vt* limpiar con una esponja. ● *vi*. **~e on** vivir a costa de. **~e-cake** *n* bizcocho *m*. **~er** /-ə(r)/ *n* gorrón *m*. **~y** *a* esponjoso
sponsor /ˈspɒnsə(r)/ *n* patrocinador *m*; (*surety*) garante *m*. ● *vt* patrocinar. **~ship** *n* patrocinio *m*
spontane|ity /spɒntəˈneɪɪtɪ/ *n* espontaneidad *f*. **~ous** /spɒnˈteɪnjəs/ *a* espontáneo. **~ously** *adv* espontáneamente
spoof /spu:f/ *n* (*sl*) parodia *f*
spooky /ˈspu:kɪ/ *a* (**-ier**, **-iest**) (*fam*) escalofriante

spool /spu:l/ *n* carrete *m*; (*of sewing-machine*) canilla *f*
spoon /spu:n/ *n* cuchara *f*. **~-fed** *a* (*fig*) mimado. **~-feed** *vt* (*pt* **-fed**) dar de comer con cuchara. **~ful** *n* (*pl* **-fuls**) cucharada *f*
sporadic /spəˈrædɪk/ *a* esporádico
sport /spɔ:t/ *n* deporte *m*; (*amusement*) pasatiempo *m*; (*person, fam*) persona *f* alegre, buen chico *m*, buena chica *f*. **be a good ~** ser buen perdedor. ● *vt* lucir. **~ing** *a* deportivo. **~ing chance** *n* probabilidad *f* de éxito. **~s car** *n* coche *m* deportivo. **~s coat** *n* chaqueta *f* de sport. **~sman** /ˈspɔ:tsmən/ *n*, (*pl* **-men**), **~swoman** /ˈspɔ:tswʊmən/ *n* (*pl* **-women**) deportista *m* & *f*
spot /spɒt/ *n* mancha *f*; (*pimple*) grano *m*; (*place*) lugar *m*; (*in pattern*) punto *m*; (*drop*) gota *f*; (*a little, fam*) poquito *m*. **in a ~** (*fam*) en un apuro. **on the ~** en el lugar; (*without delay*) en el acto. ● *vt* (*pt* **spotted**) manchar; (*notice, fam*) observar, ver. **~ check** *n* control *m* hecho al azar. **~less** *a* inmaculado. **~light** /ˈspɒtlaɪt/ *n* reflector *m*. **~ted** *a* moteado; ‹*cloth*› a puntos. **~ty** *a* (**-ier**, **-iest**) manchado; ‹*skin*› con granos
spouse /spaʊz/ *n* cónyuge *m* & *f*
spout /spaʊt/ *n* pico *m*; (*jet*) chorro *m*. **up the ~** (*ruined, sl*) perdido. ● *vi* chorrear
sprain /spreɪn/ *vt* torcer. ● *n* torcedura *f*
sprang /spræŋ/ *see* **spring**
sprat /spræt/ *n* espadín *m*
sprawl /sprɔ:l/ *vi* ‹*person*› repanchigarse; ‹*city etc*› extenderse
spray /spreɪ/ *n* (*of flowers*) ramo *m*; (*water*) rociada *f*; (*from sea*) espuma *f*; (*device*) pulverizador *m*. ● *vt* rociar. **~-gun** *n* pistola *f* pulverizadora
spread /spred/ *vt* (*pt* **spread**) (*stretch, extend*) extender; untar ‹*jam etc*›; difundir ‹*idea, news*›. ● *vi* extenderse; ‹*disease*› propagarse; ‹*idea, news*› difundirse. ● *n* extensión *f*; (*paste*) pasta *f*; (*of disease*) propagación *f*; (*feast, fam*) comilona *f*. **~-eagled** *a* con los brazos y piernas extendidos
spree /spri:/ *n*. **go on a ~** (*have fun, fam*) ir de juerga
sprig /sprɪg/ *n* ramito *m*
sprightly /ˈspraɪtlɪ/ *a* (**-ier**, **-iest**) vivo
spring /sprɪŋ/ *n* (*season*) primavera *f*; (*device*) muelle *m*; (*elasticity*) elasticidad *f*; (*water*) manantial *m*. ● *a* de primavera. ● *vt* (*pt* **sprang**, *pp* **sprung**) hacer inesperadamente. ● *vi* saltar; (*issue*) brotar. **~ from** *vi* provenir de. **~ up** *vi* surgir. **~-board** *n* trampolín *m*. **~time** *n* primavera *f*. **~y** *a* (**-ier**, **-iest**) elástico
sprinkl|e /ˈsprɪŋkl/ *vt* salpicar; (*with liquid*) rociar. ● *n* salpicadura *f*; (*of liquid*) rociada *f*. **~ed with** salpicado de. **~er** /-ə(r)/ *n* regadera *f*. **~ing** /ˈsprɪŋklɪŋ/ *n* (*fig, amount*) poco *m*
sprint /sprɪnt/ *n* carrera *f*. ● *vi* correr. **~er** *n* corredor *m*
sprite /spraɪt/ *n* duende *m*, hada *f*
sprout /spraʊt/ *vi* brotar. ● *n* brote *m*. **(Brussels) ~s** *npl* coles *fpl* de Bruselas
spruce /spru:s/ *a* elegante
sprung /sprʌŋ/ *see* **spring**. ● *a* de muelles
spry /spraɪ/ *a* (**spryer**, **spryest**) vivo
spud /spʌd/ *n* (*sl*) patata *f*, papa *f* (*LAm*)
spun /spʌn/ *see* **spin**
spur /spɜ:(r)/ *n* espuela *f*; (*stimulus*) estímulo *m*. **on the ~ of the moment** impulsivamente. ● *vt* (*pt* **spurred**). **~ (on)** espolear; (*fig*) estimular
spurious /ˈspjʊərɪəs/ *a* falso. **~ly** *adv* falsamente
spurn /spɜ:n/ *vt* despreciar; (*reject*) rechazar
spurt /spɜ:t/ *vi* chorrear; (*make sudden effort*) hacer un esfuerzo repentino. ● *n* chorro *m*; (*effort*) esfuerzo *m* repentino
spy /spaɪ/ *n* espía *m* & *f*. ● *vt* divisar. ● *vi* espiar. **~ out** *vt* reconocer. **~ing** *n* espionaje *m*
squabble /ˈskwɒbl/ *n* riña *f*. ● *vi* reñir
squad /skwɒd/ *n* (*mil*) pelotón *m*; (*of police*) brigada *f*; (*sport*) equipo *m*
squadron /ˈskwɒdrən/ *n* (*mil*) escuadrón *m*; (*naut, aviat*) escuadrilla *f*
squalid /ˈskwɒlɪd/ *a* asqueroso; (*wretched*) miserable
squall /skwɔ:l/ *n* turbión *m*. ● *vi* chillar. **~y** *a* borrascoso
squalor /ˈskwɒlə(r)/ *n* miseria *f*
squander /ˈskwɒndə(r)/ *vt* derrochar

square /skweə(r)/ *n* cuadrado *m*; (*open space in town*) plaza *f*; (*for drawing*) escuadra *f*. ● *a* cuadrado; (*not owing*) sin deudas, iguales; (*honest*) honrado; ‹*meal*› satisfactorio; (*old-fashioned, sl*) chapado a la antigua. **all ~** iguales. ● *vt* (*settle*) arreglar; (*math*) cuadrar. ● *vi* (*agree*) cuadrar. **~ up to** enfrentarse con. **~ly** *adv* directamente

squash /skwɒʃ/ *vt* aplastar; (*suppress*) suprimir. ● *n* apiñamiento *m*; (*drink*) zumo *m*; (*sport*) squash *m*. **~y** *a* blando

squat /skwɒt/ *vi* (*pt* **squatted**) ponerse en cuclillas; (*occupy illegally*) ocupar sin derecho. ● *n* casa *f* ocupada sin derecho. ● *a* (*dumpy*) achaparrado. **~ter** /-ə(r)/ *n* ocupante *m* & *f* ilegal

squawk /skwɔ:k/ *n* graznido *m*. ● *vi* graznar

squeak /skwi:k/ *n* chillido *m*; (*of door etc*) chirrido *m*. ● *vi* chillar; ‹*door etc*› chirriar. **~y** *a* chirriador

squeal /skwi:l/ *n* chillido *m*. ● *vi* chillar. **~ on** (*inform on, sl*) denunciar

squeamish /'skwi:mɪʃ/ *a* delicado; (*scrupulous*) escrupuloso. **be ~ about snakes** tener horror a las serpientes

squeeze /skwi:z/ *vt* apretar; exprimir ‹*lemon etc*›; (*extort*) extorsionar (**from** de). ● *vi* (*force one's way*) abrirse paso. ● *n* estrujón *m*; (*of hand*) apretón *m*. **credit ~** *n* restricción *f* de crédito

squelch /skweltʃ/ *vi* chapotear. ● *n* chapoteo *m*

squib /skwɪb/ *n* (*firework*) buscapiés *m invar*

squid /skwɪd/ *n* calamar *m*

squiggle /'skwɪgl/ *n* garabato *m*

squint /skwɪnt/ *vi* ser bizco; (*look sideways*) mirar de soslayo. ● *n* estrabismo *m*

squire /'skwaɪə(r)/ *n* terrateniente *m*

squirm /skwɜ:m/ *vi* retorcerse

squirrel /'skwɪrəl/ *n* ardilla *f*

squirt /skwɜ:t/ *vt* arrojar a chorros. ● *vi* salir a chorros. ● *n* chorro *m*

St *abbr* (*saint*) /sənt/ S, San(to); (*street*) C/, Calle *f*

stab /stæb/ *vt* (*pt* **stabbed**) apuñalar. ● *n* puñalada *f*; (*pain*) punzada *f*; (*attempt, fam*) tentativa *f*

stabili|ty /stə'bɪlətɪ/ *n* estabilidad *f*. **~ze** /'steɪbɪlaɪz/ *vt* estabilizar. **~zer** /-ə(r)/ *n* estabilizador *m*

stable[1] /'steɪbl/ *a* (**-er**, **-est**) estable

stable[2] /'steɪbl/ *n* cuadra *f*. ● *vt* poner en una cuadra. **~-boy** *n* mozo *m* de cuadra

stack /stæk/ *n* montón *m*. ● *vt* amontonar

stadium /'steɪdjəm/ *n* estadio *m*

staff /stɑ:f/ *n* (*stick*) palo *m*; (*employees*) personal *m*; (*mil*) estado *m* mayor; (*in school*) profesorado *m*. ● *vt* proveer de personal

stag /stæg/ *n* ciervo *m*. **~-party** *n* reunión *f* de hombres, fiesta *f* de despedida de soltero

stage /steɪdʒ/ *n* (*in theatre*) escena *f*; (*phase*) etapa *f*; (*platform*) plataforma *f*. **go on the ~** hacerse actor. ● *vt* representar; (*arrange*) organizar. **~-coach** *n* (*hist*) diligencia *f*. **~ fright** *n* miedo *m* al público. **~-manager** *n* director *m* de escena. **~ whisper** *n* aparte *m*

stagger /'stægə(r)/ *vi* tambalearse. ● *vt* asombrar; escalonar ‹*holidays etc*›. ● *n* tambaleo *m*. **~ing** *a* asombroso

stagna|nt /'stægnənt/ *a* estancado. **~te** /stæg'neɪt/ *vi* estancarse. **~tion** /-ʃn/ *n* estancamiento *m*

staid /steɪd/ *a* serio, formal

stain /steɪn/ *vt* manchar; (*colour*) teñir. ● *n* mancha *f*; (*liquid*) tinte *m*. **~ed glass window** *n* vidriera *f* de colores. **~less** /'steɪnlɪs/ *a* inmaculado. **~less steel** *n* acero *m* inoxidable. **~ remover** *n* quitamanchas *m invar*

stair /steə(r)/ *n* escalón *m*. **~s** *npl* escalera *f*. **flight of ~s** tramo *m* de escalera. **~case** /'steəkeɪs/ *n*, **~way** *n* escalera *f*

stake /steɪk/ *n* estaca *f*; (*for execution*) hoguera *f*; (*wager*) apuesta *f*; (*com*) intereses *mpl*. **at ~** en juego. ● *vt* estacar; (*wager*) apostar. **~ a claim** reclamar

stalactite /'stæləktaɪt/ *n* estalactita *f*

stalagmite /'stæləgmaɪt/ *n* estalagmita *f*

stale /steɪl/ *a* (**-er**, **-est**) no fresco; ‹*bread*› duro; ‹*smell*› viciado; ‹*news*› viejo; (*uninteresting*) gastado. **~mate** /'steɪlmeɪt/ *n* (*chess*) ahogado *m*; (*deadlock*) punto *m* muerto

stalk[1] /stɔ:k/ *n* tallo *m*

stalk[2] /stɔ:k/ *vi* andar majestuosamente. ● *vt* seguir; ‹*animal*› acechar

stall[1] /stɔ:l/ *n* (*stable*) cuadra *f*; (*in stable*) casilla *f*; (*in theatre*) butaca *f*; (*in market*) puesto *m*; (*kiosk*) quiosco *m*

stall[2] /stɔ:l/ *vt* parar ‹*engine*›. ● *vi* ‹*engine*› pararse; (*fig*) andar con rodeos

stallion /'stæljən/ *n* semental *m*

stalwart /'stɔ:lwət/ *n* partidario *m* leal

stamina /'stæmɪnə/ *n* resistencia *f*

stammer /'stæmə(r)/ *vi* tartamudear. ● *n* tartamudeo *m*

stamp /stæmp/ *vt* (*with feet*) patear; (*press*) estampar; poner un sello en ‹*envelope*›; (*with rubber stamp*) sellar; (*fig*) señalar. ● *vi* patear. ● *n* sello *m*; (*with foot*) patada *f*; (*mark*) marca *f*, señal *f*. ~ **out** (*fig*) acabar con

stampede /stæm'pi:d/ *n* desbandada *f*; (*fam*) pánico *m*. ● *vi* huir en desorden

stance /stɑ:ns/ *n* postura *f*

stand /stænd/ *vi* (*pt* **stood**) estar de pie; (*rise*) ponerse de pie; (*be*) encontrarse; (*stay firm*) permanecer; (*pol*) presentarse como candidato (**for** en). ~ **to reason** ser lógico. ● *vt* (*endure*) soportar; (*place*) poner; (*offer*) ofrecer. ~ **a chance** tener una posibilidad. ~ **one's ground** mantenerse firme. **I'll ~ you a drink** te invito a una copa.● *n* posición *f*, postura *f*; (*mil*) resistencia *f*; (*for lamp etc*) pie *m*, sostén *m*; (*at market*) puesto *m*; (*booth*) quiosco *m*; (*sport*) tribuna *f*. ~ **around** no hacer nada. ~ **back** retroceder. ~ **by** *vi* estar preparado. ● *vt* (*support*) apoyar. ~ **down** *vi* retirarse. ~ **for** *vt* representar. ~ **in for** suplir a. ~ **out** destacarse. ~ **up** *vi* ponerse de pie. ~ **up for** defender. ~ **up to** *vt* resistir a

standard /'stændəd/ *n* norma *f*; (*level*) nivel *m*; (*flag*) estandarte *m*. ● *a* normal, corriente. **~ize** *vt* uniformar. ~ **lamp** *n* lámpara *f* de pie. **~s** *npl* valores *mpl*

stand: **~-by** *n* (*person*) reserva *f*; (*at airport*) lista *f* de espera. **~-in** *n* suplente *m* & *f*. **~ing** /'stændɪŋ/ *a* de pie; (*upright*) derecho. ● *n* posición *f*; (*duration*) duración *f*. **~-offish** *a* (*fam*) frío. **~point** /'stændpɔɪnt/ *n* punto *m* de vista. **~still** /'stændstɪl/ *n*. **at a ~still** parado. **come to a ~still** pararse

stank /stæŋk/ *see* **stink**

staple[1] /'steɪpl/ *a* principal

staple[2] /'steɪpl/ *n* grapa *f*. ● *vt* sujetar con una grapa. **~r** /-ə(r)/ *n* grapadora *f*

star /stɑ:/ *n* (*incl cinema, theatre*) estrella *f*; (*asterisk*) asterisco *m*. ● *vi* (*pt* **starred**) ser el protagonista

starboard /'stɑ:bəd/ *n* estribor *m*

starch /stɑ:tʃ/ *n* almidón *m*; (*in food*) fécula *f*. ● *vt* almidonar. **~y** *a* almidonado; ‹*food*› feculento; (*fig*) formal

stardom /'stɑ:dəm/ *n* estrellato *m*

stare /steə(r)/ *n* mirada *f* fija. ● *vi*. ~ **at** mirar fijamente

starfish /'stɑ:fɪʃ/ *n* estrella *f* de mar

stark /stɑ:k/ *a* (**-er**, **-est**) rígido; (*utter*) completo. ● *adv* completamente

starlight /'stɑ:laɪt/ *n* luz *f* de las estrellas

starling /'stɑ:lɪŋ/ *n* estornino *m*

starry /'stɑ:rɪ/ *a* estrellado. **~-eyed** *a* (*fam*) ingenuo, idealista

start /stɑ:t/ *vt* empezar; poner en marcha ‹*machine*›; (*cause*) provocar. ● *vi* empezar; (*jump*) sobresaltarse; (*leave*) partir; ‹*car etc*› arrancar. ● *n* principio *m*; (*leaving*) salida *f*; (*sport*) ventaja *f*; (*jump*) susto *m*. **~er** *n* (*sport*) participante *m* & *f*; (*auto*) motor *m* de arranque; (*culin*) primer plato *m*. **~ing-point** *n* punto *m* de partida

startle /'stɑ:tl/ *vt* asustar

starv|ation /stɑ:'veɪʃn/ *n* hambre *f*. **~e** /stɑ:v/ *vt* hacer morir de hambre; (*deprive*) privar. ● *vi* morir de hambre

stash /stæʃ/ *vt* (*sl*) esconder

state /steɪt/ *n* estado *m*; (*grand style*) pompa *f*. **S~** *n* Estado *m*. **be in a ~** estar agitado. ● *vt* declarar; expresar ‹*views*›; (*fix*) fijar. ● *a* del Estado; (*schol*) público; (*with ceremony*) de gala. **~less** *a* sin patria

stately /'steɪtlɪ/ *a* (**-ier**, **-iest**) majestuoso

statement /'steɪtmənt/ *n* declaración *f*; (*account*) informe *m*. **bank ~** *n* estado *m* de cuenta

stateroom /'steɪtrʊm/ *n* (*on ship*) camarote *m*

statesman /'steɪtsmən/ *n* (*pl* **-men**) estadista *m*

static /'stætɪk/ *a* inmóvil. **~s** *n* estática *f*; (*rad*, *TV*) parásitos *mpl* atmosféricos, interferencias *fpl*
station /'steɪʃn/ *n* estación *f*; (*status*) posición *f* social. • *vt* colocar; (*mil*) estacionar
stationary /'steɪʃənərɪ/ *a* estacionario
stationer /'steɪʃənə(r)/ *n* papelero *m*. **~'s (shop)** *n* papelería *f*. **~y** *n* artículos *mpl* de escritorio
station-wagon /'steɪʃnwægən/ *n* furgoneta *f*
statistic /stə'tɪstɪk/ *n* estadística *f*. **~al** /stə'tɪstɪkl/ *a* estadístico. **~s** /stə'tɪstɪks/ *n* (*science*) estadística *f*
statue /'stætʃu:/ *n* estatua *f*. **~sque** /-ʊ'esk/ *a* escultural. **~tte** /-ʊ'et/ *n* figurilla *f*
stature /'stætʃə(r)/ *n* talla *f*, estatura *f*
status /'steɪtəs/ *n* posición *f* social; (*prestige*) categoría *f*; (*jurid*) estado *m*
statut|e /'stætʃu:t/ *n* estatuto *m*. **~ory** /-ʊtrɪ/ *a* estatutario
staunch /stɔ:nʃ/ *a* (**-er**, **-est**) leal. **~ly** *adv* lealmente
stave /'steɪv/ *n* (*mus*) pentagrama *m*. • *vt*. **~ off** evitar
stay /steɪ/ *n* soporte *m*, sostén *m*; (*of time*) estancia *f*; (*jurid*) suspensión *f*. • *vi* quedar; (*spend time*) detenerse; (*reside*) alojarse. • *vt* matar ‹*hunger*›. **~ the course** terminar. **~ in** quedar en casa. **~ put** mantenerse firme. **~ up** no acostarse. **~ing-power** *n* resistencia *f*
stays /steɪz/ *npl* (*old use*) corsé *m*
stead /sted/ *n*. **in s.o.'s ~** en lugar de uno. **stand s.o. in good ~** ser útil a uno
steadfast /'stedfɑ:st/ *a* firme
stead|ily /'stedɪlɪ/ *adv* firmemente; (*regularly*) regularmente. **~y** /'stedɪ/ *a* (**-ier**, **-iest**) firme; (*regular*) regular; (*dependable*) serio
steak /steɪk/ *n* filete *m*
steal /sti:l/ *vt* (*pt* **stole**, *pp* **stolen**) robar. **~ the show** llevarse los aplausos. **~ in** *vi* entrar a hurtadillas. **~ out** *vi* salir a hurtadillas
stealth /stelθ/ *n*. **by ~** sigilosamente. **~y** *a* sigiloso
steam /sti:m/ *n* vapor *m*; (*energy*) energía *f*. • *vt* (*cook*) cocer al vapor; empañar ‹*window*›. • *vi* echar vapor. **~ ahead** (*fam*) hacer progresos. **~ up** *vi* ‹*glass*› empañar. **~-engine** *n* máquina *f* de vapor. **~er** /'sti:mə(r)/ *n* (*ship*) barco *m* de vapor. **~-roller** /'sti:mrəʊlə(r)/ *n* apisonadora *f*. **~y** *a* húmedo
steel /sti:l/ *n* acero *m*. • *vt*. **~ o.s.** fortalecerse. **~ industry** *n* industria *f* siderúrgica. **~ wool** *n* estropajo *m* de acero. **~y** *a* acerado; (*fig*) duro, inflexible
steep /sti:p/ • *a* (**-er**, **-est**) escarpado; ‹*price*› (*fam*) exorbitante. • *vt* (*soak*) remojar. **~ed in** (*fig*) empapado de
steeple /'sti:pl/ *n* aguja *f*, campanario *m*. **~chase** /'sti:pltʃeɪs/ *n* carrera *f* de obstáculos
steep: ~ly *adv* de modo empinado. **~ness** *n* lo escarpado
steer /stɪə(r)/ *vt* guiar; gobernar ‹*ship*›. • *vi* (*in ship*) gobernar. **~ clear of** evitar. **~ing** *n* (*auto*) dirección *f*. **~ing-wheel** *n* volante *m*
stem /stem/ *n* tallo *m*; (*of glass*) pie *m*; (*of word*) raíz *f*; (*of ship*) roda *f*. • *vt* (*pt* **stemmed**) detener. • *vi*. **~ from** provenir de
stench /stentʃ/ *n* hedor *m*
stencil /'stensl/ *n* plantilla *f*; (*for typing*) cliché *m*. • *vt* (*pt* **stencilled**) estarcir
stenographer /ste'nɒgrəfə(r)/ *n* (*Amer*) estenógrafo *m*
step /step/ *vi* (*pt* **stepped**) ir. **~ down** retirarse. **~ in** entrar; (*fig*) intervenir. **~ up** *vt* aumentar. • *n* paso *m*; (*surface*) escalón *m*; (*fig*) medida *f*. **in ~** (*fig*) de acuerdo con. **out of ~** (*fig*) en desacuerdo con. **~brother** /'stepbrʌðə(r)/ *n* hermanastro *m*. **~daughter** *n* hijastra *f*. **~-father** *n* padrastro *m*. **~-ladder** *n* escalera *f* de tijeras. **~mother** *n* madrastra *f*. **~ping-stone** /'stepɪŋstəʊn/ *n* pasadera *f*; (*fig*) escalón *m*. **~sister** *n* hermanastra *f*. **~son** *n* hijastro *m*
stereo /'sterɪəʊ/ *n* (*pl* **-os**) cadena *f* estereofónica. • *a* estereofónico. **~phonic** /sterɪəʊ'fɒnɪk/ *a* estereofónico. **~type** /'sterɪəʊtaɪp/ *n* estereotipo *m*. **~typed** *a* estereotipado
steril|e /'steraɪl/ *a* estéril. **~ity** /stə'rɪlətɪ/ *n* esterilidad *f*. **~ization** /-'zeɪʃn/ *n* esterilización *f*. **~ize** /'sterɪlaɪz/ *vt* esterilizar
sterling /'stɜ:lɪŋ/ *n* libras *fpl* esterlinas. • *a* ‹*pound*› esterlina; (*fig*) excelente. **~ silver** *n* plata *f* de ley

stern[1] /stɜ:n/ *n* (*of boat*) popa *f*
stern[2] /stɜ:n/ *a* (**-er, -est**) severo. **~ly** *adv* severamente
stethoscope /ˈsteθəskəʊp/ *n* estetoscopio *m*
stew /stju:/ *vt/i* guisar. ● *n* guisado *m*. **in a ~** (*fam*) en un apuro
steward /stjʊəd/ *n* administrador *m*; (*on ship, aircraft*) camarero *m*. **~ess** /-ˈdes/ *n* camarera *f*; (*on aircraft*) azafata *f*
stick /stɪk/ *n* palo *m*; (*for walking*) bastón *m*; (*of celery etc*) tallo *m*. ● *vt* (*pt* **stuck**) (*glue*) pegar; (*put, fam*) poner; (*thrust*) clavar; (*endure, sl*) soportar. ● *vi* pegarse; (*remain, fam*) quedarse; (*jam*) bloquearse. **~ at** (*fam*) perseverar en. **~ out** sobresalir; (*catch the eye, fam*) resaltar. **~ to** aferrarse a; cumplir ‹*promise*›. **~ up for** (*fam*) defender. **~er** /ˈstɪkə(r)/ *n* pegatina *f*. **~ing-plaster** *n* esparadrapo *m*. **~-in-the-mud** *n* persona *f* chapada a la antigua
stickler /ˈstɪklə(r)/ *n*. **be a ~ for** insistir en
sticky /ˈstɪkɪ/ *a* (**-ier, -iest**) pegajoso; ‹*label*› engomado; (*sl*) difícil
stiff /stɪf/ *a* (**-er, -est**) rígido; (*difficult*) difícil; ‹*manner*› estirado; ‹*drink*› fuerte; ‹*price*› subido; ‹*joint*› tieso; ‹*muscle*› con agujetas. **~en** /ˈstɪfn/ *vt* poner tieso. **~ly** *adv* rígidamente. **~ neck** *n* tortícolis *f*. **~ness** *n* rigidez *f*
stifl|e /ˈstaɪfl/ *vt* sofocar. **~ing** *a* sofocante
stigma /ˈstɪgmə/ *n* (*pl* **-as**) estigma *m*. (*pl* **stigmata** /ˈstɪgmətə/) (*relig*) estigma *m*. **~tize** *vt* estigmatizar
stile /staɪl/ *n* portillo *m* con escalones
stiletto /stɪˈletəʊ/ *n* (*pl* **-os**) estilete *m*. **~ heels** *npl* tacones *mpl* aguja
still[1] /stɪl/ *a* inmóvil; (*peaceful*) tranquilo; ‹*drink*› sin gas. ● *n* silencio *m*. ● *adv* todavía; (*nevertheless*) sin embargo
still[2] /stɪl/ (*apparatus*) alambique *m*
still: **~born** *a* nacido muerto. **~ life** *n* (*pl* **-s**) bodegón *m*. **~ness** *n* tranquilidad *f*
stilted /ˈstɪltɪd/ *a* artificial
stilts /stɪlts/ *npl* zancos *mpl*
stimul|ant /ˈstɪmjʊlənt/ *n* estimulante *m*. **~ate** /ˈstɪmjʊleɪt/ *vt* estimular. **~ation** /-ˈleɪʃn/ *n* estímulo *m*. **~us** /ˈstɪmjʊləs/ *n* (*pl* **-li** /-laɪ/) estímulo *m*

sting /stɪŋ/ *n* picadura *f*; (*organ*) aguijón *m*. ● *vt/i* (*pt* **stung**) picar
sting|iness /ˈstɪndʒɪnɪs/ *n* tacañería *f*. **~y** /ˈstɪndʒɪ/ *a* (**-ier, -iest**) tacaño
stink /stɪŋk/ *n* hedor *m*. ● *vi* (*pt* **stank** *or* **stunk**, *pp* **stunk**) oler mal. ● *vt*. **~ out** apestar ‹*room*›; ahuyentar ‹*person*›. **~er** /-ə(r)/ *n* (*sl*) problema *m* difícil; (*person*) mal bicho *m*
stint /stɪnt/ *n* (*work*) trabajo *m*. ● *vi*. **~ on** escatimar
stipple /ˈstɪpl/ *vt* puntear
stipulat|e /ˈstɪpjʊleɪt/ *vt/i* estipular. **~ion** /-ˈleɪʃn/ *n* estipulación *f*
stir /stɜ:(r)/ *vt* (*pt* **stirred**) remover, agitar; (*mix*) mezclar; (*stimulate*) estimular. ● *vi* moverse. ● *n* agitación *f*; (*commotion*) conmoción *f*
stirrup /ˈstɪrəp/ *n* estribo *m*
stitch /stɪtʃ/ *n* (*in sewing*) puntada *f*; (*in knitting*) punto *m*; (*pain*) dolor *m* de costado; (*med*) punto *m* de sutura. **be in ~es** (*fam*) desternillarse de risa. ● *vt* coser
stoat /stəʊt/ *n* armiño *m*
stock /stɒk/ *n* (*com, supplies*) existencias *fpl*; (*com, variety*) surtido *m*; (*livestock*) ganado *m*; (*lineage*) linaje *m*; (*finance*) acciones *fpl*; (*culin*) caldo *m*; (*plant*) alhelí *m*. **out of ~** agotado. **take ~** (*fig*) evaluar. ● *a* corriente; (*fig*) trillado. ● *vt* abastecer (**with** de). ● *vi*. **~ up** abastecerse (**with** de). **~broker** /ˈstɒkbrəʊkə(r)/ *n* corredor *m* de bolsa. **S~ Exchange** *n* bolsa *f*. **well-~ed** *a* bien provisto
stocking /ˈstɒkɪŋ/ *n* media *f*
stock: **~-in-trade** /ˈstɒkɪntreɪd/ *n* existencias *fpl*. **~ist** /ˈstɒkɪst/ *n* distribuidor *m*. **~pile** /ˈstɒkpaɪl/ *n* reservas *fpl*. ● *vt* acumular. **~-still** *a* inmóvil. **~-taking** *n* (*com*) inventario *m*
stocky /ˈstɒkɪ/ *a* (**-ier, -iest**) achaparrado
stodg|e /stɒdʒ/ *n* (*fam*) comida *f* pesada. **~y** *a* pesado
stoic /ˈstəʊɪk/ *n* estoico. **~al** *a* estoico. **~ally** *adv* estoicamente. **~ism** /-sɪzəm/ *n* estoicismo *m*
stoke /stəʊk/ *vt* alimentar. **~r** /ˈstəʊkə(r)/ *n* fogonero *m*
stole[1] /stəʊl/ *see* **steal**
stole[2] /stəʊl/ *n* estola *f*
stolen /ˈstəʊlən/ *see* **steal**
stolid /ˈstɒlɪd/ *a* impasible. **~ly** *adv* impasiblemente

stomach /ˈstʌmək/ *n* estómago *m*. ● *vt* soportar. **~-ache** *n* dolor *m* de estómago

ston|e /stəʊn/ *n* piedra *f*; (*med*) cálculo *m*; (*in fruit*) hueso *m*; (*weight*, *pl* **stone**) peso *m* de 14 libras (= *6,348 kg*). ● *a* de piedra. ● *vt* apedrear; deshuesar ‹*fruit*›. **~e-deaf** *a* sordo como una tapia. **~emason** /ˈstəʊnmeɪsn/ *n* albañil *m*. **~ework** /ˈstəʊnwɜːk/ *n* cantería *f*. **~y** *a* pedregoso; (*like stone*) pétreo

stood /stʊd/ *see* **stand**

stooge /stuːdʒ/ *n* (*in theatre*) compañero *m*; (*underling*) lacayo *m*

stool /stuːl/ *n* taburete *m*

stoop /stuːp/ *vi* inclinarse; (*fig*) rebajarse. ● *n*. **have a ~** ser cargado de espaldas

stop /stɒp/ *vt* (*pt* **stopped**) parar; (*cease*) terminar; tapar ‹*a leak etc*›; (*prevent*) impedir; (*interrupt*) interrumpir. ● *vi* pararse; (*stay*, *fam*) quedarse. ● *n* (*bus etc*) parada *f*; (*gram*) punto *m*; (*mec*) tope *m*. **~ dead** *vi* pararse en seco. **~cock** /ˈstɒpkɒk/ *n* llave *f* de paso. **~gap** /ˈstɒpgæp/ *n* remedio *m* provisional. **~(-over)** *n* escala *f*. **~page** /ˈstɒpɪdʒ/ *n* parada *f*; (*of work*) paro *m*; (*interruption*) interrupción *f*. **~per** /ˈstɒpə(r)/ *n* tapón *m*. **~-press** *n* noticias *fpl* de última hora. **~ light** *n* luz *f* de freno. **~-watch** *n* cronómetro *m*

storage /ˈstɔːrɪdʒ/ *n* almacenamiento *m*. **~ heater** *n* acumulador *m*. **in cold ~** almacenaje *m* frigorífico

store /stɔː(r)/ *n* provisión *f*; (*shop*, *depot*) almacén *m*; (*fig*) reserva *f*. **in ~** en reserva. **set ~ by** dar importancia a. ● *vt* (*for future*) poner en reserva; (*in warehouse*) almacenar. **~ up** *vt* acumular

storeroom /ˈstɔːruːm/ *n* despensa *f*

storey /ˈstɔːrɪ/ *n* (*pl* **-eys**) piso *m*

stork /stɔːk/ *n* cigüeña *f*

storm /stɔːm/ *n* tempestad *f*; (*mil*) asalto *m*. ● *vi* rabiar. ● *vt* (*mil*) asaltar. **~y** *a* tempestuoso

story /ˈstɔːrɪ/ *n* historia *f*; (*in newspaper*) artículo *m*; (*fam*) mentira *f*, cuento *m*. **~-teller** *n* cuentista *m & f*

stout /staʊt/ *a* (**-er**, **-est**) (*fat*) gordo; (*brave*) valiente. ● *n* cerveza *f* negra. **~ness** *n* corpulencia *f*

stove /stəʊv/ *n* estufa *f*

stow /stəʊ/ *vt* guardar; (*hide*) esconder. ● *vi*. **~ away** viajar de polizón. **~away** /ˈstəʊəweɪ/ *n* polizón *m*

straddle /ˈstrædl/ *vt* estar a horcajadas

straggl|e /ˈstrægl/ *vi* rezagarse. **~y** *a* desordenado

straight /streɪt/ *a* (**-er**, **-est**) derecho, recto; (*tidy*) en orden; (*frank*) franco; ‹*drink*› solo, puro; ‹*hair*› lacio. ● *adv* derecho; (*direct*) directamente; (*without delay*) inmediatamente. **~ on** todo recto. **~ out** sin vacilar. **go ~** enmendarse. ● *n* recta *f*. **~ away** inmediatamente. **~en** /ˈstreɪtn/ *vt* enderezar. ● *vi* enderezarse. **~forward** /streɪtˈfɔːwəd/ *a* franco; (*easy*) sencillo. **~forwardly** *adv* francamente. **~ness** *n* rectitud *f*

strain[1] /streɪn/ *n* (*tension*) tensión *f*; (*injury*) torcedura *f*. ● *vt* estirar; (*tire*) cansar; (*injure*) torcer; (*sieve*) colar

strain[2] /streɪn/ *n* (*lineage*) linaje *m*; (*streak*) tendencia *f*

strained /streɪnd/ *a* forzado; ‹*relations*› tirante

strainer /-ə(r)/ *n* colador *m*

strains /streɪnz/ *npl* (*mus*) acordes *mpl*

strait /streɪt/ *n* estrecho *m*. **~-jacket** *n* camisa *f* de fuerza. **~-laced** *a* remilgado, gazmoño. **~s** *npl* apuro *m*

strand /strænd/ *n* (*thread*) hebra *f*; (*sand*) playa *f*. ● *vi* ‹*ship*› varar. **be ~ed** quedarse sin recursos

strange /streɪndʒ/ *a* (**-er**, **-est**) extraño, raro; (*not known*) desconocido; (*unaccustomed*) nuevo. **~ly** *adv* extrañamente. **~ness** *n* extrañeza *f*. **~r** /ˈstreɪndʒə(r)/ *n* desconocido *m*

strang|le /ˈstræŋgl/ *vt* estrangular; (*fig*) ahogar. **~lehold** /ˈstræŋglhəʊld/ *n* (*fig*) dominio *m* completo. **~ler** /-ə(r)/ *n* estrangulador *m*. **~ulation** /stræŋgjʊˈleɪʃn/ *n* estrangulación *f*

strap /stræp/ *n* correa *f*; (*of garment*) tirante *m*. ● *vt* (*pt* **strapped**) atar con correa; (*flog*) azotar

strapping /ˈstræpɪŋ/ *a* robusto

strata /ˈstrɑːtə/ *see* **stratum**

strat|agem /ˈstrætədʒəm/ *n* estratagema *f*. **~egic** /strəˈtiːdʒɪk/ *a* estratégico. **~egically** *adv* estratégicamente. **~egist** *n* estratega *m & f.* **~egy** /ˈstrætədʒɪ/ *n* estrategia *f*

stratum /ˈstrɑːtəm/ *n* (*pl* **strata**) estrato *m*

straw /strɔː/ *n* paja *f.* **the last ~** el colmo

strawberry /ˈstrɔːbərɪ/ *n* fresa *f*

stray /streɪ/ *vi* vagar; (*deviate*) desviarse (**from** de). ● *a* ‹*animal*› extraviado, callejero; (*isolated*) aislado. ● *n* animal *m* extraviado, animal *m* callejero

streak /striːk/ *n* raya *f*; (*of madness*) vena *f.* ● *vt* rayar. ● *vi* moverse como un rayo. **~y** *a* (**-ier**, **-iest**) rayado; ‹*bacon*› entreverado

stream /striːm/ *n* arroyo *m*; (*current*) corriente *f*; (*of people*) desfile *m*; (*schol*) grupo *m.* ● *vi* correr. **~ out** *vi* ‹*people*› salir en tropel

streamer /ˈstriːmə(r)/ *n* (*paper*) serpentina *f*; (*flag*) gallardete *m*

streamline /ˈstriːmlaɪn/ *vt* dar línea aerodinámica a; (*simplify*) simplificar. **~d** *a* aerodinámico

street /striːt/ *n* calle *f.* **~car** /ˈstriːtkɑː/ *n* (*Amer*) tranvía *m.* **~ lamp** *n* farol *m.* **~ map** *n*, **~ plan** *n* plano *m*

strength /streŋθ/ *n* fuerza *f*; (*of wall etc*) solidez *f.* **on the ~ of** a base de. **~en** /ˈstreŋθn/ *vt* reforzar

strenuous /ˈstrenjʊəs/ *a* enérgico; (*arduous*) arduo; (*tiring*) fatigoso. **~ly** *adv* enérgicamente

stress /stres/ *n* énfasis *f*; (*gram*) acento *m*; (*mec, med, tension*) tensión *f.* ● *vt* insistir en

stretch /stretʃ/ *vt* estirar; (*extend*) extender; (*exaggerate*) forzar. **~ a point** hacer una excepción. ● *vi* estirarse; (*extend*) extenderse. ● *n* estirón *m*; (*period*) período *m*; (*of road*) tramo *m.* **at a ~** seguido; (*in one go*) de un tirón. **~er** /ˈstretʃə(r)/ *n* camilla *f*

strew /struː/ *vt* (*pt* **strewed**, *pp* **strewn** *or* **strewed**) esparcir; (*cover*) cubrir

stricken /ˈstrɪkən/ *a.* **~ with** afectado de

strict /strɪkt/ *a* (**-er**, **-est**) severo; (*precise*) estricto, preciso. **~ly** *adv* estrictamente. **~ly speaking** en rigor

stricture /ˈstrɪktʃə(r)/ *n* crítica *f*; (*constriction*) constricción *f*

stride /straɪd/ *vi* (*pt* **strode,** *pp* **stridden**) andar a zancadas. ● *n* zancada *f.* **take sth in one's ~** hacer algo con facilidad, tomarse las cosas con calma

strident /ˈstraɪdnt/ *a* estridente

strife /straɪf/ *n* conflicto *m*

strike /straɪk/ *vt* (*pt* **struck**) golpear; encender ‹*match*›; encontrar ‹*gold etc*›; ‹*clock*› dar. ● *vi* golpear; (*go on strike*) declararse en huelga; (*be on strike*) estar en huelga; (*attack*) atacar; ‹*clock*› dar la hora. ● *n* (*of workers*) huelga *f*; (*attack*) ataque *m*; (*find*) descubrimiento *m.* **on ~** en huelga. **~ off**, **~ out** tachar. **~ up a friendship** trabar amistad. **~r** /ˈstraɪkə(r)/ *n* huelguista *m & f*

striking /ˈstraɪkɪŋ/ *a* impresionante

string /strɪŋ/ *n* cuerda *f*; (*of lies, pearls*) sarta *f.* **pull ~s** tocar todos los resortes. ● *vt* (*pt* **strung**) (*thread*) ensartar. **~ along** (*fam*) engañar. **~ out** extender(se). **~ed** *a* (*mus*) de cuerda

stringen|cy /ˈstrɪndʒənsɪ/ *n* rigor *m.* **~t** /ˈstrɪndʒənt/ *a* riguroso

stringy /ˈstrɪŋɪ/ *a* fibroso

strip /strɪp/ *vt* (*pt* **stripped**) desnudar; (*tear away, deprive*) quitar; desmontar ‹*machine*›. ● *vi* desnudarse. ● *n* tira *f.* **~ cartoon** *n* historieta *f*

stripe /straɪp/ *n* raya *f*; (*mil*) galón *m.* **~d** *a* a rayas, rayado

strip: **~ light** *n* tubo *m* fluorescente. **~per** /-ə(r)/ *n* artista *m & f* de striptease. **~-tease** *n* número *m* del desnudo, striptease *m*

strive /straɪv/ *vi* (*pt* **strove**, *pp* **striven**). **~ to** esforzarse por

strode /strəʊd/ *see* **stride**

stroke /strəʊk/ *n* golpe *m*; (*in swimming*) brazada *f*; (*med*) apoplejía *f*; (*of pen etc*) rasgo *m*; (*of clock*) campanada *f*; (*caress*) caricia *f.* ● *vt* acariciar

stroll /strəʊl/ *vi* pasearse. ● *n* paseo *m*

strong /strɒŋ/ *a* (**-er**, **-est**) fuerte. **~-box** *n* caja *f* fuerte. **~hold** /ˈstrɒŋhəʊld/ *n* fortaleza *f*; (*fig*) baluarte *m.* **~ language** *n* palabras *fpl* fuertes, palabras *fpl* subidas de tono. **~ly** *adv* (*greatly*) fuertemente; (*with energy*) enérgicamente; (*deeply*) profundamente. **~ measures** *npl* medidas *fpl* enérgicas. **~-minded** *a* resuelto. **~-room** *n* cámara *f* acorazada

stroppy /ˈstrɒpɪ/ *a* (*sl*) irascible

strove /strəʊv/ *see* **strive**

struck /strʌk/ *see* **strike**. **~ on** (*sl*) entusiasta de

structur|al /'strʌktʃərəl/ *a* estructural. **~e** /'strʌktʃə(r)/ *n* estructura *f*

struggle /'strʌgl/ *vi* luchar. **~ to one's feet** levantarse con dificultad. ● *n* lucha *f*

strum /strʌm/ *vt/i* (*pt* **strummed**) rasguear

strung /strʌŋ/ *see* **string**. ● *a*. **~ up** (*tense*) nervioso

strut /strʌt/ *n* puntal *m*; (*walk*) pavoneo *m*. ● *vi* (*pt* **strutted**) pavonearse

stub /stʌb/ *n* cabo *m*; (*counterfoil*) talón *m*; (*of cigarette*) colilla *f*; (*of tree*) tocón *m*. ● *vt* (*pt* **stubbed**). **~ out** apagar

stubble /'stʌbl/ *n* rastrojo *m*: (*beard*) barba *f* de varios días

stubborn /'stʌbən/ *a* terco. **~ly** *adv* tercamente. **~ness** *n* terquedad *f*

stubby /'stʌbɪ/ *a* (**-ier**, **-iest**) achaparrado

stucco /'stʌkəʊ/ *n* (*pl* **-oes**) estuco *m*

stuck /stʌk/ *see* **stick**. ● *a* (*jammed*) bloqueado; (*in difficulties*) en un apuro. **~ on** (*sl*) encantado con. **~-up** *a* (*sl*) presumido

stud[1] /stʌd/ *n* tachón *m*; (*for collar*) botón *m*. ● *vt* (*pt* **studded**) tachonar. **~ded with** sembrado de

stud[2] /stʌd/ *n* (*of horses*) caballeriza *f*

student /'stju:dənt/ *n* estudiante *m & f*

studied /'stʌdɪd/ *a* deliberado

studio /'stju:dɪəʊ/ *n* (*pl* **-os**) estudio *m*. **~ couch** *n* sofá *m* cama. **~ flat** *n* estudio *m* de artista

studious /'stju:dɪəs/ *a* estudioso; (*studied*) deliberado. **~ly** *adv* estudiosamente; (*carefully*) cuidadosamente

study /'stʌdɪ/ *n* estudio *m*; (*office*) despacho *m*. ● *vt/i* estudiar

stuff /stʌf/ *n* materia *f*, sustancia *f*; (*sl*) cosas *fpl*. ● *vt* rellenar; disecar ‹*animal*›; (*cram*) atiborrar; (*block up*) tapar; (*put*) meter de prisa. **~ing** *n* relleno *m*

stuffy /'stʌfɪ/ *a* (**-ier**, **-iest**) mal ventilado; (*old-fashioned*) chapado a la antigua

stumbl|e /'stʌmbl/ *vi* tropezar. **~e across**, **~e on** tropezar con. ● *n* tropezón *m*. **~ing-block** *n* tropiezo *m*, impedimento *m*

stump /stʌmp/ *n* cabo *m*; (*of limb*) muñón *m*; (*of tree*) tocón *m*. **~ed** /stʌmpt/ *a* (*fam*) perplejo. **~y** /'stʌmpɪ/ *a* (**-ier**, **-iest**) achaparrado

stun /stʌn/ *vt* (*pt* **stunned**) aturdir; (*bewilder*) pasmar. **~ning** *a* (*fabulous*, *fam*) estupendo

stung /stʌŋ/ *see* **sting**

stunk /stʌŋk/ *see* **stink**

stunt[1] /stʌnt/ *n* (*fam*) truco *m* publicitario

stunt[2] /stʌnt/ *vt* impedir el desarrollo de. **~ed** *a* enano

stupefy /'stju:pɪfaɪ/ *vt* dejar estupefacto

stupendous /stju:'pendəs/ *a* estupendo. **~ly** *adv* estupendamente

stupid /'stju:pɪd/ *a* estúpido. **~ity** /-'pɪdətɪ/ *n* estupidez *f*. **~ly** *adv* estúpidamente

stupor /'stju:pə(r)/ *n* estupor *m*

sturd|iness /'stɜ:dɪnɪs/ *n* robustez *f*. **~y** /'stɜ:dɪ/ *a* (**-ier**, **-iest**) robusto

sturgeon /'stɜ:dʒən/ *n* (*pl* **sturgeon**) esturión *m*

stutter /'stʌtə(r)/ *vi* tartamudear. ● *n* tartamudeo *m*

sty[1] /staɪ/ *n* (*pl* **sties**) pocilga *f*

sty[2] /staɪ/ *n* (*pl* **sties**) (*med*) orzuelo *m*

styl|e /staɪl/ *n* estilo *m*; (*fashion*) moda *f*. **in ~** con todo lujo. ● *vt* diseñar. **~ish** /'staɪlɪʃ/ *a* elegante. **~ishly** *adv* elegantemente. **~ist** /'staɪlɪst/ *n* estilista *m & f*. **hair ~ist** *n* peluquero *m*. **~ized** /'staɪlaɪzd/ *a* estilizado

stylus /'staɪləs/ *n* (*pl* **-uses**) aguja *f* (de tocadiscos)

suave /swɑ:v/ *a* (*pej*) zalamero

sub... /sʌb/ *pref* sub...

subaquatic /sʌbə'kwætɪk/ *a* subacuático

subconscious /sʌb'kɒnʃəs/ *a & n* subconsciente (*m*). **~ly** *adv* de modo subconsciente

subcontinent /sʌb'kɒntɪnənt/ *n* subcontinente *m*

subcontract /sʌbkən'trækt/ *vt* subcontratar. **~or** /-ə(r)/ *n* subcontratista *m & f*

subdivide /sʌbdɪ'vaɪd/ *vt* subdividir

subdue /səb'dju:/ *vt* dominar ‹*feelings*›; sojuzgar ‹*country*›. **~d** *a* (*depressed*) abatido; ‹*light*› suave

subhuman /sʌb'hju:mən/ *a* infrahumano

subject /'sʌbdʒɪkt/ *a* sometido. ~ **to** sujeto a. ● *n* súbdito *m*; (*theme*) asunto *m*; (*schol*) asignatura *f*; (*gram*) sujeto *m*; (*of painting, play, book etc*) tema *m*. /səb'dʒekt/ *vt* sojuzgar; (*submit*) someter. **~ion** /-ʃn/ *n* sometimiento *m*
subjective /səb'dʒektɪv/ *a* subjetivo. **~ly** *adv* subjetivamente
subjugate /'sʌbdʒʊgeɪt/ *vt* subyugar
subjunctive /səb'dʒʌŋktɪv/ *a & n* subjuntivo (*m*)
sublet /sʌb'let/ *vt* (*pt* **sublet**, *pres p* **subletting**) subarrendar
sublimat|e /'sʌblɪmeɪt/ *vt* sublimar. **~ion** /-'meɪʃn/ *n* sublimación *f*
sublime /sə'blaɪm/ *a* sublime. **~ly** *adv* sublimemente
submarine /sʌbmə'ri:n/ *n* submarino *m*
submerge /səb'mɜ:dʒ/ *vt* sumergir. ● *vi* sumergirse
submi|ssion /səb'mɪʃn/ *n* sumisión *f*. **~ssive** /-sɪv/ *a* sumiso. **~t** /səb'mɪt/ *vt* (*pt* **submitted**) someter. ● *vi* someterse
subordinat|e /sə'bɔ:dɪnət/ *a & n* subordinado (*m*). /sə'bɔ:dɪneɪt/ *vt* subordinar. **~ion** /-'neɪʃn/ *n* subordinación *f*
subscri|be /səb'skraɪb/ *vi* suscribir. **~be to** suscribir ‹*fund*›; (*agree*) estar de acuerdo con; abonarse a ‹*newspaper*›. **~ber** /-ə(r)/ *n* abonado *m*. **~ption** /-rɪpʃn/ *n* suscripción *f*
subsequent /'sʌbsɪkwənt/ *a* subsiguiente. **~ly** *adv* posteriormente
subservient /səb'sɜ:vjənt/ *a* servil
subside /səb'saɪd/ *vi* ‹*land*› hundirse; ‹*flood*› bajar; ‹*storm, wind*› amainar. **~nce** *n* hundimiento *m*
subsidiary /səb'sɪdɪərɪ/ *a* subsidiario. ● *n* (*com*) sucursal *m*
subsid|ize /'sʌbsɪdaɪz/ *vt* subvencionar. **~y** /'sʌbsədɪ/ *n* subvención *f*
subsist /səb'sɪst/ *vi* subsistir. **~ence** *n* subsistencia *f*
subsoil /'sʌbsɒɪl/ *n* subsuelo *m*
subsonic /sʌb'sɒnɪk/ *a* subsónico
substance /'sʌbstəns/ *n* substancia *f*
substandard /sʌb'stændəd/ *a* inferior
substantial /səb'stænʃl/ *a* sólido; ‹*meal*› substancial; (*considerable*) considerable. **~ly** *adv* considerablemente
substantiate /səb'stænʃɪeɪt/ *vt* justificar
substitut|e /'sʌbstɪtju:t/ *n* substituto *m*. ● *vt/i* substituir. **~ion** /-'tju:ʃn/ *n* substitución *f*
subterfuge /'sʌbtəfju:dʒ/ *n* subterfugio *m*
subterranean /sʌbtə'reɪnjən/ *a* subterráneo
subtitle /'sʌbtaɪtl/ *n* subtítulo *m*
subtle /'sʌtl/ *a* (**-er**, **-est**) sutil. **~ty** *n* sutileza *f*
subtract /səb'trækt/ *vt* restar. **~ion** /-ʃn/ *n* resta *f*
suburb /'sʌbɜ:b/ *n* barrio *m*. **the ~s** las afueras *fpl*. **~an** /sə'bɜ:bən/ *a* suburbano. **~ia** /sə'bɜ:bɪə/ *n* las afueras *fpl*
subvention /səb'venʃn/ *n* subvención *f*
subver|sion /səb'vɜ:ʃn/ *n* subversión *f*. **~sive** /səb'vɜ:sɪv/ *a* subversivo. **~t** /səb'vɜ:t/ *vt* subvertir
subway /'sʌbweɪ/ *n* paso *m* subterráneo; (*Amer*) metro *m*
succeed /sək'si:d/ *vi* tener éxito. ● *vt* suceder a. ~ **in doing** lograr hacer. **~ing** *a* sucesivo
success /sək'ses/ *n* éxito *m*. **~ful** *a* que tiene éxito; (*chosen*) elegido
succession /sək'seʃn/ *n* sucesión *f*. **in ~** sucesivamente, seguidos
successive /sək'sesɪv/ *a* sucesivo. **~ly** *adv* sucesivamente
successor /sək'sesə(r)/ *n* sucesor *m*
succinct /sək'sɪŋkt/ *a* sucinto
succour /'sʌkə(r)/ *vt* socorrer. ● *n* socorro *m*
succulent /'sʌkjʊlənt/ *a* suculento
succumb /sə'kʌm/ *vi* sucumbir
such /sʌtʃ/ *a* tal. ● *pron* los que, las que; (*so much*) tanto. **and ~** y tal. ● *adv* tan. ~ **a big house** una casa tan grande. ~ **and ~** tal o cual. ~ **as it is** tal como es. **~like** *a* (*fam*) semejante, de ese tipo
suck /sʌk/ *vt* chupar; sorber ‹*liquid*›. ~ **up** absorber. ~ **up to** (*sl*) dar coba a. **~er** /'sʌkə(r)/ *n* (*plant*) chupón *m*; (*person, fam*) primo *m*
suckle /sʌkl/ *vt* amamantar
suction /'sʌkʃn/ *n* succión *f*
sudden /'sʌdn/ *a* repentino. **all of a ~** de repente. **~ly** *adv* de repente. **~ness** *n* lo repentino
suds /sʌds/ *npl* espuma *f* (de jabón)
sue /su:/ *vt* (*pres p* **suing**) demandar (**for** por)
suede /sweɪd/ *n* ante *m*

suet /'su:ɪt/ *n* sebo *m*
suffer /'sʌfə(r)/ *vt* sufrir; (*tolerate*) tolerar. ● *vi* sufrir. **~ance** /'sʌfərəns/ *n*. **on ~ance** por tolerancia. **~ing** *n* sufrimiento *m*
suffic|e /sə'faɪs/ *vi* bastar. **~iency** /sə'fɪʃənsɪ/ *n* suficiencia *f*. **~ient** /sə'fɪʃnt/ *a* suficiente; (*enough*) bastante. **~iently** *adv* suficientemente, bastante
suffix /'sʌfɪks/ *n* (*pl* **-ixes**) sufijo *m*
suffocat|e /'sʌfəkeɪt/ *vt* ahogar. ● *vi* ahogarse. **~ion** /-'keɪʃn/ *n* asfixia *f*
sugar /'ʃʊgə(r)/ *n* azúcar *m* & *f*. ● *vt* azucarar. **~-bowl** *n* azucarero *m*. **~ lump** *n* terrón *m* de azúcar. **~y** *a* azucarado.
suggest /sə'dʒest/ *vt* sugerir. **~ible** /sə'dʒestɪbl/ *a* sugestionable. **~ion** /-tʃən/ *n* sugerencia *f*; (*trace*) traza *f*. **~ive** /sə'dʒestɪv/ *a* sugestivo. **be ~ive of** evocar, recordar. **~ively** *adv* sugestivamente
suicid|al /su:ɪ'saɪdl/ *a* suicida. **~e** /'su:ɪsaɪd/ *n* suicidio *m*; (*person*) suicida *m* & *f*. **commit ~e** suicidarse
suit /su:t/ *n* traje *m*; (*woman's*) traje *m* de chaqueta; (*cards*) palo *m*; (*jurid*) pleito *m*. ● *vt* convenir; ‹*clothes*› sentar bien a; (*adapt*) adaptar. **be ~ed for** ser apto para. **~ability** *n* conveniencia *f*. **~able** *a* adecuado. **~ably** *adv* convenientemente. **~case** /'su:tkeɪs/ *n* maleta *f*, valija *f* (*LAm*)
suite /swi:t/ *n* (*of furniture*) juego *m*; (*of rooms*) apartamento *m*; (*retinue*) séquito *m*
suitor /'su:tə(r)/ *n* pretendiente *m*
sulk /sʌlk/ *vi* enfurruñarse. **~s** *npl* enfurruñamiento *m*. **~y** *a* enfurruñado
sullen /'sʌlən/ *a* resentido. **~ly** *adv* con resentimiento
sully /'sʌlɪ/ *vt* manchar
sulphur /'sʌlfə(r)/ *n* azufre *m*. **~ic** /-'fjʊərɪk/ *a* sulfúrico. **~ic acid** *n* ácido *m* sulfúrico
sultan /'sʌltən/ *n* sultán *m*
sultana /sʌl'tɑ:nə/ *n* pasa *f* gorrona
sultry /'sʌltrɪ/ *a* (**-ier**, **-iest**) ‹*weather*› bochornoso; (*fig*) sensual
sum /sʌm/ *n* suma *f*. ● *vt* (*pt* **summed**). **~ up** resumir ‹*situation*›; (*assess*) evaluar
summar|ily /'sʌmərɪlɪ/ *adv* sumariamente. **~ize** *vt* resumir. **~y** /'sʌmərɪ/ *a* sumario. ● *n* resumen *m*
summer /'sʌmə(r)/ *n* verano *m*. **~-house** *n* glorieta *f*, cenador *m*. **~time** *n* verano *m*. **~ time** *n* hora *f* de verano. **~y** *a* veraniego
summit /'sʌmɪt/ *n* cumbre *f*. **~ conference** *n* conferencia *f* cumbre
summon /'sʌmən/ *vt* llamar; convocar ‹*meeting, s.o. to meeting*›; (*jurid*) citar. **~ up** armarse de. **~s** /'sʌmənz/ *n* llamada *f*; (*jurid*) citación *f*. ● *vt* citar
sump /sʌmp/ *n* (*mec*) cárter *m*
sumptuous /'sʌmptjʊəs/ *a* suntuoso. **~ly** *adv* suntuosamente
sun /sʌn/ *n* sol *m*. ● *vt* (*pt* **sunned**). **~ o.s.** tomar el sol. **~bathe** /'sʌnbeɪð/ *vi* tomar el sol. **~beam** /'sʌnbi:m/ *n* rayo *m* de sol. **~burn** /'sʌnbɜ:n/ *n* quemadura *f* de sol. **~burnt** *a* quemado por el sol
sundae /'sʌndeɪ/ *n* helado *m* con frutas y nueces
Sunday /'sʌndeɪ/ *n* domingo *m*. **~ school** *n* catequesis *f*
sun: **~dial** /'sʌndaɪl/ *n* reloj *m* de sol. **~down** /'sʌndaʊn/ *n* puesta *f* del sol
sundry /'sʌndrɪ/ *a* diversos. **all and ~** todo el mundo. **sundries** *npl* artículos *mpl* diversos
sunflower /'sʌnflaʊə(r)/ *n* girasol *m*
sung /sʌŋ/ *see* **sing**
sun-glasses /'sʌnglɑ:sɪz/ *npl* gafas *fpl* de sol
sunk /sʌŋk/ *see* **sink**. **~en** /'sʌŋkən/ ● *a* hundido
sunlight /'sʌnlaɪt/ *n* luz *f* del sol
sunny /'sʌnɪ/ *a* (**-ier**, **-iest**) ‹*day*› de sol; ‹*place*› soleado. **it is ~** hace sol
sun: **~rise** /'sʌnraɪz/ *n* amanecer *m*, salida *f* del sol. **~-roof** *n* techo *m* corredizo. **~set** /'sʌnset/ *n* puesta *f* del sol. **~-shade** /'sʌnʃeɪd/ *n* quitasol *m*, sombrilla *f*; (*awning*) toldo *m*. **~shine** /'sʌnʃaɪn/ *n* sol *m*. **~spot** /'sʌnspɒt/ *n* mancha *f* solar. **~stroke** /'sʌnstrəʊk/ *n* insolación *f*. **~tan** *n* bronceado *m*. **~tanned** *a* bronceado. **~tan lotion** *n* bronceador *m*
sup /sʌp/ *vt* (*pt* **supped**) sorber
super /'su:pə(r)/ *a* (*fam*) estupendo
superannuation /su:pərænjʊ'eɪʃn/ *n* jubilación *f*
superb /su:'pɜ:b/ *a* espléndido. **~ly** *adv* espléndidamente
supercilious /su:pə'sɪlɪəs/ *a* desdeñoso
superficial /su:pə'fɪʃl/ *a* superficial. **~ity** /-ɪ'ælətɪ/ *n* superficialidad *f*. **~ly** *adv* superficialmente

superfluous /su:'pɜ:flʊəs/ *a* superfluo

superhuman /su:pə'hju:mən/ *a* sobrehumano

superimpose /su:pərɪm'pəʊz/ *vt* sobreponer

superintend /su:pərɪn'tend/ *vt* vigilar. **~ence** *n* dirección *f*. **~ent** *n* director *m*; (*of police*) comisario *m*

superior /su:'pɪərɪə(r)/ *a* & *n* superior (*m*). **~ity** /-'ɒrətɪ/ *n* superioridad *f*

superlative /su:'pɜ:lətɪv/ *a* & *n* superlativo (*m*)

superman /'su:pəmæn/ *n* (*pl* **-men**) superhombre *m*

supermarket /'su:pəmɑ:kɪt/ *n* supermercado *m*

supernatural /su:pə'nætʃrəl/ *a* sobrenatural

superpower /'su:pəpaʊə(r)/ *n* superpotencia *f*

supersede /su:pə'si:d/ *vt* reemplazar, suplantar

supersonic /su:pə'sɒnɪk/ *a* supersónico

superstitio|n /su:pə'stɪʃn/ *n* superstición *f*. **~us** *a* supersticioso

superstructure /'su:pəstrʌktʃə(r)/ *n* superestructura *f*

supertanker /'su:pətæŋkə(r)/ *n* petrolero *m* gigante

supervene /su:pə'vi:n/ *vi* sobrevenir

supervis|e /'su:pəvaɪz/ *vt* supervisar. **~ion** /-'vɪʒn/ *n* supervisión *f*. **~or** /-zə(r)/ *n* supervisor *m*. **~ory** *a* de supervisión

supper /'sʌpə(r)/ *n* cena *f*

supplant /sə'plɑ:nt/ *vt* suplantar

supple /sʌpl/ *a* flexible. **~ness** *n* flexibilidad *f*

supplement /'sʌplɪmənt/ *n* suplemento *m*. ● *vt* completar; (*increase*) aumentar. **~ary** /-'mentərɪ/ *a* suplementario

suppl|ier /sə'plaɪə(r)/ *n* suministrador *m*; (*com*) proveedor *m*. **~y** /sə'plaɪ/ *vt* proveer; (*feed*) alimentar; satisfacer ‹*a need*›. **~y with** abastecer de. ● *n* provisión *f*, suministro *m*. **~y and demand** oferta *f* y demanda

support /sə'pɔ:t/ *vt* sostener; (*endure*) soportar, aguantar; (*fig*) apoyar. ● *n* apoyo *m*; (*tec*) soporte *m*. **~er** /-ə(r)/ *n* soporte *m*; (*sport*) seguidor *m*, hincha *m* & *f*. **~ive** *a* alentador

suppos|e /sə'pəʊz/ *vt* suponer; (*think*) creer. **be ~ed to** deber. **not be ~ed to** (*fam*) no tener permiso para, no tener derecho a. **~edly** *adv* según cabe suponer; (*before adjective*) presuntamente. **~ition** /sʌpə'zɪʃn/ *n* suposición *f*

suppository /sə'pɒzɪtərɪ/ *n* supositorio *m*

suppress /sə'pres/ *vt* suprimir. **~ion** *n* supresión *f*. **~or** /-ə(r)/ *n* supresor *m*

suprem|acy /su:'preməsɪ/ *n* supremacía *f*. **~e** /su:'pri:m/ *a* supremo

surcharge /'sɜ:tʃɑ:dʒ/ *n* sobreprecio *m*; (*tax*) recargo *m*

sure /ʃʊə(r)/ *a* (**-er**, **-est**) seguro, cierto. **make ~** asegurarse. ● *adv* (*Amer*, *fam*) ¡claro! **~ enough** efectivamente. **~-footed** *a* de pie firme. **~ly** *adv* seguramente

surety /'ʃʊərətɪ/ *n* garantía *f*

surf /sɜ:f/ *n* oleaje *m*; (*foam*) espuma *f*

surface /'sɜ:fɪs/ *n* superficie *f*. ● *a* superficial, de la superficie. ● *vt* (*smoothe*) alisar; (*cover*) recubrir (**with** de). ● *vi* salir a la superficie; (*emerge*) emerger. **~ mail** *n* por vía marítima

surfboard /'sɜ:fbɔ:d/ *n* tabla *f* de surf

surfeit /'sɜ:fɪt/ *n* exceso *m*

surfing /'sɜ:fɪŋ/ *n*, **surf-riding** /'sɜ:fraɪdɪŋ/ *n* surf *m*

surge /sɜ:dʒ/ *vi* ‹*crowd*› moverse en tropel; ‹*waves*› encresparse. ● *n* oleada *f*; (*elec*) sobretensión *f*

surgeon /'sɜ:dʒən/ *n* cirujano *m*

surgery /'sɜ:dʒərɪ/ *n* cirugía *f*; (*consulting room*) consultorio *m*; (*consulting hours*) horas *fpl* de consulta

surgical /'sɜ:rdʒɪkl/ *a* quirúrgico

surl|iness /'sɜ:lɪnɪs/ *n* aspereza *f*. **~y** /'sɜ:lɪ/ *a* (**-ier**, **-iest**) áspero

surmise /sə'maɪz/ *vt* conjeturar

surmount /sə'maʊnt/ *vt* superar

surname /'sɜ:neɪm/ *n* apellido *m*

surpass /sə'pɑ:s/ *vt* sobrepasar, exceder

surplus /'sɜ:pləs/ *a* & *n* excedente (*m*)

surpris|e /sə'praɪz/ *n* sorpresa *f*. ● *vt* sorprender. **~ing** *a* sorprendente. **~ingly** *adv* asombrosamente

surrealis|m /sə'rɪəlɪzəm/ *n* surrealismo *m*. **~t** *n* surrealista *m* & *f*

surrender /sə'rendə(r)/ *vt* entregar. ● *vi* entregarse. ● *n* entrega *f*; (*mil*) rendición *f*

surreptitious /sʌrəp'tɪʃəs/ *a* clandestino
surrogate /'sʌrəgət/ *n* substituto *m*
surround /sə'raʊnd/ *vt* rodear; (*mil*) cercar. ● *n* borde *m*. **~ing** *a* circundante. **~ings** *npl* alrededores *mpl*
surveillance /sɜː'veɪləns/ *n* vigilancia *f*
survey /'sɜːveɪ/ *n* inspección *f*; (*report*) informe *m*; (*general view*) vista *f* de conjunto. /sə'veɪ/ *vt* examinar, inspeccionar; (*inquire into*) hacer una encuesta de. **~or** *n* topógrafo *m*, agrimensor
surviv|al /sə'vaɪvl/ *n* supervivencia *f*. **~e** /sə'vaɪv/ *vt/i* sobrevivir. **~or** /-ə(r)/ *n* superviviente *m & f*
susceptib|ility /səseptə'bɪlətɪ/ *n* susceptibilidad *f*. **~le** /sə'septəbl/ *a* susceptible. **~le to** propenso a
suspect /sə'spekt/ *vt* sospechar. /'sʌspekt/ *a & n* sospechoso (*m*)
suspend /sə'spend/ *vt* suspender. **~er** /səs'pendə(r)/ *n* liga *f*. **~er belt** *n* liguero *m*. **~ers** *npl* (*Amer*) tirantes *mpl*
suspense /sə'spens/ *n* incertidumbre *f*; (*in film etc*) suspense *m*
suspension /sə'spenʃn/ *n* suspensión *f*. **~ bridge** *n* puente *m* colgante
suspicion /sə'spɪʃn/ *n* sospecha *f*; (*trace*) pizca *f*
suspicious /sə'spɪʃəs/ *a* desconfiado; (*causing suspicion*) sospechoso
sustain /sə'steɪn/ *vt* sostener; (*suffer*) sufrir
sustenance /'sʌstɪnəns/ *n* sustento *m*
svelte /svelt/ *a* esbelto
SW *abbr* (*south-west*) sudoeste *m*
swab /swɒb/ *n* (*med*) tapón *m*
swagger /'swægə(r)/ *vi* pavonearse
swallow[1] /'swɒləʊ/ *vt/i* tragar. ● *n* trago *m*. **~ up** tragar; consumir ‹*savings etc*›
swallow[2] /'swɒləʊ/ *n* (*bird*) golondrina *f*
swam /swæm/ *see* **swim**
swamp /swɒmp/ *n* pantano *m*. ● *vt* inundar; (*with work*) agobiar. **~y** *a* pantanoso
swan /swɒn/ *n* cisne *m*
swank /swæŋk/ *n* (*fam*) ostentación *f*. ● *vi* (*fam*) fanfarronear
swap /swɒp/ *vt/i* (*pt* **swapped**) (*fam*) (inter)cambiar. ● *n* (*fam*) (inter) cambio *m*
swarm /swɔːm/ *n* enjambre *m*. ● *vi* ‹*bees*› enjambrar; (*fig*) hormiguear
swarthy /'swɔːðɪ/ *a* (**-ier**, **-iest**) moreno
swastika /'swɒstɪkə/ *n* cruz *f* gamada
swat /swɒt/ *vt* (*pt* **swatted**) aplastar
sway /sweɪ/ *vi* balancearse. ● *vt* (*influence*) influir en. ● *n* balanceo *m*; (*rule*) imperio *m*
swear /sweə(r)/ *vt/i* (*pt* **swore**, *pp* **sworn**) jurar. **~ by** (*fam*) creer ciegamente en. **~-word** *n* palabrota *f*
sweat /swet/ *n* sudor *m*. ● *vi* sudar
sweat|er /'swetə(r)/ *n* jersey *m*. **~shirt** *n* sudadera *f*
swede /swiːd/ *n* naba *f*
Swede /swiːd/ *n* sueco *m*
Sweden /'swiːdn/ *n* Suecia *f*
Swedish /'swiːdɪʃ/ *a & n* sueco (*m*)
sweep /swiːp/ *vt* (*pt* **swept**) barrer; deshollinar ‹*chimney*›. **~ the board** ganar todo. ● *vi* barrer; ‹*road*› extenderse; (*go majestically*) moverse majestuosamente. ● *n* barrido *m*; (*curve*) curva *f*; (*movement*) movimiento *m*; (*person*) deshollinador *m*. **~ away** *vt* barrer. **~ing** *a* ‹*gesture*› amplio; ‹*changes etc*› radical; ‹*statement*› demasiado general. **~stake** /'swiːpsteɪk/ *n* lotería *f*
sweet /swiːt/ *a* (**-er**, **-est**) dulce; (*fragrant*) fragante; (*pleasant*) agradable. **have a ~ tooth** ser dulcero. ● *n* caramelo *m*; (*dish*) postre *m*. **~bread** /'swiːtbred/ *n* lechecillas *fpl*. **~en** /'swiːtn/ *vt* endulzar. **~ener** /-ə(r)/ *n* dulcificante *m*. **~heart** /'swiːthɑːt/ *n* amor *m*. **~ly** *adv* dulcemente. **~ness** *n* dulzura *f*. **~ pea** *n* guisante *m* de olor
swell /swel/ *vt* (*pt* **swelled**, *pp* **swollen** *or* **swelled**) hinchar; (*increase*) aumentar. ● *vi* hincharse; (*increase*) aumentarse; ‹*river*› crecer. ● *a* (*fam*) estupendo. ● *n* (*of sea*) oleaje *m*. **~ing** *n* hinchazón *m*
swelter /'sweltə(r)/ *vi* sofocarse de calor
swept /swept/ *see* **sweep**
swerve /swɜːv/ *vi* desviarse
swift /swɪft/ *a* (**-er**, **-est**) rápido. ● *n* (*bird*) vencejo *m*. **~ly** *adv* rápidamente. **~ness** *n* rapidez *f*
swig /swɪg/ *vt* (*pt* **swigged**) (*fam*) beber a grandes tragos. ● *n* (*fam*) trago *m*

swill /swɪl/ *vt* enjuagar; (*drink*) beber a grandes tragos. ● *n* (*food for pigs*) bazofia *f*

swim /swɪm/ *vi* (*pt* **swam**, *pp* **swum**) nadar; ‹*room, head*› dar vueltas. ● *n* baño *m*. **~mer** *n* nadador *m*. **~ming-bath** *n* piscina *f*. **~mingly** /'swɪmɪŋlɪ/ *adv* a las mil maravillas. **~ming-pool** *n* piscina *f*. **~ming-trunks** *npl* bañador *m*. **~suit** *n* traje *m* de baño

swindle /'swɪndl/ *vt* estafar. ● *n* estafa *f*. **~r** /-ə(r)/ *n* estafador *m*

swine /swaɪn/ *npl* cerdos *mpl*. ● *n* (*pl* **swine**) (*person, fam*) canalla *m*

swing /swɪŋ/ *vt* (*pt* **swung**) balancear. ● *vi* oscilar; ‹*person*› balancearse; (*turn round*) girar. ● *n* balanceo *m*, vaivén *m*; (*seat*) columpio *m*; (*mus*) ritmo *m*. **in full ~** en plena actividad. **~ bridge** *n* puente *m* giratorio

swingeing /'swɪndʒɪŋ/ *a* enorme

swipe /swaɪp/ *vt* golpear; (*snatch, sl*) birlar. ● *n* (*fam*) golpe *m*

swirl /swɜːl/ *vi* arremolinarse. ● *n* remolino *m*

swish /swɪʃ/ *vt* silbar. ● *a* (*fam*) elegante

Swiss /swɪs/ *a* & *n* suizo (*m*). **~ roll** *n* bizcocho *m* enrollado

switch /swɪtʃ/ *n* (*elec*) interruptor *m*; (*change*) cambio *m*. ● *vt* cambiar; (*deviate*) desviar. **~ off** (*elec*) desconectar; apagar ‹*light*›. **~ on** (*elec*) encender; arrancar ‹*engine*›. **~back** /'swɪtʃbæk/ *n* montaña *f* rusa. **~board** /'swɪtʃbɔːd/ *n* centralita *f*

Switzerland /'swɪtsələnd/ *n* Suiza *f*

swivel /'swɪvl/ ● *vi* (*pt* **swivelled**) girar

swollen /'swəʊlən/ *see* **swell**. ● *a* hinchado

swoon /swuːn/ *vi* desmayarse

swoop /swuːp/ *vi* ‹*bird*› calarse; ‹*plane*› bajar en picado. ● *n* calada *f*; (*by police*) redada *f*

sword /sɔːd/ *n* espada *f*. **~fish** /'sɔːdfɪʃ/ *n* pez *m* espada

swore /swɔː(r)/ *see* **swear**

sworn /swɔːn/ *see* **swear**. ● *a* ‹*enemy*› jurado; ‹*friend*› leal

swot /swɒt/ *vt/i* (*pt* **swotted**) (*schol, sl*) empollar. ● *n* (*schol, sl*) empollón *m*

swum /swʌm/ *see* **swim**

swung /swʌŋ/ *see* **swing**

sycamore /'sɪkəmɔː(r)/ *n* plátano *m* falso

syllable /'sɪləbl/ *n* sílaba *f*

syllabus /'sɪləbəs/ *n* (*pl* **-buses**) programa *m* (de estudios)

symbol /'sɪmbl/ *n* símbolo *m*. **~ic(al)** /-'bɒlɪk(l)/ *a* simbólico. **~ism** *n* simbolismo *m*. **~ize** *vt* simbolizar

symmetr|ical /sɪ'metrɪkl/ *a* simétrico. **~y** /'sɪmətrɪ/ *n* simetría *f*

sympath|etic /sɪmpə'θetɪk/ *a* comprensivo; (*showing pity*) compasivo. **~ize** /-aɪz/ *vi* comprender; (*pity*) compadecerse (**with** de). **~izer** *n* (*pol*) simpatizante *m* & *f*. **~y** /'sɪmpəθɪ/ *n* comprensión *f*; (*pity*) compasión *f*; (*condolences*) pésame *m*. **be in ~y with** estar de acuerdo con

symphon|ic /sɪm'fɒnɪk/ *a* sinfónico. **~y** /'sɪmfənɪ/ *n* sinfonía *f*

symposium /sɪm'pəʊzɪəm/ *n* (*pl* **-ia**) simposio *m*

symptom /'sɪmptəm/ *n* síntoma *m*. **~atic** /-'mætɪk/ *a* sintomático

synagogue /'sɪnəgɒg/ *n* sinagoga *f*

synchroniz|ation /sɪnkrənaɪ'zeɪʃn/ *n* sincronización *f*. **~e** /'sɪŋkrənaɪz/ *vt* sincronizar

syncopat|e /'sɪnkəpeɪt/ *vt* sincopar. **~ion** /-'peɪʃn/ *n* síncopa *f*

syndicate /'sɪndɪkət/ *n* sindicato *m*

syndrome /'sɪndrəʊm/ *n* síndrome *m*

synod /'sɪnəd/ *n* sínodo *m*

synonym /'sɪnənɪm/ *n* sinónimo *m*. **~ous** /-'nɒnɪməs/ *a* sinónimo

synopsis /sɪ'nɒpsɪs/ *n* (*pl* **-opses** /-siːz/) sinopsis *f*, resumen *m*

syntax /'sɪntæks/ *n* sintaxis *f invar*

synthesi|s /'sɪnθəsɪs/ *n* (*pl* **-theses** /-siːz/) síntesis *f*. **~ze** *vt* sintetizar

synthetic /sɪn'θetɪk/ *a* sintético

syphilis /'sɪfɪlɪs/ *n* sífilis *f*

Syria /'sɪrɪə/ *n* Siria *f*. **~n** *a* & *n* sirio (*m*)

syringe /'sɪrɪndʒ/ *n* jeringa *f*. ● *vt* jeringar

syrup /'sɪrəp/ *n* jarabe *m*, almíbar *m*; (*treacle*) melaza *f*. **~y** *a* almibarado

system /'sɪstəm/ *n* sistema *m*; (*body*) organismo *m*; (*order*) método *m*. **~atic** /-ə'mætɪk/ *a* sistemático. **~atically** /-ə'mætɪklɪ/ *adv* sistemáticamente. **~s analyst** *n* analista *m* & *f* de sistemas

T

tab /tæb/ *n* (*flap*) lengüeta *f*; (*label*) etiqueta *f*. **keep ~s on** (*fam*) vigilar
tabby /'tæbɪ/ *n* gato *m* atigrado
tabernacle /'tæbənækl/ *n* tabernáculo *m*
table /'teɪbl/ *n* mesa *f*; (*list*) tabla *f*. **~ of contents** índice *m*. ● *vt* presentar; (*postpone*) aplazar. **~cloth** *n* mantel *m*. **~-mat** *n* salvamanteles *m invar*. **~spoon** /'teɪblspu:n/ *n* cucharón *m*, cuchara *f* sopera. **~spoonful** *n* (*pl* **-fuls**) cucharada *f*
tablet /'tæblɪt/ *n* (*of stone*) lápida *f*; (*pill*) tableta *f*; (*of soap etc*) pastilla *f*
table tennis /'teɪbltenɪs/ *n* tenis *m* de mesa, ping-pong *m*
tabloid /'tæblɔɪd/ *n* tabloide *m*
taboo /tə'bu:/ *a & n* tabú (*m*)
tabulator /'tæbjʊleɪtə(r)/ *n* tabulador *m*
tacit /'tæsɪt/ *a* tácito
taciturn /'tæsɪtɜ:n/ *a* taciturno
tack /tæk/ *n* tachuela *f*; (*stitch*) hilván *m*; (*naut*) virada *f*; (*fig*) línea *f* de conducta. ● *vt* sujetar con tachuelas; (*sew*) hilvanar. **~ on** añadir. ● *vi* virar
tackle /'tækl/ *n* (*equipment*) equipo *m*; (*football*) placaje *m*. ● *vt* abordar ‹*problem etc*›; (*in rugby*) hacer un placaje a
tacky /'tækɪ/ *a* pegajoso; (*in poor taste*) vulgar, de pacotilla
tact /tækt/ *n* tacto *m*. **~ful** *a* discreto. **~fully** *adv* discretamente
tactic|al /'tæktɪkl/ *a* táctico. **~s** /'tæktɪks/ *npl* táctica *f*
tactile /'tæktaɪl/ *a* táctil
tact: **~less** *a* indiscreto. **~lessly** *adv* indiscretamente
tadpole /'tædpəʊl/ *n* renacuajo *m*
tag /tæg/ *n* (*on shoe-lace*) herrete *m*; (*label*) etiqueta *f*. ● *vt* (*pt* **tagged**) poner etiqueta a; (*trail*) seguir. ● *vi*. **~ along** (*fam*) seguir
tail /teɪl/ *n* cola *f*. **~s** *npl* (*tailcoat*) frac *m*; (*of coin*) cruz *f*. ● *vt* (*sl*) seguir. ● *vi*. **~ off** disminuir. **~-end** *n* extremo *m* final, cola *f*
tailor /'teɪlə(r)/ *n* sastre *m*. ● *vt* confeccionar. **~-made** *n* hecho a la medida. **~-made for** (*fig*) hecho para
tailplane /'teɪlpleɪn/ *n* plano *m* de cola
taint /teɪnt/ *n* mancha *f*. ● *vt* contaminar
take /teɪk/ *vt* (*pt* **took**, *pp* **taken**) tomar, coger (*not LAm*), agarrar (*esp LAm*); (*contain*) contener; (*capture*) capturar; (*endure*) aguantar; (*require*) requerir; tomar ‹*bath*›; dar ‹*walk*›; (*carry*) llevar; (*accompany*) acompañar; presentarse para ‹*exam*›; sacar ‹*photo*›; ganar ‹*prize*›. **~ advantage of** aprovechar. **~ after** parecerse a. **~ away** quitar. **~ back** retirar ‹*statement etc*›. **~ in** achicar ‹*garment*›; (*understand*) comprender; (*deceive*) engañar. **~ off** quitarse ‹*clothes*›; (*mimic*) imitar; (*aviat*) despegar. **~ o.s. off** marcharse. **~ on** (*undertake*) emprender; contratar ‹*employee*›. **~ out** (*remove*) sacar. **~ over** tomar posesión de; (*assume control*) tomar el poder. **~ part** participar. **~ place** tener lugar. **~ sides** tomar partido. **~ to** dedicarse a; (*like*) tomar simpatía a ‹*person*›; (*like*) aficionarse a ‹*thing*›. **~ up** dedicarse a ‹*hobby*›; (*occupy*) ocupar; (*resume*) reanudar. **~ up with** trabar amistad con. **be ~n ill** ponerse enfermo. ● *n* presa *f*; (*photo, cinema, TV*) toma *f*
takings /'teɪkɪŋz/ *npl* ingresos *mpl*
take: **~-off** *n* despegue *m*. **~-over** *n* toma *f* de posesión.
talcum /'tælkəm/ *n*. **~ powder** *n* (polvos *mpl* de) talco (*m*)
tale /teɪl/ *n* cuento *m*
talent /'tælənt/ *n* talento *m*. **~ed** *a* talentoso
talisman /'tælɪzmən/ *n* talismán *m*
talk /tɔ:k/ *vt/i* hablar. **~ about** hablar de. **~ over** discutir. ● *n* conversación *f*; (*lecture*) conferencia *f*. **small ~** charla *f*. **~ative** *a* hablador. **~er** *n* hablador *m*; (*chatterbox*) parlanchín *m*. **~ing-to** *n* reprensión *f*
tall /tɔ:l/ *a* (**-er**, **-est**) alto. **~ story** *n* (*fam*) historia *f* inverosímil. **that's a ~ order** *n* (*fam*) eso es pedir mucho
tallboy /'tɔ:lbɔɪ/ *n* cómoda *f* alta
tally /'tælɪ/ *n* tarja *f*; (*total*) total *m*. ● *vi* corresponder (**with** a)
talon /'tælən/ *n* garra *f*
tambourine /tæmbə'ri:n/ *n* pandereta *f*
tame /teɪm/ *a* (**-er**, **-est**) ‹*animal*› doméstico; ‹*person*› dócil; (*dull*) insípido. ● *vt* domesticar; domar

‹*wild animal*›. **~ly** *adv* dócilmente. **~r** /-ə(r)/ *n* domador *m*

tamper /'tæmpə(r)/ *vi*. **~ with** manosear; (*alter*) alterar, falsificar

tampon /'tæmpən/ *n* tampón *m*

tan /tæn/ *vt* (*pt* **tanned**) curtir ‹*hide*›; ‹*sun*› broncear. ● *vi* ponerse moreno. ● *n* bronceado *m*. ● *a* (*colour*) de color canela

tandem /'tændəm/ *n* tándem *m*

tang /tæŋ/ *n* sabor *m* fuerte; (*smell*) olor *m* fuerte

tangent /'tændʒənt/ *n* tangente *f*

tangerine /tændʒə'ri:n/ *n* mandarina *f*

tangibl|e /'tændʒəbl/ *a* tangible. **~y** *adv* perceptiblemente

tangle /'tæŋgl/ *vt* enredar. ● *vi* enredarse. ● *n* enredo *m*

tango /'tæŋgəʊ/ *n* (*pl* **-os**) tango *m*

tank /tæŋk/ *n* depósito *m*; (*mil*) tanque *m*

tankard /'tæŋkəd/ *n* jarra *f*, bock *m*

tanker /'tæŋkə(r)/ *n* petrolero *m*; (*truck*) camión *m* cisterna

tantaliz|e /'tæntəlaɪz/ *vt* atormentar. **~ing** *a* atormentador; (*tempting*) tentador

tantamount /'tæntəmaʊnt/ *a*. **~ to** equivalente a

tantrum /'tæntrəm/ *n* rabieta *f*

tap[1] /tæp/ *n* grifo *m*. **on ~** disponible. ● *vt* explotar ‹*resources*›; interceptar ‹*phone*›

tap[2] /tæp/ *n* (*knock*) golpe *m* ligero. ● *vt* (*pt* **tapped**) golpear ligeramente. **~-dance** *n* zapateado *m*

tape /teɪp/ *n* cinta *f*. ● *vt* atar con cinta; (*record*) grabar. **have sth ~d** (*sl*) comprender perfectamente. **~-measure** *n* cinta *f* métrica

taper /'teɪpə(r)/ *n* bujía *f*. ● *vt* ahusar. ● *vi* ahusarse. **~ off** disminuir

tape: **~ recorder** *n* magnetofón *m*, magnetófono *m*. **~ recording** *n* grabación *f*

tapestry /'tæpɪstrɪ/ *n* tapicería *f*; (*product*) tapiz *m*

tapioca /tæpɪ'əʊkə/ *n* tapioca *f*

tar /tɑ:(r)/ *n* alquitrán *m*. ● *vt* (*pt* **tarred**) alquitranar

tard|ily /'tɑ:dɪlɪ/ *adv* lentamente; (*late*) tardíamente. **~y** /'tɑ:dɪ/ *a* (**-ier**, **-iest**) (*slow*) lento; (*late*) tardío

target /'tɑ:gɪt/ *n* blanco *m*; (*fig*) objetivo *m*

tariff /'tærɪf/ *n* tarifa *f*

tarmac /'tɑ:mæk/ *n* pista *f* de aterrizaje. **T~** *n* (*P*) macadán *m*

tarnish /'tɑ:nɪʃ/ *vt* deslustrar. ● *vi* deslustrarse

tarpaulin /tɑ:'pɔ:lɪn/ *n* alquitranado *m*

tarragon /'tærəgən/ *n* estragón *m*

tart[1] /tɑ:t/ *n* pastel *m*; (*individual*) pastelillo *m*

tart[2] /tɑ:t/ *n* (*sl, woman*) prostituta *f*, fulana *f* (*fam*). ● *vt*. **~ o.s. up** (*fam*) engalanarse

tart[3] /tɑ:t/ *a* (**-er**, **-est**) ácido; (*fig*) áspero

tartan /'tɑ:tn/ *n* tartán *m*, tela *f* escocesa

tartar /'tɑ:tə(r)/ *n* tártaro *m*. **~ sauce** *n* salsa *f* tártara

task /tɑ:sk/ *n* tarea *f*. **take to ~** reprender. **~ force** *n* destacamiento *m* especial

tassel /'tæsl/ *n* borla *f*

tast|e /teɪst/ *n* sabor *m*, gusto *m*; (*small quantity*) poquito *m*. ● *vt* probar. ● *vi*. **~e of** saber a. **~eful** *a* de buen gusto. **~eless** *a* soso; (*fig*) de mal gusto. **~y** *a* (**-ier**, **-iest**) sabroso

tat /tæt/ *see* **tit**[2]

tatter|ed /'tætəd/ *a* hecho jirones. **~s** /'tætəz/ *npl* andrajos *mpl*

tattle /'tætl/ *vi* charlar. ● *n* charla *f*

tattoo[1] /tə'tu:/ (*mil*) espectáculo *m* militar

tattoo[2] /tə'tu:/ *vt* tatuar. ● *n* tatuaje *m*

tatty /'tætɪ/ *a* (**-ier**, **-iest**) gastado, en mal estado

taught /tɔ:t/ *see* **teach**

taunt /tɔ:nt/ *vt* mofarse de. **~ s.o. with sth** echar algo en cara a uno. ● *n* mofa *f*

Taurus /'tɔ:rəs/ *n* (*astr*) Tauro *m*

taut /tɔ:t/ *a* tenso

tavern /'tævən/ *n* taberna *f*

tawdry /'tɔ:drɪ/ *a* (**-ier**, **-iest**) charro

tawny /'tɔ:nɪ/ *a* bronceado

tax /tæks/ *n* impuesto *m*. ● *vt* imponer contribuciones a ‹*person*›; gravar con un impuesto ‹*thing*›; (*fig*) poner a prueba. **~able** *a* imponible. **~ation** /-'seɪʃn/ *n* impuestos *mpl*. **~-collector** *n* recaudador *m* de contribuciones. **~-free** *a* libre de impuestos

taxi /'tæksɪ/ *n* (*pl* **-is**) taxi *m*. ● *vi* (*pt* **taxied**, *pres p* **taxiing**) ‹*aircraft*› rodar por la pista. **~ rank** *n* parada *f* de taxis

taxpayer /'tækspeɪə(r)/ *n* contribuyente *m & f*
te /ti:/ *n* (*mus, seventh note of any musical scale*) si *m*
tea /ti:/ *n* té *m*. **~-bag** *n* bolsita *f* de té. **~-break** *n* descanso *m* para el té
teach /ti:tʃ/ *vt/i* (*pt* **taught**) enseñar. **~er** *n* profesor *m*; (*primary*) maestro *m*. **~-in** *n* seminario *m*. **~ing** *n* enseñanza *f*. ● *a* docente. **~ing staff** *n* profesorado *m*
teacup /'ti:kʌp/ *n* taza *f* de té
teak /ti:k/ *n* teca *f*
tea-leaf /'ti:li:f/ *n* hoja *f* de té
team /ti:m/ *n* equipo *m*; (*of horses*) tiro *m*. ● *vi*. **~ up** unirse. **~-work** *n* trabajo *m* en equipo
teapot /'ti:pɒt/ *n* tetera *f*
tear[1] /teə(r)/ *vt* (*pt* **tore**, *pp* **torn**) rasgar. ● *vi* rasgarse; (*run*) precipitarse. ● *n* rasgón *m*. **~ apart** desgarrar. **~ o.s. away** separarse
tear[2] /tɪə(r)/ *n* lágrima *f*. **in ~s** llorando
tearaway /'teərəweɪ/ *n* gamberro *m*
tear /tɪə(r)/: **~ful** *a* lloroso. **~-gas** *n* gas *m* lacrimógeno
tease /ti:z/ *vt* tomar el pelo a; cardar ‹*cloth etc*›. ● *n* guasón *m*. **~r** /-ə(r)/ *n* (*fam*) problema *m* difícil
tea: **~-set** *n* juego *m* de té. **~spoon** /'ti:spu:n/ *n* cucharilla *f*. **~spoonful** *n* (*pl* **-fuls**) (*amount*) cucharadita *f*
teat /ti:t/ *n* (*of animal*) teta *f*; (*for bottle*) tetilla *f*
tea-towel /'ti:taʊəl/ *n* paño *m* de cocina
technical /'teknɪkl/ *a* técnico. **~ity** *n* /-'kælətɪ/ *n* detalle *m* técnico. **~ly** *adv* técnicamente
technician /tek'nɪʃn/ *n* técnico *m*
technique /tek'ni:k/ *n* técnica *f*
technolog|ist /tek'nɒlədʒɪst/ *n* tecnólogo *m*. **~y** /tek'nɒlədʒɪ/ *n* tecnología *f*
teddy bear /'tedɪbeə(r)/ *n* osito *m* de felpa, osito *m* de peluche
tedious /'ti:dɪəs/ *a* pesado. **~ly** *adv* pesadamente
tedium /'ti:dɪəm/ *n* aburrimiento *m*
tee /ti:/ *n* (*golf*) tee *m*
teem /ti:m/ *vi* abundar; (*rain*) llover a cántaros
teen|age /'ti:neɪdʒ/ *a* adolescente; (*for teenagers*) para jóvenes. **~ager** /-ə(r)/ *n* adolescente *m & f*, joven *m & f*. **~s** /ti:nz/ *npl*. **the ~s** la adolescencia *f*
teeny /'ti:nɪ/ *a* (**-ier**, **-iest**) (*fam*) chiquito
teeter /'ti:tə(r)/ *vi* balancearse
teeth /ti:θ/ *see* **tooth**. **~e** /ti:ð/ *vi* echar los dientes. **~ing troubles** *npl* (*fig*) dificultades *fpl* iniciales
teetotaller /ti:'təʊtələ(r)/ *n* abstemio *m*
telecommunications /telɪkəmju:nɪ'keɪʃnz/ *npl* telecomunicaciones *fpl*
telegram /'telɪgræm/ *n* telegrama *m*
telegraph /'telɪgrɑ:f/ *n* telégrafo *m*. ● *vt* telegrafiar. **~ic** /-'græfɪk/ *a* telegráfico
telepath|ic /telɪ'pæθɪk/ *a* telepático. **~y** /tɪ'lepəθɪ/ *n* telepatía *f*
telephon|e /'telɪfəʊn/ *n* teléfono *m*. ● *vt* llamar por teléfono. **~e booth** *n* cabina *f* telefónica. **~e directory** *n* guía *f* telefónica. **~e exchange** *n* central *f* telefónica. **~ic** /-'fɒnɪk/ *a* telefónico. **~ist** /tɪ'lefənɪst/ *n* telefonista *m & f*
telephoto /telɪ'fəʊtəʊ/ *a*. **~ lens** *n* teleobjetivo *m*
teleprinter /'telɪprɪntə(r)/ *n* teleimpresor *m*
telescop|e /'telɪskəʊp/ *n* telescopio *m*. **~ic** /-'kɒpɪk/ *a* telescópico
televis|e /'telɪvaɪz/ *vt* televisar. **~ion** /'telɪvɪʒn/ *n* televisión *f*. **~ion set** *n* televisor *m*
telex /'teleks/ *n* télex *m*. ● *vt* enviar por télex
tell /tel/ *vt* (*pt* **told**) decir; contar ‹*story*›; (*distinguish*) distinguir. ● *vi* (*produce an effect*) tener efecto; (*know*) saber. **~ off** *vt* reprender. **~er** /'telə(r)/ *n* (*in bank*) cajero *m*
telling /'telɪŋ/ *a* eficaz
tell-tale /'telteɪl/ *n* soplón *m*. ● *a* revelador
telly /'telɪ/ *n* (*fam*) televisión *f*, tele *f* (*fam*)
temerity /tɪ'merətɪ/ *n* temeridad *f*
temp /temp/ *n* (*fam*) empleado *m* temporal
temper /'tempə(r)/ *n* (*disposition*) disposición *f*; (*mood*) humor *m*; (*fit of anger*) cólera *f*; (*of metal*) temple *m*. **be in a ~** estar de mal humor. **keep one's ~** contenerse. **lose one's ~** enfadarse, perder la paciencia. ● *vt* templar ‹*metal*›
temperament /'temprəmənt/ *n* temperamento *m*. **~al** /-'mentl/ *a* caprichoso

temperance /'tempərəns/ *n* moderación *f*
temperate /'tempərət/ *a* moderado; ‹*climate*› templado
temperature /'temprɪtʃə(r)/ *n* temperatura *f*. **have a ~** tener fiebre
tempest /'tempɪst/ *n* tempestad *f*. **~uous** /-'pestjʊəs/ *a* tempestuoso
temple[1] /'templ/ *n* templo *m*
temple[2] /'templ/ (*anat*) sien *f*
tempo /'tempəʊ/ *n* (*pl* **-os** *or* **tempi**) ritmo *m*
temporar|ily /'tempərərəlɪ/ *adv* temporalmente. **~y** /'tempərərɪ/ *a* temporal, provisional
tempt /tempt/ *vt* tentar. **~ s.o. to** inducir a uno a. **~ation** /-'teɪʃn/ *n* tentación *f*. **~ing** *a* tentador
ten /ten/ *a & n* diez (*m*)
tenable /'tenəbl/ *a* sostenible
tenaci|ous /tɪ'neɪʃəs/ *a* tenaz. **~ty** /-æsətɪ/ *n* tenacidad *f*
tenan|cy /'tenənsɪ/ *n* alquiler *m*. **~t** /'tenənt/ *n* inquilino *m*
tend[1] /tend/ *vi*. **~ to** tener tendencia a
tend[2] /tend/ *vt* cuidar
tendency /'tendənsɪ/ *n* tendencia *f*
tender[1] /'tendə(r)/ *a* tierno; (*painful*) dolorido
tender[2] /'tendə(r)/ *n* (*com*) oferta *f*. **legal ~** *n* curso *m* legal. • *vt* ofrecer, presentar
tender: **~ly** *adv* tiernamente. **~ness** *n* ternura *f*
tendon /'tendən/ *n* tendón *m*
tenement /'tenəmənt/ *n* vivienda *f*
tenet /'tenɪt/ *n* principio *m*
tenfold /'tenfəʊld/ *a* diez veces mayor, décuplo. • *adv* diez veces
tenner /'tenə(r)/ *n* (*fam*) billete *m* de diez libras
tennis /'tenɪs/ *n* tenis *m*
tenor /'tenə(r)/ *n* tenor *m*
tens|e /tens/ *a* (**-er**, **-est**) tieso; (*fig*) tenso. • *n* (*gram*) tiempo *m*. • *vi*. **~ up** tensarse. **~eness** *n*, **~ion** /'tenʃn/ *n* tensión *f*
tent /tent/ *n* tienda *f*, carpa *f* (*LAm*)
tentacle /'tentəkl/ *n* tentáculo *m*
tentative /'tentətɪv/ *a* provisional; (*hesitant*) indeciso. **~ly** *adv* provisionalmente; (*timidly*) tímidamente
tenterhooks /'tentəhʊks/ *npl*. **on ~** en ascuas
tenth /tenθ/ *a & n* décimo (*m*)
tenuous /'tenjʊəs/ *a* tenue
tenure /'tenjʊə(r)/ *n* posesión *f*
tepid /'tepɪd/ *a* tibio
term /tɜːm/ *n* (*of time*) período *m*; (*schol*) trimestre *m*; (*word etc*) término *m*. • *vt* llamar. **~s** *npl* condiciones *fpl*; (*com*) precio *m*. **on bad ~s** en malas relaciones. **on good ~s** en buenas relaciones
terminal /'tɜːmɪnl/ *a* terminal, final. • *n* (*rail*) estación *f* terminal; (*elec*) borne *m*. **(air) ~** *n* término *m*, terminal *m*
terminat|e /'tɜːmɪneɪt/ *vt* terminar. • *vi* terminarse. **~tion** /-'neɪʃn/ *n* terminación *f*
terminology /tɜːmɪ'nɒlədʒɪ/ *n* terminología *f*
terrace /'terəs/ *n* terraza *f*; (*houses*) hilera *f* de casas. **the ~s** *npl* (*sport*) las gradas *fpl*
terrain /tə'reɪn/ *n* terreno *m*
terrestrial /tɪ'restrɪəl/ *a* terrestre
terribl|e /'terəbl/ *a* terrible. **~y** *adv* terriblemente
terrier /'terɪə(r)/ *n* terrier *m*
terrific /tə'rɪfɪk/ *a* (*excellent*, *fam*) estupendo; (*huge*, *fam*) enorme. **~ally** *adv* (*fam*) terriblemente; (*very well*) muy bien
terrify /'terɪfaɪ/ *vt* aterrorizar. **~ing** *a* espantoso
territor|ial /terɪ'tɔːrɪəl/ *a* territorial. **~y** /'terɪtrɪ/ *n* territorio *m*
terror /'terə(r)/ *n* terror *m*. **~ism** /-zəm/ *n* terrorismo *m*. **~ist** /'terərɪst/ *n* terrorista *m & f*. **~ize** /'terəraɪz/ *vt* aterrorizar
terse /tɜːs/ *a* conciso; (*abrupt*) brusco
test /test/ *n* prueba *f*; (*exam*) examen *m*. • *vt* probar; (*examine*) examinar
testament /'testəmənt/ *n* testamento *m*. **New T~** Nuevo Testamento. **Old T~** Antiguo Testamento
testicle /'testɪkl/ *n* testículo *m*
testify /'testɪfaɪ/ *vt* atestiguar. • *vi* declarar
testimon|ial /testɪ'məʊnɪəl/ *n* certificado *m*; (*of character*) recomendación *f*. **~y** /'testɪmənɪ/ *n* testimonio *m*
test: **~ match** *n* partido *m* internacional. **~-tube** *n* tubo *m* de ensayo, probeta *f*
testy /'testɪ/ *a* irritable
tetanus /'tetənəs/ *n* tétanos *m invar*
tetchy /'tetʃɪ/ *a* irritable

tether /ˈteðə(r)/ *vt* atar. ● *n*. **be at the end of one's ~** no poder más

text /tekst/ *n* texto *m*. **~book** *n* libro *m* de texto

textile /ˈtekstaɪl/ *a & n* textil (*m*)

texture /ˈtekstʃə(r)/ *n* textura *f*

Thai /taɪ/ *a & n* tailandés (*m*). **~land** *n* Tailandia *f*

Thames /temz/ *n* Támesis *m*

than /ðæn, ðən/ *conj* que; (*with numbers*) de

thank /θæŋk/ *vt* dar las gracias a, agradecer. **~ you** gracias. **~ful** /ˈθæŋkfl/ *a* agradecido. **~fully** *adv* con gratitud; (*happily*) afortunadamente. **~less** /ˈθæŋklɪs/ *a* ingrato. **~s** *npl* gracias *fpl*. **~s!** (*fam*) ¡gracias! **~s to** gracias a

that /ðæt, ðət/ *a* (*pl* **those**) ese, aquel, esa, aquella. ● *pron* (*pl* **those**) ése, aquél, ésa, aquélla. **~ is** es decir. **~'s it!** ¡eso es! **~ is why** por eso. **is ~ you?** ¿eres tú? **like ~** así. ● *adv* tan. ● *rel pron* que; (*with prep*) el que, la que, el cual, la cual. ● *conj* que

thatch /θætʃ/ *n* techo *m* de paja. **~ed** *a* con techo de paja

thaw /θɔː/ *vt* deshelar. ● *vi* deshelarse; ‹*snow*› derretirse. ● *n* deshielo *m*

the /ðə, ðiː/ *def art* el, la, los, las. **at ~** al, a la, a los, a las. **from ~** del, de la, de los, de las. **to ~** al, a la, a los, a las. ● *adv*. **all ~ better** tanto mejor

theatr|e /ˈθɪətə(r)/ *n* teatro *m*. **~ical** /-ˈætrɪkl/ *a* teatral

theft /θeft/ *n* hurto *m*

their /ðeə(r)/ *a* su, sus

theirs /ðeəz/ *poss pron* (el) suyo, (la) suya, (los) suyos, (las) suyas

them /ðem, ðəm/ *pron* (*accusative*) los, las; (*dative*) les; (*after prep*) ellos, ellas

theme /θiːm/ *n* tema *m*. **~ song** *n* motivo *m* principal

themselves /ðəmˈselvz/ *pron* ellos mismos, ellas mismas; (*reflexive*) se; (*after prep*) sí mismos, sí mismas

then /ðen/ *adv* entonces; (*next*) luego, después. **by ~** para entonces. **now and ~** de vez en cuando. **since ~** desde entonces. ● *a* de entonces

theolog|ian /θɪəˈləʊdʒən/ *n* teólogo *m*. **~y** /θɪˈɒlədʒɪ/ *n* teología *f*

theorem /ˈθɪərəm/ *n* teorema *m*

theor|etical /θɪəˈretɪkl/ *a* teórico. **~y** /ˈθɪərɪ/ *n* teoría *f*

therap|eutic /θerəˈpjuːtɪk/ *a* terapéutico. **~ist** *n* terapeuta *m & f*. **~y** /ˈθerəpɪ/ *n* terapia *f*

there /ðeə(r)/ *adv* ahí, allí. **~ are** hay. **~ he is** ahí está. **~ is** hay. **~ it is** ahí está. **down ~** ahí abajo. **up ~** ahí arriba. ● *int* ¡vaya! **~, ~!** ¡ya, ya! **~abouts** *adv* por ahí. **~after** *adv* después. **~by** *adv* por eso. **~fore** /ˈðeəfɔː(r)/ *adv* por lo tanto.

thermal /ˈθɜːml/ *a* termal

thermometer /θəˈmɒmɪtə(r)/ *n* termómetro *m*

thermonuclear /θɜːməʊˈnjuːklɪə(r)/ *a* termonuclear

Thermos /ˈθɜːməs/ *n* (*P*) termo *m*

thermostat /ˈθɜːməstæt/ *n* termostato *m*

thesaurus /θɪˈsɔːrəs/ *n* (*pl* **-ri** /-raɪ/) diccionario *m* de sinónimos

these /ðiːz/ *a* estos, estas. ● *pron* éstos, éstas

thesis /ˈθiːsɪs/ *n* (*pl* **theses** /-siːz/) tesis *f*

they /ðeɪ/ *pron* ellos, ellas. **~ say that** se dice que

thick /θɪk/ *a* (**-er**, **-est**) espeso; (*dense*) denso; (*stupid, fam*) torpe; (*close, fam*) íntimo. ● *adv* espesamente, densamente. ● *n*. **in the ~ of** en medio de. **~en** /ˈθɪkən/ *vt* espesar. ● *vi* espesarse

thicket /ˈθɪkɪt/ *n* matorral *m*

thick: **~ly** *adv* espesamente, densamente. **~ness** *n* espesor *m*

thickset /θɪkˈset/ *a* fornido

thick-skinned /θɪkˈskɪnd/ *a* insensible

thief /θiːf/ *n* (*pl* **thieves**) ladrón *m*

thiev|e /θiːv/ *vt/i* robar. **~ing** *a* ladrón

thigh /θaɪ/ *n* muslo *m*

thimble /ˈθɪmbl/ *n* dedal *m*

thin /θɪn/ *a* (**thinner, thinnest**) delgado; ‹*person*› flaco; (*weak*) débil; (*fine*) fino; (*sparse*) escaso. ● *adv* ligeramente. ● *vt* (*pt* **thinned**) adelgazar; (*dilute*) diluir. **~ out** hacer menos denso. ● *vi* adelgazarse; (*diminish*) disminuir

thing /θɪŋ/ *n* cosa *f*. **for one ~** en primer lugar. **just the ~** exactamente lo que se necesita. **poor ~!** ¡pobrecito! **~s** *npl* (*belongings*) efectos *mpl*; (*clothing*) ropa *f*

think /θɪŋk/ *vt* (*pt* **thought**) pensar, creer. ● *vi* pensar (**about**, **of** en); (*carefully*) reflexionar; (*imagine*) imaginarse. **~ better of it** cambiar de idea. **I ~ so** creo que sí. **~ over** *vt* pensar bien. **~ up** *vt* idear, inventar. **~er** *n* pensador *m*. **~-tank** *n* grupo *m* de expertos

thin: **~ly** *adv* ligeramente. **~ness** *n* delgadez *f*; (*of person*) flaqueza *f*

third /θɜ:d/ *a* tercero. ● *n* tercio *m*, tercera parte *f*. **~-rate** *a* muy inferior. **T~ World** *n* Tercer Mundo *m*

thirst /θɜ:st/ *n* sed *f*. **~y** *a* sediento. **be ~y** tener sed

thirteen /θɜ:'ti:n/ *a* & *n* trece (*m*). **~th** *a* & *n* decimotercero (*m*)

thirt|ieth /'θɜ:tɪəθ/ *a* & *n* trigésimo (*m*). **~y** /'θɜ:tɪ/ *a* & *n* treinta (*m*)

this /ðɪs/ *a* (*pl* **these**) este, esta. **~ one** éste, ésta. ● *pron* (*pl* **these**) éste, ésta, esto. **like ~** así

thistle /'θɪsl/ *n* cardo *m*

thong /θɒŋ/ *n* correa *f*

thorn /θɔ:n/ *n* espina *f*. **~y** *a* espinoso

thorough /'θʌrə/ *a* completo; (*deep*) profundo; ‹*cleaning etc*› a fondo; ‹*person*› concienzudo

thoroughbred /'θʌrəbred/ *a* de pura sangre

thoroughfare /'θʌrəfeə(r)/ *n* calle *f*. **no ~** prohibido el paso

thoroughly /'θʌrəlɪ/ *adv* completamente

those /ðəʊz/ *a* esos, aquellos, esas, aquellas. ● *pron* ésos, aquéllos, ésas, aquéllas

though /ðəʊ/ *conj* aunque. ● *adv* sin embargo. **as ~** como si

thought /θɔ:t/ *see* **think**. ● *n* pensamiento *m*; (*idea*) idea *f*. **~ful** /'θɔ:tfl/ *a* pensativo; (*considerate*) atento. **~fully** *adv* pensativamente; (*considerately*) atentamente. **~less** /'θɔ:tlɪs/ *a* irreflexivo; (*inconsiderate*) desconsiderado

thousand /'θaʊznd/ *a* & *n* mil (*m*). **~th** *a* & *n* milésimo (*m*)

thrash /θræʃ/ *vt* azotar; (*defeat*) derrotar. **~ out** discutir a fondo

thread /θred/ *n* hilo *m*; (*of screw*) rosca *f*. ● *vt* ensartar. **~ one's way** abrirse paso. **~bare** /'θredbeə(r)/ *a* raído

threat /θret/ *n* amenaza *f*. **~en** /'θretn/ *vt/i* amenazar. **~ening** *a* amenazador. **~eningly** *adv* de modo amenazador

three /θri:/ *a* & *n* tres (*m*). **~fold** *a* triple. ● *adv* tres veces. **~some** /'θri:səm/ *n* conjunto *m* de tres personas

thresh /θreʃ/ *vt* trillar

threshold /'θreʃhəʊld/ *n* umbral *m*

threw /θru:/ *see* **throw**

thrift /θrɪft/ *n* economía *f*, ahorro *m*. **~y** *a* frugal

thrill /θrɪl/ *n* emoción *f*. ● *vt* emocionar. ● *vi* emocionarse; (*quiver*) estremecerse. **be ~ed with** estar encantado de. **~er** /'θrɪlə(r)/ *n* (*book*) libro *m* de suspense; (*film*) película *f* de suspense. **~ing** *a* emocionante

thriv|e /θraɪv/ *vi* prosperar. **~ing** *a* próspero

throat /θrəʊt/ *n* garganta *f*. **have a sore ~** dolerle la garganta

throb /θrɒb/ *vi* (*pt* **throbbed**) palpitar; (*with pain*) dar punzadas; (*fig*) vibrar. ● *n* palpitación *f*; (*pain*) punzada *f*; (*fig*) vibración *f*. **~bing** *a* ‹*pain*› punzante

throes /θrəʊz/ *npl*. **in the ~ of** en medio de

thrombosis /θrɒm'bəʊsɪs/ *n* trombosis *f*

throne /θrəʊn/ *n* trono *m*

throng /θrɒŋ/ *n* multitud *f*

throttle /'θrɒtl/ *n* (*auto*) acelerador *m*. ● *vt* ahogar

through /θru:/ *prep* por, a través de; (*during*) durante; (*by means of*) por medio de; (*thanks to*) gracias a. ● *adv* de parte a parte, de un lado a otro; (*entirely*) completamente; (*to the end*) hasta el final. **be ~** (*finished*) haber terminado. ● *a* ‹*train etc*› directo

throughout /θru:'aʊt/ *prep* por todo; (*time*) en todo. ● *adv* en todas partes; (*all the time*) todo el tiempo

throve /θrəʊv/ *see* **thrive**

throw /θrəʊ/ *vt* (*pt* **threw**, *pp* **thrown**) arrojar; (*baffle etc*) desconcertar. **~ a party** (*fam*) dar una fiesta. ● *n* tiro *m*; (*of dice*) lance *m*. **~ away** *vt* tirar. **~ over** *vt* abandonar. **~ up** *vi* (*vomit*) vomitar. **~-away** *a* desechable

thrush /θrʌʃ/ *n* tordo *m*

thrust /θrʌst/ *vt* (*pt* **thrust**) empujar; (*push in*) meter. ● *n* empuje *m*. **~ (up)on** imponer a

thud /θʌd/ *n* ruido *m* sordo

thug /θʌg/ *n* bruto *m*

thumb /θʌm/ *n* pulgar *m*. **under the ~ of** dominado por. ● *vt* hojear ‹*book*›. **~ a lift** hacer autostop. **~-index** *n* uñeros *mpl*

thump /θʌmp/ *vt* golpear. ● *vi* ‹*heart*› latir fuertemente. ● *n* porrazo *m*; (*noise*) ruido *m* sordo

thunder /'θʌndə(r)/ *n* trueno *m*. ● *vi* tronar. ~ **past** pasar con estruendo. ~**bolt** /'θʌndəbəʊlt/ *n* rayo *m*. ~**clap** /'θʌndəklæp/ *n* trueno *m*. ~**storm** /'θʌndəstɔ:m/ *n* tronada *f*. ~**y** *a* con truenos

Thursday /'θɜ:zdeɪ/ *n* jueves *m*

thus /ðʌs/ *adv* así

thwart /θwɔ:t/ *vt* frustrar

thyme /taɪm/ *n* tomillo *m*

thyroid /'θaɪrɔɪd/ *n* tiroides *m invar*

tiara /tɪ'ɑ:rə/ *n* diadema *f*

tic /tɪk/ *n* tic *m*

tick[1] /tɪk/ *n* tictac *m*; (*mark*) señal *f*, marca *f*; (*instant*, *fam*) momentito *m*. ● *vi* hacer tictac. ● *vt*. ~ **(off)** marcar. ~ **off** *vt* (*sl*) reprender. ~ **over** *vi* ‹*of engine*› marchar en vacío

tick[2] /tɪk/ *n* (*insect*) garrapata *f*

tick[3] /tɪk/ *n*. **on** ~ (*fam*) a crédito

ticket /'tɪkɪt/ *n* billete *m*, boleto *m* (*LAm*); (*label*) etiqueta *f*; (*fine*) multa *f*. ~**-collector** *n* revisor *m*. ~**-office** *n* taquilla *f*

tickl|e /'tɪkl/ *vt* hacer cosquillas a; (*amuse*) divertir. ● *n* cosquilleo *m*. ~**ish** /'tɪklɪʃ/ *a* cosquilloso; ‹*problem*› delicado. **be** ~**ish** tener cosquillas

tidal /'taɪdl/ *a* de marea. ~ **wave** *n* maremoto *m*

tiddly-winks /'tɪdlɪwɪŋks/ *n* juego *m* de pulgas

tide /taɪd/ *n* marea *f*; (*of events*) curso *m*. ● *vt*. ~ **over** ayudar a salir de un apuro

tidings /'taɪdɪŋz/ *npl* noticias *fpl*

tid|ily /'taɪdɪlɪ/ *adv* en orden; (*well*) bien. ~**iness** *n* orden *m*. ~**y** /'taɪdɪ/ *a* (**-ier**, **-iest**) ordenado; ‹*amount*, *fam*› considerable. ● *vt/i*. ~**y (up)** ordenar. ~**y o.s. up** arreglarse

tie /taɪ/ *vt* (*pres p* **tying**) atar; hacer ‹*a knot*›; (*link*) vincular. ● *vi* (*sport*) empatar. ● *n* atadura *f*; (*necktie*) corbata *f*; (*link*) lazo *m*; (*sport*) empate *m*. ~ **in with** relacionar con. ~ **up** atar; (*com*) inmovilizar. **be** ~**d up** (*busy*) estar ocupado

tier /tɪə(r)/ *n* fila *f*; (*in stadium etc*) grada *f*; (*of cake*) piso *m*

tie-up /'taɪʌp/ *n* enlace *m*

tiff /tɪf/ *n* riña *f*

tiger /'taɪgə(r)/ *n* tigre *m*

tight /taɪt/ *a* (**-er**, **-est**) ‹*clothes*› ceñido; (*taut*) tieso; ‹*control etc*› riguroso; ‹*knot*, *nut*› apretado; (*drunk*, *fam*) borracho. ● *adv* bien; (*shut*) herméticamente. ~ **corner** *n* (*fig*) apuro *m*. ~**en** /'taɪtn/ *vt* apretar. ● *vi* apretarse. ~**-fisted** *a* tacaño. ~**ly** *adv* bien; (*shut*) herméticamente. ~**ness** *n* estrechez *f*. ~**rope** /'taɪtrəʊp/ *n* cuerda *f* floja. ~**s** /taɪts/ *npl* leotardos *mpl*

tile /taɪl/ *n* (*decorative*) azulejo *m*; (*on roof*) teja *f*; (*on floor*) baldosa *f*. ● *vt* azulejar; tejar ‹*roof*›; embaldosar ‹*floor*›

till[1] /tɪl/ *prep* hasta. ● *conj* hasta que

till[2] /tɪl/ *n* caja *f*

till[3] /tɪl/ *vt* cultivar

tilt /tɪlt/ *vt* inclinar. ● *vi* inclinarse. ● *n* inclinación *f*. **at full** ~ a toda velocidad

timber /'tɪmbə(r)/ *n* madera *f* (de construcción); (*trees*) árboles *mpl*

time /taɪm/ *n* tiempo *m*; (*moment*) momento *m*; (*occasion*) ocasión *f*; (*by clock*) hora *f*; (*epoch*) época *f*; (*rhythm*) compás *m*. ~ **off** tiempo libre. **at** ~**s** a veces. **behind the** ~**s** anticuado. **behind** ~ atrasado. **for the** ~ **being** por ahora. **from** ~ **to** ~ de vez en cuando. **have a good** ~ divertirse, pasarlo bien. **in a year's** ~ dentro de un año. **in no** ~ en un abrir y cerrar de ojos. **in** ~ a tiempo; (*eventually*) con el tiempo. **on** ~ a la hora, puntual. ● *vt* elegir el momento; cronometrar ‹*race*›. ~ **bomb** *n* bomba *f* de tiempo. ~**-honoured** *a* consagrado. ~**-lag** *n* intervalo *m*

timeless /'taɪmlɪs/ *a* eterno

timely /'taɪmlɪ/ *a* oportuno

timer /'taɪmə(r)/ *n* cronómetro *m*; (*culin*) avisador *m*; (*with sand*) reloj *m* de arena; (*elec*) interruptor *m* de reloj

timetable /'taɪmteɪbl/ *n* horario *m*

time zone /'taɪmzəʊn/ *n* huso *m* horario

timid /'tɪmɪd/ *a* tímido; (*fearful*) miedoso. ~**ly** *adv* tímidamente

timing /'taɪmɪŋ/ *n* medida *f* del tiempo; (*moment*) momento *m*; (*sport*) cronometraje *m*

timorous /'tɪmərəs/ *a* tímido; (*fearful*) miedoso. ~**ly** *adv* tímidamente

tin /tɪn/ *n* estaño *m*; (*container*) lata *f*. ~ **foil** *n* papel *m* de estaño. ● *vt* (*pt* **tinned**) conservar en lata, enlatar

tinge /tɪndʒ/ *vt* teñir (**with** de); (*fig*) matizar (**with** de). ● *n* matiz *m*

tingle /'tɪŋgl/ *vi* sentir hormigueo; (*with excitement*) estremecerse

tinker /ˈtɪŋkə(r)/ *n* hojalatero *m*. ● *vi*. ~ **(with)** jugar con; (*repair*) arreglar
tinkle /ˈtɪŋkl/ *n* retintín *m*; (*phone call, fam*) llamada *f*
tin: ~**ned** *a* en lata. ~**ny** *a* metálico. ~**-opener** *n* abrelatas *m invar*. ~ **plate** *n* hojalata *f*
tinpot /ˈtɪnpɒt/ *a* (*pej*) inferior
tinsel /ˈtɪnsl/ *n* oropel *m*
tint /tɪnt/ *n* matiz *m*
tiny /ˈtaɪnɪ/ *a* (**-ier, -iest**) diminuto
tip[1] /tɪp/ *n* punta *f*
tip[2] /tɪp/ *vt* (*pt* **tipped**) (*tilt*) inclinar; (*overturn*) volcar; (*pour*) verter● *vi* inclinarse; (*overturn*) volcarse. ● *n* (*for rubbish*) vertedero *m*. ~ **out** verter
tip[3] /tɪp/ *vt* (*reward*) dar una propina a. ~ **off** advertir. ● *n* (*reward*) propina *f*; (*advice*) consejo *m*
tip-off /ˈtɪpɒf/ *n* advertencia *f*
tipped /ˈtɪpt/ *a* ‹*cigarette*› con filtro
tipple /ˈtɪpl/ *vi* beborrotear. ● *n* bebida *f* alcohólica. **have a** ~ tomar una copa
tipsy /ˈtɪpsɪ/ *a* achispado
tiptoe /ˈtɪptəʊ/ *n*. **on** ~ de puntillas
tiptop /ˈtɪptɒp/ *a* (*fam*) de primera
tirade /taɪˈreɪd/ *n* diatriba *f*
tire /ˈtaɪə(r)/ *vt* cansar. ● *vi* cansarse. ~**d** /ˈtaɪəd/ *a* cansado. ~**d of** harto de. ~**d out** agotado. ~**less** *a* incansable
tiresome /ˈtaɪəsəm/ *a* (*annoying*) fastidioso; (*boring*) pesado
tiring /ˈtaɪərɪŋ/ *a* cansado
tissue /ˈtɪʃuː/ *n* tisú *m*; (*handkerchief*) pañuelo *m* de papel. ~**-paper** *n* papel *m* de seda
tit[1] /tɪt/ *n* (*bird*) paro *m*
tit[2] /tɪt/ *n*. ~ **for tat** golpe por golpe
titbit /ˈtɪtbɪt/ *n* golosina *f*
titillate /ˈtɪtɪleɪt/ *vt* excitar
title /ˈtaɪtl/ *n* título *m*. ~**d** *a* con título nobiliario. ~**-deed** *n* título *m* de propiedad. ~**-role** *n* papel *m* principal
tittle-tattle /ˈtɪtltætl/ *n* cháchara *f*
titular /ˈtɪtjʊlə(r)/ *a* nominal
tizzy /ˈtɪzɪ/ *n* (*sl*). **get in a** ~ ponerse nervioso
to /tuː, tə/ *prep* a; (*towards*) hacia; (*in order to*) para; (*according to*) según; (*as far as*) hasta; (*with times*) menos; (*of*) de. **give it** ~ **me** dámelo. **I don't want to** no quiero. **twenty** ~ **seven** (*by clock*) las siete menos veinte. ● *adv*. **push** ~, **pull** ~ cerrar. ~ **and fro** *adv* de aquí para allá
toad /təʊd/ *n* sapo *m*
toadstool /ˈtəʊdstuːl/ *n* seta *f* venenosa
toast /təʊst/ *n* pan *m* tostado, tostada *f*; (*drink*) brindis *m*. **drink a** ~ **to** brindar por. ● *vt* brindar por. ~**er** *n* tostador *m* de pan
tobacco /təˈbækəʊ/ *n* tabaco *m*. ~**nist** *n* estanquero *m*. ~**nist's shop** *n* estanco *m*
to-be /təˈbiː/ *a* futuro
toboggan /təˈbɒgən/ *n* tobogán *m*
today /təˈdeɪ/ *n* & *adv* hoy (*m*). ~ **week** dentro de una semana
toddler /ˈtɒdlə(r)/ *n* niño *m* que empieza a andar
toddy /ˈtɒdɪ/ *n* ponche *m*
to-do /təˈduː/ *n* lío *m*
toe /təʊ/ *n* dedo *m* del pie; (*of shoe*) punta *f*. **big** ~ dedo *m* gordo (del pie). **on one's** ~**s** (*fig*) alerta. ● *vt*. ~ **the line** conformarse. ~**-hold** *n* punto *m* de apoyo
toff /tɒf/ *n* (*sl*) petimetre *m*
toffee /ˈtɒfɪ/ *n* caramelo *m*
together /təˈgeðə(r)/ *adv* junto, juntos; (*at same time*) a la vez. ~ **with** junto con. ~**ness** *n* compañerismo *m*
toil /tɔɪl/ *vi* afanarse. ● *n* trabajo *m*
toilet /ˈtɔɪlɪt/ *n* servicio *m*, retrete *m*; (*grooming*) arreglo *m*, tocado *m*. ~**-paper** *n* papel *m* higiénico. ~**ries** /ˈtɔɪlɪtrɪz/ *npl* artículos *mpl* de tocador. ~ **water** *n* agua *f* de Colonia
token /ˈtəʊkən/ *n* señal *f*; (*voucher*) vale *m*; (*coin*) ficha *f*. ● *a* simbólico
told /təʊld/ *see* **tell**. ● *a*. **all** ~ con todo
tolerabl|e /ˈtɒlərəbl/ *a* tolerable; (*not bad*) regular. ~**y** *adv* pasablemente
toleran|ce /ˈtɒlərəns/ *n* tolerancia *f*. ~**t** /ˈtɒlərənt/ *a* tolerante. ~**tly** *adv* con tolerancia
tolerate /ˈtɒləreɪt/ *vt* tolerar
toll[1] /təʊl/ *n* peaje *m*. **death** ~ número *m* de muertos. **take a heavy** ~ dejar muchas víctimas
toll[2] /təʊl/ *vi* doblar, tocar a muerto
tom /tɒm/ *n* gato *m* (macho)
tomato /təˈmɑːtəʊ/ *n* (*pl* ~**oes**) tomate *m*
tomb /tuːm/ *n* tumba *f*, sepulcro *m*
tomboy /ˈtɒmbɔɪ/ *n* marimacho *m*
tombstone /ˈtuːmstəʊn/ *n* lápida *f* sepulcral

tom-cat /'tɒmkæt/ *n* gato *m* (macho)

tome /təʊm/ *n* librote *m*

tomfoolery /tɒm'fu:lərɪ/ *n* payasadas *fpl*, tonterías *fpl*

tomorrow /tə'mɒrəʊ/ *n & adv* mañana (*f*). **see you ~!** ¡hasta mañana!

ton /tʌn/ *n* tonelada *f* (= *1,016 kg*). **~s of** (*fam*) montones de. **metric ~** tonelada *f* (métrica) (= *1,000 kg*)

tone /təʊn/ *n* tono *m*. ● *vt*. **~ down** atenuar. **~ up** tonificar ‹*muscles*›. ● *vi*. **~ in** armonizar. **~-deaf** *a* que no tiene buen oído

tongs /tɒŋz/ *npl* tenazas *fpl*; (*for hair, sugar*) tenacillas *fpl*

tongue /tʌŋ/ *n* lengua *f*. **~ in cheek** *adv* irónicamente. **~-tied** *a* mudo. **get ~-tied** trabársele la lengua. **~-twister** *n* trabalenguas *m invar*

tonic /'tɒnɪk/ *a* tónico. ● *n* (*tonic water*) tónica *f*; (*med, fig*) tónico *m*. **~ water** *n* tónica *f*

tonight /tə'naɪt/ *adv & n* esta noche (*f*); (*evening*) esta tarde (*f*)

tonne /tʌn/ *n* tonelada *f* (métrica)

tonsil /'tɒnsl/ *n* amígdala *f*. **~litis** /-'laɪtɪs/ *n* amigdalitis *f*

too /tu:/ *adv* demasiado; (*also*) también. **~ many** *a* demasiados. **~ much** *a & adv* demasiado

took /tʊk/ *see* **take**

tool /tu:l/ *n* herramienta *f*. **~-bag** *n* bolsa *f* de herramientas

toot /tu:t/ *n* bocinazo *m*. ● *vi* tocar la bocina

tooth /tu:θ/ *n* (*pl* **teeth**) diente *m*; (*molar*) muela *f*. **~ache** /'tu:θeɪk/ *n* dolor *m* de muelas. **~brush** /'tu:θbrʌʃ/ *n* cepillo *m* de dientes. **~comb** /'tu:θkəʊm/ *n* peine *m* de púa fina. **~less** *a* desdentado, sin dientes. **~paste** /'tu:θpeɪst/ *n* pasta *f* dentífrica. **~pick** /'tu:θpɪk/ *n* palillo *m* de dientes

top[1] /tɒp/ *n* cima *f*; (*upper part*) parte *f* de arriba; (*upper surface*) superficie *f*; (*lid, of bottle*) tapa *f*; (*of list*) cabeza *f*. **from ~ to bottom** de arriba abajo. **on ~ (of)** encima de; (*besides*) además. ● *a* más alto; (*in rank*) superior, principal; (*maximum*) máximo. **~ floor** *n* último piso *m*. ● *vt* (*pt* **topped**) cubrir; (*exceed*) exceder. **~ up** *vt* llenar

top[2] /tɒp/ *n* (*toy*) trompa *f*, peonza *f*

top: **~ hat** *n* chistera *f*. **~-heavy** *a* más pesado arriba que abajo

topic /'tɒpɪk/ *n* tema *m*. **~al** /'tɒpɪkl/ *a* de actualidad

top: **~less** /'tɒplɪs/ *a* ‹*bather*› con los senos desnudos. **~most** /'tɒpməʊst/ *a* (el) más alto. **~-notch** *a* (*fam*) excelente

topography /tə'pɒgrəfɪ/ *n* topografía *f*

topple /'tɒpl/ *vi* derribar; (*overturn*) volcar

top secret /tɒp'si:krɪt/ *a* sumamente secreto

topsy-turvy /tɒpsɪ'tɜ:vɪ/ *adv & a* patas arriba

torch /tɔ:tʃ/ *n* lámpara *f* de bolsillo; (*flaming*) antorcha *f*

tore /tɔ:(r)/ *see* **tear**[1]

toreador /'tɒrɪədɔ:(r)/ *n* torero *m*

torment /'tɔ:ment/ *n* tormento *m*. /tɔ:'ment/ *vt* atormentar

torn /tɔ:n/ *see* **tear**[1]

tornado /tɔ:'neɪdəʊ/ *n* (*pl* **-oes**) tornado *m*

torpedo /tɔ:'pi:dəʊ/ *n* (*pl* **-oes**) torpedo *m*. ● *vt* torpedear

torpor /'tɔ:pə(r)/ *n* apatía *f*

torrent /'tɒrənt/ *n* torrente *m*. **~ial** /tə'renʃl/ *a* torrencial

torrid /'tɒrɪd/ *a* tórrido

torso /'tɔ:səʊ/ *n* (*pl* **-os**) torso *m*

tortoise /'tɔ:təs/ *n* tortuga *f*. **~shell** *n* carey *m*

tortuous /'tɔ:tjʊəs/ *a* tortuoso

torture /'tɔ:tʃə(r)/ *n* tortura *f*, tormento *m*. ● *vt* atormentar. **~r** /-ə(r)/ *n* atormentador *m*, verdugo *m*

Tory /'tɔ:rɪ/ *a & n* (*fam*) conservador (*m*)

toss /tɒs/ *vt* echar; (*shake*) sacudir. ● *vi* agitarse. **~ and turn** (*in bed*) revolverse. **~ up** echar a cara o cruz

tot[1] /tɒt/ *n* nene *m*; (*of liquor, fam*) trago *m*

tot[2] /tɒt/ *vt* (*pt* **totted**). **~ up** (*fam*) sumar

total /'təʊtl/ *a & n* total (*m*). ● *vt* (*pt* **totalled**) sumar

totalitarian /təʊtælɪ'teərɪən/ *a* totalitario

total: **~ity** /təʊ'tælətɪ/ *n* totalidad *f*. **~ly** *adv* totalmente

totter /'tɒtə(r)/ *vi* tambalearse. **~y** *a* inseguro

touch /tʌtʃ/ *vt* tocar; (*reach*) alcanzar; (*move*) conmover. ● *vi* tocarse. ● *n* toque *m*; (*sense*) tacto *m*; (*contact*) contacto *m*; (*trace*) pizca *f*. **get in ~ with** ponerse en contacto con. **~ down** ‹*aircraft*› aterrizar. **~ off**

disparar ‹*gun*›; (*fig*) desencadenar. **~ on** tratar levemente. **~ up** retocar. **~-and-go** *a* incierto, dudoso
touching /'tʌtʃɪŋ/ *a* conmovedor
touchstone /'tʌtʃstəʊn/ *n* (*fig*) piedra *f* de toque
touchy /'tʌtʃɪ/ *a* quisquilloso
tough /tʌf/ *a* (**-er**, **-est**) duro; (*strong*) fuerte, resistente. **~en** /'tʌfn/ *vt* endurecer. **~ness** *n* dureza *f*; (*strength*) resistencia *f*
toupee /'tu:peɪ/ *n* postizo *m*, tupé *m*
tour /tʊə(r)/ *n* viaje *m*; (*visit*) visita *f*; (*excursion*) excursión *f*; (*by team etc*) gira *f*. ● *vt* recorrer; (*visit*) visitar
touris|m /'tʊərɪzəm/ *n* turismo *m*. **~t** /'tʊərɪst/ *n* turista *m* & *f*. ● *a* turístico. **~t office** *n* oficina *f* de turismo
tournament /'tɔ:nəmənt/ *n* torneo *m*
tousle /'taʊzl/ *vt* despeinar
tout /taʊt/ *vi*. **~ (for)** solicitar. ● *n* solicitador *m*
tow /təʊ/ *vt* remolcar. ● *n* remolque *m*. **on ~** a remolque. **with his family in ~** (*fam*) acompañado por su familia
toward(s) /tə'wɔ:d(z)/ *prep* hacia
towel /'taʊəl/ *n* toalla *f*. **~ling** *n* (*fabric*) toalla *f*
tower /'taʊə(r)/ *n* torre *f*. ● *vi*. **~ above** dominar. **~ block** *n* edificio *m* alto. **~ing** *a* altísimo; ‹*rage*› violento
town /taʊn/ *n* ciudad *f*, pueblo *m*. **go to ~** (*fam*) no escatimar dinero. **~ hall** *n* ayuntamiento *m*. **~ planning** *n* urbanismo *m*
tow-path /'təʊpɑ:θ/ *n* camino *m* de sirga
toxi|c /'tɒksɪk/ *a* tóxico. **~n** /'tɒksɪn/ *n* toxina *f*
toy /tɔɪ/ *n* juguete *m*. ● *vi*. **~ with** jugar con ‹*object*›; acariciar ‹*idea*›. **~shop** *n* juguetería *f*
trac|e /treɪs/ *n* huella *f*; (*small amount*) pizca *f*. ● *vt* seguir la pista de; (*draw*) dibujar; (*with tracing-paper*) calcar; (*track down*) encontrar. **~ing** /'treɪsɪŋ/ *n* calco *m*. **~ing-paper** *n* papel *m* de calcar
track /træk/ *n* huella *f*; (*path*) sendero *m*; (*sport*) pista *f*; (*of rocket etc*) trayectoria *f*; (*rail*) vía *f*. **keep ~ of** vigilar. **make ~s** (*sl*) marcharse. ● *vt* seguir la pista de. **~ down** *vt* localizar. **~ suit** *n* traje *m* de deporte, chandal *m*
tract[1] /trækt/ *n* (*land*) extensión *f*; (*anat*) aparato *m*
tract[2] /trækt/ *n* (*pamphlet*) opúsculo *m*
traction /'trækʃn/ *n* tracción *f*
tractor /'træktə(r)/ *n* tractor *m*
trade /treɪd/ *n* comercio *m*; (*occupation*) oficio *m*; (*exchange*) cambio *m*; (*industry*) industria *f*. ● *vt* cambiar. ● *vi* comerciar. **~ in** (*give in part-exchange*) dar como parte del pago. **~ on** aprovecharse de. **~ mark** *n* marca *f* registrada. **~r** /-ə(r)/ *n* comerciante *m* & *f*. **~sman** /'treɪdzmən/ *n* (*pl* **-men**) (*shopkeeper*) tendero *m*. **~ union** *n* sindicato *m*. **~ unionist** *n* sindicalista *m* & *f*. **~ wind** *n* viento *m* alisio
trading /'treɪdɪŋ/ *n* comercio *m*. **~ estate** *n* zona *f* industrial
tradition /trə'dɪʃn/ *n* tradición *f*. **~al** *a* tradicional. **~alist** *n* tradicionalista *m* & *f*. **~ally** *adv* tradicionalmente
traffic /'træfɪk/ *n* tráfico *m*. ● *vi* (*pt* **trafficked**) comerciar (**in** en). **~-lights** *npl* semáforo *m*. **~ warden** *n* guardia *m*, controlador *m* de tráfico
trag|edy /'trædʒɪdɪ/ *n* tragedia *f*. **~ic** /'trædʒɪk/ *a* trágico. **~ically** *adv* trágicamente
trail /treɪl/ *vi* arrastrarse; (*lag*) rezagarse. ● *vt* (*track*) seguir la pista de. ● *n* estela *f*; (*track*) pista *f*. (*path*) sendero *m*. **~er** /'treɪlə(r)/ *n* remolque *m*; (*film*) avance *m*
train /treɪn/ *n* tren *m*; (*of dress*) cola *f*; (*series*) sucesión *f*; (*retinue*) séquito *m*. ● *vt* adiestrar; (*sport*) entrenar; educar ‹*child*›; guiar ‹*plant*›; domar ‹*animal*›. ● *vi* adiestrarse; (*sport*) entrenarse. **~ed** *a* (*skilled*) cualificado; ‹*doctor*› diplomado. **~ee** *n* aprendiz *m*. **~er** *n* (*sport*) entrenador *m*; (*of animals*) domador *m*. **~ers** *mpl* zapatillas *fpl* de deporte. **~ing** *n* instrucción *f*; (*sport*) entrenamiento *m*
traipse /treɪps/ *vi* (*fam*) vagar
trait /treɪ(t)/ *n* característica *f*, rasgo *m*
traitor /'treɪtə(r)/ *n* traidor *m*
tram /træm/ *n* tranvía *m*
tramp /træmp/ *vt* recorrer a pie. ● *vi* andar con pasos pesados. ● *n* (*vagrant*) vagabundo *m*; (*sound*) ruido *m* de pasos; (*hike*) paseo *m* largo

trample /'træmpl/ *vt/i* pisotear. **~ (on)** pisotear
trampoline /'træmpəli:n/ *n* trampolín *m*
trance /trɑ:ns/ *n* trance *m*
tranquil /'træŋkwɪl/ *a* tranquilo. **~lity** /-'kwɪlətɪ/ *n* tranquilidad *f*
tranquillize /'træŋkwɪlaɪz/ *vt* tranquilizar. **~r** /-ə(r)/ *n* tranquilizante *m*
transact /træn'zækt/ *vt* negociar. **~ion** /-ʃn/ *n* transacción *f*
transatlantic /trænzət'læntɪk/ *a* transatlántico
transcend /træn'send/ *vt* exceder. **~ent** *a* sobresaliente
transcendental /trænsen'dentl/ *a* trascendental
transcribe /træns'kraɪb/ *vt* transcribir; grabar ‹*recorded sound*›
transcript /'trænskrɪpt/ *n* copia *f*. **~ion** /-ɪpʃn/ *n* transcripción *f*
transfer /træns'fɜ:(r)/ *vt* (*pt* **transferred**) trasladar; calcar ‹*drawing*›. ● *vi* trasladarse. **~ the charges** (*on telephone*) llamar a cobro revertido. /'trænsfɜ:(r)/ *n* traslado *m*; (*paper*) calcomanía *f*. **~able** *a* transferible
transfigur|ation /trænsfɪgjʊ'reɪʃn/ *n* transfiguración *f*. **~e** /træns'fɪgə(r)/ *vt* transfigurar
transfix /træns'fɪks/ *vt* traspasar; (*fig*) paralizar
transform /træns'fɔ:m/ *vt* transformar. **~ation** /-ə'meɪʃn/ *n* transformación *f*. **~er** /-ə(r)/ *n* transformador *m*
transfusion /træns'fju:ʒn/ *n* transfusión *f*
transgress /træns'gres/ *vt* traspasar, infringir. **~ion** /-ʃn/ *n* transgresión *f*; (*sin*) pecado *m*
transient /'trænzɪənt/ *a* pasajero
transistor /træn'zɪstə(r)/ *n* transistor *m*
transit /'trænsɪt/ *n* tránsito *m*
transition /træn'zɪʒn/ *n* transición *f*
transitive /'trænsɪtɪv/ *a* transitivo
transitory /'trænsɪtrɪ/ *a* transitorio
translat|e /trænz'leɪt/ *vt* traducir. **~ion** /-ʃn/ *n* traducción *f*. **~or** /-ə(r)/ *n* traductor *m*
translucen|ce /træns'lu:sns/ *n* traslucidez *f*. **~t** /trænz'lu:snt/ *a* traslúcido
transmission /træns'mɪʃn/ *n* transmisión *f*
transmit /trænz'mɪt/ *vt* (*pt* **transmitted**) transmitir. **~ter** /-ə(r)/ *n* transmisor *m*; (*TV, radio*) emisora *f*
transparen|cy /træns'pærənsɪ/ *n* transparencia *f*; (*photo*) diapositiva *f*. **~t** /træns'pærənt/ *a* transparente
transpire /træn'spaɪə(r)/ *vi* transpirar; (*happen, fam*) suceder, revelarse
transplant /træns'plɑ:nt/ *vt* trasplantar. /'trænsplɑ:nt/ *n* trasplante *m*
transport /træn'spɔ:t/ *vt* transportar. /'trænspɔ:t/ *n* transporte *m*. **~ation** /-'teɪʃn/ *n* transporte *m*
transpos|e /træn'spəʊz/ *vt* transponer; (*mus*) transportar. **~ition** /-pə'zɪʃn/ *n* transposición *f*; (*mus*) transporte *m*
transverse /'trænzvɜ:s/ *a* transverso
transvestite /trænz'vestaɪt/ *n* travestido *m*
trap /træp/ *n* trampa *f*. ● *vt* (*pt* **trapped**) atrapar; (*jam*) atascar; (*cut off*) bloquear. **~door** /'træpdɔ:(r)/ *n* trampa *f*; (*in theatre*) escotillón *m*
trapeze /trə'pi:z/ *n* trapecio *m*
trappings /'træpɪŋz/ *npl* (*fig*) atavíos *mpl*
trash /træʃ/ *n* pacotilla *f*; (*refuse*) basura *f*; (*nonsense*) tonterías *fpl*. **~ can** *n* (*Amer*) cubo *m* de la basura. **~y** *a* de baja calidad
trauma /'trɔ:mə/ *n* trauma *m*. **~tic** /-'mætɪk/ *a* traumático
travel /'trævl/ *vi* (*pt* **travelled**) viajar. ● *vt* recorrer. ● *n* viajar *m*. **~ler** /-ə(r)/ *n* viajero *m*. **~ler's cheque** *n* cheque *m* de viaje. **~ling** *n* viajar *m*
traverse /træ'vɜ:s/ *vt* atravesar, recorrer
travesty /'trævɪstɪ/ *n* parodia *f*
trawler /'trɔ:lə(r)/ *n* pesquero *m* de arrastre
tray /treɪ/ *n* bandeja *f*
treacher|ous *a* traidor; (*deceptive*) engañoso. **~ously** *adv* traidoramente. **~y** /'tretʃərɪ/ *n* traición *f*
treacle /'tri:kl/ *n* melaza *f*
tread /tred/ *vi* (*pt* **trod**, *pp* **trodden**) andar. **~ on** pisar. ● *vt* pisar. ● *n* (*step*) paso *m*; (*of tyre*) banda *f* de rodadura. **~le** /'tredl/ *n* pedal *m*. **~mill** /'tredmɪl/ *n* rueda *f* de molino; (*fig*) rutina *f*
treason /'tri:zn/ *n* traición *f*
treasure /'treʒə(r)/ *n* tesoro *m*. ● *vt* apreciar mucho; (*store*) guardar

treasur|er /'treʒərə(r)/ *n* tesorero *m*. **~y** /'treʒərɪ/ *n* tesorería *f*. **the T~y** *n* el Ministerio *m* de Hacienda

treat /tri:t/ *vt* tratar; (*consider*) considerar. **~ s.o.** invitar a uno. ● *n* placer *m*; (*present*) regalo *m*

treatise /'tri:tɪz/ *n* tratado *m*

treatment /'tri:tmənt/ *n* tratamiento *m*

treaty /'tri:tɪ/ *n* tratado *m*

treble /'trebl/ *a* triple; ‹*clef*› de sol; ‹*voice*› de tiple. ● *vt* triplicar. ● *vi* triplicarse. ● *n* tiple *m & f*

tree /tri:/ *n* árbol *m*

trek /trek/ *n* viaje *m* arduo, caminata *f*. ● *vi* (*pt* **trekked**) hacer un viaje arduo

trellis /'trelɪs/ *n* enrejado *m*

tremble /'trembl/ *vi* temblar

tremendous /trɪ'mendəs/ *a* tremendo; (*huge*, *fam*) enorme. **~ly** *adv* tremendamente

tremor /'tremə(r)/ *n* temblor *m*

tremulous /'tremjʊləs/ *a* tembloroso

trench /trentʃ/ *n* foso *m*, zanja *f*; (*mil*) trinchera *f*. **~ coat** *n* trinchera *f*

trend /trend/ *n* tendencia *f*; (*fashion*) moda *f*. **~-setter** *n* persona *f* que lanza la moda. **~y** *a* (**-ier**, **-iest**) (*fam*) a la última

trepidation /trepɪ'deɪʃn/ *n* inquietud *f*

trespass /'trespəs/ *vi*. **~ on** entrar sin derecho; (*fig*) abusar de. **~er** /-ə(r)/ *n* intruso *m*

tress /tres/ *n* trenza *f*

trestle /'tresl/ *n* caballete *m*. **~-table** *n* mesa *f* de caballete

trews /tru:z/ *npl* pantalón *m*

trial /'traɪəl/ *n* prueba *f*; (*jurid*) proceso *m*; (*ordeal*) prueba *f* dura. **~ and error** tanteo *m*. **be on ~** estar a prueba; (*jurid*) ser procesado

triang|le /'traɪæŋgl/ *n* triángulo *m*. **~ular** /-'æŋgjʊlə(r)/ *a* triangular

trib|al /'traɪbl/ *a* tribal. **~e** /traɪb/ *n* tribu *f*

tribulation /trɪbjʊ'leɪʃn/ *n* tribulación *f*

tribunal /traɪ'bju:nl/ *n* tribunal *m*

tributary /'trɪbjʊtrɪ/ *n* (*stream*) afluente *m*

tribute /'trɪbju:t/ *n* tributo *m*. **pay ~ to** rendir homenaje a

trice /traɪs/ *n*. **in a ~** en un abrir y cerrar de ojos

trick /trɪk/ *n* trampa *f*; engaño *m*; (*joke*) broma *f*; (*at cards*) baza *f*; (*habit*) manía *f*. **do the ~** servir. **play a ~ on** gastar una broma a. ● *vt* engañar. **~ery** /'trɪkərɪ/ *n* engaño *m*

trickle /'trɪkl/ *vi* gotear. **~ in** (*fig*) entrar poco a poco. **~ out** (*fig*) salir poco a poco

trickster /'trɪkstə(r)/ *n* estafador *m*

tricky /'trɪkɪ/ *a* delicado, difícil

tricolour /'trɪkələ(r)/ *n* bandera *f* tricolor

tricycle /'traɪsɪkl/ *n* triciclo *m*

trident /'traɪdənt/ *n* tridente *m*

tried /traɪd/ *see* **try**

trifl|e /'traɪfl/ *n* bagatela *f*; (*culin*) bizcocho *m* con natillas, jalea, frutas y nata. ● *vi*. **~e with** jugar con. **~ing** *a* insignificante

trigger /'trɪgə(r)/ *n* (*of gun*) gatillo *m*. ● *vt*. **~ (off)** desencadenar

trigonometry /trɪgə'nɒmɪtrɪ/ *n* trigonometría *f*

trilby /'trɪlbɪ/ *n* sombrero *m* de fieltro

trilogy /'trɪlədʒɪ/ *n* trilogía *f*

trim /trɪm/ *a* (**trimmer**, **trimmest**) arreglado. ● *vt* (*pt* **trimmed**) cortar; recortar ‹*hair etc*›; (*adorn*) adornar. ● *n* (*cut*) recorte *m*; (*decoration*) adorno *m*; (*state*) estado *m*. **in ~** en buen estado; (*fit*) en forma. **~ming** *n* adorno *m*. **~mings** *npl* recortes *mpl*; (*decorations*) adornos *mpl*; (*culin*) guarnición *f*

trinity /'trɪnɪtɪ/ *n* trinidad *f*. **the T~** la Trinidad

trinket /'trɪŋkɪt/ *n* chuchería *f*

trio /'tri:əʊ/ *n* (*pl* **-os**) trío *m*

trip /trɪp/ *vt* (*pt* **tripped**) hacer tropezar. ● *vi* tropezar; (*go lightly*) andar con paso ligero. ● *n* (*journey*) viaje *m*; (*outing*) excursión *f*; (*stumble*) traspié *m*. **~ up** *vi* tropezar. ● *vt* hacer tropezar

tripe /traɪp/ *n* callos *mpl*; (*nonsense*, *sl*) tonterías *fpl*

triple /'trɪpl/ *a* triple. ● *vt* triplicar. ● *vi* triplicarse. **~ts** /'trɪplɪts/ *npl* trillizos *mpl*

triplicate /'trɪplɪkət/ *a* triplicado. **in ~** por triplicado

tripod /'traɪpɒd/ *n* trípode *m*

tripper /'trɪpə(r)/ *n* (*on day trip etc*) excursionista *m & f*

triptych /'trɪptɪk/ *n* tríptico *m*

trite /traɪt/ *a* trillado

triumph /ˈtraɪʌmf/ *n* triunfo *m*. ● *vi* trinufar (**over** sobre). **~al** /-ˈʌmfl/ *a* triunfal. **~ant** /-ˈʌmfnt/ *a* triunfante

trivial /ˈtrɪvɪəl/ *a* insignificante. **~ity** /-ˈæləti/ *n* insignificancia *f*

trod, trodden /trɒd, trɒdn/ *see* **tread**

trolley /ˈtrɒlɪ/ *n* (*pl* **-eys**) carretón *m*. **tea ~** *n* mesita *f* de ruedas. **~bus** *n* trolebús *m*

trombone /trɒmˈbəʊn/ *n* trombón *m*

troop /tru:p/ *n* grupo *m*. ● *vi*. **~ in** entrar en tropel. **~ out** salir en tropel. ● *vt*. **~ing the colour** saludo *m* a la bandera. **~er** *n* soldado *m* de caballería. **~s** *npl* (*mil*) tropas *fpl*

trophy /ˈtrəʊfɪ/ *n* trofeo *m*

tropic /ˈtrɒpɪk/ *n* trópico *m*. **~al** *a* tropical. **~s** *npl* trópicos *mpl*

trot /trɒt/ *n* trote *m*. **on the ~** (*fam*) seguidos. ● *vi* (*pt* **trotted**) trotar. **~ out** (*produce, fam*) producir

trotter /ˈtrɒtə(r)/ *n* (*culin*) pie *m* de cerdo

trouble /ˈtrʌbl/ *n* problema *m*; (*awkward situation*) apuro *m*; (*inconvenience*) molestia *f*; (*conflict*) conflicto *m*; (*med*) enfermedad *f*; (*mec*) avería *f*. **be in ~** estar en un apuro. **make ~** armar un lío. **take ~** tomarse la molestia. ● *vt* (*bother*) molestar; (*worry*) preocupar. ● *vi* molestarse; (*worry*) preocuparse. **be ~d about** preocuparse por. **~-maker** *n* alborotador *m*. **~some** *a* molesto

trough /trɒf/ *n* (*for drinking*) abrevadero *m*; (*for feeding*) pesebre *m*; (*of wave*) seno *m*; (*atmospheric*) mínimo *m* de presión

trounce /traʊns/ *vt* (*defeat*) derrotar; (*thrash*) pegar

troupe /tru:p/ *n* compañía *f*

trousers /ˈtraʊzəz/ *npl* pantalón *m*; pantalones *mpl*

trousseau /ˈtru:səʊ/ *n* (*pl* **-s** /-əʊz/) ajuar *m*

trout /traʊt/ *n* (*pl* **trout**) trucha *f*

trowel /ˈtraʊəl/ *n* (*garden*) desplantador *m*; (*for mortar*) paleta *f*

truant /ˈtru:ənt/ *n*. **play ~** hacer novillos

truce /tru:s/ *n* tregua *f*

truck[1] /trʌk/ *n* carro *m*; (*rail*) vagón *m*; (*lorry*) camión *m*

truck[2] /trʌk/ *n* (*dealings*) trato *m*

truculent /ˈtrʌkjʊlənt/ *a* agresivo

trudge /trʌdʒ/ *vi* andar penosamente. ● *n* caminata *f* penosa

true /tru:/ *a* (**-er**, **-est**) verdadero; (*loyal*) leal; (*genuine*) auténtico; (*accurate*) exacto. **come ~** realizarse

truffle /ˈtrʌfl/ *n* trufa *f*; (*chocolate*) trufa *f* de chocolate

truism /ˈtru:ɪzəm/ *n* perogrullada *f*

truly /ˈtru:lɪ/ *adv* verdaderamente; (*sincerely*) sinceramente; (*faithfully*) fielmente. **yours ~** (*in letters*) le saluda atentamente

trump /trʌmp/ *n* (*cards*) triunfo *m*. ● *vt* fallar. **~ up** inventar

trumpet /ˈtrʌmpɪt/ *n* trompeta *f*. **~er** /-ə(r)/ *n* trompetero *m*, trompeta *m & f*

truncated /trʌŋˈkeɪtɪd/ *a* truncado

truncheon /ˈtrʌntʃən/ *n* porra *f*

trundle /ˈtrʌndl/ *vt* hacer rodar. ● *vi* rodar

trunk /trʌŋk/ *n* tronco *m*; (*box*) baúl *m*; (*of elephant*) trompa *f*. **~-call** *n* conferencia *f*. **~-road** *n* carretera *f* (nacional). **~s** *npl* bañador *m*

truss /trʌs/ *n* (*med*) braguero *m*. **~ up** *vt* (*culin*) espetar

trust /trʌst/ *n* confianza *f*; (*association*) trust *m*. **on ~** a ojos cerrados; (*com*) al fiado. ● *vi* confiar. **~ to** confiar en. ● *vt* confiar en; (*hope*) esperar. **~ed** *a* leal

trustee /trʌˈsti:/ *n* administrador *m*

trust: **~ful** *a* confiado. **~fully** *adv* confiadamente. **~worthy** *a*, **~y** *a* digno de confianza

truth /tru:θ/ *n* (*pl* **-s** /tru:ðz/) verdad *f*. **~ful** *a* veraz; (*true*) verídico. **~fully** *adv* sinceramente

try /traɪ/ *vt* (*pt* **tried**) probar; (*be a strain on*) poner a prueba; (*jurid*) procesar. **~ on** *vt* probarse ‹*garment*›. **~ out** *vt* probar. ● *vi* probar. **~ for** *vi* intentar conseguir. ● *n* tentativa *f*, prueba *f*; (*rugby*) ensayo *m*. **~ing** *a* difícil; (*annoying*) molesto. **~-out** *n* prueba *f*

tryst /trɪst/ *n* cita *f*

T-shirt /ˈti:ʃɜ:t/ *n* camiseta *f*

tub /tʌb/ *n* tina *f*; (*bath, fam*) baño *m*

tuba /ˈtju:bə/ *n* tuba *f*

tubby /ˈtʌbɪ/ *a* (**-ier**, **-iest**) rechoncho

tube /tju:b/ *n* tubo *m*; (*rail, fam*) metro *m*. **inner ~** *n* cámara *f* de aire

tuber /ˈtju:bə(r)/ *n* tubérculo *m*

tuberculosis /tju:bɜ:kjʊˈləʊsɪs/ *n* tuberculosis *f*

tub|ing /'tju:bɪŋ/ *n* tubería *f*, tubos *mpl*. **~ular** *a* tubular

tuck /tʌk/ *n* pliegue *m*. ● *vt* plegar; (*put*) meter; (*put away*) remeter; (*hide*) esconder. **~ up** *vt* arropar ‹*child*›. ● *vi*. **~ in(to)** (*eat, sl*) comer con buen apetito. **~-shop** *n* confitería *f*

Tuesday /'tju:zdeɪ/ *n* martes *m*

tuft /tʌft/ *n* (*of hair*) mechón *m*; (*of feathers*) penacho *m*; (*of grass*) manojo *m*

tug /tʌg/ *vt* (*pt* **tugged**) tirar de; (*tow*) remolcar. ● *vi* tirar fuerte. ● *n* tirón *m*; (*naut*) remolcador *m*. **~-of-war** *n* lucha *f* de la cuerda; (*fig*) tira *m* y afloja

tuition /tju:'ɪʃn/ *n* enseñanza *f*

tulip /'tju:lɪp/ *n* tulipán *m*

tumble /'tʌmbl/ *vi* caerse. **~ to** (*fam*) comprender. ● *n* caída *f*

tumbledown /'tʌmbldaʊn/ *a* ruinoso

tumble-drier /tʌmbl'draɪə(r)/ *n* secadora *f* (eléctrica con aire de salida)

tumbler /'tʌmblə(r)/ *n* (*glass*) vaso *m*

tummy /'tʌmɪ/ *n* (*fam*) estómago *m*

tumour /'tju:mə(r)/ *n* tumor *m*

tumult /'tju:mʌlt/ *n* tumulto *m*. **~uous** /-'mʌltjʊəs/ *a* tumultuoso

tuna /'tju:nə/ *n* (*pl* **tuna**) atún *m*

tune /tju:n/ *n* aire *m*. **be in ~** estar afinado. **be out of ~** estar desafinado. ● *vt* afinar; sintonizar ‹*radio, TV*›; (*mec*) poner a punto. ● *vi*. **~ in (to)** ‹*radio, TV*› sintonizarse. **~ up** afinar. **~ful** *a* melodioso. **~r** /-ə(r)/ *n* afinador *m*; (*radio, TV*) sintonizador *m*

tunic /'tju:nɪk/ *n* túnica *f*

tuning-fork /'tju:nɪŋfɔ:k/ *n* diapasón *m*

Tunisia /tju:'nɪzɪə/ *n* Túnez *m*. **~n** *a* & *n* tunecino (*m*)

tunnel /'tʌnl/ *n* túnel *m*. ● *vi* (*pt* **tunnelled**) construir un túnel en

turban /'tɜ:bən/ *n* turbante *m*

turbid /'tɜ:bɪd/ *a* túrbido

turbine /'tɜ:baɪn/ *n* turbina *f*

turbo-jet /'tɜ:bəʊdʒet/ *n* turborreactor *m*

turbot /'tɜ:bət/ *n* rodaballo *m*

turbulen|ce /'tɜ:bjʊləns/ *n* turbulencia *f*. **~t** /'tɜ:bjʊlənt/ *a* turbulento

tureen /tjʊ'ri:n/ *n* sopera *f*

turf /tɜ:f/ *n* (*pl* **turfs** *or* **turves**) césped *m*; (*segment*) tepe *m*. **the ~** *n* las carreras *fpl* de caballos. ● *vt*. **~ out** (*sl*) echar

turgid /'tɜ:dʒɪd/ *a* ‹*language*› pomposo

Turk /tɜ:k/ *n* turco *m*

turkey /'tɜ:kɪ/ *n* (*pl* **-eys**) pavo *m*

Turk|ey /'tɜ:kɪ/ *f* Turquía *f*. **T~ish** *a* & *n* turco (*m*)

turmoil /'tɜ:mɔɪl/ *n* confusión *f*

turn /tɜ:n/ *vt* hacer girar, dar vueltas a; volver ‹*direction, page, etc*›; cumplir ‹*age*›; dar ‹*hour*›; doblar ‹*corner*›; (*change*) cambiar; (*deflect*) desviar. **~ the tables** volver las tornas. ● *vi* girar, dar vueltas; (*become*) hacerse; (*change*) cambiar. ● *n* vuelta *f*; (*in road*) curva *f*; (*change*) cambio *m*; (*sequence*) turno *m*; (*of mind*) disposición *f*; (*in theatre*) número *m*; (*fright*) susto *m*; (*of illness, fam*) ataque *m*. **bad ~** mala jugada *f*. **good ~** favor *m*. **in ~** a su vez. **out of ~** fuera de lugar. **to a ~** (*culin*) en su punto. **~ against** *vt* volverse en contra de. **~ down** *vt* (*fold*) doblar; (*reduce*) bajar; (*reject*) rechazar. **~ in** *vt* entregar. ● *vi* (*go to bed, fam*) acostarse. **~ off** *vt* cerrar ‹*tap*›; apagar ‹*light, TV, etc*›. ● *vi* desviarse. **~ on** *vt* abrir ‹*tap*›; encender ‹*light etc*›; (*attack*) atacar; (*attract, fam*) excitar. **~ out** *vt* expulsar; apagar ‹*light etc*›; (*produce*) producir; (*empty*) vaciar. ● *vi* (*result*) resultar. **~ round** *vi* dar la vuelta. **~ up** *vi* aparecer. ● *vt* (*find*) encontrar; levantar ‹*collar*›; poner más fuerte ‹*gas*›. **~ed-up** *a* ‹*nose*› respingona. **~ing** /'tɜ:nɪŋ/ *n* vuelta *f*; (*road*) bocacalle *f*. **~ing-point** *n* punto *m* decisivo.

turnip /'tɜ:nɪp/ *n* nabo *m*

turn: ~-out *n* (*of people*) concurrencia *f*; (*of goods*) producción *f*. **~over** /'tɜ:nəʊvə(r)/ *n* (*culin*) empanada *f*; (*com*) volumen *m* de negocios; (*of staff*) rotación *f*. **~pike** /'tɜ:npaɪk/ *n* (*Amer*) autopista *f* de peaje. **~-stile** /'tɜ:nstaɪl/ *n* torniquete *m*. **~table** /'tɜ:nteɪbl/ *n* plataforma *f* giratoria; (*on record-player*) plato *m* giratorio. **~-up** *n* (*of trousers*) vuelta *f*

turpentine /'tɜ:pəntaɪn/ *n* trementina *f*

turquoise /'tɜ:kwɔɪz/ *a* & *n* turquesa (*f*)

turret /'tʌrɪt/ *n* torrecilla *f*; (*mil*) torreta *f*
turtle /'tɜ:tl/ *n* tortuga *f* de mar. **~-neck** *n* cuello *m* alto
tusk /tʌsk/ *n* colmillo *m*
tussle /'tʌsl/ *vi* pelearse. ● *n* pelea *f*
tussock /'tʌsək/ *n* montecillo *m* de hierbas
tutor /'tju:tə(r)/ *n* preceptor *m*; (*univ*) director *m* de estudios, profesor *m*. **~ial** /tju:'tɔ:rɪəl/ *n* clase *f* particular
tuxedo /tʌk'si:dəʊ/ *n* (*pl* **-os**) (*Amer*) esmoquin *m*
TV /ti:'vi:/ *n* televisión *f*
twaddle /'twɒdl/ *n* tonterías *fpl*
twang /twæŋ/ *n* tañido *m*; (*in voice*) gangueo *m*. ● *vt* hacer vibrar. ● *vi* vibrar
tweed /twi:d/ *n* tela *f* gruesa de lana
tweet /twi:t/ *n* piada *f*. ● *vi* piar
tweezers /'twi:zəz/ *npl* pinzas *fpl*
twel|fth /twelfθ/ *a* & *n* duodécimo (*m*). **~ve** /twelv/ *a* & *n* doce (*m*)
twent|ieth /'twentɪəθ/ *a* & *n* vigésimo (*m*). /'twentɪ/ *a* & *n* veinte (*m*)
twerp /twɜ:p/ *n* (*sl*) imbécil *m*
twice /twaɪs/ *adv* dos veces
twiddle /'twɪdl/ *vt* hacer girar. **~ one's thumbs** (*fig*) no tener nada que hacer. **~ with** jugar con
twig[1] /twɪg/ *n* ramita *f*
twig[2] /twɪg/ *vt*/*i* (*pt* **twigged**) (*fam*) comprender
twilight /'twaɪlaɪt/ *n* crepúsculo *m*
twin /twɪn/ *a* & *n* gemelo (*m*)
twine /twaɪn/ *n* bramante *m*. ● *vt* torcer. ● *vi* enroscarse
twinge /twɪndʒ/ *n* punzada *f*; (*fig*) remordimiento *m* (de conciencia)
twinkle /'twɪŋkl/ *vi* centellear. ● *n* centelleo *m*
twirl /twɜ:l/ *vt* dar vueltas a. ● *vi* dar vueltas. ● *n* vuelta *f*
twist /twɪst/ *vt* torcer; (*roll*) enrollar; (*distort*) deformar. ● *vi* torcerse; (*coil*) enroscarse; ‹*road*› serpentear. ● *n* torsión *f*; (*curve*) vuelta *f*; (*of character*) peculiaridad *f*
twit[1] /twɪt/ *n* (*sl*) imbécil *m*
twit[2] /twɪt/ *vt* (*pt* **twitted**) tomar el pelo a
twitch /twɪtʃ/ *vt* crispar. ● *vi* crisparse. ● *n* tic *m*; (*jerk*) tirón *m*
twitter /'twɪtə(r)/ *vi* gorjear. ● *n* gorjeo *m*
two /tu:/ *a* & *n* dos (*m*). **in ~ minds** indeciso. **~-faced** *a* falso, insincero. **~-piece (suit)** *n* traje *m* (de dos piezas). **~some** /'tu:səm/ *n* pareja *f*. **~-way** *a* ‹*traffic*› de doble sentido
tycoon /taɪ'ku:n/ *n* magnate *m*
tying /'taɪɪŋ/ *see* **tie**
type /taɪp/ *n* tipo *m*. ● *vt*/*i* escribir a máquina. **~-cast** *a* ‹*actor*› encasillado. **~script** /'taɪpskrɪpt/ *n* texto *m* escrito a máquina. **~writer** /'taɪpraɪtə(r)/ *n* máquina *f* de escribir. **~written** /-ɪtn/ *a* escrito a máquina, mecanografiado
typhoid /'taɪfɔɪd/ *n*. **~ (fever)** fiebre *f* tifoidea
typhoon /taɪ'fu:n/ *n* tifón *m*
typical /'tɪpɪkl/ *a* típico. **~ly** *adv* típicamente
typify /'tɪpɪfaɪ/ *vt* tipificar
typi|ng /'taɪpɪŋ/ *n* mecanografía *f*. **~st** *n* mecanógrafo *m*
typography /taɪ'pɒgrəfɪ/ *n* tipografía *f*
tyran|nical /tɪ'rænɪkl/ *a* tiránico. **~nize** *vi* tiranizar. **~ny** /'tɪrənɪ/ *n* tiranía *f*. **~t** /'taɪərənt/ *n* tirano *m*
tyre /'taɪə(r)/ *n* neumático *m*, llanta *f* (*Amer*)

U

ubiquitous /ju:'bɪkwɪtəs/ *a* omnipresente, ubicuo
udder /'ʌdə(r)/ *n* ubre *f*
UFO /'ju:fəʊ/ *abbr* (*unidentified flying object*) OVNI *m*, objeto *m* volante no identificado
ugl|iness /ʌglɪnɪs/ *n* fealdad *f*. **~y** /'ʌglɪ/ *a* (**-ier**, **-iest**) feo
UK /ju:'keɪ/ *abbr* (*United Kingdom*) Reino *m* Unido
ulcer /'ʌlsə(r)/ *n* úlcera *f*. **~ous** *a* ulceroso
ulterior /ʌl'tɪərɪə(r)/ *a* ulterior. **~ motive** *n* segunda intención *f*
ultimate /'ʌltɪmət/ *a* último; (*definitive*) definitivo; (*fundamental*) fundamental. **~ly** *adv* al final; (*basically*) en el fondo
ultimatum /ʌltɪ'meɪtəm/ *n* (*pl* **-ums**) ultimátum *m invar*
ultra... /'ʌltrə/ *pref* ultra...
ultramarine /ʌltrəmə'ri:n/ *n* azul *m* marino
ultrasonic /ʌltrə'sɒnɪk/ *a* ultrasónico

ultraviolet /ʌltrə'vaɪələt/ *a* ultravioleta *a invar*
umbilical /ʌm'bɪlɪkl/ *a* umbilical. **~ cord** *n* cordón *m* umbilical
umbrage /'ʌmbrɪdʒ/ *n* resentimiento *m*. **take ~** ofenderse (**at** por)
umbrella /ʌm'brelə/ *n* paraguas *m invar*
umpire /'ʌmpaɪə(r)/ *n* árbitro *m*. ● *vt* arbitrar
umpteen /'ʌmpti:n/ *a* (*sl*) muchísimos. **~th** *a* (*sl*) enésimo
UN /ju:'en/ *abbr* (*United Nations*) ONU *f*, Organización *f* de las Naciones Unidas
un... /ʌn/ *pref* in..., des..., no, poco, sin
unabated /ʌnə'beɪtɪd/ *a* no disminuido
unable /ʌn'eɪbl/ *a* incapaz (**to** de). **be ~ to** no poder
unabridged /ʌnə'brɪdʒd/ *a* íntegro
unacceptable /ʌnək'septəbl/ *a* inaceptable
unaccountabl|e /ʌnə'kaʊntəbl/ *a* inexplicable. **~y** *adv* inexplicablemente
unaccustomed /ʌnə'kʌstəmd/ *a* insólito. **be ~ to** *a* no estar acostumbrado a
unadopted /ʌnə'dɒptɪd/ *a* ‹*of road*› privado
unadulterated /ʌnə'dʌltəreɪtɪd/ *a* puro
unaffected /ʌnə'fektɪd/ *a* sin afectación, natural
unaided /ʌn'eɪdɪd/ *a* sin ayuda
unalloyed /ʌnə'lɔɪd/ *a* puro
unanimous /ju:'nænɪməs/ *a* unánime. **~ly** *adv* unánimemente
unannounced /ʌnə'naʊnst/ *a* sin previo aviso; (*unexpected*) inesperado
unarmed /ʌn'ɑ:md/ *a* desarmado
unassuming /ʌnə'sju:mɪŋ/ *a* modesto, sin pretensiones
unattached /ʌnə'tætʃt/ *a* suelto; (*unmarried*) soltero
unattended /ʌnə'tendɪd/ *a* sin vigilar
unattractive /ʌnə'træktɪv/ *a* poco atractivo
unavoidabl|e /ʌnə'vɔɪdəbl/ *a* inevitable. **~y** *adv* inevitablemente
unaware /ʌnə'weə(r)/ *a* ignorante (**of** de). **be ~ of** ignorar. **~s** /-eəz/ *adv* desprevenido
unbalanced /ʌn'bælənst/ *a* desequilibrado
unbearabl|e /ʌn'beərəbl/ *a* inaguantable. **~y** *adv* inaguantablemente
unbeat|able /ʌn'bi:təbl/ *a* insuperable. **~en** *a* no vencido
unbeknown /ʌnbɪ'nəʊn/ *a* desconocido. **~ to me** (*fam*) sin saberlo yo
unbelievable /ʌnbɪ'li:vəbl/ *a* increíble
unbend /ʌn'bend/ *vt* (*pt* **unbent**) enderezar. ● *vi* (*relax*) relajarse. **~ing** *a* inflexible
unbiased /ʌn'baɪəst/ *a* imparcial
unbidden /ʌn'bɪdn/ *a* espontáneo; (*without invitation*) sin ser invitado
unblock /ʌn'blɒk/ *vt* desatascar
unbolt /ʌn'bəʊlt/ *vt* desatrancar
unborn /ʌn'bɔ:n/ *a* no nacido todavía
unbounded /ʌn'baʊndɪd/ *a* ilimitado
unbreakable /ʌn'breɪkəbl/ *a* irrompible
unbridled /ʌn'braɪdld/ *a* desenfrenado
unbroken /ʌn'brəʊkən/ *a* (*intact*) intacto; (*continuous*) continuo
unburden /ʌn'bɜ:dn/ *vt*. **~ o.s.** desahogarse
unbutton /ʌn'bʌtn/ *vt* desabotonar, desabrochar
uncalled-for /ʌn'kɔ:ldfɔ:(r)/ *a* fuera de lugar; (*unjustified*) injustificado
uncanny /ʌn'kænɪ/ *a* (**-ier**, **-iest**) misterioso
unceasing /ʌn'si:sɪŋ/ *a* incesante
unceremonious /ʌnserɪ'məʊnɪəs/ *a* informal; (*abrupt*) brusco
uncertain /ʌn'sɜ:tn/ *a* incierto; (*changeable*) variable. **be ~ whether** no saber exactamente si. **~ty** *n* incertidumbre *f*
unchang|ed /ʌn'tʃeɪndʒd/ *a* igual. **~ing** *a* inmutable
uncharitable /ʌn'tʃærɪtəbl/ *a* severo
uncivilized /ʌn'sɪvɪlaɪzd/ *a* incivilizado
uncle /'ʌŋkl/ *n* tío *m*
unclean /ʌn'kli:n/ *a* sucio
unclear /ʌn'klɪə(r)/ *a* poco claro
uncomfortable /ʌn'kʌmfətəbl/ *a* incómodo; (*unpleasant*) desagradable. **feel ~** no estar a gusto
uncommon /ʌn'kɒmən/ *a* raro. **~ly** *adv* extraordinariamente
uncompromising /ʌn'kɒmprəmaɪzɪŋ/ *a* intransigente
unconcerned /ʌnkən'sɜ:nd/ *a* indiferente

unconditional /ʌnkən'dɪʃənl/ *a* incondicional. **~ly** *adv* incondicionalmente
unconscious /ʌn'kɒnʃəs/ *a* inconsciente; (*med*) sin sentido. **~ly** *adv* inconscientemente
unconventional /ʌnkən'venʃənl/ *a* poco convencional
uncooperative /ʌnkəʊ'ɒpərətɪv/ *a* poco servicial
uncork /ʌn'kɔ:k/ *vt* descorchar, destapar
uncouth /ʌn'ku:θ/ *a* grosero
uncover /ʌn'kʌvə(r)/ *vt* descubrir
unctuous /'ʌŋktjʊəs/ *a* untuoso; (*fig*) empalagoso
undecided /ʌndɪ'saɪdɪd/ *a* indeciso
undeniabl|e /ʌndɪ'naɪəbl/ *a* innegable. **~y** *adv* indiscutible mente
under /'ʌndə(r)/ *prep* debajo de; (*less than*) menos de; (*in the course of*) bajo, en. ● *adv* debajo, abajo. **~ age** *a* menor de edad. **~ way** *adv* en curso; (*on the way*) en marcha
under... *pref* sub...
undercarriage /'ʌndəkærɪdʒ/ *n* (*aviat*) tren *m* de aterrizaje
underclothes /'ʌndəkləʊðz/ *npl* ropa *f* interior
undercoat /'ʌndəkəʊt/ *n* (*of paint*) primera mano *f*
undercover /ʌndə'kʌvə(r)/ *a* secreto
undercurrent /'ʌndəkʌrənt/ *n* corriente *f* submarina; (*fig*) tendencia *f* oculta
undercut /'ʌndəkʌt/ *vt* (*pt* **undercut**) (*com*) vender más barato que
underdeveloped /ʌndədɪ'veləpt/ *a* subdesarrollado
underdog /'ʌndədɒg/ *n* perdedor *m*. **the ~s** *npl* los de abajo
underdone /ʌndə'dʌn/ *a* ‹*meat*› poco hecho
underestimate /ʌndər'estɪmeɪt/ *vt* subestimar
underfed /ʌndə'fed/ *a* desnutrido
underfoot /ʌndə'fʊt/ *adv* bajo los pies
undergo /'ʌndəgəʊ/ *vt* (*pt* **-went**, *pp* **-gone**) sufrir
undergraduate /ʌndə'grædjʊət/ *n* estudiante *m* & *f* universitario (no licenciado)
underground /ʌndə'graʊnd/ *adv* bajo tierra; (*in secret*) clandestinamente. /'ʌndəgraʊnd/ *a* subterráneo; (*secret*) clandestino. ● *n* metro *m*
undergrowth /'ʌndəgrəʊθ/ *n* maleza *f*
underhand /'ʌndəhænd/ *a* (*secret*) clandestino; (*deceptive*) fraudulento
underlie /ʌndə'laɪ/ *vt* (*pt* **-lay**, *pp* **-lain**, *pres p* **-lying**) estar debajo de; (*fig*) estar a la base de
underline /ʌndə'laɪn/ *vt* subrayar
underling /'ʌndəlɪŋ/ *n* subalterno *m*
underlying /ʌndə'laɪŋ/ *a* fundamental
undermine /ʌndə'maɪn/ *vt* socavar
underneath /ʌndə'ni:θ/ *prep* debajo de. ● *adv* por debajo
underpaid /ʌndə'peɪd/ *a* mal pagado
underpants /'ʌndəpænts/ *npl* calzoncillos *mpl*
underpass /'ʌndəpa:s/ *n* paso *m* subterráneo
underprivileged /ʌndə'prɪvɪlɪdʒd/ *a* desvalido
underrate /ʌndə'reɪt/ *vt* subestimar
undersell /ʌndə'sel/ *vt* (*pt* **-sold**) vender más barato que
undersigned /'ʌndəsaɪnd/ *a* abajo firmante
undersized /ʌndə'saɪzd/ *a* pequeño
understand /ʌndə'stænd/ *vt/i* (*pt* **-stood**) entender, comprender. **~able** *a* comprensible. **~ing** /ʌndə'stændɪŋ/ *a* comprensivo. ● *n* comprensión *f*; (*agreement*) acuerdo *m*
understatement /ʌndə'steɪtmənt/ *n* subestimación *f*
understudy /'ʌndəstʌdɪ/ *n* sobresaliente *m* & *f* (en el teatro)
undertake /ʌndə'teɪk/ *vt* (*pt* **-took**, *pp* **-taken**) emprender; (*assume responsibility*) encargarse de
undertaker /'ʌndəteɪkə(r)/ *n* empresario *m* de pompas fúnebres
undertaking /ʌndə'teɪkɪŋ/ *n* empresa *f*; (*promise*) promesa *f*
undertone /'ʌndətəʊn/ *n*. **in an ~** en voz baja
undertow /'ʌndətəʊ/ *n* resaca *f*
undervalue /ʌndə'vælju:/ *vt* subvalorar
underwater /ʌndə'wɔ:tə(r)/ *a* submarino. ● *adv* bajo el agua
underwear /'ʌndəweə(r)/ *n* ropa *f* interior
underweight /'ʌndəweɪt/ *a* de peso insuficiente. **be ~** estar flaco
underwent /ʌndə'went/ *see* **undergo**

underworld /ˈʌndəwɜːld/ *n* (*criminals*) hampa *f*
underwrite /ʌndəˈraɪt/ *vt* (*pt* **-wrote**, *pp* **-written**) (*com*) asegurar. **~r** /-ə(r)/ *n* asegurador *m*
undeserved /ʌndɪˈzɜːvd/ *a* inmerecido
undesirable /ʌndɪˈzaɪərəbl/ *a* indeseable
undeveloped /ʌndɪˈveləpt/ *a* sin desarrollar
undies /ˈʌndɪz/ *npl* (*fam*) ropa *f* interior
undignified /ʌnˈdɪgnɪfaɪd/ *a* indecoroso
undisputed /ʌndɪsˈpjuːtɪd/ *a* incontestable
undistinguished /ʌndɪsˈtɪŋgwɪʃt/ *a* mediocre
undo /ʌnˈduː/ *vt* (*pt* **-did**, *pp* **-done**) deshacer; (*ruin*) arruinar; reparar ‹*wrong*›. **leave ~ne** dejar sin hacer
undoubted /ʌnˈdaʊtɪd/ *a* indudable. **~ly** *adv* indudablemente
undress /ʌnˈdres/ *vt* desnudar. • *vi* desnudarse
undue /ʌnˈdjuː/ *a* excesivo
undulat|e /ˈʌndjʊleɪt/ *vi* ondular. **~ion** /-ˈleɪʃn/ *n* ondulación *f*
unduly /ʌnˈdjuːlɪ/ *adv* excesivamente
undying /ʌnˈdaɪɪŋ/ *a* eterno
unearth /ʌnˈɜːθ/ *vt* desenterrar
unearthly /ʌnˈɜːθlɪ/ *a* sobrenatural; (*impossible*, *fam*) absurdo. **~ hour** *n* hora intempestiva
uneas|ily /ʌnˈiːzɪlɪ/ *adv* inquietamente. **~y** /ʌnˈiːzɪ/ *a* incómodo; (*worrying*) inquieto
uneconomic /ʌniːkəˈnɒmɪk/ *a* poco rentable
uneducated /ʌnˈedjʊkeɪtɪd/ *a* inculto
unemploy|ed /ʌnɪmˈplɔɪd/ *a* parado, desempleado; (*not in use*) inutilizado. **~ment** *n* paro *m*, desempleo *m*
unending /ʌnˈendɪŋ/ *a* interminable, sin fin
unequal /ʌnˈiːkwəl/ *a* desigual
unequivocal /ʌnɪˈkwɪvəkl/ *a* inequívoco
unerring /ʌnˈɜːrɪŋ/ *a* infalible
unethical /ʌnˈeθɪkl/ *a* sin ética, inmoral
uneven /ʌnˈiːvn/ *a* desigual
unexceptional /ʌnɪkˈsepʃənl/ *a* corriente
unexpected /ʌnɪkˈspektɪd/ *a* inesperado
unfailing /ʌnˈfeɪlɪŋ/ *a* inagotable; (*constant*) constante; (*loyal*) leal
unfair /ʌnˈfeə(r)/ *a* injusto. **~ly** *adv* injustamente. **~ness** *n* injusticia *f*
unfaithful /ʌnˈfeɪθfl/ *a* infiel. **~ness** *n* infidelidad *f*
unfamiliar /ʌnfəˈmɪlɪə(r)/ *a* desconocido. **be ~ with** desconocer
unfasten /ʌnˈfɑːsn/ *vt* desabrochar ‹*clothes*›; (*untie*) desatar
unfavourable /ʌnˈfeɪvərəbl/ *a* desfavorable
unfeeling /ʌnˈfiːlɪŋ/ *a* insensible
unfit /ʌnˈfɪt/ *a* inadecuado, no apto; (*unwell*) en mal estado físico; (*incapable*) incapaz
unflinching /ʌnˈflɪntʃɪŋ/ *a* resuelto
unfold /ʌnˈfəʊld/ *vt* desdoblar; (*fig*) revelar. • *vi* ‹*view etc*› extenderse
unforeseen /ʌnfɔːˈsiːn/ *a* imprevisto
unforgettable /ʌnfəˈgetəbl/ *a* inolvidable
unforgivable /ʌnfəˈgɪvəbl/ *a* imperdonable
unfortunate /ʌnˈfɔːtʃənət/ *a* desgraciado; (*regrettable*) lamentable. **~ly** *adv* desgraciadamente
unfounded /ʌnˈfaʊndɪd/ *a* infundado
unfriendly /ʌnˈfrendlɪ/ *a* poco amistoso, frío
unfurl /ʌnˈfɜːl/ *vt* desplegar
ungainly /ʌnˈgeɪnlɪ/ *a* desgarbado
ungodly /ʌnˈgɒdlɪ/ *a* impío. **~ hour** *n* (*fam*) hora *f* intempestiva
ungrateful /ʌnˈgreɪtfl/ *a* desagradecido
unguarded /ʌnˈgɑːdɪd/ *a* indefenso; (*incautious*) imprudente, incauto
unhapp|ily /ʌnˈhapɪlɪ/ *adv* infelizmente; (*unfortunately*) desgraciadamente. **~iness** *n* tristeza *f*. **~y** /ʌnˈhæpɪ/ *a* (**-ier**, **-iest**) infeliz, triste; (*unsuitable*) inoportuno. **~y with** insatisfecho de ‹*plans etc*›
unharmed /ʌnˈhɑːmd/ *a* ileso, sano y salvo
unhealthy /ʌnˈhelθɪ/ *a* (**-ier**, **-iest**) enfermizo; (*insanitary*) malsano
unhinge /ʌnˈhɪndʒ/ *vt* desquiciar
unholy /ʌnˈhəʊlɪ/ *a* (**-ier**, **-iest**) impío; (*terrible*, *fam*) terrible
unhook /ʌnˈhʊk/ *vt* desenganchar
unhoped /ʌnˈhəʊpt/ *a*. **~ for** inesperado
unhurt /ʌnˈhɜːt/ *a* ileso
unicorn /ˈjuːnɪkɔːn/ *n* unicornio *m*

unification /ju:nɪfɪ'keɪʃn/ *n* unificación *f*

uniform /'ju:nɪfɔ:m/ *a & n* uniforme (*m*). **~ity** /-'fɔ:mətɪ/ *n* uniformidad *f*. **~ly** *adv* uniformemente

unify /'ju:nɪfaɪ/ *vt* unificar

unilateral /ju:nɪ'lætərəl/ *a* unilateral

unimaginable /ʌnɪ'mædʒɪnəbl/ *a* inconcebible

unimpeachable /ʌnɪm'pi:tʃəbl/ *a* irreprensible

unimportant /ʌnɪm'pɔ:tnt/ *a* insignificante

uninhabited /ʌnɪn'hæbɪtɪd/ *a* inhabitado; (*abandoned*) despoblado

unintentional /ʌnɪn'tenʃənl/ *a* involuntario

union /'ju:njən/ *n* unión *f*; (*trade union*) sindicato *m*. **~ist** *n* sindicalista *m & f*. **U~ Jack** *n* bandera *f* del Reino Unido

unique /ju:'ni:k/ *a* único. **~ly** *adv* extraordinariamente

unisex /'ju:nɪseks/ *a* unisex(o)

unison /'ju:nɪsn/ *n*. **in ~** al unísono

unit /'ju:nɪt/ *n* unidad *f*; (*of furniture etc*) elemento *m*

unite /ju:'naɪt/ *vt* unir. ● *vi* unirse. **U~d Kingdom (UK)** *n* Reino *m* Unido. **U~d Nations (UN)** *n* Organización *f* de las Naciones Unidas (ONU). **U~d States (of America) (USA)** *n* Estados *mpl* Unidos (de América) (EE.UU.)

unity /'ju:nɪtɪ/ *n* unidad *f*; (*fig*) acuerdo *m*

univers|al /ju:nɪ'vɜ:sl/ *a* universal. **~e** /'ju:nɪvɜ:s/ *n* universo *m*

university /ju:nɪ'vɜ:sətɪ/ *n* universidad *f*. ● *a* universitario

unjust /ʌn'dʒʌst/ *a* injusto

unkempt /ʌn'kempt/ *a* desaseado

unkind /ʌn'kaɪnd/ *a* poco amable; (*cruel*) cruel. **~ly** *adv* poco amablemente. **~ness** *n* falta *f* de amabilidad; (*cruelty*) crueldad *f*

unknown /ʌn'nəʊn/ *a* desconocido

unlawful /ʌn'lɔ:fl/ *a* ilegal

unleash /ʌn'li:ʃ/ *vt* soltar; (*fig*) desencadenar

unless /ʌn'les, ən'les/ *conj* a menos que, a no ser que

unlike /ʌn'laɪk/ *a* diferente; (*not typical*) impropio de. ● *prep* a diferencia de. **~lihood** *n* improbabilidad *f*. **~ly** /ʌn'laɪklɪ/ *a* improbable

unlimited /ʌn'lɪmɪtɪd/ *a* ilimitado

unload /ʌn'ləʊd/ *vt* descargar

unlock /ʌn'lɒk/ *vt* abrir (con llave)

unluck|ily /ʌn'lʌkɪlɪ/ *adv* desgraciadamente. **~y** /ʌn'lʌkɪ/ *a* (**-ier, -iest**) desgraciado; ‹*number*› de mala suerte

unmanly /ʌn'mænlɪ/ *a* poco viril

unmanned /ʌn'mænd/ *a* no tripulado

unmarried /ʌn'mærɪd/ *a* soltero. **~ mother** *n* madre *f* soltera

unmask /ʌn'mɑ:sk/ *vt* desenmascarar. ● *vi* quitarse la máscara

unmentionable /ʌn'menʃənəbl/ *a* a que no se debe aludir

unmistakabl|e /ʌnmɪ'steɪkəbl/ *a* inconfundible. **~y** *adv* claramente

unmitigated /ʌn'mɪtɪgeɪtɪd/ *a* (*absolute*) absoluto

unmoved /ʌn'mu:vd/ *a* (*fig*) indiferente (**by** a), insensible (**by** a)

unnatural /ʌn'nætʃərəl/ *a* no natural; (*not normal*) anormal

unnecessar|ily /ʌn'nesəsərɪlɪ/ *adv* innecesariamente. **~y** /ʌn'nesəsərɪ/ *a* innecesario

unnerve /ʌn'nɜ:v/ *vt* desconcertar

unnoticed /ʌn'nəʊtɪst/ *a* inadvertido

unobtainable /ʌnəb'teɪnəbl/ *a* inasequible; (*fig*) inalcanzable

unobtrusive /ʌnəb'tru:sɪv/ *a* discreto

unofficial /ʌnə'fɪʃl/ *a* no oficial. **~ly** *adv* extraoficialmente

unpack /ʌn'pæk/ *vt* desempaquetar ‹*parcel*›; deshacer ‹*suitcase*›. ● *vi* deshacer la maleta

unpalatable /ʌn'pælətəbl/ *a* desagradable

unparalleled /ʌn'pærəleld/ *a* sin par

unpick /ʌn'pɪk/ *vt* descoser

unpleasant /ʌn'pleznt/ *a* desagradable. **~ness** *n* lo desagradable

unplug /ʌn'plʌg/ *vt* (*elec*) desenchufar

unpopular /ʌn'pɒpjʊlə(r)/ *a* impopular

unprecedented /ʌn'presɪdentɪd/ *a* sin precedente

unpredictable /ʌnprɪ'dɪktəbl/ *a* imprevisible

unpremeditated /ʌnprɪ'medɪteɪtɪd/ *a* impremeditado

unprepared /ʌnprɪ'peəd/ *a* no preparado; (*unready*) desprevenido

unprepossessing /ʌnpri:pə'zesɪŋ/ *a* poco atractivo

unpretentious /ʌnprɪ'tenʃəs/ *a* sin pretensiones, modesto

unprincipled /ʌn'prɪnsɪpld/ *a* sin principios

unprofessional /ʌnprə'feʃənəl/ *a* contrario a la ética profesional

unpublished /ʌn'pʌblɪʃt/ *a* inédito

unqualified /ʌn'kwɒlɪfaɪd/ *a* sin título; (*fig*) absoluto

unquestionabl|e /ʌn'kwestʃənəbl/ *a* indiscutible. **~y** *adv* indiscutiblemente

unquote /ʌn'kwəʊt/ *vi* cerrar comillas

unravel /ʌn'rævl/ *vt* (*pt* **unravelled**) desenredar; deshacer ‹*knitting etc*›. ● *vi* desenredarse

unreal /ʌn'rɪəl/ *a* irreal. **~istic** *a* poco realista

unreasonable /ʌn'ri:zənəbl/ *a* irrazonable

unrecognizable /ʌnrekəg'naɪzəbl/ *a* irreconocible

unrelated /ʌnrɪ'leɪtɪd/ *a* ‹*facts*› inconexo, sin relación; ‹*people*› no emparentado

unreliable /ʌnrɪ'laɪəbl/ *a* ‹*person*› poco formal; ‹*machine*› poco fiable

unrelieved /ʌnrɪ'li:vd/ *a* no aliviado

unremitting /ʌnrɪ'mɪtɪŋ/ *a* incesante

unrepentant /ʌnrɪ'pentənt/ *a* impenitente

unrequited /ʌnrɪ'kwaɪtɪd/ *a* no correspondido

unreservedly /ʌnrɪ'zɜ:vɪdlɪ/ *adv* sin reserva

unrest /ʌn'rest/ *n* inquietud *f*; (*pol*) agitación *f*

unrivalled /ʌn'raɪvld/ *a* sin par

unroll /ʌn'rəʊl/ *vt* desenrollar. ● *vi* desenrollarse

unruffled /ʌn'rʌfld/ ‹*person*› imperturbable

unruly /ʌn'ru:lɪ/ *a* indisciplinado

unsafe /ʌn'seɪf/ *a* peligroso; ‹*person*› en peligro

unsaid /ʌn'sed/ *a* sin decir

unsatisfactory /ʌnsætɪs'fæktərɪ/ *a* insatisfactorio

unsavoury /ʌn'seɪvərɪ/ *a* desagradable

unscathed /ʌn'skeɪðd/ *a* ileso

unscramble /ʌn'skræmbl/ *vt* descifrar

unscrew /ʌn'skru:/ *vt* destornillar

unscrupulous /ʌn'skru:pjʊləs/ *a* sin escrúpulos

unseat /ʌn'si:t/ *vt* (*pol*) quitar el escaño a

unseemly /ʌn'si:mlɪ/ *a* indecoroso

unseen /ʌn'si:n/ *a* inadvertido. ● *n* (*translation*) traducción *f* a primera vista

unselfish /ʌn'selfɪʃ/ *a* desinteresado

unsettle /ʌn'setl/ *vt* perturbar. **~d** *a* perturbado; ‹*weather*› variable; ‹*bill*› por pagar

unshakeable /ʌn'ʃeɪkəbl/ *a* firme

unshaven /ʌn'ʃeɪvn/ *a* sin afeitar

unsightly /ʌn'saɪtlɪ/ *a* feo

unskilled /ʌn'skɪld/ *a* inexperto. **~ worker** *n* obrero *m* no cualificado

unsociable /ʌn'səʊʃəbl/ *a* insociable

unsolicited /ʌnsə'lɪsɪtɪd/ *a* no solicitado

unsophisticated /ʌnsə'fɪstɪkeɪtɪd/ *a* sencillo

unsound /ʌn'saʊnd/ *a* defectuoso, erróneo. **of ~ mind** demente

unsparing /ʌn'speərɪŋ/ *a* pródigo; (*cruel*) cruel

unspeakable /ʌn'spi:kəbl/ *a* indecible

unspecified /ʌn'spesɪfaɪd/ *a* no especificado

unstable /ʌn'steɪbl/ *a* inestable

unsteady /ʌn'stedɪ/ *a* inestable; ‹*hand*› poco firme; ‹*step*› inseguro

unstinted /ʌn'stɪntɪd/ *a* abundante

unstuck /ʌn'stʌk/ *a* suelto. **come ~** despegarse; (*fail*, *fam*) fracasar

unstudied /ʌn'stʌdɪd/ *a* natural

unsuccessful /ʌnsək'sesfʊl/ *a* fracasado. **be ~** no tener éxito, fracasar

unsuitable /ʌn'su:təbl/ *a* inadecuado; (*inconvenient*) inconveniente

unsure /ʌn'ʃʊə(r)/ *a* inseguro

unsuspecting /ʌnsə'spektɪŋ/ *a* confiado

unthinkable /ʌn'θɪŋkəbl/ *a* inconcebible

untid|ily /ʌn'taɪdɪlɪ/ *adv* desordenadamente. **~iness** *n* desorden *m*. **~y** /ʌn'taɪdɪ/ *a* (**-ier**, **-iest**) desordenado; ‹*person*› desaseado

untie /ʌn'taɪ/ *vt* desatar

until /ən'tɪl, ʌn'tɪl/ *prep* hasta. ● *conj* hasta que

untimely /ʌn'taɪmlɪ/ *a* inoportuno; (*premature*) prematuro

untiring /ʌn'taɪərɪŋ/ *a* incansable

untold /ʌn'təʊld/ *a* incalculable

untoward /ʌntə'wɔ:d/ *a* (*inconvenient*) inconveniente

untried /ʌn'traɪd/ *a* no probado

untrue /ʌn'tru:/ *a* falso

unused /ʌn'ju:zd/ *a* nuevo. /ʌn'ju:st/ *a*. **~ to** no acostumbrado a

unusual /ʌn'ju:ʒʊəl/ *a* insólito; (*exceptional*) excepcional. **~ly** *adv* excepcionalmente
unutterable /ʌn'ʌtərəbl/ *a* indecible
unveil /ʌn'veɪl/ *vt* descubrir; (*disclose*) revelar
unwanted /ʌn'wɒntɪd/ *a* superfluo; ‹*child*› no deseado
unwarranted /ʌn'wɒrəntɪd/ *a* injustificado
unwelcome /ʌn'welkəm/ *a* desagradable; ‹*guest*› inoportuno
unwell /ʌn'wel/ *a* indispuesto
unwieldy /ʌn'wi:ldɪ/ *a* difícil de manejar
unwilling /ʌn'wɪlɪŋ/ *a* no dispuesto. **be ~** no querer. **~ly** *adv* de mala gana
unwind /ʌn'waɪnd/ *vt* (*pt* **unwound**) desenvolver. ● *vi* desenvolverse; (*relax, fam*) relajarse
unwise /ʌn'waɪz/ *a* imprudente
unwitting /ʌn'wɪtɪŋ/ *a* inconsciente; (*involuntary*) involuntario. **~ly** *adv* involuntariamente
unworthy /ʌn'wɜ:ðɪ/ *a* indigno
unwrap /ʌn'ræp/ *vt* (*pt* **unwrapped**) desenvolver, deshacer
unwritten /ʌn'rɪtn/ *a* no escrito; ‹*agreement*› tácito
up /ʌp/ *adv* arriba; (*upwards*) hacia arriba; (*higher*) más arriba; (*out of bed*) levantado; (*finished*) terminado. **~ here** aquí arriba. **~ in** (*fam*) versado en, fuerte en. **~ there** allí arriba. **~ to** hasta. **be one ~ on** llevar la ventaja a. **be ~ against** enfrentarse con. **be ~ to** tramar ‹*plot*›; (*one's turn*) tocar a; a la altura de ‹*task*›; (*reach*) llegar a. **come ~** subir. **feel ~ to it** sentirse capaz. **go ~** subir. **it's ~ to you** depende de tí. **what is ~?** ¿qué pasa? ● *prep* arriba; (*on top of*) en lo alto de. ● *vt* (*pt* **upped**) aumentar. ● *n*. **~s and downs** *npl* altibajos *mpl*
upbraid /ʌp'breɪd/ *vt* reprender
upbringing /'ʌpbrɪŋɪŋ/ *n* educación *f*
update /ʌp'deɪt/ *vt* poner al día
upgrade /ʌp'greɪd/ *vt* ascender ‹*person*›; mejorar ‹*equipment*›
upheaval /ʌp'hi:vl/ *n* trastorno *m*
uphill /'ʌphɪl/ *a* ascendente; (*fig*) arduo. ● *adv* /ʌp'hɪl/ cuesta arriba. **go ~** subir
uphold /ʌp'həʊld/ *vt* (*pt* **upheld**) sostener
upholster /ʌp'həʊlstə(r)/ *vt* tapizar. **~er** /-rə(r)/ *n* tapicero *m*. **~y** *n* tapicería *f*
upkeep /'ʌpki:p/ *n* mantenimiento *m*
up-market /ʌp'mɑ:kɪt/ *a* superior
upon /ə'pɒn/ *prep* en; (*on top of*) encima de. **once ~ a time** érase una vez
upper /'ʌpə(r)/ *a* superior. **~ class** *n* clases *fpl* altas. **~ hand** *n* dominio *m*, ventaja *f*. **~most** *a* (el) más alto. ● *n* (*of shoe*) pala *f*
uppish /'ʌpɪʃ/ *a* engreído
upright /'ʌpraɪt/ *a* derecho; ‹*piano*› vertical. ● *n* montante *m*
uprising /'ʌpraɪzɪŋ/ *n* sublevación *f*
uproar /'ʌprɔ:(r)/ *n* tumulto *m*. **~ious** /-'rɔ:rɪəs/ *a* tumultuoso
uproot /ʌp'ru:t/ *vt* desarraigar
upset /ʌp'set/ *vt* (*pt* **upset**, *presp* **upsetting**) trastornar; desbaratar ‹*plan etc*›; (*distress*) alterar. /'ʌpset/ *n* trastorno *m*
upshot /'ʌpʃɒt/ *n* resultado *m*
upside-down /ʌpsaɪd'daʊn/ *adv* al revés; (*in disorder*) patas arriba. **turn ~** volver
upstairs /ʌp'steəz/ *adv* arriba. /'ʌpsteəz/ *a* de arriba
upstart /'ʌpstɑ:t/ *n* arribista *m & f*
upstream /'ʌpstri:m/ *adv* río arriba; (*against the current*) contra la corriente
upsurge /'ʌpsɜ:dʒ/ *n* aumento *m*; (*of anger etc*) arrebato *m*
uptake /'ʌpteɪk/ *n*. **quick on the ~** muy listo
uptight /'ʌptaɪt/ *a* (*fam*) nervioso
up-to-date /ʌptə'deɪt/ *a* al día; ‹*news*› de última hora; (*modern*) moderno
upturn /'ʌptɜ:n/ *n* aumento *m*; (*improvement*) mejora *f*
upward /'ʌpwəd/ *a* ascendente. ● *adv* hacia arriba. **~s** *adv* hacia arriba
uranium /jʊ'reɪnɪəm/ *n* uranio *m*
urban /'ɜ:bən/ *a* urbano
urbane /ɜ:'beɪn/ *a* cortés
urbanize /'ɜ:bənaɪz/ *vt* urbanizar
urchin /'ɜ:tʃɪn/ *n* pilluelo *m*
urge /ɜ:dʒ/ *vt* incitar, animar. ● *n* impulso *m*. **~ on** animar
urgen|cy /'ɜ:dʒənsɪ/ *n* urgencia *f*. **~t** /'ɜ:dʒənt/ *a* urgente. **~tly** *adv* urgentemente
urin|ate /'jʊərɪneɪt/ *vi* orinar. **~e** /'jʊərɪn/ *n* orina *f*

urn /ɜ:n/ *n* urna *f*

Uruguay /jʊərəgwaɪ/ *n* el Uruguay *m*. **~an** *a & n* uruguayo (*m*)

us /ʌs, əs/ *pron* nos; (*after prep*) nosotros, nosotras

US(A) /ju:es'eɪ/ *abbr* (*United States* (*of America*)) EE.UU., Estados *mpl* Unidos

usage /'ju:zɪdʒ/ *n* uso *m*

use /ju:z/ *vt* emplear. /ju:s/ *n* uso *m*, empleo *m*. **be of ~** servir. **it is no ~** es inútil, no sirve para nada. **make ~ of** servirse de. **~ up** agotar, consumir. **~d** /ju:zd/ *a* ‹*clothes*› gastado. /ju:st/ *pt*. **he ~d to say** decía, solía decir. ● *a*. **~d to** acostumbrado a. **~ful** /'ju:sfl/ *a* útil. **~fully** *adv* útilmente. **~less** *a* inútil; ‹*person*› incompetente. **~r** /-zə(r)/ *n* usuario *m*

usher /'ʌʃə(r)/ *n* ujier *m*; (*in theatre etc*) acomodador *m*. ● *vt*. **~ in** hacer entrar. **~ette** *n* acomodadora *f*

USSR *abbr* (*history*) (*Union of Soviet Socialist Republics*) URSS

usual /'ju:ʒʊəl/ *a* usual, corriente; (*habitual*) acostumbrado, habitual. **as ~** como de costumbre, como siempre. **~ly** *adv* normalmente. **he ~ly wakes up early** suele despertarse temprano

usurer /'ju:ʒərə(r)/ *n* usurero *m*

usurp /jʊ'zɜ:p/ *vt* usurpar. **~er** /-ə(r)/ *n* usurpador *m*

usury /'ju:ʒərɪ/ *n* usura *f*

utensil /ju:'tensl/ *n* utensilio *m*

uterus /'ju:tərəs/ *n* útero *m*

utilitarian /ju:tɪlɪ'teərɪən/ *a* utilitario

utility /ju:'tɪlətɪ/ *n* utilidad *f*. **public ~** *n* servicio *m* público. ● *a* utilitario

utilize /'ju:tɪlaɪz/ *vt* utilizar

utmost /'ʌtməʊst/ *a* extremo. ● *n*. **one's ~** todo lo posible

utter[1] /'ʌtə(r)/ *a* completo

utter[2] /'ʌtə(r)/ *vt* (*speak*) pronunciar; dar ‹*sigh*›; emitir ‹*sound*›. **~ance** *n* expresión *f*

utterly /'ʌtəlɪ/ *adv* totalmente

U-turn /'ju:tɜ:n/ *n* vuelta *f*

V

vacan|cy /'veɪkənsɪ/ *n* (*job*) vacante *f*; (*room*) habitación *f* libre. **~t** *a* libre; (*empty*) vacío; ‹*look*› vago

vacate /və'keɪt/ *vt* dejar

vacation /və'keɪʃn/ *n* (*Amer*) vacaciones *fpl*

vaccin|ate /'væksɪneɪt/ *vt* vacunar. **~ation** /-'neɪʃn/ *n* vacunación *f*. **~e** /'væksi:n/ *n* vacuna *f*

vacuum /'vækjʊəm/ *n* (*pl* **-cuums** *or* **-cua**) vacío *m*. **~ cleaner** *n* aspiradora *f*. **~ flask** *n* termo *m*

vagabond /'vægəbɒnd/ *n* vagabundo *m*

vagary /'veɪgərɪ/ *n* capricho *m*

vagina /və'dʒaɪnə/ *n* vagina *f*

vagrant /'veɪgrənt/ *n* vagabundo *m*

vague /veɪg/ *a* (**-er**, **-est**) vago; ‹*outline*› indistinto. **be ~ about** no precisar. **~ly** *adv* vagamente

vain /veɪn/ *a* (**-er**, **-est**) vanidoso; (*useless*) vano, inútil. **in ~** en vano. **~ly** *adv* vanamente

valance /'væləns/ *n* cenefa *f*

vale /veɪl/ *n* valle *m*

valentine /'væləntaɪn/ *n* (*card*) tarjeta *f* del día de San Valentín

valet /'vælɪt, 'væleɪ/ *n* ayuda *m* de cámara

valiant /'vælɪənt/ *a* valeroso

valid /'vælɪd/ *a* válido; ‹*ticket*› valedero. **~ate** *vt* dar validez a; (*confirm*) convalidar. **~ity** /-'ɪdətɪ/ *n* validez *f*

valley /'vælɪ/ *n* (*pl* **-eys**) valle *m*

valour /'vælə(r)/ *n* valor *m*

valuable /'væljʊəbl/ *a* valioso. **~s** *npl* objetos *mpl* de valor

valuation /væljʊ'eɪʃn/ *n* valoración *f*

value /'vælju:/ *n* valor *m*; (*usefulness*) utilidad *f*. **face ~** *n* valor *m* nominal; (*fig*) significado *m* literal. ● *vt* valorar; (*cherish*) apreciar. **~ added tax (VAT)** *n* impuesto *m* sobre el valor añadido (IVA). **~d** *a* (*appreciated*) apreciado, estimado. **~r** /-ə(r)/ *n* tasador *m*

valve /vælv/ *n* válvula *f*

vampire /'væmpaɪə(r)/ *n* vampiro *m*

van /væn/ *n* furgoneta *f*; (*rail*) furgón *m*

vandal /'vændl/ *n* vándalo *m*. **~ism** /-əlɪzəm/ *n* vandalismo *m*. **~ize** *vt* destruir

vane /veɪn/ *n* (*weathercock*) veleta *f*; (*naut, aviat*) paleta *f*

vanguard /'vængɑ:d/ *n* vanguardia *f*

vanilla /və'nɪlə/ *n* vainilla *f*

vanish /'vænɪʃ/ *vi* desaparecer

vanity /'vænɪtɪ/ *n* vanidad *f*. **~ case** *n* neceser *m*

vantage /'vɑ:ntɪdʒ/ *n* ventaja *f*. **~point** *n* posición *f* ventajosa
vapour /'veɪpə(r)/ *n* vapor *m*
variable /'veərɪəbl/ *a* variable
varian|ce /'veərɪəns/ *n*. **at ~ce** en desacuerdo. **~t** /'veərɪənt/ *a* diferente. ● *n* variante *m*
variation /veərɪ'eɪʃn/ *n* variación *f*
varicoloured /'veərɪkʌləd/ *a* multicolor
varied /'veərɪd/ *a* variado
varicose /'værɪkəʊs/ *a* varicoso. **~ veins** *npl* varices *fpl*
variety /və'raɪətɪ/ *n* variedad *f*. **~ show** *n* espectáculo *m* de variedades
various /'veərɪəs/ *a* diverso. **~ly** *adv* diversamente
varnish /'vɑ:nɪʃ/ *n* barniz *m*; (*for nails*) esmalte *m*. ● *vt* barnizar
vary /'veərɪ/ *vt/i* variar. **~ing** *a* diverso
vase /vɑ:z, *Amer* veɪs/ *n* jarrón *m*
vasectomy /və'sektəmɪ/ *n* vasectomía *f*
vast /vɑ:st/ *a* vasto, enorme. **~ly** *adv* enormemente. **~ness** *n* inmensidad *f*
vat /væt/ *n* tina *f*
VAT /vi:eɪ'ti:/ *abbr* (*value added tax*) IVA *m*, impuesto *m* sobre el valor añadido
vault /vɔ:lt/ *n* (*roof*) bóveda *f*; (*in bank*) cámara *f* acorazada; (*tomb*) cripta *f*; (*cellar*) sótano *m*; (*jump*) salto *m*. ● *vt/i* saltar
vaunt /vɔ:nt/ *vt* jactarse de
veal /vi:l/ *n* ternera *f*
veer /vɪə(r)/ *vi* cambiar de dirección; (*naut*) virar
vegetable /'vedʒɪtəbl/ *a* vegetal. ● *n* legumbre *m*; (*greens*) verduras *fpl*
vegetarian /vedʒɪ'teərɪən/ *a & n* vegetariano (*m*)
vegetate /'vedʒɪteɪt/ *vi* vegetar
vegetation /vedʒɪ'teɪʃn/ *n* vegetación *f*
vehemen|ce /'vi:əməns/ *n* vehemencia *f*. **~t** /'vi:əmənt/ *a* vehemente. **~tly** *adv* con vehemencia
vehicle /'vi:ɪkl/ *n* vehículo *m*
veil /veɪl/ *n* velo *m*. **take the ~** hacerse monja. ● *vt* velar
vein /veɪn/ *n* vena *f*; (*mood*) humor *m*. **~ed** *a* veteado
velocity /vɪ'lɒsɪtɪ/ *n* velocidad *f*
velvet /'velvɪt/ *n* terciopelo *m*. **~y** *a* aterciopelado
venal /'vi:nl/ *a* venal. **~ity** /-'nælətɪ/ *n* venalidad *f*
vendetta /ven'detə/ *n* enemistad *f* prolongada
vending-machine /'vendɪŋ məʃi:n/ *n* distribuidor *m* automático
vendor /'vendə(r)/ *n* vendedor *m*
veneer /və'nɪə(r)/ *n* chapa *f*; (*fig*) barniz *m*, apariencia *f*
venerable /'venərəbl/ *a* venerable
venereal /və'nɪərɪəl/ *a* venéreo
Venetian /və'ni:ʃn/ *a & n* veneciano (*m*). **v~ blind** *n* persiana *f* veneciana
vengeance /'vendʒəns/ *n* venganza *f*. **with a ~** (*fig*) con creces
venison /'venɪzn/ *n* carne *f* de venado
venom /'venəm/ *n* veneno *m*. **~ous** *a* venenoso
vent /vent/ *n* abertura *f*; (*for air*) respiradero *m*. **give ~ to** dar salida a. ● *vt* hacer un agujero en; (*fig*) desahogar
ventilat|e /'ventɪleɪt/ *vt* ventilar. **~ion** /-'leɪʃn/ *n* ventilación *f*. **~or** /-ə(r)/ *n* ventilador *m*
ventriloquist /ven'trɪləkwɪst/ *n* ventrílocuo *m*
venture /'ventʃə(r)/ *n* empresa *f* (arriesgada). **at a ~** a la ventura. ● *vt* arriesgar. ● *vi* atreverse
venue /'venju:/ *n* lugar *m* (de reunión)
veranda /və'rændə/ *n* terraza *f*
verb /vɜ:b/ *n* verbo *m*
verbal /'vɜ:bl/ *a* verbal. **~ly** *adv* verbalmente
verbatim /vɜ:'beɪtɪm/ *adv* palabra por palabra, al pie de la letra
verbose /vɜ:'bəʊs/ *a* prolijo
verdant /'vɜ:dənt/ *a* verde
verdict /'vɜ:dɪkt/ *n* veredicto *m*; (*opinion*) opinión *f*
verge /vɜ:dʒ/ *n* borde *m*. ● *vt*. **~ on** acercarse a
verger /'vɜ:dʒə(r)/ *n* sacristán *m*
verif|ication /verɪfɪ'keɪʃn/ *n* verificación *f*. **~y** /'verɪfaɪ/ *vt* verificar
veritable /'verɪtəbl/ *a* verdadero
vermicelli /vɜ:mɪ'tʃelɪ/ *n* fideos *mpl*
vermin /'vɜ:mɪn/ *n* sabandijas *fpl*
vermouth /'vɜ:məθ/ *n* vermut *m*
vernacular /və'nækjʊlə(r)/ *n* lengua *f*; (*regional*) dialecto *m*
versatil|e /'vɜ:sətaɪl/ *a* versátil. **~ity** /-'tɪlətɪ/ *n* versatilidad *f*
verse /vɜ:s/ *n* estrofa *f*; (*poetry*) poesías *fpl*; (*of Bible*) versículo *m*

versed /vɜ:st/ *a*. **~ in** versado en
version /'vɜ:ʃn/ *n* versión *f*
versus /'vɜ:səs/ *prep* contra
vertebra /'vɜ:tɪbrə/ *n* (*pl* **-brae** /-bri:/) vértebra *f*
vertical /'vɜ:tɪkl/ *a* & *n* vertical (*f*). **~ly** *adv* verticalmente
vertigo /'vɜ:tɪgəʊ/ *n* vértigo *m*
verve /vɜ:v/ *n* entusiasmo *m*, vigor *m*
very /'verɪ/ *adv* muy. **~ much** muchísimo. **~ well** muy bien. **the ~ first** el primero de todos. ● *a* mismo. **the ~ thing** exactamente lo que hace falta
vespers /'vespəz/ *npl* vísperas *fpl*
vessel /'vesl/ *n* (*receptacle*) recipiente *m*; (*ship*) buque *m*; (*anat*) vaso *m*
vest /vest/ *n* camiseta *f*; (*Amer*) chaleco *m*. ● *vt* conferir. **~ed interest** *n* interés *m* personal; (*jurid*) derecho *m* adquirido
vestige /'vestɪdʒ/ *n* vestigio *m*
vestment /'vestmənt/ *n* vestidura *f*
vestry /'vestrɪ/ *n* sacristía *f*
vet /vet/ *n* (*fam*) veterinario *m*. ● *vt* (*pt* **vetted**) examinar
veteran /'vetərən/ *n* veterano *m*
veterinary /'vetərɪnərɪ/ *a* veterinario. **~ surgeon** *n* veterinario *m*
veto /'vi:təʊ/ *n* (*pl* **-oes**) veto *m*. ● *vt* poner el veto a
vex /veks/ *vt* fastidiar. **~ation** /-'seɪʃn/ *n* fastidio *m*. **~ed question** *n* cuestión *f* controvertida. **~ing** *a* fastidioso
via /'vaɪə/ *prep* por, por vía de
viab|ility /vaɪə'bɪlətɪ/ *n* viabilidad *f*. **~le** /'vaɪəbl/ *a* viable
viaduct /'vaɪədʌkt/ *n* viaducto *m*
vibrant /'vaɪbrənt/ *a* vibrante
vibrat|e /vaɪ'breɪt/ *vt/i* vibrar. **~ion** /-ʃn/ *n* vibración *f*
vicar /'vɪkə(r)/ *n* párroco *m*. **~age** /-rɪdʒ/ *n* casa *f* del párroco
vicarious /vɪ'keərɪəs/ *a* indirecto
vice[1] /vaɪs/ *n* vicio *m*
vice[2] /vaɪs/ *n* (*tec*) torno *m* de banco
vice... /'vaɪs/ *pref* vice...
vice versa /vaɪsɪ'vɜ:sə/ *adv* viceversa
vicinity /vɪ'sɪnɪtɪ/ *n* vecindad *f*. **in the ~ of** cerca de
vicious /'vɪʃəs/ *a* (*spiteful*) malicioso; (*violent*) atroz. **~ circle** *n* círculo *m* vicioso. **~ly** *adv* cruelmente
vicissitudes /vɪ'sɪsɪtju:dz/ *npl* vicisitudes *fpl*
victim /'vɪktɪm/ *n* víctima *f*. **~ization** /-aɪ'zeɪʃn/ *n* persecución *f*. **~ize** *vt* victimizar
victor /'vɪktə(r)/ *n* vencedor *m*
Victorian /vɪk'tɔ:rɪən/ *a* victoriano
victor|ious /vɪk'tɔ:rɪəs/ *a* victorioso. **~y** /'vɪktərɪ/ *n* victoria *f*
video /'vɪdɪəʊ/ *a* video. ● *n* (*fam*) magnetoscopio *m*. **~ recorder** *n* magnetoscopio *m*. **~tape** *n* videocassette *f*
vie /vaɪ/ *vi* (*pres p* **vying**) rivalizar
view /vju:/ *n* vista *f*; (*mental survey*) visión *f* de conjunto; (*opinion*) opinión *f*. **in my ~** a mi juicio. **in ~ of** en vista de. **on ~** expuesto. **with a ~ to** con miras a. ● *vt* ver; (*visit*) visitar; (*consider*) considerar. **~er** /-ə(r)/ *n* espectador *m*; (*TV*) televidente *m* & *f*. **~finder** /'vju:faɪndə(r)/ *n* visor *m*. **~point** /'vju:pɔɪnt/ *n* punto *m* de vista
vigil /'vɪdʒɪl/ *n* vigilia *f*. **~ance** *n* vigilancia *f*. **~ant** *a* vigilante. **keep ~** velar
vigo|rous /'vɪgərəs/ *a* vigoroso. **~ur** /'vɪgə(r)/ *n* vigor *m*
vile /vaɪl/ *a* (*base*) vil; (*bad*) horrible; ‹*weather*, *temper*› de perros
vilif|ication /vɪlɪfɪ'keɪʃn/ *n* difamación *f*. **~y** /'vɪlɪfaɪ/ *vt* difamar
village /'vɪlɪdʒ/ *n* aldea *f*. **~r** /-ə(r)/ *n* aldeano *m*
villain /'vɪlən/ *n* malvado *m*; (*in story etc*) malo *m*. **~ous** *a* infame. **~y** *n* infamia *f*
vim /vɪm/ *n* (*fam*) energía *f*
vinaigrette /vɪnɪ'gret/ *n*. **~ sauce** *n* vinagreta *f*
vindicat|e /'vɪndɪkeɪt/ *vt* vindicar. **~ion** /-'keɪʃn/ *n* vindicación *f*
vindictive /vɪn'dɪktɪv/ *a* vengativo. **~ness** *n* carácter *m* vengativo
vine /vaɪn/ *n* vid *f*
vinegar /'vɪnɪgə(r)/ *n* vinagre *m*. **~y** *a* ‹*person*› avinagrado
vineyard /'vɪnjəd/ *n* viña *f*
vintage /'vɪntɪdʒ/ *n* (*year*) cosecha *f*. ● *a* ‹*wine*› añejo; ‹*car*› de época
vinyl /'vaɪnɪl/ *n* vinilo *m*
viola /vɪ'əʊlə/ *n* viola *f*
violat|e /'vaɪəleɪt/ *vt* violar. **~ion** /-'leɪʃn/ *n* violación *f*
violen|ce /'vaɪələns/ *n* violencia *f*. **~t** /'vaɪələnt/ *a* violento. **~tly** *adv* violentamente
violet /'vaɪələt/ *a* & *n* violeta (*f*)
violin /'vaɪəlɪn/ *n* violín *m*. **~ist** *n* violinista *m* & *f*

VIP /vi:aɪ'pi:/ *abbr* (*very important person*) personaje *m*
viper /'vaɪpə(r)/ *n* víbora *f*
virgin /'vɜ:dʒɪn/ *a & n* virgen (*f*). **~al** *a* virginal. **~ity** /və'dʒɪnətɪ/ *n* virginidad *f*
Virgo /'vɜ:gəʊ/ *n* (*astr*) Virgo *f*
viril|e /'vɪraɪl/ *a* viril. **~ity** /-'rɪlətɪ/ *n* virilidad *f*
virtual /'vɜ:tʃʊəl/ *a* verdadero. **a ~ failure** prácticamente un fracaso. **~ly** *adv* prácticamente
virtue /'vɜ:tʃu:/ *n* virtud *f*. **by ~ of, in ~ of** en virtud de
virtuoso /vɜ:tjʊ'əʊzəʊ/ *n* (*pl* **-si** /-zi:/) virtuoso *m*
virtuous /'vɜ:tʃʊəs/ *a* virtuoso
virulent /'vɪrʊlənt/ *a* virulento
virus /'vaɪərəs/ *n* (*pl* **-uses**) virus *m*
visa /'vi:zə/ *n* visado *m*, visa *f* (*LAm*)
vis-a-vis /vi:zɑ:'vi:/ *adv* frente a frente. ● *prep* respecto a; (*opposite*) en frente de
viscount /'vaɪkaʊnt/ *n* vizconde *m*. **~ess** *n* vizcondesa *f*
viscous /'vɪskəs/ *a* viscoso
visib|ility /vɪzɪ'bɪlətɪ/ *n* visibilidad *f*. **~le** /'vɪzɪbl/ *a* visible. **~ly** *adv* visiblemente
vision /'vɪʒn/ *n* visión *f*; (*sight*) vista *f*. **~ary** /'vɪʒənərɪ/ *a & n* visionario (*m*)
visit /'vɪzɪt/ *vt* visitar; hacer una visita a ‹*person*›. ● *vi* hacer visitas. ● *n* visita *f*. **~or** *n* visitante *m & f*; (*guest*) visita *f*; (*in hotel*) cliente *m & f*
visor /'vaɪzə(r)/ *n* visera *f*
vista /'vɪstə/ *n* perspectiva *f*
visual /'vɪʒʊəl/ *a* visual. **~ize** /'vɪʒʊəlaɪz/ *vt* imaginar(se); (*foresee*) prever. **~ly** *adv* visualmente
vital /'vaɪtl/ *a* vital; (*essential*) esencial
vitality /vaɪ'tælətɪ/ *n* vitalidad *f*
vital: **~ly** /'vaɪtəlɪ/ *adv* extremadamente. **~s** *npl* órganos *mpl* vitales. **~ statistics** *npl* (*fam*) medidas *fpl*
vitamin /'vɪtəmɪn/ *n* vitamina *f*
vitiate /'vɪʃɪeɪt/ *vt* viciar
vitreous /'vɪtrɪəs/ *a* vítreo
vituperat|e /vɪ'tju:pəreɪt/ *vt* vituperar. **~ion** /-'reɪʃn/ *n* vituperación *f*
vivaci|ous /vɪ'veɪʃəs/ *a* animado, vivo. **~ously** *adv* animadamente. **~ty** /-'væsətɪ/ *n* viveza *f*
vivid /'vɪvɪd/ *a* vivo. **~ly** *adv* intensamente; (*describe*) gráficamente. **~ness** *n* viveza *f*
vivisection /vɪvɪ'sekʃn/ *n* vivisección *f*
vixen /'vɪksn/ *n* zorra *f*
vocabulary /və'kæbjʊlərɪ/ *n* vocabulario *m*
vocal /'vəʊkl/ *a* vocal; (*fig*) franco. **~ist** *n* cantante *m & f*
vocation /vəʊ'keɪʃn/ *n* vocación *f*. **~al** *a* profesional
vocifer|ate /və'sɪfəreɪt/ *vt/i* vociferar. **~ous** *a* vociferador
vogue /vəʊg/ *n* boga *f*. **in ~** de moda
voice /vɔɪs/ *n* voz *f*. ● *vt* expresar
void /vɔɪd/ *a* vacío; (*not valid*) nulo. **~ of** desprovisto de. ● *n* vacío *m*. ● *vt* anular
volatile /'vɒlətaɪl/ *a* volátil; ‹*person*› voluble
volcan|ic /vɒl'kænɪk/ *a* volcánico. **~o** /vɒl'keɪnəʊ/ *n* (*pl* **-oes**) volcán *m*
volition /və'lɪʃn/ *n*. **of one's own ~** de su propia voluntad
volley /'vɒlɪ/ *n* (*pl* **-eys**) (*of blows*) lluvia *f*; (*of gunfire*) descarga *f* cerrada
volt /vəʊlt/ *n* voltio *m*. **~age** *n* voltaje *m*
voluble /'vɒljʊbl/ *a* locuaz
volume /'vɒlju:m/ *n* volumen *m*; (*book*) tomo *m*
voluminous /və'lju:mɪnəs/ *a* voluminoso
voluntar|ily /'vɒləntərəlɪ/ *adv* voluntariamente. **~y** /'vɒləntərɪ/ *a* voluntario
volunteer /vɒlən'tɪə(r)/ *n* voluntario *m*. ● *vt* ofrecer. ● *vi* ofrecerse voluntariamente; (*mil*) alistarse como voluntario
voluptuous /və'lʌptjʊəs/ *a* voluptuoso
vomit /'vɒmɪt/ *vt/i* vomitar. ● *n* vómito *m*
voracious /və'reɪʃəs/ *a* voraz
vot|e /vəʊt/ *n* voto *m*; (*right*) derecho *m* de votar. ● *vi* votar. **~er** /-ə(r)/ *n* votante *m & f*. **~ing** *n* votación *f*
vouch /vaʊtʃ/ *vi*. **~ for** garantizar
voucher /'vaʊtʃə(r)/ *n* vale *m*
vow /vaʊ/ *n* voto *m*. ● *vi* jurar
vowel /'vaʊəl/ *n* vocal *f*
voyage /'vɔɪɪdʒ/ *n* viaje *m* (en barco)
vulgar /'vʌlgə(r)/ *a* vulgar. **~ity** /-'gærətɪ/ *n* vulgaridad *f*. **~ize** *vt* vulgarizar

vulnerab|ility /vʌlnərə'bɪlətɪ/ *n* vulnerabilidad *f*. **~le** /'vʌlnərəbl/ *a* vulnerable
vulture /'vʌltʃə(r)/ *n* buitre *m*
vying /'vaɪɪŋ/ *see* **vie**

W

wad /wɒd/ *n* (*pad*) tapón *m*; (*bundle*) lío *m*; (*of notes*) fajo *m*; (*of cotton wool etc*) bolita *f*
wadding /'wɒdɪŋ/ *n* relleno *m*
waddle /'wɒdl/ *vi* contonearse
wade /weɪd/ *vt* vadear. ● *vi*. **~ through** abrirse paso entre; leer con dificultad ‹*book*›
wafer /'weɪfə(r)/ *n* barquillo *m*; (*relig*) hostia *f*
waffle[1] /'wɒfl/ *n* (*fam*) palabrería *f*. ● *vi* (*fam*) divagar
waffle[2] /'wɒfl/ *n* (*culin*) gofre *m*
waft /wɒft/ *vt* llevar por el aire. ● *vi* flotar
wag /wæg/ *vt* (*pt* **wagged**) menear. ● *vi* menearse
wage /weɪdʒ/ *n*. **~s** *npl* salario *m*. ● *vt*. **~ war** hacer la guerra. **~r** /'weɪdʒə(r)/ *n* apuesta *f*. ● *vt* apostar
waggle /'wægl/ *vt* menear. ● *vi* menearse
wagon /'wægən/ *n* carro *m*; (*rail*) vagón *m*. **be on the ~** (*sl*) no beber
waif /weɪf/ *n* niño *m* abandonado
wail /weɪl/ *vi* lamentarse. ● *n* lamento *m*
wainscot /'weɪnskət/ *n* revestimiento *m*, zócalo *m*
waist /weɪst/ *n* cintura *f*. **~band** *n* cinturón *m*
waistcoat /'weɪstkəʊt/ *n* chaleco *m*
waistline /'weɪstlaɪn/ *n* cintura *f*
wait /weɪt/ *vt/i* esperar; (*at table*) servir. **~ for** esperar. **~ on** servir. ● *n* espera *f*. **lie in ~** acechar
waiter /'weɪtə(r)/ *n* camarero *m*
wait: **~ing-list** *n* lista *f* de espera. **~ing-room** *n* sala *f* de espera
waitress /'weɪtrɪs/ *n* camarera *f*
waive /weɪv/ *vt* renunciar a
wake[1] /weɪk/ *vt* (*pt* **woke**, *pp* **woken**) despertar. ● *vi* despertarse. ● *n* velatorio *m*. **~ up** *vt* despertar. ● *vi* despertarse
wake[2] /weɪk/ *n* (*naut*) estela *f*. **in the ~ of** como resultado de, tras
waken /'weɪkən/ *vt* despertar. ● *vi* despertarse
wakeful /'weɪkfl/ *a* insomne
Wales /weɪlz/ *n* País *m* de Gales
walk /wɔ:k/ *vi* andar; (*not ride*) ir a pie; (*stroll*) pasearse. **~ out** salir; ‹*workers*› declararse en huelga. **~ out on** abandonar. ● *vt* andar por ‹*streets*›; llevar de paseo ‹*dog*›. ● *n* paseo *m*; (*gait*) modo *m* de andar; (*path*) sendero *m*. **~ of life** clase *f* social. **~about** /'wɔ:kəbaʊt/ *n* (*of royalty*) encuentro *m* con el público. **~er** /-ə(r)/ *n* paseante *m & f*
walkie-talkie /wɔ:kɪ'tɔ:kɪ/ *n* transmisor-receptor *m* portátil
walking /'wɔ:kɪŋ/ *n* paseo *m*. **~-stick** *n* bastón *m*
Walkman /'wɔ:kmən/ *n* (*P*) estereo *m* personal, Walkman *m* (*P*), magnetófono *m* de bolsillo
walk: **~-out** *n* huelga *f*. **~-over** *n* victoria *f* fácil
wall /wɔ:l/ *n* (*interior*) pared *f*; (*exterior*) muro *m*; (*in garden*) tapia *f*; (*of city*) muralla *f*. **go to the ~** fracasar. **up the ~** (*fam*) loco. ● *vt* amurallar ‹*city*›
wallet /'wɒlɪt/ *n* cartera *f*, billetera *f* (*LAm*)
wallflower /'wɔ:lflaʊə(r)/ *n* alhelí *m*
wallop /'wɒləp/ *vt* (*pt* **walloped**) (*sl*) golpear con fuerza. ● *n* (*sl*) golpe *m* fuerte
wallow /'wɒləʊ/ *vi* revolcarse
wallpaper /'wɔ:lpeɪpə(r)/ *n* papel *m* pintado
walnut /'wɔ:lnʌt/ *n* nuez *f*; (*tree*) nogal *m*
walrus /'wɔ:lrəs/ *n* morsa *f*
waltz /wɔ:ls/ *n* vals *m*. ● *vi* valsar
wan /wɒn/ *a* pálido
wand /wɒnd/ *n* varita *f*
wander /'wɒndə(r)/ *vi* vagar; (*stroll*) pasearse; (*digress*) divagar; ‹*road, river*› serpentear. ● *n* paseo *m*. **~er** /-ə(r)/ *n* vagabundo *m*. **~lust** /'wɒndəlʌst/ *n* pasión *f* por los viajes
wane /weɪn/ *vi* menguar. ● *n*. **on the ~** disminuyendo
wangle /wæŋgl/ *vt* (*sl*) agenciarse
want /wɒnt/ *vt* querer; (*need*) necesitar; (*require*) exigir. ● *vi*. **~ for** carecer de. ● *n* necesidad *f*; (*lack*) falta *f*; (*desire*) deseo *m*. **~ed** *a* ‹*criminal*› buscado. **~ing** *a* (*lacking*) falto de. **be ~ing** carecer de
wanton /'wɒntən/ *a* (*licentious*) lascivo; (*motiveless*) sin motivo
war /wɔ:(r)/ *n* guerra *f*. **at ~** en guerra

warble /ˈwɔːbl/ *vt* cantar trinando. ● *vi* gorjear. ● *n* gorjeo *m*. **~r** /-ə(r)/ *n* curruca *f*

ward /wɔːd/ *n* (*in hospital*) sala *f*; (*of town*) barrio *m*; (*child*) pupilo *m*. ● *vt*. **~ off** parar

warden /ˈwɔːdn/ *n* guarda *m*

warder /ˈwɔːdə(r)/ *n* carcelero *m*

wardrobe /ˈwɔːdrəʊb/ *n* armario *m*; (*clothes*) vestuario *m*

warehouse /ˈweəhaʊs/ *n* almacén *m*

wares /weəz/ *npl* mercancías *fpl*

war: **~fare** /ˈwɔːfeə(r)/ *n* guerra *f*. **~head** /ˈwɔːhed/ *n* cabeza *f* explosiva

warily /ˈweərɪlɪ/ *adv* cautelosamente

warlike /ˈwɔːlaɪk/ *a* belicoso

warm /wɔːm/ *a* (**-er**, **-est**) caliente; (*hearty*) caluroso. **be ~** ‹*person*› tener calor. **it is ~** hace calor. ● *vt*. **~ (up)** calentar; recalentar ‹*food*›; (*fig*) animar. ● *vi*. **~ (up)** calentarse; (*fig*) animarse. **~ to** tomar simpatía a ‹*person*›; ir entusiasmándose por ‹*idea etc*›. **~-blooded** *a* de sangre caliente. **~-hearted** *a* simpático. **~ly** *adv* (*heartily*) calurosamente

warmonger /ˈwɔːmʌŋgə(r)/ *n* belicista *m & f*

warmth /wɔːmθ/ *n* calor *m*

warn /wɔːn/ *vt* avisar, advertir. **~ing** *n* advertencia *f*; (*notice*) aviso *m*. **~ off** (*advise against*) aconsejar en contra de; (*forbid*) impedir

warp /wɔːp/ *vt* deformar; (*fig*) pervertir. ● *vi* deformarse

warpath /ˈwɔːpɑːθ/ *n*. **be on the ~** buscar camorra

warrant /ˈwɒrənt/ *n* autorización *f*; (*for arrest*) orden *f*. ● *vt* justificar. **~-officer** *n* suboficial *m*

warranty /ˈwɒrəntɪ/ *n* garantía *f*

warring /ˈwɔːrɪŋ/ *a* en guerra

warrior /ˈwɔːrɪə(r)/ *n* guerrero *m*

warship /ˈwɔːʃɪp/ *n* buque *m* de guerra

wart /wɔːt/ *n* verruga *f*

wartime /ˈwɔːtaɪm/ *n* tiempo *m* de guerra

wary /ˈweərɪ/ *a* (**-ier**, **-iest**) cauteloso

was /wəz, wɒz/ *see* **be**

wash /wɒʃ/ *vt* lavar; (*flow over*) bañar. ● *vi* lavarse. ● *n* lavado *m*; (*dirty clothes*) ropa *f* sucia; (*wet clothes*) colada *f*; (*of ship*) estela *f*. **have a ~** lavarse. **~ out** *vt* enjuagar; (*fig*) cancelar. **~ up** *vi* fregar los platos. **~able** *a* lavable. **~-basin** *n* lavabo *m*. **~ed-out** *a* (*pale*) pálido; (*tired*) rendido. **~er** /ˈwɒʃə(r)/ *n* arandela *f*; (*washing-machine*) lavadora *f*. **~ing** /ˈwɒʃɪŋ/ *n* lavado *m*; (*dirty clothes*) ropa *f* sucia; (*wet clothes*) colada *f*. **~ing-machine** *n* lavadora *f*. **~ing-powder** *n* jabón *m* en polvo. **~ing-up** *n* fregado *m*; (*dirty plates etc*) platos *mpl* para fregar. **~-out** *n* (*sl*) desastre *m*. **~-room** *n* (*Amer*) servicios *mpl*. **~-stand** *n* lavabo *m*. **~-tub** *n* tina *f* de lavar

wasp /wɒsp/ *n* avispa *f*

wastage /ˈweɪstɪdʒ/ *n* desperdicios *mpl*

waste /weɪst/ ● *a* de desecho; ‹*land*› yermo. ● *n* derroche *m*; (*rubbish*) desperdicio *m*; (*of time*) pérdida *f*. ● *vt* derrochar; (*not use*) desperdiciar; perder ‹*time*›. ● *vi*. **~ away** consumirse. **~-disposal unit** *n* trituradora *f* de basuras. **~ful** *a* dispendioso; ‹*person*› derrochador. **~-paper basket** *n* papelera *f*. **~s** *npl* tierras *fpl* baldías

watch /wɒtʃ/ *vt* mirar; (*keep an eye on*) vigilar; (*take heed*) tener cuidado con; ver ‹*TV*›. ● *vi* mirar; (*keep an eye on*) vigilar. ● *n* vigilancia *f*; (*period of duty*) guardia *f*; (*timepiece*) reloj *m*. **on the ~** alerta. **~ out** *vi* tener cuidado. **~-dog** *n* perro *m* guardián; (*fig*) guardián *m*. **~ful** *a* vigilante. **~maker** /ˈwɒtʃmeɪkə(r)/ *n* relojero *m*. **~man** /ˈwɒtʃmən/ *n* (*pl* **-men**) vigilante *m*. **~-tower** *n* atalaya *f*. **~word** /ˈwɒtʃwɜːd/ *n* santo *m* y seña

water /ˈwɔːtə(r)/ *n* agua *f*. **by ~** (*of travel*) por mar. **in hot ~** (*fam*) en un apuro. ● *vt* regar ‹*plants etc*›; (*dilute*) aguar, diluir. ● *vi* ‹*eyes*› llorar. **make s.o.'s mouth ~** hacérsele la boca agua. **~ down** *vt* diluir; (*fig*) suavizar. **~-closet** *n* wáter *m*. **~-colour** *n* acuarela *f*. **~course** /ˈwɔːtəkɔːs/ *n* arroyo *m*; (*artificial*) canal *m*. **~cress** /ˈwɔːtəkres/ *n* berro *m*. **~fall** /ˈwɔːtəfɔːl/ *n* cascada *f*. **~-ice** *n* sorbete *m*. **~ing-can** /ˈwɔːtərɪŋkæn/ *n* regadera *f*. **~-lily** *n* nenúfar *m*. **~-line** *n* línea *f* de flotación. **~logged** /ˈwɔːtəlɒgd/ *a* saturado de agua, empapado. **~ main** *n* cañería *f* principal. **~ melon** *n* sandía *f*. **~-mill** *n* molino *m* de agua. **~ polo** *n* polo *m* acuático. **~-power** *n* energía *f* hidráulica. **~proof** /'wɔ:təpru:f/ *a* & *n* impermeable (*m*); ‹*watch*› sumergible. **~shed** /'wɔ:təʃed/ *n* punto *m* decisivo. **~-skiing** *n* esquí *m* acuático. **~-softener** *n* ablandador *m* de agua. **~tight** /'wɔ:tətaɪt/ *a* hermético, estanco; (*fig*) irrecusable. **~way** *n* canal *m* navegable. **~-wheel** *n* rueda *f* hidráulica. **~-wings** *npl* flotadores *mpl*. **~works** /'wɔ:təwɜ:ks/ *n* sistema *m* de abastecimiento de agua. **~y** /'wɔ:tərɪ/ *a* acuoso; ‹*colour*› pálido; ‹*eyes*› lloroso

watt /wɒt/ *n* vatio *m*

wave /weɪv/ *n* onda *f*; (*of hand*) señal *f*; (*fig*) oleada *f*. ● *vt* agitar; ondular ‹*hair*›. ● *vi* (*signal*) hacer señales con la mano; ‹*flag*› flotar. **~band** /'weɪvbænd/ *n* banda *f* de ondas. **~length** /'weɪvleŋθ/ *n* longitud *f* de onda

waver /'weɪvə(r)/ *vi* vacilar

wavy /'weɪvɪ/ *a* (**-ier**, **-iest**) ondulado

wax[1] /wæks/ *n* cera *f*. ● *vt* encerar

wax[2] /wæks/ *vi* ‹*moon*› crecer

wax: **~en** *a* céreo. **~work** /'wækswɜ:k/ *n* figura *f* de cera. **~y** *a* céreo

way /weɪ/ *n* camino *m*; (*distance*) distancia *f*; (*manner*) manera *f*, modo *m*; (*direction*) dirección *f*; (*means*) medio *m*; (*habit*) costumbre *f*. **be in the ~** estorbar. **by the ~** a propósito. **by ~ of** a título de, por. **either ~** de cualquier modo. **in a ~** en cierta manera. **in some ~s** en ciertos modos. **lead the ~** mostrar el camino. **make ~** dejar paso a. **on the ~** en camino. **out of the ~** remoto; (*extraordinary*) fuera de lo común. **that ~** por allí. **this ~** por aquí. **under ~** en curso. **~-bill** *n* hoja *f* de ruta. **~farer** /'weɪfeərə(r)/ *n* viajero *m*. **~ in** *n* entrada *f*

waylay /weɪ'leɪ/ *vt* (*pt* **-laid**) acechar; (*detain*) detener

way: **~ out** *n* salida *f*. **~-out** *a* ultramoderno, original. **~s** *npl* costumbres *fpl*. **~side** /'weɪsaɪd/ *n* borde *m* del camino

wayward /'weɪwəd/ *a* caprichoso

we /wi:/ *pron* nosotros, nosotras

weak /wi:k/ *a* (**-er**, **-est**) débil; ‹*liquid*› aguado, acuoso; (*fig*) flojo. **~en** *vt* debilitar. **~-kneed** *a* irresoluto. **~ling** /'wi:klɪŋ/ *n* persona *f* débil. **~ly** *adv* débilmente. ● *a* enfermizo. **~ness** *n* debilidad *f*

weal /wi:l/ *n* verdugón *m*

wealth /welθ/ *n* riqueza *f*. **~y** *a* (**-ier**, **-iest**) rico

wean /wi:n/ *vt* destetar

weapon /'wepən/ *n* arma *f*

wear /weə(r)/ *vt* (*pt* **wore**, *pp* **worn**) llevar; (*put on*) ponerse; tener ‹*expression etc*›; (*damage*) desgastar. ● *vi* desgastarse; (*last*) durar. ● *n* uso *m*; (*damage*) desgaste *m*; (*clothing*) ropa *f*. **~ down** *vt* desgastar; agotar ‹*opposition etc*›. **~ off** *vi* desaparecer. **~ on** *vi* ‹*time*› pasar. **~ out** *vt* desgastar; (*tire*) agotar. **~able** *a* que se puede llevar. **~ and tear** desgaste *m*

wear|ily /'wɪərɪlɪ/ *adv* cansadamente. **~iness** *n* cansancio *m*. **~isome** /'wɪərɪsəm/ *a* cansado. **~y** /'wɪərɪ/ *a* (**-ier**, **-iest**) cansado. ● *vt* cansar. ● *vi* cansarse. **~y of** cansarse de

weasel /'wi:zl/ *n* comadreja *f*

weather /'weðə(r)/ *n* tiempo *m*. **under the ~** (*fam*) indispuesto. ● *a* meteorológico. ● *vt* curar ‹*wood*›; (*survive*) superar. **~-beaten** *a* curtido. **~cock** /'weðəkɒk/ *n*, **~-vane** *n* veleta *f*

weave /wi:v/ *vt* (*pt* **wove**, *pp* **woven**) tejer; entretejer ‹*story etc*›; entrelazar ‹*flowers etc*›. **~ one's way** abrirse paso. ● *n* tejido *m*. **~r** /-ə(r)/ *n* tejedor *m*

web /web/ *n* tela *f*; (*of spider*) telaraña *f*; (*on foot*) membrana *f*. **~bing** *n* cincha *f*

wed /wed/ *vt* (*pt* **wedded**) casarse con; ‹*priest etc*› casar. ● *vi* casarse. **~ded to** (*fig*) unido a

wedding /'wedɪŋ/ *n* boda *f*. **~-cake** *n* pastel *m* de boda. **~-ring** *n* anillo *m* de boda

wedge /wedʒ/ *n* cuña *f*; (*space filler*) calce *m*. ● *vt* acuñar; (*push*) apretar

wedlock /'wedlɒk/ *n* matrimonio *m*

Wednesday /'wenzdeɪ/ *n* miércoles *m*

wee /wi:/ *a* (*fam*) pequeñito

weed /wi:d/ *n* mala hierba *f*. ● *vt* desherbar. **~-killer** *n* herbicida *m*. **~ out** eliminar. **~y** *a* ‹*person*› débil

week /wi:k/ *n* semana *f*. **~day** /'wi:kdeɪ/ *n* día *m* laborable. **~end** *n* fin *m* de semana. **~ly** /'wi:klɪ/ *a* semanal. ● *n* semanario *m*. ● *adv* semanalmente

weep /wi:p/ *vi* (*pt* **wept**). llorar. **~ing willow** *n* sauce *m* llorón

weevil /'wi:vɪl/ *n* gorgojo *m*
weigh /weɪ/ *vt/i* pesar. ~ **anchor** levar anclas. ~ **down** *vt* (*fig*) oprimir. ~ **up** *vt* pesar; (*fig*) considerar
weight /weɪt/ *n* peso *m*. ~**less** *a* ingrávido. ~**lessness** *n* ingravidez *f*. ~**-lifting** *n* halterofilia *f*, levantamiento *m* de pesos. ~**y** *a* (**-ier**, **-iest**) pesado; (*influential*) influyente
weir /wɪə(r)/ *n* presa *f*
weird /wɪəd/ *a* (**-er**, **-est**) misterioso; (*bizarre*) extraño
welcome /'welkəm/ *a* bienvenido. ~ **to do** libre de hacer. **you're** ~**e!** (*after thank you*) ¡de nada! ● *n* bienvenida *f*; (*reception*) acogida *f*. ● *vt* dar la bienvenida a; (*appreciate*) alegrarse de
welcoming /'welkəmɪŋ/ *a* acogedor
weld /weld/ *vt* soldar. ● *n* soldadura *f*. ~**er** *n* soldador *m*
welfare /'welfeə(r)/ *n* bienestar *m*; (*aid*) asistencia *f* social. **W**~ **State** *n* estado *m* benefactor. ~ **work** *n* asistencia *f* social
well[1] /wel/ *adv* (**better**, **best**) bien. ~ **done!** ¡bravo! **as** ~ también. **as** ~ **as** tanto... como. **be** ~ estar bien. **do** ~ (*succeed*) tener éxito. **very** ~ muy bien. ● *a* bien. ● *int* bueno; (*surprise*) ¡vaya! ~ **I never!** ¡no me digas!
well[2] /wel/ *n* pozo *m*; (*of staircase*) caja *f*
well: ~**-appointed** *a* bien equipado. ~**-behaved** *a* bien educado. ~**-being** *n* bienestar *m*. ~**-bred** *a* bien educado. ~**-disposed** *a* benévolo. ~**-groomed** *a* bien aseado. ~**-heeled** *a* (*fam*) rico
wellington /'welɪŋtən/ *n* bota *f* de agua
well: ~**-knit** *a* robusto. ~**-known** *a* conocido. ~**-meaning** *a*, ~ **meant** *a* bienintencionado. ~ **off** *a* acomodado. ~**-read** *a* culto. ~**-spoken** *a* bienhablado. ~**-to-do** *a* rico. ~**-wisher** *n* bienqueriente *m & f*
Welsh /welʃ/ *a & n* galés (*m*). ~ **rabbit** *n* pan *m* tostado con queso
welsh /welʃ/ *vi*. ~ **on** no cumplir con
wench /wentʃ/ *n* (*old use*) muchacha *f*
wend /wend/ *vt*. ~ **one's way** encaminarse
went /went/ *see* **go**
wept /wept/ *see* **weep**
were /wɜ:(r), wə(r)/ *see* **be**
west /west/ *n* oeste *m*. **the** ~ el Occidente *m*. ● *a* del oeste. ● *adv* hacia el oeste, al oeste. **go** ~ (*sl*) morir. **W**~ **Germany** *n* Alemania *f* Occidental. ~**erly** *a* del oeste. ~**ern** *a* occidental. ● *n* (*film*) película *f* del Oeste. ~**erner** /-ənə(r)/ *n* occidental *m & f*. **W**~ **Indian** *a & n* antillano (*m*). **W**~ **Indies** *npl* Antillas *fpl*. ~**ward** *a*, ~**ward(s)** *adv* hacia el oeste
wet /wet/ *a* (**wetter**, **wettest**) mojado; (*rainy*) lluvioso, de lluvia; (*person*, *sl*) soso. ~ **paint** recién pintado. **get** ~ mojarse. ● *vt* (*pt* **wetted**) mojar, humedecer. ~ **blanket** *n* aguafiestas *m & f invar*. ~ **suit** *n* traje *m* de buzo
whack /wæk/ *vt* (*fam*) golpear. ● *n* (*fam*) golpe *m*. ~**ed** /wækt/ *a* (*fam*) agotado. ~**ing** *a* (*huge*, *sl*) enorme. ● *n* paliza *f*
whale /weɪl/ *n* ballena *f*. **a** ~ **of a** (*fam*) maravilloso, enorme
wham /wæm/ *int* ¡zas!
wharf /wɔ:f/ *n* (*pl* **wharves** *or* **wharfs**) muelle *m*
what /wɒt/ *a* el que, la que, lo que, los que, las que; (*in questions & exclamations*) qué. ● *pron* lo que; (*interrogative*) qué. ~ **about going?** ¿si fuésemos? ~ **about me?** ¿y yo? ~ **for?** ¿para qué? ~ **if?** ¿y si? ~ **is it?** ¿qué es? ~ **you need** lo que te haga falta. ● *int* ¡cómo! ~ **a fool!** ¡qué tonto!
whatever /wɒt'evə(r)/ *a* cualquiera. ● *pron* (todo) lo que, cualquier cosa que
whatnot /'wɒtnɒt/ *n* chisme *m*
whatsoever /wɒtsəʊ'evə(r)/ *a & pron* = **whatever**
wheat /wi:t/ *n* trigo *m*. ~**en** *a* de trigo
wheedle /'wi:dl/ *vt* engatusar
wheel /wi:l/ *n* rueda *f*. **at the** ~ al volante. **steering-**~ *n* volante *m*. ● *vt* empujar ‹*bicycle etc*›. ● *vi* girar. ~ **round** girar. ~**barrow** /'wi:lbærəʊ/ *n* carretilla *f*. ~**-chair** /'wi:ltʃeə(r)/ *n* silla *f* de ruedas
wheeze /wi:z/ *vi* resollar. ● *n* resuello *m*
when /wen/ *adv* cuándo. ● *conj* cuando
whence /wens/ *adv* de dónde
whenever /wen'evə(r)/ *adv* en cualquier momento; (*every time that*) cada vez que

where /weə(r)/ *adv & conj* donde; (*interrogative*) dónde. **~ are you going?** ¿adónde vas? **~ are you from?** ¿de dónde eres?
whereabouts /ˈweərəbaʊts/ *adv* dónde. ● *n* paradero *m*
whereas /weərˈæz/ *conj* por cuanto; (*in contrast*) mientras (que)
whereby /weəˈbaɪ/ *adv* por lo cual
whereupon /weərəˈpɒn/ *adv* después de lo cual
wherever /weərˈevə(r)/ *adv* (*in whatever place*) dónde (diablos). ● *conj* dondequiera que
whet /wet/ *vt* (*pt* **whetted**) afilar; (*fig*) aguzar
whether /ˈweðə(r)/ *conj* si. **~ you like it or not** que te guste o no te guste. **I don't know ~ she will like it** no sé si le gustará
which /wɪtʃ/ *a* (*in questions*) qué. **~ one** cuál. **~ one of you** cuál de vosotros. ● *pron* (*in questions*) cuál; (*relative*) que; (*object*) el cual, la cual, lo cual, los cuales, las cuales
whichever /wɪtʃˈevə(r)/ *a* cualquier. ● *pron* cualquiera que, el que, la que
whiff /wɪf/ *n* soplo *m*; (*of smoke*) bocanada *f*; (*smell*) olorcillo *m*
while /waɪl/ *n* rato *m*. ● *conj* mientras; (*although*) aunque. ● *vt*. **~ away** pasar ‹*time*›
whilst /waɪlst/ *conj* = **while**
whim /wɪm/ *n* capricho *m*
whimper /ˈwɪmpə(r)/ *vi* lloriquear. ● *n* lloriqueo *m*
whimsical /ˈwɪmzɪkl/ *a* caprichoso; (*odd*) extraño
whine /waɪn/ *vi* gimotear. ● *n* gimoteo *m*
whip /wɪp/ *n* látigo *m*; (*pol*) oficial *m* disciplinario. ● *vt* (*pt* **whipped**) azotar; (*culin*) batir; (*seize*) agarrar. **~-cord** *n* tralla *f*. **~ped cream** *n* nata *f* batida. **~ping-boy** /ˈwɪpɪŋbɔɪ/ *n* cabeza *f* de turco. **~-round** *n* colecta *f*. **~ up** (*incite*) estimular
whirl /wɜːl/ *vt* hacer girar rápidamente. ● *vi* girar rápidamente; (*swirl*) arremolinarse. ● *n* giro *m*; (*swirl*) remolino *m*. **~pool** /ˈwɜːlpuːl/ *n* remolino *m*. **~wind** /ˈwɜːlwɪnd/ *n* torbellino *m*
whirr /wɜː(r)/ *n* zumbido *m*. ● *vi* zumbar
whisk /wɪsk/ *vt* (*culin*) batir. ● *n* (*culin*) batidor *m*. **~ away** llevarse
whisker /ˈwɪskə(r)/ *n* pelo *m*. **~s** *npl* (*of man*) patillas *fpl*; (*of cat etc*) bigotes *mpl*
whisky /ˈwɪskɪ/ *n* güisqui *m*
whisper /ˈwɪspə(r)/ *vt* decir en voz baja. ● *vi* cuchichear; ‹*leaves etc*› susurrar. ● *n* cuchicheo *m*; (*of leaves*) susurro *m*; (*rumour*) rumor *m*
whistle /ˈwɪsl/ *n* silbido *m*; (*instrument*) silbato *m*. ● *vi* silbar. **~-stop** *n* (*pol*) breve parada *f* (en gira electoral)
white /waɪt/ *a* (**-er**, **-est**) blanco. **go ~** ponerse pálido. ● *n* blanco; (*of egg*) clara *f*. **~bait** /ˈwaɪtbeɪt/ *n* (*pl* **~bait**) chanquetes *mpl*. **~ coffee** *n* café *m* con leche. **~-collar worker** *n* empleado *m* de oficina. **~ elephant** *n* objeto *m* inútil y costoso
Whitehall /ˈwaɪthɔːl/ *n* el gobierno *m* británico
white: **~ horses** *n* cabrillas *fpl*. **~-hot** *a* ‹*metal*› candente. **~ lie** *n* mentirijilla *f*. **~n** *vt/i* blanquear. **~ness** *n* blancura *f*. **W~ Paper** *n* libro *m* blanco. **~wash** /ˈwaɪtwɒʃ/ *n* jalbegue *m*; (*fig*) encubrimiento *m*. ● *vt* enjalbegar; (*fig*) encubrir
whiting /ˈwaɪtɪŋ/ *n* (*pl* **whiting**) (*fish*) pescadilla *f*
whitlow /ˈwɪtləʊ/ *n* panadizo *m*
Whitsun /ˈwɪtsn/ *n* Pentecostés *m*
whittle /ˈwɪtl/ *vt*. **~ (down)** tallar; (*fig*) reducir
whiz /wɪz/ *vi* (*pt* **whizzed**) silbar; (*rush*) ir a gran velocidad. **~ past** pasar como un rayo. **~-kid** *n* (*fam*) joven *m* prometedor, promesa *f*
who /huː/ *pron* que, quien; (*interrogative*) quién; (*particular person*) el que, la que, los que, las que
whodunit /huːˈdʌnɪt/ *n* (*fam*) novela *f* policíaca
whoever /huːˈevə(r)/ *pron* quienquera que; (*interrogative*) quién (diablos)
whole /həʊl/ *a* entero; (*not broken*) intacto. ● *n* todo *m*, conjunto *m*; (*total*) total *m*. **as a ~** en conjunto. **on the ~** por regla general. **~-hearted** *a* sincero. **~meal** *a* integral
wholesale /ˈhəʊlseɪl/ *n* venta *f* al por mayor. ● *a & adv* al por mayor. **~r** /-ə(r)/ *n* comerciante *m & f* al por mayor
wholesome /ˈhəʊlsəm/ *a* saludable
wholly /ˈhəʊlɪ/ *adv* completamente

whom /hu:m/ *pron* que, a quien; (*interrogative*) a quién
whooping cough /'hu:pɪŋkɒf/ *n* tos *f* ferina
whore /hɔ:(r)/ *n* puta *f*
whose /hu:z/ *pron* de quién. ● *a* de quién; (*relative*) cuyo
why /waɪ/ *adv* por qué. ● *int* ¡toma!
wick /wɪk/ *n* mecha *f*
wicked /'wɪkɪd/ *a* malo; (*mischievous*) travieso; (*very bad*, *fam*) malísimo. **~ness** *n* maldad *f*
wicker /'wɪkə(r)/ *n* mimbre *m* & *f*. ● *a* de mimbre. **~work** *n* artículos *mpl* de mimbre
wicket /'wɪkɪt/ *n* (*cricket*) rastrillo *m*
wide /waɪd/ *a* (**-er**, **-est**) ancho; (*fully opened*) de par en par; (*far from target*) lejano; ‹*knowledge etc*› amplio. ● *adv* lejos. **far and ~** por todas partes. **~ awake** *a* completamente despierto; (*fig*) despabilado. **~ly** *adv* extensamente; (*believed*) generalmente; (*different*) muy. **~n** *vt* ensanchar
widespread /'waɪdspred/ *a* extendido; (*fig*) difundido
widow /'wɪdəʊ/ *n* viuda *f*. **~ed** *a* viudo. **~er** *n* viudo *m*. **~hood** *n* viudez *f*
width /wɪdθ/ *n* anchura *f*. **in ~** de ancho
wield /wi:ld/ *vt* manejar; ejercer ‹*power*›
wife /waɪf/ *n* (*pl* **wives**) mujer *f*, esposa *f*
wig /wɪg/ *n* peluca *f*
wiggle /'wɪgl/ *vt* menear. ● *vi* menearse
wild /waɪld/ *a* (**-er**, **-est**) salvaje; (*enraged*) furioso; ‹*idea*› extravagante; (*with joy*) loco; (*random*) al azar. ● *adv* en estado salvaje. **run ~** crecer en estado salvaje. **~s** *npl* regiones *fpl* salvajes
wildcat /'waɪldkæt/ *a*. **~ strike** *n* huelga *f* salvaje
wilderness /'wɪldənɪs/ *n* desierto *m*
wild: **~fire** /'waɪldfaɪl(r)/ *n*. **spread like ~fire** correr como un reguero de pólvera. **~-goose chase** *n* empresa *f* inútil. **~life** /'waɪldlaɪf/ *n* fauna *f*. **~ly** *adv* violentamente; (*fig*) locamente
wilful /'wɪlfʊl/ *a* intencionado; (*self-willed*) terco. **~ly** *adv* intencionadamente; (*obstinately*) obstinadamente
will[1] /wɪl/ *v aux*. **~ you have some wine?** ¿quieres vino? **he ~ be** será. **you ~ be back soon, won't you?** volverás pronto, ¿no?
will[2] /wɪl/ *n* voluntad *f*; (*document*) testamento *m*
willing /'wɪlɪŋ/ *a* complaciente. **~ to** dispuesto a. **~ly** *adv* de buena gana. **~ness** *n* buena voluntad *f*
willow /'wɪləʊ/ *n* sauce *m*
will-power /'wɪlpaʊə(r)/ *n* fuerza *f* de voluntad
willy-nilly /wɪlɪ'nɪlɪ/ *adv* quieras que no
wilt /wɪlt/ *vi* marchitarse
wily /'waɪlɪ/ *a* (**-ier**, **-iest**) astuto
win /wɪn/ *vt* (*pt* **won**, *pres p* **winning**) ganar; (*achieve, obtain*) conseguir. ● *vi* ganar. ● *n* victoria *f*. **~ back** *vi* reconquistar. **~ over** *vt* convencer
wince /wɪns/ *vi* hacer una mueca de dolor. **without wincing** sin pestañear. ● *n* mueca *f* de dolor
winch /wɪntʃ/ *n* cabrestante *m*. ● *vt* levantar con el cabrestante
wind[1] /wɪnd/ *n* viento *m*; (*in stomach*) flatulencia *f*. **get the ~ up** (*sl*) asustarse. **get ~ of** enterarse de. **in the ~** en el aire. ● *vt* dejar sin aliento.
wind[2] /waɪnd/ *vt* (*pt* **wound**) (*wrap around*) enrollar; dar cuerda a ‹*clock etc*›. ● *vi* ‹*road etc*› serpentear. **~ up** *vt* dar cuerda a ‹*watch, clock*›; (*provoke*) agitar, poner nervioso; (*fig*) terminar, concluir
wind /wɪnd/: **~bag** *n* charlatán *m*. **~-cheater** *n* cazadora *f*
winder /'waɪndə(r)/ *n* devanador *m*; (*of clock, watch*) llave *f*
windfall /'wɪndfɔ:l/ *n* fruta *f* caída; (*fig*) suerte *f* inesperada
winding /'waɪndɪŋ/ *a* tortuoso
wind instrument /'wɪndɪnstrəmənt/ *n* instrumento *m* de viento
windmill /'wɪndmɪl/ *n* molino *m* (de viento)
window /'wɪndəʊ/ *n* ventana *f*; (*in shop*) escaparate *m*; (*of vehicle, booking-office*) ventanilla *f*. **~-box** *n* jardinera *f*. **~-dresser** *n* escaparatista *m* & *f*. **~-shop** *vi* mirar los escaparates
windpipe /'wɪndpaɪp/ *n* tráquea *f*
windscreen /'wɪndskri:n/ *n*, **windshield** *n* (*Amer*) parabrisas *m invar*. **~ wiper** *n* limpiaparabrisas *m invar*

wind /wɪnd/: **~-swept** *a* barrido por el viento. **~y** *a* (**-ier**, **-iest**) ventoso, de mucho viento. **it is ~y** hace viento

wine /waɪn/ *n* vino *m*. **~-cellar** *n* bodega *f*. **~glass** *n* copa *f*. **~-grower** *n* vinicultor *m*. **~-growing** *n* vinicultura *f*. ● *a* vinícola. **~ list** *n* lista *f* de vinos. **~-tasting** *n* cata *f* de vinos

wing /wɪŋ/ *n* ala *f*; (*auto*) aleta *f*. **under one's ~** bajo la protección de uno. **~ed** *a* alado. **~er** /-ə(r)/ *n* (*sport*) ala *m & f*. **~s** *npl* (*in theatre*) bastidores *mpl*

wink /wɪŋk/ *vi* guiñar el ojo; ‹*light etc*› centellear. ● *n* guiño *m*. **not to sleep a ~** no pegar ojo

winkle /'wɪŋkl/ *n* bígaro *m*

win: **~ner** /-ə(r)/ *n* ganador *m*. **~ning-post** *n* poste *m* de llegada. **~ning smile** *n* sonrisa *f* encantadora. **~nings** *npl* ganancias *fpl*

winsome /'wɪnsəm/ *a* atractivo

wint|er /'wɪntə(r)/ *n* invierno *m*. ● *vi* invernar. **~ry** *a* invernal

wipe /waɪp/ *vt* limpiar; (*dry*) secar. ● *n* limpión *m*. **give sth a ~** limpiar algo. **~ out** (*cancel*) cancelar; (*destroy*) destruir; (*obliterate*) borrar. **~ up** limpiar; (*dry*) secar

wire /'waɪə(r)/ *n* alambre *m*; (*elec*) cable *m*; (*telegram, fam*) telegrama *m*

wireless /'waɪəlɪs/ *n* radio *f*

wire netting /waɪə'netɪŋ/ *n* alambrera *f*, tela *f* metálica

wiring *n* instalación *f* eléctrica

wiry /'waɪərɪ/ *a* (**-ier**, **-iest**) ‹*person*› delgado

wisdom /'wɪzdəm/ *n* sabiduría *f*. **~ tooth** *n* muela *f* del juicio

wise /waɪz/ *a* (**-er**, **-est**) sabio; (*sensible*) prudente. **~crack** /'waɪzkræk/ *n* (*fam*) salida *f*. **~ly** *adv* sabiamente; (*sensibly*) prudentemente

wish /wɪʃ/ *n* deseo *m*; (*greeting*) saludo *m*. **with best ~es** (*in letters*) un fuerte abrazo. ● *vt* desear. **~ on** (*fam*) encajar a. **~ s.o. well** desear buena suerte a uno. **~bone** *n* espoleta *f* (de las aves). **~ful** *a* deseoso. **~ful thinking** *n* ilusiones *fpl*

wishy-washy /'wɪʃɪwɒʃɪ/ *a* soso; ‹*person*› sin convicciones, falto de entereza

wisp /wɪsp/ *n* manojito *m*; (*of smoke*) voluta *f*; (*of hair*) mechón *m*

wisteria /wɪs'tɪərɪə/ *n* glicina *f*

wistful /'wɪstfl/ *a* melancólico

wit /wɪt/ *n* gracia *f*; (*person*) persona *f* chistosa; (*intelligence*) ingenio *m*. **be at one's ~s' end** no saber qué hacer. **live by one's ~s** vivir de expedientes, vivir del cuento

witch /wɪtʃ/ *n* bruja *f*. **~craft** *n* brujería *f*. **~-doctor** *n* hechicero *m*

with /wɪð/ *prep* con; (*cause, having*) de. **be ~ it** (*fam*) estar al día, estar al tanto. **the man ~ the beard** el hombre de la barba

withdraw /wɪð'drɔ:/ *vt* (*pt* **withdrew**, *pp* **withdrawn**) retirar. ● *vi* apartarse. **~al** *n* retirada *f*. **~n** *a* ‹*person*› introvertido

wither /'wɪðə(r)/ *vi* marchitarse. ● *vt* (*fig*) fulminar

withhold /wɪð'həʊld/ *vt* (*pt* **withheld**) retener; (*conceal*) ocultar (**from** a)

within /wɪð'ɪn/ *prep* dentro de. ● *adv* dentro. **~ sight** a la vista

without /wɪð'aʊt/ *prep* sin

withstand /wɪð'stænd/ *vt* (*pt* **~stood**) resistir a

witness /'wɪtnɪs/ *n* testigo *m*; (*proof*) testimonio *m*. ● *vt* presenciar; firmar como testigo ‹*document*›. **~-box** *n* tribuna *f* de los testigos

witticism /'wɪtɪsɪzəm/ *n* ocurrencia *f*

wittingly /'wɪtɪŋlɪ/ *adv* a sabiendas

witty /'wɪtɪ/ *a* (**-ier**, **-iest**) gracioso

wives /waɪvz/ *see* **wife**

wizard /'wɪzəd/ *n* hechicero *m*. **~ry** *n* hechicería *f*

wizened /'wɪznd/ *a* arrugado

wobbl|e /'wɒbl/ *vi* tambalearse; ‹*voice, jelly, hand*› temblar; ‹*chair etc*› balancearse. **~y** *a* ‹*chair etc*› cojo

woe /wəʊ/ *n* aflicción *f*. **~ful** *a* triste. **~begone** /'wəʊbɪgɒn/ *a* desconsolado

woke, woken /wəʊk, 'wəʊkən/ *see* **wake**[1]

wolf /wʊlf/ *n* (*pl* **wolves**) lobo *m*. **cry ~** gritar al lobo. ● *vt* zamparse. **~-whistle** *n* silbido *m* de admiración

woman /'wʊmən/ *n* (*pl* **women**) mujer *f*. **single ~** soltera *f*. **~ize** /'wʊmənaɪz/ *vi* ser mujeriego. **~ly** *a* femenino

womb /wu:m/ *n* matriz *f*

women /'wɪmɪn/ *npl see* **woman**. **~folk** /'wɪmɪnfəʊk/ *npl* mujeres *fpl*.

~'s lib *n* movimiento *m* de liberación de la mujer
won /wʌn/ *see* **win**
wonder /ˈwʌndə(r)/ *n* maravilla *f*; (*bewilderment*) asombro *m*. **no ~** no es de extrañarse (**that** que). ● *vi* admirarse; (*reflect*) preguntarse
wonderful /ˈwʌndəfl/ *a* maravilloso. **~ly** *adv* maravillosamente
won't /wəʊnt/ = **will not**
woo /wuː/ *vt* cortejar
wood /wʊd/ *n* madera *f*; (*for burning*) leña *f*; (*area*) bosque *m*; (*in bowls*) bola *f*. **out of the ~** (*fig*) fuera de peligro. **~cutter** /ˈwʊdkʌtə(r)/ *n* leñador *m*. **~ed** *a* poblado de árboles, boscoso. **~en** *a* de madera. **~land** *n* bosque *m*
woodlouse /ˈwʊdlaʊs/ *n* (*pl* **-lice**) cochinilla *f*
woodpecker /ˈwʊdpekə(r)/ *n* pájaro *m* carpintero
woodwind /ˈwʊdwɪnd/ *n* instrumentos *mpl* de viento de madera
woodwork /ˈwʊdwɜːk/ *n* carpintería *f*; (*in room etc*) maderaje *m*
woodworm /ˈwʊdwɜːm/ *n* carcoma *f*
woody /ˈwʊdɪ/ *a* leñoso
wool /wʊl/ *n* lana *f*. **pull the ~ over s.o.'s eyes** engañar a uno. **~len** *a* de lana. **~lens** *npl* ropa *f* de lana. **~ly** *a* (**-ier**, **-iest**) de lana; (*fig*) confuso. ● *n* jersey *m*
word /wɜːd/ *n* palabra *f*; (*news*) noticia *f*. **by ~ of mouth** de palabra. **have ~s with** reñir con. **in one ~** en una palabra. **in other ~s** es decir. ● *vt* expresar. **~ing** *n* expresión *f*, términos *mpl*. **~-perfect** *a*. **be ~-perfect** saber de memoria. **~ processor** *n* procesador *m* de textos. **~y** *a* prolijo
wore /wɔː(r)/ *see* **wear**
work /wɜːk/ *n* trabajo *m*; (*arts*) obra *f*. ● *vt* hacer trabajar; manejar ‹*machine*›. ● *vi* trabajar; ‹*machine*› funcionar; ‹*student*› estudiar; ‹*drug etc*› tener efecto; (*be successful*) tener éxito. **~ in** introducir(se). **~ off** desahogar. **~ out** *vt* resolver; (*calculate*) calcular; elaborar ‹*plan*›. ● *vi* (*succeed*) salir bien; (*sport*) entrenarse. **~ up** *vt* desarrollar. ● *vi* excitarse. **~able** /ˈwɜːkəbl/ *a* ‹*project*› factible. **~aholic** /wɜːkəˈhɒlɪk/ *n* trabajador *m* obsesivo. **~ed up** *a* agitado. **~er** /ˈwɜːkə(r)/ *n* trabajador *m*; (*manual*) obrero *m*
workhouse /ˈwɜːkhaʊs/ *n* asilo *m* de pobres
work: **~ing** /ˈwɜːkɪŋ/ *a* ‹*day*› laborable; ‹*clothes etc*› de trabajo. ● *n* (*mec*) funcionamiento *m*. **in ~ing order** en estado de funcionamiento. **~ing class** *n* clase *f* obrera. **~ing-class** *a* de la clase obrera. **~man** /ˈwɜːkmən/ *n* (*pl* **-men**) obrero *m*. **~manlike** /ˈwɜːkmənlaɪk/ *a* concienzudo. **~manship** *n* destreza *f*. **~s** *npl* (*building*) fábrica *f*; (*mec*) mecanismo *m*. **~shop** /ˈwɜːkʃɒp/ *n* taller *m*. **~-to-rule** *n* huelga *f* de celo
world /wɜːld/ *n* mundo *m*. **a ~ of** enorme. **out of this ~** maravilloso. ● *a* mundial. **~ly** *a* mundano. **~-wide** *a* universal
worm /wɜːm/ *n* lombriz *f*; (*grub*) gusano *m*. ● *vi*. **~ one's way** insinuarse. **~eaten** *a* carcomido
worn /wɔːn/ *see* **wear**. ● *a* gastado. **~-out** *a* gastado; ‹*person*› rendido
worr|ied /ˈwʌrɪd/ *a* preocupado. **~ier** /-ə(r)/ *n* aprensivo *m*. **~y** /ˈwʌrɪ/ *vt* preocupar; (*annoy*) molestar. ● *vi* preocuparse. ● *n* preocupación *f*. **~ying** *a* inquietante
worse /wɜːs/ *a* peor. ● *adv* peor; (*more*) más. ● *n* lo peor. **~n** *vt/i* empeorar
worship /ˈwɜːʃɪp/ *n* culto *m*; (*title*) señor, su señoría. ● *vt* (*pt* **worshipped**) adorar
worst /wɜːst/ *a* (el) peor. ● *adv* peor. ● *n* lo peor. **get the ~ of it** llevar la peor parte
worsted /ˈwʊstɪd/ *n* estambre *m*
worth /wɜːθ/ *n* valor *m*. ● *a*. **be ~** valer. **it is ~ trying** vale la pena probarlo. **it was ~ my while** (me) valió la pena. **~less** *a* sin valor. **~while** /ˈwɜːθwaɪl/ *a* que vale la pena
worthy /ˈwɜːðɪ/ *a* meritorio; (*respectable*) respetable; (*laudable*) loable
would /wʊd/ *v aux*. **~ you come here please?** ¿quieres venir aquí? **~ you go?** ¿irías tú? **he ~ come if he could** vendría si pudiese. **I ~ come every day** (*used to*) venía todos los días. **I ~ do it** lo haría yo. **~-be** *a* supuesto
wound[1] /wuːnd/ *n* herida *f*. ● *vt* herir
wound[2] /waʊnd/ *see* **wind**[2]

wove, woven /wəʊv, 'wəʊvn/ *see* **weave**
wow /waʊ/ *int* ¡caramba!
wrangle /'ræŋgl/ *vi* reñir. ● *n* riña *f*
wrap /ræp/ *vt* (*pt* **wrapped**) envolver. **be ~ped up in** (*fig*) estar absorto en. ● *n* bata *f*; (*shawl*) chal *m*. **~per** /-ə(r)/ *n*, **~ping** *n* envoltura *f*
wrath /rɒθ/ *n* ira *f*. **~ful** *a* iracundo
wreath /ri:θ/ *n* (*pl* **-ths** /-ðz/) guirnalda *f*; (*for funeral*) corona *f*
wreck /rek/ *n* ruina *f*; (*sinking*) naufragio *m*; (*remains of ship*) buque *m* naufragado. **be a nervous ~** tener los nervios destrozados. ● *vt* hacer naufragar; (*fig*) arruinar. **~age** *n* restos *mpl*; (*of building*) escombros *mpl*
wren /ren/ *n* troglodito *m*
wrench /rentʃ/ *vt* arrancar; (*twist*) torcer. ● *n* arranque *m*; (*tool*) llave *f* inglesa
wrest /rest/ *vt* arrancar (**from** a)
wrestl|e /'resl/ *vi* luchar. **~er** /-ə(r)/ *n* luchador *m*. **~ing** *n* lucha *f*
wretch /retʃ/ *n* desgraciado *m*; (*rascal*) tunante *m* & *f*. **~ed** *a* miserable; ‹*weather*› horrible, de perros; ‹*dog etc*› maldito
wriggle /'rɪgl/ *vi* culebrear. **~ out of** escaparse de. **~ through** deslizarse por. ● *n* serpenteo *m*
wring /rɪŋ/ *vt* (*pt* **wrung**) retorcer. **~ out of** (*obtain from*) arrancar. **~ing wet** empapado
wrinkle /'rɪŋkl/ *n* arruga *f*. ● *vt* arrugar. ● *vi* arrugarse
wrist /rɪst/ *n* muñeca *f*. **~-watch** *n* reloj *m* de pulsera
writ /rɪt/ *n* decreto *m* judicial
write /raɪt/ *vt/i* (*pt* **wrote**, *pp* **written**, *pres p* **writing**) escribir. **~ down** *vt* anotar. **~ off** *vt* cancelar; (*fig*) dar por perdido. **~ up** *vt* hacer un reportaje de; (*keep up to date*) poner al día. **~-off** *n* pérdida *f* total. **~r** /-ə(r)/ *n* escritor *m*; (*author*) autor *m*. **~-up** *n* reportaje *m*; (*review*) crítica *f*
writhe /raɪð/ *vi* retorcerse
writing /'raɪtɪŋ/ *n* escribir *m*; (*handwriting*) letra *f*. **in ~** por escrito. **~s** *npl* obras *fpl*. **~-paper** *n* papel *m* de escribir
written /'rɪtn/ *see* **write**
wrong /rɒŋ/ *a* incorrecto; (*not just*) injusto; (*mistaken*) equivocado. **be ~** no tener razón; (*be mistaken*) equivocarse. ● *adv* mal. **go ~** equivocarse; ‹*plan*› salir mal; ‹*car etc*› estropearse. ● *n* injusticia *f*; (*evil*) mal *m*. **in the ~** equivocado. ● *vt* ser injusto con. **~ful** *a* injusto. **~ly** *adv* mal; (*unfairly*) injustamente
wrote /rəʊt/ *see* **write**
wrought /rɔ:t/ *a*. **~ iron** *n* hierro *m* forjado
wrung /rʌŋ/ *see* **wring**
wry /raɪ/ *a* (**wryer, wryest**) torcido; ‹*smile*› forzado. **~ face** *n* mueca *f*

X

xenophobia /zenə'fəʊbɪə/ *n* xenofobia *f*
Xerox /'zɪərɒks/ *n* (*P*) fotocopiadora *f*. **xerox** *n* fotocopia *f*
Xmas /'krɪsməs/ *n abbr* (*Christmas*) Navidad *f*, Navidades *fpl*
X-ray /'eksreɪ/ *n* radiografía *f*. **~s** *npl* rayos *mpl* X. ● *vt* radiografiar
xylophone /'zaɪləfəʊn/ *n* xilófono *m*

Y

yacht /jɒt/ *n* yate *m*. **~ing** *n* navegación *f* a vela
yam /jæm/ *n* ñame *m*, batata *f*
yank /jæŋk/ *vt* (*fam*) arrancar violentamente
Yankee /'jæŋkɪ/ *n* (*fam*) yanqui *m* & *f*
yap /jæp/ *vi* (*pt* **yapped**) ‹*dog*› ladrar
yard[1] /jɑ:d/ *n* (*measurement*) yarda *f* (= *0.9144 metre*)
yard[2] /jɑ:d/ *n* patio *m*; (*Amer, garden*) jardín *m*
yardage /'jɑ:dɪdʒ/ *n* metraje *m*
yardstick /'jɑ:dstɪk/ *n* (*fig*) criterio *m*
yarn /jɑ:n/ *n* hilo *m*; (*tale, fam*) cuento *m*
yashmak /'jæʃmæk/ *n* velo *m*
yawn /jɔ:n/ *vi* bostezar. ● *n* bostezo *m*
year /jɪə(r)/ *n* año *m*. **be three ~s old** tener tres años. **~-book** *n* anuario *m*. **~ling** /'jɜ:lɪŋ/ *n* primal *m*. **~ly** *a* anual. ● *adv* anualmente
yearn /'jɜ:n/ *vi*. **~ for** anhelar. **~ing** *n* ansia *f*
yeast /ji:st/ *n* levadura *f*
yell /jel/ *vi* gritar. ● *n* grito *m*

yellow /ˈjeləʊ/ *a & n* amarillo (*m*). **~ish** *a* amarillento
yelp /jelp/ *n* gañido *m*. ● *vi* gañir
yen /jen/ *n* muchas ganas *fpl*
yeoman /ˈjəʊmən/ *n* (*pl* **-men**). **Y~ of the Guard** alabardero *m* de la Casa Real
yes /jes/ *adv & n* sí (*m*)
yesterday /ˈjestədeɪ/ *adv & n* ayer (*m*). **the day before ~** anteayer *m*
yet /jet/ *adv* todavía, aún; (*already*) ya. **as ~** hasta ahora. ● *conj* sin embargo
yew /juː/ *n* tejo *m*
Yiddish /ˈjɪdɪʃ/ *n* judeoalemán *m*
yield /jiːld/ *vt* producir. ● *vi* ceder. ● *n* producción *f*; (*com*) rendimiento *m*
yoga /ˈjəʊgə/ *n* yoga *m*
yoghurt /ˈjɒgət/ *n* yogur *m*
yoke /jəʊk/ *n* yugo *m*; (*of garment*) canesú *m*
yokel /ˈjəʊkl/ *n* patán *m*, palurdo *m*
yolk /jəʊk/ *n* yema *f* (de huevo)
yonder /ˈjɒndə(r)/ *adv* a lo lejos
you /juː/ *pron* (*familiar form*) tú, vos (*Arg*), (*pl*) vosotros, vosotras, ustedes (*LAm*); (*polite form*) usted, (*pl*) ustedes; (*familiar, object*) te, (*pl*) os, les (*LAm*); (*polite, object*) le, la, (*pl*) les; (*familiar, after prep*) ti, (*pl*) vosotros, vosotras, ustedes (*LAm*); (*polite, after prep*) usted, (*pl*) ustedes. **with ~** (*familiar*) contigo, (*pl*) con vosotros, con vosotras, con ustedes (*LAm*); (*polite*) con usted, (*pl*) con ustedes; (*polite reflexive*) consigo. **I know ~** te conozco, le conozco a usted. **you can't smoke here** aquí no se puede fumar
young /jʌŋ/ *a* (**-er**, **-est**) joven. **~ lady** *n* señorita *f*. **~ man** *n* joven *m*. **her ~ man** (*boyfriend*) su novio *m*. **the ~** *npl* los jóvenes *mpl*; (*of animals*) la cría *f*. **~ster** /ˈjʌŋstə(r)/ *n* joven *m*
your /jɔː(r)/ *a* (*familiar*) tu, (*pl*) vuestro; (*polite*) su
yours /jɔːz/ *poss pron* (el) tuyo, (*pl*) (el) vuestro, el de ustedes (*LAm*); (*polite*) el suyo. **a book of ~s** un libro tuyo, un libro suyo. **Y~s faithfully**, **Y~s sincerely** le saluda atentamente
yourself /jɔːˈself/ *pron* (*pl* **yourselves**) (*familiar, subject*) tú mismo, tú misma, (*pl*) vosotros mismos, vosotras mismas, ustedes mismos (*LAm*), ustedes mismas (*LAm*); (*polite, subject*) usted mismo, usted misma, (*pl*) ustedes mismos, ustedes mismas; (*familiar, object*) te, (*pl*) os, se (*LAm*); (*polite, object*) se; (*familiar, after prep*) ti, (*pl*) vosotros, vosotras, ustedes (*LAm*); (*polite, after prep*) sí
youth /juːθ/ *n* (*pl* **youths** /juːðz/) juventud *f*; (*boy*) joven *m*; (*young people*) jóvenes *mpl*. **~ful** *a* joven, juvenil. **~hostel** *n* albergue *m* para jóvenes
yowl /jaʊl/ *vi* aullar. ● *n* aullido *m*
Yugoslav /ˈjuːgəslɑːv/ *a & n* yugoslavo (*m*). **~ia** /-ˈslɑːvɪə/ *n* Yugoslavia *f*
yule /juːl/ *n*, **yule-tide** /ˈjuːltaɪd/ *n* (*old use*) Navidades *fpl*

Z

zany /ˈzeɪnɪ/ *a* (**-ier**, **-iest**) estrafalario
zeal /ziːl/ *n* celo *m*
zealot /ˈzelət/ *n* fanático *m*
zealous /ˈzeləs/ *a* entusiasta. **~ly** /ˈzeləslɪ/ *adv* con entusiasmo
zebra /ˈzebrə/ *n* cebra *f*. **~ crossing** *n* paso *m* de cebra
zenith /ˈzenɪθ/ *n* cenit *m*
zero /ˈzɪərəʊ/ *n* (*pl* **-os**) cero *m*
zest /zest/ *n* gusto *m*; (*peel*) cáscara *f*
zigzag /ˈzɪgzæg/ *n* zigzag *m*. ● *vi* (*pt* **zigzagged**) zigzaguear
zinc /zɪŋk/ *n* cinc *m*
Zionis|m /ˈzaɪənɪzəm/ *n* sionismo *m*. **~t** *n* sionista *m & f*
zip /zɪp/ *n* cremallera *f*. ● *vt*. **~ (up)** cerrar (la cremallera)
Zip code /ˈzɪpkəʊd/ *n* (*Amer*) código *m* postal
zip fastener /zɪpˈfɑːsnə(r)/ *n* cremallera *f*
zircon /ˈzɜːkən/ *n* circón *m*
zither /ˈzɪðə(r)/ *n* cítara *f*
zodiac /ˈzəʊdɪæk/ *n* zodiaco *m*
zombie /ˈzɒmbɪ/ *n* (*fam*) autómata *m & f*
zone /zəʊn/ *n* zona *f*
zoo /zuː/ *n* (*fam*) zoo *m*, jardín *m* zoológico. **~logical** /zəʊəˈlɒdʒɪkl/ *a* zoológico
zoolog|ist /zəʊˈɒlədʒɪst/ *n* zoólogo *m*. **~y** /zəʊˈɒlədʒɪ/ *n* zoología *f*
zoom /zuːm/ *vi* ir a gran velocidad. **~ in** (*photo*) acercarse rápidamente. **~ past** pasar zumbando. **~ lens** *n* zoom *m*
Zulu /ˈzuːluː/ *n* zulú *m & f*

Numbers · Números

zero	0	cero
one (first)	1	uno (primero)
two (second)	2	dos (segundo)
three (third)	3	tres (tercero)
four (fourth)	4	cuatro (cuarto)
five (fifth)	5	cinco (quinto)
six (sixth)	6	seis (sexto)
seven (seventh)	7	siete (séptimo)
eight (eighth)	8	ocho (octavo)
nine (ninth)	9	nueve (noveno)
ten (tenth)	10	diez (décimo)
eleven (eleventh)	11	once (undécimo)
twelve (twelfth)	12	doce (duodécimo)
thirteen (thirteenth)	13	trece (decimotercero)
fourteen (fourteenth)	14	catorce (decimocuarto)
fifteen (fifteenth)	15	quince (decimoquinto)
sixteen (sixteenth)	16	dieciséis (decimosexto)
seventeen (seventeenth)	17	diecisiete (decimoséptimo)
eighteen (eighteenth)	18	dieciocho (decimoctavo)
nineteen (nineteenth)	19	diecinueve (decimonoveno)
twenty (twentieth)	20	veinte (vigésimo)
twenty-one (twenty-first)	21	veintiuno (vigésimo primero)
twenty-two (twenty-second)	22	veintidós (vigésimo segundo)
twenty-three (twenty-third)	23	veintitrés (vigésimo tercero)
twenty-four (twenty-fourth)	24	veinticuatro (vigésimo cuarto)
twenty-five (twenty-fifth)	25	veinticinco (vigésimo quinto)
twenty-six (twenty-sixth)	26	veintiséis (vigésimo sexto)
thirty (thirtieth)	30	treinta (trigésimo)
thirty-one (thirty-first)	31	treinta y uno (trigésimo primero)
forty (fortieth)	40	cuarenta (cuadragésimo)
fifty (fiftieth)	50	cincuenta (quincuagésimo)
sixty (sixtieth)	60	sesenta (sexagésimo)
seventy (seventieth)	70	setenta (septuagésimo)
eighty (eightieth)	80	ochenta

		(octogésimo)
ninety (ninetieth)	90	noventa (nonagésimo)
a/one hundred (hundredth)	100	cien (centésimo)
a/one hundred and one (hundred and first)	101	ciento uno (centésimo primero)
two hundred (two hundredth)	200	doscientos (ducentésimo)
three hundred (three hundredth)	300	trescientos (tricentésimo)
four hundred (four hundredth)	400	cuatrocientos (cuadringentésimo)
five hundred (five hundredth)	500	quinientos (quingentésimo)
six hundred (six hundredth)	600	seiscientos (sexcentésimo)
seven hundred (seven hundredth)	700	setecientos (septingentésimo)
eight hundred (eight hundredth)	800	ochocientos (octingentésimo)
nine hundred (nine hundredth)	900	novecientos (noningentésimo)
a/one thousand (thousandth)	1000	mil (milésimo)
two thousand (two thousandth)	2000	dos mil (dos milésimo)
a/one million (millionth)	1,000,000	un millón (millonésimo)

Spanish Verbs · Verbos españoles

Regular verbs:
in **-ar** (*e.g.* **comprar**)
Present; compr|o, **~as, ~a, ~amos, ~áis, ~an**
Future: comprar|é, **~ás, ~á, ~emos, ~éis, ~án**
Imperfect: compr|aba, **~abas, ~aba, ~ábamos, ~abais, ~aban**
Preterite: compr|é, **~aste, ~ó, ~amos, ~asteis, ~aron**
Present subjunctive: compr|e, **~es, ~e, ~emos, ~éis, ~en**
Imperfect subjunctive: compr|ara, **~aras ~ara, ~áramos, ~arais, ~aran**
compr|ase, **~ases, ~ase, ~ásemos, ~aseis, ~asen**
Conditional: comprar|ía, **~ías, ~ía, ~íamos, ~íais, ~ían**
Present participle: comprando
Past participle: comprado
Imperative: compra, comprad

in **-er** (*e.g.* **beber**)
Present: beb|o, **~es, ~e, ~emos, ~éis, ~en**
Future: beber|é, **~ás, ~á, ~emos, ~éis, ~án**
Imperfect: beb|ía, **~ías, ~ía, ~íamos, ~íais, ~ían**
Preterite: beb|í, **~iste, ~ió, ~imos, ~isteis, ~ieron**
Present subjunctive: beb|a, **~as, ~a, ~amos, ~áis, ~an**
Imperfect subjunctive: beb|iera, **~ieras, ~iera, ~iéramos, ~ierais, ~ieran**
beb|iese, **~ieses, ~iese, ~iésemos, ~ieseis, ~iesen**
Conditional: beber|ía, **~ías, ~ía, ~íamos, ~íais, ~ían**
Present participle: bebiendo
Past participle: bebido
Imperative: bebe, bebed

in **-ir** (*e.g.* **vivir**)
Present: viv|o, **~es, ~e, ~imos, ~ís, ~en**
Future: vivir|é, **~ás, ~á, ~emos, ~éis, ~án**
Imperfect: viv|ía, **~ías, ~ía, ~íamos, ~íais, ~ían**
Preterite: viv|í, **~iste, ~ió, ~imos, ~isteis, ~ieron**
Present subjunctive: viv|a, **~as, ~a, ~amos, ~áis, ~an**
Imperfect subjunctive: viv|iera, **~ieras, ~iera, ~iéramos, ~ierais, ~ieran**
viv|iese, **~ieses, ~iese, ~iésemos, ~ieseis, ~iesen**
Conditional: vivir|ía, **~ías, ~ía, ~íamos, ~íais, ~ían**
Present participle: viviendo
Past participle: vivido
Imperative: vive, vivid

Irregular verbs:
[1] **cerrar**
Present: cierro, cierras, cierra, cerramos, cerráis, cierran
Present subjunctive: cierre, cierres, cierre, cerremos, cerréis, cierren
Imperative: cierra, cerrad

[2] **contar**, **mover**
Present: cuento, cuentas, cuenta, contamos, contáis, cuentan
muevo, mueves, mueve, movemos, movéis, mueven
Present subjunctive: cuente, cuentes, cuente, contemos, contéis, cuenten
mueva, muevas mueva, movamos, mováis, muevan
Imperative: cuenta, contad mueve, moved

[3] **jugar**
Present: juego, juegas, juega, jugamos, jugáis, juegan
Preterite: jug|ué, jugaste, jugó, jugamos, jugasteis, jugaron
Present subjunctive: juegue, juegues, juegue, juguemos, juguéis, jueguen

[4] **sentir**
Present: siento, sientes, siente, sentimos, sentís, sienten
Preterite: sentí, sentiste, sintió, sentimos, sentisteis, sintieron
Present subjunctive: sienta, sientas, sienta, sintamos, sintáis, sientan
Imperfect subjunctive: sint|iera, **~ieras, ~iera, ~iéramos, ~ierais, ~ieran**

sint|iese, **~ieses, ~iese, ~iésemos, ~ieseis, ~iesen**
Present participle: sintiendo
Imperative: siente, sentid

[5] **pedir**
Present: pido, pides, pide, pedimos, pedís, piden
Preterite: pedí, pediste, pidió, pedimos, pedisteis, pidieron
Present subjunctive: pid|a, **~as, ~a, ~amos, ~áis, ~an**
Imperfect subjunctive: pid|iera, **~ieras, ~iera, ~iéramos, ~ierais, ~ieran**
pid|iese, **~ieses, ~iese, ~iésemos, ~ieseis, ~iesen**
Present participle: pidiendo
Imperative: pide, pedid

[6] **dormir**
Present: duermo, duermes, duerme, dormimos, dormís, duermen
Preterite: dormí, dormiste, durmió, dormimos, dormisteis, durmieron
Present subjunctive: duerma, duermas, duerma, durmamos, durmáis, duerman
Imperfect subjunctive: durm|iera, **~ieras, ~iera, ~iéramos, ~ierais, ~ieran**
durm|iese, **~ieses, ~iese, ~iésemos, ~ieseis, ~iesen**
Present participle: durmiendo
Imperative: duerme, dormid

[7] **dedicar**
Preterite: dediqué, dedicaste, dedicó, dedicamos, dedicasteis, dedicaron
Present subjunctive: dediqu|e, **~ues, ~e, ~emos, ~éis, ~en**

[8] **delinquir**
Present: delinco, delinques, delinque, delinquimos, delinquís, delinquen
Present subjunctive: delinc|a, **~as, ~a, ~amos, ~áis, ~an**

[9] **vencer**, **esparcir**
Present: venzo, vences, vence, vencemos, vencéis, vencen
esparzo, esparces, esparce, esparcimos, esparcís, esparcen
Present subjunctive: venz|a, **~as, ~a, ~amos, ~áis, ~an** esparz|a, **~as, ~a, ~amos, ~áis, ~an**

[10] **rechazar**
Preterite: rechacé, rechazaste, rechazó, rechazamos, rechazasteis, rechazaron
Present subjunctive: rechac|e, **~es, ~e, ~emos, ~éis, ~en**

[11] **conocer**, **lucir**
Present: conozco, conoces, conoce, conocemos, conocéis, conocen
luzco, luces, luce, lucimos, lucís, lucen
Present subjunctive: conozc|a, **~as, ~a, ~amos, ~áis, ~an** luzc|a, **~as, ~a, ~amos, ~áis, ~an**

[12] **pagar**
Preterite: pagué, pagaste, pagó, pagamos, pagasteis, pagaron
Present subjunctive: pagu|e, **~es, ~e, ~emos, ~éis, ~en**

[13] **distinguir**
Present: distingo, distingues, distingue, distinguimos, distinguís, distinguen
Present subjunctive: disting|a, **~as, ~a, ~amos, ~áis, ~an**

[14] **acoger**, **afligir**
Present: acojo, acoges, acoge, acogemos, acogéis, acogen
aflijo, afliges, aflige, afligimos, afligís, afligen
Present subjunctive: acoj|a, **~as, ~a, ~amos, ~áis, ~an**
aflij|a, **~as, ~a, ~amos, ~áis, ~an**

[15] **averiguar**
Preterite: averigüé, averiguaste, averiguó, averiguamos, averiguasteis, averiguaron
Present subjunctive: averigü|e, **~es, ~e, ~emos, ~éis, ~en**

[16] **agorar**
Present: agüero, agüeras, agüera, agoramos, agoráis, agüeran
Present subjunctive: agüere, agüeres, agüere, agoremos, agoréis, agüeren
Imperative: agüera, agorad

[17] **huir**
Present: huyo, huyes, huye, huimos, huís, huyen

Preterite: huí, huiste, huyó, huimos, huisteis, huyeron
Present subjunctive: huy|a, **~as, ~a, ~amos, ~áis, ~an**
Imperfect subjunctive: huy|era, **~eras, ~era, ~éramos, ~erais, ~eran**
huy|ese, **~eses, ~ese, ~ésemos, ~eseis, ~esen**
Present participle: huyendo

[18] **creer**
Preterite: creí, creíste, creyó, creímos, creísteis, creyeron
Imperfect subjunctive: crey|era, **~eras, ~era, ~éramos, ~erais, ~eran**
crey|ese, **~eses, ~ese, ~ésemos, ~eseis, ~esen**
Present participle: creyendo
Past participle: creído

[19] **argüir**
Present: arguyo, arguyes, arguye, argüimos, argüís, arguyen
Preterite: argüí, argüiste, arguyó, argüimos, argüisteis, arguyeron
Present subjunctive: arguy|a, **~as, ~a, ~amos, ~áis, ~an**
Imperfect subjunctive: arguy|era, **~eras, ~era, ~éramos, ~erais, ~eran**
arguy|ese, **~eses, ~ese, ~ésemos, ~eseis, ~esen**
Present participle: arguyendo
Imperative: arguye, argüid

[20] **vaciar**
Present: vacío, vacías, vacía, vaciamos, vaciáis, vacían
Present subjunctive: vacíe, vacíes, vacíe, vaciemos, vaciéis, vacíen
Imperative: vacía, vaciad

[21] **acentuar**
Present: acentúo, acentúas, acentúa, acentuamos, acentuáis, acentúan
Present subjunctive: acentúe, acentúes, acentúe, acentuemos, acentuéis, acentúen
Imperative: acentúa, acentuad

[22] **ateñer, engullir**
Preterite: atañ|í, **~aste, ~ó, ~amos, ~asteis, ~eron** engull|í **~iste, ~ó, ~imos, ~isteis, ~eron**
Imperfect subjunctive: atañ|era, **~eras, ~era, ~éramos, ~erais, ~eran**
atañ|ese, **~eses, ~ese, ~ésemos, ~eseis, ~esen**
engull|era, **~eras, ~era, ~éramos, ~erais, ~eran**
engull|ese, **~eses, ~ese, ~ésemos, ~eseis, ~esen**
Present participle: atañendo
engullendo

[23] **aislar, aullar**
Present: aíslo, aíslas, aísla, aislamos, aisláis, aíslan
aúllo, aúllas, aúlla, aullamos aulláis, aúllan
Present subjunctive: aísle, aísles, aísle, aislemos, aisléis, aíslen
aúlle, aúlles, aúlle, aullemos, aulléis, aúllen
Imperative: aísla, aislad
aúlla, aullad

[24] **abolir, garantir**
Present: abolimos, abolís
garantimos, garantís
Present subjunctive: not used
Imperative: abolid
garantid

[25] **andar**
Preterite: anduv|e, **~iste, ~o, ~imos, ~isteis, ~ieron**
Imperfect subjunctive: anduv|iera, **~ieras, ~iera, ~iéramos, ~ierais, ~ieran**
anduv|iese, **~ieses, ~iese, ~iésemos, ~ieseis, ~iesen**

[26] **dar**
Present: doy, das, da, damos, dais, dan
Preterite: di, diste, dio, dimos, disteis, dieron
Present subjunctive: dé, des, dé, demos, deis, den
Imperfect subjunctive: diera, dieras, diera, diéramos, dierais, dieran
diese, dieses, diese, diésemos, dieseis, diesen

[27] **estar**
Present: estoy, estás, está, estamos, estáis, están
Preterite: estuv|e, **~iste, ~o, ~imos, ~isteis, ~ieron**
Present subjunctive: esté, estés, esté, estemos, estéis, estén

Imperfect subjunctive: estuv/iera, **~ieras, ~iera, ~iéramos, ~ierais, ~ieran**
estuv|iese, **~ieses, ~iese, ~iésemos, ~ieseis, ~iesen**
Imperative: está, estad

[28] **caber**
Present: quepo, cabes, cabe, cabemos, cabéis, caben
Future: cabr|é, **~ás, ~á, ~emos, ~éis, ~án**
Preterite: cup|e, **~iste, ~o, ~imos, ~isteis, ~ieron**
Present subjunctive: quep|a, **~as, ~a, ~amos, ~áis, ~an**
Imperfect subjunctive: cup|iera, **~ieras, ~iera, ~iéramos, ~ierais, ~ieran**
cup|iese, **~ieses, ~iese, ~iésemos, ~ieseis, ~iesen**
Conditional: cabr|ía, **~ías, ~ía, ~íamos, ~íais, ~ían**

[29] **caer**
Present: caigo, caes, cae, caemos, caéis, caen
Preterite: caí, caiste, cayó, caímos, caísteis, cayeron
Present subjunctive: caig|a, **~as, ~a, ~amos, ~áis, ~an**
Imperfect subjunctive: cay|era, **~eras, ~era, ~éramos, ~erais, ~eran**
cay|ese, **~eses, ~ese, ~ésemos, ~eseis, ~esen**
Present participle: cayendo
Past participle: caído

[30] **haber**
Present: he, has, ha, hemos, habéis, han
Future: habr|é **~ás, ~á, ~emos, ~éis, ~án**
Preterite: hub|e, **~iste, ~o, ~imos, ~isteis, ~ieron**
Present subjunctive: hay|a, **~as, ~a, ~amos, ~áis, ~an**
Imperfect subjunctive: hub|iera, **~ieras, ~iera, ~iéramos, ~ierais, ~ieran**
hub|iese, **~ieses, ~iese, ~iésemos, ~ieseis, ~iesen**
Conditional: habr|ía, **~ías, ~ía, ~íamos, ~íais, ~ían**
Imperative: habe, habed

[31] **hacer**
Present: hago, haces, hace, hacemos, hacéis, hacen
Future: har|é, **~ás, ~á, ~emos, ~éis, ~án**
Preterite: hice, hiciste, hizo, hicimos, hicisteis, hicieron
Present subjunctive: hag|a, **~as, ~a, ~amos, ~áis, ~an**
Imperfect subjunctive: hic|iera, **~ieras, ~iera, ~iéramos, ~ierais, ~ieran**
hic|iese, **~ieses, ~iese, ~iésemos, ~ieseis, ~iesen**
Conditional: har|ía, **~ías, ~ía, ~íamos, ~íais, ~ían**
Past participle: hecho
Imperative: haz, haced

[32] **placer**
Preterite: plació/plugo
Present subjunctive: plazca
Imperfect subjunctive: placiera/pluguiera
placiese/pluguiese

[33] **poder**
Present: puedo, puedes, puede, podemos, podéis, pueden
Future: podr|é, **~ás, ~á, ~emos, ~éis, ~án**
Preterite: pud|e, **~iste, ~o, ~imos, ~isteis, ~ieron**
Present subjunctive: pueda, puedas, pueda, podamos, podáis, puedan
Imperfect subjunctive: pud|iera, **~ieras, ~iera, ~iéramos, ~ierais, ~ieran**
pud|iese, **~ieses, ~iese, ~iésemos, ~ieseis, ~iesen**
Conditional: podr|ía, **~ías, ~ía, ~íamos, ~íais, ~ían**
Past participle: pudiendo

[34] **poner**
Present: pongo, pones, pone, ponemos, ponéis, ponen
Future: pondr|é, **~ás, ~á, ~emos, ~éis, ~án**
Preterite: pus|e, **~iste, ~o, ~imos, ~isteis, ~ieron**
Present subjunctive: pong|a, **~as, ~a, ~amos, ~áis, ~an**
Imperfect subjunctive: pus|iera, **~ieras, ~iera, ~iéramos, ~ierais, ~ieran**
pus|iese, **~ieses, ~iese, ~iésemos, ~ieseis, ~iesen**

Conditional: pondr|ía, **~ías, ~ía, ~íamos, ~íais, ~ían**
Past participle: puesto
Imperative: pon, poned

[35] **querer**
Present: quiero, quieres, quiere, queremos, queréis, quieren
Future: querr|é, **~ás, ~á, ~emos, ~éis, ~án**
Preterite: quis|e, **~iste, ~o, ~imos, ~isteis, ~ieron**
Present subjunctive: quiera, quieras, quiera, queramos, queráis, quieran
Imperfect subjunctive: quis|iera, **~ieras, ~iera, ~iéramos, ~ierais, ~ieran**
quis|iese, **~ieses, ~iese, ~iésemos, ~ieseis, ~iesen**
Conditional: querr|ía, **~ías, ~ía, ~íamos, ~íais, ~ían**
Imperative: quiere, quered

[36] **raer**
Present: raigo/rayo, raes, rae, raemos, raéis, raen
Preterite: raí, raíste, rayó, raímos, raísteis, rayeron
Present subjunctive: raig|a, **~as, ~a, ~amos, ~áis, ~an**
ray|a, **~as, ~a, ~amos, ~áis, ~an**
Imperfect subjunctive: ray|era, **~eras, ~era, ~éramos, ~erais, ~eran**
ray|ese, **~eses, ~ese, ~ésemos, ~eseis, ~esen**
Present participle: rayendo
Past participle: raído

[37] **roer**
Present: roo/roigo/royo, roes, roe, roemos, roéis, roen
Preterite: roí, roíste, royó, roímos, roísteis, royeron
Present subjunctive: roa/roiga/roya, roas, roa, roamos, roáis, roan
Imperfect subjunctive: roy|era, **~eras, ~era, ~éramos, ~erais, ~eran**
roy|ese, **~eses, ~ese, ~ésemos, ~eseis, ~esen**
Present participle: royendo
Past participle: roído

[38] **saber**
Present: sé, sabes, sabe, sabemos, sabéis, saben
Future: sabr|é, **~ás, ~á, ~emos, ~éis, ~án**
Preterite: sup|e, **~iste, ~o, ~imos, ~isteis, ~ieron**
Present subjunctive: sep|a, **~as, ~a, ~amos, ~áis, ~an**
Imperfect subjunctive: sup|iera, **~ieras, ~iera, ~iéramos, ~ierais, ~ieran**
sup|iese, **~ieses, ~iese, ~iésemos, ~ieseis, ~iesen**
Conditional: sabr|ía, **~ías, ~ía, ~íamos, ~íais, ~ían**

[39] **ser**
Present: soy, eres, es, somos, sois, son
Imperfect: era, eras, era, éramos, erais, eran
Preterite: fui, fuiste, fue, fuimos, fuisteis, fueron
Present subjunctive: se|a, **~as, ~a, ~amos, ~áis, ~an**
Imperfect subjunctive: fu|era, **~eras, ~era, ~éramos, ~erais, ~eran**
fu|ese, **~eses, ~ese, ~ésemos, ~eseis, ~esen**
Imperative: sé, sed

[40] **tener**
Present: tengo, tienes, tiene, tenemos, tenéis, tienen
Future: tendr|é, **~ás, ~á, ~emos, ~éis, ~án**
Preterite: tuv|e, **~iste, ~o, ~imos, ~isteis, ~ieron**
Present subjunctive: teng|a, **~as, ~a, ~amos, ~áis, ~an**
Imperfect subjunctive: tuv|iera, **~ieras, ~iera, ~iéramos, ~ierais, ~ieran**
tuv|iese, **~ieses, ~iese, ~iésemos, ~ieseis, ~iesen**
Conditional: tendr|ía, **~ías, ~ía, ~íamos, ~íais, ~ían**
Imperative: ten, tened

[41] **traer**
Present: traigo, traes, trae, traemos, traéis, traen
Preterite: traj|e, **~iste, ~o, ~imos, ~isteis, ~eron**
Present subjunctive: traig|a, **~as, ~a, ~amos, ~áis, ~an**
Imperfect subjunctive: traj|era, **~eras, ~era, ~éramos, ~erais, ~eran**

traj|ese, **~eses, ~ese, ~ésemos, ~eseis, ~esen**
Present participle: trayendo
Past participle: traído

[42] **valer**
Present: valgo, vales, vale, valemos, valéis, valen
Future: vald|ré, **~ás, ~á, ~emos, ~éis, ~án**
Present subjunctive: valg|a, **~as, ~a, ~amos ~áis, ~an**
Conditional: vald|ría, **~ías, ~ía, ~íamos, ~íais, ~ían**
Imperative: val/vale, valed

[43] **ver**
Present: veo, ves, ve, vemos, véis, ven
Imperfect: ve|ía, **~ías, ~ía, ~íamos, ~íais, ~ían**
Preterite: vi, viste, vio, vimos, visteis, vieron
Present subjunctive: ve|a, **~as, ~a, ~amos, ~áis, ~an**
Past participle: visto

[44] **yacer**
Present: yazco/yazgo/yago, yaces, yace, yacemos, yacéis, yacen
Present subjunctive: yazca/yazga/yaga, yazcas, yazca, yazcamos, yazcáis, yazcan
Imperative: yace/yaz, yaced

[45] **asir**
Present: asgo, ases, ase, asimos, asís, asen
Present subjunctive: asg|a, **~as, ~a, ~amos, ~áis, ~an**

[46] **decir**
Present: digo, dices, dice, decimos, decís, dicen
Future: dir|é, **~ás, ~á, ~emos, ~éis, ~án**
Preterite: dij|e, **~iste, ~o, ~imos, ~isteis, ~eron**
Present subjunctive: dig|a, **~as, ~a, ~amos, ~áis, ~an**
Imperfect subjunctive: dij|era, **~eras, ~era, ~éramos, ~erais, ~eran**
dij|ese, **~eses, ~ese, ~ésemos, ~eseis, ~esen**
Conditional: dir|ía, **~ías, ~ía, ~íamos, ~íais, ~ían**
Present participle: dicho
Imperative: di, decid

[47] **reducir**
Present: reduzco, reduces, reduce, reducimos, reducís, reducen
Preterite: reduj|e, **~iste, ~o, ~imos, ~isteis, ~eron**
Present subjunctive: reduzc|a, **~as, ~a, ~amos, ~áis, ~an**
Imperfect subjunctive: reduj|era, **~eras, ~era, ~éramos, ~erais, ~eran**
reduj|ese, **~eses, ~ese, ~ésemos, ~eseis, ~esen**

[48] **erguir**
Present: irgo, irgues, irgue, erguimos, erguís, irguen
yergo, yergues, yergue, erguimos, erguís, yerguen
Preterite: erguí, erguiste, irguió, erguimos, erguisteis, irguieron
Present subjunctive: irg|a, **~as, ~a, ~amos, ~áis, ~an**
yerg|a, **~as, ~a, ~amos, ~áis, ~an**
Imperfect subjunctive: irgu|iera, **~ieras, ~iera, ~iéramos, ~ierais, ~ieran**
irgu|iese, **~ieses, ~iese, ~iésemos, ~ieseis, ~iesen**
Present participle: irguiendo
Imperative: irgue/yergue, erguid

[49] **ir**
Present: voy, vas, va, vamos, vais, van
Imperfect: iba, ibas, iba, íbamos, ibais, iban
Preterite: fui, fuiste, fue, fuimos, fuisteis, fueron
Present subjunctive: vay|a, **~as, ~a, ~amos, ~áis, ~an**
Imperfect subjunctive: fu|era, **~eras, ~era, ~éramos, ~erais, ~eran**
fu|ese, **~eses, ~ese, ~ésemos, ~eseis, ~esen**
Present participle: yendo
Imperative: ve, id

[50] **oír**
Present: oigo, oyes, oye, oímos, oís, oyen
Preterite: oí, oíste, oyó, oímos, oísteis, oyeron
Present subjunctive: oig|a, **~as, ~a, ~amos, ~áis, ~an**

Imperfect subjunctive: oy|era, **~eras, ~era, ~éramos, ~erais, ~eran**
oy|ese, **~eses, ~ese, ~ésemos, ~eseis, ~esen**
Present participle: oyendo
Past participle: oído
Imperative: oye, oíd

[51] **reír**
Present: río, ríes, ríe, reímos, reís, ríen
Preterite: reí, reíste, rió, reímos, reísteis, rieron
Present subjunctive: ría, rías, ría, riamos, riáis, rían
Present participle: riendo
Past participle: reído
Imperative: ríe, reíd

[52] **salir**
Present: salgo, sales, sale, salimos, salís, salen
Future: saldr|é, **~ás, ~á, ~emos, ~éis, ~án**
Present subjunctive: salg|a, **~as, ~a, ~amos, ~áis, ~an**
Conditional: saldr|ía, **~ías, ~ía, ~íamos, ~íais, ~ían**
Imperative: sal, salid

[53] **venir**
Present: vengo, vienes, viene, venimos, venís, vienen
Future: vendr|é, **~ás, ~á, ~emos, ~éis, ~án**
Preterite: vin|e, **~iste, ~o, ~imos, ~isteis, ~ieron**
Present subjunctive: veng|a, **~as, ~a, ~amos, ~áis, ~an**
Imperfect subjunctive: vin|iera, **~ieras, ~iera, ~iéramos, ~ierais, ~ieran**
vin|iese, **~ieses, ~iese, ~iésemos, ~ieseis, ~iesen**
Conditional: vendr|ía, **~ías, ~ía, ~íamos, ~íais, ~ían**
Present participle: viniendo
Imperative: ven, venid

Verbos Irregulares Ingleses

Infinitivo	*Pretérito*	*Participio pasado*
arise	arose	arisen
awake	awoke	awoken
be	was	been
bear	bore	borne
beat	beat	beaten
become	became	become
befall	befell	befallen
beget	begot	begotten
begin	began	begun
behold	beheld	beheld
bend	bent	bent
beset	beset	beset
bet	bet, betted	bet, betted
bid	bade, bid	bidden, bid
bind	bound	bound
bite	bit	bitten
bleed	bled	bled
blow	blew	blown
break	broke	broken
breed	bred	bred
bring	brought	brought
broadcast	broadcast(ed)	broadcast
build	built	built
burn	burnt, burned	burnt, burned
burst	burst	burst
buy	bought	bought
cast	cast	cast
catch	caught	caught
choose	chose	chosen
cleave	clove, cleft, cleaved	cloven, cleft, cleaved
cling	clung	clung
clothe	clothed, clad	clothed, clad
come	came	come
cost	cost	cost
creep	crept	crept
crow	crowed, crew	crowed
cut	cut	cut
deal	dealt	dealt
dig	dug	dug
do	did	done
draw	drew	drawn
dream	dreamt, dreamed	dreamt, dreamed
drink	drank	drunk
drive	drove	driven
dwell	dwelt	dwelt
eat	ate	eaten
fall	fell	fallen
feed	fed	fed
feel	felt	felt
fight	fought	fought
find	found	found

Infinitivo	*Pretérito*	*Participio pasado*
flee	fled	fled
fling	flung	flung
fly	flew	flown
forbear	forbore	forborne
forbid	forbad(e)	forbidden
forecast	forecast(ed)	forecast(ed)
foresee	foresaw	foreseen
foretell	foretold	foretold
forget	forgot	forgotten
forgive	forgave	forgiven
forsake	forsook	forsaken
freeze	froze	frozen
gainsay	gainsaid	gainsaid
get	got	got
give	gave	given
go	went	gone
grind	ground	ground
grow	grew	grown
hang	hung, hanged	hung, hanged
have	had	had
hear	heard	heard
hew	hewed	hewn, hewed
hide	hid	hidden
hit	hit	hit
hold	held	held
hurt	hurt	hurt
inlay	inlaid	inlaid
keep	kept	kept
kneel	knelt	knelt
knit	knitted, knit	knitted, knit
know	knew	known
lay	laid	laid
lead	led	led
lean	leaned, leant	leaned, leant
leap	leaped, leapt	leaped, leapt
learn	learned, learnt	learned, learnt
leave	left	left
lend	lent	lent
let	let	let
lie	lay	lain
light	lit, lighted	lit, lighted
lose	lost	lost
make	made	made
mean	meant	meant
meet	met	met
mislay	mislaid	mislaid
mislead	misled	misled
misspell	misspelt	misspelt
mistake	mistook	mistaken
misunderstand	misunderstood	misunderstood
mow	mowed	mown
outbid	outbid	outbid
outdo	outdid	outdone
outgrow	outgrew	outgrown
overcome	overcame	overcome

Infinitivo	*Pretérito*	*Participio pasado*
overdo	overdid	overdone
overhang	overhung	overhung
overhear	overheard	overheard
override	overrode	overridden
overrun	overran	overrun
oversee	oversaw	overseen
overshoot	overshot	overshot
oversleep	overslept	overslept
overtake	overtook	overtaken
overthrow	overthrew	overthrown
partake	partook	partaken
pay	paid	paid
prove	proved	proved, proven
put	put	put
quit	quitted, quit	quitted, quit
read /ri:d/	read /red/	read /red/
rebuild	rebuilt	rebuilt
redo	redid	redone
rend	rent	rent
repay	repaid	repaid
rewrite	rewrote	rewritten
rid	rid	rid
ride	rode	ridden
ring	rang	rung
rise	rose	risen
run	ran	run
saw	sawed	sawn, sawed
say	said	said
see	saw	seen
seek	sought	sought
sell	sold	sold
send	sent	sent
set	set	set
sew	sewed	sewn, sewed
shake	shook	shaken
shear	sheared	shorn, sheared
shed	shed	shed
shine	shone	shone
shoe	shod	shod
shoot	shot	shot
show	showed	shown, showed
shrink	shrank	shrunk
shut	shut	shut
sing	sang	sung
sink	sank	sunk
sit	sat	sat
slay	slew	slain
sleep	slept	slept
slide	slid	slid
sling	slung	slung
slit	slit	slit
smell	smelt, smelled	smelt, smelled
smite	smote	smitten
sow	sowed	sown, sowed
speak	spoke	spoken

Infinitivo	*Pretérito*	*Participio pasado*
speed	speeded, sped	speeded, sped
spell	spelt, spelled	spelt, spelled
spend	spent	spent
spill	spilt, spilled	spilt, spilled
spin	spun	spun
spit	spat	spat
split	split	split
spoil	spoilt, spoiled	spoilt, spoiled
spread	spread	spread
spring	sprang	sprung
stand	stood	stood
steal	stole	stolen
stick	stuck	stuck
sting	stung	stung
stink	stank, stunk	stunk
strew	strewed	strewn, strewed
stride	strode	stridden
strike	struck	struck
string	strung	strung
strive	strove	striven
swear	swore	sworn
sweep	swept	swept
swell	swelled	swollen, swelled
swim	swam	swum
swing	swung	swung
take	took	taken
teach	taught	taught
tear	tore	torn
tell	told	told
think	thought	thought
thrive	thrived, throve	thrived, thriven
throw	threw	thrown
thrust	thrust	thrust
tread	trod	trodden, trod
unbend	unbent	unbent
undergo	underwent	undergone
understand	understood	understood
undertake	undertook	undertaken
undo	undid	undone
upset	upset	upset
wake	woke, waked	woken, waked
waylay	waylaid	waylaid
wear	wore	worn
weave	wove	woven
weep	wept	wept
win	won	won
wind	wound	wound
withdraw	withdrew	withdrawn
withhold	withheld	withheld
withstand	withstood	withstood
wring	wrung	wrung
write	wrote	written